Matrix Analysis

Second Edition

Linear algebra and matrix theory are fundamental tools in mathematical and physical science, as well as fertile fields for research. This new edition of the acclaimed text presents results of both classic and recent matrix analysis using canonical forms as a unifying theme, and demonstrates their importance in a variety of applications.

The authors have thoroughly revised, updated, and expanded on the first edition. The book opens with an extended summary of useful concepts and facts and includes numerous new topics and features, such as:

- New sections on the singular value and CS decompositions
- New applications of the Jordan canonical form
- A new section on the Weyr canonical form
- Expanded treatments of inverse problems and of block matrices
- A central role for the von Neumann trace theorem
- A new appendix with a modern list of canonical forms for a pair of Hermitian matrices and for a symmetric–skew symmetric pair
- Expanded index with more than 3,500 entries for easy reference
- More than 1,100 problems and exercises, many with hints, to reinforce understanding and develop auxiliary themes such as finite-dimensional quantum systems, the compound and adjugate matrices, and the Loewner ellipsoid
- A new appendix provides a collection of problem-solving hints.

Roger A. Horn is a Research Professor in the Department of Mathematics at the University of Utah. He is the author of *Topics in Matrix Analysis* (Cambridge University Press 1994).

Charles R. Johnson is the author of *Topics in Matrix Analysis* (Cambridge University Press 1994).

Matrix Analysis

Second Edition

Roger A. Horn

University of Utah

Charles R. Johnson

CAMBRIDGE
UNIVERSITY PRESS

University Printing House, Cambridge CB2 8BS, United Kingdom

One Liberty Plaza, 20th Floor, New York, NY 10006, USA

477 Williamstown Road, Port Melbourne, VIC 3207, Australia

314-321, 3rd Floor, Plot 3, Splendor Forum, Jasola District Centre, New Delhi - 110025, India

79 Anson Road, #06-04/06, Singapore 079906

Cambridge University Press is part of the University of Cambridge.

It furthers the University's mission by disseminating knowledge in the pursuit of education, learning and research at the highest international levels of excellence.

www.cambridge.org
Information on this title: www.cambridge.org/9780521548236

First published 1985
First paperback edition 1990
Second edition first published 2013
Reprinted 2013 (four times)
Corrected reprint 2018
Reprint 2019

A catalogue record for this publication is available from the British Library

Library of Congress Cataloging in Publication data
Horn, Roger A.
Matrix analysis / Roger A. Horn, Charles R. Johnson. – 2nd ed.
    p. cm.
Includes bibliographical references and index.
ISBN 978-0-521-83940-2 (hardback)
1. Matrices.   I. Johnson, Charles R.   II. Title.
QA188.H66   2012
512.9′434–dc23   2012012300

ISBN 978-0-521-83940-2 Hardback
ISBN 978-0-521-54823-6 Paperback

To the matrix theory community

Contents

Preface to the Second Edition

The basic structure of the first edition has been preserved in the second because it remains congruent with the goal of writing "a book that would be a useful modern treatment of a broad range of topics...[that] may be used as an undergraduate or graduate text and as a self-contained reference for a variety of audiences." The quotation is from the Preface to the First Edition, whose declaration of goals for the work remains unchanged.

What is different in the second edition?

The core role of canonical forms has been expanded as a unifying element in understanding similarity (complex, real, and simultaneous), unitary equivalence, unitary similarity, congruence, *congruence, unitary congruence, triangular equivalence, and other equivalence relations. More attention is paid to cases of equality in the many inequalities considered in the book. Block matrices are a ubiquitous feature of the exposition in the new edition.

Learning mathematics has never been a spectator sport, so the new edition continues to emphasize the value of exercises and problems for the active reader. Numerous 2-by-2 examples illustrate concepts throughout the book. Problem threads (some span several chapters) develop special topics as the foundation for them evolves in the text. For example, there are threads involving the adjugate matrix, the compound matrix, finite-dimensional quantum systems, the Loewner ellipsoid and the Loewner–John matrix, and normalizable matrices; see the index for page references for these threads. The first edition had about 690 problems; the second edition has more than 1,100. Many problems have hints; they may be found in an appendix that appears just before the index.

A comprehensive index is essential for a book that is intended for sustained use as a reference after initial use as a text. The index to the first edition had about 1,200 entries; the current index has more than 3,500 entries. An unfamiliar term encountered in the text should be looked up in the index, where a pointer to a definition (in Chapter 0 or elsewhere) is likely to be found.

New discoveries since 1985 have shaped the presentation of many topics and have stimulated inclusion of some new ones. A few examples of the latter are the Jordan canonical form of a rank-one perturbation, motivated by enduring student interest in the *Google* matrix; a generalization of real normal matrices (normal matrices A such that $A\bar{A}$ is real); computable block matrix criteria for simultaneous unitary similarity or simultaneous unitary congruence; G. Belitskii's discovery that a matrix commutes with a Weyr canonical form if and only if it is block upper triangular and has a special structure; the discovery by K. C. O'Meara and C. Vinsonhaler that, unlike the corresponding situation for the Jordan canonical form, a commuting family can be simultaneously upper triangularized by similarity in such a way that any one specified matrix in the family is in Weyr canonical form; and canonical forms for congruence and *congruence.

Queries from many readers have motivated changes in the way that some topics are presented. For example, discussion of Lidskii's eigenvalue majorization inequalities was moved from a section primarily devoted to singular value inequalities to the section where majorization is discussed. Fortunately, a splendid new proof of Lidskii's inequalities by C. K. Li and R. Mathias became available and was perfectly aligned with Chapter 4's new approach to eigenvalue inequalities for Hermitian matrices. A second example is a new proof of Birkhoff's theorem, which has a very different flavor from the proof in the first edition.

Instructors accustomed to the order of topics in the first edition may be interested in a chapter-by-chapter summary of what is different in the new edition:

0. Chapter 0 has been expanded by about 75% to include a more comprehensive summary of useful concepts and facts. It is intended to serve as an as-needed reference. Definitions of terms and notations used throughout the book can be found here, but it has no exercises or problems. Formal courses and reading for self-study typically begin with Chapter 1.

1. Chapter 1 contains new examples related to similarity and the characteristic polynomial, as well as an enhanced emphasis on the role of left eigenvectors in matrix analysis.

2. Chapter 2 contains a detailed presentation of real orthogonal similarity, an exposition of McCoy's theorem on simultaneous triangularization, and a rigorous treatment of continuity of eigenvalues that makes essential use of both the *unitary* and *triangular* aspects of Schur's unitary triangularization theorem. Section 2.4 (Consequences of Schur's triangularization theorem) is almost twice the length of the corresponding section in the first edition. There are two new sections, one devoted to the singular value decomposition and one devoted to the CS decomposition. Early introduction of the singular value decomposition permits this essential tool of matrix analysis to be used throughout the rest of the book.

3. Chapter 3 approaches the Jordan canonical form via the Weyr characteristic; it contains an exposition of the Weyr canonical form and its unitary variant that were not in the first edition. Section 3.2 (Consequences of the Jordan canonical form) discusses many new applications; it contains 60% more material than the corresponding section in the first edition.

4. Chapter 4 now has a modern presentation of variational principles and eigenvalue inequalities for Hermitian matrices via subspace intersections. It contains an expanded treatment of inverse problems associated with interlacing and other

classical results. Its detailed treatment of unitary congruence includes Youla's theorem (a normal form for a square complex matrix A under unitary congruence that is associated with the eigenstructure of $A\bar{A}$), as well as canonical forms for conjugate normal, congruence normal, and squared normal matrices. It also has an exposition of recently discovered canonical forms for congruence and *congruence and new algorithms to construct a basis of a coneigenspace.

5. Chapter 5 contains an expanded discussion of norm duality, many new problems, and a treatment of semi-inner products that finds application in a discussion of finite-dimensional quantum systems in Chapter 7.

6. Chapter 6 has a new treatment of the "disjoint discs" aspect of Geršgorin's theorem and a reorganized discussion of eigenvalue perturbations, including differentiability of a simple eigenvalue.

7. Chapter 7 has been reorganized now that the singular value decomposition is introduced in Chapter 2. There is a new treatment of the polar decomposition, new factorizations related to the singular value decomposition, and special emphasis on row and column inclusion. The von Neumann trace theorem (proved via Birkhoff's theorem) is now the foundation on which many applications of the singular value decomposition are built. The Loewner partial order and block matrices are treated in detail with new techniques, as are the classical determinant inequalities for positive definite matrices.

8. Chapter 8 uses facts about left eigenvectors developed in Chapter 1 to streamline its exposition of the Perron–Frobenius theory of positive and nonnegative matrices.

D. Appendix D contains new explicit perturbation bounds for the zeroes of a polynomial and the eigenvalues of a matrix.

F. Appendix F tabulates a modern list of canonical forms for a pair of Hermitian matrices, or a pair of matrices, one of which is symmetric and the other is skew symmetric. These canonical pairs are applications of the canonical forms for congruence and *congruence presented in Chapter 4.

Readers who are curious about the technology of book making may be interested to know that this book began as a set of LaTeX files created manually by a company in India from hard copy of the first edition. Those files were edited and revised using the Scientific WorkPlace® graphical user interface and typesetting system.

The cover art for the second edition was the result of a lucky encounter on a Delta flight from Salt Lake City to Los Angeles in spring 2003. The young man in the middle seat said he was an artist who paints abstract paintings that are sometimes mathematically inspired. In the course of friendly conversation, he revealed that his special area of mathematical enjoyment was linear algebra, and that he had studied *Matrix Analysis*. After mutual expressions of surprise at the chance nature of our meeting, and a pleasant discussion, we agreed that appropriate cover art would enhance the visual appeal of the second edition; he said he would send something to consider. In due course a packet arrived from Seattle. It contained a letter and a stunning 4.5- by 5-inch color photograph, identified on the back as an image of a 72- by 66-inch oil on canvas, painted in 2002. The letter said that "the painting is entitled *Surprised Again on the Diagonal* and is inspired by the recurring prevalence of the diagonal in math whether it be in geometry, analysis, algebra, set theory or logic. I think that it would be an attractive addition to your wonderful book." Thank you, Lun-Yi Tsai, for your wonderful cover art!

A great many students, instructors, and professional colleagues have contributed to the evolution of this new edition since its predecessor appeared in 1985. Special thanks are hereby acknowledged to T. Ando, Wayne Barrett, Ignat Domanov, Jim Fill, Carlos Martins da Fonseca, Tatiana Gerasimova, Geoffrey Goodson, Robert Guralnick, Thomas Hawkins, Eugene Herman, Khakim Ikramov, Ilse Ipsen, Dennis C. Jespersen, Hideki Kosaki, Zhongshan Li, Teck C. Lim, Ross A. Lippert, Roy Mathias, Dennis Merino, Arnold Neumaier, Kevin O'Meara, Peter Rosenthal, Vladimir Sergeichuk, Wasin So, Hugo Woerdeman, and Fuzhen Zhang.

R.A.H.

Preface to the First Edition

Linear algebra and matrix theory have long been fundamental tools in mathematical disciplines as well as fertile fields for research in their own right. In this book, and in the companion volume, *Topics in Matrix Analysis*, we present classical and recent results of matrix analysis that have proved to be important to applied mathematics. The book may be used as an undergraduate or graduate text and as a self-contained reference for a variety of audiences. We assume background equivalent to a one-semester elementary linear algebra course and knowledge of rudimentary analytical concepts. We begin with the notions of eigenvalues and eigenvectors; no prior knowledge of these concepts is assumed.

Facts about matrices, beyond those found in an elementary linear algebra course, are necessary to understand virtually any area of mathematical science, whether it be differential equations; probability and statistics; optimization; or applications in theoretical and applied economics, the engineering disciplines, or operations research, to name only a few. But until recently, much of the necessary material has occurred sporadically (or not at all) in the undergraduate and graduate curricula. As interest in applied mathematics has grown and more courses have been devoted to advanced matrix theory, the need for a text offering a broad selection of topics has become more apparent, as has the need for a modern reference on the subject.

There are several well-loved classics in matrix theory, but they are not well suited for general classroom use, nor for systematic individual study. A lack of problems, applications, and motivation; an inadequate index; and a dated approach are among the difficulties confronting readers of some traditional references. More recent books tend to be either elementary texts or treatises devoted to special topics. Our goal was to write a book that would be a useful modern treatment of a broad range of topics.

One view of "matrix analysis" is that it consists of those *topics* in linear algebra that have arisen out of the needs of mathematical analysis, such as multivariable calculus, complex variables, differential equations, optimization, and approximation theory. Another view is that matrix analysis is an *approach* to real and complex linear algebraic problems that does not hesitate to use notions from analysis – such as limits, continuity, and power series – when these seem more efficient or natural than a purely algebraic approach. Both views of matrix analysis are reflected in the choice and treatment of topics in this book. We prefer the term *matrix analysis* to *linear algebra* as an accurate reflection of the broad scope and methodology of the field.

For review and convenience in reference, Chapter 0 contains a summary of necessary facts from elementary linear algebra, as well as other useful, though not necessarily elementary, facts. Chapters 1, 2, and 3 contain mainly core material likely to be included in any second course in linear algebra or matrix theory: a basic treatment of eigenvalues, eigenvectors, and similarity; unitary similarity, Schur triangularization and its implications, and normal matrices; and canonical forms and factorizations, including the Jordan form, *LU* factorization, *QR* factorization, and companion matrices. Beyond this, each chapter is developed substantially independently and treats in some depth a major topic:

1. *Hermitian and complex symmetric matrices* (Chapter 4). We give special emphasis to variational methods for studying eigenvalues of Hermitian matrices and include an introduction to the notion of majorization.

2. *Norms on vectors and matrices* (Chapter 5) are essential for error analyses of numerical linear algebraic algorithms and for the study of matrix power series and iterative processes. We discuss the algebraic, geometric, and analytic properties of norms in some detail and make a careful distinction between those norm results for matrices that depend on the submultiplicativity axiom for matrix norms and those that do not.

3. *Eigenvalue location and perturbation results* (Chapter 6) for general (not necessarily Hermitian) matrices are important for many applications. We give a detailed treatment of the theory of Geršgorin regions, and some of its modern refinements, and of relevant graph theoretic concepts.

4. *Positive definite matrices* (Chapter 7) and their applications, including inequalities, are considered at some length. A discussion of the polar and singular value decompositions is included, along with applications to matrix approximation problems.

5. *Entry-wise nonnegative and positive matrices* (Chapter 8) arise in many applications in which nonnegative quantities necessarily occur (probability, economics, engineering, etc.), and their remarkable theory reflects the applications. Our development of the theory of nonnegative, positive, primitive, and irreducible matrices proceeds in elementary steps based on the use of norms.

In the companion volume, further topics of similar interest are treated: the field of values and generalizations; inertia, stable matrices, *M*-matrices and related special classes; matrix equations, Kronecker and Hadamard products; and various ways in which functions and matrices may be linked.

This book provides the basis for a variety of one- or two-semester courses through selection of chapters and sections appropriate to a particular audience. We recommend that an instructor make a careful preselection of sections and portions of sections of the book for the needs of a particular course. This would probably include Chapter 1, much of Chapters 2 and 3, and facts about Hermitian matrices and norms from Chapters 4 and 5.

Most chapters contain some relatively specialized or nontraditional material. For example, Chapter 2 includes not only Schur's basic theorem on unitary triangularization of a single matrix but also a discussion of simultaneous triangularization of families of matrices. In the section on unitary equivalence, our presentation of the usual facts is followed by a discussion of trace conditions for two matrices to be unitarily equivalent. A discussion of complex symmetric matrices in Chapter 4 provides a counterpoint to the development of the classical theory of Hermitian matrices. Basic aspects of a topic appear in the initial sections of each chapter, while more elaborate discussions occur at the ends of sections or in later sections. This strategy has the advantage of presenting topics in a sequence that enhances the book's utility as a reference. It also provides a rich variety of options to the instructor.

Many of the results discussed are valid or can be generalized to be valid for matrices over other fields or in some broader algebraic setting. However, we deliberately confine our domain to the real and complex fields where familiar methods of classical analysis as well as formal algebraic techniques may be employed.

Though we generally consider matrices to have complex entries, most examples are confined to real matrices, and no deep knowledge of complex analysis is required. Acquaintance with the arithmetic of complex numbers is necessary for an understanding of matrix analysis and is covered to the extent necessary in an appendix. Other brief appendices cover several peripheral, but essential, topics such as Weierstrass's theorem and convexity.

We have included many exercises and problems because we feel these are essential to the development of an understanding of the subject and its implications. The exercises occur throughout as part of the development of each section; they are generally elementary and of immediate use in understanding the concepts. We recommend that the reader work at least a broad selection of these. Problems are listed (in no particular order) at the end of each section; they cover a range of difficulties and types (from theoretical to computational) and they may extend the topic, develop special aspects, or suggest alternate proofs of major ideas. Significant hints are given for the more difficult problems. The results of some problems are referred to in other problems or in the text itself. We cannot overemphasize the importance of the reader's active involvement in carrying out the exercises and solving problems.

While the book itself is not about applications, we have, for motivational purposes, begun each chapter with a section outlining a few applications to introduce the topic of the chapter.

Readers who wish to consult alternative treatments of a topic for additional information are referred to the books listed in the References section following the appendices.

The list of book references is not exhaustive. As a practical concession to the limits of space in a general multitopic book, we have minimized the number of citations in the text. A small selection of references to papers – such as those we have explicitly used – does occur at the end of most sections accompanied by a brief discussion, but we have made no attempt to collect historical references to classical results. Extensive bibliographies are provided in the more specialized books we have referenced.

We appreciate the helpful suggestions of our colleagues and students who have taken the time to convey their reactions to the class notes and preliminary manuscripts that were the precursors of the book. They include Wayne Barrett, Leroy Beasley, Bryan Cain, David Carlson, Dipa Choudhury, Risana Chowdhury, Yoo Pyo Hong, Dmitry Krass, Dale Olesky, Stephen Pierce, Leiba Rodman, and Pauline van den Driessche.

R.A.H.
C.R.J.

CHAPTER 0
Review and Miscellanea

0.0 Introduction

In this initial chapter we summarize many useful concepts and facts, some of which provide a foundation for the material in the rest of the book. Some of this material is included in a typical elementary course in linear algebra, but we also include additional useful items, even though they do not arise in our subsequent exposition. The reader may use this chapter as a review before beginning the main part of the book in Chapter 1; subsequently, it can serve as a convenient reference for notation and definitions that are encountered in later chapters. We assume that the reader is familiar with the basic concepts of linear algebra and with mechanical aspects of matrix manipulations, such as matrix multiplication and addition.

0.1 Vector spaces

A finite dimensional vector space is the fundamental setting for matrix analysis.

0.1.1 Scalar field. Underlying a vector space is its *field*, or set of scalars. For our purposes, that underlying field is typically the real numbers **R** or the complex numbers **C** (see Appendix A), but it could be the rational numbers, the integers modulo a specified prime number, or some other field. When the field is unspecified, we denote it by the symbol **F**. To qualify as a field, a set must be closed under two binary operations: "addition" and "multiplication." Both operations must be associative and commutative, and each must have an identity element in the set; inverses must exist in the set for all elements under addition and for all elements except the additive identity under multiplication; multiplication must be distributive over addition.

0.1.2 Vector spaces. A *vector space* V over a field **F** is a set V of objects (called *vectors*) that is closed under a binary operation ("addition") that is associative and commutative and has an identity (the *zero vector*, denoted by 0) and additive inverses

1

in the set. The set is also closed under an operation of "scalar multiplication" of the vectors by elements of the scalar field $\mathbf{F}$, with the following properties for all $a, b \in \mathbf{F}$ and all $x, y \in V$: $a(x + y) = ax + ay$, $(a + b)x = ax + bx$, $a(bx) = (ab)x$, and $ex = x$ for the multiplicative identity $e \in \mathbf{F}$.

For a given field $\mathbf{F}$ and a given positive integer n, the set $\mathbf{F}^n$ of n-tuples with entries from $\mathbf{F}$ forms a vector space over $\mathbf{F}$ under entrywise addition in $\mathbf{F}^n$. Our convention is that *elements of $\mathbf{F}^n$ are always presented as column vectors*; we often call them *n-vectors*. The special cases $\mathbf{R}^n$ and $\mathbf{C}^n$ are the basic vector spaces of this book; $\mathbf{R}^n$ is a real vector space (that is, a vector space over the real field), while $\mathbf{C}^n$ is both a real vector space and a complex vector space (a vector space over the complex field). The set of polynomials with real or with complex coefficients (of no more than a specified degree *or* of arbitrary degree) and the set of real-valued or complex-valued functions on subsets of $\mathbf{R}$ or $\mathbf{C}$ (all with the usual notions of addition of functions and multiplication of a function by a scalar) are also examples of real or complex vector spaces.

0.1.3 Subspaces, span, and linear combinations. A *subspace* of a vector space V over a field $\mathbf{F}$ is a subset of V that is, by itself, a vector space over $\mathbf{F}$ using the same operations of vector addition and scalar multiplication as in V. A subset of V is a subspace precisely when it is closed under these two operations. For example, $\{[a, b, 0]^T : a, b \in \mathbf{R}\}$ is a subspace of $\mathbf{R}^3$; see (0.2.5) for the transpose notation. An intersection of subspaces is always a subspace; a union of subspaces need not be a subspace. The subsets $\{0\}$ and V are always subspaces of V, so they are often called *trivial subspaces*; a subspace of V is said to be *nontrivial* if it is different from both $\{0\}$ and V. A subspace of V is said to be a *proper subspace* if it is not equal to V. We call $\{0\}$ the *zero vector space*. Since a vector space always contains the zero vector, a subspace cannot be empty.

If S is a subset of a vector space V over a field $\mathbf{F}$, span S is the intersection of all subspaces of V that contain S. If S is nonempty, then span $S = \{a_1 v_1 + \cdots + a_k v_k :$ $v_1, \ldots, v_k \in S, a_1, \ldots, a_k \in \mathbf{F}$, and $k = 1, 2, \ldots\}$. If S is empty, it is contained in every subspace of V; since the intersection of every subspace of V is the subspace $\{0\}$, the definition ensures that span $S = \{0\}$. Notice that span S is always a subspace even if S is not a subspace; S is said to *span* V if span $S = V$.

A *linear combination* of vectors in a vector space V over a field $\mathbf{F}$ is any expression of the form $a_1 v_1 + \cdots + a_k v_k$ in which k is a positive integer, $a_1, \ldots, a_k \in \mathbf{F}$, and $v_1, \ldots, v_k \in V$. Thus, the span of a nonempty subset S of V consists of all linear combinations of finitely many vectors in S. A linear combination $a_1 v_1 + \cdots + a_k v_k$ is *trivial* if $a_1 = \cdots = a_k = 0$; otherwise, it is *nontrivial*. A linear combination is by definition a sum of *finitely many* elements of a vector space.

Let S_1 and S_2 be subspaces of a vector space over a field $\mathbf{F}$. The *sum* of S_1 and S_2 is the subspace

$$S_1 + S_2 = \text{span}\,\{S_1 \cup S_2\} = \{x + y : x \in S_1, y \in S_2\}$$

If $S_1 \cap S_2 = \{0\}$, we say that the sum of S_1 and S_2 is a *direct sum* and write it as $S_1 \oplus S_2$; every $z \in S_1 \oplus S_2$ can be written as $z = x + y$ with $x \in S_1$ and $y \in S_2$ in one and only one way.

0.1.4 Linear dependence and linear independence. We say that a finite list of vectors $v_1, \ldots, v_k$ in a vector space V over a field $\mathbf{F}$ is *linearly dependent* if and only if there are scalars $a_1, \ldots, a_k \in \mathbf{F}$, not all zero, such that $a_1 x_1 + \cdots + a_k x_k = 0$. Thus, a list of vectors $v_1, \ldots, v_k$ is linearly dependent if and only if some nontrivial linear combination of $v_1, \ldots, v_k$ is the zero vector. It is often convenient to say that "$v_1, \ldots, v_k$ are linearly dependent" instead of the more formal statement "the list of vectors $v_1, \ldots, v_k$ is linearly dependent." A list of vectors $v_1, \ldots, v_k$ is said to have *length k*. A list of two or more vectors is linearly dependent if one of the vectors is a linear combination of some of the others; in particular, a list of two or more vectors in which two of the vectors in the list are identical is linearly dependent. Two vectors are linearly dependent if and only if one of the vectors is a scalar multiple of the other. A list consisting only of the zero vector is linearly dependent since $a_1 0 = 0$ for $a_1 = 1$.

A finite list of vectors $v_1, \ldots, v_k$ in a vector space V over a field $\mathbf{F}$ is *linearly independent* if it is not linearly dependent. Again, it can be convenient to say that "$v_1, \ldots, v_k$ are linearly independent" instead of "the list of vectors $v_1, \ldots, v_k$ is linearly independent."

Sometimes one encounters natural lists of vectors that have infinitely many elements, for example, the monomials $1, t, t^2, t^3, \ldots$ in the vector space of all polynomials with real coefficients or the complex exponentials $1, e^{it}, e^{2it}, e^{3it}, \ldots$ in the vector space of complex-valued continuous functions that are periodic on $[0, 2\pi]$.

If certain vectors in a list (finite or infinite) are deleted, the resulting list is a *sublist* of the original list. An infinite list of vectors is said to be linearly dependent if some finite sublist is linearly dependent; it is said to be linearly independent if every finite sublist is linearly independent. Any sublist of a linearly independent list of vectors is linearly independent; any list of vectors that has a linearly dependent sublist is linearly dependent. Since a list consisting only of the zero vector is linearly dependent, any list of vectors that contains the zero vector is linearly dependent. A list of vectors can be linearly dependent, while any proper sublist is linearly independent; see (1.4.P12). An empty list of vectors is not linearly dependent, so it is linearly independent.

The *cardinality* of a finite set is the number of its (necessarily distinct) elements. For a given list of vectors $v_1, \ldots, v_k$ in a vector space V, the cardinality of the *set* $\{v_1, \ldots, v_k\}$ is less than k if and only if two or more vectors in the list are identical; if $v_1, \ldots, v_k$ are linearly independent, then the cardinality of the set $\{v_1, \ldots, v_k\}$ is k. The *span* of a list of vectors (finite or not) is the span of the set of elements of the list; a list of vectors *spans* V if V is the span of the list.

A set S of vectors is said to be linearly independent if every finite list of distinct vectors in S is linearly independent; S is said to be linearly dependent if some finite list of distinct vectors in S is linearly dependent.

0.1.5 Basis. A linearly independent list of vectors in a vector space V whose span is V is a *basis* for V. Each element of V can be represented as a linear combination of vectors in a basis in one and only one way; this is no longer true if any element whatsoever is appended to or deleted from the basis. A linearly independent list of vectors in V is a basis of V if and only if no list of vectors that properly contains it is linearly independent. A list of vectors that spans V is a basis for V if and only if no proper sublist of it spans V. The empty list is a basis for the zero vector space.

0.1.6 Extension to a basis. Any linearly independent list of vectors in a vector space V may be extended, perhaps in more than one way, to a basis of V. A vector space can have a basis that is not finite; for example, the infinite list of monomials $1, t, t^2, t^3, \ldots$ is a basis for the real vector space of all polynomials with real coefficients; each polynomial is a unique linear combination of (finitely many) elements in the basis.

0.1.7 Dimension. If there is a positive integer n such that a basis of the vector space V contains exactly n vectors, then every basis of V consists of exactly n vectors; this common cardinality of bases is the *dimension* of the vector space V and is denoted by $\dim V$. In this event, V is *finite-dimensional*; otherwise V is *infinite-dimensional*. In the infinite-dimensional case, there is a one-to-one correspondence between the elements of any two bases. The real vector space $\mathbf{R}^n$ has dimension n. The vector space $\mathbf{C}^n$ has dimension n over the field $\mathbf{C}$ but dimension $2n$ over the field $\mathbf{R}$. The basis $e_1, \ldots, e_n$ of $\mathbf{F}^n$ in which each n-vector e_i has a 1 as its ith entry and 0s elsewhere is called the *standard basis*.

It is convenient to say "V is an n-dimensional vector space" as a shorthand for "V is a finite-dimensional vector space whose dimension is n." Any subspace of an n-dimensional vector space is finite-dimensional; its dimension is strictly less than n if it is a proper subspace.

Let V be a finite-dimensional vector space and let S_1 and S_2 be two given subspaces of V. The *subspace intersection theorem* is

$$\dim (S_1 \cap S_2) + \dim (S_1 + S_2) = \dim S_1 + \dim S_2 \qquad (0.1.7.1)$$

Rewriting this identity as

$$\dim (S_1 \cap S_2) = \dim S_1 + \dim S_2 - \dim (S_1 + S_2)$$
$$\geq \dim S_1 + \dim S_2 - \dim V \qquad (0.1.7.2)$$

reveals the useful fact that if $\delta = \dim S_1 + \dim S_2 - \dim V \geq 1$, then the subspace $S_1 \cap S_2$ has dimension at least δ, and hence it contains δ linearly independent vectors, namely, any δ elements of a basis of $S_1 \cap S_2$. In particular, $S_1 \cap S_2$ contains a nonzero vector. An induction argument shows that if $S_1, \ldots, S_k$ are subspaces of V, and if $\delta = \dim S_1 + \cdots + \dim S_k - (k-1) \dim V \geq 1$, then

$$\dim (S_1 \cap \cdots \cap S_k) \geq \delta \qquad (0.1.7.3)$$

and hence $S_1 \cap \cdots \cap S_k$ contains δ linearly independent vectors; in particular, it contains a nonzero vector.

0.1.8 Isomorphism. If U and V are vector spaces over the same scalar field $\mathbf{F}$, and if $f : U \to V$ is an invertible function such that $f(ax + by) = af(x) + bf(y)$ for all $x, y \in U$ and all $a, b \in \mathbf{F}$, then f is said to be an *isomorphism* and U and V are said to be isomorphic ("same structure"). Two finite-dimensional vector spaces over the same field are isomorphic if and only if they have the same dimension; thus, any n-dimensional vector space over $\mathbf{F}$ is isomorphic to $\mathbf{F}^n$. Any n-dimensional real vector space is, therefore, isomorphic to $\mathbf{R}^n$, and any n-dimensional complex vector space is isomorphic to $\mathbf{C}^n$. Specifically, if V is an n-dimensional vector space over a field $\mathbf{F}$

with specified basis $\mathcal{B} = \{x_1, \ldots, x_n\}$, then, since any element $x \in V$ may be written uniquely as $x = a_1 x_1 + \cdots + a_n x_n$ in which each $a_i \in \mathbf{F}$, we may identify x with the n-vector $[x]_\mathcal{B} = [a_1 \ldots a_n]^T$. For any basis $\mathcal{B}$, the mapping $x \to [x]_\mathcal{B}$ is an isomorphism between V and $\mathbf{F}^n$.

0.2 Matrices

The fundamental object of study here may be thought of in two important ways: as a rectangular array of scalars and as a linear transformation between two vector spaces, given specified bases for each space.

0.2.1 Rectangular arrays. A *matrix* is an m-by-n array of scalars from a field $\mathbf{F}$. If $m = n$, the matrix is said to be *square*. The set of all m-by-n matrices over $\mathbf{F}$ is denoted by $M_{m,n}(\mathbf{F})$, and $M_{n,n}(\mathbf{F})$ is often denoted by $M_n(\mathbf{F})$. The vector spaces $M_{n,1}(\mathbf{F})$ and $\mathbf{F}^n$ are identical. If $\mathbf{F} = \mathbf{C}$, then $M_n(\mathbf{C})$ is further abbreviated to M_n, and $M_{m,n}(\mathbf{C})$ to $M_{m,n}$. Matrices are typically denoted by capital letters, and their scalar entries are typically denoted by doubly subscripted lowercase letters. For example, if

$$A = \begin{bmatrix} 2 & -\frac{3}{2} & 0 \\ -1 & \pi & 4 \end{bmatrix} = [a_{ij}]$$

then $A \in M_{2,3}(\mathbf{R})$ has entries $a_{11} = 2, a_{12} = -3/2, a_{13} = 0, a_{21} = -1, a_{22} = \pi$, $a_{23} = 4$. A *submatrix* of a given matrix is a rectangular array lying in specified subsets of the rows and columns of a given matrix. For example, $[\pi \ 4]$ is a submatrix (lying in row 2 and columns 2 and 3) of A.

Suppose that $A = [a_{ij}] \in M_{n,m}(\mathbf{F})$. The *main diagonal* of A is the list of entries $a_{11}, a_{22}, \ldots, a_{qq}$, in which $q = \min\{n, m\}$. It is sometimes convenient to express the main diagonal of A as a vector $\text{diag } A = [a_{ii}]_{i=1}^q \in \mathbf{F}^q$. The pth *superdiagonal* of A is the list $a_{1,p+1}, a_{2,p+2}, \ldots, a_{k,p+k}$, in which $k = \min\{n, m - p\}$, $p = 0, 1, 2, \ldots, m - 1$; the pth *subdiagonal* of A is the list $a_{p+1,1}, a_{p+2,2}, \ldots, a_{p+\ell,\ell}$, in which $\ell = \min\{n - p, m\}$, $p = 0, 1, 2, \ldots, n - 1$.

0.2.2 Linear transformations. Let U be an n-dimensional vector space and let V be an m-dimensional vector space, both over the same field $\mathbf{F}$; let $\mathcal{B}_U$ be a basis of U and let $\mathcal{B}_V$ be a basis of V. We may use the isomorphisms $x \to [x]_{\mathcal{B}_U}$ and $y \to [y]_{\mathcal{B}_V}$ to represent vectors in U and V as n-vectors and m-vectors over $\mathbf{F}$, respectively. A *linear transformation* is a function $T : U \to V$ such that $T(a_1 x_1 + a_2 x_2) = a_1 T(x_1) + a_2 T(x_2)$ for any scalars a_1, a_2 and vectors x_1, x_2. A matrix $A \in M_{m,n}(\mathbf{F})$ corresponds to a linear transformation $T : U \to V$ in the following way: $y = T(x)$ if and only if $[y]_{\mathcal{B}_V} = A[x]_{\mathcal{B}_U}$. The matrix A is said to *represent the linear transformation T* (relative to the bases $\mathcal{B}_U$ and $\mathcal{B}_V$); the representing matrix A depends on the bases chosen. When we study a matrix A, we realize that we are studying a linear transformation relative to a particular choice of bases, but explicit appeal to the bases is usually not necessary.

0.2.3 Vector spaces associated with a matrix or linear transformation. Any n-dimensional vector space over $\mathbf{F}$ may be identified with $\mathbf{F}^n$; we may think of

$A \in M_{m,n}(\mathbf{F})$ as a linear transformation $x \to Ax$ from $\mathbf{F}^n$ to $\mathbf{F}^m$ (and also as an array). The *domain* of this linear transformation is $\mathbf{F}^n$; its *range* is range $A = \{y \in \mathbf{F}^m : y = Ax\}$ for some $x \in \mathbf{F}^n$; its *null space* is nullspace $A = \{x \in \mathbf{F}^n : Ax = 0\}$. The range of A is a subspace of $\mathbf{F}^m$, and the null space of A is a subspace of $\mathbf{F}^n$. The dimension of nullspace A is denoted by nullity A; the dimension of range A is denoted by rank A. These numbers are related by the *rank-nullity theorem*

$$\dim(\text{range } A) + \dim(\text{nullspace } A) = \text{rank } A + \text{nullity } A = n \qquad (0.2.3.1)$$

for $A \in M_{m,n}(\mathbf{F})$. The null space of A is a set of vectors in $\mathbf{F}^n$ whose entries satisfy m homogeneous linear equations.

0.2.4 Matrix operations. Matrix addition is defined entrywise for arrays of the same dimensions and is denoted by $+$ ("$A + B$"). It corresponds to addition of linear transformations (relative to the same basis), and it inherits commutativity and associativity from the scalar field. The *zero matrix* (all entries are zero) is the additive identity, and $M_{m,n}(\mathbf{F})$ is a vector space over $\mathbf{F}$. Matrix multiplication is denoted by juxtaposition ("AB") and corresponds to the composition of linear transformations. Therefore, it is defined only when $A \in M_{m,n}(\mathbf{F})$ and $B \in M_{n,q}(\mathbf{F})$. It is associative, but not always commutative. For example,

$$\begin{bmatrix} 1 & 2 \\ 6 & 8 \end{bmatrix} = \begin{bmatrix} 1 & 0 \\ 0 & 2 \end{bmatrix} \begin{bmatrix} 1 & 2 \\ 3 & 4 \end{bmatrix} \neq \begin{bmatrix} 1 & 2 \\ 3 & 4 \end{bmatrix} \begin{bmatrix} 1 & 0 \\ 0 & 2 \end{bmatrix} = \begin{bmatrix} 1 & 4 \\ 3 & 8 \end{bmatrix}$$

The *identity matrix*

$$I = \begin{bmatrix} 1 & & \\ & \ddots & \\ & & 1 \end{bmatrix} \in M_n(\mathbf{F})$$

is the multiplicative identity in $M_n(\mathbf{F})$; its main diagonal entries are 1, and all other entries are 0. The identity matrix and any scalar multiple of it (a *scalar matrix*) commute with every matrix in $M_n(\mathbf{F})$; they are the only matrices that do so. Matrix multiplication is distributive over matrix addition.

The symbol 0 is used throughout the book to denote each of the following: the zero scalar of a field, the zero vector of a vector space, the zero n-vector in $\mathbf{F}^n$ (all entries equal to the zero scalar in $\mathbf{F}$), and the zero matrix in $M_{m,n}(\mathbf{F})$ (all entries equal to the zero scalar). The symbol I denotes the identity matrix of any size. If there is potential for confusion, we indicate the dimension of a zero or identity matrix with subscripts, for example, $0_{p,q}$, 0_k, or I_k.

0.2.5 The transpose, conjugate transpose, and trace. If $A = [a_{ij}] \in M_{m,n}(\mathbf{F})$, the *transpose* of A, denoted by A^T, is the matrix in $M_{n,m}(\mathbf{F})$ whose i, j entry is a_{ji}; that is, rows are exchanged for columns and vice versa. For example,

$$\begin{bmatrix} 1 & 2 & 3 \\ 4 & 5 & 6 \end{bmatrix}^T = \begin{bmatrix} 1 & 4 \\ 2 & 5 \\ 3 & 6 \end{bmatrix}$$

Of course, $(A^T)^T = A$. The *conjugate transpose* (sometimes called the *adjoint* or *Hermitian adjoint*) of $A \in M_{m,n}(\mathbf{C})$, is denoted by A^* and defined by $A^* = \bar{A}^T$, in which $\bar{A}$ is the entrywise conjugate. For example,

$$\begin{bmatrix} 1+i & 2-i \\ -3 & -2i \end{bmatrix}^* = \begin{bmatrix} 1-i & -3 \\ 2+i & 2i \end{bmatrix}$$

Both the transpose and the conjugate transpose obey the *reverse-order law*: $(AB)^T = B^T A^T$ and $(AB)^* = B^* A^*$. For the complex conjugate of a product, there is no reversing: $\overline{AB} = \bar{A}\bar{B}$. If x, y are real or complex vectors of the same size, then y^*x is a scalar and its conjugate transpose and complex conjugate are the same: $(y^*x)^* = \overline{y^*x} = x^*y = y^T\bar{x}$.

Many important classes of matrices are defined by identities involving the transpose or conjugate transpose. For example, $A \in M_n(\mathbf{F})$ is said to be *symmetric* if $A^T = A$, *skew symmetric* if $A^T = -A$, and *orthogonal* if $A^T A = I$; $A \in M_n(\mathbf{C})$ is said to be *Hermitian* if $A^* = A$, *skew Hermitian* if $A^* = -A$, *essentially Hermitian* if $e^{i\theta} A$ is Hermitian for some $\theta \in \mathbf{R}$, *unitary* if $A^*A = I$, and *normal* if $A^*A = AA^*$.

Each $A \in M_n(\mathbf{F})$ can be written in exactly one way as $A = S(A) + C(A)$, in which $S(A)$ is symmetric and $C(A)$ is skew symmetric: $S(A) = \frac{1}{2}(A + A^T)$ is the *symmetric part* of A; $C(A) = \frac{1}{2}(A - A^T)$ is the *skew-symmetric part* of A.

Each $A \in M_{m,n}(\mathbf{C})$ can be written in exactly one way as $A = B + iC$, in which $B, C \in M_{m,n}(\mathbf{R})$: $B = \frac{1}{2}(A + \bar{A})$ is the *real part* of A; $C = \frac{1}{2i}(A - \bar{A})$ is the *imaginary part* of A.

Each $A \in M_n(\mathbf{C})$ can be written in exactly one way as $A = H(A) + iK(A)$, in which $H(A)$ and $K(A)$ are Hermitian: $H(A) = \frac{1}{2}(A + A^*)$ is the *Hermitian part* of A; $iK(A) = \frac{1}{2}(A - A^*)$ is the *skew-Hermitian part* of A. The representation $A = H(A) + iK(A)$ of a complex or real matrix is its *Toeplitz decomposition*.

The *trace* of $A = [a_{ij}] \in M_{m,n}(\mathbf{F})$ is the sum of its main diagonal entries: $\text{tr } A = a_{11} + \cdots + a_{qq}$, in which $q = \min\{m, n\}$. For any $A = [a_{ij}] \in M_{m,n}(\mathbf{C})$, $\text{tr } AA^* = \text{tr } A^*A = \sum_{i,j} |a_{ij}|^2$, so

$$\text{tr } AA^* = 0 \text{ if and only if } A = 0 \qquad (0.2.5.1)$$

A vector $x \in \mathbf{F}^n$ is *isotropic* if $x^Tx = 0$. For example, $[1 \ i]^T \in \mathbf{C}^2$ is a nonzero isotropic vector. There are no nonzero isotropic vectors in $\mathbf{R}^n$.

0.2.6 Metamechanics of matrix multiplication.
In addition to the conventional definition of matrix-vector and matrix-matrix multiplication, several alternative viewpoints can be useful.

1. If $A \in M_{m,n}(\mathbf{F})$, $x \in \mathbf{F}^n$, and $y \in \mathbf{F}^m$, then the (column) vector Ax is a linear combination of the columns of A; the coefficients of the linear combination are the entries of x. The row vector $y^T A$ is a linear combination of the rows of A; the coefficients of the linear combination are the entries of y.
2. If b_j is the jth column of B and a_i^T is the ith row of A, then the jth column of AB is Ab_j and the ith row of AB is $a_i^T B$.

To paraphrase, in the matrix product AB, *left multiplication by A multiplies the columns of B* and *right multiplication by B multiplies the rows of A*. See (0.9.1) for an important special case of this observation when one of the factors is a diagonal matrix.

Suppose that $A \in M_{m,p}(\mathbf{F})$ and $B \in M_{n,q}(\mathbf{F})$. Let a_k be the kth column of A and let b_k be the kth column of B. Then

3. If $m = n$, then $A^T B = [a_i^T b_j]$: the i, j entry of $A^T B$ is the scalar $a_i^T b_j$.
4. If $p = q$, then $AB^T = \sum_{k=1}^{p} a_k b_k^T$: each summand is an m-by-n matrix, the *outer product* of a_k and b_k.

0.2.7 Column space and row space of a matrix. The range of $A \in M_{m,n}(\mathbf{F})$ is also called its *column space* because Ax is a linear combination of the columns of A for any $x \in \mathbf{F}^n$ (the entries of x are the coefficients in the linear combination); range A is the span of the columns of A. Analogously, $\{y^T A : y \in \mathbf{F}^m\}$ is called the *row space* of A. If the column space of $A \in M_{m,n}(\mathbf{F})$ is contained in the column space of $B \in M_{m,k}(\mathbf{F})$, then there is some $X \in M_{k,n}(\mathbf{F})$ such that $A = BX$ (and conversely); the entries in column j of X tell how to express column j of A as a linear combination of the columns of B.

If $A \in M_{m,n}(\mathbf{F})$ and $B \in M_{m,q}(\mathbf{F})$, then

$$\text{range } A + \text{range } B = \text{range} \begin{bmatrix} A & B \end{bmatrix} \tag{0.2.7.1}$$

If $A \in M_{m,n}(\mathbf{F})$ and $B \in M_{p,n}(\mathbf{F})$, then

$$\text{nullspace } A \cap \text{nullspace } B = \text{nullspace} \begin{bmatrix} A \\ B \end{bmatrix} \tag{0.2.7.2}$$

0.2.8 The all-ones matrix and vector. In $\mathbf{F}^n$, every entry of the vector $e = e_1 + \cdots + e_n$ is 1. Every entry of the matrix $J_n = ee^T$ is 1.

0.3 Determinants

Often in mathematics, it is useful to summarize a multivariate phenomenon with a single number, and the determinant function is an example of this. Its domain is $M_n(\mathbf{F})$ (square matrices only), and it may be presented in several different ways. We denote the determinant of $A \in M_n(\mathbf{F})$ by det A.

0.3.1 Laplace expansion by minors along a row or column. The determinant may be defined inductively for $A = [a_{ij}] \in M_n(\mathbf{F})$ in the following way. Assume that the determinant is defined over $M_{n-1}(\mathbf{F})$ and let $A_{ij} \in M_{n-1}(\mathbf{F})$ denote the submatrix of $A \in M_n(\mathbf{F})$ obtained by deleting row i and column j of A. Then, for any $i, j \in \{1, \ldots, n\}$, we have

$$\det A = \sum_{k=1}^{n} (-1)^{i+k} a_{ik} \det A_{ik} = \sum_{k=1}^{n} (-1)^{k+j} a_{kj} \det A_{kj} \tag{0.3.1.1}$$

The first sum is the *Laplace expansion by minors along row i*; the second sum is the *Laplace expansion by minors along column j*. This inductive presentation begins by defining the determinant of a 1-by-1 matrix to be the value of the single entry. Thus,

$$\det [\, a_{11} \,] = a_{11}$$

$$\det \begin{bmatrix} a_{11} & a_{12} \\ a_{21} & a_{22} \end{bmatrix} = a_{11}a_{22} - a_{12}a_{21}$$

$$\det \begin{bmatrix} a_{11} & a_{12} & a_{13} \\ a_{21} & a_{22} & a_{23} \\ a_{31} & a_{32} & a_{33} \end{bmatrix} = a_{11}a_{22}a_{33} + a_{12}a_{23}a_{31} + a_{13}a_{21}a_{32}$$
$$- a_{11}a_{23}a_{32} - a_{12}a_{21}a_{33} - a_{13}a_{22}a_{31}$$

and so on. Notice that $\det A^T = \det A$, $\det A^* = \overline{\det A}$ if $A \in M_n(\mathbf{C})$, and $\det I = 1$.

0.3.2 Alternating sums and permutations. A *permutation* of $\{1, \ldots, n\}$ is a one-to-one function $\sigma : \{1, \ldots, n\} \to \{1, \ldots, n\}$. The *identity permutation* satisfies $\sigma(i) = i$ for each $i = 1, \ldots, n$. There are $n!$ distinct permutations of $\{1, \ldots, n\}$, and the collection of all such permutations forms a group under composition of functions.

Consistent with the low-dimensional examples in (0.3.1), for $A = [a_{ij}] \in M_n(\mathbf{F})$ we have the alternative presentation

$$\det A = \sum_\sigma \left(\operatorname{sgn} \sigma \prod_{i=1}^n a_{i\sigma(i)} \right) \qquad (0.3.2.1)$$

in which the sum is over all $n!$ permutations of $\{1, \ldots, n\}$ and $\operatorname{sgn} \sigma$, the "sign" or "signum" of a permutation σ, is $+1$ or -1 according to whether the minimum number of transpositions (pairwise interchanges) necessary to achieve it starting from $\{1, \ldots, n\}$ is even or odd. We say that a permutation σ is *even* if $\operatorname{sgn} \sigma = +1$; σ is *odd* if $\operatorname{sgn} \sigma = -1$.

If $\operatorname{sgn} \sigma$ in (0.3.2.1) is replaced by certain other functions of σ, one obtains *generalized matrix functions* in place of $\det A$. For example, the *permanent* of A, denoted by per A, is obtained by replacing $\operatorname{sgn} \sigma$ by the function that is identically $+1$.

0.3.3 Elementary row and column operations. Three simple and fundamental operations on rows or columns, called *elementary row and column operations*, can be used to transform a matrix (square or not) into a simple form that facilitates such tasks as solving linear equations, determining rank, and calculating determinants and inverses of square matrices. We focus on *row operations*, which are implemented by matrices that act on the left. *Column operations* are defined and used in a similar fashion; the matrices that implement them act on the right.

Type 1: Interchange of two rows.

For $i \neq j$, interchange of rows i and j of A results from left multiplication of A by

$$
\begin{bmatrix}
1 & & & & & & & & \\
 & \ddots & & & & & & & \\
 & & 0 & 1 & \cdots & & 1 & & \\
 & & \vdots & & \ddots & & \vdots & & \\
 & & 1 & \cdots & \cdots & \cdots & 0 & & \\
 & & & & & & & 1 & \\
 & & & & & & & & \ddots \\
 & & & & & & & & & 1
\end{bmatrix}
$$

The two off-diagonal 1s are in the i, j and j, i positions, the two diagonal 0s are in positions i, i and j, j, and all unspecified entries are 0.

Type 2: Multiplication of a row by a nonzero scalar.

Multiplication of row i of A by a nonzero scalar c results from left multiplication of A by

$$
\begin{bmatrix}
1 & & & & & \\
 & \ddots & & & & \\
 & & 1 & & & \\
 & & & c & & \\
 & & & & 1 & \\
 & & & & & \ddots \\
 & & & & & & 1
\end{bmatrix}
$$

The i, i entry is c, all other main diagonal entries are 1, and all unspecified entries are 0.

Type 3: Addition of a scalar multiple of one row to another row.

For $i \neq j$, addition of c times row i of A to row j of A results from left multiplication of A by

$$
\begin{bmatrix}
1 & & & & & \\
 & \ddots & & & & \\
 & & 1 & & & \\
 & & & 1 & & \\
 & & c & & \ddots & \\
 & & & & & 1
\end{bmatrix}
$$

The j, i entry is c, all main diagonal entries are 1, and all unspecified entries are 0. The displayed matrix illustrates the case in which $j > i$.

The matrices of each of the three elementary row (or column) operations are just the result of applying the respective operation to the identity matrix I (on the left for a row operation; on the right for a column operation). The effect of a type 1 operation on the determinant is to multiply it by -1; the effect of a type 2 operation is to multiply it by the nonzero scalar c; a type 3 operation does not change the determinant. The determinant of a square matrix with a zero row is zero. The determinant of a square matrix is zero if and only if some subset of the rows of the matrix is linearly dependent.

0.3.4 Reduced row echelon form. To each $A = [a_{ij}] \in M_{m,n}(\mathbf{F})$ there corresponds a (unique) canonical form in $M_{m,n}(\mathbf{F})$, the *reduced row echelon form*, also known as the *Hermite normal form*.

If a row of A is nonzero, its *leading entry* is its first nonzero entry. The defining specifications of the RREF are as follows:

(a) Any zero rows occur at the bottom of the matrix.
(b) The leading entry of any nonzero row is a 1.
(c) All other entries in the column of a leading entry are zero.
(d) The leading entries occur in a stairstep pattern, left to right; that is, if row i is nonzero and a_{ik} is its leading entry, then either $i = m$, or row $i + 1$ is zero, or the leading entry in row $i + 1$ is $a_{i+1,\ell}$, in which $\ell > k$.

For example,

$$\begin{bmatrix} 0 & 1 & -1 & 0 & 0 & 2 \\ 0 & 0 & 0 & 1 & 0 & \pi \\ 0 & 0 & 0 & 0 & 1 & 4 \\ 0 & 0 & 0 & 0 & 0 & 0 \end{bmatrix}$$

is in RREF.

If $R \in M_{m,n}(\mathbf{F})$ is the RREF of A, then $R = EA$, in which the nonsingular matrix $E \in M_m(\mathbf{F})$ is a product of type 1, type 2, and type 3 elementary matrices corresponding to the sequence of elementary row operations performed to reduce A to RREF.

The determinant of $A \in M_n(\mathbf{F})$ is nonzero if and only if its RREF is I_n. The value of $\det A$ may be calculated by recording the effects on the determinant of each of the elementary operations that lead to the RREF.

For the system of linear equations $Ax = b$, with $A \in M_{m,n}(\mathbf{F})$ and $b \in \mathbf{F}^m$ given and $x \in \mathbf{F}^n$ unknown, the set of solutions is unchanged if the same sequence of elementary row operations is performed on both A and b. The solutions of $Ax = b$ are revealed by inspection of the RREF of $[A \; b]$. Since the RREF is unique, for given $A_1, A_2 \in M_{m,n}$ and given $b_1, b_2 \in \mathbf{F}^m$, consistent systems of linear equations $A_1 x = b_1$ and $A_2 x = b_2$ have the same set of solutions if and only if $[A_1 \; b_1]$ and $[A_2 \; b_2]$ have the same RREF.

0.3.5 Multiplicativity. A key property of the determinant function is that it is multiplicative: For $A, B \in M_n(\mathbf{F})$

$$\det AB = \det A \det B$$

This may be proved using elementary operations that row-reduce both A and B.

0.3.6 Functional characterization of the determinant. If we think of the determinant as a function of each row (or column) of a matrix separately with the others fixed, the Laplace expansion (0.3.1.1) reveals that the determinant is a linear function of the entries in any one given row (column). We summarize this property by saying that the function $A \rightarrow \det A$ is *multilinear* in the rows (columns) of A.

The determinant function $A \rightarrow \det A$ is the unique function $f : M_n(\mathbf{F}) \rightarrow \mathbf{F}$ that is

(a) Multilinear in the rows of its argument
(b) Alternating: any type 1 operation on A changes the sign of $f(A)$
(c) Normalized: $f(I) = 1$

The permanent function is also multilinear (as are other generalized matrix functions), and it is normalized, but it is not alternating.

0.4 Rank

0.4.1 Definition. If $A \in M_{m,n}(\mathbf{F})$, rank $A = \dim \text{range } A$ is the length of a longest linearly independent list of columns of A. There can be more than one linearly independent list of columns whose length equals the rank. It is a remarkable fact that rank A^T = rank A. Therefore, an equivalent definition of rank is the length of a longest linearly independent list of rows of A: row rank = column rank.

0.4.2 Rank and linear systems. Let $A \in M_{m,n}(\mathbf{F})$ and $b \in \mathbf{F}^m$ be given. The linear system $Ax = b$ may have no solution, exactly one solution, or infinitely many solutions; these are the only possibilities. If there is at least one solution, the linear system is *consistent*; if there is no solution, the linear system is *inconsistent*. The linear system $Ax = b$ is consistent if and only if rank$[A\ b]$ = rank A. The matrix $[A\ b] \in M_{m,n+1}(\mathbf{F})$ is the *augmented matrix*. To say that the augmented matrix and the *coefficient matrix* A of a linear system have the same rank is just to say that b is a linear combination of the columns of A. In this case, appending b to the columns of A does not increase the rank. A *solution* of the linear system $Ax = b$ is a vector x whose entries are the coefficients in a representation of b as a linear combination of the columns of A.

0.4.3 RREF and rank. Elementary operations do not change the rank of a matrix, and thus rank A is the same as the rank of the RREF of A, which is just the number of nonzero rows in the RREF. As a practical matter, however, numerical calculation of the rank by calculation of the RREF is unwise. Round-off errors in intermediate numerical calculations can make zero rows of the RREF appear to be nonzero, thereby affecting perception of the rank.

0.4.4 Characterizations of rank. The following statements about a given matrix $A \in M_{m,n}(\mathbf{F})$ are equivalent; each can be useful in a different context. Note that in (b) and (c) the key issue is linear independence of *lists* of columns or rows of a matrix:

(a) rank $A = k$.
(b) k, and no more than k, rows of A are linearly independent.
(c) k, and no more than k, columns of A are linearly independent.
(d) Some k-by-k submatrix of A has nonzero determinant, and every $(k+1)$-by-$(k+1)$ submatrix of A has zero determinant.
(e) $\dim (\text{range } A) = k$.

(f) There are k, but no more than k, linearly independent vectors $b_1, \ldots, b_k$ such that the linear system $Ax = b_j$ is consistent for each $j = 1, \ldots, k$.

(g) $k = n - \dim(\text{nullspace } A)$ (the *rank-nullity theorem*).

(h) $k = \min\{p : A = XY^T \text{ for some } X \in M_{m,p}(\mathbf{F}), Y \in M_{n,p}(\mathbf{F})\}$.

(i) $k = \min\{p : A = x_1 y_1^T + \cdots + x_p y_p^T\}$ for some $x_1, \ldots, x_p \in \mathbf{F}^m$, $y_1, \ldots, y_p \in \mathbf{F}^n$.

0.4.5 Rank inequalities. Some fundamental inequalities involving rank are:

(a) If $A \in M_{m,n}(\mathbf{F})$, then rank $A \le \min\{m, n\}$.

(b) If one or more rows and/or columns are deleted from a matrix, the rank of the resulting submatrix is not greater than the rank of the original matrix.

(c) Sylvester inequality: If $A \in M_{m,k}(\mathbf{F})$ and $B \in M_{k,n}(\mathbf{F})$, then

$$(\text{rank } A + \text{rank } B) - k \le \text{rank } AB \le \min\{\text{rank } A, \text{rank } B\}$$

(d) The rank-sum inequality: If $A, B \in M_{m,n}(\mathbf{F})$, then

$$|\text{rank } A - \text{rank } B| \le \text{rank}(A + B) \le \text{rank } A + \text{rank } B \qquad (0.4.5.1)$$

with equality in the second inequality if and only if $(\text{range } A) \cap (\text{range } B) = \{0\}$ and $(\text{range } A^T) \cap (\text{range } B^T) = \{0\}$. If rank $B = 1$ then

$$|\text{rank}(A + B) - \text{rank } A| \le 1 \qquad (0.4.5.2)$$

in particular, changing one entry of a matrix can change its rank by at most 1.

(e) Frobenius inequality: If $A \in M_{m,k}(\mathbf{F})$, $B \in M_{k,p}(\mathbf{F})$, and $C \in M_{p,n}(\mathbf{F})$, then

$$\text{rank } AB + \text{rank } BC \le \text{rank } B + \text{rank } ABC$$

with equality if and only if there are matrices X and Y such that $B = BCX + YAB$.

0.4.6 Rank equalities. Some fundamental equalities involving rank are:

(a) If $A \in M_{m,n}(\mathbf{C})$, then rank $A^* = \text{rank } A^T = \text{rank } \bar{A} = \text{rank } A$.

(b) If $A \in M_m(\mathbf{F})$ and $C \in M_n(\mathbf{F})$ are nonsingular and $B \in M_{m,n}(\mathbf{F})$, then rank $AB = \text{rank } B = \text{rank } BC = \text{rank } ABC$; that is, left or right multiplication by a non-singular matrix leaves rank unchanged.

(c) If $A, B \in M_{m,n}(\mathbf{F})$, then rank $A = \text{rank } B$ if and only if there exist a nonsingular $X \in M_m(\mathbf{F})$ and a nonsingular $Y \in M_n(\mathbf{F})$ such that $B = XAY$.

(d) If $A \in M_{m,n}(\mathbf{C})$, then rank $A^*A = \text{rank } A$.

(e) *Full-rank factorization*: If $A \in M_{m,n}(\mathbf{F})$, then rank $A = k$ if and only if $A = XY^T$ for some $X \in M_{m,k}(\mathbf{F})$ and $Y \in M_{n,k}(\mathbf{F})$ that each have independent columns. The equivalent factorization $A = XBY^T$ for some nonsingular $B \in M_k(\mathbf{F})$ can also be useful. In particular, rank $A = 1$ if and only if $A = xy^T$ for some nonzero vectors $x \in \mathbf{F}^m$ and $y \in \mathbf{F}^n$.

(f) If $A \in M_{m,n}(\mathbf{F})$, then rank $A = k$ if and only if there exist nonsingular matrices $S \in M_m(\mathbf{F})$ and $T \in M_n(\mathbf{F})$ such that $A = S \begin{bmatrix} I_k & 0 \\ 0 & 0 \end{bmatrix} T$.

(g) Let $A \in M_{m,n}(\mathbf{F})$. If $X \in M_{n,k}(\mathbf{F})$ and $Y \in M_{m,k}(\mathbf{F})$, and if $W = Y^T A X$ is nonsingular, then

$$\text{rank}(A - AXW^{-1}Y^T A) = \text{rank } A - \text{rank } AXW^{-1}Y^T A \qquad (0.4.6.1)$$

When $k = 1$, this is *Wedderburn's rank-one reduction formula*: If $x \in \mathbf{F}^n$ and $y \in \mathbf{F}^m$, and if $\omega = y^T Ax \neq 0$, then

$$\text{rank}\left(A - \omega^{-1}Axy^T A\right) = \text{rank } A - 1 \qquad (0.4.6.2)$$

Conversely, if $\sigma \in \mathbf{F}$, $u \in \mathbf{F}^n$, $v \in \mathbf{F}^m$, and $\text{rank}\left(A - \sigma u v^T\right) < \text{rank } A$, then $\text{rank}\left(A - \sigma u v^T\right) = \text{rank } A - 1$ and there are $x \in \mathbf{F}^n$ and $y \in \mathbf{F}^m$ such that $u = Ax$, $v = A^T y$, $y^T Ax \neq 0$, and $\sigma = (y^T Ax)^{-1}$.

0.5 Nonsingularity

A linear transformation or matrix is said to be *nonsingular* if it produces the output 0 only for the input 0. Otherwise, it is *singular*. If $A \in M_{m,n}(\mathbf{F})$ and $m < n$, then A is necessarily singular. An $A \in M_n(\mathbf{F})$ is *invertible* if there is a matrix $A^{-1} \in M_n(\mathbf{F})$ (the *inverse* of A) such that $A^{-1}A = I$. If $A \in M_n$ and $A^{-1}A = I$, then $AA^{-1} = I$; that is, A^{-1} is a *left inverse* if and only if it is a *right inverse*; A^{-1} is unique whenever it exists.

It is useful to be able to call on a variety of criteria for a square matrix to be nonsingular. The following are equivalent for a given $A \in M_n(\mathbf{F})$:

(a) A is nonsingular.
(b) A^{-1} exists.
(c) $\text{rank } A = n$.
(d) The rows of A are linearly independent.
(e) The columns of A are linearly independent.
(f) $\det A \neq 0$.
(g) The dimension of the range of A is n.
(h) The dimension of the null space of A is 0.
(i) $Ax = b$ is consistent for each $b \in \mathbf{F}^n$.
(j) If $Ax = b$ is consistent, then the solution is unique.
(k) $Ax = b$ has a unique solution for each $b \in \mathbf{F}^n$.
(l) The only solution to $Ax = 0$ is $x = 0$.
(m) 0 is not an eigenvalue of A (see Chapter 1).

The conditions (g) and (h) are equivalent for a linear transformation $T : V \to V$ on a finite dimensional vector space V; that is, $Tx = y$ has a solution x for every $y \in V$ if and only if the only x such that $Tx = 0$ is $x = 0$ if and only if $Tx = y$ has a *unique* solution x for every $y \in V$.

The nonsingular matrices in $M_n(\mathbf{F})$ form a group, the *general linear group*, often denoted by $GL(n, \mathbf{F})$.

If $A \in M_n(\mathbf{F})$ is nonsingular, then $((A^{-1})^T A^T)^T = A(A^{-1}) = I$, so $(A^{-1})^T A^T = I$, which means that $(A^{-1})^T = (A^T)^{-1}$. It is convenient to write either $(A^{-1})^T$ or $(A^T)^{-1}$ as A^{-T}. If $A \in M_n(\mathbf{C})$ is nonsingular, then $(A^{-1})^* = (A^*)^{-1}$, and we may safely write either as A^{-*}.

0.6 The Euclidean inner product and norm

0.6.1 Definitions. The scalar $\langle x, y \rangle = y^*x$ is the *Euclidean inner product* (*standard inner product, usual inner product, scalar product, dot product*) of $x, y \in \mathbf{C}^n$. The *Euclidean norm* (*usual norm, Euclidean length*) function on $\mathbf{C}^n$ is the real-valued function $\|x\|_2 = \langle x, x \rangle^{1/2} = (x^*x)^{1/2}$; two important properties of this function are that $\|x\|_2 > 0$ for all nonzero $x \in \mathbf{C}^n$ and $\|\alpha x\|_2 = |\alpha| \, \|x\|_2$ for all $x \in \mathbf{C}^n$ and all $\alpha \in \mathbf{C}$.

The function $\langle \cdot, \cdot \rangle : \mathbf{C}^n \times \mathbf{C}^n \to \mathbf{C}$ is linear in the first argument and *conjugate linear* in the second; that is, $\langle \alpha x_1 + \beta x_2, y \rangle = \alpha \langle x_1, y \rangle + \beta \langle x_2, y \rangle$ and $\langle x, \alpha y_1 + \beta y_2 \rangle = \bar{\alpha} \langle x, y_1 \rangle + \bar{\beta} \langle x, y_2 \rangle$ for all $\alpha, \beta \in \mathbf{C}$ and $y_1, y_2 \in \mathbf{C}^n$. If V is a real or complex vector space and $f : V \times V \to \mathbf{C}$ is a function that is linear in its first argument and conjugate linear in its second argument, we say that f is *sesquilinear* on V; f is a *semi-inner product* on V if it is sesquilinear on V and $f(x, x) \geq 0$ for every $x \in V$; f is an *inner product* on V if it is sesquilinear on V and $f(x, x) > 0$ for every nonzero $x \in V$. An *inner product space* is a pair (V, f) in which V is a real or complex vector space and f is an inner product on V.

0.6.2 Orthogonality and orthonormality. Two vectors $x, y \in \mathbf{C}^n$ are *orthogonal* if $\langle x, y \rangle = 0$. In $\mathbf{R}^2$ and $\mathbf{R}^3$, "orthogonal" has the conventional geometric interpretation of "perpendicular." A list of vectors $x_1, \ldots, x_m \in \mathbf{C}^n$ is said to be *orthogonal* if $\langle x_i, x_j \rangle = 0$ for all distinct $i, j \in \{1, \ldots, m\}$. An orthogonal list of nonzero vectors is linearly independent. A vector whose Euclidean norm is 1 is said to be *normalized* (a *unit vector*). For any nonzero $x \in \mathbf{C}^n$, $x/\|x\|_2$ is a unit vector. An orthogonal list of vectors is an *orthonormal* list if each of its elements is a unit vector. An orthonormal list of vectors is linearly independent. Each of these concepts has a straightforward generalization to the context of an inner product space.

0.6.3 The Cauchy–Schwarz inequality. The *Cauchy–Schwarz inequality* states that

$$|\langle x, y \rangle| \leq \|x\|_2 \, \|y\|_2$$

for all $x, y \in \mathbf{C}^n$, with equality if and only if one of the vectors is a scalar multiple of the other. The *angle* θ between two nonzero real vectors $x, y \in \mathbf{R}^n$ is defined by

$$\cos \theta = \frac{\langle x, y \rangle}{\|x\|_2 \, \|y\|_2}, \quad 0 \leq \theta \leq \pi \tag{0.6.3.1}$$

0.6.4 Gram–Schmidt orthonormalization. Any finite independent list of vectors in an inner product space may be replaced by an orthonormal list with the same span. This replacement may be carried out in many ways, but there is a systematic way to do so that has a useful special property. The *Gram–Schmidt process* starts with a list of vectors $x_1, \ldots, x_n$ and (if the given list is linearly independent) produces an orthonormal list of vectors $z_1, \ldots, z_n$ such that $\mathrm{span}\{z_1, \ldots, z_k\} = \mathrm{span}\{x_1, \ldots, x_k\}$ for each $k = 1, \ldots, n$. The vectors z_i may be calculated in turn as follows: Let $y_1 = x_1$ and normalize it: $z_1 = y_1/\|y_1\|_2$. Let $y_2 = x_2 - \langle x_2, z_1 \rangle z_1$ (y_2 is orthogonal to z_1) and normalize it: $z_2 = y_2/\|y_2\|_2$. Once $z_1, \ldots, z_{k-1}$ have been determined, the vector

$$y_k = x_k - \langle x_k, z_{k-1} \rangle z_{k-1} - \langle x_k, z_{k-2} \rangle z_{k-2} - \cdots - \langle x_k, z_1 \rangle z_1$$

is orthogonal to $z_1, \ldots, z_{k-1}$; normalize it: $z_k = y_k / \|y_k\|_2$. Continue until $k = n$. If we denote $Z = [z_1 \; \cdots \; z_n]$ and $X = [x_1 \; \cdots \; x_n]$, the Gram–Schmidt process gives a factorization $X = ZR$, in which the square matrix $R = [r_{ij}]$ is nonsingular and upper triangular; that is, $r_{ij} = 0$ whenever $i > j$.

If the vectors $x_1, \ldots, x_k$ are orthonormal and the vectors $x_1, \ldots, x_k, x_{k+1}, \ldots, x_n$ are linearly independent, applying the Gram–Schmidt process to the latter list produces the list $x_1, \ldots, x_k, z_{k+1}, \ldots, z_n$ of orthonormal vectors.

The Gram–Schmidt process may be applied to any finite list of vectors, independent or not. If $x_1, \ldots, x_n$ are linearly dependent, the Gram–Schmidt process produces a vector $y_k = 0$ for the least value of k for which x_k is a linear combination of $x_1, \ldots, x_{k-1}$.

0.6.5 Orthonormal bases. An *orthonormal basis* of an inner product space is a basis whose elements constitute an orthonormal list. Since any finite ordered basis may be transformed with the Gram–Schmidt process to an orthonormal basis, any finite-dimensional inner product space has an orthonormal basis, and any orthonormal list may be extended to an orthonormal basis. Such a basis is pleasant to work with, since the cross terms in inner product calculations all vanish.

0.6.6 Orthogonal complements. Given any set $S \subset \mathbf{C}^n$, its *orthogonal complement* is the set $S^\perp = \{x \in \mathbf{C}^n : x^* y = 0 \text{ for all } y \in S\}$ if S is nonempty; if S is empty, then $S^\perp = \mathbf{C}^n$. In either case, $S^\perp = (\text{span } S)^\perp$. Even if S is not a subspace, $S^\perp$ is always a subspace. We have $(S^\perp)^\perp = \text{span } S$, and $(S^\perp)^\perp = S$ if S is a subspace. It is always the case that $\dim S^\perp + \dim(S^\perp)^\perp = n$. If S_1 and S_2 are subspaces, then $(S_1 + S_2)^\perp = S_1^\perp \cap S_2^\perp$.

For a given $A \in M_{m,n}$, range A is the orthogonal complement of nullspace A^*. Therefore, for a given $b \in \mathbf{C}^m$, the linear system $Ax = b$ has a solution (not necessarily unique) if and only if $b^* z = 0$ for every $z \in \mathbf{C}^m$ such that $A^* z = 0$. This equivalence is sometimes stated as the *Fredholm alternative* (*theorem of the alternative*) — exactly one of the following two statements is true: *Either* (1) $Ax = b$ has a solution *or* (2) $A^* y = 0$ has a solution such that $y^* b \neq 0$.

If $A \in M_{m,n}$ and $B \in M_{m,q}$, if $X \in M_{m,r}$ and $Y \in M_{m,s}$, and if range $X = $ nullspace A^* and range $Y = $ nullspace B^*, then we have the following companion to (0.2.7.1) and (0.2.7.2):

$$\text{range } A \cap \text{range } B = \text{nullspace} \begin{bmatrix} X^* \\ Y^* \end{bmatrix} \tag{0.6.6.1}$$

0.7 Partitioned sets and matrices

A *partition* of a set $\mathcal{S}$ is a collection of subsets of $\mathcal{S}$ such that each element of $\mathcal{S}$ is a member of one and only one of the subsets. For example, a partition of the set $\{1, 2, \ldots, n\}$ is a collection of subsets $\alpha_1, \ldots, \alpha_t$ (called *index sets*) such that each integer between 1 and n is in one and only one of the index sets. A *sequential partition* of $\{1, 2, \ldots, n\}$ is a partition in which the index sets have the special form $\alpha_1 = \{1, \ldots, i_1\}, \alpha_2 = \{i_1 + 1, \ldots, i_2\}, \ldots, \alpha_t = \{i_{t-1} + 1, \ldots, n\}$.

A *partition* of a matrix is a decomposition of the matrix into submatrices such that each entry of the original matrix is in one and only one of the submatrices. Partitioning of matrices is often a convenient device for perception of useful structure. For example, partitioning $B = [b_1 \ldots b_n] \in M_n(\mathbf{F})$ according to its columns reveals the presentation $AB = [Ab_1 \ldots Ab_n]$ of the matrix product, partitioned according to the columns of AB.

0.7.1 Submatrices. Let $A \in M_{m,n}(\mathbf{F})$. For index sets $\alpha \subseteq \{1, \ldots, m\}$ and $\beta \subseteq \{1, \ldots, n\}$, we denote by $A[\alpha, \beta]$ the (sub)matrix of entries that lie in the rows of A indexed by α and the columns indexed by β. For example,

$$\begin{bmatrix} 1 & 2 & 3 \\ 4 & 5 & 6 \\ 7 & 8 & 9 \end{bmatrix} [\{1,3\}, \{1,2,3\}] = \begin{bmatrix} 1 & 2 & 3 \\ 7 & 8 & 9 \end{bmatrix}$$

If $\alpha = \beta$, the submatrix $A[\alpha] = A[\alpha, \alpha]$ is a *principal submatrix* of A. An n-by-n matrix has $\binom{n}{k}$ distinct principal submatrices of size k.

For $A \in M_n(\mathbf{F})$ and $k \in \{1, \ldots, n\}$, $A[\{1, \ldots, k\}]$ is a *leading principal submatrix* and $A[\{k, \ldots, n\}]$ is a *trailing principal submatrix*.

It is often convenient to indicate a submatrix or principal submatrix via deletion, rather than inclusion, of rows or columns. This may be accomplished by complementing the index sets. Let $\alpha^c = \{1, \ldots, m\} \setminus \alpha$ and $\beta^c = \{1, \ldots, n\} \setminus \beta$ denote the index sets complementary to α and β, respectively. Then $A[\alpha^c, \beta^c]$ is the submatrix obtained by *deleting* the rows indexed by α and the columns indexed by β. For example, the submatrix $A[\alpha, \varnothing^c]$ contains the rows of A indexed by α; $A[\varnothing^c, \beta]$ contains the columns of A indexed by β.

The determinant of an r-by-r submatrix of A is called a *minor*; if we wish to indicate the size of the submatrix, we call its determinant a *minor of size r*. If the r-by-r submatrix is a *principal* submatrix, then its determinant is a *principal minor (of size r)*; if the submatrix is a leading principal matrix, then its determinant is a *leading principal minor*; if the submatrix is a trailing principal submatrix, then its determinant is a *trailing principal minor*. By convention, the empty principal minor is 1; that is, $\det A[\varnothing] = 1$.

A signed minor, such as those appearing in the Laplace expansion (0.3.1.1) $[(-1)^{i+j} \det A_{ij}]$ is called a *cofactor*; if we wish to indicate the size of the submatrix, we call its signed determinant a *cofactor of size r*.

0.7.2 Partitions, block matrices, and multiplication. If $\alpha_1, \ldots, \alpha_t$ constitute a partition of $\{1, \ldots, m\}$ and $\beta_1, \ldots, \beta_s$ constitute a partition of $\{1, \ldots, n\}$, then the matrices $A[\alpha_i, \beta_j]$ form a partition of the matrix $A \in M_{m,n}(\mathbf{F})$, $1 \le i \le t, 1 \le j \le s$. If $A \in M_{m,n}(\mathbf{F})$ and $B \in M_{n,p}(\mathbf{F})$ are partitioned so that the two partitions of $\{1, \ldots, n\}$ coincide, the two matrix partitions are said to be *conformal*. In this event,

$$(AB)[\alpha_i, \gamma_j] = \sum_{k=1}^{s} A[\alpha_i, \beta_k] B[\beta_k, \gamma_j] \tag{0.7.2.1}$$

in which the respective collections of submatrices $A[\alpha_i, \beta_k]$ and $B[\beta_k, \gamma_j]$ are conformal partitions of A and B, respectively. The left-hand side of (0.7.2.1) is a submatrix of the product AB (calculated in the usual way), and each summand on the right-hand side is a standard matrix product. Thus, multiplication of conformally partitioned matrices mimics usual matrix multiplication. The sum of two partitioned matrices $A, B \in M_{m,n}(\mathbf{F})$ of the same size has a similarly pleasant representation if the partitions of their rows (respectively, of their columns) are the same:

$$(A + B)\left[\alpha_i, \beta_j\right] = A\left[\alpha_i, \beta_j\right] + B\left[\alpha_i, \beta_j\right]$$

If a matrix is partitioned by sequential partitions of its rows and columns, the resulting partitioned matrix is called a *block matrix*. For example, if the rows and columns of $A \in M_n(\mathbf{F})$ are partitioned by the same sequential partition $\alpha_1 = \{1, \ldots, k\}, \alpha_2 = \{k + 1, \ldots, n\}$, the resulting block matrix is

$$A = \begin{bmatrix} A[\alpha_1, \alpha_1] & A[\alpha_1, \alpha_2] \\ A[\alpha_2, \alpha_1] & A[\alpha_2, \alpha_2] \end{bmatrix} = \begin{bmatrix} A_{11} & A_{12} \\ A_{21} & A_{22} \end{bmatrix}$$

in which the *blocks* are $A_{ij} = A[\alpha_i, \alpha_j]$. Computations with block matrices are employed throughout the book; 2-by-2 block matrices are the most important and useful.

0.7.3 The inverse of a partitioned matrix. It can be useful to know the corresponding blocks in the inverse of a partitioned nonsingular matrix A, that is, to present the inverse of a partitioned matrix in conformally partitioned form. This may be done in a variety of apparently different, but equivalent, ways — assuming that certain submatrices of $A \in M_n(\mathbf{F})$ and A^{-1} are also nonsingular. For simplicity, let A be partitioned as a 2-by-2 block matrix

$$A = \begin{bmatrix} A_{11} & A_{12} \\ A_{21} & A_{22} \end{bmatrix}$$

with $A_{ii} \in M_{n_i}(\mathbf{F})$, $i = 1, 2$, and $n_1 + n_2 = n$. A useful expression for the correspondingly partitioned presentation of A^{-1} is

$$\begin{bmatrix} \left(A_{11} - A_{12}A_{22}^{-1}A_{21}\right)^{-1} & A_{11}^{-1}A_{12}\left(A_{21}A_{11}^{-1}A_{12} - A_{22}\right)^{-1} \\ A_{22}^{-1}A_{21}\left(A_{12}A_{22}^{-1}A_{21} - A_{11}\right)^{-1} & \left(A_{22} - A_{21}A_{11}^{-1}A_{12}\right)^{-1} \end{bmatrix} \quad (0.7.3.1)$$

assuming that all the relevant inverses exist. This expression for A^{-1} may be verified by doing a partitioned multiplication by A and then simplifying. In general index set notation, we may write

$$A^{-1}[\alpha] = \left(A[\alpha] - A[\alpha, \alpha^c] A[\alpha^c]^{-1} A[\alpha^c, \alpha]\right)^{-1}$$

and

$$A^{-1}\left[\alpha, \alpha^c\right] = A[\alpha]^{-1} A\left[\alpha, \alpha^c\right] \left(A\left[\alpha^c, \alpha\right] A[\alpha]^{-1} A\left[\alpha, \alpha^c\right] - A\left[\alpha^c\right]\right)^{-1}$$

$$= \left(A\left[\alpha, \alpha^c\right] A\left[\alpha^c\right]^{-1} A\left[\alpha^c, \alpha\right] - A[\alpha]\right)^{-1} A\left[\alpha, \alpha^c\right] A\left[\alpha^c\right]^{-1}$$

again assuming that the relevant inverses exist. There is an intimate relationship between these representations and the Schur complement; see (0.8.5). Notice that $A^{-1}[\alpha]$ is a submatrix of A^{-1}, while $A[\alpha]^{-1}$ is the inverse of a submatrix of A; these two objects are not, in general, the same.

0.7.4 The Sherman–Morrison–Woodbury formula.

Suppose that a nonsingular matrix $A \in M_n(\mathbf{F})$ has a known inverse A^{-1} and consider $B = A + XRY$, in which X is n-by-r, Y is r-by-n, and R is r-by-r and nonsingular. If B and $R^{-1} + YA^{-1}X$ are nonsingular, then

$$B^{-1} = A^{-1} - A^{-1}X(R^{-1} + YA^{-1}X)^{-1}YA^{-1} \qquad (0.7.4.1)$$

If r is much smaller than n, then R and $R^{-1} + YA^{-1}X$ may be much easier to invert than B. For example, if $x, y \in \mathbf{F}^n$ are nonzero vectors, $X = x$, $Y = y^T$, $y^T A^{-1}x \neq -1$, and $R = [1]$, then (0.7.4.1) becomes a formula for the inverse of a rank-1 adjustment to A:

$$\left(A + xy^T\right)^{-1} = A^{-1} - \left(1 + y^T A^{-1}x\right)^{-1} A^{-1}xy^T A^{-1} \qquad (0.7.4.2)$$

In particular, if $B = I + xy^T$ for $x, y \in \mathbf{F}^n$ and $y^T x \neq -1$, then $B^{-1} = I - (1 + y^T x)^{-1}xy^T$.

0.7.5 Complementary nullities.

Suppose that $A \in M_n(\mathbf{F})$ is nonsingular, let α and β be nonempty subsets of $\{1, \ldots, n\}$, and write $|\alpha| = r$ and $|\beta| = s$ for the cardinalities of α and β. The *law of complementary nullities* is

$$\text{nullity}\,(A\,[\alpha, \beta]) = \text{nullity}\,\left(A^{-1}\left[\beta^c, \alpha^c\right]\right) \qquad (0.7.5.1)$$

which is equivalent to the rank identity

$$\text{rank}\,(A\,[\alpha, \beta]) = \text{rank}\,\left(A^{-1}\left[\beta^c, \alpha^c\right]\right) + r + s - n \qquad (0.7.5.2)$$

Since we can permute rows and columns to place first the r rows indexed by α and the s columns indexed by β, it suffices to consider the presentations

$$A = \begin{bmatrix} A_{11} & A_{12} \\ A_{21} & A_{22} \end{bmatrix} \text{ and } A^{-1} = \begin{bmatrix} B_{11} & B_{12} \\ B_{21} & B_{22} \end{bmatrix}$$

in which A_{11} and B_{11}^T are r-by-s and A_{22} and B_{22}^T are $(n-r)$-by-$(n-s)$. Then (0.7.5.1) says that nullity $A_{11} = $ nullity B_{22}.

The underlying principle here is very simple. Suppose that the nullity of A_{11} is k. If $k \geq 1$, let the columns of $X \in M_{s,k}(\mathbf{F})$ be a basis for the null space of A_{11}. Since A is nonsingular,

$$A \begin{bmatrix} X \\ 0 \end{bmatrix} = \begin{bmatrix} A_{11}X \\ A_{21}X \end{bmatrix} = \begin{bmatrix} 0 \\ A_{21}X \end{bmatrix}$$

has full rank, so $A_{21}X$ has k independent columns. But

$$\begin{bmatrix} B_{12}(A_{21}X) \\ B_{22}(A_{21}X) \end{bmatrix} = A^{-1} \begin{bmatrix} 0 \\ A_{21}X \end{bmatrix} = A^{-1}A \begin{bmatrix} X \\ 0 \end{bmatrix} = \begin{bmatrix} X \\ 0 \end{bmatrix}$$

so $B_{22}(A_{21}X) = 0$ and hence nullity $B_{22} \geq k = $ nullity A_{11}, a statement that is trivially correct if $k = 0$. A similar argument starting with B_{22} shows that nullity $A_{11} \geq$ nullity B_{22}. For a different approach, see (3.5.P13).

Of course, (0.7.5.1) also tells us that nullity $A_{12} = $ nullity B_{12}, nullity $A_{21} = $ nullity B_{21}, and nullity $A_{22} = $ nullity B_{11}. If $r + s = n$, then rank $A_{11} = $ rank B_{22} and rank $A_{22} = $ rank B_{11}, while if $n = 2r = 2s$, then we also have rank $A_{12} = $ rank B_{12} and rank $A_{21} = $ rank B_{21}. Finally, (0.7.5.2) tells us that the rank of an r-by-s submatrix of an n-by-n nonsingular matrix is at least $r + s - n$.

0.7.6 Rank in a partitioned matrix and rank-principal matrices. Partition $A \in M_n(\mathbf{F})$ as

$$A = \begin{bmatrix} A_{11} & A_{12} \\ A_{21} & A_{22} \end{bmatrix}, \quad A_{11} \in M_r(\mathbf{F}), \mathbf{A}_{22} \in M_{n-r}(\mathbf{F})$$

If A_{11} is nonsingular, then of course rank $[\begin{smallmatrix} A_{11} & A_{12} \end{smallmatrix}] = r$ and rank $\begin{bmatrix} A_{11} \\ A_{21} \end{bmatrix} = r$. Remarkably, the converse is true: If $A_{11} \in M_r$, rank $A = r$, and

$$r = \text{rank}[A_{11} \ A_{12}] = \text{rank} \begin{bmatrix} A_{11} \\ A_{21} \end{bmatrix}, \text{ then } A_{11} \text{ is nonsingular} \qquad (0.7.6.1)$$

This follows from (0.4.6(c)): If A_{11} is singular, then rank $A_{11} = k < r$, and there are nonsingular $S, T \in M_r(\mathbf{F})$ such that

$$SA_{11}T = \begin{bmatrix} I_k & 0 \\ 0 & 0_{r-k} \end{bmatrix}$$

Therefore,

$$\hat{A} = \begin{bmatrix} S & 0 \\ 0 & I_{n-r} \end{bmatrix} A \begin{bmatrix} T & 0 \\ 0 & I_{n-r} \end{bmatrix} = \begin{bmatrix} \begin{bmatrix} I_k & 0 \\ 0 & 0_{r-k} \end{bmatrix} & SA_{12} \\ A_{21}T & A_{22} \end{bmatrix}$$

has rank r, as do its first block row and column. Because the rth row of the first block column of $\hat{A}$ is zero, there must be some column in SA_{12} whose rth entry is not zero, which means that $\hat{A}$ has at least $r + 1$ independent columns. This contradicts rank $\hat{A} = $ rank $A = r$, so A_{11} must be nonsingular.

Let $A \in M_{m,n}(\mathbf{F})$ and suppose that rank $A = r > 0$. Let $A = XY^T$ be a full-rank factorization with $X \in M_{m,r}(\mathbf{F})$ and $Y \in M_{n,r}(\mathbf{F})$; see (0.4.6(e)). Let $\alpha, \beta \subseteq \{1, \ldots, m\}$ and $\gamma, \delta \subseteq \{1, \ldots, n\}$ be index sets of cardinality r. Then $A[\alpha, \gamma] = X[\alpha, \varnothing^c]Y[\gamma, \varnothing^c]^T \in M_r(\mathbf{F})$, which is nonsingular whenever rank $X[\alpha, \varnothing^c] = $ rank $Y[\gamma, \varnothing^c] = r$. The multiplicativity property (0.3.5) ensures that

$$\det A[\alpha, \gamma] \det A[\beta, \delta] = \det A[\alpha, \delta] \det A[\beta, \gamma] \qquad (0.7.6.2)$$

Suppose that $A \in M_n(\mathbf{F})$ and rank $A = r$. We say that A is *rank principal* if it has a nonsingular r-by-r *principal* submatrix. It follows from (0.7.6.1) that if there is some index set $\alpha \subset \{1, \ldots, n\}$ such that

$$\text{rank } A = \text{rank } A\left[\alpha, \varnothing^c\right] = \text{rank } A\left[\varnothing^c, \alpha\right] \qquad (0.7.6.3)$$

(that is, if there are r linearly independent rows of A such that the corresponding r columns are linearly independent), then A is rank principal; moreover, $A[\alpha]$ is nonsingular.

If $A \in M_n(\mathbf{F})$ is symmetric or skew symmetric, or if $A \in M_n(\mathbf{C})$ is Hermitian or skew Hermitian, then rank $A[\alpha, \varnothing^c] = $ rank $A[\varnothing^c, \alpha]$ for every index set α, so A satisfies (0.7.6.3) and is therefore rank principal.

0.7.7 Commutativity, anticommutativity, and block diagonal matrices. Two matrices $A, B \in M_n(\mathbf{F})$ are said to *commute* if $AB = BA$. Commutativity is not typical, but one important instance is encountered frequently. Suppose that $\Lambda = [\Lambda_{ij}]_{i,j=1}^s \in M_n(\mathbf{F})$ is a block matrix in which $\Lambda_{ij} = 0$ if $i \neq j$; $\Lambda_{ii} = \lambda_i I_{n_i}$ for some $\lambda_i \in \mathbf{F}$ for each $i = 1, \ldots, s$; and $\lambda_i \neq \lambda_j$ if $i \neq j$. Partition $B = [B_{ij}]_{i,j=1}^s \in M_n(\mathbf{F})$ conformally with Λ. Then $\Lambda B = B\Lambda$ if and only if $\lambda_i B_{ij} = B_{ij}\lambda_j$ for each $i, j = 1, \ldots, s$, that is, $(\lambda_i - \lambda_j)B_{ij} = 0$ for each $i, j = 1, \ldots, s$. These identities are satisfied if and only if $B_{ij} = 0$ whenever $i \neq j$. Thus, Λ *commutes with B if and only if B is block diagonal conformal with Λ*; see (0.9.2).

Two matrices $A, B \in M_n(\mathbf{F})$ are said to *anticommute* if $AB = -BA$. For example, the matrices $\begin{bmatrix} 1 & 0 \\ 0 & -1 \end{bmatrix}$ and $\begin{bmatrix} 0 & 1 \\ 1 & 0 \end{bmatrix}$ anticommute.

0.7.8 The vec mapping. Partition a matrix $A \in M_{m,n}(\mathbf{F})$ according to its columns: $A = [a_1 \ \ldots \ a_n]$. The mapping $\text{vec} : M_{m,n}(\mathbf{F}) \to \mathbf{F}^{mn}$ is

$$\text{vec } A = [a_1^T \ \ldots \ a_n^T]^T$$

that is, vec A is the vector obtained by stacking the columns of A, left to right. The vec operator can be a convenient tool in problems involving matrix equations.

0.8 Determinants again

Some additional facts about and identities for the determinant are useful for reference.

0.8.1 Compound matrices. Let $A \in M_{m,n}(\mathbf{F})$. Let $\alpha \subseteq \{1, \ldots, m\}$ and $\beta \subseteq \{1, \ldots, n\}$ be index sets of cardinality $r \leq \min\{m, n\}$ elements. The $\binom{m}{r}$-by-$\binom{n}{r}$ matrix whose α, β entry is $\det A[\alpha, \beta]$ is called the rth *compound matrix* of A and is denoted by $C_r(A)$. In forming the rows and columns of $C_r(A)$, we arrange index sets lexicographically, that is, $\{1, 2, 4\}$ before $\{1, 2, 5\}$ before $\{1, 3, 4\}$, and so on. For example, if

$$A = \begin{bmatrix} 1 & 2 & 3 \\ 4 & 5 & 6 \\ 7 & 8 & 10 \end{bmatrix} \quad (0.8.1.0)$$

then $C_2(A) =$

$$\begin{bmatrix} \det \begin{bmatrix} 1 & 2 \\ 4 & 5 \end{bmatrix} & \det \begin{bmatrix} 1 & 3 \\ 4 & 6 \end{bmatrix} & \det \begin{bmatrix} 2 & 3 \\ 5 & 6 \end{bmatrix} \\ \det \begin{bmatrix} 1 & 2 \\ 7 & 8 \end{bmatrix} & \det \begin{bmatrix} 1 & 3 \\ 7 & 10 \end{bmatrix} & \det \begin{bmatrix} 2 & 3 \\ 8 & 10 \end{bmatrix} \\ \det \begin{bmatrix} 4 & 5 \\ 7 & 8 \end{bmatrix} & \det \begin{bmatrix} 4 & 6 \\ 7 & 10 \end{bmatrix} & \det \begin{bmatrix} 5 & 6 \\ 8 & 10 \end{bmatrix} \end{bmatrix} = \begin{bmatrix} -3 & -6 & -3 \\ -6 & -11 & -4 \\ -3 & -2 & 2 \end{bmatrix}$$

If $A \in M_{m,k}(\mathbf{F})$, $B \in M_{k,n}(\mathbf{F})$, and $r \leq \min\{m, k, n\}$, it follows from the Cauchy–Binet formula (0.8.7) that

$$C_r(AB) = C_r(A)C_r(B) \quad (0.8.1.1)$$

which is the *multiplicativity property* of the rth compound matrix.

We *define* $C_0(A) = 1$. We have $C_1(A) = A$; if $A \in M_n(\mathbf{F})$, then $C_n(A) = \det A$.

If $A \in M_{m,k}(\mathbf{F})$ and $t \in \mathbf{F}$, then $C_r(tA) = t^r C_r(A)$

If $1 \leq r \leq n$, then $C_r(I_n) = I_{\binom{n}{r}} \in M_{\binom{n}{r}}$

If $A \in M_n$ is nonsingular and $1 \leq r \leq n$, then $C_r(A)^{-1} = C_r(A^{-1})$

If $A \in M_n$ and $1 \leq r \leq n$, then $\det C_r(A) = (\det A)^{\binom{n-1}{r-1}}$

If $A \in M_{m,n}(\mathbf{F})$ and $r = \operatorname{rank} A$, then $\operatorname{rank} C_r(A) = 1$

If $A \in M_{m,n}(\mathbf{F})$ and $1 \leq r \leq \min\{m, n\}$, then $C_r(A^T) = C_r(A)^T$

If $A \in M_{m,n}(\mathbf{C})$ and $1 \leq r \leq \min\{m, n\}$, then $C_r(A^*) = C_r(A)^*$

If $\Delta = [d_{ij}] \in M_n(\mathbf{F})$ is upper (respectively, lower) triangular (see (0.9.3)), then $C_r(\Delta)$ is upper (respectively, lower) triangular; its main diagonal entries are the $\binom{n}{r}$ possible products of r entries chosen from the list $d_{11}, \ldots, d_{nn}$, that is, they are the $\binom{n}{r}$ scalars $d_{i_1 i_1} \cdots d_{i_r i_r}$ such that $1 \leq i_1 < \cdots < i_r \leq n$, arranged lexicographically. Consequently, if $D = \operatorname{diag}(d_1, \ldots, d_n) \in M_n(\mathbf{F})$ is diagonal, then so is $C_r(D)$; its main diagonal entries are the $\binom{n}{r}$ possible products of r entries chosen from the list $d_1, \ldots, d_n$, that is, they are the $\binom{n}{r}$ scalars $d_{i_1} \cdots d_{i_r}$ such that $1 \leq i_1 < \cdots < i_r \leq n$, arranged lexicographically. See chapter 6 of (Fiedler, 1986) for a detailed discussion of compound matrices.

0.8.2 The adjugate and the inverse. If $A \in M_n(\mathbf{F})$ and $n \geq 2$, the transposed matrix of cofactors of A

$$\operatorname{adj} A = \left[(-1)^{i+j} \det A \left[\{j\}^c, \{i\}^c \right] \right] \tag{0.8.2.0}$$

is the *adjugate* of A; it is also called the *classical adjoint* of A. For example, $\operatorname{adj} \begin{bmatrix} a & b \\ c & d \end{bmatrix} = \begin{bmatrix} d & -b \\ -c & a \end{bmatrix}$.

A calculation using the Laplace expansion for the determinant reveals the basic property of the adjugate:

$$(\operatorname{adj} A) A = A (\operatorname{adj} A) = (\det A) I \tag{0.8.2.1}$$

Thus, $\operatorname{adj} A$ is nonsingular if A is nonsingular, and $\det(\operatorname{adj} A) = (\det A)^{n-1}$.

If A is nonsingular, then

$$\operatorname{adj} A = (\det A) A^{-1}, \quad \text{that is,} \quad A^{-1} = (\det A)^{-1} \operatorname{adj} A \tag{0.8.2.2}$$

For example, $\begin{bmatrix} a & b \\ c & d \end{bmatrix}^{-1} = (ad - bc)^{-1} \begin{bmatrix} d & -b \\ -c & a \end{bmatrix}$ if $ad \neq bc$. In particular, $\operatorname{adj}(A^{-1}) = A/\det A = (\operatorname{adj} A)^{-1}$.

If A is singular and $\operatorname{rank} A \leq n - 2$, then every minor of A of size $n - 1$ is zero, so $\operatorname{adj} A = 0$.

If A is singular and $\operatorname{rank} A = n - 1$, then some minor of A of size $n - 1$ is nonzero, so $\operatorname{adj} A \neq 0$ and $\operatorname{rank} \operatorname{adj} A \geq 1$. Moreover, some list of $n - 1$ columns of A is linearly independent, so the identity $(\operatorname{adj} A) A = (\det A) I = 0$ ensures that the null space of $\operatorname{adj} A$ has dimension at least $n - 1$ and hence $\operatorname{rank} \operatorname{adj} A \leq 1$. We conclude that $\operatorname{rank} \operatorname{adj} A = 1$. The full-rank factorization (0.4.6(e)) ensures that $\operatorname{adj} A = xy^T$ for some nonzero $x, y \in \mathbf{F}^n$ that are determined as follows: Compute

$$(Ax)y^T = A(\operatorname{adj} A) = 0 = (\operatorname{adj} A)A = x(y^T A)$$

and conclude that $Ax = 0$ and $y^T A = 0$, that is, x (respectively, y) is determined up to a nonzero scalar factor as a nonzero element of the one-dimensional null space of A (respectively, A^T).

The function $A \to \text{adj } A$ is continuous on M_n (each entry of adj A is a multinomial in the entries of A) and every matrix in M_n is a limit of nonsingular matrices, so properties of the adjugate can be deduced from continuity and properties of the inverse function. For example, if $A, B \in M_n$ are nonsingular, then $\text{adj}(AB) = (\det AB)(AB)^{-1} = (\det A)(\det B)B^{-1}A^{-1} = (\det B)B^{-1}(\det A)A^{-1} = (\text{adj } B)(\text{adj } A)$. Continuity then ensures that

$$\text{adj}(AB) = (\text{adj } B)(\text{adj } A) \text{ for all } A, B \in M_n \qquad (0.8.2.3)$$

For any $c \in \mathbf{F}$ and any $A \in M_n(\mathbf{F})$, $\text{adj}(cA) = c^{n-1} \text{adj } A$. In particular, $\text{adj}(cI) = c^{n-1}I$ and $\text{adj } 0 = 0$.

If A is nonsingular, then

$$\text{adj}(\text{adj } A) = \text{adj}((\det A)A^{-1}) = (\det A)^{n-1} \text{adj } A^{-1}$$
$$= (\det A)^{n-1}(A/\det A) = (\det A)^{n-2} A$$

so continuity ensures that

$$\text{adj}(\text{adj } A) = (\det A)^{n-2}A \text{ for all } A \in M_n \qquad (0.8.2.4)$$

If $A + B$ is nonsingular, then $A(A + B)^{-1}B = B(A + B)^{-1}A$, so continuity ensures that

$$A \, \text{adj}(A + B) B = B \, \text{adj}(A + B) A \text{ for all } A, B \in M_n \qquad (0.8.2.5)$$

Let $A, B \in M_n$ and suppose that A commutes with B. If A is nonsingular, then $BA^{-1} = A^{-1}ABA^{-1} = A^{-1}BAA^{-1} = A^{-1}B$, so A^{-1} commutes with B. But $BA^{-1} = (\det A)^{-1}B \, \text{adj } A$ and $A^{-1}B = (\det A)^{-1}(\text{adj } A)B$, so $\text{adj } A$ commutes with B. Continuity ensures that $\text{adj } A$ commutes with B whenever A commutes with B, even if A is singular.

If $A = [a_{ij}]$ is upper triangular, then $\text{adj } A = [b_{ij}]$ is upper triangular and each $b_{ii} = \prod_{j \neq i} a_{jj}$; if A is diagonal, then so is adj A.

The adjugate is the transpose of the gradient of $\det A$:

$$(\text{adj } A) = \left[\frac{\partial}{\partial a_{ij}} \det A \right]^T \qquad (0.8.2.6)$$

If A is nonsingular, it follows from (0.8.2.6) that

$$\left[\frac{\partial}{\partial a_{ij}} \det A \right]^T = (\det A) A^{-1} \qquad (0.8.2.7)$$

If $A \in M_n$ is nonsingular, then $\text{adj } A^T = (\det A^T)A^{-T} = (\det A)A^{-T} = ((\det A)A^{-1})^T = (\text{adj } A)^T$. Continuity ensures that

$$\text{adj } A^T = (\text{adj } A)^T \text{ for all } A \in M_n(\mathbf{F}) \qquad (0.8.2.8)$$

A similar argument shows that

$$\text{adj } A^* = (\text{adj } A)^* \text{ for all } A \in M_n \qquad (0.8.2.9)$$

Let $A = [a_1 \ldots a_n] \in M_n(\mathbf{F})$ be partitioned according to its columns and let $b \in \mathbf{F}^n$. Define

$$(A \underset{i}{\leftarrow} b) = [a_1 \ldots a_{i-1} \, b \, a_{i+1} \ldots a_n]$$

that is, $(A \underset{i}{\leftarrow} b)$ denotes the matrix whose ith column is b and whose remaining columns coincide with those of A. Examination of the Laplace expansion (0.3.1.1) of $\det(A \underset{i}{\leftarrow} b)$ by minors along column i reveals that it is the ith entry of the vector $(\text{adj } A) b$, that is,

$$\left[\det(A \underset{i}{\leftarrow} b) \right]_{i=1}^n = (\text{adj } A) b \tag{0.8.2.10}$$

Applying this vector identity to each column of $C = [c_1 \ldots c_n] \in M_n(\mathbf{F})$ gives the matrix identity

$$\left[\det(A \underset{i}{\leftarrow} c_j) \right]_{i,j=1}^n = (\text{adj } A) C \tag{0.8.2.11}$$

0.8.3 Cramer's rule. Cramer's rule is a useful way to present analytically a particular entry of the solution to $Ax = b$ when $A \in M_n(\mathbf{F})$ is nonsingular. The identity

$$A \left[\det(A \underset{i}{\leftarrow} b) \right]_{i=1}^n = A (\text{adj } A) b = (\det A) b$$

follows from (0.8.2.10). If $\det A \neq 0$, we obtain Cramer's rule

$$x_i = \frac{\det(A \underset{i}{\leftarrow} b)}{\det A}$$

for the ith entry x_i of the solution vector x. Cramer's rule also follows directly from multiplicativity of the determinant. The system $Ax = b$ may be rewritten as

$$A(I \underset{i}{\leftarrow} x) = A \underset{i}{\leftarrow} b$$

and taking determinants of both sides (using multiplicativity) gives

$$(\det A) \det(I \underset{i}{\leftarrow} x) = \det(A \underset{i}{\leftarrow} b)$$

But $\det(I \underset{i}{\leftarrow} x) = x_i$, and the formula follows.

0.8.4 Minors of the inverse. *Jacobi's identity* generalizes the adjugate formula for the inverse of a nonsingular matrix and relates the minors of A^{-1} to those of $A \in M_n(\mathbf{F})$:

$$\det A^{-1} \left[\alpha^c, \beta^c \right] = (-1)^{p(\alpha,\beta)} \frac{\det A [\beta, \alpha]}{\det A} \tag{0.8.4.1}$$

in which $p(\alpha, \beta) = \sum_{i \in \alpha} i + \sum_{j \in \beta} j$. Our universal convention is that $\det A[\varnothing] = 1$. For principal submatrices, Jacobi's identity assumes the simple form

$$\det A^{-1}[\alpha^c] = \frac{\det A [\alpha]}{\det A} \tag{0.8.4.2}$$

0.8.5 Schur complements and determinantal formulae. Let $A = [a_{ij}] \in M_n(\mathbf{F})$ be given and suppose that $\alpha \subseteq \{1, \ldots, n\}$ is an index set such that $A[\alpha]$ is nonsingular. An important formula for $\det A$, based on the 2-partition of A using α and α^c, is

$$\det A = \det A [\alpha] \det \left(A [\alpha^c] - A [\alpha^c, \alpha] A [\alpha]^{-1} A [\alpha, \alpha^c] \right) \tag{0.8.5.1}$$

which generalizes the familiar formula for the determinant of a 2-by-2 matrix. The special matrix

$$A/A\,[\alpha] = A\,[\alpha^c] - A\,[\alpha^c, \alpha]\,A\,[\alpha]^{-1}\,A\,[\alpha, \alpha^c] \qquad (0.8.5.2)$$

which also appears in the partitioned form for the inverse in (0.7.3.1), is called the *Schur complement of $A\,[\alpha]$ in A*. When convenient, we take $\alpha = \{1, \ldots, k\}$ and write A as a 2-by-2 block matrix $A = [A_{ij}]$ with $A_{11} = A[\alpha]$, $A_{22} = A[\alpha^c]$, $A_{12} = A[\alpha, \alpha^c]$, and $A_{21} = A[\alpha^c, \alpha]$. The formula (0.8.5.1) may be verified by computing the determinant of both sides of the identity

$$\begin{bmatrix} I & 0 \\ -A_{21}A_{11}^{-1} & I \end{bmatrix} \begin{bmatrix} A_{11} & A_{12} \\ A_{21} & A_{22} \end{bmatrix} \begin{bmatrix} I & -A_{11}^{-1}A_{12} \\ 0 & I \end{bmatrix} \qquad (0.8.5.3)$$
$$= \begin{bmatrix} A_{11} & 0 \\ 0 & A_{22} - A_{21}A_{11}^{-1}A_{12} \end{bmatrix}$$

which contains a wealth of information about the Schur complement $S = [s_{ij}] = A/A_{11} = A_{22} - A_{21}A_{11}^{-1}A_{12}$:

(a) The Schur complement S arises (uniquely) in the lower right corner if linear combinations of the first k rows (respectively, columns) of A are added to the last $n - k$ rows (respectively, columns) in such a way as to produce a zero block in the lower left (respectively, upper right) corner; this is *block Gaussian elimination*, and it is (uniquely) possible because A_{11} is nonsingular. Any submatrix of A that includes A_{11} as a principal submatrix has the same determinant before and after the block eliminations that produce the block diagonal form in (0.8.5.3). Thus, for any index set $\beta = \{i_1, \ldots, i_m\} \subseteq \{1, \ldots, n - k\}$, if we construct the shifted index set $\tilde{\beta} = \{i_1 + k, \ldots, i_m + k\}$, then $\det A[\alpha \cup \tilde{\beta}, \alpha \cup \tilde{\gamma}]$ (before) $= \det(A_{11} \oplus S[\beta, \gamma])$ (after), so

$$\det S\,[\beta, \gamma] = \det A\,[\alpha \cup \tilde{\beta}, \alpha \cup \tilde{\gamma}]\,/\det A\,[\alpha] \qquad (0.8.5.4)$$

For example, if $\beta = \{i\}$ and $\gamma = \{j\}$, then with $\alpha = \{1, \ldots, k\}$, we have

$$\det S\,[\beta, \gamma] = s_{ij} \qquad (0.8.5.5)$$
$$= \det A\,[\{1, \ldots k, k + i\}, \{1, \ldots, k, k + j\}]\,/\det A_{11}$$

so all the entries of S are ratios of minors of A.

(b) rank A = rank A_{11} + rank $S \geq$ rank A_{11}, and rank A = rank A_{11} if and only if $A_{22} = A_{21}A_{11}^{-1}A_{12}$.

(c) A is nonsingular if and only if S is nonsingular, since $\det A = \det A_{11} \det S$. If A is nonsingular, then $\det S = \det A/\det A_{11}$.

Suppose that A is nonsingular. Then inverting both sides of (0.8.5.3) gives a presentation of the inverse different from that in (0.7.3.1):

$$A^{-1} = \begin{bmatrix} A_{11}^{-1} + A_{11}^{-1}A_{12}S^{-1}A_{21}A_{11}^{-1} & -A_{11}^{-1}A_{12}S^{-1} \\ -S^{-1}A_{21}A_{11}^{-1} & S^{-1} \end{bmatrix} \qquad (0.8.5.6)$$

Among other things, this tells us that $A^{-1}[\{k + 1, \ldots, n\}] = S^{-1}$, so

$$\det A^{-1}\,[\{k + 1, \ldots, n\}] = \det A_{11}/\det A \qquad (0.8.5.7)$$

This is a form of Jacobi's identity (0.8.4.1). Another form results from using the adjugate to write the inverse, which gives

$$\det\left((\operatorname{adj} A)\left[\{k+1,\ldots,n\}\right]\right) = (\det A)^{n-k-1}\det A_{11} \qquad (0.8.5.8)$$

When α^c consists of a single element, the Schur complement of $A[\alpha]$ in A is a scalar and (0.8.5.1) reduces to the identity

$$\det A = A\left[\alpha^c\right]\det A\left[\alpha\right] - A\left[\alpha^c,\alpha\right](\operatorname{adj} A\left[\alpha\right])A\left[\alpha,\alpha^c\right] \qquad (0.8.5.9)$$

which is valid even if $A[\alpha]$ is singular. For example, if $\alpha = \{1,\ldots,n-1\}$, then $\alpha^c = \{n\}$ and A is presented as a *bordered matrix*

$$A = \begin{bmatrix} \tilde{A} & x \\ y^T & a \end{bmatrix}$$

with $a \in \mathbf{F}$, $x, y \in \mathbf{F}^{n-1}$, and $\tilde{A} \in M_{n-1}(\mathbf{F})$; (0.8.5.9) is the *Cauchy expansion* of the determinant of a bordered matrix

$$\det\begin{bmatrix} \tilde{A} & x \\ y^T & a \end{bmatrix} = a\det\tilde{A} - y^T\left(\operatorname{adj}\tilde{A}\right)x \qquad (0.8.5.10)$$

The Cauchy expansion (0.8.5.10) involves signed minors of A of size $n-2$ (the entries of $\operatorname{adj}\tilde{A}$) and a bilinear form in the entries of a row *and* column; the Laplace expansion (0.3.1.1) involves signed minors of A of size $n-1$ and a linear form in the entries of a row *or* column. If $a \neq 0$, we can use the Schur complement of $[a]$ in A to express

$$\det\begin{bmatrix} \tilde{A} & x \\ y^T & a \end{bmatrix} = a\det(\tilde{A} - a^{-1}xy^T)$$

Equating the right-hand side of this identity to that of (0.8.5.10) and setting $a = -1$ gives *Cauchy's formula for the determinant of a rank-one perturbation*

$$\det\left(\tilde{A} + xy^T\right) = \det\tilde{A} + y^T\left(\operatorname{adj}\tilde{A}\right)x \qquad (0.8.5.11)$$

The uniqueness property of the Schur complement discussed in (a) can be used to derive an identity involving a Schur complement within a Schur complement. Suppose that the nonsingular k-by-k block A_{11} is partitioned as a 2-by-2 block matrix $A_{11} = [\mathcal{A}_{ij}]$ in which the upper left ℓ-by-ℓ block $\mathcal{A}_{11}$ is nonsingular. Write $A_{21} = [\mathcal{A}_1\ \mathcal{A}_2]$, in which $\mathcal{A}_1$ is $(n-k)$-by-ℓ, and write $A_{12}^T = [\mathcal{B}_1^T\ \mathcal{B}_2^T]$, in which $\mathcal{B}_1$ is ℓ-by-$(n-k)$; this gives the refined partition

$$A = \begin{bmatrix} \mathcal{A}_{11} & \mathcal{A}_{12} & \mathcal{B}_1 \\ \mathcal{A}_{21} & \mathcal{A}_{22} & \mathcal{B}_2 \\ \mathcal{A}_1 & \mathcal{A}_2 & A_{22} \end{bmatrix}$$

Now add linear combinations of the first ℓ rows of A to the next $k-\ell$ rows to reduce $\mathcal{A}_{21}$ to a zero block. The result is

$$A' = \begin{bmatrix} \mathcal{A}_{11} & \mathcal{A}_{12} & \mathcal{B}_1 \\ 0 & A_{11}/\mathcal{A}_{11} & \mathcal{B}_2' \\ \mathcal{A}_1 & \mathcal{A}_2 & A_{22} \end{bmatrix}$$

in which we have identified the resulting 2,2 block of A' as the (necessarily nonsingular) Schur complement of $\mathcal{A}_{11}$ in A_{11}. Now add linear combinations of the first k rows of

A' to the last $n - k$ rows to reduce $[\ \mathcal{A}_1\ \ \mathcal{A}_2\ ]$ to a zero block. The result is

$$A'' = \begin{bmatrix} \mathcal{A}_{11} & \mathcal{A}_{12} & \mathcal{B}_1 \\ 0 & A_{11}/\mathcal{A}_{11} & B_2' \\ 0 & 0 & A/A_{11} \end{bmatrix}$$

in which we have identified the resulting 3,3 block of A'' as the Schur complement of A_{11} in A. The lower right 2-by-2 block of A'' must be $A/\mathcal{A}_{11}$, the Schur complement of $\mathcal{A}_{11}$ in A. Moreover, the lower right block of $A/\mathcal{A}_{11}$ must be the Schur complement of $A_{11}/\mathcal{A}_{11}$ in $A/\mathcal{A}_{11}$. This observation is the *quotient property of Schur complements*:

$$A/A_{11} = (A/\mathcal{A}_{11}) / (A_{11}/\mathcal{A}_{11}) \tag{0.8.5.12}$$

If the four blocks A_{ij} in (0.8.5.3) are square and the same size, and if A_{11} commutes with A_{21}, then

$$\det A = \det A_{11} \det S = \det(A_{11}S)$$
$$= \det(A_{11}A_{22} - A_{11}A_{21}A_{11}^{-1}A_{12}) = \det(A_{11}A_{22} - A_{21}A_{12})$$

Alternative assumptions about commuting blocks and consideration of the Schur complement of A_{22} in A lead to other identities:

$$\det \begin{bmatrix} A_{11} & A_{12} \\ A_{21} & A_{22} \end{bmatrix} = \begin{cases} \det(A_{11}A_{22} - A_{21}A_{12}) & \text{if } A_{11}A_{21} = A_{21}A_{11} \\ \det(A_{22}A_{11} - A_{21}A_{12}) & \text{if } A_{11}A_{12} = A_{12}A_{11} \\ \det(A_{11}A_{22} - A_{12}A_{21}) & \text{if } A_{22}A_{21} = A_{21}A_{22} \\ \det(A_{22}A_{11} - A_{12}A_{21}) & \text{if } A_{22}A_{12} = A_{12}A_{22} \end{cases} \tag{0.8.5.13}$$

A continuity argument shows that these identities are valid even if $A_{11}A_{22}$ is singular.

0.8.6 Determinantal identities of Sylvester and Kronecker. We consider two consequences of (0.8.5.4). If we set

$$B = [b_{ij}] = [\det A\,[\{1, \ldots, k, k + i\}, \{1, \ldots, k, k + j\}]]_{i,j=1}^{n-k}$$

then each entry of B is the determinant of a bordered matrix of the form (0.8.5.10): $\tilde{A}$ is A_{11}, x is the jth column of A_{12}, y^T is the ith row of A_{21}, and a is the i, j entry of A_{22}. The identity (0.8.5.5) tells us that $B = (\det A_{11})S$, so

$$\det B = (\det A_{11})^{n-k} \det S$$
$$= (\det A_{11})^{n-k} (\det A/ \det A_{11}) = (\det A_{11})^{n-k-1} \det A$$

This observation about B is *Sylvester's identity for bordered determinants*:

$$\det B = (\det A\,[\alpha])^{n-k-1} \det A \tag{0.8.6.1}$$

in which $B = [\det A[\alpha \cup \{i\}, \alpha \cup \{j\}]]$ and i, j are indices *not* contained in α.

If $A_{22} = 0$, then each entry of B is the determinant of a bordered matrix of the form (0.8.5.10) with $a = 0$. In this case, the Schur complement $A/A_{11} = -A_{21}A_{11}^{-1}A_{12}$ has rank at most k, so the determinant of every $(k + 1)$-by-$(k + 1)$ submatrix of B is zero; this observation about B is *Kronecker's theorem for bordered determinants*.

0.8.7 The Cauchy–Binet formula. This useful formula can be remembered because of its similarity in appearance to the formula for matrix multiplication. This is no accident, since it is equivalent to multiplicativity of the compound matrix (0.8.1.1). Let $A \in M_{m,k}(\mathbf{F})$, $B \in M_{k,n}(\mathbf{F})$, and $C = AB$. Furthermore, let $1 \leq r \leq \min\{m, k, n\}$, and let $\alpha \subseteq \{1, \ldots, m\}$ and $\beta \subseteq \{1, \ldots, n\}$ be index sets, each of cardinality r. An expression for the α, β minor of C is

$$\det C\,[\alpha, \beta] = \sum_{\gamma} \det A\,[\alpha, \gamma] \det B\,[\gamma, \beta]$$

in which the sum is taken over all index sets $\gamma \subseteq \{1, \ldots, k\}$ of cardinality r.

0.8.8 Relations among minors. Let $A \in M_{m,n}(\mathbf{F})$ be given and let a fixed index set $\alpha \subseteq \{1, \ldots, m\}$ of cardinality k be given. The minors $\det A\,[\alpha, \omega]$, as $\omega \subseteq \{1, \ldots, n\}$ runs over *ordered* index sets of cardinality k, are not algebraically independent since there are more minors than there are distinct entries among the submatrices. Quadratic relations are known among these minors. Let $i_1, i_2, \ldots, i_k \in \{1, \ldots, n\}$ be k distinct indices, not necessarily in natural order, and let $A[\alpha; i_1, \ldots, i_k]$ denote the matrix whose rows are indicated by α and whose jth column is column i_j of $A[\alpha, \{1, \ldots, n\}]$. The difference between this and our previous notation is that columns might not occur in natural order as in $A[\{1, 3\}; 4, 2]$, whose first column has the 1, 4 and 3, 4 entries of A. We then have the relations

$$\det A\,[\alpha; i_1, \ldots, i_k] \det A\,[\alpha; j_1, \ldots, j_k]$$

$$= \sum_{t=1}^{k} \det A\,[\alpha; i_1, \ldots, i_{s-1}, j_t, i_{s+1}, \ldots, i_k] \det A\,[\alpha; j_1, \ldots, j_{t-1}, i_s, j_{t+1}, \ldots, j_k]$$

for each $s = 1, \ldots, k$ and all sequences of distinct indices $i_1, \ldots, i_k \in \{1, \ldots, n\}$ and $j_1, \ldots, j_k \in \{1, \ldots, n\}$.

0.8.9 The Laplace expansion theorem. The Laplace expansion (0.3.1.1) by minors along a given row or column is included in a natural family of expressions for the determinant. Let $A \in M_n(\mathbf{F})$, let $k \in \{1, \ldots, n\}$ be given, and let $\beta \subseteq \{1, \ldots, n\}$ be any given index set of cardinality k. Then

$$\det A = \sum_{\alpha} (-1)^{p(\alpha, \beta)} \det A\,[\alpha, \beta] \det A\,[\alpha^c, \beta^c]$$

$$= \sum_{\alpha} (-1)^{p(\alpha, \beta)} \det A\,[\beta, \alpha] \det A\,[\beta^c, \alpha^c]$$

in which the sums are over all index sets $\alpha \subseteq \{1, \ldots, n\}$ of cardinality k, and $p(\alpha, \beta) = \sum_{i \in \alpha} i + \sum_{j \in \beta} j$. Choosing $k = 1$ and $\beta = \{i\}$ or $\{j\}$ gives the expansions in (0.3.1.1).

0.8.10 Derivative of the determinant. Let $A(t) = [a_1(t) \ \ldots \ a_n(t)] = [a_{ij}(t)]$ be an n-by-n complex matrix whose entries are differentiable functions of t and define $A'(t) = [a'_{ij}(t)]$. It follows from multilinearity of the determinant (0.3.6(a)) and the definition of the derivative that

$$\frac{d}{dt} \det A(t) = \sum_{j=1}^{n} \det \left(A(t) \xleftarrow{}_{j} a'_j(t) \right) = \sum_{j=1}^{n} \sum_{i=1}^{n} ((\operatorname{adj} A(t))^T)_{ij} a'_{ij}(t)$$

$$= \operatorname{tr}((\operatorname{adj} A(t)) A'(t)) \qquad (0.8.10.1)$$

For example, if $A \in M_n$ and $A(t) = tI - A$, then $A'(t) = I$ and

$$\frac{d}{dt} \det (tI - A) = \operatorname{tr}((\operatorname{adj} A(t)) I) = \operatorname{tr} \operatorname{adj} (tI - A) \qquad (0.8.10.2)$$

0.8.11 Dodgson's identity. Let $n \geq 3$ and let $A \in M_n(\mathbf{F})$. Define $a = \det A[\{n\}^c]$, $b = \det A[\{n\}^c, \{1\}^c]$, $c = \det A[\{1\}^c, \{n\}^c]$, $d = \det A[\{1\}^c]$, and $e = \det A[\{1, n\}^c]$. Then

$$e \det A = ad - bc$$

0.8.12 Adjugates and compounds. Let $A, B \in M_n(\mathbf{F})$. Let $\alpha \subseteq \{1, \ldots, n\}$ and $\beta \subseteq \{1, \ldots, n\}$ be index sets of cardinality $r \leq n$. The α, β entry of the *rth adjugate matrix* $\operatorname{adj}_r(A) \in M_{\binom{n}{r}}(\mathbf{F})$ is

$$(-1)^{p(\alpha, \beta)} \det A[\beta^c, \alpha^c] \qquad (0.8.12.1)$$

in which $p(\alpha, \beta) = \sum_{i \in \alpha} i + \sum_{j \in \beta} j$. The rows and columns of $\operatorname{adj}_r(A)$ are formed by arranging the index sets lexicographically, just as for the rth compound matrix. For example, using the matrix A in (0.8.1.0), we have

$$\operatorname{adj}_2(A) = \begin{bmatrix} 10 & -6 & 3 \\ -8 & 5 & -2 \\ 7 & -4 & 1 \end{bmatrix}$$

The *multiplicativity property* of the rth adjugate matrix is

$$\operatorname{adj}_r(AB) = \operatorname{adj}_r(B) \operatorname{adj}_r(A) \qquad (0.8.12.2)$$

We *define* $\operatorname{adj}_n(A) = 1$. We have $\operatorname{adj}_0(A) = \det A$ and $\operatorname{adj}_1(A) = \operatorname{adj} A$. The rth adjugate and rth compound matrices are related by the identity

$$\operatorname{adj}_r(A) C_r(A) = C_r(A) \operatorname{adj}_r(A) = (\det A) I_{\binom{n}{r}}$$

of which the identities in (0.8.9) are special cases. In particular, $C_r(A)^{-1} = (\det A)^{-1} \operatorname{adj}_r(A)$ if A is nonsingular.

The determinant of a sum of matrices can be expressed using the rth adjugate and rth compound matrices:

$$\det(sA + tB) = \sum_{k=0}^{n} s^{n-k} t^k \operatorname{tr}(\operatorname{adj}_k(A) C_k(B)) \qquad (0.8.12.3)$$

In particular, $\det(A + I) = \sum_{k=0}^{n} \operatorname{tr} \operatorname{adj}_k(A) = \sum_{k=0}^{n} \operatorname{tr} C_k(A)$.

0.9 Special types of matrices

Certain matrices of special form arise frequently and have important properties. Some of these are cataloged here for reference and terminology.

0.9.1 Diagonal matrices. A matrix $D = [d_{ij}] \in M_{n,m}(\mathbf{F})$ is *diagonal* if $d_{ij} = 0$ whenever $j \neq i$. If all the diagonal entries of a diagonal matrix are positive (non-negative) real numbers, we refer to it as a *positive (nonnegative) diagonal matrix*. The term *positive diagonal matrix* means that the matrix is diagonal and has positive diagonal entries; it does not refer to a general matrix with positive diagonal entries. The identity matrix $I \in M_n$ is a positive diagonal matrix. A square diagonal matrix D is a *scalar matrix* if its diagonal entries are all equal, that is, $D = \alpha I$ for some $\alpha \in \mathbf{F}$. Left or right multiplication of a matrix by a scalar matrix has the same effect as multiplying it by the corresponding scalar.

If $A = [a_{ij}] \in M_{n,m}(\mathbf{F})$ and $q = \min\{m, n\}$, then diag $A = [a_{11}, \ldots, a_{qq}]^T \in \mathbf{F}^q$ denotes the vector of diagonal entries of A (0.2.1). Conversely, if $x \in \mathbf{F}^q$ and if m and n are positive integers such that $\min\{m, n\} = q$, then diag $x \in M_{n,m}(\mathbf{F})$ denotes the n-by-m diagonal matrix A such that diag $A = x$; for diag x to be well-defined, both m and n must be specified. For any $a_1, \ldots, a_n \in \mathbf{F}$, diag$(a_1, \ldots, a_n)$ always denotes the matrix $A = [a_{ij}] \in M_n(\mathbf{F})$ such that $a_{ii} = a_i$ for each $i = 1, \ldots, n$ and $a_{ij} = 0$ if $i \neq j$.

Suppose that $D = [d_{ij}]$, $E = [e_{ij}] \in M_n(\mathbf{F})$ are diagonal and let $A = [a_{ij}] \in M_n(\mathbf{F})$ be given. Then (a) det $D = \prod_{i=1}^n d_{ii}$; (b) D is nonsingular if and only if all $d_{ii} \neq 0$; (c) *left* multiplication of A by D multiplies the *rows* of A by the diagonal entries of D (the ith row of DA is d_{ii} times the ith row of A); (d) *right* multiplication of A by D multiplies the *columns* of A by the diagonal entries of D, that is, the jth column of AD is d_{jj} times the jth column of A; (e) $DA = AD$ if and only if $a_{ij} = 0$ whenever $d_{ii} \neq d_{jj}$; (f) if all the diagonal entries of D are distinct and $DA = AD$, then A is diagonal; (g) for any positive integer k, $D^k = \mathrm{diag}(d_{11}^k, \ldots, d_{nn}^k)$; and (h) any two diagonal matrices D and E of the same size commute: $DE = \mathrm{diag}(d_{11}e_{11}, \ldots, d_{nn}e_{nn}) = ED$.

0.9.2 Block diagonal matrices and direct sums. A matrix $A \in M_n(\mathbf{F})$ of the form

$$A = \begin{bmatrix} A_{11} & & \mathbf{0} \\ & \ddots & \\ \mathbf{0} & & A_{kk} \end{bmatrix}$$

in which $A_{ii} \in M_{n_i}(\mathbf{F})$, $i = 1, \ldots, k$, $\sum_{i=1}^k n_i = n$, and all blocks above and below the block diagonal are zero blocks, is called *block diagonal*. It is convenient to write such a matrix as

$$A = A_{11} \oplus A_{22} \oplus \cdots \oplus A_{kk} = \bigoplus_{i=1}^k A_{ii}$$

This is the *direct sum* of the matrices $A_{11}, \ldots, A_{kk}$. Many properties of block diagonal matrices generalize those of diagonal matrices. For example, $\det(\oplus_{i=1}^k A_{ii}) = \prod_{i=1}^k \det A_{ii}$, so that $A = \oplus A_{ii}$ is nonsingular if and only if each A_{ii} is nonsingular,

$i = 1, \ldots, k$. Furthermore, two direct sums $A = \oplus_{i=1}^{k} A_{ii}$ and $B = \oplus_{i=1}^{k} B_{ii}$, in which each A_{ii} is the same size as B_{ii}, commute if and only if each pair A_{ii} and B_{ii} commutes, $i = 1, \ldots, k$. Also, $\mathrm{rank}(\oplus_{i=1}^{k} A_{ii}) = \sum_{i=1}^{k} \mathrm{rank}\, A_{ii}$.

If $A \in M_n$ and $B \in M_m$ are nonsingular, then $(A \oplus B)^{-1} = A^{-1} \oplus B^{-1}$ and $(\det(A \oplus B))(A \oplus B)^{-1} = (\det A)(\det B)(A^{-1} \oplus B^{-1}) = ((\det B)(\det A)A^{-1}) \oplus ((\det A)(\det B)B^{-1})$, so a continuity argument ensures that

$$\mathrm{adj}(A \oplus B) = (\det B)\,\mathrm{adj}\, A \oplus (\det A)\,\mathrm{adj}\, B \qquad (0.9.2.1)$$

0.9.3 Triangular matrices. A matrix $T = [t_{ij}] \in M_{n,m}(\mathbf{F})$ is *upper triangular* if $t_{ij} = 0$ whenever $i > j$. If $t_{ij} = 0$ whenever $i \geq j$, then T is said to be *strictly upper triangular*. Analogously, T is *lower triangular* (or *strictly lower triangular*) if its transpose is upper triangular (or strictly upper triangular). A *triangular* matrix is either lower or upper triangular; a *strictly triangular matrix* is either strictly upper triangular or strictly lower triangular. A *unit triangular matrix* is a triangular matrix (upper or lower) that has ones on its main diagonal. Sometimes the terms *right* (in place of *upper*) and *left* (in place of *lower*) are used to describe triangular matrices.

Let $T \in M_{n,m}(\mathbf{F})$ be given. If T is upper triangular, then $T = [R \; T_2]$ if $n \leq m$, whereas $T = \begin{bmatrix} R \\ 0 \end{bmatrix}$ if $n \geq m$; $R \in M_{\min\{n,m\}}(\mathbf{F})$ is upper triangular and T_2 is arbitrary (empty if $n = m$). If T is lower triangular, then $T = [L \; 0]$ if $n \leq m$, whereas $T = \begin{bmatrix} L \\ T_2 \end{bmatrix}$ if $n \geq m$; $L \in M_{\min\{n,m\}}(\mathbf{F})$ is lower triangular and T_2 is arbitrary (empty if $n = m$).

A square triangular matrix shares with a square diagonal matrix the property that its determinant is the product of its diagonal entries. Square triangular matrices need not commute with other square triangular matrices of the same size. However, if $T \in M_n$ is triangular, has distinct diagonal entries, and commutes with $B \in M_n$, then B must be triangular of the same type as T (2.4.5.1).

For each $i = 1, \ldots, n$, left multiplication of $A \in M_n(\mathbf{F})$ by a lower triangular matrix L ($A \to LA$) replaces the ith row of A by a linear combination of the first through ith rows of A. The result of performing a finite number of type 3 row operations on A (0.3.3) is a matrix LA, in which L is a unit lower triangular matrix. Corresponding statements may be made about column operations and right multiplication by an upper triangular matrix.

The rank of a triangular matrix is at least, and can be greater than, the number of nonzero entries on the main diagonal. If a square triangular matrix is nonsingular, its inverse is a triangular matrix of the same type. A product of square triangular matrices of the same size and type is a triangular matrix of the same type; each i, i diagonal entry of such a matrix product is the product of the i, i entries of the factors.

0.9.4 Block triangular matrices. A matrix $A \in M_n(\mathbf{F})$ of the form

$$A = \begin{bmatrix} A_{11} & \bigstar & \bigstar \\ & \ddots & \bigstar \\ \mathbf{0} & & A_{kk} \end{bmatrix} \qquad (0.9.4.1)$$

in which $A_{ii} \in M_{n_i}(\mathbf{F})$, $i = 1, \ldots, k$, $\Sigma_{i=1}^{k} n_i = n$, and all blocks below the block diagonal are zero, is *block upper triangular*; it is strictly block upper triangular if, in addition, all the diagonal blocks are zero blocks. A matrix is *block lower triangular* if its transpose is block upper triangular; it is *strictly block lower triangular* if its transpose is strictly block upper triangular. We say that a matrix is *block triangular* if it is either block lower triangular or block upper triangular; a matrix is both block lower triangular and block upper triangular if and only if it is block diagonal.

A block upper triangular matrix in which all the diagonal blocks are 1-by-1 or 2-by-2 is said to be *upper quasitriangular*. A matrix is *lower quasitriangular* if its transpose is upper quasitriangular; it is *quasitriangular* if it is either upper quasitriangular or lower quasitriangular. A matrix that is both upper quasitriangular and lower quasitriangular is said to be *quasidiagonal*.

Consider the square block triangular matrix A in (0.9.4.1). We have $\det A = \det A_{11} \cdots \det A_{kk}$ and $\operatorname{rank} A \geq \operatorname{rank} A_{11} + \cdots + \operatorname{rank} A_{kk}$. If A is nonsingular (that is, if A_{ii} is nonsingular for all $i = 1, \ldots, k$), then A^{-1} is a block triangular matrix partitioned conformally to A whose diagonal blocks are $A_{11}^{-1}, \ldots, A_{kk}^{-1}$.

If $A \in M_n(\mathbf{F})$ is upper triangular, then $[A[\alpha_i, \alpha_j]]_{i,j=1}^t$ is block upper triangular for *any* sequential partition $\alpha_1, \ldots, \alpha_t$ of $\{1, \ldots, n\}$ (0.7.2).

0.9.5 Permutation matrices. A square matrix P is a *permutation matrix* if exactly one entry in each row and column is equal to 1 and all other entries are 0. Multiplication by such matrices effects a permutation of the rows or columns of the matrix multiplied. For example,

$$\begin{bmatrix} 0 & 1 & 0 \\ 1 & 0 & 0 \\ 0 & 0 & 1 \end{bmatrix} \begin{bmatrix} 1 \\ 2 \\ 3 \end{bmatrix} = \begin{bmatrix} 2 \\ 1 \\ 3 \end{bmatrix}$$

illustrates how a permutation matrix produces a permutation of the rows (entries) of a vector: it sends the first entry to the second position, sends the second entry to the first position, and leaves the third entry in the third position. Left multiplication of a matrix $A \in M_{m,n}$ by an m-by-m permutation matrix P permutes the rows of A, while right multiplication of A by an n-by-n permutation matrix P permutes the columns of A. The matrix that carries out a type 1 elementary operation (0.3.3) is an example of a special type of permutation matrix called a *transposition*. Any permutation matrix is a product of transpositions.

The determinant of a permutation matrix is ± 1, so permutation matrices are nonsingular. Although permutation matrices need not commute, the product of two permutation matrices is again a permutation matrix. Since the identity is a permutation matrix and $P^T = P^{-1}$ for every permutation matrix P, the set of n-by-n permutation matrices is a subgroup of $GL(n, C)$ with cardinality $n!$.

Since right multiplication by $P^T = P^{-1}$ permutes columns in the same way that left multiplication by P permutes rows, the transformation $A \to PAP^T$ permutes the rows and columns (and hence also the main diagonal entries) of $A \in M_n$ in the same way. In the context of linear equations with coefficient matrix A, this transformation amounts to renumbering the variables and the equations in the same way. A matrix $A \in M_n$ such that PAP^T is triangular for some permutation matrix P is called *essentially triangular*; these matrices have much in common with triangular matrices.

If $\Lambda \in M_n$ is diagonal and $P \in M_n$ is a permutation matrix, then $P \Lambda P^T$ is a diagonal matrix.

The n-by-n *reversal matrix* is the permutation matrix

$$K_n = \begin{bmatrix} & & 1 \\ & \cdot^{\cdot^{\cdot}} & \\ 1 & & \end{bmatrix} = \left[\kappa_{ij} \right] \in M_n \qquad (0.9.5.1)$$

in which $\kappa_{i,n-i+1} = 1$ for $i = 1, \ldots, n$ and all other entries are zero. The rows of $K_n A$ are the rows of A presented in reverse order; the columns of $A K_n$ are the columns of A presented in reverse order. The reversal matrix is sometimes called the *sip matrix* (*standard involutory permutation*), the *backward identity*, or the *exchange matrix*.

For any n-by-n matrix $A = [a_{ij}]$, the entries $a_{i,n-i+1}$ for $i = 1, \ldots, n$ comprise its *counterdiagonal* (sometimes called the *secondary diagonal, backward diagonal, cross diagonal, dexter-diagonal,* or *antidiagonal*).

A *generalized permutation matrix* is a matrix of the form $G = PD$, in which $P, D \in M_n$, P is a permutation matrix, and D is a nonsingular diagonal matrix. The set of n-by-n generalized permutation matrices is a subgroup of $GL(n, C)$.

0.9.6 Circulant matrices. A matrix $A \in M_n(\mathbf{F})$ of the form

$$A = \begin{bmatrix} a_1 & a_2 & \cdots & & a_n \\ a_n & a_1 & a_2 & \cdots & a_{n-1} \\ a_{n-1} & a_n & a_1 & \cdots & a_{n-2} \\ \vdots & \vdots & \ddots & \ddots & \vdots \\ a_2 & a_3 & \cdots & a_n & a_1 \end{bmatrix} \qquad (0.9.6.1)$$

is a *circulant matrix*. Each row is the previous row cycled forward one step; the entries in each row are a cyclic permutation of those in the first. The n-by-n permutation matrix

$$C_n = \begin{bmatrix} 0 & 1 & 0 & \cdots & 0 \\ \vdots & 0 & 1 & & \vdots \\ & & \ddots & \ddots & 0 \\ 0 & & & & 1 \\ 1 & 0 & & \cdots & 0 \end{bmatrix} = \begin{bmatrix} 0 & I_{n-1} \\ 1 & 0_{1,n-1} \end{bmatrix} \qquad (0.9.6.2)$$

is the *basic circulant permutation* matrix. A matrix $A \in M_n(\mathbf{F})$ can be written in the form

$$A = \sum_{k=0}^{n-1} a_{k+1} C_n^k \qquad (0.9.6.3)$$

(a polynomial in the matrix C_n) if and only if it is a circulant. We have $C_n^0 = I = C_n^n$, and the coefficients $a_1, \ldots, a_n$ are the entries of the first row of A. This representation reveals that the circulant matrices of size n are a commutative algebra: linear combinations and products of circulants are circulants; the inverse of a nonsingular circulant is a circulant; any two circulants of the same size commute.

0.9.7 Toeplitz matrices. A matrix $A = \begin{bmatrix} a_{ij} \end{bmatrix} \in M_{n+1}(\mathbf{F})$ of the form

$$
A = \begin{bmatrix}
a_0 & a_1 & a_2 & \cdots & \cdots & a_n \\
a_{-1} & a_0 & a_1 & a_2 & \cdots & a_{n-1} \\
a_{-2} & a_{-1} & a_0 & a_1 & \cdots & a_{n-2} \\
\vdots & \vdots & & \ddots & \ddots & \vdots \\
\vdots & \vdots & & & \ddots & a_1 \\
a_{-n} & a_{-n+1} & \cdots & \cdots & a_{-1} & a_0
\end{bmatrix}
$$

is a *Toeplitz matrix*. The entry a_{ij} is equal to a_{j-i} for some given sequence $a_{-n}, a_{-n+1}, \ldots, a_{-1}, a_0, a_1, a_2, \ldots, a_{n-1}, a_n \in \mathbf{C}$. The entries of A are constant down the diagonals parallel to the main diagonal. The Toeplitz matrices

$$
B = \begin{bmatrix}
0 & 1 & & \mathbf{0} \\
 & 0 & \ddots & \\
 & & \ddots & 1 \\
\mathbf{0} & & & 0
\end{bmatrix}
\quad \text{and} \quad
F = \begin{bmatrix}
0 & & & \mathbf{0} \\
1 & 0 & & \\
 & \ddots & \ddots & \\
\mathbf{0} & & 1 & 0
\end{bmatrix}
$$

are called the *backward shift* and *forward shift* because of their effect on the elements of the standard basis $\{e_1, \ldots, e_{n+1}\}$. Moreover, $F = B^T$ and $B = F^T$. A matrix $A \in M_{n+1}$ can be written in the form

$$
A = \sum_{k=1}^{n} a_{-k} F^k + \sum_{k=0}^{n} a_k B^k \tag{0.9.7.1}
$$

if and only if it is a Toeplitz matrix. Toeplitz matrices arise naturally in problems involving trigonometric moments.

Using a reversal matrix K of appropriate size (0.9.5.1), notice that the forward and backward shift matrices are related: $F = KBK = B^T$ and $B = KFK = F^T$. The representation (0.9.7.1) ensures that $KA = A^T K$ for any Toeplitz matrix A, that is, $A^T = KAK = KAK^{-1}$.

An upper triangular Toeplitz matrix $A \in M_{n+1}(\mathbf{F})$ can be represented as a polynomial in B:

$$
A = a_0 I + a_1 B + \cdots + a_n B^n
$$

This representation (and the fact that $B^{n+1} = 0$) makes it clear why the upper triangular Toeplitz matrices of size n are a commutative algebra: Linear combinations and products of upper triangular Toeplitz matrices are upper triangular Toeplitz matrices; A is nonsingular if and only if $a_0 \neq 0$, in which case $A^{-1} = b_0 I + b_1 B + \cdots + b_n B^n$ is also an upper triangular Toeplitz matrix with $b_0 = a_0^{-1}$ and $b_k = -a_0^{-1} \sum_{m=0}^{k-1} a_{k-m} b_m$ for $k = 1, \ldots, n$. Any two upper triangular Toeplitz matrices of the same size commute.

0.9.8 Hankel matrices. A matrix $A \in M_{n+1}(\mathbf{F})$ of the form

$$
A = \begin{bmatrix}
a_0 & a_1 & a_2 & \cdots & & a_n \\
a_1 & a_2 & \cdots & & \cdots & a_{n+1} \\
a_2 & & \cdots & & & \vdots \\
\vdots & a_n & & & & a_{2n-1} \\
a_n & a_{n+1} & \cdots & & a_{2n-1} & a_{2n}
\end{bmatrix}
$$

is a *Hankel matrix*. Each entry a_{ij} is equal to a_{i+j-2} for some given sequence $a_0, a_1, a_2, \ldots, a_{2n-1}, a_{2n}$. The entries of A are constant along the diagonals perpendicular to the main diagonal. Hankel matrices arise naturally in problems involving power moments. Using a reversal matrix K of appropriate size (0.9.5.1), notice that KA and AK are Hankel matrices for any Toeplitz matrix A; KH and HK are Toeplitz matrices for any Hankel matrix H. Since $K = K^T = K^{-1}$ and Hankel matrices are symmetric, this means that any Toeplitz matrix is a product of two symmetric matrices with special structure: a reversal matrix and a Hankel matrix.

0.9.9 Hessenberg matrices. A matrix $A = [a_{ij}] \in M_n(\mathbf{F})$ is said to be in *upper Hessenberg form* or to be an *upper Hessenberg matrix* if $a_{ij} = 0$ for all $i > j + 1$:

$$
A = \begin{bmatrix}
a_{11} & & & & \bigstar \\
a_{21} & a_{22} & & & \\
& a_{32} & \ddots & & \\
& & \ddots & \ddots & \\
0 & & & a_{n,n-1} & a_{nn}
\end{bmatrix}
$$

An upper Hessenberg matrix A is said to be *unreduced* if all its subdiagonal entries are nonzero, that is, if $a_{i+1,i} \neq 0$ for all $i = 1, \ldots, n-1$; the rank of such a matrix is at least $n-1$ since its first $n-1$ columns are independent.

Let $A \in M_n(\mathbf{F})$ be unreduced upper Hessenberg. Then $A - \lambda I$ is unreduced upper Hessenberg for all $\lambda \in \mathbf{F}$, so $\mathrm{rank}(A - \lambda I) \geq n-1$ for all $\lambda \in \mathbf{F}$.

A matrix $A \in M_n(\mathbf{F})$ is *lower Hessenberg* if A^T is upper Hessenberg.

0.9.10 Tridiagonal, bidiagonal, and other structured matrices. A matrix $A = [a_{ij}] \in M_n(\mathbf{F})$ that is *both* upper and lower Hessenberg is called *tridiagonal*, that is, A is tridiagonal if $a_{ij} = 0$ whenever $|i - j| > 1$:

$$
A = \begin{bmatrix}
a_1 & b_1 & & & 0 \\
c_1 & a_2 & \ddots & & \\
& \ddots & \ddots & b_{n-1} \\
0 & & c_{n-1} & a_n
\end{bmatrix}
\tag{0.9.10.1}
$$

The determinant of A can be calculated inductively starting with $\det A_1 = a_1$, $\det A_2 = a_1 a_2 - b_1 c_1$, and then computing

$$
\begin{bmatrix} \det A_{k+1} \\ \det A_k \end{bmatrix} = \begin{bmatrix} a_{k+1} & -b_k c_k \\ 1 & 0 \end{bmatrix} \begin{bmatrix} \det A_k \\ \det A_{k-1} \end{bmatrix}, \quad k = 2, \ldots, n-1
$$

A *Jacobi matrix* is a real symmetric tridiagonal matrix with positive subdiagonal entries.

An *upper bidiagonal matrix* $A \in M_n(\mathbf{F})$ is a tridiagonal matrix (0.9.10.1) in which $c_1 = \cdots = c_{n-1} = 0$. A matrix $A \in M_n(\mathbf{F})$ is *lower bidiagonal* if A^T is upper bidiagonal.

A *block tridiagonal* or *block bidiagonal* matrix has a block structure like the pattern in (0.9.10.1); the diagonal blocks are square and the sizes of the superdiagonal and subdiagonal blocks are determined by the sizes of their nearest diagonal blocks.

A matrix $A = [a_{ij}] \in M_n(\mathbf{F})$ is *persymmetric* if $a_{ij} = a_{n+1-j,n+1-i}$ for all $i, j = 1, \ldots, n$; that is, a persymmetric matrix is symmetric with respect to the counterdiagonal. An alternative, and very useful, characterization is that A is persymmetric if $K_n A = A^T K_n$, in which K_n is the reversal matrix (0.9.5.1). If A is persymmetric and invertible, then A^{-1} is also persymmetric since $K_n A^{-1} = (A K_n)^{-1} = (K_n A^T)^{-1} = A^{-T} K_n$. Toeplitz matrices are persymmetric. We say that $A \in M_n(\mathbf{F})$ is *skew persymmetric* if $K_n A = -A^T K_n$; the inverse of a nonsingular skew-persymmetric matrix is skew persymmetric.

A complex matrix $A \in M_n$ such that $K_n A = A^* K_n$ is *perhermitian*; A is *skew perhermitian* if $K_n A = -A^* K_n$. The inverse of a nonsingular perhermitian (respectively, skew perhermitian) matrix is perhermitian (respectively, skew perhermitian).

A matrix $A = [a_{ij}] \in M_n(\mathbf{F})$ is *centrosymmetric* if $a_{ij} = a_{n+1-i,n+1-j}$ for all $i, j = 1, \ldots, n$. Equivalently, A is centrosymmetric if $K_n A = A K_n$; A is *skew centrosymmetric* if $K_n A = -A K_n$. A centrosymmetric matrix is symmetric about its geometric center, as illustrated by the example

$$A = \begin{bmatrix} 1 & 2 & 3 & 4 & 5 \\ 0 & 6 & 7 & 8 & 9 \\ -1 & -2 & -3 & -2 & -1 \\ 9 & 8 & 7 & 6 & 0 \\ 5 & 4 & 3 & 2 & 1 \end{bmatrix}$$

If A is nonsingular and centrosymmetric (respectively, skew centrosymmetric), then A^{-1} is also centrosymmetric (respectively, skew centrosymmetric) since $K A^{-1} = (A K_n)^{-1} = (K_n A)^{-1} = A^{-1} K_n$. If A and B are centrosymmetric, then $A B$ is centrosymmetric since $K_n A B = A K_n B = A B K_n$. If A and B are skew centrosymmetric, then $A B$ is centrosymmetric.

A centrosymmetric matrix $A \in M_n(\mathbf{F})$ has a special block structure. If $n = 2m$, then

$$A = \begin{bmatrix} B & K_m C K_m \\ C & K_m B K_m \end{bmatrix}, \quad B, C \in M_m(\mathbf{F}) \tag{0.9.10.2}$$

If $n = 2m + 1$, then

$$A = \begin{bmatrix} B & K_m y & K_m C K_m \\ x^T & \alpha & x^T K_m \\ C & y & K_m B K_m \end{bmatrix}, \quad B, C \in M_m(\mathbf{F}), \ x, y \in \mathbf{F}^m, \ \alpha \in \mathbf{F} \tag{0.9.10.3}$$

A complex matrix $A \in M_n$ such that $K_n A = \bar{A} K_n$ is *centrohermitian*; it is *skew centrohermitian* if $K_n A = -\bar{A} K_n$. The inverse of a nonsingular centrohermitian (respectively, skew centrohermitian) matrix is centrohermitian (respectively, skew centrohermitian). A product of centrohermitian matrices is centrohermitian.

0.9.11 Vandermonde matrices and Lagrange interpolation. A *Vandermonde matrix* $A \in M_n(\mathbf{F})$ has the form

$$A = \begin{bmatrix} 1 & x_1 & x_1^2 & \cdots & x_1^{n-1} \\ 1 & x_2 & x_2^2 & \cdots & x_2^{n-1} \\ \vdots & \vdots & \vdots & \ddots & \vdots \\ 1 & x_n & x_n^2 & \cdots & x_n^{n-1} \end{bmatrix} \qquad (0.9.11.1)$$

in which $x_1, \ldots, x_n \in \mathbf{F}$; that is, $A = [a_{ij}]$ with $a_{ij} = x_i^{j-1}$. It is a fact that

$$\det A = \prod_{\substack{i,j=1 \\ i>j}}^{n} (x_i - x_j) \qquad (0.9.11.2)$$

so a Vandermonde matrix is nonsingular if and only if the parameters $x_1, \ldots, x_n$ are distinct.

If $x_1, \ldots, x_n$ are distinct, the entries of the inverse $A^{-1} = [\alpha_{ij}]$ of the Vandermonde matrix (0.9.11.1) are

$$\alpha_{ij} = (-1)^{i-1} \frac{S_{n-i}(x_1, \ldots, \hat{x}_j, \ldots, x_n)}{\prod_{k \neq j}(x_k - x_j)}, \quad i, j = 1, \ldots, n$$

in which $S_0 = 1$, and if $m > 0$, then $S_m(x_1, \ldots, \hat{x}_j, \ldots, x_n)$ is the mth elementary symmetric function of the $n-1$ variables $x_k, k = 1, \ldots, n, k \neq j$; see (1.2.14).

The Vandermonde matrix arises in the *interpolation problem* of finding a polynomial $p(x) = a_{n-1}x^{n-1} + a_{n-2}x^{n-2} + \cdots + a_1 x + a_0$ of degree at most $n-1$ with coefficients from $\mathbf{F}$ such that

$$p(x_1) = a_0 + a_1 x_1 + a_2 x_1^2 + \cdots + a_{n-1}x_1^{n-1} = y_1$$
$$p(x_2) = a_0 + a_1 x_2 + a_2 x_2^2 + \cdots + a_{n-1}x_2^{n-1} = y_2$$
$$\vdots \quad \vdots \qquad \vdots \qquad \qquad \vdots \quad \vdots \qquad (0.9.11.3)$$
$$p(x_n) = a_0 + a_1 x_n + a_2 x_n^2 + \cdots + a_{n-1}x_n^{n-1} = y_n$$

in which $x_1, \ldots, x_n$ and $y_1, \ldots, y_n$ are given elements of $\mathbf{F}$. The interpolation conditions (0.9.11.3) are a system of n equations for the n unknown coefficients $a_0, \ldots, a_{n-1}$, and they have the form $Aa = y$, in which $a = [a_0 \ldots a_{n-1}]^T \in \mathbf{F}^n$, $y = [y_1 \ldots y_n]^T \in \mathbf{F}^n$, and $A \in M_n(\mathbf{F})$ is the Vandermonde matrix (0.9.11.1). This interpolation problem always has a solution if the points $x_1, x_2, \ldots, x_n$ are distinct, since A is nonsingular in this event.

If the points $x_1, \ldots, x_n$ are distinct, the coefficients of the interpolating polynomial could in principle be obtained by solving the system (0.9.11.3), but it is usually more useful to represent the interpolating polynomial $p(x)$ as a linear combination of the

Lagrange interpolating polynomials

$$L_i(x) = \frac{\prod\limits_{j \neq i}(x - x_j)}{\prod\limits_{j \neq i}(x_i - x_j)}, \qquad i = 1, \ldots, n$$

Each polynomial $L_i(x)$ has degree $n - 1$ and has the property that $L_i(x_k) = 0$ if $k \neq i$, but $L_i(x_i) = 1$. *Lagrange's interpolation formula*

$$p(x) = y_1 L_1(x) + \cdots + y_n L_n(x) \tag{0.9.11.4}$$

provides a polynomial of degree at most $n - 1$ that satisfies the equations (0.9.11.3).

0.9.12 Cauchy matrices. A *Cauchy matrix* $A \in M_n(\mathbf{F})$ is a matrix of the form $A = [(a_i + b_j)^{-1}]_{i,j=1}^n$, in which $a_1, \ldots, a_n, b_1, \ldots, b_n$ are scalars such that $a_i + b_j \neq 0$ for all $i, j = 1, \ldots, n$. It is a fact that

$$\det A = \frac{\prod\limits_{1 \leq i < j \leq n}(a_j - a_i)(b_j - b_i)}{\prod\limits_{1 \leq i \leq j \leq n}(a_i + b_j)} \tag{0.9.12.1}$$

so A is nonsingular if and only if $a_i \neq a_j$ and $b_i \neq b_j$ for all $i \neq j$. A *Hilbert matrix* $H_n = [(i + j - 1)^{-1}]_{i,j=1}^n$ is a Cauchy matrix that is also a Hankel matrix. It is a fact that

$$\det H_n = \frac{(1!2! \cdots (n - 1)!)^4}{1!2! \cdots (2n - 1)!} \tag{0.9.12.2}$$

so a Hilbert matrix is always nonsingular. The entries of its inverse $H_n^{-1} = [h_{ij}]_{i,j=1}^n$ are

$$h_{ij} = \frac{(-1)^{i+j}(n + i - 1)!(n + j - 1)!}{((i - 1)!(j - 1)!)^2(n - i)!(n - j)!(i + j - 1)} \tag{0.9.12.3}$$

0.9.13 Involution, nilpotent, projection, coninvolution. A matrix $A \in M_n(\mathbf{F})$ is

- an *involution* if $A^2 = I$, that is, if $A = A^{-1}$ (the term *involutory* is also used)
- *nilpotent* if $A^k = 0$ for some positive integer k; the least such k is the *index of nilpotence* of A
- a *projection* if $A^2 = A$ (the term *idempotent* is also used)

Now suppose that $\mathbf{F} = \mathbf{C}$. A matrix $A \in M_n$ is

- a *Hermitian projection* if $A^* = A$ and $A^2 = A$ (the term *orthogonal projection* is also used; see (4.1.P19))
- a *coninvolution* if $A\bar{A} = I$, that is, if $\bar{A} = A^{-1}$ (the term *coninvolutory* is also used)

0.10 Change of basis

Let V be an n-dimensional vector space over the field $\mathbf{F}$, and let the list $\mathcal{B}_1 = v_1, v_2,$ $\ldots, v_n$ be a basis for V. Any vector $x \in V$ can be represented as $x = \alpha_1 v_1 + \alpha_2 v_2 + \cdots + \alpha_n v_n$ because $\mathcal{B}_1$ spans V. If there were some other representation of $x = \beta_1 v_1 + \beta_2 v_2 + \cdots + \beta_n v_n$ in the same basis, then

$$0 = x - x = (\alpha_1 - \beta_1)v_1 + (\alpha_2 - \beta_2)v_2 + \cdots + (\alpha_n - \beta_n)v_n$$

from which it follows that all $\alpha_i - \beta_i = 0$ because the list $\mathcal{B}_1$ is independent. Given the basis $\mathcal{B}_1$, the linear mapping

$$x \to [x]_{\mathcal{B}_1} = \begin{bmatrix} \alpha_1 \\ \vdots \\ \alpha_n \end{bmatrix}, \qquad \text{in which } x = \alpha_1 v_1 + \alpha_2 v_2 + \cdots + \alpha_n v_n$$

from V to $\mathbf{F}^n$ is well-defined, one-to-one, and onto. The scalars α_i are the *coordinates* of x with respect to the basis $\mathcal{B}_1$, and the column vector $[x]_{\mathcal{B}_1}$ is the unique $\mathcal{B}_1$-*coordinate representation of x*.

Let $T : V \to V$ be a given linear transformation. The action of T on any $x \in V$ is determined once one knows the n vectors $Tv_1, Tv_2, \ldots, Tv_n$, because any $x \in V$ has a unique representation $x = \alpha_1 v_1 + \cdots + \alpha_n v_n$ and $Tx = T(\alpha_1 v_1 + \cdots + \alpha_n v_n) = T(\alpha_1 v_1) + \cdots + T(\alpha_n v_n) = \alpha_1 Tv_1 + \cdots + \alpha_n Tv_n$ by linearity. Thus, the value of Tx is determined once $[x]_{\mathcal{B}_1}$ is known.

Let $\mathcal{B}_2 = \{w_1, w_2, \ldots, w_n\}$ also be a basis for V (either different from or the same as $\mathcal{B}_1$) and suppose that the $\mathcal{B}_2$-coordinate representation of Tv_j is

$$\left[Tv_j\right]_{\mathcal{B}_2} = \begin{bmatrix} t_{1j} \\ \vdots \\ t_{nj} \end{bmatrix}, \qquad j = 1, 2, \ldots, n$$

Then, for any $x \in V$, we have

$$[Tx]_{\mathcal{B}_2} = \left[\sum_{j=1}^{n} \alpha_j Tv_j\right]_{\mathcal{B}_2} = \sum_{j=1}^{n} \alpha_j \left[Tv_j\right]_{\mathcal{B}_2}$$

$$= \sum_{j=1}^{n} \alpha_j \begin{bmatrix} t_{1j} \\ \vdots \\ t_{nj} \end{bmatrix} = \begin{bmatrix} t_{11} & \cdots & t_{1n} \\ \vdots & \ddots & \vdots \\ t_{n1} & \cdots & t_{nn} \end{bmatrix} \begin{bmatrix} \alpha_1 \\ \vdots \\ \alpha_n \end{bmatrix}$$

The n-by-n array $[t_{ij}]$ depends on T and on the choice of the bases $\mathcal{B}_1$ and $\mathcal{B}_2$, but it does not depend on x. We define the $\mathcal{B}_1$-$\mathcal{B}_2$ *basis representation of T* to be

$$_{\mathcal{B}_2}[T]_{\mathcal{B}_1} = \begin{bmatrix} t_{11} & \cdots & t_{1n} \\ \vdots & \ddots & \vdots \\ t_{n1} & \cdots & t_{nn} \end{bmatrix} = \begin{bmatrix} [Tv_1]_{\mathcal{B}_2} \ldots [Tv_n]_{\mathcal{B}_2} \end{bmatrix}$$

We have just shown that $[Tx]_{\mathcal{B}_2} = {_{\mathcal{B}_2}[T]_{\mathcal{B}_1}}[x]_{\mathcal{B}_1}$ for any $x \in V$. In the important special case $\mathcal{B}_2 = \mathcal{B}_1$, we have $_{\mathcal{B}_1}[T]_{\mathcal{B}_1}$, which is called the $\mathcal{B}_1$ *basis representation of T*.

Consider the identity linear transformation $I : V \to V$ defined by $Ix = x$ for all x. Then

$$[x]_{\mathcal{B}_2} = [Ix]_{\mathcal{B}_2} = {}_{\mathcal{B}_2}[I]_{\mathcal{B}_1}[x]_{\mathcal{B}_1} = {}_{\mathcal{B}_2}[I]_{\mathcal{B}_1}[Ix]_{\mathcal{B}_1} = {}_{\mathcal{B}_2}[I]_{\mathcal{B}_1}\,{}_{\mathcal{B}_1}[I]_{\mathcal{B}_2}[x]_{\mathcal{B}_2}$$

for all $x \in V$. By successively choosing $x = w_1, w_2, \ldots, w_n$, this identity permits us to identify each column of ${}_{\mathcal{B}_2}[I]_{\mathcal{B}_1}\,{}_{\mathcal{B}_1}[I]_{\mathcal{B}_2}$ and shows that

$$_{\mathcal{B}_2}[I]_{\mathcal{B}_1}\,{}_{\mathcal{B}_1}[I]_{\mathcal{B}_2} = I_n$$

If we do the same calculation starting with $[x]_{\mathcal{B}_1} = [Ix]_{\mathcal{B}_1} = \cdots$, we find that

$$_{\mathcal{B}_1}[I]_{\mathcal{B}_2}\,{}_{\mathcal{B}_2}[I]_{\mathcal{B}_1} = I_n$$

Thus, every matrix of the form ${}_{\mathcal{B}_2}[I]_{\mathcal{B}_1}$ is invertible and ${}_{\mathcal{B}_1}[I]_{\mathcal{B}_2}$ is its inverse. Conversely, every invertible matrix $S = [s_1\ s_2\ \ldots\ s_n] \in M_n(\mathbf{F})$ has the form ${}_{\mathcal{B}_1}[I]_{\mathcal{B}}$ for some basis $\mathcal{B}$. We may take $\mathcal{B}$ to be the vectors $\{\tilde{s}_1, \tilde{s}_2, \ldots, \tilde{s}_n\}$ defined by $[\tilde{s}_i]_{\mathcal{B}_1} = s_i$, $i = 1, 2, \ldots, n$. The list $\mathcal{B}$ is independent because S is invertible.

Notice that

$$_{\mathcal{B}_2}[I]_{\mathcal{B}_1} = \big[[Iv_1]_{\mathcal{B}_2} \ldots [Iv_n]_{\mathcal{B}_2}\big] = \big[[v_1]_{\mathcal{B}_2} \ldots [v_n]_{\mathcal{B}_2}\big]$$

so ${}_{\mathcal{B}_2}[I]_{\mathcal{B}_1}$ describes how the elements of the basis $\mathcal{B}_1$ are formed from elements of the basis $\mathcal{B}_2$. Now let $x \in V$ and compute

$$\begin{aligned}
_{\mathcal{B}_2}[T]_{\mathcal{B}_2}[x]_{\mathcal{B}_2} &= [Tx]_{\mathcal{B}_2} = [I(Tx)]_{\mathcal{B}_2} = {}_{\mathcal{B}_2}[I]_{\mathcal{B}_1}[Tx]_{\mathcal{B}_1} \\
&= {}_{\mathcal{B}_2}[I]_{\mathcal{B}_1}\,{}_{\mathcal{B}_1}[T]_{\mathcal{B}_1}[x]_{\mathcal{B}_1} = {}_{\mathcal{B}_2}[I]_{\mathcal{B}_1}\,{}_{\mathcal{B}_1}[T]_{\mathcal{B}_1}[Ix]_{\mathcal{B}_1} \\
&= {}_{\mathcal{B}_2}[I]_{\mathcal{B}_1}\,{}_{\mathcal{B}_1}[T]_{\mathcal{B}_1}\,{}_{\mathcal{B}_1}[I]_{\mathcal{B}_2}[x]_{\mathcal{B}_2}
\end{aligned}$$

By choosing $x = w_1, w_2, \ldots, w_n$ successively, we conclude that

$$_{\mathcal{B}_2}[T]_{\mathcal{B}_2} = {}_{\mathcal{B}_2}[I]_{\mathcal{B}_1}\,{}_{\mathcal{B}_1}[T]_{\mathcal{B}_1}\,{}_{\mathcal{B}_1}[I]_{\mathcal{B}_2} \qquad (0.10.1.1)$$

This identity shows how the $\mathcal{B}_1$ basis representation of T changes if the basis is changed to $\mathcal{B}_2$. For this reason, the matrix ${}_{\mathcal{B}_2}[I]_{\mathcal{B}_1}$ is called the $\mathcal{B}_1 - \mathcal{B}_2$ *change of basis matrix*.

Any matrix $A \in M_n(\mathbf{F})$ is a basis representation of some linear transformation $T : V \to V$, for if $\mathcal{B}$ is any basis of V, we can determine Tx by $[Tx]_{\mathcal{B}} = A[x]_{\mathcal{B}}$. For this T, a computation reveals that ${}_{\mathcal{B}}[T]_{\mathcal{B}} = A$.

0.11 Equivalence relations

Let S be a given set and let Γ be a given subset of $S \times S = \{(a, b) : a \in S \text{ and } b \in S\}$. Then Γ defines a *relation* on S in the following way: We say that a is *related to* b, written $a \sim b$, if $(a, b) \in \Gamma$. A relation on S is said to be an *equivalence relation* if it is (a) *reflexive* ($a \sim a$ for every $a \in S$), (b) *symmetric* ($a \sim b$ whenever $b \sim a$), and (c) *transitive* ($a \sim c$ whenever $a \sim b$ and $b \sim c$). An equivalence relation on S gives a disjoint partition of S in a natural way: If we define the *equivalence class* of any $a \in S$ by $S_a = \{b \in S : b \sim a\}$, then $S = \cup_{a \in S} S_a$, and for each $a, b \in S$, either $S_a = S_b$ (if $a \sim b$) or $S_a \cap S_b = \varnothing$ (if $a \not\sim b$). Conversely, any disjoint partition of S can be used to define an equivalence relation on S.

The following table lists several equivalence relations that arise in matrix analysis. The factors $D_1, D_2, S, T, L,$ and R are square and nonsingular; U and V are unitary; L is lower triangular; R is upper triangular; D_1 and D_2 are diagonal; and A and B need not be square for equivalence, unitary equivalence, triangular equivalence, or diagonal equivalence.

Equivalence Relation $\sim$	$A \sim B$
congruence	$A = SBS^T$
unitary congruence	$A = UBU^T$
*congruence	$A = SBS^*$
consimilarity	$A = SB\bar{S}^{-1}$
equivalence	$A = SBT$
unitary equivalence	$A = UBV$
diagonal equivalence	$A = D_1BD_2$
similarity	$A = SBS^{-1}$
unitary similarity	$A = UBU^*$
triangular equivalence	$A = LBR$

Whenever an interesting equivalence relation arises in matrix analysis, it can be useful to identify a set of distinguished representatives of the equivalence classes (a *canonical form* or *normal form* for the equivalence relation). Alternatively, we often want to have effective criteria (*invariants*) that can be used to decide if two given matrices belong to the same equivalence class.

Abstractly, a *canonical form for an equivalence relation* $\sim$ on a set S is a subset C of S such that $S = \cup_{a \in C} S_a$ and $S_a \cap S_b = \varnothing$ whenever $a, b \in C$ and $a \neq b$; the *canonical form of an element* $a \in S$ is the unique element $c \in C$ such that $a \in S_c$.

For a given equivalence relation in matrix analysis, it is important to make an artful and simple choice of canonical form, and one sometimes does this in more than one way to tailor the canonical form to a specific purpose. For example, the Jordan and Weyr canonical forms are different canonical forms for similarity; the Jordan canonical form works well in problems involving powers of matrices, while the Weyr canonical form works well in problems involving commutativity.

An *invariant* for an equivalence relation $\sim$ on S is a function f on S such that $f(a) = f(b)$ whenever $a \sim b$. A family of invariants $\mathcal{F}$ for an equivalence relation $\sim$ on S is said to be *complete* if $f(a) = f(b)$ for all $f \in \mathcal{F}$ if and only if $a \sim b$; a complete family of invariants is often called a *complete system of invariants*. For example, the singular values of a matrix are a complete system of invariants for unitary equivalence.

CHAPTER 1

Eigenvalues, Eigenvectors, and Similarity

1.0 Introduction

In the initial section of each chapter, we motivate some key issues discussed in the chapter with examples of how they arise, either conceptually or in applications.

Throughout the book, we use the notation and terminology introduced in Chapter 0. Readers should consult the Index to find the definition of an unfamiliar term; unfamiliar notation can usually be identified by consulting the Notation section that follows the References.

1.0.1 Change of basis and similarity. Every invertible matrix is a change-of-basis matrix, and every change-of-basis matrix is invertible (0.10). Thus, if $\mathcal{B}$ is a given basis of a vector space V, if T is a given linear transformation on V, and if $A = {}_{\mathcal{B}}[T]_{\mathcal{B}}$ is the $\mathcal{B}$ basis representation of T, the set of all possible basis representations of T is

$$\{ {}_{\mathcal{B}_1}[I]_{\mathcal{B}} \, {}_{\mathcal{B}}[T]_{\mathcal{B}} \, {}_{\mathcal{B}}[I]_{\mathcal{B}_1} \; : \; \mathcal{B}_1 \text{ is a basis of } V \}$$
$$= \{ S^{-1}AS \colon S \in M_n(\mathbf{F}) \text{ is invertible} \}$$

This is just the set of all matrices that are *similar* to the given matrix A. Similar but not identical matrices are therefore just different basis representations of a single linear transformation.

One would expect similar matrices to share many important properties – at least, those properties that are intrinsic to the underlying linear transformation – and this is an important theme in linear algebra. It is often useful to step back from a question about a given matrix to a question about some intrinsic property of the linear transformation of which the matrix is only one of many possible representations.

The notion of similarity is a key concept in this chapter.

1.0.2 Constrained extrema and eigenvalues. A second key concept in this chapter is the notion of *eigenvector* and *eigenvalue*. Nonzero vectors x such that Ax is a scalar multiple of x play a major role in analyzing the structure of a matrix or linear transformation, but such vectors arise in the more elementary context of maximizing (or minimizing) a real symmetric quadratic form subject to a geometric constraint: For a given real symmetric $A \in M_n(\mathbf{R})$,

$$\text{maximize } x^T A x, \quad \text{subject to} \quad x \in \mathbf{R}^n, \quad x^T x = 1 \tag{1.0.3}$$

A conventional approach to such a constrained optimization problem is to introduce the Lagrangian $L = x^T A x - \lambda x^T x$. Necessary conditions for an extremum are

$$0 = \nabla L = 2(Ax - \lambda x) = 0$$

Thus, if a vector $x \in \mathbf{R}^n$ with $x^T x = 1$ (and hence $x \neq 0$) is an extremum of $x^T A x$, it must satisfy the equation $Ax = \lambda x$. A scalar λ such that $Ax = \lambda x$ for some nonzero vector x is an *eigenvalue* of A.

Problems

1.0.P1 Use Weierstrass's theorem (see Appendix E) to explain why the constrained extremum problem (1.0.3) has a solution, and conclude that every real symmetric matrix has at least one real eigenvalue.

1.0.P2 Suppose that $A \in M_n(\mathbf{R})$ is symmetric. Show that $\max\{x^T A x : x \in \mathbf{R}^n, x^T x = 1\}$ is the *largest* real eigenvalue of A.

1.1 The eigenvalue–eigenvector equation

A matrix $A \in M_n$ can be thought of as a linear transformation from $\mathbf{C}^n$ into $\mathbf{C}^n$, namely,

$$A : x \to Ax \tag{1.1.1}$$

but it is also useful to think of it as an array of numbers. The interplay between these two concepts of A, and what the array of numbers tells us about the linear transformation, is a central theme of matrix analysis and a key to applications. A fundamental concept in matrix analysis is the set of *eigenvalues* of a square complex matrix.

Definition 1.1.2. *Let $A \in M_n$. If a scalar λ and a nonzero vector x satisfy the equation*

$$Ax = \lambda x, \qquad x \in \mathbf{C}^n, x \neq 0, \lambda \in \mathbf{C} \tag{1.1.3}$$

then λ is called an eigenvalue *of A and x is called an* eigenvector *of A associated with λ. The pair λ, x is an* eigenpair *for A.*

The scalar λ and the vector x in the preceding definition occur inextricably as a pair. It is a key element of the definition that *an eigenvector can never be the zero vector.*

Exercise. Consider the diagonal matrix $D = \text{diag}(d_1, d_2, \ldots, d_n)$. Explain why the standard basis vectors e_i, $i = 1, \ldots, n$, are eigenvectors of D. With what eigenvalue is each eigenvector e_i associated?

Equation 1.1.3 can be rewritten as $\lambda x - Ax = (\lambda I - A)x = 0$, a square system of homogeneous linear equations. If this system has a nontrivial solution, then λ is an eigenvalue of A and the matrix $\lambda I - A$ is singular. Conversely, if $\lambda \in \mathbf{C}$ and if $\lambda I - A$ is singular, then there is a nonzero vector x such that $(\lambda I - A)x = 0$, so $Ax = \lambda x$, that is, λ, x is an eigenvalue–eigenvector pair for A.

Definition 1.1.4. *The* spectrum *of $A \in M_n$ is the set of all $\lambda \in \mathbf{C}$ that are eigenvalues of A; we denote this set by $\sigma(A)$.*

For a given $A \in M_n$, we do not know at this point whether $\sigma(A)$ is empty, or, if it is not empty, whether it contains finitely or infinitely many complex numbers.

Exercise. If x is an eigenvector associated with an eigenvalue λ of A, show that any nonzero scalar multiple of x is an eigenvector of A associated with λ.

If x is an eigenvector of $A \in M_n$ associated with λ, it is often convenient to normalize it, that is, to form the unit vector $\xi = x/\|x\|_2$, which is still an eigenvector of A associated with λ. Normalization does not select a unique eigenvector associated with λ, however: $\lambda, e^{i\theta}\xi$ is an eigenvalue-eigenvector pair for A for all $\theta \in \mathbf{R}$.

Exercise. If $Ax = \lambda x$, observe that $\bar{A}\bar{x} = \bar{\lambda}\bar{x}$. Explain why $\sigma(\bar{A}) = \overline{\sigma(A)}$. If $A \in M_n(\mathbf{R})$ and $\lambda \in \sigma(A)$, explain why $\bar{\lambda} \in \sigma(A)$ as well.

Even if they had no other importance, eigenvalues and eigenvectors would be interesting algebraically: according to (1.1.3), the eigenvectors are just those nonzero vectors such that multiplication by the matrix A is the same as multiplication by the scalar λ.

Exercise. Consider the matrix

$$A = \begin{bmatrix} 7 & -2 \\ 4 & 1 \end{bmatrix} \in M_2 \tag{1.1.4a}$$

Then $3 \in \sigma(A)$ and $\begin{bmatrix} 1 \\ 2 \end{bmatrix}$ is an associated eigenvector since

$$A\begin{bmatrix} 1 \\ 2 \end{bmatrix} = \begin{bmatrix} 3 \\ 6 \end{bmatrix} = 3\begin{bmatrix} 1 \\ 2 \end{bmatrix}$$

Also, $5 \in \sigma(A)$. Find an eigenvector associated with the eigenvalue 5.

Sometimes the structure of a matrix makes an eigenvector easy to perceive, so the associated eigenvalue can be computed easily.

Exercise. Let J_n be the n-by-n matrix whose entries are all equal to 1. Consider the n-vector e whose entries are all equal to 1, and let $x_k = e - ne_k$, in which $\{e_1, \ldots, e_n\}$ is the standard basis for $\mathbf{C}^n$. For $n = 2$, show that e and x_1 are linearly independent eigenvectors of J_2 and that 2 and 0, respectively, are the associated eigenvalues. For $n = 3$, show that e, x_1, and x_2 are linearly independent eigenvectors of J_3 and that 3, 0, and 0, respectively, are the associated eigenvalues. In general, show that $e, x_1, \ldots, x_{n-1}$ are linearly independent eigenvectors of J_n and that $n, 0, \ldots, 0$, respectively, are the associated eigenvalues.

Exercise. Show that 1 and 4 are eigenvalues of the matrix

$$A = \begin{bmatrix} 3 & -1 & -1 \\ -1 & 3 & -1 \\ -1 & -1 & 3 \end{bmatrix}$$

Hint: Use eigenvectors. Write $A = 4I - J_3$ and use the preceding exercise.

Evaluation of a polynomial of degree k

$$p(t) = a_k t^k + a_{k-1} t^{k-1} + \cdots + a_1 t + a_0, \quad a_k \neq 0 \qquad (1.1.5a)$$

with real or complex coefficients at a matrix $A \in M_n$ is well-defined since we may form linear combinations of integral powers of a given square matrix. We define

$$p(A) = a_k A^k + a_{k-1} A^{k-1} + \cdots + a_1 A + a_0 I \qquad (1.1.5b)$$

in which we observe the universal convention that $A^0 = I$. A polynomial (1.1.5a) of degree k is said to be *monic* if $a_k = 1$; since $a_k \neq 0$, $a_k^{-1} p(t)$ is always monic. Of course, a monic polynomial cannot be the zero polynomial.

There is an alternative way to represent $p(A)$ that has very important consequences. The *fundamental theorem of algebra* (Appendix C) ensures that any monic polynomial (1.1.5a) of degree $k \geq 1$ can be represented as a product of exactly k complex or real linear factors:

$$p(t) = (t - \alpha_1) \cdots (t - \alpha_k) \qquad (1.1.5c)$$

This representation of $p(t)$ is unique up to permutation of its factors. It tells us that $p(\alpha_j) = 0$ for each $j = 1, \ldots, k$, so that each α_j is a *root* of the equation $p(t) = 0$; one also says that each α_j is a *zero* of $p(t)$. Conversely, if β is a complex number such that $p(\beta) = 0$, then $\beta \in \{\alpha_1, \ldots, \alpha_k\}$, so a polynomial of degree $k \geq 1$ has at most k distinct zeroes. In the product (1.1.5c), some factors might be repeated, for example, $p(t) = t^2 + 2t + 1 = (t + 1)(t + 1)$. The number of times a factor $(t - \alpha_j)$ is repeated is the *multiplicity* of α_j as a zero of $p(t)$. The factorization (1.1.5c) gives a factorization of $p(A)$:

$$p(A) = (A - \alpha_1 I) \cdots (A - \alpha_k I) \qquad (1.1.5d)$$

The eigenvalues of $p(A)$ are linked to the eigenvalues of A in a simple way.

Theorem 1.1.6. *Let $p(t)$ be a given polynomial of degree k. If λ, x is an eigenvalue–eigenvector pair of $A \in M_n$, then $p(\lambda), x$ is an eigenvalue–eigenvector pair of $p(A)$. Conversely, if $k \geq 1$ and if μ is an eigenvalue of $p(A)$, then there is some eigenvalue λ of A such that $\mu = p(\lambda)$.*

Proof. We have

$$p(A)x = a_k A^k x + a_{k-1} A^{k-1} x + \cdots + a_1 A x + a_0 x, \quad a_k \neq 0$$

and $A^j x = A^{j-1} A x = A^{j-1} \lambda x = \lambda A^{j-1} x = \cdots = \lambda^j x$ by repeated application of the eigenvalue–eigenvector equation. Thus,

$$p(A)x = a_k \lambda^k x + \cdots + a_0 x = (a_k \lambda^k + \cdots + a_0)x = p(\lambda)x$$

Conversely, if μ is an eigenvalue of $p(A)$, then $p(A) - \mu I$ is singular. Since $p(t)$ has degree $k \geq 1$, the polynomial $q(t) = p(t) - \mu$ has degree $k \geq 1$, and we can factor it as $q(t) = (t - \beta_1) \cdots (t - \beta_k)$ for some complex or real $\beta_1, \ldots, \beta_k$. Since $p(A) - \mu I = q(A) = (A - \beta_1 I) \cdots (A - \beta_k I)$ is singular, some factor $A - \beta_j I$ is singular, which means that β_j is an eigenvalue of A. But $0 = q(\beta_j) = p(\beta_j) - \mu$, so $\mu = p(\beta_j)$, as claimed. $\qquad\square$

Exercise. Suppose that $A \in M_n$. If $\sigma(A) = \{-1, 1\}$, what is $\sigma(A^2)$? *Caution*: The first assertion in Theorem 1.1.6 permits you to identify a point in $\sigma(A^2)$, but you must invoke the second assertion to find out if it is the *only* point in $\sigma(A^2)$.

Exercise. Consider $A = \left[\begin{smallmatrix} 0 & 1 \\ 0 & 0 \end{smallmatrix}\right]$. What is A^2? Show that e_1 is an eigenvector of A and of A^2, both associated with the eigenvalue $\lambda = 0$. Show that e_2 is an eigenvector of A^2 but *not* of A. Explain why the "converse" part of Theorem 1.1.6 speaks only about eigenvalues of $p(A)$, not eigenvectors. Show that A has no eigenvectors other than scalar multiples of e_1, and explain why $\sigma(A) = \{0\}$.

Observation 1.1.7. *A matrix $A \in M_n$ is singular if and only if $0 \in \sigma(A)$.*

Proof. The matrix A is singular if and only if $Ax = 0$ for some $x \neq 0$. This happens if and only if $Ax = 0x$ for some $x \neq 0$, that is, if and only if $\lambda = 0$ is an eigenvalue of A. $\qquad\square$

Observation 1.1.8. *Let $A \in M_n$ and $\lambda, \mu \in \mathbf{C}$ be given. Then $\lambda \in \sigma(A)$ if and only if $\lambda + \mu \in \sigma(A + \mu I)$.*

Proof. If $\lambda \in \sigma(A)$, there is a nonzero vector x such that $Ax = \lambda x$ and hence $(A + \mu I)x = Ax + \mu x = \lambda x + \mu x = (\lambda + \mu)x$. Thus, $\lambda + \mu \in \sigma(A + \mu I)$. Conversely, if $\lambda + \mu \in \sigma(A + \mu I)$, there is a nonzero vector y such that $Ay + \mu y = (A + \mu I)y = (\lambda + \mu)y = \lambda y + \mu y$. Thus, $Ay = \lambda y$ and $\lambda \in \sigma(A)$. $\qquad\square$

We are now prepared to make a very important observation: *every complex matrix has a nonempty spectrum*, that is, for each $A \in M_n$, there is some scalar $\lambda \in \mathbf{C}$ and some nonzero $x \in \mathbf{C}^n$ such that $Ax = \lambda x$.

Theorem 1.1.9. *Let $A \in M_n$ be given. Then A has an eigenvalue. In fact, for each given nonzero $y \in \mathbf{C}^n$, there is a polynomial $g(t)$ of degree at most $n - 1$ such that $g(A)y$ is an eigenvector of A.*

Proof. Let m be the *least* integer k such that the vectors $y, Ay, A^2 y, \ldots, A^k y$ are linearly dependent. Then $m \geq 1$ since $y \neq 0$, and $m \leq n$ since *any* $n + 1$ vectors in $\mathbf{C}^n$ are linearly dependent. Let $a_0, a_1, \ldots, a_m$ be scalars, not all zero, such that

$$a_m A^m y + a_{m-1} A^{m-1} y + \cdots + a_1 Ay + a_0 y = 0 \qquad (1.1.10)$$

If $a_m = 0$, then (1.1.10) implies that the vectors $y, Ay, A^2 y, \ldots, A^{m-1} y$ are linearly dependent, contradicting the minimality of m. Thus, $a_m \neq 0$, and we may consider the polynomial $p(t) = t^m + (a_{m-1}/a_m)t^{m-1} + \cdots + (a_1/a_m)t + (a_0/a_m)$. The identity (1.1.10) ensures that $p(A)y = 0$, so $0, y$ is an eigenvalue–eigenvector pair for $p(A)$. Theorem 1.1.6 ensures that one of the m zeroes of $p(t)$ is an eigenvalue of A.

Suppose that λ is a zero of $p(t)$ that is an eigenvalue of A and factor $p(t) = (t - \lambda)g(t)$, in which $g(t)$ is a polynomial of degree $m - 1$. If $g(A)y = 0$, the minimality of m would be contradicted again, so $g(A)y \neq 0$. But $0 = p(A)y = (A - \lambda I)(g(A)y)$, so the nonzero vector $g(A)y$ is an eigenvector of A associated with the eigenvalue λ. $\qquad\square$

The preceding argument shows that for a given $A \in M_n$ we can find a polynomial of degree at most n such that *at least one* of its zeroes is an eigenvalue of A. In the next section, we introduce a polynomial $p_A(t)$ of degree exactly n such that each of its zeroes is an eigenvalue of A and each eigenvalue of A is a zero of $p_A(t)$; that is, $p_A(\lambda) = 0$ if and only if $\lambda \in \sigma(A)$.

Problems

1.1.P1 Suppose that $A \in M_n$ is nonsingular. According to (1.1.7), this is equivalent to assuming that $0 \notin \sigma(A)$. For each $\lambda \in \sigma(A)$, show that $\lambda^{-1} \in \sigma(A^{-1})$. If $Ax = \lambda x$ and $x \neq 0$, show that $A^{-1}x = \lambda^{-1}x$.

1.1.P2 Let $A \in M_n$ be given. (a) Show that the sum of the entries in each row of A is 1 if and only if $1 \in \sigma(A)$ and the vector $e = [1, 1, \ldots, 1]^T$ is an associated eigenvector, that is, $Ae = e$. (b) Suppose that the sum of the entries in each row of A is 1. If A is nonsingular, show that the sum of the entries in each row of A^{-1} is also 1. Moreover, for any given polynomial $p(t)$, show that the sums of the entries in each row $p(A)$ are equal. Equal to what?

1.1.P3 Let $A \in M_n(\mathbf{R})$. Suppose that λ is a real eigenvalue of A and that $Ax = \lambda x, x \in \mathbf{C}^n, x \neq 0$. Let $x = u + iv$, in which $u, v \in \mathbf{R}^n$ are the respective real and imaginary parts of x; see (0.2.5). Show that $Au = \lambda u$ and $Av = \lambda v$. Explain why at least one of u, v must be nonzero, and conclude that A has a real eigenvector associated with λ. Must both u and v be eigenvectors of A? Can A have a real eigenvector associated with an eigenvalue that is not real?

1.1.P4 Consider the block diagonal matrix

$$A = \begin{bmatrix} A_{11} & 0 \\ 0 & A_{22} \end{bmatrix}, \qquad A_{ii} \in M_{n_i}$$

Show that $\sigma(A) = \sigma(A_{11}) \cup \sigma(A_{22})$. You must show three things: (a) if λ is an eigenvalue of A, then it is an eigenvalue of either A_{11} or of A_{22}; (b) if λ is an eigenvalue of A_{11}, then it is an eigenvalue of A; and (c) if λ is an eigenvalue of A_{22}, then it is an eigenvalue of A.

1.1.P5 Let $A \in M_n$ be idempotent, that is, $A^2 = A$. Show that each eigenvalue of A is either 0 or 1. Explain why I is the only nonsingular idempotent matrix.

1.1.P6 Show that all eigenvalues of a nilpotent matrix are 0. Give an example of a nonzero nilpotent matrix. Explain why 0 is the only nilpotent idempotent matrix.

1.1.P7 If $A \in M_n$ is Hermitian, show that all eigenvalues of A are *real*.

1.1.P8 Explain how the argument in (1.1.9) fails if we try to use it to show that every square real matrix has a real eigenvalue.

1.1.P9 Use the definition (1.1.3) to show that the real matrix $A = \begin{bmatrix} 0 & 1 \\ -1 & 0 \end{bmatrix}$ has no real eigenvalue. However, (1.1.9) says that A has a complex eigenvalue. Actually, there are two; what are they?

1.1.P10 Provide details for the following example, which shows that a linear operator on an infinite dimensional complex vector space might have *no* eigenvalues. Let $V = \{(a_1, a_2, \ldots) : a_i \in \mathbf{C}, i = 1, 2, \ldots\}$ be the vector space of all formal infinite sequences of complex numbers, and define the *right-shift operator* S on V by $S(a_1, a_2, \ldots) = (0, a_1, a_2, \ldots)$. Verify that S is a linear transformation. If $Sx = \lambda x$, show that $x = 0$.

1.1.P11 Let $A \in M_n$ and $\lambda \in \sigma(A)$ be given. Then $A - \lambda I$ is singular, so $(A - \lambda I) \operatorname{adj}(A - \lambda I) = (\det(A - \lambda I))I = 0$; see (0.8.2). Explain why there is some $y \in \mathbf{C}^n$ ($y = 0$ is possible) such that $\operatorname{adj}(A - \lambda I) = xy^*$. Conclude that every nonzero column of $\operatorname{adj}(A - \lambda I)$ is an eigenvector of A associated with the eigenvalue λ. Why is this observation useful only if $\operatorname{rank}(A - \lambda I) = n - 1$?

1.1.P12 Suppose that λ is an eigenvalue of $A = \begin{bmatrix} a & b \\ c & d \end{bmatrix} \in M_2$. Use (1.1.P11) to show that if either column of $\begin{bmatrix} d - \lambda & -b \\ -c & a - \lambda \end{bmatrix}$ is nonzero, then it is an eigenvector of A associated with λ. Why must one of the columns be a scalar multiple of the other? Use this method to find eigenvectors of the matrix (1.1.4a) associated with the eigenvalues 3 and 5.

1.1.P13 Let $A \in M_n$ and let λ, x be an eigenvalue-eigenvector pair of A. Show that x is an eigenvector of $\operatorname{adj} A$.

1.2 The characteristic polynomial and algebraic multiplicity

How many eigenvalues does a square complex matrix have? How may they be characterized in a systematic way?

Rewrite the eigenvalue–eigenvector equation (1.1.3) as

$$(\lambda I - A)x = 0, \qquad x \neq 0 \tag{1.2.1}$$

Thus, $\lambda \in \sigma(A)$ if and only if $\lambda I - A$ is singular, that is, if and only if

$$\det(\lambda I - A) = 0 \tag{1.2.2}$$

Definition 1.2.3. *Thought of as a formal polynomial in t, the* characteristic polynomial *of $A \in M_n$ is*

$$p_A(t) = \det(tI - A)$$

We refer to the equation $p_A(t) = 0$ as the characteristic equation *of A.*

Observation 1.2.4. *The characteristic polynomial of each $A = [a_{ij}] \in M_n$ has degree n and $p_A(t) = t^n - (\operatorname{tr} A)t^{n-1} + \cdots + (-1)^n \det A$. Moreover, $p_A(\lambda) = 0$ if and only if $\lambda \in \sigma(A)$, so $\sigma(A)$ contains at most n complex numbers.*

Proof. Each summand in the presentation (0.3.2.1) of the determinant of $tI - A$ is a product of exactly n entries of $tI - A$, each from a different row and column, so each summand is a polynomial in t of degree at most n. The degree of a summand can be n

only if every factor in the product involves t, which happens only for the summand

$$(t - a_{11}) \cdots (t - a_{nn}) = t^n - (a_{11} + \cdots + a_{nn})t^{n-1} + \cdots \quad (1.2.4a)$$

that is the product of the diagonal entries. Any other summand must contain a factor $-a_{ij}$ with $i \neq j$, so the diagonal entries $(t - a_{ii})$ (in the same row as a_{ij}) and $(t - a_{jj})$ (in the same column as a_{ij}) cannot also be factors; this summand therefore cannot have degree larger than $n - 2$. Thus, the coefficients of t^n and t^{n-1} in the polynomial $p_A(t)$ arise only from the summand (1.2.4a). The constant term in $p_A(t)$ is just $p_A(0) = \det(0I - A) = \det(-A) = (-1)^n \det A$. The remaining assertion is the equivalence of (1.2.1) and (1.2.2), together with the fact that a polynomial of degree $n \geq 1$ has at most n distinct zeroes. $\qquad \square$

Exercise. Show that the roots of $\det(A - tI) = 0$ are the same as those of $\det(tI - A) = 0$ and that $\det(A - tI) = (-1)^n \det(tI - A) = (-1)^n(t^n + \cdots)$

The characteristic polynomial could alternatively be defined as $\det(A - tI) = (-1)^n t^n + \cdots$. The convention we have chosen ensures that the coefficient of t^n in the characteristic polynomial is always $+1$.

Exercise. Let $A = \begin{bmatrix} a & b \\ c & d \end{bmatrix} \in M_2$. Show that the characteristic polynomial of A is

$$p_A(t) = t^2 - (a + d)t + (ad - bc) = t^2 - (\operatorname{tr} A)t + \det A$$

Let $r = (-\operatorname{tr} A)^2 - 4 \det A = (a - d)^2 + 4bc$ (this is the *discriminant* of $p_A(t)$) and let $\sqrt{r}$ be a fixed square root of r. Deduce that each of

$$\lambda_1 = \tfrac{1}{2}\left(a + d + \sqrt{r}\right) \quad \text{and} \quad \lambda_2 = \tfrac{1}{2}\left(a + d - \sqrt{r}\right) \quad (1.2.4b)$$

is an eigenvalue of A. Verify that $\operatorname{tr} A = \lambda_1 + \lambda_2$ and $\det A = \lambda_1 \lambda_2$. Explain why $\lambda_1 \neq \lambda_2$ if and only if $r \neq 0$. If $A \in M_2(\mathbf{R})$, show that (a) the eigenvalues of A are real if and only if $r \geq 0$; (b) the eigenvalues of A are real if $bc \geq 0$; and (c) if $r < 0$, then $\lambda_1 = \overline{\lambda_2}$, that is, λ_1 is the complex conjugate of λ_2.

The preceding exercise illustrates that an eigenvalue λ of a matrix $A \in M_n$ with $n > 1$ can be a multiple zero of $p_A(t)$ (equivalently, a multiple root of its characteristic equation). Indeed, the characteristic polynomial of $I \in M_n$ is

$$p_I(t) = \det(tI - I) = \det((t - 1)I) = (t - 1)^n \det I = (t - 1)^n$$

so the eigenvalue $\lambda = 1$ has multiplicity n as a zero of $p_I(t)$. How should we account for such repetitions in an enumeration of the eigenvalues?

For a given $A \in M_n$ with $n > 1$, factor its characteristic polynomial as $p_A(t) = (t - \alpha_1) \cdots (t - \alpha_n)$. We know that each zero α_i of $p_A(t)$ (regardless of its multiplicity) is an eigenvalue of A. A computation reveals that

$$p_A(t) = t^n - (\alpha_1 + \cdots + \alpha_n)t^{n-1} + \cdots + (-1)^n \alpha_1 \cdots \alpha_n \quad (1.2.4c)$$

so a comparison of (1.2.4a) and (1.2.4c) tells us that the sum of the zeroes of $p_A(t)$ is the trace of A, and the product of the zeroes of $p_A(t)$ is the determinant of A. If each zero of $p_A(t)$ has multiplicity 1, that is, if $\alpha_i \neq \alpha_j$ whenever $i \neq j$, then $\sigma(A) = \{\alpha_1, \ldots, \alpha_n\}$, so $\operatorname{tr} A$ *is the sum of the eigenvalues* of A and $\det A$ *is the product of the eigenvalues* of A. If these two statements are to remain true even if some zeroes of $p_A(t)$ have multiplicity greater than 1, we must enumerate the eigenvalues of A according to their multiplicities as roots of the characteristic equation.

Definition 1.2.5. *Let $A \in M_n$. The* multiplicity *of an eigenvalue λ of A is its multiplicity as a zero of the characteristic polynomial $p_A(t)$. For clarity, we sometimes refer to the multiplicity of an eigenvalue as its* algebraic multiplicity.

Henceforth, *the eigenvalues of $A \in M_n$ will always mean the eigenvalues together with their respective (algebraic) multiplicities*. Thus, the zeroes of the characteristic polynomial of A (including their multiplicities) are the same as the eigenvalues of A (including their multiplicities):

$$p_A(t) = (t - \lambda_1)(t - \lambda_2) \cdots (t - \lambda_n) \tag{1.2.6}$$

in which $\lambda_1, \ldots, \lambda_n$ are the n eigenvalues of A, listed in any order. When we refer to *the distinct eigenvalues* of A, we mean the elements of the set $\sigma(A)$.

We can now say without qualification that *each matrix $A \in M_n$ has exactly n eigenvalues among the complex numbers*; the trace and determinant of A are the sum and product, respectively, of its eigenvalues. If A is real, some or all of its eigenvalues might not be real.

Exercise. Consider a real matrix $A \in M_n(\mathbf{R})$. (a) Explain why all the coefficients of $p_A(t)$ are real. (b) Suppose that A has an eigenvalue λ that is not real. Use (a) to explain why $\bar{\lambda}$ is also an eigenvalue of A and why the algebraic multiplicities of λ and $\bar{\lambda}$ are the same. If x, λ is an eigenpair for A, we know that $\bar{x}, \bar{\lambda}$ is also an eigenpair (why?). Notice that x and $\bar{x}$ are eigenvectors of A that are associated with *distinct* eigenvalues λ and $\bar{\lambda}$.

Example 1.2.7. Let $x, y \in \mathbf{C}^n$. What are the eigenvalues and determinant of $I + xy^*$? Using (0.8.5.11) and the fact that $\text{adj}(\alpha I) = \alpha^{n-1} I$, we compute

$$
\begin{aligned}
p_{I+xy^*}(t) &= \det(tI - (I + xy^*)) = \det((t - 1)I - xy^*) \\
&= \det((t - 1)I) - y^* \, \text{adj}((t - 1)I)x \\
&= (t - 1)^n - (t - 1)^{n-1} y^* x = (t - 1)^{n-1}(t - (1 + y^* x))
\end{aligned}
$$

Thus, the eigenvalues of $I + xy^*$ are $1 + y^* x$ and 1 (with multiplicity $n - 1$), so $\det(I + xy^*) = (1 + y^* x)(1)^{n-1} = 1 + y^* x$.

Example 1.2.8. Brauer's theorem. Let $x, y \in \mathbf{C}^n$, $x \neq 0$, and $A \in M_n$. Suppose that $Ax = \lambda x$ and let the eigenvalues of A be $\lambda, \lambda_2, \ldots, \lambda_n$. What are the eigenvalues of $A + xy^*$? First observe that $(t - \lambda)x = (tI - A)x$ implies that $(t - \lambda) \, \text{adj}(tI - A)x = \text{adj}(tI - A)(tI - A)x = \det(tI - A)x$, that is,

$$(t - \lambda) \, \text{adj}(tI - A)x = p_A(t)x \tag{1.2.8a}$$

Use (0.8.5.11) to compute

$$
\begin{aligned}
p_{A+xy^*}(t) &= \det(tI - (A + xy^*)) = \det((tI - A) - xy^*) \\
&= \det(tI - A) - y^* \, \text{adj}(tI - A)x
\end{aligned}
$$

Multiply both sides by $(t - \lambda)$, use (1.2.8a), and obtain

$$(t - \lambda)\, p_{A+xy^*}(t) = (t - \lambda) \det(tI - A) - y^*(t - \lambda) \operatorname{adj}(tI - A)x$$
$$= (t - \lambda)p_A(t) - p_A(t)y^*x$$

which is the polynomial identity

$$(t - \lambda)p_{A+xy^*}(t) = (t - (\lambda + y^*x))p_A(t)$$

The zeroes of the left-hand polynomial are λ together with the n eigenvalues of $A + xy^*$. The zeroes of the right-hand polynomial are $\lambda + y^*x, \lambda, \lambda_2, \ldots, \lambda_n$. It follows that the eigenvalues of $A + xy^*$ are $\lambda + y^*x, \lambda_2, \ldots, \lambda_n$.

Since we now know that each n-by-n complex matrix has finitely many eigenvalues, we may make the following definition.

Definition 1.2.9. *Let $A \in M_n$. The* spectral radius *of A is $\rho(A) = \max\{|\lambda| : \lambda \in \sigma(A)\}$.*

Exercise. Explain why every eigenvalue of $A \in M_n$ lies in the closed bounded disk $\{z : z \in \mathbf{C}$ and $|z| \le \rho(A)\}$ in the complex plane.

Exercise. Suppose that $A \in M_n$ has at least one nonzero eigenvalue. Explain why $\min\{|\lambda| : \lambda \in \sigma(A)$ and $\lambda \neq 0\} > 0$.

Exercise. Underlying both of the two preceding exercises is the fact that $\sigma(A)$ is a nonempty finite set. Explain why.

Sometimes the structure of a matrix makes the characteristic polynomial easy to calculate. This is the case for diagonal or triangular matrices.

Exercise. Consider an upper triangular matrix

$$T = \begin{bmatrix} t_{11} & \cdots & t_{1n} \\ & \ddots & \vdots \\ 0 & & t_{nn} \end{bmatrix} \in M_n$$

Show that $p_T(t) = (t - t_{11}) \cdots (t - t_{nn})$, so the eigenvalues of T are its diagonal entries $t_{11}, t_{22}, \ldots, t_{nn}$. What if T is lower triangular? What if T is diagonal?

Exercise. Suppose that $A \in M_n$ is block upper triangular

$$A = \begin{bmatrix} A_{11} & & \bigstar \\ & \ddots & \\ 0 & & A_{kk} \end{bmatrix}, \quad A_{ii} \in M_{n_i} \text{ for } i = 1, \ldots, n_k$$

Explain why $p_A(t) = p_{A_{11}}(t) \cdots p_{A_{kk}}(t)$ and the eigenvalues of A are the eigenvalues of A_{11}, together with those of $A_{22}, \ldots$, together with those of A_{kk} including all their respective algebraic multiplicities. This observation is the basis of many algorithms to compute eigenvalues. Explain why the preceding exercise is a special case of this one.

Definition 1.2.10. *Let $A \in M_n$. The sum of its principal minors of size k (there are $\binom{n}{k}$ of them) is denoted by $E_k(A)$.*

We have already encountered principal minor sums as two coefficients of the characteristic polynomial

$$p_A(t) = t^n + a_{n-1}t^{n-1} + \cdots + a_2 t^2 + a_1 t + a_0 \qquad (1.2.10a)$$

If $k = 1$, then $\binom{n}{k} = n$ and $E_1(A) = a_{11} + \cdots + a_{nn} = \operatorname{tr} A = -a_{n-1}$; if $k = n$, then $\binom{n}{k} = 1$ and $E_n(A) = \det A = (-1)^n a_0$. The broader connection between coefficients and principal minor sums is a consequence of the fact that the coefficients are explicit functions of certain derivatives of $p_A(t)$ at $t = 0$:

$$a_k = \frac{1}{k!} p_A^{(k)}(0), \quad k = 0, 1, \ldots, n-1 \qquad (1.2.11)$$

Use (0.8.10.2) to evaluate the derivative

$$p_A'(t) = \operatorname{tr} \operatorname{adj}(tI - A)$$

Observe that $\operatorname{tr} \operatorname{adj} A$ is the sum of the principal minors of A of size $n - 1$, so $\operatorname{tr} \operatorname{adj} A = E_{n-1}(A)$. Then

$$
\begin{aligned}
a_1 &= p_A'(t)\big|_{t=0} = \operatorname{tr} \operatorname{adj}(tI - A)\big|_{t=0} = \operatorname{tr} \operatorname{adj}(-A) \\
&= (-1)^{n-1} \operatorname{tr} \operatorname{adj}(A) = (-1)^{n-1} E_{n-1}(A)
\end{aligned}
$$

Now observe that $\operatorname{tr} \operatorname{adj}(tI - A) = \sum_{i=1}^{n} p_{A_{(i)}}(t)$ is the sum of the characteristic polynomials of the n principal submatrices of A of size $n - 1$, which we denote by $A_{(1)}, \ldots, A_{(n)}$. Use (0.8.10.2) again to evaluate

$$p_A''(t) = \frac{d}{dt} \operatorname{tr} \operatorname{adj}(tI - A) = \sum_{i=1}^{n} \frac{d}{dt} p_{A_{(i)}}(t) = \sum_{i=1}^{n} \operatorname{tr} \operatorname{adj}(tI - A_{(i)}) \qquad (1.2.12)$$

Each summand $\operatorname{tr} \operatorname{adj}(tI - A_{(i)})$ is the sum of the $n - 1$ principal minors of size $n - 2$ of a principal minor of $tI - A$, so each summand is a sum of certain principal minors of $tI - A$ of size $n - 2$. Each of the $\binom{n}{n-2}$ principal minors of $tI - A$ of size $n - 2$ appears twice in (1.2.12): the principal minor with rows and columns k and ℓ omitted appears when $i = k$ as well as when $i = \ell$. Thus,

$$
\begin{aligned}
a_2 &= \frac{1}{2} p_A''(t)\big|_{t=0} = \frac{1}{2} \sum_{i=1}^{n} \operatorname{tr} \operatorname{adj}(tI - A_{(i)})\big|_{t=0} = \frac{1}{2} \sum_{i=1}^{n} \operatorname{tr} \operatorname{adj}(-A_{(i)}) \\
&= \frac{1}{2}(-1)^{n-2} \sum_{i=1}^{n} \operatorname{tr} \operatorname{adj}(A_{(i)}) = \frac{1}{2}(-1)^{n-2} (2E_{n-2}(A)) \\
&= (-1)^{n-2} E_{n-2}(A)
\end{aligned}
$$

Repeating this argument reveals that $p_A^{(k)}(0) = k!(-1)^{n-k} E_{n-k}(A), \ k = 0, 1, \ldots, n - 1$, so the coefficients of the characteristic polynomial (1.2.11) are

$$a_k = \frac{1}{k!} p_A^{(k)}(0) = (-1)^{n-k} E_{n-k}(A), \quad k = 0, 1, \ldots, n - 1$$

and hence

$$p_A(t) = t^n - E_1(A)t^{n-1} + \cdots + (-1)^{n-1} E_{n-1}(A)t + (-1)^n E_n(A) \qquad (1.2.13)$$

With the identity (1.2.6) in mind, we make the following definition:

Definition 1.2.14. *The kth elementary symmetric function of n complex numbers* $\lambda_1, \ldots, \lambda_n, k \leq n,$ *is*

$$S_k(\lambda_1, \ldots, \lambda_n) = \sum_{1 \leq i_1 < \cdots < i_k \leq n} \prod_{j=1}^{k} \lambda_{i_j}$$

Notice that the sum has $\binom{n}{k}$ summands. If $A \in M_n$ and $\lambda_1, \ldots, \lambda_n$ are its eigenvalues, we define $S_k(A) = S_k(\lambda_1, \ldots, \lambda_n)$.

Exercise. What are $S_1(\lambda_1, \ldots, \lambda_n)$ and $S_n(\lambda_1, \ldots, \lambda_n)$? Explain why each of the functions $S_k(\lambda_1, \ldots, \lambda_n)$ is unchanged if the list $\lambda_1, \ldots, \lambda_n$ is reindexed and rearranged.

A calculation with (1.2.6) reveals that

$$p_A(t) = t^n - S_1(A)t^{n-1} + \cdots + (-1)^{n-1}S_{n-1}(A)t + (-1)^n S_n(A) \qquad (1.2.15)$$

Comparison of (1.2.13) and (1.2.15) gives the following identities between elementary symmetric functions of eigenvalues of a matrix and sums of its principal minors.

Theorem 1.2.16. *Let $A \in M_n$. Then $S_k(A) = E_k(A)$ for each $k = 1, \ldots, n$.*

The next theorem shows that a singular complex matrix can always be shifted slightly to become nonsingular. This important fact often permits us to use continuity arguments to deduce results about singular matrices from properties of nonsingular matrices.

Theorem 1.2.17. *Let $A \in M_n$. There is some $\delta > 0$ such that $A + \varepsilon I$ is nonsingular whenever $\varepsilon \in \mathbf{C}$ and $0 < |\varepsilon| < \delta$.*

Proof. Observation 1.1.8 ensures that $\lambda \in \sigma(A)$ if and only if $\lambda + \varepsilon \in \sigma(A + \varepsilon I)$. Therefore, $0 \in \sigma(A + \varepsilon I)$ if and only if $\lambda + \varepsilon = 0$ for some $\lambda \in \sigma(A)$, that is, if and only if $\varepsilon = -\lambda$ for some $\lambda \in \sigma(A)$. If all the eigenvalues of A are zero, take $\delta = 1$. If some eigenvalue of A is nonzero, let $\delta = \min\{|\lambda| : \lambda \in \sigma(A) \text{ and } \lambda \neq 0\}$. If we choose any ε such that $0 < |\varepsilon| < \delta$, we are assured that $-\varepsilon \notin \sigma(A)$, so $0 \notin \sigma(A + \varepsilon I)$ and $A + \varepsilon I$ is nonsingular. $\square$

There is a useful connection between the derivatives of a polynomial $p(t)$ and the multiplicity of its zeroes: α is a zero of $p(t)$ with multiplicity $k \geq 1$ if and only if we can write $p(t)$ in the form

$$p(t) = (t - \alpha)^k q(t)$$

in which $q(t)$ is a polynomial such that $q(\alpha) \neq 0$. Differentiating this identity gives $p'(t) = k(t - \alpha)^{k-1}q(t) + (t - \alpha)^k q'(t)$, which shows that $p'(\alpha) = 0$ if and only if $k > 1$. If $k \geq 2$, then $p''(t) = k(k - 1)(t - \alpha)^{k-2}q(t) +$ polynomial terms each involving a factor $(t - \alpha)^m$ with $m \geq k - 1$, so $p''(\alpha) = 0$ if and only if $k > 2$. Repetition of this calculation shows that α is a zero of $p(t)$ of multiplicity k if and only if $p(\alpha) = p'(\alpha) = \cdots = p^{(k-1)}(\alpha) = 0$ and $p^{(k)}(\alpha) \neq 0$.

Theorem 1.2.18. *Let $A \in M_n$ and suppose that $\lambda \in \sigma(A)$ has algebraic multiplicity k. Then* $\text{rank}(A - \lambda I) \geq n - k$ *with equality for $k = 1$.*

Proof. Apply the preceding observation to the characteristic polynomial $p_A(t)$ of a matrix $A \in M_n$ that has an eigenvalue λ with multiplicity $k \geq 1$. If we let $B = A - \lambda I$, then zero is an eigenvalue of B with multiplicity k and hence $p_B^{(k)}(0) \neq 0$. But $p_B^{(k)}(0) = k!(-1)^{n-k} E_{n-k}(B)$, so $E_{n-k}(B) \neq 0$. In particular, *some* principal minor of $B = A - \lambda I$ of size $n - k$ is nonzero, so $\text{rank}(A - \lambda I) \geq n - k$. If $k = 1$, we can say more: $A - \lambda I$ is singular, so $n > \text{rank}(A - \lambda I) \geq n - 1$, which means that $\text{rank}(A - \lambda I) = n - 1$ if the eigenvalue λ has algebraic multiplicity 1. $\qquad\square$

Problems

1.2.P1 Let $A \in M_n$. Use the identity $S_n(A) = E_n(A)$ to verify (1.1.7).

1.2.P2 For matrices $A \in M_{m,n}$ and $B \in M_{n,m}$, show by direct calculation that $\text{tr}(AB) = \text{tr}(BA)$. For any $A \in M_n$ and nonsingular $S \in M_n$, deduce that $\text{tr}(S^{-1}AS) = \text{tr}\,A$. For any $A \in M_n$, use multiplicativity of the determinant function to show that $\det(S^{-1}AS) = \det A$. Conclude that the determinant function on M_n is similarity invariant.

1.2.P3 Let $D \in M_n$ be a diagonal matrix. Compute the characteristic polynomial $p_D(t)$ and show that $p_D(D) = 0$.

1.2.P4 Suppose that $A \in M_n$ is idempotent. Use (1.2.15) and (1.1.P5) to show that every coefficient of $p_A(t)$ is an integer (positive, negative, or zero).

1.2.P5 Use (1.1.P6) to show that the trace of a nilpotent matrix is 0. What is the characteristic polynomial of a nilpotent matrix?

1.2.P6 If $A \in M_n$ and $\lambda \in \sigma(A)$ has multiplicity 1, we know that $\text{rank}(A - \lambda I) = n - 1$. Consider the converse: If $\text{rank}(A - \lambda I) = n - 1$, must λ be an eigenvalue of A? Must it have multiplicity 1?

1.2.P7 Use (1.2.13) to determine the characteristic polynomial of the tridiagonal matrix

$$\begin{bmatrix} 1 & 1 & 0 & 0 & 0 \\ 1 & 1 & 1 & 0 & 0 \\ 0 & 1 & 1 & 1 & 0 \\ 0 & 0 & 1 & 1 & 1 \\ 0 & 0 & 0 & 1 & 1 \end{bmatrix}$$

Consider how this procedure could be used to compute the characteristic polynomial of a general n-by-n tridiagonal matrix.

1.2.P8 Let $A \in M_n$ and $\lambda \in \mathbf{C}$ be given. Suppose that the eigenvalues of A are $\lambda_1, \dots, \lambda_n$. Explain why $p_{A + \lambda I}(t) = p_A(t - \lambda)$ and deduce from this identity that the eigenvalues of $A + \lambda I$ are $\lambda_1 + \lambda, \dots, \lambda_n + \lambda$.

1.2.P9 Explicitly compute $S_2(\lambda_1, \dots, \lambda_6)$, $S_3(\lambda_1, \dots, \lambda_6)$, $S_4(\lambda_1, \dots, \lambda_6)$, and $S_5(\lambda_1, \dots, \lambda_6)$.

1.2.P10 If $A \in M_n(\mathbf{R})$ and if n is *odd*, show that A has at least one real eigenvalue.

1.2.P11 Let V be a vector space over a field $\mathbf{F}$. An eigenvalue of a linear transformation $T : V \to V$ is a scalar $\lambda \in \mathbf{F}$ such that there is a nonzero vector $\upsilon \in V$ with $T\upsilon = \lambda\upsilon$. If $\mathbf{F} = \mathbf{C}$ and if V is finite dimensional, show that every linear transformation $T : V \to V$ has an eigenvalue. Give examples to show that if either hypothesis is weakened (V is not finite dimensional or $\mathbf{F} \neq \mathbf{C}$), then T might not have an eigenvalue.

1.2.P12 Let $x = [x_i]$, $y = [y_i] \in \mathbf{C}^n$, and $a \in \mathbf{C}$ be given and let $A = \begin{bmatrix} 0_n & x \\ y^* & a \end{bmatrix} \in M_{n+1}$. Show that $p_A(t) = t^{n-1}(t^2 - at - y^*x)$ in two ways: (a) Use Cauchy's expansion (0.8.5.10) to calculate $p_A(t)$. (b) Explain why rank $A \leq 2$ and use (1.2.13) to calculate $p_A(t)$. Why do only $E_1(A)$ and $E_2(A)$ need to be calculated and only principal submatrices of the form $\begin{bmatrix} 0 & x_i \\ \bar{y}_i & a \end{bmatrix}$ need to be considered? Show that the eigenvalues of A are $(a \pm \sqrt{a^2 + 4y^*x})/2$ together with $n - 1$ zero eigenvalues.

1.2.P13 Let $x, y \in \mathbf{C}^n$, $a \in \mathbf{C}$, and $B \in M_n$. Consider the bordered matrix $A = \begin{bmatrix} B & x \\ y^* & a \end{bmatrix} \in M_{n+1}$. (a) Use (0.8.5.10) to show that

$$p_A(t) = (t - a)p_B(t) - y^*(\mathrm{adj}(tI - B))x \qquad (1.2.19)$$

(b) If $B = \lambda I_n$, deduce that

$$p_A(t) = (t - \lambda)^{n-1}(t^2 - (a + \lambda)t + a\lambda - y^*x) \qquad (1.2.20)$$

and conclude that the eigenvalues of $\begin{bmatrix} \lambda I_n & x \\ y^* & a \end{bmatrix}$ are λ with multiplicity $n - 1$, together with $\frac{1}{2}(a + \lambda \pm ((a - \lambda)^2 + 4y^*x)^{1/2})$.

1.2.P14 Let $n \geq 3$, $B \in M_{n-2}$, and $\lambda, \mu \in \mathbf{C}$. Consider the block matrix

$$A = \begin{bmatrix} \lambda & \bigstar & \bigstar \\ 0 & \mu & 0 \\ 0 & \bigstar & B \end{bmatrix}$$

in which the $\bigstar$ entries are not necessarily zero. Show that $p_A(t) = (t - \lambda)(t - \mu)p_B(t)$.

1.2.P15 Suppose that $A(t) \in M_n$ is a given continuous matrix-valued function and each of the vector valued functions $x_1(t), \ldots, x_n(t) \in \mathbf{C}^n$ satisfies the system of ordinary differential equations $x_j'(t) = A(t)x_j(t)$. Let $X(t) = [x_1(t) \ldots x_n(t)]$ and let $W(t) = \det X(t)$. Use (0.8.10) and (0.8.2.11) and provide details for the following argument:

$$W'(t) = \sum_{j=1}^{n} \det\left(X(t) \leftarrow_j x_j'(t)\right) = \mathrm{tr}\left[\det\left(X(t) \leftarrow_i x_j'(t)\right)\right]_{i, j=1}^{n}$$

$$= \mathrm{tr}\left((\mathrm{adj}\, X(t))\, X'(t)\right) = \mathrm{tr}\left((\mathrm{adj}\, X(t))\, A(t)X(t)\right) = W(t)\, \mathrm{tr}\, A(t)$$

Thus, $W(t)$ satisfies the scalar differential equation $W'(t) = \mathrm{tr}\, A(t)W(t)$, whose solution is *Abel's formula* for the *Wronskian*

$$W(t) = W(t_0)e^{\int_{t_0}^{t} \mathrm{tr}\, A(s)\, ds}$$

Conclude that if the vectors $x_1(t), \ldots, x_n(t)$ are linearly independent for $t = t_0$, then they are linearly independent for all t. How did you use the identity $\mathrm{tr}\, BC = \mathrm{tr}\, CB$ (1.2.P2)?

1.2.P16 Let $A \in M_n$ and $x, y \in \mathbf{C}^n$ be given. Let $f(t) = \det(A + txy^T)$. Use (0.8.5.11) to show that $f(t) = \det A + \beta t$, a linear function of t. What is β? For any $t_1 \neq t_2$, show that

$\det A = (t_2 f(t_1) - t_1 f(t_2))/(t_2 - t_1)$. Now consider

$$
A = \begin{bmatrix} d_1 & b & \cdots & b \\ c & d_2 & \ddots & \vdots \\ \vdots & \ddots & \ddots & b \\ c & \cdots & c & d_n \end{bmatrix} \in M_n
$$

$x = y = e$ (the vector of all ones), $t_1 = b$, and $t_2 = c$. Let $q(t) = (d_1 - t) \cdots (d_n - t)$. Show that $\det A = (bq(c) - cq(b))/(b - c)$ if $b \neq c$, and $\det A = q(b) - bq'(b)$ if $b = c$. If $d_1 = \cdots = d_n = 0$, show that $p_A(t) = (b(t + c)^n - c(t + b)^n)/(b - c)$ if $b \neq c$, and $p_A(t) = (t + b)^{n-1}(t - (n - 1)b)$ if $b = c$.

1.2.P17 Let $A, B \in M_n$ and let $C = \begin{bmatrix} 0_n & A \\ B & 0_n \end{bmatrix}$. Use (0.8.5.13) to show that $p_C(t) = p_{AB}(t^2) = p_{BA}(t^2)$, and explain carefully why this implies that AB and BA have the same eigenvalues. Explain why this confirms that $\operatorname{tr} AB = \operatorname{tr} BA$ and $\det AB = \det BA$. Also explain why $\det(I + AB) = \det(I + BA)$.

1.2.P18 Let $A \in M_3$. Explain why $p_A(t) = t^3 - (\operatorname{tr} A)t^2 + (\operatorname{tr} \operatorname{adj} A)t - \det A$.

1.2.P19 Suppose that all the entries of $A = [a_{ij}] \in M_n$ are either zero or one, and suppose that all the eigenvalues $\lambda_1, \ldots, \lambda_n$ of A are positive real numbers. Explain why $\det A$ is a positive integer, and provide details for the following:

$$
n \geq \operatorname{tr} A = \frac{1}{n}(\lambda_1 + \cdots + \lambda_n)n \geq n(\lambda_1 \cdots \lambda_n)^{1/n}
$$
$$
= n(\det A)^{1/n} \geq n
$$

Conclude that all $\lambda_i = 1$, all $a_{ii} = 1$, and $\det A = 1$.

1.2.P20 For any $A \in M_n$, show that $\det(I + A) = 1 + E_1(A) + \cdots + E_n(A)$.

1.2.P21 Let $A \in M_n$ and nonzero vectors $x, v \in \mathbb{C}^n$ be given. Suppose that $c \in \mathbb{C}$, $v^* x = 1$, $Ax = \lambda x$, and the eigenvalues of A are $\lambda, \lambda_2, \ldots, \lambda_n$. Show that the eigenvalues of the *Google* matrix $A(c) = cA + (1 - c)\lambda x v^*$ are $\lambda, c\lambda_2, \ldots, c\lambda_n$.

1.2.P22 Consider the n-by-n circulant matrix C_n in (0.9.6.2). For a given $\varepsilon > 0$, let $C_n(\varepsilon)$ be the matrix obtained from C_n by replacing its $n, 1$ entry by ε. Show that the characteristic polynomial of $C_n(\varepsilon)$ is $p_{C_n(\varepsilon)}(t) = t^n - \varepsilon$, its spectrum is $\sigma(C_n(\varepsilon)) = \{\varepsilon^{1/n}e^{2\pi ik/n} : k = 0, 1, \ldots, n - 1\}$, and the spectral radius of $I + C_n(\varepsilon)$ is $\rho(I + C_n(\varepsilon)) = 1 + \varepsilon^{1/n}$.

1.2.P23 If $A \in M_n$ is singular and has distinct eigenvalues, show that it has a nonsingular principal minor of size $n - 1$.

Notes: Sums of principal minors arose in our discussion of the coefficients of a characteristic polynomial. *Products* of principal minors also arise in a natural way; see (7.8.11).

1.3 Similarity

We know that a similarity transformation of a matrix in M_n corresponds to representing its underlying linear transformation on $\mathbb{C}^n$ in another basis. Thus, studying similarity can be thought of as studying properties that are intrinsic to one linear transformation or the properties that are common to all its basis representations.

Definition 1.3.1. *Let $A, B \in M_n$ be given. We say that B is* similar *to A if there exists a nonsingular $S \in M_n$ such that*

$$B = S^{-1}AS$$

The transformation $A \rightarrow S^{-1}AS$ is called a similarity transformation *by the* similarity matrix S. *We say that B is* permutation similar *to A if there is a permutation matrix P such that $B = P^T AP$. The relation "B is similar to A" is sometimes abbreviated $B \sim A$.*

Observation 1.3.2. *Similarity is an* equivalence *relation on M_n; that is, similarity is reflexive, symmetric, and transitive; see (0.11).*

Like any equivalence relation, similarity partitions the set M_n into disjoint equivalence classes. Each equivalence class is the set of all matrices in M_n similar to a given matrix, a representative of the class. All matrices in an equivalence class are similar, and matrices in different classes are not similar. The crucial observation is that matrices in a similarity class share many important properties. Some of these are mentioned here; a complete description of the *similarity invariants* (for example, the Jordan or Weyr canonical forms) is in Chapter 3.

Theorem 1.3.3. *Let $A, B \in M_n$. If B is similar to A, then A and B have the same characteristic polynomial.*

Proof. Compute

$$\begin{aligned} p_B(t) &= \det(tI - B) \\ &= \det(tS^{-1}S - S^{-1}AS) = \det(S^{-1}(tI - A)S) \\ &= \det S^{-1}\det(tI - A)\det S = (\det S)^{-1}(\det S)\det(tI - A) \\ &= \det(tI - A) = p_A(t) \end{aligned}$$

$\square$

Corollary 1.3.4. *Let $A, B \in M_n$ and suppose that A is similar to B. Then*

(a) *A and B have the same eigenvalues.*
(b) *If B is a diagonal matrix, its main diagonal entries are the eigenvalues of A.*
(c) *$B = 0$ (a diagonal matrix) if and only if $A = 0$.*
(d) *$B = I$ (a diagonal matrix) if and only if $A = I$.*

Exercise. Verify the assertions in the preceding corollary.

Example 1.3.5. Having the same eigenvalues is a necessary but not sufficient condition for similarity. Consider $\begin{bmatrix} 0 & 1 \\ 0 & 0 \end{bmatrix}$ and $\begin{bmatrix} 0 & 0 \\ 0 & 0 \end{bmatrix}$, which have the same eigenvalues but are not similar (why not?).

Exercise. Suppose that $A, B \in M_n$ are similar and let $q(t)$ be a given polynomial. Show that $q(A)$ and $q(B)$ are similar. In particular, show that $A + \alpha I$ and $B + \alpha I$ are similar for any $\alpha \in \mathbf{C}$.

Exercise. Let $A, B, C, D \in M_n$. Suppose that $A \sim B$ and $C \sim D$, both via the same similarity matrix S. Show that $A + C \sim B + D$ and $AC \sim BD$.

Exercise. Let $A, S \in M_n$ and suppose that S is nonsingular. Show that $S_k(S^{-1}AS) = S_k(A)$ for all $k = 1, \ldots, n$ and explain why $E_k(S^{-1}AS) = E_k(A)$ for all $k = 1, \ldots, n$. Thus, all the principal minor sums (1.2.10) are similarity invariants, not just the determinant and trace.

Exercise. Explain why rank is a similarity invariant: If $B \in M_n$ is similar to $A \in M_n$, then rank $B =$ rank A. *Hint:* See (0.4.6).

Since diagonal matrices are especially simple and have very nice properties, we would like to know which matrices are similar to diagonal matrices.

Definition 1.3.6. *If $A \in M_n$ is similar to a diagonal matrix, then A is said to be diagonalizable.*

Theorem 1.3.7. *Let $A \in M_n$ be given. Then A is similar to a block matrix of the form*

$$\begin{bmatrix} \Lambda & C \\ 0 & D \end{bmatrix}, \quad \Lambda = \mathrm{diag}(\lambda_1, \ldots, \lambda_k), \ D \in M_{n-k}, \ 1 \le k < n \quad (1.3.7.1)$$

if and only if there are k linearly independent vectors in $\mathbf{C}^n$, each of which is an eigenvector of A. The matrix A is diagonalizable if and only if there are n linearly independent vectors, each of which is an eigenvector of A. If $x^{(1)}, \ldots, x^{(n)}$ are linearly independent eigenvectors of A and if $S = [x^{(1)} \ldots x^{(n)}]$, then $S^{-1}AS$ is a diagonal matrix. If A is similar to a matrix of the form (1.3.7.1), then the diagonal entries of Λ are eigenvalues of A; if A is similar to a diagonal matrix Λ, then the diagonal entries of Λ are all of the eigenvalues of A.

Proof. Suppose that $k < n$, the n-vectors $x^{(1)}, \ldots, x^{(k)}$ are linearly independent, and $Ax^{(i)} = \lambda_i x^{(i)}$ for each $i = 1, \ldots, k$. Let $\Lambda = \mathrm{diag}(\lambda_1, \ldots, \lambda_k)$, let $S_1 = [x^{(1)} \ldots x^{(k)}]$, and choose any $S_2 \in M_{n,n-k}$ such that $S = [S_1 \ S_2]$ is nonsingular. Calculate

$$S^{-1}AS = S^{-1}[Ax^{(1)} \ldots Ax^{(k)} \ AS_2] = S^{-1}[\lambda_1 x^{(1)} \ldots \lambda_k x^{(k)} \ AS_2]$$

$$= [\lambda_1 S^{-1}x^{(1)} \ldots \lambda_k S^{-1}x^{(k)} \ S^{-1}AS_2] = [\lambda_1 e_1 \ldots \lambda_k e_k \ S^{-1}AS_2]$$

$$= \begin{bmatrix} \Lambda & C \\ 0 & D \end{bmatrix}, \quad \Lambda = \mathrm{diag}(\lambda_1, \ldots, \lambda_k), \quad \begin{bmatrix} C \\ D \end{bmatrix} = S^{-1}AS_2$$

Conversely, if S is nonsingular, $S^{-1}AS = \begin{bmatrix} \Lambda & C \\ 0 & D \end{bmatrix}$, and we partition $S = [S_1 \ S_2]$ with $S_1 \in M_{n,k}$, then S_1 has linearly independent columns and $[AS_1 \ AS_2] = AS = S\begin{bmatrix} \Lambda & C \\ 0 & D \end{bmatrix} = [S_1\Lambda \ S_1C + S_2D]$. Thus, $AS_1 = S_1\Lambda$, so each column of S_1 is an eigenvector of A.

If $k = n$ and we have a basis $\{x^{(1)}, \ldots, x^{(n)}\}$ of $\mathbf{C}^n$ such that $Ax^{(i)} = \lambda_i x^{(i)}$ for each $i = 1, \ldots, n$, let $\Lambda = \mathrm{diag}(\lambda_1, \ldots, \lambda_n)$ and let $S = [x^{(1)} \ldots x^{(n)}]$, which is nonsingular. Our previous calculation shows that $S^{-1}AS = \Lambda$. Conversely, if S is nonsingular and $S^{-1}AS = \Lambda$, then $AS = S\Lambda$, so each column of S is an eigenvector of A.

The final assertions about the eigenvalues follow from an examination of the characteristic polynomials: $p_A(t) = p_\Lambda(t)p_D(t)$ if $k < n$ and $p_A(t) = p_\Lambda(t)$ if $k = n$. $\square$

The proof of Theorem 1.3.7 is, in principle, an algorithm for diagonalizing a diagonalizable matrix $A \in M_n$: Find all n of the eigenvalues of A; find n associated (and linearly independent!) eigenvectors; and construct the matrix S. However, except for small examples, this is *not* a practical computational procedure.

Exercise. Show that $\begin{bmatrix} 0 & 1 \\ 0 & 0 \end{bmatrix}$ is *not* diagonalizable. *Hint*: If it were diagonalizable, it would be similar to the zero matrix. Alternatively, how many linearly independent eigenvectors are associated with the eigenvalue 0?

Exercise. Let $q(t)$ be a given polynomial. If A is diagonalizable, show that $q(A)$ is diagonalizable. If $q(A)$ is diagonalizable, must A be diagonalizable? Why?

Exercise. If λ is an eigenvalue of $A \in M_n$ that has multiplicity $m \geq 1$, show that A is not diagonalizable if rank $(A - \lambda I) > n - m$.

Exercise. If there are k linearly independent vectors in $\mathbf{C}^n$, each of which is an eigenvector of $A \in M_n$ associated with a given eigenvalue λ, explain carefully why the (algebraic) multiplicity of λ is at least k.

Diagonalizability is assured if *all* the eigenvalues are distinct. The basis for this fact is the following important lemma about *some* of the eigenvalues.

Lemma 1.3.8. *Let* $\lambda_1, \ldots, \lambda_k$ *be* $k \geq 2$ *distinct eigenvalues of* $A \in M_n$ *(that is,* $\lambda_i \neq \lambda_j$ *if* $i \neq j$ *and* $1 \leq i, j \leq k$*), and suppose that* $x^{(i)}$ *is an eigenvector associated with* λ_i *for each* $i = 1, \ldots, k$. *Then the vectors* $x^{(1)}, \ldots, x^{(k)}$ *are linearly independent.*

Proof. Suppose that there are complex scalars $\alpha_1, \ldots \alpha_k$ such that $\alpha_1 x^{(1)} + \alpha_2 x^{(2)} + \cdots + \alpha_r x^{(r)} = 0$. Let $B_1 = (A - \lambda_2 I)(A - \lambda_3 I) \cdots (A - \lambda_k I)$ (the product omits $A - \lambda_1 I$). Since $x^{(i)}$ is an eigenvector associated with the eigenvalue λ_i for each $i = 1, \ldots, n$, we have $B_1 x^{(i)} = (\lambda_i - \lambda_2)(\lambda_i - \lambda_3) \cdots (\lambda_i - \lambda_k) x^{(i)}$, which is zero if $2 \leq i \leq k$ (one of the factors is zero) and nonzero if $i = 1$ (no factor is zero and $x^{(1)} \neq 0$). Thus,

$$
\begin{aligned}
0 &= B_1 \left(\alpha_1 x^{(1)} + \alpha_2 x^{(2)} + \cdots + \alpha_k x^{(k)} \right) \\
&= \alpha_1 B_1 x^{(1)} + \alpha_2 B_1 x^{(2)} + \cdots + \alpha_k B_1 x^{(k)} \\
&= \alpha_1 B_1 x^{(1)} + 0 + \cdots + 0 = \alpha_1 B_1 x^{(1)}
\end{aligned}
$$

which ensures that $\alpha_1 = 0$ since $B_1 x^{(1)} \neq 0$. Repeat this argument for each $j = 2, \ldots, k$, defining B_j by a product like that defining B_1, but in which the factor $A - \lambda_j I$ is omitted. For each j we find that $\alpha_j = 0$, so $\alpha_1 = \cdots = \alpha_k = 0$ and hence $x^{(1)}, \ldots, x^{(k)}$ are linearly independent. $\square$

Theorem 1.3.9. *If* $A \in M_n$ *has n distinct eigenvalues, then A is diagonalizable.*

Proof. Let $x^{(i)}$ be an eigenvector associated with the eigenvalue λ_i for each $i = 1, \ldots, n$. Since all the eigenvalues are distinct, Lemma 1.3.8 ensures that the vectors $x^{(1)}, \ldots, x^{(n)}$ are linearly independent. Theorem 1.3.7 then ensures that A is diagonalizable. $\square$

Having distinct eigenvalues is *sufficient* for diagonalizability, but of course, it is not *necessary*.

Exercise. Give an example of a diagonalizable matrix that does not have distinct eigenvalues.

Exercise. Let $A, P \in M_n$ and suppose that P is a permutation matrix, so every entry of P is either 0 or 1 and $P^T = P^{-1}$; see (0.9.5). Show that the permutation similarity PAP^{-1} reorders the diagonal entries of A. For any given diagonal matrix $D \in M_n$ explain why there is a permutation similarity PDP^{-1} that puts the diagonal entries of D into any given order. In particular, explain why P can be chosen so that any repeated diagonal entries occur contiguously.

In general, matrices $A, B \in M_n$ do *not* commute, but if A and B are both diagonal, they always commute. The latter observation can be generalized somewhat; the following lemma is helpful in this regard.

Lemma 1.3.10. *Let $B_1 \in M_{n_1}, \ldots, B_d \in M_{n_d}$ be given and let B be the direct sum*

$$
B = \begin{bmatrix} B_1 & & 0 \\ & \ddots & \\ 0 & & B_d \end{bmatrix} = B_1 \oplus \cdots \oplus B_d
$$

Then B is diagonalizable if and only if each of $B_1, \ldots, B_d$ is diagonalizable.

Proof. If for each $i = 1, \ldots, d$ there is a nonsingular $S_i \in M_{n_i}$ such that $S_i^{-1} B_i S_i$ is diagonal, and if we define $S = S_1 \oplus \cdots \oplus S_d$, then one checks that $S^{-1} B S$ is diagonal.

For the converse, we proceed by induction. There is nothing to prove for $d = 1$. Suppose that $d \geq 2$ and that the assertion has been established for direct sums with $d - 1$ or fewer direct summands. Let $C = B_1 \oplus \cdots \oplus B_{d-1}$, let $n = n_1 + \cdots + n_{d-1}$, and let $m = n_d$. Let $S \subset M_{n+m}$ be nonsingular and such that

$$
S^{-1} B S = S^{-1} (C \oplus B_d) S = \Lambda = \mathrm{diag}(\lambda_1, \lambda_2, \ldots, \lambda_{n+m})
$$

Rewrite this identity as $BS = S\Lambda$. Partition $S = [s_1 \, s_2 \ldots s_{n+m}]$ with

$$
s_i = \begin{bmatrix} \xi_i \\ \eta_i \end{bmatrix} \in \mathbf{C}^{n+m}, \quad \xi_i \in \mathbf{C}^n, \eta_i \in \mathbf{C}^m, i = 1, 2, \ldots, n + m
$$

Then $Bs_i = \lambda_i s_i$ implies that $C\xi_i = \lambda_i \xi_i$ and $B_d \eta_i = \lambda_i \eta_i$ for $i = 1, 2, \ldots, n + m$. The row rank of $[\xi_1 \, \cdots \, \xi_{n+m}] \in M_{n,n+m}$ is n because this matrix comprises the first n rows of the nonsingular matrix S. Thus, its column rank is also n, so the list $\xi_1, \ldots, \xi_{n+m}$ contains a linearly independent list of n vectors, each of which is an eigenvector of C. Theorem 1.3.7 ensures that C is diagonalizable and the induction hypothesis ensures that its direct summands $B_1, \ldots, B_{d-1}$ are all diagonalizable. The row rank of $[\eta_1 \, \cdots \, \eta_{n+m}] \in M_{m,n+m}$ is m, so the list $\eta_1, \ldots, \eta_{n+m}$ contains a linearly independent list of m vectors; it follows that B_d is diagonalizable as well. $\square$

Definition 1.3.11. *Two matrices $A, B \in M_n$ are said to be* simultaneously diagonalizable *if there is a single nonsingular $S \in M_n$ such that $S^{-1}AS$ and $S^{-1}BS$ are both diagonal.*

Exercise. Let $A, B, S \in M_n$ and suppose that S is nonsingular. Show that A commutes with B if and only if $S^{-1}AS$ commutes with $S^{-1}BS$.

Exercise. If $A, B \in M_n$ are simultaneously diagonalizable, show that they commute. *Hint*: Diagonal matrices commute.

Exercise. Show that if $A \in M_n$ is diagonalizable and $\lambda \in \mathbf{C}$, then A and λI are simultaneously diagonalizable.

Theorem 1.3.12. *Let $A, B \in M_n$ be diagonalizable. Then A and B commute if and only if they are simultaneously diagonalizable.*

Proof. Assume that A and B commute, perform a similarity transformation on both A and B that diagonalizes A (but not necessarily B) and groups together any repeated eigenvalues of A. If $\mu_1, \ldots, \mu_d$ are the distinct eigenvalues of A and $n_1, \ldots, n_d$ are their respective multiplicities, then we may assume that

$$A = \begin{bmatrix} \mu_1 I_{n_1} & & & 0 \\ & \mu_2 I_{n_2} & & \\ & & \ddots & \\ 0 & & & \mu_d I_{n_d} \end{bmatrix}, \quad \mu_i \neq \mu_j \text{ if } i \neq j \qquad (1.3.13)$$

Since $AB = BA$, (0.7.7) ensures that

$$B = \begin{bmatrix} B_1 & & 0 \\ & \ddots & \\ 0 & & B_d \end{bmatrix}, \quad \text{each } B_i \in M_{n_i} \qquad (1.3.14)$$

is block diagonal conformal to A. Since B is diagonalizable, (1.3.10) ensures that each B_i is diagonalizable. Let $T_i \in M_{n_i}$ be nonsingular and such that $T_i^{-1}B_iT_i$ is diagonal for each $i = 1, \ldots, d$; let

$$T = \begin{bmatrix} T_1 & & & 0 \\ & T_2 & & \\ & & \ddots & \\ 0 & & & T_d \end{bmatrix} \qquad (1.3.15)$$

Then $T_i^{-1}\mu_i I_{n_i} T_i = \mu_i I_{n_i}$, so $T^{-1}AT = A$ and $T^{-1}BT$ are both diagonal. The converse is included in an earlier exercise. $\qquad \square$

We want to have a version of Theorem 1.3.12 involving arbitrarily many commuting diagonalizable matrices. Central to our investigation is the notion of an invariant subspace and the companion notion of a block triangular matrix.

Definitions 1.3.16. *A family $\mathcal{F} \subseteq M_n$ of matrices is a nonempty finite or infinite set of matrices; a commuting family is a family of matrices in which every pair of matrices commutes. For a given $A \in M_n$, a subspace $W \subseteq \mathbf{C}^n$ is A-invariant if $Aw \in W$ for every $w \in W$. A subspace $W \subseteq \mathbf{C}^n$ is trivial if either $W = \{0\}$ or $W = \mathbf{C}^n$; otherwise, it is nontrivial. For a given family $\mathcal{F} \subseteq M_n$, a subspace $W \subseteq \mathbf{C}^n$ is $\mathcal{F}$-invariant if W is A-invariant for each $A \in \mathcal{F}$. A given family $\mathcal{F} \subseteq M_n$ is reducible if some nontrivial subspace of $\mathbf{C}^n$ is $\mathcal{F}$-invariant; otherwise, $\mathcal{F}$ is irreducible.*

Exercise. For $A \in M_n$, show that each nonzero element of a one-dimensional A-invariant subspace of $\mathbf{C}^n$ is an eigenvector of A.

Exercise. Suppose that $n \geq 2$ and $S \in M_n$ is nonsingular. Partition $S = [S_1 \ S_2]$, in which $S_1 \in M_{n,k}$ and $S_2 \in M_{n,n-k}$ with $1 \leq k < n$. Explain why

$$S^{-1}S_1 = [e_1 \ \ldots \ e_k] = \begin{bmatrix} I_k \\ 0 \end{bmatrix} \text{ and } S^{-1}S_2 = [e_{k+1} \ \ldots \ e_n] = \begin{bmatrix} 0 \\ I_{n-k} \end{bmatrix}$$

Invariant subspaces and block triangular matrices are two sides of the same valuable coin: The former is the linear algebra side, while the latter is the matrix analysis side. Let $A \in M_n$ with $n \geq 2$ and suppose that $W \subseteq \mathbf{C}^n$ is a k-dimensional subspace with $1 \leq k < n$. Choose a basis $s_1, \ldots, s_k$ of W and let $S_1 = [s_1 \ \ldots \ s_k] \in M_{n,k}$. Choose any $s_{k+1}, \ldots, s_n$ such that $s_1, \ldots, s_n$ is a basis for $\mathbf{C}^n$, let $S_2 = [s_{k+1} \ \ldots \ s_n] \in M_{n,n-k}$, and let $S = [S_1 \ S_2]$; S has linearly independent columns, so it is nonsingular. If W is A-invariant, then $As_j \in W$ for each $j = 1, \ldots, k$, so each As_j is a linear combination of $s_1, \ldots, s_k$, that is, $AS_1 = S_1 B$ for some $B \in M_k$. If $AS_1 = S_1 B$, then $AS = [AS_1 \ AS_2] = [S_1 B \ AS_2]$ and hence

$$S^{-1}AS = \begin{bmatrix} S^{-1}S_1 B & S^{-1}AS_2 \end{bmatrix} = \begin{bmatrix} \begin{bmatrix} I_k \\ 0 \end{bmatrix} B & S^{-1}AS_2 \end{bmatrix}$$

$$= \begin{bmatrix} B & C \\ 0 & D \end{bmatrix}, \quad B \in M_k, \ 1 \leq k \leq n-1 \quad (1.3.17)$$

The conclusion is that A is similar to a block triangular matrix (1.3.17) if it has a k-dimensional invariant subspace. But we can say a little more: We know that $B \in M_k$ has an eigenvalue, so suppose that $B\xi = \lambda\xi$ for some scalar λ and a nonzero $\xi \in \mathbf{C}^k$. Then $0 \neq S_1\xi \in W$ and $A(S_1\xi) = (AS_1)\xi = S_1 B\xi = \lambda(S_1\xi)$, which means that A has an eigenvector in W.

Conversely, if $S = [S_1 \ S_2] \in M_n$ is nonsingular, $S_1 \in M_{n,k}$, and $S^{-1}AS$ has the block triangular form (1.3.17), then

$$AS_1 = AS \begin{bmatrix} I_k \\ 0 \end{bmatrix} = S \begin{bmatrix} B & C \\ 0 & D \end{bmatrix} \begin{bmatrix} I_k \\ 0 \end{bmatrix} = \begin{bmatrix} S_1 & S_2 \end{bmatrix} \begin{bmatrix} B \\ 0 \end{bmatrix} = S_1 B$$

so the (k-dimensional) span of the columns of S_1 is A-invariant. We summarize the foregoing discussion in the following observation.

Observation 1.3.18. *Suppose that $n \geq 2$. A given $A \in M_n$ is similar to a block triangular matrix of the form (1.3.17) if and only if some nontrivial subspace of $\mathbf{C}^n$ is A-invariant. Moreover, if $W \subseteq \mathbf{C}^n$ is a nonzero A-invariant subspace, then some vector in W is an eigenvector of A. A given family $\mathcal{F} \subseteq M_n$ is reducible if and only if there is some $k \in \{1, \ldots, n-1\}$ and a nonsingular $S \in M_n$ such that $S^{-1}AS$ has the form (1.3.17) for every $A \in \mathcal{F}$.*

The following lemma is at the heart of many subsequent results.

Lemma 1.3.19. *Let $\mathcal{F} \subset M_n$ be a commuting family. Then some nonzero vector in $\mathbf{C}^n$ is an eigenvector of every $A \in \mathcal{F}$.*

Proof. There is always a nonzero $\mathcal{F}$-invariant subspace, namely, $\mathbf{C}^n$. Let $m = \min\{\dim V : V$ is a nonzero $\mathcal{F}$-invariant subspace of $\mathbf{C}^n\}$ and let W be any given $\mathcal{F}$-invariant subspace such that $\dim W = m$. Let any $A \in \mathcal{F}$ be given. Since W is $\mathcal{F}$-invariant, it is A-invariant, so (1.3.18) ensures that there is some nonzero $x_0 \in W$ and some $\lambda \in \mathbf{C}$ such that $Ax_0 = \lambda x_0$. Consider the subspace $W_{A,\lambda} = \{x \in W : Ax = \lambda x\}$. Then $x_0 \in W_{A,\lambda}$, so $W_{A,\lambda}$ is a nonzero subspace of W. For any $B \in \mathcal{F}$ and any $x \in W_{A,\lambda}$, $\mathcal{F}$-invariance of W ensures that $Bx \in W$. Using commutativity of $\mathcal{F}$, we compute

$$A(Bx) = (AB)x = (BA)x = B(Ax) = B(\lambda x) = \lambda(Bx)$$

which shows that $Bx \in W_{A,\lambda}$. Thus, $W_{A,\lambda}$ is $\mathcal{F}$-invariant and nonzero, so $\dim W_{A,\lambda} \geq m$. But $W_{A,\lambda} \subseteq W$, so $\dim W_{A,\lambda} \leq m$ and hence $W = W_{A,\lambda}$. We have now shown that for each $A \in \mathcal{F}$, there is some scalar λ_A such that $Ax = \lambda_A x$ for all $x \in W$, so every nonzero vector in W is an eigenvector of every matrix in $\mathcal{F}$. $\qquad\square$

Exercise. Consider the nonzero $\mathcal{F}$-invariant subspace W in the preceding proof. Explain why $m = \dim W = 1$.

Exercise. Suppose that $\mathcal{F} \subset M_n$ is a commuting family. Show that there is a nonsingular $S \in M_n$ such that for every $A \in \mathcal{F}$, $S^{-1}AS$ has the block triangular form (1.3.17) with $k = 1$.

Lemma 1.3.19 concerns commuting families of arbitrary nonzero cardinality. Our next result shows that Theorem 1.3.12 can be extended to arbitrary commuting families of diagonalizable matrices.

Definition 1.3.20. *A family $\mathcal{F} \subset M_n$ is said to be* simultaneously diagonalizable *if there is a single nonsingular $S \in M_n$ such that $S^{-1}AS$ is diagonal for every $A \in \mathcal{F}$.*

Theorem 1.3.21. *Let $\mathcal{F} \subset M_n$ be a family of diagonalizable matrices. Then $\mathcal{F}$ is a commuting family if and only if it is a simultaneously diagonalizable family. Moreover, for any given $A_0 \in \mathcal{F}$ and for any given ordering $\lambda_1, \ldots, \lambda_n$ of the eigenvalues of A_0, there is a nonsingular $S \in M_n$ such that $S^{-1}A_0S = \mathrm{diag}(\lambda_1, \ldots, \lambda_n)$ and $S^{-1}BS$ is diagonal for every $B \in \mathcal{F}$.*

Proof. If $\mathcal{F}$ is simultaneously diagonalizable, then it is a commuting family by a previous exercise. We prove the converse by induction on n. If $n = 1$, there is nothing to prove since every family is both commuting and diagonal. Let us suppose that $n \geq 2$ and that, for each $k = 1, 2, \ldots, n - 1$, any commuting family of k-by-k diagonalizable matrices is simultaneously diagonalizable. If every matrix in $\mathcal{F}$ is a scalar matrix, there is nothing to prove, so we may assume that $A \in \mathcal{F}$ is a given n-by-n diagonalizable matrix with distinct eigenvalues $\lambda_1, \lambda_2, \ldots, \lambda_k$ and $k \geq 2$, that $AB = BA$ for every $B \in \mathcal{F}$, and that each $B \in \mathcal{F}$ is diagonalizable. Using the argument in (1.3.12), we reduce to the case in which A has the form (1.3.13). Since every $B \in \mathcal{F}$ commutes with A, (0.7.7) ensures that each $B \in \mathcal{F}$ has the form (1.3.14). Let $B, \hat{B} \in \mathcal{F}$, so $B = B_1 \oplus \cdots \oplus B_k$ and $\hat{B} = \hat{B}_1 \oplus \cdots \oplus \hat{B}_k$, in which each of $B_i, \hat{B}_i$ has the same size and that size is at most $n - 1$. Commutativity and diagonalizability of B and $\hat{B}$ imply commutativity and diagonalizability of B_i and $\hat{B}_i$ for each $i = 1, \ldots, d$. By the induction hypothesis, there are k similarity matrices $T_1, T_2, \ldots, T_k$ of appropriate size, each of which diagonalizes the corresponding block of every matrix in $\mathcal{F}$. Then the direct sum (1.3.15) diagonalizes every matrix in $\mathcal{F}$.

We have shown that there is a nonsingular $T \in M_n$ such that $T^{-1}BT$ is diagonal for every $B \in \mathcal{F}$. Then $T^{-1}A_0T = P\operatorname{diag}(\lambda_1, \ldots, \lambda_n)P^T$ for some permutation matrix P, $P^T(T^{-1}A_0T)P = (TP)^{-1}A_0(TP) = \operatorname{diag}(\lambda_1, \ldots, \lambda_n)$ and $(TP)^{-1}B(TP) = P^T(T^{-1}BT)P$ is diagonal for every $B \in \mathcal{F}$ (0.9.5). $\qquad\square$

Remarks: We defer two important issues until Chapter 3: (1) Given $A, B \in M_n$, how can we determine if A is similar to B? (2) How can we tell if a given matrix is diagonalizable without knowing its eigenvectors?

Although AB and BA need not be the same (and need not be the same size even when both products are defined), their eigenvalues are as much the same as possible. Indeed, if A and B are both square, then AB and BA have exactly the same eigenvalues. These important facts follow from a simple but very useful observation.

Exercise. Let $X \in M_{m,n}$ be given. Explain why $\begin{bmatrix} I_m & X \\ 0 & I_n \end{bmatrix} \in M_{m+n}$ is nonsingular and verify that its inverse is $\begin{bmatrix} I_m & -X \\ 0 & I_n \end{bmatrix}$.

Theorem 1.3.22. *Suppose that $A \in M_{m,n}$ and $B \in M_{n,m}$ with $m \leq n$. Then the n eigenvalues of BA are the m eigenvalues of AB together with $n - m$ zeroes; that is, $p_{BA}(t) = t^{n-m}p_{AB}(t)$. If $m = n$ and at least one of A or B is nonsingular, then AB and BA are similar.*

Proof. A computation reveals that

$$\begin{bmatrix} I_m & -A \\ 0 & I_n \end{bmatrix} \begin{bmatrix} AB & 0 \\ B & 0_n \end{bmatrix} \begin{bmatrix} I_m & A \\ 0 & I_n \end{bmatrix} = \begin{bmatrix} 0_m & 0 \\ B & BA \end{bmatrix}$$

and the preceding exercise ensures that $C_1 = \begin{bmatrix} AB & 0 \\ B & 0_n \end{bmatrix}$ and $C_2 = \begin{bmatrix} 0_m & 0 \\ B & BA \end{bmatrix}$ are similar. The eigenvalues of C_1 are the eigenvalues of AB together with n zeroes. The eigenvalues of C_2 are the eigenvalues of BA together with m zeroes. Since the eigenvalues of C_1 and C_2 are the same, the first assertion of the theorem follows. The final assertion follows from the observation that $AB = A(BA)A^{-1}$ if A is nonsingular and $m = n$. $\qquad\square$

Theorem 1.3.22 has many applications, several of which emerge in the following chapters. Here are just four.

Example 1.3.23. Eigenvalues of a low-rank matrix. Suppose that $A \in M_n$ is factored as $A = XY^T$, in which $X, Y \in M_{n,r}$ and $r < n$. Then the eigenvalues of A are the same as those of the r-by-r matrix Y^TX, together with $n - r$ zeroes. For example, consider the n-by-n all-ones matrix $J_n = ee^T$ (0.2.8). Its eigenvalues are the eigenvalue of the 1-by-1 matrix $e^Te = [n]$, namely, n, together with $n - 1$ zeroes. The eigenvalues of any matrix of the form $A = xy^T$ with $x, y \in \mathbf{C}^n$ (rank A is at most 1) are y^Tx, together with $n - 1$ zeroes. The eigenvalues of any matrix of the form $A = xy^T + zw^T = [x \; z][y \; w]^T$ with $x, y, z, w \in \mathbf{C}^n$ (rank A is at most 2) are the two eigenvalues of $[y \; w]^T[x \; z] = \begin{bmatrix} y^Tx & y^Tz \\ w^Tx & w^Tz \end{bmatrix}$ (1.2.4b) together with $n - 2$ zeroes.

Example 1.3.24. Cauchy's determinant identity. Let a nonsingular $A \in M_n$ and $x, y \in \mathbf{C}^n$ be given. Then

$$\det(A + xy^T) = (\det A)\left(\det(I + A^{-1}xy^T)\right)$$

$$= (\det A)\prod_{i=1}^{n}\lambda_i(I + A^{-1}xy^T)$$

$$= (\det A)\prod_{i=1}^{n}\left(1 + \lambda_i(A^{-1}xy^T)\right)$$

$$= (\det A)\left(1 + y^T A^{-1}x\right) \quad \text{(use (1.3.23))}$$

$$= \det A + y^T\left((\det A)\,A^{-1}\right)x = \det A + y^T(\operatorname{adj} A)x$$

Cauchy's identity $\det(A + xy^T) = \det A + y^T(\operatorname{adj} A)x$, valid for any $A \in M_n$, now follows by continuity. For a different approach, see (0.8.5).

Example 1.3.25. For any $n \geq 2$, consider the n-by-n real symmetric Hankel matrix

$$A = [i + j]_{i,j=1}^{n} = \begin{bmatrix} 2 & 3 & 4 & \cdots \\ 3 & 4 & 5 & \cdots \\ 4 & 5 & 6 & \cdots \\ \vdots & & & \ddots \end{bmatrix} = ve^T + ev^T = [v\ e][e\ v]^T$$

in which every entry of $e \in \mathbf{R}^n$ is 1 and $v = [1\ 2\ \dots\ n]^T$. The eigenvalues of A are the same as those of

$$B = [e\ v]^T[v\ e] = \begin{bmatrix} e^T v & e^T e \\ v^T v & v^T e \end{bmatrix} = \begin{bmatrix} \frac{n(n+1)}{2} & n \\ \frac{n(n+1)(2n+1)}{6} & \frac{n(n+1)}{2} \end{bmatrix}$$

together with $n - 2$ zeroes. According to (1.2.4b), the eigenvalues B (one positive and one negative) are

$$n(n + 1)\left[\frac{1}{2} \pm \sqrt{\frac{2n + 1}{6(n + 1)}}\right]$$

Example 1.3.26. For any $n \geq 2$, consider the n-by-n real skew-symmetric Toeplitz matrix

$$A = [i - j]_{i,j=1}^{n} = \begin{bmatrix} 0 & -1 & -2 & \cdots \\ 1 & 0 & -1 & \cdots \\ 2 & 1 & 0 & \cdots \\ \vdots & & & \ddots \end{bmatrix} = ve^T - ev^T = [v\ -e][e\ v]^T$$

in which every entry of $e \in \mathbf{R}^n$ is 1 and $v = [1\ 2\ \dots\ n]^T$. Except for $n - 2$ zeroes, the eigenvalues of A are the same as those of

$$B = [e\ v]^T[v\ -e] = \begin{bmatrix} e^T v & -e^T e \\ v^T v & -v^T e \end{bmatrix}$$

which, using (1.2.4b) again, are $\pm\frac{ni}{2}\sqrt{\frac{n^2-1}{3}}$.

Theorem 1.3.22 on the eigenvalues of AB versus BA is only part of the story; we return to that story in (3.2.11).

If $A \in M_n$ is diagonalizable and $A = S\Lambda S^{-1}$, then aS also diagonalizes A for any $a \neq 0$. Thus, a diagonalizing similarity is never unique. Nevertheless, *every* similarity of A to a particular diagonal matrix can be obtained from just one given similarity.

Theorem 1.3.27. *Suppose that $A \in M_n$ is diagonalizable, let $\mu_1, \ldots, \mu_d$ be its distinct eigenvalues with respective multiplicities $n_1, \ldots, n_d$, let $S, T \in M_n$ be nonsingular, and suppose that $A = S\Lambda S^{-1}$, in which Λ is a diagonal matrix of the form (1.3.13). Then*

(a) *$A = T\Lambda T^{-1}$ if and only if $T = S(R_1 \oplus \cdots \oplus R_d)$ in which each $R_i \in M_{n_i}$ is nonsingular.*

(b) *If $S = [S_1 \ldots S_d]$ and $T = [T_1 \ldots T_d]$ are partitioned conformally to Λ, then $A = S\Lambda S^{-1} = T\Lambda T^{-1}$ if and only if for each $i = 1, \ldots, d$ the column space of S_i is the same as the column space of T_i.*

(c) *If A has n distinct eigenvalues and $S = [s_1 \ldots s_n]$ and $T = [t_1 \ldots t_n]$ are partitioned according to their columns, then $A = S\Lambda S^{-1} = T\Lambda T^{-1}$ if and only if there is a nonsingular diagonal matrix $R = \mathrm{diag}(r_1, \ldots, r_n)$ such that $T = SR$ if and only if, for each $i = 1, \ldots, n$, the column s_i is a nonzero scalar multiple of the corresponding column t_i.*

Proof. We have $S\Lambda S^{-1} = T\Lambda T^{-1}$ if and only if $(S^{-1}T)\Lambda = \Lambda(S^{-1}T)$ if and only if $S^{-1}T$ is block diagonal conformal to Λ (0.7.7), that is, if and only if $S^{-1}T = R_1 \oplus \cdots \oplus R_d$ and each $R_i \in M_{n_i}$ is nonsingular. For (b), observe that if $1 \leq k \leq n$, then the column space of $X \in M_{n,k}$ is contained in the column space of $Y \in M_{n,k}$ if and only if there is some $C \in M_k$ such that $X = YC$; if, in addition, rank $X =$ rank $Y = k$, then C must be nonsingular. The assertion (c) is a special case of (a) and (b). $\square$

If real matrices are similar via a complex matrix, are they similar via a real matrix? Is there a real version of (1.3.21) for commuting real matrices? The following lemma is the key to answering such questions.

Lemma 1.3.28. *Let $S \in M_n$ be nonsingular and let $S = C + iD$, in which $C, D \in M_n(\mathbf{R})$. There is a real number τ such that $T = C + \tau D$ is nonsingular.*

Proof. If C is nonsingular, take $\alpha = 0$. If C is singular, consider the polynomial $p(t) = \det(C + tD)$, which is not a constant (degree zero) polynomial since $p(0) = \det C = 0 \neq \det S = p(i)$. Since $p(t)$ has only finitely many zeroes in the complex plane, there is a real τ such that $p(\tau) \neq 0$, so $C + \tau D$ is nonsingular. $\square$

Theorem 1.3.29. *Let $\mathcal{F} = \{A_\alpha : \alpha \in \mathcal{I}\} \subset M_n(\mathbf{R})$ and $\mathcal{G} = \{B_\alpha : \alpha \in \mathcal{I}\} \subset M_n(\mathbf{R})$ be given families of real matrices. If there is a nonsingular $S \in M_n$ such that $A_\alpha = SB_\alpha S^{-1}$ for every $\alpha \in \mathcal{I}$, then there is a nonsingular $T \in M_n(\mathbf{R})$ such that $A_\alpha = TB_\alpha T^{-1}$ for every $\alpha \in \mathcal{I}$. In particular, two real matrices that are similar over $\mathbf{C}$ are similar over $\mathbf{R}$.*

Proof. Let $S = C + iD$ be nonsingular, in which $C, D \in M_n(\mathbf{R})$. The preceding lemma ensures that there is a real number τ such that $T = C + \tau D$ is nonsingular. The similarity $A_\alpha = SB_\alpha S^{-1}$ is equivalent to the identity $A_\alpha(C + iD) = A_\alpha S = SB_\alpha = (C + iD)B_\alpha$. Equating the real and imaginary parts of this identity shows that

$A_\alpha C = C B_\alpha$ and $A_\alpha D = D B_\alpha$. Consequently, $A_\alpha C = C B_\alpha$ and $A_\alpha(\tau D) = (\tau D) B_\alpha$, so $A_\alpha T = T B_\alpha$ and $A_\alpha = T B_\alpha T^{-1}$.   □

An immediate consequence of the preceding theorem is a real version of (1.3.21).

Corollary 1.3.30. *Let* $\mathcal{F} = \{A_\alpha : \alpha \in \mathcal{I}\} \subset M_n(\mathbf{R})$ *be a family of real diagonalizable matrices with real eigenvalues. Then* $\mathcal{F}$ *is a commuting family if and only if there is a nonsingular real matrix* T *such that* $T^{-1} A_\alpha T = \Lambda_\alpha$ *is diagonal for every* $A \in \mathcal{F}$. *Moreover, for any given* $\alpha_0 \in \mathcal{I}$ *and for any given ordering* $\lambda_1, \ldots, \lambda_n$ *of the eigenvalues of* A_{α_0}, *there is a nonsingular* $T \in M_n(\mathbf{R})$ *such that* $T^{-1} A_{\alpha_0} T = \mathrm{diag}(\lambda_1, \ldots, \lambda_n)$ *and* $T^{-1} A_\alpha T$ *is diagonal for every* $\alpha \in \mathcal{I}$.

Proof. For the "only if" assertion, apply the preceding theorem to the families $\mathcal{F} = \{A_\alpha : \alpha \in \mathcal{I}\}$ and $\mathcal{G} = \{\Lambda_\alpha : \alpha \in \mathcal{I}\}$. The "if" assertion follows as in (1.3.21).   □

Our final theorem about similarity shows that the *only* relationship between the eigenvalues and main diagonal entries of a complex matrix is that their respective sums be equal.

Theorem 1.3.31 (Mirsky). *Let an integer* $n \geq 2$ *and complex scalars* $\lambda_1, \ldots, \lambda_n$ *and* $d_1, \ldots, d_n$ *be given. There is an* $A \in M_n$ *with eigenvalues* $\lambda_1, \ldots, \lambda_n$ *and main diagonal entries* $d_1, \ldots, d_n$ *if and only if* $\sum_{i=1}^n \lambda_i = \sum_{i=1}^n d_i$. *If* $\lambda_1, \ldots, \lambda_n$ *and* $d_1, \ldots, d_n$ *are all real and have the same sums, there is an* $A \in M_n(\mathbf{R})$ *with eigenvalues* $\lambda_1, \ldots, \lambda_n$ *and main diagonal entries* $d_1, \ldots, d_n$.

Proof. We know that $\mathrm{tr}\, A = E_1(A) = S_1(A)$ for any $A \in M_n$ (1.2.16), which establishes the necessity of the stated condition. We must prove its sufficiency.

If $k \geq 2$ and if $\lambda_1, \ldots, \lambda_k$ and $d_1, \ldots, d_k$ are any given complex scalars such that $\sum_{i=1}^k \lambda_i = \sum_{i=1}^k d_i$, we claim that the upper bidiagonal matrix

$$T(\lambda_1, \ldots, \lambda_k) = \begin{bmatrix} \lambda_1 & 1 & & \\ & \lambda_2 & \ddots & \\ & & \ddots & 1 \\ & & & \lambda_k \end{bmatrix} \in M_k$$

is similar to a matrix with diagonal entries $d_1, \ldots, d_k$; that matrix has the property asserted. Let $L(s, t) = \begin{bmatrix} 1 & 0 \\ s-t & 1 \end{bmatrix}$, so $L(s, t)^{-1} = \begin{bmatrix} 1 & 0 \\ t-s & 1 \end{bmatrix}$.

Consider first the case $k = 2$, so $\lambda_1 + \lambda_2 = d_1 + d_2$. Compute the similarity

$$L(\lambda_1, d_1) T(\lambda_1, \lambda_2) L(\lambda_1, d_1)^{-1} = \begin{bmatrix} 1 & 0 \\ \lambda_1 - d_1 & 1 \end{bmatrix} \begin{bmatrix} \lambda_1 & 1 \\ 0 & \lambda_2 \end{bmatrix} \begin{bmatrix} 1 & 0 \\ d_1 - \lambda_1 & 1 \end{bmatrix}$$

$$= \begin{bmatrix} d_1 & \bigstar \\ \bigstar & \lambda_1 + \lambda_2 - d_1 \end{bmatrix} = \begin{bmatrix} d_1 & \bigstar \\ \bigstar & d_2 \end{bmatrix}$$

in which we use the hypothesis $\lambda_1 + \lambda_2 - d_1 = d_1 + d_2 - d_1 = d_2$. This verifies our claim for $k = 2$.

We proceed by induction. Assume that our claim has been proved for some $k \geq 2$ and that $\sum_{i=1}^{k+1} \lambda_i = \sum_{i=1}^{k+1} d_i$. Partition $T(\lambda_1, \ldots, \lambda_{k+1}) = [T_{ij}]_{i,j=1}^2$, in which $T_{11} = T(\lambda_1, \lambda_2)$, $T_{12} = E_2$, $T_{21} = 0$, and $T_{22} = T(\lambda_3, \ldots, \lambda_{k+1})$, with $E_2 = [e_2 \ 0 \ \ldots \ 0] \in M_{2,k-1}$ and $e_2 = [0 \ 1]^T \in \mathbf{C}^2$. Let $\mathcal{L} = L(\lambda_1, d_1) \oplus I_{k-1}$ and compute $\mathcal{L}T(\lambda_1, \ldots, \lambda_{k+1})\mathcal{L}^{-1}$

$$
= \begin{bmatrix} L(\lambda_1, d_1) & 0 \\ 0 & I_{k-1} \end{bmatrix} \begin{bmatrix} T(\lambda_1, \lambda_2) & E_2 \\ 0 & T(\lambda_3, \ldots, \lambda_{k+1}) \end{bmatrix} \begin{bmatrix} L(d_1, \lambda_1) & 0 \\ 0 & I_{k-1} \end{bmatrix}
$$

$$
= \begin{bmatrix} \begin{bmatrix} d_1 & \star \\ \star & \lambda_1 + \lambda_2 - d_1 \end{bmatrix} & E_2 \\ 0 & T(\lambda_3, \ldots, \lambda_{k+1}) \end{bmatrix}
$$

$$
= \begin{bmatrix} d_1 & \star \\ \star & T(\lambda_1 + \lambda_2 - d_1, \lambda_3, \ldots, \lambda_{k+1}) \end{bmatrix} = \begin{bmatrix} d_1 & \star \\ \star & D \end{bmatrix}
$$

The sum of the eigenvalues of $D = T(\lambda_1 + \lambda_2 - d_1, \lambda_3, \ldots, \lambda_{k+1}) \in M_k$ is $\sum_{i=1}^{k+1} \lambda_i - d_1 = \sum_{i=1}^{k+1} d_i - d_1 = \sum_{i=2}^{k+1} d_i$, so the induction hypothesis ensures that there is a nonsingular $S \in M_k$ such that the diagonal entries of SDS^{-1} are $d_2, \ldots, d_{k+1}$. Then $\begin{bmatrix} 1 & 0 \\ 0 & S \end{bmatrix} \begin{bmatrix} d_1 & \star \\ \star & D \end{bmatrix} \begin{bmatrix} 1 & 0 \\ 0 & S \end{bmatrix}^{-1} = \begin{bmatrix} d_1 & \star \\ \star & SDS^{-1} \end{bmatrix}$ has diagonal entries $d_1, d_2, \ldots, d_{k+1}$.

If $\lambda_1, \ldots, \lambda_n$ and $d_1, \ldots, d_n$ are all real, all of the matrices and similarities in the preceding constructions are real. $\qquad\square$

Exercise. Write out the details of the inductive step $k = 2 \Rightarrow k = 3$ in the preceding proof.

Problems

1.3.P1 Let $A, B \in M_n$. Suppose that A and B are diagonalizable and commute. Let $\lambda_1, \ldots, \lambda_n$ be the eigenvalues of A and let $\mu_1, \ldots, \mu_n$ be the eigenvalues of B. (a) Show that the eigenvalues of $A + B$ are $\lambda_1 + \mu_{i_1}, \lambda_2 + \mu_{i_2}, \ldots, \lambda_n + \mu_{i_n}$, for some permutation $i_1, \ldots, i_n$ of $1, \ldots, n$. (b) If B is nilpotent, explain why A and $A + B$ have the same eigenvalues. (c) What are the eigenvalues of AB?

1.3.P2 If $A, B \in M_n$ and if A and B commute, show that any polynomial in A commutes with any polynomial in B.

1.3.P3 If $A \in M_n$, $SAS^{-1} = \Lambda = \mathrm{diag}(\lambda_1, \ldots, \lambda_n)$, and $p(t)$ is a polynomial, show that $p(A) = S^{-1}p(\Lambda)S$ and that $p(\Lambda) = \mathrm{diag}(p(\lambda_1), \ldots, p(\lambda_n))$. This provides a simple way to evaluate $p(A)$ if one can diagonalize A.

1.3.P4 If $A \in M_n$ has distinct eigenvalues $\alpha_1, \ldots, \alpha_n$ and commutes with a given matrix $B \in M_n$, show that B is diagonalizable and that there is a polynomial $p(t)$ of degree at most $n - 1$ such that $B = p(A)$.

1.3.P5 Give an example of two commuting matrices that are not simultaneously diagonalizable. Does this contradict (1.3.12)? Why?

1.3.P6 (a) If $\Lambda = \mathrm{diag}(\lambda_1, \ldots, \lambda_n)$, show that $p_\Lambda(\Lambda)$ is the zero matrix. (b) Suppose that $A \in M_n$ is diagonalizable. Explain why $p_A(t) = p_\Lambda(t)$ and $p_\Lambda(A) = Sp_\Lambda(\Lambda)S^{-1}$. Conclude that $p_A(A)$ is the zero matrix.

1.3.P7 A matrix $A \in M_n$ is a *square root* of $B \in M_n$ if $A^2 = B$. Show that every diagonalizable $B \in M_n$ has a square root. Does $B = \begin{bmatrix} 0 & 1 \\ 0 & 0 \end{bmatrix}$ have a square root? Why?

1.3.P8 If $A, B \in M_n$ and if at least one has distinct eigenvalues (no assumption, even of diagonalizability, about the other), provide details for the following geometric argument that A and B commute if and only if they are simultaneously diagonalizable: One direction is easy; for the other, suppose that B has distinct eigenvalues and $Bx = \lambda x$ with $x \neq 0$. Then $B(Ax) = A(Bx) = A\lambda x = \lambda Ax$, so $Ax = \mu x$ for some $\mu \in \mathbf{C}$ (Why? See (1.2.18).) Thus, we can diagonalize A with the same matrix of eigenvectors that diagonalizes B. Of course, the eigenvalues of A need not be distinct.

1.3.P9 Consider the singular matrices $A = \begin{bmatrix} 1 & 0 \\ 0 & 0 \end{bmatrix}$ and $B = \begin{bmatrix} 0 & 0 \\ 1 & 0 \end{bmatrix}$. Show that AB and BA are not similar but that they do have the same eigenvalues.

1.3.P10 Let $A \in M_n$ be given, and let $\lambda_1, \ldots, \lambda_k$ be distinct eigenvalues of A. For each $i = 1, 2, \ldots, k$, suppose that $x_1^{(i)}, x_2^{(i)}, \ldots, x_{n_i}^{(i)}$ is a list of linearly independent eigenvectors of A associated with the eigenvalue λ_i. Show that the list $x_1^{(1)}, x_2^{(1)}, \ldots, x_{n_1}^{(1)}, \ldots, x_1^{(k)}, x_2^{(k)}, \ldots, x_{n_k}^{(k)}$ of all of these vectors is linearly independent.

1.3.P11 Provide details for the following alternative proof of (1.3.19): (a) Suppose that $A, B \in M_n$ commute, $x \neq 0$, and $Ax = \lambda x$. Consider the sequence of vectors $x, Bx, B^2x, B^3x, \ldots$. Suppose that k is the smallest positive integer such that B^kx is a linear combination of its predecessors; $S = \mathrm{span}\{x, Bx, B^2x, \ldots, B^{k-1}x\}$ is B-invariant and hence contains an eigenvector of B. But $AB^jx = B^jAx = B^j\lambda x = \lambda B^jx$, so every nonzero vector in S is an eigenvector for A. Conclude that A and B have a common eigenvector. (b) If $\mathcal{F} = \{A_1, A_2, \ldots, A_m\} \subset M_n$ is a finite commuting family, use induction to show that it has a common eigenvector: If $y \neq 0$ is a common eigenvector for $A_1, A_2, \ldots, A_{m-1}$, consider $y, A_my, A_m^2y, A_m^3y, \ldots$. (c) If $\mathcal{F} \subset M_n$ is an infinite commuting family, then no list of more than n^2 matrices in $\mathcal{F}$ can be linearly independent. Select a maximal linearly independent set and explain why a common eigenvector for this finite set of matrices is a common eigenvector for $\mathcal{F}$.

1.3.P12 Let $A, B \in M_n$, and suppose that either A or B is nonsingular. If AB is diagonalizable, show that BA is also diagonalizable. Consider $A = \begin{bmatrix} 0 & 1 \\ 0 & 0 \end{bmatrix}$ and $B = \begin{bmatrix} 1 & 1 \\ 0 & 0 \end{bmatrix}$ to show that this need not be true if both A and B are singular.

1.3.P13 Show that two diagonalizable matrices are similar if and only if their characteristic polynomials are the same. Is this true for two matrices that are not both diagonalizable?

1.3.P14 Suppose that $A \in M_n$ is diagonalizable. (a) Prove that the rank of A is equal to the number of its nonzero eigenvalues. (b) Prove that $\mathrm{rank}\, A = \mathrm{rank}\, A^k$ for all $k = 1, 2, \ldots$. (c) Prove that A is nilpotent if and only if $A = 0$. (d) If $\mathrm{tr}\, A = 0$, prove that $\mathrm{rank}\, A \neq 1$. (e) Use each of the four preceding results to show that $B = \begin{bmatrix} 0 & 1 \\ 0 & 0 \end{bmatrix}$ is not diagonalizable.

1.3.P15 Let $A \in M_n$ and a polynomial $p(t)$ be given. If A is diagonalizable, show that $p(A)$ is diagonalizable. What about the converse?

1.3.P16 Let $A \in M_n$ and suppose that $n > \mathrm{rank}\, A = r \geq 1$. If A is similar to $B \oplus 0_{n-r}$ (so $B \in M_r$ is nonsingular), show that A has a nonsingular r-by-r principal submatrix (that is, A is rank principal (0.7.6)). If A is rank principal, must it be similar to $B \oplus 0_{n-r}$?

1.3.P17 Let $A, B \in M_n$ be given. Prove that there is a nonsingular $T \in M_n(\mathbf{R})$ such that $A = TBT^{-1}$ if and only if there is a nonsingular $S \in M_n$ such that both $A = SBS^{-1}$ and $\bar{A} = S\bar{B}S^{-1}$.

1.3.P18 Suppose that $A, B \in M_n$ are coninvolutory, that is, $A\bar{A} = B\bar{B} = I$. Show that A and B are similar over $\mathbf{C}$ if and only if they are similar over $\mathbf{R}$.

1.3.P19 Let $B, C \in M_n$ and define $\mathcal{A} = \begin{bmatrix} B & C \\ C & B \end{bmatrix} \in M_{2n}$. Let $Q = \frac{1}{\sqrt{2}} \begin{bmatrix} I_n & I_n \\ I_n & -I_n \end{bmatrix}$ and verify that $Q^{-1} = Q = Q^T$. Let $\mathcal{K}_{2n} = \begin{bmatrix} 0_n & I_n \\ I_n & 0_n \end{bmatrix}$. (a) A matrix in M_{2n} with the block structure of $\mathcal{A}$ is said to be 2-by-2 *block centrosymmetric*. Show that $A \in M_{2n}$ is 2-by-2 block centrosymmetric if and only if $\mathcal{K}_{2n} A = A \mathcal{K}_{2n}$. Deduce from this identity that the inverse of a nonsingular 2-by-2 block centrosymmetric matrix is 2-by-2 block centrosymmetric, and that a product of 2-by-2 block centrosymmetric matrices is 2-by-2 block centrosymmetric. (b) Show that $Q^{-1} \mathcal{A} Q = (B + C) \oplus (B - C)$. (c) Explain why $\det \mathcal{A} = \det(B^2 + CB - BC - C^2)$ and $\operatorname{rank} \mathcal{A} = \operatorname{rank}(B + C) + \operatorname{rank}(B - C)$. (d) Explain why $\begin{bmatrix} 0 & C \\ C & 0 \end{bmatrix}$ is similar to $C \oplus (-C)$ and why its eigenvalues occur in $\pm$ pairs. What more can you say about the eigenvalues if C is real? For a more precise statement, see (4.6.P24).

1.3.P20 Represent any $A, B \in M_n$ as $A = A_1 + i A_2$ and $B = B_1 + i B_2$, in which $A_1, A_2, B_1, B_2 \in M_n(\mathbf{R})$. Define $R_1(A) = \begin{bmatrix} A_1 & A_2 \\ -A_2 & A_1 \end{bmatrix} \in M_{2n}(\mathbf{R})$. Show the following:
(a) $R_1(A + B) = R_1(A) + R_1(B)$, $R_1(AB) = R_1(A)R_1(B)$, and $R(I_n) = I_{2n}$.
(b) If A is nonsingular then $R_1(A)$ is nonsingular, $R_1(A)^{-1} = R_1(A^{-1})$, and $R_1(A)^{-1} = \begin{bmatrix} X & Y \\ -Y & X \end{bmatrix}$ has the same block structure as $R_1(A)$.
(c) If S is nonsingular, then $R_1(SAS^{-1}) = R_1(S)R_1(A)R_1(S)^{-1}$.
(d) If A and B are similar, then $R_1(A)$ and $R_1(B)$ are similar.
    Let the eigenvalues of A be $\lambda_1, \ldots, \lambda_n$, let $S = \begin{bmatrix} I_n & iI_n \\ 0 & I_n \end{bmatrix}$, and let $U = \frac{1}{\sqrt{2}} \begin{bmatrix} I_n & iI_n \\ iI_n & I_n \end{bmatrix}$. Show the following:
(e) $S^{-1} = \bar{S}$ and $U^{-1} = \bar{U} = U^*$.
(f) $S^{-1} R_1(A) S = \begin{bmatrix} A & 0 \\ -A_2 & \bar{A} \end{bmatrix}$ and $U^{-1} R_1(A) U = \begin{bmatrix} A & 0 \\ 0 & \bar{A} \end{bmatrix}$.
(g) The eigenvalues of $R_1(A)$ are the same as the eigenvalues of $A \oplus \bar{A}$, which are $\lambda_1, \ldots, \lambda_n, \overline{\lambda_1}, \ldots, \overline{\lambda_n}$ (see (1.3.P30) for a more precise statement).
(h) $\det R_1(A) = |\det A|^2 \geq 0$ and $\operatorname{rank} R_1(A) = 2 \operatorname{rank} A$.
(i) If $R_1(A)$ is nonsingular then A is nonsingular.
(j) iI_n is not similar to $-iI_n$, but $R_1(iI_n)$ is similar to $R_1(-iI_n)$, so the implication in (d) cannot be reversed.
(k) $p_{R_1(A)}(t) = p_A(t) p_{\bar{A}}(t)$.
(l) $R_1(A^*) = R_1(A)^T$, so A is Hermitian if and only if $R_1(A)$ is (real) symmetric and A is unitary if and only if $R_1(A)$ is real orthogonal.
(m) A commutes with A^* if and only if $R_1(A)$ commutes with $R_1(A)^T$, that is, the complex matrix A is normal if and only if the real matrix $R_1(A)$ is normal; see (2.5).
(n) A matrix in $M_{2n}(\mathbf{R})$ with the block structure of $R_1(A)$ is said to be a matrix of *complex type*. Let $S_{2n} = \begin{bmatrix} 0_n & I_n \\ -I_n & 0_n \end{bmatrix}$. Show that $A \in M_{2n}(\mathbf{R})$ is a matrix of complex type if and only if $S_{2n} A = A S_{2n}$. Deduce from this identity that the inverse of a real matrix of complex type is a matrix of complex type and that a product of real matrices of complex type is a matrix of complex type.

The block matrix $R_1(A)$ is an example of a *real representation* of A; see (4.4.P29) for a generalization to a *complex representation*, also known as a matrix of *quaternion type*.

1.3.P21 Using the same notation as in the preceding problem, define $R_2(A) = \begin{bmatrix} A_1 & A_2 \\ A_2 & -A_1 \end{bmatrix} \in M_{2n}(\mathbf{R})$. Let $V = \frac{1}{\sqrt{2}} \begin{bmatrix} -iI_n & -iI_n \\ I_n & -I_n \end{bmatrix}$, and consider $R_2(iI_n) = \begin{bmatrix} 0 & I_n \\ I_n & 0 \end{bmatrix}$ and $R_2(I_n) = \begin{bmatrix} I_n & 0 \\ 0 & -I_n \end{bmatrix}$. Show the following:
(a) $V^{-1} = V^*$, $R_2(I_n)^{-1} = R_2(I_n) = R_2(I_n)^*$, $R_2(iI_n)^{-1} = R_2(iI_n) = R_2(iI_n)^*$, and $R_2(iI_n) = V^{-1} R_2(I_n) V$.

(b) $A = B$ if and only if $R_2(A) = R_2(B)$, and $R_2(A + B) = R_2(A) + R_2(B)$.

(c) $R_2(A) = V \begin{bmatrix} 0 & \bar{A} \\ A & 0 \end{bmatrix} V^{-1}$.

(d) $\det R_2(A) = (-1)^n |\det A|^2$; see (0.8.5.13).

(e) $R_2(A)$ is nonsingular if and only if A is nonsingular.

(f) Characteristic polynomials and eigenvalues: $p_{R_2(A)}(t) = \det(t^2 I - A\bar{A}) = p_{A\bar{A}}(t^2)$ (0.8.5.13), so if $\mu_1, \ldots, \mu_n$ are the eigenvalues of $A\bar{A}$ then $\pm\mu_1^{1/2}, \ldots, \pm\mu_n^{1/2}$ are the eigenvalues of $R_2(A)$. Moreover, $p_{R_2(A)}(t)$ has real coefficients, so the non-real eigenvalues of $A\bar{A}$ occur in conjugate pairs.

(g) $R_2(AB) = R_2(A \cdot I_n \cdot B) = R_2(A)R_2(I_n)R_2(B)$.

(h) $R_2(\bar{A}) = R_2(I_n)R_2(A)R_2(I_n)$, so $R_2(\bar{A})$ is similar to $R_2(A)$ and $R_2(A\bar{B}C) = R_2(A)R_2(B)R_2(C)$.

(i) $-R_2(A) = R_2(-A) = R_2(iI_n \cdot A \cdot iI_n) = (R_2(iI_n)R_2(I_n)) \cdot R_2(A) \cdot (R_2(iI_n)R_2(I_n))^{-1}$, so $R_2(-A)$ is similar to $R_2(A)$.

(j) $R_2(A)R_2(B) = V(\bar{A}B \oplus A\bar{B})V^{-1}$.

(k) If A is nonsingular, then $R_2(A)^{-1} = R_2(\bar{A}^{-1})$.

(l) $R_2(A)^2 = R_1(\bar{A}A) = R_2(A\bar{A})R_2(I_n)$.

(m) If S is nonsingular, then $R_2(SA\bar{S}^{-1}) = (R_2(S)R_2(I_n)) \cdot R_2(A) \cdot (R_2(S)R_2(I_n))^{-1}$, so $R_2(SA\bar{S}^{-1})$ is similar to $R_2(A)$. See (4.6.P19) for the converse: if $R_2(A)$ is similar to $R_2(B)$, then there is a nonsingular S such that $B = SA\bar{S}^{-1}$.

(n) $R_2(A^T) = R_2(A)^T$, so A is (complex) symmetric if and only if $R_2(A)$ is (real) symmetric.

(o) A is unitary if and only if $R_2(A)$ is real orthogonal.

The block matrix $R_2(A)$ is a second example of a *real representation* of A.

1.3.P22 Let $A, B \in M_n$. Show that A and B are similar if and only if there are $X, Y \in M_n$, at least one of which is nonsingular, such that $A = XY$ and $B = YX$.

1.3.P23 Let $B \in M_n$ and $C \in M_{n,m}$ and define $\mathcal{A} = \begin{bmatrix} B & C \\ 0 & 0_m \end{bmatrix} \in M_{n+m}$. Show that $\mathcal{A}$ is similar to $B \oplus 0_m$ if and only if $\mathrm{rank}[B\ C] = \mathrm{rank}\,B$, that is, if and only if there is some $X \in M_{n,m}$ such that $C = BX$.

1.3.P24 For a given integer $n \geq 3$, let $\theta = 2\pi/n$ and let $A = [\cos(j\theta + k\theta)]_{j,k=1}^n \in M_n(\mathbf{R})$. Show that $A = [x\ y][x\ y]^T$, in which $x = [\alpha\ \alpha^2\ \ldots\ \alpha^n]^T$, $y = [\alpha^{-1}\ \alpha^{-2}\ \ldots\ \alpha^{-n}]^T$, and $\alpha = e^{2\pi i/n}$. Show that the eigenvalues of A are $n/2$ and $-n/2$, together with $n - 2$ zeroes.

1.3.P25 Let $x, y \in \mathbf{C}^n$ be given and suppose that $y^*x \neq -1$. (a) Verify that $(I + xy^*)^{-1} = I - cxy^*$, in which $c = (1 + y^*x)^{-1}$. (b) Let $\Lambda = \mathrm{diag}(\lambda_1, \ldots, \lambda_n)$ and suppose that $y^*x = 0$. Explain why the eigenvalues of

$$A = (I + xy^*)\Lambda(I - xy^*) = \Lambda + xy^*\Lambda - \Lambda xy^* - (y^*\Lambda x)xy^*$$

are $\lambda_1, \ldots, \lambda_n$. Notice that A has integer entries if the entries of x, y, and Λ are integers. Use this observation to construct an interesting 3-by-3 matrix with integer entries and eigenvalues 1, 2, and 7; verify that your construction has the asserted eigenvalues.

1.3.P26 Let $e_1, \ldots, e_n$ and $\varepsilon_1, \ldots, \varepsilon_m$ denote the standard orthonormal bases of $\mathbf{C}^n$ and $\mathbf{C}^m$, respectively. Consider the n-by-m block matrix $P = [P_{ij}] \in M_{mn}$ in which each block $P_{ij} \in M_{m,n}$ is given by $P_{ij} = \varepsilon_j e_i^T$. (a) Show that P is a permutation matrix. (b) Similarity of any matrix $A \in M_{mn}$ by P gives a matrix $\tilde{A} = PAP^T$ whose entries are a rearrangement of the entries of A. Appropriate partitioning of both A and $\tilde{A}$ permits us to describe this rearrangement in a simple way. Write $A = [A_{ij}] \in M_{mn}$ as an m-by-m block matrix in which each block $A_{k\ell} = [a_{ij}^{(k,\ell)}] \in M_n$, and write $\tilde{A} = [\tilde{A}_{ij}]$ as an n-by-n block matrix in which each block $\tilde{A}_{ij} \in M_m$. Explain why the i, j entry of $\tilde{A}_{pq}$ is the p, q entry of A_{ij} for all $i, j = 1, \ldots, m$ and all $p, q = 1, \ldots, n$, that is, $\tilde{A}_{pq} = [a_{pq}^{(i,j)}]$. Since A and $\tilde{A}$ are permutation similar, they have the same eigenvalues, determinant, and so forth. (c) Various special patterns in the entries of A result in special patterns in the entries of $\tilde{A}$ (and vice versa). For example, explain why (i) all of the blocks A_{ij} are upper triangular if and only if $\tilde{A}$ is block upper triangular; (ii) all of the blocks A_{ij} are upper Hessenberg if and only if $\tilde{A}$ is block upper Hessenberg; (iii) all of the blocks A_{ij} are diagonal if and only if $\tilde{A}$ is block diagonal; and (iv) A is block upper triangular and all of the blocks A_{ij} are upper triangular if and only if $\tilde{A}$ is block diagonal and all of its main diagonal blocks are upper triangular.

1.3.P27 (Continuation of 1.3.P26) Let $A = [A_{k\ell}] \in M_{mn}$ be a given m-by-m block matrix with each $A_{k\ell} = [a_{ij}^{(k,\ell)}] \in M_n$, and suppose that each block $A_{k\ell}$ is upper triangular. Explain why the eigenvalues of A are the same as those of $\tilde{A}_{11} \oplus \cdots \oplus \tilde{A}_{nn}$, in which $\tilde{A}_{pp} = [a_{pp}^{(i,j)}]$ for $p = 1, \ldots, n$. Thus, the eigenvalues of A depend only on the main diagonal entries of the blocks A_{ij}. In particular, $\det A = (\det \tilde{A}_{11}) \cdots (\det \tilde{A}_{nn})$. What can you say about the eigenvalues and determinant of A if the diagonal entries of each block A_{ij} are constant (so there are scalars $\alpha_{k\ell}$ such that $a_{ii}^{(k,\ell)} = \alpha_{k\ell}$ for all $i = 1, \ldots, n$ and all $k, \ell = 1, \ldots, m$)?

1.3.P28 Let $A \in M_{m,n}$ and $B \in M_{n,m}$ be given. Prove that $\det(I_m + AB) = \det(I_n + BA)$.

1.3.P29 Let $A = [a_{ij}] \in M_n$. Suppose that each $a_{ii} = 0$ for $i = 1, \ldots, n$ and $a_{ij} \in \{-1, 1\}$ for all $i \neq j$. Explain why $\det A$ is an integer. Use Cauchy's identity (1.3.24) to show that if any -1 entry of A is changed to $+1$, then the parity of $\det A$ is unchanged, that is, it remains even if it was even and odd if it was odd. Show that the parity of $\det A$ is the same as the parity of $\det(J_n - I)$, which is opposite to the parity of n. Conclude that A is nonsingular if n is even.

1.3.P30 Suppose that $A \in M_n$ is diagonalizable and $A = S\Lambda S^{-1}$, in which Λ has the form (1.3.13). If f is a complex valued function whose domain includes $\sigma(A)$, we define $f(A) = Sf(\Lambda)S^{-1}$, in which $f(\Lambda) = f(\mu_1)I_{n_1} \oplus \cdots \oplus f(\mu_d)I_{n_d}$. Does $f(A)$ depend on the choice of the diagonalizing similarity (which is never unique)? Use Theorem 1.3.27 to show that it does not; that is, if $A = S\Lambda S^{-1} = T\Lambda T^{-1}$, show that $Sf(\Lambda)S^{-1} = Tf(\Lambda)T^{-1}$. If A has real eigenvalues, show that $\cos^2 A + \sin^2 A = I$.

1.3.P31 Let $a, b \in \mathbf{C}$. Show that the eigenvalues of $\begin{bmatrix} a & b \\ -b & a \end{bmatrix}$ are $a \pm ib$.

1.3.P32 Let $x \in \mathbf{C}^n$ be a given nonzero vector, and write $x = u + iv$, in which $u, v \in \mathbf{R}^n$. Show that the vectors $x, \bar{x} \in \mathbf{C}^n$ arc linearly independent if and only if the vectors $u, v \in \mathbf{R}^n$ are linearly independent.

1.3.P33 Suppose that $A \in M_n(\mathbf{R})$ has a non-real eigenvalue λ and write $\lambda = a + ib$ with $a, b \in \mathbf{R}$ and $b > 0$. Let x be an eigenvector of A associated with λ and write $x = u + iv$ with $u, v \in \mathbf{R}^n$. (a) Explain why $\bar{\lambda}, \bar{x}$ is an eigenpair of A. (b) Explain why x and $\bar{x}$ are linearly independent and deduce that u and v are linearly independent. (c) Show that $Au = au - bv$ and $Av = bu + av$, so $A[u\ v] = [u\ v]B$, in which $B = \begin{bmatrix} a & b \\ -b & a \end{bmatrix}$. (d) Let $S = [u\ v\ S_1] \in M_n(\mathbf{R})$ be nonsingular. Explain why $S^{-1}[u\ v] = \begin{bmatrix} I_2 \\ 0 \end{bmatrix}$ and check that $S^{-1}AS = S^{-1}[A[u\ v]\quad AS_1] = S^{-1}[[u\ v]B\quad AS_1] = \begin{bmatrix} B & \star \\ 0 & A_1 \end{bmatrix}$, in which $A_1 \in M_{n-2}$. Thus, a real square matrix with a non-real eigenvalue λ is real similar to a 2-by-2 block upper triangular matrix whose upper left block reveals the real and imaginary parts of λ. (e) Explain why the multiplicity of each of λ and $\bar{\lambda}$ as an eigenvalue of A_1 is 1 less than its multiplicity as an eigenvalue of A.

1.3.P34 If $A, B \in M_n$ are similar, show that adj A and adj B are similar.

1.3.P35 A set $\mathcal{A} \subseteq M_n$ is an *algebra* if (i) $\mathcal{A}$ is a subspace and (ii) $AB \in \mathcal{A}$ whenever $A, B \in \mathcal{A}$. Provide details for the following assertions and assemble a proof of *Burnside's theorem on matrix algebras*: Let $n \geq 2$ and let $\mathcal{A} \subseteq M_n$ be a given algebra. Then $\mathcal{A} = M_n$ if and only if $\mathcal{A}$ is irreducible.

(a) If $n \geq 2$ and an algebra $\mathcal{A} \subseteq M_n$ is reducible, then $\mathcal{A} \neq M_n$. This is the easy implication in Burnside's theorem; some work is required to show that if $\mathcal{A}$ is irreducible, then $\mathcal{A} = M_n$. In the following, $\mathcal{A} \subseteq M_n$ is a given algebra and $\mathcal{A}^* = \{A^* : A \in \mathcal{A}\}$.
(b) If $n \geq 2$ and $\mathcal{A}$ is irreducible, then $\mathcal{A} \neq \{0\}$.
(c) If $x \in \mathbf{C}^n$ is nonzero, then $\mathcal{A}x = \{Ax : A \in \mathcal{A}\}$ is an $\mathcal{A}$-invariant subspace of $\mathbf{C}^n$.
(d) If $n \geq 2$, $x \in \mathbf{C}^n$ is nonzero, and $\mathcal{A}$ is irreducible, then $\mathcal{A}x = \mathbf{C}^n$.
(e) For any given $x \in \mathbf{C}^n$, $\mathcal{A}^*x = \{A^*x : A \in \mathcal{A}\}$ is a subspace of $\mathbf{C}^n$.
(f) If $n \geq 2$, $x \in \mathbf{C}^n$ is nonzero, and $\mathcal{A}$ is irreducible, then $\mathcal{A}^*x = \mathbf{C}^n$.
(g) If $n \geq 2$ and $\mathcal{A}$ is irreducible, there is some $A \in \mathcal{A}$ such that rank $A = 1$.
(h) If $n \geq 2$, $\mathcal{A}$ is irreducible, and there are nonzero $y, z \in \mathbf{C}^n$ such that $yz^* \in \mathcal{A}$, then $\mathcal{A}$ contains every rank-one matrix.
(i) If $\mathcal{A}$ contains every rank-one matrix, then $\mathcal{A} = M_n$; see (0.4.4(i)).

1.3.P36 Let $A, B \in M_n$ and suppose that $n \geq 2$. The *algebra generated by A and B* (denoted by $\mathcal{A}(A, B)$) is the span of the set of all words in A and B (2.2.5). (a) If $n = 2$ and if A and B have no common eigenvector, explain why $\mathcal{A}(A, B) = M_2$. (b) Let $A = \begin{bmatrix} 0 & 1 \\ 0 & 0 \end{bmatrix}$ and $B = A^T$. Show that A and B have no common eigenvector, so $\mathcal{A}(A, B) = M_2$. Give a direct proof by exhibiting a basis of M_2 consisting of words in A and B.

1.3.P37 Let $A \in M_n$ be centrosymmetric. If $n = 2m$ and A is presented in the block form (0.9.10.2), show that A is similar to $(B - K_m C) \oplus (B + K_m C)$ via the real orthogonal matrix $Q = \frac{1}{\sqrt{2}} \begin{bmatrix} I_m & I_m \\ -K_m & K_m \end{bmatrix}$. If $n = 2m + 1$ and A is presented in the block form (0.9.10.3), show that A is similar to

$$(B - K_m C) \oplus \begin{bmatrix} \alpha & \sqrt{2}x^T \\ \sqrt{2}K_m y & B + K_m C \end{bmatrix} \text{ via } Q = \frac{1}{\sqrt{2}} \begin{bmatrix} I_m & 0 & I_m \\ 0 & \sqrt{2} & 0 \\ -K_m & 0 & K_m \end{bmatrix}$$

and that Q is real orthogonal.

1.3.P38 Let J_n be the all-ones matrix (0.2.8) and let $B(t) = (1 - t)I_n + t J_n$, with $n \geq 2$. (a) Describe the entries of $B(t)$. Explain why its eigenvalues are $1 + (n - 1)t$ and $1 - t$ with multiplicity $n - 1$. (b) Verify that if $1 \neq t \neq -(n - 1)^{-1}$, then $B(t)$ is nonsingular and $B(t)^{-1} = (1 - t)^{-1}(I_n - t(1 + (n - 1)t)^{-1} J_n)$.

1.3.P39 Let $A \in M_n$ be given and suppose that tr $A = 0$. If A is diagonalizable, explain why rank $A \neq 1$. One of the matrices in (1.3.5) has rank one and trace zero. What can you conclude about it?

1.3.P40 The *Jordan product* of $A, B \in M_n$ is $]A, B[= AB + BA$. The matrices A and B anticommute if $]A, B[= 0$; see (0.7.7). (a) Give an example of a commuting family of matrices that contains infinitely many distinct matrices. (b) Let $\mathcal{F} = \{A_1, A_2, ...\}$ be a family of matrices such that $]A_i, A_j[= 0$ if $i \neq j$, but $A_i^2 \neq 0$ for all $i = 1, 2, ...$; that is, no matrix in $\mathcal{F}$ anticommutes with itself. Show that $I \notin \mathcal{F}$ and that any finite set of matrices in $\mathcal{F}$ is linearly independent. Conclude that $\mathcal{F}$ contains at most $n^2 - 1$ matrices. (c) If $\mathcal{F} = \{A_1, A_2, ...\}$ is a family of distinct pairwise anticommuting diagonalizable matrices, show that it is a finite family and that $\{A_1^2, A_2^2, ...\}$ is a (finite) commuting family of diagonalizable matrices.

1.3.P41 If $A \in M_n$ does not have distinct eigenvalues, then no matrix that is similar to A has distinct eigenvalues, but perhaps some matrix that is diagonally equivalent to A has distinct eigenvalues. (a) If $D_1, D_2 \in M_n$ are diagonal and nonsingular, and if $D_1 A D_2$ has distinct eigenvalues, explain why there is a nonsingular $D \in M_n$ such that DA has distinct eigenvalues. (b) If $A \in M_n$ is strictly triangular and $n \geq 2$, explain why no matrix that is diagonally equivalent to A has distinct eigenvalues. (c) Let $n = 2$, let $A = \begin{bmatrix} a & b \\ c & d \end{bmatrix}$, and suppose that no matrix that is diagonally equivalent to A has distinct eigenvalues. Then $A_z = \begin{bmatrix} 1 & 0 \\ 0 & z \end{bmatrix} A = \begin{bmatrix} a & b \\ zc & zd \end{bmatrix}$ has a double eigenvalue for all nonzero $z \in \mathbf{C}$. Show that the discriminant of $p_{A_z}(t)$ is $(a + dz)^2 - 4(ad - bc)z = d^2 z^2 + (2ad - 4(ad - bc))z + a^2$. Why is this discriminant equal to zero for all nonzero $z \in \mathbf{C}$? Explain why $d = 0$, $a = 0$, and $bc = 0$, and conclude that A is strictly triangular. (d) In the case $n = 2$, explain why $A \in M_n$ is not diagonally equivalent to a matrix with distinct eigenvalues if and only if A is singular and every principal minor of A of size $n - 1$ is zero. (e) The assertion in (d) is known to be correct for all $n \geq 2$.

Notes and Further Readings: Theorem 1.3.31 is due to L. Mirsky (1958); our proof is adapted from E. Carlen and E. Lieb, Short proofs of theorems of Mirsky and Horn on diagonals and eigenvalues of matrices, *Electron. J. Linear Algebra* 18 (2009) 438–441. A result complementary to Mirsky's is known: If n^2 complex numbers $\lambda_1, \ldots, \lambda_n$ and $a_{ij}, i, j = 1, \ldots, n, i \neq j$ are given, then there exist n complex numbers $a_{11}, \ldots, a_{nn}$ such that $\lambda_1, \ldots, \lambda_n$ are the eigenvalues of $A = [a_{ij}]_{i,j=1}^n$; see S. Friedland, Matrices with prescribed off-diagonal elements, *Israel J. Math.* 11 (1975) 184–189. The proof of Burnside's theorem in (1.3.P35) is adapted from I. Halperin and P. Rosenthal, Burnside's theorem on algebras of matrices, *Amer. Math. Monthly* 87 (1980) 810. For alternative approaches, see Radjavi and Rosenthal (2000) and V. Lomonosov and P. Rosenthal, The simplest proof of Burnside's theorem on matrix algebras, *Linear Algebra Appl.* 383 (2004) 45–47. For a proof of the claim in (1.3.P41(e)) see M. D. Choi, Z. Huang, C. K. Li, and N. S. Sze, Every invertible matrix is diagonally equivalent to a matrix with distinct eigenvalues, *Linear Algebra Appl.* 436 (2012) 3773–3776.

1.4 Left and right eigenvectors and geometric multiplicity

The eigenvectors of a matrix are important not only for their role in diagonalization but also for their utility in a variety of applications. We begin with an important observation about eigenvalues.

Observation 1.4.1. *Let* $A \in M_n$. *(a) The eigenvalues of* A *and* A^T *are the same.* *(b) The eigenvalues of* A^* *are the complex conjugates of the eigenvalues of* A.

Proof. Since $\det(tI - A^T) = \det(tI - A)^T = \det(tI - A)$, we have $p_{A^T}(t) = p_A(t)$, so $p_{A^T}(\lambda) = 0$ if and only if $p_A(\lambda) = 0$. Similarly, $p_{A^*}(t) = \det(tI - A^*) = \det(tI - \bar{A})^T = \det(tI - \bar{A}) = \overline{\det(\bar{t}I - A)} = \overline{p_A(\bar{t})}$. $\square$

Exercise. If $x, y \in \mathbf{C}^n$ are both eigenvectors of $A \in M_n$ associated with the eigenvalue λ, show that any nonzero linear combination of x and y is also an eigenvector associated with λ. Conclude that the set of all eigenvectors associated with a particular $\lambda \in \sigma(A)$, together with the zero vector, is a *subspace* of $\mathbf{C}^n$.

Exercise. The subspace described in the preceding exercise is the *null space* of $A - \lambda I$, that is, the solution set of the homogeneous linear system $(A - \lambda I)x = 0$. Explain why the dimension of this subspace is $n - \text{rank}(A - \lambda I)$.

Definition 1.4.2. *Let* $A \in M_n$. *For a given* $\lambda \in \sigma(A)$, *the set of all vectors* $x \in \mathbf{C}^n$ *satisfying* $Ax = \lambda x$ *is called the* eigenspace *of* A *associated with the eigenvalue* λ. *Every nonzero element of this eigenspace is an eigenvector of* A *associated with* λ.

Exercise. Show that the eigenspace of A associated with an eigenvalue λ is an A-invariant subspace, but an A-invariant subspace need not be an eigenspace of A. Explain why a *minimal* A-invariant subspace (an A-invariant subspace that contains no strictly lower-dimensional, nonzero A-invariant subspace) W is the span of a *single* eigenvector of A, that is, $\dim W = 1$.

Definition 1.4.3. *Let* $A \in M_n$ *and let* λ *be an eigenvalue of* A. *The dimension of the eigenspace of* A *associated with* λ *is the* geometric multiplicity *of* λ. *The multiplicity of* λ *as a zero of the characteristic polynomial of* A *is the* algebraic multiplicity *of* λ. *If the term* multiplicity *is used without qualification in reference to* λ, *it means the algebraic multiplicity. We say that* λ *is* simple *if its algebraic multiplicity is 1; it is* semisimple *if its algebraic and geometric multiplicities are equal.*

It can be very useful to think of the geometric multiplicity of an eigenvalue λ of $A \in M_n$ in more than one way: Since the geometric multiplicity is the dimension of the nullspace of $A - \lambda I$, it is equal to $n - \text{rank}(A - \lambda I)$. It is also the maximum number of linearly independent eigenvectors associated with λ. Theorems 1.2.18 and 1.3.7 both contain an inequality between the geometric and algebraic multiplicities of an eigenvalue but from two different viewpoints.

Exercise. Use (1.2.18) to explain why the algebraic multiplicity of an eigenvalue is greater than or equal to its geometric multiplicity. If the algebraic multiplicity is 1, why must the geometric multiplicity also be 1?

Exercise. Use (1.3.7) to explain why the geometric multiplicity of an eigenvalue is less than or equal to its algebraic multiplicity. If the algebraic multiplicity is 1, why must the geometric multiplicity also be 1?

Exercise. Verify the following statements about the respective matrices and their eigenvalue $\lambda = 1$:

(a) $A_1 = \begin{bmatrix} 1 & 0 \\ 0 & 2 \end{bmatrix}$: geometric multiplicity = algebraic multiplicity = 1; simple.

(b) $A_2 = \begin{bmatrix} 1 & 0 \\ 0 & 1 \end{bmatrix}$: geometric multiplicity = algebraic multiplicity = 2; semisimple.

(c) $A_3 = \begin{bmatrix} 1 & 1 \\ 0 & 1 \end{bmatrix}$: geometric multiplicity = 1; algebraic multiplicity = 2.

Definitions 1.4.4. *Let $A \in M_n$. We say that A is* defective *if the geometric multiplicity of some eigenvalue of A is strictly less than its algebraic multiplicity. If the geometric multiplicity of each eigenvalue of A is the same as its algebraic multiplicity, we say that A is* nondefective. *If each eigenvalue of A has geometric multiplicity 1, we say that A is* nonderogatory; *otherwise, it is* derogatory.

A matrix is diagonalizable if and only if it is nondefective; it has distinct eigenvalues if and only if it is nonderogatory and nondefective.

Exercise. Explain why, in the preceding exercise, A_1 is nondefective; A_2 is nondefective and derogatory; and A_3 is defective and nonderogatory.

Example 1.4.5. Even though A and A^T have the same eigenvalues, their eigenspaces associated with a given eigenvalue can be different. For example, let $A = \begin{bmatrix} 2 & 3 \\ 0 & 4 \end{bmatrix}$. Then the (one-dimensional) eigenspace of A associated with the eigenvalue 2 is spanned by $\begin{bmatrix} 1 \\ 0 \end{bmatrix}$, while the eigenspace of A^T associated with the eigenvalue 2 is spanned by $\begin{bmatrix} 1 \\ -3/2 \end{bmatrix}$.

Definition 1.4.6. *A nonzero vector $y \in \mathbb{C}^n$ is a* left eigenvector *of $A \in M_n$ associated with an eigenvalue λ of A if $y^* A = \lambda y^*$. If necessary for clarity, we refer to the vector x in (1.1.3) as a* right eigenvector; *when the context does not require distinction, we continue to call x an* eigenvector.

Observation 1.4.6a. *Let $x \in \mathbb{C}^n$ be nonzero, let $A \in M_n$, and suppose that $Ax = \lambda x$. If $x^* A = \mu x^*$, then $\lambda = \mu$.*

Proof. We may assume that x is a unit vector. Compute $\mu = \mu x^* x = (x^* A)x = x^* Ax = x^*(Ax) = x^*(\lambda x) = \lambda x^* x = \lambda$. $\qquad\qquad\square$

Exercise. Show that a left eigenvector y associated with an eigenvalue λ of $A \in M_n$ is a right eigenvector of A^* associated with $\bar{\lambda}$; also show that $\bar{y}$ is a right eigenvector of A^T associated with λ.

Exercise. Suppose that $A \in M_n$ is diagonalizable, S is nonsingular, and $S^{-1} A S = \Lambda = \mathrm{diag}(\lambda_1, \ldots, \lambda_n)$. Partition $S = [x_1 \ \ldots \ x_n]$ and $S^{-*} = [y_1 \ \ldots \ y_n]$ (0.5) according to their columns. The identity $AS = S\Lambda$ tells us that each column x_j of S is a right eigenvector of A associated with the eigenvalue λ_j. Explain why: $(S^{-*})^* A = \Lambda (S^{-*})^*$; each column y_j of S^{-*} is a left eigenvector of A associated with the eigenvalue λ_j; $y_j^* x_j = 1$ for each $j = 1, \ldots, n$; and $y_i^* x_j = 0$ whenever $i \neq j$.

One should not dismiss left eigenvectors as merely a parallel theoretical alternative to right eigenvectors. Each type of eigenvector can convey different information about a matrix, and it can be very useful to know how the two types of eigenvectors interact.

We next examine a version of the results in the preceding exercise for matrices that are not necessarily diagonalizable.

Theorem 1.4.7. *Let $A \in M_n$, nonzero vectors $x, y \in \mathbb{C}^n$, and scalars $\lambda, \mu \in \mathbb{C}$ be given. Suppose that $Ax = \lambda x$ and $y^*A = \mu y^*$.*

(a) *If $\lambda \neq \mu$, then $y^*x = 0$.*

(b) *If $\lambda = \mu$ and $y^*x \neq 0$, then there is a nonsingular $S \in M_n$ of the form $S = [x\ S_1]$ such that $S^{-*} = [y/(x^*y)\ Z_1]$ and*

$$A = S \begin{bmatrix} \lambda & 0 \\ 0 & B \end{bmatrix} S^{-1}, \qquad B \in M_{n-1} \qquad (1.4.8)$$

Conversely, if A is similar to a block matrix of the form (1.4.8), then it has a nonorthogonal pair of left and right eigenvectors associated with the eigenvalue λ.

Proof. (a) Let y be a left eigenvector of A associated with μ and let x be a right eigenvector of A associated with λ. Manipulate y^*Ax in two ways:

$$y^*Ax = y^*(\lambda x) = \lambda(y^*x)$$
$$= (\mu y^*)x = \mu(y^*x)$$

Since $\lambda \neq \mu$, $\lambda y^*x = \mu y^*x$ only if $y^*x = 0$.

(b) Suppose that $Ax = \lambda x$, $y^*A = \lambda y^*$, and $y^*x \neq 0$. If we replace y by $y/(x^*y)$, we may assume that $y^*x = 1$. Let the columns of $S_1 \in M_{n,n-1}$ be any basis for the orthogonal complement of y (so $y^*S_1 = 0$) and consider $S = [x\ S_1] \in M_n$. Let $z = [z_1\ \zeta^T]^T$ with $\zeta \in \mathbb{C}^{n-1}$ and suppose that $Sz = 0$. Then

$$0 = y^*Sz = y^*(z_1x + S_1\zeta) = z_1(y^*x) + (y^*S_1)\zeta = z_1$$

so $z_1 = 0$ and $0 = Sz = S_1\zeta$, which implies that $\zeta = 0$ since S_1 has full column rank. We conclude that S is nonsingular. Partition $S^{-*} = [\eta\ Z_1]$ with $\eta \in \mathbb{C}^n$ and compute

$$I_n = S^{-1}S = \begin{bmatrix} \eta^* \\ Z_1^* \end{bmatrix} [x\ \ S_1] = \begin{bmatrix} \eta^*x & \eta^*S_1 \\ Z_1^*x & Z_1^*S_1 \end{bmatrix} = \begin{bmatrix} 1 & 0 \\ 0 & I_{n-1} \end{bmatrix}$$

which contains four identities. The identity $\eta^*S_1 = 0$ implies that η is orthogonal to the orthogonal complement of y, so $\eta = \alpha y$ for some scalar α. The identity $\eta^*x = 1$ tells us that $\eta^*x = (\alpha y)^*x = \bar{\alpha}(y^*x) = \bar{\alpha} = 1$, so $\eta = y$. Using the identities $\eta^*S_1 = y^*S_1 = 0$ and $Z_1^*x = 0$ as well as the eigenvector properties of x and y, compute the similarity

$$S^{-1}AS = \begin{bmatrix} y^* \\ Z_1^* \end{bmatrix} A [x\ \ S_1] = \begin{bmatrix} y^*Ax & y^*AS_1 \\ Z_1^*Ax & Z_1^*AS_1 \end{bmatrix}$$
$$= \begin{bmatrix} (\lambda y^*)x & (\lambda y^*)S_1 \\ Z_1^*(\lambda x) & Z_1^*AS_1 \end{bmatrix} = \begin{bmatrix} \lambda(y^*x) & \lambda(y^*S_1) \\ \lambda(Z_1^*x) & Z_1^*AS_1 \end{bmatrix}$$
$$= \begin{bmatrix} \lambda & 0 \\ 0 & Z_1^*AS_1 \end{bmatrix}$$

which verifies (1.4.8).

Conversely, suppose that there is a nonsingular S such that $A = S([\lambda] \oplus B)S^{-1}$. Let x be the first column of S, let y be the first column of S^{-*}, and partition $S = [x\ S_1]$ and $S^{-*} = [y\ Z_1]$. The 1, 1 entry of the identity $S^{-1}S = I$ tells us that $y^*x = 1$; the first column of the identity

$$[Ax \quad AS_1] = AS = S([\lambda] \oplus B) = [\lambda x \quad S_1 B]$$

tells us that $Ax = \lambda x$; and the first row of the identity

$$\begin{bmatrix} y^*A \\ Z_1^*A \end{bmatrix} = S^{-1}A = ([\lambda] \oplus B)S^{-1} = \begin{bmatrix} \lambda y^* \\ BZ_1^* \end{bmatrix}$$

tells us that $y^*A = \lambda y^*$. □

The assertion in (1.4.7(a)) is the *principle of biorthogonality*. One might also ask what happens if left and right eigenvectors associated with the *same* eigenvalue are either orthogonal or linearly dependent; these cases are discussed in (2.4.11.1).

The eigenvalues of a matrix are unchanged by similarity; its eigenvectors transform under similarity in a simple way.

Theorem 1.4.9. *Let $A, B \in M_n$ and suppose that $B = S^{-1}AS$ for some nonsingular S. If $x \in \mathbf{C}^n$ is a right eigenvector of B associated with an eigenvalue λ, then Sx is a right eigenvector of A associated with λ. If $y \in \mathbf{C}^n$ is a left eigenvector of B associated with λ, then $S^{-*}y$ is a left eigenvector of A associated with λ.*

Proof. If $Bx = \lambda x$, then $S^{-1}ASx = \lambda x$, or $A(Sx) = \lambda(Sx)$. Since S is nonsingular and $x \neq 0$, $Sx \neq 0$, and hence Sx is an eigenvector of A. If $y^*B = \lambda y^*$, then $y^*S^{-1}AS = \lambda y^*$, or $(S^{-*}y)^*A = \lambda(S^{-*}y)^*$ and $S^{-*}y \neq 0$ since $y \neq 0$. □

Information about eigenvalues of principal submatrices can refine the basic observation that the algebraic multiplicity of an eigenvalue cannot be less than its geometric multiplicity.

Theorem 1.4.10. *Let $A \in M_n$ and $\lambda \in \mathbf{C}$ be given, and let $k \geq 1$ be a given positive integer. Consider the following three statements:*

(a) *λ is an eigenvalue of A with geometric multiplicity at least k.*
(b) *For each $m = n - k + 1, \ldots, n$, λ is an eigenvalue of every m-by-m principal submatrix of A.*
(c) *λ is an eigenvalue of A with algebraic multiplicity at least k.*

Then (a) implies (b), and (b) implies (c). In particular, the algebraic multiplicity of an eigenvalue is at least as great as its geometric multiplicity.

Proof. (a) $\Rightarrow$ (b): Let λ be an eigenvalue of A with geometric multiplicity at least k, which means that $\text{rank}(A - \lambda I) \leq n - k$. Suppose that $m > n - k$. Then every m-by-m minor of $A - \lambda I$ is zero. In particular, every principal m-by-m minor of $A - \lambda I$ is zero, so every m-by-m principal submatrix of $A - \lambda I$ is singular. Thus, λ is an eigenvalue of every m-by-m principal submatrix of A.
(b) $\Rightarrow$ (c): Suppose that λ is an eigenvalue of every m-by-m principal submatrix of A for each $m \geq n - k + 1$. Then every principal minor of $A - \lambda I$ of size at least $n - k + 1$ is zero, so each principal minor sum $E_j(A - \lambda I) = 0$ for all $j \geq n - k + 1$. Then

(1.2.13) and (1.2.11) ensure that $p^{(i)}_{A-\lambda I}(0) = 0$ for $i = 0, 1, \ldots, k-1$. But $p_{A-\lambda I}(t) = p_A(t+\lambda)$, so $p^{(i)}_A(\lambda) = 0$ for $i = 0, 1, \ldots, k-1$; that is, λ is a zero of $p_A(t)$ with multiplicity at least k. ☐

An eigenvalue λ with geometric multiplicity 1 can have algebraic multiplicity 2 or more, but this can happen only if the left and right eigenvectors associated with λ are orthogonal. If λ has algebraic multiplicity 1, however, then it has geometric multiplicity 1; left and right eigenvectors associated with λ can never be orthogonal. Our approach to these results relies on the following lemma.

Lemma 1.4.11. *Let $A \in M_n$, $\lambda \in \mathbf{C}$, and nonzero vectors $x, y \in \mathbf{C}^n$ be given. Suppose that λ has geometric multiplicity 1 as an eigenvalue of A, $Ax = \lambda x$, and $y^*A = \lambda y^*$. Then there is a nonzero $\gamma \in \mathbf{C}$ such that $\mathrm{adj}(\lambda I - A) = \gamma xy^*$.*

Proof. We have $\mathrm{rank}(\lambda I - A) = n-1$ and hence $\mathrm{rank}\,\mathrm{adj}(\lambda I - A) = 1$, that is, $\mathrm{adj}(\lambda I - A) = \xi\eta^*$ for some nonzero $\xi, \eta \in \mathbf{C}^n$; see (0.8.2). But $(\lambda I - A)(\mathrm{adj}(\lambda I - A)) = \det(\lambda I - A)I = 0$, so $(\lambda I - A)\xi\eta^* = 0$ and $(\lambda I - A)\xi = 0$, which implies that $\xi = \alpha x$ for some nonzero scalar α. Using the identity $(\mathrm{adj}(\lambda I - A))(\lambda I - A) = 0$ in a similar fashion, we conclude that $\eta = \beta y$ for some nonzero scalar β. Thus, $\mathrm{adj}(\lambda I - A) = \alpha\beta xy^*$. ☐

Theorem 1.4.12. *Let $A \in M_n$, $\lambda \in \mathbf{C}$, and nonzero vectors $x, y \in \mathbf{C}^n$ be given. Suppose that λ is an eigenvalue of A, $Ax = \lambda x$, and $y^*A = \lambda y^*$.*

(a) *If λ has algebraic multiplicity 1, then $y^*x \neq 0$.*
(b) *If λ has geometric multiplicity 1, then it has algebraic multiplicity 1 if and only if $y^*x \neq 0$.*

Proof. In both cases (a) and (b), λ has geometric multiplicity 1; the preceding lemma tells us that there is a nonzero $\gamma \in \mathbf{C}$ such that $\mathrm{adj}(\lambda I - A) = \gamma xy^*$. Then $p_A(\lambda) = 0$ and $p'_A(\lambda) = \mathrm{tr}\,\mathrm{adj}(\lambda I - A) = \gamma y^*x$; see (0.8.10.2). In (a) we assume that the algebraic multiplicity is 1, so $p'_A(\lambda) \neq 0$ and hence $y^*x \neq 0$. In (b) we assume that $y^*x \neq 0$, so $p'_A(\lambda) \neq 0$ and hence the algebraic multiplicity is 1. ☐

Problems

1.4.P1 Let nonzero vectors $x, y \in \mathbf{C}^n$ be given, let $A = xy^*$, and let $\lambda = y^*x$. Show that (a) λ is an eigenvalue of A; (b) x is a right and y is a left eigenvector of A associated with λ; and (c) if $\lambda \neq 0$, then it is the *only* nonzero eigenvalue of A (algebraic multiplicity = 1). Explain why any vector that is orthogonal to y is in the null space of A. What is the geometric multiplicity of the eigenvalue 0? Explain why A is diagonalizable if and only if $y^*x \neq 0$.

1.4.P2 Let $A \in M_n$ be skew symmetric. Show that $p_A(t) = (-1)^n p_A(-t)$ and deduce that if λ is an eigenvalue of A with multiplicity k, then so is $-\lambda$. If n is odd, explain why A must be singular. Explain why every principal minor of A with odd size is singular. Use the fact that a skew-symmetric matrix is rank principal (0.7.6) to show that rank A must be even.

1.4.P3 Suppose that $n \geq 2$ and let $T = [t_{ij}] \in M_n$ be upper triangular. (a) Let x be an eigenvector of T associated with the eigenvalue t_{nn}; explain why e_n is a left eigenvector associated with t_{nn}. If $t_{ii} \neq t_{nn}$ for each $i = 1, \ldots, n - 1$, show that the last entry of x *must* be nonzero. (b) Let $k \in \{1, \ldots, n - 1\}$. Show that there is an eigenvector x of T associated with the eigenvalue t_{kk} whose last $n - k$ entries of x are zero, that is, $x^T = [\xi^T \; 0]^T$ with $\xi \in \mathbf{C}^k$. If $t_{ii} \neq t_{kk}$ for all $i = 1, \ldots, k - 1$, explain why the kth entry of x *must* be nonzero.

1.4.P4 Suppose that $A \in M_n$ is tridiagonal and has a zero main diagonal. Let $S = \mathrm{diag}(-1, 1, -1, \ldots, (-1)^n)$ and show that $S^{-1}AS = -A$. If λ is an eigenvalue of A with multiplicity k, explain why $-\lambda$ is also an eigenvalue of A with multiplicity k. If n is odd, show that A is singular.

1.4.P5 Consider the block triangular matrix

$$A = \begin{bmatrix} A_{11} & A_{12} \\ 0 & A_{22} \end{bmatrix}, \qquad A_{ii} \in M_{n_i}, \quad i = 1, 2$$

If $x \in \mathbf{C}^{n_1}$ is a right eigenvector of A_{11} associated with $\lambda \in \sigma(A_{11})$, and if $y \in \mathbf{C}^{n_2}$ is a left eigenvector of A_{22} associated with $\mu \in \sigma(A_{22})$, show that $\left[\begin{smallmatrix} x \\ 0 \end{smallmatrix}\right] \in \mathbf{C}^{n_1+n_2}$ is a right eigenvector, and $\left[\begin{smallmatrix} 0 \\ y \end{smallmatrix}\right]$ is a left eigenvector, of A associated with λ and μ, respectively. Use this observation to show that the eigenvalues of A are the eigenvalues of A_{11} together with those of A_{22}.

1.4.P6 Suppose that $A \in M_n$ has an entrywise positive left eigenvector and an entrywise positive right eigenvector, both associated with an eigenvalue λ. (a) Show that A has no entrywise nonnegative left or right eigenvectors associated with any eigenvalue different from λ. (b) If λ has geometric multiplicity 1, show that it has algebraic multiplicity 1. See (8.2.2) and (8.4.4) for properties of A that are sufficient to ensure the existence of positive left and right eigenvectors associated with a special eigenvalue of A.

1.4.P7 In this problem we outline a simple version of the *power method* for finding the largest modulus eigenvalue and an associated eigenvector of $A \in M_n$. Suppose that $A \in M_n$ has distinct eigenvalues $\lambda_1, \ldots, \lambda_n$ and that there is exactly one eigenvalue λ_n of maximum modulus $\rho(A)$. If $x^{(0)} \in \mathbf{C}^n$ is *not* orthogonal to a left eigenvector associated with λ_n, show that the sequence

$$x^{(k+1)} = \frac{1}{(x^{(k)*}x^{(k)})^{1/2}} A x^{(k)}, \quad k = 0, 1, 2, \ldots$$

converges to an eigenvector of A, and the ratios of a given nonzero entry in the vectors $A x^{(k)}$ and $x^{(k)}$ converge to λ_n.

1.4.P8 Continue with the assumptions and notation of (1.4.P7). Further eigenvalues (and eigenvectors) of A can be calculated by combining the power method with a *deflation* that delivers a square matrix of size one smaller, whose spectrum (with multiplicities) contains all but one eigenvalue of A. Let $S \in M_n$ be nonsingular and have as its first column an eigenvector $y^{(n)}$ associated with the eigenvalue λ_n. Show that $S^{-1}AS = \left[\begin{smallmatrix} \lambda_n & * \\ 0 & B \end{smallmatrix}\right]$ and the eigenvalues of $B \in M_{n-1}$ are $\lambda_1, \ldots, \lambda_{n-1}$. Another eigenvalue may be calculated from B and the deflation repeated.

1.4.P9 Let $A \in M_n$ have eigenvalues $\lambda_1, \ldots, \lambda_{n-1}, 0$, so that rank $A \leq n - 1$. Suppose that the last row of A is a linear combination of the first $n - 1$ rows. Partition $A = \left[\begin{smallmatrix} B & x \\ y^T & \alpha \end{smallmatrix}\right]$, in which $B \in M_{n-1}$. (a) Explain why there is a $z \in \mathbf{C}^{n-1}$ such that $y^T = z^T B$ and $\alpha = z^T x$. Why is $\left[\begin{smallmatrix} z \\ -1 \end{smallmatrix}\right]$ a left eigenvector of A associated with the eigenvalue 0? (b) Show that $B + x z^T \in M_{n-1}$ has eigenvalues $\lambda_1, \ldots, \lambda_{n-1}$. This construction is another type of *deflation*; see (1.3.P33) for a further example of a deflation. (c) If one eigenvalue λ of A is known, explain how this construction can be applied to $P(A - \lambda I)P^{-1}$ for a suitable permutation P.

1.4.P10 Let $T \in M_n$ be a nonsingular matrix whose columns are left eigenvectors of $A \in M_n$. Show that the columns of T^{-*} are right eigenvectors of A.

1.4.P11 Suppose that $A \in M_n$ is an unreduced upper Hessenberg matrix (0.9.9). Explain why rank$(A - \lambda I) \geq n - 1$ for every $\lambda \in \mathbf{C}$ and deduce that every eigenvalue of A has geometric multiplicity 1, that is, A is nonderogatory.

1.4.P12 Let λ be an eigenvalue of $A \in M_n$. (a) Show that every list of $n - 1$ columns of $A - \lambda I$ is linearly independent if and only if no eigenvector of A associated with λ has a zero entry. (b) If no eigenvector of A associated with λ has a zero entry, why must λ have geometric multiplicity 1?

1.4.P13 Let $A \in M_n$ and nonzero vectors $x, y \in \mathbf{C}^n$ be given, and let $\lambda, \lambda_2, \ldots, \lambda_n$ be the eigenvalues of A. Suppose that $Ax = \lambda x$ and $y^*A = \lambda y^*$, and that λ has geometric multiplicity 1. Then (1.4.11) says that adj$(\lambda I - A) = \gamma x y^*$ and $\gamma \neq 0$. (a) Explain why $\gamma y^*x = \mathrm{tr}(\lambda I - A) = E_{n-1}(\lambda I - A) = S_{n-1}(\lambda I - A) = (\lambda - \lambda_2)(\lambda - \lambda_3) \cdots (\lambda - \lambda_n)$. (b) Deduce from (a) that $y^*x \neq 0$ if and only if λ is a simple eigenvalue. (c) The parameter γ is nonzero no matter what the multiplicity of λ is. If λ is simple, explain why $\gamma = (\lambda - \lambda_2) \cdots (\lambda - \lambda_n)/y^*x$. See (2.6.P12) for a different way to evaluate γ. (d) Explain why every entry of x and y is nonzero $\Leftrightarrow$ every principal minor of $\lambda I - A$ is nonzero $\Leftrightarrow$ every *main diagonal* entry of adj$(\lambda I - A)$ is nonzero $\Leftrightarrow$ *every* entry of adj$(\lambda I - A)$ is nonzero.

1.4.P14 Let $A \in M_n$ and let $t \in \mathbf{C}$. Explain why $(A - tI)\,\mathrm{adj}(A - tI) = \mathrm{adj}(A - tI)(A - tI) = p_A(t)I$. Now suppose that λ is an eigenvalue of A. Show that (a) every nonzero column of adj$(A - \lambda I)$ is an eigenvector of A associated with λ; (b) every nonzero row of adj$(A - \lambda I)$ is the conjugate transpose of a left eigenvector of A associated with λ; (c) adj$(A - \lambda I) \neq 0$ if and only if λ has geometric multiplicity one; and (d) if λ is an eigenvalue of $A = \begin{bmatrix} a & b \\ c & d \end{bmatrix}$, then each nonzero column of $\begin{bmatrix} d-\lambda & -b \\ -c & a-\lambda \end{bmatrix}$ is an eigenvector of A associated with λ; each nonzero row is the conjugate transpose of a left eigenvector of A associated with λ.

1.4.P15 Suppose that λ is a simple eigenvalue of $A \in M_n$, and suppose that $x, y, z, w \in \mathbf{C}^n$, $Ax = \lambda x$, $y^*A = \lambda y^*$, $y^*z \neq 0$, and $w^*x \neq 0$. Show that $A - \lambda I + \kappa z w^*$ is nonsingular for all $\kappa \neq 0$. Explain why it is possible to take $z = x$.

1.4.P16 Show that the complex tridiagonal Toeplitz matrix

$$A = \begin{bmatrix} a & b & & \\ c & a & \ddots & \\ & \ddots & \ddots & b \\ & & c & a \end{bmatrix} \in M_n, \quad bc \neq 0 \qquad (1.4.13)$$

has spectrum $\sigma(A) = \{a + 2\sqrt{bc}\cos(\frac{\pi\kappa}{n+1}) : \kappa = 1, \ldots, n\}$, in which Re $\sqrt{bc} \geq 0$ and Im $\sqrt{bc} > 0$ if bc is real and negative. Why is A diagonalizable?

1.4.P17 If $a = 2$ and $b = c = -1$ in (1.4.13), show that $\sigma(A) = \{4\sin^2(\frac{\pi\kappa}{2(n+1)}) : \kappa = 1, \ldots, n\}$.

CHAPTER 2

Unitary Similarity and Unitary Equivalence

2.0 Introduction

In Chapter 1, we made an initial study of similarity of $A \in M_n$ via a general nonsingular matrix S, that is, the transformation $A \to S^{-1}AS$. For certain very special nonsingular matrices, called *unitary matrices*, the inverse of S has a simple form: $S^{-1} = S^*$. Similarity via a unitary matrix U, $A \to U^*AU$, is not only conceptually simpler than general similarity (the conjugate transpose is much easier to compute than the inverse), but it also has superior stability properties in numerical computations. A fundamental property of unitary similarity is that every $A \in M_n$ is unitarily similar to an upper triangular matrix whose diagonal entries are the eigenvalues of A. This triangular form can be further refined under general similarity; we study the latter in Chapter 3.

The transformation $A \to S^*AS$, in which S is nonsingular but not necessarily unitary, is called *congruence; we study it in Chapter 4. Notice that similarity by a unitary matrix is both a similarity and a *congruence.

For $A \in M_{n,m}$, the transformation $A \to UAV$, in which $U \in M_n$ and $V \in M_m$ are both unitary, is called *unitary equivalence*. The upper triangular form achievable under unitary similarity can be greatly refined under unitary equivalence and generalized to rectangular matrices: Every $A \in M_{n,m}$ is unitarily equivalent to a nonnegative diagonal matrix whose diagonal entries (the singular values of A) are of great importance.

2.1 Unitary matrices and the QR factorization

Definition 2.1.1. *A list of vectors* $x_1, \ldots, x_k \in \mathbf{C}^n$ *is* orthogonal *if* $x_i^* x_j = 0$ *for all* $i \neq j$, $i, j \in \{1, \ldots, k\}$. *If, in addition,* $x_i^* x_i = 1$ *for all* $i = 1, \ldots, k$ *(that is, the vectors are* normalized*), then the list is* orthonormal. *It is often convenient to say that* "$x_1, \ldots, x_k$ *are orthogonal (respectively, orthonormal)" instead of the more formal statement "the list of vectors* $v_1, \ldots, v_k$ *is orthogonal (orthonormal, respectively)."*

83

Exercise. If $y_1, \ldots, y_k \in \mathbf{C}^n$ are orthogonal and nonzero, show that the vectors $x_1, \ldots, x_k$ defined by $x_i = (y_i^* y_i)^{-1/2} y_i$, $i = 1, \ldots, k$, are orthonormal.

Theorem 2.1.2. *Every orthonormal list of vectors in $\mathbf{C}^n$ is linearly independent.*

Proof. Suppose that $\{x_1, \ldots, x_k\}$ is an orthonormal set, and suppose that $0 = \alpha_1 x_1 + \cdots + \alpha_k x_k$. Then $0 = (\alpha_1 x_1 + \cdots + \alpha_k x_k)^* (\alpha_1 x_1 + \cdots + \alpha_k x_k) = \Sigma_{i,j} \bar{\alpha}_i \alpha_j x_i^* x_j = \Sigma_{i=1}^k |\alpha_i|^2 x_i^* x_i = \Sigma_{i=1}^k |\alpha_i|^2$ because the vectors x_i are orthogonal and normalized. Thus, all $\alpha_i = 0$ and hence $\{x_1, \ldots, x_k\}$ is a linearly independent set. $\square$

Exercise. Show that every orthogonal list of nonzero vectors in $\mathbf{C}^n$ is linearly independent.

Exercise. If $x_1, \ldots, x_k \in \mathbf{C}^n$ are orthogonal, show that either $k \leq n$ or at least $k - n$ of the vectors x_i are zero vectors.

A linearly independent list need not be orthonormal, of course, but one can apply the Gram–Schmidt orthonormalization procedure (0.6.4) to it and obtain an orthonormal list with the same span.

Exercise. Show that any nonzero subspace of $\mathbf{R}^n$ or $\mathbf{C}^n$ has an orthonormal basis (0.6.5).

Definition 2.1.3. *A matrix $U \in M_n$ is* unitary *if $U^* U = I$. A matrix $U \in M_n(\mathbf{R})$ is* real orthogonal *if $U^T U = I$.*

Exercise. Show that $U \in M_n$ and $V \in M_m$ are unitary if and only if $U \oplus V \in M_{n+m}$ is unitary.

Exercise. Verify that the matrices Q, U, and V in (P19, P20, and P21) in (1.3) are unitary.

The unitary matrices in M_n form a remarkable and important set. We list some of the basic equivalent conditions for U to be unitary in (2.1.4).

Theorem 2.1.4. *If $U \in M_n$, the following are equivalent:*

  (a) U is unitary.
  (b) U is nonsingular and $U^* = U^{-1}$.
  (c) $UU^* = I$.
  (d) U^* is unitary.
  (e) The columns of U are orthonormal.
  (f) The rows of U are orthonormal.
  (g) For all $x \in \mathbf{C}^n$, $\|x\|_2 = \|Ux\|_2$, that is, x and Ux have the same Euclidean norm.

Proof. (a) implies (b) since U^{-1} (when it exists) is the unique matrix, left multiplication by which produces I (0.5); the definition of unitary says that U^* is such a matrix. Since $BA = I$ if and only if $AB = I$ (for $A, B \in M_n$ (0.5)), (b) implies (c). Since $(U^*)^* = U$, (c) implies that U^* is unitary; that is, (c) implies (d). The converse of each of these implications is similarly observed, so (a)–(d) are equivalent.

Partition $U = [u_1 \ \ldots \ u_n]$ according to its columns. Then $U^* U = I$ means that $u_i^* u_i = 1$ for all $i = 1, \ldots, n$ and $u_i^* u_j = 0$ for all $i \neq j$. Thus, $U^* U = I$ is another way of saying that the columns of U are orthonormal, and hence (a) is equivalent to (e). Similarly, (d) and (f) are equivalent.

If U is unitary and $y = Ux$, then $y^*y = x^*U^*Ux = x^*Ix = x^*x$, so (a) implies (g). To prove the converse, let $U^*U = A = [a_{ij}]$, let $z, w \in \mathbf{C}$ be given, and take $x = z + w$ in (g). Then $x^*x = z^*z + w^*w + 2\operatorname{Re} z^*w$ and $y^*y = x^*Ax = z^*Az + w^*Aw + 2\operatorname{Re} z^*Aw$; (g) ensures that $z^*z = z^*Az$ and $w^*w = w^*Aw$, and hence $\operatorname{Re} z^*w = \operatorname{Re} z^*Aw$ for any z and w. Take $z = e_p$ and $w = ie_q$ and compute $\operatorname{Re} ie_p^T e_q = 0 = \operatorname{Re} ie_p^T Ae_q = \operatorname{Re} ia_{pq} = -\operatorname{Im} a_{pq}$, so every entry of A is real. Finally, take $z = e_p$ and $w = e_q$ and compute $e_p^T e_q = \operatorname{Re} e_p^T e_q = \operatorname{Re} e_p^T Ae_q = a_{pq}$, which tells us that $A = I$ and U is unitary. $\qquad\square$

Definition 2.1.5. *A linear transformation $T : \mathbf{C}^n \to \mathbf{C}^m$ is called a* Euclidean isometry *if $\|x\|_2 = \|Tx\|_2$ for all $x \in \mathbf{C}^n$. Theorem 2.1.4 says that a square complex matrix $U \in M_n$ is a Euclidean isometry (via $U : x \to Ux$) if and only if it is unitary. For other kinds of isometries, see (5.4.P11–13).*

Exercise. Let $U_\theta = \begin{bmatrix} \cos\theta & -\sin\theta \\ \sin\theta & \cos\theta \end{bmatrix}$, in which θ is a real parameter. (a) Show that a given $U \in M_2(\mathbf{R})$ is real orthogonal if and only if either $U = U_\theta$ or $U = \begin{bmatrix} 1 & 0 \\ 0 & -1 \end{bmatrix} U_\theta$ for some $\theta \in \mathbf{R}$. (b) Show that a given $U \in M_2(\mathbf{R})$ is real orthogonal if and only if either $U = U_\theta$ or $U = \begin{bmatrix} 0 & 1 \\ 1 & 0 \end{bmatrix} U_\theta$ for some $\theta \in \mathbf{R}$. These are two different presentations, involving a parameter θ, of the 2-by-2 real orthogonal matrices. Interpret them geometrically.

Observation 2.1.6. *If $U, V \in M_n$ are unitary (respectively, real orthogonal), then UV is also unitary (respectively, real orthogonal).*

Exercise. Use (b) of (2.1.4) to prove (2.1.6).

Observation 2.1.7. *The set of unitary (respectively, real orthogonal) matrices in M_n forms a group. This group is generally referred to as the n-by-n* unitary *(respectively,* real orthogonal*) group, a subgroup of $GL(n, \mathbf{C})$ (0.5).*

Exercise. A *group* is a set that is *closed* under a single *associative* binary operation ("multiplication") and is such that the *identity* for and *inverses* under the operation are contained in the set. Verify (2.1.7). *Hint*: Use (2.1.6) for closure; matrix multiplication is associative; $I \in M_n$ is unitary; and $U^* = U^{-1}$ is again unitary.

The set (group) of unitary matrices in M_n has another very important property. Notions of "convergence" and "limit" of a sequence of matrices are presented precisely in Chapter 5, but can be understood here as "convergence" and "limit" of entries. The defining identity $U^*U = I$ means that every column of U has Euclidean norm 1, and hence no entry of $U = [u_{ij}]$ can have absolute value greater than 1. If we think of the set of unitary matrices as a subset of $\mathbf{C}^{n^2}$, this says it is a *bounded* subset. If $U_k = [u_{ij}^{(k)}]$ is an infinite sequence of unitary matrices, $k = 1, 2, \ldots$, such that $\lim_{k\to\infty} u_{ij}^{(k)} = u_{ij}$ exists for all $i, j = 1, 2, \ldots, n$, then from the identity $U_k^*U_k = I$ for all $k = 1, 2, \ldots$, we see that $\lim_{k\to\infty} U_k^*U_k = U^*U = I$, in which $U = [u_{ij}]$. Thus, the limit matrix U is also unitary. This says that the set of unitary matrices is a *closed* subset of $\mathbf{C}^{n^2}$.

Since a closed and bounded subset of a finite dimensional Euclidean space is a *compact* set (see Appendix E), we conclude that the set (group) of unitary matrices in M_n is compact. For our purposes, the most important consequence of this observation is the following *selection principle* for unitary matrices.

Lemma 2.1.8. *Let $U_1, U_2, \ldots \in M_n$ be a given infinite sequence of unitary matrices. There exists an infinite subsequence $U_{k_1}, U_{k_2}, \ldots, 1 \leq k_1 < k_2 < \cdots$, such that all of the entries of U_{k_i} converge (as sequences of complex numbers) to the entries of a unitary matrix as $i \to \infty$.*

Proof. All that is required here is the fact that from any infinite sequence in a compact set, one may always select a convergent subsequence. We have already observed that if a sequence of unitary matrices converges to some matrix, then the limit matrix must be unitary.   □

The unitary limit guaranteed by the lemma need not be unique; it can depend on the subsequence chosen.

Exercise. Consider the sequence of unitary matrices $U_k = \begin{bmatrix} 0 & 1 \\ 1 & 0 \end{bmatrix}^k$, $k = 1, 2, \ldots$. Show that there are two possible limits of subsequences.

Exercise. Explain why the selection principle (2.1.8) applies as well to the (real) orthogonal group; that is, an infinite sequence of real orthogonal matrices has an infinite subsequence that converges to a real orthogonal matrix.

A unitary matrix U has the property that U^{-1} *equals* U^*. One way to generalize the notion of a unitary matrix is to require that U^{-1} be *similar* to U^*. The set of such matrices is easily characterized as the range of the mapping $A \to A^{-1}A^*$ for all nonsingular $A \in M_n$.

Theorem 2.1.9. *Let $A \in M_n$ be nonsingular. Then A^{-1} is similar to A^* if and only if there is a nonsingular $B \in M_n$ such that $A = B^{-1}B^*$.*

Proof. If $A = B^{-1}B^*$ for some nonsingular $B \in M_n$, then $A^{-1} = (B^*)^{-1}B$ and $B^*A^{-1}(B^*)^{-1} = B(B^*)^{-1} = (B^{-1}B^*)^* = A^*$, so A^{-1} is similar to A^* via the similarity matrix B^*. Conversely, if A^{-1} is similar to A^*, then there is a nonsingular $S \in M_n$ such that $SA^{-1}S^{-1} = A^*$ and hence $S = A^*SA$. Set $S_\theta = e^{i\theta}S$ for $\theta \in \mathbf{R}$ so that $S_\theta = A^*S_\theta A$ and $S_\theta^* = A^*S_\theta^* A$. Adding these two identities gives $H_\theta = A^*H_\theta A$, in which $H_\theta = S_\theta + S_\theta^*$ is Hermitian. If H_θ were singular, there would be a nonzero $x \in \mathbf{C}^n$ such that $0 = H_\theta x = S_\theta x + S_\theta^* x$, so $-x = S_\theta^{-1}S_\theta^* x = e^{-2i\theta}S^{-1}S^*x$ and $S^{-1}S^*x = -e^{2i\theta}x$. Choose a value of $\theta = \theta_0 \in [0, 2\pi)$ such that $-e^{2i\theta_0}$ is not an eigenvalue of $S^{-1}S^*$; the resulting Hermitian matrix $H = H_{\theta_0}$ is nonsingular and has the property that $H = A^*HA$.

Now choose any complex α such that $|\alpha| = 1$ and α is not an eigenvalue of A^*. Set $B = \beta(\alpha I - A^*)H$, in which the complex parameter $\beta \neq 0$ is to be chosen, and observe that B is nonsingular. We want to have $A = B^{-1}B^*$, that is, $BA = B^*$. Compute $B^* = H(\bar{\beta}\bar{\alpha}I - \bar{\beta}A)$, and $BA = \beta(\alpha I - A^*)HA = \beta(\alpha HA - A^*HA) = \beta(\alpha HA - H) = H(\alpha\beta A - \beta I)$. We are done if we can select a nonzero β such that $\beta = -\bar{\beta}\bar{\alpha}$, but if $\alpha = e^{i\psi}$, then $\beta = e^{i(\pi-\psi)/2}$ will do.   □

If a unitary matrix is presented as a 2-by-2 block matrix, then the ranks of its off-diagonal blocks are equal; the ranks of its diagonal blocks are related by a simple formula.

Lemma 2.1.10. *Let a unitary $U \in M_n$ be partitioned as $U = \begin{bmatrix} U_{11} & U_{12} \\ U_{21} & U_{22} \end{bmatrix}$, in which $U_{11} \in M_k$. Then* rank $U_{12} =$ rank U_{21} *and* rank $U_{22} =$ rank $U_{11} + n - 2k$. *In particular, $U_{12} = 0$ if and only if $U_{21} = 0$, in which case U_{11} and U_{22} are unitary.*

Proof. The two assertions about rank follow immediately from the law of complementary nullities (0.7.5) using the fact that $U^{-1} = \begin{bmatrix} U_{11}^* & U_{21}^* \\ U_{12}^* & U_{22}^* \end{bmatrix}$. $\qquad\square$

Exercise. Use the preceding lemma to show that a unitary matrix is upper triangular if and only if it is diagonal.

Plane rotations and Householder matrices are special (and very simple) unitary matrices that play an important role in establishing some basic matrix factorizations.

Example 2.1.11. Plane rotations. Let $1 \leq i < j \leq n$ and let

$$U(\theta; i, j) = \begin{bmatrix} 1 \\ & \ddots \\ & & 1 \\ & & & \cos\theta & & & & -\sin\theta \\ & & & & 1 \\ & & & & & \ddots \\ & & & & & & 1 \\ & & & \sin\theta & & & & \cos\theta \\ & & & & & & & & 1 \\ & & & & & & & & & \ddots \\ & & & & & & & & & & 1 \end{bmatrix}$$

denote the result of replacing the i, i and j, j entries of the n-by-n identity matrix by $\cos\theta$, replacing its i, j entry by $-\sin\theta$ and replacing its j, i entry by $\sin\theta$. The matrix $U(\theta; i, j)$ is called a *plane rotation* or *Givens rotation*.

Exercise. Verify that $U(\theta; i, j) \in M_n(\mathbf{R})$ is real orthogonal for any pair of indices i, j with $1 \leq i < j \leq n$ and any parameter $\theta \in [0, 2\pi)$. The matrix $U(\theta; i, j)$ carries out a rotation (through an angle θ) in the i, j coordinate plane of $\mathbf{R}^n$. Left multiplication by $U(\theta; i, j)$ affects only rows i and j of the matrix multiplied; right multiplication by $U(\theta; i, j)$ affects only columns i and j of the matrix multiplied.

Exercise. Verify that $U(\theta; i, j)^{-1} = U(-\theta; i, j)$.

Example 2.1.12. Householder matrices. Let $w \in \mathbf{C}^n$ be a nonzero vector. The *Householder matrix* $U_w \in M_n$ is defined by $U_w = I - 2(w^*w)^{-1}ww^*$. If w is a unit vector, then $U_w = I - 2ww^*$.

Exercise. Show that a Householder matrix U_w is both unitary and Hermitian, so $U_w^{-1} = U_w$.

Exercise. Let $w \in \mathbf{R}^n$ be a nonzero vector. Show that the Householder matrix U_w is real orthogonal and symmetric. Why is every eigenvalue of U_w either $+1$ or -1?

Exercise. Show that a Householder matrix U_w acts as the identity on the subspace $w^\perp$ and that it acts as a reflection on the one-dimensional subspace spanned by w; that is, $U_w x = x$ if $x \perp w$ and $U_w w = -w$.

Exercise. Use (0.8.5.11) to show that $\det U_w = -1$ for all n. Thus, for all n and every nonzero $w \in \mathbf{R}^n$, the Householder matrix $U_w \in M_n(\mathbf{R})$ is a real orthogonal matrix that is never a *proper rotation matrix* (a real orthogonal matrix whose determinant is $+1$).

Exercise. Use (1.2.8) to show that the eigenvalues of a Householder matrix are always $-1, 1, \ldots, 1$ and explain why its determinant is always -1.

Exercise. Let $n \geq 2$ and let $x, y \in \mathbf{R}^n$ be unit vectors. If $x = y$, let w be any real unit vector that is orthogonal to x. If $x \neq y$, let $w = x - y$. Show that $U_w x = y$. Conclude that any $x \in \mathbf{R}^n$ can be transformed by a real Householder matrix into any $y \in \mathbf{R}^n$ such that $\|x\|_2 = \|y\|_2$.

Exercise. The situation is different in $\mathbf{C}^n$. Show that there is no $w \in \mathbf{C}^n$ such that $U_w e_1 = i e_1$.

Householder matrices and unitary scalar matrices can be used to construct a unitary matrix that takes any given vector in $\mathbf{C}^n$ into any other vector in $\mathbf{C}^n$ that has the same Euclidean norm.

Theorem 2.1.13. *Let $x, y \in \mathbf{C}^n$ be given and suppose that $\|x\|_2 = \|y\|_2 > 0$. If $y = e^{i\theta} x$ for some real θ, let $U(y, x) = e^{i\theta} I_n$; otherwise, let $\phi \in [0, 2\pi)$ be such that $x^* y = e^{i\phi} |x^* y|$ (take $\phi = 0$ if $x^* y = 0$); let $w = e^{i\phi} x - y$; and let $U(y, x) = e^{i\phi} U_w$, in which $U_w = I - 2(w^* w)^{-1} w w^*$ is a Householder matrix. Then $U(y, x)$ is unitary and essentially Hermitian, $U(y, x) x = y$, and $U(y, x) z \perp y$ whenever $z \perp x$. If x and y are real, then $U(y, x)$ is real orthogonal: $U(y, x) = I$ if $y = x$, and $U(y, x)$ is the real Householder matrix U_{x-y} otherwise.*

Proof. The assertions are readily verified if x and y are linearly dependent, that is, if $y = e^{i\theta} x$ for some real θ. If x and y are linearly independent, the Cauchy–Schwarz inequality (0.6.3) ensures that $x^* x \neq |x^* y|$. Compute

$$w^* w = (e^{i\phi} x - y)^*(e^{i\phi} x - y) = x^* x - e^{-i\phi} x^* y - e^{i\phi} y^* x + y^* y$$
$$= 2(x^* x - \operatorname{Re}(e^{-i\phi} x^* y)) = 2(x^* x - |x^* y|)$$

and

$$w^* x = e^{-i\phi} x^* x - y^* x = e^{-i\phi} x^* x - e^{-i\phi} |y^* x| = e^{-i\phi} (x^* x - |x^* y|).$$

and, finally,

$$e^{i\phi}U_w x = e^{i\phi}(x - 2(w^*w)^{-1}ww^*x) = e^{i\phi}(x - (e^{i\phi}x - y)e^{-i\phi}) = y$$

If z is orthogonal to x, then $w^*z = -y^*z$ and

$$y^*U(y,x)z = e^{i\phi}\left(y^*z - \frac{1}{\|x\|_2^2 - |x^*y|}\left(e^{i\phi}y^*x - \|y\|_2^2\right)\left(-y^*x\right)\right)$$
$$= e^{i\phi}\left(y^*z + (-y^*x)\right) = 0$$

Since U_w is unitary and Hermitian, $U(y,x) = (e^{i\phi}I)U_w$ is unitary (as a product of two unitary matrices) and essentially Hermitian; see (0.2.5). $\qquad\square$

Exercise. Let $y \in \mathbf{C}^n$ be a given unit vector and let e_1 be the first column of the n-by-n identity matrix. Construct $U(y,e_1)$ using the recipe in the preceding theorem and verify that its first column is y (which it should be, since $y = U(y,e_1)e_1$).

Exercise. Let $x \in \mathbf{C}^n$ be a given nonzero vector. Explain why the matrix $U(\|x\|_2 e_1, x)$ constructed in the preceding theorem is an essentially Hermitian unitary matrix that takes x into $\|x\|_2 e_1$.

The following *QR factorization* of a complex or real matrix is of considerable theoretical and computational importance.

Theorem 2.1.14 (QR factorization). *Let $A \in M_{n,m}$ be given.*

(a) *If $n \geq m$, there is a $Q \in M_{n,m}$ with orthonormal columns and an upper triangular $R \in M_m$ with nonnegative main diagonal entries such that $A = QR$.*

(b) *If rank $A = m$, then the factors Q and R in (a) are uniquely determined and the main diagonal entries of R are all positive.*

(c) *If $m = n$, then the factor Q in (a) is unitary.*

(d) *There is a unitary $Q \in M_n$ and an upper triangular $R \in M_{n,m}$ with nonnegative diagonal entries such that $A = QR$.*

(e) *If A is real, then the factors Q and R in (a), (b), (c), and (d) may be taken to be real.*

Proof. Let $a_1 \in \mathbf{C}^n$ be the first column of A, let $r_1 = \|a_1\|_2$, and let U_1 be a unitary matrix such that $U_1 a_1 = r_1 e_1$. Theorem 2.1.13 gives an explicit construction for such a matrix, which is either a unitary scalar matrix or the product of a unitary scalar matrix and a Householder matrix. Partition

$$U_1 A = \begin{bmatrix} r_1 & \bigstar \\ 0 & A_2 \end{bmatrix}$$

in which $A_2 \in M_{n-1,m-1}$. Let $a_2 \in \mathbf{C}^{n-1}$ be the first column of A_2 and let $r_2 = \|a_2\|_2$. Use (2.1.13) again to construct a unitary $V_2 \in M_{n-1}$ such that $V_2 a_2 = r_2 e_1$ and let $U_2 = I_1 \oplus V_2$. Then

$$U_2 U_1 A = \begin{bmatrix} r_1 & & \bigstar \\ 0 & r_2 & \\ 0 & 0 & A_3 \end{bmatrix}$$

Repeat this construction m times to obtain

$$U_m U_{m-1} \cdots U_2 U_1 A = \begin{bmatrix} R \\ 0 \end{bmatrix}$$

in which $R \in M_m$ is upper triangular. Its main diagonal entries are $r_1, \ldots, r_m$; they are all nonnegative. Let $U = U_m U_{m-1} \cdots U_2 U_1$. Partition $U^* = U_1^* U_2^* \cdots U_{m-1}^* U_m^* = [Q \ Q_2]$, in which $Q \in M_{n,m}$ has orthonormal columns (it contains the first m columns of a unitary matrix). Then $A = QR$, as desired. If A has full column rank, then R is nonsingular, so its main diagonal entries are all positive.

Suppose that rank $A = m$ and $A = QR = \tilde{Q}\tilde{R}$, in which R and $\tilde{R}$ are upper triangular and have positive main diagonal entries, and Q and $\tilde{Q}$ have orthonormal columns. Then $A^*A = R^*(Q^*Q)R = R^*IR = R^*R$ and also $A^*A = \tilde{R}^*\tilde{R}$, so $R^*R = \tilde{R}^*\tilde{R}$ and $\tilde{R}^{-*}R^* = \tilde{R}R^{-1}$. This says that a lower triangular matrix equals an upper triangular matrix, so both must be diagonal: $\tilde{R}R^{-1} = D$ is diagonal, and it must have positive main diagonal entries because the main diagonal entries of both $\tilde{R}$ and R^{-1} are positive. But $\tilde{R} = DR$ implies that $D = \tilde{R}R^{-1} = \tilde{R}^{-*}R^* = (DR)^{-*}R^* = D^{-1}R^{-*}R^* = D^{-1}$, so $D^2 = I$ and hence $D = I$. We conclude that $\tilde{R} = R$ and hence $\tilde{Q} = Q$.

The assertion in (c) follows from the fact that a square matrix with orthonormal columns is unitary.

If $n \geq m$ in (d), we may start with the factorization in (a), let $\tilde{Q} = [Q \ Q_2] \in M_n$ be unitary, let $\tilde{R} = \begin{bmatrix} R \\ 0 \end{bmatrix} \in M_{n,m}$, and observe that $A = QR = \tilde{Q}\tilde{R}$. If $n < m$, we may undertake the construction in (a) (left multiplication by a sequence of scalar multiples of Householder transformations) and stop after n steps, when the factorization $U_n \cdots U_1 A = [R \ \bigstar]$ is achieved and R is upper triangular. Entries in the $\bigstar$ block need not be zero.

The final assertion (e) follows from the assurance in (2.1.13) that the unitary matrices U_i involved in the constructions in (a) and (d) may all be chosen to be real. $\square$

Exercise. Show that any $B \in M_n$ of the form $B = A^*A$, $A \in M_n$, may be written as $B = LL^*$, in which $L \in M_n$ is lower triangular and has non-negative diagonal entries. Explain why this factorization is unique if A is nonsingular. This is the *Cholesky factorization* of B; every positive definite or semidefinite matrix may be factored in this way; see (7.2.9).

Some easy variants of the QR factorization of $A \in M_{n,m}$ can be useful. First, suppose that $n \leq m$ and let $A^* = QR$, in which $Q \in M_{n,m}$ has orthonormal columns and $R \in M_m$ is upper triangular. Then $A = R^*Q^*$ is a factorization of the form

$$A = LQ \tag{2.1.15a}$$

in which $Q \in M_{n,m}$ has orthonormal rows and $L \in M_n$ is lower triangular. If $\tilde{Q} = \begin{bmatrix} Q \\ \tilde{Q}_2 \end{bmatrix}$ is unitary, we have a factorization of the form

$$A = [L \ 0]\tilde{Q} \tag{2.1.15b}$$

Now let K_p be the (real orthogonal and symmetric) p-by-p reversal matrix (0.9.5.1), which has the pleasant property that $K_p^2 = I_p$. For square matrices $R \in M_p$, the matrix

$L = K_p R K_p$ is lower triangular if R is upper triangular; the main diagonal entries of L are those of R, with the order reversed.

If $n \geq m$ and $A K_m = QR$ as in (2.1.14a), then $A = (Q K_m)(K_m R K_m)$, which is a factorization of the form with a unitary $Q \in M_n$ and an upper triangular $R \in M_n'$, then

$$A = QL \tag{2.1.17a}$$

in which $Q \in M_{n,m}$ has orthonormal columns and $L \in M_m$ is lower triangular. If $\tilde{Q} = [Q \ Q_2]$ is unitary, we have a factorization of the form

$$A = \tilde{Q} \begin{bmatrix} L \\ 0 \end{bmatrix} \tag{2.1.17b}$$

If $n \leq m$, we can apply (2.1.17a) and (2.1.17b) to A^* and obtain factorizations of the form

$$A = RQ = [R \ 0]\tilde{Q} \tag{2.1.17c}$$

in which $R \in M_n$ is upper triangular, $Q \in M_{n,m}$ has orthonormal rows, and $\tilde{Q} \in M_m$ is unitary. If $n \leq m$ and we apply (2.1.14d) to $A K_m$, we obtain $A = (Q K_n)(K_n [R \ \bigstar] K_m)$, which is a factorization of the form

$$A = \tilde{Q} L \tag{2.1.17d}$$

in which $\tilde{Q} \in M_n$ is unitary and $L \in M_{n,m}$ is lower triangular.

An important geometrical fact is that any two lists containing equal numbers of orthonormal vectors are related via a unitary transformation.

Theorem 2.1.18. *If $X = [x_1 \ \ldots \ x_k] \in M_{n,k}$ and $Y = [y_1 \ \ldots \ y_k] \in M_{n,k}$ have orthonormal columns, then there is a unitary $U \in M_n$ such that $Y = UX$. If X and Y are real, then U may be taken to be real.*

Proof. Extend each of the orthonormal lists $x_1, \ldots, x_k$ and $y_1, \ldots, y_k$ to orthonormal bases of $\mathbf{C}^n$; see (0.6.4–5). That is, construct unitary matrices $V = [X \ X_2]$ and $W = [Y \ Y_2] \in M_n$. Then $U = WV^*$ is unitary and $[Y \ Y_2] = W = UV = [UX \ UX_2]$, so $Y = UX$. If X and Y are real, the matrices $[X \ X_2]$ and $[Y \ Y_2]$ may be chosen to be real orthogonal (their columns are orthonormal bases of $\mathbf{R}^n$). □

Problems

2.1.P1 If $U \in M_n$ is unitary, show that $|\det U| = 1$.

2.1.P2 Let $U \in M_n$ be unitary and let λ be a given eigenvalue of U. Show that (a) $|\lambda| = 1$ and (b) x is a (right) eigenvector of U associated with λ if and only if x is a left eigenvector of U associated with λ.

2.1.P3 Given real parameters $\theta_1, \theta_2, \ldots, \theta_n$, show that $U = \mathrm{diag}(e^{i\theta_1}, e^{i\theta_2}, \ldots, e^{i\theta_n})$ is unitary. Show that every diagonal unitary matrix has this form.

2.1.P4 Characterize the diagonal real orthogonal matrices.

2.1.P5 Show that the permutation matrices (0.9.5) in M_n are a subgroup (a subset that is itself a group) of the group of real orthogonal matrices. How many different permutation matrices are there in M_n?

2.1.P6 Give a parametric presentation of the 3-by-3 orthogonal group. Two presentations of the 2-by-2 orthogonal group are given in the exercise following (2.1.5).

2.1.P7 Suppose that $A, B \in M_n$ and $AB = I$. Provide details for the following argument that $BA = I$: Every $y \in \mathbb{C}^n$ can be represented as $y = A(By)$, so rank $A = n$ and hence dim(nullspace(A)) = 0 (0.2.3.1). Compute $A(AB - BA) = A(I - BA) = A - (AB)A = A - A = 0$, so $AB - BA = 0$.

2.1.P8 A matrix $A \in M_n$ is *complex orthogonal* if $A^T A = I$. (a) Show that a complex orthogonal matrix is unitary if and only if it is real. (b) Let $S = \begin{bmatrix} 0 & 1 \\ -1 & 0 \end{bmatrix} \in M_2(\mathbb{R})$. Show that $A(t) = (\cosh t)I + (i \sinh t)S \in M_2$ is complex orthogonal for all $t \in \mathbb{R}$, but that $A(t)$ is unitary only for $t = 0$. The hyperbolic functions are defined by $\cosh t = (e^t + e^{-t})/2$, $\sinh t = (e^t - e^{-t})/2$. (c) Show that, unlike the unitary matrices, the set of complex orthogonal matrices is not a bounded set, and it is therefore not a compact set. (d) Show that the set of complex orthogonal matrices of a given size forms a group. The smaller (and compact) group of *real* orthogonal matrices of a given size is often called the *orthogonal group*. (e) If $A \in M_n$ is complex orthogonal, show that $|\det A| = 1$; consider $A(t)$ in (b) to show that A can have eigenvalues λ with $|\lambda| \neq 1$. (f) If $A \in M_n$ is complex orthogonal, show that $\bar{A}$, A^T, and A^* are all complex orthogonal and nonsingular. Are the rows (respectively, columns) of A orthogonal? (g) Characterize the diagonal complex orthogonal matrices. Compare with 2.1.P4 (h) Show that $A \in M_n$ is both complex orthogonal and unitary if and only if it is real orthogonal.

2.1.P9 If $U \in M_n$ is unitary, show that $\bar{U}$, U^T, and U^* are all unitary.

2.1.P10 If $U \in M_n$ is unitary, show that $x, y \in \mathbb{C}^n$ are orthogonal if and only if Ux and Uy are orthogonal.

2.1.P11 A nonsingular matrix $A \in M_n$ is *skew orthogonal* if $A^{-1} = -A^T$. Show that A is skew orthogonal if and only if $\pm i A$ is orthogonal. More generally, if $\theta \in \mathbb{R}$, show that $A^{-1} = e^{i\theta} A^T$ if and only if $e^{i\theta/2}A$ is orthogonal. What is this for $\theta = 0$ and π?

2.1.P12 Show that if $A \in M_n$ is similar to a unitary matrix, then A^{-1} is similar to A^*.

2.1.P13 Consider $\text{diag}(2, \frac{1}{2}) \in M_2$ and show that the set of matrices that are similar to unitary matrices is a proper subset of the set of matrices A for which A^{-1} is similar to A^*.

2.1.P14 Show that the intersection of the group of unitary matrices in M_n with the group of complex orthogonal matrices in M_n is the group of real orthogonal matrices in M_n.

2.1.P15 If $U \in M_n$ is unitary, $\alpha \subset \{1, \ldots, n\}$, and $U[\alpha, \alpha^c] = 0$, (0.7.1) show that $U[\alpha^c, \alpha] = 0$, and $U[\alpha]$ and $U[\alpha^c]$ are unitary.

2.1.P16 Let $x, y \in \mathbb{R}^n$ be given linearly independent unit vectors and let $w = x + y$. Consider the *Palais matrix* $P_{x,y} = I - 2(w^T w)^{-1} ww^T + 2yx^T$. Show that: (a) $P_{x,y} = (I - 2(w^T w)^{-1} ww^T)(I - 2xx^T) = U_w U_x$ is a product of two real Householder matrices, so it is a real orthogonal matrix; (b) $\det P_{x,y} = +1$, so $P_{x,y}$ is always a proper rotation matrix; (c) $P_{x,y}x = y$ and $P_{x,y}y = -x + 2(x^T y)y$; (d) $P_{x,y}z = z$ if $z \in \mathbb{R}^n$, $z \perp x$, and $z \perp y$; (e) $P_{x,y}$ acts as the identity on the $(n - 2)$-dimensional subspace $(\text{span}\{x, y\})^\perp$ and it is a proper rotation on the 2-dimensional subspace $\text{span}\{x, y\}$ that takes x into y; (f) If $n = 3$, explain why $P_{x,y}$ is the unique proper rotation that takes x into y and leaves fixed their vector cross product $x \times y$; (g) the eigenvalues of $P_{x,y}$ are $x^T y \pm i(1 - (x^T y)^2)^{1/2} = e^{\pm i\theta}, 1, \ldots, 1$, in which $\cos\theta = x^T y$.

2.1.P17 Suppose that $A \in M_{n,m}$, $n \geq m$, and rank $A = m$. Describe the steps of the Gram–Schmidt process applied to the columns of A, proceeding from left to right. Explain why this process produces, column–by–column, an explicit matrix $Q \in M_{n,m}$ with orthonormal columns and an explicit upper triangular matrix $R \in M_m$ such that $Q = AR$. How is this factorization related to the one in (2.1.14)?

2.1.P18 Let $A \in M_n$ be factored as $A = QR$ as in (2.1.14), partition $A = [a_1 \ \ldots \ a_n]$ and $Q = [q_1 \ \ldots \ q_n]$ according to their columns and let $R = [r_{ij}]_{i,j=1}^n$. (a) Explain why $\{q_1, \ldots, q_k\}$ is an orthonormal basis for span$\{a_1, \ldots, a_k\}$ for each $k = 1, \ldots, n$. (b) Show that r_{kk} is the Euclidean distance from a_k to span$\{a_1, \ldots, a_{k-1}\}$ for each $k = 2, \ldots, n$.

2.1.P19 Let $X = [x_1 \ \ldots \ x_m] \in M_{n,m}$, suppose that rank $X = m$, and factor $X = QR$ as in (2.1.14). Let $Y = QR^{-*} = [y_1 \ \ldots \ y_m]$. (a) Show that the columns of Y are a basis for the subspace $S = $ span$\{x_1, \ldots, x_m\}$ and that $Y^* X = I_m$, so $y_i^* x_j = 0$ if $i \neq j$ and each $y_i^* x_i = 1$. Given the basis $x_1, \ldots, x_m$ of S, its *dual basis* (*reciprocal basis*) is the list $y_1, \ldots, y_m$. (b) Explain why the dual basis of $x_1, \ldots, x_m$ is unique, that is, if the columns of $Z \in M_{n,m}$ are in S and $Z^* X = I$, then $Z = Y$. (c) Show that the dual basis of the list $y_1, \ldots, y_m$ is $x_1, \ldots, x_m$. (d) If $n = m$, show that the columns of X^{-*} are a basis of $\mathbf{C}^n$ that is dual to the basis $x_1, \ldots, x_n$.

2.1.P20 If $U \in M_n$ is unitary, show that adj $U = (\det U)U^*$ and conclude that adj U is unitary.

2.1.P21 Explain why (2.1.10) remains true if *unitary* is replaced with *complex orthogonal*. Deduce that a complex orthogonal matrix is upper triangular if and only if it is diagonal. What does a diagonal complex orthogonal matrix look like?

2.1.P22 Suppose that $X, Y \in M_{n,m}$ have orthonormal columns. Show that X and Y have the same range (column space) if and only if there is a unitary $U \in M_m$ such that $X = YU$.

2.1.P23 Let $A \in M_n$, let $A = QR$ be a QR factorization, let $R = [r_{ij}]$, and partition A, Q, and R according to their columns: $A = [a_1 \ \ldots \ a_n]$, $Q = [q_1 \ \ldots \ q_n]$, and $R = [r_1 \ldots r_n]$. Explain why $|\det A| = \det R = r_{11} \cdots r_{nn}$ and why $\|a_i\|_2 = \|r_i\|_2 \geq r_{ii}$ for each $i = 1, \ldots, n$, with equality for some i if and only if $a_i = r_{ii}q_i$. Conclude that $|\det A| \leq \prod_{i=1}^n \|a_i\|_2$, with equality if and only if either (a) some $a_i = 0$ or (b) A has orthogonal columns ($A^* A = $ diag$(\|a_1\|_2^2, \ldots, \|a_n\|_2^2)$). This *Hadamard's inequality*.

2.1.P24 Let $E = [e_{ij}] \in M_3$, in which each $e_{ij} = +1$. (a) Show that the permanent of E (0.3.2) is per $E = 6$. (b) Let $B = [b_{ij}] \in M_3$, in which each $b_{ij} = \pm 1$. Use Hadamard's inequality to show that there is no choice of $\pm$ signs such that per $E = \det B$.

2.1.P25 If $U \in M_n$ is unitary and $r \in \{1, \ldots, n\}$, explain why the compound matrix $C_r(U)$ is unitary.

2.1.P26 Explain why (a) each $A \in M_n$ can be factored as $A = H_1 \cdots H_{n-1} R$, in which each H_i is a Householder matrix and R is upper triangular; (b) each unitary $U \in M_n$ can be factored as $U = H_1 \cdots H_{n-1} D$, in which each H_i is a Householder matrix and D is a diagonal unitary matrix; and (c) each real orthogonal $Q \in M_n(\mathbf{R})$ can be factored as $Q = H_1 \cdots H_{n-1} D$, in which each H_i is a real Householder matrix and $D = $ diag$(1, \ldots, 1, \pm 1) = $ diag$(1, \ldots, 1, (-1)^{n-1} \det Q)$.

The following three problems provide an analog of the preceding problem, in which plane rotations are used instead of Householder matrices.

2.1.P27 Let $n \geq 2$ and let $x = [x_i] \in \mathbf{R}^n$. If $x_n = x_{n-1} = 0$, let $\theta_1 = 0$; otherwise, choose $\theta_1 \in [0, 2\pi)$ such that $\cos\theta_1 = x_{n-1}/\sqrt{x_n^2 + x_{n-1}^2}$ and $\sin\theta_1 = -x_n/\sqrt{x_n^2 + x_{n-1}^2}$. Let $x^{(1)} = [x_i^{(1)}] = U(\theta_1; n-1, n)x$. Show that $x_n^{(1)} = 0$ and $x_{n-1}^{(1)} \geq 0$. Let $x^{(2)} = [x_i^{(2)}] = U(\theta_2; n-2, n-1)U(\theta_1; n-1, n)x$. How can you choose θ_2 so that $x_n^{(2)} = x_{n-1}^{(2)} = 0$ and $x_{n-2}^{(2)} \geq 0$? If $1 \leq k < n$, explain how to construct a sequence of k plane rotations $U_1, \ldots, U_k$ such that the vector $x^{(k)} = [x_i^{(k)}] = U_k \cdots U_1 x$ has $x_n^{(k)} = \cdots = x_{n-k+1}^{(k)} = 0$ and $x_{n-k}^{(k)} \geq 0$. Why is $\|x\|_2 = \|x^{(k)}\|_2$?

2.1.P28 Let $A \in M_{n,m}(\mathbf{R})$ with $n \geq m$. (a) Explain how to construct a finite sequence of plane rotations $U_1, \ldots, U_N$ such that $U_N \cdots U_1 A = \begin{bmatrix} B \\ 0 \end{bmatrix}$, in which $B = [b_{ij}] \in M_m(\mathbf{R})$ is upper triangular and each of $b_{11}, \ldots, b_{m-1,m-1}$ is nonnegative. (b) Explain why this reduction to upper triangular form can be achieved with a sequence of $N = m(n - \frac{m+1}{2})$ plane rotations; some of them may be the identity, so fewer than N nontrivial plane rotations are possible. (c) Use (a) to prove that each $A \in M_n(\mathbf{R})$ can be factored as $A = U_1 \cdots U_N R$, in which $N = n(n-1)/2$, each U_i is a plane rotation, $R = [r_{ij}]$ is upper triangular, and each $r_{11}, \ldots, r_{n-1,n-1}$ (but not necessarily r_{nn}) is nonnegative.

2.1.P29 Explain why each real orthogonal matrix $Q \in M_n(\mathbf{R})$ can be factored as $Q = U_1 \cdots U_N D$, in which $N = n(n-1)/2$, each U_i is a plane rotation, and $D = \text{diag}(1, \ldots, 1, \det Q) = \text{diag}(1, \ldots, 1, \pm 1) \in M_n(\mathbf{R})$.

Further Reading. For more information about matrices that satisfy the conditions of (2.1.9), see C. R. DePrima and C. R. Johnson, The range of $A^{-1}A^*$ in $GL(n, \mathbf{C})$, *Linear Algebra Appl.* 9 (1974) 209–222.

2.2 Unitary similarity

Since $U^* = U^{-1}$ for a unitary U, the transformation on M_n given by $A \rightarrow U^*AU$ is a similarity transformation if U is unitary. This special type of similarity is called *unitary similarity*.

Definition 2.2.1. *Let $A, B \in M_n$ be given. We say that A is* unitarily similar *to B if there is a unitary $U \in M_n$ such that $A = UBU^*$. If U may be taken to be real (and hence is real orthogonal), then A is said to be* real *orthogonally similar to B. We say that A is* unitarily diagonalizable *if it is unitarily similar to a diagonal matrix; A is* real orthogonally diagonalizable *if it is real orthogonally similar to a diagonal matrix.*

Exercise. Show that unitary similarity is an equivalence relation.

Theorem 2.2.2. *Let $U \in M_n$ and $V \in M_m$ be unitary, let $A = [a_{ij}] \in M_{n,m}$ and $B = [b_{ij}] \in M_{n,m}$, and suppose that $A = UBV$. Then $\sum_{i,j=1}^{n,m} |b_{ij}|^2 = \sum_{i,j=1}^{n,m} |a_{ij}|^2$. In particular, this identity is satisfied if $m = n$ and $V = U^*$, that is, if A is unitarily similar to B.*

Proof. It suffices to check that $\text{tr } B^*B = \text{tr } A^*A$; see (0.2.5). Compute $\text{tr } A^*A = \text{tr}(UBV)^*(UBV) = \text{tr}(V^*B^*U^*UBV) = \text{tr } V^*B^*BV = \text{tr } B^*BVV^* = \text{tr } B^*B$. $\quad\square$

Exercise. Show that the matrices $\begin{bmatrix} 3 & 1 \\ -2 & 0 \end{bmatrix}$ and $\begin{bmatrix} 1 & 1 \\ 0 & 2 \end{bmatrix}$ are similar but not unitarily similar.

Unitary similarity implies similarity but not conversely. The unitary similarity equivalence relation partitions M_n into *finer* equivalence classes than the similarity equivalence relation. Unitary similarity, like similarity, corresponds to a change of basis, but of a special type—it corresponds to a change from one *orthonormal* basis to another.

Exercise. Using the notation of (2.1.11), explain why only rows and columns i and j are changed under real orthogonal similarity via the plane rotation $U(\theta; i, j)$.

Exercise. Using the notation of (2.1.13), explain why $U(y, x)^* A U(y, x) = U_w^* A U_w$ for any $A \in M_n$; that is, a unitary similarity via an essentially Hermitian unitary matrix of the form $U(y, x)$ is a unitary similarity via a Householder matrix. Unitary (or real orthogonal) similarity via a Householder matrix is often called a *Householder transformation*.

For computational or theoretical reasons, it is often convenient to transform a given matrix by unitary similarity into another matrix with a special form. Here are two examples.

Example 2.2.3. Unitary similarity to a matrix with equal diagonal entries.
Let $A = [a_{ij}] \in M_n$ be given. We claim that there is a unitary $U \in M_n$ such that all the main diagonal entries of $U^* A U = B = [b_{ij}]$ are equal; if A is real, then U may be taken to be real orthogonal. If this claim is true, then $\operatorname{tr} A = \operatorname{tr} B = n b_{11}$, so every main diagonal entry of B is equal to the average of the main diagonal entries of A.

Begin by considering the complex case and $n = 2$. Since we can replace $A \in M_2$ by $A - (\frac{1}{2} \operatorname{tr} A) I$, there is no loss of generality to assume that $\operatorname{tr} A = 0$, in which case the two eigenvalues of A are $\pm \lambda$ for some $\lambda \in \mathbf{C}$. We wish to determine a unit vector u such that $u^* A u = 0$. If $\lambda = 0$, let u be any unit vector such that $Au = 0$. If $\lambda \neq 0$, let w and z be any unit eigenvectors associated with the distinct eigenvalues $\pm \lambda$. Let $x(\theta) = e^{i\theta} w + z$, which is nonzero for all $\theta \in \mathbf{R}$ since w and z are linearly independent. Compute $x(\theta)^* A x(\theta) = \lambda (e^{i\theta} w + z)^* (e^{i\theta} w - z) = 2i\lambda \operatorname{Im}(e^{i\theta} z^* w)$. If $z^* w = e^{i\phi} |z^* w|$, then $x(-\phi)^* A x(-\phi) = 0$. Let $u = x(-\phi) / \|x(-\phi)\|_2$. Now let $v \in \mathbf{C}^2$ be any unit vector that is orthogonal to u and let $U = [u \ v]$. Then U is unitary and $(U^* A U)_{11} = u^* A u = 0$. But $\operatorname{tr}(U^* A U) = 0$, so $(U^* A U)_{22} = 0$ as well.

Now suppose that $n = 2$ and A is real. If the diagonal entries of $A = [a_{ij}]$ are not equal, consider the plane rotation matrix $U_\theta = \begin{bmatrix} \cos\theta & -\sin\theta \\ \sin\theta & \cos\theta \end{bmatrix}$. A calculation reveals that the diagonal entries of $U_\theta A U_\theta^T$ are equal if $(\cos^2\theta - \sin^2\theta)(a_{11} - a_{22}) = 2\sin\theta \cos\theta (a_{12} + a_{21})$, so equal diagonal entries are achieved if $\theta \in (0, \pi/2)$ is chosen so that $\cot 2\theta = (a_{12} + a_{21})/(a_{11} - a_{22})$.

We have now shown that any 2-by-2 complex matrix A is unitarily similar to a matrix with both diagonal entries equal to the average of the diagonal entries of A; if A is real, the similarity may be taken to be real orthogonal.

Now suppose that $n > 2$ and define $f(A) = \max\{|a_{ii} - a_{jj}| : i, j = 1, 2, \ldots, n\}$. If $f(A) > 0$, let $A_2 = \begin{bmatrix} a_{ii} & a_{ij} \\ a_{ji} & a_{jj} \end{bmatrix}$ for a pair of indices i, j for which $f(A) = |a_{ii} - a_{jj}|$ (there could be several pairs of indices for which this maximum positive separation is attained; choose any one of them). Let $U_2 \in M_2$ be unitary, real if A is real, and such that $U_2^* A_2 U_2$ has both main diagonal entries equal to $\frac{1}{2}(a_{ii} + a_{jj})$. Construct $U(i, j) \in M_n$ from U_2 in the same way that $U(\theta; i, j)$ was constructed from a 2-by-2 plane rotation in (2.1.11). The unitary similarity $U(i, j)^* A U(i, j)$ affects only entries in rows and columns i and j, so it leaves unchanged every main diagonal entry of A except the entries in positions i and j, which it replaces with the average $\frac{1}{2}(a_{ii} + a_{jj})$. For any $k \neq i, j$ the triangle inequality ensures that

$$
\begin{aligned}
\left| a_{kk} - \frac{1}{2}(a_{ii} + a_{jj}) \right| &= \left| \frac{1}{2}(a_{kk} - a_{ii}) + \frac{1}{2}(a_{kk} - a_{jj}) \right| \\
&\leq \frac{1}{2}|a_{kk} - a_{ii}| + \frac{1}{2}|a_{kk} - a_{jj}| \\
&\leq \frac{1}{2}f(A) + \frac{1}{2}f(A) = f(A)
\end{aligned}
$$

with equality only if the scalars $a_{kk} - a_{ii}$ and $a_{kk} - a_{jj}$ both lie on the same ray in the complex plane and $|a_{kk} - a_{ii}| = |a_{kk} - a_{jj}|$. These two conditions imply that $a_{ii} = a_{jj}$, so it follows that $|a_{kk} - \frac{1}{2}(a_{ii} + a_{jj})| < f(A)$ for all $k \neq i, j$. Thus, the unitary similarity we have just constructed reduces by one the finitely many pairs of indices k, ℓ for which $f(A) = |a_{kk} - a_{\ell\ell}|$. Repeat the construction, if necessary, to deal with any such remaining pairs and achieve a unitary U (real if A is real) such that $f(U^*AU) < f(A)$.

Finally, consider the compact set $R(A) = \{U^*AU : U \in M_n \text{ is unitary}\}$. Since f is a continuous nonnegative-valued function on $R(A)$, it achieves its minimum value there, that is, there is some $B \in R(A)$ such that $f(A) \geq f(B) \geq 0$ for all $A \in R(A)$. If $f(B) > 0$, we have just seen that there is a unitary U (real if A is real) such that $f(B) > f(U^*BU)$. This contradiction shows that $f(B) = 0$, so all the diagonal entries of B are equal.

Example 2.2.4. Unitary similarity to an upper Hessenberg matrix. Let $A = [a_{ij}] \in M_n$ be given. The following construction shows that A is unitarily similar to an upper Hessenberg matrix with nonnegative entries in its first subdiagonal. Let a_1 be the first column of A, partitioned as $a_1^T = [a_{11}\ \xi^T]$ with $\xi \in \mathbf{C}^{n-1}$. Let $U_1 = I_{n-1}$ if $\xi = 0$; otherwise, use (2.1.13) to construct $U_1 = U(\|\xi\|_2 e_1, \xi) \in M_{n-1}$, a unitary matrix that takes ξ into a positive multiple of e_1. Form the unitary matrix $V_1 = I_1 \oplus U_1$ and observe that the first column of $V_1 A$ is the vector $[a_{11}\ \|\xi\|_2\ 0]^T$. Moreover, $\mathcal{A}_1 = (V_1 A)V_1^*$ has the same first column as $V_1 A$ and is unitarily similar to A. Partition it as

$$
\mathcal{A}_1 = \begin{bmatrix} \begin{bmatrix} a_{11} \\ \|\xi\|_2 \\ 0 \end{bmatrix} & \bigstar \\ & A_2 \end{bmatrix}, \qquad A_2 \in M_{n-1}
$$

Use (2.1.13) again to form a unitary matrix $(I_1 \oplus U_2)$ that takes the first column of A_2 into a vector whose entries below the second are all zero and whose second entry is nonnegative. Let $V_2 = I_2 \oplus U_2$ and let $\mathcal{A}_2 = V_2 \mathcal{A}_1 V_2^*$; the first column of $\mathcal{A}_1$ is undisturbed. After $n - 2$ of these reductions, we obtain an upper Hessenberg matrix $\mathcal{A}_{n-2}$ that is unitarily similar to A and has nonnegative subdiagonal entries except perhaps for the entry in position $(n, n - 1)$; a final unitary similarity via $I_{n-1} \oplus [e^{i\theta}]$ may be necessary to rotate it to be nonnegative.

Exercise. If A is Hermitian or skew Hermitian, explain why the construction in the preceding example produces a tridiagonal Hermitian or skew-Hermitian matrix that is unitarily similar to A.

Theorem 2.2.2 provides a necessary but not sufficient condition for two given matrices to be unitarily similar. It can be augmented with additional identities that collectively do provide necessary and sufficient conditions. A key role is played by the following simple notion. Let s, t be two given noncommuting variables. Any finite formal product of nonnegative powers of s and t

$$W(s, t) = s^{m_1} t^{n_1} s^{m_2} t^{n_2} \cdots s^{m_k} t^{n_k}, \quad m_1, n_1, \ldots, m_k, n_k \geq 0 \qquad (2.1)$$

is called a *word in s and t*. The *length* of the word $W(s, t)$ is the nonnegative integer $m_1 + n_1 + m_2 + n_2 + \cdots + m_k + n_k$, that is, the sum of all the exponents in the word. If $A \in M_n$ is given, we define a *word in A and A^** as

$$W(A, A^*) = A^{m_1} (A^*)^{n_1} A^{m_2} (A^*)^{n_2} \cdots A^{m_k} (A^*)^{n_k}$$

Since the powers of A and A^* need not commute, it may not be possible to simplify the expression of $W(A, A^*)$ by rearranging the terms in the product.

Suppose that A is unitarily similar to $B \in M_n$, that is, $A = UBU^*$ for some unitary $U \in M_n$. For any word $W(s, t)$ we have

$$\begin{aligned}
W(A, A^*) &= (UBU^*)^{m_1} (UB^*U^*)^{n_1} \cdots (UBU^*)^{m_k} (UB^*U^*)^{n_k} \\
&= UB^{m_1} U^* U(B^*)^{n_1} U^* \cdots UB^{m_k} U^* U(B^*)^{n_k} U^* \\
&= UB^{m_1} (B^*)^{n_1} \cdots B^{m_k} (B^*)^{n_k} U^* \\
&= UW(B, B^*) U^*
\end{aligned}$$

so $W(A, A^*)$ is unitarily similar to $W(B, B^*)$. Thus, $\operatorname{tr} W(A, A^*) = \operatorname{tr} W(B, B^*)$. If we take the word $W(s, t) = ts$, we obtain the identity in (2.2.2).

If one considers all possible words $W(s, t)$, this observation gives infinitely many necessary conditions for two matrices to be unitarily similar. A theorem of W. Specht, which we state without proof, guarantees that these necessary conditions are also sufficient.

Theorem 2.2.6. *Two matrices $A, B \in M_n$ are unitarily similar if and only if*

$$\operatorname{tr} W(A, A^*) = \operatorname{tr} W(B, B^*) \qquad (2.2.7)$$

for every word $W(s, t)$ in two noncommuting variables.

Specht's theorem can be used to show that two matrices are *not* unitarily similar by exhibiting a specific word that violates (2.2.7). However, except in special situations (see 2.2.P6) it may be useless in showing that two given matrices *are* unitarily similar because infinitely many conditions must be verified. Fortunately, a refinement of Specht's theorem ensures that it suffices to check the trace identities (2.2.7) for only finitely many words, which can provide a practical criterion to assess unitary similarity of matrices of small size.

Theorem 2.2.8. *Let* $A, B \in M_n$ *be given.*

(a) *A and B are unitarily similar if and only if (2.2.7) is satisfied for every word* $W(s, t)$ *in two noncommuting variables whose length is at most*

$$n\sqrt{\frac{2n^2}{n-1} + \frac{1}{4}} + \frac{n}{2} - 2$$

(b) *If* $n = 2$, *A and B are unitarily similar if and only if (2.2.7) is satisfied for the three words* $W(s, t) = s; s^2$, *and* st.

(c) *If* $n = 3$, *A and B are unitarily similar if and only if (2.2.7) is satisfied for the seven words* $W(s, t) = s; s^2$, $st; s^3$, $s^2t; s^2t^2$; *and* s^2t^2st.

(d) *If* $n = 4$, *A and B are unitarily similar if and only if (2.2.7) is satisfied for the 20 words* $W(s, t)$ *in the following table:*

s	s^2, st
s^3, s^2t	s^4, s^3t, s^2t^2, $stst$
s^3t^2	s^2ts^2t, s^2t^2st, t^2s^2ts
s^3t^2st	$s^3t^2s^2t$, s^3t^3st, t^3s^3ts
s^3ts^2tst, $s^2t^2sts^2t$	$s^3t^3s^2t^2$

Two *real* matrices are unitarily similar if and only if they are real orthogonally similar; see (2.5.21). Thus, the criteria in (2.2.8) are necessary and sufficient for two real matrices A and B to be real orthogonally similar.

Problems

2.2.P1 Let $A = [a_{ij}] \in M_n(\mathbf{R})$ be symmetric but *not* diagonal, and choose indices i, j with $i < j$ so that $|a_{ij}| = \max\{|a_{pq}| : p < q\}$. Define θ by $\cot 2\theta = (a_{ii} - a_{jj})/2a_{ij}$, let $U(\theta; i, j)$ be the plane rotation (2.1.11), and let $B = U(\theta; i, j)^T A U(\theta; i, j) = [b_{pq}]$. Show that $b_{ij} = 0$, $\sum_{p,q=1}^n |b_{pq}|^2 = \sum_{p,q=1}^n |a_{pq}|^2$, and

$$\sum_{p \neq q} |b_{pq}|^2 = \sum_{p \neq q} |a_{pq}|^2 - 2|a_{ij}|^2 \leq \left(1 - \frac{2}{n^2 - n}\right) \sum_{p \neq q} |a_{pq}|^2$$

Explain why a sequence of real orthogonal similarities via plane rotations chosen in this way (at each step, do a plane rotation that annihilates a largest-magnitude off-diagonal entry) converges to a diagonal matrix whose diagonal entries are the eigenvalues of A. How can corresponding eigenvectors be obtained as a by-product of this process? This is *Jacobi's method* for calculating the eigenvalues of a real symmetric matrix. In practice, one implements Jacobi's method with an algorithm that avoids calculation of any trigonometric functions or their inverses; see Golub and VanLoan (1996).

2.2.P2 The eigenvalue calculation method of Givens for real matrices also uses plane rotations, but in a different way. For $n \geq 3$, provide details for the following argument to show that every $A = [a_{ij}] \in M_n(\mathbf{R})$ is real orthogonally similar to a real lower Hessenberg matrix, which is necessarily tridiagonal if A is symmetric; see (0.9.9) and (0.9.10). Choose a plane rotation $U_{1,3}$ of the form $U(\theta; 1, 3)$, as in the preceding problem, so that the 1,3 entry of $U_{1,3}^* A U_{1,3}$ is 0. Choose another plane rotation of the form $U_{1,4} = U(\theta; 1, 4)$ so that the 1,4 entry of $U_{1,4}^*(U_{1,3}^* A U_{1,3})U_{1,4}$ is 0; continue in this way to zero out the rest of the first row with a sequence of real orthogonal similarities. Then start on the second row beginning with the 2,4 entry and zero out the 2,4, 2,5, ..., 2,n entries. Explain why this process does not disturb previously manufactured zero entries and why it preserves symmetry if A is symmetric. Proceeding in this way through row $n - 3$ produces a lower Hessenberg matrix after finitely many real orthogonal similarities via plane rotations; that matrix is tridiagonal if A is symmetric. However, the eigenvalues of A are not displayed as in Jacobi's method; they must be obtained from a further calculation.

2.2.P3 Let $A \in M_2$. (a) Show that tr $W(A, A^*) =$ tr $W(A^T, \bar{A})$ for each of the three words in (2.2.8b). (b) Explain why every 2-by-2 complex matrix is unitarily similar to its transpose.

2.2.P4 Let $A \in M_3$. (a) Show that tr $W(A, A^*) =$ tr $W(A^T, \bar{A})$ for each of the first six words of the list in (2.2.8c), and conclude that A is unitarily similar to A^T if and only if $\text{tr}(A^2(A^{*2})AA^*) = \text{tr}((A^T)^2\bar{A}^2 A^T \bar{A})$. (b) Explain why A is unitarily similar to A^T if and only if $\text{tr}(AA^*(A^*A - AA^*)A^*A) = 0$. (c) Use the criterion in either (a) or (b) to show that the matrix

$$\begin{bmatrix} 1 & 1 & 1 \\ -1 & 0 & 1 \\ -1 & -1 & -1 \end{bmatrix}$$

is not unitarily similar to its transpose. Note, however that *every* square complex matrix is *similar* to its transpose (3.2.3).

2.2.P5 If $A \in M_n$ and there is a unitary $U \in M_n$ such that $A^* = UAU^*$, show that U commutes with $A + A^*$. Apply this observation to the 3-by-3 matrix in the preceding problem and conclude that if it is unitarily similar to its transpose, then any such unitary similarity must be diagonal. Show that no diagonal unitary similarity can take this matrix into its transpose, so it is not unitarily similar to its transpose.

2.2.P6 Let $A \in M_n$ and $B, C \in M_m$ be given. Use either (2.2.6) or (2.2.8) to show that B and C are unitarily similar if and only if any one of the following conditions is satisfied:

(a) $\begin{bmatrix} A & 0 \\ 0 & B \end{bmatrix}$ and $\begin{bmatrix} A & 0 \\ 0 & C \end{bmatrix}$ are unitarily similar.

(b) $B \oplus \cdots \oplus B$ and $C \oplus \cdots \oplus C$ are unitarily similar if both direct sums contain the same number of direct summands.

(c) $A \oplus B \oplus \cdots \oplus B$ and $A \oplus C \oplus \cdots \oplus C$ are unitarily similar if both direct sums contain the same number of direct summands.

2.2.P7 Give an example of two 2-by-2 matrices that satisfy the identity (2.2.2) but are not unitarily similar. Explain why.

2.2.P8 Let $A, B \in M_2$ and let $C = AB - BA$. Use Example 2.2.3 to show that $C^2 = \lambda I$ for some scalar λ.

2.2.P9 Let $A \in M_n$ and suppose that $\operatorname{tr} A = 0$. Use (2.2.3) to show that A can be written as a sum of two nilpotent matrices. Conversely, if A can be written as a sum of nilpotent matrices, explain why $\operatorname{tr} A = 0$.

2.2.P10 Let $n \geq 2$ be a given integer and define $\omega = e^{2\pi i/n}$. (a) Explain why $\sum_{k=0}^{n-1} \omega^{k\ell} = 0$ unless $\ell = mn$ for some $m = 0, \pm 1, \pm 2, \ldots$, in which case the sum is equal to n. (b) Let $F_n = n^{-1/2}[\omega^{(i-1)(j-1)}]_{i,j=1}^n$ denote the n-by-n *Fourier matrix*. Show that F_n is symmetric, unitary, and coninvolutory: $F_n F_n^* = F_n \overline{F_n} = I$. (c) Let C_n denote the basic circulant permutation matrix (0.9.6.2). Explain why C_n is unitary (real orthogonal). (d) Let $D = \operatorname{diag}(1, \omega, \omega^2, \ldots, \omega^{n-1})$ and show that $C_n F_n = F_n D$, so $C_n = F_n D F_n^*$ and $C_n^k = F_n D^k F_n^*$ for all $k = 1, 2, \ldots$. (e) Let A denote the circulant matrix (0.9.6.1) whose first row is $[a_1 \ \ldots \ a_n]$, expressed as the sum in (0.9.6.3). Explain why $A = F_n \Lambda F_n^*$, in which $\Lambda = \operatorname{diag}(\lambda_1, \ldots, \lambda_n)$, the eigenvalues of A are

$$\lambda_\ell = \sum_{k=0}^{n-1} a_{k+1} \omega^{k(\ell-1)}, \quad \ell = 1, \ldots, n \tag{2.2.9}$$

and $\lambda_1, \ldots, \lambda_n$ are the entries of the vector $n^{1/2} F_n^* A e_1$. Thus, the Fourier matrix provides an explicit unitary diagonalization for every circulant matrix. (f) If there is some $i \in \{1, \ldots, n\}$ such that $|a_i| > \sum_{j \neq i} |a_j|$, deduce from (2.2.9) that A is nonsingular. We can restate this criterion as follows: If a circulant matrix is singular and has first row $[a_1 \ \ldots \ a_n]$, then that row vector is *balanced*; see (7.2.P28). (g) Write $F_n = \mathcal{C}_n + i\mathcal{S}_n$, in which $\mathcal{C}_n$ and $\mathcal{S}_n$ are real. What are the entries of $\mathcal{C}_n$ and $\mathcal{S}_n$? The matrix $H_n = \mathcal{C}_n + \mathcal{S}_n$ is the n-by-n *Hartley matrix*. (h) Show that $\mathcal{C}_n^2 + \mathcal{S}_n^2 = I$, $\mathcal{C}_n \mathcal{S}_n = \mathcal{S}_n \mathcal{C}_n = 0$, H_n is symmetric, and H_n is real orthogonal. (i) Let K_n denote the reversal matrix (0.9.5.1). Show that $\mathcal{C}_n K_n = K_n \mathcal{C}_n = \mathcal{C}_n$, $\mathcal{S}_n K_n = K_n \mathcal{S}_n = -\mathcal{S}_n$, and $H_n K_n = K_n H_n$, so $\mathcal{C}_n$, $\mathcal{S}_n$, and H_n are centrosymmetric. It is known that $H_n A H_n = \Lambda$ is diagonal for any matrix of the form $A = E + K_n F$, in which E and F are real circulant matrices, $E = E^T$, and $F = -F^T$; the diagonal entries of Λ (the eigenvalues of such a matrix A) are the entries of the vector $n^{1/2} H_n A e_1$. In particular, the Hartley matrix provides an explicit real orthogonal diagonalization for every real symmetric circulant matrix.

Notes and Further Readings. For the original proof of (2.2.6), see W. Specht, Zur Theorie der Matrizen II, *Jahresber. Deutsch. Math.-Verein.* 50 (1940) 19–23; there is a modern proof in [Kap]. For a survey of the issues addressed in (2.2.8), see D. Djoković and C. R. Johnson, Unitarily achievable zero patterns and traces of words in A and A^*, *Linear Algebra Appl.* 421 (2007) 63–68. The list of words in (2.2.8d) is in Theorem 4.4 of D. Djoković, Poincaré series of some pure and mixed trace algebras of two generic matrices, *J. Algebra* 309 (2007) 654–671. A 4-by-4 complex matrix is unitarily similar to its transpose if and only if *seven* zero-trace identities of the type in (2.2.P4(b)) are satisfied; see Theorem 1 in S. R. Garcia, D. E. Poore, and J. E. Tener, Unitary equivalence to a complex symmetric matrix: low dimensions, *Linear Algebra Appl.* 437 (2012) 271–284. There is an approximate version of (2.2.6) for two *nonsingular* matrices $A, B \in M_n$: A and B are unitarily similar if and only if $|\operatorname{tr} W(A, A^*) - \operatorname{tr} W(B, B^*)| \leq 1$ and $|\operatorname{tr} W((A, A^*)^{-1}) - \operatorname{tr} W((B, B^*)^{-1})| \leq 1$ for every word $W(s, t)$ in two noncommuting variables; see L. W. Marcoux, M. Mastnak, and H. Radjavi, An approximate, multivariable version of Specht's theorem, *Linear Multilinear Algebra* 55 (2007) 159–173.

2.3 Unitary and real orthogonal triangularizations

Perhaps the most fundamentally useful fact of elementary matrix theory is a theorem attributed to I. Schur: Any square complex matrix A is unitarily similar to a triangular matrix whose diagonal entries are the eigenvalues of A, in any prescribed order. Our proof involves a sequence of deflations by unitary similarities.

Theorem 2.3.1 (Schur form; Schur triangularization). *Let $A \in M_n$ have eigenvalues $\lambda_1, \ldots, \lambda_n$ in any prescribed order and let $x \in \mathbf{C}^n$ be a unit vector such that $Ax = \lambda_1 x$.*

(a) *There is a unitary $U = [x\, u_2\, \ldots\, u_n] \in M_n$ such that $U^* A U = T = [t_{ij}]$ is upper triangular with diagonal entries $t_{ii} = \lambda_i$, $i = 1, \ldots, n$.*

(b) *If $A \in M_n(\mathbf{R})$ has only real eigenvalues, then x may be chosen to be real and there is a real orthogonal $Q = [x\, q_2\, \ldots\, q_n] \in M_n(\mathbf{R})$ such that $Q^T A Q = T = [t_{ij}]$ is upper triangular with diagonal entries $t_{ii} = \lambda_i$, $i = 1, \ldots, n$.*

Proof. Let x be a normalized eigenvector of A associated with the eigenvalue λ_1, that is, $x^* x = 1$ and $Ax = \lambda_1 x$. Let $U_1 = [x\, u_2\, \ldots\, u_n]$ be any unitary matrix whose first column is x. For example, one may take $U_1 = U(x, e_1)$ as in (2.1.13), or one can proceed as in 2.3.P1. Then

$$U_1^* A U_1 = U_1^* \begin{bmatrix} Ax & Au_2 & \ldots & Au_n \end{bmatrix} = U_1^* \begin{bmatrix} \lambda_1 x & Au_2 & \ldots & Au_n \end{bmatrix}$$

$$= \begin{bmatrix} x^* \\ u_2^* \\ \vdots \\ u_n^* \end{bmatrix} \begin{bmatrix} \lambda_1 x & Au_2 & \ldots & Au_n \end{bmatrix}$$

$$= \begin{bmatrix} \lambda_1 x^* x & x^* A u_2 & \ldots & x^* A u_n \\ \lambda_1 u_2^* x & & & \\ \vdots & & A_1 & \\ \lambda_1 u_n^* x & & & \end{bmatrix} = \begin{bmatrix} \lambda_1 & \bigstar \\ 0 & A_1 \end{bmatrix}$$

because the columns of U_1 are orthonormal. The eigenvalues of the submatrix $A_1 = [u_i^* A u_j]_{i,j=2}^n \in M_{n-1}$ are $\lambda_2, \ldots, \lambda_n$. If $n = 2$, we have achieved the desired unitary triangularization. If not, let $\xi \in \mathbf{C}^{n-1}$ be a unit eigenvector of A_1 associated with λ_2, and perform the preceding reduction on A_1. If $U_2 \in M_{n-1}$ is any unitary matrix whose first column is ξ, then we have seen that

$$U_2^* A_1 U_2 = \begin{bmatrix} \lambda_2 & \bigstar \\ 0 & A_2 \end{bmatrix}$$

Let $V_2 = [1] \oplus U_2$ and compute the unitary similarity

$$(U_1 V_2)^* A U_1 V_2 = V_2^* U_1^* A U_1 V_2 = \begin{bmatrix} \lambda_1 & \bigstar & \bigstar \\ 0 & \lambda_2 & \bigstar \\ 0 & 0 & A_2 \end{bmatrix}$$

Continue this reduction to produce unitary matrices $U_i \in M_{n-i+1}, i = 1, \ldots, n-1$ and unitary matrices $V_i \in M_n, i = 2, \ldots, n-2$. The matrix $U = U_1 V_2 V_3 \cdots V_{n-2}$ is unitary and $U^* A U$ is upper triangular.

If all the eigenvalues of $A \in M_n(\mathbf{R})$ are real, then all of the eigenvectors and unitary matrices in the preceding algorithm can be chosen to be real (1.1.P3 and (2.1.13)). $\qquad\square$

Exercise. Using the notation of (2.3.1), let $U^* A^T U$ be upper triangular. Let $V = \bar{U}$ and explain why $V^* A V$ is lower triangular.

Example 2.3.2. If the eigenvalues of A are reordered and the corresponding upper triangularization (2.3.1) is performed, the entries of T above the main diagonal can be different. Consider

$$T_1 = \begin{bmatrix} 1 & 1 & 4 \\ 0 & 2 & 2 \\ 0 & 0 & 3 \end{bmatrix}, \quad T_2 = \begin{bmatrix} 2 & -1 & 3\sqrt{2} \\ 0 & 1 & \sqrt{2} \\ 0 & 0 & 3 \end{bmatrix}, \quad U = \frac{1}{\sqrt{2}} \begin{bmatrix} 1 & 1 & 0 \\ 1 & -1 & 0 \\ 0 & 0 & \sqrt{2} \end{bmatrix}$$

Verify that U is unitary and $T_2 = U T_1 U^*$.

Exercise (Schur's inequality; defect from normality). If $A = [a_{ij}] \in M_n$ has eigenvalues $\lambda_1, \ldots, \lambda_n$ and is unitarily similar to an upper triangular matrix $T = [t_{ij}] \in M_n$, the diagonal entries of T are the eigenvalues of A in some order. Apply (2.2.2) to A and T to show that

$$\sum_{i=1}^{n} |\lambda_i|^2 = \sum_{i,j=1}^{n} |a_{ij}|^2 - \sum_{i<j} |t_{ij}|^2 \leq \sum_{i,j=1}^{n} |a_{ij}|^2 = \mathrm{tr}(AA^*) \qquad (2.3.2a)$$

with equality if and only if T is diagonal.

Exercise. If $A = [a_{ij}]$ and $B = [b_{ij}] \in M_2$ have the same eigenvalues and if $\sum_{i,j=1}^{2} |a_{ij}|^2 = \sum_{i,j=1}^{2} |b_{ij}|^2$, use the criterion in (2.2.8) to show that A and B are unitarily similar. However, consider

$$A = \begin{bmatrix} 1 & 3 & 0 \\ 0 & 2 & 4 \\ 0 & 0 & 3 \end{bmatrix} \quad \text{and} \quad B = \begin{bmatrix} 1 & 0 & 0 \\ 0 & 2 & 5 \\ 0 & 0 & 3 \end{bmatrix} \qquad (2.3.2b)$$

which have the same eigenvalues and the same sums of squared entries. Use the criterion in (2.2.8) or the exercise following (2.4.5.1) to show that A and B are not unitarily similar. Nevertheless, A and B are similar. Why?

There is a useful extension of (2.3.1): A commuting family of complex matrices can be reduced simultaneously to upper triangular form by a single unitary similarity.

Theorem 2.3.3. *Let* $\mathcal{F} \subseteq M_n$ *be a nonempty commuting family. There is a unitary* $U \in M_n$ *such that* U^*AU *is upper triangular for every* $A \in \mathcal{F}$.

Proof. Return to the proof of (2.3.1). Exploiting (1.3.19) at each step of the proof in which a choice of an eigenvector (and unitary matrix) is made, choose a unit eigenvector that is common to every $A \in \mathcal{F}$ and construct a unitary matrix that has this common eigenvector as its first column; it deflates (via unitary similarity) every matrix in $\mathcal{F}$ in the same way. Similarity preserves commutativity, and a partitioned multiplication calculation reveals that, if two matrices of the form $\begin{bmatrix} A_{11} & A_{12} \\ 0 & A_{22} \end{bmatrix}$ and $\begin{bmatrix} B_{11} & B_{12} \\ 0 & B_{22} \end{bmatrix}$ commute, then A_{22} and B_{22} commute also. We conclude that all ingredients in the U of (2.3.1) may be chosen in the same way for all members of a commuting family. $\square$

In (2.3.1) we may specify the main diagonal of T (that is, we may specify in advance the order in which the eigenvalues of A appear as the deflation progresses), but (2.3.3) makes no such claim. At each stage of the deflation, the common eigenvector used is associated with *some* eigenvalue of *each* matrix in $\mathcal{F}$, but we may not be able to specify which one. We must take the eigenvalues as they come, according to the common eigenvectors guaranteed by (1.3.19).

The following exercise illuminates why quasitriangular and quasidiagonal matrices arise when one looks for triangular forms that a real matrix can achieve under real similarity; see (0.9.4).

Exercise. Show that $a \pm ib$ are the eigenvalues of the real 2-by-2 matrix $\begin{bmatrix} a & b \\ -b & a \end{bmatrix}$.

If a real matrix A has any non-real eigenvalues, there is no hope of reducing it to upper triangular form T by a *real* similarity because some main diagonal entries of T (eigenvalues of A) would be non-real. However, we can always reduce A to a real upper quasitriangular form by a real orthogonal similarity; conjugate pairs of non-real eigenvalues are associated with 2-by-2 blocks.

Theorem 2.3.4 (real Schur form). *Let* $A \in M_n(\mathbf{R})$ *be given.*

(a) *There is a real nonsingular* $S \in M_n(\mathbf{R})$ *such that* $S^{-1}AS$ *is a real upper quasi-triangular matrix*

$$\begin{bmatrix} A_1 & & & \bigstar \\ & A_2 & & \\ & & \ddots & \\ 0 & & & A_m \end{bmatrix}, \text{ each } A_i \text{ is 1-by-1 or 2-by-2} \qquad (2.3.5)$$

with the following properties: (i) its 1-by-1 diagonal blocks display the real eigenvalues of A; (ii) each of its 2-by-2 diagonal blocks has a special form that displays a conjugate pair of non-real eigenvalues of A:

$$\begin{bmatrix} a & b \\ -b & a \end{bmatrix}, a, b \in \mathbf{R}, b > 0, \text{ and } a \pm ib \text{ are eigenvalues of } A \qquad (2.3.5a)$$

(iii) its diagonal blocks are completely determined by the eigenvalues of A; they may appear in any prescribed order.

(b) There is a real orthogonal $Q \in M_n(\mathbf{R})$ such that $Q^T A Q$ is a real upper qua-
sitriangular matrix with the following properties: (i) its 1-by-1 diagonal blocks
display the real eigenvalues of A; (ii) each of its 2-by-2 diagonal blocks has a
conjugate pair of non-real eigenvalues (but no special form); (iii) the ordering of
its diagonal blocks may be prescribed in the following sense: If the real eigenval-
ues and conjugate pairs of non-real eigenvalues of A are listed in a prescribed
order, then the real eigenvalues and conjugate pairs of non-real eigenvalues of
the respective diagonal blocks $A_1, \ldots, A_m$ of $Q^T A Q$ are in the same order.

Proof. (a) The proof of (2.3.1) shows how to deflate A by a real orthogonal similarity
corresponding to any given real eigenpair; that deflation produces a real 1-by-1 diagonal
block and a deflated matrix of the form $\begin{bmatrix} \lambda & * \\ 0 & \mathcal{A} \end{bmatrix}$. Problem 1.3.P33 describes how to
deflate A via a real similarity corresponding to an eigenpair λ, x in which λ is not
real; that deflation produces a real 2-by-2 diagonal block B of the special form (2.3.5a)
and a deflated matrix of the form $\begin{bmatrix} B & * \\ 0 & \mathcal{A} \end{bmatrix}$. Only finitely many deflations are needed
to construct a nonsingular S such that $S^{-1}AS$ has the asserted upper quasitriangular
form. We can control the order in which the diagonal blocks appear by choosing, in
each deflation step, a particular eigenvalue and corresponding eigenvector.
(b) Suppose that an ordering of the real and conjugate non-real pairs of eigenvalues
of A has been given, and let S be a nonsingular real matrix such that $S^{-1}AS$ has the
form (2.3.5) with diagonal blocks in the prescribed order. Use (2.1.14) to factor S as
$S = QR$, in which Q is real orthogonal and R is real and upper triangular. Partition
$R = [R_{ij}]$ conformally to (2.3.5) and compute $S^{-1}AS = R^{-1}Q^T A Q R$, so

$$Q^T A Q = R \begin{bmatrix} A_1 & & & \bigstar \\ & A_2 & & \\ & & \ddots & \\ 0 & & & A_m \end{bmatrix} R^{-1}$$

$$= \begin{bmatrix} R_{11} A_1 R_{11}^{-1} & & & \bigstar \\ & R_{22} A_2 R_{22}^{-1} & & \\ & & \ddots & \\ 0 & & & R_{mm} A_m R_{mm}^{-1} \end{bmatrix}$$

is upper quasitriangular, its 1-by-1 diagonal blocks are identical to those of (2.3.5), and
its 2-by-2 diagonal blocks are similar to the corresponding blocks of (2.3.5). $\square$

There is a commuting families version of the preceding theorem: A commuting
family of real matrices may be reduced simultaneously to a common upper quasitriangular form by a single real or real orthogonal similarity. It is convenient to describe the
partitioned structure of (2.3.5) by saying that it is partitioned conformally to a given
quasidiagonal matrix $D = J_{n_1} \oplus \cdots \oplus J_{n_m} \in M_n$, in which J_k denotes the k-by-k all-
ones matrix (0.2.8) and each n_j is either 1 or 2.

Theorem 2.3.6. *Let $\mathcal{F} \subseteq M_n(\mathbf{R})$ be a nonempty commuting family.*

(a) *There is a nonsingular $S \in M_n(\mathbf{R})$ and a quasidiagonal $D = J_{n_1} \oplus \cdots \oplus J_{n_m} \in M_n$ such that: (i) for each $A \in \mathcal{F}$, $S^{-1}AS$ is a real upper quasitriangular matrix of the form*

$$\begin{bmatrix} A_1(A) & & & \bigstar \\ & A_2(A) & & \\ & & \ddots & \\ 0 & & & A_m(A) \end{bmatrix} \tag{2.3.6.1}$$

that is partitioned conformally to D; (ii) if $n_j = 2$, then for each $A \in \mathcal{F}$ we have

$$A_j(A) = \begin{bmatrix} a_j(A) & b_j(A) \\ -b_j(A) & a_j(A) \end{bmatrix} \in M_2(\mathbf{R}) \tag{2.3.6.2}$$

and $a_j(A) \pm i b_j(A)$ are eigenvalues of A; and (iii) for each $j \in \{1, \ldots, m\}$ such that $n_j = 2$, there is some $A \in \mathcal{F}$ such that $b_j(A) \neq 0$. If every matrix in $\mathcal{F}$ has only real eigenvalues, then $S^{-1}AS$ is upper triangular for every $A \in \mathcal{F}$.

(b) *There is a real orthogonal $Q \in M_n(\mathbf{R})$ and a quasidiagonal $D = J_{n_1} \oplus \cdots \oplus J_{n_m} \in M_n$ such that (i) for each $A \in \mathcal{F}$, $Q^T A Q$ is an upper quasitriangular matrix of the form (2.3.6.1) that is partitioned conformally to D, and (ii) for each $j \in \{1, \ldots, m\}$ such that $n_j = 2$, there is some $A \in \mathcal{F}$ such that $A_j(A)$ has a conjugate pair of non-real eigenvalues. If every matrix in $\mathcal{F}$ has only real eigenvalues, then $Q^T A Q$ is upper triangular for every $A \in \mathcal{F}$.*

Proof. (a) Following the inductive pattern of the proof of (2.3.3), it suffices to construct a nonsingular real matrix that deflates (via similarity) each matrix in $\mathcal{F}$ in the same way. Use (1.3.19) to choose a common unit eigenvector $x \in \mathbf{C}^n$ of every $A \in \mathcal{F}$. Write $x = u + iv$, in which $u, v \in \mathbf{R}^n$. There are two possibilities, the first of which is (i) $\{u, v\}$ is linearly dependent. In this event, there is a real unit vector $w \in \mathbf{R}^n$ and real scalars α and β, not both zero, such that $u = \alpha w$ and $v = \beta w$. Then $x = (\alpha + i\beta)w$ and $w = (\alpha + i\beta)^{-1}x$ is a real unit eigenvector of every $A \in \mathcal{F}$. Let Q be a real orthogonal matrix whose first column is w and observe that for every $A \in \mathcal{F}$, $Q^T A Q = \begin{bmatrix} \lambda(A) & * \\ 0 & * \end{bmatrix}$, in which $\lambda(A)$ is a real eigenvalue of A. The second possibility is (ii) $\{u, v\}$ is linearly independent. In this event (1.3.P3) shows how to construct a real nonsingular matrix S such that for every $A \in \mathcal{F}$, $S^{-1}AS = \begin{bmatrix} A_1(A) & * \\ 0 & * \end{bmatrix}$, in which $A_1(A)$ has the form (2.3.6.2). If $b_1(A) \neq 0$, then $a_1(A) \pm i b_1(A)$ is a conjugate pair of non-real eigenvalues of A. If $b_1(A) = 0$, however, then $a_1(A)$ is a double real eigenvalue of A. If $b_1(A) = 0$ for every $A \in \mathcal{F}$ (for example, if every matrix in $\mathcal{F}$ has only real eigenvalues), then split the 2-by-2 block into two 1-by-1 blocks.

(b) Let S be a nonsingular real matrix that has the properties asserted in (a), and let $S = QR$ be a QR factorization (2.1.14). In the same way as in the proof of (2.3.4), one shows that Q has the asserted properties. $\qquad\square$

Just as in (2.3.3), we cannot control the order of appearance of the eigenvalues corresponding to the diagonal blocks in the preceding theorem; we have to take the eigenvalues as they come, according to the common eigenvectors guaranteed by (1.3.19).

Exercise. Let $A \in M_n$. Explain why A commutes with $\bar{A}$ if and only if $A\bar{A}$ is real.

Exercise. Let $A = \begin{bmatrix} 1 & i \\ -i & 1 \end{bmatrix}$. Show that $A\bar{A}$ is real and that Re A commutes with Im A.

Exercise. Let $A \in M_n$ and write $A = B + iC$, in which B and C are real. Show that $A\bar{A} = \bar{A}A$ if and only if $BC = CB$.

The set $\mathcal{S} = \{A \in M_n : A\bar{A} = \bar{A}A\}$ of matrices such that $A\bar{A}$ is real is larger than the set $M_n(\mathbf{R})$ of real matrices, but they have an important property in common: Any real square matrix is real orthogonally similar to a *real* upper quasitriangular matrix, while any matrix in $\mathcal{S}$ is real orthogonally similar to a *complex* upper quasitriangular matrix.

Corollary 2.3.7. *Let $A \in M_n$ and suppose that $A\bar{A} = \bar{A}A$. There is a real orthogonal $Q \in M_n(\mathbf{R})$ and a quasidiagonal $D = J_{n_1} \oplus \cdots \oplus J_{n_m} \in M_n$ such that $Q^T A Q \in M_n$ is a complex upper quasitriangular matrix of the form (2.3.6.1) that is partitioned conformally to D and has the following property: For each $j \in \{1, \ldots, m\}$ such that $n_j = 2$, at least one of Re A_j or Im A_j has a conjugate pair of non-real eigenvalues. If each of Re A and Im A has only real eigenvalues, then $Q^T A Q \in M_n$ is upper triangular.*

Proof. Write $A = B + iC$, in which B and C are real. The hypothesis and the preceding exercise ensure that B and C commute. It follows from (2.3.6b) that there is a real orthogonal $Q \in M_n(\mathbf{R})$ and a quasidiagonal $D = J_{n_1} \oplus \cdots \oplus J_{n_m} \in M_n$ such that each of $Q^T B Q$ and $Q^T C Q$ is a real upper quasitriangular matrix of the form (2.3.6.1) that is partitioned conformally to D. Moreover, for each $j \in \{1, \ldots, m\}$ such that $n_j = 2$, at least one of $A_j(B)$ or $A_j(C)$ has a conjugate pair of non-real eigenvalues. It follows that $Q^T A Q = Q^T (B + iC)Q = Q^T B Q + i Q^T C Q$ is a complex upper quasitriangular matrix that is partitioned conformally to D. If each of B and C has only real eigenvalues, then every $n_j = 1$ and each of $Q^T B Q$ and $Q^T C Q$ is upper triangular. $\square$

Problems

2.3.P1 Let $x \in \mathbf{C}^n$ be a given unit vector and write $x = [x_1 \ y^T]^T$, in which $x_1 \in \mathbf{C}$ and $y \in \mathbf{C}^{n-1}$. Choose $\theta \in \mathbf{R}$ such that $e^{i\theta} x_1 \geq 0$ and define $z = e^{i\theta} x = [z_1 \ \zeta^T]^T$, in which $z_1 \in \mathbf{R}$ is nonnegative and $\zeta \in \mathbf{C}^{n-1}$. Consider the Hermitian matrix

$$V_x = \begin{bmatrix} z_1 & \zeta^* \\ \hline \zeta & -I + \frac{1}{1+z_1}\zeta\zeta^* \end{bmatrix} \qquad (2.3.8)$$

Use partitioned multiplication to compute $V_x^* V_x = V_x^2$. Conclude that $U = e^{-i\theta} V_x = [x \ u_2 \ \ldots \ u_n]$ is a unitary matrix whose first column is the given vector x.

2.3.P2 If $x \in \mathbf{R}^n$ is a given unit vector, show how to streamline the construction described in (2.3.P1) to produce a real orthogonal matrix $Q \in M_n(\mathbf{R})$ whose first column is x. Prove that your construction works.

2.3.P3 Let $A \in M_n(\mathbf{R})$. Explain why the non-real eigenvalues of A (if any) must occur in conjugate pairs.

2.3.P4 Consider the family $\mathcal{F} = \left\{ \begin{bmatrix} 0 & -1 \\ 0 & -1 \end{bmatrix}, \begin{bmatrix} 1 & 1 \\ 0 & -1 \end{bmatrix} \right\}$ and show that the hypothesis of commutativity in (2.3.3), while sufficient to imply simultaneous unitary upper triangularizability of $\mathcal{F}$, is not necessary.

2.3.P5 Let $\mathcal{F} = \{A_1, \ldots, A_k\} \subset M_n$ be a given family, and let $\mathcal{G} = \{A_i A_j : i, j = 1, 2, \ldots, k\}$ be the family of all pairwise products of matrices in $\mathcal{F}$. If $\mathcal{G}$ is commutative, it is known that $\mathcal{F}$ can be simultaneously unitarily upper triangularized if and only if every eigenvalue of every commutator $A_i A_j - A_j A_i$ is zero. Show that assuming commutativity of $\mathcal{G}$ is a weaker hypothesis than assuming commutativity of $\mathcal{F}$. Show that the family $\mathcal{F}$ in (2.3.P4) has a corresponding $\mathcal{G}$ that is commutative and that it also satisfies the zero eigenvalue condition.

2.3.P6 Let $A, B \in M_n$ be given, and suppose A and B are simultaneously similar to upper triangular matrices; that is, $S^{-1}AS$ and $S^{-1}BS$ are both upper triangular for some nonsingular $S \in M_n$. Show that every eigenvalue of $AB - BA$ must be zero.

2.3.P7 If a given $A \in M_n$ can be written as $A = Q \Delta Q^T$, in which $Q \in M_n$ is complex orthogonal and $\Delta \in M_n$ is upper triangular, show that A has at least one eigenvector $x \in \mathbf{C}^n$ such that $x^T x \neq 0$. Consider $A = \begin{bmatrix} 1 & i \\ i & -1 \end{bmatrix}$ to show that not every $A \in M_n$ can be upper triangularized by a complex orthogonal similarity.

2.3.P8 Let $Q \in M_n$ be complex orthogonal, and suppose that $x \in \mathbf{C}^n$ is an eigenvector of Q associated with an eigenvalue $\lambda \neq \pm 1$. Show that $x^T x = 0$. See (2.1.P8a) for an example of a family of 2-by-2 complex orthogonal matrices with both eigenvalues different from ± 1. Show that none of these matrices can be reduced to upper triangular form by complex orthogonal similarity.

2.3.P9 Let $\lambda, \lambda_2, \ldots, \lambda_n$ be the eigenvalues of $A \in M_n$, suppose that x is a nonzero vector such that $Ax = \lambda x$, and let $y \in \mathbf{C}^n$ and $\alpha \in \mathbf{C}$ be given. Provide details for the following argument to show that the eigenvalues of the bordered matrix $\mathcal{A} = \begin{bmatrix} \alpha & y^* \\ x & A \end{bmatrix} \in M_{n+1}$ are the two eigenvalues of $\begin{bmatrix} \alpha & y^*x \\ 1 & \lambda \end{bmatrix}$ together with $\lambda_2, \ldots, \lambda_n$: Form a unitary U whose first column is $x / \|x\|_2$, let $V = [1] \oplus U$, and show that $V^* \mathcal{A} V = \begin{bmatrix} B & \star \\ 0 & C \end{bmatrix}$, in which $B = \begin{bmatrix} \alpha & y^*x/\|x\|_2 \\ \|x\|_2 & \lambda \end{bmatrix} \in M_2$ and $C \in M_{n-2}$ has eigenvalues $\lambda_2, \ldots, \lambda_n$. Consider a similarity of B via $\mathrm{diag}(1, \|x\|_2^{-1})$. If $y \perp x$, conclude that the eigenvalues of $\mathcal{A}$ are $\alpha, \lambda, \lambda_2, \ldots, \lambda_n$. Explain why the eigenvalues of $\begin{bmatrix} \alpha & y^* \\ x & A \end{bmatrix}$ and $\begin{bmatrix} A & x \\ y^* & \alpha \end{bmatrix}$ are the same.

2.3.P10 Let $A = [a_{ij}] \in M_n$ and let $c = \max\{|a_{ij}| : 1 \leq i, j \leq n\}$. Show that $|\det A| \leq c^n n^{n/2}$ in two ways: (a) Let $\lambda_1, \ldots, \lambda_n$ be the eigenvalues of A. Use the arithmetic-geometric mean inequality and (2.3.2a) to explain why $|\det A|^2 = |\lambda_1 \cdots \lambda_n|^2 \leq ((|\lambda_1|^2 + \cdots + |\lambda_n|^2)/n)^n \leq (\sum_{i,j=1}^n |a_{ij}|^2/n)^n \leq (nc^2)^n$. (b) Use Hadamard's inequality in (2.1.P23).

2.3.P11 Use (2.3.1) to prove that if all the eigenvalues of $A \in M_n$ are zero, then $A^n = 0$.

2.3.P12 Let $A \in M_n$, let $\lambda_1, \ldots, \lambda_n$ be its eigenvalues, and let $r \in \{1, \ldots, n\}$. (a) Use (2.3.1) to show that the eigenvalues of the compound matrix $C_r(A)$ are the $\binom{n}{r}$ possible products $\lambda_{i_1} \cdots \lambda_{i_r}$ such that $1 \leq i_1 < i_2 < \cdots < i_r \leq n$. (b) Explain why $\mathrm{tr}\, C_r(A) = S_r(\lambda_1, \ldots, \lambda_n) = E_r(A)$; see (1.2.14) and (1.2.16). (c) If the eigenvalues of

A are arranged so that $|\lambda_1| \geq \cdots \geq |\lambda_n|$, explain why the spectral radius of $C_r(A)$ is $\rho(C_r(A)) = |\lambda_1 \cdots \lambda_r|$. (d) Explain why $p_A(t) = \sum_{k=0}^{n}(-1)^k t^{n-k} \operatorname{tr} C_r(A)$ and hence $\det(I + A) = \sum_{k=0}^{n} \operatorname{tr} C_r(A)$. (e) Provide details for the following: If A is nonsingular, then

$$\det(A + B) = \det A \det(I + A^{-1}B) = \det A \sum_{k=0}^{n} \operatorname{tr} C_r(A^{-1}B)$$

$$= \det A \sum_{k=0}^{n} \operatorname{tr}(C_r(A^{-1})C_r(B)) = \det A \sum_{k=0}^{n} \operatorname{tr}(C_r(A)^{-1}C_r(B))$$

$$= \det A \sum_{k=0}^{n} \operatorname{tr}(\det A^{-1} \operatorname{adj}_k(A)C_r(B)) = \sum_{k=0}^{n} \operatorname{tr}(\operatorname{adj}_k(A)C_r(B)).$$

(f) Prove the identity (0.8.12.3).

2.3.P13 Consider $A = \begin{bmatrix} -2 & 5 \\ -1 & 2 \end{bmatrix}$. (a) Show that $\pm i$ are the eigenvalues of A and explain why A is real similar to $B = \begin{bmatrix} 0 & 1 \\ -1 & 0 \end{bmatrix}$. (b) Explain why A is not real orthogonally similar to B.

2.3.P14 Let $A = [a_{ij}] \in M_n$. (a) Let $V = [v_{ij}] \in M_n$ be unitary. Explain why $|\operatorname{tr} VA| = |\sum_{i,j} v_{ij}a_{ji}| \leq \sum_{i,j} |a_{ji}|$. (b) Let $\lambda_1, \ldots, \lambda_n$ be the eigenvalues of A. Show that $\sum_i |\lambda_i| \leq \sum_{i,j} |a_{ji}|$.

Further Readings. See (3.4.3.1) for a refinement of the upper triangularization (2.3.1). For a proof of the stronger form of (2.3.3) asserted in (2.3.P5) see Y. P. Hong and R. A. Horn, On simultaneous reduction of families of matrices to triangular or diagonal form by unitary congruences, *Linear Multilinear Algebra* 17 (1985) 271–288.

2.4 Consequences of Schur's triangularization theorem

A bounty of results can be harvested from Schur's unitary triangularization theorem. We investigate several of them in this section.

2.4.1 The trace and determinant. Suppose that $A \in M_n$ has eigenvalues $\lambda_1, \ldots, \lambda_n$. In (1.2) we used the characteristic polynomial to show that $\sum_{i=1}^{n} \lambda_i = \operatorname{tr} A$, $\sum_{i=1}^{n} \Pi_{j \neq i}^{n} \lambda_j = \operatorname{tr}(\operatorname{adj} A)$, and $\det A = \Pi_{i=1}^{n} \lambda_i$, but these identities and others follow simply from inspection of the triangular form in (2.3.1).

For any nonsingular $S \in M_n$ we have $\operatorname{tr}(S^{-1}AS) = \operatorname{tr}(ASS^{-1}) = \operatorname{tr} A$; $\operatorname{tr}(\operatorname{adj}(S^{-1}AS)) = \operatorname{tr}((\operatorname{adj} S)(\operatorname{adj} A)(\operatorname{adj} S^{-1})) = \operatorname{tr}((\operatorname{adj} S)(\operatorname{adj} A)(\operatorname{adj} S)^{-1}) = \operatorname{tr}(\operatorname{adj} A)$; and $\det(S^{-1}AS) = (\det S^{-1})(\det A)(\det S) = (\det S)^{-1}(\det A)(\det S) = \det A$. Thus, $\operatorname{tr} A$, $\operatorname{tr}(\operatorname{adj} A)$, and $\det A$ can be evaluated using any matrix that is similar to A. The upper triangular matrix $T = [t_{ij}]$ in (2.3.1) is convenient for this purpose, since its main diagonal entries $t_{11}, \ldots, t_{nn}$ are the eigenvalues of A, $\operatorname{tr} T = \sum_{i=1}^{n} t_{ii}$, $\det T = \Pi_{i=1}^{n} t_{ii}$, and the main diagonal entries of $\operatorname{adj} T$ are $\Pi_{j \neq 1}^{n} t_{jj}, \ldots, \Pi_{j \neq n}^{n} t_{jj}$.

2.4.2 The eigenvalues of a polynomial in A. Suppose that $A \in M_n$ has eigenvalues $\lambda_1, \ldots, \lambda_n$ and let $p(t)$ be a given polynomial. We showed in (1.1.6) that $p(\lambda_i)$ is an eigenvalue of $p(A)$ for *each* $i = 1, \ldots, n$ and that if μ is an eigenvalue of $p(A)$, then there is *some* $i \in \{1, \ldots, n\}$ such that $\mu = p(\lambda_i)$. These observations identify the distinct eigenvalues of $p(A)$ (that is, its *spectrum* (1.1.4)) but not their multiplicities. Schur's theorem 2.3.1 reveals the multiplicities.

Let $A = UTU^*$, in which U is unitary and $T = [t_{ij}]$ is upper triangular with main diagonal entries $t_{11} = \lambda_1, t_{22} = \lambda_2, \ldots, t_{nn} = \lambda_n$. Then $p(A) = p(UTU^*) = Up(T)U^*$ (1.3.P2). The main diagonal entries of $p(T)$ are $p(\lambda_1), p(\lambda_2), \ldots, p(\lambda_n)$, so these are the eigenvalues (including multiplicities) of $p(T)$ and hence also of $p(A)$. In particular, for each $k = 1, 2, \ldots$, the eigenvalues of A^k are $\lambda_1^k, \ldots, \lambda_n^k$ and

$$\operatorname{tr} A^k = \lambda_1^k + \cdots + \lambda_n^k \qquad (2.4.2.1)$$

Exercise. If $T \in M_n$ is strictly upper triangular, show that all of the entries in the main diagonal and the first $p - 1$ superdiagonals of T^p are zero, $p = 1, \ldots, n$; in particular, $T^n = 0$.

Suppose that $A \in M_n$. We know (1.1.P6) that if $A^k = 0$ for some positive integer k, then $\sigma(A) = \{0\}$, so the characteristic polynomial of A is $p_A(t) = t^n$. We can now prove the converse, and a little more. If $\sigma(A) = \{0\}$, then there is a unitary U and a *strictly* upper triangular T such that $A = UTU^*$; the preceding exercise tells us that $T^n = 0$, so $A^n = UT^nU^* = 0$. Thus, the following are equivalent for $A \in M_n$: (a) A is nilpotent; (b) $A^n = 0$; and (c) $\sigma(A) = \{0\}$.

2.4.3 The Cayley–Hamilton theorem. The fact that every square complex matrix satisfies its own characteristic equation follows from Schur's theorem and an observation about multiplication of triangular matrices with special patterns of zero entries.

Lemma 2.4.3.1. *Suppose that $R = [r_{ij}]$, $T = [t_{ij}] \in M_n$ are upper triangular and that $r_{ij} = 0$, $1 \le i, j \le k < n$, and $t_{k+1,k+1} = 0$. Let $S = [s_{ij}] = RT$. Then $s_{ij} = 0$, $1 \le i, j \le k + 1$.*

Proof. The hypotheses describe block matrices R and T of the form

$$R = \begin{bmatrix} 0_k & R_{12} \\ 0 & R_{22} \end{bmatrix}, \qquad T = \begin{bmatrix} T_{11} & T_{12} \\ 0 & T_{22} \end{bmatrix}, \qquad T_{11} \in M_k$$

in which R_{22}, T_{11}, and T_{22} are upper triangular and the first column of T_{22} is zero. The product RT is necessarily upper triangular. We must show that it has a zero upper-left principal submatrix of size $k + 1$. Partition $T_{22} = [0 \ Z]$ to reveal its first column and perform a block multiplication

$$RT = \begin{bmatrix} 0_k T_{11} + R_{12}0 & 0_k T_{12} + R_{12}[0 \ Z] \\ 0 T_{11} + R_{22}0 & 0 T_{12} + R_{22}[0 \ Z] \end{bmatrix} = \begin{bmatrix} 0_k & [0 \ R_{12}Z] \\ 0 & [0 \ R_{22}Z] \end{bmatrix}$$

which reveals the desired zero upper-left principal submatrix of size $k + 1$. $\square$

Theorem 2.4.3.2 (Cayley–Hamilton). Let $p_A(t)$ be the characteristic polynomial of $A \in M_n$. Then $p_A(A) = 0$

Proof. Factor $p_A(t) = (t - \lambda_1)(t - \lambda_2) \cdots (t - \lambda_n)$ as in (1.2.6) and use (2.3.1) to write A as $A = UTU^*$, in which U is unitary, T is upper triangular, and the main diagonal entries of T are $\lambda_1, \ldots, \lambda_n$. Compute

$$p_A(A) = p_A(UTU^*) = Up_A(T)U^*$$
$$= U[(T - \lambda_1 I)(T - \lambda_2 I) \cdots (T - \lambda_n I)]U^*$$

It suffices to show that $p_A(T) = 0$. The upper left 1-by-1 block of $T - \lambda_1 I$ is 0, and the 2, 2 entry of $T - \lambda_2 I$ is 0, so the preceding lemma ensures that the upper left 2-by-2 principal submatrix of $(T - \lambda_1 I)(T - \lambda_2 I)$ is 0. Suppose that the upper left k-by-k principal submatrix of $(T - \lambda_1 I) \cdots (T - \lambda_k I)$ is zero. The $k + 1, k + 1$ entry of $(T - \lambda_{k+1} I)$ is 0, so invoking the lemma again, we know that the upper left principal submatrix of $(T - \lambda_1 I) \cdots (T - \lambda_{k+1} I)$ of size $k + 1$ is 0. By induction, we conclude that $((T - \lambda_1 I) \cdots (T - \lambda_{n-1} I))(T - \lambda_n I) = 0$. $\square$

Exercise. What is wrong with the following argument? "Since $p_A(\lambda_i) = 0$ for every eigenvalue λ_i of $A \in M_n$, and since the eigenvalues of $p_A(A)$ are $p_A(\lambda_1), \ldots, p_A(\lambda_n)$, all eigenvalues of $p_A(A)$ are 0. Therefore, $p_A(A) = 0$." Give an example to illustrate the fallacy in the argument.

Exercise. What is wrong with the following argument? "Since $p_A(t) = \det(tI - A)$, we have $p_A(A) = \det(AI - A) = \det(A - A) = \det 0 = 0$. Therefore, $p_A(A) = 0$."

The Cayley–Hamilton theorem is often paraphrased as "every square matrix satisfies its own characteristic equation" (1.2.3), but this must be understood carefully: The scalar polynomial $p_A(t)$ is first computed as $p_A(t) = \det(tI - A)$; one then computes the matrix $p_A(A)$ by substituting $t \to A$.

We have proved the Cayley–Hamilton theorem for matrices with complex entries, and hence it must hold for matrices whose entries come from any subfield of the complex numbers (the reals or the rationals, for example). In fact, the Cayley–Hamilton theorem is a completely formal result that is valid for matrices whose entries come from any field or, more generally, any commutative ring; see (2.4.P3).

One important use of the Cayley–Hamilton theorem is to write powers A^k of $A \in M_n$, for $k \geq n$, as linear combinations of $I, A, A^2, \ldots, A^{n-1}$.

Example 2.4.3.3. Let $A = \begin{bmatrix} 3 & 1 \\ -2 & 0 \end{bmatrix}$. Then $p_A(t) = t^2 - 3t + 2$, so $A^2 - 3A + 2I = 0$. Thus, $A^2 = 3A - 2I$; $A^3 = A(A^2) = 3A^2 - 2A = 3(3A - 2I) - 2A = 7A - 6I$; $A^4 = 7A^2 - 6A = 15A - 14I$, and so on. We can also express negative powers of the nonsingular matrix A as linear combinations of A and I. Write $A^2 - 3A + 2I = 0$ as $2I = -A^2 + 3A = A(-A + 3I)$, or $I = A[\frac{1}{2}(-A + 3I)]$. Thus, $A^{-1} = -\frac{1}{2}A + \frac{3}{2}I = \begin{bmatrix} 0 & -1/2 \\ 1 & 3/2 \end{bmatrix}$, $A^{-2} = (-\frac{1}{2}A + \frac{3}{2}I)^2 = \frac{1}{4}A^2 - \frac{3}{2}A + \frac{9}{4}I = \frac{1}{4}(3A - 2I) - \frac{3}{2}A + \frac{9}{4}I = -\frac{3}{4}A + \frac{7}{4}I$, and so on.

Corollary 2.4.3.4. *Suppose $A \in M_n$ is nonsingular and let $p_A(t) = t^n + a_{n-1}t^{n-1} + \cdots + a_1 t + a_0$. Let $q(t) = -(t^{n-1} + a_{n-1}t^{n-2} + \cdots + a_2 t + a_1)/a_0$. Then $A^{-1} = q(A)$ is a polynomial in A.*

Proof. Write $p_A(A) = 0$ as $A(A^{n-1} + a_{n-1}A^{n-2} + \cdots + a_2A + a_1I) = -a_0I$, that is, $Aq(A) = I$.   □

Exercise. If $A, B \in M_n$ are similar and $g(t)$ is any given polynomial, show that $g(A)$ is similar to $g(B)$, and that any polynomial equation satisfied by A is satisfied by B. Give some thought to the converse: Satisfaction of the same polynomial equations implies similarity – true or false?

Example 2.4.3.5. We have shown that each $A \in M_n$ satisfies a polynomial equation of degree n, for example, its characteristic equation. It is possible for $A \in M_n$ to satisfy a polynomial equation of degree less than n, however. Consider

$$A = \begin{bmatrix} 1 & 0 & 0 \\ 0 & 1 & 1 \\ 0 & 0 & 1 \end{bmatrix} \in M_3$$

The characteristic polynomial is $p_A(t) = (t - 1)^3$ and indeed $(A - I)^3 = 0$. But $(A - I)^2 = 0$ so A satisfies a polynomial equation of degree 2. There is no polynomial $h(t) = t + a_0$ of degree 1 such that $h(A) = 0$ since $h(A) = A + a_0I \neq 0$ for all $a_0 \in \mathbf{C}$.

Exercise. Suppose that a diagonalizable matrix $A \in M_n$ has $d \leq n$ distinct eigenvalues $\lambda_1, \ldots, \lambda_d$. Let $q(t) = (t - \lambda_1) \cdots (t - \lambda_d)$. Show that $q(A) = 0$, so A satisfies a polynomial equation of degree d. Why is there no polynomial $g(t)$ of degree strictly less than d such that $g(A) = 0$? Consider the matrix in the preceding example to show that the minimum degree of a polynomial equation satisfied by a nondiagonalizable matrix can be strictly larger than the number of its distinct eigenvalues.

2.4.4 Sylvester's theorem on linear matrix equations. The equation $AX - XA = 0$ associated with commutativity is a special case of the linear matrix equation $AX - XB = C$, often called *Sylvester's equation*. The following theorem gives a necessary and sufficient condition for Sylvester's equation to have a unique solution X for every given C. It relies on the Cayley–Hamilton theorem and on the observation that if $AX = XB$, then $A^2X = A(AX) = A(XB) = (AX)B = (XB)B = XB^2$, $A^3X = A(A^2X) = A(XB^2) = (AX)B^2 = XB^3$, and so on. Thus, with the standard understanding that A^0 denotes the identity matrix, we have

$$\left(\sum_{k=0}^{m} a_k A^k \right) X = \sum_{k=0}^{m} a_k A^k X = \sum_{k=0}^{m} a_k X B^k = X \left(\sum_{k=0}^{m} a_k B^k \right)$$

We formalize this observation in the following lemma.

Lemma 2.4.4.0. *Let* $A \in M_n$, $B \in M_m$, *and* $X \in M_{n,m}$. *If* $AX - XB = 0$, *then* $g(A)X - Xg(B) = 0$ *for any polynomial* $g(t)$.

Theorem 2.4.4.1 (Sylvester). Let $A \in M_n$ and $B \in M_m$ be given. The equation $AX - XB = C$ has a unique solution $X \in M_{n,m}$ for each given $C \in M_{n,m}$ if and only if

$\sigma(A) \cap \sigma(B) = \varnothing$, that is, if and only if A and B have no eigenvalue in common. In particular, if $\sigma(A) \cap \sigma(B) = \varnothing$ then the only X such that $AX - XB = 0$ is $X = 0$. If A and B are real, then $AX - XB = C$ has a unique solution $X \in M_{n,m}(\mathbf{R})$ for each given $C \in M_{n,m}(\mathbf{R})$.

Proof. Consider the linear transformation $T : M_{n,m} \to M_{n,m}$ defined by $T(X) = AX - XB$. To ensure that the equation $T(X) = C$ has a unique solution X for every given $C \in M_{n,m}$ it suffices to show that the only solution of $T(X) = 0$ is $X = 0$; see (0.5). If $AX - XB = 0$, we know from the preceding discussion that $p_B(A)X - Xp_B(B) = 0$. The Cayley–Hamilton theorem ensures that $p_B(B) = 0$, so $p_B(A)X = 0$. Let $\lambda_1, \ldots, \lambda_m$ be the eigenvalues of B, so $p_B(t) = (t - \lambda_1) \cdots (t - \lambda_m)$ and $p_B(A) = (A - \lambda_1 I) \cdots (A - \lambda_m I)$.

If $\sigma(A) \cap \sigma(B) = \varnothing$, then each factor $A - \lambda_j I$ is nonsingular, $p_B(A)$ is nonsingular, and the only solution of $p_B(A)X = 0$ is $X = 0$.

If $\lambda \in \sigma(A) \cap \sigma(B)$, let $y \in \mathbf{C}^m$ and $x \in \mathbf{C}^n$ be nonzero vectors such that $y^*B = \lambda y^*$ and $Ax = \lambda x$ (left and right eigenvectors). Then $\hat{X} = xy^* \neq 0$ and $A\hat{X} - \hat{X}B = \lambda \hat{X} - \lambda \hat{X} = 0$.

If A and B are real, consider the linear transformation $T : M_{n,m}(\mathbf{R}) \to M_{n,m}(\mathbf{R})$ defined by $T(X) = AX - XB$. The same argument shows that: (a) If $\sigma(A) \cap \sigma(B) = \varnothing$ then $AX - XB = 0 \Rightarrow p_B(A)X = 0 \Rightarrow X = 0$, and (b) if $\sigma(A) \cap \sigma(B) \neq \varnothing$ then $\mathrm{Re}(A\hat{X} - \hat{X}B) = A(\mathrm{Re}\hat{X}) - (\mathrm{Re}\hat{X})B = 0$, $\mathrm{Im}(A\hat{X} - \hat{X}B) = A(\mathrm{Im}\hat{X}) - (\mathrm{Im}\hat{X})B = 0$, and either $\mathrm{Re}\hat{X}$ or $\mathrm{Im}\hat{X}$ is nonzero. $\square$

A matrix identity of the form $AX = XB$ is known as an *intertwining relation*. The commutativity equation $AB = BA$ is perhaps the most familiar intertwining relation; other examples are the anticommutativity equation $AB = -BA$, $AB = BA^T$, $AB = B\bar{A}$, and $AB = BA^*$. The following consequence of Sylvester's theorem is often used to show that a matrix is block diagonal if it satisfies a certain type of intertwining relation.

Corollary 2.4.4.2. *Let $B, C \in M_n$ be block diagonal and partitioned conformally as $B = B_1 \oplus \cdots \oplus B_k$ and $C = C_1 \oplus \cdots \oplus C_k$. Suppose that $\sigma(B_i) \cap \sigma(C_j) = \varnothing$ whenever $i \neq j$. If $A \in M_n$ and $AB = CA$, then A is block diagonal conformal to B and C, that is, $A = A_1 \oplus \cdots \oplus A_k$, and $A_i B_i = C_i A_i$ for each $i = 1, \ldots, k$.*

Proof. Partition $A = [A_{ij}]$ conformally to B and C. Then $AB = CA$ if and only if $A_{ij}B_j = C_i A_{ij}$. If $i \neq j$, then (2.4.4.1) ensures that $A_{ij} = 0$. $\square$

A fundamental principle worth keeping in mind is that if $AX = XB$ and if there is something special about the structure of A and B, then there is likely to be something special about the structure of X. One may be able to discover what that special structure is by replacing A and B by canonical forms and studying the resulting intertwining relation involving the canonical forms and a transformed X. The following corollary is an example of a result of this type.

Corollary 2.4.4.3. *Let $A, B \in M_n$. Suppose that there is a nonsingular $S \in M_n$ such that $A = S(A_1 \oplus \cdots \oplus A_d)S^{-1}$, in which each $A_j \in M_{n_j}$, $j = 1, \ldots, d$, and $\sigma(A_i) \cap$*

$\sigma(A_j) = \emptyset$ *whenever* $i \neq j$. *Then* $AB = BA$ *if and only if* $B = S(B_1 \oplus \cdots \oplus B_d)S^{-1}$, *in which each* $B_j \in M_{n_j}$, $j = 1, \ldots, d$, *and* $A_i B_i = B_i A_i$ *for each* $i = 1, \ldots, d$.

Proof. If A commutes with B, then $(S^{-1}AS)(S^{-1}BS) = (S^{-1}BS)(S^{-1}AS)$, so the asserted direct sum decomposition of $S^{-1}BS$ follows from the preceding corollary. The converse follows from a computation. $\qquad\square$

In a common application of the preceding result, each matrix A_i has a single eigenvalue and is often a scalar matrix: $A_i = \lambda_i I_{n_i}$.

2.4.5 Uniqueness in Schur's triangularization theorem. For a given $A \in M_n$, an upper triangular form T described in (2.3.1) that can be achieved by unitary similarity need not be unique. That is, different upper triangular matrices with the same main diagonals can be unitarily similar.

If $T, T' \in M_n$ are upper triangular and have identical main diagonals in which equal entries are grouped together, what can be said about a unitary $W \in M_n$ such that $T' = WTW^*$, that is, $WT = T'W$? The following theorem says that W must be block diagonal, and under certain assumptions about the superdiagonal entries of T, W must be a diagonal matrix or even a scalar matrix. In the latter case, $T = T'$.

Theorem 2.4.5.1. *Let* $n, d, n_1, \ldots, n_d$ *be positive integers such that* $n_1 + \cdots + n_d = n$. *Let* $\Lambda = \lambda_1 I_{n_1} \oplus \cdots \oplus \lambda_d I_{n_d} \in M_n$, *in which* $\lambda_i \neq \lambda_j$ *if* $i \neq j$. *Let* $T = [t_{ij}] \in M_n$ *and* $T' = [t'_{ij}] \in M_n$ *be upper triangular matrices that have the same main diagonal as* Λ. *Partition* $T = [T_{ij}]_{i,j=1}^d$, $T' = [T'_{ij}]_{i,j=1}^d$, *and* $W = [W_{ij}]_{i,j=1}^d \in M_n$ *conformally to* Λ. *Suppose that* $WT = T'W$. *Then*

(a) *$W_{ij} = 0$ if $i > j$, that is, W is block upper triangular conformal to Λ.*

(b) *If W is unitary, then it is block diagonal conformal to Λ: $W = W_{11} \oplus \cdots \oplus W_{dd}$.*

(c) *Suppose that every entry in the first superdiagonal of each block $T_{11}, \ldots, T_{dd}$ is nonzero. Then W is upper triangular. If W is unitary, then it is diagonal: $W = \mathrm{diag}(w_1, \ldots, w_n)$.*

(d) *If W is unitary, and if $t_{i,i+1} > 0$ and $t'_{i,i+1} > 0$ for each $i = 1, \ldots, n-1$, then W is a scalar unitary matrix: $W = wI$. In this event, $T = T'$.*

Proof. (a) If $d = 1$, there is nothing to prove, so assume that $d \geq 2$. Our strategy is to exploit the equality of corresponding blocks of both sides of the identity $WT = T'W$. The $d, 1$ block of WT is $W_{d1}T_{11}$ and the $d, 1$ block of $T'W$ is $T'_{dd}W_{d1}$. Since $\sigma(T_{11})$ and $\sigma(T'_{dd})$ are disjoint, (2.4.4.1) ensures that $W_{d1} = 0$ is the only solution to $W_{d1}T_{11} = T'_{dd}W_{d1}$. If $d = 2$, we stop at this point. If $d > 2$, then the $d, 2$ block of WT is $W_{d2}T_{22}$ (since $W_{d1} = 0$), and the $d, 2$ block of $T'W$ is $T'_{d,d}W_{d2}$; we have $W_{d2}T_{22} = T'_{d,d}W_{d2}$. Again, (2.4.4.1) ensures that $W_{d2} = 0$ since $\sigma(T_{22})$ and $\sigma(T'_{dd})$ are disjoint. Proceeding in this way across the dth block row of $TW = WT'$, we find that $W_{d1}, \ldots, W_{d,d-1}$ are all zero. Now equate the blocks of $WT = T'W$ in positions $(d-1), k$ for $k = 1, \ldots, d-2$ and conclude in the same way that $W_{d-1,1}, \ldots, W_{d-1,d-2}$

are all zero. Working our way up the block rows of $WT = T'W$, left to right, we conclude that $W_{ij} = 0$ for all $i > j$.

(b) Now assume that W is unitary. Partition $W = \begin{bmatrix} W_{11} & X \\ 0 & \hat{W} \end{bmatrix}$ and conclude from (2.1.10) that $X = 0$. Since $\hat{W}$ is also block upper triangular and unitary, an induction leads to the conclusion that $W = W_{11} \oplus \cdots \oplus W_{dd}$; see (2.5.2).

(c) We have d identities $W_{ii}T_{ii} = T'_{ii}W_{ii}$, $i = 1, \ldots, d$, and we assume that all the entries in the first superdiagonal of each T_{ii} are nonzero. Thus, it suffices to consider the case $d = 1$: $T = [t_{ij}] \in M_n$ and $T' = [t'_{ij}] \in M_n$ are upper triangular, $t_{ii} = t'_{ii} = \lambda$ for all $i = 1, \ldots, n$, $t_{i,i+1}$ is nonzero for all $i = 1, \ldots, n-1$, and $WT = T'W$. As in (a), we equate corresponding entries of the identity $WT = T'W$: In position $n, 2$ we have $w_{n1}t_{12} + w_{n2}\lambda = \lambda w_{n2}$ or $w_{n1}t_{12} = 0$; since $t_{12} \neq 0$ it follows that $w_{n1} = 0$. Proceeding across the nth row of $WT = T'W$ we obtain a sequence of identities $w_{ni}t_{i,i+1} + w_{n,i+1}\lambda = \lambda w_{n,i+1}$, $i = 1, \ldots, n-1$, from which it follows that $w_{ni} = 0$ for all $i = 1, \ldots, n-1$. Working our way up the rows of $WT = T'W$ in this fashion, left to right, we find that $w_{ij} = 0$ for all $i > j$. Thus, W is upper triangular; if it is unitary, the argument in (b) ensures that it is diagonal.

(d) The hypotheses and (c) ensure that $W = \text{diag}(w_1, \ldots, w_n)$ is diagonal and unitary. Equating the entries of $WT = T'W$ in position $i, i+1$, we have $w_i t_{i,i+1} = t'_{i,i+1}w_{i+1}$, so $t_{i,i+1}/t'_{i,i+1} = w_{i+1}/w_i$, which is real, positive, and has modulus one. We conclude that $w_{i+1}/w_i = 1$ for each $i = 1, \ldots n-1$, and hence $w_1 = \cdots = w_n$ and $W = w_{11}I$. $\square$

Exercise. Suppose that $A, B \in M_n$ are unitarily similar via a scalar unitary matrix. Explain why $A = B$.

Exercise. Suppose that $T, T' \in M_n$ are upper triangular, have identical main diagonals with distinct entries, and are similar via a unitary $U \in M_n$. Explain why U must be diagonal. If all the entries of the first superdiagonals of T and T' are real and positive, explain why $T = T'$.

Exercise. If $A = [a_{ij}] \in M_n$ and $a_{i,i+1} \neq 0$ for each $i = 1, \ldots, n-1$, show that there is a diagonal unitary matrix D such that DAD^* has real positive entries in its first superdiagonal. *Hint*: Consider $D = \text{diag}(1, a_{12}/|a_{12}|, a_{12}a_{23}/|a_{12}a_{23}|, \ldots)$.

2.4.6 Every square matrix is block diagonalizable. The following application and extension of (2.3.1) is an important step toward the Jordan canonical form, which we discuss in the next chapter.

Theorem 2.4.6.1. *Let the distinct eigenvalues of $A \in M_n$ be $\lambda_1, \ldots, \lambda_d$, with respective multiplicities $n_1, \ldots, n_d$. Theorem 2.3.1 ensures that A is unitarily similar to a d-by-d block upper triangular matrix $T = [T_{ij}]_{i,j=1}^d$ in which each block T_{ij} is n_i-by-n_j, $T_{ij} = 0$ if $i > j$, and each diagonal block T_{ii} is upper triangular with diagonal entries λ_i, that is, each $T_{ii} = \lambda_i I_{n_i} + R_i$ and $R_i \in M_{n_i}$ is strictly upper triangular. Then A is* *similar to*

$$\begin{bmatrix} T_{11} & & & 0 \\ & T_{22} & & \\ & & \ddots & \\ 0 & & & T_{dd} \end{bmatrix} \qquad (2.4.6.2)$$

If $A \in M_n(R)$ and if all its eigenvalues are real, then the unitary similarity that reduces A to the special upper triangular form T and the similarity matrix that reduces T to the block diagonal form (2.4.6.2) may both be taken to be real.

Proof. Partition T as

$$T = \begin{bmatrix} T_{11} & Y \\ 0 & S_2 \end{bmatrix}$$

in which $S_2 = [T_{ij}]_{i,j=2}^d$. Notice that the only eigenvalue of T_{11} is λ_1 and that the eigenvalues of S_2 are $\lambda_2, \ldots, \lambda_d$. Sylvester's theorem 2.4.4.1 ensures that the equation $T_{11}X - XS_2 = -Y$ has a solution X; use it to construct

$$M = \begin{bmatrix} I_{n_1} & X \\ 0 & I \end{bmatrix} \text{ and its inverse } M^{-1} = \begin{bmatrix} I_{n_1} & -X \\ 0 & I \end{bmatrix}$$

Then

$$\begin{aligned} M^{-1}TM &= \begin{bmatrix} I_{n_1} & -X \\ 0 & I \end{bmatrix} \begin{bmatrix} T_{11} & Y \\ 0 & S_2 \end{bmatrix} \begin{bmatrix} I_{n_1} & X \\ 0 & I \end{bmatrix} \\ &= \begin{bmatrix} T_{11} & T_{11}X - XS_2 + Y \\ 0 & S_2 \end{bmatrix} = \begin{bmatrix} T_{11} & 0 \\ 0 & S_2 \end{bmatrix} \end{aligned}$$

If $d = 2$, this is the desired block diagonalization. If $d > 2$, repeat this reduction process to show that S_2 is similar to $T_{22} \oplus S_3$ in which $S_3 = [T_{ij}]_{i,j=3}^d$. After $d - 1$ reductions, we find that T is similar to $T_{11} \oplus \cdots \oplus T_{dd}$.

If A is real and has real eigenvalues, then it is real orthogonally similar to a real block upper triangular matrix of the form just considered. Each of the reduction steps can be carried out with a real similarity. □

Exercise. Suppose that $A \in M_n$ is unitarily similar to a d-by-d block upper triangular matrix $T = [T_{ij}]_{i,j=1}^d$. If any block T_{ij} with $j > i$ is nonzero, use (2.2.2) to explain why T is not *unitarily* similar to $T_{11} \oplus \cdots \oplus T_{dd}$.

There are two extensions of the preceding theorem that, for commuting families and simultaneous (but not necessarily unitary) similarity, significantly refine the block structure achieved in (2.3.3).

Theorem 2.4.6.3. *Let $\mathcal{F} \subset M_n$ be a commuting family, let A_0 be any given matrix in $\mathcal{F}$, and suppose that A_0 has d distinct eigenvalues $\lambda_1, \ldots, \lambda_d$, with respective multiplicities $n_1, \ldots, n_d$. Then there is a nonsingular $S \in M_n$ such that*

(a) $\hat{A}_0 = S^{-1}A_0S = T_1 \oplus \cdots \oplus T_d$, in which each $T_i \in M_{n_i}$ is upper triangular and all its diagonal entries are λ_i; and

(b) for every $A \in \mathcal{F}$, $S^{-1}AS$ is upper triangular and block diagonal conformal to $\hat{A}_0$.

Proof. First use (2.4.6.1) to choose a nonsingular S_0 such that $S_0^{-1} A_0 S_0 = R_1 \oplus$ $\cdots \oplus R_d = \tilde{A}_0$, in which each $R_i \in M_{n_i}$ has λ_i as its only eigenvalue. Let $S_0^{-1} \mathcal{F} S_0 = \{S_0^{-1} A S_0 : A \in \mathcal{F}\}$, which is also a commuting family. Partition any given $B \in S_0^{-1} \mathcal{F} S_0$ as $B = [B_{ij}]_{i,j=1}^{d}$, conformal to $\tilde{A}_0$. Then $[R_i B_{ij}] = \tilde{A}_0 B = B \tilde{A}_0 = [B_{ij} R_j]$, so $R_i B_{ij} = B_{ij} R_j$ for all $i, j = 1, \ldots, d$. Sylvester's theorem 2.4.4.1 now ensures that $B_{ij} = 0$ for all $i \neq j$ since R_i and R_j have no eigenvalues in common. Thus, $S_0^{-1} \mathcal{F} S_0$ is a commuting family of block diagonal matrices that are all conformal to $\tilde{A}_0$. For each $i = 1, \ldots, d$, consider the family $\mathcal{F}_i \subset M_{n_i}$ consisting of the ith diagonal block of every matrix in $S_0^{-1} \mathcal{F} S_0$; notice that $R_i \in \mathcal{F}_i$ for each $i = 1, \ldots, d$. Each $\mathcal{F}_i$ is a commuting family, so (2.3.3) ensures that there is a unitary $U_i \in M_{n_i}$ such that $U_i^* \mathcal{F}_i U_i$ is an upper triangular family. The main diagonal entries of $U_i^* R_i U_i$ are its eigenvalues, which are all equal to λ_i. Let $U = U_1 \oplus \cdots \oplus U_d$ and observe that $S = S_0 U$ accomplishes the asserted reduction, in which $T_i = U_i^* R_i U_i$.   □

Corollary 2.4.6.4. *Let $\mathcal{F} \subset M_n$ be a commuting family. There is a nonsingular $S \in M_n$ and positive integers $k, n_1, \ldots, n_k$ such that $n_1 + \cdots + n_k = n$ and, for every $A \in \mathcal{F}$, $S^{-1} A S = A_1 \oplus \cdots \oplus A_k$ is block diagonal with $A_i \in M_{n_i}$ for each $i = 1, \ldots, k$. Moreover, each diagonal block A_i is upper triangular and has exactly one eigenvalue.*

Proof. If every matrix in $\mathcal{F}$ has only one eigenvalue, apply (2.3.3) and stop. If some matrix in $\mathcal{F}$ has at least two distinct eigenvalues, let $A_0 \in \mathcal{F}$ be any matrix that has the maximum number of distinct eigenvalues among all matrices in $\mathcal{F}$. Construct a simultaneous block diagonal upper triangularization as in the preceding theorem, and observe that the size of every diagonal block obtained is strictly smaller than the size of A_0. Associated with each diagonal block of the reduced form of A_0 is a commuting family of matrices. Among the members of that family, either (a) each matrix has only one eigenvalue (no further reduction required) or (b) some matrix has at least two distinct eigenvalues, in which case, we choose any matrix that has the maximum number of distinct eigenvalues and reduce again to obtain a set of strictly smaller diagonal blocks. Recursively repeat this reduction, which must terminate in finitely many steps, until no member of any commuting family has more than one eigenvalue.   □

2.4.7 Every square matrix is almost diagonalizable. Another use of Schur's result is to make it clear that every square complex matrix is "almost diagonalizable" in two possible interpretations of the phrase. The first says that arbitrarily close to a given matrix, there is a diagonalizable matrix; the second says that any given matrix is similar to an upper triangular matrix whose off-diagonal entries are arbitrarily small.

Theorem 2.4.7.1. *Let $A = [a_{ij}] \in M_n$. For each $\epsilon > 0$, there exists a matrix $A(\epsilon) = [a_{ij}(\epsilon)] \in M_n$ that has n distinct eigenvalues (and is therefore diagonalizable) and is such that $\sum_{i,j=1}^{n} |a_{ij} - a_{ij}(\epsilon)|^2 < \epsilon$.*

Proof. Let $U \in M_n$ be unitary and such that $U^* A U = T$ is upper triangular. Let $E = \text{diag}(\varepsilon_1, \varepsilon_2, \ldots, \varepsilon_n)$, in which $\varepsilon_1, \ldots, \varepsilon_n$ are chosen so that $|\varepsilon_i| < \left(\frac{\epsilon}{n}\right)^{1/2}$ and so that $t_{ii} + \varepsilon_i \neq t_{jj} + \varepsilon_j$ for all $i \neq j$. (Reflect for a moment to see that this can be done.)

Then $T + E$ has n distinct eigenvalues: $t_{11} + \varepsilon_1, \ldots, t_{nn} + \varepsilon_n$, and so does $A + UEU^*$, which is similar to $T + E$. Let $A(\epsilon) = A + UEU^*$, so that $A - A(\epsilon) = -UEU^*$, and hence (2.2.2) ensures that $\sum_{i,j} |a_{ij} - a_{ij}(\epsilon)|^2 = \sum_{i=1}^n |\varepsilon_i|^2 < n \left(\frac{\epsilon}{n} \right) = \epsilon$. □

Exercise. Show that the condition $\sum_{i,j} |a_{ij} - a_{ij}(\epsilon)|^2 < \epsilon$ in (2.4.6) could be replaced by $\max_{i,j} |a_{ij} - a_{ij}(\epsilon)| < \epsilon$. *Hint*: Apply the theorem with ϵ^2 in place of ϵ and realize that, if a sum of squares is less than ϵ^2, each of the items must be less than ϵ in absolute value.

Theorem 2.4.7.2. *Let $A \in M_n$. For each $\epsilon > 0$ there is a nonsingular matrix $S_\epsilon \in M_n$ such that $S_\epsilon^{-1} A S_\epsilon = T_\epsilon = [t_{ij}(\epsilon)]$ is upper triangular and $|t_{ij}(\epsilon)| \leq \epsilon$ for all $i, j \in \{1, \ldots, n\}$ such that $i < j$.*

Proof. First apply Schur's theorem to produce a unitary matrix $U \in M_n$ and an upper triangular matrix $T \in M_n$ such that $U^* A U = T$. Define $D_\alpha = \mathrm{diag}(1, \alpha, \alpha^2, \ldots, \alpha^{n-1})$ for a nonzero scalar α and set $t = \max_{i<j} |t_{ij}|$. Assume that $\epsilon < 1$, since it certainly suffices to prove the statement in this case. If $t \leq 1$, let $S_\epsilon = UD_\epsilon$; if $t > 1$, let $S_\epsilon = UD_{1/t}D_\epsilon$. In either case, the appropriate S_ϵ substantiates the claim of the theorem. If $t \leq 1$, a calculation reveals that $t_{ij}(\epsilon) = t_{ij}\epsilon^{-i}\epsilon^j = t_{ij}\epsilon^{j-i}$, whose absolute value is no more than ϵ^{j-i}, which is, in turn, no more than ϵ if $i < j$. If $t > 1$, the similarity by $D_{1/t}$ preprocesses the matrix, producing one in which all off-diagonal entries are no more than 1 in absolute value. □

Exercise. Prove the following variant of (2.4.7.2): If $A \subset M_n$ and $c > 0$, there is a nonsingular $S_\epsilon \in M_n$ such that $S_\epsilon^{-1} A S_\epsilon = T_\epsilon = [t_{ij}(\epsilon)]$ is upper triangular and $\sum_{j>i} |t_{ij}(\epsilon)| \leq \epsilon$. *Hint*: Apply (2.4.7) with $[2/n(n-1)]\epsilon$ in place of ϵ.

2.4.8 Commuting families and simultaneous triangularization. We now use the commuting families version (2.3.3) of Schur's theorem to show that the eigenvalues "add" and "multiply" – in some order – for commuting matrices.

Theorem 2.4.8.1. *Suppose that $A, B \in M_n$ commute. Then there is an ordering $\alpha_1, \ldots, \alpha_n$ of the eigenvalues of A and an ordering $\beta_1, \ldots, \beta_n$ of the eigenvalues of B such that the eigenvalues of $A + B$ are $\alpha_1 + \beta_1, \alpha_2 + \beta_2, \ldots, \alpha_n + \beta_n$ and the eigenvalues of AB are $\alpha_1\beta_1, \alpha_2\beta_2, \ldots, \alpha_n\beta_n$. In particular, $\sigma(A + B) \subseteq \sigma(A) + \sigma(B)$, $\sigma(AB) \subseteq \sigma(A)\sigma(B)$.*

Proof. Since A and B commute, (2.3.3) ensures that there is a unitary $U \in M_n$ such that $U^* A U = T = [t_{ij}]$ and $U^* B U = R = [r_{ij}]$ are both upper triangular. The main diagonal entries (and hence also the eigenvalues) of the upper triangular matrix $T + R = U^*(A + B)U$ are $t_{11} + r_{11}, \ldots, t_{nn} + r_{nn}$; these are the eigenvalues of $A + B$ since $A + B$ is similar to $T + R$. The main diagonal entries (and hence also the eigenvalues) of the upper triangular matrix $TR = U^*(AB)U$ are $t_{11}r_{11}, \ldots, t_{nn}r_{nn}$; these are the eigenvalues of AB, which is similar to TR. □

Exercise. Suppose that $A, B \in M_n$ commute. Explain why $\rho(A + B) \leq \rho(A) + \rho(B)$ and $\rho(AB) \leq \rho(A)\rho(B)$, so the spectral radius function is subadditive and submultiplicative for commuting matrices.

Example 2.4.8.2. Even if A and B commute, not every sum of their respective eigenvalues need be an eigenvalue of $A + B$. Consider the diagonal matrices

$$A = \begin{bmatrix} 1 & 0 \\ 0 & 2 \end{bmatrix} \quad \text{and} \quad B = \begin{bmatrix} 3 & 0 \\ 0 & 4 \end{bmatrix}$$

Since $1 + 4 = 5 \notin \{4, 6\} = \sigma(A + B)$, we see that $\sigma(A + B)$ is contained in, but is not equal to, $\sigma(A) + \sigma(B)$.

Example 2.4.8.3. If A and B do *not* commute, it is difficult to say how $\sigma(A + B)$ is related to $\sigma(A)$ and $\sigma(B)$. In particular, $\sigma(A + B)$ need not be contained in $\sigma(A) + \sigma(B)$. Let

$$A = \begin{bmatrix} 0 & 1 \\ 0 & 0 \end{bmatrix} \quad \text{and} \quad B = \begin{bmatrix} 0 & 0 \\ 1 & 0 \end{bmatrix}$$

Then $\sigma(A + B) = \{-1, 1\}$, while $\sigma(A) = \sigma(B) = \{0\}$.

Exercise. Consider the matrices in the preceding example. Explain why $\rho(A + B) > \rho(A) + \rho(B)$, so the spectral radius function is not subadditive on M_n.

Example 2.4.8.4. Is there a converse of (2.4.8.1)? If the eigenvalues of A and B add, in some order, must A and B commute? The answer is no, even if the eigenvalues of αA and βB add, in some order, for all scalars α and β. This is an interesting phenomenon, and the characterization of such pairs of matrices is an unsolved problem! Consider the noncommuting matrices

$$A = \begin{bmatrix} 0 & 1 & 0 \\ 0 & 0 & -1 \\ 0 & 0 & 0 \end{bmatrix} \quad \text{and} \quad B = \begin{bmatrix} 0 & 0 & 0 \\ 1 & 0 & 0 \\ 0 & 1 & 0 \end{bmatrix}$$

for which $\sigma(A) = \sigma(B) = \{0\}$. Moreover, $p_{\alpha A + \beta B}(t) = t^3$, so $\sigma(\alpha A + \beta B) = \{0\}$ for all $\alpha, \beta \in \mathbf{C}$ and the eigenvalues add. If A and B were simultaneously upper triangularizable, the proof of (2.4.8.1) shows that the eigenvalues of AB would be products, in some order, of the eigenvalues of A and B. However, $\sigma(AB) = \{-1, 0, 1\}$ is not contained in $\sigma(A) \cdot \sigma(B) = \{0\}$, so A and B are not simultaneously triangularizable.

Corollary 2.4.8.5. *Suppose that $A, B \in M_n$ commute, $\sigma(A) = \{\alpha_1, \dots, \alpha_{d_1}\}$, and $\sigma(B) = \{\beta_1, \dots, \beta_{d_2}\}$. If $\alpha_i \neq -\beta_j$ for all i, j, then $A + B$ is nonsingular.*

Exercise. Verify (2.4.8.5) using (2.4.8.1).

Exercise. Suppose that $T = [t_{ij}]$ and $R = [r_{ij}]$ are n-by-n upper triangular matrices of the same size and let $p(s, t)$ be a polynomial in two noncommuting variables, that is, any linear combination of words in two noncommuting variables. Explain why $p(T, R)$ is upper triangular and its main diagonal entries (its eigenvalues) are $p(t_{11}, r_{11}), \dots, p(t_{nn}, r_{nn})$.

For complex matrices, simultaneous triangularization and simultaneous unitary triangularization are equivalent concepts.

Theorem 2.4.8.6. *Let $A_1, \ldots, A_m \in M_n$ be given. There is a nonsingular $S \in M_n$ such that $S^{-1} A_i S$ is upper triangular for all $i = 1, \ldots, m$ if and only if there is a unitary $U \in M_n$ such that $U^* A_i U$ is upper triangular for all $i = 1, \ldots, m$.*

Proof. Use (2.1.14) to write $S = QR$, in which Q is unitary and R is upper triangular. Then $T_i = S^{-1} A_i S = (QR)^{-1} A_i (QR) = R^{-1}(Q^* A_i Q) R$ is upper triangular, so $Q^* A_i Q = R T_i R^{-1}$ is upper triangular, as the product of three upper triangular matrices. $\square$

Simultaneous upper triangularizability of m matrices by similarity is completely characterized by the following theorem of McCoy. It involves a polynomial $p(t_1, \ldots, t_m)$ in m noncommuting variables, which is a linear combination of products of powers of the variables, that is, a linear combination of words in m noncommuting variables. The key observation is captured in the preceding exercise: If $T_1, \ldots, T_m$ are upper triangular, then so is $p(T_1, \ldots, T_m)$, and the main diagonals of $T_1, \ldots, T_m$ and $p(T_1, \ldots, T_m)$ exhibit specific orderings of their eigenvalues. For each $k = 1, \ldots, n$, the kth main diagonal entry of $p(T_1, \ldots, T_m)$ (an eigenvalue of $p(T_1, \ldots, T_m)$) is the same polynomial in the respective kth main diagonal entries of $T_1, \ldots, T_m$.

Theorem 2.4.8.7 (McCoy). *Let $m \geq 2$ and let $A_1, \ldots, A_m \in M_n$ be given. The following statements are equivalent:*

(a) *For every polynomial $p(t_1, \ldots, t_m)$ in m noncommuting variables and every $k, \ell = 1, \ldots, m$, $p(A_1, \ldots, A_m)(A_k A_\ell - A_\ell A_k)$ is nilpotent.*

(b) *There is a unitary $U \in M_n$ such that $U^* A_i U$ is upper triangular for each $i = 1, \ldots, m$.*

(c) *There is an ordering $\lambda_1^{(i)}, \ldots, \lambda_n^{(i)}$ of the eigenvalues of each of the matrices A_i, $i = 1, \ldots, m$ such that for any polynomial $p(t_1, \ldots, t_m)$ in m noncommuting variables, the eigenvalues of $p(A_1, \ldots, A_m)$ are $p(\lambda_i^{(1)}, \ldots, \lambda_i^{(m)})$, $i = 1, \ldots, n$.*

Proof. (b) $\Rightarrow$ (c): Let $T_k = U^* A_k U = [t_{ij}^{(k)}]$ be upper triangular and let $\lambda_1^{(k)} = t_{11}^{(k)}, \ldots, \lambda_n^{(k)} = t_{nn}^{(k)}$. Then the eigenvalues of $p(A_1, \ldots, A_m) = p(U T_1 U^*, \ldots, U T_m U^*) = U p(T_1, \ldots, T_m) U^*$ are the main diagonal entries of $p(T_1, \ldots, T_m)$, which are $p(\lambda_i^{(1)}, \ldots, \lambda_i^{(m)})$, $i = 1, \ldots, m$.

(c) $\Rightarrow$ (a) For any given polynomial $p(t_1, \ldots, t_m)$ in m noncommuting variables, consider the polynomials $q_{k\ell}(t_1, \ldots, t_m) = p(t_1, \ldots, t_m)(t_k t_\ell - t_\ell t_k)$, $k, \ell = 1, \ldots, m$ in m noncommuting variables. The eigenvalues of $q_{k\ell}(A_1, \ldots, A_m)$ are, according to (c), $q_{k\ell}(\lambda_i^{(1)}, \ldots, \lambda_i^{(m)}) = p(\lambda_i^{(1)}, \ldots, \lambda_i^{(m)})(\lambda_i^{(k)} \lambda_i^{(\ell)} - \lambda_i^{(\ell)} \lambda_i^{(k)}) = p(\lambda_i^{(1)}, \ldots, \lambda_i^{(m)}) \cdot 0 = 0$ for all $i = 1, \ldots, n$. Thus, each matrix $p(A_1, \ldots, A_m)(A_k A_\ell - A_\ell A_k)$ is nilpotent; see (2.4.2).

(a) $\Rightarrow$ (b): Suppose (see the following lemma) that $A_1, \ldots, A_m$ have a common unit eigenvector x. Subject to this assumption, we proceed by induction as in the proof of (2.3.3). Let U_1 be any unitary matrix that has x as its first column. Use U_1 to deflate each A_i in the same way:

$$\mathcal{A}_i = U_1^* A_i U_1 = \begin{bmatrix} \lambda_1^{(i)} & \bigstar \\ 0 & \tilde{A}_i \end{bmatrix}, \ \tilde{A}_i \in M_{n-1}, \ i = 1, \ldots, m \qquad (2.4.8.8)$$

Let $p(t_1, \ldots, t_m)$ be any given polynomial in m noncommuting variables. Then (a) ensures that the matrix

$$U^* p(A_1, \ldots, A_m)(A_k A_\ell - A_\ell A_k)U = p(\mathcal{A}_1, \ldots, \mathcal{A}_m)(\mathcal{A}_k \mathcal{A}_\ell - \mathcal{A}_\ell \mathcal{A}_k) \quad (2.4.8.9)$$

is nilpotent for each $k, \ell = 1, \ldots, m$. Partition each of the matrices (2.4.8.9) conformally to (2.4.8.8) and observe that its $1, 1$ entry is zero and its lower right block is $p(\tilde{A}_1, \ldots, \tilde{A}_m)(\tilde{A}_k \tilde{A}_\ell - \tilde{A}_\ell \tilde{A}_k)$, which is necessarily nilpotent. Thus, the matrices $\tilde{A}_1, \ldots, \tilde{A}_m \in M_{n-1}$ inherit property (a), and hence (b) follows by induction, as in (2.3.3). $\qquad\square$

We know that commuting matrices always have a common eigenvector (1.3.19). If the matrices $A_1, \ldots, A_m$ in the preceding theorem commute, then the condition (a) is trivially satisfied since $p(A_1, \ldots, A_m)(A_k A_\ell - A_\ell A_k) = 0$ for all $k, \ell = 1, \ldots, m$. The following lemma shows that the condition (a), weaker than commutativity, is sufficient to ensure existence of a common eigenvector.

Lemma 2.4.8.10. *Let $A_1, \ldots, A_m \in M_n$ be given. Suppose that for every polynomial $p(t_1, \ldots, t_m)$ in $m \geq 2$ noncommuting variables and every $k, \ell = 1, \ldots, m$, each of the matrices $p(A_1, \ldots, A_m)(A_k A_\ell - A_\ell A_k)$ is nilpotent. Then, for each given nonzero vector $x \in \mathbf{C}^n$, there is a polynomial $q(t_1, \ldots, t_m)$ in m noncommuting variables such that $q(A_1, \ldots, A_m)x$ is a common eigenvector of $A_1, \ldots, A_m$.*

Proof. We consider only the case $m = 2$, which illustrates all the features of the general case. Let $A, B \in M_n$, let $C = AB - BA$, and assume that $p(A, B)C$ is nilpotent for every polynomial $p(s, t)$ in two noncommuting variables. Let $x \in \mathbf{C}^n$ be any given nonzero vector. We claim that there is a polynomial $q(s, t)$ in two noncommuting variables such that $q(A, B)x$ is a common eigenvector of A and B.

Begin with (1.1.9) and let $g_1(t)$ be a polynomial such that $\xi_1 = g_1(A)x$ is an eigenvector of A: $A\xi_1 = \lambda \xi_1$.

Case I: Suppose that $Cp(B)\xi_1 = 0$ for *every* polynomial $p(t)$, that is,

$$ABp(B)\xi_1 = BAp(B)\xi_1 \text{ for every polynomial } p(t) \quad (2.4.8.11)$$

Using this identity with $p(t) = 1$ shows that $AB\xi_1 = BA\xi_1$. Now proceed by induction: Suppose $AB^k\xi_1 = B^k A\xi_1$ for some $k \geq 1$. Using (2.4.8.11) and the induction hypothesis, we compute

$$AB^{k+1}\xi_1 = AB \cdot B^k\xi_1 = BA \cdot B^k\xi_1 = B \cdot AB^k\xi_1$$
$$= B \cdot B^k A\xi_1 = B^{k+1}A\xi_1$$

We conclude that $AB^k\xi_1 = B^k A\xi_1$ for every $k \geq 1$, and hence $Ap(B)\xi_1 = p(B)A\xi_1 = p(B)\lambda\xi_1 = \lambda(p(B)\xi_1)$ for every polynomial $p(t)$. Thus, $p(B)\xi_1$ is an eigenvector of A if it is nonzero. Use (1.1.9) again to choose a polynomial $g_2(t)$ such that $g_2(B)\xi_1 = g_2(B)g_1(A)x$ is an eigenvector of B (necessarily nonzero). Let $q(s, t) = g_2(t)g_1(s)$. We have shown that $q(A, B)x$ is a common eigenvector of A and B, as claimed.

Case II: Suppose that there is *some* polynomial $f_1(t)$ such that $Cf_1(B)\xi_1 \neq 0$. Use (1.1.9) to find a polynomial $q_1(t)$ such that $\xi_2 = q_1(A)Cf_1(B)\xi_1$ is an eigenvector of A. If $Cp(B)\xi_2 = 0$ for every polynomial $p(t)$, then Case I permits us to construct the desired common eigenvector; otherwise, let $f_2(t)$ be a polynomial such that $Cf_2(B)\xi_2 \neq 0$ and let $q_2(t)$ be a polynomial such that $\xi_3 = q_2(A)Cf_2(B)\xi_2$ is an eigenvector of A.

Continue in this fashion to construct a sequence of eigenvectors

$$\xi_k = q_{k-1}(A)Cf_{k-1}(B)\xi_{k-1}, k = 2, 3, \ldots \tag{2.4.8.12}$$

of A until either (i) $Cp(B)\xi_k = 0$ for every polynomial $p(t)$ or (ii) $k = n + 1$. If (i) occurs for some $k \le n$, Case I permits us to construct the desired common eigenvector of A and B. If (i) is false for each $k = 1, 2, \ldots, n$, our construction produces $n + 1$ vectors $\xi_1, \ldots, \xi_{n+1}$ that must be linearly dependent, so there are $n + 1$ scalars $c_1, \ldots, c_{n+1}$, not all zero, such that $c_1\xi_1 + \cdots + c_{n+1}\xi_{n+1} = 0$. Let $r = \min\{i : c_i \ne 0\}$. Then

$$-c_r\xi_r = \sum_{i=r}^{n} c_{i+1}\xi_{i+1} = \sum_{i=r}^{n} c_{i+1}q_i(A)Cf_i(B)\xi_i$$

$$= c_{r+1}q_r(A)Cf_r(B)\xi_r$$

$$+ \sum_{i=r}^{n-1} c_{i+2}q_{i+1}(A)Cf_{i+1}(B)\xi_{i+1} \tag{2.4.8.13}$$

Using (2.4.8.12), the summand in (2.4.8.13) in which $i = r$ can be expanded to the expression

$$c_{r+2}q_{r+1}(A)Cf_{r+1}(B)q_r(A)Cf_r(B)\xi_r$$

In the same fashion, we can use (2.4.8.12) to expand each of the summands in (2.4.8.13) with $i = r + 1, r + 2, \ldots, n - 1$ to an expression of the form $h_i(A, B)Cf_r(B)\xi_r$, in which each $h_i(A, B)$ is a polynomial in A and B. We obtain in this way an identity of the form $-c_r\xi_r = p(A, B)Cf_r(B)\xi_r$, in which $p(s, t)$ is a polynomial in two noncommuting variables. This means that $f_r(B)\xi_r$ is an eigenvector of $p(A, B)C$ associated with the nonzero eigenvalue $-c_r$, in contradiction to the hypothesis that $p(A, B)C$ is nilpotent. This contradiction shows that (i) is true for some $k \le n$, and hence A and B have a common eigenvector of the asserted form. □

We have stated McCoy's theorem 2.4.8.7 for complex matrices, but if we restate (b) to assert only simultaneous similarity (not simultaneous unitary similarity), then the theorem is valid for matrices and polynomials over any subfield of $\mathbf{C}$ that contains the eigenvalues of all the matrices $A_1, \ldots, A_m$.

2.4.9 Continuity of eigenvalues. Schur's unitary triangularization theorem can be used to prove a basic and widely useful fact: The eigenvalues of a square real or complex matrix depend continuously on its entries. Both aspects of Schur's theorem – *unitary* and *triangular* – play key roles in the proof. The following lemma encapsulates the fundamental principle involved.

Lemma 2.4.9.1. *Let an infinite sequence of matrices* $A_1, A_2, \ldots \in M_n$ *be given and suppose that* $\lim_{k\to\infty} A_k = A$ *(entrywise convergence). Then there is an infinite sequence of positive integers* $k_1 < k_2 < \cdots$ *and unitary matrices* $U_{k_i} \in M_n$ *for* $i = 1, 2, \ldots$ *such that*

(a) $T_i = U_{k_i}^* A_{k_i} U_{k_i}$ *is upper triangular for all* $i = 1, 2, \ldots$;
(b) $U = \lim_{i\to\infty} U_{k_i}$ *exists and is unitary;*
(c) $T = U^* A U$ *is upper triangular; and*
(d) $\lim_{i\to\infty} T_i = T$.

Proof. Using (2.3.1), for each $k = 1, 2, \ldots$, let $U_k \in M_n$ be unitary and such that $U_k^* A_k U_k$ is upper triangular. Lemma 2.1.8 ensures that there is an infinite subsequence $U_{k_1}, U_{k_2}, \ldots$ and a unitary U such that $U_{k_i} \to U$ as $i \to \infty$. Then convergence of each of its three factors ensures that the product $T_i = U_{k_i}^* A_{k_i} U_{k_i}$ converges to a limit $T = U^* A U$, which is upper triangular because each T_i is upper triangular.   □

In the preceding argument, the main diagonal of each upper triangular matrix $T, T_1, T_2, \ldots$ is a particular presentation (think of it as an n-vector) of the eigenvalues of $A, A_{k_1}, A_{k_2}, \ldots$, respectively. The entrywise convergence $T_i \to T$ ensures that among the up to $n!$ different ways of presenting the eigenvalues of each of the matrices $A, A_{k_1}, A_{k_2}, \ldots$ as an n-vector, there is at least one presentation for each matrix such that the respective vectors of eigenvalues converge to a vector whose entries comprise all the eigenvalues of A. It is in this sense, formalized in the following theorem, that the eigenvalues of a square real or complex matrix depend continuously on its entries.

Theorem 2.4.9.2. *Let an infinite sequence $A_1, A_2, \ldots \in M_n$ be given and suppose that $\lim_{k \to \infty} A_k = A$ (entrywise convergence). Let $\lambda(A) = [\lambda_1(A) \ \ldots \ \lambda_n(A)]^T$ and $\lambda(A_k) = [\lambda_1(A_k) \ \ldots \ \lambda_n(A_k)]^T$ be given presentations of the eigenvalues of A and A_k, respectively, for $k = 1, 2, \ldots$. Let $S_n = \{\pi : \pi \text{ is a permutation of } \{1, 2, \ldots, n\}\}$. Then for each given $\varepsilon > 0$ there exists a positive integer $N = N(\varepsilon)$ such that*

$$\min_{\pi \in S_n} \max_{i=1,\ldots,n} \left|\lambda_{\pi(i)}(A_k) - \lambda_i(A)\right| \leq \varepsilon \text{ for all } k \geq N \qquad (2.4.9.3)$$

Proof. If the assertion (2.4.9.3) is false, then there is some $\varepsilon_0 > 0$ and an infinite sequence of positive integers $k_1 < k_2 < \cdots$ such that for every $j = 1, 2, \ldots$ we have

$$\max_{i=1,\ldots,n} \left|\lambda_{\pi(i)}(A_{k_j}) - \lambda_i(A)\right| > \varepsilon_0 \text{ for every } \pi \in S_n \qquad (2.4.9.4)$$

However, (2.4.9.1) ensures that there is an infinite sub-subsequence $k_1 \leq k_{j_1} < k_{j_2} < \cdots$, unitary matrices $U, U_{k_{j_1}}, U_{k_{j_2}}, \ldots$, and upper triangular matrices $T = U^* A U$ and $T_p = U_{k_{j_p}}^* A_{k_{j_p}} U_{k_{j_p}}$ for $p = 1, 2, \ldots$ such that all of the entries of T_p (in particular, the main diagonal entries) converge to the corresponding entries of T as $p \to \infty$. Since the vectors of main diagonal entries of $T, T_1, T_2, \ldots$ are obtained, respectively, from the given presentations of eigenvalues $\lambda(A), \lambda(A_{k_{j_1}}), \lambda(A_{k_{j_2}}), \ldots$ by permuting their entries, the entrywise convergence we have observed contradicts (2.4.9.4) and proves the theorem.   □

The existential assertion "for each given $\varepsilon > 0$ there exists a positive integer $N = N(\varepsilon)$" in the preceding theorem can be replaced by an explicit bound; see (D2) in Appendix D.

2.4.10 Eigenvalues of a rank-one perturbation. It is often useful to know that any one eigenvalue of a matrix can be shifted arbitrarily by a rank-one perturbation, without disturbing the rest of the eigenvalues.

Theorem 2.4.10.1 (A. Brauer). *Suppose that $A \in M_n$ has eigenvalues $\lambda, \lambda_2, \ldots, \lambda_n$, and let x be a nonzero vector such that $Ax = \lambda x$. Then for any $v \in \mathbf{C}^n$ the eigenvalues of $A + xv^*$ are $\lambda + v^* x, \lambda_2, \ldots, \lambda_n$.*

Proof. Let $\xi = x/\|x\|_2$ and let $U = [\xi \; u_2 \; \ldots \; u_n]$ be unitary. Then the proof of (2.3.1) shows that

$$U^*AU = \begin{bmatrix} \lambda & \bigstar \\ 0 & A_1 \end{bmatrix}$$

in which $A_1 \in M_{n-1}$ has eigenvalues $\lambda_2, \ldots, \lambda_n$. Also,

$$U^*xv^*U = \begin{bmatrix} \xi^*x \\ u_2^*x \\ \vdots \\ u_n^*x \end{bmatrix} v^*U = \begin{bmatrix} \|x\|_2 \\ 0 \\ \vdots \\ 0 \end{bmatrix} \begin{bmatrix} v^*\xi & v^*u_2 & \cdots & v^*u_n \end{bmatrix}$$

$$= \begin{bmatrix} \|x\|_2 \, v^*\xi & \bigstar \\ 0 & 0 \end{bmatrix} = \begin{bmatrix} v^*x & \bigstar \\ 0 & 0 \end{bmatrix}$$

Therefore,

$$U^*(A + xv^*)U = \begin{bmatrix} \lambda + v^*x & \bigstar \\ 0 & A_1 \end{bmatrix}$$

has eigenvalues $\lambda + v^*x, \lambda_2, \ldots, \lambda_n$. $\qquad\square$

For a different approach to this result, see (1.2.8).

2.4.11 The complete principle of biorthogonality. The *principle of biorthogonality* says that left and right eigenvectors associated with different eigenvalues are orthogonal; see (1.4.7(a)). We now address all the possibilities for left and right eigenvectors.

Theorem 2.4.11.1. *Let $A \in M_n$, unit vectors $x, y \in \mathbf{C}^n$, and $\lambda, \mu \in \mathbf{C}$ be given.*

(a) *If $Ax = \lambda x$, $y^*A = \mu y^*$, and $\lambda \neq \mu$, then $y^*x = 0$. Let $U = [x \; y \; u_3 \; \ldots \; u_n] \in M_n$ be unitary. Then*

$$U^*AU = \begin{bmatrix} \lambda & \bigstar & \bigstar \\ 0 & \mu & 0 \\ 0 & \bigstar & A_{n-2} \end{bmatrix}, \quad A_{n-2} \in M_{n-2} \qquad (2.4.11.2)$$

(b) *Suppose that $Ax = \lambda x$, $y^*A = \lambda y^*$, and $y^*x = 0$. Let $U = [x \; y \; u_3 \; \ldots \; u_n] \in M_n$ be unitary. Then*

$$U^*AU = \begin{bmatrix} \lambda & \bigstar & \bigstar \\ 0 & \lambda & 0 \\ 0 & \bigstar & A_{n-2} \end{bmatrix}, \quad A_{n-2} \in M_{n-2} \qquad (2.4.11.3)$$

and the algebraic multiplicity of λ is at least two.

(c) *Suppose that $Ax = \lambda x$, $y^*A = \lambda y^*$, and $y^*x \neq 0$. Let $S = [x \; S_1] \in M_n$, in which the columns of S_1 are any given basis for the orthogonal complement of y. Then S is nonsingular, the first column of S^{-*} is a nonzero scalar multiple of y, and $S^{-1}AS$ has the block form*

$$\begin{bmatrix} \lambda & 0 \\ 0 & A_{n-1} \end{bmatrix}, \quad A_{n-1} \in M_{n-1} \qquad (2.4.11.4)$$

If the geometric multiplicity of λ is 1, then its algebraic multiplicity is also 1. Conversely, if A is similar to a block matrix of the form (2.4.11.4), then it has a nonorthogonal pair of left and right eigenvectors associated with the eigenvalue λ.

(d) *Suppose that* $Ax = \lambda x$, $y^*A = \lambda y^*$, *and* $x = y$ *(such an x is called a* normal eigenvector*). Let* $U = [x \ U_1] \in M_n$ *be unitary. Then* U^*AU *has the block form* *(2.4.11.4).*

Proof. (a) Compared with the reduction in (2.3.1), the extra zeroes in the second row of (2.4.11.2) come from the left eigenvector: $y^*Au_i = \mu y^*u_i = 0$ for $i = 3, \ldots, n$. (b) The zero pattern in (2.4.11.3) is the same as that in (2.4.11.2), and for the same reason. For the assertion about the algebraic multiplicity, see (1.2.P14). (c) See (1.4.7) and (1.4.12). (d) Compared with the reduction in (2.3.1), the extra zeroes in the first row of (2.4.11.4) appear because x is also a left eigenvector: $x^*AU_1 = \lambda x^*U_1 = 0$. $\qquad\square$

Problems

2.4.P1 Suppose that $A = [a_{ij}] \in M_n$ has n distinct eigenvalues. Use (2.4.9.2) to show that there is an $\epsilon > 0$ such that every $B = [b_{ij}] \in M_n$ with $\sum_{i,j=1}^n |a_{ij} - b_{ij}|^2 < \epsilon$ has n distinct eigenvalues. Conclude that the set of matrices with distinct eigenvalues is an *open* subset of M_n.

2.4.P2 Why is the rank of an upper triangular matrix at least as large as the number of its nonzero main diagonal entries? Let $A = [a_{ij}] \in M_n$, and suppose A has exactly $k \geq 1$ nonzero eigenvalues $\lambda_1, \ldots, \lambda_k$. Write $A = UTU^*$, in which U is unitary, $T = [t_{ij}]$ is upper triangular, and $|t_{11}| \geq \cdots \geq |t_{nn}|$. Show that rank $A \geq k$, with equality if A is diagonalizable. Explain why

$$|\sum_{i=1}^k \lambda_i|^2 \leq k \sum_{i=1}^k |\lambda_i|^2 = k \sum_{i=1}^n |t_{ii}|^2 \leq k \sum_{i,j=1}^n |t_{ij}|^2 = k \sum_{i,j=1}^k |a_{ij}|^2$$

and conclude that rank $A \geq |\operatorname{tr} A|^2/(\operatorname{tr} A^*A)$, with equality if and only if $T = aI_k \oplus 0_{n-k}$ for some nonzero $a \in \mathbf{C}$.

2.4.P3 Our proof of (2.4.3.2) relies on the fact that complex matrices have eigenvalues, but neither the definition of the characteristic polynomial nor the substitution $p_A(t) \to p_A(A)$ involves eigenvalues or any special properties of the complex field. In fact, the Cayley–Hamilton theorem is valid for matrices whose entries come from a *commutative ring with unit*, examples of which are the ring of integers modulo some integer k (which is a field if and only if k is prime) and the ring of polynomials in one or more formal indeterminants with complex coefficients. Provide details for the following proof of (2.4.3.2). Observe that the algebraic operations used in the proof involve addition and multiplication, but no division operations or roots of polynomial equations are involved.

(a) Start with the fundamental identity $(tI - A)(\operatorname{adj}(tI - A)) = \det(tI - A)I = p_A(t)I$ (0.8.2) and write

$$p_A(t)I = It^n + a_{n-1}It^{n-1} + a_{n-2}It^{n-2} + \cdots + a_1It + a_0I \qquad (2.4.12)$$

a polynomial in t of degree n with matrix coefficients; each coefficient is a scalar matrix.

(b) Explain why $\operatorname{adj}(tI - A)$ is a matrix whose entries are polynomials in t of degree at most $n - 1$, and hence it can be written as

$$\operatorname{adj}(tI - A) = A_{n-1}t^{n-1} + A_{n-2}t^{n-2} + \cdots + A_1t + A_0 \qquad (2.4.13)$$

in which $A_0 = (-1)^{n-1} \operatorname{adj} A$ and each A_k is an n-by-n matrix whose entries are polynomial functions of the entries of A.

(c) Use (2.4.13) to compute the product $(tI - A)(\operatorname{adj}(tI - A))$ as

$$A_{n-1}t^n + (A_{n-2} - AA_{n-1})t^{n-1} + \cdots + (A_0 - AA_1)t - AA_0 \qquad (2.4.14)$$

(d) Equate corresponding coefficients of (2.4.12) and (2.4.14) to obtain $n + 1$ equations

$$A_{n-1} = I$$
$$A_{n-2} - AA_{n-1} = a_{n-1}I$$
$$\vdots \qquad\qquad (2.4.15)$$
$$A_0 - AA_1 = a_1 I$$
$$-AA_0 = a_0 I$$

(e) For each $k = 1, \ldots, n$, left-multiply the kth equation in (2.4.15) by A^{n-k+1}, add all $n + 1$ equations, and obtain the Cayley–Hamilton theorem $0 = p_A(A)$.

(f) For each $k = 1, \ldots, n - 1$, left-multiply the kth equation in (2.4.15) by A^{n-k}, add only the first n equations, and obtain the identity

$$\operatorname{adj} A = (-1)^{n-1}(A^{n-1} + a_{n-1}A^{n-2} + \cdots + a_2 A + a_1 I) \qquad (2.4.16)$$

Thus, $\operatorname{adj} A$ is a polynomial in A whose coefficients (except for $a_0 = (-1)^n \det A$) are the same as the coefficients in $p_A(t)$, but in reversed order.

(g) Use (2.4.15) to show that the matrix coefficients in the right-hand side of (2.4.13) are $A_{n-1} = I$ and

$$A_{n-k-1} = A^k + a_{n-1}A^{k-1} + \cdots + a_{n-k+1}A + a_{n-k}I \qquad (2.4.17)$$

for $k = 1, \ldots, n - 1$.

2.4.P4 Let $A, B \in M_n$ and suppose that A commutes with B. Explain why B commutes with $\operatorname{adj} A$ and why $\operatorname{adj} A$ commutes with $\operatorname{adj} B$. If A is nonsingular, deduce that B commutes with A^{-1}.

2.4.P5 Consider the matrices $\begin{bmatrix} 0 & \epsilon \\ 0 & 0 \end{bmatrix}$ and explain why there can be nondiagonalizable matrices arbitrarily close to a given diagonalizable matrix. Use (2.4.P1) to explain why this cannot happen if the given matrix has distinct eigenvalues.

2.4.P6 Show that for

$$A = \begin{bmatrix} 1 & 0 & 0 \\ 0 & 2 & 0 \\ 0 & 0 & 3 \end{bmatrix} \quad \text{and} \quad B = \begin{bmatrix} -2 & 1 & 2 \\ -1 & -2 & -1 \\ 1 & 1 & 1 \end{bmatrix}$$

$\sigma(aA + bB) = \{a - 2b, 2a - 2b, 3a + b\}$ for all scalars $a, b \in \mathbf{C}$, but A and B are not simultaneously similar to upper triangular matrices. What are the eigenvalues of AB?

2.4.P7 Use the criterion in (2.3.P6) to show that the two matrices in (2.4.8.4) cannot be simultaneously upper triangularized. Apply the same test to the two matrices in (2.4.P6).

2.4.P8 An observation in the spirit of McCoy's theorem can sometimes be useful in showing that two matrices are *not* unitarily similar. Let $p(t, s)$ be a polynomial with complex coefficients in two noncommuting variables, and let $A, B \in M_n$ be unitarily similar with

$A = UBU^*$ for some unitary $U \in M_n$. Explain why $p(A, A^*) = Up(B, B^*)U^*$. Conclude that if A and B are unitarily similar, then $\operatorname{tr} p(A, A^*) = \operatorname{tr} p(B, B^*)$ for every complex polynomial $p(t, s)$ in two noncommuting variables. How is this related to (2.2.6)?

2.4.P9 Let $p(t) = t^n + a_{n-1}t^{n-1} + \cdots + a_1 t + a_0$ be a given monic polynomial of degree n with zeroes $\lambda_1, \ldots, \lambda_n$. Let $\mu_k = \lambda_1^k + \cdots + \lambda_n^k$ denote the kth moments of the zeroes, $k = 0, 1, \ldots$ (take $\mu_0 = n$). Provide details for the following proof of *Newton's identities*:

$$ka_{n-k} + a_{n-k+1}\mu_1 + a_{n-k+2}\mu_2 + \cdots + a_{n-1}\mu_{k-1} + \mu_k = 0 \qquad (2.4.18)$$

for $k = 1, 2, \ldots, n-1$ and

$$a_0\mu_k + a_1\mu_{k+1} + \cdots + a_{n-1}\mu_{n+k-1} + \mu_{n+k} = 0 \qquad (2.4.19)$$

for $k = 0, 1, 2, \ldots$. First show that if $|t| > R = \max\{|\lambda_i| : i = 1, \ldots, n\}$, then $(t - \lambda_i)^{-1} = t^{-1} + \lambda_i t^{-2} + \lambda_i^2 t^{-3} + \cdots$ and hence

$$f(t) = \sum_{i=1}^{n} (t - \lambda_i)^{-1} = nt^{-1} + \mu_1 t^{-2} + \mu_2 t^{-3} + \cdots \quad \text{for} \quad |t| > R$$

Now show that $p'(t) = p(t)f(t)$ and compare coefficients. Newton's identities show that the first n moments of the zeroes of a monic polynomial of degree n uniquely determine its coefficients. See (3.3.P18) for a matrix-analytic approach to Newton's identities.

2.4.P10 Show that $A, B \in M_n$ have the same characteristic polynomials, and hence the same eigenvalues, if and only if $\operatorname{tr} A^k = \operatorname{tr} B^k$ for all $k = 1, 2, \ldots, n$. Deduce that A is nilpotent if and only if $\operatorname{tr} A^k = 0$ for all $k = 1, 2, \ldots, n$.

2.4.P11 Let $A, B \in M_n$ be given and consider their *commutator* $C = AB - BA$. Show that (a) $\operatorname{tr} C = 0$. (b) Consider $A = \begin{bmatrix} 0 & 0 \\ 1 & 0 \end{bmatrix}$ and $B = \begin{bmatrix} 0 & 1 \\ 0 & 0 \end{bmatrix}$ and show that a commutator need not be nilpotent; that is, a commutator can have some nonzero eigenvalues, but their sum must be zero. (c) If rank $C \leq 1$, show that C is nilpotent. (d) If rank $C = 0$, explain why A and B are simultaneously unitarily triangularizable. (e) If rank $C = 1$, *Laffey's theorem* says that A and B are simultaneously triangularizable by similarity. Provide details for the following sketch of a proof of Laffey's theorem: We may assume that A is singular (replace A by $A - \lambda I$, if necessary). If the null space of A is B-invariant, then it is a common nontrivial invariant subspace, so A and B are simultaneously similar to a block matrix of the form (1.3.17). If the null space of A is *not* B-invariant, let $x \neq 0$ be such that $Ax = 0$ and $ABx \neq 0$. Then $Cx = ABx$ so there is a $z \neq 0$ such that $C = ABxz^T$. For any y, $(z^T y)ABx = Cy = ABy - BAy$, $BAy = AB(y - (z^T y)x)$, and hence range $BA \subset$ range $AB \subset$ range A, range A is B-invariant, and A and B are simultaneously similar to a block matrix of the form (1.3.17). Now assume that $A = \begin{bmatrix} A_{11} & A_{12} \\ 0 & A_{22} \end{bmatrix}$, $B = \begin{bmatrix} B_{11} & B_{12} \\ 0 & B_{22} \end{bmatrix}$, $A_{11}, B_{11} \in M_k$, $1 \leq k < n$, and $C = \begin{bmatrix} A_{11}B_{11} - B_{11}A_{11} & X \\ 0 & A_{22}B_{22} - B_{22}A_{22} \end{bmatrix}$ has rank one. At least one of the diagonal blocks of C is zero, so we may invoke (2.3.3). If one diagonal block has rank one and size greater than one, repeat the reduction. A 1-by-1 diagonal block cannot have rank one.

2.4.P12 Let $A, B \in M_n$ and let $C = AB - BA$. This problem examines some consequences of assuming that C commutes with either A or B, or both. (a) If C commutes with A, explain why $\operatorname{tr}(C^k) = \operatorname{tr}(C^{k-1}(AB - BA)) = \operatorname{tr}(AC^{k-1}B - C^{k-1}BA) = 0$ for all $k = 2, \ldots, n$. Deduce *Jacobson's Lemma* from (2.4.P10): *C is nilpotent if it commutes* *with either* A *or* B. (b) If $n = 2$, show that C commutes with both A and B if and only if $C = 0$, that is, if and only if A commutes with B. (c) If A is diagonalizable, show that C commutes with A if and only if $C = 0$. (d) A and B are said to *quasicommute* if they both commute with C. If A and B quasicommute and $p(s, t)$ is any polynomial in two noncommuting variables, show that $p(A, B)$ commutes with C, invoke (2.4.8.1), and use (a) to show that $p(A, B)C$ is nilpotent. (e) If A and B quasicommute, use (2.4.8.7) to show that A and B are simultaneously triangularizable. This is known as the *little McCoy theorem*. (f) Let $n = 2$. If C commutes with A, (3.2.P32) ensures that A and B (and hence also B and C) are simultaneously triangularizable. Show that A and B are simultaneously triangularizable if and only if $C^2 = 0$. (g) The situation is different for $n = 3$. Consider

$$A = \begin{bmatrix} 0 & 1 & 0 \\ 0 & 0 & 0 \\ 0 & 0 & 0 \end{bmatrix} \quad \text{and} \quad B = \begin{bmatrix} 0 & 0 & 0 \\ 0 & 0 & 1 \\ 1 & 0 & 0 \end{bmatrix}$$

Show that (i) A commutes with C, so A and C are simultaneously triangularizable; (ii) B does not commute with C; and (iii) B and C (and hence also A and B) are not simultaneously triangularizable. (h) Let $n = 3$. Another theorem of Laffey says that A and B are simultaneously triangularizable if and only if C, AC^2, BC^2, and at least one of A^2C^2, ABC^2, and B^2C^2 is nilpotent. Deduce from this theorem that A and B are simultaneously triangularizable if $C^2 = 0$. Give an example to show that the condition $C^3 = 0$ does *not* imply that A and B are simultaneously triangularizable.

2.4.P13 Provide details for the following alternative proof of (2.4.4.1) on the linear matrix equation $AX - XB = C$: Suppose that $A \in M_n$ and $B \in M_m$ have no eigenvalues in common. Consider the linear transformations $T_1, T_2 : M_{n,m} \to M_{n,m}$ defined by $T_1(X) = AX$ and $T_2(X) = XB$. Show that T_1 and T_2 commute, and deduce from (2.4.8.1) that the eigenvalues of $T = T_1 - T_2$ are differences of eigenvalues of T_1 and T_2. Argue that λ is an eigenvalue of T_1 if and only if there is a nonzero $X \in M_{n,m}$ such that $AX - \lambda X = 0$, which can happen if and only if λ is an eigenvalue of A (and every nonzero column of X is a corresponding eigenvector). The spectra of T_1 and A are therefore the same, and similarly for T_2 and B. Thus, T is nonsingular if A and B have no eigenvalues in common. If x is an eigenvector of A associated with the eigenvalue λ and y is a left eigenvector of B associated with the eigenvalue μ, consider $X = xy^*$; show that $T(X) = (\lambda - \mu)X$, and conclude that the spectrum of T consists of all possible differences of eigenvalues of A and B.

2.4.P14 Let $A \in M_n$ and suppose that rank $A = r$. Show that A is unitarily similar to an upper triangular matrix whose first r rows are linearly independent and whose last $n - r$ rows are zero.

2.4.P15 Let $A, B \in M_n$ and consider the polynomial in two complex variables defined by $p_{A,B}(s, t) = \det(tB - sA)$. (a) Suppose that A and B are simultaneously triangularizable, with $A = S\mathcal{A}S^{-1}$, $B = S\mathcal{B}S^{-1}$, $\mathcal{A}$ and $\mathcal{B}$ upper triangular, diag $\mathcal{A} = (\alpha_1, \ldots, \alpha_n)$, and diag $\mathcal{B} = (\beta_1, \ldots, \beta_n)$. Show that $p_{A,B}(s, t) = \det(t\mathcal{B} - s\mathcal{A}) = \prod_{i=1}^{n}(t\beta_i - s\alpha_i)$. (b) Now suppose that A and B commute. Deduce that

$$p_{A,B}(B, A) = \prod_{i=1}^{n}(\beta_i A - \alpha_i B) = S(\prod_{i=1}^{n}(\beta_i \mathcal{A} - \alpha_i \mathcal{B}))S^{-1}$$

Explain why the i, i entry of the upper triangular matrix $\beta_i \mathcal{A} - \alpha_i \mathcal{B}$ is zero. (c) Use Lemma 2.4.3.1 to show that $p_{A,B}(B, A) = 0$ if A and B commute. Explain why this identity is a two-variable generalization of the Cayley–Hamilton theorem. (d) Suppose that $A, B \in M_n$ commute. For $n = 2$, show that $p_{A,B}(B, A) = (\det B)A^2 - (\operatorname{tr}(A \operatorname{adj} B))AB + (\det A)B^2$. For $n = 3$, show that $p_{A,B}(B, A) = (\det B)A^3 - (\operatorname{tr}(A \operatorname{adj} B))A^2B + (\operatorname{tr}(B \operatorname{adj} A))AB^2 - (\det A)B^3$. What are these identities for $B = I$? (e) Calculate $\det(tB - sA)$ for the matrices in Examples 2.4.8.3 and 2.4.8.4; discuss. (f) Why did we assume commutativity in (b) but not in (a)?

2.4.P16 Let λ be an eigenvalue of $A = \begin{bmatrix} a & b \\ c & d \end{bmatrix} \in M_2$. (a) Explain why $\mu = a + d - \lambda$ is an eigenvalue of A. (b) Explain why $(A - \lambda I)(A - \mu I) = (A - \mu I)(A - \lambda I) = 0$. (c) Deduce that any nonzero column of $\begin{bmatrix} a - \lambda & b \\ c & d - \lambda \end{bmatrix}$ is an eigenvector of A associated with μ, and any nonzero row is the conjugate transpose of a left eigenvector associated with λ. (d) Deduce that any nonzero column of $\begin{bmatrix} \lambda - d & b \\ c & \lambda - a \end{bmatrix}$ is an eigenvector of A associated with μ and any nonzero row is the conjugate transpose of a left eigenvector associated with λ.

2.4.P17 Let $A, B \in M_n$ be given and consider $\mathcal{A}(A, B)$, the subalgebra of M_n generated by A and B (see (1.3.P36)). Then $\mathcal{A}(A, B)$ is a subspace of M_n, so $\dim \mathcal{A}(A, B) \leq n^2$. Consider $n = 2$, $A = \begin{bmatrix} 0 & 1 \\ 0 & 0 \end{bmatrix}$, and $B = A^T$; show in this case that $\dim \mathcal{A}(A, B) = n^2$. Use the Cayley–Hamilton theorem to show that $\dim \mathcal{A}(A, I) \leq n$ for any $A \in M_n$. *Gerstenhaber's theorem* says that if $A, B \in M_n$ commute, then $\dim \mathcal{A}(A, B) \leq n$.

2.4.P18 Suppose that $A = \begin{bmatrix} A_{11} & A_{12} \\ 0 & A_{22} \end{bmatrix} \in M_n$, $A_{11} \in M_k$, $1 \leq k < n$, $A_{22} \in M_{n-k}$. Show that A is nilpotent if and only if both A_{11} and A_{22} are nilpotent.

2.4.P19 Let $n \geq 3$ and $k \in \{1, \ldots, n-1\}$ be given. (a) Suppose that $A = \begin{bmatrix} A_{11} & A_{12} \\ 0 & A_{22} \end{bmatrix} \in M_n$, $B = \begin{bmatrix} B_{11} & B_{12} \\ 0 & B_{22} \end{bmatrix} \in M_n$, $A_{11}, B_{11} \in M_k$, and $A_{22}, B_{22} \in M_{n-k}$. Show that A and B are simultaneously upper triangularizable if and only if both (i) A_{11} and B_{11} are simultaneously upper triangularizable and (ii) A_{22} and B_{22} are simultaneously upper triangularizable. (b) If $m \geq 3$, $\mathcal{F} = \{A_1, \ldots, A_m\} \subset M_n$, and each $A_j = \begin{bmatrix} A_{j1} & A_{j2} \\ 0 & A_{j3} \end{bmatrix}$ with $A_{j1} \in M_k$, show that $\mathcal{F}$ is simultaneously upper triangularizable if and only if each of $\{A_{11}, \ldots, A_{m1}\}$ and $\{A_{13}, \ldots, A_{m3}\}$ is (separately) simultaneously upper triangularizable.

2.4.P20 Suppose that $A, B \in M_n$ and $AB = 0$, so $C = AB - BA = -BA$. Let $p(s, t)$ be a polynomial in two noncommuting variables. (a) If $p(0, 0) = 0$, show that $Ap(A, B)B = 0$ and hence $(p(A, B)C)^2 = 0$. (b) Show that $C^2 = 0$. (c) Use (2.4.8.7) to show that A and B are simultaneously upper triangularizable. (d) Are $\begin{bmatrix} -3 & 3 \\ -4 & 4 \end{bmatrix}$ and $\begin{bmatrix} 2 & -1 \\ 2 & -1 \end{bmatrix}$ simultaneously upper triangularizable?

2.4.P21 Let $A \in M_n$ have eigenvalues $\lambda_1, \ldots, \lambda_n$. The Hankel matrix $K = [\operatorname{tr} A^{i+j-2}]_{i,j=1}^n$ is the *moment matrix* associated with A. We always take $A^0 = I$, so $\operatorname{tr} A^0 = n$. (a) Show that $K = VV^T$, in which $V \in M_n$ is the Vandermonde matrix (0.9.11.1) whose jth column is $[1 \ \lambda_j \ \lambda_j^2 \ \ldots \ \lambda_j^{m-1}]^T$, $j = 1, \ldots, n$. (b) Explain why $\det K = (\det V)^2 = \prod_{i<j}(\lambda_j - \lambda_i)^2$; this product is the *discriminant* of A. (c) Conclude that the eigenvalues of A are distinct if and only if its moment matrix is nonsingular. (d) Explain why K (and hence the discriminant of A) is invariant under similarity of A. (e) Calculate the determinant of the moment matrix

of $A = \begin{bmatrix} a & b \\ c & d \end{bmatrix} \in M_2$; verify that it is the discriminant of A, as computed in the exercise containing (1.2.4b). (f) Consider the real matrix

$$A = \begin{bmatrix} a & b & 0 \\ 0 & 0 & c \\ d & -e & 0 \end{bmatrix}, \quad a, b, c, d, e \text{ are positive} \tag{2.4.20}$$

whose zero entries are specified, but only the sign pattern of the remaining entries is specified. The moment matrix of A is

$$K = \begin{bmatrix} 3 & a & a^2 - 2ce \\ a & a^2 - 2ce & a^3 + 3bcd \\ a^2 - 2ce & a^3 + 3bcd & a^4 + 4bdac + 2e^2c^2 \end{bmatrix}$$

and $\det K = -27b^2c^2d^2 - 4c^3e^3 - 4a^4ce - 8a^2c^2e^2 - 4a^3bcd - 36abc^2de$. Explain why A always has three distinct eigenvalues.

2.4.P22 Suppose that $A \in M_n$ has d *distinct* eigenvalues $\mu_1, \ldots, \mu_d$ with respective multiplicities $\nu_1, \ldots, \nu_d$. The matrix $K_m = [\text{tr } A^{i+j-2}]_{i,j=1}^m$ is the *moment matrix of order m* associated with A, $m = 1, 2, \ldots$; if $m \leq n$, it is a leading principal submatrix of the moment matrix K in the preceding problem. Let $v_j^{(m)} = [1 \; \mu_j \; \mu_j^2 \; \ldots \; \mu_j^{m-1}]^T$, $j = 1, \ldots, d$ and form the m-by-d matrix $V_m = [v_1^{(m)} \; \ldots \; v_d^{(m)}]$. Let $D = \text{diag}(\nu_1, \ldots, \nu_d) \in M_d$. Show that (a) V_m has row rank m if $m \leq d$ and has column rank d if $m \geq d$; (b) $K_m = V_m D V_m^T$; (c) if $1 \leq p < q$, K_p is a leading principal submatrix of K_q; (d) K_d is nonsingular; (e) rank $K_m = d$ if $m \geq d$; (f) $d = \max\{m \geq 1 : K_m \text{ is nonsingular}\}$ but K_p can be singular for some $p < d$; (g) K_d is nonsingular and each of $K_{d+1}, \ldots, K_n, K_{n+1}$ are all singular; (h) $K_n = K$, the moment matrix in the preceding problem; (i) rank K is exactly the number of distinct eigenvalues of A.

2.4.P23 Suppose that $T = [t_{ij}] \in M_n$ is upper triangular. Show that adj $T = [\tau_{ij}]$ is upper triangular and has main diagonal entries $\tau_{ii} = \prod_{j \neq i} t_{jj}$.

2.4.P24 Let $A \in M_n$ have eigenvalues $\lambda_1, \ldots, \lambda_n$. Show that the eigenvalues of adj A are $\prod_{j \neq i} \lambda_j$, $i = 1, \ldots, n$.

2.4.P25 Let $A, B \in M_2$ and suppose that λ_1, λ_2 are the eigenvalues of A. (a) Show that A is unitarily similar to $\begin{bmatrix} \lambda_1 & x \\ 0 & \lambda_2 \end{bmatrix}$ in which $x \geq 0$ and $x^2 = \text{tr } AA^* - |\lambda_1|^2 - |\lambda_2|^2$. (b) Show that A is unitarily similar to B if and only if $\text{tr } A = \text{tr } B$, $\text{tr } A^2 = \text{tr } B^2$, and $\text{tr } AA^* = \text{tr } BB^*$.

2.4.P26 Let $B \in M_{n,k}$ and $C \in M_{k,n}$. Show that $BCp(BC) = Bp(CB)C$ for any polynomial $p(t)$.

2.4.P27 Let $A \in M_n$ be given. (a) If $A = BC$ and $B, C^T \in M_{n,k}$, use (2.4.3.2) to show that there is a polynomial $q(t)$ of degree at most $k + 1$ such that $q(A) = 0$.

2.4.P28 Suppose $A \in M_n$ is singular and let $r = \text{rank } A$. Show that there is a polynomial $p(t)$ of degree at most $r + 1$ such that $p(A) = 0$.

2.4.P29 Let $A \in M_n$ and suppose that $x, y \in \mathbf{C}^n$ are nonzero vectors such that $Ax = \lambda x$ and $y^*A = \lambda y^*$. If λ is a simple eigenvalue of A, show that $A - \lambda I + \kappa xy^*$ is nonsingular for all $\kappa \neq 0$.

2.4.P30 There is a systematic approach to the calculations illustrated in (2.4.3.3). Let $A \in M_n$ be given and suppose that $p(t)$ is a polynomial of degree greater than n. Use the Euclidean algorithm (polynomial long division) to express $p(t) = h(t)p_A(t) + r(t)$, in which the degree of $r(t)$ is strictly less than n (possibly zero). Explain why $p(A) = r(A)$.

2.4.P31 Use (2.4.3.2) to prove that if all the eigenvalues of $A \in M_n$ are zero, then $A^n = 0$.

2.4.P32 Let $A, B \in M_n$ and let $C = AB - BA$. Explain why $\operatorname{tr} C \neq 0$ is impossible. In particular, $C = cI$ is impossible if $c \neq 0$.

2.4.P33 Let $A, B \in M_n$ and let p be a positive integer. Suppose that $A = \begin{bmatrix} A_{11} & A_{12} \\ 0 & A_{22} \end{bmatrix}$, in which $A_{11} \in M_k$ and $A_{22} \in M_{n-k}$ have no eigenvalues in common. If $B^p = A$, show that B is block upper triangular conformal to A, $B = \begin{bmatrix} B_{11} & B_{12} \\ 0 & B_{22} \end{bmatrix}$, $B_{11}^p = A_{11}$, and $B_{22}^p = A_{22}$.

2.4.P34 Let $A = \begin{bmatrix} a & b \\ c & d \end{bmatrix} \in M_2$. Verify the Cayley–Hamilton theorem for A by an explicit computation, that is, verify that $A^2 - (a+d)A + (ad - bc)I_2 = 0$.

2.4.P35 Let $A \in M_n(\mathbf{F})$ ($\mathbf{F} = \mathbf{R}$ or $\mathbf{C}$). Use (2.4.4.2) to show that A commutes with every unitary matrix in $M_n(\mathbf{F})$ if and only if A is a scalar matrix.

Notes and Further Readings. See Radjavi and P. Rosenthal (2000) for a detailed exposition of simultaneous triangularization. Theorem 2.4.8.7 and its generalizations were proved by N. McCoy, On the characteristic roots of matric polynomials, *Bull. Amer. Math. Soc.* 42 (1936) 592–600. Our proof for (2.4.8.7) is adapted from M. P. Drazin, J. W. Dungey, and K. W. Gruenberg, Some theorems on commutative matrices, *J. Lond. Math. Soc.* 26 (1951) 221–228, which contains a proof of (2.4.8.10) in the general case $m \geq 2$. The relationship between eigenvalues and linear combinations is discussed in T. Motzkin and O. Taussky, Pairs of matrices with property L, *Trans. Amer. Math. Soc.* 73 (1952) 108–114. A pair $A, B \in M_n$ such that $\sigma(aA + bB) = \{a\alpha_j + b\beta_{i_j} : j = 1, \dots, n\}$ for all $a, b \in \mathbf{C}$ is said to have *property L*; the condition (2.4.8.7(c)) is called *property P*. Property P implies property L for all $n = 2, 3, \dots$; property L implies property P only for $n = 2$. Property L is not fully understood, but it is known that a pair of normal matrices has property L if and only if they commute; see N. A. Wiegmann, A note on pairs of normal matrices with property L, *Proc. Amer. Math. Soc.* 4 (1953) 35–36. There is a remarkable approximate version of the assertion in (2.4.P10): Nonsingular matrices $A, B \in M_n$ have the same characteristic polynomial (and hence have the same eigenvalues) if and only if $|\operatorname{tr} A^k - \operatorname{tr} B^k| \leq 1$ for all $k = \pm 1, \pm 2, \dots$; see the paper by Marcoux, Mastnak, and Radjavi cited at the end of (2.2). Jacobson's lemma (2.4.P12) is Lemma 2 in N. Jacobson, Rational methods in the theory of Lie algebras, *Ann. of Math.* (2) 36 (1935) 875–881. The notion of quasicommutativity (2.4.P12) arises in quantum mechanics: The position and momentum operators x and p_x (linear operators, but not finite dimensional; see (2.4.P32)) satisfy the identity $xp_x - p_x x = i\hbar I$. This identity, which implies the Heisenberg uncertainty principle for position and momentum, ensures that both x and p_x commute with their commutator. The example in (2.4.P12(g)) is due to Gérald Bourgeois. The Laffey theorems mentioned in (2.4.P11 and 2.4.P12) are proved in T. J. Laffey, Simultaneous triangularization of a pair of matrices – low rank cases and the nonderogatory case, *Linear Multilinear Algebra* 6 (1978) 269–306, and T. J. Laffey, Simultaneous quasidiagonalization of a pair of 3×3 complex matrices, *Rev. Roumaine Math. Pures Appl.* 23 (1978) 1047–1052). The matrix (2.4.20) is an example of a sign pattern matrix that *requires* distinct eigenvalues.

For a discussion of this fascinating property, see Z. Li and L. Harris, Sign patterns that require all distinct eigenvalues, *JP J. Algebra Number Theory Appl.* 2 (2002) 161–179; the paper contains a list of all the irreducible 3-by-3 sign pattern matrices with this property. The explicit calculation required in (2.4.P34) was published in A. Cayley, A memoir on the theory of matrices, *Philos. Trans. R. Soc. London* 148 (1858) 17–37; see p. 23. On p. 24, Cayley says that he had also verified the 3-by-3 case (presumably by another explicit calculation, which he does not present in the paper), but he had "not thought it necessary to undertake the labour of a formal proof of the theorem in the general case of a matrix of any degree." In a paper published in 1878, F. G. Frobenius gave a rigorous proof that every square complex matrix satisfies its own characteristic equation, but his approach was very different from our proof of (2.4.3.2). Frobenius first defined the minimal polynomial of a matrix (a new concept of his own invention; see (3.3)) and then showed that it divides the characteristic polynomial; see p. 355 in Vol. I of *Ferdinand Georg Frobenius: Gesammelte Abhandlungen*, ed. by J-P. Serre, Springer, Berlin, 1968. The proof of the Cayley–Hamilton theorem outlined in (2.4.P3) is taken from A. Buchheim, Mathematical notes, *Messenger Math.* 13 (1884) 62–66.

2.5 Normal matrices

The class of normal matrices, which arises naturally in the context of unitary similarity, is important throughout matrix analysis; it includes the unitary, Hermitian, skew Hermitian, real orthogonal, real symmetric, and real skew-symmetric matrices.

Definition 2.5.1. *A matrix $A \in M_n$ is normal if $AA^* = A^*A$, that is, if A commutes with its conjugate transpose.*

> *Exercise.* If $A \in M_n$ is normal and $\alpha \in \mathbf{C}$, show that αA is normal. *The class of normal matrices of a given size is closed under multiplication by complex scalars.*

> *Exercise.* If $A \in M_n$ is normal, and if B is unitarily similar to A, show that B is normal. *The class of normal matrices of a given size is closed under unitary similarity.*

> *Exercise.* If $A \in M_n$ and $B \in M_m$ are normal, show that $A \oplus B \in M_{n+m}$ is normal. *The class of normal matrices is closed under direct sums.*

> *Exercise.* If $A \in M_n$ and $B \in M_m$, and if $A \oplus B \in M_{n+m}$ is normal, show that A and B are normal.

> *Exercise.* Let $a, b \in \mathbf{C}$ be given. Show that $\begin{bmatrix} a & b \\ -b & a \end{bmatrix}$ is normal and has eigenvalues $a \pm ib$.

> *Exercise.* Show that every unitary matrix is normal.

Exercise. Show that every Hermitian or skew-Hermitian matrix is normal.

Exercise. Verify that $A = \begin{bmatrix} 1 & e^{i\pi/4} \\ -e^{i\pi/4} & 1 \end{bmatrix}$ is normal but that no scalar multiple of A is unitary, Hermitian, or skew Hermitian.

Exercise. Explain why every diagonal matrix is normal. If a diagonal matrix is Hermitian, why must it be real?

Exercise. Show that each of the classes of unitary, Hermitian, and skew-Hermitian matrices is closed under unitary similarity. If A is unitary and $|\alpha| = 1$, show that αA is unitary. If A is Hermitian and α is real, show that αA is Hermitian. If A is skew Hermitian and α is real, show that αA is skew Hermitian.

Exercise. Show that a Hermitian matrix has real main diagonal entries. Show that a skew-Hermitian matrix has pure imaginary main diagonal entries. What are the main diagonal entries of a real skew-symmetric matrix?

Exercise. Review the proof of (1.3.7) and conclude that $A \in M_n$ is unitarily diagonalizable if and only if there is a set of n orthonormal vectors in $\mathbf{C}^n$, each of which is an eigenvector of A.

In understanding and using the defining identity for normal matrices $- AA^* = A^*A -$ it can be helpful to keep a geometric interpretation in mind. Partition $A = [c_1 \ldots c_n]$ and $A^T = [r_1 \ldots r_n]$ according to their columns; the vectors c_j are the columns of A and the vectors r_i^T are the rows of A. Inspection of the defining identity $A^*A = AA^*$ reveals that A is normal if and only if $c_i^*c_j = \overline{r_i^*r_j}$ for all $i, j = 1, \ldots, n$. In particular, $c_i^*c_i = \|c_i\|_2^2 = \|r_i\|_2^2 = r_i^*r_i$, so each column of A has the same Euclidean norm as its corresponding row; a column is zero if and only if the corresponding row is zero.

If $A \in M_n(\mathbf{R})$ is a real normal matrix, then $c_i^Tc_j = \langle c_i, c_j \rangle = \langle r_i, r_j \rangle = r_i^Tr_j$ for all i and j. If columns i and j are nonzero, then rows i and j are nonzero and the identity

$$\frac{\langle c_i, c_j \rangle}{\|c_i\|_2 \|c_j\|_2} = \frac{\langle r_i, r_j \rangle}{\|r_i\|_2 \|r_j\|_2}$$

tells us that the angle between the vectors in columns i and j of A is the same as the angle between the vectors in rows i and j of A; see (0.6.3.1).

There is something special not only about zero rows or columns of a normal matrix, but also about certain zero blocks.

Lemma 2.5.2. *Let $A \in M_n$ be partitioned as $A = \begin{bmatrix} A_{11} & A_{12} \\ 0 & A_{22} \end{bmatrix}$, in which A_{11} and A_{22} are square. Then A is normal if and only if A_{11} and A_{22} are normal and $A_{12} = 0$. A block upper triangular matrix is normal if and only if each of its off-diagonal blocks is zero and each of its diagonal blocks is normal; in particular, an upper triangular matrix is normal if and only if it is diagonal.*

Proof. If A_{11} and A_{22} are normal and $A_{12} = 0$, then $A = A_{11} \oplus A_{22}$ is a direct sum of normal matrices, so it is normal.

Conversely, if A is normal, then

$$AA^* = \begin{bmatrix} A_{11}A_{11}^* + A_{12}A_{12}^* & \bigstar \\ \bigstar & \bigstar \end{bmatrix} = \begin{bmatrix} A_{11}^*A_{11} & \bigstar \\ \bigstar & \bigstar \end{bmatrix} = A^*A$$

so $A_{11}^*A_{11} = A_{11}A_{11}^* + A_{12}A_{12}^*$, which implies that

$$\text{tr } A_{11}^*A_{11} = \text{tr}(A_{11}A_{11}^* + A_{12}A_{12}^*)$$
$$= \text{tr } A_{11}A_{11}^* + \text{tr } A_{12}A_{12}^* = \text{tr } A_{11}^*A_{11} + \text{tr } A_{12}A_{12}^*$$

and hence $\text{tr } A_{12}A_{12}^* = 0$. Since $\text{tr } A_{12}A_{12}^*$ is the sum of squares of the absolute values of the entries of A_{12} (0.2.5.1), it follows that $A_{12} = 0$. Then $A = A_{11} \oplus A_{22}$ is normal, so A_{11} and A_{22} are normal.

Suppose that $B = [B_{ij}]_{i,j=1}^k \in M_n$ is normal and block upper triangular, that is, $B_{ii} \in M_{n_i}$ for $i = 1, \ldots, k$ and $B_{ij} = 0$ if $i > j$. Partition it as $B = \begin{bmatrix} B_{11} & X \\ 0 & \tilde{B} \end{bmatrix}$, in which $X = [B_{12} \ \ldots \ B_{1k}]$ and $\tilde{B} = [B_{ij}]_{i,j=2}^k$ is block upper triangular. Then $X = 0$ and $\tilde{B}$ is normal, so a finite induction permits us to conclude that B is block diagonal. For the converse, we have observed in a preceding exercise that a direct sum of normal matrices is normal. $\qquad\square$

Exercise. Let $A \in M_n$ be normal and let $\alpha \in \{1, \ldots, n\}$ be a given index set. If $A[\alpha, \alpha^c] = 0$, show that $A[\alpha^c, \alpha] = 0$.

We next catalog the most fundamental facts about normal matrices. The equivalence of (a) and (b) in the following theorem is often called the *spectral theorem for normal matrices.*

Theorem 2.5.3. *Let $A = [a_{ij}] \in M_n$ have eigenvalues $\lambda_1, \ldots, \lambda_n$. The following statements are equivalent:*

(a) A is normal.
(b) A is unitarily diagonalizable.
(c) $\sum_{i,j=1}^n |a_{ij}|^2 = \sum_{i=1}^n |\lambda_i|^2$.
(d) A has n orthonormal eigenvectors.

Proof. Use (2.3.1) to write $A = UTU^*$, in which $U = [u_1 \ \ldots \ u_n]$ is unitary and $T = [t_{ij}] \in M_n$ is upper triangular.

If A is normal, then so is T (as is every matrix that is unitarily similar to A). The preceding lemma ensures that T is actually a diagonal matrix, so A is unitarily diagonalizable.

If there is a unitary V such that $A = V\Lambda V^*$ and $\Lambda = \text{diag}(\lambda_1, \ldots, \lambda_n)$, then $\text{tr } A^*A = \text{tr } \Lambda^*\Lambda$ by (2.2.2), which is the assertion in (c).

The diagonal entries of T are $\lambda_1, \ldots, \lambda_n$ in some order, and hence $\text{tr } A^*A = \text{tr } T^*T = \sum_{i=1}^n |\lambda_i|^2 + \sum_{i<j} |t_{ij}|^2$. Thus, (c) implies that $\sum_{i<j} |t_{ij}|^2 = 0$, so T is diagonal. The factorization $A = UTU^*$ is equivalent to the identity $AU = UT$, which says that $Au_i = t_{ii}u_i$ for each $i = 1, \ldots, n$. Thus, the n columns of U are orthonormal eigenvectors of A.

Finally, an orthonormal list is linearly independent, so (d) ensures that A is diagonalizable and that a diagonalizing similarity can be chosen with orthonormal columns

(1.3.7). This means that A is unitarily similar to a diagonal (and hence normal) matrix, so A is normal.   □

A representation of a normal matrix $A \in M_n$ as $A = U \Lambda U^*$, in which U is unitary and Λ is diagonal, is called a *spectral decomposition* of A.

Exercise. Explain why a normal matrix is nondefective, that is, the geometric multiplicity of every eigenvalue is the same as its algebraic multiplicity.

Exercise. If $A \in M_n$ is normal, show that $x \in \mathbf{C}^n$ is a right eigenvector of A associated with the eigenvalue λ of A if and only if x is a left eigenvector of A associated with $\bar{\lambda}$; that is, $Ax = \lambda x$ is equivalent to $x^*A = \lambda x^*$ if A is normal. *Hint*: Normalize x and write $A = U \Lambda U^*$ with x as the first column of U. Then what is A^*? A^*x? See (2.5.P20) for another proof.

Exercise. If $A \in M_n$ is normal, and if x and y are eigenvectors of A associated with distinct eigenvalues, use the preceding exercise and the principle of biorthogonality to show that x and y are orthogonal.

Once the distinct eigenvalues $\lambda_1, \ldots, \lambda_d$ of a normal matrix $A \in M_n$ are known, it can be unitarily diagonalized via the following conceptual prescription: For each eigenspace $\{x \in \mathbf{C}^n : Ax = \lambda x\}$, determine a basis and orthonormalize it to obtain an orthonormal basis. The eigenspaces are mutually orthogonal and the dimension of each eigenspace is equal to the multiplicity of the corresponding eigenvalue (normality of A is the reason for both), so the union of these bases is an orthonormal basis for $\mathbf{C}^n$. Arraying these basis vectors as the columns of a matrix U produces a unitary matrix such that U^*AU is diagonal.

However, an eigenspace always has more than one orthonormal basis, so the diagonalizing unitary matrix constructed in the preceding conceptual prescription is never unique. If $X, Y \in M_{n,k}$ have orthonormal columns ($X^*X = I_k = Y^*Y$), and if range $X = $ range Y, then each column of X is a linear combination of the columns of Y, that is, $X = YG$ for some $G \in M_k$. Then $I_k = X^*X = (YG)^*(YG) = G^*(Y^*Y)G = G^*G$, so G must be unitary. This observation gives a geometric interpretation for the first part of the following uniqueness theorem.

Theorem 2.5.4. *Let $A \in M_n$ be normal and have distinct eigenvalues $\lambda_1, \ldots, \lambda_d$, with respective multiplicities $n_1, \ldots, n_d$. Let $\Lambda = \lambda_1 I_{n_1} \oplus \cdots \oplus \lambda_d I_{n_d}$, and suppose that $U \in M_n$ is unitary and $A = U \Lambda U^*$.*

(a) *$A = V \Lambda V^*$ for some unitary $V \in M_n$ if and only if there are unitary matrices $W_1, \ldots, W_d$ with each $W_i \in M_{n_i}$ such that $U = V(W_1 \oplus \cdots \oplus W_d)$.*

(b) *Two normal matrices are unitarily similar if and only if they have the same eigenvalues.*

Proof. (a) If $U \Lambda U^* = V \Lambda V^*$, then $\Lambda U^*V = U^*V \Lambda$, so $W = U^*V$ is unitary and commutes with Λ; (2.4.4.2) ensures that W is block diagonal conformal to Λ. Conversely, if $U = VW$ and $W = W_1 \oplus \cdots \oplus W_d$ with each $W_i \in M_{n_i}$, then W commutes with Λ and $U \Lambda U^* = VW \Lambda W^*V^* = V \Lambda WW^*V^* = V \Lambda V^*$.

(b)   If $B = V \Lambda V^*$ for   some   unitary V,   then $(UV^*)B(UV^*)^* =$ $(UV^*)V \Lambda V^*(UV^*)^* = U \Lambda U^* = A$.   Conversely, if B is similar to A, then they have the same eigenvalues; if B is unitarily similar to a normal matrix, then it is normal. $\qquad \square$

We next note that commuting normal matrices may be simultaneously unitarily diagonalized.

Theorem 2.5.5. *Let $\mathcal{N} \subseteq M_n$ be a nonempty family of normal matrices. Then $\mathcal{N}$ is a commuting family if and only if it is a simultaneously unitarily diagonalizable family. For any given $A_0 \in \mathcal{N}$ and for any given ordering $\lambda_1, \ldots, \lambda_n$ of the eigenvalues of A_0, there is a unitary $U \in M_n$ such that $U^* A_0 U = \mathrm{diag}(\lambda_1, \ldots, \lambda_n)$ and $U^* B U$ is diagonal for every $B \in \mathcal{N}$.*

Exercise. Use (2.3.3) and the fact that a triangular normal matrix must be diagonal to prove (2.5.5). The final assertion about A_0 follows as in the proof of (1.3.21) since every permutation matrix is unitary.

Application of (2.5.3) to the special case of Hermitian matrices yields a fundamental result called the *spectral theorem for Hermitian matrices.*

Theorem 2.5.6. *Let $A \in M_n$ be Hermitian and have eigenvalues $\lambda_1, \ldots, \lambda_n$. Let $\Lambda = \mathrm{diag}(\lambda_1, \ldots, \lambda_n)$. Then*

(a) $\lambda_1, \ldots, \lambda_n$ *are real.*
(b) A *is unitarily diagonalizable.*
(c) *There is a unitary $U \in M_n$ such that $A = U \Lambda U^*$.*

Proof. A diagonal Hermitian matrix must have real diagonal entries, so (a) follows from (b) and the fact that the set of Hermitian matrices is closed under unitary similarity. Statement (b) follows from (2.5.3) because Hermitian matrices are normal. Statement (c) restates (b) and incorporates the information that the diagonal entries of Λ are necessarily the eigenvalues of A. $\qquad \square$

In contrast to the discussion of diagonalizability in Chapter 1, there is no reason to assume distinctness of eigenvalues in (2.5.4) and (2.5.6), and diagonalizability need not be assumed in (2.5.5). A basis of eigenvectors (in fact, an orthonormal basis) is structurally guaranteed by normality. This is one reason why Hermitian and normal matrices are so important and have such pleasant properties.

We now turn to a discussion of *real* normal matrices. They can be diagonalized by a *complex unitary* similarity, but what special form can be achieved by a *real orthogonal* similarity? Since a real normal matrix can have non-real eigenvalues, it might not be possible to diagonalize it with a real similarity. However, each real matrix is real orthogonally similar to a real quasitriangular matrix, which must be quasidiagonal if it is normal.

Lemma 2.5.7. *Suppose that $A = \begin{bmatrix} a & b \\ c & d \end{bmatrix} \in M_2(\mathbf{R})$ is normal and has a conjugate pair of non-real eigenvalues. Then $c = -b \neq 0$ and $d = a$.*

Proof. A computation reveals that $AA^T = A^T A$ if and only if $b^2 = c^2$ and $ac + bd = ab + cd$. If $b = c$, then A is Hermitian (because it is real symmetric), so the preceding theorem ensures that it has two real eigenvalues. Therefore, we must have $b = -c \neq 0$ and $b(d - a) = b(a - d)$, which implies that $a = d$.   □

Theorem 2.5.8. *Let $A \in M_n(\mathbf{R})$ be normal.*

(a) *There is a real orthogonal $Q \in M_n(\mathbf{R})$ such that $Q^T A Q$ is a real quasidiagonal matrix*

$$A_1 \oplus \cdots \oplus A_m \in M_n(\mathbf{R}), \ \text{each } A_i \text{ is } 1\text{-by-}1 \text{ or } 2\text{-by-}2 \qquad (2.5.9)$$

with the following properties: The 1-by-1 direct summands in (2.5.9) display all the real eigenvalues of A. Each 2-by-2 direct summand in (2.5.9) has the special form

$$\begin{bmatrix} a & b \\ -b & a \end{bmatrix} \qquad (2.5.10)$$

in which $b > 0$; it is normal and has eigenvalues $a \pm ib$.

(b) *The direct summands in (2.5.9) are completely determined by the eigenvalues of A; they may appear in any prescribed order.*

(c) *Two real normal n-by-n matrices are real orthogonally similar if and only if they have the same eigenvalues.*

Proof. (a) Theorem 2.3.4b ensures that A is real orthogonally similar to a real upper quasitriangular matrix, each of whose 2-by-2 diagonal blocks has a conjugate pair of non-real eigenvalues. Since this upper quasitriangular matrix is normal, (2.5.2) ensures that it is actually quasidiagonal, and each of its 2-by-2 direct summands is normal and has a conjugate pair of non-real eigenvalues. The preceding lemma tells us that each of these 2-by-2 direct summands has the special form (2.5.10) in which $b \neq 0$. If necessary, we can ensure that $b > 0$ by performing a similarity via the matrix $\begin{bmatrix} 1 & 0 \\ 0 & -1 \end{bmatrix}$.

(b) The direct summands in (2.5.9) display all the eigenvalues of A, and any desired ordering of these direct summands may be achieved via a permutation similarity.

(c) Two real normal n-by-n matrices with the same eigenvalues are real orthogonally similar to the same direct sum of the form (2.5.9).   □

The preceding theorem reveals a canonical form for real normal matrices under real orthogonal similarity. It leads to canonical forms for real symmetric, real skew-symmetric, and real orthogonal matrices under real orthogonal similarity.

Corollary 2.5.11. *Let $A \in M_n(\mathbf{R})$. Then*

(a) *$A = A^T$ if and only if there is a real orthogonal $Q \in M_n(\mathbf{R})$ such that*

$$Q^T A Q = \mathrm{diag}(\lambda_1, \ldots, \lambda_n) \in M_n(\mathbf{R}) \qquad (2.5.12)$$

The eigenvalues of A are $\lambda_1, \ldots, \lambda_n$. Two real symmetric matrices are real orthogonally similar if and only if they have the same eigenvalues.

(b) *$A = -A^T$ if and only if there is a real orthogonal $Q \in M_n(\mathbf{R})$ and a nonnegative integer p such that $Q^T A Q$ has the form*

$$0_{n-2p} \oplus b_1 \begin{bmatrix} 0 & 1 \\ -1 & 0 \end{bmatrix} \oplus \cdots \oplus b_p \begin{bmatrix} 0 & 1 \\ -1 & 0 \end{bmatrix}, \quad \text{all } b_j > 0 \qquad (2.5.13)$$

If $A \neq 0$, its nonzero eigenvalues are $\pm ib_1, \ldots, \pm ib_p$. Two real skew-symmetric matrices are real orthogonally similar if and only if they have the same eigenvalues.

(c) $AA^T = I$ if and only if there is a real orthogonal $Q \in M_n(\mathbf{R})$ and a nonnegative integer p such that $Q^T A Q$ has the form

$$\Lambda_{n-2p} \oplus \begin{bmatrix} \cos\theta_1 & \sin\theta_1 \\ -\sin\theta_1 & \cos\theta_1 \end{bmatrix} \oplus \cdots \oplus \begin{bmatrix} \cos\theta_p & \sin\theta_p \\ -\sin\theta_p & \cos\theta_p \end{bmatrix} \tag{2.5.14}$$

in which $\Lambda_{n-2p} = \mathrm{diag}(\pm 1, \ldots, \pm 1) \in M_{n-2p}(\mathbf{R})$ and each $\theta_j \in (0, \pi)$. The eigenvalues of A are the diagonal entries of Λ_{n-2p} together with $e^{\pm i\theta_1}, \ldots, e^{\pm i\theta_p}$. Two real orthogonal matrices are real orthogonally similar if and only if they have the same eigenvalues.

Proof. Each of the hypotheses ensures that A is real and normal, so it is real orthogonally similar to a quasidiagonal matrix of the form (2.5.9). It suffices to consider what each of the hypotheses implies about the direct summands in (2.5.9). If $A = A^T$, there can be no direct summands of the form (2.5.10). If $A = -A^T$, then every 1-by-1 direct summand is zero and any 2-by-2 direct summand has zero diagonal entries. If $AA^T = I$, then each 1-by-1 direct summand has the form $[\pm 1]$ and any 2-by-2 block (2.5.10) has determinant ± 1, so $a^2 + b^2 = 1$ and there is some $\theta \in (0, \pi)$ such that $a = \cos\theta$ and $b = \sin\theta$, that is, $a \pm ib = e^{\pm i\theta}$. $\square$

Exercise. Let $A_1 = \begin{bmatrix} a & b \\ -b & a \end{bmatrix} \in M_2$, $A_2 = \begin{bmatrix} \alpha & \beta \\ \gamma & \delta \end{bmatrix} \in M_2$, and suppose that $b \neq 0$. Show that A_1 and A_2 commute if and only if $\alpha = \delta$ and $\gamma = -\beta$.

Exercise. Let $a, b \in \mathbf{C}$. Explain why $\begin{bmatrix} a & b \\ -b & a \end{bmatrix}$ and $\begin{bmatrix} a & -b \\ b & a \end{bmatrix}$ are real orthogonally similar. *Hint:* Consider a similarity via $\begin{bmatrix} 1 & 0 \\ 0 & -1 \end{bmatrix}$.

The following theorem is the real normal version of (2.5.5).

Theorem 2.5.15. *Let $\mathcal{N} \subseteq M_n(\mathbf{R})$ be a nonempty commuting family of real normal matrices. There is a real orthogonal matrix Q and a nonnegative integer q such that, for each $A \in \mathcal{N}$, $Q^T A Q$ is a real quasidiagonal matrix of the form*

$$\Lambda(A) \oplus \begin{bmatrix} a_1(A) & b_1(A) \\ -b_1(A) & a_1(A) \end{bmatrix} \oplus \cdots \oplus \begin{bmatrix} a_q(A) & b_q(A) \\ -b_q(A) & a_q(A) \end{bmatrix} \tag{2.5.15a}$$

in which each $\Lambda(A) \in M_{n-2q}(\mathbf{R})$ is diagonal; the parameters $a_j(A)$ and $b_j(A)$ are real for all $A \in \mathcal{N}$ and all $j = 1, \ldots, q$; and for each $j \in \{1, \ldots, q\}$ there is some $A \in \mathcal{N}$ for which $b_j(A) > 0$.

Proof. Theorem 2.3.6b ensures that there is a real orthogonal Q and a quasidiagonal matrix $D = J_{n_1} \oplus \cdots \oplus J_{n_m}$ such that for every $A \in \mathcal{N}$, $Q^T A Q$ is an upper quasitriangular matrix of the form (2.3.6.1) that is partitioned conformally to D. Moreover, if $n_j = 2$, then for some $A \in \mathcal{N}$, $A_j(A)$ has a conjugate pair of non-real eigenvalues. Because each upper quasitriangular matrix $Q^T A Q$ is normal, (2.5.2) ensures that it is actually quasidiagonal, that is, every $A \in \mathcal{N}$, $Q^T A Q = A_1(A) \oplus \cdots \oplus A_m(A)$ is partitioned conformally to D and each direct summand $A_j(A)$ is normal. If every $n_j = 1$, there is nothing further to prove. Suppose that $n_j = 2$ and consider the commuting family $\mathcal{F} = \{A_j(A) : A \in \mathcal{N}\}$. Since some matrix in $\mathcal{F}$ has a conjugate pair of non-real eigenvalues, (2.5.7) tells us that it has the special form (2.5.10) with $b \neq 0$; if necessary, we can ensure that $b > 0$ by performing a similarity via the matrix $\begin{bmatrix} 1 & 0 \\ 0 & -1 \end{bmatrix}$. The preceding exercise now tells us that *every* matrix in $\mathcal{F}$ has the special form (2.5.10). Perform a final simultaneous permutation similarity to achieve the ordering of direct summands displayed in the partitioned matrix (2.5.15a). $\qquad\square$

If $A, B \in M_n$ are normal (either complex or real) and satisfy an intertwining relation, the *Fuglede–Putnam theorem* says that A^* and B^* satisfy the same intertwining relation. The key to our proof of this result is the fact that, for $a, b \in \mathbf{C}$, $ab = 0$ if and only if $a\bar{b} = 0$; for a different proof, see (2.5.P26).

Theorem 2.5.16 (Fuglede–Putnam). *Let $A \in M_n$ and $B \in M_m$ be normal and let $X \in M_{n,m}$ be given. Then $AX = XB$ if and only if $A^*X = XB^*$.*

Proof. Let $A = U\Lambda U^*$ and $B = VMV^*$ be spectral decompositions in which $\Lambda = \mathrm{diag}(\lambda_1, \ldots, \lambda_n)$ and $M = \mathrm{diag}(\mu_1, \ldots, \mu_m)$. Let $U^*XV = [\xi_{ij}]$. Then $AX = XB \iff U\Lambda U^*X = XVMV^* \iff \Lambda(U^*XV) = (U^*XV)M \iff \lambda_i\xi_{ij} = \xi_{ij}\mu_j$ for all $i, j \iff \xi_{ij}(\lambda_i - \mu_j) = 0$ for all $i, j \iff \xi_{ij}(\overline{\lambda_i - \mu_j}) = 0$ for all $i, j \iff \bar{\lambda}_i\xi_{ij} = \xi_{ij}\bar{\mu}_j$ for all $i, j \iff \bar{\Lambda}(U^*XV) = (U^*XV)\bar{M} \iff U\bar{\Lambda}U^*X = XV\bar{M}V^* \iff A^*X = XB^*$. $\qquad\square$

The preceding two theorems lead to a useful representation for normal matrices that commute with their transpose or, equivalently, with their complex conjugate.

Exercise. Suppose that $A \in M_n$ is normal, $\bar{A}A = A\bar{A}$, and $A = B + iC$, in which B and C are real. Explain why B and C are normal and commute. *Hint*: $B = (A + \bar{A})/2$.

Exercise. Let $A = \begin{bmatrix} a & b \\ -b & a \end{bmatrix} \in M_2(\mathbf{C})$ be given with $b \neq 0$, let $A_0 = \begin{bmatrix} 1 & i \\ -i & 1 \end{bmatrix}$, and let $Q = \begin{bmatrix} 1 & 0 \\ 0 & -1 \end{bmatrix}$. Show that (a) A is nonsingular if and only if $A = c\begin{bmatrix} \alpha & \beta \\ -\beta & \alpha \end{bmatrix}$, $c, \beta \neq 0$, and $\alpha^2 + \beta^2 = 1$; (b) A is singular and nonzero if and only A is a nonzero scalar multiple of A_0 or $\bar{A}_0$; and (c) Q is real orthogonal and $\bar{A}_0 = QA_0Q^T$.

Theorem 2.5.17. *Let $A \in M_n$ be normal. The following three statements are equivalent:*

(a) $\bar{A}A = A\bar{A}$.

(b) $A^TA = AA^T$.

(c) *There is a real orthogonal Q such that Q^TAQ is a direct sum of blocks, in any prescribed order, each of which is either a zero block or a nonzero scalar multiple of*

$$[1], \begin{bmatrix} 0 & 1 \\ -1 & 0 \end{bmatrix}, \begin{bmatrix} a & b \\ -b & a \end{bmatrix}, \text{ or } \begin{bmatrix} 1 & i \\ -i & 1 \end{bmatrix}, \quad a, b \in \mathbf{C} \qquad (2.5.17.1)$$

in which $a \neq 0 \neq b$ and $a^2 + b^2 = 1$.

Conversely, if A is real orthogonally similar to a direct sum of complex scalar multiples of blocks of the form (2.5.17.1), then A is normal and $A\bar{A} = \bar{A}A$.

Proof. Equivalence of (a) and (b) follows from the preceding theorem: $\bar{A}A = A\bar{A}$ if and only if $A^T A = (\bar{A})^* A = A(\bar{A})^* = AA^T$.

Let $A = B + iC$, in which B and C are real. The exercise following (2.5.16) shows that $\{B, C\}$ is a commuting real normal family, so (2.5.15) ensures that there is a real orthogonal Q and a nonnegative integer q such that

$$Q^T B Q = \Lambda(B) \oplus \begin{bmatrix} a_1(B) & b_1(B) \\ -b_1(B) & a_1(B) \end{bmatrix} \oplus \cdots \oplus \begin{bmatrix} a_q(B) & b_q(B) \\ -b_q(B) & a_q(B) \end{bmatrix}$$

and

$$Q^T C Q = \Lambda(C) \oplus \begin{bmatrix} a_1(C) & b_1(C) \\ -b_1(C) & a_1(C) \end{bmatrix} \oplus \cdots \oplus \begin{bmatrix} a_q(C) & b_q(C) \\ -b_q(C) & a_q(C) \end{bmatrix}$$

in which each of $\Lambda(B), \Lambda(C) \in M_{n-2q}$ is diagonal and, for each $j \in \{1, \ldots, q\}$, at least one of $b_j(B)$ or $b_j(C)$ is positive. Therefore,

$$Q^T A Q = Q^T(B + iC)Q \qquad (2.5.17.2)$$
$$= \Lambda(A) \oplus \begin{bmatrix} \alpha_1(A) & \beta_1(A) \\ -\beta_1(A) & \alpha_1(A) \end{bmatrix} \oplus \cdots \oplus \begin{bmatrix} \alpha_q(A) & \beta_q(A) \\ -\beta_q(A) & \alpha_q(A) \end{bmatrix}$$

in which $\Lambda(A) = \Lambda(B) + i\Lambda(C)$, each $\alpha_j(A) = a_j(B) + ia_j(C)$, and each $\beta_j(A) = b_j(B) + ib_j(C) \neq 0$. The preceding exercise shows that every nonsingular 2-by-2 block in (2.5.17.2) is a nonzero scalar multiple of either $\begin{bmatrix} 0 & 1 \\ -1 & 0 \end{bmatrix}$ or $\begin{bmatrix} a & b \\ -b & a \end{bmatrix}$, in which $a \neq 0 \neq b$ and $a^2 + b^2 = 1$. It also shows that every singular 2-by-2 block in (2.5.17.2) is either a nonzero scalar multiple of $\begin{bmatrix} 1 & i \\ -i & 1 \end{bmatrix}$ or is real orthogonally similar to a nonzero scalar multiple of it. $\qquad \square$

Two special cases of the preceding theorem play an important role in the next section: unitary matrices that are either symmetric or skew symmetric.

Exercise. Show that the first two blocks in (2.5.17.1) are unitary; the third block is complex orthogonal but not unitary; the fourth block is singular, so it is neither unitary nor complex orthogonal.

Corollary 2.5.18. *Let $U \in M_n$ be unitary.*

(a) *If U is symmetric, then there is a real orthogonal $Q \in M_n(\mathbf{R})$ and real $\theta_1, \ldots, \theta_n \in [0, 2\pi)$ such that*

$$U = Q \operatorname{diag}(e^{i\theta_1}, \ldots, e^{i\theta_n}) Q^T \qquad (2.5.19.1)$$

(b) *If U is skew symmetric, then n is even and there is a real orthogonal $Q \in M_n(\mathbf{R})$ and real $\theta_1, \ldots, \theta_{n/2} \in [0, 2\pi)$ such that*

$$U = Q \left(e^{i\theta_1} \begin{bmatrix} 0 & 1 \\ -1 & 0 \end{bmatrix} \oplus \cdots \oplus e^{i\theta_{n/2}} \begin{bmatrix} 0 & 1 \\ -1 & 0 \end{bmatrix} \right) Q^T \qquad (2.5.19.2)$$

Conversely, any matrix of the form (2.5.19.1) is unitary and symmetric; any matrix of the form (2.5.19.2) is unitary and skew symmetric.

Proof. A unitary matrix U that is either symmetric or skew symmetric satisfies the identity $UU^T = U^TU$, so (2.5.17) ensures that there is a real orthogonal Q such that Q^TAQ is a direct sum of nonzero scalar multiples of blocks selected from the four types in (2.5.17.1).

(a) If U is symmetric, only symmetric blocks may be selected from (2.5.17.1), so Q^TUQ is a direct sum of blocks of the form $c[1]$, in which $|c| = 1$ because U is unitary.

(b) If U is skew symmetric, only skew-symmetric blocks may be selected from (2.5.17.1), so Q^TUQ is a direct sum of blocks of the form $c\begin{bmatrix} 0 & 1 \\ -1 & 0 \end{bmatrix}$, in which $|c| = 1$ because U is unitary. It follows that n is even. $\qquad\square$

Our final theorem is an analog of (1.3.29) in the setting of unitary similarity: Real matrices are unitarily similar if and only if they are real orthogonally similar. We prepare for it with the following exercise and corollary.

Exercise. Let $U \in M_n$ be given. Explain why U is both unitary and complex orthogonal if and only if it is real orthogonal. *Hint:* $U^{-1} = U^* = U^T$.

Corollary 2.5.20. *Let $U \in M_n$ be unitary.*

(a) *If U is symmetric, there is a unitary symmetric V such that $V^2 = U$ and V is a polynomial in U. Consequently, V commutes with any matrix that commutes with U.*

(b) *(QS factorization of a unitary matrix) There is a real orthogonal Q and a symmetric unitary S such that $U = QS$ and S is a polynomial in U^TU. Consequently, S commutes with any matrix that commutes with U^TU.*

Proof. (a) Use the preceding corollary to factor $U = P\,\text{diag}(e^{i\theta_1}, \ldots, e^{i\theta_n})P^T$, in which P is real orthogonal and $\theta_1, \ldots, \theta_n \in [0, 2\pi)$ are real. Let $p(t)$ be a polynomial such that $p(e^{i\theta_j}) = e^{i\theta_j/2}$ for each $j = 1, \ldots, n$ (0.9.11.4), and let $V = p(U)$. Then

$$V = p(U) = p(P\,\text{diag}(e^{i\theta_1}, \ldots, e^{i\theta_n})P^T)$$
$$= Pp(\text{diag}(e^{i\theta_1}, \ldots, e^{i\theta_n}))P^T = P\,\text{diag}(p(e^{i\theta_1}), \ldots, p(e^{i\theta_n}))P^T$$
$$= P\,\text{diag}(e^{i\theta_1/2}, \ldots, e^{i\theta_n/2})P^T$$

so V is unitary and symmetric, and $V^2 = P(\text{diag}(e^{i\theta_1/2}, \ldots, e^{i\theta_n/2})^2 P^T = P\,\text{diag}(e^{i\theta_1}, \ldots, e^{i\theta_n})P^T = U$. The final assertion follows from (2.4.4.0).

(b) Part (a) ensures that there is a symmetric unitary matrix S such that $S^2 = U^TU$ and S is a polynomial in U^TU. Consider the unitary matrix $Q = US^*$, which has the property that $QS = US^*S = U$. Compute $Q^TQ = S^*U^TUS^* = S^*S^2S^* = I$. Thus, Q is both orthogonal and unitary, so the preceding exercise ensures that it is real orthogonal. $\qquad\square$

Exercise. If $U \in M_n$ is unitary, explain why there is a real orthogonal matrix Q and a symmetric unitary matrix S such that $U = SQ$ and S is a polynomial in UU^T.

Theorem 2.5.21. *Let $\mathcal{F} = \{A_\alpha : \alpha \in \mathcal{I}\} \subset M_n(\mathbf{R})$ and $\mathcal{G} = \{B_\alpha : \alpha \in \mathcal{I}\} \subset M_n(\mathbf{R})$ be given families of real matrices. If there is a unitary $U \in M_n$ such that $A_\alpha = U B_\alpha U^*$ for every $\alpha \in \mathcal{I}$, then there is a real orthogonal $Q \in M_n(\mathbf{R})$ such that $A_\alpha = Q B_\alpha Q^T$ for every $\alpha \in \mathcal{I}$. In particular, two real matrices that are unitarily similar are real orthogonally similar.*

Proof. Since each A_α and B_α is real, $A_\alpha = U B_\alpha U^* = \bar{U} B_\alpha U^T = \bar{A}_\alpha$ and hence $U^T U B_\alpha = B_\alpha U^T U$ for every $\alpha \in \mathcal{I}$. The preceding corollary ensures that there is a symmetric unitary matrix S and a real orthogonal matrix Q such that $U = QS$ and S commutes with B_α. Thus, $A_\alpha = U B_\alpha U^* = Q S B_\alpha S^* Q^T = Q B_\alpha S S^* Q^T = Q B_\alpha Q^T$ for every $\alpha \in \mathcal{I}$. $\qquad \square$

Problems

2.5.P1 Show that $A \in M_n$ is normal if and only if $(Ax)^*(Ax) = (A^*x)^*(A^*x)$ for all $x \in \mathbf{C}^n$, that is, $\|Ax\|_2 = \|A^*x\|_2$ for all $x \in \mathbf{C}^n$.

2.5.P2 Show that a normal matrix is unitary if and only if all its eigenvalues have absolute value 1.

2.5.P3 Show that a normal matrix is Hermitian if and only if all its eigenvalues are real.

2.5.P4 Show that a normal matrix is skew Hermitian if and only if all its eigenvalues are pure imaginary (have real part equal to 0).

2.5.P5 If $A \in M_n$ is skew Hermitian (respectively, Hermitian), show that iA is Hermitian (respectively, skew Hermitian).

2.5.P6 Show that $A \in M_n$ is normal if and only if it commutes with some normal matrix with distinct eigenvalues.

2.5.P7 Consider matrices $A \in M_n$ of the form $A = B^{-1}B^*$ for a nonsingular $B \in M_n$, as in (2.1.9). (a) Show that A is unitary if and only if B is normal. (b) If B has the form $B = HNH$, in which N is normal and H is Hermitian (and both are nonsingular), show that A is similar to a unitary matrix.

2.5.P8 Write $A \in M_n$ as $A = H(A) + i K(A)$ in which $H(A)$ and $K(A)$ are Hermitian; see (0.2.5). Show that A is normal if and only if $H(A)$ and $K(A)$ commute.

2.5.P9 Write $A \in M_n$ as $A = H(A) + i K(A)$ in which $H(A)$ and $K(A)$ are Hermitian. If every eigenvector of $H(A)$ is an eigenvector of $K(A)$, show that A is normal. What about the converse? Consider $A = \begin{bmatrix} 1 & i \\ -i & 1 \end{bmatrix}$.

2.5.P10 Suppose $A, B \in M_n$ are both normal. If A and B commute, show that AB and $A \pm B$ are all normal. What about the converse? Verify that $A = \begin{bmatrix} 1 & -1 \\ 1 & 1 \end{bmatrix}$, $B = \begin{bmatrix} -1 & 1 \\ 1 & 1 \end{bmatrix}$, AB, and BA are all normal, but A and B do not commute.

2.5.P11 For any complex number $z \in \mathbf{C}$, show that there are $\theta, \tau \in \mathbf{R}$ such that $\bar{z} = e^{i\theta}z$ and $|z| = e^{i\tau}z$. Notice that $[e^{i\theta}] \in M_1$ is a unitary matrix. What do diagonal unitary matrices $U \in M_n$ look like?

2.5.P12 Generalize (2.5.P11) to show that if $\Lambda = \mathrm{diag}(\lambda_1, \ldots, \lambda_n) \in M_n$, then there are diagonal unitary matrices U and V such that $\bar{\Lambda} = U\Lambda = \Lambda U$ and $|\Lambda| = \mathrm{diag}(|\lambda_1|, \ldots, |\lambda_n|) = V\Lambda = \Lambda V$.

2.5.P13 Use (2.5.P12) to show that $A \in M_n$ is normal if and only if there is a unitary $V \in M_n$ such that $A^* = AV$. Deduce that range A = range A^* if A is normal.

2.5.P14 Let $A \in M_n(\mathbf{R})$ be given. Explain why A is normal and all its eigenvalues are real if and only if A is symmetric.

2.5.P15 Show that two normal matrices are similar if and only if they have the same characteristic polynomial. Is this true if we omit the assumption that both matrices are normal? Consider $\begin{bmatrix} 0 & 0 \\ 0 & 0 \end{bmatrix}$ and $\begin{bmatrix} 0 & 1 \\ 0 & 0 \end{bmatrix}$.

2.5.P16 If $U, V, \Lambda \in M_n$ and U, V are unitary, show that $U\Lambda U^*$ and $V\Lambda V^*$ are unitarily similar. Deduce that two normal matrices are similar if and only if they are unitarily similar. Give an example of two diagonalizable matrices that are similar but not unitarily similar.

2.5.P17 If $A \in M_n$ is normal and $p(t)$ is a given polynomial, use (2.5.1) to show that $p(A)$ is normal. Give another proof of this fact using (2.5.3).

2.5.P18 If $A \in M_n$ and there is nonzero polynomial $p(t)$ such that $p(A)$ is normal, does it follow that A is normal?

2.5.P19 Let $A \in M_n$ and $a \in \mathbf{C}$ be given. Use the definition (2.5.1) to show that A is normal if and only if $A + aI$ is normal; do not invoke the spectral theorem (2.5.3).

2.5.P20 Let $A \in M_n$ be normal and suppose that $x \in \mathbf{C}^n$ is a right eigenvector of A associated with the eigenvalue λ. Use (2.5.P1) and (2.5.P19) to show that x is a left eigenvector of A associated with the same eigenvalue λ.

2.5.P21 Suppose that $A \in M_n$ is normal. Use the preceding problem to show that $Ax = 0$ if and only if $A^*x = 0$, that is, the null space of A is the same as that of A^*. Consider $B = \begin{bmatrix} 0 & 1 \\ 0 & 1 \end{bmatrix}$ to show that the null space of a non-normal matrix B need not be the same as the null space B^*, even if B is diagonalizable.

2.5.P22 Use (2.5.6) to show that the characteristic polynomial of a complex Hermitian matrix has real coefficients.

2.5.P23 The matrices $\begin{bmatrix} 1 & i \\ i & 1 \end{bmatrix}$ and $\begin{bmatrix} i & i \\ i & -1 \end{bmatrix}$ are both symmetric. Show that one is normal and the other is not. This is an important difference between real symmetric matrices and complex symmetric matrices.

2.5.P24 If $A \in M_n$ is both normal and nilpotent, show that $A = 0$.

2.5.P25 Suppose that $A \in M_n$ and $B \in M_m$ are normal and let $X \in M_{n,m}$ be given. Explain why $\bar{B}$ is normal and deduce that $AX = X\bar{B}$ if and only if $A^*X = XB^T$.

2.5.P26 Let $A \in M_n$ be given. (a) If there is a polynomial $p(t)$ such that $A^* = p(A)$, show that $A \in M_n$ is normal. (b) If A is normal, show that there is a polynomial $p(t)$ of degree at most $n - 1$ such that $A^* = p(A)$. (c) If A is real and normal, show that there is a polynomial $p(t)$ with real coefficients and degree at most $n - 1$ such that $A^T = p(A)$. (d) If A is normal, show that there is a polynomial $p(t)$ with real coefficients and degree at most $2n - 1$ such that $A^* = p(A)$. (e) If A is normal and $B \in M_m$ is normal, show that there is a polynomial $p(t)$ of degree at most $n + m - 1$ such that $A^* = p(A)$ and $B^* = p(B)$.

(f) If A is normal and $B \in M_m$ is normal, show that there is a polynomial $p(t)$ with real coefficients and degree at most $2n + 2m - 1$ such that $A^* = p(A)$ and $B^* = p(B)$. (g) Use (e) and (2.4.4.0) to prove the Fuglede–Putnam theorem (2.5.16). (h) Use (f) to prove the assertion in (2.5.P25).

2.5.P27 (a) Let $A, B \in M_{n,m}$. If AB^* and B^*A are both normal, show that $BA^*A = AA^*B$. (b) Let $A \in M_n$. Prove that $A\bar{A}$ is normal (such a matrix is said to be *congruence normal*) if and only if $AA^*A^T = A^T A^* A$. (c) If $A \in M_n$ is congruence normal, show that the three normal matrices $A\bar{A}$, $\overline{A^*A}$, and AA^* commute and hence are simultaneously unitarily diagonalizable.

2.5.P28 Let Hermitian matrices $A, B \in M_n$ be given and assume that AB is normal. (a) Why is BA normal? (b) Show that A commutes with B^2 and B commutes with A^2. (c) If there is a polynomial $p(t)$ such that *either* $A = p(A^2)$ or $B = p(B^2)$, show that A commutes with B and AB is actually Hermitian. (d) Explain why the condition in (c) is met if *either* A or B has the property that whenever λ is a nonzero eigenvalue, then $-\lambda$ is not also an eigenvalue. For example, if either A or B has all nonnegative eigenvalues, this condition is met. (d) Discuss the example $A = \begin{bmatrix} 0 & 1 \\ 1 & 0 \end{bmatrix}$, $B = \begin{bmatrix} 0 & i \\ -i & 0 \end{bmatrix}$.

2.5.P29 Let $A = \begin{bmatrix} a & b \\ c & d \end{bmatrix} \in M_2$ and assume that $bc \neq 0$. (a) Show that A is normal if and only if there is some $\theta \in \mathbf{R}$ such that $c = e^{i\theta}b$ and $a - d = e^{i\theta}b(\bar{a} - \bar{d})/\bar{b}$. In particular, if A is normal, it is necessary that $|c| = |b|$. (b) Let $b = |b|e^{i\phi}$. If A is normal and $c = be^{i\theta}$, show that $e^{-i(\phi+\theta/2)}(A - aI)$ is Hermitian. Conversely, if there is a $\gamma \in \mathbf{C}$ such that $A - \gamma I$ is essentially Hermitian, show that A is normal. (c) If A is real, deduce from part (a) that it is normal if and only if either $c = b$ ($A = A^T$) or $c = -b$ and $a = d$ ($AA^T = (a^2 + b^2)I$ and $A = -A^T$ if $a = 0$).

2.5.P30 Show that a given $A \in M_n$ is normal if and only if $(Ax)^*(Ay) = (A^*x)^*(A^*y)$ for all $x, y \in \mathbf{C}^n$. If A, x, and y are real, this means that the angle between Ax and Ay is always the same as the angle between A^Tx and A^Ty. Compare with (2.5.P1). What does this condition say if we take $x = e_i$ and $y = e_j$ (the standard Euclidean basis vectors)? If $(Ae_i)^*(Ae_j) = (A^*e_i)^*(A^*e_j)$ for all $i, j = 1, \ldots, n$, show that A is normal.

2.5.P31 Let $A \in M_n(\mathbf{R})$ be a real normal matrix, that is, $AA^T = A^T A$. If AA^T has n distinct eigenvalues, show that A is symmetric.

2.5.P32 If $A \in M_3(\mathbf{R})$ is real orthogonal, observe that A has either one or three real eigenvalues. If it has a positive determinant, use (2.5.11) to show that it is orthogonally similar to the direct sum of $[1] \in M_1$ and a plane rotation. Discuss the geometrical interpretation of this as a rotation by an angle θ around some fixed axis passing through the origin in $\mathbf{R}^3$. This is part of Euler's theorem in mechanics: Every motion of a rigid body is the composition of a translation and a rotation about some axis.

2.5.P33 If $\mathcal{F} \subseteq M_n$ is a commuting family of normal matrices, show that there exists a single Hermitian matrix B such that for each $A_\alpha \in \mathcal{F}$ there is a polynomial $p_\alpha(t)$ of degree at most $n - 1$ such that $A_\alpha = p_\alpha(B)$. Notice that B is fixed for all of $\mathcal{F}$ but the polynomial may depend on the element of $\mathcal{F}$.

2.5.P34 Let $A \in M_n$, and let $x \in \mathbf{C}^n$ be nonzero. We say that x is a *normal eigenvector* of A if it is both a right and left eigenvector of A. (a) If $Ax = \lambda x$ and $x^*A = \mu x^*$, show that

$\lambda = \mu$. (b) If x is a normal eigenvector of A associated with an eigenvalue λ, show that A is unitarily similar to $[\lambda] \oplus A_1$, in which $A_1 \in M_{n-1}$ is upper triangular. (c) Show that A is normal if and only if each of its eigenvectors is a normal eigenvector.

2.5.P35 Let $x, y \in \mathbf{C}^n$ be given nonzero vectors. (a) Show that $xx^* = yy^*$ if and only if there is some real θ such that $x = e^{i\theta} y$. (b) Show that the following are equivalent for a rank-one matrix $A = xy^*$: (i) A is normal; (ii) there is a positive real number r and a real $\theta \in [0, 2\pi)$ such that $x = re^{i\theta} y$; (iii) A is essentially Hermitian.

2.5.P36 For any $A \in M_n$, show that $\begin{bmatrix} A & A^* \\ A^* & A \end{bmatrix} \in M_{2n}$ is normal. Thus, any square matrix can be a principal submatrix of a normal matrix (every $A \in M_n$ has a *dilation* to a normal matrix). Can any square matrix be a principal submatrix of a Hermitian matrix? Of a unitary matrix?

2.5.P37 Let $n \geq 2$ and suppose that $A = \begin{bmatrix} a & x^* \\ y & B \end{bmatrix} \in M_n$ is normal with $B \in M_{n-1}$ and $x, y \in \mathbf{C}^{n-1}$. (a) Show that $\|x\|_2 = \|y\|_2$ and $xx^* - yy^* = BB^* - B^*B$. (b) Explain why rank$(FF^* - F^*F) \neq 1$ for every square complex matrix F. (c) Explain why there are two mutually exclusive possibilities: either (i) the principal submatrix B is normal or (ii) rank$(BB^* - B^*B) = 2$. (d) Explain why B is normal if and only if $x = e^{i\theta} y$ for some real θ. (e) Discuss the example $B = \begin{bmatrix} 0 & 1 \\ 0 & 0 \end{bmatrix}$, $x = [-\sqrt{2} \ 1]^T$, $y = [1 \ -\sqrt{2}]^T$, and $a = 1 - \sqrt{2}$.

2.5.P38 Let $A = [a_{ij}] \in M_n$ and let $C = AA^* - A^*A$. (a) Explain why C is Hermitian and why C is nilpotent if and only if $C = 0$. (b) Show that A is normal if and only if it commutes with C. (c) Show that rank $C \neq 1$. (d) Explain why A is normal if and only if rank $C \leq 1$, that is, there are only two possibilities: rank $C = 0$ (A is normal) and rank $C \geq 2$ (A is not normal). We say that A is *nearly normal* if rank $C = 2$. (d) Suppose that A is a tridiagonal Toeplitz matrix. Show that $C = \text{diag}(\alpha, 0, \ldots, 0, -\alpha)$, in which $\alpha = |a_{12}|^2 - |a_{21}|^2$. Conclude that A is normal if and only if $|a_{12}| = |a_{21}|$; otherwise, A is nearly normal.

2.5.P39 Suppose that $U \in M_n$ is unitary, so all its eigenvalues have modulus one. (a) If U is symmetric, show that its eigenvalues uniquely determine its representation (2.5.19.1), up to permutation of the diagonal entries. (b) If U is skew symmetric, explain how the scalars $e^{i\theta_j}$ in (2.5.19.2) are related to its eigenvalues. Why must the eigenvalues of U occur in $\pm$ pairs? Show that the eigenvalues of U uniquely determine its representation (2.5.19.2), up to permutation of direct summands.

2.5.P40 Let $A = \begin{bmatrix} 0 & B \\ 0 & 0 \end{bmatrix} \in M_4$, in which $B = \begin{bmatrix} 1 & i \\ -i & 1 \end{bmatrix}$. Verify that A commutes with A^T, A commutes with $\bar{A}$, but A does not commute with A^*, that is, A is not normal.

2.5.P41 Let $z \in \mathbf{C}^n$ be nonzero and write $z = x + iy$ with $x, y \in \mathbf{R}^n$. (a) Show that the following three statements are equivalent: (1) $\{z, \bar{z}\}$ is linearly dependent; (2) $\{x, y\}$ is linearly dependent; (3) there is a unit vector $u \in \mathbf{R}^n$ and a nonzero $c \in \mathbf{C}$ such that $z = cu$. (b) Show that the following are equivalent: (1) $\{z, \bar{z}\}$ is linearly independent; (2) $\{x, y\}$ is linearly independent; (3) there are real orthonormal vectors $v, w \in \mathbf{R}^n$ such that span$\{z, \bar{z}\} = \text{span}\{v, w\}$ (over $\mathbf{C}$).

2.5.P42 Let $A, B \in M_n$ and suppose that $\lambda_1, \ldots, \lambda_n$ are the eigenvalues of A. The function $\Delta(A) = \text{tr} \, A^*A - \sum_{i=1}^{n} |\lambda_i|^2$ is the *defect of A from normality*. Schur's inequality (2.3.2a) says that $\Delta(A) \geq 0$, and 2.5.3(c) ensures that A is normal if and only if $\Delta(A) = 0$. (a) If A,

B, and AB are normal, show that $\operatorname{tr}((AB)^*(AB)) = \operatorname{tr}((BA)^*(BA))$ and explain why BA is normal. (b) Suppose that A is normal, A and B have the same characteristic polynomials, and $\operatorname{tr} A^*A = \operatorname{tr} B^*B$. Show that B is normal and unitarily similar to A. Compare with (2.5.P15).

2.5.P43 Let $A = [a_{ij}] \in M_n$ be normal. (a) Partition $A = [A_{ij}]_{i,j=1}^{k}$, in which each A_{ii} is square. Suppose that the eigenvalues of A are the eigenvalues of $A_{11}, A_{22}, \ldots,$ and A_{kk} (including their multiplicities); for example, we could assume that $p_A(t) = p_{A_{11}}(t) \cdots p_{A_{kk}}(t)$. Show that A is block diagonal, that is, $A_{ij} = 0$ for all $i \neq j$, and each diagonal block A_{ii} is normal. (b) If each diagonal entry a_{ij} of A is an eigenvalue of A, explain why A is diagonal. (c) If $n = 2$ and one of the main diagonal entries of A is an eigenvalue of A, explain why it is diagonal.

2.5.P44 (a) Show that $A \in M_n$ is Hermitian if and only if $\operatorname{tr} A^2 = \operatorname{tr} A^*A$. (b) Show that Hermitian matrices $A, B \in M_n$ commute if and only if $\operatorname{tr}(AB)^2 = \operatorname{tr}(A^2B^2)$.

2.5.P45 Let $\mathcal{N} \subseteq M_n(\mathbf{R})$ be a commuting family of real symmetric matrices. Show that there is a single real orthogonal matrix Q such that $Q^T A Q$ is diagonal for every $A \in \mathcal{N}$.

2.5.P46 Use (2.3.1) to show that any non-real eigenvalues of a real matrix must occur in complex conjugate pairs.

2.5.P47 Suppose $A \in M_n$ is normal and has eigenvalues $\lambda_1, \ldots, \lambda_n$. Show that (a) adj A is normal and has eigenvalues $\prod_{j \neq i} \lambda_j, i = 1, \ldots, n$; (b) adj A is Hermitian if A is Hermitian; (c) adj A has positive (respectively, nonnegative) eigenvalues if A has positive (respectively, nonnegative) eigenvalues; (d) adj A is unitary if A is unitary.

2.5.P48 Let $A \in M_n$ be normal and suppose that rank $A = r > 0$. Use (2.5.3) to write $A = U \Lambda U^*$, in which $U \in M_n$ is unitary and $\Lambda = \Lambda_r \oplus 0_{n-r}$ is diagonal. (a) Explain why $\det \Lambda_r \neq 0$. Let $\det \Lambda_r = |\det \Lambda_r| e^{i\theta}$ with $\theta \in [0, 2\pi)$. (b) Partition $U = [V \ U_2]$, in which $V \in M_{n,r}$. Explain why $A = V \Lambda_r V^*$; this is a full-rank factorization of A. (c) Let $\alpha, \beta \subseteq \{1, \ldots, n\}$ be index sets of cardinality r, and let $V[\alpha, \varnothing^c] = V_\alpha$. Explain why $A[\alpha, \beta] = V_\alpha \Lambda_r V_\beta^*$; $\det A[\alpha] \det A[\beta] = \det A[\alpha, \beta] \det A[\beta, \alpha]$; and $\det A[\alpha] = |\det V_\alpha|^2 \det \Lambda_r$. (d) Explain why every principal minor of A of size r lies on the ray $\{s e^{i\theta} : s \geq 0\}$ in the complex plane, and at least one of those principal minors is nonzero. (e) Conclude that A is rank principal. For a version of this result if A is Hermitian, see (4.2.P30); see also (3.1.P20).

2.5.P49 Suppose that $A \in M_n$ is upper triangular and diagonalizable. Show that it can be diagonalized via an upper triangular similarity.

2.5.P50 The reversal matrix K_n (0.9.5.1) is real symmetric. Check that it is also real orthogonal, and explain why its eigenvalues can only be ± 1. Check that $\operatorname{tr} K_n = 0$ if n is even and $\operatorname{tr} K_n = 1$ if n is odd. Explain why if n is even, the eigenvalues of K_n are ± 1, each with multiplicity $n/2$; if n is odd, the eigenvalues of K_n are $+1$ with multiplicity $(n + 1)/2$ and -1 with multiplicity $(n - 1)/2$.

2.5.P51 Let $A \in M_n$ be normal, let $A = U \Lambda U^*$ be a spectral decomposition in which $\Lambda = \operatorname{diag}(\lambda_1, \ldots, \lambda_n)$, let $x \in \mathbf{C}^n$ be any given unit vector, and let $\xi = [\xi_i] = U^*x$. Explain why $x^*Ax = \sum_{i=1}^{n} |\xi_i|^2 \lambda_i$, why the x^*Ax lies in the convex hull of the eigenvalues of A, and why each complex number in the convex hull of the eigenvalues of A equals x^*Ax for some unit vector x. Thus, if A is normal, $x^*Ax \neq 0$ for every unit vector x if and only if 0 is not in the convex hull of the eigenvalues of A.

2.5.P52 Let $A, B \in M_n$ be nonsingular. The matrix $C = ABA^{-1}B^{-1}$ is the *multiplicative commutator* of A and B. Explain why $C = I$ if and only if A commutes with B. Suppose that A and C are normal and 0 is not in the convex hull of the eigenvalues of B. Provide details for the following sketch of a proof that A commutes with C if and only if A commutes with B (this is the *Marcus-Thompson theorem*): Let $A = U\Lambda U^*$ and $C = UMU^*$ be spectral decompositions in which $\Lambda = \mathrm{diag}(\lambda_1, \ldots, \lambda_n)$ and $M = \mathrm{diag}(\mu_1, \ldots, \mu_n)$. Let $\mathcal{B} = U^*BU = [\beta_{ij}]$. Then all $\beta_{ii} \neq 0$ and $M = U^*CU = \Lambda \mathcal{B}\Lambda^{-1}\mathcal{B}^{-1} \Rightarrow M\mathcal{B} = \Lambda \mathcal{B}\Lambda^{-1} \Rightarrow \mu_i\beta_{ii} = \beta_{ii} \Rightarrow M = I \Rightarrow C = I$. Compare with (2.4.P12(c)).

2.5.P53 Let $U, V \in M_n$ be unitary and suppose that all of the eigenvalues of V lie on an open arc of the unit circle of length π; such a matrix is called a *cramped unitary matrix*. Let $C = UVU^*V^*$ be the multiplicative commutator of U and V. Use the preceding problem to prove *Frobenius's theorem*: U commutes with C if and only if U commutes with V.

2.5.P54 If $A, B \in M_n$ are normal, show that (a) the null space of A is orthogonal to the range of A; (b) the ranges of A and A^* are the same; (c) the null space of A is contained in the null space of B if and only if the range of A contains the range of B.

2.5.P55 Verify the following improvement of (2.2.8) for normal matrices: If $A, B \in M_n$ are normal, then A is unitarily similar to B if and only if $\mathrm{tr}\, A^k = \mathrm{tr}\, B^k$, $k = 1, 2, \ldots, n$.

2.5.P56 Let $A \in M_n$ and an integer $k \geq 2$ be given, and let $\omega = e^{2\pi i/(k+1)}$. Show that $A^k = A^*$ if and only if A is normal and its spectrum is contained in the set $\{0, 1, \omega, \omega^2, \ldots, \omega^k\}$. If $A^k = A^*$ and A is nonsingular, explain why it is unitary.

2.5.P57 Let $A \in M_n$ be given. Show that (a) A is normal and symmetric if and only if there is a real orthogonal $Q \in M_n$ and a diagonal $\Lambda \in M_n$ such that $A = Q\Lambda Q^T$; (b) A is normal and skew symmetric if and only if there is a real orthogonal $Q \in M_n$ such that $Q^T A Q$ is a direct sum of a zero block and blocks of the form $\begin{bmatrix} 0 & z_j \\ -z_j & 0 \end{bmatrix}$, $z_j \in \mathbf{C}$.

2.5.P58 Let $A \in M_n$ be normal. Then $A\bar{A} = 0$ if and only if $AA^T = A^T A = 0$. (a) Use (2.5.17) to prove this. (b) Provide details for an alternative proof: $A\bar{A} = 0 \Rightarrow 0 = A^*A\bar{A} = AA^*\bar{A} \Rightarrow \bar{A}A^T A = 0 \Rightarrow (A^T A)^*(A^T A) = 0 \Rightarrow A^T A = 0$ (0.2.5.1).

2.5.P59 Let $A, B \in M_n$. Suppose that A is normal and has distinct eigenvalues. If $AB = BA$, show that B is normal. Compare with (1.3.P3).

2.5.P60 Let $x = [x_i] \in \mathbf{C}^n$ be given. (a) Explain why $\max_i |x_i| \leq \|x\|_2$ (0.6.1). (b) Let $e = e_1 + \cdots + e_n \in \mathbf{C}^n$ be the vector all of whose entries are $+1$. If $x^T e = 0$, show that $\max_i |x_i| \leq \sqrt{\frac{n-1}{n}} \|x\|_2$ with equality if and only if, for some $c \in M_n$ and some index j, $x = c(ne_j - e)$.

2.5.P61 Let the eigenvalues of a given $A \in M_n$ be $\lambda_1, \ldots, \lambda_n$. (a) Show that

$$\max_{i=1,\ldots,n} \left| \lambda_i - \frac{\mathrm{tr}\, A}{n} \right| \leq \sqrt{\frac{n-1}{n}} \left(\sum_{i=1}^n |\lambda_i|^2 - \frac{|\mathrm{tr}\, A|^2}{n} \right)^{1/2}$$

and deduce that

$$\max_{i=1,\ldots,n} \left| \lambda_i - \frac{\mathrm{tr}\, A}{n} \right| \leq \sqrt{\frac{n-1}{n}} \left(\mathrm{tr}\, A^*A - \frac{|\mathrm{tr}\, A|^2}{n} \right)^{1/2}$$

with equality if and only if A is normal and has eigenvalues $(n-1)c, -c, \ldots, -c$ for some $c \in \mathbf{C}$. (b) What does this say geometrically about the eigenvalues of A? What if A is

Hermitian? (c) The quantity spread $A = \max\{|\lambda - \mu| : \lambda, \mu \in \sigma(A)\}$ is the maximum distance between any two eigenvalues of A. Explain why spread $A \leq 2\sqrt{\frac{n-1}{n}}\left(\operatorname{tr} A^*A - \frac{|\operatorname{tr} A|^2}{n}\right)^{1/2}$ and, if A is Hermitian, spread $A \leq 2\sqrt{\frac{n-1}{n}}\left(\operatorname{tr} A^2 - \frac{|\operatorname{tr} A|^2}{n}\right)^{1/2}$. For a lower bound on spread A, see (4.3.P16).

2.5.P62 If $A \in M_n$ has exactly k nonzero eigenvalues, we know that rank $A \geq k$. If A is normal, why is rank $A = k$?

2.5.P63 Suppose that $A = [a_{ij}] \in M_n$ is tridiagonal. If A is normal, show that $|a_{i,i+1}| = |a_{i+1,i}|$ for each $i = 1, \ldots, n - 1$. What about the converse? Compare with (2.5.P38(d)).

2.5.P64 Let $A \in M_n$ be given. Show that $\operatorname{rank}(AA^* - A^*A) \neq 1$.

2.5.P65 Suppose that $A \in M_n$ is normal and let $r \in \{1, \ldots, n\}$. Explain why the compound matrix $C_r(A)$ is normal.

2.5.P66 Let $A \in M_n$. We say that A is *squared normal* if A^2 is normal. It is known that A is squared normal if and only if A is unitarily similar to a direct sum of blocks, each of which is

$$[\lambda] \text{ or } \tau\begin{bmatrix} 0 & 1 \\ \mu & 0 \end{bmatrix}, \text{ in which } \tau \in \mathbf{R}, \lambda, \mu \in \mathbf{C}, \tau > 0, \text{ and } |\mu| < 1 \qquad (2.5.22a)$$

This direct sum is uniquely determined by A, up to permutation of its blocks. Use (2.2.8) to show that each 2-by-2 block in (2.5.22a) is unitarily similar to a block of the form $\begin{bmatrix} \nu & r \\ 0 & -\nu \end{bmatrix}$, in which $\nu = \tau\sqrt{\mu} \in \mathcal{D}_+$, $r = \tau(1 - |\mu|)$, and $\mathcal{D}_+ = \{z \in \mathbf{C} : \operatorname{Re} z > 0\} \cup \{it : t \in \mathbf{R} \text{ and } t \geq 0\}$. Conclude that A^2 is normal if and only if A is unitarily similar to a direct sum of blocks, each of which is

$$[\lambda] \text{ or } \begin{bmatrix} \nu & r \\ 0 & -\nu \end{bmatrix}, \text{ in which } \lambda, \nu \in \mathbf{C}, r \in \mathbf{R}, r > 0, \text{ and } \nu \in \mathcal{D}_+ \qquad (2.5.22b)$$

Explain why this direct sum is uniquely determined by A, up to permutation of its blocks.

2.5.P67 Let $A, B \in M_n$ be normal. Show that $AB = 0$ if and only if $BA = 0$.

2.5.P68 Let $A, B \in M_n$. Suppose that B is normal and that every null vector of A is a normal eigenvector. Show that $AB = 0$ if and only if $BA = 0$.

2.5.P69 Consider a k-by-k block matrix $M_A = [A_{ij}]_{i,j=1}^k \in M_{kn}$ in which $A_{ij} = 0$ if $i \geq j$ and $A_{ij} = I_n$ if $j = i + 1$. Define $M_B = [B_{ij}]_{i,j=1}^k$ similarly. Let $W = [W_{ij}]_{i,j=1}^k \in M_{kn}$ be partitioned conformally to M_A and M_B. (a) If $M_A W = W M_B$, show that W is block upper triangular and $W_{11} = \cdots = W_{kk}$. (b) If W is unitary and $M_A W = W M_B$ (that is, $M_A = W M_B W^*$, so M_A is unitarily similar to M_B via W), show that W is block diagonal, $W_{11} = U$ is unitary, $W = U \oplus \cdots \oplus U$, and $A_{ij} = U B_{ij} U^*$ for all i, j. For further properties of the block matrices M_A and M_B, see (4.4.P46) and (4.4.P47).

2.5.P70 Let pairs $(A_1, B_1), \ldots, (A_m, B_m)$ of n-by-n complex matrices be given. We say that these pairs are *simultaneously unitarily similar* if there is a unitary $U \in M_n$ such that $A_j = U B_j U^*$ for each $j = 1, \ldots, m$. Consider an $(m + 2)$-by-$(m + 2)$ block matrix $N_A = [N_{ij}]_{i,j=1}^k$ in which $N_{ij} = I_n$ if $j = i + 1$, $N_{ij} = A_i$ if $j = i + 2$, and $N_{ij} = 0$ if $j - i \notin \{1, 2\}$. Define N_B in a similar fashion. (a) Explain why N_A is unitarily similar to N_B if and only if the pairs $(A_1, B_1), \ldots, (A_m, B_m)$ are simultaneously unitarily similar. (b) Describe some other block matrices with this pleasant property. (c) Explain how simultaneous unitary similarity of finitely many pairs of complex matrices of the same size can be verified or refuted with finitely many computations.

2.5.P71 The matrix $\begin{bmatrix} a & b \\ -b & a \end{bmatrix} \in M_2(\mathbf{R})$ plays an important role in our discussion of real normal matrices. Discuss properties of this matrix as a special case of the real representation $R_1(A)$ studied in (3.1.P20).

2.5.P72 Consider the matrices $A_1 = \begin{bmatrix} i & 0 \\ 0 & -i \end{bmatrix}$ and $A_2 = \begin{bmatrix} 0 & 1 \\ -1 & 0 \end{bmatrix}$. (a) Show that each matrix is normal and commutes with its complex conjugate. (b) What are the eigenvalues of each matrix? (c) Explain why A_1 is unitarily similar to A_2. (d) Show that A_1 is not real orthogonally similar to A_2. (e) Let $A \in M_n$ be normal and satisfy either of the conditions (a) or (b) in (2.5.17). Then A is real orthogonally similar to the direct sum of a zero matrix and nonzero scalar multiples of one or more of the four types of blocks in (2.5.17.1). To determine which blocks and scalar multiples can occur in that direct sum, explain why one must know more about A than just its eigenvalues.

2.5.P73 Let $A \in M_n(\mathbf{R})$ be normal and let λ, x be an eigenpair of A in which $\lambda = a + ib$ is not real. (a) Show that $\bar{\lambda}, \bar{x}$ is an eigenpair of A and that $x^T x = 0$. (b) Let $x = u + iv$, in which u and v are real vectors. Show that $u^T u = v^T v \neq 0$ and $u^T v = 0$. (c) Let $q_1 = u/\sqrt{u^T u}$ and $q_2 = v/\sqrt{v^T v}$, and let $Q = [q_1 \ q_2 \ Q_1] \in M_n(\mathbf{R})$ be real orthogonal. Show that $Q^T A Q = \begin{bmatrix} B & * \\ 0 & * \end{bmatrix}$ in which $B = \begin{bmatrix} a & b \\ -b & a \end{bmatrix}$. (d) Give an alternative proof of (2.5.8) that does not rely on (2.3.4b).

2.5.P74 Let $A, B, X \in M_n$. If $AX = XB$ and X is normal, is it correct that $AX^* = X^* B$? Compare with the Fuglede–Putnam theorem (2.5.16).

2.5.P75 Let $A, B, X \in M_n$. (a) Show that $AX = XB$ and $XA = BX$ if and only if $\begin{bmatrix} 0 & X \\ X & 0 \end{bmatrix}$ commutes with $\begin{bmatrix} A & 0 \\ 0 & B \end{bmatrix}$. (b) If X is normal, $AX = XB$, and $XA = BX$, show that $AX^* = X^* B$ and $X^* A = BX^*$.

2.5.P76 Suppose that every entry of $A \in M_n(\mathbf{R})$ is either zero or one, let $e \in \mathbf{R}^n$ be the all-ones vector, and let $J \in M_n(\mathbf{R})$ be the all-ones matrix. Let $A = [c_1 \ \ldots \ c_n]$ and $A^T = [r_1 \ \ldots \ r_n]$. (a) Explain why the entries of Ae are both the row sums and the squares of the Euclidean norms of the rows of A. Interpret the entries of $A^T e$ in a similar manner. (b) Show that A is normal if and only if $Ae = A^T e$ and $c_i^T c_j = r_i^T r_j$ for all $i \neq j$. (c) Show that A is normal if and only if the complementary zero-one matrix $J - A$ is normal.

Notes and Further Readings. For a discussion of 89 characterizations of normality, see R. Grone, C. R. Johnson, E. Sa, and H. Wolkowicz, Normal matrices, *Linear Algebra Appl.* 87 (1987) 213–225 as well as L. Elsner and Kh. Ikramov, Normal matrices: an update, *Linear Algebra Appl.* 285 (1998) 291–303. Despite the issue raised in (2.5.P72), the representation described in (2.5.17) is actually a canonical form under real orthogonal similarity, up to permutation of direct summands and replacement of any direct summand by its transpose (there are real orthogonal similarity invariants other than eigenvalues); see G. Goodson and R. A. Horn, Canonical forms for normal matrices that commute with their complex conjugate, *Linear Algebra Appl.* 430 (2009) 1025–1038. For a proof of the canonical forms (2.5.22a,b) for squared-normal matrices, see R. A. Horn and V. V. Sergeichuk, Canonical forms for unitary congruence and *congruence, *Linear Multilinear Algebra* 57 (2009) 777–815. Problem 4.4.P38 contains a far-reaching generalization of (2.5.20(b)): Every nonsingular square complex matrix (and some singular matrices) has a QS factorization.

2.6 Unitary equivalence and the singular value decomposition

Suppose that a given matrix A is the basis representation of a linear transformation T : $V \to V$ on an n-dimensional complex vector space, with respect to a given orthonormal basis. A unitary similarity $A \to UAU^*$ corresponds to changing the basis from the given one to another orthonormal basis; the unitary matrix U is the change of basis matrix.

If $T : V_1 \to V_2$ is a linear transformation from an n-dimensional complex vector space into an m-dimensional one, and if $A \in M_{m,n}$ is its basis representation with respect to given orthonormal bases of V_1 and V_2, then the unitary equivalence $A \to UAW^*$ corresponds to changing the bases of V_1 and V_2 from the given ones to other orthonormal bases.

A unitary equivalence $A \to UAV$ involves *two* unitary matrices that can be selected independently. This additional flexibility permits us to achieve some reductions to special forms that may be unattainable with unitary similarity.

To ensure that we can reduce $A, B \in M_n$ to upper triangular form by the same unitary similarity, some condition (commutativity, for example) must be imposed on them. However, we can reduce *any* two given matrices to upper triangular form by the same unitary equivalence.

Theorem 2.6.1. *Let $A, B \in M_n$. There are unitary $V, W \in M_n$ such that $A = VT_AW^*$, $B = VT_BW^*$, and both T_A and T_B are upper triangular. If B is nonsingular, the main diagonal entries of $T_B^{-1}T_A$ are the eigenvalues of $B^{-1}A$.*

Proof. Suppose that B is nonsingular, and use (2.3.1) to write $B^{-1}A = UTU^*$, in which U is unitary and T is upper triangular. Use the QR factorization (2.1.14) to write $BU = QR$, in which Q is unitary and R is upper triangular. Then $A = BUTU^* = Q(RT)U^*$, RT is upper triangular, and $B = QRU^*$. Moreover, the eigenvalues of $B^{-1}A = UR^{-1}Q^*QRTU^* = UTU^*$ are the main diagonal entries of T.

If both A and B are singular, there is a $\delta > 0$ such that $B_\varepsilon = B + \varepsilon I$ is nonsingular whenever $0 < \varepsilon < \delta$; see (1.2.17). For any ε satisfying this constraint, we have shown that there are unitary $V_\varepsilon, W_\varepsilon \in M_n$ such that $V_\varepsilon^*AW_\varepsilon$ and $V_\varepsilon^*BW_\varepsilon$ are both upper triangular. Choose a sequence of nonzero scalars ε_k such that $\varepsilon_k \to 0$ and both $\lim_{k\to\infty} V_{\varepsilon_k} = V$ and $\lim_{k\to\infty} W_{\varepsilon_k} = W$ exist; each of the limits V and W is unitary; see (2.1.8). Then each of $\lim_{k\to\infty} V_{\varepsilon_k}^*AW_{\varepsilon_k} = V^*AW = T_A$ and $\lim_{k\to\infty} V_{\varepsilon_k}^*BW_{\varepsilon_k} = V^*BW = T_B$ is upper triangular. We conclude that $A = VT_AW^*$ and $B = VT_BW^*$, as asserted. $\square$

There is also a real version of this theorem, which uses the following fact.

Exercise. Suppose that $A, B \in M_n$, A is upper triangular, and B is upper quasi-triangular. Show that AB is upper quasitriangular conformal to B.

Theorem 2.6.2. *Let $A, B \in M_n(\mathbf{R})$. There are real orthogonal $V, W \in M_n$ such that $A = VT_AW^T$, $B = VT_BW^T$, T_A is real and upper quasitriangular, and T_B is real and upper triangular.*

Proof. If B is nonsingular, one uses (2.3.4) to write $B^{-1}A = UTU^T$, in which U is real orthogonal and T is real and upper quasitriangular. Use (2.1.14(d)) to write $BU = QR$, in which Q is real orthogonal and R is real and upper triangular. Then RU is upper quasitriangular, $A = Q(RT)U^T$, and $B = QRU^T$. If both A and B are singular, one can use a real version of the limit argument in the preceding proof. $\square$

Although only square matrices that are normal can be diagonalized by unitary similarity, any complex matrix can be diagonalized by unitary equivalence.

Theorem 2.6.3 (Singular value decomposition). *Let $A \in M_{n,m}$ be given, let $q = \min\{m, n\}$, and suppose that* rank $A = r$.

(a) *There are unitary matrices $V \in M_n$ and $W \in M_m$, and a square diagonal matrix*

$$\Sigma_q = \begin{bmatrix} \sigma_1 & & 0 \\ & \ddots & \\ 0 & & \sigma_q \end{bmatrix} \qquad (2.6.3.1)$$

such that $\sigma_1 \geq \sigma_2 \geq \cdots \geq \sigma_r > 0 = \sigma_{r+1} = \cdots = \sigma_q$ and $A = V\Sigma W^*$, in which

$$\Sigma = \Sigma_q \text{ if } m = n,$$
$$\Sigma = \begin{bmatrix} \Sigma_q & 0 \end{bmatrix} \in M_{n,m} \text{ if } m > n, \text{ and} \qquad (2.6.3.2)$$
$$\Sigma = \begin{bmatrix} \Sigma_q \\ 0 \end{bmatrix} \in M_{n,m} \text{ if } n > m$$

(b) *The parameters $\sigma_1, \ldots, \sigma_r$ are the positive square roots of the decreasingly ordered nonzero eigenvalues of AA^*, which are the same as the decreasingly ordered nonzero eigenvalues of A^*A.*

Proof. First suppose that $m = n$. The Hermitian matrices $AA^* \in M_n$ and $A^*A \in M_n$ have the same eigenvalues (1.3.22), so they are unitarily similar (2.5.4(d)), and hence there is a unitary U such that $A^*A = U(AA^*)U^*$. Then

$$(UA)^*(UA) = A^*U^*UA = A^*A = UAA^*U^* = (UA)(UA)^*$$

so UA is normal. Let $\lambda_1 = |\lambda_1|e^{i\theta_1}, \ldots, \lambda_n = |\lambda_n|e^{i\theta_n}$ be the eigenvalues of UA, ordered so that $|\lambda_1| \geq \cdots \geq |\lambda_n|$. Then $r = $ rank $A = $ rank UA is the number of nonzero eigenvalues of the normal matrix UA, so $|\lambda_r| > 0$ and $\lambda_{r+1} = \cdots = \lambda_n = 0$. Let $\Lambda = \text{diag}(\lambda_1, \ldots, \lambda_n)$, let $D = \text{diag}(e^{i\theta_1}, \ldots, e^{i\theta_n})$, let $\Sigma_q = \text{diag}(|\lambda_1|, \ldots, |\lambda_n|)$, and let X be a unitary matrix such that $UA = X\Lambda X^*$. Then D is unitary and $A = U^*X\Lambda X^* = U^*X\Sigma_q DX^* = (U^*X)\Sigma_q(DX^*)$ exhibits the desired factorization, in which $V = U^*X$ and $W = XD^*$ are unitary and $\sigma_j = |\lambda_j|$, $j = 1, \ldots, n$.

Now suppose that $m > n$. Then $r \leq n$, so the null space of A has dimension $m - r \geq m - n$. Let $x_1, \ldots, x_{m-n}$ be any orthonormal list of vectors in the null space of A, let $X_2 = [x_1 \cdots x_{m-n}] \in M_{m,m-n}$, and let $X = [X_1 \ X_2] \in M_m$ be unitary, that is, extend the given orthonormal list to a basis of $\mathbf{C}^m$. Then $AX = [AX_1 \ AX_2] = [AX_1 \ 0]$ and $AX_1 \in M_n$. Using the preceding case, write $AX_1 = V\Sigma_q W^*$, in which $V, W \in M_n$ are unitary and Σ_q has the form (2.6.3.1). This gives

$$A = [AX_1 \; 0]X^* = [V\Sigma_q W^* \; 0]X^* = V[\Sigma_q \; 0]\left(\begin{bmatrix} W^* & 0 \\ 0 & I_{m-n} \end{bmatrix} X^*\right)$$

which is a factorization of the asserted form.

If $n > m$, apply the preceding case to A^*.

Using the factorization $A = V\Sigma W^*$, notice that rank A = rank Σ since V and W are nonsingular. But rank Σ equals the number of nonzero (and hence positive) diagonal entries of Σ, as asserted. Now compute $AA^* = V\Sigma W^* W\Sigma^T V^* = V\Sigma\Sigma^T V^*$, which is unitarily similar to $\Sigma\Sigma^T$. If $n = m$, then $\Sigma\Sigma^T = \Sigma_q^2 = \text{diag}(\sigma_1^2, \ldots, \sigma_n^2)$. If $m > n$, then $\Sigma\Sigma^T = [\Sigma_q \; 0][\Sigma_q \; 0]^T = \Sigma_q^2 + 0_n = \Sigma_q^2$. Finally, if $n > m$, then

$$\Sigma\Sigma^T = \begin{bmatrix} \Sigma_q \\ 0 \end{bmatrix}[\Sigma_q \; 0] = \begin{bmatrix} \Sigma_q^2 & 0 \\ 0 & 0_{n-m} \end{bmatrix}$$

In each case, the nonzero eigenvalues of AA^* are $\sigma_1^2, \ldots, \sigma_r^2$, as asserted. $\square$

The diagonal entries of the matrix Σ in (2.6.3.2) (that is, the scalars $\sigma_1, \ldots, \sigma_q$ that are the diagonal entries of the square matrix Σ_q) are the *singular values* of A. The *multiplicity* of a singular value σ of A is the multiplicity of σ^2 as an eigenvalue of AA^* or, equivalently, of A^*A. A singular value σ of A is said to be *simple* if σ^2 is a simple eigenvalue of AA^* or, equivalently, of A^*A. The rank of A is *equal* to the number of its nonzero singular values, while rank A is *not less than* (and can be greater than) the number of its nonzero eigenvalues.

The singular values of A are uniquely determined by the eigenvalues of A^*A (equivalently, by the eigenvalues of AA^*), so the diagonal factor Σ in the singular value decomposition of A is determined up to permutation of its diagonal entries; a conventional choice that makes Σ unique is to require that the singular values be arranged in nonincreasing order, but other choices are possible.

Exercise. Let $A \in M_{m,n}$. Explain why A, $\bar{A}$, A^T, and A^* have the same singular values.

Let $\sigma_1, \ldots, \sigma_n$ be the singular values of $A \in M_n$. Explain why

$$\sigma_1 \cdots \sigma_n = |\det A| \quad \text{and} \quad \sigma_1^2 + \cdots + \sigma_n^2 = \text{tr } A^*A \qquad (2.6.3.3)$$

Exercise. Show that the two squared singular values of $A \in M_2$ are

$$\sigma_1^2, \sigma_2^2 = \frac{1}{2}\left((\text{tr } A^*A) \pm \sqrt{(\text{tr } A^*A)^2 - 4|\det A|^2}\right). \qquad (2.6.3.4)$$

Exercise. Explain why the singular values of the nilpotent matrix

$$A = \begin{bmatrix} 0 & a_{12} & & 0 \\ & \ddots & \ddots & \\ & & \ddots & a_{n-1,n} \\ 0 & & & 0 \end{bmatrix} \in M_n$$

(zero entries everywhere except for some nonzero entries in the first superdiagonal) are 0, $|a_{12}|, \ldots, |a_{n-1,n}|$.

The following theorem gives a precise formulation of the assertion that the singular values of a matrix depend continuously on its entries.

Theorem 2.6.4. *Let an infinite sequence* $A_1, A_2, \ldots \in M_{n,m}$ *be given, suppose that* $\lim_{k\to\infty} A_k = A$ *(entrywise convergence), and let* $q = \min\{m, n\}$. *Let* $\sigma_1(A) \geq \cdots \geq \sigma_q(A)$ *and* $\sigma_1(A_k) \geq \cdots \geq \sigma_q(A_k)$ *be the nonincreasingly ordered singular values of* A *and* A_k, *respectively, for* $k = 1, 2, \ldots$. *Then* $\lim_{k\to\infty} \sigma_i(A_k) = \sigma_i(A)$ *for each* $i = 1, \ldots, q$.

Proof. If the assertion of the theorem is false, then there is some $\varepsilon_0 > 0$ and an infinite sequence of positive integers $k_1 < k_2 < \cdots$ such that for every $j = 1, 2, \ldots$ we have

$$\max_{i=1,\ldots,q} \left|\sigma_i(A_{k_j}) - \sigma_i(A)\right| > \varepsilon_0 \tag{2.6.4.1}$$

For each $j = 1, 2, \ldots$ let $A_{k_j} = V_{k_j} \Sigma_{k_j} W_{k_j}^*$, in which $V_{k_j} \in M_n$ and $W_{k_j} \in M_m$ are unitary and $\Sigma_{k_j} \in M_{n,m}$ is the nonnegative diagonal matrix such that diag $\Sigma_{k_j} = [\sigma_1(A_{k_j}) \ldots \sigma_q(A_{k_j})]^T$. Lemma 2.1.8 ensures that there is an infinite sub-subsequence $k_{j_1} < k_{j_2} < \cdots$ and unitary matrices V and W such that $\lim_{\ell\to\infty} V_{k_{j_\ell}} = V$ and $\lim_{\ell\to\infty} W_{k_{j_\ell}} = W$. Then

$$\lim_{\ell\to\infty} \Sigma_{k_{j_\ell}} = \lim_{\ell\to\infty} V_{k_{j_\ell}}^* A_{k_{j_\ell}} W_{k_{j_\ell}} = \left(\lim_{\ell\to\infty} V_{k_{j_\ell}}^*\right)\left(\lim_{\ell\to\infty} A_{k_{j_\ell}}\right)\left(\lim_{\ell\to\infty} W_{k_{j_\ell}}\right) = V^* A W$$

exists and is a nonnegative diagonal matrix with nonincreasingly ordered diagonal entries; we denote it by Σ and observe that $A = V\Sigma W^*$. Uniqueness of the singular values of A ensures that diag $\Sigma = [\sigma_1(A) \ldots \sigma_q(A)]^T$, contradicts (2.6.4.1), and proves the theorem. $\quad\square$

The unitary factors in a singular value decomposition are never unique. For example, if $A = V\Sigma W^*$, we may replace V by $-V$ and W by $-W$. The following theorem describes in an explicit and very useful fashion how, given one pair of unitary factors in a singular value decomposition, all possible pairs of unitary factors can be obtained.

Theorem 2.6.5 (Autonne's uniqueness theorem). *Let* $A \in M_{n,m}$ *be given with* rank $A = r$. *Let* $s_1, \ldots, s_d$ *be the distinct positive singular values of* A, *in any order, with respective multiplicities* $n_1, \ldots, n_d$, *and let* $\Sigma_d = s_1 I_{n_1} \oplus \cdots \oplus s_d I_{n_d} \in M_r$. *Let* $A = V\Sigma W^*$ *be a singular value decomposition with* $\Sigma = \begin{bmatrix} \Sigma_d & 0 \\ 0 & 0 \end{bmatrix} \in M_{n,m}$ *as in (2.6.3.2), so that* $\Sigma^T\Sigma = s_1^2 I_{n_1} \oplus \cdots \oplus s_d^2 I_{n_d} \oplus 0_{n-r}$ *and* $\Sigma\Sigma^T = s_1^2 I_{n_1} \oplus \cdots \oplus s_d^2 I_{n_d} \oplus 0_{m-r}$ *(one zero direct summand is absent if* A *has full rank; both are absent if* A *is square and nonsingular). Let* $\hat{V} \in M_n$ *and* $\hat{W} \in M_m$ *be unitary. Then* $A = \hat{V}\Sigma\hat{W}^*$ *if and only if there are unitary matrices* $U_1 \in M_{n_1}, \ldots, U_d \in M_{n_d}$, $\tilde{V} \in M_{n-r}$, *and* $\tilde{W} \in M_{m-r}$ *such that*

$$\hat{V} = V(U_1 \oplus \cdots \oplus U_d \oplus \tilde{V}) \text{ and } \hat{W} = W(U_1 \oplus \cdots \oplus U_d \oplus \tilde{W}) \tag{2.6.5.1}$$

If A *is real and the factors* $V, W, \hat{V}, \hat{W}$ *are real orthogonal, then the matrices* $U_1, \ldots, U_d, \tilde{V}$, *and* $\tilde{W}$ *may be taken to be real orthogonal.*

Proof. The Hermitian matrix A^*A is represented as $A^*A = (V\Sigma W^*)^*(V\Sigma W^*) = W\Sigma^T\Sigma W^*$ and also as $A^*A = \hat{W}\Sigma^T\Sigma\hat{W}^*$. Theorem 2.5.4 ensures that there are unitary matrices $W_1, \ldots, W_d, W_{d+1}$ with $W_i \in M_{n_i}$ for $i = 1, \ldots, d$ such that $\hat{W} = W(W_1 \oplus \cdots \oplus W_d \oplus W_{d+1})$. We also have $AA^* = V\Sigma\Sigma^T V^* = \hat{V}\Sigma\Sigma^T\hat{V}^*$, so (2.5.4) again tells us that there are unitary matrices $V_1, \ldots, V_d, V_{d+1}$ with $V_i \in M_{n_i}$ for $i = 1, \ldots, d$ such that $\hat{V} = V(V_1 \oplus \cdots \oplus V_d \oplus V_{d+1})$. Since $A = V\Sigma W^* = \hat{V}\Sigma\hat{W}^*$, we have $\Sigma = (V^*\hat{V})\Sigma(\hat{W}^*W)$, that is, $s_i I_{n_i} = V_i(s_i I_{n_i})W_i^*$ for $i = 1, \ldots, d+1$, or $V_i W_i^* = I_{n_i}$ for each $i = 1, \ldots, d$. The matrices $\tilde{V}$ and $\tilde{W}$, if present, are arbitrary. It follows that $V_i = W_i$ for each $i = 1, \ldots, d$. The final assertion follows from the preceding argument and the fact that $V^T\hat{V}$ and $W^T\hat{W}$ are real. $\square$

The singular value decomposition is a very important tool in matrix analysis, with myriad applications in engineering, numerical computation, statistics, image compression, and many other areas; for more details, see Chapter 7 and chapter 3 of Horn and Johnson (1991).

We close this chapter with three applications of the preceding uniqueness theorem: Singular value decompositions of symmetric or skew-symmetric matrices can be chosen to be unitary congruences, and a real matrix has a singular value decomposition in which all three factors are real.

Corollary 2.6.6. *Let $A \in M_n$ and let $r = \text{rank } A$.*

(a) *(Autonne) $A = A^T$ if and only if there is a unitary $U \in M_n$ and a nonnegative diagonal matrix Σ such that $A = U\Sigma U^T$. The diagonal entries of Σ are the singular values of A.*

(b) *If $A = -A^T$, then r is even and there is a unitary $U \in M_n$ and positive real scalars $s_1, \ldots, s_{r/2}$ such that*

$$A = U\left(\begin{bmatrix} 0 & s_1 \\ -s_1 & 0 \end{bmatrix} \oplus \cdots \oplus \begin{bmatrix} 0 & s_{r/2} \\ -s_{r/2} & 0 \end{bmatrix} \oplus 0_{n-r}\right)U^T \qquad (2.6.6.1)$$

The nonzero singular values of A are $s_1, s_1, \ldots, s_{r/2}, s_{r/2}$. Conversely, any matrix of the form (2.6.6.1) is skew symmetric.

Proof. (a) If $A = U\Sigma U^T$ for a unitary $U \in M_n$ and a nonnegative diagonal matrix Σ, then A is symmetric and the diagonal entries of Σ are its singular values. Conversely, let $s_1, \ldots, s_d$ be the distinct positive singular values of A, in any order, with respective multiplicities $n_1, \ldots, n_d$, and let $A = V\Sigma W^*$ be a singular value decomposition in which $V, W \in M_n$ are unitary and $\Sigma = s_1 I_{n_1} \oplus \cdots \oplus s_d I_{n_d} \oplus 0_{n-r}$; the zero block is missing if A is nonsingular. We have $A = V\Sigma W^* = \bar{W}\Sigma\bar{V}^* = A$, so the preceding theorem ensures that there are unitary matrices U_W and U_V such that $\bar{V} = WU_W, \bar{W} = VU_V, U_V = U_1 \oplus \cdots \oplus U_d \oplus \tilde{V}, U_W = U_1 \oplus \cdots \oplus U_d \oplus \tilde{W}$, and each $U_i \in M_{n_i}, i = 1, \ldots, d$. Then $U_W = W^*\bar{V} = (V^*\bar{W})^T = U_V^T$, which implies that each $U_j = U_j^T$, that is, each U_j is unitary and symmetric. Corollary 2.5.20a ensures that there are symmetric unitary matrices R_j such that $R_j^2 = U_j$ for each $j = 1, \ldots, d$. Let $R = R_1 \oplus \cdots \oplus R_d \oplus I_{n-r}$. Then R is symmetric and unitary, and $U_V\Sigma = R^2\Sigma = R\Sigma R$, so $A = \bar{W}\Sigma V^T = VU_V\Sigma V^T = VR\Sigma RV^T = (VR)\Sigma(VR)^T$ is a factorization of the asserted form.

(b) Starting with the identity $V\Sigma W^* = -\bar{W}\Sigma V^T = -\bar{W}\Sigma\bar{V}^*$ and proceeding exactly as in (a), we have $\bar{V} = WU_W, \bar{W} = -VU_V$, that is, $U_W = W^*\bar{V}$ and $U_V = -V^*\bar{W} = -U_W^T$. In particular, $U_j = -U_j^T$ for $j = 1, \ldots, d$, that is, each U_j is unitary and skew symmetric. Corollary 2.5.18b ensures that, for each $j = 1, \ldots, d$, n_j is even and there are real orthogonal matrices Q_j and real parameters $\theta_1^{(j)}, \ldots, \theta_{n_j/2}^{(j)} \in [0, 2\pi)$ such that

$$U_j = Q_j \left(e^{i\theta_1^{(j)}} \begin{bmatrix} 0 & 1 \\ -1 & 0 \end{bmatrix} \oplus \cdots \oplus e^{i\theta_{n_j/2}^{(j)}} \begin{bmatrix} 0 & 1 \\ -1 & 0 \end{bmatrix} \right) Q_j^T$$

Define the real orthogonal matrix $Q = Q_1 \oplus \cdots \oplus Q_d \oplus I_{n-r}$ and the skew-symmetric unitary matrices

$$S_j = e^{i\theta_1^{(j)}} \begin{bmatrix} 0 & 1 \\ -1 & 0 \end{bmatrix} \oplus \cdots \oplus e^{i\theta_{n_j/2}^{(j)}} \begin{bmatrix} 0 & 1 \\ -1 & 0 \end{bmatrix}, \; j = 1, \ldots, d$$

Let $S = S_1 \oplus \cdots \oplus S_d \oplus 0_{n-r}$. Then $U_V \Sigma = QSQ^T \Sigma = QS\Sigma Q^T$, so $A = -\bar{W}\Sigma V^T = VU_V \Sigma V^T = VQS\Sigma Q^T V^T = (VQ)S\Sigma(VQ)^T$ is a factorization of the asserted form and rank $A = n_1 + \cdots + n_d$ is even. $\qquad\square$

Corollary 2.6.7. *Let $A \in M_{n,m}(\mathbf{R})$ and suppose that* rank $A = r$. *Then $A = P\Sigma Q^T$, in which $P \in M_n(\mathbf{R})$ and $Q \in M_m(\mathbf{R})$ are real orthogonal, and $\Sigma \in M_{n,m}(\mathbf{R})$ is nonnegative diagonal and has the form (2.6.3.1) or (2.6.3.2).*

Proof. Using the notation of (2.6.4), let $A = V\Sigma W^*$ be a given singular value decomposition; the unitary matrices V and W need not be real. We have $V\Sigma W^* = A = \bar{A} = \bar{V}\Sigma\bar{W}^*$, so $V^T V \Sigma = \Sigma W^T W$. Theorem 2.6.5 ensures that there are unitary matrices $U_V = U_1 \oplus \cdots \oplus U_d \oplus \tilde{V} \in M_n$ and $U_W = U_1 \oplus \cdots \oplus U_d \oplus \tilde{W} \in M_m$ such that $\bar{V} = VU_V$ and $\bar{W} = WU_W$. Then $U_V = V^*\bar{V} = \bar{V}^T V$ and $U_W = \bar{W}^T W$ are unitary and symmetric, so $\tilde{V}$, $\tilde{W}$, and each U_i is unitary and symmetric. Corollary 2.5.20(a) ensures that there are symmetric unitary matrices $R_{\tilde{V}}$, $R_{\tilde{W}}$, and $R_1, \ldots, R_d$ such that $R_{\tilde{V}}^2 = \tilde{V}$, $R_{\tilde{W}}^2 = \tilde{W}$, and $R_i^2 = U_i$ for each $i = 1, \ldots, d$. Let $R_V = R_1 \oplus \cdots \oplus R_d \oplus R_{\tilde{V}}$ and $R_W = R_1 \oplus \cdots \oplus R_d \oplus R_{\tilde{W}}$. Then R_V and R_W are symmetric and unitary, $R_V^{-1} = R_V^* = \overline{R_V}$, $R_W^{-1} = R_W^* = \overline{R_W}$, $R_V^2 = U_V$, $R_W^2 = U_W$, and $R_V \Sigma \overline{R_W} = \Sigma$, so

$$A = \bar{V}\Sigma\bar{W}^* = VU_V\Sigma(WU_W)^* = VR_V^2\Sigma(WR_W^2)^*$$
$$= (VR_V)(R_V\Sigma\overline{R_W})(WR_W)^* = (VR_V)\Sigma(WR_W)^*$$

We conclude the argument by observing that $\bar{V} = VU_V = VR_V^2$ and $\bar{W} = WU_W = WR_W^2$, so $\overline{VR_V} = \bar{V}R_V^* = VR_V$ and $\overline{WR_W} = \bar{W}R_W^* = WR_W$. That is, both VR_V and WR_W are unitary and real, so they are both real orthogonal. $\qquad\square$

Problems

2.6.P1 Let $A \in M_{n,m}$ with $n \geq m$. Show that A has full column rank if and only if all of its singular values are positive.

2.6.P2 Suppose that $A, B \in M_{n,m}$ can be simultaneously diagonalized by unitary equivalence, that is, suppose that there are unitary $X \in M_n$ and $Y \in M_m$ such that each of $X^*AY = \Lambda$ and $X^*BY = M$ is diagonal (0.9.1). Show that both AB^* and B^*A are normal.

2.6.P3 When are $A, B \in M_{n,m}$ simultaneously unitarily equivalent to diagonal matrices? Show that AB^* and B^*A are both normal *if and only if* there are unitary $X \in M_n$ and $Y \in M_m$ such that $A = X\Sigma Y^*$; $B = X\Lambda Y^*$; $\Sigma, \Lambda \in M_{n,m}$ are diagonal; and $\Sigma \in M_{n,m}$ has the form (2.6.3.1,2).

2.6.P4 When are $A, B \in M_{n,m}$ simultaneously unitarily equivalent to *real* or *nonnegative real* diagonal matrices? (a) Show that AB^* and B^*A are both Hermitian if and only if there are unitary $X \in M_n$ and $Y \in M_m$ such that $A = X\Sigma Y^*$; $B = X\Lambda Y^*$; $\Sigma, \Lambda \in M_{n,m}(\mathbf{R})$ are diagonal; and Σ has the form (2.6.3.1,2). (b) If A and B are real, show that AB^T and $B^T A$ are both real symmetric if and only if there are real orthogonal matrices $X \in M_n(\mathbf{R})$ and $Y \in M_m(\mathbf{R})$ such that $A = X\Sigma Y^T$; $B = X\Lambda Y^T$; $\Sigma, \Lambda \in M_{n,m}(\mathbf{R})$ are diagonal; and Σ has the form (2.6.3.1,2). (c) In both (a) and (b), show that Λ can be chosen to have nonnegative diagonal entries if and only if all the eigenvalues of the Hermitian matrices AB^* and B^*A are nonnegative.

2.6.P5 Let $A \in M_{n,m}$ be given and write $A = B + iC$, in which $B, C \in M_{n,m}(\mathbf{R})$. Show that there are real orthogonal matrices $X \in M_n(\mathbf{R})$ and $Y \in M_m(\mathbf{R})$ such that $A = X\Lambda Y^T$ and $\Lambda \in M_{n,m}(\mathbf{C})$ is diagonal if and only if both BC^T and $C^T B$ are real symmetric.

2.6.P6 Let $A \in M_n$ be given and let $A = QR$ be a QR factorization (2.1.14). (a) Explain why QR is normal if and only if RQ is normal. (b) Show that A is normal if and only if Q and R^* can be simultaneously diagonalized by unitary equivalence.

2.6.P7 Show that two complex matrices of the same size are unitarily equivalent if and only if they have the same singular values.

2.6.P8 Let $A \in M_{n,k}$ and $B \in M_{k,m}$ be given. Use the singular value decomposition to show that rank $AB \leq \min\{\text{rank } A, \text{rank } B\}$.

2.6.P9 Let $A \in M_n$ be given. Suppose that rank $A = r$, form $\Sigma_1 = \text{diag}(\sigma_1, \ldots, \sigma_r)$ from its decreasingly ordered positive singular values, and let $\Sigma = \Sigma_1 \oplus 0_{n-r}$. Suppose that $W \in M_n$ is unitary and $A^*A = W\Sigma^2 W^*$. Show that there is a unitary $V \in M_n$ such that $A = V\Sigma W^*$.

2.6.P10 Let $A, B \in M_n$ be given, let $\sigma_1 \geq \cdots \geq \sigma_n \geq 0$ be the singular values of A, and let $\Sigma = \text{diag}(\sigma_1, \ldots, \sigma_n)$. Show that the following three statements are equivalent: (a) $A^*A = B^*B$; (b) there are unitary matrices $W, X, Y \in M_n$ such that $A = X\Sigma W^*$ and $B = Y\Sigma W^*$; (c) there is a unitary $U \in M_n$ such that $B = UA$. See (7.3.11) for a generalization.

2.6.P11 Let $A \in M_{n,m}$ and a normal $B \in M_m$ be given. Show that A^*A commutes with B if and only if there are unitary matrices $V \in M_n$ and $W \in M_m$, and diagonal matrices $\Sigma \in M_{n,m}$ and $\Lambda \in M_m$, such that $A = V\Sigma W^*$ and $B = W\Lambda W^*$.

2.6.P12 Let $A \in M_n$ have a singular value decomposition $A = V\Sigma W^*$, in which $\Sigma = \text{diag}(\sigma_1, \ldots, \sigma_n)$ and $\sigma_1 \geq \cdots \geq \sigma_n$. (a) Show that adj A has a singular value decomposition adj $A = X^*SY$ in which $X = (\det W)(\text{adj } W)$, $Y = (\det V)(\text{adj } V)$, and $S = \text{diag}(s_1, \ldots, s_n)$, in which each $s_i = \prod_{j \neq i} \sigma_j$. (b) Use (a) to explain why adj $A = 0$ if rank $A \leq n - 2$. (c) If rank $A = n - 1$ and $v_n, w_n \in \mathbf{C}^n$ are the last columns of V and W, respectively, show that adj $A = \sigma_1 \cdots \sigma_{n-1} e^{i\theta} w_n v_n^*$, in which $\det(VW^*) = e^{i\theta}$, $\theta \in \mathbf{R}$.

2.6.P13 Let $A \in M_n$ and let $A = V\Sigma W^*$ be a singular value decomposition. (a) Show that A is unitary if and only if $\Sigma = I$. (b) Show that A is a scalar multiple of a unitary matrix if and only if Ax is orthogonal to Ay whenever $x, y \in \mathbf{C}^n$ are orthogonal.

2.6.P14 Let $A \in M_n$ be given. (a) Suppose that A is normal and let $A = U\Lambda U^*$ be a spectral decomposition, in which U is unitary and $\Lambda = \mathrm{diag}(\lambda_1, \ldots, \lambda_n) = \mathrm{diag}(e^{i\theta_1}|\lambda_1|, \ldots, e^{i\theta_n}|\lambda_n|)$. Let $D = \mathrm{diag}(e^{i\theta_1}, \ldots, e^{i\theta_n})$ and $\Sigma = \mathrm{diag}(|\lambda_1|, \ldots, |\lambda_n|)$. Explain why $A = (UD)\Sigma U^*$ is a singular value decomposition of A and why the singular values of A are the absolute values of its eigenvalues. (b) Let $s_1, \ldots, s_d$ be the distinct singular values of A and let $A = V\Sigma W^*$ be a singular value decomposition, in which $V, W \in M_n$ are unitary and $\Sigma = s_1 I_{n_1} \oplus \cdots \oplus s_d I_{n_d}$. Show that A is normal if and only if there is a block diagonal unitary matrix $U = U_1 \oplus \cdots \oplus U_d$, partitioned conformally to Σ, such that $V = WU$. (c) If A is normal and has distinct singular values, and if $A = V\Sigma W^*$ is a singular value decomposition, explain why $V = WD$, in which D is a diagonal unitary matrix. What does the hypothesis of distinct singular values say about the eigenvalues of A?

2.6.P15 Let $A = [a_{ij}] \in M_n$ have eigenvalues $\lambda_1, \ldots, \lambda_n$ ordered so that $|\lambda_1| \geq \cdots \geq |\lambda_n|$ and singular values $\sigma_1, \ldots, \sigma_n$ ordered so that $\sigma_1 \geq \cdots \geq \sigma_n$. Show that (a) $\sum_{i,j=1}^n |a_{ij}|^2 = \mathrm{tr}\, A^*A = \sum_{i=1}^n \sigma_i^2$; (b) $\sum_{i=1}^n |\lambda_i|^2 \leq \sum_{i=1}^n \sigma_i^2$ with equality if and only if A is normal (Schur's inequality); (c) $\sigma_i = |\lambda_i|$ for all $i = 1, \ldots, n$ if and only if A is normal; (d) if $|a_{ii}| = \sigma_i$ for all $i = 1, \ldots, n$, then A is diagonal; (e) if A is normal and $|a_{ii}| = |\lambda_i|$ for all $i = 1, \ldots, n$, then A is diagonal.

2.6.P16 Let $U, V \in M_n$ be unitary. (a) Show that there are always unitary $X, Y \in M_n$ and a diagonal unitary $D \in M_n$ such that $U = XDY$ and $V = Y^*DX^*$. (b) Explain why the unitary equivalence map $A \to UAV = XDYAY^*DX^*$ on M_n is the composition of a unitary similarity, a diagonal unitary congruence, and a unitary similarity.

2.6.P17 Let $A \in M_{n,m}$. Use the singular value decomposition to explain why rank $A = $ rank $AA^* = $ rank A^*A.

2.6.P18 Let $A \in M_n$ be a projection and suppose that rank $A = r$. (a) Show that A is unitarily similar to $\begin{bmatrix} I_r & X \\ 0 & 0_{n-r} \end{bmatrix}$; see (1.1.P5). (b) Let $X = V\Sigma W^*$ be a singular value decomposition. Show that A is unitarily similar to $\begin{bmatrix} I_r & \Sigma \\ 0 & 0_{n-r} \end{bmatrix}$ via $V \oplus W$, and hence the singular values of A are the diagonal entries of $(I_r + \Sigma\Sigma^T) \oplus 0_{n-r}$; let $\sigma_1, \ldots, \sigma_g$ be the singular values of A that are greater than 1. (c) Show that A is unitarily similar to $0_{n-r-g} \oplus I_{r-g} \oplus \begin{bmatrix} 1 & (\sigma_1^2-1)^{1/2} \\ 0 & 0 \end{bmatrix} \oplus \cdots \oplus \begin{bmatrix} 1 & (\sigma_g^2-1)^{1/2} \\ 0 & 0 \end{bmatrix}$.

2.6.P19 Let $U = \begin{bmatrix} U_{11} & U_{12} \\ U_{21} & U_{22} \end{bmatrix} \in M_{k+\ell}$ be unitary with $U_{11} \in M_k$, $U_{22} \in M_\ell$, and $k \leq \ell$. Show that the nonincreasingly ordered singular values of the blocks of U are related by the identities $\sigma_i(U_{11}) = \sigma_i(U_{22})$ and $\sigma_i(U_{12}) = \sigma_i(U_{21}) = (1 - \sigma_{k-i+1}^2(U_{11}))^{1/2}$ for each $i = 1, \ldots, k$; and $\sigma_i(U_{22}) = 1$ for each $i = k+1, \ldots, \ell$. In particular, $|\det U_{11}| = |\det U_{22}|$ and $\det U_{12}U_{12}^* = \det U_{21}^*U_{21}$. Explain why these results imply (2.1.10).

2.6.P20 Let $A \in M_n$ be symmetric. Suppose that the special singular value decomposition in (2.6.6(a)) is known if A is *nonsingular*. Provide details for the following two approaches to showing that it is valid even if A is singular. (a) Consider $A_\varepsilon = A + \varepsilon I$; use (2.1.8) and (2.6.4). (b) Let the columns of $U_1 \in M_{n,\nu}$ be an orthonormal basis for the null space of A and let $U = [U_1 \ U_2] \in M_n$ be unitary. Let $U^T A U = [A_{ij}]_{i,j=1}^2$ (partitioned conformally to U). Explain why A_{11}, A_{12}, and A_{21} are zero matrices, while A_{22} is nonsingular and symmetric.

2.6.P21 Let $A, B \in M_n$ be symmetric. Show that $A\bar{B}$ is normal if and only if there is a unitary $U \in M_n$ such that $A = U\Sigma U^T$, $B = U\Lambda U^T$, $\Sigma, \Lambda \in M_n$ are diagonal and the diagonal entries of Σ are nonnegative.

2.6.P22 Let $A, B \in M_n$ be symmetric. (a) Show that $A\bar{B}$ is Hermitian if and only if there is a unitary $U \in M_n$ such that $A = U\Sigma U^T$, $B = U\Lambda U^T$, $\Sigma, \Lambda \in M_n(\mathbf{R})$ are diagonal and the diagonal entries of Σ are nonnegative. (b) Show that $A\bar{B}$ is Hermitian and has nonnegative eigenvalues if and only if there is a unitary $U \in M_n$ such that $A = U\Sigma U^T$, $B = U\Lambda U^T$, $\Sigma, \Lambda \in M_n(\mathbf{R})$ are diagonal and the diagonal entries of Σ and Λ are nonnegative.

2.6.P23 Let $A \in M_n$ be given. Suppose that rank $A = r \geq 1$ and suppose that A is *self-annihilating*, that is, $A^2 = 0$. Provide details for the following outline of a proof that A is unitarily *similar* to

$$\sigma_1 \begin{bmatrix} 0 & 1 \\ 0 & 0 \end{bmatrix} \oplus \cdots \oplus \sigma_r \begin{bmatrix} 0 & 1 \\ 0 & 0 \end{bmatrix} \oplus 0_{n-2r} \tag{2.6.8}$$

in which $\sigma_1 \geq \cdots \geq \sigma_r > 0$ are the positive singular values of A. (a) range $A \subseteq$ nullspace A and hence $2r \leq n$. (b) Let the columns of $U_2 \in M_{n,n-r}$ be an orthonormal basis for the null space of A^*, so $U_2^*A = 0$. Let $U = [U_1 \ U_2] \in M_n$ be unitary. Explain why the columns of $U_1 \in M_{n,r}$ are an orthonormal basis for the range of A and $AU_1 = 0$. (c) $U^*AU = \begin{bmatrix} 0 & B \\ 0 & 0 \end{bmatrix}$, in which $B \in M_{r,n-r}$ and rank $B = r$. (d) $B = V[\Sigma_r \ 0_{r,n-2r}]W^*$, in which $V \in M_r$ and $W \in M_{n-r}$ are unitary, and $\Sigma_r = \text{diag}(\sigma_1, \ldots, \sigma_r)$. (e) Let $Z = V \oplus W$. Then $Z^*(U^*AU)Z = \begin{bmatrix} 0 & \Sigma_r \\ 0 & 0 \end{bmatrix} \oplus 0_{n-2r}$, which is similar to (2.6.8) via a permutation matrix.

2.6.P24 Let $A \in M_n$ be given. Suppose that rank $A = r \geq 1$ and that A is *conjugate self-annihilating*: $A\bar{A} = 0$. Provide details for the following outline of a proof that A is unitarily *congruent* to (2.6.8), in which $\sigma_1 \geq \cdots \geq \sigma_r > 0$ are the positive singular values of A. (a) range $\bar{A} \subseteq$ nullspace A and hence $2r \leq n$. (b) Let the columns of $U_2 \in M_{n,n-r}$ be an orthonormal basis for the null space of A^T, so $U_2^T A = 0$. Let $U = [U_1 \ U_2] \in M_n$ be unitary. Explain why the columns of $U_1 \in M_{n,r}$ are an orthonormal basis for the range of $\bar{A}$ and $AU_1 = 0$. (c) $U^T AU = \begin{bmatrix} 0 & B \\ 0 & 0 \end{bmatrix}$, in which $B \in M_{r,n-r}$ and rank $B = r$. (d) $B = V[\Sigma_r \ 0_{r,n-2r}]W^*$, in which $V \in M_r$ and $W \in M_{n-r}$ are unitary, and $\Sigma_r = \text{diag}(\sigma_1, \ldots, \sigma_r)$. (e) Let $Z = \bar{V} \oplus W$. Then $Z^T(U^T AU)Z = \begin{bmatrix} 0 & \Sigma_r \\ 0 & 0 \end{bmatrix} \oplus 0_{n-2r}$, which is unitarily congruent to (2.6.8) via a permutation matrix. For a different approach, see (3.4.P5).

2.6.P25 Let $A \in M_n$ and suppose that rank $A = r < n$. Let $\sigma_1 \geq \cdots \geq \sigma_r > 0$ be the positive singular values of A and let $\Sigma_r = \text{diag}(\sigma_1, \ldots, \sigma_r)$. Show that there is a unitary $U \in M_n$, and matrices $K \in M_r$ and $L \in M_{r,n-r}$, such that

$$A = U \begin{bmatrix} \Sigma_r K & \Sigma_r L \\ 0 & 0_{n-r} \end{bmatrix} U^*, \quad KK^* + LL^* = I_r \tag{2.6.9}$$

2.6.P26 Let $A \in M_n$, suppose that $1 \leq \text{rank } A = r < n$, and consider the representation (2.6.9). Show that (a) A is normal if and only if $L = 0$ and $\Sigma_r K = K\Sigma_r$; (b) $A^2 = 0$ if and only if $K = 0$ (in which case $LL^* = I_r$); (c) $A^2 = 0$ if and only if A is unitarily similar to a direct sum of the form (2.6.8).

2.6.P27 Let $A \in M_n$ be skew symmetric. If rank $A \leq 1$, explain why $A = 0$.

2.6.P28 A matrix $A \in M_n$ is an *EP matrix* if the ranges of A and A^* are the same. Every normal matrix is an EP matrix (2.5.P54(b)) and every nonsingular matrix (normal or not) is an EP matrix. (a) Show that A is an EP matrix and rank $A = r$ if and only if there is a nonsingular $B \in M_r$ and a unitary $V \in M_n$ such that $A = V \begin{bmatrix} B & 0 \\ 0 & 0 \end{bmatrix} V^*$. (b) Explain why an EP matrix is rank principal.

2.6.P29 If $x \in \mathbb{C}^n$ is a normal eigenvector of $A \in M_n$ with associated eigenvalue λ, show that $|\lambda|$ is a singular value of A.

2.6.P30 Use the singular value decomposition to verify (0.4.6(f)) for complex matrices: $A \in M_{m,n}$ has rank r if and only if there are nonsingular matrices $S \in M_m$ and $T \in M_n$ such that $A = S \begin{bmatrix} I_r & 0 \\ 0 & 0 \end{bmatrix} T$.

2.6.P31 Let $A \in M_{m,n}$. (a) Use the singular value decomposition $A = V \Sigma W^*$ to show that the Hermitian matrix $\mathcal{A} = \begin{bmatrix} 0 & A \\ A^* & 0 \end{bmatrix} \in M_{m+n}$ is unitarily similar to the real matrix $\begin{bmatrix} 0 & \Sigma \\ \Sigma^T & 0 \end{bmatrix}$. (b) If $m = n$ and $\Sigma = \mathrm{diag}(\sigma_1, \ldots, \sigma_n)$, explain why the eigenvalues of $\mathcal{A}$ are $\pm \sigma_1, \ldots, \pm \sigma_n$.

2.6.P32 Let $A \in M_n$ and let $\mathcal{A} = \begin{bmatrix} 0 & A \\ A^T & 0 \end{bmatrix} \in M_{2n}$. If $\sigma_1, \ldots, \sigma_n$ are the singular values of A, show that $\sigma_1, \sigma_1, \ldots, \sigma_n, \sigma_n$ are the singular values of $\mathcal{A}$.

2.6.P33 Let $\sigma_1 \geq \cdots \geq \sigma_n$ be the ordered singular values of $A \in M_n$ and let $r \in \{1, \ldots, n\}$. Show that the singular values of the compound matrix $C_r(A)$ are the $\binom{n}{r}$ possible products $\sigma_{i_1} \cdots \sigma_{i_r}$ in which $1 \leq i_1 < i_2 < \cdots < i_r \leq n$. Explain why $\mathrm{tr}(C_r(A) C_r(A)^*) = \mathrm{tr}\, C_r(AA^*) = S_r(\sigma_1^2, \ldots, \sigma_n^2)$ is the sum of the squares of the singular values of $C_r(A)$; see (1.2.14). In particular, $\mathrm{tr}\, C_2(AA^*) = \sum_{1 \leq i < j \leq n} \sigma_i^2(A) \sigma_j^2(A)$. Explain why $\sigma_1 \cdots \sigma_r$ is the largest singular value of $C_r(A)$. See (2.3.P12) for related results about eigenvalues of $C_r(A)$.

2.6.P34 Denote the eigenvalues of $A \in M_n$ and A^2 by $\lambda_1(A), \ldots, \lambda_n(A)$ and $\lambda_1(A^2), \ldots, \lambda_n(A^2)$, respectively; denote their singular values by $\sigma_1(A), \ldots, \sigma_n(A)$ and $\sigma_1(A^2), \ldots, \sigma_n(A^2)$, respectively. (a) Derive the inequality $\sum_{i=1}^{n} |\lambda_i(A)|^4 \leq \sum_{i=1}^{n} \sigma_i^2(A^2)$ by applying Schur's inequality (2.3.2a) to A^2. (b) Derive the inequality $\sum_{1 \leq i < j \leq n} |\lambda_i(A) \lambda_j(A)|^2 \leq \sum_{1 \leq i < j \leq n} \sigma_i^2(A) \sigma_j^2(A)$ by applying Schur's inequality to the compound matrix $C_2(A)$. (c) Show that $(\mathrm{tr}\, AA^*)^2 = \sum_{i=1}^{n} \sigma_i^4(A) + 2 \sum_{1 \leq i < j \leq n} \sigma_i^2(A) \sigma_j^2(A)$. (d) Show that $\mathrm{tr}((AA^* - A^*A)^2) = 2 \sum_{i=1}^{n} \sigma_i^4(A) - 2 \sum_{i=1}^{n} \sigma_i^2(A^2)$. (e) Show that $(\sum_{i=1}^{n} |\lambda_i(A)|^2)^2 = \sum_{i=1}^{n} |\lambda_i(A)|^4 + 2 \sum_{1 \leq i < j \leq n} |\lambda_i(A) \lambda_j(A)|^2$. (f) Conclude that

$$\sum_{i=1}^{n} |\lambda_i(A)|^2 \leq \sqrt{(\mathrm{tr}\, AA^*)^2 - \frac{1}{2} \mathrm{tr}((AA^* - A^*A)^2)} \tag{2.6.9}$$

which strengthens Schur's inequality (2.3.2a). Why is (2.6.9) an equality if A is normal? Explain why (2.6.9) is an equality if and only if both A^2 and $C_2(A)$ are normal.

2.6.P35 Using the notation of the preceding problem, show that

$$\sum_{i=1}^{n} |\lambda_i(A)|^2 \leq \sqrt{(\mathrm{tr}\, AA^* - \frac{1}{n} |\mathrm{tr}\, A|^2)^2 - \frac{1}{2} \mathrm{tr}((AA^* - A^*A)^2)} + \frac{1}{n} |\mathrm{tr}\, A|^2 \tag{2.6.10}$$

In addition, show that the upper bound in (2.6.10) is less than or equal to the upper bound in (2.6.9), with equality if and only if either $\mathrm{tr}\, A = 0$ or A is normal.

2.6.P36 Let $A \in M_n$ have rank r, let $s_1, \ldots, s_d$ be the distinct positive singular values of A, in any order, with respective multiplicities $n_1, \ldots, n_d$, and let $A = V\Sigma W^*$ be a singular value decomposition in which $V, W \in M_n$ are unitary and $\Sigma = s_1 I_{n_1} \oplus \cdots \oplus s_d I_{n_d} \oplus 0_{n-r}$. (a) Explain why A is symmetric if and only if $V = \bar{W}(S_1 \oplus \cdots \oplus S_d \oplus \tilde{W})$, in which $\tilde{W} \in M_{n-r}$ is unitary and each $S_j \in M_{n_j}$ is unitary and symmetric. (b) If the singular values of A are distinct (that is, if $d \geq n - 1$), explain why A is symmetric if and only if $V = \bar{W}D$, in which $D \in M_n$ is a diagonal unitary matrix.

2.6.P37 Suppose that $A \in M_n$ has distinct singular values. Let $A = V\Sigma W^*$ and $A = \hat{V}\Sigma\hat{W}^*$ be singular value decompositions. (a) If A is nonsingular, explain why there is a diagonal unitary matrix D such that $\hat{V} = VD$ and $\hat{W} = WD$. (b) If A is singular, explain why there are diagonal unitary matrices D and $\tilde{D}$ that differ in at most one diagonal entry such that $\hat{V} = VD$ and $\hat{W} = W\tilde{D}$.

2.6.P38 Let $A \in M_n$ be nonsingular and let σ_n be the smallest singular value of $A + A^{-*}$. Show that $\sigma_n \geq 2$. What can you say about the case of equality?

2.6.P39 Let $A \in M_n$ be coninvolutory, so A is nonsingular and $A = \bar{A}^{-1}$. Explain why the singular values of A that are not equal to 1 occur in reciprocal pairs.

2.6.P40 Using the notation of (2.4.5.1), suppose that T and T' are unitarily similar. (a) Explain why, for each $i, j = 1, \ldots, d$, it is necessary that the singular values of T_{ij} and T'_{ij} be the same. (b) What does this necessary condition say when $n = 2$? Why is it both necessary and sufficient in this case? (c) Let $n = 4$ and $d = 2$. Consider the example $T_{11} = T'_{11} = \begin{bmatrix} 1 & 1 \\ 0 & 1 \end{bmatrix}$, $T_{22} = T'_{22} = \begin{bmatrix} 2 & 2 \\ 0 & 2 \end{bmatrix}$, $T_{12} = \begin{bmatrix} 3 & 0 \\ 0 & 4 \end{bmatrix}$, and $T'_{11} = \begin{bmatrix} 0 & 4 \\ 3 & 0 \end{bmatrix}$; explain why the necessary condition in (a) need not be sufficient.

Notes and Further Readings. The special singular value decomposition (2.6.6a) for complex symmetric matrices was published by L. Autonne in 1915; it has been rediscovered many times since then. Autonne's proof used a version of (2.6.4), but his approach required that the matrix be nonsingular; (2.6.P20) shows how to deduce the singular case from the nonsingular case. See section 3.0 of Horn and Johnson (1991) for a history of the singular value decomposition, including an account of Autonne's contributions.

2.7 The *CS* decomposition

The *CS* decomposition is a canonical form for partitioned unitary matrices under partitioned unitary equivalence. Its proof involves the singular value decomposition, the QR factorization, and the observation in the following exercise.

Exercise. Let $\Gamma, L \in M_p$. Suppose that $\Gamma = \text{diag}(\gamma_1, \ldots, \gamma_p)$ with $0 \leq \gamma_1 \leq \cdots \leq \gamma_p \leq 1$, $L = [\ell_{ij}]$ is lower triangular, and the rows of $[\Gamma \ L \ 0] \in M_{p,2p+k}$ are orthonormal. Explain why L is diagonal, $L = \text{diag}(\lambda_1, \ldots, \lambda_p)$, and $|\lambda_j|^2 = 1 - \gamma_j^2$, $j = 1, \ldots, p$. *Hint*: If $\gamma_1 = 1$, why is $L = 0$? If $\gamma_1 < 1$, then $|\ell_{11}|^2 = 1 - \gamma_1^2$. Why does orthogonality ensure that $\ell_{21} = \cdots = \ell_{p1} = 0$? Work down the rows.

Theorem 2.7.1 (CS decomposition). *Let p, q, and n be given integers with $1 < p \le q < n$ and $p + q = n$. Let $U = \begin{bmatrix} U_{11} & U_{12} \\ U_{21} & U_{22} \end{bmatrix} \in M_n$ be unitary, with $U_{11} \in M_p$ and $U_{22} \in M_q$. There are unitary $V_1, W_1 \in M_p$ and $V_2, W_2 \in M_q$ such that*

$$
\begin{bmatrix} V_1 & 0 \\ 0 & W_1 \end{bmatrix} \begin{bmatrix} U_{11} & U_{12} \\ U_{21} & U_{22} \end{bmatrix} \begin{bmatrix} V_2 & 0 \\ 0 & W_2 \end{bmatrix} = \begin{bmatrix} C & S & 0 \\ -S & C & 0 \\ 0 & 0 & I_{q-p} \end{bmatrix} \tag{2.7.1.2}
$$

in which $C = \mathrm{diag}(\sigma_1, \ldots, \sigma_p)$, $\sigma_1 \ge \cdots \ge \sigma_p$ are the nonincreasingly ordered singular values of U_{11} and $S = \mathrm{diag}((1 - \sigma_1^2)^{1/2}, \ldots, (1 - \sigma_p^2)^{1/2})$.

Proof. Our strategy is to perform a sequence of partitioned unitary equivalences that, step by step, reduce U to a block matrix that has the asserted form. The first step is to use the singular value decomposition: Write $U_{11} = V \Sigma W = (V K_p)(K_p \Sigma K_p)(K_p W) = \tilde{V} \Gamma \tilde{W}$, in which $V, W \in M_p$ are unitary, K_p is the p-by-p reversal matrix (0.9.5.1), $\tilde{V} = V K_p$, $\tilde{W} = K_p W$, $\Sigma = \mathrm{diag}(\sigma_1, \ldots, \sigma_p)$ with $\sigma_1 \ge \cdots \ge \sigma_p$, and $\Gamma = K_p \Sigma K_p = \mathrm{diag}(\sigma_p, \ldots, \sigma_1)$. Compute

$$
\begin{bmatrix} \tilde{V}^* & 0 \\ 0 & I_q \end{bmatrix} \begin{bmatrix} U_{11} & U_{12} \\ U_{21} & U_{22} \end{bmatrix} \begin{bmatrix} \tilde{W}^* & 0 \\ 0 & I_q \end{bmatrix} = \begin{bmatrix} \Gamma & \tilde{V}^* U_{12} \\ U_{21} \tilde{W}^* & U_{22} \end{bmatrix}
$$

This matrix is unitary (a product of three unitary matrices), so each column has unit Euclidean norm, which means that $\sigma_1 = \gamma_p \le 1$. Now invoke the QR factorization (2.1.14) and its variant (2.1.15b) to write $\tilde{V}^* U_{12} = [L \ \ 0]\tilde{Q}$ and $U_{21} \tilde{W}^* = Q \begin{bmatrix} R \\ 0 \end{bmatrix}$, in which $\tilde{Q}, Q \in M_q$ are unitary, $L = [\ell_{ij}] \in M_p$ is lower triangular, and $R = [r_{ij}] \in M_p$ is upper triangular. Compute

$$
\begin{bmatrix} I_p & 0 \\ 0 & Q^* \end{bmatrix} \begin{bmatrix} \Gamma & \tilde{V}^* U_{12} \\ U_{21} \tilde{W}^* & U_{22} \end{bmatrix} \begin{bmatrix} I_p & 0 \\ 0 & \tilde{Q}^* \end{bmatrix} = \begin{bmatrix} \Gamma & [L \ \ 0] \\ \begin{bmatrix} R \\ 0 \end{bmatrix} & Q^* U_{22} \tilde{Q}^* \end{bmatrix}
$$

The argument in the preceding exercise shows that both L and R are diagonal and that $|r_{ii}| = |\ell_{ii}| = \sqrt{1 - \gamma_i^2}$ for each $i = 1, \ldots, p$. Let $M = \mathrm{diag}(\sqrt{1 - \gamma_1^2}, \ldots, \sqrt{1 - \gamma_p^2})$ and let $t = \max\{i : \gamma_i < 1\}$. (We may assume that $t \ge 1$, for if $\gamma_1 = 1$, then $\Gamma = I_p$ and $M = 0$, so we have (2.7.1.2) with $C = I_p$ and $S = 0_p$.) There are diagonal unitary matrices $D_1, D_2 \in M_p$ such that $D_1 R = -M$ and $L D_2 = M$, so a unitary congruence via $I_p \oplus D_1 \oplus I_{n-2p}$ on the left and $I_p \oplus D_2 \oplus I_{n-2p}$ on the right results in a unitary matrix of the form

$$
\begin{bmatrix} \Gamma & [M \ \ 0] \\ \begin{bmatrix} -M \\ 0 \end{bmatrix} & Z \end{bmatrix} = \begin{bmatrix} \Gamma_1 & 0 & M_1 & 0 & 0 \\ 0 & I_{p-t} & 0 & 0_{p-t} & 0 \\ -M_1 & 0 & Z_{11} & Z_{12} & Z_{13} \\ 0 & 0_{p-t} & Z_{21} & Z_{22} & Z_{23} \\ 0 & 0 & Z_{31} & Z_{32} & Z_{33} \end{bmatrix}
$$

in which we have partitioned $\Gamma = \Gamma_1 \oplus I_{p-t}$ and $M = M_1 \oplus 0_{p-t}$, so M_1 is nonsingular. Orthogonality of the first and third block columns (and nonsingularity of M_1) implies that $Z_{11} = \Gamma_1$, whereupon the requirement that each row and each column be a unit vector ensures that Z_{12}, Z_{13}, Z_{21}, and Z_{31} are all zero blocks. Thus, we have

$$
\begin{bmatrix}
\Gamma_1 & 0 & M_1 & 0 & 0 \\
0 & I_{p-t} & 0 & 0_{p-t} & 0 \\
-M_1 & 0 & \Gamma_1 & 0 & 0 \\
0 & 0_{p-t} & 0 & Z_{22} & Z_{23} \\
0 & 0 & 0 & Z_{32} & Z_{33}
\end{bmatrix}
$$

The lower-right block $\tilde{Z} = \begin{bmatrix} Z_{22} & Z_{23} \\ Z_{32} & Z_{33} \end{bmatrix} \in M_{q-t}$ is a direct summand of a unitary matrix, so it is unitary, and hence $\tilde{Z} = \hat{V} I_{q-t} \hat{W}$ for some unitary $\hat{V}$, $\hat{W} \in M_{q-t}$. A unitary equivalence via $I_{p+t} \oplus \hat{V}^*$ on the left and $I_{p+t} \oplus \hat{W}^*$ on the right produces the block matrix

$$
\begin{bmatrix}
\Gamma_1 & 0 & M_1 & 0 & 0 \\
0 & I_{p-t} & 0 & 0_{p-t} & 0 \\
-M_1 & 0 & \Gamma_1 & 0 & 0 \\
0 & 0_{p-t} & 0 & I_{p-t} & 0 \\
0 & 0 & 0 & 0 & I_{q-p}
\end{bmatrix}
$$

Finally, a unitary similarity via $K_p \oplus K_p \oplus I_{q-p}$ produces a unitary matrix with the required structure (2.7.1.2). $\qquad\square$

The *CS* decomposition is a parametric description of the set of all unitary matrices $U = \begin{bmatrix} U_{11} & U_{12} \\ U_{21} & U_{22} \end{bmatrix} \in M_n$ of size $n - p + q$ (by convention, $p \le q$, but that is not essential) that are partitioned conformally to $I_p \oplus I_q$. The parameters are four smaller arbitrary unitary matrices V_1, $W_1 \in M_p$ and V_2, $W_2 \in M_q$, and any p real numbers $\sigma_1, \ldots, \sigma_p$ such that $1 \ge \sigma_1 \ge \cdots \ge \sigma_p \ge 0$. The parametrization of the four blocks is

$$
U_{11} = V_1 C W_1, \quad U_{12} = V_1 [S \ 0] W_2, \tag{2.7.1.3}
$$

$$
U_{21} = V_2 \begin{bmatrix} -S \\ 0 \end{bmatrix} W_1, \quad \text{and } U_{22} = V_2 \begin{bmatrix} C & 0 \\ 0 & I_{q-p} \end{bmatrix} W_2
$$

in which $C = \mathrm{diag}(\sigma_1, \ldots, \sigma_p)$ and $S = \mathrm{diag}((1 - \sigma_1^2)^{1/2}, \ldots, (1 - \sigma_p^2)^{1/2})$. The *CS* decomposition is a widely useful tool, notably in problems involving distances and angles between subspaces.

Problems

Use the *CS* decomposition to solve each of the following problems, even though other approaches are possible. A given $A \in M_{n,m}$ is a *contraction* if its largest singular value is less than or equal to 1.

2.7.P1 Show that every submatrix of a unitary matrix is a contraction; the submatrix need not be principal and need not be square.

2.7.P2 There is an interesting converse to the preceding problem. Let $A \in M_{m,n}$ be a contraction and suppose that exactly ν of its singular values are strictly less than 1. Show that there are matrices B, C, and D such that $U = \begin{bmatrix} D & B \\ C & A \end{bmatrix} \in M_{\max\{m,n\}+|m-n|+\nu}$ is unitary. Such a matrix U is called a *unitary dilation* of the contraction A.

2.7.P3 If A is a k-by-k submatrix of an n-by-n unitary matrix, and if $2k > n$, show that some singular value of A is equal to 1.

2.7.P4 Let $U = \begin{bmatrix} U_{11} & U_{12} \\ U_{21} & U_{22} \end{bmatrix}$ be unitary, with $U_{11} \in M_p$ and $U_{22} \in M_q$. Show that the nullities of U_{11} and U_{22} are equal and the nullities of U_{12} and U_{21}^* are equal. Compare with (0.7.5).

2.7.P5 Prove the assertions in (2.6.P19).

2.7.P6 Let $U = \begin{bmatrix} u_{11} & u_{12} \\ u_{21} & u_{22} \end{bmatrix} \in M_2$ be unitary. (a) Show that $|u_{11}| = |u_{22}|$ and $|u_{21}| = |u_{12}| = (1 - |u_{11}|^2)^{1/2}$. (b) Show that U is diagonally unitarily similar to a complex symmetric matrix.

Further Reading. For a historical survey and a version of the CS decomposition that embraces an arbitrary 2-by-2 partition of a unitary $U = [U_{ij}] \in M_n$ ($U_{ij} \in M_{r_i, c_j}$, $i, j = 1, 2$ with $r_1 + r_2 = n = c_1 + c_2$), see C. Paige and M. Wei, History and generality of the CS decomposition, *Linear Algebra Appl.* 208/209 (1994) 303–326.

CHAPTER 3

Canonical Forms for Similarity and Triangular Factorizations

3.0 Introduction

How can we tell if two given matrices are similar? The matrices

$$A = \begin{bmatrix} 0 & 1 & 0 & 0 \\ 0 & 0 & 0 & 0 \\ 0 & 0 & 0 & 1 \\ 0 & 0 & 0 & 0 \end{bmatrix} \quad \text{and} \quad B = \begin{bmatrix} 0 & 1 & 0 & 0 \\ 0 & 0 & 1 & 0 \\ 0 & 0 & 0 & 0 \\ 0 & 0 & 0 & 0 \end{bmatrix} \tag{3.0.0}$$

have the same eigenvalues, and hence they have the same characteristic polynomial, trace, and determinant. They also have the same rank, but $A^2 = 0$ and $B^2 \neq 0$, so A and B are not similar.

One approach to determining whether given square complex matrices A and B are similar would be to have in hand a set of special matrices of prescribed form and see if both given matrices can be reduced by similarity to the same special matrix. If so, then A and B must be similar because the similarity relation is transitive and reflexive. If not, then we would like to be able to conclude that A and B are not similar. What sets of special matrices would be suitable for this purpose?

Every square complex matrix is similar to an upper triangular matrix. However, two upper triangular matrices with the same main diagonals but some different off-diagonal entries can still be similar (2.3.2b). Thus, we have a uniqueness problem: If we reduce A and B to two unequal upper triangular matrices with the same main diagonal, we cannot conclude from this fact alone that A and B are not similar.

The class of upper triangular matrices is too large for our purposes, but what about the smaller class of diagonal matrices? Uniqueness is no longer an issue, but now we have an existence problem: Some similarity equivalence classes contain no diagonal matrices.

One approach to finding a suitable set of special matrices turns out to be a deft compromise between diagonal matrices and upper triangular matrices: A *Jordan matrix* is a special block upper triangular form that can be achieved by similarity for every complex matrix. Two Jordan matrices are similar if and only if they have the same

163

diagonal blocks, without regard to their ordering. Moreover, no other matrix in the similarity equivalence class of a Jordan matrix J has strictly fewer nonzero off-diagonal entries than J.

Similarity is only one of many equivalence relations of interest in matrix theory; several others are listed in (0.11). Whenever we have an equivalence relation on a set of matrices, we want to be able to decide whether given matrices A and B are in the same equivalence class. A classical and broadly successful approach to this decision problem is to identify a set of representative matrices for the given equivalence relation such that (a) there is a representative in each equivalence class and (b) distinct representatives are not equivalent. The test for equivalence of A and B is to reduce each via the given equivalence to a representative matrix and see if the two representative matrices are the same. Such a set of representatives is a *canonical form* for the equivalence relation.

For example, (2.5.3) provides a canonical form for the set of normal matrices under unitary similarity: The diagonal matrices are a set of representative matrices (we identify two diagonal matrices if one is a permutation similarity of the other). Another example is the singular value decomposition (2.6.3), which provides a canonical form for M_n under unitary equivalence: The diagonal matrices $\Sigma = \mathrm{diag}(\sigma_1, \ldots, \sigma_n)$ with $\sigma_1 \geq \cdots \geq \sigma_n \geq 0$ are the representative matrices.

3.1 The Jordan canonical form theorem

Definition 3.1.1. *A Jordan block $J_k(\lambda)$ is a k-by-k upper triangular matrix of the form*

$$J_k(\lambda) = \begin{bmatrix} \lambda & 1 & & & \\ & \lambda & 1 & & \\ & & \ddots & \ddots & \\ & & & \lambda & 1 \\ & & & & \lambda \end{bmatrix}; J_1(\lambda) = [\lambda], J_2(\lambda) = \begin{bmatrix} \lambda & 1 \\ 0 & \lambda \end{bmatrix} \qquad (3.1.2)$$

The scalar λ appears k times on the main diagonal; if $k > 1$, there are $k - 1$ entries "+1" in the superdiagonal; all other entries are zero. A Jordan matrix $J \in M_n$ is a direct sum of Jordan blocks

$$J = J_{n_1}(\lambda_1) \oplus J_{n_2}(\lambda_2) \oplus \cdots \oplus J_{n_q}(\lambda_q), \qquad n_1 + n_2 + \cdots + n_q = n \qquad (3.1.3)$$

Neither the block sizes n_i nor the scalars λ_i need be distinct.

The main result of this section is that every complex matrix is similar to an essentially unique Jordan matrix. We proceed to this conclusion in three steps, two of which have already been taken:

Step 1. Theorem 2.3.1 ensures that every complex matrix is similar to an upper triangular matrix whose eigenvalues appear on the main diagonal, and equal eigenvalues are grouped together.

Step 2. Theorem 2.4.6.1 ensures that a matrix of the form described in Step 1 is similar to a block diagonal upper triangular matrix (2.4.6.2) in which each diagonal block has equal diagonal entries.

Step 3. In this section, we show that an upper triangular matrix with equal diagonal entries is similar to a Jordan matrix.

We are also interested in concluding that if a matrix is real and has only real eigenvalues, then it can be reduced to a Jordan matrix via a *real* similarity. If a real matrix A has only real eigenvalues, then (2.3.1) and (2.4.6.1) ensure that there is a real similarity matrix S such that $S^{-1}AS$ is a (real) block diagonal upper triangular matrix of the form (2.4.6.2). Thus, it suffices to show that a real upper triangular matrix with equal main diagonal entries can be reduced to a direct sum of Jordan blocks via a real similarity.

The following lemma is helpful in taking Step 3; its proof is an entirely straightforward computation. The k-by-k Jordan block with eigenvalue zero is called a *nilpotent Jordan block*.

Lemma 3.1.4. *Let $k \geq 2$ be given. Let $e_i \in \mathbf{C}^k$ denote the ith standard unit basis vector, and let $x \in \mathbf{C}^k$ be given. Then*

$$J_k^T(0)J_k(0) = \begin{bmatrix} 0 & 0 \\ 0 & I_{k-1} \end{bmatrix} \quad and \quad J_k(0)^p = 0 \quad if \quad p \geq k$$

Moreover, $J_k(0)e_{i+1} = e_i$ for $i = 1, 2, \ldots, k-1$ and $[I_k - J_k^T(0)J_k(0)]x = (x^T e_1)e_1$.

We now address the issue in Step 3.

Theorem 3.1.5. *Let $A \in M_n$ be strictly upper triangular. There is a nonsingular $S \in M_n$ and there are integers $n_1, n_2, \ldots, n_m$ with $n_1 \geq n_2 \geq \cdots \geq n_m \geq 1$ and $n_1 + n_2 + \cdots + n_m = n$ such that*

$$A = S\left(J_{n_1}(0) \oplus J_{n_2}(0) \oplus \cdots \oplus J_{n_m}(0)\right)S^{-1} \tag{3.1.6}$$

If A is real, the similarity matrix S may be chosen to be real.

Proof. If $n = 1$, $A = [0]$ and the result is trivial. We proceed by induction on n. Assume that $n > 1$ and that the result has been proved for all strictly upper triangular matrices of size less than n. Partition $A = \begin{bmatrix} 0 & a^T \\ 0 & A_1 \end{bmatrix}$, in which $a \in \mathbf{C}^{n-1}$ and $A_1 \in M_{n-1}$ is strictly upper triangular. By the induction hypothesis, there is a nonsingular $S_1 \in M_{n-1}$ such that $S_1^{-1}A_1S_1$ has the desired form (3.1.6); that is,

$$S_1^{-1}A_1S_1 = \begin{bmatrix} J_{k_1} & & \\ & \ddots & \\ & & J_{k_s} \end{bmatrix} = \begin{bmatrix} J_{k_1} & 0 \\ 0 & J \end{bmatrix} \tag{3.1.7}$$

in which $k_1 \geq k_2 \geq \cdots \geq k_s \geq 1$, $k_1 + k_2 + \cdots + k_s = n - 1$, $J_{k_i} = J_{k_i}(0)$, and $J = J_{k_2} \oplus \cdots \oplus J_{k_s} \in M_{n-k_1-1}$. No diagonal Jordan block in J has size greater than k_1, so $J^{k_1} = 0$. A computation reveals that

$$\begin{bmatrix} 1 & 0 \\ 0 & S_1^{-1} \end{bmatrix} A \begin{bmatrix} 1 & 0 \\ 0 & S_1 \end{bmatrix} = \begin{bmatrix} 0 & a^T S_1 \\ 0 & S_1^{-1}A_1S_1 \end{bmatrix} \tag{3.1.8}$$

Partition $a^T S_1 = [a_1^T a_2^T]$ with $a_1 \in \mathbf{C}^{k_1}$ and $a_2 \in \mathbf{C}^{n-k_1-1}$, and write (3.1.8) as

$$
\begin{bmatrix} 1 & 0 \\ 0 & S_1^{-1} \end{bmatrix} A \begin{bmatrix} 1 & 0 \\ 0 & S_1 \end{bmatrix} = \begin{bmatrix} 0 & a_1^T & a_2^T \\ 0 & J_{k_1} & 0 \\ 0 & 0 & J \end{bmatrix}
$$

Now consider the similarity

$$
\begin{bmatrix} 1 & -a_1^T J_{k_1}^T & 0 \\ 0 & I & 0 \\ 0 & 0 & I \end{bmatrix} \begin{bmatrix} 0 & a_1^T & a_2^T \\ 0 & J_{k_1} & 0 \\ 0 & 0 & J \end{bmatrix} \begin{bmatrix} 1 & a_1^T J_{k_1}^T & 0 \\ 0 & I & 0 \\ 0 & 0 & I \end{bmatrix}
$$

$$
= \begin{bmatrix} 0 & a_1^T(I - J_{k_1}^T J_{k_1}) & a_2^T \\ 0 & J_{k_1} & 0 \\ 0 & 0 & J \end{bmatrix} = \begin{bmatrix} 0 & (a_1^T e_1)e_1^T & a_2^T \\ 0 & J_{k_1} & 0 \\ 0 & 0 & J \end{bmatrix} \qquad (3.1.9)
$$

in which we use the identity $(I - J_k^T J_k)x = (x^T e_1)e_1$. There are now two possibilities, depending on whether $a_1^T e_1 \neq 0$ or $a_1^T e_1 = 0$.

If $a_1^T e_1 \neq 0$, then

$$
\begin{bmatrix} 1/a_1^T e_1 & 0 & 0 \\ 0 & I & 0 \\ 0 & 0 & (1/a_1^T e_1)I \end{bmatrix} \begin{bmatrix} 0 & (a_1^T e_1)e_1^T & a_2^T \\ 0 & J_{k_1} & 0 \\ 0 & 0 & J \end{bmatrix} \begin{bmatrix} a_1^T e_1 & 0 & 0 \\ 0 & I & 0 \\ 0 & 0 & a_1^T e_1 I \end{bmatrix}
$$

$$
= \begin{bmatrix} 0 & e_1^T & a_2^T \\ 0 & J_{k_1} & 0 \\ 0 & 0 & J \end{bmatrix} = \begin{bmatrix} \tilde{J} & e_1 a_2^T \\ 0 & J \end{bmatrix}
$$

Notice that $\tilde{J} = \begin{bmatrix} 0 & e_1^T \\ 0 & J_{k_1} \end{bmatrix} = J_{k_1+1}(0)$. Since $\tilde{J} e_{i+1} = e_i$ for $i = 1, 2, \ldots, k_1$, a computation reveals that

$$
\begin{bmatrix} I & e_2 a_2^T \\ 0 & I \end{bmatrix} \begin{bmatrix} \tilde{J} & e_1 a_2^T \\ 0 & J \end{bmatrix} \begin{bmatrix} I & -e_2 a_2^T \\ 0 & I \end{bmatrix} = \begin{bmatrix} \tilde{J} & -\tilde{J} e_2 a_2^T + e_1 a_2^T + e_2 a_2^T J \\ 0 & J \end{bmatrix}
$$

$$
= \begin{bmatrix} \tilde{J} & e_2 a_2^T J \\ 0 & J \end{bmatrix}
$$

We can proceed recursively to compute the sequence of similarities

$$
\begin{bmatrix} I & e_{i+1} a_2^T J^{i-1} \\ 0 & I \end{bmatrix} \begin{bmatrix} \tilde{J} & e_i a_2^T J^{i-1} \\ 0 & J \end{bmatrix} \begin{bmatrix} I & -e_{i+1} a_2^T J^{i-1} \\ 0 & I \end{bmatrix} = \begin{bmatrix} \tilde{J} & e_{i+1} a_2^T J^i \\ 0 & J \end{bmatrix},
$$

for $i = 2, 3, \ldots$ Since $J^{k_1} = 0$, after at most k_1 steps in this sequence of similarities, the off-diagonal term finally vanishes. We conclude that A is similar to $\begin{bmatrix} \tilde{J} & 0 \\ 0 & J \end{bmatrix}$, which is a strictly upper triangular Jordan matrix of the required form.

If $a_1^T e_1 = 0$, then (3.1.9) shows that A is similar to

$$
\begin{bmatrix} 0 & 0 & a_2^T \\ 0 & J_{k_1} & 0 \\ 0 & 0 & J \end{bmatrix}
$$

which is permutation similar to

$$\begin{bmatrix} J_{k_1} & 0 & 0 \\ 0 & 0 & a_2^T \\ 0 & 0 & J \end{bmatrix} \tag{3.1.10}$$

By the induction hypothesis, there is a nonsingular $S_2 \in M_{n-k_1}$ such that $S_2^{-1} \begin{bmatrix} 0 & a_2^T \\ 0 & J \end{bmatrix} S_2 = \hat{J} \in M_{n-k_1}$ is a Jordan matrix with zero main diagonal. Thus, the matrix (3.1.10), and therefore A itself, is similar to $\begin{bmatrix} J_{k_1} & 0 \\ 0 & \hat{J} \end{bmatrix}$, which is a Jordan matrix of the required form, except that the diagonal Jordan blocks might not be arranged in nonincreasing order of their size. A block permutation similarity, if necessary, produces the required form.

Finally, observe that if A is real, then all the similarities in this proof are real, so A is similar via a real similarity to a Jordan matrix of the required form. $\square$

Theorem 3.1.5 essentially completes Step 3, as the general case is an easy consequence of the nilpotent case. If $A \in M_n$ is an upper triangular matrix with all diagonal entries equal to λ, then $A_0 = A - \lambda I$ is strictly upper triangular. If $S \in M_n$ is nonsingular and $S^{-1} A_0 S$ is a direct sum of nilpotent Jordan blocks $J_{n_i}(0)$, as guaranteed by (3.1.5), then $S^{-1} A S = S^{-1} A_0 S + \lambda I$ is a direct sum of Jordan blocks $J_{n_i}(\lambda)$ with eigenvalue λ. We have now established the existence assertion of the *Jordan canonical form theorem*.

Theorem 3.1.11. *Let $A \in M_n$ be given. There is a nonsingular $S \in M_n$, positive integers q and $n_1, \ldots, n_q$ with $n_1 + n_2 + \cdots + n_q = n$, and scalars $\lambda_1, \ldots, \lambda_q \in \mathbf{C}$ such that*

$$A = S \begin{bmatrix} J_{n_1}(\lambda_1) & & \\ & \ddots & \\ & & J_{n_q}(\lambda_q) \end{bmatrix} S^{-1} \tag{3.1.12}$$

The Jordan matrix $J_A = J_{n_1}(\lambda_1) \oplus \cdots \oplus J_{n_q}(\lambda_q)$ is uniquely determined by A up to permutation of its direct summands. If A is real and has only real eigenvalues, then S can be chosen to be real.

The Jordan matrix J_A in the preceding theorem is (up to permutation of its direct summands) the *Jordan canonical form* of A. The matrices $J_k(\lambda)$, $\lambda \in \mathbf{C}$, $k = 1, 2, \ldots$ are *canonical blocks for similarity*.

Two facts provide the key to understanding the uniqueness assertion in the Jordan canonical form theorem: (1) similarity of two matrices is preserved if they are both translated by the same scalar matrix and (2) rank is a similarity invariant.

If $A, B, S \in M_n$, S is nonsingular, and $A = SBS^{-1}$, then for any $\lambda \in \mathbf{C}$, $A - \lambda I = SBS^{-1} - \lambda SS^{-1} = S(B - \lambda I)S^{-1}$. Moreover, for every $k = 1, 2, \ldots$, the matrices $(A - \lambda I)^k$ and $(B - \lambda I)^k$ are similar; in particular, their ranks are equal. We focus on this assertion when $B = J = J_{n_1}(\lambda_1) \oplus \cdots \oplus J_{n_q}(\lambda_q)$ is a Jordan matrix that is similar to A (the existence assertion of (3.1.11)) and λ is an eigenvalue of A. After a permutation of the diagonal blocks of J (a permutation similarity), we may assume that $J = J_{m_1}(\lambda) \oplus \cdots \oplus J_{m_p}(\lambda) \oplus \hat{J}$, in which the Jordan matrix $\hat{J}$ is a direct sum of

Jordan blocks with eigenvalues different from λ. Then $A - \lambda I$ is similar to

$$J - \lambda I = (J_{m_1}(\lambda) - \lambda I) \oplus \cdots \oplus (J_{m_p}(\lambda) - \lambda I) \oplus (\hat{J} - \lambda I)$$
$$= J_{m_1}(0) \oplus \cdots \oplus J_{m_p}(0) \oplus (\hat{J} - \lambda I)$$

which is a direct sum of p nilpotent Jordan blocks of various sizes and a nonsingular Jordan matrix $\hat{J} - \lambda I \in M_m$, in which $m = n - (m_1 + \cdots + m_p)$. Moreover, $(A - \lambda I)^k$ is similar to $(J - \lambda I)^k = J_{m_1}(0)^k \oplus \cdots \oplus J_{m_p}(0)^k \oplus (\hat{J} - \lambda I)^k$ for each $k = 1, 2, \ldots$. Since the rank of a direct sum is the sum of the ranks of the summands (0.9.2), we have

$$\begin{aligned} \operatorname{rank}(A - \lambda I)^k &= \operatorname{rank}(J - \lambda I)^k \\ &= \operatorname{rank} J_{m_1}(0)^k + \cdots + \operatorname{rank} J_{m_p}(0)^k + \operatorname{rank}(\hat{J} - \lambda I)^k \\ &= \operatorname{rank} J_{m_1}(0)^k + \cdots + \operatorname{rank} J_{m_p}(0)^k + m \end{aligned} \qquad (3.1.13)$$

for each $k = 1, 2, \ldots$.

What is the rank of a power of a nilpotent Jordan block? Inspection of (3.1.2) reveals that the first column of $J_\ell(0)$ is zero and its last $\ell - 1$ columns are linearly independent (the only nonzero entries are ones in the first superdiagonal), so $\operatorname{rank} J_\ell(0) = \ell - 1$. The only nonzero entries in $J_\ell(0)^2$ are ones in the second superdiagonal, so its first two columns are zero, its last $\ell - 2$ columns are linearly independent, and $\operatorname{rank} J_\ell(0)^2 = \ell - 2$. The ones move up one superdiagonal (so the number of zero columns increases by one and the rank drops by one) with each successive power until $J_\ell(0)^{\ell-1}$ has just one nonzero entry (in position $1, \ell$) and $\operatorname{rank} J_\ell(0)^{\ell-1} = 1 = \ell - (\ell - 1)$. Of course, $J_\ell(0)^k = 0$ for all $k = \ell, \ell + 1, \ldots$. In general, we have $\operatorname{rank} J_\ell(0)^k = \max\{\ell - k, 0\}$ for each $k = 1, 2, \ldots$, and so

$$\operatorname{rank} J_\ell(0)^{k-1} - \operatorname{rank} J_\ell(0)^k = \begin{cases} 1 \text{ if } k \le \ell \\ 0 \text{ if } k > \ell \end{cases}, \quad k = 1, 2, \ldots \qquad (3.1.14)$$

in which we observe the standard convention that $\operatorname{rank} J_\ell(0)^0 = \ell$.

Now let $A \in M_n$, let $\lambda \in \mathbf{C}$, let k be a positive integer, let

$$r_k(A, \lambda) = \operatorname{rank}(A - \lambda I)^k, \quad r_0(A, \lambda) := n \qquad (3.1.15)$$

and define

$$w_k(A, \lambda) = r_{k-1}(A, \lambda) - r_k(A, \lambda), \quad w_1(A, \lambda) := n - r_1(A, \lambda) \qquad (3.1.16)$$

Exercise. If $A \in M_n$ and $\lambda \in \mathbf{C}$ is *not* an eigenvalue of A, explain why $w_k(A, \lambda) = 0$ for all $k = 1, 2, \ldots$.

Exercise. Consider the Jordan matrix

$$J = J_3(0) \oplus J_3(0) \oplus J_2(0) \oplus J_2(0) \oplus J_2(0) \oplus J_1(0) \qquad (3.1.16a)$$

Verify that $r_1(J, 0) = 7$, $r_2(J, 0) = 2$, and $r_3(J, 0) = r_4(J, 0) = 0$. Also verify that $w_1(J, 0) = 6$ is the number of blocks of size at least 1, $w_2(J, 0) = 5$ is the number of blocks of size at least 2, $w_3(J, 0) = 2$ is the number of blocks of size at least 3, and $w_4(J, 0) = 0$ is the number of blocks of size at least 4. Observe that $w_1(J, 0) - w_2(J, 0) = 1$ is the number of blocks of size 1,

$w_2(J, 0) - w_3(J, 0) = 3$ is the number of blocks of size 2, and $w_3(J, 0) - w_4(J, 0) = 2$ is the number of blocks of size 3. This is not an accident.

Exercise. Use (3.1.13) and (3.1.14) to explain the algebraic meaning of $w_k(A, \lambda)$:

$$
\begin{aligned}
w_k(A, \lambda) &= \left(\operatorname{rank} J_{m_1}(0)^{k-1} - \operatorname{rank} J_{m_1}(0)^k\right) + \\
&\quad \cdots + \left(\operatorname{rank} J_{m_p}(0)^{k-1} - \operatorname{rank} J_{m_p}(0)^k\right) \\
&= (1 \text{ if } m_1 \geq k) + \cdots + (1 \text{ if } m_p \geq k) \qquad (3.1.17) \\
&= \text{number of blocks with eigenvalue } \lambda \text{ that have size at least } k
\end{aligned}
$$

In particular, $w_1(A, \lambda)$ is the number of Jordan blocks of A of all sizes that have eigenvalue λ, which is the geometric multiplicity of λ as an eigenvalue of A.

Using the characterization (3.1.17), we see that $w_k(A, \lambda) - w_{k+1}(A, \lambda)$ is the number of blocks with eigenvalue λ that have size at least k but do not have size at least $k + 1$; this is the number of blocks with eigenvalue λ that have size exactly k.

Exercise. Let $A, B \in M_n$ and $\lambda \in \mathbf{C}$ be given. If A and B are similar, explain why $w_k(A, \lambda) = w_k(B, \lambda)$ for all $k = 1, 2, \ldots$.

Exercise. Let $A \in M_n$ and $\lambda \in \mathbf{C}$ be given. Explain why $w_1(A, \lambda) \geq w_2(A, \lambda) \geq w_3(A, \lambda) \geq \cdots$, that is, the sequence $w_1(A, \lambda), w_2(A, \lambda), \ldots$ is nonincreasing. *Hint*: $w_k(A, \lambda) - w_{k+1}(A, \lambda)$ is always a nonnegative integer. Why?

Exercise. Let $A \in M_n$ and $\lambda \in \mathbf{C}$ be given. Let q denote the size of the largest Jordan block of A with eigenvalue λ, and consider the rank identity (3.1.13). Explain why (a) $\operatorname{rank}(A - \lambda I)^k = \operatorname{rank}(A - \lambda I)^{k+1}$ for all $k \geq q$, (b) $w_q(A, \lambda)$ is the number of Jordan blocks of A with eigenvalue λ and maximum size q, and (c) $w_k(A, \lambda) = 0$ for all $k > q$. This integer q is called the *index* of λ as an eigenvalue of A.

Exercise. Let $A \in M_n$ and $\lambda \in \mathbf{C}$ be given. Explain why the index of λ as an eigenvalue of A may (equivalently) be defined as the smallest integer $k \geq 0$ such that $\operatorname{rank}(A - \lambda I)^k = \operatorname{rank}(A - \lambda I)^{k+1}$. Delicate point: Why must such a smallest integer exist?

Exercise. Let $A \in M_n$ and $\lambda \in \mathbf{C}$ be given. If λ is not an eigenvalue of A, explain why its index as an eigenvalue of A is zero.

The preceding exercises ensure that only finitely many terms of the sequence $w_1(A, \lambda), w_2(A, \lambda), \ldots$ defined by (3.1.16) are nonzero. The number of nonzero terms is the index of λ as an eigenvalue of A.

Exercise. Let $A \in M_n$ and let λ have index q as an eigenvalue of A. Explain why (a) $w_1(A, \lambda)$ is the geometric multiplicity of λ (the number of Jordan blocks with eigenvalue λ in the Jordan canonical form of A); (b) $w_1(A, \lambda) + w_2(A, \lambda) + \cdots + w_q(A, \lambda)$ is the algebraic multiplicity of λ (the sum of the sizes of all the Jordan blocks of A with eigenvalue λ); (c) for each $p = 2, 3, \ldots, q$, $w_p(A, \lambda) + w_{p+1}(A, \lambda) + \cdots + w_q(A, \lambda) = \operatorname{rank}(A - \lambda I)^{p-1}$.

The *Weyr characteristic* of $A \in M_n$ associated with $\lambda \in \mathbf{C}$ is

$$w(A, \lambda) = (w_1(A, \lambda), \ldots, w_q(A, \lambda))$$

in which the sequence of integers $w_1(A, \lambda)$, $w_2(A, \lambda)$, ... is defined by (3.1.16) and q is the index of λ as an eigenvalue of A. It is sometimes convenient to refer to the sequence $w_1(A, \lambda)$, $w_2(A, \lambda)$, ... itself as the Weyr characteristic of A associated with λ. We have just seen that the structure of a Jordan matrix J that is similar to A is completely determined by the Weyr characteristics of A associated with its distinct eigenvalues: If λ is an eigenvalue of A, and if J is a Jordan matrix that is similar to A, then the number of Jordan blocks $J_k(\lambda)$ in J is exactly $w_k(A, \lambda) - w_{k+1}(A, \lambda)$, $k = 1, 2, \ldots$. This means that two essentially different Jordan matrices (that is, for some eigenvalue, their respective lists of nonincreasingly ordered block sizes associated with that eigenvalue are not identical) cannot both be similar to A because their Weyr characteristics must be different. We have now proved the uniqueness portion of (3.1.11) and a little more.

Lemma 3.1.18. *Let λ be a given eigenvalue of $A \in M_n$ and let $w_1(A, \lambda)$, $w_2(A, \lambda)$, ... be the Weyr characteristic of A associated with λ. The number of blocks of the form $J_k(\lambda)$ in the Jordan canonical form of A is $w_k(A, \lambda) - w_{k+1}(A, \lambda)$, $k = 1, 2, \ldots$. Two square complex matrices $A, B \in M_n$ are similar if and only if (a) they have the same distinct eigenvalues $\lambda_1, \ldots, \lambda_d$, and (b) for each $i = 1, \ldots, d$, $w_k(A, \lambda_i) = w_k(B, \lambda_i)$ for all $k = 1, 2, \ldots$, that is, the same Weyr characteristics are associated with each eigenvalue.*

The Jordan structure of a given $A \in M_n$ can be completely specified by giving, for each distinct eigenvalue λ of A, a list of the sizes of all the Jordan blocks of A that have eigenvalue λ. The *nonincreasingly ordered* list of sizes of Jordan blocks of A with eigenvalue λ,

$$s_1(A, \lambda) \geq s_2(A, \lambda) \geq \cdots \geq s_{w_1(A,\lambda)}(A, \lambda) > 0$$
$$= 0 = s_{w_1(A,\lambda)+1}(A, \lambda) = \cdots \qquad (3.1.19)$$

is called the *Segre characteristic* of A associated with the eigenvalue λ. It is convenient to define $s_k(A, \lambda) = 0$ for all $k > w_1(A, \lambda)$. Observe that $s_1(A, \lambda)$ is the index of λ as an eigenvalue of A (the size of the largest Jordan block of A with eigenvalue λ) and $s_{w_1(A,\lambda)}(A, \lambda)$ is the size of the smallest Jordan block of A with eigenvalue λ. For example, the Segre characteristic of the matrix (3.1.16a) associated with the zero eigenvalue is 3, 3, 2, 2, 2, 1 ($s_1(J, 0) = 3$ and $s_6(J, 0) = 1$).

If $s_k = s_k(A, \lambda)$, $k = 1, 2, \ldots$ is the Segre characteristic of $A \in M_n$ associated with the eigenvalue λ and $w_1 = w_1(A, \lambda)$, the part of the Jordan canonical form that contains all the Jordan blocks of A with eigenvalue λ is

$$\begin{bmatrix} J_{s_1}(\lambda) & & & \\ & J_{s_2}(\lambda) & & \\ & & \ddots & \\ & & & J_{s_{w_1}}(\lambda) \end{bmatrix} \qquad (3.1.20)$$

It is easy to derive the Weyr characteristic if the Segre characteristic is known, and vice versa. For example, from the Segre characteristic 3, 3, 2, 2, 2, 1 we see that there are 6 blocks of size 1 or greater, 5 blocks of size 2 or greater, and 2 blocks of size 3 or greater: The Weyr characteristic is 6,5,2. Conversely, from the Weyr characteristic 6,5,2, we see that there are $6 - 5 = 1$ blocks of size 1, $5 - 2 = 3$ blocks of size 2, and $2 - 0 = 2$ blocks of size 3: The Segre characteristic is 3, 3, 2, 2, 2, 1.

Our derivation of the Jordan canonical form is based on an explicit algorithm, but it cannot be recommended for implementation in a software package to compute Jordan canonical forms. A simple example illustrates the difficulty: If $A_\epsilon = \begin{bmatrix} \epsilon & 0 \\ 1 & 0 \end{bmatrix}$ and $\epsilon \neq 0$, then $A_\epsilon = S_\epsilon J_\epsilon S_\epsilon^{-1}$ with $S_\epsilon = \begin{bmatrix} 0 & \epsilon \\ 1 & 1 \end{bmatrix}$ and $J_\epsilon = \begin{bmatrix} 0 & 0 \\ 0 & \epsilon \end{bmatrix}$. If we let $\epsilon \to 0$, then $J_\epsilon \to \begin{bmatrix} 0 & 0 \\ 0 & 0 \end{bmatrix} = J_1(0) \oplus J_1(0)$, but $A_\epsilon \to A_0 = \begin{bmatrix} 0 & 0 \\ 1 & 0 \end{bmatrix}$, whose Jordan canonical form is $J_2(1)$. Small variations in the entries of a matrix can result in major changes in its Jordan canonical form. The root of the difficulty is that rank A is not a continuous function of the entries of A.

It is sometimes useful to know that every matrix is similar to a matrix of the form (3.1.12) in which all the "+1" entries in the Jordan blocks are replaced by any $\epsilon \neq 0$.

Corollary 3.1.21. *Let $A \in M_n$ and a nonzero $\epsilon \in \mathbf{C}$ be given. Then there exists a nonsingular $S(\epsilon) \in M_n$ such that*

$$A = S(\epsilon) \begin{bmatrix} J_{n_1}(\lambda_1, \epsilon) & & \\ & \ddots & \\ & & J_{n_k}(\lambda_k, \epsilon) \end{bmatrix} S(\epsilon)^{-1} \qquad (3.1.22)$$

in which $n_1 + n_2 + \cdots + n_k = n$ and

$$J_m(\lambda, \epsilon) = \begin{bmatrix} \lambda & \epsilon & & & \\ & \ddots & \ddots & & \\ & & \ddots & \ddots & \\ & & & \ddots & \epsilon \\ & & & & \lambda \end{bmatrix} \in M_m$$

If A is real and has real eigenvalues, and if $\epsilon \in \mathbf{R}$, then $S(\epsilon)$ may be taken to be real.

Proof. First find a nonsingular matrix $S_1 \in M_n$ such that $S_1^{-1} A S_1$ is a Jordan matrix of the form (3.1.3) (with a real S_1 if A is real and has real eigenvalues). Let $D_{\epsilon,i} = \mathrm{diag}(1, \epsilon, \epsilon^2, \ldots, \epsilon^{n_i-1})$, define $D_\epsilon = D_{\epsilon,1} \oplus \cdots \oplus D_{\epsilon,k}$, and compute $D_\epsilon^{-1}(S_1^{-1} A S_1) D_\epsilon$. This matrix has the form (3.1.22), so $S(\epsilon) = S_1 D_\epsilon$ meets the stated requirements. $\qquad \square$

Problems

3.1.P1 Supply the computational details to prove (3.1.4).

3.1.P2 What are the Jordan canonical forms of the two matrices in (3.0.0)?

3.1.P3 Suppose that $A \in M_n$ has some non-real entries, but only real eigenvalues. Show that A is similar to a real matrix. Can the similarity matrix ever be chosen to be real?

3.1.P4 Let $A \in M_n$ be given. If A is similar to cA for some complex scalar c with $|c| \neq 1$, show that $\sigma(A) = \{0\}$ and hence A is nilpotent. Conversely, if A is nilpotent, show that A is similar to cA for all nonzero $c \in \mathbf{C}$.

3.1.P5 Explain why every Jordan block $J_k(\lambda)$ has a one-dimensional eigenspace associated with the eigenvalue λ. Conclude that λ has geometric multiplicity 1 and algebraic multiplicity k as an eigenvalue of $J_k(\lambda)$.

3.1.P6 Carry out the three steps in the proof of (3.1.11) to find the Jordan canonical forms of

$$\begin{bmatrix} 1 & 1 \\ 1 & 1 \end{bmatrix} \quad \text{and} \quad \begin{bmatrix} 3 & 1 & 2 \\ 0 & 3 & 0 \\ 0 & 0 & 3 \end{bmatrix}$$

Confirm your answers by using (3.1.18).

3.1.P7 Let $A \in M_n$, let λ be an eigenvalue of A, and let $k \in \{1, \ldots, n\}$. Explain why $r_{k-1}(A, \lambda) - 2r_k(A, \lambda) + r_{k+1}(A, \lambda)$ is the number of Jordan blocks of A that have size k and eigenvalue λ.

3.1.P8 Let $A \in M_n$ be given. Suppose that rank $A = r \geq 1$ and $A^2 = 0$. Use the preceding problem or (3.1.18) to show that the Jordan canonical form of A is $J_2(0) \oplus \cdots \oplus J_2(0) \oplus 0_{n-2r}$ (there are r 2-by-2 blocks). Compare with (2.6.P23).

3.1.P9 Let $n \geq 3$. Show that the Jordan canonical form of $J_n(0)^2$ is $J_m(0) \oplus J_m(0)$ if $n = 2m$ is even, and it is $J_{m+1}(0) \oplus J_m(0)$ if $n = 2m + 1$ is odd.

3.1.P10 For any $\lambda \in \mathbf{C}$ and any positive integer k, show that the Jordan canonical form of $-J_k(\lambda)$ is $J_k(-\lambda)$. In particular, the Jordan canonical form of $-J_k(0)$ is $J_k(0)$.

3.1.P11 The information contained in the Weyr characteristic of a matrix associated with a given eigenvalue can be presented as a *dot diagram*, sometimes called a *Ferrers diagram* or *Young diagram*. For example, consider the Jordan matrix J in (3.1.16a) and its Weyr characteristic $w(J, 0) = (w_1, w_2, w_3)$. Construct the dot diagram

$$
\begin{array}{cccccc}
w_1 & \bullet & \bullet & \bullet & \bullet & \bullet & \bullet \\
w_2 & \bullet & \bullet & \bullet & \bullet & \bullet \\
w_3 & \bullet & \bullet \\
& s_1 & s_2 & s_3 & s_4 & s_5 & s_6
\end{array}
$$

by putting w_1 dots in the first row, w_2 dots in the second row, and w_3 dots in the third row. We stop with the third row since $w_k = 0$ for all $k \geq 4$. Proceeding from the left, the respective column lengths are $3, 3, 2, 2, 2, 1$, which is the Segre characteristic $s_k = s_k(J, 0)$, $k = 1, 2, \ldots, 6$. That is, J has 2 Jordan blocks of the form $J_3(0)$, 3 blocks of the form $J_2(0)$, and 1 block of the form $J_1(0)$. Conversely, if one first constructs a dot diagram by putting s_1 dots in the first column, s_2 dots in the second column, and so forth, then there are w_1 dots in the first row, w_2 dots in the second row, and w_3 dots in the third row. In this sense, the Segre and Weyr characteristics are *conjugate partitions* of their common sum n; either characteristic can be derived from the other via a dot diagram. In general, for $A \in M_n$ and a given eigenvalue λ of A, use the Weyr characteristic to construct a dot diagram with $w_k(A, \lambda)$ dots in row $k = 1, 2, \ldots$ so long as $w_k(A, \lambda) > 0$. (a) Explain why there are $s_j(A, \lambda)$ dots in column j for each $j = 1, 2, \ldots$. (b) Explain why one can also start with the Segre characteristic, construct the columns of a dot diagram from it, and then read off the Weyr characteristic from the rows.

3.1.P12 Let $A \in M_n$, and let k and p be given positive integers. Let $w_k = w_k(A, \lambda)$, $k = 1, 2, \ldots$ and $s_k = s_k(A, \lambda)$, $k = 1, 2, \ldots$ denote the Weyr and Segre characteristics of A, respectively, both associated with a given eigenvalue λ. Show that (a) $s_{w_k} \geq k$ if $w_k > 0$; (b) $k > s_{w_k+1}$ for all $k = 1, 2, \ldots$; (c) $w_{s_k} \geq k$ if $s_k > 0$; and (d) $k > w_{s_k+1}$ for all $k = 1, 2, \ldots$. (e) Explain why $s_k \geq p > s_{k+1}$ if and only if $w_p = k$. (f) Show that the following three statements are equivalent: (i) $s_k \geq p > p - 1 > s_{k+1}$; (ii) $s_k \geq p > s_{k+1}$ and $s_k \geq p - 1 > s_{k+1}$; (iii) $p \geq 2$ and $w_p = w_{p-1} = k$. (g) Explain why $s_k > s_k - 1 > s_{k+1}$ if and only if there is no block $J_\ell(\lambda)$ of size $\ell = s_k - 1$ in the Jordan canonical form of A. (h) Show that the following four statements are equivalent: (i) $s_k - s_{k+1} \geq 2$; (ii) $s_k = s_k \geq s_k - 1 > s_{k+1}$; (iii) $p = s_k \geq 2$ and there is no block $J_{p-1}(\lambda)$ in the Jordan canonical form of A; (iv) $p = s_k \geq 2$ and $w_p = w_{p-1} = k$.

3.1.P13 Let k and m be given positive integers and consider the *block Jordan matrix*

$$J_k^+(\lambda I_m) := \begin{bmatrix} \lambda I_m & I_m & & & \\ & \lambda I_m & \ddots & & \\ & & \ddots & I_m & \\ & & & \lambda I_m \end{bmatrix} \in M_{km}$$

(a block k-by-k matrix). Compute the Weyr characteristic of $J_k^+(\lambda I_m)$ and use it to show that the Jordan canonical form of $J_k^+(\lambda I_m)$ is $J_k(\lambda) \oplus \cdots \oplus J_k(\lambda)$ (m summands).

3.1.P14 Let $A \in M_n$. Use (3.1.18) to show that A and A^T are similar. Are A and A^* similar?

3.1.P15 Let $n \geq 2$, let $x, y \in \mathbf{C}^n$ be given nonzero vectors, and let $A = xy^*$. (a) Show that the Jordan canonical form of A is $B \oplus 0_{n-2}$, in which $B = \begin{bmatrix} y^*x & 0 \\ 0 & 0 \end{bmatrix}$ if $y^*x \neq 0$ and $B = J_2(0)$ if $y^*x = 0$. (b) Explain why a rank-one matrix is diagonalizable if and only if its trace is nonzero.

3.1.P16 Suppose that $\lambda \neq 0$ and $k \geq 2$. Then $J_k(\lambda)^{-1}$ is a polynomial in $J_k(\lambda)$ (2.4.3.4). (a) Explain why $J_k(\lambda)^{-1}$ is an upper triangular Toeplitz matrix, all of whose main diagonal entries are λ^{-1}. (b) Let $[\lambda^{-1} \ a_2 \ \ldots \ a_n]$ be the first row of $J_k(\lambda)^{-1}$. Verify that the 1, 2 entry of $J_k(\lambda)J_k(\lambda)^{-1}$ is $\lambda a_2 + \lambda^{-1}$ and explain why all the entries in the first superdiagonal of $J_k(\lambda)^{-1}$ are $-\lambda^{-2}$; in particular, these entries are all nonzero. (c) Show that rank$(J_k(\lambda)^{-1} - \lambda^{-1}I)^k = n - k$ for $k = 1, \ldots, n$ and explain why the Jordan canonical form of $J_k(\lambda)^{-1}$ is $J_k(\lambda^{-1})$.

3.1.P17 Suppose that $A \in M_n$ is nonsingular. Show that A is similar to A^{-1} if and only if for each eigenvalue λ of A with $\lambda \neq \pm 1$, the number of Jordan blocks of the form $J_k(\lambda)$ in the Jordan canonical form of A is equal to the number of blocks of the form $J_k(\lambda^{-1})$, that is, the blocks $J_k(\lambda)$ and $J_k(\lambda^{-1})$ occur in pairs if $\lambda \neq \pm 1$ (there is no restriction on the blocks with eigenvalues ± 1).

3.1.P18 Suppose that $A \in M_n$ is nonsingular. (a) If each eigenvalue of A is either $+1$ or -1, explain why A is similar to A^{-1}. (b) Suppose that there are nonsingular $B, C, S \in M_n$ such that $A = BC$, $B^{-1} = SBS^{-1}$, and $C^{-1} = SCS^{-1}$. Show that A is similar to A^{-1}.

3.1.P19 Let $x, y \in \mathbf{R}^n$ and $t \in \mathbf{R}$ be given. Define the upper triangular matrix

$$A_{x,y,t} = \begin{bmatrix} 1 & x^T & t \\ & I_n & y \\ & & 1 \end{bmatrix} \in M_{n+2}(\mathbf{R})$$

and let $\mathcal{H}_n(\mathbf{R}) = \{A_{x,y,t} : x, y \in \mathbf{R}^n \text{ and } t \in \mathbf{R}\}$. (a) Show that $A_{x,y,t} A_{\xi,\eta,\tau} = A_{x+\xi,y+\eta,t+\tau}$ and $(A_{x,y,t})^{-1} = A_{-x,-y,-t}$. (b) Explain why $\mathcal{H}_n(\mathbf{R})$ is a subgroup (called the nth *Heisenberg group*) of the group of upper triangular matrices in $M_{n+2}(\mathbf{R})$ that have all main diagonal entries equal to $+1$. (c) Explain why the Jordan canonical form of $A_{x,y,t}$ is $J_3(1) \oplus I_{n-1}$ if $x^T y \neq 0$; if $x^T y = 0$, it is either $J_2(1) \oplus J_2(1) \oplus I_{n-2}$ ($x \neq 0 \neq y$), or $J_2(1) \oplus I_n$ ($x = 0$ or $y = 0$ but not both), or I_{n+2} ($x = y = 0$). (d) Explain why $A_{x,y,t}$ is always similar to its inverse.

3.1.P20 Let $A \in M_n$ and suppose that $n > \operatorname{rank} A \geq 1$. If $\operatorname{rank} A = \operatorname{rank} A^2$, that is, if 0 is a semisimple eigenvalue of A, show that A is rank principal; see (0.7.6). For special cases of this result, see (2.5.P48) and (4.1.P30).

3.1.P21 Let $A \in M_n$ be an unreduced upper Hessenberg matrix; see (0.9.9). (a) For each eigenvalue λ of A, explain why $w_1(A, \lambda) = 1$ and A is nonderogatory. (b) Suppose that A is diagonalizable (for example, A might be Hermitian and tridiagonal). Explain why A has n distinct eigenvalues.

3.1.P22 Let $A \in M_n(\mathbf{R})$ be tridiagonal. (a) If $a_{i,i+1} a_{i+1,i} > 0$ for all $i = 1, \ldots, n-1$, show that A has n distinct real eigenvalues. (b) If $a_{i,i+1} a_{i+1,i} \geq 0$ for all $i = 1, \ldots, n-1$, show that all the eigenvalues of A are real.

3.1.P23 Let $A = [a_{ij}] \in M_n$ be tridiagonal with a_{ii} real for all $i = 1, \ldots, n$. (a) If $a_{i,i+1} a_{i+1,i}$ is real and positive for $i = 1, \ldots, n-1$, show that A has n distinct real eigenvalues. (b) If $a_{i,i+1} a_{i+1,i}$ is real and nonnegative for all $i = 1, \ldots, n-1$, show that all the eigenvalues of A are real. (c) If each $a_{ii} = 0$ and $a_{i,i+1} a_{i+1,i}$ is real and negative for $i = 1, \ldots, n-1$, deduce from (a) that A has n distinct pure imaginary eigenvalues. In addition, show that those eigenvalues occur in $\pm$ pairs, so A is singular if n is odd.

3.1.P24 Consider the 4-by-4 matrices $A = [A_{ij}]_{i,j=1}^2$ and $B = [B_{ij}]_{i,j=1}^2$, in which $A_{11} = A_{22} = B_{11} = B_{22} = J_2(0)$, $A_{21} = B_{21} = 0_2$, $A_{12} = \begin{bmatrix} 0 & 1 \\ 1 & 1 \end{bmatrix}$, and $B_{12} = \begin{bmatrix} 1 & 1 \\ 1 & 0 \end{bmatrix}$. (a) For all $k = 1, 2, \ldots$, show that A^k and B^k are $0 - 1$ matrices (that is, every entry is 0 or 1) that have the same number of entries equal to 1. (b) Explain why A and B are nilpotent and similar. What is their Jordan canonical form? (c) Explain why two permutation-similar $0 - 1$ matrices have the same number of entries equal to 1. (d) Show that A and B are not permutation similar.

3.1.P25 Use (3.1.11) to show that $A \in M_n$ is diagonalizable if and only if the following condition is satisfied for each eigenvalue λ of A: If $x \in \mathbf{C}^n$ and $(A - \lambda I)^2 x = 0$, then $(A - \lambda I)x = 0$.

3.1.P26 Let $A \in M_n$ be normal. Use the preceding problem to deduce from the definition (2.5.1) that A is diagonalizable; do not invoke the spectral theorem (2.5.3).

3.1.P27 Let $A \in M_n$ be normal. Use the preceding problem and the QR factorization (2.1.14) to show that A is unitarily diagonalizable.

3.1.P28 Let $A, B \in M_n$. Show that A and B are similar if and only if $r_k(A, \lambda) = r_k(B, \lambda)$ for every eigenvalue λ of A and every $k = 1, \ldots, n$.

3.1.P29 Let $A \in M_k$ be upper triangular. Suppose that $a_{ii} = 1$ for each $i = 1, \ldots, n$ and $a_{i,i+1} \neq 0$ for each $i = 1, \ldots, n-1$. Show that A is similar to $J_k(1)$.

3.1.P30 Suppose that the only eigenvalue of $A \in M_n$ is $\lambda = 1$. Show that A is similar to A^k for each $k = 1, 2, \ldots$.

Notes and Further Readings. Camille Jordan published his eponymous canonical form in C. Jordan, *Traité des Substitutions et des Équations Algébriques*, Gauthier-Villars, Paris, 1870; see section 157, pp. 125–126. Our proof of (3.1.11) is in the spirit of R. Fletcher and D. Sorensen, An algorithmic derivation of the Jordan canonical form, *Amer. Math. Monthly* 90 (1983) 12–16. For a combinatorial approach, see R. Brualdi, The Jordan canonical form: An old proof, *Amer. Math. Monthly* 94 (1987) 257–267.

3.2 Consequences of the Jordan canonical form

3.2.1 The structure of a Jordan matrix. The Jordan matrix

$$
J = \begin{bmatrix} J_{n_1}(\lambda_1) & & \\ & \ddots & \\ & & J_{n_k}(\lambda_k) \end{bmatrix}, \quad n_1 + n_2 + \cdots + n_k = n \tag{3.2.1.1}
$$

has a definite structure that makes apparent certain basic properties of any matrix that is similar to it.

1. The number k of Jordan blocks (counting multiple occurrences of the same block) is the maximum number of linearly independent eigenvectors of J.
2. The matrix J is diagonalizable if and only if $k = n$, that is, if and only if all the Jordan blocks are 1-by-1.
3. The number of Jordan blocks corresponding to a given eigenvalue is the geometric multiplicity of the eigenvalue, which is the dimension of the associated eigenspace. The sum of the sizes of all the Jordan blocks corresponding to a given eigenvalue is its algebraic multiplicity.
4. Let $A \in M_n$ be a given nonzero matrix, and suppose that λ is an eigenvalue of A. Using (3.1.14) and the notation of (3.1.15), we know that there is some positive integer q such that

$$
r_1(A, \lambda) > r_2(A, \lambda) > \cdots > r_{q-1}(A, \lambda) > r_q(A, \lambda) = r_{q+1}(A, \lambda)
$$

This integer q is the index of λ as an eigenvalue of A; it is also the size of the largest Jordan block of A with eigenvalue λ.

3.2.2 Linear systems of ordinary differential equations. One application of the Jordan canonical form that is of considerable theoretical importance is the analysis

of solutions of a system of first order linear ordinary differential equations with constant coefficients. Let $A \in M_n$ be given, and consider the first-order initial value problem

$$x'(t) = Ax(t)$$
$$x(0) = x_0 \text{ is given} \qquad (3.2.2.1)$$

in which $x(t) = [x_1(t), x_2(t), \ldots, x_n(t)]^T$, and the prime ($'$) denotes differentiation with respect to t. If A is not a diagonal matrix, this system of equations is *coupled*; that is, $x_i'(t)$ is related not only to $x_i(t)$ but to the other entries of the vector $x(t)$ as well. This coupling makes the problem hard to solve, but if A can be transformed to diagonal (or almost diagonal) form, the amount of coupling can be reduced or even eliminated and the problem may be easier to solve. If $A = SJS^{-1}$ and J is the Jordan canonical form of A, then (3.2.2.1) becomes

$$y'(t) = Jy(t)$$
$$y(0) = y_0 \text{ is given} \qquad (3.2.2.2)$$

in which $x(t) = Sy(t)$ and $y_0 = S^{-1}x_0$. If the problem (3.2.2.2) can be solved, then each entry of the solution $x(t)$ to (3.2.2.1) is just a linear combination of the entries of the solution to (3.2.2.2), and the linear combinations are given by S.

If A is diagonalizable, then J is a diagonal matrix, and (3.2.2.2) is just an uncoupled set of equations of the form $y_k'(t) = \lambda_k y_k(t)$, which have the solutions $y_k(t) = y_k(0)e^{\lambda_k t}$. If the eigenvalue λ_k is real, this is a simple exponential, and if $\lambda_k = a_k + ib_k$ is not real, $y_k(t) = y_k(0)e^{a_k t}[\cos(b_k t) + i \sin(b_k t)]$ is an oscillatory term with a real exponential factor if $a_k \neq 0$.

If J is not diagonal, the solution is more complicated but can be described explicitly. The entries of $y(t)$ that correspond to distinct Jordan blocks in J are not coupled, so it suffices to consider the case in which $J = J_m(\lambda)$ is a single Jordan block. The system (3.2.2.2) is

$$y_1'(t) = \lambda y_1(t) + y_2(t)$$
$$\vdots \qquad \vdots$$
$$y_{m-1}'(t) = \lambda y_{m-1}(t) + y_m(t)$$
$$y_m'(t) = \lambda y_m(t)$$

which can be solved in a straightforward way from the bottom up. Starting with the last equation, we obtain

$$y_m(t) = y_m(0)e^{\lambda t}$$

so that

$$y_{m-1}'(t) = \lambda y_{m-1}(t) + y_m(0)e^{\lambda t}$$

This has the solution

$$y_{m-1}(t) = e^{\lambda t}[y_m(0)t + y_{m-1}(0)]$$

which can now be used in the next equation. It becomes

$$y'_{m-2}(t) = \lambda y_{m-2}(t) + y_m(0)te^{\lambda t} + y_{m-1}(0)e^{\lambda t}$$

which has the solution

$$y_{m-2}(t) = e^{\lambda t}[y_m(0)\frac{t^2}{2} + y_{m-1}(0)t + y_{m-2}(0)]$$

and so forth. Each entry of the solution has the form

$$y_k(t) = e^{\lambda t}q_k(t) = e^{\lambda t}\sum_{i=k}^{m} y_i(0)\frac{t^{i-k}}{(i-k)!}$$

so $q_k(t)$ is an explicitly determined polynomial of degree at most $m - k$, $k = 1, \ldots, m$.

From this analysis, we conclude that the entries of the solution $x(t)$ of the problem (3.2.2.1) have the form

$$x_j(t) = e^{\lambda_1 t}p_1(t) + e^{\lambda_2 t}p_2(t) + \cdots + e^{\lambda_k t}p_k(t)$$

in which $\lambda_1, \lambda_2, \ldots, \lambda_k$ are the distinct eigenvalues of A and each $p_j(t)$ is a polynomial whose degree is strictly less than the size of the largest Jordan block corresponding to the eigenvalue λ_j (that is, strictly less than the index of λ_j). Real eigenvalues are associated with terms that contain a real exponential factor, while non-real eigenvalues are associated with terms that contain an oscillatory factor and possibly also a real exponential factor.

3.2.3 Similarity of a matrix and its transpose.

Let K_m be the m-by-m reversal matrix (0.9.5.1), which is symmetric and involutory: $K_m = K_m^T = K_m^{-1}$.

Exercise. Verify that $K_m J_m(\lambda) = J_m(\lambda)^T K_m$ and $J_m(\lambda)K_m = K_m J_m(\lambda)^T$. Deduce that $K_m J_m(\lambda)$ and $J_m(\lambda)K_m$ are symmetric, and $J_m(\lambda) = K_m^{-1} J_m(\lambda)^T K_m = K_m J_m(\lambda)^T K_m$. *Hint*: See (0.9.7).

The preceding exercise shows that each Jordan block is similar to its transpose via a reversal matrix. Therefore, if J is a given Jordan matrix (3.2.1.1), then J^T is similar to J via the symmetric involutory matrix $K = K_{n_1} \oplus \cdots \oplus K_{n_k}$: $J^T = KJK$. If $S \in M_n$ is nonsingular (but not necessarily symmetric) and $A = SJS^{-1}$, then $J = S^{-1}AS$,

$$A^T = S^{-T}J^T S^T = S^{-T}KJKS^T = S^{-T}K(S^{-1}AS)KS^T$$
$$= (S^{-T}KS^{-1})A(SKS^T) = (SKS^T)^{-1}A(SKS^T)$$

and the similarity matrix SKS^T between A and A^T is symmetric. We have proved the following theorem.

Theorem 3.2.3.1. *Let $A \in M_n$. There is a nonsingular complex symmetric matrix S such that $A^T = SAS^{-1}$.*

If A is nonderogatory, we can say a little more: *every* similarity between A and A^T must be via a symmetric matrix; see (3.2.4.4).

Returning to the similarity between A and its Jordan canonical form, we can write

$$A = SJS^{-1} = (SKS^T)(S^{-T}KJS^{-1}) = (SJKS^T)(S^{-T}KS^{-1})$$

in which KJ and JK are symmetric. This observation proves the following theorem.

Theorem 3.2.3.2. *Each square complex matrix is a product of two complex symmetric matrices, in which either factor may be chosen to be nonsingular.*

For *any* field $\mathbf{F}$, it is known that every matrix in $M_n(\mathbf{F})$ is similar, via some symmetric matrix in $M_n(\mathbf{F})$, to its transpose. In particular, each real square matrix is similar, via some real symmetric matrix, to its transpose.

3.2.4 Commutativity and nonderogatory matrices.

For any polynomial $p(t)$ and any $A \in M_n$, $p(A)$ always commutes with A. What about the converse? If $A, B \in M_n$ are given and if A commutes with B, is there some polynomial $p(t)$ such that $B = p(A)$? Not always, for if we take $A = I$, then A commutes with every matrix and $p(I) = p(1)I$ is a scalar matrix; no nonscalar matrix can be a polynomial in I. The problem is that the form of A permits it to commute with many matrices, but to generate only a limited set of matrices of the form $p(A)$.

What can we say if $A = J_m(\lambda)$ is a single Jordan block of size 2 or greater?

Exercise. Let $\lambda \in \mathbf{C}$ and an integer $m \geq 2$ be given. Show that $B \in M_m$ commutes with $J_m(\lambda)$ if and only if it commutes with $J_m(0)$. *Hint:* $J_m(\lambda) = \lambda I_m + J_m(0)$.

Exercise. Show that $B = \begin{bmatrix} b_{11} & b_{12} \\ b_{21} & b_{22} \end{bmatrix} \in M_2$ commutes with $J_2(0)$ if and only if $b_{21} = 0$ and $b_{11} = b_{22}$; this is the case if and only if $B = b_{11}I_2 + b_{12}J_2(0)$, which is a polynomial in $J_2(0)$.

Exercise. Show that $B = [b_{ij}] \in M_3$ commutes with $J_3(0)$ if and only if B is upper triangular, $b_{11} = b_{22} = b_{33}$, and $b_{12} = b_{23}$, that is, if and only if B is an upper triangular Toeplitz matrix (0.9.7). This is the case if and only if $B = b_{11}I_3 + b_{12}J_3(0) + b_{13}J_3(0)^2$, which is a polynomial in $J_3(0)$.

Exercise. What can you say about $B = [b_{ij}] \in M_4$ if it commutes with $J_4(0)$?

Definition 3.2.4.1. *A square complex matrix is* nonderogatory *if each of its eigenvalues has geometric multiplicity 1.*

Since the geometric multiplicity of a given eigenvalue of a Jordan matrix is equal to the number of Jordan blocks corresponding to that eigenvalue, a matrix is nonderogatory if and only if each of its distinct eigenvalues corresponds to exactly one block in its Jordan canonical form. Examples of nonderogatory matrices $A \in M_n$ are any matrix with n distinct eigenvalues or any matrix with only one eigenvalue, which has geometric multiplicity 1 (that is, A is similar to a single Jordan block). A scalar matrix is the antithesis of a nonderogatory matrix.

Exercise. If $A \in M_n$ is nonderogatory, why is rank $A \geq n - 1$?

Theorem 3.2.4.2. *Suppose that $A \in M_n$ is nonderogatory. If $B \in M_n$ commutes with A, then there is a polynomial $p(t)$ of degree at most $n - 1$ such that $B = p(A)$.*

Proof. Let $A = S J_A S^{-1}$ be the Jordan canonical form of A. If $BA = AB$, then $BSJ_AS^{-1} = SJ_AS^{-1}B$ and hence $(S^{-1}BS)J_A = J_A(S^{-1}BS)$. If we can show that

$S^{-1}BS = p(J_A)$, then $B = Sp(J_A)S^{-1} = p(SJ_AS^{-1}) = p(A)$ is a polynomial in A. Thus, it suffices to assume that A is itself a Jordan matrix.

Assume that (a) $A = J_{n_1}(\lambda_1) \oplus \cdots \oplus J_{n_k}(\lambda_k)$, in which $\lambda_1, \lambda_2, \ldots, \lambda_k$ are distinct, and (b) A commutes with B. If we partition $B = [B_{ij}]_{i,j=1}^k$ conformally with J, then (2.4.4.2) ensures that $B = B_{11} \oplus \cdots \oplus B_{kk}$ is block diagonal. Moreover, $B_{ii}J_{n_i}(0) = J_{n_i}(0)B_{ii}$ for each $i = 1, 2, \ldots, k$. A computation reveals that each B_{ii} must be an upper triangular Toeplitz matrix (0.9.7), that is,

$$B_{ii} = \begin{bmatrix} b_1^{(i)} & b_2^{(i)} & \cdots & b_{n_i}^{(i)} \\ & \ddots & \ddots & \vdots \\ & & \ddots & b_2^{(i)} \\ & & & b_1^{(i)} \end{bmatrix} \tag{3.2.4.3}$$

which is a polynomial in $J_{n_i}(0)$ and hence also a polynomial in $J_{n_i}(\lambda)$:

$$B_{ii} = b_1^{(i)} I_{n_i} + b_2^{(i)} J_{n_i}(0) + \cdots + b_{n_i}^{(i)} J_{n_i}(0)^{n_i-1}$$
$$= b_1^{(i)}(J_{n_i}(\lambda) - \lambda_i I_{n_i})^0 + b_2^{(i)}(J_{n_i}(\lambda) - \lambda_i I_{n_i})^1 + \cdots + b_{n_i}^{(i)}(J_{n_i}(\lambda) - \lambda_i I_{n_i})^{n_i-1}$$

If we can construct polynomials $p_i(t)$ of degree at most $n - 1$ with the property that $p_i(J_{n_j}(\lambda_j)) = 0$ for all $i \neq j$, and $p_i(J_{n_i}(\lambda_i)) = B_{ii}$, then

$$p(t) = p_1(t) + \cdots + p_k(t)$$

fulfills the assertions of the theorem. Define

$$q_i(t) = \prod_{\substack{j=1 \\ j \neq i}}^k (t - \lambda_j)^{n_j}, \qquad \text{degree } q_i(t) = n - n_i$$

and observe that $q_i(J_{n_j}(\lambda_j)) = 0$ whenever $i \neq j$ because $(J_{n_j}(\lambda_j) - \lambda_j I)^{n_j} = 0$. The upper triangular Toeplitz matrix $q_i(J_{n_i}(\lambda_i))$ is nonsingular because its main diagonal entries $q_i(\lambda_i)$ are nonzero.

The key to our construction of the polynomials $p_i(t)$ is observing that the product of two upper triangular Toeplitz matrices is upper triangular Toeplitz, and the inverse of a nonsingular upper triangular Toeplitz matrix has the same form (0.9.7). Thus, $[q_i(J_{n_i}(\lambda_i))]^{-1}B_{ii}$ is an upper triangular Toeplitz matrix, which is therefore a polynomial in $J_{n_i}(\lambda_i)$:

$$[q_i(J_{n_i}(\lambda_i))]^{-1}B_{ii} = r_i(J_{n_i}(\lambda_i))$$

in which $r_i(t)$ is a polynomial of degree at most $n_i - 1$. The polynomial $p_i(t) = q_i(t)r_i(t)$ has degree at most $n - 1$,

$$p_i(J_{n_j}(\lambda_j)) = q_i(J_{n_j}(\lambda_j))r_i(J_{n_j}(\lambda_j)) = 0 \text{ whenever } i \neq j$$

and

$$p_i(J_{n_i}(\lambda_i)) = q_i(J_{n_i}(\lambda_i))r_i(J_{n_i}(\lambda_i))$$
$$= q_i(J_{n_i}(\lambda_i))(q_i(J_{n_i}(\lambda_i))^{-1}B_{ii}) = B_{ii}$$

$\square$

There is a converse to the preceding theorem; see (3.2.P2).

An illustrative application of (3.2.4.2) is the following strengthening of (3.2.3.1) in a special case.

Corollary 3.2.4.4. *Let $A, B, S \in M_n$ be given and suppose that A is nonderogatory.*

(a) *If $AB = BA^T$, then B is symmetric.*
(b) *If S is nonsingular and $A^T = S^{-1}AS$, then S is symmetric.*

Proof. (a) There is a symmetric nonsingular $R \in M_n$ such that $A^T = RAR^{-1}$ (3.2.3.1), so $AB = BA^T = BRAR^{-1}$ and hence $A(BR) = (BR)A$. Then (3.2.4.2) ensures that there is a polynomial $p(t)$ such that $BR = p(A)$. Compute $RB^T = (BR)^T = p(A)^T = p(A^T) = p(RAR^{-1}) = Rp(A)R^{-1} = R(BR)R^{-1} = RB$. Since R is nonsingular, it follows that $B^T = B$.

(b) If $A^T = S^{-1}AS$, then $SA^T = AS$, so (a) ensures that S is symmetric. $\square$

3.2.5 Convergent and power-bounded matrices. A matrix $A \in M_n$ is *convergent* if each entry of A^m tends to zero as $m \to \infty$; it is is *power bounded* if all of the entries of the family $\{A^m : m = 1, 2, \ldots\}$ are contained in a bounded subset of **C**. A convergent matrix is power bounded; the identity matrix is an example of a power-bounded matrix that is not convergent. Convergent matrices play an important role in the analysis of algorithms in numerical linear algebra.

A diagonal matrix (and hence also a diagonalizable matrix) is convergent if and only if all its eigenvalues have modulus strictly less than one. The same is true of nondiagonalizable matrices, but a careful analysis is required to come to this conclusion.

Let $A = SJ_AS^{-1}$ be the Jordan canonical form of A, so $A^m = SJ_A^mS^{-1}$ and $A^m \to 0$ as $m \to \infty$ if and only if $J_A^m \to 0$ as $m \to \infty$. Since J_A is a direct sum of Jordan blocks, it suffices to consider the behavior of powers of a single Jordan block. For 1-by-1 blocks, $J_1(\lambda)^m = [\lambda^m] \to 0$ as $m \to \infty$ if and only if $|\lambda| < 1$. For blocks of size 2 or greater, $J_k(\lambda)^m = (\lambda I_k + J_k(0))^m$, which we can compute using the binomial theorem. We have $J_k(0)^m = 0$ for all $m \geq k$, so

$$J_k(\lambda)^m = (\lambda I + J_k(0))^m = \sum_{j=0}^{m} \binom{m}{m-j} \lambda^{m-j} J_k(0)^j$$

$$= \sum_{j=0}^{k-1} \binom{m}{m-j} \lambda^{m-j} J_k(0)^j$$

for all $m \geq k$. The diagonal entries of $J_k(\lambda)^m$ are all equal to λ^m, so $J_k(\lambda)^m \to 0$ implies that $\lambda^m \to 0$, which means that $|\lambda| < 1$. Conversely, if $|\lambda| < 1$, it suffices to prove that

$$\binom{m}{m-j} \lambda^{m-j} \to 0 \text{ as } m \to \infty \text{ for each } j = 0, 1, 2, \ldots, k-1$$

There is nothing to prove if $\lambda = 0$ or $j = 0$, so suppose that $0 < |\lambda| < 1$ and $j \geq 1$; compute

$$\left| \binom{m}{m-j} \lambda^{m-j} \right| = \left| \frac{m(m-1)(m-2)\cdots(m-j+1)\lambda^m}{j!\lambda^j} \right|$$

$$\leq \left| \frac{m^j \lambda^m}{j!\lambda^j} \right| \tag{3.2.5.1}$$

It suffices to show that $m^j |\lambda|^m \to 0$ as $m \to \infty$. One way to see this is to take logarithms and observe that $j \log m + m \log |\lambda| \to -\infty$ as $m \to \infty$ because $\log |\lambda| < 0$ and l'Hôpital's rule ensures that $(\log m)/m \to 0$ as $m \to \infty$. For an alternative approach that is independent of the Jordan canonical form, see (5.6.12).

To analyze the power-bounded case, we need to examine what happens for eigenvalues with unit modulus. For 1-by-1 blocks and $|\lambda| = 1$, $J_1(\lambda)^m = [\lambda^m]$ remains bounded as $m \to \infty$. The identity in (3.2.5.1) reveals that superdiagonal entries of $J_k(\lambda)^m$ do not remain bounded if $k \geq 2$ and $|\lambda| = 1$. We summarize our observations in the following theorem.

Theorem 3.2.5.2. *Let $A \in M_n$ be given. Then A is convergent if and only if every eigenvalue of A has modulus strictly less than one; A is power bounded if and only if every eigenvalue of A has modulus at most one and every Jordan block associated with an eigenvalue of modulus one is 1-by-1, that is, every eigenvalue of modulus one is semisimple.*

3.2.6 The geometric multiplicity–algebraic multiplicity inequality. The *geometric multiplicity* of an eigenvalue λ of a given $A \in M_n$ is the number of Jordan blocks of A corresponding to λ. This number is less than or equal to the sum of the sizes of all the Jordan blocks corresponding to λ; this sum is the *algebraic multiplicity* of λ. Thus, the geometric multiplicity of an eigenvalue is less than or equal to its algebraic multiplicity. The geometric and algebraic multiplicities of an eigenvalue λ are equal (that is, λ is a *semisimple eigenvalue*) if and only if every Jordan block corresponding to λ is 1-by-1. We have previously discussed the inequality between the algebraic and geometric multiplicities of an eigenvalue from very different points of view: see (1.2.18), (1.3.7), and (1.4.10).

3.2.7 Diagonalizable + nilpotent: the Jordan decomposition. For any Jordan block, we have the identity $J_k(\lambda) = \lambda I_k + J_k(0)$, and $J_k(0)^k = 0$. Thus, any Jordan block is the sum of a diagonal matrix and a nilpotent matrix.

More generally, a Jordan matrix (3.2.1.1) can be written as $J = D + N$, in which D is a diagonal matrix whose main diagonal is the same as that of J, and $N = J - D$. The matrix N is nilpotent, and $N^k = 0$ if k is the size of the largest Jordan block in J, which is the index of 0 as an eigenvalue of N.

Finally, if $A \in M_n$ and $A = SJ_AS^{-1}$ is its Jordan canonical form, then $A = S(D + N)S^{-1} = SDS^{-1} + SNS^{-1} = A_D + A_N$, in which A_D is diagonalizable and A_N is nilpotent. Moreover, $A_D A_N = A_N A_D$ because both D and N are conformal block diagonal matrices, and the diagonal blocks in D are scalar matrices. Of course, A_D and A_N also commute with $A = A_D + A_N$.

The preceding discussion establishes the existence of a *Jordan decomposition*: Any square complex matrix is a sum of two commuting matrices, one of which is diagonalizable and the other of which is nilpotent. For uniqueness of the Jordan decomposition, see (3.2.P18).

3.2.8 The Jordan canonical form of a direct sum. Let $A_i \in M_{n_i}$ be given for $i = 1, \ldots, m$ and suppose that each $A_i = S_i J_i S_i^{-1}$, in which each J_i is a Jordan matrix. Then the direct sum $A = A_1 \oplus \cdots \oplus A_m$ is similar to the direct sum $J = J_1 \oplus \cdots \oplus J_m$ via $S = S_1 \oplus \cdots \oplus S_m$. Moreover, J is a direct sum of direct sums of Jordan blocks, so it is a Jordan matrix and hence uniqueness of the Jordan canonical form ensures that it is the Jordan canonical form of A.

3.2.9 An optimality property of the Jordan canonical form. The Jordan canonical form of a matrix is a direct sum of upper triangular matrices that have nonzero off-diagonal entries only in the first superdiagonal, so it has many zero entries. However, among all the matrices that are similar to a given matrix, the Jordan canonical form need not have the least number of nonzero entries. For example,

$$A = \begin{bmatrix} 0 & 0 & 0 & -1 \\ 1 & 0 & 0 & 0 \\ 0 & 1 & 0 & 2 \\ 0 & 0 & 1 & 0 \end{bmatrix} \tag{3.2.9.1}$$

has five nonzero entries, but its Jordan canonical form $J = J_2(1) \oplus J_2(-1)$ has six nonzero entries. However, A has five nonzero *off-diagonal* entries, while J has only two nonzero off-diagonal entries. We now explain why no matrix similar to A can have fewer than two nonzero off-diagonal entries.

Observation 3.2.9.2. *Suppose that $B = [b_{ij}] \in M_m$ has fewer than $m - 1$ nonzero off-diagonal entries. Then there exists a permutation matrix P such that $P^T B P = B_1 \oplus B_2$ in which each $B_i \in M_{n_i}$ and each $n_i \geq 1$.*

Why is this? Here is an informal argument that can be made precise: Consider m islands $C_1, \ldots, C_m$ located near each other in the sea. There is a footbridge between two different islands C_i and C_j if and only if $i \neq j$ and either $b_{ij} \neq 0$ or $b_{ji} \neq 0$. Suppose that $C_1, C_{j_2}, \ldots, C_{j_\nu}$ are all the different islands that one can walk to starting from C_1. The minimum number of bridges required to link up all the islands is $m - 1$. We are assuming that there are fewer than $m - 1$ bridges, so $\nu < m$. Relabel all the islands (1 through m again) in any way that gives the new labels $1, 2, \ldots, \nu$ to $C_1, C_{j_2}, \ldots, C_{j_\nu}$. Let $P \in M_m$ be the permutation matrix corresponding to the relabeling. Then $P^T B P = B_1 \oplus B_2$, in which $B_1 \in M_\nu$. The direct sum structure reflects the fact that no bridge joins any of the first (relabeled) ν islands to any of the remaining $n - \nu$ islands.

We say that a given $B \in M_m$ is *indecomposable under permutation similarity* if there is no permutation matrix P such that $P^T B P = B_1 \oplus B_2$, in which each $B_i \in M_{n_i}$ and each $n_i \geq 1$. Then (3.2.9.2) says that if $B \in M_m$ is indecomposable under permutation similarity, it has at least $m - 1$ nonzero off-diagonal entries.

Observation 3.2.9.3. *Any given $B \in M_n$ is permutation similar to a direct sum of matrices that are indecomposable under permutation similarity.*

Proof. Consider the finite set $\mathcal{S} = \{P^T B P : P \in M_n \text{ is a permutation matrix}\}$. Some of the elements of $\mathcal{S}$ are block diagonal (take $P = I_n$, for example). Let q be the largest positive integer such that B is permutation similar to $B_1 \oplus \cdots \oplus B_q$, each $B_i \in M_{n_i}$, and each $n_i \geq 1$; maximality of q ensures that no direct summand B_i is decomposable under permutation similarity. $\qquad\Box$

The number of nonzero off-diagonal entries in a square matrix is not changed by a permutation similarity, so we can combine the two preceding observations to obtain a lower bound on the number of Jordan blocks in the Jordan canonical form of a matrix.

Observation 3.2.9.4. *Suppose that a given $B \in M_n$ has p nonzero off-diagonal entries and that its Jordan canonical form J_B contains r Jordan blocks. Then $r \geq n - p$.*

Proof. Suppose that B is permutation similar to $B_1 \oplus \cdots \oplus B_q$, in which each $B_i \in M_{n_i}$ is indecomposable under permutation similarity, and each $n_i \geq 1$. The number of nonzero off-diagonal entries in B_i is at least $n_i - 1$, so the number of nonzero off-diagonal entries in B is at least $(n_1 - 1) + \cdots + (n_q - 1) = n - q$. That is, $p \geq n - q$, so $q \geq n - p$. But (3.2.8) ensures that J_B contains at least q Jordan blocks, so $r \geq q \geq n - p$. $\qquad\Box$

Our final observation is that the number of nonzero off-diagonal entries in an n-by-n Jordan matrix $J = J_{n_1}(\lambda_1) \oplus \cdots \oplus J_{n_r}(\lambda_r)$ is exactly $(n_1 - 1) + \cdots + (n_r - 1) = n - r$.

Theorem 3.2.9.5. *Let $A, B \in M_n$ be given. Suppose that B has exactly p nonzero off-diagonal entries and is similar to A. Let J_A be the Jordan canonical form of A and suppose that J_A consists of r Jordan blocks. Then $p \geq n - r$, which is the number of nonzero off-diagonal entries of J_A.*

Proof. Since B is similar to A, J_A is also the Jordan canonical form of B, and (3.2.9.4) ensures that $r \geq n - p$, so $p \geq n - r$. $\qquad\Box$

3.2.10 The index of an eigenvalue of a block upper triangular matrix. If the indices of λ as an eigenvalue of $A_{11} \in M_{n_1}$ and $A_{22} \in M_{n_2}$ are ν_1 and ν_2, respectively, then inspection of the Jordan canonical forms of A_{11}, A_{22}, and their direct sum reveals that the index of λ as an eigenvalue of $A_{11} \oplus A_{22}$ is $\max\{\nu_1, \nu_2\}$.

The situation is different for block triangular matrices. If $A_{11}^{\nu_1} = 0$ and $A_{22}^{\nu_2} = 0$, then $A = \begin{bmatrix} A_{11} & A_{12} \\ 0 & A_{22} \end{bmatrix}$ is nilpotent. Compute

$$A^m = \begin{bmatrix} A_{11}^m & \sum_{k=0}^{m-1} A_{11}^k A_{12} A_{22}^{m-k-1} \\ 0 & A_{22}^m \end{bmatrix}, \quad m = 1, 2, \ldots$$

and let $m = \nu_1 + \nu_2$. Then $A_{11}^m = A_{11}^{\nu_1} A_{11}^{\nu_2} = 0$ and $A_{22}^m = A_{22}^{\nu_1} A_{22}^{\nu_2} = 0$. If $0 \leq k \leq \nu_1 - 1$, then $A_{22}^{m-k-1} = A_{22}^{\nu_1 - k - 1} A_{22}^{\nu_2} = 0$. If $\nu_1 \leq k \leq m - 1$, then $A_{11}^k = A_{11}^{\nu_1} A_{11}^{k - \nu_1} = 0$. This shows that $A_{11}^k A_{12} A_{22}^{m-k-1} = 0$ for each $k = 0, 1, \ldots, m - 1$. Therefore, $A^{\nu_1 + \nu_2} = 0$, so the index of 0 as an eigenvalue of A is at most $\nu_1 + \nu_2$. The example $A_{11} = J_2(0)$, $A_{22} = J_2(0)^T$, and $A_{12} = I_2$ shows that this upper bound on the index can be achieved: $A^4 = 0$ but $A^3 \neq 0$.

If not both of A_{11} and A_{22} are nilpotent and the indices of 0 as an eigenvalue of A_{11} and A_{22} are ν_1 and ν_2, respectively, Theorem 3.1.11 ensures that A_{11} is similar to

$B \oplus N_1$ and A_{22} is similar to $N_2 \oplus C$, in which B and C are nonsingular, $N_1^{\nu_1} = 0$, and $N_2^{\nu_2} = 0$. Then $A = \begin{bmatrix} A_{11} & A_{12} \\ 0 & A_{22} \end{bmatrix}$ is similar to $\begin{bmatrix} B \oplus N_1 & * \\ 0 & N_2 \oplus C \end{bmatrix}$, which is similar (see the proof of Theorem 2.4.6.1) to

$$\begin{bmatrix} B & 0 & 0 & * \\ 0 & N_1 & * & 0 \\ & & N_2 & 0 \\ & & 0 & C \end{bmatrix}, \text{ which is permutation similar to } \begin{bmatrix} B & * & 0 & 0 \\ 0 & C & 0 & 0 \\ & & N_1 & * \\ & & 0 & N_2 \end{bmatrix}$$

We have shown that the index of 0 as an eigenvalue of $\begin{bmatrix} N_1 & * \\ 0 & N_2 \end{bmatrix}$ (and hence of A itself) is at most $\nu_1 + \nu_2$. An induction permits us to extend this conclusion to any block triangular matrix and any of its eigenvalues.

Theorem 3.2.10.1. *Let $A = [A_{ij}]_{i,j=1}^{p} \in M_n$ be block upper triangular. If the index of λ as an eigenvalue of each A_{ii} is ν_i, then the index of λ as an eigenvalue of A is at most $\nu_1 + \cdots + \nu_p$.*

3.2.11 AB versus BA. If $A \in M_{m,n}$ and $B \in M_{n,m}$, (1.3.22) ensures that the nonzero eigenvalues of AB and BA are the same, including their multiplicities. In fact, we can make a much stronger statement: the nonsingular parts of the Jordan canonical forms of AB and BA are identical.

Theorem 3.2.11.1. *Suppose that $A \in M_{m,n}$ and $B \in M_{n,m}$. For each nonzero eigenvalue λ of AB and for each $k = 1, 2, \ldots$, the respective Jordan canonical forms of AB and BA contain the same number of Jordan blocks $J_k(\lambda)$.*

Proof. In the proof of (1.3.22), we found that $C_1 = \begin{bmatrix} AB & 0 \\ B & 0_n \end{bmatrix}$ and $C_2 = \begin{bmatrix} 0_m & 0 \\ B & BA \end{bmatrix}$ are similar. Let $\lambda \neq 0$ be given and let k be any given positive integer. First observe that the row rank of

$$(C_1 - \lambda I_{m+n})^k = \begin{bmatrix} (AB - \lambda I_m)^k & 0 \\ \bigstar & (-\lambda I_n)^k \end{bmatrix}$$

is $n + \text{rank}((AB - \lambda I_m)^k)$, then observe that the column rank of

$$(C_2 - \lambda I_{m+n})^k = \begin{bmatrix} (-\lambda I_m)^k & 0 \\ \bigstar & (BA - \lambda I_n)^k \end{bmatrix}$$

is $m + \text{rank}((BA - \lambda I_n)^k)$. But $(C_1 - \lambda I_{m+n})^k$ is similar to $(C_2 - \lambda I_{m+n})^k$, so their ranks are equal, that is,

$$\text{rank}((AB - \lambda I_m)^k) = \text{rank}((BA - \lambda I_n)^k) + m - n$$

for each $k = 1, 2, \ldots$, which implies that

$$\text{rank}((AB - \lambda I_m)^{k-1}) - \text{rank}((AB - \lambda I_m)^k)$$
$$= \text{rank}((BA - \lambda I_n)^{k-1}) - \text{rank}((BA - \lambda I_n)^k)$$

for each $k = 1, 2, \ldots$. Thus, the respective Weyr characteristics of AB and BA associated with any given nonzero eigenvalue λ of AB are identical, so (3.1.18) ensures that their respective Jordan canonical forms contain exactly the same number of blocks $J_k(\lambda)$ for each $k = 1, 2, \ldots$. $\square$

3.2.12 The Drazin inverse. For a given $A \in M_n$, any $X \in M_n$ such that $AXA = A$ is called a *generalized inverse* of A. Several types of generalized inverse are available, each of which has some features of the ordinary inverse. The generalized inverse that we consider in this section is the Drazin inverse.

Definition 3.2.12.1. *Let $A \in M_n$ and suppose that*

$$A = S \begin{bmatrix} B & 0 \\ 0 & N \end{bmatrix} S^{-1} \tag{3.2.12.2}$$

in which S and B are square and nonsingular and N is nilpotent. The direct summand B is absent if A is nilpotent; N is absent if A is nonsingular. The Drazin inverse *of A is*

$$A^D = S \begin{bmatrix} B^{-1} & 0 \\ 0 & 0 \end{bmatrix} S^{-1} \tag{3.2.12.3}$$

Every $A \in M_n$ has a representation of the form (3.2.12.2): Use the Jordan canonical form (3.1.12), in which B is a direct sum of all the nonsingular Jordan blocks of A and N is a direct sum of all the nilpotent blocks.

In addition to (3.2.12.2), suppose that A is represented as

$$A = T \begin{bmatrix} C & 0 \\ 0 & N' \end{bmatrix} T^{-1} \tag{3.2.12.4}$$

in which T and C are square and nonsingular and N' is nilpotent. Then $A^n = S \begin{bmatrix} B^n & 0 \\ 0 & 0 \end{bmatrix} S^{-1} = T \begin{bmatrix} C^n & 0 \\ 0 & 0 \end{bmatrix} T^{-1}$, so rank $A^n =$ rank $B^n =$ rank B is the size of B since it is nonsingular; for the same reason, it is also the size of C. We conclude that B and C have the same size, and hence N and N' have the same size. Since $A = S \begin{bmatrix} B & 0 \\ 0 & N \end{bmatrix} S^{-1} = T \begin{bmatrix} C & 0 \\ 0 & N' \end{bmatrix} T^{-1}$, it follows that $R \begin{bmatrix} B & 0 \\ 0 & N \end{bmatrix} = \begin{bmatrix} C & 0 \\ 0 & N' \end{bmatrix} R$, in which $R = T^{-1}S$. Partition $R = [R_{ij}]_{i,j=1}^2$ conformally with $\begin{bmatrix} B & 0 \\ 0 & N \end{bmatrix}$. Then (2.4.4.2) ensures that $R_{12} = 0$ and $R_{21} = 0$, so $R = R_{11} \oplus R_{22}$, R_{11} and R_{22} are nonsingular, $C = R_{11}BR_{11}^{-1}$, $N' = R_{22}NR_{22}^{-1}$, and $T = SR^{-1}$. Finally, compute the Drazin inverse using (3.2.12.4):

$$T \begin{bmatrix} C^{-1} & 0 \\ 0 & 0 \end{bmatrix} T^{-1} = SR^{-1} \begin{bmatrix} (R_{11}BR_{11}^{-1})^{-1} & 0 \\ 0 & 0 \end{bmatrix} RS^{-1}$$

$$= S \begin{bmatrix} R_{11}^{-1} & 0 \\ 0 & R_{22}^{-1} \end{bmatrix} \begin{bmatrix} R_{11}B^{-1}R_{11}^{-1} & 0 \\ 0 & 0 \end{bmatrix} \begin{bmatrix} R_{11} & 0 \\ 0 & R_{22} \end{bmatrix} S^{-1}$$

$$= S \begin{bmatrix} B^{-1} & 0 \\ 0 & 0 \end{bmatrix} S^{-1} = A^D$$

We conclude that the Drazin inverse is well-defined by (3.2.12.3).

Exercise. Explain why $A^D = A^{-1}$ if A is nonsingular.

Exercise. If $A \in M_n$, show that $AA^DA = A$ if and only if rank $A =$ rank A^2.

Let q be the index of the eigenvalue 0 of A and consider the three identities

$$AX = XA \tag{3.2.12.5}$$

$$A^{q+1}X = A^q \tag{3.2.12.6}$$

$$XAX = X \tag{3.2.12.7}$$

Exercise. Use (3.2.12.2) and (3.2.12.3) to explain why A and $X = A^D$ satisfy the preceding three identities if and only if $A = \begin{bmatrix} B & 0 \\ 0 & N \end{bmatrix}$ and $X = \begin{bmatrix} B^{-1} & 0 \\ 0 & 0 \end{bmatrix}$ satisfy them. Verify that they do.

There is a converse to the result in the preceding exercise: *If X satisfies (3.2.12.5–7), then $X = A^D$.* To verify this assertion, proceed as in the preceding exercise to replace A by $\begin{bmatrix} B & 0 \\ 0 & N \end{bmatrix}$ and partition the unknown matrix $X = [X_{ij}]_{i,j=1}^2$ conformally. We must show that $X_{11} = B^{-1}$ and that X_{12}, X_{21}, and X_{22} are zero blocks. Combining the first identity (3.2.12.5) with (2.4.4.2) ensures that $X_{12} = 0$ and $X_{21} = 0$; in addition, $NX_{22} = X_{22}N$. The second identity (3.2.12.6) says that $\begin{bmatrix} B^{q+1} & 0 \\ 0 & 0 \end{bmatrix}\begin{bmatrix} X_{11} & 0 \\ 0 & X_{22} \end{bmatrix} = \begin{bmatrix} B^q & 0 \\ 0 & 0 \end{bmatrix}$, so $B^{q+1}X_{11} = B^q$, $BX_{11} = I$, and $X_{11} = B^{-1}$. The third identity (3.2.12.7) ensures that

$$X_{22} = X_{22}NX_{22} = NX_{22}^2 \tag{3.2.12.8}$$

which implies that $N^{q-1}X_{22} = N^{q-1}NX_{22}^2 = N^q X_{22}^2 = 0$, so $N^{q-1}X_{22} = 0$. Using (3.2.12.8) again, we see that $N^{q-2}X_{22} = N^{q-2}NX_{22}^2 = (N^{q-1}X_{22})X_{22} = 0$, so $N^{q-2}X_{22} = 0$. Continuing this argument reveals that $N^{q-3}X_{22} = 0, \ldots, NX_{22} = 0$, and finally, $X_{22} = 0$.

Our final observation is that *the Drazin inverse A^D is a polynomial in A.*

Exercise. Represent A as in (3.2.12.2). According to (2.4.3.4), there is a polynomial $p(t)$ such that $p(B^{q+1}) = (B^{q+1})^{-1}$. Let $g(t) = t^q p(t^{q+1})$. Verify that $g(A) = A^D$.

Exercise. Let $A \in M_n$ and suppose that λ is a nonzero eigenvalue of A. If $x \neq 0$ and $Ax = \lambda x$, explain why $A^D x = \lambda^{-1} x$.

3.2.13 The Jordan canonical form of a rank-one perturbation. Brauer's theorem about eigenvalues of rank-one perturbations ((1.2.8) and (2.4.10.1)) has an analog for Jordan blocks: Under certain conditions, one eigenvalue of a square complex matrix can be shifted almost arbitrarily by a rank-one perturbation without disturbing the rest of its Jordan structure.

Theorem 3.2.13.1. *Let $n \geq 2$ and let $\lambda, \lambda_2, \ldots, \lambda_n$ be the eigenvalues of $A \in M_n$. Suppose that there are nonzero vectors $x, y \in \mathbf{C}^n$ such that $Ax = \lambda x$, $y^*A = \lambda y^*$, and $y^*x \neq 0$. Then*

(a) *the Jordan canonical form of A is*

$$[\lambda] \oplus J_{n_1}(\nu_1) \oplus \cdots \oplus J_{n_k}(\nu_k) \tag{3.2.13.2}$$

for some positive integers $k, n_1, \ldots, n_k$ and some $\{\nu_1, \ldots, \nu_k\} \subset \{\lambda_2, \ldots, \lambda_n\}$.

(b) *for any $v \in \mathbf{C}^n$ such that $\lambda + v^*x \neq \lambda_j$, $j = 2, \ldots, n$, the Jordan canonical form of $A + xv^*$ is*

$$\left[\lambda + v^*x\right] \oplus J_{n_1}(\nu_1) \oplus \cdots \oplus J_{n_k}(\nu_k)$$

Proof. The assertions in (a) follow from (1.4.7), which ensures that there is a nonsingular $S = [x \ S_1]$ such that $S^{-1}AS = [\lambda] \oplus B$ for some $B \in M_{n-1}$. The direct sum $J_{n_1}(\nu_1) \oplus \cdots \oplus J_{n_k}(\nu_k)$ is the Jordan canonical form of B. Compute $S^{-1}(xv^*)S = (S^{-1}x)(v^*S) = e_1(v^*S) = \begin{bmatrix} v^*x & w^* \\ 0 & 0 \end{bmatrix}$, in which $w^* = v^*S_1$. Combining the preceding similarities of A and xv^* gives $S^{-1}(A + xv^*)S = \begin{bmatrix} \lambda + v^*x & w^* \\ 0 & B \end{bmatrix}$. It suffices to show that this block matrix is similar to $[\lambda + v^*x] \oplus B$. For any $\xi \in \mathbf{C}^{n-1}$, we have $\begin{bmatrix} 1 & \xi^* \\ 0 & I \end{bmatrix}^{-1} = \begin{bmatrix} 1 & -\xi^* \\ 0 & I \end{bmatrix}$, so

$$\begin{bmatrix} 1 & \xi^* \\ 0 & I \end{bmatrix}^{-1} \begin{bmatrix} \lambda + v^*x & w^* \\ 0 & B \end{bmatrix} \begin{bmatrix} 1 & \xi^* \\ 0 & I \end{bmatrix} = \begin{bmatrix} \lambda + v^*x & w^* + \xi^*((\lambda + v^*x)I - B) \\ 0 & B \end{bmatrix}$$

We have assumed that $\lambda + v^*x$ is not an eigenvalue of B, so we may take $\xi^* = -w^*((\lambda + v^*x)I - B)^{-1}$, which reveals that $A + xv^*$ is similar to $[\lambda + v^*x] \oplus B$. $\square$

Problems

3.2.P1 Let $\mathcal{F} = \{A_\alpha : \alpha \in \mathcal{I}\} \subset M_n$ be a given family of matrices, indexed by the index set $\mathcal{I}$, and suppose that there is a nonderogatory matrix $A_0 \in \mathcal{F}$ such that $A_\alpha A_0 = A_0 A_\alpha$ for all $\alpha \in \mathcal{I}$. Show that for every $\alpha \in \mathcal{I}$, there is a polynomial $p_\alpha(t)$ of degree at most $n - 1$ such that $A_\alpha = p_\alpha(A_0)$, and hence $\mathcal{F}$ is a commuting family.

3.2.P2 Let $A \in M_n$. If every matrix that commutes with A is a polynomial in A, show that A is nonderogatory.

3.2.P3 Let $A = B + iC \in M_n$, in which B and C are real (0.2.5), and let J be the Jordan canonical form of A. Consider the real representation $R_1(A) = \begin{bmatrix} B & C \\ -C & B \end{bmatrix} \in M_{2n}$ discussed in (1.3.P20). Explain why $J \oplus \bar{J}$ is the Jordan canonical form of $R_1(A)$ and why it is a direct sum of pairs of the form $J_k(\lambda) \oplus J_k(\bar{\lambda})$, even if λ is real.

3.2.P4 Suppose that $A \in M_n$ is singular and let $r = \operatorname{rank} A$. In (2.4.P28) we learned that there is a polynomial of degree $r + 1$ that annihilates A. Provide details for the following argument to show that $h(t) = p_A(t)/t^{n-r-1}$ is such a polynomial: Let the Jordan canonical form of A be $J \oplus J_{n_1}(0) \oplus \cdots \oplus J_{n_k}(0)$, in which the Jordan matrix J is nonsingular. Let $\nu = n_1 + \cdots + n_k$ and let $n_{\max} = \max_i n_i$ be the index of the eigenvalue zero. (a) Explain why $p_A(t) = p_1(t)t^\nu$, in which $p_1(t)$ is a polynomial and $p_1(0) \neq 0$. (b) Show that $p(t) = p_1(t)t^{n_{\max}}$ annihilates A, so $p_A(t) = (p_1(t)t^{n_{\max}})t^{\nu-n_{\max}}$. (c) Explain why $k = n - r$, $\nu - n_{\max} \geq k - 1 = n - r - 1$, and $h(A) = 0$.

3.2.P5 What is the Jordan canonical form of $A = \begin{bmatrix} i & 1 \\ 1 & -i \end{bmatrix}$?

3.2.P6 The linear transformation $d/dt : p(t) \to p'(t)$ acting on the vector space of all polynomials with degree at most 3 has the basis representation

$$\begin{bmatrix} 0 & 1 & 0 & 0 \\ 0 & 0 & 2 & 0 \\ 0 & 0 & 0 & 3 \\ 0 & 0 & 0 & 0 \end{bmatrix}$$

in the basis $B = \{1, t, t^2, t^3\}$. What is the Jordan canonical form of this matrix?

3.2.P7 What are the possible Jordan forms of a matrix $A \in M_n$ such that $A^3 = I$?

3.2.P8 What are the possible Jordan canonical forms for a matrix $A \in M_6$ with characteristic polynomial $p_A(t) = (t + 3)^4 (t - 4)^2$?

3.2.P9 Suppose that $k \geq 2$. Explain why the Jordan canonical form of adj $J_k(\lambda)$ is $J_k(\lambda^{k-1})$ if $\lambda \neq 0$, and it is $J_2(0) \oplus 0_{k-2}$ if $\lambda = 0$.

3.2.P10 Suppose that the Jordan canonical form of a given nonsingular $A \in M_n$ is $J_{n_1}(\lambda_1) \oplus \cdots \oplus J_{n_k}(\lambda_k)$. Explain why the Jordan canonical form of adj A is $J_{n_1}(\mu_1) \oplus \cdots \oplus J_{n_k}(\mu_k)$, in which each $\mu_i = \lambda_i^{n_i-1} \prod_{j \neq i} \lambda_j^{n_j}$, $i = 1, \ldots, k$.

3.2.P11 Suppose that the Jordan canonical form of a given singular $A \in M_n$ is $J_{n_1}(\lambda_1) \oplus \cdots \oplus J_{n_{k-1}}(\lambda_{k-1}) \oplus J_{n_k}(0)$. Explain why the Jordan canonical form of adj A is $J_2(0) \oplus 0_{n-2}$ if $n_k \geq 2$, and it is $\prod_{i=1}^{k-1} \lambda_i^{n_i} \oplus 0_{n-1}$ if $n_k = 1$; the former case is characterized by rank $A < n - 1$ and the latter case is characterized by rank $A = n - 1$.

3.2.P12 Explain why adj $A = 0$ if the Jordan canonical form of A contains two or more singular Jordan blocks.

3.2.P13 (Cancellation theorem for similarity) Let $A \in M_n$ and $B, C \in M_m$ be given. Show that $\begin{bmatrix} A & 0 \\ 0 & B \end{bmatrix} \in M_{n+m}$ is similar to $\begin{bmatrix} A & 0 \\ 0 & C \end{bmatrix}$ if and only if B is similar to C.

3.2.P14 Let $B, C \in M_m$ and a positive integer k be given. Show that

$$\underbrace{B \oplus \cdots \oplus B}_{k \text{ summands}} \quad \text{and} \quad \underbrace{C \oplus \cdots \oplus C}_{k \text{ summands}}$$

are similar if and only if B and C are similar.

3.2.P15 Let $A \in M_n$ and $B, C \in M_m$ be given. Show that

$$A \oplus \underbrace{B \oplus \cdots \oplus B}_{k \text{ summands}} \quad \text{and} \quad A \oplus \underbrace{C \oplus \cdots \oplus C}_{k \text{ summands}}$$

are similar if and only if B and C are similar

3.2.P16 Let $A \in M_n$ have Jordan canonical form $J_{n_1}(\lambda_1) \oplus \cdots \oplus J_{n_k}(\lambda_k)$. If A is nonsingular, show that the Jordan canonical form of A^2 is $J_{n_1}(\lambda_1^2) \oplus \cdots \oplus J_{n_k}(\lambda_k^2)$; that is, the Jordan canonical form of A^2 is composed of precisely the same collection of Jordan blocks as A, but the respective eigenvalues are squared. However, the Jordan canonical form of $J_m(0)^2$ is not $J_m(0^2)$ if $m \geq 2$; explain.

3.2.P17 Let $A \in M_n$ be given. Show that rank $A = $ rank A^2 if and only if the geometric and algebraic multiplicities of the eigenvalue $\lambda = 0$ are equal, that is, if and only if all the Jordan blocks corresponding to $\lambda = 0$ (if any) in the Jordan canonical form of A are

1-by-1. Explain why A is diagonalizable if and only if $\operatorname{rank}(A - \lambda I) = \operatorname{rank}(A - \lambda I)^2$ for all $\lambda \in \sigma(A)$.

3.2.P18 Let $A \in M_n$ be given. In (3.2.7) we used the Jordan canonical form to write A as a sum of two commuting matrices, one of which is diagonalizable and the other of which is nilpotent: the Jordan decomposition $A = A_D + A_N$. The goal of this problem is to show that *the Jordan decomposition is unique*. That is, suppose that (a) $A = B + C$, (b) B commutes with C, (c) B is diagonalizable, and (d) C is nilpotent; we claim that $B = A_D$ and $C = A_N$. It is helpful to use the fact that there are polynomials $p(t)$ and $q(t)$ such that $A_D = p(A)$ and $A_N = q(A)$; see Problem 14(d) in section 6.1 of Horn and Johnson (1991). Provide details for the following: (a) B and C commute with A; (b) B and C commute with A_D and A_N; (c) B and A_D are simultaneously diagonalizable, so $A_D - B$ is diagonalizable; (d) C and A_N are simultaneously upper triangularizable, so $C - A_N$ is nilpotent; (e) $A_D - B = C - A_N$ is both diagonalizable and nilpotent, so it is a zero matrix. The (uniquely determined) matrix A_D is called the *diagonalizable part* of A; A_N is the *nilpotent part* of A.

3.2.P19 Let $A \in M_n$ be given and let λ be an eigenvalue of A. (a) Prove that the following two assertions are equivalent: (i) every Jordan block of A with eigenvalue λ has size two or greater; (ii) every eigenvector of A associated with λ is in the range of $A - \lambda I$. (b) Prove that the following five assertions are equivalent: (i) some Jordan block of A is 1-by-1; (ii) there is a nonzero vector x such that $Ax = \lambda x$ but x is not in the range of $A - \lambda I$; (iii) there is a nonzero vector x such that $Ax = \lambda x$ but x is not orthogonal to the null space of $A^* - \bar{\lambda} I$; (iv) there are nonzero vectors x and y such that $Ax = \lambda x$, $y^* A = \lambda y^*$, and $x^* y \neq 0$; (v) A is similar to $[\lambda] \oplus B$ for some $B \in M_{n-1}$.

3.2.P20 Let $A, B \in M_n$ be given. (a) Show that AB is similar to BA if and only $\operatorname{rank}(AB)^k = \operatorname{rank}(BA)^k$ for each $k = 1, 2, \ldots, n$. (b) If $r = \operatorname{rank} A = \operatorname{rank} AB = \operatorname{rank} BA$, show that AB is similar to BA. Explain why we may replace A by SAT and B by $T^{-1} B S^{-1}$ for any nonsingular $S, T \in M_n$. Choose S and T so that $SAT = I_r \oplus 0_{n-r}$. Consider $A = I_r \oplus 0_{n-r}$ and $B = [B_{ij}]_{i,j=1}^2$. Compute AB and BA; explain why each of $X = [B_{11} \ B_{12}]$ and $Y^T = [B_{11}^T \ B_{21}^T]$ has full rank. Explain why $\operatorname{rank} CX = \operatorname{rank} C = \operatorname{rank} YC$ for any $C \in M_r$. Explain why $\operatorname{rank}((AB)^{k+1}) = \operatorname{rank}((B_{11})^k X) = \operatorname{rank}((B_{11})^k) = \operatorname{rank}(Y(B_{11})^k) = \operatorname{rank}((BA)^{k+1})$ for each $k = 1, 2, \ldots, n$.

3.2.P21 Let $A = \begin{bmatrix} J_2(0) & 0 \\ x^T & 0 \end{bmatrix} \in M_3$ with $x^T = [1 \ 0]$, and let $B = I_2 \oplus [0] \in M_3$. Show that the Jordan canonical form of AB is $J_3(0)$, while that of BA is $J_2(0) \oplus J_1(0)$.

3.2.P22 Let $A \in M_n$. Show that both AA^D and $I - AA^D$ are projections and that $AA^D(I - AA^D) = 0$.

3.2.P23 Let $A \in M_n$, let q be the index of 0 as an eigenvalue of A, and let $k \geq q$ be a given integer. Show that $A^D = \lim_{t \to 0} (A^{k+1} + tI)^{-1} A^k$.

3.2.P24 This problem is an analog of (2.4.P12). Let $A, B \in M_n$, let $\lambda_1, \ldots, \lambda_d$ be the distinct eigenvalues of A, let $D = AB - BA^T$, and suppose that $AD = DA^T$. (a) Show that D is singular. (b) If A is diagonalizable, show that $D = 0$, that is, $AB = BA^T$. (c) If $DA = A^T D$ as well as $AD = DA^T$, show that D is nilpotent. (d) Suppose that A is nonderogatory. Then (3.2.4.4) ensures that D is symmetric. In addition, show that $\operatorname{rank} D \leq n - d$, so the geometric multiplicity of 0 as an eigenvalue of D is at least d.

3.2.P25 Let $A \in M_n$ be given and suppose that A^2 is nonderogatory. Explain why (a) A is nonderogatory; (b) if λ is a nonzero eigenvalue of A, then $-\lambda$ is not an eigenvalue of A;

(c) if A is singular, then 0 has algebraic multiplicity 1 as an eigenvalue of A; (d) rank $A \geq n - 1$; (e) there is a polynomial $p(t)$ such that $A = p(A^2)$.

3.2.P26 Let $A, B \in M_n$ be given and suppose that A^2 is nonderogatory. If $AB = B^T A$ and $BA = AB^T$, show that B is symmetric.

3.2.P27 (a) For each $k = 1, 2, \ldots$ show that adj $J_k(0)$ is similar to $J_2(0) \oplus 0_{k-2}$. (b) If $A \in M_n$ is nilpotent and rank $A = n - 1$, explain why A is similar to $J_n(0)$. (c) If $A \in M_n$ is nilpotent, show that $(\text{adj } A)^2 = 0$.

3.2.P28 Let A, x, y, and λ satisfy the hypotheses of (3.2.13.1) so that (3.2.13.2) is the Jordan canonical form of A. Let $v \in \mathbf{C}^n$ be any vector such that $v^* x = 1$ and consider the *Google* matrix $A(c) = cA + (1 - c)\lambda x v^*$. If c is nonzero and $c\lambda_j \neq \lambda$ for each $j = 2, \ldots, n$, show that the Jordan canonical form of $A(c)$ is $[\lambda] \oplus J_{n_1}(cv_1) \oplus \cdots \oplus J_{n_k}(cv_k)$. Compare with (1.2.P21).

3.2.P29 Let $\lambda \in \mathbf{C}$, $A = J_k(\lambda)$, and $B = [b_{ij}] \in M_k$, and let $C = AB - BA$. If we were to assume that $C = 0$, then (3.2.4.2) ensures that B is upper triangular and Toeplitz, so all of the eigenvalues of B are the same. Instead, make the weaker assumption that $AC = CA$. (a) Explain why C is upper triangular, Toeplitz, and nilpotent, that is, $C = [\gamma_{j-i}]_{i,j=1}^k$ in which $\gamma_{-k+1} = \cdots = \gamma_0 = 0$ and $\gamma_1, \ldots, \gamma_{k-1} \in \mathbf{C}$. (b) Use the form of C to show that B is upper triangular (but not Toeplitz) and its eigenvalues $b_{11}, b_{11} + \gamma_1, b_{11} + 2\gamma_1, \ldots, b_{11} + (k - 1)\gamma_1$ are in arithmetic progression.

3.2.P30 Let $A \in M_n$ and a subspace $\mathcal{S} \subset \mathbf{C}^n$ be given. Provide details for the following outline of a proof that $\mathcal{S}$ is an invariant subspace of A if and only if there is a $B \in M_n$ such that $AB = BA$ and $\mathcal{S}$ is the null space of B. *Only if:* $B(A\mathcal{S}) = A(B\mathcal{S}) = A\{0\} = \{0\}$, so $A\mathcal{S} \subset \mathcal{S}$. *If:* (a) If $\mathcal{S} = \{0\}$ or $\mathbf{C}^n$, take $B = I$ or $B = 0$, so we may assume that $1 \leq \dim \mathcal{S} \leq n - 1$. (b) It suffices to prove the implication for some matrix that is similar to A (why?), so we may assume that $A = \begin{bmatrix} A_{11} & A_{12} \\ 0 & A_{22} \end{bmatrix}$ with $A_{11} \in M_k$; see (1.3.17(c)). There is a nonsingular $X \in M_n$ such that $AX = XA^T$; see (3.2.3.1). (d) There is a nonsingular $Y \in M_{n-k}$ such that $Y A_{22} = A_{22}^T Y$. Let $C = 0_k \oplus Y$. (e) $CA = A^T C$. (f) Let $B = XC$. Then $AB = AXC = XA^T C = XCA = BA$.

3.2.P31 Let $A \in M_n$ and a subspace $\mathcal{S} \subset \mathbf{C}^n$ be given. Show that $\mathcal{S}$ is an invariant subspace of A if and only if there is a $B \in M_n$ such that $AB = BA$ and $\mathcal{S}$ is the range of B.

3.2.P32 Let $A, B \in M_n$, let $C = AB - BA$, and suppose that A commutes with C. If $n = 2$, show that A and B are simultaneously upper triangularizable. Problem (2.4.P12(f)) shows that A and B need *not* be simultaneously triangularizable if $n > 2$.

3.2.P33 Let $A \in M_n$. Explain why A^* is nonderogatory if and only if A is nonderogatory.

3.2.P34 Let $A, B \in M_n$. Suppose that A is nonderogatory and that B commutes with both A and A^*. Show that B is normal.

3.2.P35 This problem considers a partial converse to (2.5.17). (a) Let $A \in M_n$ be non-derogatory. If $A\bar{A} = \bar{A}A$ and $AA^T = A^T A$, show that $AA^* = A^* A$, that is, A is normal. (b) If $A \in M_2$, $A\bar{A} = \bar{A}A$, and $AA^T = A^T A$, show that A is normal. (c) The implication in (b) is correct for $n = 3$, but known proofs are technical and tedious. Can you find a simple proof? (c) Let $B = \begin{bmatrix} 1 & 1 \\ -1 & 1 \end{bmatrix}$ and $C = \begin{bmatrix} 1 & i \\ -i & 1 \end{bmatrix}$ and define $A = \begin{bmatrix} B & C \\ 0 & B \end{bmatrix} \in M_4$. Show that $A\bar{A} = \bar{A}A$ and $AA^T = A^T A$, but A is not normal.

3.2.P36 Let $A \in M_n$ be coninvolutory, so A is nonsingular and $A = \bar{A}^{-1}$. (a) Explain why the Jordan canonical form of A is a direct sum of blocks of the form $J_k(e^{i\theta})$ with $\theta \in [0, 2\pi)$, and blocks of the form $J_k(\lambda) \oplus J_k(1/\bar{\lambda})$ with $0 \neq |\lambda| \neq 1$. (b) If A is diagonalizable, explain why its Jordan canonical form is a direct sum of blocks of the form $[e^{i\theta}]$ with $\theta_1, \ldots, \theta_n \in [0, 2\pi)$ and blocks of the form $[\lambda] \oplus [1/\bar{\lambda}]$ with $0 \neq |\lambda| \neq 1$.

3.2.P37 A matrix $A \in M_n$ is said to be *semiconvergent* if $\lim_{k \to \infty} A^k$ exists. (a) Explain why A is semiconvergent if and only if $\rho(A) \leq 1$ and, if λ is an eigenvalue of A and $|\lambda| = 1$, then $\lambda = 1$ and λ is semisimple. (b) If $A \in M_n$ is semiconvergent, show that $\lim_{k \to \infty} A^k = I - (I - A)(I - A)^D$.

Notes and Further Readings. For a detailed discussion of the optimality property (3.2.9.4) and a characterization of the case of equality, see R. Brualdi, P. Pei, and X. Zhan, An extremal sparsity property of the Jordan canonical form, *Linear Algebra Appl.* 429 (2008) 2367–2372. Problem 3.2.P21 illustrates that the nilpotent Jordan structures of AB and BA need not be the same, but in the following sense, they cannot differ by much: If $m_1 \geq m_2 \geq \cdots$ are the sizes of the nilpotent Jordan blocks of AB while $n_1 \geq n_2 \geq \cdots$ are the sizes of the nilpotent Jordan blocks of BA (append zero sizes to one list or the other, if necessary. to achieve lists of equal length), then $|m_i - n_i| \leq 1$ for all i. For a discussion and proof, see C. R. Johnson and E. Schreiner, The relationship between AB and BA, *Amer. Math. Monthly* 103 (1996) 578–582. For a very different proof that uses the Weyr characteristic, see R. Lippert and G. Strang, The Jordan forms of AB and BA, *Electron. J. Linear Algebra* 18 (2009) 281–288. The argument involving similarity of a matrix and its transpose in Problem 3.2.P30 is due to Ignat Domanov; the assertions of this problem are theorem 3 in P. Halmos, Eigenvectors and adjoints, *Linear Algebra Appl.* 4 (1971) 11–15. Problem 3.2.P35 is due to G. Goodson.

3.3 The minimal polynomial and the companion matrix

A polynomial $p(t)$ is said to *annihilate* $A \in M_n$ if $p(A) = 0$. The Cayley–Hamilton theorem 2.4.2 guarantees that for each $A \in M_n$ there is a monic polynomial $p_A(t)$ of degree n (the characteristic polynomial) such that $p_A(A) = 0$. Of course, there may be a monic polynomial of degree $n - 1$ that annihilates A, or one of degree $n - 2$ or less. Of special interest is a monic polynomial of minimum degree that annihilates A. It is clear that such a polynomial exists; the following theorem says that it is unique.

Theorem 3.3.1. *Let $A \in M_n$ be given. There exists a unique monic polynomial $q_A(t)$ of minimum degree that annihilates A. The degree of $q_A(t)$ is at most n. If $p(t)$ is any monic polynomial such that $p(A) = 0$, then $q_A(t)$ divides $p(t)$, that is, $p(t) = h(t)q_A(t)$ for some monic polynomial $h(t)$.*

Proof. The set of monic polynomials that annihilate A contains $p_A(t)$, which has degree n. Let $m = \min\{k : p(t)$ is a monic polynomial of degree k and $p(A) = 0\}$; necessarily $m \leq n$. If $p(t)$ is any monic polynomial that annihilates A, and if $q(t)$ is a monic polynomial of degree m that annihilates A, then the degree of $p(t)$ is m or greater. The Euclidean algorithm ensures that there is a monic polynomial $h(t)$ and a polynomial $r(t)$ of degree strictly less than m such that $p(t) = q(t)h(t) + r(t)$. But $0 = p(A) = q(A)h(A) + r(A) = 0h(A) + r(A)$, so $r(A) = 0$. If $r(t)$ is not the zero polynomial, we could normalize it and obtain a monic annihilating polynomial of degree less than m, which would be a contradiction. We conclude that $r(t)$ is the zero polynomial, so $q(t)$ divides $p(t)$ with quotient $h(t)$. If there are two monic polynomials of minimum degree that annihilate A, this argument shows that each divides the other; since the degrees are the same, one must be a scalar multiple of the other. But since both are monic, the scalar factor must be $+1$ and they are identical. $\qquad\square$

Definition 3.3.2. *Let $A \in M_n$ be given. The unique monic polynomial $q_A(t)$ of minimum degree that annihilates A is called the* minimal polynomial *of A.*

Corollary 3.3.3. *Similar matrices have the same minimal polynomial.*

Proof. If $A, B, S \in M_n$ and if $A = SBS^{-1}$, then $q_B(A) = q_B(SBS^{-1}) = Sq_B(B)S^{-1} = 0$, so $q_B(t)$ is a monic polynomial that annihilates A and hence the degree of $q_A(t)$ is less than or equal to the degree of $q_B(t)$. But $B = S^{-1}AS$, so the same argument shows that the degree of $q_B(t)$ is less than or equal to the degree of $q_A(t)$. Thus, $q_A(t)$ and $q_B(t)$ are monic polynomials of minimum degree that annihilate A, so (3.3.1) ensures that they are identical. $\qquad\square$

> **Exercise.** Consider $A = J_2(0) \oplus J_2(0) \in M_4$ and $B = J_2(0) \oplus 0_2 \in M_4$. Explain why A and B have the same minimal polynomial but are not similar.

Corollary 3.3.4. *For each $A \in M_n$, the minimal polynomial $q_A(t)$ divides the characteristic polynomial $p_A(t)$. Moreover, $q_A(\lambda) = 0$ if and only if λ is an eigenvalue of A, so every root of $p_A(t) = 0$ is a root of $q_A(t) = 0$.*

Proof. Since $p_A(A) = 0$, the fact that there is a polynomial $h(t)$ such that $p_A(t) = h(t)q_A(t)$ follows from (3.2.1). This factorization makes it clear that every root of $q_A(t) = 0$ is a root of $p_A(t) = 0$, and hence every root of $q_A(t) = 0$ is an eigenvalue of A. If λ is an eigenvalue of A, and if x is an associated eigenvector, then $Ax = \lambda x$ and $0 = q_A(A)x = q_A(\lambda)x$, so $q_A(\lambda) = 0$ since $x \neq 0$. $\qquad\square$

The preceding corollary shows that if the characteristic polynomial $p_A(t)$ has been completely factored as

$$p_A(t) = \prod_{i=1}^{d}(t - \lambda_i)^{s_i}, \quad 1 \leq s_i \leq n, \quad s_1 + s_2 + \cdots + s_d = n \quad (3.3.5a)$$

with $\lambda_1, \lambda_2, \ldots, \lambda_d$ distinct, then the minimal polynomial $q_A(t)$ must have the form

$$q_A(t) = \prod_{i=1}^{d}(t - \lambda_i)^{r_i}, \quad 1 \leq r_i \leq s_i \quad (3.3.5b)$$

In principle, this gives an algorithm for finding the minimal polynomial of a given matrix A:

1. First compute the eigenvalues of A, together with their algebraic multiplicities, perhaps by finding the characteristic polynomial and factoring it completely. By some means, determine the factorization (3.3.5a).
2. There are finitely many polynomials of the form (3.3.5b). Starting with the product in which all $r_i = 1$, determine by explicit calculation the product of minimal degree that annihilates A; it is the minimal polynomial.

Numerically, this is not a good algorithm if it involves factoring the characteristic polynomial of a large matrix, but it can be very effective for hand calculations involving small matrices of simple form. Another approach to computing the minimal polynomial that does not involve knowing either the characteristic polynomial or the eigenvalues is outlined in (3.3.P5).

There is an intimate connection between the Jordan canonical form of $A \in M_n$ and the minimal polynomial of A. Suppose that $A = SJS^{-1}$ is the Jordan canonical form of A, and suppose first that $J = J_n(\lambda)$ is a single Jordan block. The characteristic polynomial of A is $(t - \lambda)^n$, and since $(J - \lambda I)^k \neq 0$ if $k < n$, the minimal polynomial of J is also $(t - \lambda)^n$. However, if $J = J_{n_1}(\lambda) \oplus J_{n_2}(\lambda) \in M_n$ with $n_1 \geq n_2$, then the characteristic polynomial of J is still $(t - \lambda)^n$, but now $(J - \lambda I)^{n_1} = 0$ and no lower power vanishes. The minimal polynomial of J is therefore $(t - \lambda)^{n_1}$. If there are more Jordan blocks with eigenvalue λ, the conclusion is the same: The minimal polynomial of J is $(t - \lambda)^r$, in which r is the size of the largest Jordan block corresponding to λ. If J is a general Jordan matrix, the minimal polynomial must contain a factor $(t - \lambda_i)^{r_i}$ for each distinct eigenvalue λ_i, and r_i must be the size of the largest Jordan block corresponding to λ_i; no smaller power annihilates all the Jordan blocks corresponding to λ_i, and no greater power is needed. Since similar matrices have the same minimal polynomial, we have proved the following theorem.

Theorem 3.3.6. *Let $A \in M_n$ be a given matrix whose distinct eigenvalues are $\lambda_1, \ldots, \lambda_d$. The minimal polynomial of A is*

$$q_A(t) = \prod_{i=1}^{d}(t - \lambda_i)^{r_i} \tag{3.3.7}$$

in which r_i is the size of the largest Jordan block of A corresponding to the eigenvalue λ_i.

In practice, this result is not very helpful in computing the minimal polynomial since it is usually harder to determine the Jordan canonical form of a matrix than it is to determine its minimal polynomial. Indeed, if only the eigenvalues of a matrix are known, its minimal polynomial can be determined by simple trial and error. There are important theoretical consequences, however. Since a matrix is diagonalizable if and only if all its Jordan blocks have size 1, a necessary and sufficient condition for diagonalizability is that all $r_i = 1$ in (3.3.7).

Corollary 3.3.8. *Let $A \in M_n$ have distinct eigenvalues $\lambda_1, \lambda_2, \ldots, \lambda_d$ and let*

$$q(t) = (t - \lambda_1)(t - \lambda_2) \cdots (t - \lambda_d) \tag{3.3.9}$$

Then A is diagonalizable if and only if $q(A) = 0$.

This criterion is actually useful for determining if a given matrix is diagonalizable, provided that we know its distinct eigenvalues: Form the polynomial (3.3.9) and see if it annihilates A. If it does, it must be the minimal polynomial of A, since no lower-order polynomial could have as zeroes all the distinct eigenvalues of A. If it does not annihilate A, then A is not diagonalizable. It can be useful to have this result formulated in several equivalent ways:

Corollary 3.3.10. *Let $A \in M_n$ and let $q_A(t)$ be its minimal polynomial. The following are equivalent:*

(a) *$q_A(t)$ is a product of distinct linear factors.*
(b) *Every eigenvalue of A has multiplicity 1 as a root of $q_A(t) = 0$.*
(c) *$q'_A(\lambda) \neq 0$ for every eigenvalue λ of A.*
(d) *A is diagonalizable.*

We have been considering the problem of finding, for a given $A \in M_n$, a monic polynomial of minimum degree that annihilates A. But what about the converse? Given a monic polynomial

$$p(t) = t^n + a_{n-1}t^{n-1} + a_{n-2}t^{n-2} + \cdots + a_1 t + a_0 \qquad (3.3.11)$$

is there a matrix A for which $p(t)$ is the minimal polynomial? If so, the size of A must be at least n-by-n. Consider

$$A = \begin{bmatrix} 0 & & & & -a_0 \\ 1 & 0 & & & -a_1 \\ & 1 & \ddots & & \vdots \\ & & \ddots & 0 & -a_{n-2} \\ 0 & & & 1 & -a_{n-1} \end{bmatrix} \in M_n \qquad (3.3.12)$$

and observe that

$$\begin{array}{rcl}
Ie_1 & = & e_1 & = & A^0 e_1 \\
Ae_1 & = & e_2 & = & Ae_1 \\
Ae_2 & = & e_3 & = & A^2 e_1 \\
Ae_3 & = & e_4 & = & A^3 e_1 \\
\vdots & & \vdots & & \vdots \\
Ae_{n-1} & = & e_n & = & A^{n-1} e_1
\end{array}$$

In addition,

$$\begin{aligned}
Ae_n &= -a_{n-1}e_n - a_{n-2}e_{n-1} - \cdots - a_1 e_2 - a_0 e_1 \\
&= -a_{n-1}A^{n-1}e_1 - a_{n-2}A^{n-2}e_1 - \cdots - a_1 Ae_1 - a_0 e_1 = A^n e_1 \\
&= (A^n - p(A))e_1
\end{aligned}$$

Thus,

$$\begin{aligned}
p(A)e_1 &= (a_0 e_1 + a_1 Ae_1 + a_2 A^2 e_1 + \cdots + a_{n-1}A^{n-1}e_1) + A^n e_1 \\
&= (p(A) - A^n)e_1 + (A^n - p(A))e_1 = 0
\end{aligned}$$

Furthermore, $p(A)e_k = p(A)A^{k-1}e_1 = A^{k-1}p(A)e_1 = A^{k-1}0 = 0$ for each $k = 1, 2, \ldots, n$. Since $p(A)e_k = 0$ for every basis vector e_k, we conclude that $p(A) = 0$. Thus $p(t)$ is a monic polynomial of degree n that annihilates A. If there were a polynomial $q(t) = t^m + b_{m-1}t^{m-1} + \cdots + b_1 t + b_0$ of lower degree $m < n$ that annihilates A, then

$$0 = q(A)e_1 = A^m e_1 + b_{m-1}A^{m-1}e_1 + \cdots + b_1 Ae_1 + b_0 e_1$$
$$= e_{m+1} + b_{m-1}e_m + \cdots + b_1 e_2 + b_0 e_1 = 0$$

which is impossible since $e_1, , \ldots, e_{m+1}$ are linearly independent. We conclude that nth degree polynomial $p(t)$ is a monic polynomial of *minimum* degree that annihilates A, so it is the minimal polynomial of A. The characteristic polynomial $p_A(t)$ is also a monic polynomial of degree n that annihilates A, so (3.3.1) ensures that $p(t)$ is also the characteristic polynomial of the matrix (3.3.12).

Definition 3.3.13. *The matrix (3.3.12) is the* companion matrix *of the polynomial (3.3.11).*

We have proved the following.

Theorem 3.3.14. *Every monic polynomial is both the minimal polynomial and the characteristic polynomial of its companion matrix.*

If the minimal polynomial of $A \in M_n$ has degree n, then the exponents in (3.3.7) satisfy $r_1 + \cdots + r_d = n$; that is, the *largest* Jordan block corresponding to each eigenvalue is the *only* Jordan block corresponding to each eigenvalue. Such a matrix is nonderogatory. In particular, every companion matrix is nonderogatory. A nonderogatory matrix $A \in M_n$ need not be a companion matrix, of course, but A and the companion matrix C of the characteristic polynomial of A have the same Jordan canonical form (one block $J_{r_i}(\lambda_i)$ corresponding to each distinct eigenvalue λ_i), so A is similar to C.

Exercise. Provide details for a proof of the following theorem.

Theorem 3.3.15. *Let $A \in M_n$ have minimal polynomial $q_A(t)$ and characteristic polynomial $p_A(t)$. The following are equivalent:*

(a) $q_A(t)$ *has degree n.*
(b) $p_A(t) = q_A(t)$.
(c) A *is nonderogatory.*
(d) A *is similar to the companion matrix of $p_A(t)$.*

Problems

3.3.P1 Let $A, B \in M_3$ be nilpotent. Show that A and B are similar if and only if A and B have the same minimal polynomial. Is this true in M_4?

3.3.P2 Suppose that $\lambda_1, \ldots, \lambda_d$ are the distinct eigenvalues of $A \in M_n$. Explain why the minimal polynomial of A (3.3.7) is determined by the following algorithm: For each

$i = 1, \ldots, d$ compute $(A - \lambda_i I)^k$ for $k = 1, \ldots, n$. Let r_i be the smallest value of k for which $\mathrm{rank}(A - \lambda_i I)^k = \mathrm{rank}(A - \lambda_i I)^{k+1}$.

3.3.P3 Use (3.3.10) to show that every projection matrix (idempotent matrix) is diagonalizable. What is the minimal polynomial of A? What can you say if A is *tripotent* ($A^3 = A$)? What if $A^k = A$?

3.3.P4 If $A \in M_n$ and $A^k = 0$ for some $k > n$, use properties of the minimal polynomial to explain why $A^r = 0$ for some $r \leq n$.

3.3.P5 Show that the following application of the Gram–Schmidt process permits the minimal polynomial of a given $A \in M_n$ to be computed without knowing either the characteristic polynomial of A or any of its eigenvalues.

(a) Let the mapping $T : M_n \to \mathbf{C}^{n^2}$ be defined as follows: For any $A \in M_n$ partitioned according to its columns as $A = [a_1 \ldots a_n]$, let $T(A)$ denote the unique vector in $\mathbf{C}^{n^2}$ whose first n entries are the entries of the first column a_1, whose entries from $n + 1$ to $2n$ are the entries of the second column a_2, and so forth. Show that this mapping T is an isomorphism (linear, one-to-one, and onto) of the vector spaces M_n and $\mathbf{C}^{n^2}$.

(b) Consider the vectors

$$v_0 = T(I), v_1 = T(A), v_2 = T(A^2), \ldots, v_k = T(A^k), \ldots$$

in $\mathbf{C}^{n^2}$ for $k = 0, 1, 2, \ldots, n$. Use the Cayley–Hamilton theorem to show that the vectors $v_0, v_1, \ldots, v_n$ are linearly dependent.

(c) Apply the Gram–Schmidt process to the list $v_0, v_1, \ldots, v_n$ until it stops by producing a first zero vector. Why must a zero vector be produced?

(d) If the Gram–Schmidt process produces a first zero vector at the kth step, argue that $k - 1$ is the degree of the minimal polynomial of A.

(e) If the kth step of the Gram–Schmidt process produces the vector $\alpha_0 v_0 + \alpha_1 v_1 + \cdots + \alpha_{k-1} v_{k-1} = 0$, show that

$$T^{-1}(\alpha_0 v_0 + \alpha_1 v_1 + \cdots + \alpha_{k-1} v_{k-1})$$
$$= \alpha_0 I + \alpha_1 A + \alpha_2 A^2 + \cdots + \alpha_{k-1} A^{k-1} = 0$$

and conclude that $q_A(t) = (\alpha_{k-1} t^{k-1} + \cdots + \alpha_2 t^2 + \alpha_1 t + \alpha_0)/\alpha_{k-1}$ is the minimal polynomial of A. Why is $\alpha_{k-1} \neq 0$?

3.3.P6 Carry out the computations required by the algorithm in (3.3.P5) to determine the minimal polynomials of $\begin{bmatrix} 1 & 1 \\ 0 & 2 \end{bmatrix}$, $\begin{bmatrix} 1 & 1 \\ 0 & 1 \end{bmatrix}$, and $\begin{bmatrix} 1 & 0 \\ 0 & 1 \end{bmatrix}$.

3.3.P7 Consider $A = \begin{bmatrix} 0 & 1 \\ 0 & 0 \end{bmatrix}$ and $B = \begin{bmatrix} 0 & 0 \\ 0 & 1 \end{bmatrix}$ to show that the minimal polynomials of AB and BA need not be the same. However, if $C, D \in M_n$, why must the characteristic polynomials of CD and DC be the same?

3.3.P8 Let $A_i \in M_{n_i}, i = 1, \ldots, k$ and let $q_{A_i}(t)$ be the minimal polynomial of each A_i. Show that the minimal polynomial of $A = A_1 \oplus \cdots \oplus A_k$ is the least common multiple of $q_{A_1}(t), \ldots, q_{A_k}(t)$. This is the unique monic polynomial of minimum degree that is divisible by each $q_i(t)$. Use this result to give a different proof for (1.3.10).

3.3.P9 If $A \in M_5$ has characteristic polynomial $p_A(t) = (t - 4)^3(t + 6)^2$ and minimal polynomial $q_A(t) = (t - 4)^2(t + 6)$, what is the Jordan canonical form of A?

3.3.P10 Show by direct computation that the polynomial (3.3.11) is the characteristic polynomial of the companion matrix (3.3.12).

3.3.P11 Let $A \in M_n$ be the companion matrix (3.3.12) of the polynomial $p(t)$ in (3.3.11). Let K_n be the n-by-n reversal matrix. Let $A_2 = K_n A K_n$, $A_3 = A^T$, and $A_4 = K_n A^T K_n$. (a) Write A_2, A_3, and A_4 as explicit arrays like the one in (3.3.12). (b) Explain why $p(t)$ is both the minimal and characteristic polynomial of A_2, A_3, and A_4, each of which is encountered in the literature as an alternative definition of *companion matrix*.

3.3.P12 Let $A, B \in M_n$. Suppose that $p_A(t) = p_B(t) = q_A(t) = q_B(t)$. Explain why A and B are similar. Use this fact to show that the alternative forms for the companion matrix noted in the preceding problem are all similar to (3.3.12).

3.3.P13 Explain why any n complex numbers can be the eigenvalues of an n-by-n companion matrix. However, the singular values of a companion matrix are subject to some very strong restrictions. Write the companion matrix (3.3.12) as a block matrix $A = \begin{bmatrix} 0 & -a_0 \\ I_{n-1} & \xi \end{bmatrix}$, in which $\xi = [-a_1 \ \ldots \ -a_{n-1}]^T \in \mathbf{C}^{n-1}$. Verify that $A^*A = \begin{bmatrix} I_{n-1} & \xi \\ \xi^* & s \end{bmatrix}$, in which $s = |a_0|^2 + \|\xi\|_2^2$. Let $\sigma_1 \geq \cdots \geq \sigma_n$ denote the ordered singular values of A. (a) Show that $\sigma_2 = \cdots = \sigma_{n-1} = 1$ and

$$\sigma_1^2, \sigma_n^2 = \frac{1}{2}\left(s + 1 \pm \sqrt{(s+1)^2 - 4|a_0|^2}\right) \tag{3.3.16}$$

Another approach is to use interlacing (4.3.18) to show that the multiplicity of the eigenvalue 1 of A^*A is at least $n - 2$; determine the other two eigenvalues from the trace and determinant of A^*A. (b) Verify that $\sigma_1 \sigma_n = |a_0|$, $\sigma_1^2 + \sigma_n^2 = s + 1$, and $\sigma_1 \geq 1 \geq \sigma_n$, in which both inequalities are strict if $\xi \neq 0$. (c) The formulae (3.3.16) show that the singular values of a companion matrix depend only on the absolute values of its entries. Show this in a different way by applying a suitable diagonal unitary equivalence to A. Problems 5.6.P28 and 5.6.P31 use (3.3.16) to give bounds on the zeroes of a polynomial.

3.3.P14 Let $A \in M_n$ be a companion matrix (3.3.12). Show that (a) if $n = 2$, then A is normal if and only if $|a_0| = 1$ and $a_1 = -a_0 \bar{a}_1$; it is unitary if and only if $|a_0| = 1$ and $a_1 = 0$; (b) if $n \geq 3$, then A is normal if and only if $|a_0| = 1$ and $a_1 = \cdots = a_{n-1} = 0$, that is, if and only if $p_A(t) = t^n - c$ and $|c| = 1$; (c) if $n \geq 3$ and A is normal, then A is unitary and there is a $\varphi \in [0, 2\pi/n)$ such that the eigenvalues of A are $e^{i\varphi} e^{2\pi i k/n}$, $k = 0, 1, \ldots, n - 1$.

3.3.P15 Let $A \in M_n$ be given, and let $P(A) = \{p(A) : p(t) \text{ is a polynomial}\}$. Show that $P(A)$ is a subalgebra of M_n: *the subalgebra generated by A*. Explain why the dimension of $P(A)$ is the degree of the minimal polynomial of A, and hence $\dim P(A) \leq n$.

3.3.P16 If $A, B, C \in M_n$ and if there are polynomials $p_1(t)$ and $p_2(t)$ such that $A = p_1(C)$ and $B = p_2(C)$, then A and B commute. Does every pair of commuting matrices arise in this way? Provide details for the following construction of two commuting 3-by-3 matrices that are *not* polynomials in a third matrix: (a) Let $A = J_2(0) \oplus J_1(0)$ and $B = J_3(0)^2$. Show that $AB = BA = A^2 = B^2 = 0$; $\{I, A, B\}$ is a basis for $\mathcal{A}(A, B)$, the algebra generated by A and B; and $\dim \mathcal{A}(A, B) = 3$. (b) If there is a $C \in M_3$ and polynomials $p_1(t)$ and $p_2(t)$ such that $A = p_1(C)$ and $B = p_2(C)$, then $\mathcal{A}(A, B) \subset P(C)$ so $\dim P(C) \geq 3$; $\dim P(C) = 3$; and $\mathcal{A}(A, B) = P(C)$. (c) Let $C = \gamma I + \alpha A + \beta B$. Then $(C - \gamma I)^2 = 0$; the minimal polynomial of C has degree at most 2; and $\dim P(C) \leq 2$. Contradiction.

3.3.P17 Explain why any matrix that commutes with a companion matrix C must be a polynomial in C.

3.3.P18 Newton's identities (2.4.18–19) can be proved by applying standard matrix analytic identities to the companion matrix. Adopt the notation of (2.4.P3) and (2.4.P9) and let $A \in M_n$ be the companion matrix of $p(t) = t^n + a_{n-1}t^{n-1} + \cdots + a_1 t + a_0$. Provide details for the following: (a) Since $p(t) = p_A(t)$, we have $p(A) = 0$ and $0 = \text{tr}(A^k p(A)) = \mu_{n+k} + a_{n-1}\mu_{n+k-1} + \cdots + a_1\mu_{k+1} + a_0\mu_k$ for $k = 0, 1, 2, \ldots$, which is (2.4.19). (b) Use (2.4.13) to show that

$$\text{tr}(\text{adj}(tI - A)) = nt^{n-1} + \text{tr}\, A_{n-2}t^{n-2} + \cdots + \text{tr}\, A_1 t + \text{tr}\, A_0 \qquad (3.3.17)$$

and use (2.4.17) to show that $\text{tr}\, A_{n-k-1} = \mu_k + a_{n-1}\mu_{k-1} + \cdots + a_{n-k+1}\mu_1 + na_{n-k}$, which is the coefficient of t^{n-k-1} in the right-hand side of (3.3.17) for $k = 1, \ldots, n-1$. Use (0.8.10.2) to show that $\text{tr}(\text{adj}(tI - A)) = nt^{n-1} + (n-1)a_{n-1}t^{n-2} + \cdots + 2a_2 t + a_1$, so $(n-k)a_{n-k}$ is the coefficient of t^{n-k-1} in the left-hand side of (3.3.17) for $k = 1, \ldots, n-1$. Conclude that $(n-k)a_{n-k} = \mu_k + a_{n-1}\mu_{k-1} + \cdots + a_{n-k+1}\mu_1 + na_{n-k}$ for $k = 1, \ldots, n-1$, which is equivalent to (2.4.17).

3.3.P19 Let $A, B \in M_n$ and let $C = AB - BA$ be their commutator. In (2.4.P12) we learned that if C commutes with *either* A or B, then $C^n = 0$. If C commutes with *both* A and B, show that $C^{n-1} = 0$. What does this say if $n = 2$?

3.3.P20 Let $A, B \in M_n$ be companion matrices (3.3.12) and let $\lambda \in \mathbf{C}$. (a) Show that λ is an eigenvalue of A if and only if $x_\lambda = [1 \ \lambda \ \lambda^2 \ \ldots \ \lambda^{n-1}]^T$ is an eigenvector of A^T. (b) If λ is an eigenvalue of A, show that every eigenvector of A^T associated with λ is a scalar multiple of x_λ. Deduce that every eigenvalue of A has geometric multiplicity 1. (c) Explain why A^T and B^T have a common eigenvector if and only if they have a common eigenvalue. (d) If A commutes with B, show that $A = B$.

3.3.P21 Let $n \geq 2$, let C_n be the companion matrix (3.3.12) of $p(t) = t^n + 1$, let $L_n \in M_n$ be the strictly lower triangular matrix whose entries below the main diagonal are all equal to $+1$, let $E_n = L_n - L_n^T$, and let $\theta_k = \frac{\pi}{n}(2k+1)$, $k = 0, 1, \ldots, n-1$. Provide details for the following proof that the spectral radius of E_n is $\cot \frac{\pi}{2n}$. (a) The eigenvalues of C_n are $\lambda_k = e^{i\theta_k}$, $k = 0, 1, \ldots, n-1$ with respective associated eigenvectors $x_k = [1 \ \lambda_k \ \ldots \ \lambda_k^{n-1}]^T$. (b) $E_n = C_n + C_n^2 + \cdots + C_n^{n-1}$ has eigenvectors x_k, $k = 0, 1, \ldots, n-1$ with respective associated eigenvalues

$$\lambda_k + \lambda_k^2 + \cdots + \lambda_k^{n-1} = \frac{\lambda_k - \lambda_k^n}{1 - \lambda_k} = \frac{1 + \lambda_k}{1 - \lambda_k}$$

$$= \frac{e^{-i\theta_k/2} + e^{i\theta_k/2}}{e^{-i\theta_k/2} - e^{i\theta_k/2}} = i \cot \frac{\theta_k}{2}$$

for $k = 0, 1, \ldots, n-1$. (c) $\rho(E_n) = \cot \frac{\pi}{2n}$.

3.3.P22 Let $A \in M_n$. Explain why the degree of the minimal polynomial of A is at most rank $A + 1$, and show by example that this upper bound on the degree is best possible for singular matrices: For each $r = 1, \ldots, n-1$ there is some $A \in M_n$ such that rank $A = r$ and the degree of $q_A(t)$ is $r + 1$.

3.3.P23 Show that a companion matrix is diagonalizable if and only if it has distinct eigenvalues.

3.3.P24 Use the example in the exercise preceding (3.3.4) to show that there are nonsimilar $A, B \in M_n$ such that for every polynomial $p(t)$, $p(A) = 0$ if and only if $p(B) = 0$.

3.3.P25 If $a_0 \neq 0$, show that the inverse of the companion matrix A in (3.3.12) is

$$A^{-1} = \begin{bmatrix} \frac{-a_1}{a_0} & 1 & 0 & \cdots & 0 \\ \frac{-a_2}{a_0} & 0 & 1 & & 0 \\ \vdots & \vdots & & \ddots & \ddots & \vdots \\ \frac{-a_{n-1}}{a_0} & 0 & & & \ddots & 1 \\ \frac{-1}{a_0} & 0 & \cdots & & \cdots & 0 \end{bmatrix} \tag{3.3.18}$$

and that its characteristic polynomial is

$$t^n + \frac{a_1}{a_0}t^{n-1} + \cdots + \frac{a_{n-1}}{a_0}t + \frac{1}{a_0} = \frac{t^n}{a_0}p_A(t^{-1}) \tag{3.3.19}$$

3.3.P26 This problem is a generalization of (2.4.P16). Let $\lambda_1, \ldots, \lambda_d$ be the distinct eigenvalues of $A \in M_n$, and let $q_A(t) = (t - \lambda_1)^{\mu_1} \cdots (t - \lambda_d)^{\mu_d}$ be the minimal polynomial of A. For $i = 1, \ldots, d$, let $q_i(t) = q_A(t)/(t - \lambda_i)$ and let ν_i denote the number of blocks $J_{\mu_i}(\lambda_i)$ in the Jordan canonical form of A. Show that (a) for each $i = 1, \ldots, d$, $q_i(A) \neq 0$, each of its nonzero columns is an eigenvector of A associated with λ_i, and each of its nonzero rows is the complex conjugate of a left eigenvector of A associated with λ_i; (b) for each $i = 1, \ldots, d$, $q_i(A) = X_i Y_i^*$, in which $X_i, Y_i \in M_{n,\nu_i}$ each have rank ν_i, $A X_i = \lambda_i X_i$, and $Y_i^* A = \lambda_i Y_i^*$; (c) $\operatorname{rank} q_i(A) = \nu_i$, $i = 1, \ldots, d$; (d) if $\nu_i = 1$ for some $i = 1, \ldots, d$, then there exists a polynomial $p(t)$ such that $\operatorname{rank} p(A) = 1$; (e) if A is nonderogatory, then there is a polynomial $p(t)$ such that $\operatorname{rank} p(A) = 1$; (f) the converse of the assertion in (d) is correct as well – can you prove it?

3.3.P27 The nth-order linear homogeneous ordinary differential equation

$$y^{(n)} + a_{n-1}y^{(n-1)} + a_{n-2}y^{(n-2)} + \cdots + a_1 y' + a_0 y = 0$$

for a complex-valued function $y(t)$ of a real parameter t can be transformed into a first-order homogeneous system of ordinary differential equations $x' = Ax$, $A \in M_n$, $x = [x_1 \ldots x_n]^T$ by introducing auxiliary variables $x_1 = y$, $x_2 = y', \ldots, x_n = y^{(n-1)}$. Perform this transformation and show that A^T is the companion matrix (3.3.12).

3.3.P28 Suppose that $K \in M_n$ is an involution. Explain why K is diagonalizable, and why K is similar to $I_m \oplus (-I_{n-m})$ for some $m \in \{0, 1, \ldots, n\}$.

3.3.P29 Suppose that $A, K \in M_n$, K is an involution, and $A = KAK$. Show that (a) there is some $m \in \{0, 1, \ldots, n\}$ and matrices $A_{11} \in M_m$, $A_{22} \in M_{n-m}$ such that A is similar to $A_{11} \oplus A_{22}$ and KA is similar to $A_{11} \oplus (-A_{22})$; (b) λ is an eigenvalue of A if and only if either $+\lambda$ or $-\lambda$ is an eigenvalue of KA; (c) if $A \in M_n$ is centrosymmetric (0.9.10) and $K = K_n$ is the reversal matrix (0.9.5.1), then λ is an eigenvalue of A if and only if either $+\lambda$ or $-\lambda$ is an eigenvalue of $K_n A$, which presents the rows of A in reverse order.

3.3.P30 Suppose that $A, K \in M_n$, K is an involution, and $A = -KAK$. Show that (a) there is some $m \in \{0, 1, \ldots, n\}$ and matrices $A_{12} \in M_{m,n-m}$, $A_{21} \in M_{n-m,m}$ such that A is similar to $\mathcal{B} = \begin{bmatrix} 0_m & A_{12} \\ A_{21} & 0_{n-m} \end{bmatrix}$ and KA is similar to $\begin{bmatrix} 0_m & A_{12} \\ -A_{21} & 0_{n-m} \end{bmatrix}$; (b) A is similar to iKA, so λ is an eigenvalue of A if and only if $i\lambda$ is an eigenvalue of KA; (c) if $A \in M_n$ is skew centrosymmetric (0.9.10) and K_n is the reversal matrix (0.9.5.1), then A is similar to $i K_n A$ (thus, λ is an eigenvalue of A if and only if $i\lambda$ is an eigenvalue of $K_n A$, which presents the rows of A in reverse order).

3.3.P31 Show that there is no real 3-by-3 matrix whose minimal polynomial is $x^2 + 1$, but that there is a real 2-by-2 matrix as well as a complex 3-by-3 matrix with this property.

3.3.P32 Let $\lambda_1, \ldots, \lambda_d$ be the distinct eigenvalues of a given $A \in M_n$. Make a list of the $N = w_1(A, \lambda_1) + \cdots + w_1(A, \lambda_d)$ blocks in the Jordan canonical form of A. For $j = 1, 2, \ldots$ until no blocks remain on the list (which must occur for some $j = r \le N$), perform the following two steps (i) for each $k = 1, \ldots, d$, if there is a block with eigenvalue λ_k in the list, remove one of largest size from the list; (ii) let J_j denote the direct sum of the (at most d) blocks removed from the list in (i), let $p_j(t)$ be the characteristic polynomial of J_j, and let C_j be the companion matrix of $p_j(t)$. Explain why (a) each matrix J_j is nonderogatory; (b) each J_j is similar to C_j; (c) A is similar to $F = C_1 \oplus \cdots \oplus C_r$; (d) $p_1(t)$ is the minimal polynomial of A and $p_1(t) \cdots p_r(t)$ is the characteristic polynomial of A; (e) F is real if A is real; (f) $p_{j+1}(t)$ divides $p_j(t)$ for each $j = 1, \ldots, r - 1$; (g) if $F' = C_1' \oplus \cdots \oplus C_s'$ is a direct sum of companion matrices, if F is similar to A, and if $p_{C_{j+1}'}(t)$ divides $p_{C_j'}(t)$ for each $j = 1, \ldots, s$, then $F' = F$. The polynomials $p_1(t), \ldots, p_r(t)$ are the *invariant factors* of A. Although we have used the Jordan canonical form of A (and hence its eigenvalues) to construct F, the eigenvalues of A do not appear explicitly in F. In fact, one can compute the invariant factors of A (and hence the companion matrices $C_1, \ldots, C_r$) solely by means of finitely many rational operations on the entries of A, without knowing its eigenvalues. If A is real, those rational operations involve only real numbers; if the entries of A are in a field $\mathbf{F} \subset \mathbf{C}$, those rational operations involve only elements of $\mathbf{F}$. The matrix F is the *rational canonical form* of A.

3.3.P33 Let $z_1, \ldots, z_n$ be the zeroes of the polynomial p in (3.3.11). Show that

$$\frac{1}{n} \sum_{i=1}^{n} |z_i|^2 \le 1 - \frac{1}{n} + \frac{1}{n} \sum_{i=0}^{n-1} |a_i|^2 < 1 + \max_{0 \le i \le n-1} |a_i|^2$$

3.3.P34 Let $A, B \in M_n$, let $C = AB - BA$, consider the minimal polynomial (3.3.5b) of A, and let $m = 2\max\{r_1, \ldots, r_d\} - 1$. If A commutes with C, it is known that $C^m = 0$. Deduce the assertions in (2.4.P12 (a,c)) from this fact.

3.3.P35 Let $A \in M_n$ and suppose that rank $A = 1$. Show that the minimal polynomial of A is $q_A(t) = t(t - \operatorname{tr} A)$ and conclude that A is diagonalizable if and only if $\operatorname{tr} A \ne 0$.

Further Readings. The first proof of (3.3.16) is in F. Kittaneh, Singular values of companion matrices and bounds on zeroes of polynomials, *SIAM J. Matrix Anal. Appl.* 16 (1995) 333–340. For a discussion of the rational canonical form of a matrix over any field, see section 7.2 of Hoffman and Kunze (1971) or section V.4 of Turnbull and Aitken (1945). The result mentioned in (3.3.P34) is proved in J. Bračič and B. Kuzma, Localizations of the Kleinecke–Shirokov theorem, *Oper. Matrices* 1 (2007) 385–389.

3.4 The real Jordan and Weyr canonical forms

In this section we discuss a real version of the Jordan canonical form for real matrices, as well as an alternative to the Jordan canonical form for complex matrices that is especially useful in problems involving commutativity.

3.4.1 The real Jordan canonical form. Suppose that $A \in M_n(\mathbf{R})$, so any non-real eigenvalues must occur in complex conjugate pairs. We have $\operatorname{rank}(A - \lambda I)^k = \operatorname{rank}\overline{(A - \lambda I)^k} = \operatorname{rank}(\overline{A - \lambda I})^k = \operatorname{rank}(A - \bar{\lambda}I)^k$ for any $\lambda \in \mathbf{C}$ and all $k = 1, 2, \ldots$, so the Weyr characteristics of A associated with any complex conjugate pair of eigenvalues are the same (that is, $w_k(A, \lambda) = w_k(A, \bar{\lambda})$ for all $k = 1, 2, \ldots$). Lemma 3.1.18 ensures that the Jordan structure of A corresponding to any eigenvalue λ is the same as the Jordan structure of A corresponding to the eigenvalue $\bar{\lambda}$ (that is, $s_k(A, \lambda) = s_k(A, \bar{\lambda})$ for all $k = 1, 2, \ldots$). Thus, all the Jordan blocks of A of all sizes with non-real eigenvalues occur in conjugate pairs of equal size.

For example, if λ is a non-real eigenvalue of $A \in M_n(\mathbf{R})$, and if k blocks $J_2(\lambda)$ are in the Jordan canonical form of A, then there are k blocks $J_2(\bar{\lambda})$ as well. The block diagonal matrix

$$\begin{bmatrix} J_2(\lambda) & \\ & J_2(\bar{\lambda}) \end{bmatrix} = \begin{bmatrix} \lambda & 1 & & \\ 0 & \lambda & & \\ \hline & & \bar{\lambda} & 1 \\ & & 0 & \bar{\lambda} \end{bmatrix}$$

is permutation similar (interchange rows and columns 2 and 3) to the block upper triangular matrix

$$\begin{bmatrix} \lambda & 0 & 1 & 0 \\ 0 & \bar{\lambda} & 0 & 1 \\ \hline & & \lambda & 0 \\ & & 0 & \bar{\lambda} \end{bmatrix} = \begin{bmatrix} D(\lambda) & I_2 \\ & D(\lambda) \end{bmatrix}$$

in which $D(\lambda) = \begin{bmatrix} \lambda & 0 \\ 0 & \bar{\lambda} \end{bmatrix} \in M_2$.

In general, any Jordan matrix of the form

$$\begin{bmatrix} J_k(\lambda) & \\ & J_k(\bar{\lambda}) \end{bmatrix} \in M_{2k} \tag{3.4.1.1}$$

is permutation similar to the block upper triangular (block bidiagonal) matrix

$$\begin{bmatrix} D(\lambda) & I_2 & & & \\ & D(\lambda) & I_2 & & \\ & & \ddots & \ddots & \\ & & & \ddots & I_2 \\ & & & & D(\lambda) \end{bmatrix} \in M_{2k} \tag{3.4.1.2}$$

which has k 2-by-2 blocks $D(\lambda)$ on the main block diagonal and $k - 1$ blocks I_2 on the block superdiagonal.

Let $\lambda = a + ib, a, b \in \mathbf{R}$. A computation reveals that $D(\lambda)$ is similar to a real matrix

$$C(a, b) := \begin{bmatrix} a & b \\ -b & a \end{bmatrix} = SD(\lambda)S^{-1} \tag{3.4.1.3}$$

in which $S = \begin{bmatrix} -i & -i \\ 1 & -1 \end{bmatrix}$ and $S^{-1} = \frac{1}{2i}\begin{bmatrix} -1 & i \\ -1 & -i \end{bmatrix}$. Moreover, every block matrix of the form (3.4.1.2) with a non-real λ is similar to a real block matrix of the form

$$C_k(a, b) := \begin{bmatrix} C(a, b) & I_2 & & & \\ & C(a, b) & I_2 & & \\ & & \ddots & \ddots & \\ & & & \ddots & I_2 \\ & & & & C(a, b) \end{bmatrix} \in M_{2k} \tag{3.4.1.4}$$

via the similarity matrix $S \oplus \cdots \oplus S$ (k direct summands). Thus, every block matrix of the form (3.4.1.1) is similar to the matrix $C_k(a, b)$ in (3.4.1.4). These observations lead us to the *real Jordan canonical form theorem*.

Theorem 3.4.1.5. *Each $A \in M_n(\mathbf{R})$ is similar via a real similarity to a real block diagonal matrix of the form*

$$C_{n_1}(a_1, b_1) \oplus \cdots \oplus C_{n_p}(a_p, b_p) \oplus J_{m_1}(\mu_1) \oplus \cdots \oplus J_{m_r}(\mu_r) \tag{3.4.1.6}$$

in which $\lambda_k = a_k + ib_k$, $k = 1, 2, \ldots, p$, are non-real eigenvalues of A, each a_k and b_k is real and $b_k > 0$, and $\mu_1, \ldots, \mu_r$ are real eigenvalues of A. Each real block triangular matrix $C_{n_k}(a_k, b_k) \in M_{2n_k}$ is of the form (3.4.1.4) and corresponds to a pair of conjugate Jordan blocks $J_{n_k}(\lambda_k), J_{n_k}(\bar{\lambda}_k) \in M_{n_k}$ with non-real λ_k in the Jordan canonical form (3.1.12) of A. The real Jordan blocks $J_{m_k}(\mu_k)$ in (3.4.6) are the Jordan blocks in (3.1.12) that have real eigenvalues.

Proof. We have shown that A is similar to (3.4.1.6) over $\mathbf{C}$. Theorem 1.3.28 ensures that A is similar to (3.4.6) over $\mathbf{R}$. $\qquad\square$

The block matrix (3.4.1.6) is the *real Jordan canonical form* of A. The following corollary formulates several useful alternative criteria for similarity to a real matrix.

Corollary 3.4.1.7. *Let $A \in M_n$ be given. The following are equivalent:*

(a) *A is similar to a real matrix.*

(b) *For each nonzero eigenvalue λ of A and each $k = 1, 2, \ldots$ the respective numbers of blocks $J_k(\lambda)$ and $J_k(\bar{\lambda})$ are equal.*

(c) *For each non-real eigenvalue λ of A and each $k = 1, 2, \ldots$ the respective numbers of blocks $J_k(\lambda)$ and $J_k(\bar{\lambda})$ are equal.*

(d) *For each non-real eigenvalue λ of A and each $k = 1, 2, \ldots$ $\operatorname{rank}(A - \lambda I)^k = \operatorname{rank}(A - \bar{\lambda}I)^k$.*

(e) *For each non-real eigenvalue λ of A and each $k = 1, 2, \ldots$ $\operatorname{rank}(A - \lambda I)^k = \operatorname{rank}(\bar{A} - \lambda I)^k$.*

(f) *For each non-real eigenvalue λ of A the Weyr characteristics of A associated with λ and $\bar{\lambda}$ are the same.*

(g) *A is similar to $\bar{A}$.*

Corollary 3.4.1.8. *If $A = \begin{bmatrix} B & C \\ 0 & 0 \end{bmatrix} \in M_n$ and $B \in M_m$ is similar to a real matrix, then A is similar to a real matrix.*

Proof. Suppose that $S \in M_m$ is nonsingular and $SBS^{-1} = R$ is real. Then $\mathcal{A} = (S \oplus I_{n-m})A(S \oplus I_{n-m})^{-1} = \begin{bmatrix} R & \star \\ 0 & 0 \end{bmatrix}$ is similar to A. If $\lambda \neq 0$ then the column ranks of

$$(\mathcal{A} - \lambda I)^k = \begin{bmatrix} (R - \lambda I)^k & \star \\ & (-\lambda)^k I_{n-m} \end{bmatrix}$$

and

$$(\overline{\mathcal{A}} - \lambda I)^k = \begin{bmatrix} (R - \lambda I)^k & \star \\ & (-\lambda)^k I_{n-m} \end{bmatrix}$$

are the same: $n - m + \operatorname{rank}(R - \lambda I)^k$. We conclude that $\mathcal{A}$ is similar to $\overline{\mathcal{A}}$, so $\mathcal{A}$ (and hence also A) is similar to a real matrix. $\qquad\square$

Corollary 3.4.1.9. *For each $A \in M_n$, $A\overline{A}$ is similar to $\overline{A}A$ as well as to a real matrix.*

Proof. Theorem 3.2.11.1 ensures that the nonsingular Jordan structures of $A\overline{A}$ and $\overline{A}A$ are the same. Since a matrix and its complex conjugate have the same rank, $\operatorname{rank}(A\overline{A})^k = \operatorname{rank}\overline{(A\overline{A})^k} = \operatorname{rank}(\overline{A}A)^k = \operatorname{rank}(\overline{A}A)^k$ for each $k = 1, 2, \ldots$. Thus, the nilpotent Jordan structures of $A\overline{A}$ and $\overline{A}A$ are also the same, so $A\overline{A}$ and $\overline{A}A$ are similar. Since $\overline{A}A = \overline{A\overline{A}}$, (3.4.1.7) ensures that $A\overline{A}$ is similar to a real matrix. $\qquad\square$

Each complex square matrix A is similar, via a complex similarity, to a complex upper triangular matrix T (2.3.1). If A is diagonalizable, then it is similar, via a complex similarity, to a diagonal matrix whose diagonal entries are the same as those of T; these entries are the eigenvalues of A. What is the real analog of this observation?

Each real square matrix A is similar, via a real similarity, to a real upper quasitriangular matrix T of the form (2.3.5) in which any 2-by-2 diagonal blocks have the special form (2.3.5a), which is the same as (3.4.1.3). If A is diagonalizable, the following corollary of (3.4.1.5) ensures that A is similar, via a real similarity, to a real quasidiagonal matrix whose diagonal blocks are the same as those of T.

Corollary 3.4.1.10. *Let $A \in M_n(\mathbf{R})$ be given and suppose that it is diagonalizable. Let $\mu_1, \ldots, \mu_q$ be the real eigenvalues of A and let $a_1 \pm ib_1, \ldots, a_\ell \pm ib_\ell$ be the non-real eigenvalues of A, in which each $b_j > 0$. Then A is similar, via a real similarity, to*

$$C_1(a_1, b_1) \oplus \cdots \oplus C_1(a_\ell, b_\ell) \oplus [\mu_1] \oplus \cdots \oplus [\mu_q]$$

Proof. This is the case $n_1 = \cdots = n_p = m_1 = \cdots = m_r = 1$ in (3.4.1.6). $\qquad\square$

3.4.2 The Weyr canonical form. The Weyr characteristic (3.1.16) played a key role in our discussion of uniqueness of the Jordan canonical form. It can also be used to define a canonical form for similarity that has certain advantages over the Jordan form. We begin by defining a *Weyr block*.

Let $\lambda \in \mathbf{C}$ be given, let $q \geq 1$ be a given positive integer, let $w_1 \geq \cdots \geq w_q \geq 1$ be a given nonincreasing sequence of positive integers, and let $w = (w_1, \ldots, w_q)$. The

Weyr block $W(w, \lambda)$ associated with λ and w is the upper triangular q-by-q block bidiagonal matrix

$$W(w, \lambda) = \begin{bmatrix} \lambda I_{w_1} & G_{w_1, w_2} & & & \\ & \lambda I_{w_2} & G_{w_2, w_3} & & \\ & & \ddots & \ddots & \\ & & & \ddots & G_{w_{q-1}, w_q} \\ & & & & \lambda I_{w_q} \end{bmatrix} \tag{3.4.2.1}$$

in which

$$G_{w_i, w_j} = \begin{bmatrix} I_{w_j} \\ 0 \end{bmatrix} \in M_{w_i, w_j}, \quad 1 \leq i < j$$

Notice that rank $G_{w_i, w_j} = w_j$ and that if $w_i = w_{i+1}$, then $G_{w_i, w_{i+1}} = I_{w_i}$.

A Weyr block $W(w, \lambda)$ may be thought of as a q-by-q block matrix analog of a Jordan block. The diagonal blocks are *scalar matrices* λI with nonincreasingly ordered sizes, and the superdiagonal blocks are *full-column-rank blocks* $\begin{bmatrix} I \\ 0 \end{bmatrix}$ whose sizes are dictated by the sizes of the diagonal blocks.

Exercise. The size of the Weyr block $W(w, \lambda)$ in (3.4.2.1) is $w_1 + \cdots + w_q$. Explain why rank$(W(w, \lambda) - \lambda I) = w_2 + \cdots + w_q$.

Exercise. Verify that $G_{w_{k-1}, w_k} G_{w_k, w_{k+1}} = G_{w_{k-1}, w_{k+1}}$, that is,

$$\begin{bmatrix} I_{w_k} \\ 0_{w_{k-1} - w_k, w_k} \end{bmatrix} \begin{bmatrix} I_{w_{k+1}} \\ 0_{w_k - w_{k+1}, w_{k+1}} \end{bmatrix} = \begin{bmatrix} I_{w_{k+1}} \\ 0_{w_{k-1} - w_{k+1}, w_{k+1}} \end{bmatrix}$$

Using the preceding exercise, we find that $(W(w, \lambda) - \lambda I)^2 =$

$$\begin{bmatrix} 0_{w_1} & 0 & G_{w_1, w_3} & & & \\ & 0_{w_2} & 0 & \ddots & & \\ & & 0_{w_3} & \ddots & G_{w_{q-2}, w_q} \\ & & & \ddots & 0 \\ & & & & 0_{w_q} \end{bmatrix}$$

so rank$(W(w, \lambda) - \lambda I)^2 = w_3 + \cdots + w_q$. Moving from one power to the next, each block $G_{w_1, w_{p+1}}, \ldots, G_{w_{q-p}, w_q}$ in the nonzero superdiagonal of $(W(w, \lambda) - \lambda I)^p$ moves up one block row into the next higher superdiagonal of $(W(w, \lambda) - \lambda I)^{p+1}$, whose blocks are $G_{w_1, w_{p+2}}, \ldots, G_{w_{q-p-1}, w_q}$. In particular, rank$(W(w, \lambda) - \lambda I)^p = w_{p+1} + \cdots + w_q$ for each $p = 1, 2, \ldots$. Consequently,

$$\text{rank}(W(w, \lambda) - \lambda I)^{p-1} - \text{rank}(W(w, \lambda) - \lambda I)^p = w_p, \, p = 1, \ldots, q$$

so the Weyr characteristic of $W(w, \lambda)$ associated with the eigenvalue λ is w.

Exercise. Explain why the number of diagonal blocks in (3.4.2.1) (the parameter q) is the index of λ as an eigenvalue of $W(w, \lambda)$.

A *Weyr matrix* is a direct sum of Weyr blocks with *distinct* eigenvalues.

For any given $A \in M_n$, let q be the index of an eigenvalue λ of A, let $w_k = w_k(A, \lambda)$, $k = 1, 2, \ldots$ be the Weyr characteristic of A associated with λ, and define the *Weyr block of A associated with the eigenvalue λ* to be

$$W_A(\lambda) = W(w(A, \lambda), \lambda)$$

For example, the Weyr characteristic of the Jordan matrix J in (3.1.16a) associated with the eigenvalue 0 is $w_1(J, 0) = 6$, $w_2(J, 0) = 5$, $w_3(J, 0) = 2$, so

$$W_J(0) = \begin{bmatrix} 0_6 & G_{6,5} & \\ & 0_5 & G_{5,2} \\ & & 0_2 \end{bmatrix} \tag{3.4.2.2}$$

Exercise. Let λ be an eigenvalue of $A \in M_n$. Explain why the size of the Weyr block $W_A(\lambda)$ is the algebraic multiplicity of λ.

Exercise. For the Weyr block (3.4.2.2), show by explicit calculation that

$$W_J(0)^2 = \begin{bmatrix} 0_6 & 0_{6,5} & G_{6,2} \\ & 0_5 & 0_{5,2} \\ & & 0_2 \end{bmatrix}$$

and $W_J(0)^3 = 0$. Explain why rank $W_J(0) = 7 = w_2 + w_3$ and rank $W_J(0)^2 = 2 = w_3$, and why the Weyr characteristic of $W_J(0)$ associated with its (only) eigenvalue 0 is 6, 5, 2. Deduce that $W_J(0)$ is similar to J.

We can now state the *Weyr canonical form theorem*.

Theorem 3.4.2.3. *Let $A \in M_n$ be given and let $\lambda_1, \ldots, \lambda_d$ be its distinct eigenvalues in any given order. There is a nonsingular $S \in M_n$ and there are Weyr blocks $W_1, \ldots, W_d$, each of the form (3.4.2.1), such that (a) the (only) eigenvalue of W_j is λ_j for each $j = 1, \ldots, d$ and (b) $A = S(W_1 \oplus \cdots \oplus W_d)S^{-1}$. The Weyr matrix $W_1 \oplus \cdots \oplus W_d$ to which A is similar is uniquely determined by A and the given enumeration of its distinct eigenvalues: $W_j = W_A(\lambda_j)$ for each $j = 1, \ldots, d$, so*

$$A = S \begin{bmatrix} W_A(\lambda_1) & & \\ & \ddots & \\ & & W_A(\lambda_d) \end{bmatrix} S^{-1}$$

If A is similar to a Weyr matrix, then that matrix is obtained from $W_A = W_A(\lambda_1) \oplus \cdots \oplus W_A(\lambda_d)$ by a permutation of its direct summands. If A is real and has only real eigenvalues, then S can be chosen to be real.

Proof. The preceding observations show that $W_A = W_A(\lambda_1) \oplus \cdots \oplus W_A(\lambda_d)$ and A have identical Weyr characteristics associated with each of their distinct eigenvalues. Lemma 3.1.18 ensures that W_A and A are similar since they are both similar to the same Jordan canonical form. If two Weyr matrices are similar, then they must have the same distinct eigenvalues and the same Weyr characteristics associated with each eigenvalue; it follows that they contain the same Weyr blocks, perhaps permuted in the respective direct sums. If A and all its eigenvalues are real, then W_A is real and (1.3.29) ensures that A is similar to W_A via a real similarity. $\square$

The Weyr matrix $W_A = W_A(\lambda_1) \oplus \cdots \oplus W_A(\lambda_d)$ in the preceding theorem is (up to permutation of its direct summands) the *Weyr canonical form* of A. The Weyr and Jordan canonical forms W_A and J_A contain the same information about A, but each presents that information differently. The Weyr form explicitly displays the Weyr characteristics of A, while the Jordan form explicitly displays its Segre characteristics. A dot diagram (3.1.P11) can be used to construct one form from the other. Moreover, W_A and J_A are permutation similar; see (3.4.P8).

Exercise. For a given $\lambda \in \mathbf{C}$, consider the Jordan matrix $J = J_3(\lambda) \oplus J_2(\lambda)$. Explain why $w(J, \lambda) = 2, 2, 1$,

$$
J = \begin{bmatrix} \lambda & 1 & 0 & & \\ 0 & \lambda & 1 & & \\ 0 & 0 & \lambda & & \\ & & & \lambda & 1 \\ & & & 0 & \lambda \end{bmatrix}, \text{ and } W_J(\lambda) = \begin{bmatrix} \lambda & 0 & 1 & 0 & \\ 0 & \lambda & 0 & 1 & \\ & & \lambda & 0 & 1 \\ & & 0 & \lambda & 0 \\ & & & & \lambda \end{bmatrix}
$$

Exercise. Observe that the matrices J and W_J in the preceding exercise have the same number of off-diagonal nonzero entries as well as the same number of nonzero entries.

Exercise. Let $\lambda_1, \ldots, \lambda_d$ be the distinct eigenvalues of $A \in M_n$. (a) If A is nonderogatory, explain why there are d positive integers $p_1, \ldots, p_d$ such that (i) $w_1(A, \lambda_i) = \cdots = w_{p_i}(A, \lambda_i) = 1$ and $w_{p_i+1}(A, \lambda_i) = 0$ for each $i = 1, \ldots, d$; (ii) the Weyr canonical form of A is its Jordan canonical form. (b) If $w_1(A, \lambda_i) = 1$ for each $i = 1, \ldots, d$, why must A be nonderogatory?

Exercise. Let $\lambda_1, \ldots, \lambda_d$ be the distinct eigenvalues of $A \in M_n$. (a) If A is diagonalizable, explain why (i) $w_2(A, \lambda_i) = 0$ for each $i = 1, \ldots, d$; (ii) $W_A(\lambda_i) = \lambda_i I_{w_1(A, \lambda_i)}, i = 1, \ldots, d$; (iii) the Weyr canonical form of A is its Jordan canonical form. (b) If $w_2(A, \lambda_i) = 0$ for some i, why is $w_1(A, \lambda_i)$ equal to the algebraic multiplicity of λ_i (it is *always* equal to the geometric multiplicity)? (c) If $w_2(A, \lambda_i) = 0$ for all $i = 1, \ldots, d$, why must A be diagonalizable?

Exercise. Let $\lambda_1, \ldots, \lambda_d$ be the distinct eigenvalues of $A \in M_n$. Explain why for each $i = 1, \ldots, d$ there are at most p Jordan blocks of A with eigenvalue λ_i if and only if $w_1(A, \lambda_i) \leq p$ for each $i = 1, \ldots, d$, which is equivalent to requiring that *every* diagonal block of *every* Weyr block $W_A(\lambda_i)$ (3.4.2.1) is at most p-by-p.

In (3.2.4) we investigated the set of matrices that commute with a single given nonderogatory matrix. The key to understanding the structure of this set is knowing that $A \in M_k$ commutes with a single Jordan block $J_k(\lambda)$ if and only if A is an upper triangular Toeplitz matrix (3.2.4.3). Thus, a matrix commutes with a *nonderogatory* Jordan matrix J if and only if it is a direct sum (conformal to J) of upper triangular Toeplitz matrices; in particular, it is upper triangular. The Jordan and Weyr canonical forms of a nonderogatory matrix are identical. The Jordan and Weyr canonical forms of a *derogatory* matrix need not be the same, and if they are not, there is a very important difference in the structures of the matrices that commute with them.

Exercise. Let $J = J_2(\lambda) \oplus J_2(\lambda)$ and $A \in M_4$. Show that (a) $W_J = \begin{bmatrix} \lambda I_2 & I_2 \\ 0_2 & \lambda I_2 \end{bmatrix}$; (b) A commutes with J if and only if $A = \begin{bmatrix} B & C \\ D & E \end{bmatrix}$ in which each block $B, C, D, E \in M_2$ is upper triangular Toeplitz; (c) A commutes with W_J if and only if $A = \begin{bmatrix} B & C \\ 0 & B \end{bmatrix}$, which is block upper triangular.

The following lemma identifies the feature of a Weyr block that forces any matrix that commutes with it to be block upper triangular.

Lemma 3.4.2.4. *Let $\lambda \in \mathbb{C}$ and positive integers $n_1 \geq n_2 \geq \cdots \geq n_k \geq 1$ be given. Consider the upper triangular and identically partitioned matrices*

$$F = [F_{ij}]_{i,j=1}^k = \begin{bmatrix} \lambda I_{n_1} & F_{12} & & \bigstar \\ & \lambda I_{n_2} & \ddots & \\ & & \ddots & F_{k-1,k} \\ & & & \lambda I_{n_k} \end{bmatrix} \in M_n$$

and

$$F' = [F'_{ij}]_{i,j=1}^k = \begin{bmatrix} \lambda I_{n_1} & F'_{12} & & \bigstar \\ & \lambda I_{n_2} & \ddots & \\ & & \ddots & F'_{k-1,k} \\ & & & \lambda I_{n_k} \end{bmatrix} \in M_n$$

Assume that all of the superdiagonal blocks $F'_{i,i+1}$ have full column rank. If $A \in M_n$ and $AF = F'A$, then A is block upper triangular conformal to F and F'. If, in addition, A is normal, then A is block diagonal conformal to F and F'.

Proof. Partition $A = [A_{ij}]_{i,j=1}^k$ conformally to F and F'. Our strategy is to inspect corresponding blocks of the identity $AF = F'A$ in a particular order. In block position $k - 1, 1$ we have $\lambda A_{k-1,1} = \lambda A_{k-1,1} + F'_{k-1,k} A_{k1}$, so $F'_{k-1,k} A_{k1} = 0$ and hence $A_{k1} = 0$ since $F'_{k-1,k}$ has full column rank. In block position $k - 2, 1$ we have $\lambda A_{k-2,1} = \lambda A_{k-2,1} + F'_{k-2,k-1} A_{k-1,1}$ (since $A_{k1} = 0$), so $F'_{k-2,k-1} A_{k-1,1} = 0$ and $A_{k-1,1} = 0$. Proceeding upward in the first block column of A and using at each step the fact that the lower blocks in that block column have been shown to be zero blocks, we find that $A_{i1} = 0$ for each $i = k, k-1, \ldots, 2$. Now inspect block position $k - 1, 2$ and proceed upward in the same fashion to show that $A_{i2} = 0$ for each $i = k, k-1, \ldots, 3$. Continuing this process left to right and bottom to top, we find that A is block upper triangular conformal to F and F'. If A is normal and block triangular, then (2.5.2) ensures that it is block diagonal. $\square$

Using the preceding lemma, we now show that if $A, B \in M_n$ commute, then there is a simultaneous similarity that takes A into its Weyr form W_A and takes B into a block upper triangular matrix whose block structure is determined by the block structure of W_A.

Theorem 3.4.2.5 (Belitskii). *Let $A \in M_n$ be given, let $\lambda_1, \ldots, \lambda_d$ be its distinct eigenvalues in any prescribed order, let $w_k(A, \lambda_j), k = 1, 2, \ldots$, be the Weyr* *characteristic of A associated with the eigenvalue λ_j, $j = 1, \ldots, d$, and let $W_A(\lambda_j)$ be a Weyr block for $j = 1, 2, \ldots, d$. Let $S \in M_n$ be nonsingular and such that $A = S(W_A(\lambda_1) \oplus \cdots \oplus W_A(\lambda_d))S^{-1}$. Suppose that $B \in M_n$ and $AB = BA$. Then (1) $S^{-1}BS = B^{(1)} \oplus \cdots \oplus B^{(k)}$ is block diagonal conformal to $W_A(\lambda_1) \oplus \cdots \oplus W_A(\lambda_d)$, and (2) each matrix $B^{(\ell)}$ is block upper triangular conformal to the partition (3.4.2.1) of $W_A(\lambda_\ell)$.*

Proof. The assertion (1) follows from the basic result (2.4.4.2); the assertion (2) follows from the preceding lemma. $\square$

Any matrix that commutes with a Weyr matrix is block upper triangular, but we can say a little more. Consider once again the Jordan matrix J in (3.1.16a), whose Weyr canonical form $W_J = W_J(0)$ is (3.4.2.2). To expose certain identities among the blocks of a (necessarily block upper triangular) matrix that commutes with W_J, we impose a finer partition on W_J. Let $m_k = w_k - w_{k+1}, k = 1, 2, 3$, so each m_k is the number of Jordan blocks of size k in J: $m_3 = 2$, $m_2 = 3$, and $m_1 = 1$. We have $w_1 = m_3 + m_2 + m_1 = 6$, $w_2 = m_3 + m_2 = 5$, and $w_3 = m_3 = 2$. Now repartition W_J (3.4.2.2) with diagonal block sizes $m_3, m_2, m_1; m_2, m_1; m_1$, that is, $2, 3, 1; 2, 3; 2$ – this is known as the *standard partition*: the coarsest partition of a Weyr block such that every diagonal block is a scalar matrix (square) and every off-diagonal block is either an identity matrix (square) or a zero matrix (not necessarily square). In the standard partition, W_J has the form

$$W_J = \left[\begin{array}{ccc|cc|c} 0_2 & 0 & 0 & I_2 & 0 & \\ & 0_3 & 0 & 0 & I_3 & \\ & & 0_1 & 0 & 0 & \\ \hline & & & 0_2 & 0 & I_2 \\ & & & & 0_3 & 0 \\ \hline & & & & & 0_2 \end{array}\right] \qquad (3.4.2.6)$$

Although the diagonal blocks in a Weyr block are arranged in nonincreasing order of size, after imposing the standard partition, the new, smaller, diagonal blocks need not occur in nonincreasing order of size. A computation reveals that N commutes with W_J if and only if it has the following block structure, conformal to that of (3.4.2.6):

$$N = \left[\begin{array}{ccc|cc|c} B & C & \bigstar & D & \bigstar & \bigstar \\ & F & \bigstar & E & \bigstar & \bigstar \\ & & G & 0 & \bigstar & \bigstar \\ \hline & & & B & C & D \\ & & & & F & E \\ \hline & & & & & B \end{array}\right] \qquad (3.4.2.7)$$

There are no constraints on the entries of the $\bigstar$ blocks. It may be easier to see how the equalities among the blocks of (3.4.2.7) are structured if we collapse its standard partition to the coarser partition of (3.4.2.2): $N = [N_{ij}]_{i,j=1}^3$ with $N_{11} \in M_{w_1} = M_6$,

$N_{22} \in M_{w_2} = M_5$, and $N_{33} \in M_{w_3} = M_2$. Then

$$N_{33} = [B], \quad N_{23} = \begin{bmatrix} D \\ E \end{bmatrix}, \quad N_{22} = \begin{bmatrix} B & C \\ 0 & F \end{bmatrix},$$

$$N_{12} = \begin{bmatrix} D & \bigstar \\ E & \bigstar \\ 0 & \bigstar \end{bmatrix}, \quad N_{11} = \begin{bmatrix} B & C & \bigstar \\ 0 & F & \bigstar \\ 0 & 0 & G \end{bmatrix},$$

that is,

$$N_{22} = \begin{bmatrix} N_{33} & \bigstar \\ 0 & \bigstar \end{bmatrix}, \quad N_{11} = \begin{bmatrix} N_{22} & \bigstar \\ 0 & \bigstar \end{bmatrix}, \quad N_{12} = \begin{bmatrix} N_{23} & \bigstar \\ 0 & \bigstar \end{bmatrix}$$

The pattern

$$N_{i-1,j-1} = \begin{bmatrix} N_{ij} & \bigstar \\ 0 & \bigstar \end{bmatrix} \tag{3.4.2.8}$$

permits us to determine all the equalities among the blocks in the standard partition (including the positions of the off-diagonal zero block(s)) starting with the blocks in the last block column and working backward up their block diagonals.

Exercise. Consider a Jordan matrix $J \in M_{4n_1+2n_2}$ that is a direct sum of n_1 copies of $J_4(\lambda)$ and n_2 copies of $J_2(\lambda)$. Explain why (a) $w(J, \lambda) = n_1 + n_2, n_1 + n_2, n_1, n_1$; (b) the block sizes $m_k = w_k - w_{k+1}$ in the standard partition of $W_J(\lambda)$ are $n_2, n_1; n_2, n_1; n_1; n_1$ (zero values of m_k are removed); (c) $W_J(\lambda)$ and its presentation according to a standard partition are

$$W_J(\lambda) = \begin{bmatrix} \lambda I_{n_1+n_2} & I_{n_1+n_2} & & \\ & \lambda I_{n_1+n_2} & G_{n_1+n_2,n_1} & \\ & & \lambda I_{n_1} & I_{n_1} \\ & & & \lambda I_{n_1} \end{bmatrix}$$

$$= \left[\begin{array}{cc|cc|cc} \lambda I_{n_1} & 0 & I_{n_1} & 0 & & \\ & \lambda I_{n_2} & 0 & I_{n_2} & & \\ \hline & & \lambda I_{n_1} & 0 & I_{n_1} & \\ & & & \lambda I_{n_2} & 0 & \\ \hline & & & & \lambda I_{n_1} & I_{n_1} \\ & & & & & \lambda I_{n_1} \end{array}\right]$$

(d) $N W_J = W_J N$ if and only if, partitioned conformally to the preceding matrix, N has the form

$$N = \left[\begin{array}{cc|cc|cc} B & D & E & \bigstar & F & \bigstar \\ & C & 0 & \bigstar & G & \bigstar \\ \hline & & B & D & E & F \\ & & & C & 0 & G \\ \hline & & & & B & E \\ & & & & & B \end{array}\right]$$

One final structural simplification of (3.4.2.7) is available to us. Let $U_3, \Delta_3 \in M_{m_3}$, $U_2, \Delta_2 \in M_{m_2}$, and $U_1, \Delta_1 \in M_{m_1}$ be unitary and upper triangular matrices (2.3.1) such that $B = U_3\Delta_3 U_3^*$, $F = U_2\Delta_2 U_2^*$, and $G = U_1\Delta_1 U_1^*$ (a trivial factorization in this case). Let

$$U = U_3 \oplus U_2 \oplus U_1 \oplus U_3 \oplus U_2 \oplus U_3$$

Then

$$N' := U^*NU = \left[\begin{array}{ccc|ccc}
\Delta_3 & C' & \star & D' & \star & \star \\
 & \Delta_2 & \star & E' & \star & \star \\
 & & \Delta_1 & 0 & \star & \star \\
\hline
 & & & \Delta_3 & C' & D' \\
 & & & & \Delta_2 & E' \\
 & & & & & \Delta_3
\end{array}\right] \tag{3.4.2.9}$$

is upper triangular; we have $C' = U_3^*CU_2$, $D' = U_3^*DU_3$, and $E' = U_2^*EU_3$. The equalities among the blocks of N' on and above the block diagonal are the same as those of N. Moreover, W_J is unchanged after a similarity via U: $U^*W_JU = W_J$.

We can draw a remarkable conclusion from the preceding example. Suppose that $A \in M_{13}$ has the Jordan canonical form (3.1.16a); $\mathcal{F} = \{A, B_1, B_2, \ldots\}$ is a commuting family; and $S \in M_{13}$ is nonsingular and $S^{-1}AS = W_A$ is the Weyr canonical form (3.4.2.11). Then $S^{-1}\mathcal{F}S = \{W_A, S^{-1}B_1S, S^{-1}B_2S, \ldots\}$ is a commuting family. Since each matrix $S^{-1}B_iS$ commutes with W_A, it has the block upper triangular form (3.4.2.7) in the standard partition. Thus, for each $j = 1, \ldots, 6$ the diagonal blocks in position j, j of all the matrices $S^{-1}B_iS$ constitute a commuting family, which can be upper triangularized by a single unitary matrix U_j (2.3.3). For each $i = 1, 2, \ldots$ the diagonal blocks of $S^{-1}B_iS$ in positions $(1, 1)$, $(4, 4)$, and $(6, 6)$ are constrained to be the same, so we may (and do) insist that $U_1 = U_4 = U_6$. For the same reason, we insist that $U_2 = U_5$. Let $U = U_1 \oplus \cdots \oplus U_6$. Then each $U^*(S^{-1}B_iS)U$ is upper triangular and has the form (3.4.2.9), and $U^*S^{-1}ASU = U^*W_AU = W_A$. The conclusion is that *there is a simultaneous similarity of the commuting family* $\{A, B_1, B_2, \ldots\}$ *that reduces A to Weyr canonical form and reduces every B_i to the upper triangular form* (3.4.2.9).

All the essential features of the general case are captured in the preceding example, and by following its development, one can prove the following theorem.

Theorem 3.4.2.10. *Let $\lambda_1, \ldots, \lambda_d$ be the distinct eigenvalues of a given $A \in M_n$ in any prescribed order, let their respective indices as eigenvalues of A be $q_1, \ldots, q_d$, and let their respective algebraic multiplicities be $p_1, \ldots, p_d$. For each $i = 1, \ldots, d$, let $w(A, \lambda_i) = (w_1(A, \lambda_i), \ldots, w_{q_i}(A, \lambda_i))$ be the Weyr characteristic of A associated with λ_i and let $W_A(\lambda_i)$ be the Weyr block of A associated with λ_i. Let*

$$W_A = W_A(\lambda_1) \oplus \cdots \oplus W_A(\lambda_d) \tag{3.4.2.11}$$

be the Weyr canonical form of A, and let $A = SW_AS^{-1}$.

(a) *(Belitskii) Suppose that $B \in M_n$ commutes with A. Then $S^{-1}BS = B^{(1)} \oplus \cdots \oplus B^{(d)}$ is block diagonal conformal to W_A. For each $\ell = 1, \ldots, d$, partition $B^{(\ell)} = [B_{ij}^{(\ell)}]_{i,j=1}^{q_\ell} \in M_{p_\ell}$, in which each $B_{jj}^{(\ell)} \in M_{w_j(A,\lambda_\ell)}$, $j = 1, \ldots, q_\ell$. In this partition, $B^{(\ell)}$ is block upper triangular conformal to $W_A(\lambda_\ell)$, and its blocks along the kth* *block superdiagonal are related by the identities*

$$B_{j-k-1,j-1}^{(\ell)} = \begin{bmatrix} B_{j-k,j}^{(\ell)} & \star \\ 0 & \star \end{bmatrix}, \qquad \begin{array}{l} k = 0, 1, \ldots, q_\ell - 1; \\ j = q_\ell, q_\ell - 1, \ldots, k + 1 \end{array} \qquad (3.4.2.12)$$

(b) *(O'Meara and Vinsonhaler) Let $\mathcal{F} = \{A, A_1, A_2, \ldots\} \subset M_n$ be a commuting family. There is a nonsingular $T \in M_n$ such that $T^{-1}\mathcal{F}T = \{W_A, T^{-1}A_1T, T^{-1}A_2T, \ldots\}$ is an upper triangular family. Each matrix $T^{-1}A_iT$ is block diagonal conformal to (3.4.2.11). If the diagonal block of $T^{-1}A_iT$ corresponding to $W_A(\lambda_\ell)$ is partitioned with diagonal block sizes $w_1(A, \lambda_\ell)$, $w_2(A, \lambda_\ell), \ldots, w_{q_\ell}(A, \lambda_\ell)$, then its blocks along its kth block superdiagonal are related by identities of the form (3.4.2.12).*

3.4.3 The unitary Weyr form.
Theorem 3.4.2.3 and the QR factorization imply a refinement of (2.3.1) that incorporates the block structure of the Weyr canonical form.

Theorem 3.4.3.1 (Littlewood). *Let $\lambda_1, \ldots, \lambda_d$ be the distinct eigenvalues of a given $A \in M_n$ in any prescribed order, let $q_1, \ldots, q_d$ be their respective indices, and let $q = q_1 + \cdots + q_d$. Then A is unitarily similar to an upper triangular matrix of the form*

$$F = \begin{bmatrix} \mu_1 I_{n_1} & F_{12} & F_{13} & \cdots & & F_{1p} \\ & \mu_2 I_{n_2} & F_{23} & \cdots & & F_{2p} \\ & & \mu_3 I_{n_3} & \ddots & & \vdots \\ & & & \ddots & F_{p-1,p} \\ & & & & \mu_p I_{n_p} \end{bmatrix} \qquad (3.4.3.2)$$

in which

(a) $\mu_1 = \cdots = \mu_{q_1} = \lambda_1; \mu_{q_1+1} = \cdots = \mu_{q_1+q_2} = \lambda_2; \ldots; \mu_{p-q_d+1} = \cdots = \mu_p = \lambda_d$

(b) *For each $j = 1, \ldots, d$ the q_j integers $n_i, \ldots, n_{i+q_j-1}$ for which $\mu_i = \cdots = \mu_{i+q_j-1} = \lambda_j$ are the Weyr characteristic of λ_j as an eigenvalue of A, that is, $n_i = w_1(A, \lambda_j) \geq \cdots \geq n_{i+q_j-1} = w_{q_j}(A, \lambda_j)$*

(c) *if $\mu_i = \mu_{i+1}$ then $n_i \geq n_{i+1}$, $F_{i,i+1} \in M_{n_i,n_{i+1}}$ is upper triangular, and its diagonal entries are real and positive*

If $A \in M_n(\mathbf{R})$ and if $\lambda_1, \ldots, \lambda_d \in \mathbf{R}$, then A is real orthogonally similar to a real matrix F of the form (3.4.3.2) that satisfies conditions (a), (b), and (c).

The matrix F in (3.4.3.2) is determined by A up to the following equivalence: If A is unitarily similar to a matrix F' of the form (3.4.3.2) that satisfies the conditions (a), (b), and (c), then there is a block diagonal unitary matrix $U = U_1 \oplus \cdots \oplus U_p$ conformal to F such that $F' = UFU^*$, that is, $F'_{ij} = U_i^* F_{ij}U_j, i \leq j, i, j = 1, \ldots, p$.

Proof. Let $S \in M_n$ be nonsingular and such that

$$A = SW_AS^{-1} = S(W_A(\lambda_1) \oplus \cdots \oplus W_A(\lambda_d))S^{-1}$$

Let $S = QR$ be a QR factorization (2.1.14), so Q is unitary, R is upper triangular with positive diagonal entries, and $A = Q(RW_AR^{-1})Q^*$ is unitarily similar to the upper triangular matrix $RW_A R^{-1}$. Partition $R = [R_{ij}]_{i,j=1}^d$ conformally to W_A and compute

$$RW_A R^{-1} = \begin{bmatrix} R_{11} W(A, \lambda_1) R_{11}^{-1} & & \bigstar \\ & \ddots & \\ & & R_{dd} W(A, \lambda_d) R_{dd}^{-1} \end{bmatrix}$$

It suffices to consider only the diagonal blocks, that is, matrices of the form $TW(A, \lambda)T^{-1}$. The matrix T is upper triangular with positive diagonal entries; we partition $T = [T_{ij}]_{i,j=1}^q$ and $T^{-1} = [T^{ij}]_{i,j=1}^q$ conformally to $W(A, \lambda)$, whose diagonal block sizes are $w_1 \geq \cdots \geq w_q \geq 1$. The diagonal blocks of $TW(A, \lambda)T^{-1}$ are $T_{ii} \lambda I_{w_i} T^{ii} = \lambda I_{w_i}$ since $T^{ii} = T_{ii}^{-1}$ (0.9.10); the superdiagonal blocks are $T_{ii} G_{i,i+1} T^{i+1,i+1} + \lambda(T_{ii} T^{i,i+1} + T_{i,i+1} T^{i+1,i+1}) = T_{ii} G_{i,i+1} T^{i+1,i+1}$ (the term in parentheses is the $(i, i+1)$ block entry of $TT^{-1} = I$). If we partition $T_{ii} = \begin{bmatrix} C & \bigstar \\ 0 & D \end{bmatrix}$ with $C \in M_{w_i}$ (C is upper triangular with positive diagonal entries), then

$$T_{ii} G_{i,i+1} T^{i+1,i+1} = \begin{bmatrix} C & \bigstar \\ 0 & D \end{bmatrix} \begin{bmatrix} I_{w_{i+1}} \\ 0 \end{bmatrix} T_{i+1,i+1}^{-1} = \begin{bmatrix} CT_{i+1,i+1}^{-1} \\ 0 \end{bmatrix}$$

is upper triangular and has positive diagonal entries, as asserted.

If A is real and has real eigenvalues, (2.3.1), (3.4.2.3), and (2.1.14) ensure that the reductions in the preceding argument (as well as the QR factorization) can be achieved with real matrices.

Finally, suppose that $V_1, V_2 \in M_n$ are unitary, $A = V_1 F V_1^* = V_2 F' V_2^*$, and both F and F' satisfy the conditions (a), (b), and (c). Then $(V_2^* V_1)F = F'(V_2^* V_1)$, so (3.4.2.4) ensures that $V_2^* V_1 = U_1 \oplus \cdots \oplus U_p$ is block diagonal conformal with F and F', that is, $V_1 = V_2(U_1 \oplus \cdots \oplus U_p)$ and $F' = UFU^*$. $\square$

The following corollary illustrates how (3.4.3.1) can be used.

Corollary 3.4.3.3. *Let $A \in M_n$ be a projection: $A^2 = A$. Let*

$$\sigma_1 \geq \cdots \geq \sigma_g > 1 \geq \sigma_{g+1} \geq \cdots \geq \sigma_r > 0 = \sigma_{r+1} = \cdots$$

be the singular values of A, so $r = $ rank A and g is the number of singular values of A that are greater than 1. Then A is unitarily similar to

$$\begin{bmatrix} 1 & (\sigma_1^2 - 1)^{1/2} \\ 0 & 0 \end{bmatrix} \oplus \cdots \oplus \begin{bmatrix} 1 & (\sigma_g^2 - 1)^{1/2} \\ 0 & 0 \end{bmatrix} \oplus I_{r-g} \oplus 0_{n-r-g}$$

Proof. The minimal polynomial of A is $q_A(t) = t(t - 1)$, so A is diagonalizable; its distinct eigenvalues are $\lambda_1 = 1$ and $\lambda_2 = 0$; their respective indices are $q_1 = q_2 = 1$; and their respective Weyr characteristics are $w_1(A, 1) = r = $ tr A and $w_1(A, 0) = n - r$. Theorem 3.4.3.1 ensures that A is unitarily similar to $F = \begin{bmatrix} I_r & F_{12} \\ 0 & 0_{n-r} \end{bmatrix}$ and that F_{12} is determined up to unitary equivalence. Let $h = $ rank F_{12} and let $F_{12} = V\Sigma W^*$ be a singular value decomposition: $V \in M_r$ and $W \in M_{n-r}$ are unitary, and $\Sigma \in M_{r,n-r}$ is diagonal with diagonal entries $s_1 \geq \cdots \geq s_h > 0 = s_{h+1} = \cdots$. Then F is unitarily similar (via $V \oplus W$) to $\begin{bmatrix} I_r & \Sigma \\ 0 & 0_{n-r} \end{bmatrix}$, which is permutation similar to

$$C = \begin{bmatrix} 1 & s_1 \\ 0 & 0 \end{bmatrix} \oplus \cdots \oplus \begin{bmatrix} 1 & s_h \\ 0 & 0 \end{bmatrix} \oplus I_{r-h} \oplus 0_{n-r-h}$$

The singular values of C (and hence also of A) are $(s_1^2 + 1)^{1/2}, \ldots, (s_h^2 + 1)^{1/2}$ together with $r - h$ ones and $n - r - h$ zeroes. It follows that $h = g$ and $s_i = (\sigma_i^2 - 1)^{1/2}, i = 1, \ldots, g$.   □

Exercise. Provide details for the preceding proof. Explain why two projections of the same size are unitarily similar if and only if they are unitarily equivalent, that is, if and only if they have the same singular values. Compare the preceding proof with the approach in (2.6.P18).

Problems

3.4.P1 Suppose that $A \in M_n(\mathbf{R})$ and $A^2 = -I_n$. Show that n must be even and that there is a nonsingular $S \in M_n(\mathbf{R})$ such that

$$S^{-1}AS = \begin{bmatrix} 0 & -I_{n/2} \\ I_{n/2} & 0 \end{bmatrix}$$

In the following three problems, for a given $A \in M_n$, $\mathcal{C}(A) = \{B \in M_n : AB = BA\}$ denotes the *centralizer* of A: the set of matrices that commute with A.

3.4.P2 Explain why $\mathcal{C}(A)$ is an algebra.

3.4.P3 Let $J \in M_{13}$ be the matrix in (3.1.16a). (a) Use (3.4.2.7) to show that $\dim \mathcal{C}(J) = 65$. (b) Show that $w_1(J, 0)^2 + w_2(J, 0)^2 + w_3(J, 0)^2 = 65$.

3.4.P4 Let the distinct eigenvalues of $A \in M_n$ be $\lambda_1, \ldots, \lambda_d$ with respective indices $q_1, \ldots, q_d$. (a) Show that $\dim \mathcal{C}(A) = \sum_{j=1}^{d} \sum_{i=1}^{q_j} w_i(A, \lambda_j)^2$. (b) Show that $\dim \mathcal{C}(A) \geq n$ with equality if and only if A is nonderogatory. (c) Let the Segre characteristic of each eigenvalue λ_j of A be $s_i(A, \lambda_j), i = 1, \ldots, w_1(A, \lambda_j)$. It is known that $\dim \mathcal{C}(A) = \sum_{j=1}^{d} \sum_{i=1}^{w_1(A,\lambda)} (2i - 1)s_i(A, \lambda_j)$; see Problem 9 in section 4.4 of Horn and Johnson (1991). Explain why

$$\sum_{j=1}^{d} \sum_{i=1}^{q} w_i(A, \lambda)^2 = \sum_{j=1}^{d} \sum_{i=1}^{w_1(A,\lambda_j)} (2i - 1)s_i(A, \lambda_j)$$

Verify this identity for the matrix in (3.1.16a).

3.4.P5 Let $A \in M_n$ be given and suppose that $A^2 = 0$. Let $r = \operatorname{rank} A$ and let $\sigma_1 \geq \cdots \geq \sigma_r$ be the positive singular values of A. Show that A is unitarily similar to

$$\begin{bmatrix} 0 & \sigma_1 \\ 0 & 0 \end{bmatrix} \oplus \cdots \oplus \begin{bmatrix} 0 & \sigma_r \\ 0 & 0 \end{bmatrix} \oplus 0_{n-2r}$$

Explain why two self-annihilating matrices of the same size are unitarily similar if and only if they have the same singular values, that is, if and only if they are unitarily equivalent. For a different approach, see (2.6.P24).

3.4.P6 Show that $A \in M_2(\mathbf{R})$ is similar to $\begin{bmatrix} 1 & 1 \\ -1 & 1 \end{bmatrix}$ if and only if $A = \begin{bmatrix} 1+\alpha & (1+\alpha^2)/\beta \\ -\beta & 1-\alpha \end{bmatrix}$ for some $\alpha, \beta \in \mathbf{R}$ with $\beta \neq 0$.

3.4.P7 Provide details for the following example, which shows that the simultaneous similarity described in (3.4.2.10b) need not be possible if "Weyr" is replaced by "Jordan." Define $J = \begin{bmatrix} J_2(0) & 0 \\ 0 & J_2(0) \end{bmatrix}$ and $A = \begin{bmatrix} 0 & I_2 \\ J_2(0) & 0 \end{bmatrix}$, and refer to the exercise preceding (3.4.2.4). (a) Notice that the blocks of A are upper triangular Toeplitz, and explain (no computations!) why A must commute with J. (b) Suppose that there is a simultaneous similarity of the commuting family $\{J, A\}$ that puts J into Jordan canonical form and puts A into upper triangular form, that is, suppose that there is a nonsingular $S = [s_{ij}] \in M_4$ such that $S^{-1}JS = J$, and $S^{-1}AS = T = [t_{ij}]$ is upper triangular. Verify that

$$S = \begin{bmatrix} s_{11} & s_{12} & s_{13} & s_{14} \\ 0 & s_{11} & 0 & s_{13} \\ s_{31} & s_{32} & s_{33} & s_{34} \\ 0 & s_{31} & 0 & s_{33} \end{bmatrix}, \quad AS = \begin{bmatrix} s_{31} & * & * & * \\ * & s_{31} & * & * \\ 0 & s_{11} & * & * \\ * & * & * & * \end{bmatrix}$$

and

$$ST = \begin{bmatrix} s_{11}t_{11} & * & * & * \\ * & s_{11}t_{22} & * & * \\ s_{31}t_{11} & s_{31}t_{12} + s_{32}t_{22} & * & * \\ * & * & * & * \end{bmatrix}$$

(the $*$ entries are not relevant to the argument). (c) Deduce that $s_{31} = 0$ and $s_{11} = 0$. Why is this a contradiction? (d) Conclude that there is no simultaneous similarity of $\{J, A\}$ that has the properties asserted in (b).

3.4.P8 An algorithm to construct a permutation similarity between the Weyr and Jordan forms of a matrix involves an interesting mathematical object known as a (*standard*) *Young tableau*. For example, consider $J = J_3(0) \oplus J_2(0) \in M_5$, whose Weyr characteristic is $w_1 = 2$, $w_2 = 2$, $w_3 = 1$. (a) Verify that its associated dot diagram (3.1.P11) and Weyr canonical form are

$$\begin{matrix} \bullet & \bullet \\ \bullet & \bullet \\ \bullet & \end{matrix} \quad \text{and} \quad W = \begin{bmatrix} 0_2 & I_2 & \\ & 0_2 & G_{2,1} \\ & & 0 \end{bmatrix} \in M_5$$

Label the dot diagram with consecutive integers $1, \ldots, 5$ left to right across the rows, top to bottom (this labeled dot diagram is the Young tableau); then read the labeled diagram top to bottom down the columns, left to right; use the sequence obtained to construct a permutation σ:

$$\text{The Young tableau} \quad \begin{matrix} 1 & 2 \\ 3 & 4 \\ 5 & \end{matrix} \quad \text{leads to the permutation } \sigma = \begin{pmatrix} 1 & 2 & 3 & 4 & 5 \\ 1 & 3 & 5 & 2 & 4 \end{pmatrix}$$

Construct the permutation matrix $P = [e_1 \; e_3 \; e_5 \; e_2 \; e_4] \in M_5$ whose columns are the permutation of the columns of I_n that is specified by σ. Verify that $J = P^T W P$, so $W = P J P^T$. In general, to construct a permutation similarity between a given Weyr form $W \in M_n$ and its Jordan form J, first construct a Young tableau by labeling the dot diagram for the Weyr characteristic of J and W with consecutive integers $1, 2, \ldots, n$ left to right across each successive row, top to bottom. Then construct a matrix $\sigma = [\sigma_{ij}] \in M_{2,n}$ whose first row entries are $1, 2, \ldots, n$; its second row entries are obtained by reading the Young tableau top to bottom down each successive column, left to right. Construct the permutation matrix $P = [e_{\sigma_{2,1}} \, e_{\sigma_{2,2}} \, \cdots \, e_{\sigma_{2,n}}] \in M_n$ by permuting the columns of I_n as specified by σ. Then $J = P^T W P$ and $W = P J P^T$. (b) Explain why. (c) Use this algorithm to construct a permutation similarity between the Jordan matrix (3.1.16a) and its Weyr canonical form; verify that it does so.

3.4.P9 Explain why the Weyr canonical form of a given square matrix A, like its Jordan canonical form (3.2.9), contains the least number of nonzero off-diagonal entries among all matrices in the similarity class of A.

3.4.P10 Show that the Weyr canonical form of a Jordan matrix J is J if and only if, for each eigenvalue λ of J, either there is exactly one Jordan block in J with eigenvalue λ or every Jordan block in J with eigenvalue λ is 1-by-1.

3.4.P11 Let $A \in M_n$ be given. Show that the Weyr and Jordan canonical forms of A are the same if and only if A is nonderogatory or diagonalizable, or if there are matrices B and C such that (a) B is nonderogatory, (b) C is diagonalizable, (c) B and C have no eigenvalues in common, and (d) A is similar to $B \oplus C$.

Notes and Further Readings. Eduard Weyr announced his eponymous characteristic and canonical form in E. Weyr, Répartition des matrices en espèces et formation de toutes les espèces, *C. R. Acad. Sci. Paris* 100 (1885) 966–969; his paper was submitted to the Paris Academy by Charles Hermite. Weyr later published a detailed exposition and several applications in E. Weyr, Zur Theorie der bilinearen Formen, *Monatsh. Math. und Physik* 1 (1890) 163–236. For modern expositions, including derivations of the Weyr canonical form that do not rely on prior knowledge of the Jordan canonical form, see H. Shapiro, The Weyr characteristic, *Amer. Math. Monthly* 196 (1999) 919–929, and the monograph Clark, O'Meara, Vinsonhaler (2011), which contains numerous applications of the Weyr form. The Weyr form (in its standard partition) was rediscovered by G. Belitskii, whose motivation was to find a canonical form for similarity with the property that every matrix commuting with it is block upper triangular. For accounts of Belitskii's work in English, which also describe identities among the blocks of the standard partition of a matrix that commutes with a Weyr block, see G. Belitskii, Normal forms in matrix spaces, *Integral Equations Operator Theory* 38 (2000) 251–283, and V. V. Sergeichuk, Canonical matrices for linear matrix problems, *Linear Algebra Appl.* 317 (2000) 53–102. Sergeichuk's paper also discusses the permutation similarity described in (3.4.P8). The remarkable Theorem 3.4.2.10b about commuting families is in K. C. O'Meara and C. Vinsonhaler, On approximately simultaneously diagonalizable matrices, *Linear Algebra Appl.* 412 (2006) 39–74, which contains yet another rediscovery of the Weyr canonical form as well as the efficient formulation (3.4.2.12) of the identities among the blocks of matrices that commute with a Weyr block. Theorem 3.4.3.1 has been rediscovered repeatedly; the original source is D. E. Littlewood, On unitary equivalence, *J. London Math. Soc.* 28 (1953) 314–322. The illuminating example in (3.4.P7) is taken from Clark, O'Meara, and Vinsonhaler (2011).

3.5 Triangular factorizations and canonical forms

If a linear system $Ax = b$ has a nonsingular triangular (0.9.3) coefficient matrix $A \in M_n$, computation of the unique solution x is remarkably easy. If, for example, $A = [a_{ij}]$ is upper triangular and nonsingular, then all $a_{ii} \neq 0$ and one can employ *back substitution*: $a_{nn}x_n = b_n$ determines x_n; $a_{n-1,n-1}x_{n-1} + a_{n-1,n}x_n = b_{n-1}$ then determines x_{n-1} since x_n is known and $a_{n-1,n-1} \neq 0$; proceeding in the same fashion upward through successive rows of A, one determines $x_{n-2}, x_{n-3}, \ldots, x_2, x_1$.

Exercise. Describe *forward substitution* as a solution technique for $Ax = b$ if $A \in M_n$ is nonsingular and lower triangular.

If $A \in M_n$ is not triangular, one can still use forward and back substitution to solve $Ax = b$ provided that A is nonsingular and can be factored as $A = LU$, in which L is lower triangular and U is upper triangular: First use forward substitution to solve $Ly = b$, and then use back substitution to solve $Ux = y$.

Definition 3.5.1. *Let $A \in M_n$. A presentation $A = LU$, in which $L \in M_n$ is lower triangular and $U \in M_n$ is upper triangular, is called an LU factorization of A.*

Exercise. Explain why $A \in M_n$ has an LU factorization in which L (respectively, U) is nonsingular if and only if it has an LU factorization in which L (respectively, U) is unit lower (respectively, unit upper) triangular. *Hint*: If L is nonsingular, write $L = L'D$, in which L' is unit lower triangular and D is diagonal.

Lemma 3.5.2. *Let $A \in M_n$ and suppose that $A = LU$ is an LU factorization. For any block 2-by-2 partition*

$$A = \begin{bmatrix} A_{11} & A_{12} \\ A_{21} & A_{22} \end{bmatrix}, \quad L = \begin{bmatrix} L_{11} & 0 \\ L_{21} & L_{22} \end{bmatrix}, \quad U = \begin{bmatrix} U_{11} & U_{12} \\ 0 & U_{22} \end{bmatrix}$$

with $A_{11}, L_{11}, U_{11} \in M_k$ and $k \leq n$, we have $A_{11} = L_{11}U_{11}$. Consequently, each leading principal submatrix of A has an LU factorization in which the factors are the corresponding leading principal submatrices of L and U.

Theorem 3.5.3. *Let $A \in M_n$ be given. Then*

(a) *A has an LU factorization in which L is nonsingular if and only if A has the* row inclusion property: *For each $i = 1, \ldots, n - 1$, $A[\{i + 1; 1, \ldots, i\}]$ is a linear combination of the rows of $A[\{1, \ldots, i\}]$*

(b) *A has an LU factorization in which U is nonsingular if and only if A has the* column inclusion property: *For each $j = 1, \ldots, n - 1$, $A[\{1, \ldots, j; j + 1\}]$ is a linear combination of the columns of $A[\{1, \ldots, j\}]$*

Proof. If $A = LU$, then $A[\{1, \ldots, i + 1\}] = L[\{1, \ldots, i + 1\}]U[\{1, \ldots, i + 1\}]$. Thus, to verify the necessity of the row inclusion property, it suffices to take $i = k = n - 1$ in the partitioned presentation given in (3.5.2). Since L is nonsingular and triangular, L_{11} is also nonsingular, and we have $A_{21} = L_{21}U_{11} = L_{21}L_{11}^{-1}L_{11}U_{11} = (L_{21}L_{11}^{-1}) A_{11}$, which verifies the row inclusion property.

Conversely, if A has the row inclusion property, we may construct inductively an LU factorization with nonsingular L as follows (the cases $n = 1, 2$ are easily verified): Suppose that $A_{11} = L_{11}U_{11}$, L_{11} is nonsingular, and the row vector A_{21} is a linear combination of the rows of A_{11}. Then there is a vector y such that $A_{21} = y^T A_{11} = y^T L_{11}U_{11}$, and we may take $U_{12} = L_{11}^{-1}A_{12}$, $L_{21} = y^T L_{11}$, $L_{22} = 1$, and $U_{22} = A_{22} - L_{21}U_{12}$ to obtain an LU factorization of A in which L is nonsingular.

The assertions about the column inclusion property follow from considering an LU factorization of A^T. $\qquad\square$

Exercise. Consider the matrix $J_n \in M_n$, all of whose entries are 1. Find an LU factorization of J_n in which L is unit lower triangular. With this factorization in hand, $J_n = J_n^T = U^T L^T$ is an LU factorization of J_n with a unit upper triangular factor.

Exercise. Show that the row inclusion property is equivalent to the following formally stronger property: For each $i = 1, ..., n - 1$, every row of $A[\{i + 1, ..., n\}; \{1, ..., i\}]$ is a linear combination of the rows of $A[\{1, ..., i\}]$. What is the corresponding statement for column inclusion?

If $A \in M_n$, rank $A = k$, and det $A[\{1, ..., j\}] \neq 0$, $j = 1, ..., k$, then A has both the row inclusion and column inclusion properties. The following result follows from (3.5.3).

Corollary 3.5.4. *Suppose that $A \in M_n$ and rank $A = k$. If $A[\{1, \ldots, j\}]$ is nonsingular for all $j = 1, \ldots, k$, then A has an LU factorization. Furthermore, either factor may be chosen to be unit triangular; both L and U are nonsingular if and only if $k = n$, that is, if and only if A and all of its leading principal submatrices are nonsingular.*

Example 3.5.5. Not every matrix has an LU factorization. If $A = \begin{bmatrix} 0 & 1 \\ 1 & 0 \end{bmatrix}$ could be written as $A = LU = \begin{bmatrix} \ell_{11} & 0 \\ \ell_{21} & \ell_{22} \end{bmatrix}\begin{bmatrix} u_{11} & u_{12} \\ 0 & u_{22} \end{bmatrix}$, then $\ell_{11}u_{11} = 0$ implies that one of L or U is singular; but $LU = A$ is nonsingular.

Exercise. Explain why a nonsingular matrix that has a singular leading principal submatrix cannot have an LU factorization.

Exercise. Verify that

$$A = \begin{bmatrix} 0 & 0 & 0 \\ 0 & 0 & 1 \\ 0 & 1 & 0 \end{bmatrix} = \begin{bmatrix} 0 & 0 & 0 \\ 1 & 0 & 0 \\ 0 & 1 & 1 \end{bmatrix}\begin{bmatrix} 0 & 0 & 1 \\ 0 & 1 & 0 \\ 0 & 0 & 0 \end{bmatrix}$$

has an LU factorization even though A has neither the row nor column inclusion property. However, A is a principal submatrix of a 4-by-4 matrix

$$\hat{A} = \begin{bmatrix} A & e_1 \\ 0 & 0 \end{bmatrix} = \begin{bmatrix} 0 & \hat{A}_{12} \\ \hat{A}_{21} & 0 \end{bmatrix}, \quad \hat{A}_{12} = \begin{bmatrix} 0 & 1 \\ 1 & 0 \end{bmatrix}, \hat{A}_{21} = \begin{bmatrix} 0 & 1 \\ 0 & 0 \end{bmatrix}$$

that *does not* have an LU factorization. Verify this by considering the block factorization in (3.5.2) with $k = 2$: $\hat{A}_{12} = L_{11}U_{12}$ implies that L_{11} is nonsingular, and hence $0 = L_{11}U_{11}$ implies that $U_{11} = 0$, which is inconsistent with $L_{21}U_{11} = \hat{A}_{21} \neq 0$.

Exercise. Consider $A = \begin{bmatrix} 1 & 0 \\ a & 1 \end{bmatrix} \begin{bmatrix} 0 & 1 \\ 0 & 2-a \end{bmatrix}$ and explain why an LU factorization need not be unique even if the L is required to be unit lower triangular.

It is now clear that an LU factorization of a given matrix may or may not exist, and if it exists, it need not be unique. Much of the trouble arises from singularity, either of A or of its leading principal submatrices. Using the tools of (3.5.2) and (3.5.3), however, we can give a full description in the nonsingular case, and we can impose a normalization that makes the factorization unique.

Corollary 3.5.6 (*LDU factorization*). *Let $A = [a_{ij}] \in M_n$ be given.*

(a) *Suppose that A is nonsingular. Then A has an LU factorization $A = LU$ if and only if $A[\{1, \ldots, i\}]$ is nonsingular for all $i = 1, \ldots, n$.*

(b) *Suppose that $A[\{1, \ldots, i\}]$ is nonsingular for all $i = 1, \ldots, n$. Then $A = LDU$, in which $L, D, U \in M_n$, L is unit lower triangular, U is unit upper triangular, $D = \mathrm{diag}(d_1, \ldots, d_n)$ is diagonal, $d_1 = a_{11}$, and*

$$d_i = \det A\,[\{1, \ldots, i\}]/\det A\,[\{1, \ldots, i-1\}], \quad i = 2, \ldots, n$$

The factors L, U, and D are uniquely determined.

Exercise. Use (3.5.2), (3.5.3), and prior exercises to provide details for a proof of the preceding corollary.

Exercise. If $A \in M_n$ has an LU factorization with $L = [\ell_{ij}]$ and $U = [u_{ij}]$, show that $\ell_{11}u_{11} = \det[A\{1\}]$ and $\ell_{ii}u_{ii} \det A[\{1, \ldots, i-1\}] = \det A[\{1, \ldots, i\}]$, $i = 2, \ldots, n$.

Returning to the solution of the nonsingular linear system $Ax = b$, suppose that $A \in M_n$ cannot be factored as LU but can be factored as PLU, in which $P \in M_n$ is a permutation matrix and L and U are lower and upper triangular, respectively. This amounts to a reordering of the equations in the linear system prior to factorization. In this event, solution of $Ax = b$ is still quite simple via $Ly = P^T b$ and $Ux = y$. It is worth knowing that any $A \in M_n$ may be so factored and that L may be taken to be nonsingular. The solutions of $Ax = b$ are the same as those of $Ux = L^{-1}P^T b$.

Lemma 3.5.7. *Let $A \in M_k$ be nonsingular. Then there is a permutation matrix $P \in M_k$ such that $\det(P^T A)[\{1, \ldots, j\}] \neq 0$, $j = 1, \ldots, k$.*

Proof. The proof is by induction on k. If $k = 1$ or 2, the result is clear by inspection. Suppose that it is valid up to and including $k - 1$. Consider a nonsingular $A \in M_k$ and delete its last column. The remaining $k - 1$ columns are linearly independent and hence they contain $k - 1$ linearly independent rows. Permute these rows to the first $k - 1$ positions and apply the induction hypothesis to the nonsingular upper $(k - 1)$-by-$(k - 1)$ submatrix. This determines a desired overall permutation P, and $P^T A$ is nonsingular. □

The factorization in the following theorem is known as a PLU factorization; the factors need not be unique.

Theorem 3.5.8 (PLU factorization). *For each $A \in M_n$ there is a permutation matrix $P \in M_n$, a unit lower triangular $L \in M_n$, and an upper triangular $U \in M_n$ such that $A = PLU$.*

Proof. If we show that there is a permutation matrix Q such that QA has the row inclusion property, then (3.5.3) and the exercise following it ensure that $QA = LU$ with a unit lower triangular factor L, so $A = PLU$ for $P = Q^T$.

If A is nonsingular, the desired permutation matrix is guaranteed by (3.5.7).

If rank $A = k < n$, first permute the rows of A so that the first k rows are linearly independent. It follows that $A[\{i + 1\}; \{1, ..., i\}]$ is a linear combination of the rows of $A[\{1, ..., i\}], i = k, ..., n - 1$. If $A[\{1, ..., k\}]$ is nonsingular, apply (3.5.7) again to further permute the rows so that $A[\{1, ..., k\}]$, and thus A, has the row inclusion property. If rank $A[\{1, ..., k\}] = \ell < k$, treat it in the same way that we have just treated A, and obtain row inclusion for the indices $i = \ell, ..., n - 1$. Continue in this manner until either the upper left block is 0, in which case we have row inclusion for all indices, or it is nonsingular, in which case one further permutation completes the argument. □

Exercise. Show that each $A \in M_n$ may be factored as $A = LUP$, in which L is lower triangular, U is unit upper triangular, and P is a permutation matrix.

Exercise. For a given $X \in M_n$ and $k, \ell \in \{1, ..., n\}$, define

$$X_{[p,q]} = X[\{1, ..., p\}, \{1, ..., q\}] \tag{3.5.9}$$

Let $A = LBU$, in which $L, B, U \in M_n$, L is lower triangular, and U is upper triangular. Explain why $A_{[p,q]} = L_{[p,p]} B_{[p,q]} U_{[q,q]}$ for all $p, q \in \{1, ..., n\}$. If L and U are nonsingular, explain why

$$\text{rank } A_{[p,q]} = \text{rank } B_{[p,q]} \text{ for all } p, q \in \{1, ..., n\} \tag{3.5.10}$$

The following theorem, like the preceding one, describes a particular triangular factorization (the LPU factorization) that is valid for every square complex matrix. Uniqueness of the P factor in the nonsingular case has important consequences.

Theorem 3.5.11 (LPU factorization). *For each $A \in M_n$ there is a permutation matrix $P \in M_n$, a unit lower triangular $L \in M_n$, and an upper triangular $U \in M_n$ such that $A = LPU$. Moreover, the factor P is uniquely determined if A is nonsingular.*

Proof. Construct inductively permutations $\pi_1, ..., \pi_n$ of the integers $1, ..., n$ as follows: Let $A^{(0)} = [a_{ij}^{(0)}] = A$ and define the index set $\mathcal{I}_1 = \{i \in \{1, ..., n\} : a_{i1}^{(0)} \neq 0\}$. If $\mathcal{I}_1$ is nonempty, let π_1 be the smallest integer in $\mathcal{I}_1$; otherwise, let π_1 be any $i \in \{1, ..., n\}$ and proceed to the next step. If $\mathcal{I}_1$ is nonempty, use type 3 elementary row operations based on row π_1 of $A^{(0)}$ to eliminate all the nonzero entries $a_{i,1}^{(0)}$ in column 1 of $A^{(0)}$ other than $a_{\pi_1,1}^{(0)}$ (for all such entries, $i > \pi_1$); denote the resulting matrix by $A^{(1)}$. Observe that $A^{(1)} = \mathsf{L}_1 A^{(0)}$ for some unit lower triangular matrix L_1 (0.3.3).

Suppose that $2 \leq k \leq n$ and that $\pi_1, ..., \pi_{k-1}$ and $A^{(k-1)} = [a_{ij}^{(k-1)}]$ have been constructed. Let $\mathcal{I}_k = \{i \in \{1, ..., n\} : i \neq \pi_1, ..., \pi_{k-1} \text{ and } a_{ik}^{(k-1)} \neq 0\}$. If $\mathcal{I}_k$ is nonempty, let π_k be the smallest integer in $\mathcal{I}_k$; otherwise, let π_k be any $i \in \{1, ..., n\}$ such that $i \neq \pi_1, ..., \pi_{k-1}$ and proceed to the next step. If $\mathcal{I}_k$ is nonempty, use type 3 elementary row operations based on row π_k of $A^{(k-1)}$ to eliminate every nonzero entry $a_{i,k}^{(k-1)}$ in column k of $A^{(k-1)}$ below the entry $a_{\pi_k,k}^{(k-1)}$ (for all such entries, $i > \pi_k$); denote the resulting matrix by $A^{(k)}$. Observe that these eliminations do not change any entries in columns $1, \ldots, k-1$ of $A^{(k-1)}$ (because $a_{\pi_k,j}^{(k-1)} = 0$ for $j = 1, \ldots, k-1$) and that $A^{(k)} = \mathsf{L}_k A^{(k-1)}$ for some unit lower triangular matrix L_k.

After n steps, our construction produces a permutation $\pi_1, \ldots, \pi_n$ of the integers $1, \ldots, n$ and a matrix $A^{(n)} = [a_{ij}^{(n)}] = \mathsf{L}A$, in which $\mathsf{L} = \mathsf{L}_n \cdots \mathsf{L}_1$ is unit lower triangular. Moreover, $a_{ij}^{(n)} = 0$ whenever $i > \pi_j$ or $i < \pi_j$ and $i \notin \{\pi_1, \ldots, \pi_{j-1}\}$. Let $L = \mathsf{L}^{-1}$ so that $A = L A^{(n)}$ and L is unit lower triangular. Let $P = [p_{ij}] \in M_n$ in which $p_{\pi_j,j} = 1$ for $j = 1, \ldots, n$ and all other entries are zero. Then P is a permutation matrix, $P^T A^{(n)} = U$ is upper triangular, and $A = LPU$.

If A is nonsingular, then both L and U are nonsingular. The preceding exercise ensures that rank $A_{[p,q]} =$ rank $P_{[p,q]}$ for all $p, q \in \{1, \ldots, n\}$, and these ranks uniquely determine the permutation matrix P (see (3.5.P11)). $\square$

Exercise. Explain how the construction in the preceding proof ensures that $P^T A^{(n)}$ is upper triangular.

Definition 3.5.12. *Matrices $A, B \in M_n$ are said to be* triangularly equivalent *if there are nonsingular matrices $L, U \in M_n$ such that L is lower triangular, U is upper triangular, and $A = LBU$.*

Exercise. Verify that triangular equivalence is an equivalence relation on M_n.

Theorem 3.5.11 provides a canonical form for triangular equivalence of nonsingular matrices. The canonical matrices are the permutation matrices; the set of ranks of submatrices described in (3.5.10) is a complete set of invariants.

Theorem 3.5.13. *Let $A, B \in M_n$ be nonsingular. The following are equivalent:*

(a) *There is a unique permutation matrix $P \in M_n$ such that both A and B are triangularly equivalent to P.*

(b) *A and B are triangularly equivalent.*

(c) *The rank equalities (3.5.10) are satisfied.*

Proof. The implication (a) $\Rightarrow$ (b) is clear, and the implication (b) $\Rightarrow$ (c) is the content of the exercise preceding (3.5.11). If $A = L_1 P U_1$ and $B = L_2 P' U_2$ are LPU factorizations and if the hypothesis (c) is assumed, then (using the notation in (3.5.9)) rank $P_{[p,q]} =$ rank $A_{[p,q]} =$ rank $B_{[p,q]} =$ rank $P'_{[p,q]}$ for all $p, q \in \{1, \ldots, n\}$. Problem 3.5.P11 ensures that $P = P'$, which implies (a). $\square$

Exercise. Let $A, P \in M_n$. Suppose that P is a permutation matrix and that all the main diagonal entries of A are ones. Explain why all the main diagonal entries of $P^T A P$ are ones.

Our final theorem concerns triangular equivalence via unit triangular matrices. It uses the facts that (a) the inverse of a unit lower triangular matrix is unit lower triangular, and (b) a product of unit lower triangular matrices is unit lower triangular, with corresponding assertions about unit upper triangular matrices.

Theorem 3.5.14 ($LPDU$ factorization). *For each nonsingular $A \in M_n$ there is a unique permutation matrix P, a unique nonsingular diagonal matrix D, a unit lower triangular matrix L, and a unit upper triangular matrix U such that $A = LPDU$.*

Proof. Theorem 3.5.11 ensures that there is a unit lower triangular matrix L, a unique permutation matrix P, and a nonsingular upper triangular matrix U' such that $A = LPU'$. Let D denote the diagonal matrix whose respective diagonal entries are the same as those of U', that is, $D = \text{diag}(\text{diag}(U'))$, and let $U = D^{-1}U'$. Then U is unit upper triangular and $A = LPDU$. Suppose that D_2 is a diagonal matrix such that $A = L_2 P D_2 U_2$, in which L_2 is unit lower triangular and U_2 is unit triangular. Then

$$(P^T(L_2^{-1}L)P)D = D_2(U_2 U^{-1}) \tag{3.5.15}$$

The main diagonal entries of $U_2 U^{-1}$ and $L_2^{-1}L$ are, of course, all ones; the preceding exercise ensures that the main diagonal entries of $P^T(L_2^{-1}L)P$ are also all ones. Thus, the main diagonal of the left-hand side of (3.5.15) is the same as that of D, while the main diagonal of the right-hand side of (3.5.15) is the same as that of D_2. It follows that $D = D_2$. $\square$

Problems

3.5.P1 We have discussed the factorization $A = LU$, in which L is lower triangular and U is upper triangular. Discuss a parallel theory of $A = UL$ factorization, noting that the factors may be different.

3.5.P2 Describe how $Ax = b$ may be solved if A is presented as $A = QR$, in which Q is unitary and R is upper triangular (2.1.14).

3.5.P3 Matrices $A, B \in M_n$ are said to be *unit triangularly equivalent* if $A = LBU$ for some unit lower triangular matrix L and some unit upper triangular matrix U. Explain why (a) unit triangular equivalence is an equivalence relation on both M_n and $GL(n, \mathbf{C})$; (b) if $P, D, P', D' \in M_n$, P and P' are permutation matrices, and D and D' are nonsingular diagonal matrices, then $PD = P'D'$ if and only if $P = P'$ and $D = D'$; (c) each nonsingular matrix in M_n is unit triangularly equivalent to a unique generalized permutation matrix (0.9.5); (d) two generalized permutation matrices in M_n are unit triangularly equivalent if and only if they are identical; (e) the n-by-n generalized permutation matrices are a set of canonical matrices for the equivalence relation of unit triangular equivalence on $GL(n, \mathbf{C})$.

3.5.P4 If the leading principal minors of $A \in M_n$ are all nonzero, describe how an LU factorization of A may be obtained by using type 3 elementary row operations to zero out entries below the diagonal.

3.5.P5 (Lanczos tridiagonalization algorithm) Let $A \in M_n$ and $x \in \mathbf{C}^n$ be given. Define $X = [x \; Ax \; A^2 x \; \ldots \; A^{n-1}x]$. The columns of X are said to form a *Krylov sequence*. Assume that X is nonsingular. (a) Show that $X^{-1}AX$ is a companion matrix (3.3.12) for the characteristic polynomial of A. (b) If $R \in M_n$ is any given nonsingular upper triangular matrix and $S = XR$, show that $S^{-1}AS$ is in upper Hessenberg form. (c) Let $y \in \mathbf{C}^n$ and define $Y = [y \; A^*y \; (A^*)^2 y \; \ldots \; (A^*)^{n-1}y]$. Suppose that Y is nonsingular and that Y^*X can be written as LDU, in which L is lower triangular and U is upper triangular and nonsingular, and D is diagonal and nonsingular. Show that there exist nonsingular upper triangular matrices R and T such that $(XR)^{-1} = T^*Y^*$ and such that T^*Y^*AXR is tridiagonal and similar to A. (d) If $A \in M_n$ is Hermitian, use these ideas to describe an algorithm that produces a tridiagonal Hermitian matrix that is similar to A.

3.5.P6 Explain why the n, n entry of a given matrix in M_n has no influence on whether it has an LU factorization, or has one with L nonsingular, or has one with U nonsingular.

3.5.P7 Show that $C_n = [1/\max\{i, j\}] \in M_n(\mathbf{R})$ has an LU decomposition of the form $C_n = L_n L_n^T$, in which the entries of the lower triangular matrix L_n are $\ell_{ij} = 1/\max\{i, j\}$ for $i \geq j$. Conclude that $\det L_n = (1/n!)^2$.

3.5.P8 Show that the condition "$A[\{1, \ldots, j\}]$ is nonsingular for all $j = 1, \ldots, n$" in (3.5.6) may be replaced with the condition "$A[\{j, \ldots, n\}]$ is nonsingular for all $j = 1, \ldots, n$."

3.5.P9 Let $A \in M_n(\mathbf{R})$ be the symmetric tridiagonal matrix (0.9.10) with all main diagonal entries equal to $+2$ and all entries in the first superdiagonal and subdiagonal equal to -1. Consider

$$
L = \begin{bmatrix} 1 \\ -\frac{1}{2} & 1 \\ & -\frac{2}{3} & \ddots \\ & & \ddots & 1 \\ & & & -\frac{n-1}{n} & 1 \end{bmatrix}, \quad U = \begin{bmatrix} 2 & -1 \\ & \frac{3}{2} & -1 \\ & & \ddots & \ddots \\ & & & \frac{n}{n-1} & -1 \\ & & & & \frac{n+1}{n} \end{bmatrix}
$$

Show that $A = LU$ and $\det A = n + 1$. The eigenvalues of A are $\lambda_k = 4\sin^2\frac{k\pi}{2(n+1)}$, $k = 1, \ldots, n$ (see (1.4.P17)). Notice that $\lambda_1(A) \to 0$ and $\lambda_n(A) \to 4$ as $n \to \infty$, and $\det A = \lambda_1 \cdots \lambda_n \to \infty$.

3.5.P10 Suppose that $A \in M_n$ is symmetric and that all its leading principal submatrices are nonsingular. Show that there is a nonsingular lower triangular L such that $A = LL^T$, that is, A has an LU factorization in which $U = L^T$.

3.5.P11 Consider a permutation matrix $P = [p_{ij}] \in M_n$ corresponding to a permutation $\pi_1, \ldots, \pi_n$ of $1, \ldots, n$, that is, $p_{\pi_j, j} = 1$ for $j = 1, \ldots, n$ and all other entries are zero. Use the notation in (3.5.9) and define rank $P_{[\ell, 0]} = 0$, $\ell = 1, \ldots, n$. Show that $\pi_j = \min\{k \in \{1, \ldots, n\} : \text{rank } P_{[k, j]} = \text{rank } P_{[k, j-1]} + 1\}$, $j = 1, \ldots, n$. Conclude that the n^2 numbers rank $P_{[k, j]}$, $k, j \in \{1, \ldots, n\}$, uniquely determine P.

3.5.P12 Let $P \in M_n$ be a permutation matrix, partitioned as $P = \begin{bmatrix} P_{11} & P_{12} \\ P_{21} & P_{22} \end{bmatrix}$, so $P^{-1} = \begin{bmatrix} P_{11}^T & P_{12}^T \\ P_{21}^T & P_{22}^T \end{bmatrix}$. Provide details for the following argument to prove the law of complementary nullities (0.7.5) for P: nullity $P_{11} = $ number of zero columns in $P_{11} = $ number of ones in $P_{21} = $ number of zero rows in $P_{22} = $ nullity P_{22}^T.

3.5.P13 Provide details for the following approach to the law of complementary nullities (0.7.5), which deduces the general case from the (easy) permutation matrix case via an LPU factorization. (a) Let $A \in M_n$ be nonsingular. Partition A and A^{-1} conformally as $A = \begin{bmatrix} A_{11} & A_{12} \\ A_{21} & A_{22} \end{bmatrix}$ and $A^{-1} = \begin{bmatrix} B_{11} & B_{12} \\ B_{21} & B_{22} \end{bmatrix}$. The law of complementary nullities asserts that nullity $A_{11} = $ nullity B_{22}; this is what we seek to prove. (b) Let $A = LPU$ be an LPU factorization, so $A^{-1} = U^{-1}P^T L^{-1}$ is an LPU factorization. The permutation matrix factors in both factorizations are uniquely determined. Partition P as in the preceding problem, conformal to the partition of A. (c) nullity $A_{11} = $ nullity $P_{11} = $ nullity $P_{22}^T = $ nullity B_{22}.

Further Readings. Problem 3.5.P5 is adapted from [Ste], where additional information about numerical applications of LU factorizations may be found. Our discussion of LPU and $LPDU$ factorizations and triangular equivalence is adapted from L. Elsner, On some algebraic problems in connection with general eigenvalue algorithms, *Linear Algebra Appl.* 26 (1979) 123–138, which also discusses *lower triangular congruence* ($A = LBL^T$, in which L is lower triangular and nonsingular) of symmetric or skew-symmetric matrices.

CHAPTER 4
Hermitian Matrices, Symmetric Matrices, and Congruences

4.0 Introduction

Example 4.0.1. If $f: D \to \mathbf{R}$ is a twice continuously differentiable function on some domain $D \subset \mathbf{R}^n$, the real matrix

$$H(x) = [h_{ij}(x)] = \left[\frac{\partial^2 f(x)}{\partial x_i \partial x_j} \right] \in M_n$$

is the *Hessian* of f. It is a function of x and plays an important role in the theory of optimization because it can be used to determine if a critical point is a relative maximum or minimum; see (7.0). The only property of $H = H(x)$ that interests us here follows from the fact that the mixed partials are equal; that is,

$$\frac{\partial^2 f}{\partial x_i \, \partial x_j} = \frac{\partial^2 f}{\partial x_j \, \partial x_i} \quad \text{for all} \quad i, j = 1, \dots, n$$

Thus, the Hessian matrix of a real-valued twice continuously differentiable function is always a real symmetric matrix.

Example 4.0.2. Let $A = [a_{ij}] \in M_n$ have real or complex entries, and consider the quadratic form on $\mathbf{R}^n$ or $\mathbf{C}^n$ generated by A:

$$Q(x) = x^T A x = \sum_{i,j=1}^{n} a_{ij} x_i x_j = \sum_{i,j=1}^{n} \frac{1}{2}(a_{ij} + a_{ji}) x_i x_j = x^T \left[\frac{1}{2}(A + A^T) \right] x$$

Thus, A and $\frac{1}{2}(A + A^T)$ both generate the same quadratic form, and the latter matrix is symmetric. To study real or complex quadratic forms, therefore, it suffices to study only those forms generated by symmetric matrices. Real quadratic forms arise naturally in physics, for example, as an expression for the inertia of a physical body.

Example 4.0.3. Consider a second-order linear partial differential operator L defined by

$$Lf(x) = \sum_{i,j=1}^{n} a_{ij}(x) \frac{\partial^2 f(x)}{\partial x_i \, \partial x_j} \qquad (4.0.4)$$

The coefficient functions a_{ij} and the function f are assumed to be defined on the same domain $D \subset \mathbf{R}^n$, and f should be twice continuously differentiable on D. The operator L is associated in a natural way with a matrix $A(x) = [a_{ij}(x)]$, which need not be symmetric, but since the mixed partial derivatives of f are equal, we have

$$Lf = \sum_{i,j=1}^{n} a_{ij}(x) \frac{\partial^2 f}{\partial x_i \, \partial x_j} = \sum_{i,j=1}^{n} \frac{1}{2} \left[a_{ij}(x) \frac{\partial^2 f}{\partial x_i \, \partial x_j} + a_{ji}(x) \frac{\partial^2 f}{\partial x_j \, \partial x_i} \right]$$

$$= \sum_{i,j=1}^{n} \frac{1}{2} [a_{ij}(x) + a_{ij}(x)] \frac{\partial^2 f}{\partial x_i \, \partial x_j}$$

Thus, the symmetric matrix $\frac{1}{2}(A(x) + A(x)^T)$ yields the same operator L as the matrix $A(x)$. For the study of real or complex linear partial differential operators of the form (4.0.4), it suffices to consider only symmetric coefficient matrices.

Example 4.0.5. Consider an undirected graph Γ: a collection N of *nodes* $\{P_1, P_2, \ldots, P_n\}$ and a collection E of unordered pairs of nodes called *edges*, $E = \{\{P_{i_1}, P_{j_1}\}, \{P_{i_2}, P_{j_2}\}, \ldots\}$. Associated with the graph Γ is its *adjacency matrix* $A = [a_{ij}]$, in which

$$a_{ij} = \begin{cases} 1 & \text{if } \{P_i, P_j\} \in E \\ 0 & \text{otherwise} \end{cases}$$

Since Γ is undirected, its adjacency matrix is symmetric.

Example 4.0.6. Let $A = [a_{ij}] \in M_n(\mathbf{R})$ and consider the real bilinear form

$$Q(x, y) = y^T A x = \sum_{i,j=1}^{n} a_{ij} y_i x_j, \qquad x, y \in \mathbf{R}^n \qquad (4.0.7)$$

which reduces to the ordinary inner product when $A = I$. If we want to have $Q(x, y) = Q(y, x)$ for all x, y, then it is necessary and sufficient that $a_{ij} = a_{ji}$ for all $i, j = 1, \ldots, n$. To show this, it suffices to observe that if $x = e_j$ and $y = e_i$, then $Q(e_j, e_i) = a_{ij}$ and $Q(e_i, e_j) = a_{ji}$. Thus, symmetric real bilinear forms are naturally associated with symmetric real matrices.

Now let $A = [a_{ij}] \in M_n$ be a real or complex matrix, and consider the complex form

$$H(x, y) = y^* A x = \sum_{i,j=1}^{n} a_{ij} \bar{y}_i x_j, \qquad x, y \in \mathbf{C}^n \qquad (4.0.8)$$

which, like (4.0.7), reduces to the ordinary inner product when $A = I$. This form is no longer bilinear but is linear in the first variable and conjugate linear in the second variable: $H(ax, by) = a\bar{b}H(x, y)$. Such forms are called *sesquilinear*. If we want to have $H(x, y) = \overline{H(y, x)}$, then the same argument as in the previous case shows that it is necessary and sufficient to have $a_{ij} = \bar{a}_{ji}$; that is, $A = \bar{A}^T = A^*$, so A must be Hermitian.

The class of n-by-n complex Hermitian matrices is in many respects the natural generalization of the class of n-by-n real symmetric matrices. Of course, a real Hermitian matrix is a real symmetric matrix. The class of complex non-real symmetric matrices – although interesting in its own right – fails to have many important properties of the class of real symmetric matrices. In this chapter we study complex Hermitian and symmetric matrices; we indicate by specialization what happens in the real symmetric case.

4.1 Properties and characterizations of Hermitian matrices

Definition 4.1.1. *A matrix $A = [a_{ij}] \in M_n$ is Hermitian if $A = A^*$; it is skew Hermitian if $A = -A^*$.*

Some observations for $A, B \in M_n$ follow:

1. $A + A^*$, AA^*, and A^*A are Hermitian.
2. If A is Hermitian, then A^k is Hermitian for all $k = 1, 2, 3, \ldots$. If A is nonsingular as well, then A^{-1} is Hermitian.
3. If A and B are Hermitian, then $aA + bB$ is Hermitian for all real scalars a, b.
4. $A - A^*$ is skew Hermitian.
5. If A and B are skew Hermitian, then $aA + bB$ is skew Hermitian for all real scalars a, b.
6. If A is Hermitian, then iA is skew Hermitian.
7. If A is skew Hermitian, then iA is Hermitian.
8. $A = \frac{1}{2}(A + A^*) + \frac{1}{2}(A - A^*) = H(A) + S(A) = H(A) + iK(A)$, in which $H(A) = \frac{1}{2}(A + A^*)$ is the *Hermitian part* of A, $S(A) = \frac{1}{2}(A - A^*)$ is the *skew-Hermitian part* of A, and $K(A) = \frac{1}{2i}(A - A^*)$.
9. If A is Hermitian, the main diagonal entries of A are all real. To specify the n^2 elements of A, one may choose freely any n real numbers (for the main diagonal entries) and any $\frac{1}{2}n(n-1)$ complex numbers (for the off-diagonal entries).
10. If we write $A = C + iD$ with $C, D \in M_n(\mathbf{R})$ (real and imaginary parts of A), then A is Hermitian if and only if C is symmetric and D is skew symmetric.
11. A real symmetric matrix is a complex Hermitian matrix.

Theorem 4.1.2 (Toeplitz decomposition). *Each $A \in M_n$ can be written uniquely as $A = H + iK$, in which both H and K are Hermitian. It can also be written uniquely as $A = H + S$, in which H is Hermitian and S is skew Hermitian.*

Proof. Write $A = \frac{1}{2}(A + A^*) + i[\frac{1}{2i}(A - A^*)]$ and observe that both $H = \frac{1}{2}(A + A^*)$ and $K = \frac{1}{2i}(A - A^*)$ are Hermitian. For the uniqueness assertion, observe that if

$A = E + iF$ with both E and F Hermitian, then

$$2H = A + A^* = (E + iF) + (E + iF)^* = E + iF + E^* - iF^* = 2E$$

so $E = H$. Similarly, one shows that $F = K$. The assertions about the representation $A = H + S$ are proved in the same way. $\square$

The foregoing observations suggest that if one thinks of M_n as being analogous to the complex numbers, then the Hermitian matrices are analogous to the real numbers. The analog of the operation of complex conjugation in $\mathbf{C}$ is the *operation (conjugate transpose) on M_n. A real number is a complex number z such that $z = \bar{z}$; a Hermitian matrix is a matrix $A \in M_n$ such that $A = A^*$. Just as every complex number z can be written uniquely as $z = s + it$ with $s, t \in \mathbf{R}$, every complex matrix A can be written uniquely as $A = H + iK$ with H and K Hermitian. There are some further properties that strengthen this analogy.

Theorem 4.1.3. *Let $A \in M_n$ be Hermitian. Then*

(a) *$x^* A x$ is real for all $x \in \mathbf{C}^n$*
(b) *the eigenvalues of A are real*
(c) *$S^* A S$ is Hermitian for all $S \in M_n$*

Proof. Compute $\overline{(x^* A x)} = (x^* A x)^* = x^* A^* x = x^* A x$, so $x^* A x$ equals its complex conjugate and hence is real. If $A x = \lambda x$ and $x^* x = 1$, then $\lambda = \lambda x^* x = x^* \lambda x = x^* A x$ is real by (a). Finally, $(S^* A S)^* = S^* A^* S = S^* A S$, so $S^* A S$ is always Hermitian. $\square$

Exercise. What does each of the properties of a Hermitian matrix $A \in M_n$ in the preceding theorem say when $n = 1$?

Each of the properties in (4.1.3) is actually (almost) a characterization of Hermitian matrices.

Theorem 4.1.4. *Let $A = [a_{ij}] \in M_n$ be given. Then A is Hermitian if and only if at least one of the following conditions is satisfied:*

(a) *$x^* A x$ is real for all $x \in \mathbf{C}^n$*
(b) *A is normal and has only real eigenvalues*
(c) *$S^* A S$ is Hermitian for all $S \in M_n$*

Proof. It suffices to prove only the sufficiency of each condition. If $x^* A x$ is real for all $x \in \mathbf{C}^n$, then $(x + y)^* A (x + y) = (x^* A x + y^* A y) + (x^* A y + y^* A x)$ is real for all $x, y \in \mathbf{C}^n$. Since $x^* A x$ and $y^* A y$ are real by assumption, we conclude that $x^* A y + y^* A x$ is real for all $x, y \in \mathbf{C}^n$. If we choose $x = e_k$ and $y = e_j$, then $x^* A y + y^* A x = a_{kj} + a_{jk}$ is real, so $\operatorname{Im} a_{kj} = - \operatorname{Im} a_{jk}$. If we choose $x = i e_k$ and $y = e_j$, then $x^* A y + y^* A x = -i a_{kj} + i a_{jk}$ is real, so $\operatorname{Re} a_{kj} = \operatorname{Re} a_{jk}$. Combining the identities for the real and imaginary parts of a_{kj} and a_{jk} leads to the identity $a_{kj} = \bar{a}_{jk}$, and since j, k are arbitrary, we conclude that $A = A^*$.

If A is normal, it is unitarily diagonalizable, so $A = U \Lambda U^*$ with $\Lambda = \operatorname{diag}(\lambda_1, \lambda_2, \ldots, \lambda_n)$. In general, we have $A^* = U \bar{\Lambda} U^*$, but if Λ is real, we have $A^* = U \Lambda U^* = A$.

Condition (c) implies that A is Hermitian by choosing $S = I$. $\square$

Since a Hermitian matrix is normal ($AA^* = A^2 = A^*A$), all the results about normal matrices in Chapter 2 apply to Hermitian matrices. For example, eigenvectors associated with distinct eigenvalues are orthogonal, there is an orthonormal basis of eigenvectors, and Hermitian matrices are unitarily diagonalizable.

For convenient reference, we restate the spectral theorem for Hermitian matrices (2.5.6).

Theorem 4.1.5. *A matrix $A \in M_n$ is Hermitian if and only if there is a unitary $U \in M_n$ and a real diagonal $\Lambda \in M_n$ such that $A = U \Lambda U^*$. Moreover, A is real and Hermitian (that is, real symmetric) if and only if there is a real orthogonal $P \in M_n$ and a real diagonal $\Lambda \in M_n$ such that $A = P \Lambda P^T$.*

Although a real linear combination of Hermitian matrices is always Hermitian, a complex linear combination of Hermitian matrices need not be Hermitian. For example, if A is Hermitian, iA is Hermitian only if $A = 0$. Furthermore, if A and B are Hermitian, then $(AB)^* = B^*A^* = BA$, so AB is Hermitian if and only if A and B commute.

One of the most famous results about commuting Hermitian matrices (there is an important generalization to operators in quantum mechanics) is the following special case of (2.5.5).

Theorem 4.1.6. *Let $\mathcal{F}$ be a given nonempty family of Hermitian matrices. There exists a unitary U such that $U A U^*$ is diagonal for all $A \in \mathcal{F}$ if and only if $AB = BA$ for all $A, B \in \mathcal{F}$.*

A Hermitian matrix A has the property that A is *equal* to A^*. One way to generalize the notion of a Hermitian matrix is to consider the class of matrices such that A is *similar* to A^*. The following theorem extends (3.4.1.7) and characterizes this class in several ways, the first of which says that such matrices must be similar, but not necessarily unitarily similar, to a real, but not necessarily diagonal, matrix.

Theorem 4.1.7. *Let $A \in M_n$ be given. The following statements are equivalent:*

(a) *A is similar to a real matrix.*
(b) *A is similar to A^*.*
(c) *A is similar to A^* via a Hermitian similarity transformation.*
(d) *$A = HK$, in which $H, K \in M_n$ are Hermitian and at least one factor is nonsingular.*
(e) *$A = HK$, in which $H, K \in M_n$ are Hermitian.*

Proof. First note that (a) and (b) are equivalent: Every complex matrix is similar to its transpose (3.2.3.1), so A is similar to $A^* = \bar{A}^T$ if and only if A is similar to $\bar{A}$ if and only if A is similar to a real matrix (3.4.1.7).

To verify that (b) implies (c), suppose that there is a nonsingular $S \in M_n$ such that $S^{-1}AS = A^*$. Let $\theta \in \mathbf{R}$ and let $T = e^{i\theta}S$. Observe that $T^{-1}AT = A^*$. Thus, $AT = TA^*$ or, equivalently, $AT^* = T^*A^*$. Adding these two identities produces the identity $A(T + T^*) = (T + T^*)A^*$. If $T + T^*$ were nonsingular, we could conclude that A is similar to A^* via the Hermitian matrix $T + T^*$, so it suffices to show that there is some θ such that $T + T^*$ is nonsingular. The matrix $T + T^*$ is nonsingular if and only if $T^{-1}(T + T^*) = I + T^{-1}T^*$ is nonsingular, if and only if $-1 \notin \sigma(T^{-1}T^*)$. However, $T^{-1}T^* = e^{-2i\theta} S^{-1}S^*$, so we may choose any θ such that $-e^{2i\theta} \notin \sigma(S^{-1}S^*)$.

Now assume (c) and write $R^{-1}AR = A^*$ with $R \in M_n$ nonsingular and Hermitian. Then $R^{-1}A = A^*R^{-1}$ and $A = R(A^*R^{-1})$. But $(A^*R^{-1})^* = R^{-1}A = A^*R^{-1}$, so A is the product of the two Hermitian matrices R and A^*R^{-1}, and R is nonsingular.

If $A = HK$ with H and K Hermitian and H nonsingular, then $H^{-1}AH = KH = (HK)^* = A^*$. The argument is similar if K is nonsingular. Thus, (d) is equivalent to (b).

Certainly (d) implies (e); we now show that (e) implies (a). If $A = HK$ with H and K Hermitian and both singular, consider $U^*AU = (U^*HU)(U^*KU)$, in which $U \in M_n$ is unitary, $U^*HU = \begin{bmatrix} D & 0 \\ 0 & 0 \end{bmatrix}$, and $D \in M_k$ is nonsingular, real, and diagonal. Partition $U^*KU = \begin{bmatrix} K' & \star \\ \star & \star \end{bmatrix}$ conformally to U^*HU and compute

$$U^*AU = (U^*HU)(U^*KU) = \begin{bmatrix} D & 0 \\ 0 & 0 \end{bmatrix}\begin{bmatrix} K' & \star \\ \star & \star \end{bmatrix} = \begin{bmatrix} DK' & \star \\ 0 & 0 \end{bmatrix}$$

The block $DK' \in M_k$ is the product of two Hermitian matrices, one of which is nonsingular, so the equivalence of (d), (b), and (a) ensures that it is similar to a real matrix. Corollary 3.4.1.8 now tells us that U^*AU (and hence also A) is similar to a real matrix. $\qquad\Box$

The characterization (4.1.4(a)) can be refined by considering Hermitian forms that take only positive (or nonnegative) values.

Theorem 4.1.8. *Let $A \in M_n$ be given. Then x^*Ax is real and positive (respectively, x^*Ax is real and nonnegative) for all nonzero $x \in \mathbb{C}^n$ if and only if A is Hermitian and all of its eigenvalues are positive (respectively, nonnegative).*

Proof. If x^*Ax is real and positive (respectively, real and nonnegative) whenever $x \neq 0$, then x^*Ax is real for all $x \in \mathbb{C}^n$, so (4.1.4(a)) ensures that A is Hermitian. Moreover, $\lambda = u^*(\lambda u) = u^*Au$ if $u \in \mathbb{C}^n$ is a unit eigenvector of A associated with an eigenvalue λ, so the hypothesis ensures that $\lambda > 0$ (respectively, $\lambda \geq 0$). Conversely, if A is Hermitian and has only positive (respectively, nonnegative) eigenvalues, then (4.2.5) ensures that $A = U\Lambda U^*$, in which the columns of the unitary matrix $U = [u_1 \ \ldots \ u_n]$ are eigenvectors of A associated with the positive (respectively, nonnegative) diagonal entries of $\Lambda = \text{diag}(\lambda_1, \ldots, \lambda_n)$. Then $x^*Ax = x^*U\Lambda U^*x = (U^*x)^*\Lambda(U^*x) = \sum_{k=1}^{n} \lambda_k |u_k^*x|^2$ is always nonnegative; it is positive if all $\lambda_k > 0$ and some $u_k^*x \neq 0$, which is certainly the case if $x \neq 0$. $\qquad\Box$

Definition 4.1.9. *A matrix $A \in M_n$ is* positive definite *if x^*Ax is real and positive for all nonzero $x \in \mathbb{C}^n$; it is* positive semidefinite *if x^*Ax is real and nonnegative for all nonzero $x \in \mathbb{C}^n$; it is* indefinite *if x^*Ax is real for all $x \in \mathbb{C}^n$ and there are vectors $y, z \in \mathbb{C}^n$ such that $y^*Ay < 0 < z^*Az$.*

Exercise. Let $A \in M_n$ and let $B = A^*A$. Show that B is positive semidefinite in two ways: (a) replace A by its singular value decomposition; and (b) observe that $x^*Bx = \|Ax\|_2^2$.

The preceding theorem says that a complex matrix is positive definite (respectively, semidefinite) if and only if it is Hermitian and all its eigenvalues are positive (respectively, nonnegative). Some authors include in the definition of positive definiteness or semidefiniteness an assumption that the matrix is Hermitian. The preceding theorem shows that this assumption is unnecessary for complex matrices and vectors, but it is harmless. However, the situation is different if one considers real matrices and the real quadratic forms that they generate. If $A \in M_n(\mathbf{R})$ and $x \in \mathbf{R}^n$, then $x^T A x = \frac{1}{2} x^T (A + A^T) x$, so assuming that $x^T A x > 0$ or $x^T A x \geq 0$ for all nonzero $x \in \mathbf{R}^n$ imposes a condition only on the symmetric part of A; its skew-symmetric part is unrestricted. The real analog of the preceding theorem must incorporate a symmetry assumption.

Theorem 4.1.10. *Let* $A \in M_n(\mathbf{R})$ *be symmetric. Then* $x^T A x > 0$ *(respectively,* $x^T A x \geq 0$*) for all nonzero* $x \in \mathbf{R}^n$ *if and only if every eigenvalue of* A *is positive (respectively, nonnegative).*

Proof. Since A is Hermitian, it suffices to show that $z^* A z > 0$ (respectively, $z^* A z \geq 0$) whenever $z = x + iy \in \mathbf{C}^n$ with $x, y \in \mathbf{R}^n$ and at least one is nonzero. Since $= (y^T A x)^T = x^T A y$, we have

$$z^* A z = (x + iy)^* A (x + iy) = x^T A x + y^T A y + i(x^T A y - y^T A x)$$
$$= x^T A x + y^T A y$$

which is positive (respectively, nonnegative) if at least one of x and y is nonzero. $\square$

Definition 4.1.11. *A symmetric matrix* $A \in M_n(\mathbf{R})$ *is* positive definite *if* $x^T A x > 0$ *for all nonzero* $x \in \mathbf{R}^n$*; it is* positive semidefinite *if* $x^T A x \geq 0$ *for all nonzero* $x \in \mathbf{R}^n$*; it is* indefinite *if there are vectors* $y, z \in \mathbf{R}^n$ *such that* $y^T A y < 0 < z^T A z$*.*

Exercise. Explain why a positive semidefinite matrix is positive definite if and only if it is nonsingular.

Our final general observation about Hermitian matrices is that $A \in M_n$ is Hermitian if and only if it can be written as $A = B - C$, in which $B, C \in M_n$ are positive semidefinite. Half of this assertion is evident; the other half relies on the following definition.

Definition 4.1.12. *Let* $A \in M_n$ *be Hermitian and let* $\lambda_1 \geq \cdots \geq \lambda_n$ *be its nonincreasingly ordered eigenvalues. Let* $\Lambda = \mathrm{diag}(\lambda_1, \ldots, \lambda_n)$*, and let* $U \in M_n$ *be unitary and such that* $A = U \Lambda U^*$*. Let* $\lambda_i^+ = \max\{\lambda_i, 0\}$ *and let* $\lambda_i^- = \min\{\lambda_i, 0\}$*, both for* $i = 1, \ldots, n$*. Let* $\Lambda_+ = \mathrm{diag}(\lambda_1^+, \ldots, \lambda_n^+)$ *and let* $A_+ = U \Lambda_+ U^*$*; let* $\Lambda_- = \mathrm{diag}(\lambda_1^-, \ldots, \lambda_n^-)$ *and let* $A_- = -U \Lambda_- U^*$*. The matrix* A_+ *is called the* positive semidefinite part *of* A*.*

Proposition 4.1.13. *Let* $A \in M_n$ *be Hermitian. Then* $A = A_+ - A_-$*; each of* A_+ *and* A_- *is positive semidefinite;* A_+ *and* A_- *commute;* $\mathrm{rank}\, A = \mathrm{rank}\, A_+ + \mathrm{rank}\, A_-$*;* $A_+ A_- = A_- A_+ = 0$*; and* A_- *is the positive semidefinite part of* $-A$*.*

Exercise. Verify the statements in the preceding proposition.

Problems

4.1.P1 Show that every principal submatrix of a Hermitian matrix is Hermitian. Is this true for skew-Hermitian matrices? For normal matrices? Prove or give a counterexample.

4.1.P2 If $A \in M_n$ is Hermitian and $S \in M_n$, show that SAS^* is Hermitian. What about SAS^{-1} (if S is nonsingular)?

4.1.P3 Let $A, B \in M_n$ be Hermitian. Show that A and B are similar if and only if they are unitarily similar.

4.1.P4 Verify the statements 1–9 following (4.1.1).

4.1.P5 Sometimes one can show that a matrix has only real eigenvalues by showing that it is similar to a Hermitian matrix. Let $A = [a_{ij}] \in M_n(\mathbf{R})$ be tridiagonal. Suppose that $a_{i,i+1}a_{i+1,i} > 0$ for all $i = 1, 2, \ldots, n - 1$. Show that there is a real diagonal matrix D with positive diagonal entries such that DAD^{-1} is symmetric, and conclude that A has only real eigenvalues. Consider $\begin{bmatrix} 0 & 1 \\ -1 & 0 \end{bmatrix}$ and explain why the assumption on the signs of the off-diagonal entries is necessary. Use a limit argument to show that the conclusion that the eigenvalues are real continues to hold if $a_{i,i+1}a_{i+1,i} \geq 0$ for all $i = 1, 2, \ldots, n - 1$.

4.1.P6 Let $A = [a_{ij}]$, $B = [b_{ij}] \in M_n$ be given. (a) Show that $x^*Ax = x^*Bx$ for all $x \in \mathbf{C}^n$ if and only if $A = B$, that is, a complex matrix is determined by the sesquilinear form that it generates.

4.1.P7 Let $A, B \in M_n(\mathbf{F})$ be given, with $n \geq 2$ and $\mathbf{F} = \mathbf{R}$ or $\mathbf{C}$. Show that $x^T Ax = 0$ for all $x \in \mathbf{F}^n$ if and only if $A^T = -A$. Give an example to show that A and B need not be equal if $x^T Ax = x^T Bx$ for all $x \in \mathbf{F}^n$, so a real or complex matrix is not determined by the quadratic form that it generates.

4.1.P8 Let $A = \begin{bmatrix} 1 & 1 \\ 0 & 1 \end{bmatrix}$ and show that $|x^*Ax| = |x^*A^Tx|$ for all $x \in \mathbf{C}^2$. Conclude that $A \in M_n$ is not determined by the absolute value of the Hermitian form that it generates.

4.1.P9 Show that $A \in M_n$ is almost determined by the absolute value of the sesquilinear form that it generates, in the following sense: If $A, B \in M_n$ are given, show that $|x^*Ay| = |x^*By|$ for all $x, y \in \mathbf{C}^n$ if and only if $A = e^{i\theta} B$ for some $\theta \in \mathbf{R}$.

4.1.P10 Show that $A \in M_n$ is Hermitian if and only if iA is skew Hermitian. Let $B \in M_n$ be skew Hermitian. Show that (a) the eigenvalues of B are pure imaginary; (b) the eigenvalues of B^2 are real and nonpositive and are all zero if and only if $B = 0$.

4.1.P11 Let $A, B \in M_n$ be Hermitian. Explain why $AB - BA$ is skew Hermitian and deduce from (4.1.P10) that $\mathrm{tr}(AB)^2 \leq \mathrm{tr}(A^2 B^2)$ with equality if and only if $AB = BA$.

4.1.P12 Let $A \in M_n$ be given. If A is Hermitian, explain why the rank of A is equal to the number of its nonzero eigenvalues, but that this need not be true for non-Hermitian matrices. If A is normal, show that $\mathrm{rank}\, A \geq \mathrm{rank}\, H(A)$ with equality if and only if A has no nonzero imaginary eigenvalues. Can the normality hypothesis be dropped?

4.1.P13 Suppose that $A \in M_n$ is nonzero. (a) Show that $\mathrm{rank}\, A \geq |\mathrm{tr}\, A|^2/(\mathrm{tr}\, A^*A)$ with equality if and only if $A = aH$ for some nonzero $a \in \mathbf{C}$ and some Hermitian projection H. (b) If A is normal, explain why $\mathrm{rank}\, A \geq |\mathrm{tr}\, H(A)|^2/(\mathrm{tr}\, H(A)^2)$, so $\mathrm{rank}\, A \geq |\mathrm{tr}\, A|^2/(\mathrm{tr}\, A^2)$ if A is Hermitian.

4.1.P14 Show that $A = e^{i\theta} A^*$ for some $\theta \in \mathbf{R}$ if and only if $e^{-i\theta/2} A$ is Hermitian. What does this say for $\theta = \pi$? For $\theta = 0$? Explain why the class of skew-Hermitian matrices may be thought of as one of infinitely many classes of *essentially Hermitian* matrices (0.2.5), and describe the structure of each such class.

4.1.P15 Explain why $A \in M_n$ is similar to a Hermitian matrix if and only if it is diagonalizable and has real eigenvalues. See (7.6.P1) for additional equivalent conditions.

4.1.P16 For any $s, t \in \mathbf{R}$, show that $\max\{|s|, |t|\} = \frac{1}{2}(|s + t| + |s - t|)$. For any Hermitian $A \in M_2$, deduce that $\rho(A) = \frac{1}{2}|\operatorname{tr} A| + \frac{1}{2}(\operatorname{tr} A^2 - 2 \det A)^{1/2}$.

4.1.P17 Let $A = [a_{ij}] \in M_2$ be Hermitian and have eigenvalues λ_1 and λ_2. Show that $(\lambda_1 - \lambda_2)^2 = (a_{11} - a_{22})^2 + 4|a_{12}|^2$ and deduce that spread $A \geq 2|a_{12}|$ with equality if and only if $a_{11} = a_{22}$.

4.1.P18 Let $A \in M_n$ be given. Show that A is Hermitian if and only if $A^2 = A^* A$.

4.1.P19 Let $A \in M_n$ be a projection ($A^2 = A$). One says that A is a *Hermitian projection* if A is Hermitian; A is an *orthogonal projection* if the range of A is orthogonal to its null space. Use (4.1.5) and (4.1.4) to show that A is a Hermitian projection if and only if it is an orthogonal projection.

4.1.P20 Let $A \in M_n$ be a projection. Show that A is Hermitian if and only if $AA^*A = A$.

4.1.P21 Let $n \geq 2$, and let $x, y \in \mathbf{C}^n$ and $z_1, \ldots, z_n \in \mathbf{C}$ be given. Consider the Hermitian matrix $A = xy^* + yx^* = [x \ y][y \ x]^*$ and the skew-Hermitian matrix $B = xy^* - yx^* = [x \ -y][y \ x]^*$. (a) Show that the eigenvalues of A are $\operatorname{Re} y^*x \pm (\|x\|^2 \|y\|^2 - (\operatorname{Im} y^*x)^2)^{1/2}$ (one positive and one negative if x and y are linearly independent) together with $n - 2$ zero eigenvalues. (b) Show that the eigenvalues of B are $i(\operatorname{Im} y^*x \pm (\|x\|^2 \|y\|^2 - (\operatorname{Re} y^*x)^2)^{1/2})$. (c) What are the eigenvalues of A and B if x and y are real? (d) Show that the eigenvalues of $C = [z_i + \bar{z}_j] \in M_n$ are $\operatorname{Re} \sum_{i=1}^n z_i \pm (n^2 \sum_{i=1}^n |z_i|^2 - (\operatorname{Im} \sum_{i=1}^n z_i)^2)^{1/2}$ (one positive and one negative if not all z_i are equal) together with $n - 2$ zero eigenvalues. (e) What are the eigenvalues of $D = [z_i - \bar{z}_j] \in M_n$? (f) What are the eigenvalues of C and D if all z_i are real? Check your answers using the special cases (1.3.25) and (1.3.26).

4.1.P22 In the definition (4.1.12) of the positive semidefinite part A_+ of a Hermitian matrix A, the diagonal factor Λ_+ is uniquely determined, but the unitary factor U is not. Why? Use the uniqueness portion of (2.5.4) to explain why A_+ (and hence also A_-) is nevertheless well-defined.

4.1.P23 Let $A, B \in M_n$ be Hermitian. Show that (a) AB is Hermitian if and only if A commutes with B; (b) $\operatorname{tr} AB$ is real.

4.1.P24 Let $A \in M_n$ be Hermitian. Explain why (a) adj A is Hermitian; (b) adj A is positive semidefinite if A positive semidefinite; (c) adj A is positive definite if A is positive definite.

4.1.P25 Let $A \in M_n$ be Hermitian and let $r \in \{1, \ldots, n\}$. Explain why the compound matrix $C_r(A)$ is Hermitian. If A is positive definite (respectively, positive semidefinite), explain why $C_r(A)$ is positive definite (respectively, positive semidefinite).

4.1.P26 Show that a Hermitian $P \in M_n$ is a projection if and only if there is a unitary $U \in M_n$ such that $P = U(I_k \oplus 0_{n-k})U^*$, in which $0 \leq k \leq n$.

4.1.P27 Let $A, P \in M_n$ and suppose that P is a Hermitian projection that is neither 0 nor I. Show that A commutes with P if and only if A is unitarily similar to $B \oplus C$, in which $B \in M_k$, $C \in M_{n-k}$, and $1 \leq k \leq n - 1$.

4.1.P28 If $A \in M_n$ is unitarily similar to $B \oplus C$, in which $B \in M_k$, $C \in M_{n-k}$, and $1 \leq k \leq n - 1$, then A is said to be *unitarily reducible*; otherwise, A is *unitarily irreducible*.

Explain why A is unitarily irreducible if and only if the only Hermitian projections that commute with A are the zero and identity matrices.

4.1.P29 Let $A \in M_n$ be either Hermitian or real symmetric. Explain why A is indefinite if and only if it has at least one positive eigenvalue and at least one negative eigenvalue.

4.1.P30 Let $A \in M_n$ be Hermitian and suppose that rank $A = r > 0$. We know that A is rank principal, so it has a nonzero principal minor of size r; see (0.7.6). But we can say more: Use (4.1.5) to write $A = U \Lambda U^*$, in which $U \in M_n$ is unitary and $\Lambda = \Lambda_r \oplus 0_{n-r}$ is real diagonal. (a) Explain why $\det \Lambda_r$ is real and nonzero. (b) Partition $U = [V \; U_2]$, in which $V \in M_{n,r}$. Explain why $A = V \Lambda_r V^*$; this is a full-rank factorization of A. (c) Let $\alpha, \beta \subseteq \{1, \ldots, n\}$ be index sets of cardinality r, and let $V[\alpha, \varnothing^c] = V_\alpha$. Explain why $A[\alpha, \beta] = V_\alpha \Lambda_r V_\beta^*$; $\det A[\alpha] = |\det V_\alpha|^2 \det \Lambda_r$; and $\det A[\alpha] \det A[\beta] = \det A[\alpha, \beta] \det A[\beta, \alpha]$. (d) Why does the factorization $A[\alpha] = V_\alpha \Lambda_r V_\alpha^*$ ensure that A has at least one nonzero principal minor of size r, and why does every such minor have the same sign? See (2.5.P48) for a version of this result if A is normal.

Notes. Perhaps the earliest version of (4.1.5) is due to A. Cauchy in 1829. He showed that the eigenvalues of a real symmetric matrix A are real and that the real quadratic form $Q(x_1, \ldots, x_n) = x^T A x$ can be transformed by a real orthogonal change of variables into a sum of real multiples of squares of real linear forms.

4.2 Variational characterizations and subspace intersections

Since the eigenvalues of a Hermitian matrix $A \in M_n$ are real, we may (and do) adopt the convention that they are always arranged in algebraically nondecreasing order:

$$\lambda_{\min} = \lambda_1 \le \lambda_2 \le \cdots \le \lambda_{n-1} \le \lambda_n = \lambda_{\max} \tag{4.2.1}$$

When several Hermitian matrices are under discussion, it is convenient to denote their respective eigenvalues (always algebraically ordered as in (4.2.1)) by $\{\lambda_i(A)\}_{i=1}^n$, $\{\lambda_i(B)\}_{i=1}^n$, and so on.

The smallest and largest eigenvalues of a Hermitian matrix A can be characterized as the solutions to minimum and maximum problems involving the *Rayleigh quotient* $x^* A x / x^* x$. The basic facts supporting the following *Rayleigh quotient theorem* are as follows: for a Hermitian $A \in M_n$, eigenvectors of A associated with distinct eigenvalues are *automatically orthogonal*, the span of any nonempty list of eigenvectors of A associated with a single eigenvalue λ contains an *orthonormal* basis of eigenvectors associated with λ, and there is an orthonormal basis of $\mathbf{C}^n$ consisting of eigenvectors of A.

Theorem 4.2.2 (Rayleigh). *Let $A \in M_n$ be Hermitian, let the eigenvalues of A be ordered as in (4.2.1), let $i_1, \ldots, i_k$ be given integers with $1 \le i_1 < \cdots < i_k \le n$, let $x_{i_1}, \ldots, x_{i_k}$ be orthonormal and such that $A x_{i_p} = \lambda_{i_p} x_{i_p}$ for each $p = 1, \ldots, k$, and let $S = \mathrm{span}\{x_{i_1}, \ldots, x_{i_k}\}$. Then*

(a)

$$\lambda_{i_1} = \min_{\{x:0\neq x\in S\}} \frac{x^*Ax}{x^*x} = \min_{\{x:x\in S \text{ and } \|x\|_2=1\}} x^*Ax$$

$$\leq \max_{\{x:x\in S \text{ and } \|x\|_2=1\}} x^*Ax = \max_{\{x:0\neq x\in S\}} \frac{x^*Ax}{x^*x} = \lambda_{i_k}$$

(b) $\lambda_{i_1} \leq x^*Ax \leq \lambda_{i_k}$ *for any unit vector* $x \in S$, *with equality in the right-hand (respectively, left-hand) inequality if and only if* $Ax = \lambda_{i_k}x$ *(respectively,* $Ax = \lambda_{i_1}x$*)*

(c) $\lambda_{\min} \leq x^*Ax \leq \lambda_{\max}$ *for any unit vector* $x \in \mathbf{C}^n$, *with equality in the right-hand (respectively, left-hand) inequality if and only if* $Ax = \lambda_{\max}x$ *(respectively,* $Ax = \lambda_{\min}x$*); moreover,*

$$\lambda_{\max} = \max_{x\neq 0} \frac{x^*Ax}{x^*x} \text{ and } \lambda_{\min} = \min_{x\neq 0} \frac{x^*Ax}{x^*x}$$

Proof. If $x \in S$ is nonzero, then $\xi = x/\|x\|_2$ is a unit vector and $x^*Ax/x^*x = x^*Ax/\|x\|_2^2 = \xi^*A\xi$. For any given unit vector $x \in S$, there are scalars $\alpha_1, \ldots, \alpha_k$ such that $x = \alpha_1 x_{i_1} + \cdots + \alpha_k x_{i_k}$; orthonormality ensures that $1 = x^*x = \sum_{p,q=1}^k \bar{\alpha}_p \alpha_q x_{i_p}^* x_{i_q} = |\alpha_1|^2 + \cdots + |\alpha_k|^2$. Then

$$x^*Ax = (\alpha_1 x_{i_1} + \cdots + \alpha_k x_{i_k})^*(\alpha_1\lambda_{i_1} x_{i_1} + \cdots + \alpha_k\lambda_{i_k} x_{i_k})$$

$$= |\alpha_1|^2\lambda_{i_1} + \cdots + |\alpha_k|^2\lambda_{i_k}$$

is a convex combination of the real numbers $\lambda_{i_1}, \ldots, \lambda_{i_k}$, so it lies between the smallest of these numbers (λ_{i_1}) and the largest (λ_{i_k}); see Appendix B. Moreover, $x^*Ax = |\alpha_1|^2\lambda_{i_1} + \cdots + |\alpha_k|^2\lambda_{i_k} = \lambda_{i_k}$ if and only if $\alpha_p = 0$ whenever $\lambda_{i_p} \neq \lambda_{i_k}$, if and only if $x = \sum_{\{p:\lambda_{i_p}=\lambda_{i_k}\}} \alpha_p x_{i_p}$, if and only if $x \in S$ is an eigenvector of A associated with the eigenvalue λ_{i_k}. A similar argument establishes the case of equality for $x^*Ax = \lambda_{i_1}$. The assertions in (c) follow from those in (b) since $S = \mathbf{C}^n$ if $k = n$. $\square$

The geometrical interpretation of (4.2.2(c)) is that $\lambda_{\max}$ is the maximum (and $\lambda_{\min}$ is the minimum) of the continuous real-valued function $f(x) = x^*Ax$ on the unit sphere in $\mathbf{C}^n$, a compact set.

Exercise. Let $A \in M_n$ be Hermitian, let $x \in \mathbf{C}^n$ be nonzero, and let $\alpha = x^*Ax/x^*x$. Explain why there is at least one eigenvalue of A in the interval $(-\infty, \alpha]$ and at least one eigenvalue of A in the interval $[\alpha, \infty)$.

In our discussion of eigenvalues of Hermitian matrices, we have several opportunities to invoke the following basic observation about subspace intersections.

Lemma 4.2.3 (Subspace intersection). *Let* $S_1, \ldots, S_k$ *be given subspaces of* $\mathbf{C}^n$. *If* $\delta = \dim S_1 + \cdots + \dim S_k - (k-1)n \geq 1$, *there are orthonormal vectors* $x_1, \ldots, x_\delta$ *such that* $x_1, \ldots, x_\delta \in S_i$ *for every* $i = 1, \ldots, k$. *In particular,* $S_1 \cap \cdots \cap S_k$ *contains a unit vector.*

Proof. See (0.1.7). The set $S_1 \cap \cdots \cap S_k$ is a subspace, and the stated inequality ensures that $\dim(S_1 \cap \cdots \cap S_k) \geq \delta \geq 1$. Let $x_1, \ldots, x_\delta$ be any δ elements of an orthonormal basis of $S_1 \cap \cdots \cap S_k$. $\square$

Exercise. Prove (0.1.7.3) by induction, starting with (0.1.7.2).

Exercise. Let S_1, S_2, and S_3 be given subspaces of $\mathbf{C}^n$. If $\dim S_1 + \dim S_2 \geq n + 1$, explain why $S_1 \cap S_2$ contains a unit vector. If $\dim S_1 + \dim S_2 + \dim S_3 \geq 2n + 2$, explain why there are two unit vectors $x, y \in S_1 \cap S_2 \cap S_3$ such that $x^* y = 0$.

Inequalities resulting from variational characterizations are often the result of the simple observation that, for a suitable real-valued function f and a nonempty set S, $\sup\{f(x) : x \in S\}$ does not decrease (and $\inf\{f(x) : x \in S\}$ does not increase) if S is replaced by a larger set $S' \supset S$.

Lemma 4.2.4. *Let f be a bounded real-valued function on a set S, and suppose that S_1 and S_2 are sets such that S_1 is nonempty and $S_1 \subset S_2 \subset S$. Then*

$$\sup_{x \in S_2} f(x) \geq \sup_{x \in S_1} f(x) \geq \inf_{x \in S_1} f(x) \geq \inf_{x \in S_2} f(x)$$

In many eigenvalue inequalities for a Hermitian matrix A, lower bounds on the eigenvalues of A follow from applying upper bounds to the eigenvalues of $-A$. The following observation is useful in this regard.

Observation 4.2.5. *Let $A \in M_n$ be Hermitian and have eigenvalues $\lambda_1(A) \leq \cdots \leq \lambda_n(A)$, ordered as in (4.2.1). Then the ordered eigenvalues of $-A$ are $-\lambda_n(A) \leq \cdots \leq -\lambda_1(A)$, that is, $\lambda_k(-A) = -\lambda_{n-k+1}(A)$, $k = 1, \ldots, n$.*

Our proof of the celebrated *Courant–Fischer min-max theorem* uses the preceding two lemmas in concert with the Rayleigh quotient theorem.

Theorem 4.2.6 (Courant–Fischer). *Let $A \in M_n$ be Hermitian and let $\lambda_1 \leq \cdots \leq \lambda_n$ be its algebraically ordered eigenvalues. Let $k \in \{1, \ldots, n\}$ and let S denote a subspace of $\mathbf{C}^n$. Then*

$$\lambda_k = \min_{\{S : \dim S = k\}} \max_{\{x : 0 \neq x \in S\}} \frac{x^* A x}{x^* x} \tag{4.2.7}$$

and

$$\lambda_k = \max_{\{S : \dim S = n-k+1\}} \min_{\{x : 0 \neq x \in S\}} \frac{x^* A x}{x^* x} \tag{4.2.8}$$

Proof. Let $x_1, \ldots, x_n \in \mathbf{C}^n$ be orthonormal and such that $A x_i = \lambda_i x_i$ for each $i = 1, \ldots, n$. Let S be any k-dimensional subspace of $\mathbf{C}^n$ and let $S' = \text{span}\{x_k, \ldots, x_n\}$. Then

$$\dim S + \dim S' = k + (n - k + 1) = n + 1$$

so (4.2.3) ensures that $\{x : 0 \neq x \in S \cap S'\}$ is nonempty. Invoking (4.2.4) and (4.2.2), we see that

$$\sup_{\{x : 0 \neq x \in S\}} \frac{x^* A x}{x^* x} \geq \sup_{\{x : 0 \neq x \in S \cap S'\}} \frac{x^* A x}{x^* x} \geq \inf_{\{x : 0 \neq x \in S \cap S'\}} \frac{x^* A x}{x^* x}$$

$$\geq \inf_{\{x : 0 \neq x \in S'\}} \frac{x^* A x}{x^* x} = \min_{\{x : 0 \neq x \in S'\}} \frac{x^* A x}{x^* x} = \lambda_k$$

which implies that

$$\inf_{\{S:\dim S=k\}} \sup_{\{x:0\neq x\in S\}} \frac{x^*Ax}{x^*x} \geq \lambda_k \qquad (4.2.9)$$

However, span$\{x_1, \ldots, x_k\}$ contains the eigenvector x_k, span$\{x_1, \ldots, x_k\}$ is one of the choices for the subspace S, and $x_k^*Ax_k/x_k^*x_k = \lambda_k$, so the inequality (4.2.9) is actually an equality in which the infimum and supremum are attained:

$$\inf_{\{S:\dim S=k\}} \sup_{\{x:0\neq x\in S\}} \frac{x^*Ax}{x^*x} = \min_{\{S:\dim S=k\}} \max_{\{x:0\neq x\in S\}} \frac{x^*Ax}{x^*x} = \lambda_k$$

The assertion (4.2.8) follows from applying (4.2.7) and (4.2.5) to $-A$:

$$-\lambda_k = \min_{\{S:\dim S=n-k+1\}} \max_{\{x:0\neq x\in S\}} \frac{x^*(-A)x}{x^*x}$$

$$= \min_{\{S:\dim S=n-k+1\}} \max_{\{x:0\neq x\in S\}} \left(-\frac{x^*Ax}{x^*x}\right)$$

$$= \min_{\{S:\dim S=n-k+1\}} \left(-\min_{\{x:0\neq x\in S\}} \frac{x^*Ax}{x^*x}\right)$$

$$= -\left(\max_{\{S:\dim S=n-k+1\}} \min_{\{x:0\neq x\in S\}} \frac{x^*Ax}{x^*x}\right)$$

from which (4.2.8) follows. □

If $k = n$ in (4.2.7) or $k = 1$ in (4.2.8), we may omit the outer optimization and set $S = \mathbf{C}^n$ since this is the only n-dimensional subspace. In these two cases, the assertions reduce to (4.2.2(c)).

If one has a Hermitian $A \in M_n$ and bounds on its Hermitian form x^*Ax on a subspace, something can be said about its eigenvalues.

Theorem 4.2.10. *Let $A \in M_n$ be Hermitian, let the eigenvalues of A be arranged in increasing order (4.2.1), let S be a given k-dimensional subspace of $\mathbf{C}^n$, and let $c \in \mathbf{R}$ be given.*

(a) *If $x^*Ax \geq c$ (respectively, $x^*Ax > c$) for every unit vector $x \in S$, then $\lambda_{n-k+1}(A) \geq c$ (respectively, $\lambda_{n-k+1}(A) > c$).*

(b) *If $x^*Ax \leq c$ (respectively, $x^*Ax < c$) for every unit vector $x \in S$, then $\lambda_k(A) \leq c$ (respectively, $\lambda_k(A) < c$).*

Proof. Let $x_1, \ldots, x_n \in \mathbf{C}^n$ be orthonormal and such that $Ax_i = \lambda_i(A)x_i$ for each $i = 1, \ldots, n$ and let $S_1 = \text{span}\{x_1, \ldots, x_{n-k+1}\}$. Then $\dim S + \dim S_1 = k + (n - k + 1) = n + 1$, so (4.2.3) ensures that there is a unit vector $x \in S \cap S_1$. Our assumption in (a) that $x^*Ax \geq c$ ($x \in S$), together with (4.2.2) ($x \in S_1$), ensures that

$$c \leq x^*Ax \leq \lambda_{n-k+1}(A) \qquad (4.2.11)$$

so $\lambda_{n-k+1}(A) \geq c$, with strict inequality if $x^*Ax > c$. The assertions in (b) about upper bounds on eigenvalues of A follow from applying (a) to $-A$. □

Corollary 4.2.12. *Let $A \in M_n$ be Hermitian. If $x^*Ax \geq 0$ for all x in a k-dimensional subspace, then A has at least k nonnegative eigenvalues. If $x^*Ax > 0$ for all nonzero x in a k-dimensional subspace, then A' has at least k positive eigenvalues.*

Proof. The preceding theorem ensures that $\lambda_{n-k+1}(A) \geq 0$ (respectively, $\lambda_{n-k+1}(A) > 0$), and $\lambda_n(A) \geq \cdots \geq \lambda_{n-k+1}(A)$. □

Problems

4.2.P1 Explain why the assertions (4.2.7-8) are equivalent to

$$\lambda_k = \min_{\{S:\dim S=k\}} \max_{\{x:x\in S \text{ and } \|x\|_2=1\}} x^*Ax$$

and

$$\lambda_k = \max_{\{S:\dim S=n-k+1\}} \min_{\{x:x\in S \text{ and } \|x\|_2=1\}} x^*Ax$$

4.2.P2 Let $A \in M_n$ be Hermitian and suppose that at least one eigenvalue of A is positive. Show that $\lambda_{\max}(A) = \max\{1/x^*x : x^*Ax = 1\}$.

4.2.P3 If $A = [a_{ij}] \in M_n$ is Hermitian, use (4.2.2(c)) to show that $\lambda_{\max}(A) \geq a_{ii} \geq \lambda_{\min}(A)$ for all $i = 1, \ldots, n$, with equality in one of the inequalities for some i only if $a_{ij} = a_{ji} = 0$ for all $j = 1, \ldots, n$, $j \neq i$. Consider $A = \mathrm{diag}(1, 2, 3)$ and explain why the condition $a_{ij} = a_{ji} = 0$ for all $j = 1, \ldots, n$, $j \neq i$ does not imply that a_{ii} is equal to either $\lambda_{\max}(A)$ or $\lambda_{\min}(A)$.

4.2.P4 Let $A = [a_{ij}] = [a_1 \ \ldots \ a_n] \in M_n$ and let σ_1 be the largest singular value of A. Apply the preceding problem to the Hermitian matrix A^*A and show that $\sigma_1 \geq \|a_j\|_2 \geq |a_{ij}|$ for all $i, j = 1, \ldots, n$.

4.2.P5 Let $A = \begin{bmatrix} 1 & 2 \\ 0 & 1 \end{bmatrix}$. What are the eigenvalues of A? What is $\max\{x^TAx/x^Tx : 0 \neq x \in \mathbf{R}^2\}$? What is $\max \mathrm{Re}\{x^*Ax/x^*x : 0 \neq x \in \mathbf{C}^2\}$? Does this contradict (4.2.2)?

4.2.P6 Let $\lambda_1, \ldots, \lambda_n$ be the eigenvalues of $A \in M_n$, which we do not assume to be Hermitian. Show that

$$\min_{x\neq 0} \left| \frac{x^*Ax}{x^*x} \right| \leq |\lambda_i| \leq \max_{x\neq 0} \left| \frac{x^*Ax}{x^*x} \right|, \quad i = 1, 2, \ldots, n$$

and that strict inequality is possible in either inequality.

4.2.P7 Show that the rank-nullity theorem 0.2.3.1 implies the subspace intersection theorem 0.1.7.1 and (4.2.3).

4.2.P8 Suppose that $A, B \in M_n$ are Hermitian, B is positive semidefinite, and the eigenvalues $\{\lambda_i(A)\}_{i=1}^n$ and $\{\lambda_i(B)\}_{i=1}^n$ are ordered as in (4.2.1). Use (4.2.6) to show that $\lambda_k(A + B) \geq \lambda_k(A)$ for all $k = 1, \ldots, n$.

4.2.P9 Provide details for an alternative proof of (4.2.10) that first derives (b) from (4.2.7) and then deduces (a) from (b) applied to $-A$.

Notes and Further Readings. The variational characterization (4.2.2) was discovered by John William Strutt, 3rd Baron Rayleigh and 1904 Nobel laureate for his discovery of the element argon; see section 89 of Rayleigh (1945). The min-max characterization (4.2.6) for eigenvalues of real symmetric matrices appeared in E. Fischer, Über quadratische Formen mit reelen Koeffizienten, *Monatsh. Math. und Physik* 16 (1905) 234–249; it was extended to infinite dimensional operators by Richard Courant and incorporated into his classic text Courant and Hilbert (1937).

4.3 Eigenvalue inequalities for Hermitian matrices

The following theorem of Hermann Weyl is the source of a great many inequalities involving either a sum of two Hermitian matrices or bordered Hermitian matrices.

Theorem 4.3.1 (Weyl). *Let $A, B \in M_n$ be Hermitian and let the respective eigenvalues of A, B, and $A + B$ be $\{\lambda_i(A)\}_{i=1}^n$, $\{\lambda_i(B)\}_{i=1}^n$, and $\{\lambda_i(A + B)\}_{i=1}^n$, each algebraically ordered as in (4.2.1). Then*

$$\lambda_i(A + B) \leq \lambda_{i+j}(A) + \lambda_{n-j}(B), \quad j = 0, 1, \ldots, n - i \qquad (4.3.2a)$$

for each $i = 1, \ldots, n$, with equality for some pair i, j if and only if there is a nonzero vector x such that $Ax = \lambda_{i+j}(A)x$, $Bx = \lambda_{n-j}(B)x$, and $(A + B)x = \lambda_i(A + B)x$. Also,

$$\lambda_{i-j+1}(A) + \lambda_j(B) \leq \lambda_i(A + B), \quad j = 1, \ldots, i \qquad (4.3.2b)$$

for each $i = 1, \ldots, n$, with equality for some pair i, j if and only if there is a nonzero vector x such that $Ax = \lambda_{i-j+1}(A)x$, $Bx = \lambda_j(B)x$, and $(A + B)x = \lambda_i(A + B)x$. If A and B have no common eigenvector, then every inequality in (4.3.2a,b) is a strict inequality.

Proof. Let $x_1, \ldots, x_n$, $y_1, \ldots, y_n$, and $z_1, \ldots, z_n$ be orthonormal lists of eigenvectors of A, B, and $A + B$, respectively, such that $Ax_i = \lambda_i(A)x_i$, $By_i = \lambda_i(B)y_i$, and $(A + B)z_i = \lambda_i(A + B)z_i$ for each $i = 1, \ldots, n$. For a given $i \in \{1, \ldots, n\}$ and any $j \in \{0, \ldots, n - i\}$, let $S_1 = \text{span}\{x_1, \ldots, x_{i+j}\}$, $S_2 = \text{span}\{y_1, \ldots, y_{n-j}\}$, and $S_3 = \text{span}\{z_i, \ldots, z_n\}$. Then

$$\dim S_1 + \dim S_2 + \dim S_3 = (i + j) + (n - j) + (n - i + 1) = 2n + 1$$

so (4.2.3) ensures that there is a unit vector $x \in S_1 \cap S_2 \cap S_3$. Now invoke (4.2.2) three times to obtain the two inequalities

$$\lambda_i(A + B) \leq x^*(A + B)x = x^*Ax + x^*Bx \leq \lambda_{i+j}(A) + \lambda_{n-j}(B)$$

The first inequality follows from $x \in S_3$ and the second inequality follows from $x \in S_1$ and $x \in S_2$, respectively. The statements about the cases of equality in (4.3.2a) follow from the cases of equality in (4.2.2) for the unit vector x and the inequalities $x^*Ax \leq \lambda_{i+j}(A)$, $x \in S_1$; $x^*Bx \leq \lambda_{n-j}(B)$, $x \in S_2$; and $\lambda_i(A + B) \leq x^*(A + B)x$, $x \in S_3$.

The inequalities (4.3.2b) and their cases of equality follow from applying (4.3.2a) to $-A$, $-B$, and $-(A + B)$ and using (4.2.5):

$$-\lambda_{n-i+1}(A + B) = \lambda_i(-A - B) \leq \lambda_{i+j}(-A) + \lambda_{n-j}(-B)$$
$$= -\lambda_{n-i-j+1}(A) - \lambda_{j+1}(B)$$

If we set $i' = n - i + 1$ and $j' = j + 1$, the preceding inequality becomes

$$\lambda_{i'}(A + B) \geq \lambda_{i'-j'+1}(A) + \lambda_{j'}(B), \quad j' = 1, \ldots, i'$$

which is (4.3.2b).

If A and B have no common eigenvector, then the necessary conditions for equality in (4.3.2a,b) cannot be met.   □

Weyl's theorem describes what can happen to the eigenvalues of a Hermitian matrix A if it is additively perturbed by a Hermitian matrix B. Various assumptions about the perturbing matrix B lead to inequalities that are special cases of (4.3.2a,b). In each of the following corollaries, we continue to use the same notation as in (4.3.1), and we continue to insist on the algebraic ordering (4.2.1) for all lists of eigenvalues.

Exercise. Let $B \in M_n$ be Hermitian. If B has exactly π positive eigenvalues and exactly ν negative eigenvalues, explain why $\lambda_{n-\pi}(B) \leq 0$ and $\lambda_{\nu+1}(B) \geq 0$ with equality if and only if $n > \pi + \nu$, that is, if and only if B is singular.

Corollary 4.3.3. *Let $A, B \in M_n$ be Hermitian. Suppose that B has exactly π positive eigenvalues and exactly ν negative eigenvalues. Then*

$$\lambda_i(A + B) \leq \lambda_{i+\pi}(A), \quad i = 1, \ldots, n - \pi \tag{4.3.4a}$$

with equality for some i if and only if B is singular and there is a nonzero vector x such that $Ax = \lambda_{i+\pi}(A)x$, $Bx = 0$, and $(A + B)x = \lambda_i(A + B)x$. Also,

$$\lambda_{i-\nu}(A) \leq \lambda_i(A + B), \quad i = \nu + 1, \ldots, n \tag{4.3.4b}$$

with equality for some i if and only if B is singular and there is a nonzero vector x such that $Ax = \lambda_{i-\nu}(A)x$, $Bx = 0$, and $(A + B)x = \lambda_i(A + B)x$. Every inequality in (4.3.4a,b) is a strict inequality if either (a) B is nonsingular or (b) $Bx \neq 0$ for every eigenvector x of A.

Proof. Take $j = n - \pi$ in (4.3.2a) and use the preceding exercise to obtain $\lambda_i(A + B) \leq \lambda_{i+\pi}(A) + \lambda_{n-\pi}(B) \leq \lambda_{i+\pi}(A)$ with equality if and only if B is singular and there is a nonzero vector x such that $Ax = \lambda_{i+\pi}(A)x$, $Bx = 0$, and $(A + B)x = \lambda_i(A + B)x$. A similar argument shows that (4.3.4b) follows from (4.3.2b) with $j = \nu + 1$.   □

Exercise. Let $B \in M_n$ be Hermitian. If B is singular and rank $B = r$, explain why $\lambda_{n-r}(B) \leq 0$ and $\lambda_{r+1}(B) \geq 0$.

Corollary 4.3.5. *Let $A, B \in M_n$ be Hermitian. Suppose that B is singular and rank $B = r$. Then*

$$\lambda_i(A + B) \leq \lambda_{i+r}(A), \quad i = 1, \ldots, n - r \tag{4.3.6a}$$

with equality for some i if and only if $\lambda_{n-r}(B) = 0$ and there is a nonzero vector x such that $Ax = \lambda_{i+r}(A)x$, $Bx = 0$, and $(A + B)x = \lambda_i(A + B)x$. Also,

$$\lambda_{i-r}(A) \leq \lambda_i(A + B), \quad i = r + 1, \ldots, n \tag{4.3.6b}$$

with equality for some i if and only if $\lambda_{r+1}(B) = 0$ and there is a nonzero vector x such that $Ax = \lambda_{i-r}(A)x$, $Bx = 0$, and $(A + B)x = \lambda_i(A + B)x$. If $Bx \neq 0$ for every eigenvector x of A, then every inequality in (4.3.6a,b) is a strict inequality.

Proof. To verify (4.3.6a), take $j = r$ in (4.3.2a) and use the preceding exercise to obtain $\lambda_i(A + B) \leq \lambda_{i+r}(A) + \lambda_{n-r}(B) \leq \lambda_{i+r}(A)$ with equality if and only if $\lambda_{n-r}(B) = 0$ and there is equality in (4.3.2a) with $j = r$. A similar argument shows that (4.3.6b) follows from (4.3.4b) with $j = r + 1$. □

Exercise. Let $B \in M_n$ be Hermitian. If B has exactly one positive eigenvalue and exactly one negative eigenvalue, explain why $\lambda_2(B) \geq 0$ and $\lambda_{n-1}(B) \leq 0$ with equality if and only if $n > 2$.

Corollary 4.3.7. *Let $A, B \in M_n$ be Hermitian. Suppose that B has exactly one positive eigenvalue and exactly one negative eigenvalue. Then*

$$\lambda_1(A + B) \leq \lambda_2(A)$$
$$\lambda_{i-1}(A) \leq \lambda_i(A + B) \leq \lambda_{i+1}(A), \quad i = 2, \ldots, n - 1 \tag{4.3.8}$$
$$\lambda_{n-1}(A) \leq \lambda_n(A + B)$$

The cases of equality are as described in (4.3.3) with $\pi = \nu = 1$, for example, $\lambda_i(A + B) = \lambda_{i+1}(A)$ if and only if $n > 2$ and there is a nonzero vector x such that $Ax = \lambda_{i+1}(A)x$, $Bx = 0$, and $(A + B)x = \lambda_i(A + B)x$. Every inequality in (4.3.8) is a strict inequality if either (a) $n = 2$ or (b) $Bx \neq 0$ for every eigenvector x of A.

Proof. Take $\pi = \nu = 1$ in (4.3.4a,b) and use the preceding exercise. □

Exercise. Suppose that $z \in \mathbf{C}^n$ is nonzero and $n \geq 2$. Explain why $\lambda_{n-1}(zz^*) = 0 = \lambda_2(zz^*)$.

The following corollary is the *interlacing theorem* for a rank-one Hermitian perturbation of a Hermitian matrix.

Corollary 4.3.9. *Let $n \geq 2$, let $A \in M_n$ be Hermitian, and let $z \in \mathbf{C}^n$ be nonzero. Then*

$$\lambda_i(A) \leq \lambda_i(A + zz^*) \leq \lambda_{i+1}(A), \quad i = 1, \ldots, n - 1 \tag{4.3.10}$$
$$\lambda_n(A) \leq \lambda_n(A + zz^*)$$

The cases of equality in (4.3.10) are as described in (4.3.3) with $\pi = 1$ and $\nu = 0$, for example, $\lambda_i(A + zz^*) = \lambda_{i+1}(A)$ if and only if there is a nonzero vector x such that $Ax = \lambda_{i+1}(A)x$, $z^*x = 0$, and $(A + zz^*)x = \lambda_i(A + zz^*)x$. Also,

$$\lambda_1(A - zz^*) \leq \lambda_1(A) \tag{4.3.11}$$
$$\lambda_{i-1}(A) \leq \lambda_i(A - zz^*) \leq \lambda_i(A), \quad i = 2, \ldots, n$$

The cases of equality in (4.3.11) are as described in (4.3.3) with $\pi = 0$ and $\nu = 1$. If no eigenvector of A is orthogonal to z, then every inequality in (4.3.10,11) is a strict inequality.

Proof. In (4.3.4a), take $\pi = 1$ and $\nu = 0$; in (4.3.4b), take $\pi = 0$ and $\nu = 1$. Use the preceding exercise. □

Exercise. Let $B \in M_n$ be positive semidefinite. Explain why $\lambda_1(B) = 0$ if and only if B is singular.

The following corollary is known as the *monotonicity theorem.*

Corollary 4.3.12. *Let $A, B \in M_n$ be Hermitian and suppose that B is positive semi-definite. Then*

$$\lambda_i(A) \leq \lambda_i(A + B), \quad i = 1, \ldots, n \tag{4.3.13}$$

with equality for some i if and only if B is singular and there is a nonzero vector x such that $Ax = \lambda_i(A)x$, $Bx = 0$, and $(A + B)x = \lambda_i(A + B)x$. If B is positive definite, then

$$\lambda_i(A) < \lambda_i(A + B), \quad i = 1, \ldots, n \tag{4.3.14}$$

Proof. Use (4.3.4b) with $\nu = 0$ and use the preceding exercise. If B is nonsingular, equality cannot occur in (4.3.13). $\square$

Corollary 4.3.15. *Let $A, B \in M_n$ be Hermitian. Then*

$$\lambda_i(A) + \lambda_1(B) \leq \lambda_i(A + B) \leq \lambda_i(A) + \lambda_n(B), \quad i = 1, \ldots, n \tag{4.3.16}$$

with equality in the upper bound if and only if there is nonzero vector x such that $Ax = \lambda_i(A)x$, $Bx = \lambda_n(B)x$, and $(A + B)x = \lambda_i(A + B)x$; equality in the lower bound occurs if and only if there is nonzero vector x such that $Ax = \lambda_i(A)x$, $Bx = \lambda_1(B)x$, and $(A + B)x = \lambda_i(A + B)x$. If A and B have no common eigenvector, then every inequality in (4.3.16) is a strict inequality.

Proof. Take $j = 0$ in (4.3.2a) and $j = 1$ in (4.3.2b). $\square$

Exercise. Let $y \in \mathbf{C}^n$ and $a \in \mathbf{R}$ be given, and let $\mathcal{K} = \begin{bmatrix} 0_n & y \\ y^* & a \end{bmatrix} \in M_{n+1}$. Show that the eigenvalues of $\mathcal{K}$ are $(a \pm \sqrt{a^2 + 4y^*y})/2$ together with $n - 1$ zero eigenvalues. If $y \neq 0$, conclude that $\mathcal{K}$ has exactly one positive eigenvalue and exactly one negative eigenvalue. *Hint*: (1.2.P13(b)).

Weyl's inequalities and their corollaries are concerned with additive Hermitian perturbations of a Hermitian matrix. Additional eigenvalue inequalities arise from extracting a principal submatrix from a Hermitian matrix, or from bordering it to form a larger Hermitian matrix. The following result is *Cauchy's interlacing theorem* for a bordered Hermitian matrix, sometimes called the *separation theorem*.

Theorem 4.3.17 (Cauchy). *Let $B \in M_n$ be Hermitian, let $y \in \mathbf{C}^n$ and $a \in \mathbf{R}$ be a given, and let $A = \begin{bmatrix} B & y \\ y^* & a \end{bmatrix} \in M_{n+1}$. Then*

$$\lambda_1(A) \leq \lambda_1(B) \leq \lambda_2(A) \leq \cdots \leq \lambda_n(A) \leq \lambda_n(B) \leq \lambda_{n+1}(A) \tag{4.3.18}$$

in which $\lambda_i(A) = \lambda_i(B)$ if and only if there is a nonzero $z \in \mathbf{C}^n$ such that $Bz = \lambda_i(B)z$, $y^*z = 0$, and $Bz = \lambda_i(A)z$; $\lambda_i(B) = \lambda_{i+1}(A)$ if and only if there is a nonzero $z \in \mathbf{C}^n$ such that $Bz = \lambda_i(B)z$, $y^*z = 0$, and $Bz = \lambda_{i+1}(A)z$. If no eigenvector of B is orthogonal to y, then every inequality in (4.3.18) is a strict inequality.

Proof. The asserted interlacing of ordered eigenvalues is unchanged if we replace A with $A + \mu I_{n+1}$, which replaces B with $B + \mu I_n$. Thus, there is no loss of generality to assume that B and A are positive definite. Consider the Hermitian matrices $\mathcal{H} = \begin{bmatrix} B & 0 \\ 0 & 0_1 \end{bmatrix}$ and $\mathcal{K} = \begin{bmatrix} 0_n & y \\ y^* & a \end{bmatrix}$, for which $A = \mathcal{H} + \mathcal{K}$. The ordered eigenvalues of $\mathcal{H} = B \oplus [0]$ are $\lambda_1(\mathcal{H}) = 0 < \lambda_1(B) = \lambda_2(\mathcal{H}) \leq \lambda_2(B) = \lambda_3(\mathcal{H}) \leq \cdots$, that is, $\lambda_{i+1}(\mathcal{H}) = \lambda_i(B)$

for all $i = 1, \ldots, n$. The preceding exercise shows that $\mathcal{K}$ has exactly one positive eigenvalue and one negative eigenvalue, so the inequalities (4.3.8) ensure that

$$\lambda_i(A) = \lambda_i(\mathcal{H} + \mathcal{K}) \leq \lambda_{i+1}(\mathcal{H}) = \lambda_i(B), \quad i = 1, \ldots, n \qquad (4.3.19)$$

Necessary and sufficient conditions for equality in (4.3.19) for a given i are described in (4.3.7): there is a nonzero $x \in \mathbf{C}^{n+1}$ such that $\mathcal{H}x = \lambda_{i+1}(\mathcal{H})x$, $\mathcal{K}x = 0$, and $Ax = \lambda_i(A)x$. If we partition $x = \begin{bmatrix} z \\ \zeta \end{bmatrix}$ with $z \in \mathbf{C}^n$ and use the identity $\lambda_{i+1}(\mathcal{H}) = \lambda_i(B)$, a computation reveals that these conditions are equivalent to the existence of a nonzero $z \in \mathbf{C}^n$ such that $Bz = \lambda_i(B)z$, $y^*z = 0$, and $Bz = \lambda_i(A)z$. In particular, if no eigenvector of B is orthogonal to y, then there is no i for which the necessary conditions $z \neq 0$, $Bz = \lambda_i(B)z$, and $y^*z = 0$ can be met.

The inequalities $\lambda_i(B) \leq \lambda_{i+1}(A)$ for $i = 1, \ldots, n$ follow from applying (4.3.19) to $-A$ and using (4.2.5):

$$-\lambda_{(n+1)-i+1}(A) = \lambda_i(-A) \leq \lambda_i(-B) = -\lambda_{n-i+1}(B) \qquad (4.3.20)$$

If we set $i' = n - i + 1$, we obtain the equivalent inequalities $\lambda_{i'+1}(A) \geq \lambda_{i'}(B)$ for $i' = 1, \ldots, n$. The case of equality for (4.3.20) follows again from (4.3.7). $\qquad \square$

We have discussed two examples of interlacing theorems for eigenvalues: If a given Hermitian matrix is modified either by adding a rank-one Hermitian matrix or by bordering, then the new and old eigenvalues must interlace. In fact, each of (4.3.9) and (4.3.17) implies the other; see (7.2.P15). What about converses of these theorems? If two interlacing sets of real numbers are given, are they the eigenvalues of a Hermitian matrix and a bordering of it? Are they the eigenvalues of a Hermitian matrix and a rank-one additive perturbation of it? The following two theorems provide affirmative answers to both questions.

Theorem 4.3.21. *Let* $\lambda_1, \ldots, \lambda_n$ *and* $\mu_1, \ldots, \mu_{n+1}$ *be real numbers that satisfy the interlacing inequalities*

$$\mu_1 \leq \lambda_1 \leq \mu_2 \leq \lambda_2 \leq \cdots \leq \lambda_{n-1} \leq \mu_n \leq \lambda_n \leq \mu_{n+1} \qquad (4.3.22)$$

Let $\Lambda = \mathrm{diag}(\lambda_1, \lambda_2, \ldots, \lambda_n)$. *A real number* a *and a real vector* $y = [y_i] \in \mathbf{R}^n$ *may be chosen such that the eigenvalues of*

$$A = \begin{bmatrix} \Lambda & y \\ y^T & a \end{bmatrix} \in M_{n+1}(\mathbf{R}) \qquad (4.3.23)$$

are $\mu_1, \ldots, \mu_{n+1}$.

Proof. We want the eigenvalues of A to be $\mu_1, \ldots, \mu_{n+1}$, so a is determined by the identity

$$\mu_1 + \cdots + \mu_{n+1} = \mathrm{tr}\, A = \mathrm{tr}\, \Lambda + a = \lambda_1 + \cdots + \lambda_n + a$$

The characteristic polynomial of A is $p_A(t) = (t - \mu_1) \cdots (t - \mu_{n+1})$; it is also the determinant of the bordered matrix $tI - A$, which we can evaluate with

Cauchy's expansion (0.8.5.10):

$$p_A(t) = \det(tI - A) = (t - a)\det(tI - \Lambda) - y^T \operatorname{adj}(tI - \Lambda)y$$

$$= (t - a)\prod_{i=1}^{n}(t - \lambda_i) - \sum_{i=1}^{n}\left(y_i^2 \prod_{j\neq i}(t - \lambda_j)\right)$$

If we combine these two representations for $p_A(t)$ and introduce the variables $\eta_i = y_i^2$, $i = 1, \ldots, n$, we obtain the identity

$$\sum_{i=1}^{n}\left(\eta_i \prod_{j\neq i}(t - \lambda_j)\right) = (t - a)\prod_{i=1}^{n}(t - \lambda_i) - \prod_{i=1}^{n+1}(t - \mu_i) \qquad (4.3.24)$$

We must show that there are *nonnegative* $\eta_1, \ldots, \eta_n$ that satisfy (4.3.24).

Choose any $\lambda \in \{\lambda_1, \ldots, \lambda_n\}$; suppose that λ has multiplicity (exactly) $m \geq 1$ as an eigenvalue of Λ and that $\lambda = \lambda_k = \cdots = \lambda_{k+m-1}$, so $\lambda_{k-1} < \lambda < \lambda_{k+m}$. To focus the exposition on the generic case, we assume that $1 < k < n + 1 - m$ and leave it to the reader to modify the following argument in the special (and easier) cases $k = 1$ or $k + m - 1 = n$.

The interlacing inequalities (4.3.22) *require* that $\mu_{k+1} = \cdots = \mu_{k+m-1} = \lambda$; they *permit* $\mu_k = \lambda$ and/or $\lambda = \mu_{k+m}$. Let

$$f_{\lambda+}(t) = \prod_{i=1}^{k-1}(t - \lambda_i), \quad f_{\lambda-}(t) = \prod_{i=k+m}^{n}(t - \lambda_i)$$

and

$$g_{\lambda+}(t) = \prod_{i=1}^{k}(t - \mu_i), \quad g_{\lambda-}(t) = \prod_{i=k+m}^{n+1}(t - \mu_i)$$

Observe that $f_{\lambda+}(\lambda) > 0$, $f_{\lambda-}(\lambda) \neq 0$, and $\operatorname{sign}(f_{\lambda-}(\lambda)) = (-1)^{n-(k+m)+1}$. Also, $g_{\lambda+}(\lambda) \geq 0$, $g_{\lambda-}(\lambda) \leq 0$, $g_{\lambda+}(\lambda) = 0$ if and only if $\mu_k = \lambda$, and $g_{\lambda-}(\lambda) = 0$ if and only if $\lambda = \mu_{k+m}$. If $\lambda < \mu_{k+m}$, then $\operatorname{sign}(g_{\lambda-}(\lambda)) = (-1)^{n+1-(k+m)+1}$.

Now examine (4.3.24) carefully: If $i \leq k - 1$ or if $i \geq k + m$, the coefficient of η_i contains a factor $(t - \lambda)^m$ since $\lambda = \lambda_k = \cdots = \lambda_{k+m-1}$. However, for each $i = k, \ldots, k + m - 1$ the coefficient of η_i is $f_{\lambda+}(t)(t - \lambda)^{m-1} f_{\lambda-}(t)$. On the right-hand side of (4.3.24), the first term contains a factor $(t - \lambda)^m$ and the second term is equal to $g_{\lambda+}(t)(t - \lambda)^{m-1}g_{\lambda-}(t)$. These observations permit us to divide both sides of (4.3.24) by $(t - \lambda)^{m-1}$, set $t = \lambda$, and obtain the identity

$$(\eta_k + \cdots + \eta_{k+m-1}) f_{\lambda+}(\lambda) f_{\lambda-}(\lambda) = -g_{\lambda+}(\lambda)g_{\lambda-}(\lambda)$$

that is,

$$(\eta_k + \cdots + \eta_{k+m-1}) = \left(\frac{g_{\lambda+}(\lambda)}{f_{\lambda+}(\lambda)}\right)\left(\frac{-g_{\lambda-}(\lambda)}{f_{\lambda-}(\lambda)}\right) \qquad (4.3.25)$$

If the right-hand side of (4.3.25) is zero (that is, if $\mu_k = \lambda$ or $\lambda = \mu_{k+m}$), we take $\eta_k = \cdots = \eta_{k+m-1} = 0$. Otherwise (that is, if $\mu_k < \lambda < \mu_{k+m}$), we know that

$(g_{\lambda+}(\lambda)/f_{\lambda+}(\lambda)) > 0$, so it suffices to check that

$$\text{sign}\left(\frac{-g_{\lambda-}(\lambda)}{f_{\lambda-}(\lambda)}\right) = \frac{-(-1)^{n-k-m+2}}{(-1)^{n-k-m+1}} = +1$$

and then choose *any* nonnegative $\eta_k, \ldots, \eta_{k+m-1}$ whose sum is the positive value in (4.3.25). □

Exercise. Give details for a proof of the preceding theorem if (a) $m = n$, or (b) $k = 1$ and $m < n$.

Theorem 4.3.26. *Let $\lambda_1, \ldots, \lambda_n$ and $\mu_1, \ldots, \mu_n$ be real numbers that satisfy the interlacing inequalities*

$$\lambda_1 \leq \mu_1 \leq \lambda_2 \leq \mu_2 \leq \cdots \leq \lambda_n \leq \mu_n \qquad (4.3.27)$$

Let $\Lambda = \text{diag}(\lambda_1, \ldots, \lambda_n)$. Then there is a real vector $z \in \mathbf{R}^n$ such that the eigenvalues of $\Lambda + zz^*$ are $\mu_1, \ldots, \mu_n$.

Proof. There is no loss of generality to assume that $\lambda_1 > 0$, for if $\lambda_1 \leq 0$ let $c > -\lambda_1$ and replace each λ_i by $\lambda_i + c$ and each μ_i by $\mu_i + c$. This shift does not disturb the interlacing inequalities (4.3.27). If there is a vector z such that the eigenvalues of $\Lambda + cI + zz^*$ are $\mu_1 + c, \ldots, \mu_n + c$, then the eigenvalues of $\Lambda + zz^*$ are $\mu_1, \ldots, \mu_n$. Let $\mu_0 = 0$ and suppose that

$$0 = \mu_0 < \lambda_1 \leq \mu_1 \leq \lambda_2 \leq \mu_2 \leq \cdots \leq \lambda_n \leq \mu_n$$

Theorem 4.3.21 ensures that there is a real number a and a real vector y such that $\mu_0, \mu_1, \ldots, \mu_n$ are the eigenvalues of the singular matrix $A = \begin{bmatrix} \Lambda & y \\ y^T & a \end{bmatrix}$ (its smallest eigenvalue is zero). Let $R = \Lambda^{1/2} = \text{diag}(\lambda_1^{1/2}, \ldots, \lambda_n^{1/2})$. The first n columns of A are linearly independent (Λ is nonsingular), so the last column of A must be a linear combination of its first n columns, that is, there is a real vector w such that $\begin{bmatrix} y \\ a \end{bmatrix} = \begin{bmatrix} \Lambda \\ y^T \end{bmatrix} w = \begin{bmatrix} R^2 w \\ y^T w \end{bmatrix}$. We conclude that $y = R^2 w$, $w = R^{-2}y$, and $a = y^T w = w^T R^2 w = (Rw)^T(Rw)$. Let $z = Rw = R^{-1}y$. Since the eigenvalues of

$$A = \begin{bmatrix} R^2 & R(Rw) \\ (Rw)^T R & (Rw)^T(Rw) \end{bmatrix} = \begin{bmatrix} R \\ z^T \end{bmatrix} \begin{bmatrix} R & z \end{bmatrix}$$

are $0, \mu_1, \ldots, \mu_n$, (1.3.22) ensures that $\mu_1, \ldots, \mu_n$ are the eigenvalues of

$$\begin{bmatrix} R & z \end{bmatrix} \begin{bmatrix} R \\ z^T \end{bmatrix} = R^2 + zz^T = \Lambda + zz^T \qquad \square$$

Theorem 4.3.17 can be appreciated from two points of view: On the one hand, it considers the eigenvalues of a matrix that is obtained by *bordering* a given Hermitian matrix by appending to it a new last row and column; on the other hand, it considers the behavior of the eigenvalues of a matrix that is obtained by *deleting* the last row and column of a given Hermitian matrix. With regard to eigenvalue interlacing, there is, of course, nothing special about deleting the *last* row and column: The eigenvalues of a matrix obtained by deleting *any* row and the corresponding column of a Hermitian matrix A are the same as the eigenvalues of a matrix obtained by deleting the last row and column of a matrix that is a certain permutation similarity of A.

One may wish to delete several rows and the corresponding columns from a Hermitian matrix. The remaining matrix is a principal submatrix of the original matrix. The following result – sometimes called the *inclusion principle* – can be obtained by repeated application of the interlacing inequalities (4.3.18); we prove it using the Rayleigh quotient theorem and the subspace intersection lemma, which permit us to clarify the cases of equality and multiple eigenvalues.

Theorem 4.3.28. *Let $A \in M_n$ be Hermitian, partitioned as*

$$A = \begin{bmatrix} B & C \\ C^* & D \end{bmatrix}, \quad B \in M_m, D \in M_{n-m}, C \in M_{m,n-m} \quad (4.3.29)$$

Let the eigenvalues of A and B be ordered as in (4.2.1). Then

$$\lambda_i(A) \le \lambda_i(B) \le \lambda_{i+n-m}(A), \quad i = 1, \dots, m \quad (4.3.30)$$

with equality in the lower bound for some i if and only if there is a nonzero $\xi \in M_m$ such that $B\xi = \lambda_i(B)\xi$ and $C^*\xi = 0$; equality in the upper bound occurs for some i if and only if there is a nonzero $\xi \in M_m$ such that $B\xi = \lambda_{i+n-m}(A)\xi$ and $C^*\xi = 0$.

If $i \in \{1, \dots, m\}$, $1 \le r \le i$, and

$$\lambda_{i-r+1}(A) = \cdots = \lambda_i(A) = \lambda_i(B) \quad (4.3.31)$$

then $\lambda_{i-r+1}(B) = \cdots = \lambda_i(B)$ and there are orthonormal vectors $\xi_1, \dots, \xi_r \in \mathbf{C}^m$ such that $B\xi_j = \lambda_i(B)\xi_j$ and $C^*\xi_j = 0$ for each $j = 1, \dots, r$.

If $i \in \{1, \dots, m\}$, $1 \le r \le m - i + 1$, and

$$\lambda_i(B) = \lambda_{i+n-m}(A) = \cdots = \lambda_{i+n-m+r-1}(A) \quad (4.3.32)$$

then $\lambda_i(B) = \cdots = \lambda_{i+n-m+r-1}(B)$ and there are orthonormal vectors $\xi_1, \dots, \xi_r \in \mathbf{C}^m$ such that $B\xi_j = \lambda_i(B)\xi_j$ and $C^*\xi_j = 0$ for each $j = 1, \dots, r$.

Proof. Let $x_1, \dots, x_n \in \mathbf{C}^n$ and $y_1, \dots, y_n \in \mathbf{C}^m$ be orthonormal lists of eigenvectors of A and B, respectively, such that $Ax_i = \lambda_i(A)x_i$ for each $i = 1, \dots, n$ and $By_i = \lambda_i(B)y_i$ for each $i = 1, \dots, m$. Let $\hat{y}_i = \begin{bmatrix} y_i \\ 0 \end{bmatrix} \in \mathbf{C}^n$ for each $i = 1, \dots, m$. For a given $i \in \{1, \dots, m\}$, let $S_1 = \mathrm{span}\{x_1, \dots, x_{i+n-m}\}$ and $S_2 = \mathrm{span}\{\hat{y}_i, \dots, \hat{y}_m\}$. Then

$$\dim S_1 + \dim S_2 = (i + n - m) + (m - i + 1) = n + 1$$

so (4.2.3) ensures that there is a unit vector $x \in S_1 \cap S_2$. Since $x \in S_2$, it has the form $x = \begin{bmatrix} \xi \\ 0 \end{bmatrix}$ for some unit vector $\xi \in \mathrm{span}\{y_i, \dots, y_m\} \subset \mathbf{C}^m$. Observe that

$$x^*Ax = \begin{bmatrix} \xi^* & 0 \end{bmatrix} \begin{bmatrix} B & C \\ C^* & D \end{bmatrix} \begin{bmatrix} \xi \\ 0 \end{bmatrix} = \begin{bmatrix} \xi^* & 0 \end{bmatrix} \begin{bmatrix} B\xi \\ C^*\xi \end{bmatrix} = \xi^*B\xi$$

Now invoke (4.2.2) twice to obtain the two inequalities

$$\lambda_i(B) \le \xi^*B\xi = x^*Ax \le \lambda_{i+n-m}(A) \quad (4.3.33)$$

The first inequality follows from $\xi \in \mathrm{span}\{y_i, \dots, y_m\}$ and the second inequality follows from $x \in S_1$. The statements about the cases of equality in (4.3.33) follow from the cases of equality in (4.2.2) for the unit vector x, and the inequalities $\lambda_i(B) \leq \xi^* B \xi, \xi \in \text{span}\{y_i, \ldots, y_m\}$ and $x^* A x \leq \lambda_{i+n-m}(A), x = \begin{bmatrix} \xi \\ 0 \end{bmatrix} \in S_1 \cap S_2$.

If the eigenvalues of A and B satisfy (4.3.31), then (4.3.30) ensures that $\lambda_{i-r+1}(A) \leq \lambda_{i-r+1}(B) \leq \cdots \leq \lambda_{i-1}(B) \leq \lambda_i(B) = \lambda_{i-r+1}(A)$, so $\lambda_{i-r+1}(B) = \cdots = \lambda_i(B)$. Let $S_1 = \text{span}\{x_1, \ldots, x_{i+n-m}\}$ and $S_2 = \text{span}\{\hat{y}_{i-r+1}, \ldots, \hat{y}_m\}$. Then $\dim S_1 + \dim S_2 = (i+n-m) + (m-i+r) = n+r$, so (4.2.3) tells us that $\dim(S_1 \cap S_2) \geq r$. It follows that there are orthonormal vectors $x_1, \ldots, x_r \in S_1 \cap S_2$ such that (just as in the preceding cases of equality) each $x_j = \begin{bmatrix} \xi_j \\ 0 \end{bmatrix}$, $\xi_1, \ldots, \xi_r$ are orthonormal vectors in $\text{span}\{y_{i-r+1}, \ldots, y_m\}$, $B\xi_j = \lambda_i(B)\xi_j$, and $C^*\xi_j = 0$ for each $= 1, \ldots, r$. The assertions (4.3.31) are verified in a similar fashion. $\square$

Exercise. For $r = 1$, explain why the assertions following (4.3.31,32) reduce to the assertions following (4.3.30).

Exercise. For $m = 1$ and $i = 1$, explain why the assertions following (4.3.30) are equivalent to the assertions in (4.2.P3).

Exercise. Explain why the inequalities (4.3.30) are all strict inequalities if either (a) C has full row rank or (b) $C^*x \neq 0$ for every eigenvector x of B.

Corollary 4.3.34. *Let $A = [a_{ij}] \in M_n$ be Hermitian, partitioned as in (4.3.29), and let the eigenvalues of A be ordered as in (4.2.1). Then*

$$a_{11} + a_{22} + \cdots + a_{mm} \geq \lambda_1(A) + \cdots + \lambda_m(A) \tag{4.3.35a}$$

and

$$a_{11} + a_{22} + \cdots + a_{mm} < \lambda_{n-m+1}(A) + \cdots + \lambda_n(A) \tag{4.3.35b}$$

If either inequality (4.3.35a,b) is an equality, then $C = 0$ and $A = B \oplus D$. More generally, suppose that $k \geq 2$ and partition $A = [A_{ij}]_{i,j=1}^k$ so that each $A_{ii} \in M_{n_i}$. If

$$\text{tr } A_{11} + \cdots + \text{tr } A_{pp} = \sum_{i=1}^{n_1+\cdots+n_p} \lambda_i(A) \tag{4.3.36a}$$

for each $p = 1, \ldots, k-1$, then $A = A_{11} \oplus \cdots \oplus A_{kk}$; the eigenvalues of A_{11} are $\lambda_1(A), \ldots, \lambda_{n_1}(A)$, the eigenvalues of A_{22} are $\lambda_{n_1+1}(A), \ldots, \lambda_{n_1+n_2}(A)$, and so on. If

$$\text{tr } A_{11} + \cdots + \text{tr } A_{pp} = \sum_{i=n-n_1-\cdots-n_p+1}^{n} \lambda_i(A) \tag{4.3.36b}$$

for each $p = 1, \ldots, k-1$, then $A = A_{11} \oplus \cdots \oplus A_{kk}$; the eigenvalues of A_{11} are $\lambda_{n-n_1+1}(A), \ldots, \lambda_n(A)$, the eigenvalues of A_{22} are $\lambda_{n-n_1-n_2+1}(A), \ldots, \lambda_{n-n_1}(A)$, and so on.

Proof. The left-hand inequalities in (4.3.30) ensure that $\lambda_i(B) \geq \lambda_i(A)$ for each $i = 1, \ldots, m$, so $\text{tr } B = \lambda_1(B) + \cdots + \lambda_m(B) \geq \lambda_1(A) + \cdots + \lambda_m(A) = \text{tr } B$ implies that $\lambda_i(A) = \lambda_i(B)$ for each $i = 1, \ldots, m$. Similarly, the right-hand inequalities in (4.3.30) imply that $\lambda_i(B) = \lambda_{i+n-m}(A)$ for each $i = 1, \ldots, m$. The equality cases of (4.3.30,31,32) ensure that there is are orthonormal eigenvectors $\xi_1, \ldots, \xi_m$ of B for which $C^*\xi_j = 0$ for each $j = 1, \ldots, m$. Since $\operatorname{rank} C = \operatorname{rank} C^* \leq m - \dim(\operatorname{nullspace} C^*) \leq m - m = 0$, we conclude that $C = 0$.

The assertion that the equalities (4.3.36a) imply that $A = A_{11} \oplus \cdots \oplus A_{kk}$ follows from the case of equality in (4.3.35a) by induction. Write $A = \begin{bmatrix} A_{11} & C_2 \\ C_2^* & D_2 \end{bmatrix}$ in which $D_2 = [A_{ij}]_{i,j=2}^k$. Since $\operatorname{tr} A_{11} = \sum_{i=1}^{n_1} \lambda_i(A)$, we know that $C_2 = 0$, $A = A_{11} \oplus D_2$, and the ordered eigenvalues of D_2 are $\lambda_{n_1+1}(A) \leq \cdots \leq \lambda_n(A)$. Write $D_2 = \begin{bmatrix} A_{22} & C_3 \\ C_3^* & D_3 \end{bmatrix}$ in which $D_3 = [A_{ij}]_{i,j=3}^k$. The hypothesis (4.3.36a) for $p = 2$ ensures that $\operatorname{tr} A_{22} = \lambda_{n_1+1}(A) + \cdots + \lambda_{n_1+n_2}(A)$ is the sum of the n_2 smallest eigenvalues of D_2, so $C_3 = 0$ and so forth. The assertion about (4.3.36b) follows in a similar fashion from (4.3.35b) by induction. □

The following consequence of (4.3.28) is known as the *Poincaré separation theorem.*

Corollary 4.3.37. *Let $A \in M_n$ be Hermitian, suppose that $1 \leq m \leq n$, and let $u_1, \ldots, u_m \in \mathbf{C}^n$ be orthonormal. Let $B_m = [u_i^* A u_j]_{i,j=1}^m \in M_m$ and let the eigenvalues of A and B_m be arranged as in (4.2.1). Then*

$$\lambda_i(A) \leq \lambda_i(B_m) \leq \lambda_{i+n-m}(A), \quad i = 1, \ldots, m \tag{4.3.38}$$

Proof. If $m < n$, choose $n - m$ additional vectors $u_{m+1}, \ldots, u_n$ so that $U = [u_1 \ \cdots \ u_n] \in M_n$ is unitary. Then U^*AU has the same eigenvalues as A, and B_m is a principal submatrix of U^*AU obtained by deleting its last $n - m$ rows and columns. The assertion now follows from (4.3.28). □

The matrix B_m in the preceding corollary can be written as $B_m = V^*AV$, in which $V \in M_{n,m}$ has orthonormal columns. Since $\operatorname{tr} B_m = \lambda_1(B_m) + \cdots + \lambda_m(B_m)$, the following two variational characterizations follow from summing the inequalities (4.3.38) and making suitable choices of V. They are generalizations of (4.2.2).

Corollary 4.3.39. *Let $A \in M_n$ be Hermitian and suppose that $1 \leq m \leq n$. Then*

$$\begin{aligned} \lambda_1(A) + \cdots + \lambda_m(A) &= \min_{\substack{V \in M_{n,m} \\ V^*V = I_m}} \operatorname{tr} V^*AV \\ \lambda_{n-m+1}(A) + \cdots + \lambda_n(A) &= \max_{\substack{V \in M_{n,m} \\ V^*V = I_m}} \operatorname{tr} V^*AV \end{aligned} \tag{4.3.40}$$

For each $m = 1, \ldots, n - 1$ the minimum or maximum in (4.3.40) is achieved for a matrix V whose columns are orthonormal eigenvectors associated with the m smallest or largest eigenvalues of A; for $m = n$ we have $\operatorname{tr} V^*AV = \operatorname{tr} AVV^* = \operatorname{tr} A$ for any unitary V.

The eigenvalues and main diagonal elements of a Hermitian matrix are real numbers whose respective sums are equal. The precise relationship between the main diagonal entries and the eigenvalues of a Hermitian matrix involves the notion of *majorization*, which is motivated by the variational identities (4.4.40).

Definition 4.3.41. *Let* $x = [x_i] \in \mathbf{R}^n$ *and* $y = [y_i] \in \mathbf{R}^n$ *be given. We say that* x *majorizes* y *if*

$$\max_{1 \leq i_1 < \cdots < i_k \leq n} \sum_{j=1}^{k} x_{i_j} \geq \max_{1 \leq i_1 < \cdots < i_k \leq n} \sum_{j=1}^{k} y_{i_j} \qquad (4.3.42)$$

for each $k = 1, \ldots, n$, *with equality for* $k = n$.

Definition 4.3.43. *Let* $z = [z_i] \in \mathbf{R}^n$ *be given. The* nonincreasing rearrangement *of* z *is the vector* $z^{\downarrow} = [z_i^{\downarrow}] \in \mathbf{R}^n$ *whose list of entries is the same as that of* z *(including multiplicities) but rearranged in nonincreasing order* $z_{i_1} = z_1^{\downarrow} \geq \cdots \geq z_{i_n} = z_n^{\downarrow}$. *The* nondecreasing rearrangement *of* z *is the vector* $z^{\uparrow} = [z_i^{\uparrow}] \in \mathbf{R}^n$ *whose list of entries is the same as that of* z *(including multiplicities) but rearranged in nondecreasing order* $z_{j_1} = z_1^{\uparrow} \leq \cdots \leq z_{j_n} = z_n^{\uparrow}$.

Exercise. Explain why the majorization inequalities (4.3.42) are equivalent to the "top-down" inequalities

$$\sum_{i=1}^{k} x_i^{\downarrow} \geq \sum_{i=1}^{k} y_i^{\downarrow} \qquad (4.3.44a)$$

for each $k = 1, 2, \ldots, n$, with equality for $k = n$, as well as to the "bottom-up" inequalities

$$\sum_{i=1}^{k} y_i^{\uparrow} \geq \sum_{i=1}^{k} x_i^{\uparrow} \qquad (4.3.44b)$$

for each $k = 1, 2, \ldots, n$, with equality for $k = n$. *Hint*: Let $s = \sum_{l=1}^{n} y_i = \sum_{i=1}^{n} x_i$. Then $\sum_{i=1}^{k} y_i^{\uparrow} = s - \sum_{i=1}^{n-k+1} y_i^{\downarrow}$ and $\sum_{i=1}^{k} x_i^{\uparrow} = s - \sum_{i=1}^{n-k+1} x_i^{\downarrow}$.

Exercise. Let $x, y \in \mathbf{R}^n$ and let $P, Q \in M_n$ be permutation matrices. Explain why x majorizes y if and only if Px majorizes Qy.

Exercise. Let $x = [x_i]$, $y = [y_i] \in \mathbf{R}^n$ and suppose that x majorizes y. Explain why $x_1^{\downarrow} \geq y_1^{\downarrow} \geq y_n^{\downarrow} \geq x_n^{\downarrow}$.

The following two theorems demonstrate how the notion of majorization arises in matrix analysis.

Theorem 4.3.45 (Schur). *Let* $A = [a_{ij}] \in M_n$ *be Hermitian. Its vector of eigenvalues* $\lambda(A) = [\lambda_i(A)]_{i=1}^n$ *majorizes its vector of main diagonal entries* $d(A) = [a_{ii}]_{i=1}^n$, *that is,*

$$\sum_{i=1}^{k} \lambda_i(A)^{\downarrow} \geq \sum_{i=1}^{k} d_i(A)^{\downarrow} \qquad (4.3.46)$$

for each $k = 1, 2, \ldots, n$, *with equality for* $k = n$. *If the inequality (4.3.46) is an equality for some* $k \in \{1, \ldots, n-1\}$, *then* A *is permutation similar to* $B \oplus D$ *with* $B \in M_k$.

Proof. Let $P \in M_n$ be a permutation matrix such that the i, i entry of PAP^T is $d_i(A)^{\downarrow}$ for each $i = 1, \ldots, n$ (0.9.5). The vector of eigenvalues PAP^T (as well as of A) is $\lambda(A)$ of A. Partition PAP^T as in (4.3.29). Let $k \in \{1, \ldots, n-1\}$ be given. Then (4.3.36) ensures that $\lambda_1(A)^\downarrow + \cdots + \lambda_k(A)^\downarrow \geq d_1(A)^\downarrow + \cdots + d_k(A)^\downarrow$; of course, $\lambda_n(A)^\downarrow + \cdots + \lambda_1(A)^\downarrow = \operatorname{tr} A = d_1(A)^\downarrow + \cdots + d_n(A)^\downarrow$. If (4.3.46) is an equality for some $k \in \{1, \ldots, n-1\}$, then (4.3.34) ensures that $PAP^T = B \oplus D$ with $B \in M_k$. $\qquad\square$

Exercise. Let $x, y \in \mathbf{R}^n$. Explain why $(x^\downarrow + y^\downarrow)^\downarrow = x^\downarrow + y^\downarrow$.

Exercise. Let $x \in \mathbf{R}^n$. Explain why $(-x)^\downarrow = -(x^\uparrow)$.

Theorem 4.3.47. *Let $A, B \in M_n$ be Hermitian. Let $\lambda(A)$, $\lambda(B)$, and $\lambda(A+B)$, respectively, denote the real n-vectors of eigenvalues of A, B, and $A+B$, respectively. Then*

(a) (Fan) $\lambda(A)^\downarrow + \lambda(B)^\downarrow$ majorizes $\lambda(A+B)$

(b) (Lidskii) $\lambda(A+B)$ majorizes $\lambda(A)^\downarrow + \lambda(B)^\uparrow$

Proof. (a) For any $k \in \{1, \ldots, n-1\}$, use (4.3.40) to write the sum of the k largest eigenvalues of $A+B$ as

$$\sum_{i=1}^{k} \lambda_i(A+B)^\downarrow = \max_{\substack{V \in M_{n,k} \\ V^*V = I_k}} \operatorname{tr} V^*(A+B)V$$

$$= \max_{\substack{V \in M_{n,k} \\ V^*V = I_k}} \left(\operatorname{tr} V^*AV + V^*BV\right)$$

$$\leq \max_{\substack{V \in M_{n,k} \\ V^*V = I_k}} \operatorname{tr} V^*AV + \max_{\substack{V \in M_{n,k} \\ V^*V = I_k}} \operatorname{tr} V^*BV$$

$$= \sum_{i=1}^{k} \lambda_i(A)^\downarrow + \sum_{i=1}^{k} \lambda_i(B)^\downarrow = \sum_{i=1}^{k} (\lambda_i(A)^\downarrow + \lambda_i(B)^\downarrow)$$

Since $\operatorname{tr}(A+B) = \operatorname{tr} A + \operatorname{tr} B$, we have equality for $k = n$.

(b) It is convenient to restate Lidskii's inequalities in an equivalent form. If we replace A by $A' + B'$ and B by $-A'$, then we obtain the equivalent assertion "$\lambda(A+B) = \lambda(B')$ majorizes $\lambda(A)^\downarrow + \lambda(B)^\uparrow = \lambda(A'+B')^\downarrow + \lambda(-A')^\uparrow = \lambda(A'+B')^\downarrow - \lambda(A')^\downarrow$." Thus, it suffices to show that $\lambda(B)$ majorizes $\lambda(A+B)^\downarrow - \lambda(A)^\downarrow$ for any Hermitian $A, B \in M_n$. The case $k = n$ in the majorization inequalities follows as in (a): $\operatorname{tr} B = \operatorname{tr}(A+B) - \operatorname{tr} A$. Let $k \in \{1, \ldots, n-1\}$ be given. We must show that

$$\sum_{i=1}^{k} (\lambda_i(A+B)^\downarrow - \lambda_i(A)^\downarrow)^\downarrow \leq \sum_{i=1}^{k} \lambda_i(B)^\downarrow$$

We may assume that $\lambda_k(B)^\downarrow = 0$; otherwise, replace B by $B - \lambda_k(B)^\downarrow I$, which diminishes both sides of the preceding inequality by $k\lambda_k(B)$. Express B as a difference of two positive semidefinite matrices as in (4.1.13): $B = B_+ - B_-$. Then $\lambda_i(B)^\downarrow = \lambda_i(B_+)^\downarrow$ for each $i = 1, \ldots, k$ and so $\sum_{i=1}^{k} \lambda_i(B)^\downarrow = \sum_{i=1}^{k} \lambda_i(B_+)^\downarrow = \sum_{i=1}^{n} \lambda_i(B_+) = \operatorname{tr} B_+$

since $\lambda_i(B)^\downarrow \leq 0$ for all $i \geq k$. Thus, we must prove that

$$\sum_{i=1}^{k} (\lambda_i(A + B_+ - B_-)^\downarrow - \lambda_i(A)^\downarrow)^\downarrow \leq \operatorname{tr} B_+$$

Since $-B_-$ is positive semidefinite, (4.3.12) ensures that $\lambda_i(A + B_+ - B_-)^\downarrow - \lambda_i(A)^\downarrow \leq \lambda_i(A + B_+) - \lambda_i(A)^\downarrow$ for each $i = 1, \ldots, n$ and also that $\lambda_i(A + B_+)^\downarrow \geq \lambda_i(A)^\downarrow$ for each $i = 1, \ldots, n$. Therefore,

$$\sum_{i=1}^{k} (\lambda_i(A + B_+ - B_-)^\downarrow - \lambda_i(A)^\downarrow)^\downarrow \leq \sum_{i=1}^{k} (\lambda_i(A + B_+)^\downarrow - \lambda_i(A)^\downarrow)^\downarrow$$

$$\leq \sum_{i=1}^{n} (\lambda_i(A + B_+)^\downarrow - \lambda_i(A)^\downarrow)^\downarrow = \sum_{i=1}^{n} (\lambda_i(A + B_+)^\downarrow - \lambda_i(A)^\downarrow)$$

$$= \operatorname{tr}(A + B_+) - \operatorname{tr} A = \operatorname{tr} A + \operatorname{tr} B_+ - \operatorname{tr} A = \operatorname{tr} B_+$$

$\square$

Exercise. Under the assumptions of the preceding theorem, explain why $\lambda(A)^\uparrow + \lambda(B)^\uparrow$ majorizes $\lambda(A + B)$ and why $\lambda(B)$ majorizes $\lambda(A + B)^\uparrow - \lambda(A)^\uparrow$.

The following converse of (4.3.45) shows that majorization is the *precise* relationship between the diagonal entries and eigenvalues of a Hermitian matrix.

Theorem 4.3.48. *Let $n \geq 1$, let $x = [x_i] \in \mathbf{R}^n$ and $y = [y_i] \in \mathbf{R}^n$ be given, and suppose that x majorizes y. Let $\Lambda = \operatorname{diag} x \in M_n(\mathbf{R})$. There exists a real orthogonal matrix Q such that $\operatorname{diag}(Q^T \Lambda Q) = y$, that is, there is a real symmetric matrix whose eigenvalues are $x_1, \ldots, x_n$ and whose main diagonal entries are $y_1, \ldots, y_n$.*

Proof. There is no loss of generality if we assume that the entries of the vectors x and y are in nonincreasing order: $x_1 \geq x_2 \geq \cdots$ and $y_1 \geq y_2 \geq \cdots$.

The assertion is trivial for $n = 1$: $x_1 = y_1$, $Q = [1]$, and $A = [x_1]$, so we may assume that $n \geq 2$.

The inequalities (4.3.44a,b) ensure that $x_1 \geq y_1 \geq y_n \geq x_n$, so if $x_1 = x_n$ it follows that all the entries of x and y are equal, $Q = I$, and $A = x_1 I$. We may therefore assume that $x_1 > x_n$.

For $n = 2$, we have $x_1 > x_2$ and $x_1 \geq y_1 \geq y_2 = (x_1 - y_1) + x_2 \geq x_2$. Consider the real matrix

$$P = \frac{1}{\sqrt{x_1 - x_2}} \begin{bmatrix} \sqrt{x_1 - y_2} & -\sqrt{y_2 - x_2} \\ \sqrt{y_2 - x_2} & \sqrt{x_1 - y_2} \end{bmatrix}$$

A computation reveals that $PP^T = I$, so P is real orthogonal. A further computation reveals that the 1, 1 entry of $P^T \Lambda P$ is $x_1 + x_2 - y_2 = y_1$, and the 2, 2 entry is y_2, that is, $\operatorname{diag}(P^T \Lambda P) = [y_1 \ y_2]^T$.

We now proceed by induction. Suppose that $n \geq 3$ and assume that the theorem is true if the vectors x and y have size at most $n - 1$.

Let $k \in \{1, \ldots, n\}$ be the largest integer such that $x_k \geq y_1$. Since $x_1 \geq y_1$, we know that $k \geq 1$, and we may assume that $k \leq n - 1$ and hence that $x_k \geq y_1 > x_{k+1} \geq x_n$.

Why? If $k = n$, then $x_n \geq y_1 \geq y_n \geq x_n$, so $x_n = y_1$ and $y_i = y_1$ for each $i = 1, \ldots, n$. Moreover, the condition $\sum_{i=1}^{n} x_i = \sum_{i=1}^{n} y_i$ implies that

$$0 = \sum_{i=1}^{n} (x_i - y_i) = \sum_{i=1}^{n} (x_i - y_1) \geq \sum_{i=1}^{n} (x_n - y_1) = \sum_{i=1}^{n} 0 = 0$$

Since each $x_i - y_1 \geq x_n - y_1 \geq 0$, we conclude that each $x_i = y_1$, which violates our general assumption that $x_1 > x_n$.

Let $\eta = x_k + x_{k+1} - y_1$ and observe that $\eta = (x_k - y_1) + x_{k+1} \geq x_{k+1}$. Then $x_k \geq y_1 > x_{k+1}$, and $x_k + x_{k+1} = y_1 + \eta$, so the vector $[x_k \ x_{k+1}]^T$ majorizes the vector $[y_1 \ \eta]^T$ and $x_k > x_{k+1}$. Let $D_1 = \begin{bmatrix} x_k & 0 \\ 0 & x_{k+1} \end{bmatrix}$. Our construction in the case $n = 2$ shows how to obtain a real orthogonal matrix P_1 such that $\mathrm{diag}(P_1^T D_1 P_1) = [y_1 \ \eta]^T$. Since $x_k = \eta + (y_1 - x_{k+1}) > \eta$, we have $x_1 > \eta \geq x_2 \geq \cdots \geq x_n$ if $k = 1$ and we have $x_1 \geq \cdots \geq x_{k-1} \geq x_k > \eta \geq x_{k+1} \geq \cdots \geq x_n$ if $k > 1$. Let $D_2 = \mathrm{diag}(x_3, \ldots, x_n)$ if $k = 1$ and let $D_2 = \mathrm{diag}(x_1, \ldots, x_{k-1}, x_{k+2}, \ldots, x_n)$ if $k > 1$. Then

$$\begin{bmatrix} P_1 & 0 \\ 0 & I_{n-2} \end{bmatrix}^T \begin{bmatrix} D_1 & 0 \\ 0 & D_2 \end{bmatrix} \begin{bmatrix} P_1 & 0 \\ 0 & I_{n-2} \end{bmatrix} = \begin{bmatrix} y_1 & z^T \\ z & [\eta] \oplus D_2 \end{bmatrix}$$

for some $z \in \mathbf{R}^{n-1}$. It suffices to show that there is a real orthogonal $P_2 \in M_{n-1}$ such that $\mathrm{diag}(P_2^T([\eta] \oplus D_2)P_2) = [y_2 \ \ldots \ y_n]^T$. According to the induction hypothesis, such a P_2 exists if the vector $\hat{x} = \mathrm{diag}([\eta] \oplus D_2)$ majorizes the vector $\hat{y} = [y_2 \ \ldots \ y_n]^T$.

Observe that $\hat{y} = \hat{y}^{\downarrow}$. If $k = 1$, then $\hat{x} = \hat{x}^{\downarrow} = [\eta \ x_3 \ \ldots \ x_n]^T$; if $k > 1$, then $\hat{x}^{\downarrow} = [x_1 \ \ldots \ x_{k-1} \ \eta \ x_{k+2} \ \ldots \ x_n]^T$ because $x_{k-1} \geq x_k > \eta \geq x_{k+1} \geq x_{k+2}$.

Suppose that $k = 1$. Then

$$\sum_{i=1}^{m} \hat{x}_i^{\downarrow} = \eta + \sum_{i=2}^{m} x_{i+1} = x_1 + x_2 - y_1 + \sum_{i=3}^{m+1} x_i$$

$$= \sum_{i=1}^{m+1} x_i - y_1 \geq \sum_{i=1}^{m+1} y_i - y_1 = \sum_{i=2}^{m+1} y_i = \sum_{i=1}^{m} \hat{y}_i^{\downarrow}$$

for each $m = 1, \ldots, n - 1$, with equality for $m = n - 1$.

Now suppose that $k > 1$. For each $m \in \{1, \ldots, k - 1\}$ we have

$$\sum_{i=1}^{m} \hat{x}_i^{\downarrow} = \sum_{i=1}^{m} x_i \geq \sum_{i=1}^{m} y_i \geq \sum_{i=1}^{m} y_{i+1} = \sum_{i=2}^{m+1} y_i = \sum_{i=1}^{m} \hat{y}_i^{\downarrow}$$

For $m = k$ we have

$$\sum_{i=1}^{k} \hat{x}_i^{\downarrow} = \sum_{i=1}^{k-1} x_i + \eta = \sum_{i=1}^{k-1} x_i + x_k + x_{k+1} - y_1 = \sum_{i=1}^{k+1} x_i - y_1$$

$$\geq \sum_{i=1}^{k+1} y_i - y_1 = \sum_{i=2}^{k+1} y_i = \sum_{i=1}^{k} \hat{y}_i^{\downarrow}$$

with equality if $k = n - 1$. Finally, if $k \le n - 2$ and $m \in \{k + 1, \ldots, n - 1\}$, we have

$$
\sum_{i=1}^{m} \hat{x}_i^{\downarrow} = \sum_{i=1}^{k} \hat{x}_i^{\downarrow} + \sum_{i=k+1}^{m} \hat{x}_i^{\downarrow} = \sum_{i=1}^{k+1} x_i - y_1 + \sum_{i=k+1}^{m} x_{i+1}
$$

$$
= \sum_{i=1}^{k+1} x_i - y_1 + \sum_{i=k+2}^{m+1} x_i = \sum_{i=1}^{m+1} x_i - y_1
$$

$$
\ge \sum_{i=1}^{m+1} y_i - y_1 = \sum_{i=2}^{m+1} y_i = \sum_{i=1}^{m} \hat{y}_i^{\downarrow}
$$

with equality for $m = n - 1$. $\qquad\square$

The preceding theorem permits us to give a geometric characterization of the majorization relation. A matrix $A = [a_{ij}] \in M_n$ is *doubly stochastic* if it has nonnegative entries and the sum of the entries in every row and every column is equal to $+1$. Theorem 8.7.2 (Birkhoff's theorem) says that an n-by-n matrix is doubly stochastic if and only if it is a convex combination of (at most $n!$) permutation matrices.

Exercise. If $S, P_1, P_2 \in M_n$, S is doubly stochastic, and P_1 and P_2 are permutation matrices, explain why $P_1 S P_2$ is doubly stochastic.

Exercise. Let $A \in M_n$ and let $e \in \mathbf{R}^n$ be the vector whose entries are all $+1$. Explain why the sum of the entries in every row and column of A is equal to $+1$ if and only if $Ae = A^T e = e$.

Theorem 4.3.49. *Let $n \ge 2$, let $x = [x_i] \in \mathbf{R}^n$, and let $y = [y_i] \in \mathbf{R}^n$ be given. The following are equivalent:*

(a) x majorizes y.
(b) There is a doubly stochastic $S = [s_{ij}] \in M_n$ such that $y = Sx$.
(c) $y \in \{\sum_{i=1}^{n!} \alpha_i P_i x : \alpha_i \ge 0, \sum_{i=1}^{n!} \alpha_i = 1,$ and each P_i is a permutation matrix$\}$.

Proof. If x majorizes y, then the preceding theorem ensures that there is a real orthogonal matrix $Q = [q_{ij}] \in M_n$ such that $y = \operatorname{diag}(Q \operatorname{diag}(x) Q^T)$. A computation reveals that $y_i = \sum_{j=1}^{n} q_{ij}^2 x_j$ for each $i = 1, \ldots, n$, that is, $y = Sx$ with $S = [q_{ij}^2] \in M_n$. The entries of S are nonnegative, and its row and column sums are all equal to $+1$ since every row and column of Q is a unit vector.

Theorem 8.7.2 asserts the equivalence of (b) and (c), so it suffices to show that (b) implies (a).

Suppose that $y = Sx$ and that S is doubly stochastic. Let P_1 and P_2 be permutation matrices such that $x = P_1 x^{\downarrow}$ and $y = P_2 y^{\downarrow}$. Then $y^{\downarrow} = (P_2^T S P_1) x^{\downarrow}$. Invoking the preceding exercise, we see that there is no loss of generality to assume that $x_1 \ge \cdots \ge x_n$, $y_1 \ge \cdots \ge y_n$, $y = Sx$, and S is doubly stochastic. Let $w_j^{(k)} = \sum_{i=1}^{k} s_{ij}$ and observe that $0 \le w_j^{(k)} \le 1$, $w_j^{(n)} = 1$, and $\sum_{j=1}^{n} w_j^{(k)} = k$. Since $y_i = \sum_{j=1}^{n} s_{ij} x_j$, we have $\sum_{i=1}^{k} y_i = \sum_{i=1}^{k} \sum_{j=1}^{n} s_{ij} x_j = \sum_{j=1}^{n} w_j^{(k)} x_j$ for each $k = 1, \ldots, n$. In particular,

$\sum_{i=1}^{n} y_i = \sum_{j=1}^{n} w_j^{(n)} x_j = \sum_{j=1}^{n} x_j$. For $k \in \{1, \ldots, n-1\}$ compute

$$\sum_{i=1}^{k} x_i - \sum_{i=1}^{k} y_i = \sum_{i=1}^{k} x_i - \sum_{i=1}^{n} w_i^{(k)} x_i$$

$$= \sum_{i=1}^{k} (x_i - x_k) + k x_k - \sum_{i=1}^{n} w_i^{(k)}(x_i - x_k) - k x_k$$

$$= \sum_{i=1}^{k} (x_i - x_k) - \sum_{i=1}^{k} w_i^{(k)}(x_i - x_k) - \sum_{i=k+1}^{n} w_i^{(k)}(x_i - x_k)$$

$$= \sum_{i=1}^{k} (x_i - x_k)(1 - w_i^{(k)}) + \sum_{i=k+1}^{n} w_i^{(k)}(x_k - x_i) \geq 0$$

We conclude that $\sum_{i=1}^{k} x_i \geq \sum_{i=1}^{k} y_i$ for each $k = 1, \ldots, n$ with equality for $k = n$. $\qquad\qquad\square$

Thus, the set of all vectors that are majorized by a given vector x is the convex hull of the (at most $n!$ distinct) vectors obtained by permuting the entries of x.

The following characterization of the majorization relationship tells us that the eigenvalues of the Hermitian part of a matrix A majorize the Hermitian parts of the eigenvalues of A.

Theorem 4.3.50. *Let $x = [x_i] \in \mathbf{R}^n$ and $z = [z_i] \in \mathbf{C}^n$ be given. Then x majorizes $\operatorname{Re} z = [\operatorname{Re} z_i]_{i=1}^n$ if and only if there is an $A \in M_n$ such that $z_1, \ldots, z_n$ are the eigenvalues of A and $x_1, \ldots, x_n$ are the eigenvalues of $H(A) = (A + A^*)/2$.*

Proof. Let $\lambda_1, \ldots, \lambda_n$ be the eigenvalues of $A \in M_n$, and use (2.3.1) to write $A = UTU^*$, in which $T = [t_{ij}] \in M_n$ is upper triangular and $t_{ii} = \lambda_i$ for $i = 1, \ldots, n$. A computation reveals that $H(A) = U H(T) U^*$ and $\operatorname{diag} H(T) = [\operatorname{Re} \lambda_1 \ \cdots \ \operatorname{Re} \lambda_n]^T$, which is majorized by the eigenvalues of $H(T)$, which are the same as the eigenvalues of $H(A)$ (4.3.45).

Conversely, if x majorizes $\operatorname{Re} z$, there is a Hermitian $B = [b_{ij}] \in M_n$ such that the entries of x are the eigenvalues of B and $\operatorname{diag} B = \operatorname{Re} z$ (4.3.48). Let $T = [t_{ij}] \in M_n$ be the upper triangular matrix such that $\operatorname{diag} T = z$ and $t_{ij} = 2b_{ij}$ for $1 \leq i < j \leq n$. Then $z_1, \ldots, z_n$ are the eigenvalues of T, and $H(T) = B$, whose eigenvalues are $x_1, \ldots, x_n$. $\qquad\square$

Our final result involving majorization concerns bounds on $\operatorname{tr} AB$, in which each of A and B (but not necessarily their product) is Hermitian.

Exercise. Let $x = [x_i]$, $y = [y_i] \in \mathbf{R}^n$ and suppose that x majorizes y. Explain why $\sum_{i=1}^{k} x_i^{\downarrow} \geq \sum_{i=1}^{k} y_i$ for each $k = 1, \ldots, n$, with equality for $k = n$.

Exercise. Let $x = [x_i]$, $y = [y_i] \in \mathbf{R}^n$. Explain why x majorizes y if and only if $-x$ majorizes $-y$.

Lemma 4.3.51. *Let* $x = [x_i]$, $y = [y_i]$, $w = [w_i] \in \mathbf{R}^n$. *Suppose that* x *majorizes* y. *Then*

$$\sum_{i=1}^n w_i^\downarrow x_i^\uparrow \leq \sum_{i=1}^n w_i^\downarrow y_i \leq \sum_{i=1}^n w_i^\downarrow x_i^\downarrow \qquad (4.3.52)$$

Let $\hat{X}_k = \sum_{i=1}^k x_i^\downarrow$, $\check{X}_k = \sum_{i=1}^k x_i^\uparrow$, *and* $Y_k = \sum_{i=1}^k y_i$ *for each* $k = 1, \dots, n$. *The right-hand inequality in (4.3.52) is an equality if and only if*

$$(w_i^\downarrow - w_{i+1}^\downarrow)(\hat{X}_i - Y_i) = 0 \text{ for each } i = 1, \dots, n-1 \qquad (4.3.52a)$$

The left-hand inequality in (4.3.52) is an equality if and only if

$$(w_i^\downarrow - w_{i+1}^\downarrow)(\check{X}_i - Y_i) = 0 \text{ for each } i = 1, \dots, n-1 \qquad (4.3.52b)$$

Proof. Since x majorizes y, we have $\hat{X}_k \geq Y_k$ for each $k = 1, \dots, n-1$ and $\hat{X}_n = Y_n$. Use partial summation (twice) and the assumption that each $w_i^\downarrow - w_{i+1}^\downarrow$ is nonnegative to compute

$$\sum_{i=1}^k w_i^\downarrow y_i = \sum_{i=1}^{n-1}(w_i^\downarrow - w_{i+1}^\downarrow)Y_i + w_n^\downarrow Y_n \leq \sum_{i=1}^{n-1}(w_i^\downarrow - w_{i+1}^\downarrow)\hat{X}_i + w_n^\downarrow Y_n$$

$$= \sum_{i=1}^{n-1}(w_i^\downarrow - w_{i+1}^\downarrow)\hat{X}_i + w_n^\downarrow \hat{X}_n = \sum_{i=1}^k w_i^\downarrow x_i^\downarrow$$

This proves the asserted upper bound, which is an equality if and only if $(w_i^\downarrow - w_{i+1}^\downarrow)Y_i = (w_i^\downarrow - w_{i+1}^\downarrow)\hat{X}_i$ for each $i = 1, \dots, n-1$, that is, if and only if $(w_i^\downarrow - w_{i+1}^\downarrow)(\hat{X}_i - Y_i) = 0$ for each $i = 1, \dots, n-1$. Since $-x$ majorizes $-y$, we may apply the upper bound to the vectors $-x$, $-y$, and w:

$$\sum_{i=1}^n w_i^\downarrow(-y_i) \leq \sum_{i=1}^n w_i^\downarrow(-x)_i^\downarrow - \sum_{i=1}^n w_i^\downarrow(-(x)_i^\uparrow) = -\sum_{i=1}^n w_i^\downarrow x_i^\uparrow$$

that is, $\sum_{i=1}^n w_i^\downarrow x_i^\uparrow \leq \sum_{i=1}^n w_i^\downarrow y_i$, which is the asserted lower bound. The case of equality follows in the same way, using the inequality $-Y_k \leq -\check{X}_k$. $\qquad \square$

Theorem 4.3.53. *Let* $A, B \in M_n$ *be Hermitian and have respective vectors of eigenvalues* $\lambda(A) = [\lambda_i(A)]_{i=1}^n$ *and* $\lambda(B) = [\lambda_i(B)]_{i=1}^n$. *Then*

$$\sum_{i=1}^n \lambda_i(A)^\downarrow \lambda_i(B)^\uparrow \leq \operatorname{tr} AB \leq \sum_{i=1}^n \lambda_i(A)^\downarrow \lambda_i(B)^\downarrow \qquad (4.3.54)$$

If either inequality in (4.3.54) is an equality, then A and B commute. If the right-hand inequality in (4.3.54) is an equality, then there is a unitary $U \in M_n$ such that $A = U \operatorname{diag}(\lambda(A)^\downarrow)U^*$ and $B = U \operatorname{diag}(\lambda(B)^\downarrow)U^*$. If the left-hand inequality in (4.3.54) is an equality, then there is a unitary $U \in M_n$ such that $A = U \operatorname{diag}(\lambda(A)^\downarrow)U^*$ and $B = U \operatorname{diag}(\lambda(B)^\uparrow)U^*$.

Proof. Let $A = U\Lambda U^*$, in which $U \in M_n$ is unitary and $\Lambda = \operatorname{diag} \lambda(A)^\downarrow$. Let $\tilde{B} = [\beta_{ij}] = U^* BU$. Then $\operatorname{tr} AB = \operatorname{tr} U\Lambda U^* B = \operatorname{tr} \Lambda U^* BU = \operatorname{tr} \Lambda \tilde{B} = \sum_{i=1}^n \lambda_i(A)^\downarrow \beta_{ii}$.

Since the vector of eigenvalues of $\tilde{B}$ (which is the vector of eigenvalues of B) majorizes the vector of main diagonal entries of $\tilde{B}$ (4.3.48), the asserted inequalities follow from the preceding lemma applied to $x = \lambda(B)$, $y = \text{diag}\,\tilde{B}$, and $w = \lambda(A)$.

Suppose that the right-hand inequality in (4.3.54) is an equality. Suppose that A has k distinct eigenvalues $\alpha_1 > \alpha_2 > \cdots > \alpha_k$ with respective multiplicities n_1, $n_2, \ldots, n_k$, and partition $\tilde{B} = [\tilde{B}_{ij}]_{i,j=1}^k$ so that each $\tilde{B}_{ii} \in M_{n_i}$. The case of equality in (4.3.52) ensures that each $(\lambda_i(A)^\downarrow - \lambda_{i+1}(A)^\downarrow)(\check{X}_i - Y_i) = 0$, which is interesting only for $i = n_1, n_1 + n_2, \ldots$, in which cases $\lambda_{n_i}(A)^\downarrow - \lambda_{n_i+1}(A)^\downarrow = \alpha_i - \alpha_{i+1} > 0$ and necessarily $\check{X}_{n_1+\cdots+n_i} - Y_{n_1+\cdots+n_i} = 0$. Thus, equality in the right-hand inequality in (4.3.54) implies that

$$Y_{n_1+\cdots+n_p} = \text{tr}\,\tilde{B}_{11} + \cdots + \text{tr}\,\tilde{B}_{pp} = \check{X}_{n_1+\cdots+n_p} = \sum_{i=1}^{n_1+\cdots+n_p} \lambda_i(\tilde{B})^\downarrow$$

for each $p = 1, \ldots, k-1$. It now follows from Corollary 4.3.34 (the equalities (4.3.36b)) that $\tilde{B} = \tilde{B}_{11} \oplus + \cdots + \tilde{B}_{kk}$; the eigenvalues of $\tilde{B}_{11}$ are $\lambda_1(B)^\downarrow, \ldots, \lambda_{n_1}(B)^\downarrow$, the eigenvalues of $\tilde{B}_{22}$ are $\lambda_{n_1+1}(B)^\downarrow, \ldots, \lambda_{n_1+n_2}(B)^\downarrow$, and so forth. Since $\Lambda = \alpha_1 I_{n_1} \oplus \cdots \oplus \alpha_k I_{n_k}$ is conformal to the block diagonal matrix $\tilde{B}$, we have $\Lambda\tilde{B} = \tilde{B}\Lambda$, which is $\Lambda U^* B U = U^* B U \Lambda$ or $AB = U\Lambda U^* B = BU\Lambda U^* = BA$. Finally, each $\tilde{B}_{ii} = \tilde{U}_i \tilde{\Lambda}_i \tilde{U}_i^*$, in which $\tilde{U}_i \in M_{n_i}$ is unitary and $\tilde{\Lambda}_i$ is a diagonal matrix whose main diagonal entries are the nonincreasingly ordered eigenvalues of $\tilde{B}_{ii}$. Let $\tilde{U} = \tilde{U}_1 \oplus \cdots \oplus \tilde{U}_k$ and observe that $\tilde{\Lambda} = \tilde{\Lambda}_1 \oplus \cdots \oplus \tilde{\Lambda}_k = \text{diag}(\lambda(B)^\downarrow)$. Moreover, $\tilde{B} = \tilde{U}\tilde{\Lambda}\tilde{U}^*$ and $\tilde{U}\Lambda\tilde{U}^* = \Lambda$. Then $A = U\Lambda U^* = (U\tilde{U})\Lambda(U\tilde{U})^* = (U\tilde{U})\,\text{diag}(\lambda(A)^\downarrow)(U\tilde{U})^*$ and $B = U\,\tilde{B}U^* = (U\,\tilde{U})\tilde{\Lambda}(U\tilde{U})^* = (U\,\tilde{U})\,\text{diag}(\lambda(B)^\downarrow)(U\tilde{U})^*$.

The case of equality in the left-hand inequality in (4.3.54) follows from replacing B by $-B$ in the right-hand inequality. $\qquad\square$

Problems

4.3.P1 Let $A, B \in M_n$ be Hermitian. Use (4.3.1) to show that $\lambda_1(B) \le \lambda_i(A+B) - \lambda_i(A) \le \lambda_n(B)$ and conclude that $|\lambda_i(A+B) - \lambda_i(A)| \le \rho(B)$ for all $i = 1, \ldots, n$. This is a simple example of a *perturbation theorem* for the eigenvalues of a Hermitian matrix; see (6.3) for more perturbation theorems.

4.3.P2 Consider $A = \begin{bmatrix} 0 & 1 \\ 0 & 0 \end{bmatrix}$ and $B = \begin{bmatrix} 0 & 0 \\ 1 & 0 \end{bmatrix}$ and show that Weyl's inequalities (4.3.2a,b) need not hold if A and B are not Hermitian.

4.3.P3 If $A, B \in M_n$ are Hermitian and their eigenvalues are arranged as in (4.2.1), explain why $\lambda_i(A+B) \le \min\{\lambda_j(A) + \lambda_k(B) : j + k = i + n\}$, $i \in \{1, \ldots, n\}$.

4.3.P4 If $A, B \in M_n$ are Hermitian and $A - B$ has only nonnegative eigenvalues, explain why $\lambda_i(A) \ge \lambda_i(B)$ for all $i = 1, 2, \ldots, n$.

4.3.P5 Let $A \in M_n$ be Hermitian, let $a_k = \det A[\{1, \ldots, k\}]$ be the leading principal minor of A of size k, $k = 1, \ldots, n$, and suppose that all $a_k \ne 0$. Show that the number of negative eigenvalues of A is equal to the number of sign changes in the sequence $+1, a_1, a_2, \ldots, a_n$. Explain why A is positive definite if and only if every principal minor of A is positive. What happens if some $a_i = 0$?

4.3.P6 Suppose that $A = [a_{ij}] \in M_n$ is Hermitian, has smallest and largest eigenvalues λ_1 and λ_n, and for some $i \in \{1, \ldots, n\}$, either $a_{ii} = \lambda_1$ or $a_{ii} = \lambda_n$. Use (4.3.34) to show that $a_{ik} = a_{ki} = 0$ for all $k = 1, \ldots, n, k \neq i$. Does anything special happen if a main diagonal entry of A is an eigenvalue of A different from λ_1 and λ_n?

4.3.P7 Provide details for the following sketch of a proof of (4.3.45) that proceeds by induction on the dimension and uses Cauchy's interlacing theorem. For $n = 1$, there is nothing to show; suppose that the asserted majorization is valid for Hermitian matrices of size $n - 1$. Let $\hat{A} \in M_{n-1}$ be the principal submatrix obtained by deleting the row and column of A corresponding to its algebraically smallest diagonal entry $d_n^{\downarrow}$. Let $\hat{\lambda}^{\downarrow}$ be the vector of nonincreasingly ordered eigenvalues of $\hat{A}$. The induction hypothesis ensures that $\sum_{i=1}^{k} \hat{\lambda}_i^{\downarrow} \geq \sum_{i=1}^{k} d_i^{\downarrow}$, and (4.3.17) ensures that $\sum_{i=1}^{k} \lambda_i^{\downarrow} \geq \sum_{i=1}^{k} \hat{\lambda}_i^{\downarrow}$, both for $k = 1, \ldots, n - 1$. Thus, $\sum_{i=1}^{k} \lambda_i^{\downarrow} \geq \sum_{i=1}^{k} d_i^{\downarrow}$ for $k = 1, \ldots, n - 1$. Why is there equality for $k = n$?

4.3.P8 Let $e \in \mathbf{R}^n$ be the vector all of whose entries are one, let $e_i \in \mathbf{R}^n$ be one of the standard Euclidean basis vectors, and let $y \in \mathbf{R}^n$. (a) If e majorizes y, show that $y = e$. (b) If e_i majorizes y, show that all the entries of y lie between zero and one.

4.3.P9 Let $A \in M_n(\mathbf{R})$ and suppose that x majorizes Ax for every $x \in \mathbf{R}^n$. Show that A is doubly stochastic.

4.3.P10 If $A = [a_{ij}] \in M_n$ is normal, then $A = U\Lambda U^*$, in which $U = [u_{ij}] \in M_n$ is unitary and $\Lambda = \text{diag}(\lambda_1, \ldots, \lambda_n) \in M_n$, which need not be real. Show that $S = [|u_{ij}|^2]$ is doubly stochastic and that $\text{diag } A = S(\text{diag } \Lambda)$. A doubly stochastic matrix S that arises from a unitary matrix U in this way is called *unistochastic*; if U is real (as in the proof of (4.3.52)), S is called *orthostochastic*.

4.3.P11 Let $A = [a_{ij}] \in M_n$. Provide details for the following argument to show that if A has some "small" columns or rows, then it must also have some "small" singular values. Let $\sigma_1^2 \geq \cdots \geq \sigma_n^2$ be the ordered squares of the singular values of A (the nonincreasingly ordered eigenvalues of AA^*). Let $R_1^2 \geq \cdots \geq R_n^2$ be the ordered squared Euclidean lengths of the rows of A (the nonincreasingly ordered main diagonal entries of AA^*). Explain why $\sum_{i=n-k+1}^{n} R_i^2 \geq \sum_{i=n-k+1}^{n} \sigma_i^2$ for $k = 1, \ldots, n$, with a similar set of inequalities involving the Euclidean lengths of the columns. If A is normal, what can you conclude about the eigenvalues of A?

4.3.P12 Let $A \in M_n$ be partitioned as in (4.3.29) with $B = [b_{ij}] \in M_m$ and $C = [c_{ij}] \in M_{m,n-m}$. Continue to use the notation of the preceding problem. If the m largest singular values of A are the singular values of B, show that $C = 0$ and $A = B \oplus D$.

4.3.P13 Provide details for the following sketch of a proof that Cauchy's interlacing theorem for a bordered Hermitian matrix (4.3.17) implies the interlacing theorem for a rank-one perturbation of a Hermitian matrix (4.3.9): Let $z \in \mathbf{C}^n$ and let $A \in M_n$ be Hermitian. We wish to prove (4.3.10). As in the proof of (4.3.26), we may assume that $A = \Lambda = \text{diag}(\lambda_1, \ldots, \lambda_n)$ is diagonal and positive definite. Why? Let $R = \text{diag}(\lambda_1^{1/2}, \ldots, \lambda_n^{1/2})$. Then $\Lambda + zz^* = \begin{bmatrix} R & z \end{bmatrix} \begin{bmatrix} R \\ z^* \end{bmatrix}$ has the same eigenvalues (except for one extra zero) as $\begin{bmatrix} R \\ z^* \end{bmatrix} \begin{bmatrix} R & z \end{bmatrix} = \begin{bmatrix} \Lambda & Rz \\ z^*R & z^*z \end{bmatrix}$, which are $0 \leq \lambda_1(\Lambda + zz^*) \leq \cdots \leq \lambda_n(\Lambda + zz^*)$. Cauchy's theorem says that these eigenvalues are interlaced by those of Λ.

4.3.P14 Let $r \in \{1, \ldots, n\}$ and let H_n denote the real vector space of n-by-n Hermitian matrices. For a given $A \in H_n$, order its eigenvalues as in (4.2.1). Let $f_r(A) = \lambda_1(A) +$

$\cdots + \lambda_r(A)$ and let $g_r(A) = \lambda_{n-r+1}(A) + \cdots + \lambda_n(A)$. Show that f_r is a concave function on H_n and that g_r is a convex function on H_n.

4.3.P15 Let $A \in M_n$ be positive semidefinite and let $m \in \{1, \ldots, n\}$. (a) Let $V \in M_{n,m}$ have orthonormal columns. Use (4.3.37) to show that

$$\lambda_1(A) \cdots \lambda_m(A) \le \det V^* A V \le \lambda_{n-m+1}(A) \cdots \lambda_n(A)$$

(b) If $X \in M_{n,m}$, show that

$$\lambda_1(A) \cdots \lambda_m(A) \det X^* X \le \det X^* A X \le \lambda_{n-m+1}(A) \cdots \lambda_n(A) \det X^* X$$

4.3.P16 (a) If $A \in M_2$ is normal, show that spread $A \ge 2|a_{12}|$ and give an example to show that this bound is sharp. Why is spread $A \ge 2|a_{21}|$ as well? (b) If $A \in M_n$ is Hermitian and $\hat{A}$ is a principal submatrix of A, show that spread $A \ge$ spread $\hat{A}$ and use (a) to conclude that spread $A \ge 2 \max\{|a_{ij}| : i, j = 1, \ldots, n, i \ne j\}$. For an upper bound on spread A, see (2.5.P61).

4.3.P17 Let $A = [a_{ij}] \in M_n$ be Hermitian and tridiagonal, and assume that $a_{i,i+1} \ne 0$ for each $i = 1, \ldots, n-1$ (A is *unreduced* (0.9.9); it is also *irreducible* (6.2.22) since it is Hermitian). This problem presents two proofs that the interlacing inequalities (4.3.18) are strict inequalities and three proofs that A has distinct eigenvalues. (a) Let $x = [x_i]$ be an eigenvector of A. Show that $x_n \ne 0$. (b) Let $\hat{A}$ be any principal submatrix of A of size $n - 1$. Use the conditions in (4.3.17) for equality to show that the interlacing inequalities (4.3.18) between the ordered eigenvalues of A and $\hat{A}$ are all strict inequalities. (c) Deduce from (b) that A has distinct eigenvalues. (d) Use (1.4.P11) to show in a different way that A has distinct eigenvalues. (e) Let $p_k(t)$ be the characteristic polynomial of the leading k-by-k principal submatrix of A, and let $p_0(t) = 1$. Show that $p_1(t) = t - a_{11}$ and that $p_k(t) = (t - a_{kk})p_{k-1}(t) - |a_{k-1,k}|^2 p_{k-2}(t)$ for $k = 2, \ldots, n$. (f) Use (e) to prove that the interlacing inequalities (4.3.18) between the ordered eigenvalues of A and $\hat{A}$ are all strict inequalities, and deduce that A has distinct eigenvalues.

4.3.P18 Let $\Lambda = \mathrm{diag}(\lambda_1, \ldots, \lambda_n) \in M_n(\mathbf{R})$ and suppose that that $\lambda_1 < \cdots < \lambda_n$ (distinct eigenvalues). If no entry of $z \in \mathbf{C}^n$ is zero, explain why $\lambda_1 < \lambda_1(\Lambda + zz^*) < \cdots < \lambda_{n-1}(\Lambda + zz^*) < \lambda_n < \lambda_n(\Lambda + zz^*)$ (interlacing with all inequalities strict).

4.3.P19 Let $A \in M_n$ be Hermitian and let $z \in \mathbf{C}^n$. Using the notation of (4.3.9), explain why $\lambda_i(A + zz^*) = \lambda_i(A) + \mu_i$ for each $i = 1, \ldots, n$, and each $\mu_i \ge 0$. Show that $\sum_{i=1}^n \mu_i = z^* z = \|z\|_2^2$.

4.3.P20 Let $\lambda \in \mathbf{R}$, $a \in \mathbf{R}$, $y \in \mathbf{C}^n$, and $A = \begin{bmatrix} \lambda I_n & y \\ y^* & a \end{bmatrix} \in M_{n+1}$. Use (4.3.17) to show that λ is an eigenvalue of A with multiplicity at least $n - 1$. What are the other two eigenvalues?

4.3.P21 Let $A \in M_n$ be Hermitian, let $a \in \mathbf{R}$, and let $y \in \mathbf{C}^n$. (a) Let $\hat{A} = \begin{bmatrix} A & y \\ y^* & a \end{bmatrix} \in M_{n+1}$. Explain why rank $\hat{A}$ − rank A can take only the values 0, 1, or 2. (b) Let $\hat{A} = A \pm yy^*$. Explain why rank $\hat{A}$ − rank A can take only the values −1, 0, or +1. See (4.3.P27) for a generalization and refinement.

4.3.P22 Let $a_1, \ldots, a_n$ be n given positive real numbers that are not all equal. Let $A = [a_i + a_j]_{i,j=1}^n \in M_n(\mathbf{R})$. Explain from general principles why A has exactly one positive and one negative eigenvalue.

4.3.P23 Let $x, y \in \mathbf{R}^n$ be given. Use (4.3.47a,b) to give one-line proofs that (a) $x + y$ is majorized by $x^{\downarrow} + y^{\downarrow}$ and (b) $x + y$ majorizes $x^{\downarrow} + y^{\uparrow}$. Can you prove these two majorizations without invoking (4.3.47)?

4.3.P24 Let $A, B \in M_n$ be Hermitian. In (4.3.47) we encounter two equivalent versions of the Lidskii inequalities: "$\lambda(A + B)$ majorizes $\lambda(A)^{\downarrow} + \lambda(B)^{\uparrow}$" and "$\lambda(B)$ majorizes $\lambda(A + B)^{\downarrow} - \lambda(A)^{\downarrow}$." Show that they are equivalent to a third version: "$\lambda(A - B)$ majorizes $\lambda(A)^{\downarrow} - \lambda(B)^{\downarrow}$."

4.3.P25 Let $A \in M_n$ be upper bidiagonal (0.9.10). (a) Show that its singular values depend only on the absolute values of its entries. (b) Assume that $a_{ii} \neq 0$ for each $i = 1, \ldots, n$ and $a_{i,i+1} \neq 0$ for each $i = 1, \ldots, n - 1$. Show that A has n distinct singular values.

4.3.P26 Let $A, B \in M_n$ be tridiagonal. (a) If A and B^* are unreduced (0.9.9) and $\lambda \in \mathbf{C}$, explain why the first $n - 2$ columns of $AB^* - \lambda I$ are linearly independent, so rank$(AB - \lambda I) \geq n - 2$ for every $\lambda \in \mathbf{C}$. Why does every eigenvalue of AB have geometric multiplicity at most 2? (b) If every superdiagonal and subdiagonal entry of A is nonzero (that is, if A is irreducible), explain why every singular value of A has multiplicity at most 2. (c) What are the eigenvalues and singular values of the irreducible tridiagonal matrix $A = \begin{bmatrix} 0 & 1 \\ 1 & 0 \end{bmatrix}$?

4.3.P27 Let $A \in M_n$, $y, z \in \mathbf{C}^n$, and $a \in \mathbf{C}$ be given. Suppose that $1 \leq \text{rank } A = r < n$, let $\hat{A} = \begin{bmatrix} A & y \\ z^T & a \end{bmatrix}$, and let $\delta = \text{rank } \hat{A} - \text{rank } A$. (a) Explain why rank $\begin{bmatrix} 0_n & y \\ z^T & a \end{bmatrix} \leq 2$ and conclude that $1 \leq \delta \leq 2$. (b) Provide details for the following sketch of a proof that $\delta = 2$ if and only if y is not in the column space of A and z^T is not in the row space of A: Use (0.4.6(f)) to write $A = S \begin{bmatrix} I_r & 0 \\ 0 & 0_{n-r} \end{bmatrix} R$, in which $S, R \in M_n$ are nonsingular. Partition $S^{-1}y = \begin{bmatrix} y_1 \\ y_2 \end{bmatrix}$ and $R^{-T}z = \begin{bmatrix} z_1 \\ z_2 \end{bmatrix}$, in which $y_1, z_1 \in \mathbf{C}^r$. Then y is in the column space of A if and only if $y_2 = 0$, and z is in the column space of A^T if and only if $z_2 = 0$. Moreover,

$$\text{rank } \hat{A} = \text{rank} \begin{bmatrix} I_r & 0 & y_1 \\ 0 & 0_{n-r} & y_2 \\ z_1^T & z_2^T & a \end{bmatrix} = r + 2$$

if and only if $y_2 \neq 0 \neq z_2$. (c) If A is Hermitian and $z = \bar{y}$, explain why $\delta = 2$ if and only if y is not in the column space of A. Compare with (4.3.P21(a)).

4.3.P28 Under the hypotheses of (4.3.47), show that the assertion "$\lambda(A + B)^{\downarrow} + \lambda(A - B)^{\downarrow}$ majorizes $2\lambda(A)$" is equivalent to Fan's inequality (4.3.47a).

4.3.P29 Let $A = [A_{ij}]_{i,j=1}^m \in M_n$ be Hermitian and partitioned so that $A_{ii} \in M_{n_i}$ for each $i = 1, \ldots, m$ and $n_1 + \cdots + n_m = n$. Show that the vector of eigenvalues of A majorizes the vector of eigenvalues of $A_{11} \oplus \cdots \oplus A_{mm}$. Explain why this assertion is a generalization of (4.3.45).

4.3.P30 Schur's majorization theorem (4.3.45) has a block matrix generalization that puts an intermediate term in the inequalities (4.3.46). Let $A = [A_{ij}]_{i,j=1}^k$ be a partitioned Hermitian matrix and let $d(A) = [a_{ii}]_{i=1}^n$ be its vector of main diagonal entries. Prove that $d(A)$ is majorized by $\lambda(A_{11} \oplus \cdots \oplus A_{kk})$, which in turn is majorized by $\lambda(A)$.

Notes and Further Readings. For more information about majorization, see Marshall and Olkin (1979). The Lidskii inequalities (4.3.47b) are the basis for many important perturbation bounds; see (6.3) and (7.4). Many proofs of these famous inequalities are in the literature; the proof in the text is from C. K. Li and R. Mathias, The Lidskii–Mirsky–Wielandt theorem – additive and multiplicative versions, *Numer. Math.* 81 (1999) 377–413. For a comprehensive discussion of a general set of eigenvalue inequalities that includes as special cases the Weyl inequalities (4.3.1), the Fan inequalities (4.3.47a), and the Lidskii inequalities (4.3.47b), see R. Bhatia, Linear algebra to quantum cohomology: The story of Alfred Horn's inequalities, *Amer. Math. Monthly* 108 (2001) 289–318. Problem (4.3.P17) provides two proofs that the interlacing inequalities (4.3.18) are strict for a class of matrices that includes the (real) Jacobi matrices. Conversely, it is known that if $2n - 1$ real numbers satisfy the inequalities $\lambda_1 < \mu_1 < \lambda_2 < \mu_2 < \cdots < \lambda_{n-1} < \mu_{n-1} < \lambda_n$, then there is a unique Jacobi matrix A such that $\lambda_1, \ldots, \lambda_n$ are its eigenvalues and $\mu_1, \ldots, \mu_{n-1}$ are the eigenvalues of its leading principal submatrix of size $n - 1$; see O. H. Hald, Inverse eigenvalue problems for Jacobi matrices, *Linear Algebra Appl.* 14 (1976) 63–85.

4.4 Unitary congruence and complex symmetric matrices

Complex Hermitian and symmetric matrices both arise in the study of analytic mappings of the unit disc in the complex plane. If f is a complex analytic function on the unit disc that is normalized so that $f(0) = 0$ and $f'(0) = 1$, then $f(z)$ is one-to-one (sometimes called *univalent* or *schlicht*) if and only if it satisfies *Grunsky's inequalities*

$$\sum_{i,j=1}^{n} x_i \bar{x}_j \log \frac{1}{1 - z_i \bar{z}_j} \geq \left| \sum_{i,j=1}^{n} x_i x_j \log \left(\frac{z_i z_j}{f(z_i) f(z_j)} \frac{f(z_i) - f(z_j)}{z_i - z_j} \right) \right|$$

for all $z_1, \ldots, z_n \in \mathbf{C}$ with $|z_i| < 1$, all $x_1, \ldots, x_n \in \mathbf{C}$, and all $n = 1, 2, \ldots$ If $z_i = z_j$, then the difference quotient on the right-hand side of Grunsky's inequalities is to be interpreted as $f'(z_i)$; if $z_i = 0$, then we interpret $z_i / f(z_i)$ as $1/f'(0)$. These formidable inequalities have the very simple algebraic form

$$x^* A x \geq |x^T B x| \tag{4.4.1}$$

in which $x = [x_i] \in \mathbf{C}^n$, $A = [a_{ij}] \in M_n$, $B = [b_{ij}] \in M_n$,

$$a_{ij} = \log \frac{1}{1 - z_i \bar{z}_j}, \quad b_{ij} = \log \left(\frac{z_i z_j}{f(z_i) f(z_j)} \frac{f(z_i) - f(z_j)}{z_i - z_j} \right)$$

The matrix A is Hermitian, while B is complex symmetric. See (7.7.P19) for a simpler equivalent form of Grunsky's inequalities.

If we make a unitary change of variables $x \to Ux$ in (4.4.1), then $A \to U^*AU$ is transformed by unitary similarity and $B \to U^T BU$ is transformed by unitary congruence.

Complex symmetric matrices appear in moment problems of various kinds. For example, let $a_0, a_1, a_2, \ldots$ be a given sequence of complex numbers, let $n \geq 1$ be a given positive integer, and define $A_{n+1} = [a_{i+j}]_{i,j=0}^{n} \in M_{n+1}$, a complex symmetric Hankel matrix (0.9.8). We consider the quadratic form $x^T A_{n+1} x$ for $x \in \mathbf{C}^{n+1}$ and ask whether there is some fixed constant $c > 0$ such that

$$|x^T A_{n+1} x| \leq c x^* x \quad \text{for all } x \in \mathbf{C}^{n+1} \text{ and each } n = 0, 1, 2, \ldots$$

According to a theorem of Nehari, this condition is satisfied if and only if there is a Lebesgue measurable and almost everywhere bounded function $f : \mathbf{R} \to \mathbf{C}$ whose Fourier coefficients are the given numbers $a_0, a_1, a_2, \ldots$; the constant c in the preceding inequalities is the essential supremum of $|f|$.

Complex symmetric matrices arise in the study of damped vibrations of linear systems, in classical theories of wave propagation in continuous media, and in general relativity.

Unitary *similarity* is a natural equivalence relation in the study of normal or Hermitian matrices: $U^* A U$ is normal (respectively, Hermitian) if U is unitary and A is normal (respectively, Hermitian). Unitary *congruence* is a natural equivalence relation in the study of complex symmetric or skew-symmetric matrices: $U^T A U$ is symmetric (respectively, skew symmetric) if U is unitary and A is symmetric (respectively, skew symmetric). In our study of unitary congruence, we make frequent use of the fact that if $A, B \in M_n$ are unitarily congruent, then $A\bar{A}$ and $B\bar{B}$ are unitarily similar and hence have the same eigenvalues.

Exercise. Suppose that $A, B \in M_n$ are unitarily congruent, that is, $A = U B U^T$ for some unitary $U \in M_n$. Explain why $A\bar{A}$ and $B\bar{B}$ are unitarily similar; AA^* and BB^* are unitarily similar; and $A^T \bar{A}$ and $B^T \bar{B}$ are unitarily similar. Moreover, all three unitary similarities can be accomplished via the same unitary matrix.

Exercise. Let $A = \begin{bmatrix} 0 & 1 \\ 0 & 0 \end{bmatrix}$ and $B = 0_2$. Show that $A\bar{A} = B\bar{B}$ and explain why A and B are not unitarily congruent.

Exercise. Suppose that $\Delta \in M_n$ is upper triangular. Explain why all the eigenvalues of $\Delta \bar{\Delta}$ are real and nonnegative. *Hint*: What are the main diagonal entries of $\Delta \bar{\Delta}$?

Exercise. Consider $A = [a] \in M_1$, in which $a = |a| e^{i\theta}$ and $\theta \in \mathbf{R}$. Show that A is unitarily congruent to the real matrix $B = [|a|]$. *Hint*: Consider $U = [e^{-i\theta/2}]$.

We already know (see (2.6.6a)) that a complex symmetric matrix A is unitarily congruent to a nonnegative diagonal matrix whose diagonal entries are the singular values of A. In this section, we show that this basic result is a consequence of a factorization involving unitary congruence that is an analog of (2.3.1): Every complex matrix is unitarily congruent to a block upper triangular matrix in which the diagonal blocks are 1-by-1 or 2-by-2. Our first step is to show how the nonnegative eigenvalues of $A\bar{A}$ can be used to achieve a partial triangularization by unitary congruence.

Lemma 4.4.2. *Let $A \in M_n$ be given, let λ be an eigenvalue of $A\bar{A}$, and let $x \in \mathbf{C}^n$ be a unit eigenvector of $A\bar{A}$ associated with λ. Let $\mathcal{S} = \mathrm{span}\{A\bar{x}, x\}$, which has dimension one or two.*

(a) *If $\dim \mathcal{S} = 1$, then λ is real and nonnegative, and there is a unit vector $z \in \mathcal{S}$ such that $A\bar{z} = \sigma z$, in which $\sigma \geq 0$ and $\sigma^2 = \lambda$.*

(b) *Suppose that* $\dim S = 2$. *If* λ *is real and nonnegative, then there is a unit vector* $z \in S$ *such that* $A\bar{z} = \sigma z$, *in which* $\sigma \geq 0$ *and* $\sigma^2 = \lambda$. *If* λ *is not real or is real and negative, then* $A\bar{y} \in S$ *for every* $y \in S$.

Proof. (a) If $\dim S = 1$, then $\{A\bar{x}, x\}$ is linearly dependent and $A\bar{x} = \mu x$ for some $\mu \in \mathbb{C}$. Compute $\lambda x = A\bar{A}x = A(\overline{A\bar{x}}) = A\bar{\mu}\bar{x} = \bar{\mu}A\bar{x} = \bar{\mu}\mu x = |\mu|^2 x$, so $|\mu|^2 = \lambda$. Choose $\theta \in \mathbb{R}$ so that $e^{-2i\theta}\mu = |\mu|$, and let $\sigma = |\lambda|$. Then

$$A(\overline{e^{i\theta}x}) = e^{-i\theta}A\bar{x} = e^{-i\theta}\mu x = (e^{-2i\theta}\mu)(e^{i\theta}x) = |\mu|(e^{i\theta}x) = \sigma(e^{i\theta}x)$$

so $z = e^{i\theta}x$ is a unit vector in S such that $A\bar{z} = \sigma z$, in which $\sigma \geq 0$ and $\sigma^2 = \lambda$.
(b) If $\dim S = 2$, then $\{A\bar{x}, x\}$ is linearly independent and hence is a basis for S. Any $y \in S$ can be expressed as $y = \alpha A\bar{x} + \beta x$ for some $\alpha, \beta \in \mathbb{C}$, and $A\bar{y} = A(\bar{\alpha}\bar{A}x + \bar{\beta}\bar{x}) = \bar{\alpha}A\bar{A}x + \bar{\beta}A\bar{x} = \bar{\alpha}\lambda x + \bar{\beta}A\bar{x} \in S$. If λ is real and nonnegative, let $\sigma = \sqrt{\lambda} \geq 0$ and let $y = A\bar{x} + \sigma x$, which is nonzero since it is a nontrivial linear combination of basis vectors. Then

$$A\bar{y} = A(\bar{A}x + \sigma\bar{x}) = A\bar{A}x + \sigma A\bar{x} = \lambda x + \sigma A\bar{x}$$
$$= \sigma^2 x + \sigma A\bar{x} = \sigma(A\bar{x} + \sigma\bar{x}) = \sigma y$$

so $z = y/\|y\|_2$ is a unit vector in S such that $A\bar{z} = \sigma z$, $\sigma \geq 0$, and $\sigma^2 = \lambda$. $\qquad\square$

A subspace $S \subset \mathbb{C}^n$ is said to be A-*coninvariant* if $A \in M_n$ and $A\bar{x} \in S$ for every $x \in S$. The concept of A-coninvariance is a natural analog of A-invariance (1.3.16). For each $A \in M_n$ there is always a one-dimensional A-invariant subspace: the span of any eigenvector. The preceding lemma ensures that for each $A \in M_n$ there is always an A-coninvariant subspace of dimension one or two: If $A\bar{A}$ has a nonnegative eigenvalue, there is an A-coninvariant subspace of dimension one (the span of the vector z constructed in (4.4.2a,b); otherwise, there is an A-coninvariant subspace of dimension two (the subspace S constructed in (4.4.2)).

Theorem 4.4.3. *Let* $A \in M_n$ *and* $p \in \{0, 1, \ldots, n\}$ *be given. Suppose that* $A\bar{A}$ *has at least* p *real nonnegative eigenvalues, including* $\lambda_1, \ldots, \lambda_p$. *Then there is a unitary* $U \in M_n$ *such that*

$$A = U \begin{bmatrix} \Delta & \bigstar \\ 0 & C \end{bmatrix} U^T$$

in which $\Delta = [d_{ij}] \in M_p$ *is upper triangular,* $d_{ii} = \sqrt{\lambda_i} \geq 0$ *for* $i = 1, \ldots, p$, *and* $C \in M_{n-p}$. *If* $A\bar{A}$ *has exactly* p *real nonnegative eigenvalues, then* $C\bar{C}$ *has no real nonnegative eigenvalues.*

Proof. The case $n = 1$ is trivial (see the preceding exercise), as is the case $p = 0$, so we assume that $n \geq 2$ and $p \geq 1$.

Consider the following reduction: Let x be a unit eigenvector of $A\bar{A}$ associated with a real nonnegative eigenvalue λ and let $\sigma = \sqrt{\lambda} \geq 0$. The preceding lemma ensures that there is a unit vector z such that $A\bar{z} = \sigma z$. Let $V = [z\, v_2\, \ldots\, v_n] \in M_n$ be unitary and consider the unitary congruence $\bar{V}^T A\bar{V}$. Its 1, 1 entry is $z^*A\bar{z} = \sigma z^*z = \sigma$. Orthogonality of the columns of V ensures that the other entries in the first column of

$\bar{V}^T A \bar{V}$ are zero: $v_i^* A \bar{z} = \sigma v_i^* z = 0$ for $i = 2, \ldots, n$. Thus,

$$A = V \begin{bmatrix} \sigma & \bigstar \\ 0 & A_2 \end{bmatrix} V^T, \quad A_2 \in M_{n-1}, \quad \sigma = \sqrt{\lambda} \geq 0$$

and

$$A\bar{A} = V \begin{bmatrix} \sigma^2 & \bigstar \\ 0 & A_2\bar{A}_2 \end{bmatrix} V^* = V \begin{bmatrix} \lambda & \bigstar \\ 0 & A_2\bar{A}_2 \end{bmatrix} V^*$$

If $A_2 \in M_{n-p}$ or if $A_2\bar{A}_2$ has no real nonnegative eigenvalues, we stop.

If $A_2\bar{A}_2$ has a real nonnegative eigenvalue, apply the preceding reduction to A_2, as in the proof of (2.3.1). In at most p reduction steps, we obtain the asserted block form. $\square$

Exercise. Explain why an upper triangular matrix is symmetric if and only if it is diagonal.

Corollary 4.4.4. *Let $A \in M_n$ be given.*

(a) *If there is a unitary $U \in M_n$ such that $A = U \Delta U^T$ and Δ is upper triangular, then every eigenvalue of $A\bar{A}$ is nonnegative.*

(b) *If $A\bar{A}$ has at least $n - 1$ real nonnegative eigenvalues, there is a unitary $U \in M_n$ such that $A = U \Delta U^T$, in which $\Delta = [d_{ij}]$ is upper triangular, each $d_{ii} \geq 0$, and $d_{11}^2, \ldots, d_{nn}^2$ are the eigenvalues of $A\bar{A}$, all of which are real and nonnegative.*

(c) *(Autonne) If A is symmetric, there is a unitary $U \in M_n$ such that $A = U \Sigma U^T$, in which Σ is a nonnegative diagonal matrix whose diagonal entries are the singular values of A, in any desired order.*

(d) *(Uniqueness) Suppose that A is symmetric and rank $A = r$. Let $s_1, \ldots, s_d$ be the distinct positive singular values of A, in any given order, with respective multiplicities $n_1, \ldots, n_d$. Let $\Sigma = s_1 I_{n_1} \oplus \cdots \oplus s_d I_{n_d} \oplus 0_{n-r}$; the zero block is missing if A is nonsingular. Let $U, V \in M_n$ be unitary. Then $A = U \Sigma U^T = V \Sigma V^T$ if and only if $V = UZ$, $Z = Q_1 \oplus \cdots \oplus Q_d \oplus \tilde{Z}$, $\tilde{Z} \in M_{n-r}$ is unitary, and each $Q_j \in M_{n_j}$ is real orthogonal. If the singular values of A are distinct (that is, if $d \geq n - 1$), then $V = UD$, in which $D = \text{diag}(d_1, \ldots, d_n)$, $d_i = \pm 1$ for each $i = 1, \ldots, n - 1$, $d_n = \pm 1$ if A is nonsingular, and $d_n \in \mathbb{C}$ with $|d_n| = 1$ if A is singular.*

Proof. (a) If $A = U \Delta U^T$ and Δ is upper triangular, then the eigenvalues of $A\bar{A} = U \Delta U^T \bar{U} \bar{\Delta} U^* = U \Delta \bar{\Delta} U^*$ are the main diagonal entries of $\Delta \bar{\Delta}$, which are nonnegative.

(b) If $A\bar{A}$ has at least $n - 1$ real nonnegative eigenvalues, we have the case $p \geq n - 1$ in (4.4.3), so A is unitarily congruent to an upper triangular matrix. Part (a) ensures that *every* eigenvalue of $A\bar{A}$ is nonnegative, and the asserted factorization now follows from a final invocation of (4.4.3) with $p = n$.

(c) If A is symmetric, the eigenvalues of $A\bar{A} = A\bar{A}^T = AA^*$ are the squares of the singular values of A (2.6.3b), so (a) ensures that $A = U \Delta U^T$, in which $\Delta = [d_{ij}]$ is upper triangular and $d_{11}, \ldots, d_{nn}$ are the singular values of A. Since Δ is unitarily congruent to the symmetric matrix A, it is itself symmetric and hence is diagonal. For any permutation matrix P, we have $A = (UP)(P^T \Delta P)(UP)^T$, so the singular values can be presented in any desired order.

(d) The two factorizations $A = U\Sigma\bar{U}^* = V\Sigma\bar{V}^*$ are singular value decomposition, so (2.6.5) ensures that $U = VX$ and $\bar{U} = \bar{V}Y$, in which $X = Z_1 \oplus \cdots \oplus Z_d \oplus \tilde{Z}$ and $Y = Z_1 \oplus \cdots \oplus Z_d \oplus \tilde{Y}$ are unitary and each $Z_j \in M_{n_j}$. Then $X = V^*U = \overline{V^T\bar{U}} = \bar{Y}$, so each $Z_j = \overline{Z_j}$ is real and unitary, that is, real orthogonal. The assertions in the case of distinct singular values follow by specialization.   □

Theorem 4.4.3 stimulates us to consider a form to which $A \in M_n$ can be reduced via unitary congruence if $A\bar{A}$ has no real nonnegative eigenvalues. Corollary 3.4.1.9 ensures that $A\bar{A}$ is always similar to a real matrix, so tr $A\bar{A}$ is always real and any non-real eigenvalues of $A\bar{A}$ must occur in conjugate pairs.

Suppose that $A \in M_2$. If neither eigenvalue of $A\bar{A}$ is real and nonnegative, then because $A\bar{A}$ is similar to a real matrix, there are only two possibilities for its eigenvalues: They are either a non-real conjugate pair or they are both real and negative; in the latter case, the following proposition says something remarkable: they must be equal. The characteristic polynomial of $A\bar{A}$ is $p_{A\bar{A}}(t) = t^2 - (\text{tr } A\bar{A})t + \det A\bar{A} = t^2 - (\text{tr } A\bar{A})t + |\det A|^2$, which has real zeroes if and only if its discriminant is nonnegative, that is, if and only if $(\text{tr } A\bar{A})^2 - 4|\det A|^2 \geq 0$. If $A\bar{A}$ has two negative eigenvalues, then tr $A\bar{A}$, their sum, must be negative.

Exercise. If $A \in M_n$ is skew symmetric, explain why every eigenvalue of $A\bar{A}$ is real and nonpositive. *Hint*: $A\bar{A} = -AA^*$.

Exercise. Let $A \in M_n$. Explain why $f(A) = \text{tr}(A\bar{A})$ and $g(A) = \det(A\bar{A}) = |\det A|^2$ are unitary congruence invariant functions of A, that is, $f(UAU^T) = f(A)$ and $g(UAU^T) = g(A)$ for every unitary $U \in M_n$. If $n = 2$, why is the discriminant of the characteristic polynomial of $A\bar{A}$ a unitary congruence invariant function of A?

Proposition 4.4.5. *Let $A \in M_2$ be given and let $\sigma_1 \geq \sigma_2 \geq 0$ be the singular values of $S(A) = \frac{1}{2}(A + A^T)$, the symmetric part of A. If $\sigma_1 = \sigma_2$, let $\sigma = \sigma_1$.*

(a) A is unitarily congruent to

$$\begin{bmatrix} \sigma_1 & \zeta \\ -\zeta & \sigma_2 \end{bmatrix}, \zeta \in \mathbf{C} \tag{4.4.6}$$

(b) $A\bar{A}$ has a non-real pair of conjugate eigenvalues if and only if A is unitarily congruent to

$$\begin{bmatrix} \sigma_1 & \zeta \\ -\zeta & \sigma_2 \end{bmatrix}, \zeta \in \mathbf{C}, \, 2|\sigma_1\bar{\zeta} + \sigma_2\zeta| > \sigma_1^2 - \sigma_2^2 \tag{4.4.7}$$

If $\sigma_1 = \sigma_2$, the conditions on σ_1, σ_2, and ζ in (4.4.7) are equivalent to the conditions $\sigma > 0$ and $\text{Re }\zeta \neq 0$.

(c) $A\bar{A}$ has two real negative eigenvalues if and only if $\sigma_1 = \sigma_2$ and A is unitarily congruent to

$$\begin{bmatrix} \sigma & i\xi \\ -i\xi & \sigma \end{bmatrix}, \xi \in \mathbf{R}, \, \xi > \sigma \geq 0 \tag{4.4.8a}$$

in which case $\sigma^2 - \xi^2$ is a double negative eigenvalue of $A\bar{A}$. If $\sigma = 0$ in (4.4.8a), then A is unitarily congruent to

$$\begin{bmatrix} 0 & \xi \\ -\xi & 0 \end{bmatrix}, \xi \in \mathbf{R}, \xi > 0 \qquad (4.4.8b)$$

in which case $-\xi^2$ is a double negative eigenvalue of $A\bar{A}$ and ξ is a double singular value of A.

(d) *Let λ_1, λ_2 be the eigenvalues of $A\bar{A}$. If λ_1 is not real, then $\lambda_2 = \bar{\lambda}_1$. If λ_1 is real and nonnegative, then so is λ_2. If λ_1 is real and negative, then $\lambda_2 = \lambda_1$.*

Proof. (a) Write $A = S(A) + C(A)$ as the sum of its symmetric and skew-symmetric parts (0.2.5). The preceding corollary ensures that there is a unitary $U \in M_2$ such that $S(A) = U \begin{bmatrix} \sigma_1 & 0 \\ 0 & \sigma_2 \end{bmatrix} U^T$, so $A = U(\begin{bmatrix} \sigma_1 & 0 \\ 0 & \sigma_2 \end{bmatrix} + U^* C(A) \bar{U}) U^T$. The matrix $U^* C(A) \bar{U}$ is skew symmetric, so it has the form $\begin{bmatrix} 0 & \zeta \\ -\zeta & 0 \end{bmatrix}$ for some $\zeta \in \mathbf{C}$.

(b) If we wish to compute the trace or determinant of $A\bar{A}$, the preceding exercise permits us to assume that A has the form (4.4.6). In this case, a calculation reveals that

$$\operatorname{tr} A\bar{A} = \sigma_1^2 + \sigma_2^2 - 2|\zeta|^2 \quad \text{and} \quad |\det A|^2 = \sigma_1^2 \sigma_2^2 + 2\sigma_1 \sigma_2 \operatorname{Re} \zeta^2 + |\zeta|^4$$

and the discriminant of $p_{A\bar{A}}(t)$ is

$$r(A) = (\operatorname{tr} A\bar{A})^2 - 4|\det A|^2 = (\sigma_1^2 - \sigma_2^2)^2 - 4|\sigma_1 \bar{\zeta} + \sigma_2 \zeta|^2$$

Then $A\bar{A}$ has a pair of non-real conjugate eigenvalues if and only if $r(A) < 0$ if and only if $2|\sigma_1 \bar{\zeta} + \sigma_2 \zeta| > \sigma_1^2 - \sigma_2^2$.

(c) Now suppose that $A\bar{A}$ has two real negative eigenvalues, so $r(A) \geq 0$ and $\operatorname{tr} A\bar{A} < 0$, that is, $2|\zeta|^2 > \sigma_1^2 + \sigma_2^2$. Let $\zeta = |\zeta| e^{i\theta}$ with $\theta \in \mathbf{R}$ and compute

$$\begin{aligned} 0 \leq r(A) &= (\sigma_1^2 - \sigma_2^2)^2 - 4|\sigma_1 \bar{\zeta} + \sigma_2 \zeta|^2 \\ &= (\sigma_1^2 - \sigma_2^2)^2 - 4|\zeta|^2 |\sigma_1 e^{-i\theta} + \sigma_2 e^{i\theta}|^2 \\ &= (\sigma_1^2 - \sigma_2^2)^2 - 4|\zeta|^2 (\sigma_1^2 + 2\sigma_1 \sigma_2 \cos 2\theta + \sigma_2^2) \\ &\leq (\sigma_1^2 - \sigma_2^2)^2 - 2(\sigma_1^2 + \sigma_2^2)(\sigma_1^2 - 2\sigma_1 \sigma_2 + \sigma_2^2) \\ &= (\sigma_1 - \sigma_2)^2 (\sigma_1 + \sigma_2)^2 - 2(\sigma_1^2 + \sigma_2^2)(\sigma_1 - \sigma_2)^2 \\ &= (\sigma_1 - \sigma_2)^2 (\sigma_1^2 + 2\sigma_1 \sigma_2 + \sigma_2^2 - 2\sigma_1^2 - 2\sigma_2^2) \\ &= -(\sigma_1 - \sigma_2)^4 \end{aligned}$$

which implies that $\sigma_1 = \sigma_2$. In this case, $r(A) = -4\sigma^2 |\bar{\zeta} + \zeta|^2 = -8\sigma^2 |\operatorname{Re} \zeta| \geq 0$, so either $\operatorname{Re} \zeta = 0$ or $\sigma = 0$. If $\operatorname{Re} \zeta = 0$, let $\zeta = i\xi$, in which ξ is real and nonzero. Since $\operatorname{tr} A\bar{A} = \sigma_1^2 + \sigma_2^2 - 2|\zeta|^2 = 2(\sigma^2 - \xi^2) < 0$, it follows that $|\xi| > \sigma$. Thus, A has the form (4.4.8a) if $\xi > 0$; it has the form of the transpose of (4.4.8a) if $\xi < 0$. In the latter case, perform a unitary congruence via the reversal matrix $\begin{bmatrix} 0 & 1 \\ 1 & 0 \end{bmatrix}$ to obtain the form (4.4.8a). If $\sigma = 0$ then $S(A) = 0$, so $A = C(A)$ has the form $\begin{bmatrix} 0 & \zeta \\ -\zeta & 0 \end{bmatrix}$ for some nonzero $\zeta \in \mathbf{C}$. Let $\zeta = |\zeta| e^{i\theta}$ for some $\theta \in \mathbf{R}$ and compute the unitary congruence $(e^{-i\theta/2} I) A (e^{-i\theta/2} I) = \begin{bmatrix} 0 & |\zeta| \\ -|\zeta| & 0 \end{bmatrix}$, which has the form (4.4.8b). For the assertions about the double negative eigenvalues, see the following exercise.

(d) Since $A\bar{A}$ is similar to a real matrix, it has a non-real eigenvalue if and only if it has a pair of conjugate non-real eigenvalues. Therefore, $\lambda_2 = \bar{\lambda}_1$ if λ_1 is not real; if λ_1 is real, then λ_2 must also be real. If λ_1 is real and nonnegative, (4.4.4b) ensures that λ_2 is real and nonnegative. Finally, if λ_1 is real and negative, the previous two cases ensure that λ_2 is real and cannot be nonnegative, that is, λ_1 and λ_2 are both negative; (c) ensures that they are equal. □

Exercise. If A is unitarily congruent to a matrix of the form (4.4.8a), show that $\sigma^2 - \xi^2$ is a double negative eigenvalue of $A\bar{A}$. If A is unitarily congruent to a matrix of the form (4.4.8b), show that $-\xi^2$ is a double negative eigenvalue of $A\bar{A}$ and ξ is a double singular value of A.

Exercise. Let $A \in M_2$ be given, so it is unitarily congruent to a matrix of the form (4.4.6). Explain why $\sigma_1 = \sigma_2$ if and only if the symmetric part of A is a scalar multiple of a unitary matrix.

We now have all the tools necessary to show that each square complex matrix A is unitarily congruent to a block upper triangular matrix in which every diagonal block is either 1-by-1 or 2-by-2 and has a special form.

Theorem 4.4.9 (Youla). *Let $A \in M_n$ be given. Let $p \in \{0, 1, \ldots, n\}$ and suppose that $A\bar{A}$ has exactly p real nonnegative eigenvalues. Then there is a unitary $U \in M_n$ such that*

$$A = U \begin{bmatrix} \Delta & \bigstar \\ 0 & \Gamma \end{bmatrix} U^T \qquad (4.4.10)$$

in which either Δ is missing (if $p = 0$), or $\Delta = [d_{ij}] \in M_p$ is upper triangular, $d_{ii} \geq 0$ for $i = 1, \ldots, p$, and $d_{11}^2, \ldots, d_{pp}^2$ are the nonnegative eigenvalues of $A\bar{A}$. Either Γ is missing (if $p = n$), or $q = n - p \geq 2$ is even, $\Gamma = [\Gamma_{ij}]_{i,j=1}^{q/2} \in M_q$ is block 2-by-2 upper triangular, and for each $j = 1, \ldots, q/2$ the eigenvalues of $\Gamma_{jj}\bar{\Gamma}_{jj}$ are either a non-real conjugate pair or a real negative equal pair.

(a) *If $\Gamma_{jj}\bar{\Gamma}_{jj}$ has a non-real conjugate pair of eigenvalues, then Γ_{jj} may be chosen to have either the form*

$$\Gamma_{jj} = \begin{bmatrix} \sigma_1 & \zeta \\ -\zeta & \sigma_2 \end{bmatrix}, \quad \begin{cases} \sigma_1, \sigma_2 \in \mathbf{R}, \ \zeta \in \mathbf{C}, \ \sigma_1 > \sigma_2 \geq 0, \\ 2|\sigma_1\bar{\zeta} + \sigma_2\zeta| > \sigma_1^2 - \sigma_2^2 \end{cases} \qquad (4.4.11a)$$

or the form

$$\Gamma_{jj} = \begin{bmatrix} \sigma & \zeta \\ -\zeta & \sigma \end{bmatrix}, \ \sigma \in \mathbf{R}, \ \zeta \in \mathbf{C}, \ \sigma > 0, \ \mathrm{Re}\,\zeta \neq 0 \qquad (4.4.11b)$$

(b) *If $\Gamma_{jj}\bar{\Gamma}_{jj}$ has a real negative equal pair of eigenvalues, then Γ_{jj} may be chosen to have either the form*

$$\Gamma_{jj} = \begin{bmatrix} \sigma & i\xi \\ -i\xi & \sigma \end{bmatrix}, \ \sigma, \xi \in \mathbf{R}, \ \xi > \sigma > 0 \qquad (4.4.12a)$$

or the form

$$\Gamma_{jj} = \begin{bmatrix} 0 & \xi \\ -\xi & 0 \end{bmatrix}, \ \xi \in \mathbf{R}, \ \xi > 0 \qquad (4.4.12b)$$

Proof. Theorem 4.4.3 ensures that A is unitarily congruent to a block upper triangular matrix of the form $\begin{bmatrix} \Delta & \star \\ 0 & C \end{bmatrix}$, in which $\Delta \in M_p$ has the stated properties. It suffices to consider a matrix $C \in M_q$ such that $q = n - p > 0$ and $C\bar{C}$ has no nonnegative eigenvalues. If $q = 1$ and $C = [c]$, then $C\bar{C} = [|c|^2]$ has a nonnegative eigenvalue, so $q \geq 2$.

Consider the following reduction: Let λ be an eigenvalue of $C\bar{C}$, so λ is either not real or it is real and negative. Let x be a unit eigenvector of $C\bar{C}$ associated with λ, and consider the subspace $\mathcal{S} = \text{span}\{C\bar{x}, x\} \subset \mathbb{C}^q$. Lemma 4.2.2 ensures that $\dim \mathcal{S} = 2$ and that $\mathcal{S}$ is C-coninvariant, that is, $C\bar{\mathcal{S}} \subset \mathcal{S}$. Let $\{u, v\}$ be an orthonormal basis for $\mathcal{S}$ and let $V = [u \; v \; v_3 \; \dots \; v_n] \in M_n$ be unitary, so each of $v_3, \dots, v_n$ is orthogonal to $\mathcal{S}$. The first two columns of $C\bar{V} = [C\bar{u} \; C\bar{v} \; C\bar{v}_3 \; \dots \; C\bar{v}_n]$ are vectors in $\mathcal{S}$, so they are orthogonal to each of $v_3, \dots, v_n$. This means that

$$V^* C\bar{V} = \begin{bmatrix} C_{11} & \star \\ 0 & D \end{bmatrix}$$

in which $C_{11} \in M_2$, $D \in M_{q-2}$, and each of $C_{11}\bar{C}_{11}$ and $D\bar{D}$ has no real nonnegative eigenvalues. Indeed, (4.4.5d) ensures that if λ is not real, then $C_{11}\bar{C}_{11}$ has eigenvalues λ and $\bar{\lambda}$; if λ is real, it is a double negative eigenvalue of $C_{11}\bar{C}_{11}$.

If $q - 2 = 0$, we stop; otherwise, $q - 2 \geq 2$ ($q - 2 = 1$ is forbidden since $D\bar{D}$ has no real nonnegative eigenvalues). Thus, we may apply the reduction algorithm to D. In finitely many reduction steps, we find that C is unitarily congruent to a 2-by-2 block upper triangular matrix $\hat{C} = [C_{ij}]_{i,j=1}^{q/2}$.

Each 2-by-2 matrix $C_{jj}\bar{C}_{jj}$ has either a non-real conjugate pair of eigenvalues or a negative equal pair of eigenvalues, so (4.4.5) ensures that there are unitary matrices $U_j \in M_2$ such that each matrix $U_j^* C_{jj}\bar{U}_j$ has the form (4.4.7), (4.4.8a), or (4.4.8b). Let $U = U_1 \oplus \cdots \oplus U_{q/2}$. Then $\Gamma = U^*\hat{C}\bar{U}$ is unitarily congruent to C and has the asserted block upper triangular structure. $\square$

Corollary 4.4.13. *Let $A \in M_n$ be given. The non-real eigenvalues of $A\bar{A}$ occur in conjugate pairs. The real negative eigenvalues of $A\bar{A}$ occur in equal pairs.*

Proof. Factor A as in (4.4.10). The eigenvalues of $A\bar{A}$ are the eigenvalues of $\Delta\bar{\Delta} \oplus \Gamma\bar{\Gamma}$, which are the squares of the absolute values of the main diagonal entries of Δ together with the eigenvalues of all the 2-by-2 diagonal blocks of $\Gamma\bar{\Gamma}$. The latter eigenvalues occur as non-real conjugate pairs or as equal negative real pairs. $\square$

The assertion in the preceding corollary can be significantly strengthened: If λ is a real negative eigenvalue of $A\bar{A}$, then for each $k = 1, 2, \dots$ there are an even number of blocks $J_k(\lambda)$ in the Jordan canonical form of $A\bar{A}$; see (4.6.16).

Youla's factorization (4.4.10) is useful in the context of unitary congruence, just as Schur's factorization described in (2.3.1) is useful in the context of unitary similarity. But neither factorization provides a canonical form for the respective equivalence relation. The eigenvalues of $A\bar{A}$ determine the main diagonal entries of Δ in (4.4.10), but not its off-diagonal entries; they determine the eigenvalues of the diagonal blocks of Γ, but not the specific form of those blocks, let alone the off-diagonal blocks.

We next study a set of square complex matrices for which (4.4.10) *does* provide a canonical form under unitary congruence. For matrices A in this set (which includes all complex symmetric, complex skew-symmetric, unitary, and real normal matrices) the eigenvalues of $A\bar{A}$ completely determine the unitary congruence equivalence class of A.

Definition 4.4.14. *A matrix $A \in M_n$ is conjugate normal if $AA^* = \overline{A^*A}$.*

> *Exercise.* Verify that complex symmetric, complex skew-symmetric, unitary, and real normal matrices are conjugate normal.

The following analog of (2.5.2) plays a key role in the study of conjugate-normal matrices.

Lemma 4.4.15. *Suppose that $A \in M_n$ is partitioned as*

$$A = \begin{bmatrix} A_{11} & A_{12} \\ 0 & A_{22} \end{bmatrix}$$

in which A_{11} and A_{22} are square. Then A is conjugate normal if and only if A_{11} and A_{22} are conjugate normal and $A_{12} = 0$. A block upper triangular matrix is conjugate normal if and only if each of its off-diagonal blocks is zero and each of its diagonal blocks is conjugate normal. In particular, an upper triangular matrix is conjugate normal if and only if it is diagonal.

Proof. Proceed as in the proof of (2.5.2) by equating the $1, 1$ blocks of the identity $\overline{A^*A} = AA^*$: $\overline{A_{11}^*A_{11}} = A_{11}A_{11}^* + A_{12}A_{12}^*$. However, $\operatorname{tr}\overline{A_{11}^*A_{11}} = \overline{\operatorname{tr} A_{11}^*A_{11}} = \operatorname{tr} A_{11}^*A_{11}$ since the trace of a Hermitian matrix is real. The rest of the proof is identical to that of (2.5.2). □

> *Exercise.* If $A \in M_n$ and $U \in M_n$ is unitary, show that A is conjugate normal if and only if UAU^T is conjugate normal, that is, conjugate normality is a unitary congruence invariant.

> *Exercise.* If $A \in M_n$ is conjugate normal and $c \in \mathbf{C}$, show that cA is conjugate normal.

> *Exercise.* If $A \in M_n$ and $B \in M_m$, show that A and B are conjugate normal if and only if $A \oplus B$ is conjugate normal.

> *Exercise.* If $A = [a] \in M_1$ or $A = \begin{bmatrix} a & b \\ -b & a \end{bmatrix} \in M_2(\mathbf{R})$, show that A is conjugate normal.

The following canonical form for conjugate normal matrices is an analog of the spectral theorem for normal matrices that has much in common with the canonical form (2.5.8) for real normal matrices.

Theorem 4.4.16. *A matrix $A \in M_n$ is conjugate normal if and only if it is unitarily congruent to a direct sum of the form*

$$\Sigma \oplus \tau_1 \begin{bmatrix} a_1 & b_1 \\ -b_1 & a_1 \end{bmatrix} \oplus \cdots \oplus \tau_q \begin{bmatrix} a_q & b_q \\ -b_q & a_q \end{bmatrix} \qquad (4.4.17)$$

in which $2q \leq n$, $\Sigma \in M_{n-2q}$ *is a nonnegative diagonal matrix, and* a_j, b_j, τ_j *are real scalars such that* $a_j \geq 0$, $0 < b_j \leq 1$, $a_j^2 + b_j^2 = 1$, *and* $\tau_j > 0$ *for each* $j = 1, \ldots, q$. *The parameters in* (4.4.17) *are uniquely determined by the eigenvalues of* $A\bar{A}$: *The diagonal entries of* Σ^2 *are the real nonnegative eigenvalues of* $A\bar{A}$; *if the eigenvalues of* $A\bar{A}$ *that are not real and nonnegative are expressed as* $r_j e^{\pm 2i\theta_j}$, $j = 1, \ldots, m$, $r_j > 0$, $0 < \theta_j \leq \pi/2$, *then* $\tau_j = \sqrt{r_j} > 0$, $a_j = \cos\theta_j$, *and* $b_j = \sin\theta_j$, $j = 1, \ldots, q$.

Proof. Suppose that A is conjugate normal and factor it as in (4.4.10). Unitary congruence invariance of conjugate normality ensures that $\begin{bmatrix} \Delta & \star \\ 0 & \Gamma \end{bmatrix}$ is conjugate normal, and (4.4.15) tells us that it is block diagonal, Δ is diagonal, $\Gamma = \Gamma_{11} \oplus \cdots \oplus \Gamma_{qq}$ is 2-by-2 block diagonal, and each Γ_{jj} is conjugate normal. Then $\Sigma = \Delta$ and the blocks Γ_{jj} are specializations of the block types described in (4.4.9).

If a block Γ_{jj} has the form (4.4.11a) and is conjugate normal, a calculation reveals that $\sigma_1 \bar{\zeta} = \sigma_2 \zeta$, which implies that $\sigma_1 |\zeta| = \sigma_2 |\zeta|$. Since $\sigma_1 > 0$ and $\zeta \neq 0$ are required by the conditions on the parameters in (4.4.11a), it follows that $\sigma_1 = \sigma_2 > 0$ and $\zeta = \bar{\zeta}$ is real. Thus, Γ_{jj} has the form (4.4.11b) in which ζ is real and positive, that is, a block of the form $\begin{bmatrix} \alpha & \beta \\ -\beta & \alpha \end{bmatrix}$ with $\alpha, \beta > 0$. Let $\tau = (\alpha^2 + \beta^2)^{1/2}$, $a = \alpha/\tau$, and $b = \beta/\tau$. We have shown that a conjugate-normal block of the form (4.4.11a) is unitarily congruent to $\tau \begin{bmatrix} a & b \\ -b & a \end{bmatrix}$, in which $\tau, a > 0$, $0 < b < 1$, and $a^2 + b^2 = 1$.

One verifies that no block of the form (4.4.12a) is conjugate normal $(2i\sigma\xi \neq 0)$; however, any block of the form (4.4.12b) is real normal, so it is conjugate normal. If we let $\tau = \xi$, we have a block of the form $\tau \begin{bmatrix} a & b \\ -b & a \end{bmatrix}$ in which $\tau > 0$, $a = 0$, and $b = 1$.

We have now established a factorization of A that has the asserted form. Each direct summand of (4.4.17) is conjugate normal, so their direct sum and any unitary congruence of it are conjugate normal.

The block diagonal matrix (4.4.17) is real, so its *square* has the same eigenvalues as $A\bar{A}$. Each 2-by-2 block in (4.4.17) is a positive scalar multiple of a real orthogonal matrix and has a pair of eigenvalues of the form $\tau_j(a_j \pm ib_j) = \tau_j e^{\pm i\theta_j} = \tau_j(\cos\theta_j \pm i\sin\theta_j)$ in which $\theta_j \in (0, \pi/2]$ (because $a_j \geq 0$ and $b_j > 0$); the eigenvalues of its square are $\tau_j^2 e^{\pm 2i\theta_j}$, $\theta_j \in (0, \pi/2]$. The eigenvalues of $A\bar{A}$ are either real and nonnegative (the nonnegative square roots of these eigenvalues determine Σ in (4.4.17)) or are pairs of the form $\tau_j^2 e^{\pm 2i\theta_j}$, $\tau_j > 0$, $\theta_j \in (0, \pi/2]$ (these determine the parameters in the 2-by-2 blocks in (4.4.17)). $\qquad\square$

Corollary 4.4.18. *A square complex matrix is conjugate normal if and only if it is unitarily congruent to a direct sum of nonnegative scalar multiples of real orthogonal matrices.*

Proof. The previous theorem states that a conjugate-normal matrix is unitarily congruent to a direct sum of nonnegative scalar multiples of 1-by-1 and 2-by-2 real orthogonal matrices. Conversely, if $A = UZU^T$, in which U is unitary, $Z = \sigma_1 Q_1 \oplus \cdots \oplus \sigma_m Q_m$, each $\sigma_j \geq 0$, and each $Q_j \in M_{n_j}$ is real orthogonal, then $AA^* = UZZ^T U^* = U(\sigma_1^2 Q_1 Q_1^T \oplus \cdots \oplus \sigma_m^2 Q_m Q_m^T)U^* = U(\sigma_1^2 I_{n_1} \oplus \cdots \oplus \sigma_m^2 I_{n_m})U^*$ and $\overline{A^*A} = U(Z^T Z)U^* = U(\sigma_1^2 Q_1^T Q_1 \oplus \cdots \oplus \sigma_m^2 Q_m^T Q_m)U^* = U(\sigma_1^2 I_{n_1} \oplus \cdots \oplus \sigma_m^2 I_{n_m})U^*$. $\qquad\square$

Canonical forms for skew-symmetric and unitary matrices under unitary congruence are consequences of the canonical form for conjugate-normal matrices.

Corollary 4.4.19. *Let $A \in M_n$ be skew symmetric. Then $r = \text{rank}\, A$ is even, the nonzero singular values of A occur in pairs $\sigma_1 = \sigma_2 = s_1 \geq \sigma_3 = \sigma_4 = s_2 \geq \cdots \geq \sigma_{r-1} = \sigma_r = s_{r/2} \geq 0$, and A is unitarily congruent to*

$$0_{n-r} \oplus \begin{bmatrix} 0 & s_1 \\ -s_1 & 0 \end{bmatrix} \oplus \cdots \oplus \begin{bmatrix} 0 & s_{r/2} \\ -s_{r/2} & 0 \end{bmatrix} \qquad (4.4.20)$$

Proof. Since A is skew symmetric, it is conjugate normal and its canonical form (4.4.17) must be skew symmetric. This implies that $\Sigma = 0$ and every $a_j = 0$. $\square$

Corollary 4.4.21. *Let $V \in M_n$ be unitary. Then V is unitarily congruent to*

$$I_{n-2q} \oplus \begin{bmatrix} a_1 & b_1 \\ -b_1 & a_1 \end{bmatrix} \oplus \cdots \oplus \begin{bmatrix} a_q & b_q \\ -b_q & a_q \end{bmatrix} \qquad (4.4.22)$$

in which $2q \leq n$; a_j, b_j are real scalars such that $a_j \geq 0, 0 < b_j \leq 1$; and $a_j^2 + b_j^2 = 1$ for each $j = 1, \ldots, q$. The parameters in (4.4.22) are uniquely determined by the eigenvalues of $V\bar{V}$: $n - 2q$ is the multiplicity of $+1$ as an eigenvalue of $V\bar{V}$; if $e^{\pm 2i\theta_j}$, $j = 1, \ldots, q$, $0 < \theta_j \leq \pi/2$, are the eigenvalues of $V\bar{V}$ that are not real and nonnegative, then $a_j = \cos\theta_j$ and $b_j = \sin\theta_j$, $j = 1, \ldots, q$.

Proof. Since V is unitary, it is conjugate normal and its canonical form (4.4.17) must be unitary. This implies that Σ is unitary, so $\Sigma = I$. It also implies that each 2-by-2 direct summand in (4.4.17) is unitary. The blocks $\begin{bmatrix} a_j & b_j \\ -b_j & a_j \end{bmatrix}$ in (4.4.17) are real orthogonal, so each $\tau_j = 1$.

The parameters in (4.4.22) are determined in the same way as in (4.4.17). $\square$

Is there anything special about the Jordan canonical form of a complex symmetric matrix? As a first step in answering this question, consider the symmetric matrix

$$S_m = \frac{1}{\sqrt{2}}(I_m + iK_m) \qquad (4.4.23)$$

in which $K_m \in M_m$ is a reversal matrix (0.9.5.1). Note that K_m is symmetric and $K_m^2 = I_m$.

Exercise. Verify that the matrix S_m defined in (4.4.23) is unitary.

Exercise. Let $J_m(0)$ be the nilpotent Jordan block of size m. Verify that (a) $K_m J_m(0)$ is symmetric; (b) $J_m(0)K_m$ is symmetric; (c) $K_m J_m(0) K_m = J_m(0)^T$. *Hint*: See (3.2.3).

Exercise. Let $J_m(\lambda)$ be the Jordan block of size m with eigenvalue λ. Verify that
$S_m J_m(\lambda) S_m^{-1} =$

$$= S_m J_m(\lambda) S_m^* = S_m(\lambda I_m + J_m(0)) S_m^*$$

$$= \lambda I_m + S_m J_m(0) S_m^* = \lambda I_m + \frac{1}{2}(I + iK_m) J_m(0)(I - iK_m)$$

$$= \lambda I_m + \frac{1}{2}(J_m(0) + K_m J_m(0) K_m) + \frac{i}{2} K_m J_m(0) - \frac{i}{2} J_m(0) K_m$$

and explain why $S_m J_m(\lambda) S_m^{-1}$ is symmetric.

Theorem 4.4.24. *Each $A \in M_n$ is similar to a complex symmetric matrix.*

Proof. Each $A \in M_n$ is similar to a direct sum of Jordan blocks, and the preceding exercises show that each Jordan block is unitarily similar to a symmetric matrix. Thus, each $A \in M_n$ is similar to a direct sum of symmetric matrices. ☐

The preceding theorem shows that there is nothing special about the Jordan canonical form of a symmetric complex matrix: Each Jordan matrix is similar to a complex symmetric matrix.

Theorem 4.4.24 also implies that every complex matrix is similar to its transpose and can be written as a product of two complex symmetric matrices; for a different approach to this result, see (3.2.3.2).

Corollary 4.4.25. *Let $A \in M_n$ be given. There are symmetric matrices $B, C \in M_n$ such that $A = BC$. Either B or C may be chosen to be nonsingular.*

Proof. Use the preceding theorem to write $A = SES^{-1}$, in which $E = E^T$ and S is nonsingular. Then $A = (SES^T)S^{-T}S^{-1} = (SES^T)(SS^T)^{-1} = (SS^T)(S^{-T}ES^{-1})$. ☐

The following lemma is useful in discussing diagonalization of a complex symmetric matrix.

Lemma 4.4.26. *Let $X \in M_{n,k}$ with $k \leq n$. Then $X^T X$ is nonsingular if and only if $X = YB$, in which $Y \in M_{n,k}$, $Y^T Y = I_k$, and $B \in M_k$ is nonsingular.*

Proof. If $X = YB$ and $Y^T Y = I_k$, then $X^T X = B^T Y^T Y B = B^T B$ is nonsingular if and only if B is nonsingular. Conversely, use (4.4.4c) to factor $X^T X = U\Sigma U^T$, in which $U \in M_k$ is unitary and $U^* X^T X \bar{U} = (X\bar{U})^T(X\bar{U}) = \Sigma = \text{diag}(\sigma_1, \ldots, \sigma_k)$ is nonnegative diagonal. If $X^T X$ is nonsingular, then so is $U^* X^T X \bar{U} = \Sigma$. Let $R = \text{diag}(\sigma_1^{1/2}, \ldots, \sigma_k^{1/2})$ and observe that $R^{-1}(X\bar{U})^T(X\bar{U})R^{-1} = (X\bar{U}R^{-1})^T(X\bar{U}R^{-1}) = R^{-1}\Sigma R^{-1} = I_k$. Thus, with $Y = (X\bar{U}R^{-1})$ and $B = RU^T$, we have $Y^T Y = I_k$, $X = YB$, and B nonsingular. ☐

Example. For $X \in M_{n,k}$ and $k < n$, explain why $\text{rank } X = k$ if $X^T X$ is nonsingular, but $X^T X$ can be singular even if $\text{rank } X = k$. Hint: Consider $X = [1\ i]^T$.

Example. Let λ be an eigenvalue of a symmetric matrix $A \in M_n$. Explain why x is a right eigenvector of A associated with λ if and only if $\bar{x}$ is a left eigenvector of A associated with λ.

If $A \in M_n$ is symmetric and if $A = S \Lambda S^{-1}$ for a diagonal $\Lambda \in M_n$ and a nonsingular $S \in M_n$, then Λ is symmetric, but it is not evident from this factorization that A is symmetric. If S is a complex orthogonal matrix, however, then $S^{-1} = S^T$ and the complex orthogonal diagonalization $A = S \Lambda S^{-1} = S \Lambda S^T$ is evidently symmetric.

Theorem 4.4.27. *Let $A \in M_n$ be symmetric. Then A is diagonalizable if and only if it is complex orthogonally diagonalizable.*

Proof. If there is a complex orthogonal Q such that $Q^T A Q$ is diagonal, then of course A is diagonalizable; only the converse assertion is interesting. Suppose that A is diagonalizable and let $x, y \in \mathbf{C}^n$ be eigenvectors of A with $Ax = \lambda x$ and $Ay = \mu y$, so $\bar{y}^* A = (Ay)^T = \mu y^T = \mu \bar{y}^*$. If $\lambda \neq \mu$, then the biorthogonality principle (1.4.7) ensures that x is orthogonal to $\bar{y}$, that is, $\bar{y}^* x = y^T x = 0$. Let $\lambda_1, \ldots, \lambda_d$ be the distinct eigenvalues of A with respective multiplicities $n_1, \ldots, n_d$ and let $A = S \Lambda S^{-1}$, in which S is nonsingular and $\Lambda = \lambda_1 I_{n_1} \oplus \cdots \oplus \lambda_d I_{n_d}$. Partition the columns of $S = [S_1 \ S_2 \ \ldots \ S_d]$ conformally to Λ and observe that $A S_i = \lambda_i S_i$ for $i = 1, 2, \ldots, d$. Biorthogonality ensures that $S_i^T S_j = 0$ if $i \neq j$. It follows that $S^T S = S_1^T S_1 \oplus \cdots \oplus S_d^T S_d$ is block diagonal. Since $S^T S$ is nonsingular, each diagonal block $S_i^T S_i$ is nonsingular, $i = 1, 2, \ldots, d$. Therefore, (4.4.26) ensures that each $S_i = Y_i B_i$ in which $Y_i^T Y_i = I_{n_i}$ and B_i is nonsingular. Moreover, $0 = S_i^T S_j = B_i^T Y_i^T Y_j B_j$ for $i \neq j$ implies that $Y_i^T Y_j = 0$ for $i \neq j$. Let $Y = [Y_1 \ \ldots \ Y_d]$ and $B = B_1 \oplus \cdots \oplus B_d$. Then Y is complex orthogonal, B is nonsingular, $S = YB$, and $A = S \Lambda S^{-1} = Y B \Lambda B^{-1} Y^T = Y(\lambda_1 B_1 B_1^{-1} \oplus \cdots \oplus \lambda_d B_d B_d^{-1}) Y^T = Y \Lambda Y^T.$ $\quad\square$

Theorem 4.4.27 has an important generalization: If $A, B \in M_n$ and there is a single polynomial $p(t)$ such that $A^T = p(A)$ and $B^T = p(B)$ (in particular, if A and B are symmetric), then A and B are similar if and only if they are similar via a complex orthogonal similarity; see corollary 6.4.18 in Horn and Johnson (1991).

The exercise preceding (4.4.24) shows how to construct symmetric canonical blocks that can be pieced together to obtain a symmetric canonical form under similarity for any square complex matrix.

Problems

4.4.P1 Let $A \in M_n$. Show that (a) A is symmetric if and only if there exists an $S \in M_n$ such that rank S = rank A and $A = SS^T$; (b) A is symmetric and unitary if and only if there exists a unitary $V \in M_n$ such that $A = VV^T$.

4.4.P2 Provide details for the following approach to (4.4.4c) that uses real representations. Let $A \in M_n$ be symmetric. If A is singular and rank $A = r$, it is unitarily congruent to $A' \oplus 0_{n-r}$, in which $A' \in M_r$ is nonsingular and symmetric; see (2.6.P20(b)). Suppose that A is symmetric and nonsingular. Let $A = A_1 + i A_2$ with A_1, A_2 real and let $x, y \in \mathbf{R}^n$. Consider the real representation $R_2(A) = \begin{bmatrix} A_1 & A_2 \\ A_2 & -A_1 \end{bmatrix}$ (see (1.3.P21)), in which A_1, A_2, and $R_2(A)$ are real symmetric. (a) $R_2(A)$ is nonsingular. (b) $R_2(A) \begin{bmatrix} x \\ -y \end{bmatrix} = \lambda \begin{bmatrix} x \\ -y \end{bmatrix}$ if and only if $R_2(A) \begin{bmatrix} y \\ x \end{bmatrix} = -\lambda \begin{bmatrix} y \\ x \end{bmatrix}$, so the eigenvalues of $R_2(A)$ occur in $\pm$ pairs. (c) Let $\begin{bmatrix} x_1 \\ -y_1 \end{bmatrix}, \ldots, \begin{bmatrix} x_n \\ -y_n \end{bmatrix}$ be orthonormal eigenvectors of $R_2(A)$ associated with its positive eigenvalues $\lambda_1, \ldots, \lambda_n$, let $X = [x_1 \ \ldots \ x_n]$, $Y = [y_1 \ \ldots \ y_n]$, $\Sigma = \text{diag}(\lambda_1, \ldots, \lambda_n)$,

$V = \begin{bmatrix} X & Y \\ -Y & X \end{bmatrix}$, and $\Lambda = \Sigma \oplus (-\Sigma)$. Then V is real orthogonal and $R_2(A) = V \Lambda V^T$. Let $U = X - iY$, so $V = R_1(\bar{U})$ (see (1.3.P20)). Explain why U is unitary and show that $U \Sigma U^T = A$.

4.4.P3 Provide details for the following approach to (4.4.4c). Let $A \in M_n$ be symmetric. (a) $A\bar{A}$ is Hermitian, so $A\bar{A} = V \Lambda_1 V^*$ in which V is unitary and Λ_1 is real diagonal. (b) $V^* A \bar{V} = B$ is symmetric and normal, so (2.5.P57) ensures that $B = Q \Lambda Q^T$ in which Λ is diagonal and Q is real orthogonal. (c) $A = (VQ)\Lambda(VQ)^T$. Let $\Lambda = E \Sigma E^T$ with diagonal E and Σ, E unitary, and Σ nonnegative to get $A = U \Sigma U^T$ with $U = VQE$.

4.4.P4 What does (4.4.4c) say when A is real symmetric? How is it related to the spectral decomposition (2.5.11a) of a real symmetric matrix?

4.4.P5 Let $A \in M_n$. (a) Use (2.5.20(a)) to show that A is unitarily similar to a complex symmetric matrix if and only if A is similar to A^T via a symmetric unitary matrix. (b) If A is unitarily similar to A^T and $n \in \{2, \dots, 7\}$, then A must be unitarily similar to a complex symmetric matrix, but not if $n = 8$! See S. R. Garcia and J. E. Tener, Unitary equivalence of a matrix to its transpose, *J. Operator Theory* 68 (2012) 179–203. (c) However, $A \oplus A^T$ is always unitarily similar to its transpose; prove this using $\begin{bmatrix} 0 & I \\ I & 0 \end{bmatrix}$.

4.4.P6 Let $A \in M_2$ be given and adopt the notation in (4.4.5). Show that (a) $A\bar{A}$ has two non-real conjugate eigenvalues if and only if $-2|\det A| < \operatorname{tr} A\bar{A} < 2|\det A|$; (b) $A\bar{A}$ has two real negative eigenvalues if and only if $\operatorname{tr} A\bar{A} \le -2|\det A| < 0$.

4.4.P7 Apply the reduction algorithm in the proof of (4.4.3) to $A = \begin{bmatrix} 1 & i \\ -i & 1 \end{bmatrix}$. Show that $A = U \Delta U^T$ with $\Delta = \begin{bmatrix} 0 & 2 \\ 0 & 0 \end{bmatrix}$ and $U = \frac{1}{\sqrt{2}} \begin{bmatrix} 1 & 1 \\ -i & i \end{bmatrix}$.

4.4.P8 Apply the reduction algorithm in the proof of (4.4.3) to $A = \begin{bmatrix} 1 & i \\ i & 1 \end{bmatrix}$. Show that it is unitarily congruent to $\operatorname{diag}(\sqrt{2}, \sqrt{2})$.

4.4.P9 Let $A \in M_n$. (a) Show that there is a unitary $U \in M_n$ such that UAU^* is real if and only if there is a symmetric unitary $W \in M_n$ such that $\bar{A} = WAW^* = WA\bar{W}$. (b) Show that there is a unitary $U \in M_n$ such that UAU^T is real if and only if there is a symmetric unitary $W \in M_n$ such that $\bar{A} = WAW^T = WAW$.

4.4.P10 If $n > 1$ and $v \in \mathbf{C}^n$ is a nonzero isotropic vector, why is the symmetric matrix $A = vv^T$ not diagonalizable? What is its Jordan canonical form?

4.4.P11 If $A \in M_n$ is symmetric and nonsingular, show that A^{-1} is symmetric.

4.4.P12 Deduce from (4.4.24) that every square complex matrix is similar to its transpose.

4.4.P13 Is every real square matrix similar to a real symmetric matrix? to a complex symmetric matrix? Via a real similarity matrix? Why?

4.4.P14 Let $z = [z_1 \, z_2 \, \dots \, z_n]^T$ be a vector of n complex variables and let $f(z)$ be a complex analytic function on some domain $D \subset \mathbf{C}^n$. Then $H = [\partial^2 f / \partial z_i \, \partial z_j] \in M_n$ is symmetric at every point $z \in D$. The discussion in (4.0.3) shows that one may assume that the coefficient matrix $A = [a_{ij}]$ in the general linear partial differential operator $Lf = \sum_{i,j=1}^n a_{ij}(z) \frac{\partial^2 f}{\partial z_i \partial z_j}$ is symmetric. At each point $z_0 \in D$, explain why there is a unitary change of variables $z \to U\zeta$ such that L is diagonal at z_0 in the new coordinates, that is, $Lf = \sum_{i=1}^n \sigma_i \frac{\partial^2 f}{\partial \zeta_i^2}$, $\sigma_1 \ge \sigma_2 \ge \dots \ge \sigma_n \ge 0$ at $z = z_0$.

4.4.P15 Let $A, B \in M_n$ be conjugate normal. Explain why A and B are unitarily congruent if and only if $A\bar{A}$ and $B\bar{B}$ have the same eigenvalues.

4.4.P16 The 2-by-2 real orthogonal blocks in (4.4.17) are determined ($a_j = \cos\theta_j$, $b_j = \sin\theta_j$) by angles obtained from the pairs of eigenvalues $\tau_j^2 e^{\pm 2i\theta_j}$, $\theta_j \in (0, \pi/2]$, of $A\bar{A}$ that are not real and nonnegative. Use the preceding problem to show that each such block is unitarily congruent to a unitary (and hence conjugate normal) block $\begin{bmatrix} 0 & 1 \\ e^{2i\theta_j} & 0 \end{bmatrix}$ and explain how this observation leads to an alternative to (4.4.17) as a canonical form for conjugate normal matrices: $A \in M_n$ is conjugate normal if and only if it is unitarily congruent to a direct sum of blocks, each of which is

$$[\sigma] \text{ or } \tau \begin{bmatrix} 0 & 1 \\ e^{i\theta} & 0 \end{bmatrix}, \sigma, \tau, \theta \in \mathbf{R}, \sigma \geq 0, \tau > 0, 0 < \theta \leq \pi \qquad (4.4.28)$$

Explain why the parameters in the blocks of this direct sum are uniquely determined by the eigenvalues of $A\bar{A}$.

4.4.P17 Let σ_1 be the largest singular value of a symmetric matrix $A \in M_n$. Show that $\{x^T A x : \|x\|_2 = 1\} = \{z \in \mathbf{C} : |z| \leq \sigma_1\}$. Compare this result with (4.2.2). What can you say if A is not symmetric?

4.4.P18 Explain why $A \in M_n$ is conjugate normal if and only if it is unitarily congruent to a real normal matrix.

4.4.P19 Let $A \in M_n$. Explain why $\text{tr } A\bar{A}$ is real (but not necessarily nonnegative), and show that $\text{tr } AA^* \geq \text{tr } A\bar{A}$.

4.4.P20 Let $J_m(\lambda)$ be a Jordan block (3.1.2) and let K_m be a reversal matrix (0.9.5.1). Explain why $\hat{J}_m(\lambda) = K_m J_m(\lambda)$ is symmetric and is real symmetric if λ is real. Suppose that $A \in M_n$ has the Jordan canonical form (3.1.12), let $\hat{J} = K_{n_1} J_{n_1}(\lambda_1) \oplus \cdots \oplus K_{n_q} J_{n_q}(\lambda_q)$, and let $\hat{K} = K_{n_1} \oplus \cdots \oplus K_{n_q}$. Explain why $A = (S\hat{K}S^T)(S^{-T}\hat{J}S^{-1})$ is a product of two complex symmetric matrices. This is a proof of (4.4.25) that does not rely on (4.4.24).

4.4.P21 Let $C_m(a, b)$ be a real Jordan block (3.4.1.4) and let K_{2m} be a reversal matrix (0.9.5.1). Explain why $\hat{C}_m(a, b) = K_{2m} C_m(a, b)$ is real symmetric. Let $A \in M_n(\mathbf{R})$, suppose that $S \in M_n(\mathbf{R})$ is nonsingular and $S^{-1}AS$ equals the real Jordan matrix (3.4.1.6), let $\hat{J} = K_{2n_1} C_{n_1}(a_1, b_1) \oplus \cdots \oplus K_{2n_p} C_{n_p}(a_p, b_p) \oplus K_{m_1} J_{m_1}(\mu_1) \oplus \cdots \oplus K_{m_r} J_{m_r}(\mu_r)$, and let $\hat{K} = K_{2n_1} \oplus \cdots \oplus K_{2n_p} \oplus K_{m_1} \oplus \cdots \oplus K_{m_r}$. Explain why $A = (S\hat{K}S^T)(S^{-T}\hat{J}S^{-1})$ is a product of two real symmetric matrices.

4.4.P22 Let $A \in M_n$ be symmetric and suppose that $A^2 = I$. Explain why there is a complex orthogonal $Q \in M_n$ and $k \in \{0, 1, \ldots, n\}$ such that $A = Q(-I_k \oplus I_{n-k})Q^T$.

4.4.P23 Let $A \in M_n$ be a Toeplitz matrix and let $K_n \in M_n$ be a reversal matrix. Explain why A has a singular value decomposition of the form $A = (K_n U)\Sigma U^T$ for some unitary $U \in M_n$.

4.4.P24 Let λ be an eigenvalue of $A \in M_n$. Suppose that x is a (right) λ-eigenvector of A and $\bar{x}$ is a left λ-eigenvector of A. (a) If x is isotropic (0.2.5), show that λ cannot be a simple eigenvalue. (b) If λ has geometric multiplicity 1 and is not a simple eigenvalue, show that x is isotropic.

4.4.P25 Let λ, x be an eigenpair of a symmetric matrix $A \in M_n$. (a) Explain why $\bar{x}$ is a left λ-eigenvector of A. (b) If x is isotropic, use the preceding problem to show that λ is not a simple eigenvalue of A; in particular, A cannot have n distinct eigenvalues. (c) If λ has geometric multiplicity one and is not a simple eigenvalue, explain why x is isotropic. (d) Explain why the symmetric block $S_m J_m(\lambda) S_m^{-1}$ constructed in the exercise preceding (4.4.24) has an isotropic eigenvector if $m > 1$.

4.4.P26 Show that real matrices $A, B \in M_n(\mathbf{R})$ are complex orthogonally similar if and only if they are real orthogonally similar.

4.4.P27 Let $A = [a_{ij}] \in M_n$ be symmetric, and let $A = U\Sigma U^T$ with a unitary $U = [u_{ij}]$ and a nonnegative diagonal $\Sigma = \mathrm{diag}(\sigma_1, \ldots, \sigma_n) \in M_n$ in which $\sigma_1 \geq \cdots \geq \sigma_n \geq 0$. (a) Explain why $\mathrm{diag}\, A = S\,\mathrm{diag}\,\Sigma = \sum_{j=1}^n \sigma_j s_j$, in which every absolute row and column sum of the complex matrix $S = [u_{ij}^2] = [s_1 \ \cdots \ s_n] \in M_n$ is equal to 1. Compare with (4.3.P10). (b) Choose real $\theta_1, \ldots, \theta_n$ such that $e^{-i\theta_j} u_{j1}^2 = |u_{j1}^2|$ for $j = 1, \ldots, n$ and let $z = [e^{i\theta_1} \ \cdots \ e^{i\theta_n}]^T$. Explain why $\sigma_1 = \sigma_1 z^* s_1 = z^* \mathrm{diag}\, A - \sigma_2 z^* s_2 - \cdots - \sigma_n z^* s_n \leq |a_{11}| + \cdots + |a_{nn}| + \sigma_2 + \cdots + \sigma_n$. (c) If A has zero main diagonal entries, explain why its singular values must satisfy the inequality $\sigma_1 \leq \sigma_2 + \cdots + \sigma_n$.

4.4.P28 Let $A \in M_n$ be given. Show that $\det(I + A\bar{A})$ is real and nonnegative.

4.4.P29 Let $A_{ij} \in M_n$, $i, j = 1, 2$, and let $A = \begin{bmatrix} A_{11} & A_{12} \\ A_{21} & A_{22} \end{bmatrix} \subset M_{2n}$. We say that A is a *matrix of quaternion type* if $A_{21} = -\bar{A}_{12}$ and $A_{22} = \bar{A}_{11}$. A matrix $A = [A_{ij}]_{i,j=1}^2$ (with each $A_{ij} \in M_n$) of quaternion type is also called a *complex representation* (of the quaternion matrix $A_{11} + A_{12}j$). (a) Explain why a real matrix $A = [A_{ij}]_{i,j=1}^2$ of quaternion type is the *real representation* $R_1(A_{11} + iA_{12})$ discussed in (1.3.P20). (b) If $A \in M_{2n}$ is a matrix of quaternion type, show that $\det A$ is real and nonnegative. (c) Let $S_{2n} = \begin{bmatrix} 0_n & I_n \\ -I_n & 0_n \end{bmatrix}$. Show that $A = [A_{ij}]_{i,j=1}^2$ (with each $A_{ij} \in M_n$) is a matrix of quaternion type if and only if $S_{2n} A = \bar{A} S_{2n}$. (d) Let $A, B \in M_{2n}$ be matrices of quaternion type, let $\alpha, \beta \in \mathbf{R}$, and let $p(s, t)$ be a polynomial in two noncommuting variables with real coefficients. Use the identity in (c) to show that $\bar{A}, A^T, A^*, AB, \alpha A + \beta B$, and $p(A, B)$ are matrices of quaternion type. (e) Use (c) to show that a matrix $A \in M_{2n}$ of quaternion type is similar to $\bar{A}$ via S_{2n}, so the non-real blocks in the Jordan canonical form of A occur in conjugate pairs. Why is A similar to a real matrix? (f) If $A = [A_{ij}]_{i,j=1}^2 \in M_{2n}$ is a real matrix of quaternion type, explain why its Jordan canonical form consists only of pairs of blocks of the form $J_k(\lambda) \oplus J_k(\bar{\lambda})$ (λ can be either real or non real), and why A is similar to $F \oplus \bar{F}$, in which $F = A_{11} + iA_{12}$. (g) It is known that the assertion in (f) remains true even if A is complex. The Jordan canonical form of a matrix $A \in M_n$ of quaternion type consists only of pairs of blocks of the form $J_k(\lambda) \oplus J_k(\bar{\lambda})$, that is, A is similar to $F \oplus \bar{F}$ for some $F \in M_n$. Why does this confirm that $\det A$ is real and nonnegative?

4.4.P30 Explain why the following property of $A \in M_n$ is a similarity invariant (that is, if one matrix in a similarity equivalence class has the property, then every matrix in that similarity equivalence class has it): A can be written as $A = BC$, in which one factor is symmetric and the other factor is skew symmetric (respectively, both factors are skew symmetric, or both factors are symmetric). Why should one expect that there are three sets of explicit properties of the Jordan canonical form of A that are, respectively, necessary and sufficient for A to be factorable in each of these three ways?

4.4.P31 What property of the Jordan canonical form of $A \in M_n$ is necessary and sufficient for it to be written as $A = BC$, in which both factors are symmetric?

4.4.P32 It is known that $A \in M_n$ can be written as a product of a symmetric matrix and a skew-symmetric matrix if and only if A is similar to $-A$, that is, if and only if the nonsingular part of the Jordan canonical form of A consists only of pairs of the form $J_k(\lambda) \oplus J_k(-\lambda)$. Prove the "only if" part of this assertion.

4.4.P33 It is known that $A \in M_n$ is a product of two skew-symmetric matrices if and only if the nonsingular part of the Jordan canonical form of A consists only of pairs of the form $J_k(\lambda) \oplus J_k(\lambda)$ *and* the Segre characteristic of A (3.1.19) associated with the eigenvalue zero satisfies the inequalities $s_{2k-1}(A, 0) - s_{2k}(A, 0) \leq 1$ for $k = 1, 2, \ldots$. (a) Prove a special case of half of this assertion: If $A \in M_n$ is nonsingular and its Jordan canonical form consists only of pairs of the form $J_k(\lambda) \oplus J_k(\lambda)$, then n is even and A is similar to a block matrix of the form $\begin{bmatrix} F & 0 \\ 0 & F \end{bmatrix}$, which is similar to $\begin{bmatrix} F & 0 \\ 0 & F^T \end{bmatrix} = \begin{bmatrix} 0 & I \\ -I & 0 \end{bmatrix} \begin{bmatrix} 0 & -F^T \\ F & 0 \end{bmatrix} = \begin{bmatrix} 0 & -F \\ F^T & 0 \end{bmatrix} \begin{bmatrix} 0 & I \\ -I & 0 \end{bmatrix}$. (b) Let $w_p(A, 0)$, $p = 1, 2, \ldots$ be the Weyr characteristic of A (3.1.17) associated with the eigenvalue zero. Show that $s_{2k-1}(A, 0) - s_{2k}(A, 0) \leq 1$ for $k = 1, 2, \ldots$ if and only if $w_{p-1}(A, 0) > w_p(A, 0)$ for each $p = 2, 3, \ldots$ such that $w_p(A, 0)$ is odd (that is, the Jordan canonical form of A contains at least one nilpotent block of size $p - 1$ whenever $p \geq 2$ and $w_p(A, 0)$ is odd).

4.4.P34 Although a symmetric complex matrix can have any given Jordan canonical form (4.4.24), the Jordan canonical form of a skew-symmetric complex matrix has a special form. It consists of only the following three types of direct summands: (a) pairs of the form $J_k(\lambda) \oplus J_k(-\lambda)$, in which $\lambda \neq 0$; (b) pairs of the form $J_k(0) \oplus J_k(0)$, in which k is even; and (c) $J_k(0)$, in which k is odd. Explain why the Jordan canonical form of a complex skew-symmetric matrix A ensures that A is similar to $-A$; also deduce this fact from (3.2.3.1).

4.4.P35 Why is a real orthogonal matrix diagonalizable? What is its Jordan canonical form? What is its real Jordan form?

4.4.P36 Let a nonsingular $A \in M_n$ be given, and suppose that there is a nonsingular complex symmetric $S \in M_n$ such that $A^T = SA^{-1}S^{-1}$. Show that A is similar to a complex orthogonal matrix as follows: Choose a $Y \in M_n$ such that $S = Y^T Y$ (4.4.P1) and explain why YAY^{-1} is complex orthogonal.

4.4.P37 Let a nonzero $\lambda \in \mathbb{C}$ and an integer $m \geq 2$ be given. Let $B \in M_n$ be a complex symmetric matrix to which the Jordan block $J_m(\lambda)$ is similar (4.4.24), and let $A = B \oplus B^{-1}$. (a) Explain why A^{-1} is similar to A^T via the reversal matrix $K_{2m} = \begin{bmatrix} 0 & I_m \\ I_m & 0 \end{bmatrix}$. (b) Use the preceding problem to explain why A is similar to a complex orthogonal matrix. (c) Explain why there is a complex orthogonal matrix $Q \in M_4$ whose Jordan canonical form is $J_2(2) \oplus J_2(\frac{1}{2})$; in particular, and in contrast to (4.4.P35) Q is not diagonalizable. It is known that the Jordan canonical form of a complex orthogonal matrix is a direct sum of only the following five types of summands: $J_k(\lambda) \oplus J_k(\lambda^{-1})$, in which $0 \neq \lambda \neq \pm 1$; $J_k(1) \oplus J_k(1)$, in which k is even; $J_k(-1) \oplus J_k(-1)$, in which k is even; $J_k(1)$, in which k is odd; and $J_k(-1)$, in which k is odd.

4.4.P38 Let $A \in M_n$ and let $\mathcal{A} = \begin{bmatrix} 0 & A \\ A^T & 0 \end{bmatrix} \in M_{2n}$. One says that A has a *QS factorization* if there is a complex orthogonal $Q \in M_n$ and a complex symmetric $S \in M_n$ such that $A = QS$. It is known that A has a QS factorization if and only if $\mathrm{rank}(AA^T)^k = \mathrm{rank}(A^T A)^k$ for each $k = 1, \ldots, n$. (a) Use this rank condition to show that A has a QS factorization if and only if AA^T is similar to $A^T A$. (b) Does $\begin{bmatrix} 1 & i \\ 0 & 0 \end{bmatrix}$ have a QS factorization? (c) Suppose that A is nonsingular. Why does it have a QS factorization? In this case it is known that there is a polynomial $p(t)$ such that $S = p(A^T A)$, and if A is real, then both Q and S may be chosen to be real; see theorem 6.4.16 in Horn and Johnson (1991). (d) If A has a QS factorization, show that $\mathcal{A}$ is similar to $\begin{bmatrix} 0 & S \\ S & 0 \end{bmatrix}$ via the similarity matrix $Q \oplus I$. Explain why the Jordan canonical form of $\mathcal{A}$ consists only of direct summands of the form $J_k(\lambda) \oplus J_k(-\lambda)$. See (2.5.20(b)) for a special QS factorization of a unitary matrix.

4.4.P39 Use the QS factorization in the preceding problem to prove a slightly stronger version of (4.4.27): Suppose that $A \in M_n$ is symmetric and $A = B\Lambda B^{-1}$ for some nonsingular B and diagonal Λ. Write $B = QS$, in which Q is complex orthogonal and S is symmetric. Then $A = B\Lambda B^{-1} = Q\Lambda Q^T$.

4.4.P40 Let $A \in M_n$ and let $\mathcal{A} = \begin{bmatrix} 0 & A \\ \bar{A} & 0 \end{bmatrix} \in M_{2n}$. Show that (a) A is normal if and only if $\mathcal{A}$ is conjugate normal; (b) A is conjugate normal if and only if $\mathcal{A}$ is normal.

4.4.P41 Let $A \in M_n$. It is known that $A\bar{A}$ is normal (that is, A is congruence normal) if and only if A is unitarily congruent to a direct sum of blocks, each of which is

$$[\sigma] \text{ or } \tau \begin{bmatrix} 0 & 1 \\ \mu & 0 \end{bmatrix}, \text{ in which } \sigma, \tau \in \mathbf{R}, \sigma \geq 0, \tau > 0, \mu \in \mathbf{C}, \text{ and } \mu \neq 1 \quad (4.4.29)$$

This direct sum is uniquely determined by A, up to permutation of its blocks and replacement of any nonzero parameter μ by μ^{-1} with a corresponding replacement of τ by $\tau|\mu|$. (a) Use the canonical forms (4.4.28) and (4.4.29) to show that every conjugate-normal matrix is congruence normal. (b) Use the definition of conjugate normality and the characterization of congruence normal matrices in (2.5.P27) to give a different proof that every conjugate-normal matrix is congruence normal.

4.4.P42 Let $A \in M_n$ and suppose that $A\bar{A}$ is Hermitian. Deduce from the canonical form (4.4.29) that A is unitarily congruent to a direct sum of blocks, each of which is

$$[\sigma] \text{ or } \tau \begin{bmatrix} 0 & 1 \\ \mu & 0 \end{bmatrix}, \text{ in which } \sigma, \tau, \mu \in \mathbf{R}, \sigma \geq 0, \tau > 0, \text{ and } \mu \in [-1, 1) \quad (4.4.30)$$

Explain why this direct sum is uniquely determined by A, up to permutation of its blocks.

4.4.P43 Let $A \in M_n$ and suppose that $A\bar{A}$ is positive semidefinite. Deduce from the canonical form (4.4.30) that A is unitarily congruent to a direct sum of blocks, each of which is

$$[\sigma] \text{ or } \tau \begin{bmatrix} 0 & 1 \\ \mu & 0 \end{bmatrix}, \text{ in which } \sigma, \tau, \mu \in \mathbf{R}, \sigma \geq 0, \tau > 0, \text{ and } \mu \in [0, 1) \quad (4.4.31)$$

Explain why this direct sum is uniquely determined by A, up to permutation of its blocks.

4.4.P44 Let $A \in M_n$. (a) Show that $A\bar{A} = AA^*$ if and only if A is symmetric. (b) Show that $A\bar{A} = -AA^*$ if and only if A is skew symmetric.

4.4.P45 If $U, V \in M_n$ are unitary and symmetric, show that they are unitarily congruent.

4.4.P46 This problem builds on (2.5.P69) and (2.5.P70). (a) If $A, B \in M_n$ are unitarily congruent, show that the three pairs (AA^*, BB^*), $(A\bar{A}, B\bar{B})$, and $(A^T\bar{A}, B^T\bar{B})$ are simultaneously unitarily similar. It is known that this necessary condition is also sufficient for A

and B to be unitarily congruent. (b) Define the $4n$-by-$4n$ block upper triangular matrices

$$K_A = \begin{bmatrix} 0 & I & AA^* & A\bar{A} \\ & 0 & I & A^T\bar{A} \\ & & 0 & I \\ & & & 0 \end{bmatrix} \text{ and } K_B = \begin{bmatrix} 0 & I & BB^* & B\bar{B} \\ & 0 & I & B^T\bar{B} \\ & & 0 & I \\ & & & 0 \end{bmatrix} \qquad (4.4.32)$$

Explain why A and B are unitarily *congruent* if and only if K_A and K_B are unitarily *similar*.

4.4.P47 Use the definitions and notation of (2.5.P69). (a) If $M_A\bar{W} = WM_B$, show that W is block upper triangular, $W_{ii} = W_{11}$ if i is odd, and $W_{ii} = \overline{W_{11}}$ if i is even. (b) Suppose that W is unitary and $M_A\bar{W} = WM_B$ (that is, $M_A = WM_BW^T$, so M_A is unitarily congruent to M_B via W). Show that $W_{11} = U$ is unitary, $W_{ii} = U$ if i is odd, $W_{ii} = \bar{U}$ if i is even, and $W = U \oplus \bar{U} \oplus U \oplus \cdots$ is block diagonal. Moreover, $A_{ij} = UB_{ij}U^*$ if i is odd and j is even (simultaneous unitary similarity via U); $A_{ij} = UB_{ij}U^T$ if i and j are both odd (simultaneous unitary congruence via U); $A_{ij} = \bar{U}B_{ij}U^*$ if i and j are both even (simultaneous unitary congruence via $\bar{U}$); and $A_{ij} = \bar{U}B_{ij}U^T$ if i is even and j is odd (simultaneous unitary similarity via $\bar{U}$). (c) Describe how the ideas in (a) and (b) can be used in an algorithm to decide if given pairs of matrices are simultaneously unitarily similar/congruent.

4.4.P48 Suppose that $A \in M_n$ has distinct singular values. Use (4.4.16) to show that A is conjugate normal if and only if it is symmetric.

4.4.P49 Let $\Sigma \in M_n$ be a nonnegative quasidiagonal matrix that is a direct sum of an identity matrix and blocks of the form $\begin{bmatrix} 0 & \sigma^{-1} \\ \sigma & 0 \end{bmatrix}$, in which $\sigma > 1$. If $U \in M_n$ is unitary, show that $A = U\Sigma U^T$ is coninvolutory. It is known that every coninvolutory matrix has a special singular value decomposition of this form, which is an analog of the special singular value decomposition (4.4.4c) for a complex symmetric matrix.

Notes and Further Readings. A block upper triangular form that can be achieved for any square complex matrix under unitary congruence is in D. C. Youla, A normal form for a matrix under the unitary congruence group, *Canad. J. Math.* 13 (1961) 694–704; Youla's 2-by-2 diagonal blocks are different from (but of course are unitarily congruent to) those in (4.4.9). For more information about unitary congruence, conjugate-normal matrices, congruence-normal matrices, and a proof of the canonical form in (4.4.P41), see R. A. Horn and V. V. Sergeichuk, Canonical forms for unitary congruence and *congruence, *Linear Multilinear Algebra* 57 (2009) 777–815. For an exposition of analogies between normal and conjugate-normal matrices, and a list of 45 criteria for conjugate normality, see H. Faßbender and Kh. Ikramov, Conjugate-normal matrices: A survey, *Linear Algebra Appl.* 429 (2008) 1425–1441. Leon Autonne (1915) seems to have discovered the canonical form (4.4.4c) for a complex symmetric matrix (the *Autonne–Takagi factorization*; see the *Notes and Further Readings* in (2.6)); there have been many subsequent independent rediscoveries and different proofs, for example, Takagi (1925), Jacobson (1939), Siegel (1943; see (4.4.P3)), Hua (1944), Schur (1945; see (4.4.P2)) and Benedetti and Cragnolini (1984). For a canonical form for complex symmetric matrices under complex orthogonal similarity, see N. H. Scott, A new canonical form for complex symmetric matrices, *Proc. R. Soc. Lond. Ser. A* 440 (1993) 431–442; it makes use of the information in (4.4.P24) and (4.4.P25). For a proof of the assertion in (4.4.P29(g)) about the Jordan canonical form of a matrix of quaternion type, see F. Zhang and Y. Wei, Jordan canonical form of a partitioned complex matrix and its application to real quaternion matrices, *Comm. Algebra* 29 (2001) 2363–2375. Characterizations of the matrix products considered in (4.4.P31 to P33) were known as early as 1922 (H. Stenzel); modern proofs may be found in L. Rodman, Products of symmetric and skew-symmetric matrices, *Linear Multilinear Algebra* 43 (1997) 19–34. The equivalence asserted in (4.4.P33(b)) is due to Ross Lippert. The assertions about Jordan canonical forms in (4.4.P34) and (4.4.P38) are proved in chapter XI of Gantmacher (1959) and in R. A. Horn and D. I. Merino, The Jordan canonical forms of complex orthogonal and skew-symmetric matrices, *Linear Algebra Appl.* 302– 303 (1999) 411–421. For more about the QS factorization in (4.4.P36), see theorem 6.4.16 in Horn and Johnson (1991); a proof of the rank condition is in I. Kaplansky, Algebraic polar decomposition, *SIAM J. Matrix Analysis Appl.* 11 (1990) 213–217; also see theorem 13 in R. A. Horn and D. I. Merino, Contragredient equivalence: A canonical form and some applications, *Linear Algebra Appl.* 214 (1995) 43–92, where two additional equivalent conditions may be found: $A = P A^T Q$ or $A = Q A^T Q$ for some complex orthogonal matrices P and Q. For a derivation of the canonical form in (4.4.P41), see theorem 7.1 in R. A. Horn and V. V. Sergeichuk, Canonical forms for unitary congruence and *congruence, *Linear Multilinear Algebra* 57 (2009) 777–815. The necessary and sufficient condition for unitary congruence mentioned in (4.4.P46) is proved in R. A. Horn and Y. P. Hong, A characterization of unitary congruence, *Linear Multilinear Algebra* 25 (1989) 105–119. For more information about unitary congruence, simultaneous unitary congruences, simultaneous unitary similarities and the block matrix M_A in (2.5.P69), see T. G. Gerasimova, R. A. Horn, and V. V. Sergeichuk, Simultaneous unitary equivalences, *Linear Algebra Appl.* (in press). The special singular value decomposition in (4.4.P49) was discovered by L. Autonne; for a proof, see theorem 1.5 in R. A. Horn and D. I. Merino, A real-coninvolutory analog of the polar decomposition, *Linear Algebra Appl.* 190 (1993) 209–227.

4.5 Congruences and diagonalizations

A real second-order linear partial differential operator has the form

$$Lf = \sum_{i,j=1}^{n} a_{ij}(x) \frac{\partial^2 f(x)}{\partial x_i \, \partial x_j} + \text{lower order terms} \tag{4.5.1}$$

in which the coefficients $a_{ij}(x)$ are defined on a domain $D \subset \mathbf{R}^n$ and f is twice continuously differentiable on D. As in (4.0.3), we may assume without loss of generality that the matrix of coefficients $A(x) = [a_{ij}(x)]$ is real symmetric for all $x \in D$. By *lower-order terms* we mean terms involving f and its first partial derivatives only.

If we make a nonsingular change of independent variables to new variables $s = [s_i] \in D \subset \mathbf{R}^n$, then each $s_i = s_i[x] = s_i(x_1, \ldots, x_n)$, and non-singularity means that the Jacobian matrix

$$S(x) = \left[\frac{\partial s_i(x)}{\partial x_j} \right] \in M_n$$

is nonsingular at each point of D. This assumption guarantees that the inverse change of variables $x = x(s)$ exists locally. In these new coordinates, the operator L has the form

$$Lf = \sum_{i,j=1}^{n} \left[\sum_{p,q=1}^{n} \frac{\partial s_i}{\partial x_p} a_{pq} \frac{\partial s_j}{\partial x_q} \right] \frac{\partial^2 f}{\partial s_i \, \partial s_j} + \text{lower order terms}$$

$$= \sum_{i,j=1}^{n} b_{ij} \frac{\partial^2 f}{\partial s_i \, \partial s_j} + \text{lower order terms} \tag{4.5.2}$$

Thus, the new matrix of coefficients B (in the coordinates $s = [s_i]$) is related to the old matrix of coefficients A (in the coordinates $x = [x_i]$) by the relation

$$B = SAS^T \tag{$4.5.3^T$}$$

in which S is a real nonsingular matrix.

If the differential operator L is associated with some physical law (for example, the Laplacian $L = \nabla^2$ and electrostatic potentials), the choice of coordinates for the independent variable should not affect the law, even though it affects the form of L. What are the invariants of the set of all matrices B that are related to a given matrix A by the relation $(4.5.3^T)$?

Another example comes from probability and statistics. Suppose that $X_1, X_2, \ldots,$ X_n are complex random variables with finite second moments on some probability space with expectation operator E, and let $\mu_i = E(X_i)$ denote the respective means. The Hermitian matrix $A = [a_{ij}] = (E[(X_i - \mu_i)(\overline{X_j - \mu_j})]) = \text{Cov}(X)$ is the *covariance matrix* of the random vector $X = [X_1 \ \ldots \ X_n]^T$. If $S = [s_{ij}] \in M_n$, then SX is a random vector whose entries are linear combinations of the entries of X. The means of the entries of SX are

$$E((SX)_i) = E\left(\sum_{k=1}^{n} s_{ik} X_k \right) = \sum_{k=1}^{n} s_{ik} E(X_k) = \sum_{k=1}^{n} s_{ik} \mu_k$$

and the covariance matrix of SX is

$$\text{Cov}(SX) = (E[((SX)_i - E((SX)_i))((\overline{SX})_j - E((\overline{SX})_j))])$$

$$= \left(E\left[\left(\sum_{p=1}^{n} s_{ip}(X_p - \mu_p) \right) \left(\sum_{q=1}^{n} \bar{s}_{jq}(\bar{X}_q - \bar{\mu}_q) \right) \right] \right)$$

$$= \left(\sum_{p,q=1}^{n} s_{ip} E[(X_p - \mu_p)(\bar{X}_q - \bar{\mu}_q)] \bar{s}_{jq} \right) = \left(\sum_{p,q=1}^{n} s_{ip} a_{pq} \bar{s}_{jq} \right)$$

$$= SAS^*$$

This shows that

$$\text{Cov}(SX) = S \, \text{Cov}(X) S^* \tag{$4.5.3^*$}$$

As a final example, consider the general quadratic form

$$Q_A(x) = \sum_{i,j=1}^{n} a_{ij} x_i x_j = x^T A x, \quad x = [x_i] \in \mathbf{C}^n$$

and the Hermitian form

$$H_B(x) = \sum_{i,j=1}^{n} b_{ij} \bar{x}_i x_j = x^* B x, \quad x = [x_i] \in \mathbf{C}^n$$

in which $A = [a_{ij}]$ and $B = [b_{ij}]$. If $S \in M_n$, then

$$Q_A(Sx) = (Sx)^T A(Sx) = x^T (S^T A S)x = Q_{S^T A S}(x)$$
$$H_B(Sx) = (Sx)^* B(Sx) = x^* (S^* B S)x = H_{S^* B S}(x)$$

Definition 4.5.4. *Let $A, B \in M_n$ be given. If there exists a nonsingular matrix S such that*

(a) $B = SAS^*$, *then B is said to be* *congruent *("star-congruent") or* conjunctive *to A*

(b) $B = SAS^T$, *then B is said to be* congruent *or* ^{T}congruent *("tee-congruent") to A.*

Exercise. Explain why congruent (respectively, *congruent) matrices have the same rank.

If A is Hermitian, then so is SAS^*, even if S is singular; if A is symmetric, then so is SAS^T, even if S is singular. Usually, one is interested in congruences that preserve the type of the matrix: *congruence for Hermitian matrices and ^{T}congruence for symmetric matrices.

Both types of congruence share an important property with similarity.

Theorem 4.5.5. *Both* *congruence *and congruence are equivalence relations.*

Proof. Reflexivity: $A = IAI^*$. Symmetry: If $A = SBS^*$ and S is nonsingular, then $B = S^{-1}A(S^{-1})^*$. Transitivity: If $A = S_1 B S_1^*$ and $B = S_2 C S_2^*$, then $A = (S_1 S_2)C(S_1 S_2)^*$. Verification of reflexivity, symmetry, and transitivity for ^{T}congruence can be verified in the same fashion. $\qquad\square$

What canonical forms are available for *congruence and ^{T}congruence? That is, if we partition M_n into *congruence (respectively, into ^{T}congruence) equivalence classes, what choice can we make for a canonical representative from each equivalence class? We begin by considering the simplest cases first: Canonical forms for Hermitian matrices under *congruence and complex symmetric matrices under ^{T}congruence.

Definition 4.5.6. *Let $A \in M_n$ be Hermitian. The* inertia *of A is the ordered triple*

$$i(A) = (i_+(A), i_-(A), i_0(A))$$

in which $i_+(A)$ is the number of positive eigenvalues of A, $i_-(A)$ is the number of negative eigenvalues of A, and $i_0(A)$ is the number of zero eigenvalues of A. The signature *of A is the quantity $i_+(A) - i_-(A)$.*

Exercise. Explain why rank $A = i_+(A) + i_-(A)$.

Exercise. Explain why the inertia of a Hermitian matrix is uniquely determined by its rank and signature.

Let $A \in M_n$ be Hermitian, and write $A = U\Lambda U^*$, in which $\Lambda = \mathrm{diag}(\lambda_1, \ldots, \lambda_n)$ and U is unitary. It is convenient to assume that the positive eigenvalues occur first among the diagonal entries of Λ, then the negative eigenvalues, and then the zero eigenvalues (if any). Thus $\lambda_1, \lambda_2, \ldots, \lambda_{i_+(A)} > 0$, $\lambda_{i_+(A)+1}, \ldots, \lambda_{i_+(A)+i_-(A)} < 0$, and $\lambda_{i_+(A)+i_-(A)+1} = \cdots = \lambda_n = 0$. Define the real diagonal nonsingular matrix

$$D = \mathrm{diag}(\underbrace{\lambda_1^{1/2}, \ldots, \lambda_{i_+(A)}^{1/2}}_{i_+(A) \text{ entries}}, \underbrace{(-\lambda_{i_+(A)+1})^{1/2}, \ldots, (-\lambda_{i_+(A)+i_-(A)})^{1/2}}_{i_-(A) \text{ entries}}, \underbrace{1, \ldots, 1}_{i_0(A) \text{ entries}})$$

Then $\Lambda = DI(A)D$, in which the real matrix

$$I(A) = I_{i_+(A)} \oplus (-I_{i_-(A)}) \oplus 0_{i_0(A)}$$

is the *inertia matrix* of A. Finally, $A = U\Lambda U^* = UDI(A)DU^* = SI(A)S^*$, in which $S = UD$ is nonsingular. We have proved the following theorem.

Theorem 4.5.7. *Each Hermitian matrix is *congruent to its inertia matrix.*

Exercise. If $A \in M_n(\mathbf{R})$ is symmetric, modify the preceding argument to show that A is congruent via a real matrix to its inertia matrix.

The inertia matrix would be a very pleasant canonical representative of the equivalence class of matrices that are *congruent to A if we knew that *congruent Hermitian matrices have the same inertia. This is the content of the following theorem, *Sylvester's law of inertia.*

Theorem 4.5.8 (Sylvester). *Hermitian matrices A, $B \in M_n$ are *congruent if and only if they have the same inertia, that is, if and only if they have the same number of positive eigenvalues and the same number of negative eigenvalues.*

Proof. Since each of A and B is *congruent to its inertia matrix, if they have the same inertia, they must be *congruent. The converse assertion is more interesting.

Suppose that $S \in M_n$ is nonsingular and that $A = SBS^*$. Congruent matrices have the same rank, so it follows that $i_0(A) = i_0(B)$ and hence it suffices to show that $i_+(A) = i_+(B)$. Let $v_1, v_2, \ldots, v_{i_+(A)}$ be orthonormal eigenvectors of A associated with the positive eigenvalues $\lambda_1(A), \ldots, \lambda_{i_+(A)}(A)$, and let $\mathcal{S}_+(A) = \mathrm{span}\{v_1, \ldots, v_{i_+(A)}\}$. If $x = \alpha_1 v_1 + \cdots + \alpha_{i_+(A)} v_{i_+(A)} \neq 0$ then $x^*Ax = \lambda_1(A)|\alpha_1|^2 + \cdots + \lambda_{i_+(A)}(A)|\alpha_{i_+(A)}|^2 > 0$; that is, $x^*Ax > 0$ for all nonzero x in the subspace $\mathcal{S}_+(A)$, whose dimension is $i_+(A)$. The subspace $S^*\mathcal{S}_+(A) = \{y : y = S^*x \text{ and } x \in \mathcal{S}_+(A)\}$ also has dimension $i_+(A)$. If $y = S^*x \neq 0$ and $x \in \mathcal{S}_+(A)$, then $y^*By = x^*(SBS^*)x = x^*Ax > 0$, so (4.2.12) ensures that $i_+(B) \geq i_+(A)$. If we reverse the roles of A and B in the preceding argument, it shows that $i_+(A) \geq i_+(B)$. We conclude that $i_+(B) = i_+(A)$. $\square$

Exercise. Explain why a Hermitian $A \in M_n$ is *congruent to the identity matrix if and only if it is positive definite.

Exercise. Let $A, B \in M_n(\mathbf{R})$ be symmetric. Explain why A and B are *congruent via a complex matrix if and only if they are congruent via a real matrix.

Exercise. Let $A, S \in M_n$ with A Hermitian and S nonsingular. Let $\lambda_1 \leq \cdots \leq \lambda_n$ be the nondecreasingly ordered eigenvalues of A and let $\mu_1 \leq \cdots \leq \mu_n$ be the nondecreasingly ordered eigenvalues of SAS^*. Explain why, for each $j = 1, \ldots, n$, λ_j and μ_j are both negative, both zero, or both positive.

Although the respective *signs* of the nonincreasingly ordered eigenvalues of a Hermitian matrix do not change under *congruence, their *magnitudes* can change. Bounds on the change in magnitude are given in the following quantitative form of Sylvester's theorem.

Theorem 4.5.9 (Ostrowski). *Let $A, S \in M_n$ with A Hermitian and S nonsingular. Let the eigenvalues of A, SAS^*, and SS^* be arranged in nondecreasing order (4.2.1). Let $\sigma_1 \geq \cdots \geq \sigma_n > 0$ be the singular values of S. For each $k = 1, \ldots, n$ there is a positive real number $\theta_k \in [\sigma_n^2, \sigma_1^2]$ such that*

$$\lambda_k(SAS^*) = \theta_k \lambda_k(A) \tag{4.5.10}$$

Proof. First observe that $\sigma_n^2 = \lambda_1(SS^*) \leq \cdots \leq \sigma_1^2 = \lambda_n(SS^*)$. Let $1 \leq k \leq n$ and consider the Hermitian matrix $A - \lambda_k(A)I$, whose kth nondecreasingly ordered eigenvalue is zero. According to the preceding exercise and theorem, for each $j = 1, \ldots, n$, the respective jth nondecreasingly ordered eigenvalues of $A - \lambda_k(A)I$ and $S(A - \lambda_k(A)I)S^* = SAS^* - \lambda_k(A)SS^*$ have the same sign: negative, zero, or positive. Since the kth eigenvalue of $A - \lambda_k(A)I$ is zero, (4.3.2a,b) ensure that

$$0 = \lambda_k(SAS^* - \lambda_k(A)SS^*) \leq \lambda_k(SAS^*) + \lambda_n(-\lambda_k(A)SS^*)$$
$$= \lambda_k(SAS^*) - \lambda_1(\lambda_k(A)SS^*)$$

and that

$$0 = \lambda_k(SAS^* - \lambda_k(A)SS^*) \geq \lambda_k(SAS^*) + \lambda_1(-\lambda_k(A)SS^*)$$
$$= \lambda_k(SAS^*) - \lambda_n(\lambda_k(A)SS^*)$$

Combining these two inequalities gives the bounds

$$\lambda_1(\lambda_k(A)SS^*) \leq \lambda_k(SAS^*) \leq \lambda_n(\lambda_k(A)SS^*)$$

If $\lambda_k(A) < 0$, then $\lambda_1(\lambda_k(A)SS^*) = \lambda_k(A)\lambda_n(SS^*) = \lambda_k(A)\sigma_1^2$, $\lambda_n(\lambda_k(A)SS^*) = \lambda_k(A)\lambda_1(SS^*) = \lambda_k(A)\sigma_n^2$, and

$$\sigma_1^2 \lambda_k(A) \leq \lambda_k(SAS^*) \leq \sigma_n^2 \lambda_k(A)$$

If $\lambda_k(A) > 0$, it follows in the same way that

$$\sigma_n^2 \lambda_k(A) \leq \lambda_k(SAS^*) \leq \sigma_1^2 \lambda_k(A)$$

In either case (or in the trivial case $\lambda_k(A) = \lambda_k(SAS^*) = 0$), we have $\lambda_k(SAS^*) = \theta_k \lambda_k(A)$ for some $\theta_k \in [\sigma_n, \sigma_1]$. $\square$

If $A = I \in M_n$ in Ostrowski's theorem, then all $\lambda_k(A) = 1$ and $\theta_k = \lambda_k(SS^*) = \sigma_{n-k+1}$. If $S \in M_n$ is unitary, then $\sigma_1 = \sigma_n = 1$ and all $\theta_k = 1$; this expresses the invariance of the eigenvalues under a unitary similarity.

A continuity argument can be used to extend the preceding theorem to the case in which S is singular. In this case, let $\delta > 0$ be such that $S + \epsilon I$ is nonsingular for all $\epsilon \in (0, \delta)$. Apply the theorem to A and $S + \epsilon I$, and conclude that $\lambda_k((S + \epsilon I)A(S + \epsilon I)^*) = \theta_k\lambda_k(A)$ with $\lambda_1((S + \epsilon I)(S + \epsilon I)^*) \leq \theta_k \leq \lambda_n((S + \epsilon I)(S + \epsilon I)^*)$. Now let $\epsilon \to 0$ to obtain the bound $0 \leq \theta_k \leq \lambda_n(SS^*) = \sigma_1^2$. This result may be thought of as an extension of Sylvester's law of inertia to singular *congruences.

Corollary 4.5.11. *Let $A, S \in M_n$ and let A be Hermitian. Let the eigenvalues of A be arranged in nonincreasing order (4.2.1); let σ_n and σ_1 be the smallest and largest singular values of S. For each $k = 1, 2, \ldots, n$ there is a nonnegative real number θ_k such that $\sigma_n^2 \leq \theta_k \leq \sigma_1^2$ and $\lambda_k(SAS^*) = \theta_k\lambda_k(A)$. In particular, the number of positive (respectively, negative) eigenvalues of SAS^* is at most the number of positive (respectively, negative) eigenvalues of A.*

The problem of finding a canonical representative for each equivalence class of complex symmetric matrices under ^{T}congruence has a very simple solution: Just compute the rank.

Theorem 4.5.12. *Let $A, B \in M_n$ be symmetric. There is a nonsingular $S \in M_n$ such that $A = SBS^T$ if and only if rank A = rank B.*

Proof. If $A = SBS^T$ and S is nonsingular, then rank A = rank B (0.4.6b). Conversely, use (4.4.4c) to write

$$A = U_1\Sigma_1U_1^T = U_1I(\Sigma_1)D_1^2U_1^T = (U_1D_1)I(\Sigma_1)(U_1D_1)^T$$

in which the inertia matrix $I(\Sigma_1)$ is determined solely by the rank of A, U_1 is unitary, $\Sigma_1 = \text{diag}(\sigma_1, \sigma_2, \ldots, \sigma_n)$ with all $\sigma_i \geq 0$, and $D_1 = \text{diag}(d_1, d_2, \ldots, d_n)$, in which

$$d_i = \begin{cases} \sqrt{\sigma_i} & \text{if } \sigma_i > 0 \\ 1 & \text{if } \sigma_i = 0 \end{cases}$$

Notice that D_1 is nonsingular. In the same way, we can also write $B = (U_2D_2)I(\Sigma_2)(U_2D_2)^T$ with similar definitions. If rank A = rank B, then $I(\Sigma_1) = I(\Sigma_2)$ and

$$I(\Sigma_1) = (U_1D_1)^{-1}A(U_1D_1)^{-T} = I(\Sigma_2) = (U_2D_2)^{-1}B(U_2D_2)^{-T}$$

and hence $A = SBS^T$, in which $S = (U_1D_1)(U_2D_2)^{-1}$. $\square$

Exercise. Let $A, B \in M_n$ be symmetric. Show that there are nonsingular matrices $X, Y \in M_n$ such that $A = XBY$ if and only if there is a nonsingular $S \in M_n$ such that $A = SBS^T$. *Hint:* (0.4.6c).

The preceding theorem is an analog of Sylvester's law of inertia (4.5.8) for ^{T}congruence of complex matrices. The following result is an analog of (4.5.9) and (4.5.11).

Theorem 4.5.13. *Let $A, S \in M_n$ and suppose that A is symmetric. Let $A = U\Sigma U^T$ and $SAS^T = VMV^T$ be factorizations (4.4.4c) of A and SAS^T in which U and V are unitary, $\Sigma = \mathrm{diag}(\sigma_1, \sigma_2, \ldots, \sigma_n)$, and $M = \mathrm{diag}(\mu_1, \mu_2, \ldots, \mu_n)$ with all $\sigma_i, \mu_i \geq 0$. Let $\lambda_i(SS^*)$ denote the eigenvalues of SS^*. Suppose that σ_i, μ_i, and $\lambda_i(SS^*)$ are all arranged in nondecreasing order (4.2.1). For each $k = 1, 2, \ldots, n$ there exists a nonnegative real θ_k with $\lambda_1(SS^*) \leq \theta_k \leq \lambda_n(SS^*)$ such that $\mu_k = \theta_k \sigma_k$. If S is nonsingular, all $\theta_k > 0$.*

Proof. We have $\mu_k^2 = \lambda_k(SAS^T \bar{S}\bar{A}S^*) = \lambda_k(S(AS^T \bar{S}\bar{A})S^*) = \hat{\theta}_k \lambda_k(AS^T \bar{S}\bar{A})$ in which (4.5.11) ensures that $\lambda_1(SS^*) \leq \hat{\theta}_k \leq \lambda_n(SS^*)$. Invoking (1.3.22), we also have $\mu_k^2 = \hat{\theta}_k \lambda_k(AS^T \bar{S}\bar{A}) = \hat{\theta}_k \lambda_k(\bar{S}\bar{A}AS^T) = \hat{\theta}_k \lambda_k(S\bar{A}A S^*)$ because $S\bar{A}AS^*$ has real eigenvalues (it is Hermitian). Applying (4.5.11) again, we obtain $\mu_k^2 = \hat{\theta}_k \tilde{\theta}_k \lambda_k(A\bar{A}) = \hat{\theta}_k \tilde{\theta}_k \sigma_k^2$ for some $\tilde{\theta}_k$ with $\lambda_1(SS^*) \leq \tilde{\theta}_k \leq \lambda_n(SS^*)$. Thus, $\mu_k = (\hat{\theta}_k \tilde{\theta}_k)^{1/2}\sigma_k = \theta_k \sigma_k$, in which $\theta_k = (\hat{\theta}_k \tilde{\theta}_k)^{1/2}$ satisfies the asserted bounds. $\square$

We know from (1.3.19) that two diagonalizable (by similarity) matrices can be simultaneously diagonalized by the same similarity if and only if they commute. What is the corresponding result for simultaneous diagonalization by congruence?

Perhaps the earliest motivation for results about simultaneous diagonalization by congruence came from mechanics in the study of small oscillations about a stable equilibrium. If the configuration of a dynamical system is specified by generalized (Lagrangian) coordinates $q_1, q_2, \ldots, q_n$ in which the origin is a point of stable equilibrium, then near the origin the potential energy function V and the kinetic energy T can be approximated by real quadratic forms

$$V = \sum_{i,j=1}^n a_{ij}q_i q_j \quad \text{and} \quad T = \sum_{i,j=1}^n b_{ij}\dot{q}_i \dot{q}_j$$

in the generalized coordinates q_i and generalized velocities $\dot{q}_i$. The behavior of the system is governed by Lagrange's equations

$$\frac{d}{dt}\left(\frac{\partial T}{\partial \dot{q}_i}\right) - \frac{\partial T}{\partial q_i} + \frac{\partial V}{\partial q_i} = 0$$

a system of second-order linear ordinary differential equations with constant coefficients that is *coupled* (and hence is difficult to solve) if the two quadratic forms T and V are not diagonal. The real matrices $A = [a_{ij}]$ and $B = [b_{ij}]$ are symmetric.

If a real nonsingular transformation $S = [s_{ij}] \in M_n$ can be found such that SAS^T and SBS^T are both diagonal, then with respect to new generalized coordinates p_i with

$$q_i = \sum_{j=1}^n s_{ij}p_j \tag{4.5.14}$$

the kinetic and potential energy quadratic forms are both diagonal. In this event, Lagrange's equations are an *uncoupled* set of n separate second-order linear ordinary differential equations with constant coefficients. These equations have standard solutions involving exponentials and trigonometric functions; the solution to the original problem can be obtained by using (4.5.14).

Thus, a substantial simplification in an important class of mechanics problems can be achieved if we can simultaneously diagonalize two real symmetric matrices by congruence. On physical grounds, the kinetic energy quadratic form is positive definite, and it turns out that this is a sufficient (but not necessary) condition for simultaneous diagonalization by congruence.

We are interested in several types of simultaneous diagonalizations of matrices $A, B \in M_n$. If A and B are Hermitian, we might wish to have UAU^* and UBU^* be diagonal for some unitary matrix U, or we might be satisfied with having SAS^* and SBS^* be diagonal for some nonsingular matrix S. If A and B are symmetric, we might want UAU^T and UBU^T (or SAS^T and SBS^T) to be diagonal. We might have a mixed problem (for example, the Grunsky inequalities (4.4.1)) in which A is Hermitian and B is symmetric, and we want UAU^* and UBU^T (or SAS^* and SBS^T) to be diagonal. The following theorem addresses the unitary cases.

Theorem 4.5.15. *Let $A, B \in M_n$ be given.*

(a) *Suppose that A and B are Hermitian. There is a unitary $U \in M_n$ and real diagonal $\Lambda, M \in M_n(\mathbf{R})$ such that $A = U\Lambda U^*$ and $B = UMU^*$ if and only if AB is Hermitian, that is, $AB = BA$.*

(b) *Suppose that A and B are symmetric. There is a unitary $U \in M_n$ and diagonal $\Lambda, M \in M_n$ such that $A = U\Lambda U^T$ and $B = UMU^T$ if and only if $A\bar{B}$ is normal. There is a unitary $U \in M_n$ and real diagonal $\Lambda, M \in M_n(\mathbf{R})$ such that $A = U\Lambda U^T$ and $B = UMU^T$ if and only if $A\bar{B}$ is Hermitian, that is, $A\bar{B} = B\bar{A}$.*

(c) *Suppose that A is Hermitian and B is symmetric. There is a unitary $U \in M_n$ and diagonal $\Lambda, M \in M_n$ such that $A = U\Lambda U^*$ and $B = UMU^T$ if and only if AB is symmetric, that is, $AB = B\bar{A}$.*

Proof. (a) See (4.1.6).

(b) See (2.6.P21) and (2.6.P22).

(c) If $A = U\Lambda U^*$ and $B = UMU^T$, then $AB = U\Lambda U^*UMU^T = U\Lambda MU^T$ is symmetric. Moreover, $AB = (AB)^T = B^T A^T = B\bar{A}$. Conversely, suppose that $AB = B\bar{A}$ and A has d distinct eigenvalues $\lambda_1, \ldots, \lambda_d$. Let $A = U\Lambda U^*$, in which U is unitary and $\Lambda = \lambda_1 I_{n_1} \oplus \cdots \oplus \lambda_d I_{n_d}$. Then $AB = U\Lambda U^*B = B\bar{U}\Lambda U^T = B\bar{A}$, so $\Lambda U^*B\bar{U} = U^*B\bar{U}\Lambda$, which means that $U^*B\bar{U} = B_1 \oplus \cdots \oplus B_d$ is block diagonal conformal to Λ (2.4.4.2). Moreover, each block $B_j \in M_{n_j}$ is symmetric, so (4.4.4c) ensures that there are unitary matrices $V_j \in M_{n_j}$ and nonnegative diagonal matrices $\Sigma_j \in M_{n_j}$ such that $B_j = V_j\Sigma_j V_j^T$, $j = 1, \ldots, d$. Let $V = V_1 \oplus \cdots \oplus V_d$, $\Sigma = \Sigma_1 \oplus \cdots \oplus \Sigma_d$, and $W = UV$; observe that V commutes with Λ. Then $B = U(B_1 \oplus \cdots \oplus B_d)U^T = UV\Sigma V^T U^T = W\Sigma W^T$ and $W\Lambda W^* = UV\Lambda V^*U^* = U\Lambda VV^*U^* = U\Lambda U^* = A$. $\qquad\square$

We now enlarge the class of congruences considered from unitary congruences to nonsingular congruences, but we add an assumption that one of A or B is nonsingular. Part (c) in the following theorem requires the following new concept, which we explore in more detail in the next section.

Definition 4.5.16. *A matrix $A \in M_n$ is said to be condiagonalizable if there is a nonsingular $S \in M_n$ and a diagonal $\Lambda \in M_n$ such that $A = S\Lambda \bar{S}^{-1}$.*

Three facts about a condiagonalizable matrix are used in the proof of the next theorem. The first is that the scalars in the diagonal matrix Λ in the preceding definition may be assumed to appear in any desired order: If P is a permutation matrix, then $A = S\Lambda\bar{S}^{-1} = SP^T P\Lambda P^T P\bar{S}^{-1} = (SP^T)(P\Lambda P^T)(\overline{SP^T})^{-1}$. The second is that we may assume that Λ is real and nonnegative diagonal: If $\Lambda = \mathrm{diag}(|\lambda_1|e^{i\theta_1}, \ldots, |\lambda_n|e^{i\theta_n})$, let $|\Lambda| = \mathrm{diag}(|\lambda_1|, \ldots, |\lambda_n|)$ and let $D = \mathrm{diag}(|\lambda_1|e^{i\theta_1/2}, \ldots, |\lambda_n|e^{i\theta_n/2})$, which is equal to $\bar{D}^{-1}$. Then $\Lambda = D|\Lambda|D$ and $A = S\Lambda\bar{S}^{-1} = SD|\Lambda|D\bar{S}^{-1} = (SD)|\Lambda|(\overline{SD})^{-1}$. The third is that if A is nonsingular, then it is condiagonalizable if and only if A^{-1} is condiagonalizable: $A = S\Lambda\bar{S}^{-1}$ if and only if $A^{-1} = \bar{S}\Lambda^{-1}S^{-1}$.

Theorem 4.5.17. *Let $A, B \in M_n$ be given.*

(a) *Suppose that A and B are Hermitian and A is nonsingular. Let $C = A^{-1}B$. There is a nonsingular $S \in M_n$ and real diagonal matrices Λ and M such that $A = S\Lambda S^*$ and $B = SMS^*$ if and only if C is diagonalizable and has real eigenvalues.*

(b) *Suppose that A and B are symmetric and A is nonsingular. Let $C = A^{-1}B$. There is a nonsingular $S \in M_n$ and complex diagonal matrices Λ and M such that $A = S\Lambda S^T$ and $B = SMS^T$ if and only if C is diagonalizable.*

(c) *Suppose that A is Hermitian, B is symmetric, and at least one of A or B is nonsingular. If A is nonsingular, let $C = A^{-1}B$; if B is nonsingular, let $C = B^{-1}A$. There is a nonsingular $S \in M_n$ and real diagonal matrices Λ and M such that $A = S\Lambda S^*$ and $B = SMS^T$ if and only if C is condiagonalizable.*

Proof. In each case, a computation verifies necessity of the stated conditions for simultaneous diagonalization by congruence, so we discuss only their sufficiency. The first two cases can be proved with parallel arguments, but the third case is a bit different.

(a) Assume that A and B are Hermitian, A is nonsingular, and there is a nonsingular S such that $C = A^{-1}B = S\Lambda S^{-1}$, $\Lambda = \lambda_1 I_{n_1} \oplus \cdots \oplus \lambda_d I_{n_d}$ is real diagonal, and $\lambda_1 < \cdots < \lambda_d$. Then $BS = AS\Lambda$ and hence $S^*BS = S^*AS\Lambda$. If we partition $S^*BS = [B_{ij}]_{i,j=1}^d$ and $S^*AS = [A_{ij}]_{i,j=1}^d$ conformally to Λ, we have the identities $B_{ij} = \lambda_j A_{ij}$ (equivalent to $B_{ij}^* = \lambda_j A_{ij}^*$ since λ_j is real) and $B_{ji} = \lambda_i A_{ji}$ for all $i, j = 1, \ldots, d$. Both S^*BS and S^*AS are Hermitian, so $B_{ji} = B_{ij}^*$, $A_{ji} = A_{ij}^*$, and $B_{ij}^* = \lambda_i A_{ij}^*$. Combining these identities, we conclude that $(\lambda_i - \lambda_j)A_{ij}^* = 0$. Thus, $A_{ij} = 0$ for all $i \neq j$, so $S^*AS = A_{11} \oplus \cdots \oplus A_{dd}$ and $S^*BS = S^*AS\Lambda$. For each $i = 1, \ldots, d$ let $V_i \in M_{n_i}$ be unitary and such that $A_{ii} = V_i^* D_i V_i$, in which D_i is diagonal and real (4.1.5). Let $V = V_1 \oplus \cdots \oplus V_d$ and $D = D_1 \oplus \cdots \oplus D_d$; observe that V commutes with Λ. Then $S^*AS = V^*DV$ and $S^*BS = S^*AS\Lambda = V^*DV\Lambda = V^*D\Lambda V$. We conclude that $A = S^{-*}V^*DVS^{-1} = RDR^*$ and $B = S^{-*}V^*D\Lambda VS^{-1} = R(D\Lambda)R^*$, in which $R = S^{-*}V^*$.

(b) Assume that A and B are symmetric, A is nonsingular, and there is a nonsingular S such that $C = A^{-1}B = S\Lambda S^{-1}$, $\Lambda = \lambda_1 I_{n_1} \oplus \cdots \oplus \lambda_d I_{n_d}$ is complex diagonal, and $\lambda_i \neq \lambda_j$ for all $i \neq j$. Then $BS = AS\Lambda$ and hence $S^TBS = S^TAS\Lambda$. If we partition $S^TBS = [B_{ij}]_{i,j=1}^d$ and $S^TAS = [A_{ij}]_{i,j=1}^d$ conformally to Λ, we have the identities $B_{ij} = \lambda_j A_{ij}$ (equivalent to $B_{ij}^T = \lambda_j A_{ij}^T$) and $B_{ji} = \lambda_i A_{ji}$ for all $i, j = 1, \ldots, d$. Both S^TBS and S^TAS are symmetric, so $B_{ji} = B_{ij}^T$, $A_{ji} = A_{ij}^T$, and

$B_{ij}^T = \lambda_i A_{ij}^T$. Combining these identities, we conclude that $(\lambda_i - \lambda_j) A_{ij}^T = 0$. Thus, $A_{ij} = 0$ for all $i \neq j$, so $S^T A S = A_{11} \oplus \cdots \oplus A_{dd}$ and $S^T B S = S^T A S \Lambda$. For each $i = 1, \ldots, d$ let $V_i \in M_{n_i}$ be unitary and such that $A_{ii} = V_i^T D_i V_i$, in which D_i is diagonal and nonnegative (4.4.4c). Let $V = V_1 \oplus \cdots \oplus V_d$ and $D = D_1 \oplus \cdots \oplus D_d$; observe that V commutes with Λ. Then $S^T A S = V^T D V$ and $S^T B S = S^T A S \Lambda = V^T D V \Lambda = V^T D \Lambda V$. We conclude that $A = S^{-T} V^T D V S^{-1} = R D R^T$ and $B = S^{-T} V^T D \Lambda V S^{-1} = R(D\Lambda)R^T$, in which $R = S^{-T} V^T$.

(c) Assume that A is Hermitian, B is symmetric, and at least one of them is nonsingular. If A is nonsingular, let $C = A^{-1} B$; if B is nonsingular, let $C = B^{-1} A$. We also assume that there is a nonsingular S such that $C = S \Lambda \bar{S}^{-1}$, in which $\Lambda = \lambda_1 I_{n_1} \oplus \cdots \oplus \lambda_d I_{n_d} \in M_n$ is real and nonnegative diagonal, $0 \leq \lambda_1 < \cdots < \lambda_d$. If both A and B are nonsingular, it does not matter which choice is made for C since it is condiagonalizable if and only if its inverse is condiagonalizable.

First suppose that A is nonsingular. Then $A^{-1} B = S \Lambda \bar{S}^{-1}$, so $B \bar{S} = A S \Lambda$ and hence $S^* B \bar{S} = S^* A S \Lambda$. If we partition the symmetric matrix $S^* B \bar{S} = \bar{S}^T B \bar{S} = [B_{ij}]_{i,j=1}^d$ and the Hermitian matrix $S^* A S = [A_{ij}]_{i,j=1}^d$ conformally to Λ, we have the identities $B_{ij} = \lambda_j A_{ij}$ and $B_{ji} = \lambda_i A_{ji}$ (equivalent to $B_{ij}^T = \lambda_i A_{ij}^*$ and to $B_{ij} = \lambda_i \bar{A}_{ij}$) for all $i, j = 1, \ldots, d$. Combining these identities, we obtain $\lambda_j A_{ij} = \lambda_i \bar{A}_{ij}$, which implies that $A_{ij} = 0$ if $i \neq j$ (look at the entries of A_{ij}: $\lambda_j a = \lambda_i \bar{a} \Rightarrow \lambda_j |a| = \lambda_i |a| \Rightarrow a = 0$ if $i \neq j$). Thus, $S^* A S = A_{11} \oplus \cdots \oplus A_{dd}$ is block diagonal and Hermitian. In addition, each block $B_{ii} = \lambda_i A_{ii}$ is both symmetric and Hermitian, so A_{ii} is real symmetric if $\lambda_i \neq 0$. If $\lambda_i \neq 0$, let $V_i \in M_{n_i}$ be real orthogonal and such that $A_{ii} = V_i^T D_i V_i$, in which D_i is diagonal and real (4.1.5). If $\lambda_1 = 0$, let V_1 be unitary and such that $A_{11} = V_1^* D_1 V_1$, in which D_1 is diagonal and real. Let $V = V_1 \oplus \cdots \oplus V_d$ and $D = D_1 \oplus \cdots \oplus D_d$. Observe that D is real, V commutes with Λ, Λ is real, and ΛV is real (V_i is real for all $i > 1$ and $\lambda_1 V_1 = 0$ if V_1 is not real). Then $S^* A S = V^* D V$ and $S^* B \bar{S} = S^* A S \Lambda = V^* D V \Lambda = V^* D \Lambda V = V^* D \overline{\Lambda V} = V^* D \Lambda \bar{V}$. We conclude that $A = S^{-*} V^* D V S^{-1} = R D R^*$ and $B = S^{-*} V^* \Lambda D \bar{V} \bar{S}^{-1} = R(\Lambda D)R^T$, in which $R = S^{-*} V^*$.

Finally, if B is nonsingular, then $B^{-1} A = S \Lambda \bar{S}^{-1}$ and $S^T A \bar{S} = S^T B S \Lambda$. From this point, the argument proceeds just as in the case in which A is nonsingular: One finds that the symmetric matrix $S^T B S = B_{11} \oplus \cdots \oplus B_{dd}$ is block diagonal, and B_{ii} is real symmetric if $\lambda_i \neq 0$. If $\lambda_1 = 0$, use (4.4.4c) to diagonalize B_{11} by unitary congruence and assemble the respective congruences that diagonalize A and B. □

Exercise. Provide details for the second part of the proof of part (c) of the preceding theorem.

Exercise. Revisit (4.4.25) and explain why the criterion in (4.5.17b) significantly restricts A and B.

In parts (a) and (b) of the preceding theorem, there is a familiar condition on the matrix $C = A^{-1} B$ that is equivalent to simultaneous diagonalizability by the respective congruence: C is diagonalizable (perhaps with real eigenvalues). In part (c), we require that C be condiagonalizable, which is equivalent to requiring that rank $C = \text{rank } C\bar{C}$, every eigenvalue of $C\bar{C}$ is real and nonnegative, and $C\bar{C}$ is diagonalizable; see (4.6.11).

To study the problem of diagonalizing a pair of nonzero singular Hermitian matrices by simultaneous *congruence, we step back and take a new path. Any $A \in M_n$ can be represented uniquely as $A = H + iK$ (its Toeplitz decomposition; see (4.1.2)), in which H and K are Hermitian. The matrix $H = \frac{1}{2}(A + A^*)$ is the *Hermitian part* of A, and $K = \frac{1}{2i}(A - A^*)$ is the *skew-Hermitian* part of A.

Lemma 4.5.18. *Let $A \in M_n$ be given, and let $A = H + iK$, in which H and K are Hermitian. Then A is diagonalizable by *congruence if and only if H and K are simultaneously diagonalizable by *congruence.*

Proof. If there is a nonsingular $S \in M_n$ such that $SHS^* = \Lambda$ and $SKS^* = M$ are both diagonal, then $SAS^* = SHS^* + iSKS^* = \Lambda + iM$ is diagonal. To prove the converse, it suffices to show that if $B = [b_{jk}]$ and $C = [c_{jk}]$ are n-by-n Hermitian matrices and $B + iC = [b_{jk} + ic_{jk}]$ is diagonal, then both B and C are diagonal. For any $j \neq k$ we have $b_{jk} + ic_{jk} = 0$ and $b_{kj} + ic_{kj} = \bar{b}_{jk} + i\bar{c}_{kj} = 0$, so $\overline{b}_{jk} + i\overline{c}_{kj} = b_{jk} - ic_{jk} = 0$. The pair of equations $b_{jk} + ic_{jk} = 0$ and $b_{jk} - ic_{jk} = 0$ has only the trivial solution $b_{jk} = c_{jk} = 0$. $\square$

The preceding lemma shows that the problem of simultaneously diagonalizing a pair of Hermitian matrices of the same size by *congruence is equivalent to the problem of diagonalizing a square complex matrix by *congruence. One way to approach the latter problem is via a canonical form for *congruence, which involves three types of canonical blocks. The first type is the family of singular Jordan blocks $J_k(0)$, $k = 1, 2, \ldots$; the smallest such block is $J_1(0) = [0]$. The second type is a family of nonsingular Hankel matrices

$$\Delta_k = \begin{bmatrix} & & & & 1 \\ & & & \cdots & i \\ & & 1 & \cdots & \\ & \cdot & \cdot & & \\ 1 & i & & & \end{bmatrix} \in M_k, \quad k = 1, 2, \ldots \quad (4.5.19)$$

The blocks of this type with sizes one and two are $\Delta_1 = [1]$ and $\Delta_2 = \begin{bmatrix} 0 & 1 \\ 1 & i \end{bmatrix}$. The third type is a family of nonsingular complex blocks of even size that incorporate nonsingular Jordan blocks

$$H_{2k}(\mu) = \begin{bmatrix} 0 & I_k \\ J_k(\mu) & 0 \end{bmatrix} \in M_{2k}, \mu \neq 0, \quad k = 1, 2, \ldots \quad (4.5.20)$$

The smallest block of this type is $H_2 = \begin{bmatrix} 0 & 1 \\ \mu & 0 \end{bmatrix}$.

We can now state the *congruence canonical form theorem.

Theorem 4.5.21. *Each square complex matrix is *congruent to a direct sum, uniquely determined up to permutation of direct summands, of matrices of the following three types:*

Type 0: $J_k(0)$, $k = 1, 2, \ldots$;

Type I: $\lambda\Delta_k$, $k = 1, 2, \ldots$, in which $\lambda = e^{i\theta}$, $0 \le \theta < 2\pi$;

Type II: $H_{2k}(\mu)$, $k = 1, 2, \ldots$, in which $|\mu| > 1$.

Alternatively, instead of the symmetric Type I matrices Δ_k, one may use the real *matrices Γ_k defined in (4.5.24.1) or any other nonsingular matrices $F_k \in M_k$ for which there exists a real ϕ_k such that $F_k^{-*}F_k$ is similar to the Jordan block $J_k(e^{i\phi_k})$.*

Just as with the Jordan canonical form, the uniqueness assertion of the *congruence canonical form theorem is perhaps its most useful feature for applications.

Exercise. Let $A, B, S \in M_n$ be nonsingular and suppose that $A = SBS^*$. Explain why $A^{-*}A = S^{-*}(B^{-*}B)S^*$ and why $A^{-*}A$ has the same Jordan canonical form as $B^{-*}B$.

Exercise. Let $A = [i] \in M_1$ and $B = [-i] \in M_1$. Explain why $A^{-*}A = B^{-*}B$, but A is not *congruent to B. *Hint*: If $S = [s]$, $SAS^* = ?$.

The matrices of Type 0, Type I, and Type II in (4.5.21) are *canonical blocks for *congruence*. A direct sum of canonical blocks that is *congruent to a given $A \in M_n$ is its *congruence canonical form*. Two *congruence canonical forms are *the same* if one can be obtained from the other by permuting its canonical blocks.

If, for a given $\theta \in [0, 2\pi)$, the *congruence canonical form of a given $A \in M_n$ contains exactly m blocks of the form $e^{i\theta}\Delta_k$, one says that θ is a *canonical angle of A of order k and multiplicity m*; alternatively, one says that the ray $\{re^{i\theta} : 0 < r < \infty\}$ in the complex plane is a *canonical ray of A with order k and multiplicity m*. If all the Type I blocks of A are known to be 1-by-1 (for example, if A is normal; see (4.5.P11)), it is customary to speak only of the canonical angle (ray) θ and its multiplicity without mentioning the order.

A direct sum of all of the Type 0 blocks of a given $A \in M_n$ is its *singular part* (with respect to *congruence); any matrix that is *congruent to the direct sum of all of the Type I and Type II blocks of A is a *regular part* (again, with respect to *congruence). The singular part of A is uniquely determined (up to permutation of its direct summands, of course); the *congruence equivalence class of a regular part of A is uniquely determined.

If $A \in M_n$ is nonsingular, the matrix $A^{-*}A$ is the *cosquare* of A. The preceding exercises show that nonsingular *congruent matrices have similar *cosquares, but matrices with similar *cosquares need not be *congruent. Although many different matrices can be regular parts of A, they must all be in the same *congruence equivalence class.

The Jordan canonical form of a *cosquare is subject to a constraint that is revealed by a computation: $(A^{-*}A)^{-*} = AA^{-*}$, which is similar to $A^{-*}A$ (1.3.22). Therefore, if μ is an eigenvalue of $A^{-*}A$ (necessarily nonzero) and $J_k(\mu)$ is a block in the Jordan canonical form of AA^{-*}, then a Jordan block similar to $J_k(\mu)^{-*}$ (namely, $J_k(\bar{\mu}^{-1})$) must also be present. If $|\mu| = 1$, this observation does not yield any useful information since $\bar{\mu}^{-1} = \mu$ in this case. However, it tells us that if $|\mu| \neq 1$, then any block $J_k(\mu)$ in the Jordan canonical form of $A^{-*}A$ is paired with a block $J_k(\bar{\mu}^{-1})$. Thus, the Jordan canonical form of a *cosquare contains *only* blocks of the form $J_k(e^{i\theta})$ for some real θ and pairs of the form $J_k(\mu) \oplus J_k(\bar{\mu}^{-1})$ with $0 \neq |\mu| \neq 1$.

The blocks $\lambda \Delta_k$ and $H_{2k}(\mu)$ in the *congruence canonical form (4.5.21) of a nonsingular $A \in M_n$ arise from the special Jordan canonical form of a *cosquare. If $\mu \neq 0$, then

$$
\begin{aligned}
H_{2k}(\mu)^{-*} H_{2k}(\mu) &= \begin{bmatrix} 0 & J_k(\mu)^{-1} \\ I_k & 0 \end{bmatrix}^* \begin{bmatrix} 0 & I_k \\ J_k(\mu) & 0 \end{bmatrix} \\
&= \begin{bmatrix} 0 & I_k \\ J_k(\mu)^{-*} & 0 \end{bmatrix} \begin{bmatrix} 0 & I_k \\ J_k(\mu) & 0 \end{bmatrix} \\
&= \begin{bmatrix} J_k(\mu) & 0 \\ 0 & J_k(\mu)^{-*} \end{bmatrix}
\end{aligned}
$$

which is similar to $J_k(\mu) \oplus J_k(\bar{\mu}^{-1})$. There is a one-to-one correspondence between blocks $H_{2k}(\mu)$ and pairs of the form $J_k(\mu) \oplus J_k(\bar{\mu}^{-1})$ with $|\mu| \neq 1$ in the Jordan canonical form of the *cosquare of A.

If $|\lambda| = 1$, a computation reveals that $(\lambda \Delta_k)^{-*}(\lambda \Delta_k)$ is similar to $J_k(\lambda^2)$; see (4.5.P15). If the Jordan canonical form of the *cosquare of a nonsingular $A \in M_n$ is $J_{k_1}(e^{i\theta}) \oplus \cdots \oplus J_{k_p}(e^{i\theta}) \oplus J$, in which $\theta \in [0, 2\pi)$ and $e^{i\theta}$ is not an eigenvalue of J, then the *congruence canonical form of A is $\pm e^{i\theta/2} \Delta_{k_1} \oplus \cdots \oplus \pm e^{i\theta/2} \Delta_{k_p} \oplus C$, in which a particular choice of $\pm$ signs is made and no block of the form $\pm e^{i\theta/2} \Delta_k$ appears in C. The $\pm$ signs cannot be determined from the *cosquare of A, but they can be determined by using other information about A.

Exercise. Explain why the Jordan canonical form of the *cosquare of a nonsingular $A \in M_n$ determines only the *lines* in the complex plane that contain its canonical rays; it determines the *orders* of the canonical rays (angles), but not their *multiplicities*.

Our first application of the *congruence canonical form is to obtain the following *cancellation theorem*:

Theorem 4.5.22. *Let A, $B \in M_p$ and $C \in M_q$ be given. Then $A \oplus C$ and $B \oplus C$ are *congruent if and only if A and B are *congruent.*

Proof. If there is a nonsingular $S \in M_p$ such that $A = SBS^*$, then $S \oplus I_q$ is nonsingular and $(S \oplus I_q)(B \oplus C)(S \oplus I_q)^* = SBS^* \oplus C = A \oplus C$. Conversely, suppose that $A \oplus C$ and $B \oplus C$ are *congruent. Let L_A, L_B, and L_C denote the respective *congruence canonical forms of A, B, and C; each is a direct sum of certain blocks of types 0, I, and II. Let S_A, S_B, and S_C be nonsingular matrices such that $A = S_A L_A S_A^*$, $B = S_B L_B S_B^*$, and $C = S_C L_C S_C^*$. Then $L_A \oplus L_C$ is *congruent to $A \oplus C$ (via $S_A \oplus S_C$), which by assumption is *congruent to $B \oplus C$, which is *congruent to $L_B \oplus L_C$ (via $S_B \oplus S_C$). Therefore, $L_A \oplus L_C$ and $L_B \oplus L_C$ are *congruent and each is a direct sum of canonical blocks, so uniqueness of the *congruence canonical form ensures that one can be obtained from the other by permuting its direct summands; this statement remains true after the direct summands of L_C are removed. The direct sum of the remaining canonical blocks is the *congruence canonical form of A; it is also the *congruence canonical form of B. Since A and B have the same *congruence canonical forms, they are *congruent. $\square$

Determination of the *congruence canonical form of a given $A \in M_n$ typically proceeds in three steps:

Step 1. Construct a nonsingular $S \in M_n$ such that $A = S(B \oplus N)S^*$, in which $N = J_{r_1}(0) \oplus \cdots \oplus J_{r_p}(0)$ is a direct sum of nilpotent Jordan blocks and B is nonsingular. Such a construction is called a *regularization* of A. A regularization can be performed in an ad hoc fashion (perhaps some special information about the matrix facilitates the construction), or one can employ a known *regularization algorithm*. Since the direct sum $B \oplus N$ produced by a regularization of A is *congruent to the *congruence canonical form of A, the uniqueness assertion in (4.5.21) ensures that N is the singular part of A; the cancellation theorem then ensures that B is a regular part of A for *congruence.

Step 2. Compute the Jordan canonical form of the *cosquare of a regular part of A. It determines completely the Type II blocks of A, and it determines the Type I blocks up to sign.

Step 3. Determine the signs of the Type I blocks of A using a known algorithm or an ad hoc method.

Exercise. Let $A, B, S \in M_n$ be given. Suppose that S is nonsingular and $A = SBS^*$. Let $\nu = \dim\text{nullspace } A$, $\delta = \dim((\text{nullspace } A) \cap (\text{nullspace } A^*))$, $\nu' = \dim\text{nullspace } B$, and $\delta' = \dim((\text{nullspace } B) \cap (\text{nullspace } B^*))$. Explain why $\nu = \nu'$ and $\delta = \delta'$, that is, $\dim\text{nullspace } A$ and $\dim((\text{nullspace } A) \cap (\text{nullspace } A^*))$ are *congruence invariants. Explain why $\nu = \delta$ if and only if the null spaces of A and A^* are the same.

The regularization algorithm for a given $A \in M_n$ begins by computing the two invariants ν and δ described in the preceding exercise. The number of 1-by-1 blocks $J_1(0)$ in the *congruence canonical form of A is δ. If $\nu = \delta$ (that is, if the null spaces of A and A^* are the same), the algorithm terminates and 0_d is the singular part of A. If $\nu > d$, the algorithm determines a *congruence that reduces A to a special block form, and the algorithm is then repeated on a specific block of the reduced matrix. The output of the algorithm is a sequence of integer invariants that determine the number of blocks $J_k(0)$ in the singular part of A, for each $k = 1, \ldots, n$.

A given $A \in M_n$ is diagonalizable by *congruence if and only if (a) its *congruence canonical form contains no Type II blocks (the smallest Type II blocks are 2-by-2), (b) its Type 0 blocks are $J_1(0) = [0]$, and (c) its Type I blocks are $\lambda \Delta_1 = [\lambda]$ for some λ with $|\lambda| = 1$. Thus, A is diagonalizable by *congruence and rank $A = r$ if and only if there is a nonsingular $S \in M_n$ such that $A = S(\Lambda \oplus 0_{n-r})S^*$, in which $\Lambda = \text{diag}(\lambda_1, \ldots, \lambda_r)$ and $|\lambda_j| = 1$ for all $j = 1, \ldots, r$. If we partition $S = [S_1 \ S_2]$ with $S_1 \in M_r$, then $A = [S_1 \ S_2](\Lambda \oplus 0_{n-r})[S_1 \ S_2]^* = S_1 \Lambda S_1^*$. Let $S_1 = U_1 R$ be a QR factorization (2.1.14) and let $U = [U_1 \ U_2] \in M_n$ be unitary. Then $A = S_1 \Lambda S_1^* = U_1 R \Lambda R^* U_1^*$, so

$$U^* A U = \begin{bmatrix} U_1^* \\ U_2^* \end{bmatrix} U_1 R \Lambda R^* U_1^* [U_1 \ U_2] = \begin{bmatrix} R \Lambda R^* & 0 \\ 0 & 0_{n-r} \end{bmatrix}$$

Thus, A is (unitarily) *congruent to $R \Lambda R^* \oplus 0_{n-r}$, and hence $R \Lambda R^*$ is a regular part of A.

Suppose that a nonsingular $B \in M_r$ is diagonalizable by *congruence. Since there are no Type II blocks in the *congruence canonical form of B, the Jordan canonical form of the *cosquare $B^{-*}B$ contains only blocks of the form $J_1(\lambda)$ with $|\lambda| = 1$, that is, $B^{-*}B$ is diagonalizable and all its eigenvalues have modulus one. Let $S \in M_n$ be nonsingular and such that $B^{-*}B = S\Lambda S^{-1}$, in which $\Lambda = \mathrm{diag}(e^{i\theta_1}I_{n_1} \oplus \cdots \oplus e^{i\theta_d}I_{n_d})$ with each $\theta_j \in [0, 2\pi)$ and $\theta_j \neq \theta_k$ if $j \neq k$. Then $B = B^*S\Lambda S^{-1}$, $BS = B^*S\Lambda$, and $S^*BS = S^*B^*S\Lambda$. Let $\mathcal{B} = S^*BS$, and observe that $\mathcal{B} = \mathcal{B}^*\Lambda \Rightarrow \mathcal{B} = (\mathcal{B}^*\Lambda)^*\Lambda = \Lambda^*\mathcal{B}\Lambda \Rightarrow \Lambda\mathcal{B} = \mathcal{B}\Lambda$ since Λ is unitary. If we partition $[B_{jk}]_{j,k=1}^d$ conformally to Λ, commutativity of $\mathcal{B}$ and Λ implies (2.4.4.2) that $\mathcal{B}$ is block diagonal conformal to Λ: $\mathcal{B} = B_1 \oplus \cdots \oplus B_d$. Moreover, the identity $\mathcal{B} = \mathcal{B}^*\Lambda$ implies that $B_j = e^{i\theta_j}B_j^*$ and $e^{-i\theta_j/2}B_j = e^{i\theta_j/2}B_j^* = (e^{-i\theta_j/2}B_j)^*$, so $e^{-i\theta_j/2}B_j$ is Hermitian for each $j = 1, \ldots, d$. Every Hermitian matrix is *congruent to its inertia matrix (4.5.7), so for each $j = 1, \ldots, d$ there is a nonsingular $S_j \in M_{n_j}$ and nonnegative integers n_j^+ and n_j^- such that $n_j^+ + n_j^- = n_j$ and $e^{-i\theta_j/2}B_j = S_j(I_{n_j^+} \oplus (-I_{n_j^-}))S_j^*$, that is, $B_j = e^{i\theta_j/2}S_j(I_{n_j^+} \oplus (-I_{n_j^-}))S_j^* = e^{i\theta_j/2}S_j(e^{i\theta_j/2}I_{n_j^+} \oplus e^{i(\pi+\theta_j/2)}I_{n_j^-})S_j^*$. We conclude that B is *congruent to

$$e^{i\theta_1/2}I_{n_1^+} \oplus e^{i(\pi+\theta_1/2)}I_{n_1^-} \oplus \cdots \oplus e^{i\theta_d/2}I_{n_d^+} \oplus e^{i(\pi+\theta_d/2)}I_{n_d^-} \qquad (4.5.23)$$

which is a direct sum of Type I blocks and is therefore the *congruence canonical form of B. The canonical angles (rays) of B are $\frac{1}{2}\theta_1, \ldots, \frac{1}{2}\theta_d$ with respective multiplicities $n_1^+, \ldots, n_d^+$, together with $\pi + \frac{1}{2}\theta_1, \ldots, \pi + \frac{1}{2}\theta_d$, with respective multiplicities $n_1^-, \ldots, n_d^-$.

The preceding analysis leads to an algorithm to decide whether a given $A \in M_n$ is diagonalizable by *congruence and, if it is, to determine its *congruence canonical form:

Step 1. Check whether A and A^* have the same null space. If they do not, stop; A is not diagonalizable by *congruence. If they do, let $U_2 \in M_{n,n-r}$ have orthonormal columns that comprise a basis of the null space of A, and let $U = [U_1\ U_2] \in M_n$ be unitary. Then AU_2 and U_2^*A are both zero matrices, so

$$U^*AU = \begin{bmatrix} U_1^* \\ U_2^* \end{bmatrix} A[U_1\ U_2] = \begin{bmatrix} U_1^*AU_1 & U_1^*AU_2 \\ U_2^*AU_1 & U_2^*AU_2 \end{bmatrix}$$
$$= \begin{bmatrix} U_1^*AU_1 & 0 \\ 0 & 0_{n-r} \end{bmatrix}$$

and $B = U_1^*AU_1$ is a regular part of A.

Step 2. Check whether (a) $B^{-*}B$ is diagonalizable, and (b) every eigenvalue of $B^{-*}B$ has modulus one. If either of these conditions is not satisfied, stop; A is not diagonalizable by *congruence. Otherwise, A is diagonalizable by *congruence.

Step 3. Diagonalize $B^{-*}B$, that is, construct a nonsingular $S \in M_n$ such that $B^{-*}B = S\Lambda S^{-1}$, in which $\Lambda = e^{i\theta_1}I_{n_1} \oplus \cdots \oplus e^{i\theta_d}I_{n_d}$ with each $\theta_j \in [0, 2\pi)$ and $\theta_j \neq \theta_k$ if $j \neq k$. Then $S^*BS = B_1 \oplus \cdots \oplus B_d$ is block diagonal conformal to Λ and $e^{-i\theta_j/2}B_j$ is Hermitian, $j = 1, \ldots, d$. Determine the number n_j^+ of positive eigenvalues of $e^{-i\theta_j/2}B_j$ and let $n_j^- = n - n_j^+$, $j = 1, \ldots, d$. The *congruence canonical form of A is

$$e^{i\theta_1/2}I_{n_1^+} \oplus e^{i(\pi+\theta_j/2)}I_{n_j^-} \oplus \cdots \oplus e^{i\theta_d/2}I_{n_d^+} \oplus e^{i(\pi+\theta_d/2)}I_{n_d^-} \oplus 0_{n-r}$$

We summarize part of the foregoing discussion in the following theorem.

Theorem 4.5.24. *Let $A \in M_n$ be given and let $A = H + iK$, in which H and K are Hermitian. Let B be a regular part of A and let $\mathcal{B} = B^{-*}B$. The following are equivalent:*

(a) *H and K are simultaneously diagonalizable by *congruence.*
(b) *A is diagonalizable by *congruence.*
(c) *A and A^* have the same null space, $\mathcal{B}$ is diagonalizable, and every eigenvalue of $\mathcal{B}$ has modulus one.*

Exercise. Let $A \in M_n$ be given, let $r = \text{rank } A$, and suppose that A and A^* have the same null space. Let $A = V\Sigma W^*$ be a singular value decomposition (2.6.3), in which $V = [V_1 \ V_2]$ and $V_1 \in M_{n,r}$. Explain why $V_1^* A V_1$ is a regular part of A with respect to *congruence.

There is also a simple canonical form for congruence (^{T}congruence) of matrices. It involves a new family of nonsingular canonical blocks

$$\Gamma_k = \begin{bmatrix} & & & & & (-1)^{k+1} \\ & & & & (-1)^k & \\ & & & -1 & \ddots & \\ & & 1 & 1 & & \\ & -1 & -1 & & & \\ 1 & 1 & & & & \end{bmatrix} \in M_k, \ k = 1, 2, \ldots \qquad (4.5.24.1)$$

The blocks in this family with sizes one and two are $\Gamma_1 = [1]$ and $\Gamma_2 = \begin{bmatrix} 0 & -1 \\ 1 & 1 \end{bmatrix}$. The *congruence canonical form theorem* is as follows:

Theorem 4.5.25. *Each square complex matrix is congruent to a direct sum, uniquely determined up to permutation of direct summands, of matrices of the following three types:*

Type 0: $J_k(0)$, $k = 1, 2, \ldots$;
Type I: Γ_k, $k = 1, 2, \ldots$;
Type II: $H_{2k}(\mu)$, $k = 1, 2, \ldots$, in which $0 \neq \mu \neq (-1)^{k+1}$ and μ is determined up to replacement by μ^{-1}.

Exercise. Let $A, B, S \in M_n$ be nonsingular and suppose that $A = SBS^T$. Explain why $A^{-T}A = S^{-T}(B^{-T}B)S^T$ and why $A^{-T}A$ has the same Jordan canonical form as $B^{-T}B$.

Exercise. Let $A \in M_n$ be nonsingular. Explain why $A^{-T}A$ is similar to $(A^{-T}A)^{-T}$, which is similar to $(A^{-T}A)^{-1}$. Why must any block $J_k(\mu)$ in the Jordan canonical form of $A^{-T}A$ be paired with a block $J_k(\mu^{-1})$ if $\mu \neq \pm 1$?

The *cosquare* of a nonsingular $A \in M_n$ is the matrix $A^{-T}A$, whose Jordan canonical form has a very special form: It contains *only* blocks of the form $J_k((-1)^{k+1})$ and pairs of blocks of the form $J_k(\mu) \oplus J_k(\mu^{-1})$, in which $0 \neq \mu \neq (-1)^{k+1}$. The matrix Γ_k appears in the canonical form for congruence because its cosquare is similar to $J_k((-1)^{k+1})$; see (4.5.P25).

Exercise. If $\mu \neq 0$, explain why the cosquare of $H_{2k}(\mu)$ is similar to $J_k(\mu) \oplus J_k(\mu^{-1})$.

The canonical form theorem for congruence implies a *cancellation theorem*, which is proved in the same way as (4.5.22) and again relies heavily on the uniqueness assertion in (4.5.25).

Theorem 4.5.26. *Let $A, B \in M_p$ and $C \in M_q$ be given. Then $A \oplus C$ and $B \oplus C$ are congruent if and only if A and B are congruent.*

Determination of the congruence canonical form of a given $A \in M_n$ has only two steps:

Step 1. Regularize A by constructing a nonsingular $S \in M_n$ such that $A = S(B \oplus N)S^T$, in which $N = J_{r_1}(0) \oplus \cdots \oplus J_{r_p}(0)$ (the uniquely determined *singular part* of A for congruence) and the *regular part* B is nonsingular. One can proceed in an ad hoc fashion to determine N, or one can employ a known *regularization algorithm*. The congruence equivalence class of the nonsingular summand B (but not B itself) is uniquely determined by A; B is a *regular part* of A for congruence.

Step 2. Compute the Jordan canonical form of the cosquare $B^{-T}B$. It determines the Type I and Type II blocks of A, as follows: Each block $J_k((-1)^{k+1})$ corresponds to a Type I block Γ_k; each pair $J_k(\mu) \oplus J_k(\mu^{-1})$ corresponds to a Type II block $H_{2k}(\mu)$, in which μ may be replaced by μ^{-1} (the two variants are congruent).

The vexing issue of signs of the Type I blocks that required resolution in the algorithm to determine the *congruence canonical form of a matrix does not arise in the congruence canonical form.

Theorem 4.5.27. *Let $A, B \in M_n$ be nonsingular. Then A is congruent to B if and only if $A^{-T}A$ is similar to $B^{-T}B$.*

As a consequence of the canonical form theorems for *congruence and congruence, one can derive *canonical pairs* for an arbitrary pair of Hermitian matrices of the same size, as well as for a pair of matrices of the same size, one of which is an arbitrary complex symmetric matrix and the other of which is an arbitrary complex skew-symmetric matrix; see Appendix F.

Problems

4.5.P1 Let $A, B \in M_n$ and suppose that B is nonsingular. Show that there is a $C \in M_n$ such that $A = BC$. Moreover, for any nonsingular $S \in M_n$, we have $SAS^* = (SBS^*)C'$, in which C' is similar to C.

4.5.P2 Let $A, B \in M_n$ be skew symmetric. Show that there is a nonsingular $S \in M_n$ such that $A = SBS^T$ if and only if rank A = rank B.

4.5.P3 Let $A, B \in M_n$ be Hermitian. (a) If A is *congruent to B, show that A^k is *congruent to B^k for all $k = 2, 3, \ldots$. (b) If A^2 is *congruent to B^2, is A *congruent to B? Why?

(c) Show that $C = \begin{bmatrix} 0 & 1 \\ 0 & 0 \end{bmatrix}$ and $D = \begin{bmatrix} 0 & 1 \\ 0 & 1 \end{bmatrix}$ are *congruent, but C^2 is not *congruent to D^2. Does this contradict (a)?

4.5.P4 Prove the following generalization of (4.5.17(a)): Let $A_1, A_2, \ldots, A_k \in M_n$ be Hermitian and suppose that A_1 is nonsingular. There is a nonsingular $T \in M_n$ such that $T^* A_i T$ is diagonal for all $i = 1, \ldots, k$ if and only if $\{A_1^{-1} A_i : i = 2, \ldots, n\}$ is a commuting family of diagonalizable matrices with real eigenvalues. What is the corresponding generalization of (4.5.17(b))?

4.5.P5 A differential operator L (4.0.4) with a real symmetric coefficient matrix $A(x) = [a_{ij}(x)]$ is *elliptic* at a point $x \in D \subset \mathbf{R}^n$ if $A(x)$ is nonsingular and all its eigenvalues have the same sign; L is *hyperbolic* at x if $A(x)$ is nonsingular, $n - 1$ of its eigenvalues have the same sign, and one eigenvalue has the opposite sign. Explain why a differential operator that is elliptic (or hyperbolic) at a point with respect to one coordinate system is elliptic (or hyperbolic) at that point with respect to every other coordinate system. Laplace's equation $\frac{\partial^2 f}{\partial x^2} + \frac{\partial^2 f}{\partial y^2} + \frac{\partial^2 f}{\partial z^2} = 0$ gives an example of an elliptic differential operator; the wave equation $\frac{\partial^2 f}{\partial x^2} + \frac{\partial^2 f}{\partial y^2} + \frac{\partial^2 f}{\partial z^2} - \frac{\partial^2 f}{\partial t^2} = 0$ gives an example of a hyperbolic one. Both are presented in Cartesian coordinates. Even though both look very different in spherical polar or cylindrical coordinates, they remain, respectively, elliptic and hyperbolic.

4.5.P6 Show that $\begin{bmatrix} 0 & 1 \\ 1 & 0 \end{bmatrix}$ and $\begin{bmatrix} 1 & 0 \\ 0 & -1 \end{bmatrix}$ can be reduced simultaneously to diagonal form by a unitary congruence but cannot be reduced simultaneously to diagonal form by *congruence. Follow the proof of (4.5.17b) to find a unitary matrix that achieves a simultaneous diagonalization by congruence.

4.5.P7 Show that $\begin{bmatrix} 1 & 1 \\ 1 & 0 \end{bmatrix}$ and $\begin{bmatrix} 0 & 1 \\ 1 & 0 \end{bmatrix}$ cannot be reduced simultaneously to diagonal form by either *congruence or congruence.

4.5.P8 Let $A, S \in M_n$ with A Hermitian and S nonsingular. Let the eigenvalues of A and SAS^* be arranged in nondecreasing order, as in (4.2.1). Let $\lambda_k(A)$ be a nonzero eigenvalue. Deduce the relative eigenvalue perturbation bound $|\lambda_k(SAS^*) - \lambda_k(A)|/|\lambda_k(A)| \leq \rho(I - SS^*)$ from (4.5.9). What does this say if S is unitary? If S is "close to unitary"?

4.5.P9 Let $A \in M_n$ and suppose that rank $A = r$. Explain why the following are equivalent: (a) the *congruence regularization algorithm terminates after the first step; (b) nullspace $A =$ nullspace A^*; (c) range $A =$ range A^*; (d) there is a nonsingular $B \in M_r$ and a unitary $U \in M_n$ such that $A = U(B \oplus 0_{n-r})U^*$; (e) A is an EP matrix (2.6.P28).

4.5.P10 Let $A \in M_n$. Suppose that rank $A = r$ and nullspace $A =$ nullspace A^*. Show that A is rank principal (0.7.6.2), that is, A has a nonsingular r-by-r principal submatrix.

4.5.P11 Let $A \in M_n$ be nonzero and normal, let $\lambda_1 = |\lambda_1|e^{i\theta_1}, \ldots, \lambda_r = |\lambda_r|e^{i\theta_r}$ be its nonzero eigenvalues, in which each $\theta_j \in [0, 2\pi)$; let $\Lambda = \text{diag}(\lambda_1, \ldots, \lambda_r)$. Explain carefully why (a) A satisfies each of the three stated conditions in (4.5.23). (Justify each one; don't just prove one and invoke their equivalence.) (b) The *congruence canonical form of A is $[e^{i\theta_1}] \oplus \cdots \oplus [e^{i\theta_r}] \oplus 0_{n-r}$. (c) If among the angles $\theta_1, \ldots, \theta_r$ there are d distinct angles $\phi_1, \ldots, \phi_d$ with respective multiplicities $n_1, \ldots, n_d$ and $n_1 + \cdots + n_d = r$, then each ϕ_j is a canonical angle of A with multiplicity n_j. (d) On each ray $\{re^{i\phi_j} : 0 < r < \infty\}$ there are exactly n_j eigenvalues of A, $j = 1, \ldots, d$. (e) If $B \in M_n$ is normal, then B is *congruent to A if and only if rank $B =$ rank A and B has exactly n_j eigenvalues on each ray $\{re^{i\phi_j} : 0 < r < \infty\}$, $j = 1, \ldots, d$.

4.5.P12 Reconsider the preceding problem under the assumption that A is Hermitian. Why are 1 and 2 the only possible values of d? Why are $\theta_1 = 0$ or $\theta_2 = \pi$ the only possible canonical angles? How are the multiplicities n_1 and n_2 related to the inertia of A? Explain why the *congruence canonical form theorem may be thought of as a generalization to arbitrary square complex matrices of Sylvester's inertia theorem about Hermitian matrices.

4.5.P13 Let $U, V \in M_n$ be unitary. Show that U and V are *congruent if and only if they are similar if and only if they have the same eigenvalues.

4.5.P14 Let $A \in M_n$ and let $A = H + iK$, in which H and K are Hermitian and H is nonsingular. (a) Use the statements of (4.5.17), (4.5.18), and (4.5.24) to explain why $H^{-1}K$ is diagonalizable and has real eigenvalues if and only if A is nonsingular and diagonalizable by *congruence . No computations! (b) Now do the computations: Suppose that $S \in M_n$ is nonsingular and $H^{-1}K = S\Lambda S^{-1}$, in which $\Lambda = \mathrm{diag}(\lambda_1, \ldots, \lambda_n)$ is real. Show that A is nonsingular and $A^{-*}A = SMS^*$, $M = \mathrm{diag}(\mu_1, \ldots, \mu_n)$, and each $\mu_j = (1 + i\lambda_j)/(1 - i\lambda_j)$ has modulus one. Conclude that H and K are simultaneously diagonalizable by *congruence. (c) Suppose that A is nonsingular and $A^{-*}A = S\Lambda S^*$, in which $S \in M_n$ is nonsingular and Λ is diagonal and unitary. Write $\Lambda = \mathrm{diag}(e^{i\theta_1}, \ldots, e^{i\theta_n})$ with all $\theta_j \in [0, 2\pi)$. Explain why $H = S\,\mathrm{diag}(\cos\theta_1, \ldots, \cos\theta_n)S^*$, $K = S\,\mathrm{diag}(\sin\theta_1, \ldots, \sin\theta_n)S^*$, and $\frac{3}{2}\pi \neq \theta_j \neq \frac{1}{2}\pi$ for all $j = 1, \ldots, n$.

4.5.P15 (a) Explain why a Hankel matrix with zero entries in every position below its counterdiagonal is completely determined by the entries in its first row. (b) The inverse of the canonical block Δ_k (4.5.19) is a Hankel matrix with zero entries in every position below its counterdiagonal, whose first row is constructed from right to left by entering successive elements of the sequence $1, -i, -1, i, 1, -i, -1, i, 1, \ldots$ until the row is filled. For example, the first row of Δ_3^{-1} is $[-1 \ -i \ 1]$, the first row of Δ_4^{-1} is $[i \ -1 \ -i \ 1]$, and the first row of Δ_5^{-1} is $[1 \ i \ -1 \ -i \ 1]$. Verify this assertion by using the stated form of Δ_k^{-1} to compute $\Delta_k^{-1}\Delta_k$. (c) Show that $\Delta_k^{-*}\Delta_k$ is an upper triangular matrix (it is actually a Toeplitz matrix) whose main diagonal entries are all $+1$ and whose first superdiagonal entries are all $2i$. (d) Explain why the Jordan canonical form of $\Delta_k^{-*}\Delta_k$ is $J_k(1)$.

4.5.P16 How many disjoint equivalence classes under *congruence are there in the set of n-by-n complex Hermitian matrices? In the set of n-by-n real symmetric matrices?

4.5.P17 How many disjoint equivalence classes under congruence are there in the set of n-by-n complex symmetric matrices? In the set of n-by-n real symmetric matrices?

4.5.P18 Let $A, B \in M_n$ with A and B symmetric and A nonsingular. Show that if the generalized characteristic polynomial $p_{A,B}(t) = \det(tA - B)$ has n distinct zeroes, then A and B are simultaneously diagonalizable by congruence.

4.5.P19 Provide details for the following outline of an alternative proof of Sylvester's law of inertia (4.5.8). Let $A, S \in M_n$ be nonsingular and suppose that A is Hermitian. Let $S = QR$ be a QR factorization (2.1.14) in which $Q \in M_n$ is unitary and $R \in M_n$ is upper triangular with positive main diagonal entries. Show that $S(t) = tQ + (1 - t)QR$ is nonsingular if $0 \leq t \leq 1$ and let $A(t) = S(t)AS(t)^*$. What is $A(0)$? $A(1)$? Explain why $A(0)$ has the same number of positive (negative) eigenvalues as $A(1)$. Treat the general case by considering $A \pm \epsilon I$ for small $\epsilon > 0$.

4.5.P20 Let $A \in M_n$ and let $p_A(t) = t^n + a_{n-1}(A)t^{n-1} + \cdots + a_1(A)t + a_0(A)$ be its characteristic polynomial. (a) Recall that the coefficients $a_i(A)$, $i = 0, 1, \ldots, n - 1$, are elementary symmetric functions of the eigenvalues of A (1.2.15). Why are these coefficients continuous functions of A? (a) If A is normal, explain why rank $A = r$ implies that A has exactly r nonzero eigenvalues (denote them by $\lambda_1, \ldots, \lambda_r$), which implies that $a_{n-r+1}(A) = a_{n-r+2}(A) = \cdots = a_0(A) = 0$ and $a_{n-r}(A) = \lambda_1 \cdots \lambda_r$. (c) Let $\mathcal{S} \subset M_n$ be a connected set of Hermitian matrices, all of which have the same rank r. Show that every matrix in $\mathcal{S}$ has the same inertia. (d) Show by example that the assertion in (c) need not be correct if $\mathcal{S}$ is not connected.

4.5.P21 Let $A \in M_n$ be Hermitian and partitioned as $A = \begin{bmatrix} B & C \\ C^* & D \end{bmatrix}$, in which B is nonsingular. Let $S = D - C^* B^{-1} C$ denote the Schur complement of B in A. (a) Explain why the identity (0.8.5.3) exhibits a *congruence between A and $B \oplus S$. (b) Prove *Haynsworth's theorem*: The inertias of A, B, and S (4.5.6) are related by the identities

$$i_+(A) = i_+(B) + i_+(S)$$
$$i_-(A) = i_-(B) + i_-(S) \tag{4.5.28}$$
$$i_0(A) = i_0(S)$$

See (7.1.P28) for a related result.

4.5.P22 Let $B \in M_n$ be Hermitian, let $y \in \mathbf{C}^n$ and $a \in \mathbf{R}$ be given, and let $A = \begin{bmatrix} B & y \\ y^* & a \end{bmatrix} \in M_{n+1}$. Use Haynsworth's theorem in the preceding problem to prove Cauchy's interlacing inequalities (4.3.18).

4.5.P23 Let $\{A_1, \ldots, A_k\} \subset M_n$ be a given family of complex symmetric matrices and let $\mathcal{G} = \{A_i \bar{A}_j : i, j = 1, \ldots, k\}$. If there is a unitary $U \in M_n$ such that $U A_i U^T$ is diagonal for all $i = 1, \ldots, k$, explain why $\mathcal{G}$ is a commuting family. What does this reduce to when $k = 2$, and what is the connection with (4.5.15b)? In fact, commutativity of $\mathcal{G}$ is also sufficient to ensure the simultaneous diagonalizability of $\mathcal{F}$ by unitary congruence.

4.5.P24 Let $\mathcal{F} = \{A_1, \ldots, A_k\} \subset M_n$ be a family of complex symmetric matrices, let $\mathcal{H} = \{B_1, \ldots, B_m\} \subset M_n$ be a family of Hermitian matrices, and let $\mathcal{G} = \{A_i \bar{A}_j : i, j = 1, \ldots, k\}$. If there is a unitary $U \in M_n$ such that every $U A_i U^T$ and every $U B_j U^*$ is diagonal, show that each of $\mathcal{G}$ and $\mathcal{H}$ is a commuting family and $B_j A_i$ symmetric for all $i = 1, \ldots, k$ and all $j = 1, \ldots, m$. What does this reduce to when $k = m = l$, and what is the connection with (4.5.15(c))? In fact, these conditions are also sufficient to ensure the simultaneous diagonalizability of $\mathcal{F}$ and $\mathcal{H}$ by the respective congruences.

4.5.P25 Show that the cosquare of the canonical block Γ_k (4.5.24.1) is similar to $J_k((-1)^{k+1})$ by verifying that

$$\Gamma_k^{-T} \Gamma_k = \begin{bmatrix} & \vdots & \vdots & \vdots & \vdots & \ddots \\ & -1 & -1 & -1 & -1 & \\ & 1 & 1 & 1 & & \\ & -1 & -1 & & & \\ & 1 & & & & \end{bmatrix} \quad \Gamma_k = (-1)^{k+1} \begin{bmatrix} 1 & 2 & & \bigstar \\ & 1 & \ddots & \\ & & \ddots & 2 \\ & & & 1 \end{bmatrix}$$

4.5.P26 Suppose that $\mu \in \mathbf{C}$ is nonzero. Show that $H_{2k}(\mu)$ is congruent to $H_{2k}(\bar{\mu}) = \overline{H_{2k}(\mu)}$ if and only if either μ is real or $|\mu| = 1$.

4.5.P27 Let $B \in M_n$, let $C = B \oplus \bar{B}$, and define $S = \frac{e^{i\pi/4}}{\sqrt{2}} \begin{bmatrix} 0_n & iI_n \\ -iI_n & 0_n \end{bmatrix}$. (a) Explain why S is unitary, symmetric, and coninvolutory. (b) Show that SCS^T and SCS^* are both real, that is,

C is both congruent and *congruent to a real matrix. (c) Explain why C is both congruent and *congruent to $\bar{C}$.

4.5.P28 Let $A \in M_n$ and suppose that A is congruent to $\bar{A}$. (a) Use (4.5.25) and (4.5.P26) to show that A is congruent to a direct sum of (i) real blocks of the form $J_k(0)$, Γ_k, or $H_{2k}(r)$ in which r is real and either $r = (-1)^k$ or $|r| > 1$; (ii) blocks of the form $H_{2k}(\mu)$ in which $|\mu| = 1$, $\mu \neq \pm 1$, and μ is determined up to replacement by $\bar{\mu}$ (that is, $H_{2k}(\mu)$ is congruent to $H_{2k}(\bar{\mu})$); and (iii) pairs of blocks of the form $H_{2k}(\mu) \oplus H_{2k}(\bar{\mu})$ in which μ is not real and $|\mu| > 1$. (b) Use the preceding problem to show that A is congruent to a real matrix.

4.5.P29 Let $A \in M_n$ and suppose that A is *congruent to $\bar{A}$. (a) Use (4.5.21) to show that A is *congruent to a direct sum of (i) real blocks of the form $J_k(0)$, $\pm\Gamma_k$, or $H_{2k}(r)$ in which r is real and $|r| > 1$, and (ii) pairs of blocks of the form $\lambda\Gamma_k \oplus \bar{\lambda}\Gamma_k$ in which $|\lambda| = 1$ and $\lambda \neq \pm 1$, or of the form $H_{2k}(\mu) \oplus H_{2k}(\bar{\mu})$ in which μ is not real and $|\mu| > 1$. (b) Use (4.5.P27) to show that A is *congruent to a real matrix.

4.5.P30 Let $A \in M_n$. Explain why A is congruent (respectively, *congruent) to $\bar{A}$ if and only if A is congruent (respectively, *congruent) to a real matrix.

4.5.P31 Explain why (a) $\begin{bmatrix} 0 & 1 \\ i & 0 \end{bmatrix}$ is congruent to a real matrix but $\begin{bmatrix} 0 & 1 \\ 2i & 0 \end{bmatrix}$ is not; (b) neither matrix in (a) is *congruent to a real matrix.

4.5.P32 Suppose that $A \in M_{2n}$ is a matrix of quaternion type (see (4.4.P29)). Explain why (a) A is congruent to $\bar{A}$ via $S_{2n} = \begin{bmatrix} 0_n & I_n \\ -I_n & 0_n \end{bmatrix}$; (b) any matrix of quaternion type is congruent to a real matrix.

4.5.P33 Let $A \in M_n$. Use (4.5.25) to show that A is congruent to A^T.

4.5.P34 Let $A \in M_n$. Use (4.5.21) to show that A is *congruent to A^T.

4.5.P35 Let $A = \begin{bmatrix} 1 & -1 \\ -1 & 1 \end{bmatrix}$ and $B = \begin{bmatrix} 1 & 0 \\ 0 & -1 \end{bmatrix}$. (a) Use both (4.5.17) and (4.5.24) to show that A and B are not simultaneously diagonalizable by *congruence. (b) Use (4.5.17) to show that A and B are not simultaneously diagonalizable by congruence. (c) Show that $x^* B x = 0$ whenever $x \in \mathbf{C}^2$ and $x^* A x = 0$.

4.5.P36 Let $A, B \in M_n$ be Hermitian. Suppose that A is indefinite, and $x^* B x = 0$ whenever $x \in \mathbf{C}^n$ and $x^* A x = 0$. (a) Show that there is a real scalar κ such that $B = \kappa A$; in particular, A and B are simultaneously diagonalizable by *congruence. (b) Show that the assertion (a) is still correct if we assume that $A, B \in M_n(\mathbf{R})$ are symmetric, A is indefinite, and $x^T B x = 0$ whenever $x \in \mathbf{R}^n$ and $x^T A x = 0$. (c) Use the preceding problem to explain why we cannot omit the hypothesis that A is indefinite.

4.5.P37 Let $A \in M_n$ be nonsingular. Show that the following statements are equivalent (equivalence of (a) and (b) is in (4.5.24)):

(a) A is diagonalizable by *congruence, that is, there is a nonsingular $S \in M_n$ and a diagonal unitary $D = \mathrm{diag}(e^{i\theta_1}, \ldots, e^{i\theta_n})$ such that $A = SDS^*$.

(b) $A^{-*}A$ is diagonalizable and each of its eigenvalues has modulus one.

(c) There is a nonsingular $S \in M_n$ and a nonsingular diagonal matrix Λ such that $A = S\Lambda S^*$.

(d) There are two bases of $\mathbf{C}^n$, given by the columns of $X = [x_1 \ \ldots \ x_n]$ and $Y = [y_1 \ \ldots \ y_n]$, and a nonsingular diagonal matrix $\Lambda = \mathrm{diag}(\lambda_1, \ldots, \lambda_n)$ such that $X^* Y = I$ and $Ax_j = \lambda_j y_j$ for each $j = 1, \ldots, n$.

(e) There is a positive definite $B \in M_n$ such that $A^*BA = ABA^*$.
(f) There is a positive definite $B \in M_n$ and a nonsingular normal $C \in M_n$ such that $A = BCB$.

Explain how each of these six statements expresses a property of normal matrices. For this reason, matrices that are diagonalizable by *congruence are said to be *normalizable*.

Notes and Further Readings. For results about simultaneous diagonalization of more than two matrices (and proofs of the assertion in (4.5.P23 and P24)), see Y. P. Hong and R. A. Horn, On simultaneous reduction of families of matrices to triangular or diagonal form by unitary congruence, *Linear Multilinear Algebra* 17 (1985), 271–288. For proofs of the *congruence and congruence canonical form theorems, details for the algorithms following (4.5.22) and (4.5.26), two algorithms to determine the signs of the Type I blocks for *congruence, and extensions to matrices over fields other than **C**, see R. A. Horn and V. V. Sergeichuk, (a) A regularization algorithm for matrices of bilinear and sesquilinear forms, *Linear Algebra Appl.* 412 (2006) 380–395, (b) Canonical forms for complex matrix congruence and *congruence, *Linear Algebra Appl.* 416 (2006) 1010–1032, and (c) Canonical matrices of bilinear and sesquilinear forms, *Linear Algebra Appl.* 428 (2008) 193–223. Problem (4.5.P36) arises in the theory of special relativity, with $A = \mathrm{diag}(1, 1, 1, -c)$, in which c is the speed of light. It implies that Lorentz transformations are the only linear changes of coordinates in four-dimensional space-time that are consistent with Einstein's postulate of the universality of the speed of light. For an exposition, see J. H. Elton, Indefinite quadratic forms and the invariance of the interval in special relativity, *Amer. Math. Monthly* 117 (2010) 540–547. The term *normalizable matrix* (see (4.5.P37)) seems to have been coined in K. Fan, Normalizable operators, *Linear Algebra Appl.* 52/53 (1983) 253–263.

4.6 Consimilarity and condiagonalization

In this section, we study and broaden the notion of condiagonalization, which arose naturally in (4.5.17(c)).

Definition 4.6.1. *Matrices $A, B \in M_n$ are* consimilar *if there exists a nonsingular $S \in M_n$ such that $A = SB\bar{S}^{-1}$.*

If U is unitary, then $\bar{U}^{-1} = \bar{U}^* = U^T$, so unitary congruence $(A = UBU^T)$ and unitary consimilarity $(A = UB\bar{U}^{-1})$ are the same; if Q is complex orthogonal, then $\bar{Q}^{-1} = \bar{Q}^T = Q^*$, so complex orthogonal *congruence $(A = QBQ^*)$ and complex orthogonal consimilarity $(A = QB\bar{Q}^{-1})$ are the same; if R is real and nonsingular, then $\bar{R}^{-1} = R^{-1}$, so real similarity $(A = RBR^{-1})$ and real consimilarity $(A = RB\bar{R}^{-1})$ are the same.

For matrices of size one, similarity is trivial $(sas^{-1} = a)$, but consimilarity is a rotation: $sa\bar{s}^{-1} = |s|e^{i\theta}a|s|^{-1}e^{i\theta} = e^{2i\theta}a$ if $s = |s|e^{i\theta}$.

Exercise. Explain why each 1-by-1 complex matrix $[a]$ is consimilar to $[\bar{a}]$, to $[-a]$, and to the real matrix $[|a|]$.

Exercise. How are A, $B \in M_n$ related if they are consimilar via a coninvolutory matrix?

Consimilarity is an equivalence relation on M_n (0.11), and we may ask which equivalence classes contain block triangular, triangular, or diagonal representatives.

Definition 4.6.2. *A matrix $A \in M_n$ is* contriangularizable *(respectively,* block contriangularizable*) if there exists a nonsingular $S \in M_n$ such that $S^{-1}A\bar{S}$ is upper triangular (respectively, block upper triangular); it is* condiagonalizable *if S can be chosen so that $S^{-1}A\bar{S}$ is diagonal. It is* unitarily contriangularizable *or* unitarily condiagonalizable *if it is unitarily congruent to a matrix of the required form.*

We encountered unitary contriangularization (triangularization by unitary congruence) and unitary condiagonalization (diagonalization by unitary congruence) in (4.4.4). If $A \in M_n$ is contriangularizable, and if $S^{-1}A\bar{S} = \Delta$ is upper triangular, a computation reveals that the main diagonal entries of $\Delta\bar{\Delta} = S^{-1}(A\bar{A})S$ are nonnegative. Conversely, if every eigenvalue of $A\bar{A}$ is nonnegative, (4.4.4) ensures that A is unitarily contriangularizable. If some eigenvalue of $A\bar{A}$ is not real or is real and negative, then A is not contriangularizable, but (4.4.9) ensures that it is block contriangularizable with diagonal blocks of sizes one and two; the diagonal blocks are associated with pairs of eigenvalues of $A\bar{A}$ that are either non-real and conjugate, or are real, negative, and equal.

Theorem 4.6.3. *Let $A \in M_n$ be given. The following are equivalent:*

(a) A is contriangularizable.
(b) A is unitarily contriangularizable.
(c) Every eigenvalue of $A\bar{A}$ is real and nonnegative.

If $A \in M_n$ is unitarily condiagonalizable, then there is a unitary U such that $A = U\Lambda\bar{U}^{-1} = U\Lambda U^T$, in which $\Lambda = \mathrm{diag}(\lambda_1, \dots, \lambda_n)$; consequently, A is symmetric. Conversely, if A is symmetric, then (4.4.4c) ensures that A is unitarily condiagonalizable.

Theorem 4.6.4. *A matrix $A \in M_n$ is unitarily condiagonalizable if and only if it is symmetric.*

How can we decide whether a given square nonsymmetric matrix can be condiagonalized by a (necessarily nonunitary) consimilarity? If $S = [s_1 \ \dots \ s_n]$ is nonsingular and partitioned according to its columns, and if $S^{-1}A\bar{S} = \Lambda = \mathrm{diag}(\lambda_1, \dots, \lambda_n)$, then $A\bar{S} = S\Lambda$, so $A\bar{s}_i = \lambda_i s_i$ for $i = 1, \dots, n$.

Definition 4.6.5. *Let $A \in M_n$ be given. A nonzero vector $x \in \mathbf{C}^n$ such that $A\bar{x} = \lambda x$ for some $\lambda \in \mathbf{C}$ is a* coneigenvector *of A; the scalar λ is a* coneigenvalue *of A. We say that the coneigenvector x is associated with the coneigenvalue λ. The pair λ, x is a* coneigenpair *for A.*

Exercise. Let $A \in M_n$ be singular and let $\mathcal{N}$ denote its null space. Explain why the coneigenvectors of A associated with the coneigenvalue 0 are the nonzero vectors in the complex subspace $\overline{\mathcal{N}}$ (not very interesting).

The span of a coneigenvector of $A \in M_n$ is a one-dimensional coninvariant subspace. Lemma 4.4.2 ensures that every $A \in M_n$ has a coninvariant subspace of dimension one or two.

If $S^{-1}A\bar{S} = \Lambda$ is diagonal, the identity $A\bar{S} = S\Lambda$ ensures that every column of S is a coneigenvector of A, so there is a basis of $\mathbf{C}^n$ consisting of coneigenvectors of A. Conversely, if there is a basis $\{s_1, \ldots, s_n\}$ of $\mathbf{C}^n$ consisting of coneigenvectors of A, then $S = [s_1 \ \ldots \ s_n]$ is nonsingular, $A\bar{S} = S\Lambda$ for some diagonal matrix Λ, and $S^{-1}A\bar{S} = \Lambda$. Just as in the case of ordinary diagonalization, we conclude that $A \in M_n$ is condiagonalizable if and only if it has n linearly independent coneigenvectors.

If $A\bar{x} = \lambda x$, then $A(\overline{e^{i\theta}x}) = e^{-i\theta}A\bar{x} = e^{-i\theta}\lambda x = (e^{-2i\theta}\lambda)(e^{i\theta}x)$ for any $\theta \in \mathbf{R}$. Thus, if λ is a coneigenvalue of A, then so is $e^{-2i\theta}\lambda$ for all $\theta \in \mathbf{R}$; $e^{i\theta}x$ is an associated coneigenvector. It is often convenient to select from the coneigenvalues $e^{-2i\theta}\lambda$ of equal modulus the unique nonnegative representative $|\lambda|$, and to use a coneigenvector associated with it.

Exercise. Let $A \in M_n$ be given, suppose that λ, x is a coneigenpair for A, and let $\mathcal{S} = \{x : A\bar{x} = \lambda x\}$. If $\lambda \neq 0$, explain why $\mathcal{S}$ is *not* a vector space over $\mathbf{C}$ (and hence it is not a subspace of $\mathbf{C}^n$), but it *is* a vector space over $\mathbf{R}$. The vector space $\mathcal{S}$ (only a real vector space if $\lambda > 0$, but a complex vector space if $\lambda = 0$) is the *coneigenspace of A associated with the coneigenvalue λ.*

Moreover, if $A\bar{x} = \lambda x$, then $A\bar{A}x = A(\overline{A\bar{x}}) = A(\overline{\lambda x}) = \bar{\lambda}A\bar{x} = \bar{\lambda}\lambda x = |\lambda|^2 x$, so λ is a coneigenvalue of A only if $|\lambda|^2$ is an eigenvalue (necessarily nonnegative) of $A\bar{A}$.

Exercise. For $A = \begin{bmatrix} 0 & -1 \\ 1 & 0 \end{bmatrix}$, show that $A\bar{A}$ has no nonnegative eigenvalues and explain why A has no coneigenvectors (and hence no coneigenvalues).

The necessary condition that we have just observed for existence of a coneigenvalue is also sufficient.

Proposition 4.6.6. *Let $A \in M_n$, let $\lambda \geq 0$ be given, let $\sigma = \sqrt{\lambda} \geq 0$, and suppose that there is a nonzero vector x such that $A\bar{A}x = \lambda x$. There is a nonzero vector y such that $A\bar{y} = \sigma y$.*

(a) *If $\lambda = 0$, then $\bar{A}$ is singular and one may take $y = \bar{z}$ for any nonzero vector z in the null space of $\bar{A}$.*

(b) *If $\lambda > 0$, one may take $y = e^{-i\theta}A\bar{x} + e^{i\theta}\sigma x$ for any $\theta \in [0, \pi)$ such that $A\bar{x} \neq -e^{2i\theta}\sigma x$; at most one value of $\theta \in [0, \pi)$ is excluded.*

(c) *If λ has geometric multiplicity 1 as an eigenvalue of $A\bar{A}$, then x is a coneigenvector of A associated with a coneigenvalue $e^{2i\theta}\sigma$ for some $\theta \in [0, \pi)$ and $y = e^{i\theta}x$ satisfies $A\bar{y} = \sigma y$.*

Proof. (a) If $A\bar{A}x = 0$, then $A\bar{A}$ is singular, so $0 = \det A\bar{A} = |\det \bar{A}|^2$, $\bar{A}$ is singular, there is a nonzero vector z such that $\bar{A}z = 0$ and $\overline{\bar{A}z} = A\bar{z} = 0$.

(b) Suppose that $\sigma > 0$, so $\sigma x \neq 0$. If the vectors $A\bar{x}, \sigma x$ are linearly dependent, then $A\bar{x} = c\sigma x$ for a unique complex scalar c; if they are linearly independent, then $A\bar{x} \neq c\sigma x$ for every complex scalar c. In particular, $A\bar{x} = -e^{2i\theta}\sigma x$ for at most one $\theta \in [0, \pi)$. If $\theta \in [0, \pi)$ is such that $A\bar{x} \neq -e^{2i\theta}\sigma x$, then $y \neq 0$ and $A\bar{y} = e^{i\theta}A\bar{A}x + e^{-i\theta}\sigma A\bar{x} = e^{i\theta}\lambda x + e^{-i\theta}\sigma A\bar{x} = \sigma(e^{-i\theta}A\bar{x} + e^{i\theta}\sigma x) = \sigma y$.

(c) $A\bar{A}(A\bar{x}) = A(\overline{A\bar{A}x}) = A(\overline{\lambda x}) = \lambda(A\bar{x})$, so $A\bar{x} = cx$ for some scalar c ($c = 0$ is possible) since λ has geometric multiplicity 1 as an eigenvalue of $A\bar{A}$ Then $\lambda x = A\bar{A}x = A(\overline{A\bar{x}}) = A(\overline{cx}) = \bar{c}A\bar{x} = \bar{c}cx = |c|^2 x$, which ensures that $|c| = \sigma$. Let $c = e^{2i\theta}\sigma$. Then $A\bar{x} = cx = e^{2i\theta}\sigma x$ and $A(\overline{e^{i\theta}x}) = \sigma(e^{i\theta}\sigma x)$. $\qquad\square$

Exercise. Let σ be real and nonnegative, let $A = \begin{bmatrix} \sigma & i \\ 0 & \sigma \end{bmatrix}$, and let $x = [1 \; i]^T$. Explain why x is an eigenvector of $A\bar{A}$ associated with the eigenvalue σ^2, but it is not a coneigenvector of A. If $\sigma = 0$, verify that the standard basis vector e_1 is a coneigenvector of A associated with the coneigenvalue zero. If $\sigma > 0$, verify that $y = A\bar{x} + \sigma x$ is a coneigenvector of A associated with the coneigenvalue σ.

Although an eigenvector of $A\bar{A}$ associated with a nonnegative eigenvalue λ need not be a coneigenvector of A, the proof of (4.6.6) shows how such a vector can be used to construct a coneigenvector of A associated with the nonnegative coneigenvalue $\sqrt{\lambda}$. If λ is positive and has geometric multiplicity g as an eigenvalue of $A\bar{A}$, then the coneigenspace of A associated with the positive coneigenvalue $\sqrt{\lambda}$ is a g-dimensional real vector space, and there is a generalization of (4.6.6) that shows how to construct a basis for it from any given basis for the eigenspace of $A\bar{A}$ associated with the eigenvalue λ; see (4.6.P16 to P18).

The following result is an analog of (1.3.8).

Proposition 4.6.7. *Let $A \in M_n$ be given, and let $x_1, x_2, \ldots, x_k$ be coneigenvectors of A with corresponding coneigenvalues $\lambda_1, \lambda_2, \ldots, \lambda_k$. If $|\lambda_i| \neq |\lambda_j|$ whenever $1 \leq i, j \leq k$ and $i \neq j$, then the vectors $x_1, \ldots, x_k$ are linearly independent.*

Proof. Each x_i is an eigenvector of $A\bar{A}$ with associated eigenvalue $|\lambda_i|^2$. Lemma 1.3.8 ensures that the vectors $x_1, \ldots, x_k$ are linearly independent. $\qquad\square$

This result, together with (4.6.6), gives a lower bound on the number of linearly independent coneigenvectors of a given matrix and yields a sufficient condition for condiagonalizability that is an analog of (1.3.9). Perhaps surprisingly, if $A\bar{A}$ has distinct eigenvalues, any nonsingular matrix that diagonalizes $A\bar{A}$ (by similarity) also condiagonalizes A.

Corollary 4.6.8. *Let $A \in M_n$ be given and suppose that $A\bar{A}$ has k distinct nonnegative eigenvalues.*

(a) *The matrix A has at least k linearly independent coneigenvectors.*

(b) *If $k = 0$, then A has no coneigenvectors.*

(c) *Suppose that $k = n$. Then A is condiagonalizable. Moreover, if $S \in M_n$ is nonsingular, $A\bar{A} = S\Lambda S^{-1}$, and Λ is nonnegative diagonal, then $S^{-1}A\bar{S} = D$ is diagonal and there is a diagonal unitary matrix Θ such that $A = Y\Sigma\bar{Y}^{-1}$, in which $Y = S\Theta$, Σ is nonnegative diagonal, and $\Sigma^2 = \Lambda$.*

Proof. Only the second assertion in (d) requires justification. Let $\Lambda = \text{diag}(\lambda_1, \ldots, \lambda_n)$, let $\sigma_j^2 = \lambda_j$ and $\sigma_j \geq 0$ for each $j = 1, \ldots, n$, let $\Sigma = \text{diag}(\sigma_1, \ldots, \sigma_n)$, and let $S = [s_1 \; \ldots \; s_n]$ be partitioned according to its columns. Each column s_j is an eigenvector of $A\bar{A}$ associated with a nonnegative eigenvalue λ_j that has algebraic (and hence geometric) multiplicity 1. Thus, (4.4.6(d)) ensures that each s_j is a coneigenvector of A, that $A\overline{s_j} = e^{2i\theta_j}\sigma_j s_j$ for some $\theta_j \in [0, \pi)$, and that $A\overline{y_j} = \sigma_j y_j$, in which $y_j = e^{i\theta}s_j$. Let $Y = [y_1 \ \cdots \ y_n]$ and $\Theta = \text{diag}(e^{i\theta_1}, \ldots, e^{i\theta_n})$. Then $Y = S\Theta$, $D = \Sigma\Theta^2$, and $A = (S\Theta)\Sigma(\overline{S\Theta})^{-1} = Y\Sigma\bar{Y}^{-1}$. $\square$

Our objective is to give a simple condition for a given matrix to be condiagonalizable, and as a first step, we prove that a coninvolutory matrix (0.9.13) is consimilar to the identity matrix.

Lemma 4.6.9. *Let $A \in M_n$ be given. Then $A\bar{A} = I$ if and only if there is a nonsingular $S \in M_n$ such that $A = S\bar{S}^{-1}$.*

Proof. If $A = S\bar{S}^{-1}$, then $A\bar{A} = S\bar{S}^{-1}\bar{S}S^{-1} = I$. Conversely, suppose that $A\bar{A} = I$, let $S_\theta = e^{i\theta}A + e^{-i\theta}I$, $\theta \in \mathbf{R}$, and compute

$$A\bar{S}_\theta = A(e^{-i\theta}\bar{A} + e^{i\theta}I) = e^{-i\theta}A\bar{A} + e^{i\theta}A = e^{i\theta}A + e^{-i\theta}I = S_\theta \qquad (4.6.10)$$

There is some $\theta_0 \in [0, \pi)$ such that $-e^{2i\theta_0}$ is not an eigenvalue of A (at most n values are excluded), and S_{θ_0} is nonsingular; (4.6.10) ensures that $A = S_{\theta_0}\bar{S}_{\theta_0}^{-1}$. $\square$

We can now state and prove a necessary and sufficient condition for condiago-nalizability; for an algorithm to compute a condiagonalization, see (4.6.P21), which generalizes the construction in (4.6.8(c)).

Theorem 4.6.11. *A given $A \in M_n$ is condiagonalizable if and only if $A\bar{A}$ is diago-nalizable (by similarity), every eigenvalue of $A\bar{A}$ is real and nonnegative, and $\text{rank } A = \text{rank } A\bar{A}$.*

Proof. If $A = SD\bar{S}^{-1}$ and $D \in M_n$ is diagonal, then $A\bar{A} = SD\bar{S}^{-1}\bar{S}\bar{D}S^{-1} = SD\bar{D}S^{-1}$ and the rank of both $A\bar{A}$ and A is the number of nonzero diagonal entries in D.

Conversely, suppose that $\text{rank } A = \text{rank } A\bar{A}$ and there is a nonsingular S such that $A\bar{A} = S\Lambda S^{-1}$, in which $\Lambda = \lambda_1 I_{n_1} \oplus \cdots \oplus \lambda_d I_{n_d}$ is real diagonal, $0 \le \lambda_1 < \cdots < \lambda_d$, and $\lambda_i \ne \lambda_j$ whenever $i \ne j$. Then

$$S^{-1}A\bar{A}S = S^{-1}A\bar{S}\bar{S}^{-1}\bar{A}S = (S^{-1}A\bar{S})(\overline{S^{-1}A\bar{S}}) = \Lambda$$

Let $S^{-1}A\bar{S} = B = [B_{ij}]_{i,j=1}^d$, partitioned conformally to Λ. We have $B\bar{B} = \Lambda = \bar{\Lambda} = \overline{B\bar{B}} = \bar{B}B$, so B and $\bar{B}$ commute. Moreover, $B\Lambda = B(B\bar{B}) = B(\bar{B}B) = (B\bar{B})B = \Lambda B$, so B and Λ commute and (2.4.4.2) ensures that $B = B_{11} \oplus \cdots \oplus B_{dd}$ is block diagonal and conformal to Λ. The identity $B\bar{B} = \Lambda$ tells us that $B_{ii}\bar{B}_{ii} = \lambda_i I_{n_i}$ for each $i = 1, \ldots, k$. Suppose that $\sigma_i \ge 0$ and $\sigma_i^2 = \lambda_i$ for each $i = 1, \ldots, d$. If $\lambda_i > 0$, then B_{ii} is nonsingular and $(\sigma_i^{-1}B_{ii})(\overline{\sigma_i^{-1}B_{ii}}) = I_{n_i}$. The preceding lemma ensures that there is a nonsingular $R_i \in M_{n_i}$ such that $\sigma_i^{-1}B_{ii} = R_i\bar{R}_i^{-1}$, that is, $B_{ii} = \sigma_i R_i\bar{R}_i^{-1}$. On one hand, we have $\text{rank } A = \text{rank } B = \sum_{i=1}^d \text{rank } B_{ii} = \text{rank } B_{11} + \sum_{i=2}^d n_i$. On the other hand, we have $\text{rank } A = \text{rank } A\bar{A} = \text{rank } \Lambda = \beta_1 + \sum_{i=2}^d n_i$, in which $\beta_1 = 0$ if $\lambda_1 = 0$ and $\beta_1 = n_1$ if $\lambda_1 > 0$. We conclude that $\text{rank } B_{11} = \beta_1$, so $B_{11} = 0$ if $\lambda_1 = 0$. Thus, $B_{11} = \sigma_1 I_{n_1}$ in either case, $\lambda_1 = 0$ or $\lambda_1 > 0$, so we may take $R_1 = I_{n_1}$ if $\lambda_1 = 0$. If we let $R = R_1 \oplus \cdots \oplus R_d$ and $\Sigma = \sigma_1 I_{n_1} \oplus \cdots \oplus \sigma_d I_{n_d}$ we conclude that $B = R(\sigma_1 I_{n_1} \oplus \cdots \oplus \sigma_d I_{n_d})\bar{R}^{-1} = R\Sigma\bar{R}^{-1}$ and $A = SB\bar{S}^{-1} = (SR)\Sigma(\overline{SR})^{-1}$. $\square$

See (4.6.P26 and P27) for versions of the preceding lemma and theorem in the context of unitary congruence.

The theory of ordinary similarity arises from studying linear transformations referred to different bases. In its general context, consimilarity arises from studying antilinear transformations referred to different bases. A *semilinear transformation* (sometimes called an *antilinear transformation*) is a mapping $T: V \to W$ from one complex vector space into another that is additive ($T(x + y) = Tx + Ty$ for all $x, y \in V$) and *conjugate homogeneous* ($T(ax) = \bar{a}Tx$ for all $a \in \mathbf{C}$ and all $x \in V$, sometimes called *antihomogeneous*). Time reversal in quantum mechanics is an example of a semilinear transformation.

Of course, not every matrix is condiagonalizable, but there is a standard form to which each square complex matrix is consimilar. The *consimilarity canonical form theorem* is as follows:

Theorem 4.6.12. *Each square complex matrix is consimilar to a direct sum, uniquely determined up to permutation of direct summands, of matrices of the following three types:*

Type 0: $J_k(0)$, $k = 1, 2, \ldots$;

Type I: $J_k(\sigma)$, $k = 1, 2, \ldots$, *in which σ is real and positive;*

Type II: $H_{2k}(\mu)$, $k = 1, 2, \ldots$, *in which $H_{2k}(\mu)$ has the form (4.5.20) and μ is either not real or is real and negative.*

Exercise. If $\mu \in \mathbf{C}$, explain why $H_{2k}(\mu)H_{2k}(\bar{\mu})$ is similar to $J_k(\mu) \oplus J_k(\bar{\mu})$.

A *cancellation theorem* for consimilarity follows from (4.6.12), as in the proof of (4.5.22).

Theorem 4.6.13. *Let $A, B \in M_p$ and $C \in M_q$ be given. Then $A \oplus C$ is consimilar to $B \oplus C$ if and only if A is consimilar to B.*

The Jordan canonical form of $A\bar{A}$ has a special form: its nonsingular part contains *only* blocks of the form $J_k(\lambda)$ with λ real and positive, and pairs of blocks of the form $J_k(\mu) \oplus J_k(\bar{\mu})$, in which μ is not both real and positive (μ is either real and negative, or it is not real); its singular part is similar to the square of a nilpotent matrix (see 4.6.P22).

The *concanonical form* of a given $A \in M_n$ is a direct sum of Type 0, Type I, and Type II blocks to which it is consimilar, as described in (4.6.12). It can be determined in two steps:

Step 1. Let $r_0 = n$, $r_1 = \text{rank } A$, $r_{2k} = \text{rank}(A\bar{A})^k$, and $r_{2k+1} = \text{rank}((A\bar{A})^k A)$, for $k = 1, 2, \ldots$. The integers $w_j = r_{j-1} - r_j$, $j = 1, \ldots, n+1$, determine the Type 0 blocks in the concanonical form of A as follows: There are $w_k - w_{k+1}$ blocks of the form $J_k(0)$, $k = 1, \ldots, n$.

Step 2. Compute the nonsingular part of the Jordan canonical form of $A\bar{A}$. It determines the Type I and Type II blocks in the concanonical form of A as follows: Each block $J_k(\lambda)$ with $\lambda > 0$ corresponds to a Type I block $J_k(\sigma)$ with $\sigma > 0$ and $\sigma^2 = \lambda$; each pair $J_k(\mu) \oplus J_k(\bar{\mu})$ with a non-real or a real negative μ corresponds to a Type II block $H_{2k}(\mu)$.

Corollary 4.6.14. *Let $A, B \in M_n$. Then A is consimilar to B if and only if $A\bar{A}$ is similar to $B\bar{B}$, rank A = rank B, and rank$(A\bar{A})^k A$ = rank$(B\bar{B})^k B$ for $k = 1, \ldots, [n/2]$. If A and B are nonsingular, then A is consimilar to B if and only if $A\bar{A}$ is similar to $B\bar{B}$.*

Proof. Necessity of the stated conditions is clear, so we consider only their sufficiency. The Type I and Type II blocks of the concanonical forms of A and B are determined by the Jordan canonical form of $A\bar{A}$, since it is the same as the Jordan canonical form of $B\bar{B}$. The stated rank conditions (together with the conditions rank$(A\bar{A})^k$ = rank$(B\bar{B})^k$, $k = 1, 2, \ldots, [n/2]$, which are a consequence of similarity of $A\bar{A}$ and $B\bar{B}$) ensure that the Type 0 blocks of the concanonical forms of A and B are the same. $\square$

Exercise. Use the preceding corollary to show that each square complex matrix is consimilar to its negative, its conjugate, its transpose, and its conjugate transpose.

One can show that each of the three types of concanonical blocks is consimilar to a Hermitian matrix, as well as to a real matrix; it then follows from (4.6.12) that each square complex matrix is consimilar to a Hermitian matrix, as well as to a real matrix.

Corollary 4.6.15. *Let $A \in M_n$ be given. Then A is consimilar to $-A$, to $\bar{A}$, to A^T, to A^*, to a Hermitian matrix, and to a real matrix.*

Corollary 4.6.16. *Let $A \in M_n$ be given. Then $A\bar{A}$ is similar to the square of a real matrix.*

Proof. Corollary 4.6.15 ensures that there is a nonsingular $S \in M_n$ and a real matrix $R \in M_n(\mathbf{R})$ such that $A = SR\bar{S}^{-1}$. Then $A\bar{A} = SR\bar{S}^{-1}\overline{SR\bar{S}^{-1}} = SR^2S^{-1}$. $\square$

The preceding corollary and its proof provide a complete explanation for the phenomenon identified in (4.4.13): Suppose that λ is a real negative eigenvalue of $A\bar{A}$. Let μ be the pure imaginary scalar such that $\mu^2 = \lambda$ and Im $\mu > 0$. A block $J_k(\lambda)$ is present in the Jordan canonical form of $A\bar{A}$ if and only if it is present in the Jordan canonical form of R^2 if and only if either $J_k(\mu)$ or $J_k(-\mu)$ is present in the Jordan canonical form of R. Because R is real and $\bar{\mu} = -\mu$, however, *both* blocks must be present in the Jordan canonical form of R, and indeed, there must be the same number of each. Thus, the number of blocks $J_k(\lambda)$ in the Jordan canonical form of R^2 (and hence also of $A\bar{A}$) is even.

Corollary 4.6.17. *Let $A \in M_n$ be given.*

(a) *$A = HS$ (as well as $A = SH$), in which H is Hermitian, S is symmetric, and either factor may be chosen to be nonsingular.*

(b) *$A = BE$ (as well as $A = EB$), in which B is similar to a real matrix and E is coninvolutory.*

Proof. (a) Use (4.6.15) to write $A = SH\bar{S}^{-1}$, in which S is nonsingular and H is Hermitian. Then $A = (SHS^*)(S^{-*}\bar{S}^{-1}) = (SHS^*)(\bar{S}^{-T}\bar{S}^{-1})$ is a product of a Hermitian matrix and a nonsingular symmetric matrix. Now write $A = SB\bar{S}^{-1}$, in which B is symmetric. Then $A = (SS^*)(S^{-*}B\bar{S}^{-1}) = (SS^*)(\bar{S}^{-T}B\bar{S}^{-1})$ is a product of a nonsingular Hermitian matrix and a symmetric matrix. To reverse the order of the factors, write $A = (SS^T)(S^{-T}H\bar{S}^{-1})$ or $A = (SBS^T)(S^{-T}\bar{S}^{-1})$.

(b) Use (4.6.15) to write $A = SR\bar{S}^{-1}$, in which S is nonsingular and R is real. Then $A = (SRS^{-1})(S\bar{S}^{-1}) = (S\bar{S}^{-1})(\bar{S}R\bar{S}^{-1})$ is a product of a coninvolutory matrix and a matrix similar to a real matrix, in both orders. $\qquad\square$

Our final result about consimilarity is a criterion that can be a useful alternative to (4.6.14); see (4.6.P17) for another criterion.

Theorem 4.6.18. *Let $A, B \in M_n$ be given. The following are equivalent:*

   (a) A and B are consimilar.
   (b) $\begin{bmatrix} 0 & A \\ \bar{A} & 0 \end{bmatrix}$ is similar to $\begin{bmatrix} 0 & B \\ \bar{B} & 0 \end{bmatrix}$.
   (c) $\begin{bmatrix} 0 & A \\ -\bar{A} & 0 \end{bmatrix}$ is similar to $\begin{bmatrix} 0 & B \\ -\bar{B} & 0 \end{bmatrix}$.

Exercise. If $A = SB\bar{S}^{-1}$, use the similarity matrix $\begin{bmatrix} S & 0 \\ 0 & \bar{S} \end{bmatrix}$ to show that (a) implies (b) and (c) in the preceding theorem.

Problems

4.6.P1 Explain why consimilarity is an equivalence relation on M_n.

4.6.P2 Show that (a) $\begin{bmatrix} i & 1 \\ 0 & i \end{bmatrix}$ is not diagonalizable (by similarity) but is condiagonalizable.
(b) $\begin{bmatrix} 1 & -1 \\ 1 & 1 \end{bmatrix}$ is diagonalizable but not condiagonalizable. (c) $\begin{bmatrix} 0 & 1 \\ 0 & 0 \end{bmatrix}$ is neither diagonalizable nor condiagonalizable.

4.6.P3 Let $A \in M_n$ be given, suppose that λ is a positive coneigenvalue of A, and let $x_1, \ldots, x_k$ be coneigenvectors of A associated with λ. Show that the vectors $x_1, \ldots, x_k$ are linearly independent over $\mathbf{C}$ if and only if they are linearly independent over $\mathbf{R}$.

4.6.P4 Theorem 4.6.11 gives necessary and sufficient conditions for a single matrix to be condiagonalizable, but what if one has several matrices that are to be condiagonalized simultaneously? Let $\{A_1, A_2, \ldots, A_k\} \subset M_n$ be given and suppose that there is a nonsingular $S \in M_n$ such that $A_i = S\Lambda_i\bar{S}^{-1}$ for $i = 1, \ldots, k$ and each Λ_i is diagonal. Show that (a) each A_i is condiagonalizable; (b) each $A_i\bar{A}_j$ is diagonalizable; (c) the family of products $\{A_i\bar{A}_j : i, j = 1, \ldots, k\}$ commutes; and (d) every eigenvalue of $A_i\bar{A}_j + A_j\bar{A}_i$ is real, and every eigenvalue of $A_i\bar{A}_j - A_j\bar{A}_i$ is imaginary for all $i, j = 1, \ldots, k$. What does this say when $k = 1$?

4.6.P5 If $A \in M_n$ is such that $A\bar{A} = \Lambda = \lambda_1 I_{n_1} \oplus \cdots \oplus \lambda_k I_{n_k}$ with $\lambda_i \neq \lambda_j$ if $i \neq j$, and all $\lambda_i \geq 0$, show that there is a unitary $U \in M_n$ such that $A = U\Delta U^T$ and $\Delta = \Delta_1 \oplus \cdots \oplus \Delta_k$, in which each $\Delta_i \in M_{n_i}$ is upper triangular.

4.6.P6 Lemma 4.6.9 says that $A \in M_n$ has a factorization $A = S\bar{S}^{-1}$ for some nonsingular $S \in M_n$ if and only if $A\bar{A} = I$. Show that $A = U\bar{U}^{-1} = UU^T$ for some unitary $U \in M_n$ if and only if $A^{-1} = \bar{A}$ and A is symmetric.

4.6.P7 Let $A \in M_n$ be coninvolutory. Show that there is a single nonsingular $S \in M_n$ such that SXS^{-1} is real for any $X \in M_n$ such that $A\bar{X} = XA$.

4.6.P8 If $A \in M_n$ is diagonal or upper triangular, show that the eigenvalues and coneigenvalues of A are related in the following way: If λ is an eigenvalue of A, then $e^{i\theta}\lambda$ is a coneigenvalue of A for *all* $\theta \in \mathbf{R}$ and if μ is a coneigenvalue of A, then $e^{i\theta}\mu$ is an eigenvalue of A for *some* $\theta \in \mathbf{R}$.

4.6.P9 Let $A \in M_n$ be given and suppose that n is odd. Explain why A has a coneigenpair.

4.6.P10 Let $A \in M_n$ be symmetric. Deduce from the statement of (4.6.11) that A is condiagonalizable (not necessarily unitarily). Modify the proof of (4.6.11) in the following three steps to show that the condiagonalization can be achieved via a unitary matrix: (a) Explain why S may be taken to be unitary. (b) Why is each block B_{ii} symmetric? Use (2.5.18) to show that $\sigma_j^{-1} B_{jj} = R_j^2 = R_j \bar{R}_j^{-1}$ with $R_j = Q_j D_j Q_j^T$, in which Q_j is real orthogonal and D_j is diagonal and unitary. (c) Explain why R may be taken to be unitary. Put all this together to obtain yet another proof of (4.4.4c).

4.6.P11 If $A \in M_n(\mathbf{R})$, explain why the singular part of its concanonical form is the same as the singular part of its Jordan canonical form.

4.6.P12 Let $A = \begin{bmatrix} 1 & i \\ i & -1 \end{bmatrix}$. Explain why the Jordan canonical form of A is $J_2(0)$. (a) Use the algorithm following (4.6.13) to verify that the concanonical form of A is $J_1(2) \oplus J_1(0)$. (b) Use (4.4.4c) to arrive at the same conclusion.

4.6.P13 How does the factorization in (4.6.17b) generalize the fact that each complex number z can be written as $z = re^{i\theta}$ with r and θ real? If $A \in M_{m,n}$ can be factored as $A = RE$, in which $R \in M_{m,n}(\mathbf{R})$ is real and $E \in M_n$ is coninvolutory, explain why range $A =$ range $\bar{A}$. This necessary condition for a factorization $A = RE$ is sufficient as well; see theorem 6.4.23 in Horn and Johnson (1991).

4.6.P14 If $\mu \in \mathbf{C}$, show that $H_{2k}(\mu)$ is consimilar to $H_{2k}(\bar{\mu})$. Now use (4.6.12) to show that each $A \in M_n$ is consimilar to $\bar{A}$.

4.6.P15 Use (4.6.14) to show that each $A \in M_n$ is consimilar to $e^{i\theta}A$ for any $\theta \in \mathbf{R}$.

4.6.P16 Let $A \in M_n$ be given, suppose that λ is a positive eigenvalue of $A\bar{A}$ with geometric multiplicity $g \geq 1$, and let $\sigma = \sqrt{\lambda} > 0$. (a) If $z_1, \ldots, z_k$ are linearly independent (over $\mathbf{C}$) vectors such that $A\bar{z}_j = \sigma z_j$ for each $j = 1, \ldots, k$, explain why $k \leq g$. (b) Use (4.6.12) to show that there exist g linearly independent (over $\mathbf{C}$) vectors $z_1, \ldots, z_g$ such that $A\bar{z}_j = \sigma z_j$ for each $j = 1, \ldots, g$. (c) Explain why (b) ensures that the dimension of the coneigenspace of A associated with the coneigenvalue σ (a real vector space) is *at least g*. (d) Explain why (a) and (4.6.P3) ensure that the dimension of the coneigenspace of A associated with the coneigenvalue σ (a real vector space) is *at most g*. (e) Conclude that *the coneigenspace of A associated with the coneigenvalue σ is a g-dimensional real vector space*.

The following two problems present four algorithms to determine a basis for the coneigenspace of $A \in M_n$ associated with a given positive coneigenvalue σ, given a basis for the eigenspace of $A\bar{A}$ associated with the eigenvalue σ^2.

4.6.P17 Let $A \in M_n$ be given, suppose that λ is a positive eigenvalue of $A\bar{A}$ with geometric multiplicity $g \geq 1$, and let $\sigma = \sqrt{\lambda} > 0$. Let $x_1, \ldots, x_g$ be linearly independent eigenvectors of $A\bar{A}$ associated with the eigenvalue λ and let $X = [x_1 \ldots x_g] \in M_{n,g}$, so rank $X = g$ and $A\bar{A} = \lambda X$. The problem is to construct a matrix $Y = [y_1 \ \ldots \ y_g] \in M_{n,g}$ such that rank $Y = g$ and $A\bar{Y} = \sigma Y$, which ensure that the columns of Y are a basis for the coneigenspace of A (a g-dimensional real vector space) associated with the coneigenvalue σ. Explain why the identity $A\bar{A}X = \lambda X$ implies that rank $A\bar{X} = g$. Why is the column space of X equal to the eigenspace of $A\bar{A}$ associated with its eigenvalue λ? Explain why $A\bar{A}(A\bar{X}) = \lambda(A\bar{X})$ and why this identity implies that there is a some matrix $B \in M_g$ such that $A\bar{X} = XB$. Why is B unique? Why is B nonsingular?

Algorithm I. Determine the matrix B by solving the linear system $A\bar{X} = XB$. Why is it possible to do so? Since $\operatorname{rank}(B + e^{2i\theta}\sigma I) < g$ if and only if $-e^{2i\theta}\sigma$ is an eigenvalue of B, there are at most g values of $\theta \in [0, \pi)$ such that $B + e^{2i\theta}\sigma I$ does not have full column rank. Choose any $\theta \in [0, \pi)$ such that $\operatorname{rank}(B + e^{2i\theta}\sigma I) = g$, and let $Y = e^{-i\theta}X(B + e^{2i\theta}\sigma I)$. Verify that rank $Y = g$ and $A\bar{Y} = \sigma Y$. Notice that each column of Y (that is, each coneigenvector produced by the algorithm) is a rotation of a linear combination of all the eigenvectors $x_1, \ldots, x_g$.

Algorithm II. Use the identity $XB = A\bar{X}$ to write $e^{-i\theta}X(B + e^{2i\theta}\sigma I) = e^{-i\theta}A\bar{X} + e^{i\theta}\sigma X$, which fails to have full column rank for at most g values of $\theta \in [0, \pi)$. Choose any $\theta \in [0, \pi)$ such that $\operatorname{rank}(e^{-i\theta}A\bar{X} + e^{i\theta}\sigma X) = g$, and let $Y = e^{-i\theta}A\bar{X} + e^{i\theta}\sigma X$. Verify that $A\bar{Y} = \sigma Y$. Explain how a suitable θ can be identified by trial and error (guess and check). Notice that each column of Y has the form $y_j = e^{-i\theta}A\overline{x_j} + e^{i\theta}x_j$, $j = 1, \ldots, g$; each y_j depends only on x_j, just as in (4.6.6).

Algorithm III. Verify that $\lambda X = A\bar{A}X = A(\overline{A\bar{X}}) = A(\overline{XB}) = XB\bar{B}$, and explain why $B\bar{B} = \lambda I$, that is, $\sigma^{-1}B$ is coninvolutory. If C is a coninvolutory matrix, it is known that there is a coninvolutory matrix E such that $C = E^2$; see (6.4.22) in Horn and Johnson (1991). Let $\sigma^{-1}B = E^2$, in which E is coninvolutory, and let $Y = XE$. Verify that rank $Y = g$ and $A\bar{Y} = \sigma Y$. Notice that each column of Y is a linear combination of all the eigenvectors $x_1, \ldots, x_g$ via coefficients whose matrix comprises a sort of rotation ($E\bar{E} = I$).

What do these three algorithms produce when $g = 1$? Compare with (4.6.6).

4.6.P18 Write $A \in M_n$ as $A = A_1 + iA_2$ with $A_1, A_2 \in M_n(\mathbf{R})$ and consider its real representation $R_2(A) = \left[\begin{smallmatrix} A_1 & A_2 \\ A_2 & -A_1 \end{smallmatrix}\right] \in M_{2n}(\mathbf{R})$; see (1.3.P21). Let $x = u + iv \neq 0$, $u, v \in \mathbf{R}^n$, $w = \left[\begin{smallmatrix} u \\ v \end{smallmatrix}\right]$, $T = \left[\begin{smallmatrix} 0 & -I_n \\ I_n & 0 \end{smallmatrix}\right]$, and $z = Tw$. (a) Show that $A\bar{x} = \sigma x, \sigma \in \mathbf{R} \Leftrightarrow$ $\left[\begin{smallmatrix} \operatorname{Re} A\bar{x} \\ \operatorname{Im} A\bar{x} \end{smallmatrix}\right] = \sigma \left[\begin{smallmatrix} \operatorname{Re} x \\ \operatorname{Im} x \end{smallmatrix}\right], \sigma \in \mathbf{R} \Leftrightarrow R_2(A)w = \sigma w, \sigma \in \mathbf{R}$, which is an ordinary real eigenpair problem. (b) Show that $A\bar{x} = \sigma x, \sigma \in \mathbf{R} \Leftrightarrow A(\overline{ix}) = -\sigma(ix), \sigma \in \mathbf{R} \Leftrightarrow R_2(A)z = -\sigma z, \sigma \in \mathbf{R}$. (c) Use (1.3.P21(f)) to show that the nonzero eigenvalues of $R_2(A)$ occur in $\pm$ pairs, and the non-real eigenvalues occur in conjugate pairs. (d) Explain why A has a coneigenvalue if and only if $R_2(A)$ has a real eigenvalue if and only if $A\bar{A}$ has a real nonnegative eigenvalue. (e) Suppose that σ is a positive eigenvalue of $R_2(A)$ with geometric multiplicity $g \geq 1$; let $w_1, \ldots, w_g$ be linearly independent (over $\mathbf{R}$) eigenvectors of $R_2(A)$ associated with σ; and let $w_j = \left[\begin{smallmatrix} u_j \\ v_j \end{smallmatrix}\right]$, $z_j = Tw_j$, $y_j = u_j + iv_j$, $\alpha_j, \beta_j \in \mathbf{R}$, and $c_j = \alpha_j + i\beta_j$ for $j = 1, \ldots, g$. Explain why $z_1, \ldots, z_g$ are linearly independent (over $\mathbf{R}$) eigenvectors of $R_2(A)$ associated with the eigenvalue $-\sigma$. Show that $\sum_{j=1}^g c_j y_j = 0 \Rightarrow \sum_{j=1}^g \alpha_j u_j = \sum_{j=1}^g \beta_j v_j$ and $\sum_{j=1}^g \alpha_j v_j = -\sum_{j=1}^g \beta_j u_j \Rightarrow$ $\sum_{j=1}^g \alpha_j w_j = -\sum_{j=1}^g \beta_j z_j \Rightarrow \sum_{j=1}^g \alpha_j \sigma w_j = \sum_{j=1}^g \beta_j \sigma z_j \Rightarrow \sum_{j=1}^g \alpha_j w_j = 0$ and $\sum_{j=1}^g \beta_j z_j = 0 \Rightarrow \alpha_1 = \cdots = \alpha_g = 0$ and $\beta_1 = \cdots = \beta_g = 0 \Rightarrow c_1 = \cdots = c_g = 0$. Conclude that $y_1, \ldots, y_g$ are linearly independent (over $\mathbf{C}$) coneigenvectors of A associated with the positive coneigenvalue σ. (f) Use (1.3.P21(c)) to show that g is equal to the geometric multiplicity of σ^2 as an eigenvalue of $A\bar{A}$.

4.6.P19 Let $A, B \in M_n$. Show that the real representations $R_2(A)$ and $R_2(B)$ are similar if and only if A and B are consimilar.

4.6.P20 Let $A \in M_n$ be singular. Explain why $\mathcal{N} = \{x \in \mathbf{C}^n : A\bar{x} = 0\}$, the coneigenspace of A associated with its zero coneigenvalue is a subspace of $\mathcal{S} = \{x \in \mathbf{C}^n : A\bar{A}x = 0\}$, the null space of $A\bar{A}$. If rank $A = $ rank $A\bar{A}$, explain why $\mathcal{N} = \mathcal{S}$.

4.6.P21 Suppose that $A \in M_n$ is condiagonalizable. Provide details to justify the following algorithm for constructing a condiagonalization of A, given an ordinary diagonalization of $A\bar{A}$. Let $A\bar{A} = S\Lambda S^{-1}$, in which $\Lambda = \lambda_1 I_{n_1} \oplus \cdots \oplus \lambda_d I_{n_d}$, $\lambda_1 > \cdots > \lambda_d \geq 0$ are the distinct eigenvalues of $A\bar{A}$, and the nonsingular matrix S is partitioned conformally to Λ as $S = [S_1 \ldots S_d]$. For each $j = 1, \ldots, d$ such that $\lambda_j > 0$, let $\sigma_j = \sqrt{\lambda} > 0$; if $\lambda_d = 0$, let $\sigma_d = 1$. Let $\Sigma = \sigma_1 I_{n_1} \oplus \cdots \oplus \sigma_d I_{n_d}$. For each $j = 1, \ldots, d$ let $Y_j = e^{-i\theta_j} A\bar{S}_j + e^{i\theta_j}\sigma_j S_j$, in which θ_j is any value in the real interval $[0, \pi)$ such that rank $Y_j = n_j$ (at most n_j values are excluded); if $\lambda_d = 0$, let $\theta_d = 0$. Let $Y = [Y_1 \ldots Y_d]$. Then $A\bar{Y} = Y\Sigma$ and Y is nonsingular, so $A = Y\Sigma\bar{Y}^{-1}$ is a condiagonalization of A. Let $\Theta = e^{i\theta_1} I_{n_1} \oplus \cdots \oplus e^{i\theta_d} I_{n_d}$ and observe that $Y = A\bar{S}\Theta + S\Sigma\Theta$. What happens if each $n_j = 1$? Compare with (4.6.8).

4.6.P22 Consider the nilpotent Jordan matrix $J = J_{n_1}(0) \oplus \cdots \oplus J_{n_k}(0)$, let $q = \max\{n_1, \ldots, n_k\}$, and let $w_1, \ldots, w_q$ be the Weyr characteristic of J, so $w_1 = k$. It is known that J is the Jordan canonical form of the square of a nilpotent matrix if and only if the sequence $w_1, \ldots, w_q$ does not contain two successive occurrences of the same odd integer and, if k is odd, then $w_1 - w_2 > 0$ (that is, if the number of blocks in J is odd, then there is at least one block of size one); see corollary 6.4.13 in Horn and Johnson (1991). Which of the following Jordan matrices is the Jordan canonical form of the square of a nilpotent matrix? If so, what is that matrix? $J = J_2(0)$; $J = J_2(0) \oplus J_2(0)$; $J = J_2(0) \oplus J_2(0) \oplus J_2(0)$; $J = J_3(0) \oplus J_1(0)$; $J = J_5(0) \oplus J_2(0) \oplus J_1(0)$.

4.6.P23 Let $A \in M_n$. Let $J = B \oplus N$ be the Jordan canonical form of $A\bar{A}$, in which B is nonsingular and N is nilpotent. Deduce from (4.6.16) that: (a) B is a direct sum of blocks of only two types: $J_k(\lambda)$ with λ real and positive, and pairs of blocks of the form $J_k(\mu) \oplus J_k(\bar{\mu})$, in which μ is not real and positive; and (b) N is similar to the square of a nilpotent matrix. Can $J_1(1) \oplus J_2(0)$ be the Jordan canonical form of $A\bar{A}$ for some $A \in M_3$?

4.6.P24 Let $A \in M_n$, and let $\mathcal{A} = \begin{bmatrix} 0 & A \\ \bar{A} & 0 \end{bmatrix} \in M_{2n}$. Corollary 4.6.15 ensures that there is a nonsingular $S \in M_n$ and a real $R \in M_n(\mathbf{R})$ such that $A = SR\bar{S}^{-1}$. (a) Show that $\mathcal{A}$ is similar to $\begin{bmatrix} 0 & R \\ R & 0 \end{bmatrix}$ via the similarity matrix $S \oplus \bar{S}$. (b) Explain why the Jordan canonical form of $\mathcal{A}$ consists only of the following two types of direct summands: $J_k(\lambda) \oplus J_k(-\lambda)$ with λ real and nonnegative, and $J_k(\lambda) \oplus J_k(-\lambda) \oplus J_k(\bar{\lambda}) \oplus J_k(-\bar{\lambda})$ with λ not real.

4.6.P25 Revisit (4.5.P35). Use (4.5.17) to show that there is no nonsingular $S \in M_2$ such that S^*AS and S^TBS are both diagonal.

4.6.P26 Let $\Delta = [d_{ij}] \in M_n$ be upper triangular. Suppose that Δ and $D = d_1 I_{n_1} \oplus \cdots \oplus d_k I_{n_k}$ have the same main diagonals and that $d_1, \ldots, d_k$ are real, nonnegative, and distinct. If $\Delta\bar{\Delta}$ is normal, show that $\Delta = \Delta_1 \oplus \cdots \oplus \Delta_k$, in which each $\Delta_j \in M_{n_j}$ is upper triangular and has the same main diagonal as $d_j I_{n_j}$.

4.6.P27 There is a unitary consimilarity analog of (4.6.9): If $B \in M_m$ and $B\bar{B} = I$, then there is a unitary $U \in M_m$ such that

$$B = U \left(I_{n-2q} \oplus \begin{bmatrix} 0 & \sigma_1^{-1} \\ \sigma_1 & 0 \end{bmatrix} \oplus \cdots \oplus \begin{bmatrix} 0 & \sigma_q^{-1} \\ \sigma_q & 0 \end{bmatrix} \right) U^T \qquad (4.6.19)$$

in which $\sigma_1, \sigma_1^{-1}, \ldots, \sigma_q, \sigma_q^{-1}$ are the singular values of B that are different from 1. (a) Use this fact and (3.4.P5) to prove a unitary consimilarity analog of (4.6.11): A given $A \in M_n$ is unitarily congruent to a direct sum of blocks of the form

$$[\sigma], \begin{bmatrix} 0 & s \\ 0 & 0 \end{bmatrix}, \text{ and } \tau \begin{bmatrix} 0 & t \\ t^{-1} & 0 \end{bmatrix} = \tau t \begin{bmatrix} 0 & 1 \\ t^{-2} & 0 \end{bmatrix}, \qquad (4.6.20)$$
$$\sigma, \tau, s, t \in \mathbf{R}, \sigma \geq 0, \tau > 0, s > 0, 0 < t < 1$$

if and only if $A\bar{A}$ is positive semidefinite (that is, $A\bar{A}$ is *unitarily* diagonalizable and has real nonnegative eigenvalues). (b) Explain why two coninvolutions are unitarily congruent if and only if they have the same singular values. The factorization (4.6.19) may be thought of as a special singular value decomposition of a coninvolutory matrix that is analogous to the special singular value decomposition (2.6.6.1) for a skew-symmetric matrix. (c) Explain why the factorization in (a) is uniquely determined by A, up to permutation of its blocks. (d) Compare the canonical form (a) with the one in (4.4.P43).

4.6.P28 Let $A = \begin{bmatrix} 0 & -1 \\ 1 & 0 \end{bmatrix}$, regarded as a real, complex, or quaternion matrix. Verify the following statements: (a) A has no real eigenvectors, and hence no real eigenvalues. That is, there is no nonzero real vector x and real scalar λ such that $Ax = \lambda x$. However, it has complex eigenvectors $x_\pm = \begin{bmatrix} \pm i \\ 1 \end{bmatrix}$ and associated complex eigenvalues $\lambda_\pm = \pm i$. (b) A has no complex coneigenvectors, and hence no complex coneigenvalues. That is, there is no nonzero complex vector x and complex scalar λ such that $A\bar{x} = \lambda x$ (equivalently, $A\bar{x} = x\lambda$). However, it has quaternion coneigenvectors $x_\pm = \begin{bmatrix} \pm j \\ k \end{bmatrix}$ associated with quaternion right coneigenvalues $\lambda_\pm = \pm i$: $A\bar{x}_\pm = x_\pm \lambda_\pm$. But it has no quaternion coneigenvector associated with a quaternion left coneigenvalue: there is no nonzero quaternion vector x and quaternion scalar λ such that $A\bar{x} = \lambda x$. In your verifications, be sure to observe the reversing rule for quaternion conjugation of products: $\overline{ab} = \bar{b}\bar{a}$.

Notes and Further Readings. For more information about consimilarity and a proof of the assertion in the last sentence of (4.6.P4), see the paper of Hong and Horn cited at the end of (4.5). For proofs of (4.6.12) and the assertions in (4.6.15) about consimilarity to a Hermitian or real matrix, see Y. P. Hong and R. A. Horn, A canonical form for matrices under consimilarity, *Linear Algebra Appl.* 102 (1988) 143–168. The assertion in (4.6.16) can be proved without invoking the concanonical form (4.6.12); see Satz 20 in K. Asano and T. Nakayama, Über halblineare Transformationen, *Math. Ann.* 115 (1938) 87–114. For a proof of (4.6.17) and another proof of (4.6.16), see P. L. Hsu, On a kind of transformations of matrices, *Acta Math. Sinica* 5 (1955) 333–346. A different approach to (4.6.17) is in Theorem 30 of R. A. Horn and D. I. Merino, Contragredient equivalence: A canonical form and some applications, *Linear Algebra Appl.* 214 (1995) 43–92. For a proof of the canonical form (4.6.19), see corollary 8.4 in R. A. Horn and V. V. Sergeichuk, Canonical forms for unitary congruence and *congruence, *Linear Multilinear Algebra* 57 (2009) 777–815. For more information about the ideas in (4.6.P28), see Huang Liping, Consimilarity of quaternion matrices and complex matrices, *Linear Algebra Appl.* 331 (2001) 21–30.

CHAPTER 5
Norms for Vectors and Matrices

5.0 Introduction

Euclidean length (0.6.1) is the most familiar measure of "size" and "proximity" in $\mathbf{R}^2$ or $\mathbf{R}^3$. A real vector x is thought of as "small" if $\|x\|_2 = (x^T x)^{1/2}$ is small. Two real vectors x and y are "close" if $\|x - y\|_2$ is small.

Are there useful ways other than Euclidean length to measure the "size" of real or complex vectors? What may be said about the "size" of matrices that reflects the algebraic structure of M_n?

We address these questions by studying *norms* of vectors and matrices. Norms may be thought of as generalizations of Euclidean length, but the study of norms is more than an exercise in mathematical generalization. Norms arise naturally in the study of power series of matrices and in the analysis and assessment of algorithms for numerical computations.

Example 5.0.1 (Convergence). If x is a complex number such that $|x| < 1$, we know that

$$(1 - x)^{-1} = 1 + x + x^2 + x^3 + \cdots$$

This suggests the formula

$$(I - A)^{-1} = I + A + A^2 + A^3 + \cdots$$

for calculating the inverse of the square matrix $I - A$, but when is it valid? It turns out that it is sufficient that *any* matrix norm of A be less than 1. Many other power series that are plausible ways to define matrix-valued functions of a matrix, such as

$$e^A = \sum_{k=0}^{\infty} \frac{1}{k!} A^k$$

313

can be shown to be convergent using norms. Norms may also be useful in determining the number of terms required in a truncated power series that must calculate a particular function value to a desired degree of accuracy.

Example 5.0.2 (Accuracy). Suppose that we wish to compute A^{-1} (or e^A or some other function of A), but the entries of A are not known exactly. Perhaps they have been obtained from an experiment, by analysis of other data, or from prior calculations that have introduced roundoff errors. We may think of $A = A_0 + E$ as being composed of the "true" A_0 plus an error E, and we would like to assess the potential error in computing $A^{-1} = (A_0 + E)^{-1}$ instead of the true A_0^{-1}. Bounds for $(A_0 + E)^{-1} - A_0^{-1}$ may be as important to know as the exact value of the inverse, and norms provide a systematic way to deal with such questions.

Example 5.0.3 (Bounds). Bounds for eigenvalues and singular values often involve norms, as do bounds for changes in these quantities resulting from perturbation of the matrix.

Example 5.0.4 (Continuity). The standard definition of continuity of a real- or complex-valued function f on $\mathbf{F}^n$ ($\mathbf{F} = \mathbf{R}$ or $\mathbf{C}$) at a point x_0 in the domain $\mathcal{D}$ of f is that for each given $\varepsilon > 0$ there is a $\delta > 0$ such that $|f(x) - f(x_0)| < \varepsilon$ whenever $x \in \mathcal{D}$ and $\|x - x_0\|_2 < \delta$. A natural generalization is to consider continuity of functions defined on vector spaces endowed with norms other than the Euclidean norm.

5.1 Definitions of norms and inner products

The four axioms for a norm on a real or complex vector space are as follows:

Definition 5.1.1. *Let* V *be a vector space over the field* $\mathbf{F}$ *(*$\mathbf{F} = \mathbf{R}$ *or* $\mathbf{C}$*). A function* $\|\cdot\| : V \to \mathbf{R}$ *is a* norm *(sometimes one says* vector norm*) if, for all* $x, y \in V$ *and all* $c \in \mathbf{F}$,

(1)	$\|x\| \geq 0$	*Nonnegativity*
(1a)	$\|x\| = 0$ *if and only if* $x = 0$	*Positivity*
(2)	$\|cx\| = |c|\|x\|$	*Homogeneity*
(3)	$\|x + y\| \leq \|x\| + \|y\|$	*Triangle Inequality*

These four axioms express some of the familiar properties of Euclidean length in the plane. Euclidean length possesses additional properties that cannot be deduced from these four axioms; an example is the parallelogram identity (5.1.9). The triangle inequality expresses the *subadditivity* of a norm.

If $\|\cdot\|$ is a norm on a real or complex vector space V, the positivity and homogeneity axioms (1a) and (2) ensure that any nonzero vector x can be normalized to produce a *unit vector* $u = \|x\|^{-1}x$: $\|u\| = \|\|x\|^{-1}x\| = \|x\|^{-1}\|x\| = 1$. A real or complex vector space V, together with a given norm $\|\cdot\|$, is called a *normed linear space* (*normed vector space*).

A function $\|\cdot\| : V \to \mathbf{R}$ that satisfies axioms (1), (2), and (3) of (5.1.1) is called a *seminorm*. The seminorm of a nonzero vector can be zero.

Lemma 5.1.2. *If $\| \cdot \|$ is a vector seminorm on a real or complex vector space V, then $|\,\|x\| - \|y\|\,| \le \|x - y\|$ for all $x, y \in V$.*

Proof. Since $y = x + (y - x)$, the inequality

$$\|y\| \le \|x\| + \|y - x\| = \|x\| + \|x - y\|$$

follows from the triangle inequality (3) and the homogeneity axiom (2). It follows that

$$\|y\| - \|x\| \le \|x - y\|$$

But $x = y + (x - y)$ as well, so invoking the triangle inequality (3) again ensures that $\|x\| \le \|y\| + \|x - y\|$, and hence

$$\|x\| - \|y\| \le \|x - y\|$$

Thus, we have shown that $\pm(\|x\| - \|y\|) \le \|x - y\|$, which is equivalent to the assertion of the lemma. $\Box$

Associated with Euclidean length on $\mathbf{R}^n$ or $\mathbf{C}^n$ is the usual Euclidean inner product y^*x of vectors y and x (0.6.1), which has something to do with the "angle" between two vectors: x and y are orthogonal if $y^*x = 0$. We formulate the axioms for an inner product by selecting the most basic properties of the Euclidean inner product.

Definition 5.1.3. *Let V be a vector space over the field $\mathbf{F}$ ($\mathbf{F} = \mathbf{R}$ or $\mathbf{C}$). A function $\langle \cdot, \cdot \rangle : V \times V \to \mathbf{F}$ is an inner product if for all $x, y, z \in V$ and all $c \in \mathbf{F}$,*

(1)	$\langle x, x \rangle \ge 0$	*Nonnegativity*
(1a)	$\langle x, x \rangle = 0$ if and only if $x = 0$	*Positivity*
(2)	$\langle x + y, z \rangle = \langle x, z \rangle + \langle y, z \rangle$	*Additivity*
(3)	$\langle cx, y \rangle = c \langle x, y \rangle$	*Homogeneity*
(4)	$\langle x, y \rangle = \overline{\langle y, x \rangle}$	*Hermitian Property*

The axioms (2), (3), and (4) say that $\langle \cdot, \cdot \rangle$ is a sesquilinear function; the axioms (1a) and (1) require that $\langle x, x \rangle > 0$ if $x \ne 0$.

Exercise. Show that the Euclidean inner product $\langle x, y \rangle = y^*x$ on $\mathbf{C}^n$ satisfies the five axioms for an inner product.

Exercise. Let $D = \mathrm{diag}(d_1, \ldots, d_n) \in M_n(\mathbf{F})$ and consider the function $(\cdot, \cdot) : V \times V \to \mathbf{F}$ defined by $(x, y) = y^*Dx$. Which of the five axioms for an inner product does $(\cdot, \cdot)$ satisfy? Under what conditions on D is $(\cdot, \cdot)$ an inner product?

Exercise. Let $a, b, c, d \in \mathbf{F}$ and let $x, y, w, z \in \mathbf{F}^n$. Deduce the following properties of an inner product from the five axioms in (5.1.3):

(a) $\langle x, cy \rangle = \bar{c} \langle x, y \rangle$
(b) $\langle x, y + z \rangle = \langle x, y \rangle + \langle x, z \rangle$
(c) $\langle ax + by, cw + dz \rangle = a\bar{c} \langle x, w \rangle + b\bar{c} \langle y, w \rangle + a\bar{d} \langle x, z \rangle + b\bar{d} \langle y, z \rangle$
(d) $\langle x, \langle x, y \rangle y \rangle = |\langle x, y \rangle|^2$
(e) $\langle x, y \rangle = 0$ for all $y \in V$ if and only if $x = 0$

Properties (a)–(d) are shared by all sesquilinear functions; only property (e) relies on axioms (1) and (1a).

The Cauchy–Schwarz inequality is an important property of all inner products.

Theorem 5.1.4 (Cauchy–Schwarz inequality). *Let* $\langle \cdot, \cdot \rangle$ *be an inner product on a vector space* V *over the field* $\mathbf{F}$ ($\mathbf{F} = \mathbf{R}$ *or* $\mathbf{C}$). *Then*

$$|\langle x, y \rangle|^2 \leq \langle x, x \rangle \langle y, y \rangle \quad \text{for all} \quad x, y \in V \qquad (5.1.5)$$

with equality if and only if x and y are linearly dependent, that is, if and only if $x = \alpha y$ or $y = \alpha x$ for some $\alpha \in \mathbf{F}$.

Proof. Let $x, y \in V$ be given. If $x = y = 0$, there is nothing to prove, so we may assume that $y \neq 0$. Let $v = \langle y, y \rangle x - \langle x, y \rangle y$ and compute

$$\begin{aligned}
0 \leq \langle v, v \rangle &= \langle \langle y, y \rangle x - \langle x, y \rangle y, \langle y, y \rangle x - \langle x, y \rangle y \rangle \\
&= \langle y, y \rangle^2 \langle x, x \rangle - \langle y, y \rangle \overline{\langle x, y \rangle} \langle x, y \rangle - \langle x, y \rangle \langle y, x \rangle \langle y, y \rangle + \langle y, y \rangle \overline{\langle x, y \rangle} \langle x, y \rangle \\
&= \langle y, y \rangle^2 \langle x, x \rangle - \langle y, y \rangle |\langle x, y \rangle|^2 \\
&= \langle y, y \rangle \left(\langle x, x \rangle \langle y, y \rangle - |\langle x, y \rangle|^2 \right) \qquad (5.1.6)
\end{aligned}$$

Since $\langle y, y \rangle > 0$, we conclude that $\langle x, x \rangle \langle y, y \rangle \geq |\langle x, y \rangle|^2$, with equality if and only if $\langle v, v \rangle = 0$ if and only if $v = \langle y, y \rangle x - \langle x, y \rangle y = 0$, which is a nontrivial linear combination of x and y. This confirms the inequality (5.1.5), with equality if and only if x and y are linearly dependent. $\square$

Corollary 5.1.7. *If $\langle \cdot, \cdot \rangle$ is an inner product on a real or complex vector space V, then the function $\|\cdot\| : V \to [0, \infty)$ defined by $\|x\| = \langle x, x \rangle^{1/2}$ is a norm on V.*

> *Exercise.* Prove (5.1.7). *Hint*: To verify the triangle inequality, compute $\|x + y\|^2 = \langle x + y, x + y \rangle$ and use the Cauchy–Schwarz inequality.

If $\langle \cdot, \cdot \rangle$ is an inner product on a real or complex vector space V, the function $\|x\| = \langle x, x \rangle^{1/2}$ on V is said to be *derived from an inner product* (namely, from $\langle \cdot, \cdot \rangle$); (5.1.7) ensures that $\|\cdot\|$ is a norm on V. A real or complex vector space V, together with a given inner product $\langle \cdot, \cdot \rangle$, is called an *inner product space*; endowed with its derived norm, any inner product space is also a normed linear space.

A function $\langle \cdot, \cdot \rangle : V \times V \to \mathbf{F}$ that satisfies the inner product axioms (1), (2), (3), and (4) in (5.1.3), but not necessarily axiom (1a), is called a *semi-inner product*; it is a sesquilinear function such that $\langle x, x \rangle \geq 0$ for all $x \in V$. An important fact about semi-inner products is that they, like inner products, satisfy the Cauchy–Schwarz inequality.

Theorem 5.1.8. *Let $\langle \cdot, \cdot \rangle$ be a semi-inner product on a vector space V over the field $\mathbf{F}$ ($\mathbf{F} = \mathbf{R}$ or $\mathbf{C}$). Then $|\langle x, y \rangle|^2 \leq \langle x, x \rangle \langle y, y \rangle$ for all $x, y \in V$ and the function $\|\cdot\| : V \to [0, \infty)$ defined by $\|x\| = \langle x, x \rangle^{1/2}$ is a seminorm on V.*

Proof. Let $x, y \in V$ be given. Consider the polynomial $p(t) = \langle tx - e^{i\theta} y, tx - e^{i\theta} y \rangle$. Then $p(t) = t^2 \|x\|^2 - 2t \operatorname{Re}(e^{-i\theta} \langle x, y \rangle) + \|y\|^2 \geq 0$ for all $t, \theta \in \mathbf{R}$. Choose any θ such that $\operatorname{Re}(e^{-i\theta} \langle x, y \rangle) = |\langle x, y \rangle|$. If $\|x\| = 0$ and $\langle x, y \rangle \neq 0$, then $p(t) = 2t|\langle x, y \rangle| + \|y\|^2$ would be negative for sufficiently large negative values of t. We conclude that $\langle x, y \rangle = 0$ if $\|x\| = 0$, so the inequality $|\langle x, y \rangle|^2 \leq \|x\|^2 \|y\|^2$ is valid in this case. Now suppose that $\|x\| \neq 0$ and define $t_0 = |\langle x, y \rangle| / \|x\|^2$. Then $p(t_0) = -|\langle x, y \rangle|^2 / \|x\|^2 + \|y\|^2 \geq 0$, so $|\langle x, y \rangle|^2 \leq \|x\|^2 \|y\|^2$. The assertion that $\|\cdot\|$ is a seminorm follows as in the proof of (5.1.7), since the Cauchy–Schwarz inequality for $\langle \cdot, \cdot \rangle$ implies the triangle inequality for $\|\cdot\|$. $\qquad\square$

Exercise. Let $A = \mathrm{diag}(1, 0) \in M_2$. Show that $\langle x, y \rangle = y^* A x$ defines a semi-inner product on $\mathbf{C}^2$. Consider the independent vectors $x = [1\,0]^T$ and $y = [1\,1]^T$. Show that $|\langle x, y \rangle|^2 = \langle x, x \rangle \langle y, y \rangle \neq 0$, so the very useful characterization of the case of equality in the Cauchy–Schwarz inequality for inner products is lost when we generalize to semi-inner products.

Problems

In each of the following problems, V is a given vector space over $\mathbf{F} = \mathbf{R}$ or $\mathbf{C}$.

5.1.P1 Let e_i denote the ith standard basis vector in $\mathbf{F}^n$ and suppose that $\|\cdot\|$ is a seminorm on $\mathbf{F}^n$. Show that $\|x\| \leq \sum_{i=1}^{n} |x_i| \, \|e_i\|$.

5.1.P2 If $\|\cdot\|$ is a seminorm on V, show that $V_0 = \{v \in V : \|v\| = 0\}$ is a subspace of V (called the *null space of* $\|\cdot\|$). (a) If S is any subspace of V such that $V_0 \cap S = \{0\}$, show that $\|\cdot\|$ is a norm on S. (b) Consider the relation $x \sim y$ defined by $x \sim y$ if and only if $\|x - y\| = 0$. Show that: $\sim$ is an equivalence relation on V; the equivalence classes of this equivalence relation are of the form $\hat{x} = \{x + y \in V : y \in V_0\}$; the set of these equivalence classes forms a vector space in a natural way. Show that the function $\|\hat{x}\| = \{\|x\| : x \in \hat{x}\}$ is well-defined and is a norm on the vector space of equivalence classes. (c) Explain why there is a natural norm associated with every vector seminorm. (d) Is the zero function ($f(x) = 0$ for all x) a seminorm? (e) Let $n \geq 1$ and let $z \in \mathbf{C}^n$ be a given nonzero vector. Explain why the function $\|x\| = |z^*x|$ is a seminorm on $\mathbf{C}^n$ that is not a norm. What is the null space of $\|\cdot\|$? Describe the equivalence relation $\sim$ geometrically.

5.1.P3 Let x and y be given nonzero vectors in V. Define the *angle θ between the subspaces* span$\{x\}$ *and* span$\{y\}$ by

$$\cos \theta = \frac{|\langle x, y \rangle|}{\langle x, x \rangle^{1/2} \langle y, y \rangle^{1/2}}, \quad 0 \leq \theta \leq \frac{\pi}{2}$$

Why is θ is well-defined, that is, why is the stated fraction between 0 and 1? The terminology is justified by observing that θ is unchanged if x and y are replaced, respectively, by cx and dy, for any nonzero $c, d \in \mathbf{C}$; explain why.

5.1.P4 Let $\|\cdot\|$ be a norm on V that is derived from an inner product. (a) Show that it satisfies the *parallelogram identity*

$$\frac{1}{2}(\|x + y\|^2 + \|x - y\|^2) = \|x\|^2 + \|y\|^2 \tag{5.1.9}$$

for all $x, y \in V$. Why is this identity so named? Validity of the parallelogram identity is necessary and sufficient for a given norm to be derived from an inner product; see (5.1.P12). (b) For any $m \in \{2, 3, \ldots\}$ and any given vectors $x_1, \ldots, x_m \in V$, show that $\sum_{1 \leq i < j \leq m} \|x_i - x_j\|^2 + \|\sum_{i=1}^{m} x_i\|^2 = m \sum_{i=1}^{m} \|x_i\|^2$ and explain why this identity reduces to (5.1.9) for $m = 2$.

5.1.P5 Consider the function $\|x\|_\infty = \max_{1 \le i \le n} |x_i|$ on $\mathbf{C}^n$. Show that $\|\cdot\|_\infty$ is a norm that is not derived from an inner product. You might consider $x = e_1$ and $y = e_2$.

5.1.P6 If $\|\cdot\|$ is a norm on V that is derived from an inner product, show that

$$\operatorname{Re}\langle x, y\rangle = \frac{1}{4}(\|x + y\|^2 - \|x - y\|^2) \tag{5.1.10}$$

for all $x, y \in V$. This is known as the *polarization identity*. Show also that

$$\operatorname{Re}\langle x, y\rangle = \frac{1}{2}(\|x + y\|^2 - \|x\|^2 - \|y\|^2)$$

5.1.P7 Show that the function $\|x\|_1 = |x_1| + \cdots + |x_n|$ is a norm on $\mathbf{C}^n$ that does not satisfy the parallelogram identity (5.1.9). It is not, therefore, derived from any inner product.

5.1.P8 If $\|\cdot\|$ is a norm on V that is derived from an inner product $\langle \cdot, \cdot\rangle$, show that

$$\|x + y\|\,\|x - y\| \le \|x\|^2 + \|y\|^2 \tag{5.1.11}$$

for all $x, y \in V$, with equality if and only if $\operatorname{Re}\langle x, y\rangle = 0$. What is the geometrical meaning of this inequality in $\mathbf{R}^2$ with the Euclidean inner product? Using the norm defined in the preceding problem, show that (5.1.11) is not satisfied with $x = e_1$ and $y = e_2$.

5.1.P9 Let $\|\cdot\|$ be a norm on V that is derived from an inner product, let $x, y \in V$, and suppose that $y \ne 0$. Show that: (a) the scalar α_0 that minimizes the value of $\|x - \alpha y\|$ is $\alpha_0 = \langle x, y\rangle / \|y\|^2$, (b) $\|x - \alpha_0 y\|^2 = \|x\|^2 - |\langle x, y\rangle|^2 / \|y\|^2$, and (c) $x - \alpha_0 y$ is orthogonal to y.

5.1.P10 Show that the nonnegativity axiom (1) in (5.1.1) is implied by axioms (2) and (3).

5.1.P11 Let $\|\cdot\|$ be a norm on V that is derived from an inner product $\langle \cdot, \cdot\rangle$, and let $x, y \in V$. Show that $\|x + y\|^2 = \|x\|^2 + \|y\|^2$ if and only if $\operatorname{Re}\langle x, y\rangle = 0$. What does this say if $V = \mathbf{R}^2$?

5.1.P12 Provide details for the following sketch of a proof that the parallelogram identity (5.1.9) is a sufficient condition for a given norm on a real or complex vector space to be derived from an inner product. First consider the case of a vector space V over $\mathbf{R}$ with a given norm $\|\cdot\|$. (a) Define

$$\langle x, y\rangle = \frac{1}{2}(\|x + y\|^2 - \|x\|^2 - \|y\|^2) \tag{5.1.12}$$

Show that $\langle \cdot, \cdot\rangle$ defined in this way satisfies axioms (1), (1a), and (4) in (5.1.3) and that $\langle x, x\rangle = \|x\|^2$. (b) Use (5.1.9) to show that

$$4\langle x, y\rangle + 4\langle z, y\rangle = 2\|x + y\|^2 + 2\|z + y\|^2 - 2\|x\|^2 - 2\|z\|^2 - 4\|y\|^2$$
$$= \|x + 2y + z\|^2 - \|x + z\|^2 - 4\|y\|^2 = 4\langle x + z, y\rangle$$

and conclude that the additivity axiom (2) in (5.1.3) is satisfied. (c) Use the additivity axiom to show that $\langle nx, y\rangle = n\langle x, y\rangle$ and $m\langle m^{-1}nx, y\rangle = \langle nx, y\rangle = n\langle x, y\rangle$ whenever m and n are nonnegative integers. Use (5.1.9) and (5.1.12) to show that $\langle -x, y\rangle = -\langle x, y\rangle$ and conclude that $\langle ax, y\rangle = a\langle x, y\rangle$ whenever $a \in \mathbf{R}$ is rational. (d) Let $p(t) = t^2\|x\|^2 + 2t\langle x, y\rangle + \|y\|^2$, $t \in \mathbf{R}$, and show that $p(t) = \|tx + y\|^2$ if t is rational. Conclude from the continuity of $p(t)$ that $p(t) \ge 0$ for all $t \in \mathbf{R}$. Deduce the Cauchy–Schwarz inequality $|\langle x, y\rangle|^2 \le \|x\|^2\|y\|^2$ from the fact that the discriminant of $p(t)$ must be nonpositive.

(e) Now let $a \in \mathbf{R}$ and show that

$$|\langle ax, y \rangle - a\langle x, y \rangle| = |\langle (a-b)x, y \rangle + (b-a)\langle x, y \rangle|$$
$$\leq |\langle (a-b)x, y \rangle| + |(b-a)\langle x, y \rangle| \leq 2|a-b| \, \|x\| \, \|y\|$$

for any rational b; observe that the upper bound can be made arbitrarily small. Conclude that the homogeneity axiom (3) in (5.1.3) is satisfied. This shows that $\langle \cdot, \cdot \rangle$ is an inner product on V.

The triangle inequality for the function $\| \cdot \|$ on V, that is, axiom (3) in (5.1.1), was not used in the preceding argument. Thus, the axioms (1), (1a), and (2) in (5.1.1) for a function $\| \cdot \|$ on V, together with (5.1.9), imply that it is derived from an inner product, is therefore a norm, and hence must satisfy the triangle inequality. (f) Now suppose that V is a complex vector space. Define

$$\langle x, y \rangle = \frac{1}{2}(\|x+y\|^2 - \|x\|^2 - \|y\|^2) + \frac{i}{2}(\|x+iy\|^2 - \|x\|^2 - \|y\|^2)$$

Why is $\mathrm{Re}\langle x, y \rangle$ an inner product on V considered as a vector space over $\mathbf{R}$? Use this fact and (5.1.9) to show that $\langle \cdot, \cdot \rangle$ is an inner product on V as a vector space over $\mathbf{C}$.

5.1.P13 Let $\| \cdot \|$ be a norm on V that is derived from an inner product. Provide details for the following sketch of a proof of *Hlawka's inequality*:

$$\|x+y\| + \|x+z\| + \|y+z\| \leq \|x+y+z\| + \|x\| + \|y\| + \|z\| \quad (5.1.13)$$

for all $x, y, z \in V$. (a) Let s denote the left-hand side of (5.1.13) and let h denote its right-hand side. To show that $s \leq h$ it suffices to show that $h^2 - hs \geq 0$. (b) Compute $h^2 - hs =$

$$\|x+y+z\|^2 + \|x\|^2 + \|y\|^2 + \|z\|^2 - \|x+y\|^2 - \|x+z\|^2 - \|y+z\|^2$$
$$+ (\|x\| + \|y\| - \|x+y\|)(\|z\| - \|x+y\| + \|x+y+z\|)$$
$$+ (\|y\| + \|z\| - \|y+z\|)(\|x\| - \|y+z\| + \|x+y+z\|)$$
$$+ (\|z\| + \|x\| - \|z+x\|)(\|y\| - \|z+x\| + \|x+y+z\|)$$

and show that the first line is zero. (c) Is (5.1.13) correct for the norm $\| \cdot \|_\infty$ on $\mathbf{R}^3$, $x = [1\ 1\ -1]^T$, $y = [1\ -1\ 1]^T$, and $z = [-1\ 1\ 1]^T$?

5.1.P14 Let $x_1, \ldots, x_n$ be n given real numbers with *mean* $\mu = n^{-1} \sum_{i=1}^n x_i$ and *variance* $\sigma = (n^{-1} \sum_{i=1}^n (x_i - \mu)^2)^{1/2}$. Use the Cauchy–Schwarz inequality to show that $(x_j - \mu)^2 \leq (n-1)\sigma^2$ for any $j \in \{1, \ldots, n\}$, with equality for some j if and only if $x_p = x_q$ for all $p, q \neq j$. The resulting sharp bounds

$$\mu - \sigma\sqrt{n-1} \leq x_j \leq \mu + \sigma\sqrt{n-1} \quad (5.1.14)$$

for the elements of a list of real numbers are associated with the names of E. N. Laguerre (1880) and P. N. Samuelson (1968).

5.1.P15 Let $\| \cdot \|$ be a norm on V that is derived from an inner product. Let m be a positive integer, let $x_1, \ldots, x_m, z \in V$, and let $y = m^{-1}(x_1 + \cdots + x_m)$. Show that $\|z - y\|^2 = m^{-1} \sum_{i=1}^m (\|z - x_i\|^2 - \|y - x_i\|^2)$.

Further Readings. The first proof that the parallelogram identity is both necessary and sufficient for a given norm to be derived from an inner product seems to be due to P. Jordan and J. von Neumann, On inner products in linear, metric spaces, *Ann. of Math. (2)* 36 (1935), 719–723. The proof outline in (5.1.P12) follows D. Fearnley-Sander and J. S. V. Symons, Apollonius and inner products, *Amer. Math. Monthly* 81 (1974), 990–993.

5.2 Examples of norms and inner products

The Euclidean norm (l_2-norm) of a vector $x = [x_1 \ \ldots \ x_n]^T \in \mathbf{C}^n$,

$$\|x\|_2 = (|x_1|^2 + \cdots + |x_n|^2)^{1/2} \tag{5.2.1}$$

is perhaps the most familiar norm, since $\|x - y\|_2$ measures the standard Euclidean distance between two points $x, y \in \mathbf{C}^n$. It is derived from the Euclidean inner product (that is, $\|x\|_2 = \langle x, x \rangle^{1/2} = (x^*x)^{1/2}$) and it is *unitarily invariant*: $\|Ux\|_2 = \|x\|_2$ for all $x \in \mathbf{C}^n$ and every unitary $U \in M_n$ (2.1.4). In fact, positive scalar multiples of the Euclidean norm are the *only* unitarily invariant norms on $\mathbf{C}^n$; see (5.2.P6).

The *sum norm* (l_1-norm) on $\mathbf{C}^n$ is

$$\|x\|_1 = |x_1| + \cdots + |x_n| \tag{5.2.2}$$

This norm is also called the *Manhattan norm* (or *taxicab norm*) because it models the distance traveled by a taxi on a network of perpendicular streets and avenues.

Exercise. Verify that $\|\cdot\|_1$ is a norm on $\mathbf{C}^n$. Problem 5.1.P7 shows that $\|\cdot\|_1$ does not satisfy the parallelogram identity, so it is not derived from an inner product.

The *max norm* (l_∞-norm) on $\mathbf{C}^n$ is

$$\|x\|_\infty = \max\{|x_1|, \ldots, |x_n|\} \tag{5.2.3}$$

Problem 5.2.P5 shows that $\|\cdot\|_\infty$ is not derived from an inner product.
The l_p-*norm* on $\mathbf{C}^n$ is

$$\|x\|_p = (|x_1|^p + \cdots + |x_n|^p)^{1/p}, \quad p \geq 1 \tag{5.2.4}$$

Exercise. Verify that $\| \cdot \|_p$ is a norm on $\mathbf{C}^n$ for $p \geq 1$. *Hint*: The triangle inequality for the l_p-norms is *Minkowski's sum inequality*; see (B9).

An important discrete family of norms on $\mathbf{C}^n$ bridges the gap between the sum norm and the max norm. For each $k = 1, \ldots, n$ the k-*norm* of a vector x is obtained by nonincreasingly ordering the absolute values of the entries of x and adding the k largest values, that is,

$$\|x\|_{[k]} = |x_{i_1}| + \cdots + |x_{i_k}|, \text{ in which } |x_{i_1}| \geq \cdots \geq |x_{i_n}| \tag{5.2.5}$$

The k-norms play an important role in the theory of unitarily invariant matrix norms; see (7.4.7).

Exercise. Verify that $\| \cdot \|_{[k]}$ is a norm on $\mathbf{C}^n$ for each $k = 1, 2, \ldots$, and that $\|\cdot\|_\infty = \| \cdot \|_{[1]} \leq \| \cdot \|_{[2]} \leq \cdots \leq \| \cdot \|_{[n]} = \|\cdot\|_1$.

Any norm on $\mathbf{C}^n$ can be used to define a norm on an n-dimensional real or complex vector space V via a basis. If $\mathcal{B} = \{b^{(1)}, \ldots, b^{(n)}\}$ is a basis for V and if we express $x = \sum_{i=1}^{n} x_i b^{(i)}$ as a (unique) linear combination of basis vectors, then the mapping $x \to [x]_{\mathcal{B}} = [x_1 \ldots x_n]^T \in \mathbf{C}^n$ is an isomorphism of V onto $\mathbf{C}^n$. If $\|\cdot\|$ is any given norm on $\mathbf{C}^n$, then $\|x\|_{\mathcal{B}} = \|[x]_{\mathcal{B}}\|$ is a norm on V.

Exercise. Verify the preceding assertion.

Exercise. Verify that the l_p-norms and the k-norms are absolute norms that are *permutation invariant*, that is, the norms of x and Px are equal for all $x \in \mathbf{C}^n$ and every permutation matrix $P \in M_n$. Which of these norms are unitarily invariant?

Let $S \in M_{m,n}$ have full column rank, so $m \geq n$. Let $\|\cdot\|$ be a given norm on $\mathbf{C}^m$ and define

$$\|x\|_S = \|Sx\| \qquad (5.2.6)$$

for $x \in \mathbf{C}^n$. Then $\|\cdot\|_S$ is a norm on $\mathbf{C}^n$.

Exercise. Verify the preceding assertion. What happens if S does not have full column rank?

Exercise. For what nonsingular $S \in M_2$ and norm $\|\cdot\|$ on $\mathbf{C}^2$ is the function $(|2x_1 - 3x_2|^2 + |x_2|^2)^{1/2}$ a norm of the form $\|Sx\|$?

Consider the complex vector space $V = M_{m,n}$ with the *Frobenius inner product*:

$$\langle A, B \rangle_F = \operatorname{tr} B^* A \qquad (5.2.7)$$

The norm derived from the Frobenius inner product is the l_2-norm (*Frobenius norm*) on $M_{m,n}$: $\|A\|_2 = (\operatorname{tr} A^* A)^{1/2}$; it played a role in the proof of (2.5.2).

Exercise. Verify that the Frobenius inner product on $M_{m,n}$ satisfies the axioms for an inner product.

Exercise. What does the Frobenius inner product on $M_{m,1}$ look like?

The definitions of a norm and inner product do not require that the underlying vector space be finite dimensional. Here are four examples of norms on the vector space $C[a, b]$ of all continuous real- or complex-valued functions on the real interval $[a, b]$:

$$\|f\|_2 = \left[\int_a^b |f(t)|^2 dt \right]^{1/2} \qquad \qquad L_2\text{-norm}$$

$$\|f\|_1 = \int_a^b |f(t)| dt \qquad \qquad L_1\text{-norm}$$

$$\|f\|_p = \left[\int_a^b |f(t)|^p dt \right]^{1/p}, \ p \geq 1 \qquad \qquad L_p\text{-norm}$$

$$\|f\|_\infty = \max\{|f(x)| : x \in [a, b]\} \qquad \qquad L_\infty\text{-norm}$$

Exercise. Verify that

$$\langle f, g \rangle = \int_a^b f(t)\overline{g(t)} dt \qquad (5.2.8)$$

is an inner product on $C[a, b]$ and that the L_2-norm is derived from it.

Problems

5.2.P1 If $0 < p < 1$, then $\|x\|_p = (|x_1|^p + \cdots + |x_n|^p)^{1/p}$ defines a function on $\mathbf{C}^n$ that satisfies all but one of the axioms for a norm. Which one fails? Give an example.

5.2.P2 Show that $\|x\|_\infty = \lim_{p\to\infty} \|x\|_p$ for each $x \in \mathbf{C}^n$.

5.2.P3 Show that any seminorm on $\mathbf{C}^n$ is of the form $\| \cdot \|_S$ for some norm $\| \cdot \|$ and some $S \in M_n$.

5.2.P4 Let $w_1, \ldots, w_n$ be given positive real numbers and let $p \geq 1$. For what $S \in M_n$ is the *weighted l_p-norm* $\|x\| = (w_1|x_1|^p + \cdots + w_n|x_n|^p)^{1/p}$ a norm of the form $\|Sx\|_p$?

5.2.P5 Let $x_0 \in [a, b] \in \mathbf{R}$ be a given point. Show that the function $\|f\|_{x_0} = |f(x_0)|$ is a seminorm on $C[a, b]$ that is not a norm if $a < b$. What is its null space? What is an analogous seminorm on $\mathbf{C}^n$?

5.2.P6 If $\| \cdot \|$ is a unitarily invariant norm on $\mathbf{C}^n$, show that $\|x\| = \|x\|_2 \|e_1\|$ for every $x \in \mathbf{C}^n$. Explain why the Euclidean norm is the only unitarily invariant norm on $\mathbf{C}^n$ for which $\|e_1\| = 1$.

5.2.P7 Suppose that $\| \cdot \|$ is a norm on a real or complex vector space V. (a) Show that, for all nonzero $x, y \in V$

$$\left\| \frac{x}{\|x\|} - \frac{y}{\|y\|} \right\| \leq \frac{c\|x - y\|}{\|x\| + \|y\|} \tag{5.2.9}$$

in which $c = 4$. (b) Consider the sum norm $\|x\|_1$ on $\mathbf{R}^2$ and the vectors $x = [1\ \varepsilon]^T$ and $y = [1\ 0]^T$, in which $\varepsilon > 0$. Show that the inequality (5.2.9) in this case is $2\varepsilon(1 + \varepsilon)^{-1} \leq c\varepsilon(2 + \varepsilon)^{-1}$ and explain why (5.2.9) is correct for every norm on every real or complex vector space if and only if $c \geq 4$. (c) If the norm $\| \cdot \|$ is derived from an inner product, show that the assertion in (a) is correct with $c = 2$.

5.2.P8 Let V be a real or complex inner product space and let $u \in V$ be a unit vector (with respect to the derived norm). For any $x \in V$, define $x_{\perp u} = x - \langle x, u\rangle u$. Show that: (a) $x_{\perp u}$ is orthogonal to u and $\|x_{\perp u}\|^2 = \|x\|^2 - |\langle x, u\rangle|^2 \leq \|x\|^2$; (b) $\|x_{\perp u}\| = \|(x - \lambda u)_{\perp u}\|$ for any scalar λ; and (c) $\langle x, y\rangle - \langle x, u\rangle\langle u, y\rangle = \langle x_{\perp u}, y_{\perp u}\rangle$. Conclude that

$$|\langle x, y\rangle - \langle x, u\rangle\langle u, y\rangle| \leq \|x - \lambda u\|\|y - \mu u\| \tag{5.2.10}$$

for any $x, y, u \in V$ such that u is a unit vector, and any scalars λ, μ. Explain why the optimal choice of λ and μ in (5.2.10) is $\lambda = \langle x, u\rangle$, $\mu = \langle y, u\rangle$; these may not be the most illuminating or convenient choices in particular cases.

5.2.P9 Suppose that $-\infty < a < b < \infty$ and let V be the real inner product space of continuous real-valued functions on $[a, b]$ with the inner product (5.2.8). For given $f, g \in V$, suppose that $-\infty < \alpha \leq f(t) \leq \beta < \infty$ and $-\infty < \gamma \leq g(t) \leq \delta < \infty$ for all $t \in [a, b]$. Deduce the *Grüss inequality*

$$\left| \frac{1}{b - a} \int_a^b f(t)g(t)dt - \frac{1}{(b - a)^2} \int_a^b f(t)dt \int_a^b g(t)dt \right| \leq \frac{(\beta - \alpha)(\delta - \gamma)}{4} \tag{5.2.11}$$

from the inequality (5.2.10).

5.2.P10 Let $\lambda_1, \ldots, \lambda_n$ be the eigenvalues of $A \in M_n$. Explain why Schur's inequality (2.3.2a) can be written as

$$\sum_{i=1}^{n} |\lambda_i|^2 \leq \|A\|_2^2 \tag{5.2.12}$$

Explain why the better inequality (2.6.9) can be written as

$$\sum_{i=1}^{n} |\lambda_i|^2 \leq \sqrt{\|A\|_2^4 - \|AA^* - A^*A\|_2^2} \tag{5.2.13}$$

and the even better inequality (2.6.10) can be written as

$$\sum_{i=1}^{n} |\lambda_i|^2 \leq \sqrt{\left(\|A\|_2^2 - \frac{1}{n}|\langle A, I\rangle_F|^2\right)^2 - \|AA^* - A^*A\|_2^2} + \frac{1}{n}|\langle A, I\rangle_F|^2 \tag{5.2.14}$$

5.2.P11 Suppose that $\|\cdot\|$ is a norm on a real or complex vector space V, and let x and y be given nonzero vectors in V. Prove that

$$\|x + y\| \leq \|x\| + \|y\| - \left(2 - \left\|\frac{x}{\|x\|} + \frac{y}{\|y\|}\right\|\right) \min\{\|x\|, \|y\|\} \tag{5.2.15}$$

and

$$\|x + y\| \geq \|x\| + \|y\| - \left(2 - \left\|\frac{x}{\|x\|} + \frac{y}{\|y\|}\right\|\right) \max\{\|x\|, \|y\|\} \tag{5.2.16}$$

with equality if either $\|x\| = \|y\|$ or $x = cy$ with c real and positive.

5.2.P12 Using the notation of the preceding problem, deduce from (5.2.15) and (5.2.16) that

$$\frac{\|x - y\| - |\|x\| - \|y\||}{\min\{\|x\|, \|y\|\}} \leq \left\|\frac{x}{\|x\|} - \frac{y}{\|y\|}\right\|$$
$$\leq \frac{\|x - y\| + |\|x\| - \|y\||}{\max\{\|x\|, \|y\|\}} \tag{5.2.17}$$

5.2.P13 Using the notation of the preceding problem, show that the upper bound in (5.2.17) is less than or equal to the upper bound in (5.2.9) with the optimal value $c = 4$:

$$\left\|\frac{x}{\|x\|} - \frac{y}{\|y\|}\right\| \leq \frac{\|x - y\| + |\|x\| - \|y\||}{\max\{\|x\|, \|y\|\}}$$
$$\leq \frac{2\|x - y\|}{\max\{\|x\|, \|y\|\}} \leq \frac{4\|x - y\|}{\|x\| + \|y\|} \tag{5.2.18}$$

5.2.P14 Let $A \in M_n$ and write $A = H + iK$, in which H and K are Hermitian; see (0.2.5). What is the best Hermitian approximation to A in the Frobenius norm, that is, for what Hermitian X_0 is $\|A - X_0\|_2^2 \leq \|A - X\|_2^2$ for every Hermitian $X \in M_n$? What is the best positive semidefinite approximation? (a) Show that $\|A\|_2^2 = \|H\|_2^2 + \|K\|_2^2$. (b) If $X \in M_n$ is Hermitian, show that $\|A - X\|_2^2 = \|H - X\|_2^2 + \|K\|_2^2 \geq \|K\|_2^2$ with equality for $X = X_0 = H$. (c) If $H = U\Lambda U^*$, in which U is unitary and $\Lambda = \text{diag}(\lambda_1, \ldots, \lambda_n)$, X is positive semidefinite, and $U^*XU = Y = [y_{ij}]$, show that $\|H - X\|_2^2 = \|\Lambda - Y\|_2^2 = \sum_{i=1}^{n}(\lambda_i - y_{ii})^2 + \sum_{i \neq j}|y_{ij}|^2$. Why is this minimized for $X = X_0 = H_+$, the positive semidefinite part of H; see (4.1.12).

Further Readings. For a detailed discussion of Minkowski's inequality and other classical inequalities, see Beckenbach and Bellman (1965). For a proof that (a) equality occurs in (5.2.1) with $c = 4$ if and only if $x = y$, and (b) validity of (5.2.1) with $c = 2$ for all nonzero x and y is necessary and sufficient for the norm to be derived from an inner product, see W. A. Kirk and M. F. Smiley, Another characterization of inner product spaces, *Amer. Math. Monthly* 71 (1964) 890–891.

5.3 Algebraic properties of norms

New norms may be constructed from given norms in several ways. For example, the sum of two norms is a norm and any positive multiple of a norm is a norm. Also, if $\| \cdot \|_\alpha$ and $\| \cdot \|_\beta$ are norms, then the function $\| \cdot \|$ defined by $\|x\| \equiv \max\{\|x\|_\alpha, \|x\|_\beta\}$ is a norm. These observations are all special cases of the following result.

Theorem 5.3.1. *Let $\| \cdot \|_{\alpha_1}, \ldots, \| \cdot \|_{\alpha_m}$ be given norms on a vector space V over the field $\mathbf{F}$ ($\mathbf{F} = \mathbf{R}$ or $\mathbf{C}$), and let $\| \cdot \|$ be a norm on $\mathbf{R}^m$ such that $\|y\| \leq \|y + z\|$ for all vectors $y, z \in \mathbf{R}^m$ that have nonnegative entries. Then the function $f : V \to \mathbf{R}$ defined by $f(x) = \|[\|x\|_{\alpha_1}, \ldots, \|x\|_{\alpha_m}]^T\|$ is a norm on V.*

The monotonicity assumption on the norm $\| \cdot \|$ in the preceding theorem is needed to ensure that the constructed function f satisfies the triangle inequality. Every l_p-norm has this monotonicity property, as does any norm $\|x\|_\beta$ on $\mathbf{R}^m$ that is a function only of the absolute values of the entries of x; see (5.4.19(c)) and (5.6.P42). Some norms do not have this property, however.

Exercise. Prove (5.3.1).

Problems

5.3.P1 Deduce from (5.3.1) that the sum or max of two norms is a norm. What about the min?

5.3.P2 Let $m = 2$. Show that $\|x\| = |x_1 - x_2| + |x_2|$ is a norm on $\mathbf{R}^2$ that does not satisfy the monotonicity condition in (5.3.1). Show that $f(x) = \|[\|x\|_\infty, \|x\|_1]^T\| = \min\{|x_1|, |x_2|\} + |x_1| + |x_2|$ satisfies the nonnegativity, positivity, and homogeneity axiom for a norm on $\mathbf{R}^2$ but not the triangle inequality.

5.4 Analytic properties of norms

The examples in the preceding two sections show that many different real-valued functions on a real or complex vector space can satisfy the axioms for a norm. This is a good thing, because one norm may be more convenient or more appropriate than another for a given purpose. For example, the l_2-norm is convenient for optimization problems because it is continuously differentiable, except at the origin. On the other hand, the l_1-norm, while differentiable on a smaller set, is popular in statistics because it leads to estimators that can be more robust than classical regression estimators. The l_∞-norm is often the most natural one to use, since it directly monitors entrywise convergence, but it can be analytically and algebraically awkward to use.

In actual applications, the norm on which a theory is most naturally based and the norm that is most easily calculated in a given situation may not coincide. It is important, therefore, to know what relationships there may be between two different norms. Fortunately, in the finite-dimensional case all norms are "equivalent" in a certain strong sense.

A basic notion in analysis is that of *convergence of a sequence*. In a normed linear space we have the following definition of convergence:

Definition 5.4.1. *Let V be a real or complex vector space with a given norm $\| \cdot \|$. We say that a sequence $\{x^{(k)}\}$ of vectors in V converges to a vector $x \in V$ with respect to $\| \cdot \|$ if and only if $\lim_{k\to\infty}\|x^{(k)} - x\| = 0$. If $\{x^{(k)}\}$ converges to x with respect to $\| \cdot \|$, we write $\lim_{k\to\infty}x^{(k)} = x$ with respect to $\|\cdot\|$.*

Can a sequence of vectors converge to two different limits with respect to a given norm?

Exercise. If $\lim_{k\to\infty} x^{(k)} = x$ and $\lim_{k\to\infty} x^{(k)} = y$ with respect to $\|\cdot\|$, consider $\|x - y\| = \|x - x_k + x_k - y\|$ and the triangle inequality to show that $x = y$. Thus, the limit of a sequence (with respect to a given norm) is unique if it exists.

Can a sequence of vectors converge with respect to one norm but not with respect to another?

Example 5.4.2. Consider the sequence $\{f_k\}$ of functions in $C[0, 1]$ (the vector space of all real-valued or complex-valued continuous functions on $[0, 1]$) defined by

$$
\begin{aligned}
f_k(x) &= 0, & 0 \leq x \leq \tfrac{1}{k} \\
f_k(x) &= 2(k^{3/2}x - k^{1/2}), & \tfrac{1}{k} \leq x \leq \tfrac{3}{2k} \\
f_k(x) &= 2(-k^{3/2}x + 2k^{1/2}), & \tfrac{3}{2k} \leq x \leq \tfrac{2}{k} \\
f_k(x) &= 0, & \tfrac{2}{k} \leq x \leq 1
\end{aligned}
$$

for $k = 2, 3, 4, \ldots$. A calculation reveals that

$$\|f_k\|_1 = \frac{1}{2}k^{-1/2} \to 0 \text{ as } k \to \infty$$

$$\|f_k\|_2 = \frac{1}{\sqrt{3}} \text{ for all } k = 1, 2, \ldots$$

$$\|f_k\|_\infty = k^{1/2} \to \infty \text{ as } k \to \infty$$

Thus, $\lim_{k\to\infty} f_k = 0$ with respect to the L_1-norm, but not with respect to the L_2-norm or the L_∞-norm. The sequence $\{f_k\}$ is bounded in the L_2-norm but is unbounded in the L_∞-norm.

Exercise. Sketch the functions described in the preceding example and verify the assertions made about the L_1-, L_2-, and L_∞-norms.

Fortunately, the strange phenomena in (5.4.2) cannot occur in a finite-dimensional normed linear space. Underlying this fact are some basic results about continuous functions on a normed linear space; see Appendix E.

Lemma 5.4.3. *Let $\| \cdot \|$ be a norm on a vector space V over the field $\mathbf{F}$ ($\mathbf{F} = \mathbf{R}$ or $\mathbf{C}$), let $m \geq 1$ be a given positive integer, let $x^{(1)}, x^{(2)}, \ldots, x^{(m)} \in V$ be given vectors, and define $x(z) = z_1 x^{(1)} + z_2 x^{(2)} + \cdots + z_m x^{(m)}$ for any $z = [z_1 \; \ldots \; z_m]^T \in \mathbf{F}^m$. The function $g: \mathbf{F}^m \to \mathbf{R}$ defined by*

$$g(z) = \|x(z)\| = \|z_1 x^{(1)} + z_2 x^{(2)} + \cdots + z_m x^{(m)}\|$$

is a uniformly continuous function on $\mathbf{F}^m$ with respect to the Euclidean norm.

Proof. Let $u = [u_1 \; \ldots \; u_m]^T$ and $v = [v_1 \; \ldots \; v_m]^T$. Use (5.2.1) and the Cauchy–Schwarz inequality to calculate

$$|g(x(u)) - g(x(v))| = |\|x(u)\| - \|x(v)\|| \leq \|x(u) - x(v)\|$$

$$= \left\| \sum_{i=1}^{m} (u_i - v_i) x^{(i)} \right\| \leq \sum_{i=1}^{m} |u_i - v_i| \|x^{(i)}\|$$

$$\leq \left(\sum_{i=1}^{m} |u_i - v_i|^2 \right)^{1/2} \left(\sum_{i=1}^{m} \|x^{(i)}\|^2 \right)^{1/2} = C \|u - v\|_2$$

in which the finite constant $C = (\sum_{i=1}^{m} \|x^{(i)}\|^2)^{1/2}$ depends only on the norm $\| \cdot \|$ and the m vectors $x^{(1)}, \ldots, x^{(m)}$. If every $x^{(i)} = 0$, then $g(z) = 0$ for every z so g is certainly uniformly continuous. If some $x^{(i)} \neq 0$, then $C > 0$ and $|g(x(u)) - g(x(v))| < \epsilon$ whenever $\|u - v\|_2 < \epsilon / C$. $\quad\square$

The normed linear space V need not be finite dimensional in the preceding lemma. However, finite dimensionality of V is essential for the following fundamental result.

Theorem 5.4.4. *Let f_1 and f_2 be real-valued functions on a finite-dimensional vector space V over the field $\mathbf{F}$ ($\mathbf{F} = \mathbf{R}$ or $\mathbf{C}$), let $\mathcal{B} = \{x^{(1)}, \ldots, x^{(n)}\}$ be a basis for V, and let $x(z) = z_1 x^{(1)} + \cdots + z_n x^{(n)}$ for all $z = [z_1 \; \ldots \; z_n]^T \in \mathbf{F}^n$. Assume that f_1 and f_2 are*

(a) *Positive: $f_i(x) \geq 0$ for all $x \in V$, and $f_i(x) = 0$ if and only if $x = 0$*
(b) *Homogeneous: $f_i(\alpha x) = |\alpha| f_i(x)$ for all $\alpha \in \mathbf{F}$ and all $x \in V$*
(c) *Continuous: $f_i(x(z))$ is continuous on $\mathbf{F}^n$ with respect to the Euclidean norm*

Then there exist finite positive constants C_m and C_M such that

$$C_m f_1(x) \leq f_2(x) \leq C_M f_1(x) \text{ for all } x \in V$$

Proof. Define $h(z) = f_2(x(z))/f_1(x(z))$ on the Euclidean unit sphere $S = \{z \in \mathbf{F}^n : \|z\|_2 = 1\}$, a compact set in $\mathbf{F}^n$ with respect to the Euclidean norm. The positivity hypothesis (a) ensures that each $f_i(x(z)) > 0$ for all $z \in S$, and therefore $h(z)$, a product of continuous functions (invoke the continuity hypothesis (c) here) is continuous on S. The Weierstrass theorem on $\mathbf{F}^n$ with the Euclidean norm (see Appendix E) ensures that h achieves a finite positive maximum C_M and a positive minimum C_m on S, so $C_m \leq f_2(x(z))/f_1(x(z)) \leq C_M$ and $C_m f_1(x(z)) \leq f_2(x(z)) \leq C_M f_1(x(z))$

for all $z \in S$. Because $z/\|z\|_2 \in S$ for every nonzero $z \in \mathbf{F}^n$, the homogeneity hypothesis (b) ensures that each $f_i(x(\frac{z}{\|z\|_2})) = f_i(\|z\|_2^{-1} x(z)) = \|z\|_2^{-1} f_i(x(z))$, so $C_m f_1(x(z)) \leq f_2(x(z)) \leq C_M f_1(x(z))$ for all nonzero $z \in \mathbf{F}^n$; these inequalities are also valid for $z = 0$ since $f_1(0) = f_2(0) = 0$. But every $x \in V$ can be expressed as $x = x(z)$ for some $z \in \mathbf{F}^n$ because $\mathcal{B}$ is a basis, so the asserted inequalities hold for all $x \in V$. $\qquad\square$

If a real-valued function on a finite-dimensional real or complex vector space satisfies the three hypotheses of positivity, homogeneity, and continuity stated in (5.4.4), it is called a *pre-norm*.

The most important example of a pre-norm is, of course, a norm; (5.4.3) says that every norm satisfies the continuity assumption (c) of (5.4.4). A pre-norm that satisfies the triangle inequality is a norm.

Corollary 5.4.5. *Let* $\|\cdot\|_\alpha$ *and* $\|\cdot\|_\beta$ *be given norms on a finite-dimensional real or complex vector space* V. *Then there exist finite positive constants* C_m *and* C_M *such that* $C_m\|x\|_\alpha \leq \|x\|_\beta \leq C_M\|x\|_\alpha$ *for all* $x \in V$.

Exercise. Let $x = [x_1\ x_2]^T \in \mathbf{R}^2$ and consider the following norms on $\mathbf{R}^2$: $\|x\|_\alpha \equiv \|[10x_1\ x_2]^T\|_\infty$ and $\|x\|_\beta \equiv \|[x_1\ 10x_2]^T\|_\infty$. Show that $f(x) = (\|x\|_\alpha\|x\|_\beta)^{1/2}$ is a pre-norm on $\mathbf{R}^2$ that is not a norm; see (5.4.P15). *Hint*: Consider $f([1\ 1]^T)$, $f([0\ 1]^T)$, and $f([1\ 0]^T)$.

Exercise. If $\|\cdot\|_{\alpha_1}, \ldots, \|\cdot\|_{\alpha_k}$ are norms on V, show that $f(x) = (\|x\|_{\alpha_1} \cdots \|x\|_{\alpha_k})^{1/k}$ and $h(x) = \min\{\|x\|_{\alpha_1}, \ldots, \|x\|_{\alpha_k}\}$ are pre-norms on V that are not necessarily norms.

An important consequence of (5.4.5) is the fact that convergence of a sequence of vectors in a finite-dimensional complex vector space is independent of the norm used.

Corollary 5.4.6. *If* $\|\cdot\|_\alpha$ *and* $\|\cdot\|_\beta$ *are norms on a finite-dimensional real or complex vector space* V, *and if* $\{x^{(k)}\}$ *is a given sequence of vectors in* V, *then* $\lim_{k\to\infty} x^{(k)} = x$ *with respect to* $\|\cdot\|_\alpha$ *if and only if* $\lim_{k\to\infty} x^{(k)} = x$ *with respect to* $\|\cdot\|_\beta$.

Proof. Since $C_m\|x^{(k)} - x\|_\alpha \leq \|x^{(k)} - x\|_\beta \leq C_M\|x^{(k)} - x\|_\alpha$ for all k, it follows that $\|x^{(k)} - x\|_\alpha \to 0$ as $k \to \infty$ if and only if $\|x^{(k)} - x\|_\beta \to 0$ as $k \to \infty$. $\qquad\square$

Definition 5.4.7. *Two given norms on a real or complex vector space are* equivalent *if whenever a sequence* $\{x^{(k)}\}$ *of vectors converges to a vector* x *with respect to the one of the norms, then it converges to* x *with respect to the other norm.*

Corollary 5.4.6 ensures that *for finite-dimensional real or complex vector spaces, all norms are equivalent*. Example 5.4.2 illustrates that the situation is very different for an infinite-dimensional space.

Since all norms on $\mathbf{R}^n$ or $\mathbf{C}^n$ are equivalent to $\|\cdot\|_\infty$, for a given sequence of vectors $x^{(k)} = [x_i^{(k)}]_{i=1}^n$ we have $\lim_{k\to\infty} x^{(k)} = x$ with respect to any norm if and only if $\lim_{k\to\infty} x_i^{(k)} = x_i$ for each $i = 1, \ldots, n$.

Another important fact is that the unit ball and unit sphere with respect to any pre-norm or norm on $\mathbf{R}^n$ or $\mathbf{C}^n$ are always compact. Consequently, a continuous real- or complex-valued function on such a unit ball or unit sphere is bounded; it achieves its maximum and minimum if it is real-valued.

Corollary 5.4.8. *Let $V = \mathbf{F}^n$ ($\mathbf{F} = \mathbf{R}$ or $\mathbf{C}$) and let $f(\cdot)$ be a pre-norm or norm on V. The sets $\{x : f(x) \leq 1\}$ and $\{x : f(x) = 1\}$ are compact.*

Proof. It suffices to show that the respective sets are closed and bounded with respect to the Euclidean norm. Theorem 5.4.4 ensures that there is some finite $C > 0$ such that $\|x\|_2 \leq Cf(x)$ for all $x \in V$, so both of the sets $\{x : f(x) \leq 1\}$ and $\{x : f(x) = 1\}$ are contained in a Euclidean ball with radius C centered at the origin. Both of the sets $\{x : f(x) = 1\}$ and $\{x : f(x) \leq 1\}$ are closed because $f(\cdot)$ is continuous. $\qquad\square$

We are sometimes confronted with the problem of determining whether a given sequence $\{x^{(k)}\}$ converges to anything at all. For this reason, it is important to have a convergence criterion that does not explicitly involve the limit of the sequence (if any). If there were such a limit x, then

$$\|x^{(k)} - x^{(j)}\| = \|x^{(k)} - x + x - x^{(j)}\| \leq \|x^{(k)} - x\| + \|x - x^{(j)}\| \to 0$$

as $k, j \to \infty$. This is the motivation for the following.

Definition 5.4.9. *A sequence $\{x^{(k)}\}$ in a vector space V is a* Cauchy sequence *with respect to a norm $\|\cdot\|$ if for each $\epsilon > 0$ there is a positive integer $N(\epsilon)$ such that $\|x^{(k_1)} - x^{(k_2)}\| \leq \epsilon$ whenever $k_1, k_2 \geq N(\epsilon)$.*

Theorem 5.4.10. *Let $\|\cdot\|$ be a given norm on a finite-dimensional real or complex vector space V, and let $\{x^{(k)}\}$ be a given sequence of vectors in V. The sequence $\{x^{(k)}\}$ converges to a vector in V if and only if it is a Cauchy sequence with respect to the norm $\|\cdot\|$.*

Proof. By choosing a basis $\mathcal{B}$ of V and considering the equivalent norm $\|[x]_{\mathcal{B}}\|_\infty$, we see that there is no loss of generality if we assume that $V = \mathbf{R}^n$ or $\mathbf{C}^n$ for some integer n and if we assume that the norm is $\|\cdot\|_\infty$. If $\{x^{(k)}\}$ is a Cauchy sequence, then so is each real or complex sequence $\{x_i^{(k)}\}$ of entries for each $i = 1, \ldots, n$. Since a Cauchy sequence of real or complex numbers must have a limit, this means that for each $i = 1, \ldots, n$ there is a scalar x_i such that $\lim_{k\to\infty} x_i^{(k)} = x_i$; one checks that $\lim_{k\to\infty} x^{(k)} = x = [x_1 \ldots x_n]^T$. Conversely, if there is a vector x such that $\lim_{k\to\infty} x^{(k)} = x$, then $\|x^{(k_1)} - x^{(k_2)}\| \leq \|x^{(k_1)} - x\| + \|x - x^{(k_2)}\|$ and the given sequence is a Cauchy sequence. $\qquad\square$

It is a fundamental property of the real and complex fields (used in the proof of the preceding theorem) that a sequence is a Cauchy sequence if and only if it converges to some (real or complex, respectively) scalar. This is known as the *completeness property* of the real and complex fields. We have just shown that the completeness property extends to finite-dimensional real and complex vector spaces with respect to any norm. Unfortunately, an infinite-dimensional normed linear space might not have the completeness property.

Definition 5.4.11. *A normed linear space* V *is said to be* complete *with respect to its norm* $\| \cdot \|$ *if every sequence in* V *that is a Cauchy sequence with respect to* $\| \cdot \|$ *converges to a point of* V.

Exercise. Consider the vector space $C[0, 1]$ with the L_1-norm, and consider the sequence of functions $\{f_k\}$ defined by

$$
\begin{aligned}
f_k(t) &= 0, & 0 \leq t \leq \tfrac{1}{2} - \tfrac{1}{k} \\
f_k(t) &= \tfrac{k}{2}(t - \tfrac{1}{2} + \tfrac{1}{k}), & \tfrac{1}{2} - \tfrac{1}{k} \leq t \leq \tfrac{1}{2} + \tfrac{1}{k} \\
f_k(t) &= 1, & \tfrac{1}{2} + \tfrac{1}{k} \leq t \leq 1
\end{aligned}
$$

Sketch the functions f_k. Show that $\{f_k\}$ is a Cauchy sequence but there is no function $f \in C[0, 1]$ for which $\lim_{k \to \infty} f_k = f$ with respect to the L_1-norm.

Using the fact that the unit ball of any norm or prenorm on $\mathbf{R}^n$ or $\mathbf{C}^n$ is compact, we can introduce another useful method to generate new norms from old ones using the Euclidean inner product.

Definition 5.4.12. *Let* $f(\cdot)$ *be a pre-norm on* $V = \mathbf{F}^n$ *(*$\mathbf{F} = \mathbf{R}$ *or* $\mathbf{C}$*). The function*

$$
f^D(y) = \max_{f(x)=1} \operatorname{Re} \langle x, y \rangle = \max_{f(x)=1} \operatorname{Re} y^* x
$$

is the dual norm *of* f.

Observe first that the dual norm is a well-defined function on V because $\operatorname{Re} y^* x$ is a continuous function of x for each fixed $y \in V$, and the set $\{x : f(x) = 1\}$ is compact. The Weierstrass theorem ensures that the maximum value of $\operatorname{Re} y^* x$ is attained at some point in this set. If c is a scalar such that $|c| = 1$, then the homogeneity of f permits us to compute

$$
\max_{f(x)=1} |y^* x| = \max_{f(x)=1} \max_{|c|=1} \operatorname{Re}(c y^* x) = \max_{f(x)=1} \max_{|c|=1} \operatorname{Re} y^*(cx)
$$

$$
= \max_{|c|=1} \max_{f(x/c)=1} \operatorname{Re} y^* x = \max_{f(x)=1} \operatorname{Re} y^* x
$$

Moreover,

$$
\max_{f(x)=1} |y^* x| = \max_{x \neq 0} \left| y^* \frac{x}{f(x)} \right| = \max_{x \neq 0} \frac{|y^* x|}{f(x)}
$$

so equivalent and sometimes convenient alternative expressions for the dual norm are

$$
f^D(y) = \max_{f(x)=1} |y^* x| = \max_{x \neq 0} \frac{|y^* x|}{f(x)} \tag{5.4.12a}
$$

Finally, we observe that the name *dual norm* for the function f^D is well deserved. The function $f^D(\cdot)$ is evidently homogeneous. It is positive, for if $y \neq 0$, homogeneity of $f(\cdot)$ ensures that

$$
f^D(y) = \max_{f(x)=1} |y^* x| \geq \left| y^* \frac{y}{f(y)} \right| = \frac{\|y\|_2^2}{f(y)} > 0
$$

It is noteworthy that even if $f(\cdot)$ does not obey the triangle inequality, $f^D(\cdot)$ always does:

$$f^D(y + z) = \max_{f(x)=1} |(y + z)^* x| \leq \max_{f(x)=1} (|y^* x| + |z^* x|)$$

$$\leq \max_{f(x)=1} |y^* x| + \max_{f(x)=1} |z^* x| = f^D(y) + f^D(z)$$

The dual norm of a pre-norm is positive, homogeneous, and satisfies the triangle inequality, so it is a norm. In particular, the dual norm of a norm is always a norm.

A simple inequality for the dual norm is given in the following lemma, which is a natural generalization of the Cauchy–Schwarz inequality.

Lemma 5.4.13. *Let $f(\cdot)$ be a pre-norm on $V = \mathbf{F}^n$ ($\mathbf{F} = \mathbf{R}$ or $\mathbf{C}$). Then for all $x, y \in V$ we have*

$$|y^* x| \leq f(x) f^D(y)$$

and

$$|y^* x| \leq f^D(x) f(y)$$

Proof. If $x \neq 0$, then

$$\left| y^* \frac{x}{f(x)} \right| \leq \max_{f(z)=1} |y^* z| = f^D(y)$$

and hence $|y^* x| \leq f(x) f^D(y)$. Of course, this inequality is also valid for $x = 0$. The second inequality follows from the first since $|y^* x| = |x^* y|$. $\qquad\square$

It is instructive to identify the duals of some familiar norms. For example, if $\|\cdot\|$ is a norm on $\mathbf{C}^n$ and $S \in M_n$ is nonsingular, what is the dual of the norm $\|\cdot\|_S$ defined by (5.2.6)? We compute

$$\|y\|_S^D = \max_{x \neq 0} \frac{|y^* x|}{\|x\|_S} = \max_{x \neq 0} \frac{|y^* x|}{\|Sx\|} = \max_{z \neq 0} \frac{|y^* S^{-1} z|}{\|z\|}$$

$$= \max_{z \neq 0} \frac{|(S^{-*} y)^* z|}{\|z\|} = \|S^{-*} y\|^D \tag{5.4.14}$$

and conclude that $(\|\cdot\|_S)^D = (\|\cdot\|^D)_{S^{-*}}$.

If $x, y \in \mathbf{C}^n$, then

$$|y^* x| = \left| \sum_{i=1}^{n} \bar{y}_i x_i \right| \leq \sum_{i=1}^{n} |\bar{y}_i x_i|$$

$$\leq \begin{cases} (\max_{1 \leq i \leq n} |y_i|) \sum_{j=1}^{n} |x_j| = \|y\|_\infty \|x\|_1 \\ (\max_{1 \leq i \leq n} |x_i|) \sum_{j=1}^{n} |y_j| = \|x\|_\infty \|y\|_1 \end{cases} \tag{5.4.15}$$

Let $y \in \mathbf{C}^n$ be a given nonzero vector. The top inequality in (5.4.15) is an equality for a vector x such that $x_i = 1$ for a single i such that $|y_i| = \|y\|_\infty$, and $x_j = 0$ for all $j \neq i$. The bottom inequality in (5.4.15) is an equality for a vector x such that $x_i = y_i/|y_i|$

for all i such that $y_i \neq 0$, and $x_j = 0$ otherwise. Thus,

$$\|y\|_1^D = \max_{\|x\|_1=1} |y^*x| = \max_{\|x\|_1=1} \|y\|_\infty \|x\|_1 = \|y\|_\infty$$

$$\|y\|_\infty^D = \max_{\|x\|_\infty=1} |y^*x| = \max_{\|x\|_\infty=1} \|y\|_1 \|x\|_\infty = \|y\|_1$$

We conclude that

$$\| \cdot \|_1^D = \| \cdot \|_\infty \text{ and } \| \cdot \|_\infty^D = \| \cdot \|_1 \tag{5.4.15a}$$

Now consider the Euclidean norm, a given nonzero vector y, and an arbitrary vector x. The Cauchy–Schwarz inequality says that

$$|y^*x| = \left| \sum_{i=1}^n \bar{y}_i x_i \right| \leq \|y\|_2 \|x\|_2$$

with equality if $x = y/\|y\|_2$. Using the preceding argument for the l_1- and l_∞-norms, we find that $\|y\|_2^D = \|y\|_2$, so the Euclidean norm is its own dual.

Exercise. Explain why the inequalities in (5.4.13) are a generalization of the Cauchy–Schwarz inequality (5.1.4).

For any $p \geq 1$, consider the l_p- and l_q-norms, with q defined by the relation $1/p + 1/q = 1$. Note that $1 < p < \infty$ if and only if $1 < q < \infty$. Hölder's inequality (see Appendix B) permits us to replace (5.4.15) with the inequality $|y^*x| \leq \|x\|_p \|y\|_q$. Thus, for a given vector $y = [y_i]$, $\|y\|_p^D = \max_{\|x\|_p=1} |y^*x| \leq \|y\|_q$, which is an equality for all x if $y = 0$; if $y \neq 0$, it is an equality for $x = [x_i]$ defined by

$$x_i = \begin{cases} 0 & \text{if } y_i = 0 \\ \dfrac{|y_i|^q}{\bar{y}_i \|y\|_q^{q-1}} & \text{if } y_i \neq 0 \end{cases}$$

It follows that $\|y\|_p^D = \|y\|_q$, and hence $\| \cdot \|_q^D = \| \cdot \|_p$.

For all of the l_p-norms, therefore, the dual of the dual norm is the original norm. This is no accident; see (5.5.9). Moreover, the only l_p-norm that is its own dual is the Euclidean norm. This is also no accident, as we explain after noting two useful properties of dual norms.

Lemma 5.4.16. *Let $f(\cdot)$ and $g(\cdot)$ be pre-norms on $V = \mathbf{F}^n$ ($\mathbf{F} = \mathbf{R}$ or $\mathbf{C}$) and let $c > 0$ be given. Then*

(a) *$cf(\cdot)$ is a pre-norm on V, and its dual norm is $c^{-1}f^D(\cdot)$*
(b) *if $f(x) \leq g(x)$ for all $x \in V$, then $f^D(y) \geq g^D(y)$ for all $y \in V$*

Proof. The function $cf(\cdot)$ is positive, homogeneous, and continuous, so it is a pre-norm. The remaining assertions follow from (5.4.12a). $\square$

Exercise. Provide details for the proof of the preceding lemma.

Theorem 5.4.17. *Let $\| \cdot \|$ be a norm on $V = \mathbf{F}^n$ ($\mathbf{F} = \mathbf{R}$ or $\mathbf{C}$) and let $c > 0$ be given. Then $\|x\| = c\|x\|^D$ for all $x \in V$ if and only if $\| \cdot \| = \sqrt{c}\| \cdot \|_2$. In particular, $\| \cdot \| = \| \cdot \|^D$ if and only if $\| \cdot \| = \| \cdot \|_2$.*

Proof. If $\| \cdot \| = \sqrt{c}\| \cdot \|_2$ and $x \in V$, then (5.4.16(a)) ensures that $\| \cdot \|^D = \frac{1}{\sqrt{c}}\| \cdot \|_2^D = \frac{1}{\sqrt{c}}\| \cdot \|_2 = c^{-1}\| \cdot \|$. For the converse assertion, consider the norm $N(x) = c^{-1/2}\|x\|$. The hypothesis $\| \cdot \| = c\| \cdot \|^D$ ensures that $N^D(\cdot) = c^{1/2}\| \cdot \|^D = c^{1/2}c^{-1}\| \cdot \| = c^{-1/2}\| \cdot \| = N(\cdot)$, so $N(\cdot)$ is self-dual. It now follows from (5.4.13) that $\|x\|_2^2 = |x^*x| \leq N(x)N^D(x)$ for every $x \in V$, that is, $\|x\|_2 \leq N(x)$ for all $x \in V$. But (5.4.16(b)) ensures that $\|x\|_2 \geq N(x)$ for all $x \in V$, so $\|\cdot\|_2 = N(\cdot)$. $\qquad\square$

Every k-norm and every l_p-norm on $\mathbf{R}^n$ or $\mathbf{C}^n$ has the property that the norm of a vector depends only on the *absolute values* of its entries and it is a *nondecreasing function* of the absolute values of those entries of x. These two properties are not unrelated.

Definition 5.4.18. *If $x = [x_i] \in V = \mathbf{F}^n$ ($\mathbf{F} = \mathbf{R}$ or $\mathbf{C}$), let $|x| = [|x_i|]$ denote the entrywise absolute value of x. We say that $|x| \leq |y|$ if $|x_i| \leq |y_i|$ for all $i = 1, \ldots, n$. A norm $\| \cdot \|$ on V is*

(a) *monotone if $|x| \leq |y|$ implies $\|x\| \leq \|y\|$ for all $x, y \in V$*
(b) *absolute if $\|x\| = \| |x| \|$ for all $x \in V$*

Theorem 5.4.19. *Let $\| \cdot \|$ be a norm on $V = \mathbf{F}^n$ ($\mathbf{F} = \mathbf{R}$ or $\mathbf{C}$).*

(a) *If $\| \cdot \|$ is absolute, then*

$$\|y\|^D = \max_{x \neq 0} \frac{|y|^T |x|}{\|x\|} \qquad (5.4.20)$$

for all $y \in V$.

(b) *If $\| \cdot \|$ is absolute, then $\| \cdot \|^D$ is absolute and monotone.*
(c) *The norm $\| \cdot \|$ is absolute if and only if it is monotone.*

Proof. Suppose that $\mathbf{F} = \mathbf{C}$.
(a) Suppose that $\| \cdot \|$ is absolute. For a given $y = [y_k] \in \mathbf{C}^n$, any $x = [x_k] \in \mathbf{C}^n$, and any $z = [z_k] \in \mathbf{C}^n$ such that $|z| = |x|$ we have $|y^*z| = |\sum_{k=1}^n \bar{y}_k z_k| \leq \sum_{k=1}^n |y_k||z_k| = |y|^T|z| = |y|^T|x|$, with equality for $z_k = e^{i\theta_k}x_k$ if we choose the real parameters $\theta_1, \ldots, \theta_n$ so that $e^{i\theta_k}\bar{y}_k x_k$ is real and nonnegative. Thus,

$$\|y\|^D = \max_{x \neq 0} \frac{|y^*x|}{\|x\|} = \max_{x \neq 0} \max_{|z|=|x|} \frac{|y^*z|}{\|z\|} = \max_{x \neq 0} \frac{|y|^T|x|}{\|x\|}$$

(b) Suppose that $\| \cdot \|$ is absolute. The representation (5.4.20) shows that $\|y\|^D = \| |y| \|^D$ for all $y \in \mathbf{C}^n$. Moreover, if $|z| \leq |y|$, then

$$\|z\|^D = \max_{x \neq 0} \frac{|z|^T|x|}{\|x\|} \leq \max_{x \neq 0} \frac{|y|^T|x|}{\|x\|} = \|y\|^D$$

so $\| \cdot \|^D$ is monotone.
(c) If $\| \cdot \|$ is monotone and $|y| = |x|$, then $|y| \leq |x|$ and $|y| \geq |x|$, so $\|y\| \leq \|x\|$, $\|y\| \geq \|x\|$, and $\|y\| = \|x\|$. Conversely, suppose that $\|\cdot\|$ is absolute. Let $k \in \{1, \ldots, n\}$ and

$\alpha \in [0, 1]$. Then

$$\|[x_1 \ \ldots \ x_{k-1} \ \alpha x_k \ x_{k+1} \ \ldots \ x_n]^T\|$$

$$= \|\frac{1}{2}(1 - \alpha)[x_1 \ \ldots \ x_{k-1} \ -x_k \ x_{k+1} \ \ldots \ x_n]^T + \frac{1}{2}(1 - \alpha)x + \alpha x\|$$

$$\leq \frac{1}{2}(1 - \alpha)\|[x_1 \ \ldots \ x_{k-1} \ -x_k \ x_{k+1} \ \ldots \ x_n]^T\| + \frac{1}{2}(1 - \alpha)\|x\| + \alpha\|x\|$$

$$= \frac{1}{2}(1 - \alpha)\|x\| + \frac{1}{2}(1 - \alpha)\|x\| + \alpha\|x\| = \|x\|$$

It follows that $\|[\alpha_1 x_1 \ \ldots \ \alpha_n x_n]^T\| \leq \|x\|$ for every $x \in \mathbf{C}^n$ and all choices of $\alpha_k \in [0, 1]$, $k = 1, \ldots, n$. If $|y| \leq |x|$ then there are $\alpha_k \in [0, 1]$ such that $|y_k| = \alpha_k |x_k|$, $k = 1, \ldots, n$, so $\|y\| \leq \|x\|$. $\square$

Exercise. Prove the preceding theorem for $\mathbf{F} = \mathbf{R}$.

For a conceptual proof that an absolute norm is monotone, see (5.5.11).

Problems

5.4.P1 Explain why (5.4.5) may be stated equivalently as

$$C_m\left(\|\cdot\|_\alpha, \|\cdot\|_\beta\right) \leq \frac{\|x\|_\beta}{\|x\|_\alpha} \leq C_M(\|\cdot\|_\alpha, \|\cdot\|_\beta), \text{ for all } x \neq 0$$

in which $C_m(\cdot, \cdot)$ and $C_M(\cdot, \cdot)$ denote the best possible constants relating the respective norms in (5.4.5). Show that $C_m(\|\cdot\|_\beta, \|\cdot\|_\alpha) = C_M(\|\cdot\|_\alpha, \|\cdot\|_\beta)^{-1}$.

5.4.P2 Give a bound for $C_m(\|\cdot\|_\alpha, \|\cdot\|_\gamma)$ that involves $C_m(\|\cdot\|_\alpha, \|\cdot\|_\beta)$ and $C_m(\|\cdot\|_\beta, \|\cdot\|_\gamma)$. Do likewise for C_M.

5.4.P3 If $1 \leq p_1 < p_2 < \infty$, show that the best bounds between the corresponding l_p-norms on $\mathbf{C}^n$ or $\mathbf{R}^n$ are

$$\|x\|_{p_2} \leq \|x\|_{p_1} \leq n^{\left(\frac{1}{p_1} - \frac{1}{p_2}\right)}\|x\|_{p_2} \tag{5.4.21}$$

Verify the entries in the following table of bounds $\|x\|_\alpha \leq C_{\alpha\beta}\|x\|_\beta$.

$$[C_{\alpha\beta}] = \begin{array}{c|ccc} \alpha \backslash \beta & 1 & 2 & \infty \\ \hline 1 & 1 & \sqrt{n} & n \\ 2 & 1 & 1 & \sqrt{n} \\ \infty & 1 & 1 & 1 \end{array}$$

For each entry, exhibit a nonzero vector x for which the asserted bound is attained.

5.4.P4 Show that two norms on a real or complex vector space are equivalent if and only if they are related by two constants and an inequality as in (5.4.5).

5.4.P5 Show that the functions f_k in (5.4.2) have the property that $f(x) \to 0$ for each x, $\|f_k - f_j\|_1 \to 0$ as $k, j \to \infty$, and for each $k \leq 2$ there is some $J > k$ for which $\|f_k - f_j\|_\infty > k^{1/2}$ for all $j > J$. Thus, a sequence in an infinite dimensional normed linear space can be convergent in one sense (pointwise), Cauchy in a norm, and not Cauchy in another norm.

5.4.P6 Let V be a complete real or complex vector space, let $\{x^{(k)}\}$ be a given sequence in V, and let $\|\cdot\|$ be a given norm on V. If there is an $M \geq 0$ such that $\sum_{k=1}^{n} \|x^{(k)}\| \leq M$ for all $n = 1, 2, \ldots$, show that the sequence of partial sums $\{y^{(n)}\}$ defined by $y^{(n)} = \sum_{k=1}^{n} x^{(k)}$ converges to a point of V. What theorem about convergence of infinite series of real numbers does this generalize?

5.4.P7 Show that $\|x\|_\infty = \lim_{p \to \infty} \|x\|_p$ for every $x \in \mathbf{C}^n$. If $|x| > 0$, what is $\lim_{p \to -\infty} \|x\|_p$?

5.4.P8 Show that the dual norm of the k-norm on $\mathbf{R}^n$ or $\mathbf{C}^n$ is

$$\|y\|_{[k]}^D = \max \left\{ \frac{1}{k} \|y\|_1, \|y\|_\infty \right\} \qquad (5.4.22)$$

What does this say if $k = 1$ or $k = n$?

5.4.P9 Let $\|\cdot\|$ be a norm on $\mathbf{R}^n$ or $\mathbf{C}^n$ and let e_i be a standard basis vector (0.1.7). Explain why $\|e_i\| \|e_i\|^D \geq 1$. Can you find a norm for which $\|e_1\| \|e_1\|^D > 1$?

5.4.P10 Let $\|\cdot\|_\alpha$ and $\|\cdot\|_\beta$ be two given norms on $\mathbf{C}^n$, and suppose that there is some $C > 0$ such that $\|x\|_\alpha \leq C \|x\|_\beta$ for all $x \in \mathbf{C}^n$. Explain why $\|x\|_\beta^D \leq C \|x\|_\alpha^D$ for all $x \in \mathbf{C}^n$.

5.4.P11 Let $\|\cdot\|$ be a norm on $\mathbf{F}^n$ ($\mathbf{F} = \mathbf{R}$ or $\mathbf{C}$). A matrix $A \in M_n(\mathbf{F})$ is an *isometry* for $\|\cdot\|$ if $\|Ax\| = \|x\|$ for all $x \in \mathbf{F}^n$. For example, any unitary matrix is an isometry for the Euclidean norm, and the identity matrix is an isometry for every norm. Show the following: (a) Every isometry for $\|\cdot\|$ is nonsingular. (b) If $A, B \in M_n(\mathbf{F})$ are isometries for $\|\cdot\|$, then so are A^{-1} and AB. Consequently, the set of isometries for $\|\cdot\|$ is a subgroup of the general linear group. This subgroup is known as the *isometry group of* $\|\cdot\|$. (c) If $A \in M_n$ is an isometry for $\|\cdot\|$, then every eigenvalue of A has modulus one. It is known that the isometry group of $\|\cdot\|$ is similar to a group of unitary matrices in $M_n(\mathbf{F})$ (Auerbach's theorem), so A is similar to a unitary matrix; see (7.6.P21 to P23). (d) If $A \in M_n(\mathbf{F})$ is an isometry for $\|\cdot\|$, then $|\det A| = 1$. (e) Any unitary generalized permutation matrix is an isometry for every k-norm and every l_p-norm with $1 \leq p \leq \infty$. Describe a typical unitary generalized permutation matrix.

5.4.P12 Let $\|\cdot\|$ be a norm on $\mathbf{F}^n$ ($\mathbf{F} = \mathbf{R}$ or $\mathbf{C}$). If $A \in M_n$ is an isometry for $\|\cdot\|$, show that A^* is an isometry for $\|\cdot\|^D$. Now explain why the group of isometries of $\|\cdot\|^D$ is *exactly* the set of conjugate transposes of the elements of the group of isometries of $\|\cdot\|$. When do $\|\cdot\|$ and $\|\cdot\|^D$ have the same isometry groups?

5.4.P13 Let $A \in M_n(\mathbf{F})$ ($\mathbf{F} = \mathbf{R}$ or $\mathbf{C}$) and suppose that $1 \leq p \leq \infty$ but $p \neq 2$. Prove that A is an isometry for the l_p-norm on $\mathbf{F}^n$ if and only if it is a unitary generalized permutation matrix.

5.4.P14 Consider the function $f : \mathbf{R}^2 \to \mathbf{R}$ given by $f(x) = |x_1 x_2|^{1/2}$. Show that the set $\{x : f(x) = 1\}$ is not compact. Does this contradict (5.4.8)?

5.4.P15 Consider the example of a pre-norm $f(x) = (\|x\|_\alpha \|x\|_\beta)^{1/2}$ on $\mathbf{R}^2$ given in the text, with $\|x\|_\alpha = \|[10x_1 \ x_2]^T\|_\infty$ and $\|x\|_\beta = \|[x_1 \ 10x_2]^T\|_\infty$. Show that the portion of the unit ball $\{x \in \mathbf{R}^2 : f(x) \leq 1\}$ in the first quadrant is bounded by segments of the lines $x_2 = 1/\sqrt{10}$ and $x_1 = 1/\sqrt{10}$ and an arc of the hyperbola $x_1 x_2 = 1/100$. Sketch this set and show that it is not convex. Why is the rest of the unit ball in the three remaining quadrants obtained by successive reflections of this set across the axes? Show that the unit ball of the dual norm $\{x \in \mathbf{R}^2 : f^D(x) \leq 1\}$ is bounded in the first quadrant by segments of the lines $x_1/10 + x_2 = \sqrt{10}$ and $x_1 + x_2/10 = \sqrt{10}$, that the whole unit ball of f^D is obtained by successive reflections of the portion in the first quadrant, and that it is convex. Show that the portion of the unit ball of f^{DD} in the first quadrant is bounded by segments of the lines $x_2 = 1/\sqrt{10}$, $x_1 = 1/\sqrt{10}$, and $x_1 + x_2 = 11/(10\sqrt{10})$; that the rest of the unit ball is obtained by successive reflections of this set across the axes; and that it is convex. Finally, compare the unit ball of f^{DD} with that of f and show that the former is exactly the closed convex hull of the latter.

5.4.P16 Let $\|\cdot\|$ be a norm on $V = \mathbf{R}^n$ or $\mathbf{C}^n$. Show that $\max_{\|x\| \neq 0}(\|x\|^D / \|x\|) = \max_{\|x\|=1} \max_{\|y\|=1}(\frac{x}{\|x\|_2})^*(\frac{y}{\|y\|_2})\|x\|_2\|y\|_2 \leq \max_{\|x\|=1} \|x\|_2^2$ (call this constant C_M) and that $\min_{x \neq 0}(\|x\|^D / \|x\|) \geq \min_{\|x\|=1} \|x\|_2^2$ (call this constant C_m). Deduce that $C_m|x| \leq \|x\|^D \leq C_M\|x\|$ for all $x \in V$, so geometrical constants give bounds between every norm and its dual.

5.4.P17 Let $f(\cdot)$ be a pre-norm on $\mathbf{R}^n$ or $\mathbf{C}^n$. Show that $f^D(y) = \max_{f(x)\leq 1} \operatorname{Re} y^*x = \max_{f(x)\leq 1} |y^*x| = \max_{x \neq 0} \frac{\operatorname{Re} y^*x}{f(x)}$.

5.4.P18 Let $\|\cdot\|$ be a norm on $V = \mathbf{R}^n$ or $\mathbf{C}^n$, and let $x_1, \ldots, x_n \in V$ be linearly independent. Explain why there is some $\varepsilon > 0$ such that $y_1, \ldots, y_n \subset V$ are linearly independent whenever $\|x_i - y_i\| < \varepsilon$ for all $i = 1, \ldots, n$.

Further Readings. See Householder (1964) for more information about dual norms. The idea that the dual of a pre-norm is a norm seems to be due to J. von Neumann, who discussed *gauge functions* (what we now call *symmetric absolute norms*) in Some matrix-inequalities and metrization of matric-space, *Tomsk Univ. Rev.* 1 (1937) 205–218. A more readily available source for this paper may be vol. 4 of von Neumann's *Collected Works*, ed. A. H. Taub, Macmillan, New York, 1962.

5.5 Duality and geometric properties of norms

The primary geometric feature of a norm is its unit ball, through which considerable insight into properties of the norm may be gained.

Definition 5.5.1. *Let $\|\cdot\|$ be a norm on a real or complex vector space V, let x be a point of V, and let $r > 0$ be given. The ball of radius r around x is the set*

$$B_{\|\cdot\|}(r;x) = \{y \in V : \|y - x\| \leq r\}$$

The unit ball *of $\|\cdot\|$ is the set*

$$B_{\|\cdot\|} = B_{\|\cdot\|}(1;0) = \{y \in V : \|y\| \leq 1\}$$

Exercise. Show that for every $r > 0$ and for every $x \in V$, $B(r;x) = \{y + x : y \in B(r;0)\} = x + B(r;0)$.

A ball of given radius around any point x looks the same as a ball of the same radius around zero; it is just translated to the point x. The unit ball is a geometric summary of a norm, which, because of the homogeneity property, characterizes the norm (actually only the boundary of $B_{\|\cdot\|}$ is needed). Our goal is to determine exactly which subsets of $\mathbf{C}^n$ can be the unit ball of some norm.

Exercise. Sketch the unit balls for the l_1, l_2, and l_∞ norms on $\mathbf{R}^2$ and identify their extreme points. Are there any containment relationships among these unit balls? Which points must be on the boundary of the unit ball of every l_p-norm on $\mathbf{R}^2$? Sketch the unit ball of some other l_p-norms.

Exercise. If $\|\cdot\|_\alpha$ and $\|\cdot\|_\beta$ are norms on V, explain why $\|x\|_\alpha \leq \|x\|_\beta$ for all $x \in V$ if and only if $B_{\|\cdot\|_\beta} \subset B_{\|\cdot\|_\alpha}$. The natural partial order on norms is reflected in geometric containment of their unit balls. What happens to the unit ball when a norm is multiplied by a positive constant?

Exercise. If $\|\cdot\|$ is a norm on V, if $x \in V$, and if α is a scalar such that $\|\alpha x\| = \|x\|$, show that either $x = 0$ or $|\alpha| = 1$. If $x \neq 0$, conclude that each ray $\{\alpha x : \alpha > 0\}$ intersects the boundary of the unit ball of $\|\cdot\|$ exactly once.

Definition 5.5.2. *A norm is* polyhedral *if its unit ball is a polyhedron.*

Exercise. Which of the l_p-norms are polyhedral?

Exercise. If $\|\cdot\|$ is a polyhedral norm and if $S \in M_n$ is nonsingular, is $\|\cdot\|_S$ polyhedral?

In a vector space that has a norm, the basic topological notions of open and closed sets are defined in the same way as in the Euclidean space $\mathbf{R}^n$.

Definition 5.5.3. *Let* $\|\cdot\|$ *be a norm on a real or complex vector space V, and let S be a subset of V. A point $x \in S$ is an* interior point *of S if there is some $\epsilon > 0$ such that $B(\epsilon; x) \subset S$. The set S is* open *if every point of S is an interior point; S is* closed *if its complement is open. A* limit point *of S is a point $x \in V$ such that $\lim_{k \to \infty} x^{(k)} = x$ with respect to $\|\cdot\|$ for some sequence $\{x^{(k)}\} \subset S$. The* closure *of S is the union of S with the set of its limit points. The* boundary *of S is the intersection of the closure of S with the closure of the complement of S. The set S is* bounded *if there exists some $M > 0$ such that $S \subset B_{\|\cdot\|}(M; 0)$. The set S is* compact *if from every covering $\cup_\alpha S_\alpha \supset S$ by open sets S_α one can extract finitely many open sets $S_{\alpha_1}, \ldots, S_{\alpha_N}$ such that $\cup_{i=1}^N S_{\alpha_i} \supset S$.*

Observation 5.5.4. *If $\|\cdot\|$ is a norm on a real or complex vector space V with positive dimension, then 0 is an interior point of the unit ball $B_{\|\cdot\|}$. This follows from homogeneity and positivity of the norm $\|\cdot\|$: $B_{\|\cdot\|}(\frac{1}{2}; 0) \subset B_{\|\cdot\|}(1; 0)$, with the boundary of the former in the interior of the latter.*

Observation 5.5.5. *The unit ball of a norm is* equilibrated; *that is, if x is in the unit ball, then so is αx for all scalars α such that $|\alpha| = 1$. This follows from the homogeneity property of the norm.*

Observation 5.5.6. *The unit ball of a norm on a finite-dimensional vector space is* compact: *It is bounded, and it is closed because the norm is always a continuous function. In the finite-dimensional case, a closed bounded set is compact, although this need not be true in the infinite-dimensional case. A basic property of compact sets is the Weierstrass theorem (see Appendix E): A continuous real-valued function on a compact set is bounded and achieves both its supremum and infimum on the set. For* *this reason, we usually refer to the "max" or "min" of such a function on a compact set.*

Exercise. Consider the complex vector space l_2 of vectors $x = [x_i]$ with countably many entries, endowed with the norm $\|x\|_2 = (\sum_{k=1}^{\infty} |x_k|^2)^{1/2}$. Show that $\|e_k - e_j\|_2 = \sqrt{2}$ for every pair of distinct unit basis vectors e_k and e_j, $k, j = 1, 2, \ldots$. Thus, no infinite subsequence of $e_1, e_2, e_3, \ldots$ can be a Cauchy sequence, so there can be no convergent subsequence. Conclude that the unit ball of l_2 cannot be compact.

Observation 5.5.7. *The unit ball of a norm is* convex*: If $\|x\| \leq 1, \|y\| \leq 1$, and $\alpha \in [0, 1]$, then*

$$\|\alpha x + (1 - \alpha)y\| \leq \|\alpha x\| + \|(1 - \alpha)y\|$$
$$= \alpha\|x\| + (1 - \alpha)\|y\| \leq \alpha + (1 - \alpha) \leq 1$$

and hence the convex combination $\alpha x + (1 - \alpha)y$ is in the unit ball.

The foregoing necessary conditions on the unit ball of a norm are also sufficient to characterize a norm.

Theorem 5.5.8. *A set B in a finite-dimensional real or complex vector space V with positive dimension is the unit ball of a norm if and only if B (i) is compact, (ii) is convex, (iii) is equilibrated, and (iv) has 0 as an interior point.*

Proof. The necessity of conditions (i)–(iv) has already been observed. To establish their sufficiency, consider any nonzero point $x \in V$. Construct a ray segment $\{\alpha x : 0 \leq \alpha \leq 1\}$ from the origin through x and define the "length" of x by the proportional distance along this ray from the origin to x, with the length of the interval of the ray from the origin to the unique point on the boundary of the unit ball serving as one unit. That is, we define $\|x\|$ by

$$\|x\| = \begin{cases} 0 & \text{if } x = 0 \\ \min\left\{\frac{1}{t} : t > 0 \text{ and } tx \in B\right\} & \text{if } x \neq 0 \end{cases}$$

This function is well-defined, finite, and positive for each nonzero vector x because B is compact and has 0 as an interior point. Using the equilibration assumption, it follows that $\|\cdot\|$ is a homogeneous function, so it remains only to check that it satisfies the triangle inequality. If x and y are given nonzero vectors, then $x/\|x\|$ and $y/\|y\|$ are unit vectors that are in B. By convexity, the vector

$$z = \frac{\|x\|}{\|x\| + \|y\|} \frac{x}{\|x\|} + \frac{\|y\|}{\|x\| + \|y\|} \frac{y}{\|y\|}$$

is also in B. Therefore, $\|z\| \leq 1$, and hence $\|x + y\| \leq \|x\| + \|y\|$. $\qquad \square$

Exercise. Provide details for the proof of (5.5.8), noting carefully where each of the four hypotheses is used.

Convexity of the unit ball of a norm is a fact with many deep and sometimes startling implications. One of these is the following duality theorem, which we state in the context of pre-norms. The key ideas involved are very natural (see Appendix B):

(a) $Co(S)$, the convex hull of a given set S in $\mathbf{R}^n$ or $\mathbf{C}^n$, is the smallest convex set containing S, namely, the intersection of all convex sets that contain S; (b) $\overline{Co(S)}$, the closure of the convex hull of S, is the intersection of all closed half-spaces (everything on one side of a hyperplane) that contain S, and (c) if x is a point that is in every closed half-space that contains S, then $x \in \overline{Co(S)}$. These geometric notions lead directly to the important fact that any norm is the dual of its dual norm.

Theorem 5.5.9 (Duality theorem). *Let f be a pre-norm on $V = \mathbf{R}^n$ or $\mathbf{C}^n$, let f^D denote the dual norm of f, let f^{DD} denote the dual norm of f^D, let $B = \{x \in V : f(x) \leq 1\}$, and let $B'' = \{x \in V : f^{DD}(x) \leq 1\}$. Then*

> *(a) $f^{DD}(x) \leq f(x)$ for all $x \in V$, so $B \subset B''$*
> *(b) $B'' = \overline{Co(S)}$, the closure of the convex hull of B*
> *(c) If f is a norm, then $B = B''$ and $f^{DD} = f$*
> *(d) If f is a norm and $x_0 \in V$ is given, then there is some $z \in V$ (not necessarily unique) such that $f^D(z) = 1$ and $f(x_0) = z^*x_0$, that is, $|z^*x| \leq f(x)$ for all $x \in V$ and $f(x_0) = z^*x_0$*

Proof. (a) If $x \in V$ is a given vector, then (5.4.13) ensures that $|y^*x| \leq f(x)f^D(y)$ for any $y \in V$, and hence

$$f^{DD}(x) = \max_{f^D(y)=1} |y^*x| \leq \max_{f^D(y)=1} f(x)f^D(y) = f(x)$$

Thus, $f^{DD}(x) \leq f(x)$ for all $x \in V$, an inequality that is equivalent to the geometric statement $B \subset B''$.

(b) The set $\{t \in V : \operatorname{Re} t^*v \leq 1\}$ is a closed half-space that contains the origin, and any such half-space can be represented in this way. Using the definition of the dual norm, let $u \in B''$ be a given point and observe that

$$u \in \{t : \operatorname{Re} t^*v \leq 1 \text{ for every } v \text{ such that } f^D(v) \leq 1\}$$
$$= \{t : \operatorname{Re} t^*v \leq 1 \text{ for every } v \text{ such that } v^*w \leq 1$$
$$\text{for every } w \text{ such that } f(w) \leq 1\}$$
$$= \{t : \operatorname{Re} t^*v \leq 1 \text{ for every } v \text{ such that } w^*v \leq 1 \text{ for all } w \in B\}$$

Thus, u lies in every closed half-space that contains B. Since the intersection of all such closed half-spaces is $\overline{Co(S)}$, we conclude that $u \in \overline{Co(S)}$. But the point $u \in B''$ is arbitrary, so $B'' \subset \overline{Co(S)}$. Since $Co(B)$ is the intersection of all convex sets containing B and B'' is a convex set that contains B, we have $Co(B) \subset B''$. The set B'' is the unit ball of a norm, so it is compact and hence is closed. We conclude that $\overline{Co(S)} \subset \overline{B''} = B''$ and hence $B'' = \overline{Co(S)}$.

(c) If f is a norm, then its unit ball B is convex and closed, so $B = \overline{Co(B)} = B''$. Since their unit balls are identical, the norms f and f^{DD} are the same.

(d) For each given $x_0 \in V$, (c) ensures that $f(x_0) = \max_{f^D(y)=1} \operatorname{Re} y^*x_0$, and compactness of the unit sphere of the norm f^D ensures that there is some z such that $f^D(z) = 1$ and $\max_{f^D(y)=1} \operatorname{Re} y^*x_0 = \operatorname{Re} z^*x_0$. If z^*x_0 were not real and nonnegative, there would be a real number θ such that $\operatorname{Re}(e^{-i\theta}z^*x_0) > 0 > \operatorname{Re} z^*x_0$ (of course,

$f^D(e^{i\theta}z) = f^D(z) = 1$), which would contradict maximality: $\mathrm{Re}\, z^*x_0 \geq \mathrm{Re}\, y^*x_0$ for all y in the unit sphere of f^D. $\qquad\square$

The assertion (c) in the preceding theorem – $f^{DD} = f$ for any norm f – is arguably the most important and widely useful part of the duality theorem. For example, it permits us to represent *any* norm f as

$$f(x) = \max_{f^D(y)=1} \mathrm{Re}\, y^*x \qquad (5.5.10)$$

This representation is an example of a *quasilinearization*.

The following corollary illustrates how duality can be used to give a very short conceptual proof of (5.4.19(c)).

Corollary 5.5.11. *An absolute norm on $\mathbf{R}^n$ or $\mathbf{C}^n$ is monotone.*

Proof. Suppose that $\|\cdot\|$ is an absolute norm on $\mathbf{F}^n$. Theorem 5.4.19(b) ensures that its dual $\|\cdot\|^D$ is absolute. The duality theorem tells us that $\|\cdot\|$ is the dual of the absolute norm $\|\cdot\|^D$, so it follows from (5.4.19(b)) that $\|\cdot\|$ is monotone. $\qquad\square$

Problems

5.5.P1 Show that a set in a normed linear space is closed if and only if it contains all its limit points.

5.5.P2 Show that every point of a set S in a normed linear space is a limit point of S, so that the closure of S is just the set of limit points of S.

5.5.P3 Give an example of a set in a normed linear space that is both open and closed. Give an example of a set that is neither open nor closed.

5.5.P4 Let S be a compact set in a real or complex vector space V with norm $\|\cdot\|$. Show that S is closed and bounded. If $\{x_\alpha\} \subset S$ is a given infinite sequence, show that there is a countable subsequence $\{x_{\alpha_i}\} \subset \{x_\alpha\}$ and a point $x \in S$ such that $\lim_{i\to\infty} x_{\alpha_i} = x$ with respect to $\|\cdot\|$. Show that any closed subset of a compact set is compact.

5.5.P5 What happens in (5.5.8) if $\dim V = 0$?

5.5.P6 For $x = [x_i] \in \mathbf{R}^2$ define $f(x) = |x_2|$. Show that f is a seminorm on $\mathbf{R}^2$ and describe the set $B = \{x \in \mathbf{R}^2 : f(x) \leq 1\}$. Which of the properties (i)-(iv) in (5.5.8) does B *not* have?

5.5.P7 If $\|\cdot\|_\alpha$ and $\|\cdot\|_\beta$ are norms on a vector space and if $\|\cdot\|$ is the norm defined by $\|x\| = \max\{\|x\|_\alpha, \|x\|_\beta\}$, show that $B_{\|\cdot\|} = B_{\|\cdot\|_\alpha} \cap B_{\|\cdot\|_\beta}$.

5.5.P8 Let $f(\cdot)$ be a pre-norm on $\mathbf{R}^n$ or $\mathbf{C}^n$. Show that $f^{DD}(\cdot)$ is the greatest norm that is uniformly less than or equal to $f(\cdot)$, that is, if $\|\cdot\|$ is a norm such that $\|x\| \leq f(x)$ for all x, show that $\|x\| \leq f^{DD}(x)$ for all x.

5.5.P9 Let $\|\cdot\|$ be an absolute norm on $\mathbf{F}^n$ ($\mathbf{R}^n$ or $\mathbf{C}^n$), let $z = [z_i] \in \mathbf{F}^n$ be a given nonzero vector, and let e_i be a standard basis vector in $\mathbf{F}^n$ (0.1.7) for some $i \in \{1, \ldots, n\}$. (a) Why is $\|e_i\| \|e_i\|^D \geq 1$? (b) Explain why $|z_i| \|e_i\| = \| |z_i| e_i \| \leq \| |z| \| = \|z\|$ and why $\|e_i\|^D = \max_{\|y\|=1} |y_i| \leq 1/\|e_i\|$. (c) Conclude that $\|e_i\| \|e_i\|^D = 1$ for each $i = 1, \ldots, n$

and revisit (5.4.P9). (d) A norm $\nu(\cdot)$ on $\mathbf{F}^n$ is said to be *standardized* if $\nu(e_i) = 1$ for each $i = 1, \ldots, n$. Explain why the dual of an absolute standardized norm is an absolute standardized norm.

5.5.P10 Let V be $\mathbf{R}^n$ or $\mathbf{C}^n$ and let $k \in \{1, \ldots, n\}$. Explain why $\|\cdot\|_{(k)} = \max\{\frac{1}{k}\|\cdot\|_1, \|\cdot\|_\infty\}$ is a norm on V and why its dual is the k-norm, that is, $\|\cdot\|_{(k)}^D = \|\cdot\|_{[k]}$. What does (5.5.P7) tell us about the unit ball of the norm $\|\cdot\|_{(k)}$? Draw a picture to illustrate the intersection property for the two k-norms on $\mathbf{R}^2$.

5.5.P11 Suppose that a norm $\|\cdot\|$ on $\mathbf{F}^n$ ($\mathbf{R}^n$ or $\mathbf{C}^n$) is *weakly monotone*:

$$\|[x_1 \ \ldots \ x_{k-1} \ 0 \ x_{k+1} \ \ldots \ x_n]^T\| \leq \|[x_1 \ \ldots \ x_{k-1} \ x_k \ x_{k+1} \ \ldots \ x_n]^T\|$$

for all $x \in \mathbf{F}^n$ and all $k = 1, \ldots, n$. (a) Explain why it satisfies the apparently stronger condition that arose in the proof of (5.4.19(c)): $\|[\alpha_1 x_1 \ \ldots \ \alpha_n x_n]^T\| \leq \|x\|$ for every $x \in \mathbf{C}^n$ and all choices of $\alpha_k \in [0, 1]$, $k = 1, \ldots, n$. Thus, if a point on the unit sphere of a weakly monotone norm is given, and if one of its coordinates is shrunk to zero, the entire line segment thus produced must be in the unit ball. Explain why a monotone norm is weakly monotone.

(b) Show that the parallelogram with vertices at $\pm[2 \ 2]^T$ and $\pm[1 \ -1]^T$ is the unit ball of a norm on $\mathbf{R}^2$ that is not weakly monotone. (c) Is the function $f(x) = |x_1 - x_2| + |x_2|$ a norm on $\mathbf{R}^2$? Is it monotone? Is it weakly monotone? Sketch its unit ball. (d) If $x = [x_1 \ x_2]^T$ is a point on the boundary of the unit ball of an absolute norm, then so are the points $[\pm x_1 \ \pm x_2]^T$ (all four possible choices). Illustrate this geometric property with a sketch and exhibit a unit ball of a norm on $\mathbf{R}^2$ that is not absolute. What happens in $\mathbf{R}^n$? (e) Sketch the polygon in $\mathbf{R}^2$ with vertices at $\pm[0 \ 1]^T$, $\pm[1 \ 0]^T$, and $\pm[1 \ 1]^T$. Explain why it is the unit ball of a norm on $\mathbf{R}^n$ that is weakly monotone but not monotone (and hence not absolute).

Further Readings. See Householder (1964) for more discussion of geometrical aspects of norms. The key idea in our proof of the duality theorem (identification of the unit ball of the second dual of a norm or pre-norm with the intersection of all the half-spaces containing its the unit ball) is used by von Neumann in the paper cited at the end of (5.4). See Valentine (1964) for a detailed discussion of convex sets, convex hulls, and half-spaces.

5.6 Matrix norms

Since M_n is itself a vector space of dimension n^2, one can measure the "size" of a matrix by using any norm on $\mathbf{C}^{n^2}$. However, M_n is not just a high-dimensional vector space; it has a natural multiplication operation, and it is often useful in making estimates to relate the "size" of a product AB to the "sizes" of A and B.

A function $\|\|\cdot\|\| : M_n \to \mathbf{R}$ is a *matrix norm* if, for all $A, B \in M_n$, it satisfies the following five axioms:

    (1) $\|\|A\|\| \geq 0$ Nonnegative

    (1a) $\|\|A\|\| = 0$ if and only if $A = 0$ Positive

    (2) $\|\|cA\|\| = |c| \ \|\|A\|\|$ for all $c \in \mathbf{C}$ Homogeneous

(3) $\||A + B\|| \leq \||A\|| + \||B\||$ Triangle Inequality

(4) $\||AB\|| \leq \||A\|| \, \||B\||$ Submultiplicativity

A matrix norm is sometimes called a *ring norm*. The first four properties of a matrix norm are identical to the axioms for a norm (5.1.1). A norm on matrices that does not satisfy property (4) for all A and B is a *vector norm on matrices*, sometimes called a *generalized matrix norm*. The notions of a matrix seminorm and a generalized matrix seminorm may also be defined via omission of axiom (1a).

Since $\||A^2\|| = \||AA\|| \leq \||A\|| \, \||A\|| = \||A\||^2$ for any matrix norm, it follows that $\||A\|| \geq 1$ for any nonzero matrix A for which $A^2 = A$. In particular, $\||I\|| \geq 1$ for any matrix norm. If A is nonsingular, then $I = AA^{-1}$, so $\||I\|| = \||AA^{-1}\|| \leq \||A\|| \, \||A^{-1}\||$, and we have the lower bound

$$\||A^{-1}\|| \geq \frac{\||I\||}{\||A\||}$$

valid for any matrix norm $\|| \cdot \||$.

Exercise. If $\|| \cdot \||$ is a matrix norm, show that $\||A^k\|| \leq \||A\||^k$ for every $k = 1, 2, \ldots$, and all $A \in M_n$. Give an example of a norm on matrices for which this inequality is not valid.

Some of the norms introduced in (5.2) are matrix norms when applied to the vector space M_n and some are not. The most familiar examples are the l_p-norms for $p = 1, 2, \infty$. They are already known to be norms, so only axiom (4) requires verification.

Example. The l_1-*norm* defined for $A \in M_n$ by

$$\|A\|_1 = \sum_{i,j=1}^{n} |a_{ij}| \tag{5.6.0.1}$$

is a matrix norm because

$$\|AB\|_1 = \sum_{i,j=1}^{n} \left| \sum_{k=1}^{n} a_{ik} b_{kj} \right| \leq \sum_{i,j,k=1}^{n} |a_{ik} b_{kj}|$$

$$\leq \sum_{i,j,k,m=1}^{n} |a_{ik} b_{mj}| = \left(\sum_{i,k=1}^{n} |a_{ik}| \right) \left(\sum_{j,m=1}^{n} |b_{mj}| \right)$$

$$= \|A\|_1 \|B\|_1$$

The first inequality comes from the triangle inequality, while the second comes from adding additional terms to the sum.

Example. The l_2-norm (*Frobenius norm, Schur norm*, or *Hilbert–Schmidt norm*) defined for $A \in M_n$ by

$$\|A\|_2 = |\operatorname{tr} AA^*|^{1/2} = \left(\sum_{i,j=1}^{n} |a_{ij}|^2 \right)^{1/2} \tag{5.6.0.2}$$

is a matrix norm because

$$\|AB\|_2 = \left(\sum_{i,j=1}^n \left|\sum_{k=1}^n a_{ik}b_{kj}\right|^2\right)^{1/2} \le \left(\sum_{i,j=1}^n \left(\sum_{k=1}^n |a_{ik}|^2\right)\left(\sum_{m=1}^n |b_{mj}|^2\right)\right)^{1/2}$$

$$= \left(\sum_{i,k=1}^n |a_{ik}|^2\right)^{1/2}\left(\sum_{m,j=1}^n |b_{mj}|^2\right)^{1/2} = \|A\|_2\|B\|_2$$

Notice that the Frobenius norm is an absolute norm; it is just the Euclidean norm of A thought of as a vector in $\mathbf{C}^{n^2}$. Since tr AA^* is the sum of the eigenvalues of AA^*, and these eigenvalues are just the squares of the singular values of A, we have an alternative characterization of the Frobenius norm (2.6.3.3):

$$\|A\|_2 = \sqrt{\sigma_1(A)^2 + \cdots + \sigma_n(A)^2}$$

The singular values of A are the same as those of A^*, and they are invariant under unitary equivalence transformations of A (2.6), so

$$\|A\|_2 = \|A^*\|_2 \text{ and } \|A\|_2 = \|UAV\|_2$$

for all unitary $U, V \in M_n$.

Exercise. Prove the two identities in the preceding display from the definition $\|A\|_2 = |\operatorname{tr} AA^*|^{1/2}$, without using properties of singular values.

Example. The l_∞-*norm* defined for $A \in M_n$ by

$$\|A\|_\infty = \max_{1 \le i,j \le n} |a_{ij}| \tag{5.6.0.3}$$

is a norm on the vector space M_n but is not a matrix norm. Consider the matrix $J = \left[\begin{smallmatrix} 1 & 1 \\ 1 & 1 \end{smallmatrix}\right] \in M_2$ and compute $J^2 = 2J$, $\|J\|_\infty = 1$, $\|J^2\|_\infty = \|2J\|_\infty = 2\|J\|_\infty = 2$. Since $\|J^2\|_\infty > \|J\|_\infty^2$, $\|\cdot\|_\infty$ is not submultiplicative. However, if we define

$$N(A) = n\|A\|_\infty, \qquad A \in M_n \tag{5.6.0.4}$$

then we have

$$N(AB) = n \max_{1 \le i,j \le n} \left|\sum_{k=1}^n a_{ik}b_{kj}\right| \le n \max_{1 \le i,j \le n} \sum_{k=1}^n |a_{ik}b_{kj}|$$

$$\le n \max_{1 \le i,j \le n} \sum_{k=1}^n \|A\|_\infty\|B\|_\infty = n\|A\|_\infty n\|B\|_\infty$$

$$= N(A)N(B)$$

Thus, a scalar multiple of the l_∞-norm on matrices is a matrix norm. This is not an accident; see (5.7.11).

The following example exhibits a matrix norm that lies between $\|A\|_\infty$ and $n\|A\|_\infty$.

Example. Let $A = [a_1 \; \cdots \; a_n] \in M_n$ be partitioned according to its columns and define

$$N_\infty(A) = \sum_{j=1}^n \|a_j\|_\infty \qquad (5.6.0.5)$$

One checks that $N_\infty(\cdot)$ is a norm, and if we let $B = [b_{ij}] = [b_1 \; \cdots \; b_n] \in M_n$, the following computation demonstrates that $N_\infty(\cdot)$ is a matrix norm:

$$N_\infty(AB) = \sum_{j=1}^n \|Ab_j\|_\infty = \sum_{j=1}^n \left\| \sum_{k=1}^n a_k b_{kj} \right\|_\infty \le \sum_{j=1}^n \sum_{k=1}^n \|a_k b_{kj}\|_\infty$$

$$= \sum_{j=1}^n \sum_{k=1}^n \|a_k\|_\infty |b_{kj}| \le \sum_{j=1}^n \sum_{k=1}^n \|a_k\|_\infty \|b_j\|_\infty$$

$$= \left(\sum_{k=1}^n \|a_k\|_\infty \right) \left(\sum_{j=1}^n \|b_j\|_\infty \right) = N_\infty(A) N_\infty(B)$$

For a structural proof that $N_\infty(\cdot)$ is a matrix norm, see (5.6.40) and the exercise that follows it.

Associated with each norm $\| \cdot \|$ on $\mathbf{C}^n$ is a matrix norm $\|\| \cdot \|\|$ that is "induced" by $\| \cdot \|$ on M_n according to the following definition.

Definition 5.6.1. *Let* $\| \cdot \|$ *be a norm on* $\mathbf{C}^n$. *Define* $\|\| \cdot \|\|$ *on* M_n *by*

$$\|\|A\|\| = \max_{\|x\|=1} \|Ax\|$$

Exercise. Show that the function defined in (5.6.1) may be computed in the following alternative ways:

$$\|\|A\|\| = \max_{\|x\| \le 1} \|Ax\| = \max_{x \ne 0} \frac{\|Ax\|}{\|x\|}$$

$$= \max_{\|x\|_\alpha = 1} \frac{\|Ax\|}{\|x\|}, \quad \text{in which } \| \cdot \|_\alpha \text{ is any given norm on } \mathbf{C}^n$$

Theorem 5.6.2. *The function* $\|\| \cdot \|\|$ *defined in (5.6.1) has the following properties:*

(a) $\|\|I\|\| = 1$
(b) $\|Ay\| \le \|\|A\|\| \|y\|$ *for any* $A \in M_n$ *and any* $y \in \mathbf{C}^n$
(c) $\|\| \cdot \|\|$ *is a matrix norm on* M_n
(d) $\|\|A\|\| = \max_{\|x\|=\|y\|^D=1} |y^* A x|$

Proof. (a) $\|\|I\|\| = \max_{\|x\|=1} \|Ix\| = \max_{\|x\|=1} \|x\| = 1$.
(b) The asserted inequality is correct for $y = 0$, so let $y \ne 0$ be given and consider the unit vector $y/\|y\|$. We have $\|\|A\|\| = \max_{\|x\|=1} \|Ax\| \ge \|A\frac{y}{\|y\|}\| = \|Ay\|/\|y\|$, so $\|\|A\|\| \|y\| \ge \|Ay\|$.
(c) We verify the five axioms:

Axiom (1): $\|\|A\|\|$ is the maximum of a nonnegative-valued function, so it is nonnegative.

Axiom (1a): If $A \neq 0$, there is a unit vector y such that $Ay \neq 0$, so $\||A\|| \geq \|Ay\| > 0$. If $A = 0$, then $Ax = 0$ for all x and hence $\||A\|| = 0$.

Axiom (2):

$$\||cA\|| = \max_{\|x\|=1} \|cAx\| = \max_{\|x\|=1}(|c| \, \|Ax\|)$$

$$= |c| \max_{\|x\|=1} \|Ax\| = |c| \, \||A\||$$

Axiom (3): For any unit vector x we have

$$\|(A + B)x\| = \|Ax + Bx\| \leq \|Ax\| + \|Bx\|$$

$$\leq \||A\|| + \||B\||$$

so $\||A + B\|| = \max_{\|x\|=1} \|(A + B)x\| \leq \||A\|| + \||B\||$.

Axiom (4): For any unit vector x we have

$$\|ABx\| = \|A(Bx)\| \leq \||A\|| \, \|Bx\|$$

$$\leq \||A\|| \, \||B\||$$

so $\||AB\|| = \max_{\|x\|=1} \|ABx\| \leq \||A\|| \, \||B\||$.

(d) Use the duality theorem (5.5.9(c)) to compute

$$\max_{\|x\|=\|y\|^D=1} |y^*Ax| = \max_{\|x\|=1}\left(\max_{\|y\|^D=1} |y^*Ax|\right) = \max_{\|x\|=1} \|Ax\|^{DD}$$

$$= \max_{\|x\|=1} \|Ax\| = \||A\||$$

$\square$

Definition 5.6.3. *The function* $\|| \cdot \||$ *defined in (5.6.1) is the* matrix norm induced by the norm $\| \cdot \|$. *It is sometimes called the* operator norm *or* lub norm *(least upper bound norm) associated with the vector norm* $\| \cdot \|$.

The inequality in (5.6.2(b)) says that the vector norm $\| \cdot \|$ is *compatible* with the matrix norm $\|| \cdot \||$. Theorem 5.6.2 shows that *given any norm on* $\mathbf{C}^n$, *there is a compatible matrix norm on* M_n.

A norm $\|| \cdot \||$ on matrices such that $\||I\|| = 1$ is said to be *unital*. The preceding theorem says that every induced matrix norm is unital. The l_∞ norm on matrices is unital but is not a matrix norm; (5.6.33.1) exhibits a unital matrix norm that is not induced.

An *induced* norm on matrices is *always* a matrix norm. Therefore, one way to prove that a nonnegative-valued function on M_n is a matrix norm is to show that it arises from some vector norm according to the prescription in (5.6.1). In each of the following examples of this principle, we take $A = [a_{ij}] \in M_n$.

Example 5.6.4. The *maximum column sum matrix norm* $\|| \cdot \||_1$ is defined on M_n by

$$\||A\||_1 = \max_{1 \leq j \leq n} \sum_{i=1}^{n} |a_{ij}|$$

We claim that $\|| \cdot \||_1$ is induced by the l_1-norm on $\mathbf{C}^n$ and hence is a matrix norm. To show this, partition A according to its columns as $A = [a_1 \ \ldots \ a_n]$.

Then $\||A\||_1 = \max_{1 \le i \le n} \|a_i\|_1$. If $x = [x_i]$, then

$$\|Ax\|_1 = \|x_1 a_1 + \cdots + x_n a_n\|_1 \le \sum_{i=1}^{n} \|x_i a_i\|_1 = \sum_{i=1}^{n} |x_i| \|a_i\|_1$$

$$\le \sum_{i=1}^{n} |x_i| \left(\max_{1 \le k \le n} \|a_k\|_1 \right) = \sum_{i=1}^{n} |x_i| \||A\||_1 = \|x\|_1 \||A\||_1$$

Thus, $\max_{\|x\|_1 = 1} \|Ax\|_1 \le \||A\||_1$. If we now choose $x = e_k$ (the kth unit basis vector), then for any $k = 1, \ldots, n$ we have

$$\max_{\|x\|_1 = 1} \|Ax\|_1 \ge \|1 a_k\|_1 = \|a_k\|_1$$

and hence

$$\max_{\|x\|_1 = 1} \|Ax\|_1 \ge \max_{1 \le k \le n} \|a_k\|_1 = \||A\||_1$$

Exercise. Prove directly from the definition that $\||\cdot\||_1$ is a matrix norm.

Example 5.6.5. The *maximum row sum matrix norm* $\||\cdot\||_\infty$ is defined on M_n by

$$\||A\||_\infty = \max_{1 \le i \le n} \sum_{j=1}^{n} |a_{ij}|$$

We claim that $\||\cdot\||_\infty$ is induced by the l_∞-norm on $\mathbf{C}^n$ and hence is a matrix norm. Compute

$$\|Ax\|_\infty = \max_{1 \le i \le n} \left| \sum_{j=1}^{n} a_{ij} x_j \right| \le \max_{1 \le i \le n} \sum_{j=1}^{n} |a_{ij} x_j|$$

$$\le \max_{1 \le i \le n} \sum_{j=1}^{n} |a_{ij}| \|x\|_\infty = \||A\||_\infty \|x\|_\infty$$

and hence $\max_{\|x\|_\infty = 1} \|Ax\|_\infty \le \||A\||_\infty$. If $A = 0$, there is nothing to prove, so we may assume that $A \ne 0$. Suppose that the kth row of A is nonzero and define the vector $z = [z_i] \in \mathbf{C}^n$ by

$$z_i = \frac{\bar{a}_{ki}}{|a_{ki}|} \quad \text{if } a_{ki} \ne 0$$

$$z_i = 1 \quad \text{if } a_{ki} = 0$$

Then $\|z\|_\infty = 1$, $a_{kj} z_j = |a_{kj}|$ for all $j = 1, 2, \ldots, n$, and

$$\max_{\|x\|_\infty = 1} \|Ax\|_\infty \ge \|Az\|_\infty = \max_{1 \le i \le n} \left| \sum_{j=1}^{n} a_{ij} z_j \right| \ge \left| \sum_{j=1}^{n} a_{kj} z_j \right| = \sum_{j=1}^{n} |a_{kj}|$$

Thus,

$$\max_{\|x\|_\infty = 1} \|Ax\|_\infty \ge \max_{1 \le k \le n} \sum_{j=1}^{n} |a_{kj}| = \||A\||_\infty$$

Exercise. Verify directly from the definition that $\||\cdot\||_\infty$ is a matrix norm on M_n.

Example 5.6.6. The *spectral norm* $\||\cdot\||_2$ is defined on M_n by

$$\||A\||_2 = \sigma_1(A), \text{ the largest singular value of } A$$

We claim that $\||\cdot\||_2$ is induced by the l_2-norm on $\mathbf{C}^n$ and hence is a matrix norm. Let $A = V\Sigma W^*$ be a singular value decomposition of A, in which V and W are unitary, $\Sigma = \text{diag}(\sigma_1, \ldots, \sigma_n)$, and $\sigma_1 \geq \cdots \geq \sigma_n \geq 0$; see (2.6.3). Use unitary invariance and monotonicity of the Euclidean norm (5.4.19) to compute

$$\max_{\|x\|_2=1} \|Ax\|_2 = \max_{\|x\|_2=1} \|V\Sigma W^*x\|_2 = \max_{\|x\|_2=1} \|\Sigma W^*x\|_2$$

$$= \max_{\|Wy\|_2=1} \|\Sigma y\|_2 = \max_{\|y\|_2=1} \|\Sigma y\|_2$$

$$\leq \max_{\|y\|_2=1} \|\sigma_1 y\|_2 = \sigma_1 \max_{\|y\|_2=1} \|y\|_2 = \sigma_1$$

However, $\|\Sigma y\|_2 = \sigma_1$ for $y = e_1$, so we conclude that $\max_{\|x\|_2=1} \|Ax\|_2 = \sigma_1(A)$.

Exercise. Use (4.2.2) to give an alternative proof that the spectral norm is induced by the Euclidean vector norm: $\max_{\|x\|_2=1} \|Ax\|_2^2 = \max_{\|x\|_2=1} x^*A^*Ax = \lambda_{\max}(A^*A) = \sigma_1(A)^2$.

Exercise. Give details for the following proof that the representation (5.6.2(d)) is correct for the spectral norm:

$$\max_{\|x\|_2=\|y\|_2=1} |y^*V\Sigma W^*x| = \max_{\|W\xi\|_2=\|V\eta\|_2=1} |\eta^*\Sigma\xi|$$

$$= \max_{\|\xi\|_2=\|\eta\|_2=1} |\eta^*\Sigma\xi| = \sigma_1(A)$$

Exercise. Explain why $\||UAV\||_2 = \||A\||_2$ for any $A \in M_n$ and any unitary $U, V \in M_n$.

We next show that new matrix norms may be created by inserting a fixed similarity into any matrix norm.

Theorem 5.6.7. *Suppose that $\||\cdot\||$ is a matrix norm on M_n and $S \in M_n$ is nonsingular. Then the function*

$$\||A\||_S = \||SAS^{-1}\|| \quad \text{for all } A \in M_n$$

is a matrix norm. Moreover, if $\||\cdot\||$ is induced by the norm $\|\cdot\|$ on $\mathbf{C}^n$, then the matrix norm $\||\cdot\||_S$ is induced by the norm $\|\cdot\|_S$ on $\mathbf{C}^n$ defined in (5.2.6).

Proof. The axioms (1), (1a), (2), and (3) are verified in a straightforward manner for $\||\cdot\||_S$. Submultiplicativity of $\||\cdot\||_S$ follows from a calculation:

$$\||AB\||_S = \||SABS^{-1}\|| = \||(SAS^{-1})(SBS^{-1})\||$$

$$\leq \||SAS^{-1}\|| \, \||SBS^{-1}\|| = \||A\||_S \, \||B\||_S$$

The final assertion follows from the computation

$$\max_{\|x\|_S=1} \|Ax\|_S = \max_{\|Sx\|=1} \|SAx\| = \max_{\|y\|=1} \|SAS^{-1}y\| = \||SAS^{-1}\||$$

$\square$

See (5.6.10) for an example of how (5.6.7) can be used to tailor a matrix norm for a specific purpose.

One important application of matrix norms is to provide bounds for the spectral radius of a matrix (1.2.9). If λ is *any* eigenvalue of A, $Ax = \lambda x$, and $x \neq 0$, consider the rank-one matrix $X = xe^T = [x \; \ldots \; x] \in M_n$ and observe that $AX = \lambda X$. If $||| \cdot |||$ is any matrix norm, then

$$|\lambda| \, |||X||| = |||\lambda X||| = |||AX||| \leq |||A||| \, |||X||| \tag{5.6.8}$$

and therefore $|\lambda| \leq |||A|||$. Since there is some eigenvalue λ for which $|\lambda| = \rho(A)$, it follows that $\rho(A) \leq |||A|||$. Now suppose that A is nonsingular and let λ be any eigenvalue of A. We know that λ^{-1} is an eigenvalue of A^{-1} and hence $|\lambda^{-1}| \leq |||A^{-1}|||$. We have proved the following theorem.

Theorem 5.6.9. *Let $||| \cdot |||$ be a matrix norm on M_n, let $A \in M_n$, and let λ be an eigenvalue of A. Then*

(a) $|\lambda| \leq \rho(A) \leq |||A|||$

If A is nonsingular, then

(b) $\rho(A) \geq |\lambda| \geq 1/|||A^{-1}|||$

Exercise. If $A, B \in M_n$ are normal, explain why $\rho(A) = |||A|||_2$ and $\rho(AB) \leq |||AB|||_2 \leq |||A|||_2 \, |||B|||_2 = \rho(A)\rho(B)$. Give an example of $C, D \in M_n$ such that $\rho(CD) > \rho(C)\rho(D)$.

Exercise. Let $A \in M_n$ have singular values $\sigma_1 \geq \cdots \geq \sigma_n \geq 0$ and absolute eigenvalues $|\lambda_1| \geq \cdots \geq |\lambda_n|$. Using the spectral norm and the bounds in the preceding theorem, show that $|\lambda_1| \leq \sigma_1$ and, if A is nonsingular, $|\lambda_n| \geq \sigma_n > 0$. *Hint*: What is the largest singular value of A^{-1}?

Exercise. Let $||| \cdot |||$ be a matrix norm on M_n. Show that (a) the function $|| \cdot ||$ defined on $\mathbf{C}^n$ by $||x|| = |||xe^T||| = |||[x \; \ldots \; x]|||$ is a norm on $\mathbf{C}^n$; and (b) $||Ax|| \leq |||A||| \, ||x||$ for all $x \in \mathbf{C}^n$ and all $A \subset M_n$, that is, the norm $|| \cdot ||$ on $\mathbf{C}^n$ is compatible with the matrix norm $||| \cdot |||$. Thus, *given any matrix norm on M_n, there is a compatible vector norm on $\mathbf{C}^n$.*

Exercise. Let $N(\cdot)$ be a norm on M_n, not necessarily a matrix norm, and suppose that there is a norm $|| \cdot ||$ on $\mathbf{C}^n$ that is compatible with it, that is, $||Ax|| \leq N(A) \, ||x||$ for all $A \in M_n$ and all $x \in \mathbf{C}^n$. Deduce that $N(A) \geq \rho(A)$ for all $A \in M_n$. *Hint*: Consider a nonzero x such that $Ax = \lambda x$ and $|\lambda| = \rho(A)$.

Although the spectral radius function is not itself a norm on M_n (see (5.6.P19)), for each $A \in M_n$, it is the greatest lower bound for the values of all matrix norms of A.

Lemma 5.6.10. *Let $A \in M_n$ and $\epsilon > 0$ be given. There is a matrix norm $||| \cdot |||$ such that $\rho(A) \leq |||A||| \leq \rho(A) + \epsilon$.*

Proof. Theorem 2.3.1 ensures that there is a unitary $U \in M_n$ and an upper triangular $\Delta \in M_n$ such that $A = U\Delta U^*$. Set $D_t = \text{diag}(t, t^2, t^3, \ldots, t^n)$ and compute

$$
D_t \Delta D_t^{-1} = \begin{bmatrix}
\lambda_1 & t^{-1}d_{12} & t^{-2}d_{13} & \cdots & t^{-n+1}d_{1n} \\
0 & \lambda_2 & t^{-1}d_{23} & \cdots & t^{-n+2}d_{2n} \\
0 & 0 & \lambda_3 & \cdots & t^{-n+3}d_{3n} \\
\cdot & \cdot & \cdot & \cdots & \cdot \\
0 & 0 & 0 & \cdots & t^{-1}d_{n-1,n} \\
0 & 0 & 0 & 0 & \lambda_n
\end{bmatrix}
$$

Thus, for $t > 0$ large enough, the sum of all the absolute values of the off-diagonal entries of $D_t \Delta D_t^{-1}$ is less than ϵ. In particular, $\||D_t \Delta D_t^{-1}\||_1 \le \rho(A) + \epsilon$ for all large enough t. Thus, if we define the matrix norm $\|| \cdot \||$ by

$$
\||B\|| = \||D_t U^* B U D_t^{-1}\||_1 = \||(D_t U^*)B(D_t U^*)^{-1}\||_1
$$

for any $B \in M_n$, and if we choose t large enough, then (5.6.7) ensures that we have constructed a matrix norm such that $\||A\|| \le \rho(A) + \epsilon$. Of course, the lower bound $\||A\|| \ge \rho(A)$ is valid for any matrix norm. $\qquad\square$

Exercise. Let $A \in M_n$ be given. Use the preceding results to show that $\rho(A) = \inf\{\||A\|| : \|| \cdot \|| \text{ is an induced matrix norm}\}$. *Hint:* (5.6.10) and (5.6.7). See (5.6.P38 and P39) for a characterization of the matrices A for which $\rho(A) = \||A\||$ for some matrix norm $\|| \cdot \||$.

We are interested in characterizing matrices A such that $A^k \to 0$ as $k \to \infty$. The following result is the final tool we need to attack this problem.

Lemma 5.6.11. *Let $A \in M_n$ be given. If there is a matrix norm $\|| \cdot \||$ such that $\||A\|| < 1$, then $\lim_{k\to\infty} A^k = 0$, that is, each entry of A^k tends to zero as $k \to \infty$.*

Proof. If $\||A\|| < 1$, then $\||A^k\|| \le \||A\||^k \to 0$ as $k \to \infty$. This says that $A^k \to 0$ with respect to the norm $\|| \cdot \||$, but since all norms on the n^2 dimensional normed linear space M_n are equivalent, it follows that $A^k \to 0$ with respect to the vector norm $\| \cdot \|_\infty$ on M_n. $\qquad\square$

Exercise. Compare the proof of the preceding theorem with the approach in (3.2.5).

Exercise. Give an example of a matrix $A \in M_n$ and two matrix norms $\||A\||_\alpha$ and $\|| \cdot \||_\beta$ such that $\||A\||_\alpha < 1$ and $\||A\||_\beta > 1$. Conclusion? Is $\lim_{k\to\infty} A^k = 0$ or not?

Matrices $A \in M_n$ such that $\lim_{k\to\infty} A^k = 0$ are called *convergent*; they are important in the analysis of iterative processes and many other applications. We are fortunate that they can be characterized by a spectral radius inequality.

Theorem 5.6.12. *Let $A \in M_n$. Then $\lim_{k\to\infty} A^k = 0$ if and only if $\rho(A) < 1$.*

Proof. If $A^k \to 0$ and if $x \neq 0$ is a vector such that $Ax = \lambda x$, then $A^k x = \lambda^k x \to 0$ only if $|\lambda| < 1$. Since this inequality must hold for every eigenvalue of A, we conclude that $\rho(A) < 1$. Conversely, if $\rho(A) < 1$, then (5.6.10) ensures that there is some matrix norm $||| \cdot |||$ such that $|||A||| < 1$. Thus, (5.6.11) ensures that $A^k \to 0$ as $k \to \infty$. □

Exercise. Consider the matrix $A = \begin{bmatrix} .5 & 1 \\ 0 & .5 \end{bmatrix} \in M_2$. Compute A^k and $\rho(A^k)$ explicitly for $k = 2, 3, \ldots$. Show that $\rho(A^k) = \rho(A)^k$. How do the following behave as $k \to \infty$? The entries of A^k; $|||A^k|||_1$; $|||A^k|||_\infty$; $|||A^k|||_2$.

Exercise. Let $A = \begin{bmatrix} .5 & 1 \\ -.125 & .5 \end{bmatrix}$, and define a sequence of vectors $x^{(0)}, x^{(1)}, x^{(2)}, \ldots$ by the recursion $x^{(k+1)} = Ax^{(k)}$, $k = 0, 1, \ldots$. Show that, regardless of the initial vector $x^{(0)}$ chosen, $x^{(k)} \to 0$ as $k \to \infty$. *Hint*: $x^{(k)} = A^k x^{(0)}$; select suitable norms and use the bound $\|x^{(k)}\| \le |||A^k||| \, \|x^{(0)}\|$.

Sometimes one needs bounds on the size of the entries of A^k as $k \to \infty$. One useful bound is an immediate consequence of the previous theorem.

Corollary 5.6.13. *Let $A \in M_n$ and $\epsilon > 0$ be given. There is a constant $C = C(A, \epsilon)$ such that $|(A^k)_{ij}| \le C(\rho(A) + \epsilon)^k$ for all $k = 1, 2, \ldots$ and all $i, j = 1, \ldots, n$.*

Proof. Consider the matrix $\tilde{A} = [\rho(A) + \epsilon]^{-1} A$, whose spectral radius is strictly less than 1. We know that $\tilde{A}^k \to 0$ as $k \to \infty$. In particular, the sequence $\{\tilde{A}^k\}$ is bounded, so there is some finite $C > 0$ such that $|(\tilde{A}^k)_{ij}| \le C$ for all $k = 1, 2, \ldots$ and all $i, j = 1, \ldots, n$. This is the asserted bound. □

Exercise. Let $A = \begin{bmatrix} a & 1 \\ 0 & a \end{bmatrix}$, compute A^k explicitly, and show that one may not always take $\epsilon = 0$ in (5.6.13).

Even though it is not accurate to say that individual entries of A^k behave like $\rho(A)^k$ as $k \to \infty$, the sequence $\{|||A^k|||\}$ does have this asymptotic behavior for any matrix norm $||| \cdot |||$.

Corollary 5.6.14 (Gelfand formula). *Let $||| \cdot |||$ be a matrix norm on M_n and let $A \in M_n$. Then $\rho(A) = \lim_{k \to \infty} |||A^k|||^{1/k}$.*

Proof. Since $\rho(A)^k = \rho(A^k) \le |||A^k|||$, we have $\rho(A) \le |||A^k|||^{1/k}$ for all $k = 1, 2, \ldots$. If $\epsilon > 0$ is given, the matrix $\tilde{A} = [\rho(A) + \epsilon]^{-1} A$ has spectral radius strictly less than 1 and hence is convergent. Thus, $|||\tilde{A}^k||| \to 0$ as $k \to \infty$ and there is some $N = N(\epsilon, A)$ such that $|||\tilde{A}^k||| \le 1$ for all $k \ge N$. This is just the statement that $|||\tilde{A}^k||| \le (\rho(A) + \epsilon)^k$ for all $k \ge N$, or that $|||A^k|||^{1/k} \le \rho(A) + \epsilon$ for all $k \ge N$. Since $\epsilon > 0$ is arbitrary and $\rho(A) \le |||A^k|||^{1/k}$ for all k, we conclude that $\lim_{k \to \infty} |||A^k|||^{1/k}$ exists and equals $\rho(A)$. □

Many questions about the convergence of infinite sequences or series of matrices can be answered using norms.

Exercise. Let $\{A_k\} \subset M_n$ be a given infinite sequence of matrices. Show that the series $\sum_{k=0}^{\infty} A_k$ converges to some matrix in M_n if there is a norm $\| \cdot \|$ on M_n such that the numerical series $\sum_{k=0}^{\infty} \|A_k\|$ is convergent (or even if its partial sums are bounded). *Hint*: Show that the partial sums form a Cauchy sequence.

Matrix norms are ideally suited to dealing with power series of matrices. The key fact from analysis is that a complex scalar power series $\sum_{k=0}^{\infty} a_k z^k$ has a *radius of convergence* $R \geq 0$: the power series is absolutely convergent if $|z| < R$ and it is divergent if $|z| > R$; $R = \infty$ and $R = 0$ are both possible, and either convergence or divergence may occur if $|z| = R$. The radius of convergence can be computed as $R = (\limsup_{k\to\infty} \sqrt[k]{|a_k|})^{-1}$, which is equal to $\lim_{k\to\infty} |\frac{a_k}{a_{k+1}}|$ if the limit exists (the ratio test). For a given $A \in M_n$ and any matrix norm $||| \cdot |||$, the computation

$$||| \sum a_k A^k ||| \leq \sum |||a_k A^k||| = \sum |a_k| \, |||A^k|||$$
$$\leq \sum |a_k| \, |||A|||^k$$

reveals that a matrix power series $\sum_{k=0}^{\infty} a_k A^k$ is convergent if $|||A||| < R$, the radius of convergence of the corresponding scalar power series. However, $||| \cdot |||$ can be *any* matrix norm, and (5.6.10) ensures that such a matrix norm exists if and only if $\rho(A) < R$. We summarize these observations in the following theorem.

Theorem 5.6.15. *Let R be the radius of convergence of a scalar power series $\sum_{k=0}^{\infty} a_k z^k$, and let $A \in M_n$ be given. The matrix power series $\sum_{k=0}^{\infty} a_k A^k$ converges if $\rho(A) < R$. This condition is satisfied if there is a matrix norm $||| \cdot |||$ on M_n such that $|||A||| < R$.*

Exercise. Suppose that an analytic function $f(z)$ is defined in a neighborhood of zero by a power series $f(z) = \sum_{k=0}^{\infty} a_k z^k$ that has radius of convergence $R > 0$, and let $||| \cdot |||$ be a matrix norm on M_n. Explain why $f(A) = \sum_{k=0}^{\infty} a_k A^k$ is well-defined for all $A \in M_n$ such that $|||A||| < R$. More generally, explain why $f(A)$ is well-defined for all $A \in M_n$ such that $\rho(A) < R$.

Exercise. What is the radius of convergence of the power series for the exponential function $e^z = \sum_{k=0}^{\infty} \frac{1}{k!} z^k$? Explain why the matrix exponential function given by the power series $e^A = \sum_{k=0}^{\infty} \frac{1}{k!} A^k$ is well-defined for every $A \in M_n$.

Exercise. How would you define $\cos A$? $\sin A$? $\log(I - A)$? For what matrices are they defined?

If $A \in M_n$ is diagonalizable, $A = S\Lambda S^{-1}$, $\Lambda = \text{diag}(\lambda_1, \ldots, \lambda_n)$, and the domain of a given complex-valued function f includes the set $\{\lambda_1, \ldots, \lambda_n\}$, the *primary matrix function* $f(A)$ is defined by $f(A) = Sf(\Lambda)S^{-1}$, in which $f(\Lambda) = \text{diag}(f(\lambda_1), \ldots, f(\lambda_n))$. This definition seems to depend on the choice of the (always nonunique) diagonalizing matrix S, but it actually doesn't. To see why, it is convenient to assume that any equal eigenvalues are grouped together in Λ, as in (1.3.13). If $A = T\Lambda T^{-1}$, then (1.3.27) ensures that $T = SR$, in which R is a block diagonal matrix conformal to Λ; its essential feature is that $R\Lambda = \Lambda R$ and hence also $Rf(\Lambda) = f(\Lambda)R$, since $f(\Lambda)$ is a block diagonal direct sum of scalar matrices that is conformal with R. Then $f(A) = f(T\Lambda T^{-1}) = Tf(\Lambda)T^{-1} = SRf(\Lambda)R^{-1}S = Sf(\Lambda)RR^{-1}S = Sf(\Lambda)S$, which shows that a primary matrix function of a diagonalizable matrix is well-defined.

In the preceding definition of $f(A)$ as a primary matrix function, rather than as a power series, we demand less of the function f (it need not be analytic), but we demand more of the matrix (it must be diagonalizable). Primary matrix functions of nondiagonalizable matrices can be defined, but one must require something about their differentiability; see chapter 6 of Horn and Johnson (1991).

Exercise. If $A \in M_n$ is diagonalizable, and if an analytic function $f(z) = \sum_{k=0}^{\infty} a_k z^k$ is defined by a power series with radius of convergence greater than $\rho(A)$, show that the primary matrix function definition of $f(A)$ agrees with its power series definition. *Hint*: Consider $\sum_{k=0}^{\infty} a_k (S \Lambda S^{-1})^k = S(\sum_{k=0}^{\infty} a_k \Lambda^k) S^{-1}$.

Corollary 5.6.16. *A matrix* $A \in M_n$ *is nonsingular if there is a matrix norm* $||| \cdot |||$ *such that* $||| I - A ||| < 1$. *If this condition is satisfied,*

$$A^{-1} = \sum_{k=0}^{\infty} (I - A)^k$$

Proof. If $||| I - A ||| < 1$, then the series $\sum_{k=0}^{\infty} (I - A)^k$ converges to some matrix C because the radius of convergence of the series $\sum z^k$ is 1. But since

$$A \sum_{k=0}^{N} (I - A)^k = (I - (I - A)) \sum_{k=0}^{N} (I - A)^k = I - (I - A)^{N+1} \to I$$

as $N \to \infty$, we conclude that $C = A^{-1}$. □

Exercise. Explain why the statement in the preceding corollary is equivalent to the following statement: If $||| \cdot |||$ is a matrix norm, and if $||| A ||| < 1$, then $I - A$ is nonsingular and $(I - A)^{-1} = \sum_{k=0}^{\infty} A^k$.

Exercise. Let $||| \cdot |||$ be a matrix norm on M_n. Suppose that $A, B \in M_n$ satisfy the inequality $||| BA - I ||| < 1$. Show that A and B are both nonsingular. One may think of B as an *approximate inverse* of A.

Exercise. If a matrix norm $||| \cdot |||$ has the property that $||| I ||| = 1$ (which would be the case if it were an induced norm), and if $A \in M_n$ is such that $||| A ||| < 1$, show that

$$\frac{1}{1 + ||| A |||} \leq ||| (I - A)^{-1} ||| \leq \frac{1}{1 - ||| A |||}$$

Hint: Use the inequality $||| (I - A)^{-1} ||| \leq \sum_{k=0}^{\infty} ||| A |||^k$ for the upper bound. Use the general inequality $||| B^{-1} ||| \geq 1/||| B |||$ and the triangle inequality for the lower bound.

Exercise. Let $||| \cdot |||$ be a matrix norm, so that $||| I ||| \geq 1$. Show that

$$\frac{||| I |||}{||| I ||| + ||| A |||} \leq ||| (I - A)^{-1} ||| \leq \frac{||| I ||| - (||| I ||| - 1) ||| A |||}{1 - ||| A |||}$$

whenever $||| A ||| < 1$.

Exercise. If $A, B \in M_n$, if A is nonsingular, and if $A + B$ is singular, show that $||| B ||| \geq 1/||| A^{-1} |||$ for any matrix norm $||| \cdot |||$. Thus, there is an intrinsic limit to how well a nonsingular matrix can be approximated by a singular one. *Hint*: $A + B = A(I + A^{-1}B)$. If $||| A^{-1}B ||| < 1$, then $I + A^{-1}B$ would be nonsingular.

One useful and easily computed criterion for nonsingularity is a consequence of the preceding corollary.

Corollary 5.6.17. *Let $A = [a_{ij}] \in M_n$. If $|a_{ii}| > \sum_{j \neq i} |a_{ij}|$ for all $i = 1, \ldots, n$, then A is nonsingular.*

Proof. The hypothesis ensures that every main diagonal entry of A is nonzero. Set $D = \text{diag}(a_{11}, \ldots, a_{nn})$ and check that $D^{-1}A$ has all 1s on the main diagonal, $B = [b_{ij}] = I - D^{-1}A$ has all 0s on the main diagonal, and $b_{ij} = -a_{ij}/a_{ii}$ if $i \neq j$. Consider the maximum row sum matrix norm $||| \cdot |||_\infty$. The hypothesis guarantees that $|||B|||_\infty < 1$, so (5.6.16) ensures that $I - B = D^{-1}A$ is nonsingular, and hence A is nonsingular. $\quad\square$

A matrix that satisfies the hypothesis of (5.6.17) is said to be *strictly diagonally dominant*. This sufficient condition for nonsingularity is known as the *Levy–Desplanques theorem*, and it can be improved somewhat; see (6.1), (6.2), and (6.4).

We now focus on the induced matrix norms defined in (5.6.1), which have an important minimality property. Because one often wishes to establish that a given matrix A is convergent by using the test $|||A||| < 1$, it is natural to prefer matrix norms that are uniformly as small as possible. All induced matrix norms have this desirable property, which actually characterizes them.

Any two norms on a finite-dimensional space are equivalent, so for any given pair of matrix norms $||| \cdot |||_\alpha$ and $||| \cdot |||_\beta$ there is a least finite positive constant $C_{\alpha\beta}$ such that $|||A|||_\alpha \leq C_{\alpha\beta}|||A|||_\beta$ for all $A \in M_n$. This constant can be computed as

$$C_{\alpha\beta} = \max_{A \neq 0} \frac{|||A|||_\alpha}{|||A|||_\beta}$$

If the roles of α and β are reversed, there must be a similarly defined least finite positive constant $C_{\beta\alpha}$ such that $|||A|||_\beta \leq C_{\beta\alpha}|||A|||_\alpha$ for all $A \in M_n$. In general, there is no obvious relation between $C_{\alpha\beta}$ and $C_{\beta\alpha}$, but if we examine the table in (5.6.P23), we see that its upper left 3×3 corner is symmetric; that is, $C_{\alpha\beta} = C_{\beta\alpha}$ for each pair of the three matrix norms $||| \cdot |||_1$, $||| \cdot |||_2$, and $||| \cdot |||_\infty$. All three of these matrix norms are induced norms, and the following theorem shows that this symmetry reflects a property of all induced norms: If $|||A|||_\alpha \leq C|||A|||_\beta$ for all $A \in M_n$, then $|||A|||_\beta \leq C|||A|||_\alpha$ for all $A \in M_n$, that is,

$$\frac{1}{C}|||A|||_\beta \leq |||A|||_\alpha \leq C|||A|||_\beta \text{ for all } A \in M_n$$

Theorem 5.6.18. *Let $|| \cdot ||_\alpha$ and $|| \cdot ||_\beta$ be given norms on $\mathbf{C}^n$. Let $||| \cdot |||_\alpha$ and $||| \cdot |||_\beta$ denote their respective induced matrix norms on M_n, that is,*

$$|||A|||_\alpha = \max_{x \neq 0} \frac{||Ax||_\alpha}{||x||_\alpha} \quad and \quad |||A|||_\beta = \max_{x \neq 0} \frac{||Ax||_\beta}{||x||_\beta}$$

Define

$$R_{\alpha\beta} = \max_{x \neq 0} \frac{||x||_\alpha}{||x||_\beta} \quad and \quad R_{\beta\alpha} = \max_{x \neq 0} \frac{||x||_\beta}{||x||_\alpha} \tag{5.6.19}$$

Then

$$\max_{A \neq 0} \frac{\|\|A\|\|_\alpha}{\|\|A\|\|_\beta} = R_{\alpha\beta} R_{\beta\alpha} \tag{5.6.20}$$

and

$$\max_{A \neq 0} \frac{\|\|A\|\|_\alpha}{\|\|A\|\|_\beta} = \max_{A \neq 0} \frac{\|\|A\|\|_\beta}{\|\|A\|\|_\alpha} = R_{\alpha\beta} R_{\beta\alpha} \tag{5.6.21}$$

Proof. The inequalities (5.6.10) say that $\|x\|_\alpha \leq R_{\alpha\beta} \|x\|_\beta$ and $\|y\|_\beta \leq R_{\beta\alpha} \|y\|_\alpha$ for all $x, y \in \mathbf{C}^n$, with equality possible in both cases for some nonzero vectors. Let a nonzero $A \in M_n$ be given, and let $\xi \in \mathbf{C}^n$ be a nonzero vector such that $\|A\xi\|_\alpha = \|\|A\|\|_\alpha \|\xi\|_\alpha$. Then

$$\|\|A\|\|_\alpha = \frac{\|A\xi\|_\alpha}{\|\xi\|_\alpha} = \frac{\|\xi\|_\beta \|A\xi\|_\alpha}{\|\xi\|_\alpha \|\xi\|_\beta} \leq \frac{\|\xi\|_\beta}{\|\xi\|_\alpha} \frac{R_{\alpha\beta} \|A\xi\|_\beta}{\|\xi\|_\beta}$$

$$\leq R_{\beta\alpha} R_{\alpha\beta} \frac{\|A\xi\|_\beta}{\|\xi\|_\beta} \leq R_{\beta\alpha} R_{\alpha\beta} \|\|A\|\|_\beta \tag{5.6.22}$$

for *all* nonzero A. Thus,

$$\max_{A \neq 0} \frac{\|\|A\|\|_\alpha}{\|\|A\|\|_\beta} \leq R_{\alpha\beta} R_{\beta\alpha}$$

We claim that there is *some* nonzero $B \in M_n$ for which the inequality (5.6.22) can be reversed, in which case we would have

$$R_{\alpha\beta} R_{\beta\alpha} \leq \frac{\|\|B\|\|_\alpha}{\|\|B\|\|_\beta} \leq \max_{A \neq 0} \frac{\|\|A\|\|_\alpha}{\|\|A\|\|_\beta} \leq R_{\alpha\beta} R_{\beta\alpha}$$

and (5.6.20) would be proved.

To verify our claim, let y_0 and z_0 be nonzero vectors such that $\|y_0\|_\alpha = R_{\alpha\beta} \|y_0\|_\beta$ and $\|z_0\|_\beta = R_{\beta\alpha} \|z_0\|_\alpha$. Theorem 5.5.9(d) ensures that there is a $w \in \mathbf{C}^n$ such that

(a) $|w^*x| \leq \|x\|_\beta$ for all $x \subset \mathbf{C}^n$
(b) $w^*z_0 = \|z_0\|_\beta$

Consider the matrix $B = y_0 w^*$. Using (b), we have

$$\frac{\|Bz_0\|_\alpha}{\|z_0\|_\alpha} = \frac{\|y_0 w^* z_0\|_\alpha}{\|z_0\|_\alpha} = \frac{|w^* z_0| \|y_0\|_\alpha}{\|z_0\|_\alpha} = \frac{\|z_0\|_\beta}{\|z_0\|_\alpha} \frac{\|y_0\|_\alpha}{\|y_0\|_\beta}$$

so we have the lower bound

$$\|\|B\|\|_\alpha \geq \frac{\|z_0\|_\beta \|y_0\|_\alpha}{\|z_0\|_\alpha} = \frac{\|z_0\|_\beta}{\|z_0\|_\alpha} \frac{\|y_0\|_\alpha}{\|y_0\|_\beta} \|y_0\|_\beta = R_{\beta\alpha} R_{\alpha\beta} \|y_0\|_\beta$$

On the other hand, we can use (a) to obtain

$$\frac{\|By_0\|_\beta}{\|y_0\|_\beta} = \frac{\|y_0 w^* y_0\|_\beta}{\|y_0\|_\beta} = \frac{|w^* y_0| \|y_0\|_\beta}{\|y_0\|_\beta} = |w^* y_0| \leq \|y_0\|_\beta$$

and hence we have the upper bound $\|\|B\|\|_\beta \leq \|y_0\|_\beta$. Combining these two bounds, we have

$$\|\|B\|\|_\alpha \geq R_{\beta\alpha} R_{\alpha\beta} \|y_0\|_\beta \geq R_{\alpha\beta} R_{\beta\alpha} \|\|B\|\|_\beta$$

as desired.

The assertion (5.6.21) follows from symmetry in α and β of the right-hand side of the identity (5.6.20). $\qquad\qquad\square$

When do two given norms on $\mathbf{C}^n$ induce the same matrix norm on M_n? The answer is that one of the norms must be a scalar multiple of the other.

Lemma 5.6.23. *Let* $\|\cdot\|_\alpha$ *and* $\|\cdot\|_\beta$ *be norms on* $\mathbf{C}^n$, *and let* $\|\|\cdot\|\|_\alpha$ *and* $\|\|\cdot\|\|_\beta$ *denote their respective induced matrix norms on* M_n. *Then*

$$R_{\alpha\beta}R_{\beta\alpha} \geq 1 \qquad\qquad (5.6.24)$$

Moreover, the following are equivalent:

(a) $R_{\alpha\beta}R_{\beta\alpha} = 1$.
(b) *There is some* $c > 0$ *such that* $\|x\|_\alpha = c\|x\|_\beta$ *for all* $x \in \mathbf{C}^n$.
(c) $\|\|\cdot\|\|_\alpha = \|\|\cdot\|\|_\beta$.

Proof. Observe that

$$R_{\beta\alpha} = \max_{x\neq 0} \frac{\|x\|_\beta}{\|x\|_\alpha} = \left(\min_{x\neq 0}\frac{\|x\|_\alpha}{\|x\|_\beta}\right)^{-1} \geq \left(\max_{x\neq 0}\frac{\|x\|_\alpha}{\|x\|_\beta}\right)^{-1} = \frac{1}{R_{\alpha\beta}}$$

with equality if and only if

$$\min_{x\neq 0}\frac{\|x\|_\alpha}{\|x\|_\beta} = \max_{x\neq 0}\frac{\|x\|_\alpha}{\|x\|_\beta}$$

which occurs if and only if the function $\|x\|_\alpha/\|x\|_\beta$ is constant for all $x \neq 0$. Thus, (a) and (b) are equivalent. If $\|\cdot\|_\alpha = c\|\cdot\|_\beta$, then for any $A \in M_n$, we have

$$\|\|A\|\|_\alpha = \max_{x\neq 0}\frac{\|Ax\|_\alpha}{\|x\|_\alpha} = \max_{x\neq 0}\frac{c\|Ax\|_\beta}{c\|x\|_\beta}$$
$$= \max_{x\neq 0}\frac{\|Ax\|_\beta}{\|x\|_\beta} = \|\|A\|\|_\beta$$

so (b) $\Rightarrow$ (c). Finally, if $\|\|\cdot\|\|_\alpha = \|\|\cdot\|\|_\beta$, then (5.6.20) shows that $R_{\alpha\beta}R_{\beta\alpha} = 1$, and hence (c) $\Rightarrow$ (a). $\qquad\qquad\square$

Corollary 5.6.25. *Let* $\|\|\cdot\|\|_\alpha$ *and* $\|\|\cdot\|\|_\beta$ *be induced matrix norms on* M_n. *Then* $\|\|A\|\|_\alpha \leq \|\|A\|\|_\beta$ *for all* $A \in M_n$ *if and only if* $\|\|A\|\|_\alpha = \|\|A\|\|_\beta$ *for all* $A \in M_n$.

Proof. If $\|\|A\|\|_\alpha \leq \|\|A\|\|_\beta$ for all $A \in M_n$, then (5.6.21) ensures that $\|\|A\|\|_\beta \leq \|\|A\|\|_\alpha$ for all $A \in M_n$. $\qquad\qquad\square$

The preceding corollary says that no *induced matrix norm* is uniformly less than an induced matrix norm that is different from it. The following theorem says more: No *matrix norm* is uniformly less than an induced matrix norm that is different from it.

Theorem 5.6.26. *Let* $\|\|\cdot\|\|$ *be a given matrix norm on* M_n, *let* $\|\|\cdot\|\|_\alpha$ *be a given induced matrix norm on* M_n, *let a nonzero* $z \in \mathbf{C}^n$ *be given, and define*

$$\|x\|_z = \|\|xz^*\|\| \text{ for any } x \in \mathbf{C}^n \qquad\qquad (5.6.27)$$

Then

(a) $\|\cdot\|_z$ *is a norm on* $\mathbf{C}^n$

(b) *The induced matrix norm*

$$N_z(A) = \max_{x \neq 0} \frac{\|Ax\|_z}{\|x\|_z} = \max_{x \neq 0} \frac{\||Axz^*\||}{\||xz^*\||} \qquad (5.6.28)$$

satisfies the inequality $N_z(A) \leq \||A\||$ *for every* $A \in M_n$

(c) $\||A\|| \leq \||A\||_\alpha$ *for every* $A \in M_n$ *if and only if* $N_z(A) = \||A\|| = \||A\||_\alpha$ *for every*
$A \in M_n$.

Proof. (a) One verifies that $\|\cdot\|_z$ satisfies the four axioms in (5.1.1); submultiplicativity
of $\|| \cdot \||$ is not necessary for this purpose.
(b) Use submultiplicativity of $\|| \cdot \||$ to compute

$$N_z(A) = \max_{x \neq 0} \frac{\|Ax\|_z}{\|x\|_z} = \max_{x \neq 0} \frac{\||Axz^*\||}{\||xz^*\||} \leq \max_{x \neq 0} \frac{\||A\|| \, \||xz^*\||}{\||xz^*\||}$$

$$= \||A\||$$

for all $A \in M_n$.
(c) Suppose that $\||A\|| \leq \||A\||_\alpha$ for all $A \in M_n$. Then (b) ensures that $N_z(A) \leq \||A\|| \leq$
$\||A\||_\alpha$ for all $A \in M_n$. But $N_z(\cdot)$ and $\|| \cdot \||_\alpha$ are both induced norms, so (5.6.25) ensures
that $N_z(A) = \||A\||_\alpha$ for all $A \in M_n$. $\qquad \square$

Exercise. Let $\|| \cdot \||$ be a given induced matrix norm on M_n and let $N_z(\cdot)$ be the
induced matrix norm defined in (5.6.28). Deduce from the preceding theorem
that $N_z(\cdot) = \|| \cdot \||$ for each nonzero $z \in \mathbf{C}^n$.

The result in the preceding exercise can be approached in an instructively different
fashion. For a given induced matrix norm $\|| \cdot \||$ we use (5.6.2(d)) and (5.5.9(d)) to
compute

$$\||Axz^*\|| = \max_{\|\xi\| = \|\eta\|^D = 1} |\eta^* Axz^* \xi| = \max_{\|\eta\|^D = 1} |\eta^* Ax| \max_{\|\xi\| = 1} |\xi^* z|$$

$$= \|Ax\|^{DD} \|z\|^D = \|Ax\| \, \|z\|^D \qquad (5.6.29)$$

The special case $A = I$ gives the identity

$$\||xz^*\|| = \|x\|_z = \|x\| \, \|z\|^D \qquad (5.6.30)$$

If $z \neq 0$ we then have

$$N_z(\cdot) = \max_{x \neq 0} \frac{\||Axz^*\||}{\||xz^*\||} = \max_{x \neq 0} \frac{\|Ax\| \, \|z\|^D}{\|x\| \, \|z\|^D}$$

$$= \max_{x \neq 0} \frac{\|Ax\|}{\|x\|} = \||A\||$$

The preceding results motivate the following definition and a further property of
induced/minimal matrix norms.

Definition 5.6.31. *A matrix norm* $\|| \cdot \||$ *on* M_n *is a* minimal matrix norm *(or just
minimal) if the only matrix norm* $N(\cdot)$ *on* M_n *such that* $N(A) \leq \||A\||$ *for all* $A \in M_n$
is $N(\cdot) = \|| \cdot \||$.

Theorem 5.6.32. *Let $\vert\vert\vert \cdot \vert\vert\vert$ be a matrix norm on M_n. For a nonzero $z \in \mathbf{C}^n$ let $N_z(\cdot)$ be the induced matrix norm defined by (5.6.27) and (5.6.28). The following are equivalent:*

(a) *$\vert\vert\vert \cdot \vert\vert\vert$ is an induced matrix norm.*
(b) *$\vert\vert\vert \cdot \vert\vert\vert$ is a minimal matrix norm.*
(c) *$\vert\vert\vert \cdot \vert\vert\vert = N_z(\cdot)$ for all nonzero $z \in \mathbf{C}^n$.*
(d) *$\vert\vert\vert \cdot \vert\vert\vert = N_z(\cdot)$ for some nonzero $z \in \mathbf{C}^n$.*

Proof. The implication (a) $\Rightarrow$ (b) follows from (5.6.26(c)). The implication (b) $\Rightarrow$ (c) is (5.6.26(b)). The implications (c) $\Rightarrow$ (d) $\Rightarrow$ (a) are straightforward. $\square$

Somewhat more can be gleaned from these observations. If $\vert\vert\vert \cdot \vert\vert\vert$ is a matrix norm and $N_y(\cdot) = N_z(\cdot)$ for all nonzero $y, z \in \mathbf{C}^n$, then the implication (c) $\Rightarrow$ (b) in (5.6.23) ensures that there is a positive constant c_{yz} such that $\|x\|_y = c_{yz}\|x\|_z$ for all $x \in \mathbf{C}^n$. For example, if $\vert\vert\vert \cdot \vert\vert\vert$ is *induced*, the preceding theorem ensures that $N_z(\cdot)$ is independent of z and the following exercise identifies the constant c_{yz}.

Exercise. If $\vert\vert\vert \cdot \vert\vert\vert$ is induced by a norm $\| \cdot \|$ on $\mathbf{C}^n$, show that $c_{yz} = \|y\|^D/\|z\|^D$. *Hint*: (5.6.30).

Theorem 5.6.33. *Let $\vert\vert\vert \cdot \vert\vert\vert$ be a matrix norm on M_n and let $\| \cdot \|_z$ be the norm on $\mathbf{C}^n$ defined by (5.6.27). The following two statements are equivalent:*

(a) *For each pair of nonzero vectors $y, z \in \mathbf{C}^n$ there is a positive constant c_{yz} such that $\|x\|_y = c_{yz}\|x\|_z$ for all $x \in \mathbf{C}^n$.*
(b) *$\vert\vert\vert xy^* \vert\vert\vert \, \vert\vert\vert zz^* \vert\vert\vert = \vert\vert\vert xz^* \vert\vert\vert \, \vert\vert\vert zy^* \vert\vert\vert$ for all $x, y, z \in \mathbf{C}^n$.*

Suppose that $\vert\vert\vert \cdot \vert\vert\vert$ is induced by a norm $\| \cdot \|$ on $\mathbf{C}^n$. Then

(c) *$\vert\vert\vert \cdot \vert\vert\vert$ satisfies (a) and (b) with $c_{yz} = \|y\|^D / \|z\|^D$, and for any nonzero $z \in \mathbf{C}^n$ we have*

$$\|x\|_y = \vert\vert\vert xy^* \vert\vert\vert = \frac{\vert\vert\vert xz^* \vert\vert\vert \, \vert\vert\vert zy^* \vert\vert\vert}{\vert\vert\vert zz^* \vert\vert\vert} = \frac{\|x\|_z \|z\|_y}{\|z\|_z}$$

Proof. Assume (a). Since (b) is correct if either $y = 0$ or $z = 0$, we may assume that $y \neq 0 \neq z$. Then

$$\vert\vert\vert xz^* \vert\vert\vert \, \vert\vert\vert zy^* \vert\vert\vert = \|x\|_z \|z\|_y = c_{yz}^{-1}\|x\|_y \, c_{yz}\|z\|_z$$
$$= \|x\|_y \|z\|_z = \vert\vert\vert xy^* \vert\vert\vert \, \vert\vert\vert zz^* \vert\vert\vert$$

Conversely, if we assume (b) and if $y \neq 0 \neq z$, then (a) follows with $c_{yz} = \vert\vert\vert zy^* \vert\vert\vert / \vert\vert\vert zz^* \vert\vert\vert$.

If $\vert\vert\vert \cdot \vert\vert\vert$ is induced, then (5.6.30) identifies $\|x\|_y = \|x\| \, \|y\|^D$ and $\|x\|_z = \|x\| \, \|z\|^D$. A calculation reveals that (a) is satisfied with $\|x\|_y = \|y\|^D \|x\|_z / \|z\|^D$ and hence (b) is also satisfied. $\square$

Exercise. Any positive scalar multiple of an induced norm satisfies the identity in (5.6.33(b)). Show that the matrix norms $\| \cdot \|_1$ and $\| \cdot \|_2$ both satisfy this identity, but that neither norm is a scalar multiple of an induced norm.

Exercise. Explain why the function

$$\||A\|| = \max\{\||A\||_1, \||A\||_\infty\} \tag{5.6.33.1}$$

is a unital matrix norm on M_n.

We saw in (5.6.2) that every induced matrix norm is unital, but (5.6.33.1) is a unital matrix norm that is not induced: $\||A\||_1 \leq \||A\||$ for all $A \in M_n$ and $\||A_0\||_1 < \||A_0\||$ for $A_0 = \begin{bmatrix} 1 & 0 \\ 1 & 3 \end{bmatrix}$, so $\|| \cdot \||$ is not minimal and hence cannot be induced. See (5.6.P7) for a generalization of the construction in (5.6.33.1).

A norm $\|\cdot\|$ on M_n (not necessarily a matrix norm) is *unitarily invariant* if $\|A\| = \|UAV\|$ for all $A \in M_n$ and all unitary $U, V \in M_n$; a *unitarily invariant matrix norm* is a unitarily invariant norm on M_n that is submultiplicative. We have seen that the Frobenius and spectral norms are unitarily invariant matrix norms, but the Frobenius norm is not an induced norm.

Theorem 5.6.34. *Let $\|| \cdot \||$ be a unitarily invariant matrix norm on M_n, and suppose that $z \in \mathbf{C}^n$ is nonzero. Then*

(a) *the vector norm $\|\cdot\|_z$ defined by (5.6.27) is unitarily invariant*

(b) *$\|\cdot\|_z$ is a scalar multiple of the Euclidean vector norm, that is, there is a positive scalar c_z such that $\|\cdot\|_z = c_z \|\cdot\|_2$*

(c) *the induced matrix norm $N_z(\cdot)$ defined by (5.6.28) is the spectral norm*

(d) *$\||A\||_2 \leq \||A\||$ for all $A \in M_n$*

(e) *if $\|| \cdot \||$ is induced (as well as unitarily invariant) then it is the spectral norm*

Proof. (a) If $U \in M_n$ is unitary, then $\|Ux\|_z = \||Uxz^*\|| = \||xz^*\|| = \|x\|_z$.

(b) For each $x \in \mathbf{C}^n$ there is a unitary $U \in M_n$ such that $Ux = \|x\|_2 e_1$ (2.1.13), so $\|x\|_z = \|Ux\|_z = \||Uxz^*\|| = \|| \|x\|_2 e_1 z^* \|| = \|x\|_2 \||e_1 z^*\||$ for all $x \in \mathbf{C}^n$.

(c) $N_z(A) = \max_{x \neq 0} \|Ax\|_z / \|x\|_z = \max_{x \neq 0}(c_z \|Ax\|_2 / (c_z \|x\|_2)) = \max_{x \neq 0} \|Ax\|_2 / \|x\|_2 = \||A\||_2$.

(d) This assertion is (5.6.26(b)).

(e) This assertion is (5.6.26(c)). $\qquad\qquad\square$

If $\|\cdot\|$ is a norm on M_n, a calculation reveals that the function $\|\cdot\|'$ defined by

$$\|A\|' = \|A^*\|$$

is a norm on M_n and $(\|A\|')' = \|A\|$. The norm $\|\cdot\|'$ is the *adjoint* of $\|\cdot\|$.

Exercise. If $\|| \cdot \||$ is a matrix norm on M_n, show that its adjoint is also a matrix norm.

A calculation also shows that $\|A\|'_2 = \|A^*\|_2 = \|A\|_2$ and $\|A\|'_1 = \|A^*\|_1 = \|A\|_1$ for all $A \in M_n$, but not every norm or matrix norm has this property: $\|| \cdot \||'_1 = \|| \cdot \||_\infty$. A norm $\|\cdot\|$ on M_n such that $\|A\| = \|A\|'$ for all $A \in M_n$ is said to be *self-adjoint*. The l_1 matrix norm is self-adjoint, as are the Frobenius norm and the spectral norm.

Exercise. Explain why every unitarily invariant norm on M_n is self-adjoint, and give an example of a self-adjoint norm on M_n that is not unitarily invariant.

Hint: If $A = V\Sigma W^*$ is a singular value decomposition and if $\|\cdot\|$ is unitarily invariant, then $\|A\| = \|\Sigma\|$.

The spectral norm is distinguished as the only induced matrix norm that is self-adjoint:

Theorem 5.6.35. *Let* $\|\|\cdot\|\|$ *be the matrix norm on* M_n *that is induced by a norm* $\|\cdot\|$ *on* $\mathbf{C}^n$. *Then*

   (a) $\|\|\cdot\|\|'$ *is induced by the norm* $\|\cdot\|^D$

   (b) if $\|\|\cdot\|\|$ *is self-adjoint (as well as induced) then it is the spectral norm*

Proof. (a) Use (5.6.2(d)) to compute

$$\|\|A\|\|' = \|\|A^*\|\| = \max_{\|x\|=\|y\|^D=1} |y^*A^*x| = \max_{\|x\|=\|y\|^D=1} |x^*Ay|$$

$$= \max_{\|y\|^D=1} \max_{\|x\|=1} |x^*Ay| = \max_{\|y\|^D=1} \|Ay\|^D$$

(b) If $\|\|\cdot\|\| = \|\|\cdot\|\|'$, then (a) ensures that $\|\|\cdot\|\|$ is induced by both $\|\cdot\|$ and $\|\cdot\|^D$. Lemma 5.6.23 tells us that $\|\cdot\|$ is a scalar multiple of $\|\cdot\|^D$, and (5.4.17) ensures that $\|\cdot\| = \|\cdot\|_2$. Since $\|\|\cdot\|\|$ is induced by the Euclidean vector norm, it is the spectral norm. $\qquad\square$

Absolute and monotone norms were introduced in (5.4.18); there is a useful characterization of the matrix norms that they induce.

Theorem 5.6.36. *Let* $\|\|\cdot\|\|$ *be the matrix norm on* M_n *induced by a norm* $\|\cdot\|$ *on* $\mathbf{C}^n$. *The following are equivalent:*

   (a) $\|\cdot\|$ *is an absolute norm.*

   (b) $\|\cdot\|$ *is a monotone norm.*

   (c) If $\Lambda = \text{diag}(\lambda_1, \ldots, \lambda_n) \in M_n$ *then* $\|\|\Lambda\|\| = \max_{1\leq i\leq n} |\lambda_i|$.

Proof. The equivalence of (a) and (b) is the assertion in (5.4.19(c)). To prove that (b) implies (c), suppose that $\|\cdot\|$ is monotone, let $\Lambda = \text{diag}(\lambda_1, \ldots, \lambda_n)$, and let $L = \max\{|\lambda_1|, \ldots, |\lambda_n|\} = |\lambda_k|$. Then $|\Lambda x| \leq |Lx|$ and hence $\|\Lambda x\| \leq \|Lx\| = L\|x\|$ with equality for $x = e_k$. Thus,

$$\|\|\Lambda\|\| = \max_{x\neq 0} \frac{\|\Lambda x\|}{\|x\|} \leq \max_{x\neq 0} \frac{L\|x\|}{\|x\|} = L \qquad\qquad (5.6.37)$$

with equality for $x = e_k$. To show that (c) implies (b), let $x, y \in \mathbf{C}^n$ be given with $|x| \leq |y|$. Choose complex numbers λ_k such that $x_k = \lambda_k y_k$ and $|\lambda_k| \leq 1, k = 1, \ldots, n$, and let $\Lambda = \text{diag}(\lambda_1, \ldots, \lambda_n)$. Then $\|x\| = \|\Lambda y\| \leq \|\|\Lambda\|\| \|y\| = \max_{1\leq i\leq n} |\lambda_i| \|y\| \leq \|y\|$, so $\|\cdot\|$ is monotone. $\qquad\square$

Because the vector space M_n is an inner product space with the Frobenius inner product (5.2.7), we can define a *dual norm* (5.4.12) for any norm on M_n.

Definition 5.6.38. *Let* $\|\cdot\|$ *be a norm on* M_n. *Its dual norm is*

$$\|A\|^D = \max_{\|B\|=1} \text{Re} \langle A, B\rangle_F = \max_{\|B\|=1} \text{Re} \,\text{tr}\, B^*A \quad \text{for each } A \in M_n$$

Exercise. Explain why analogs of (5.4.12a) are available as alternative representations for the dual of a norm $\|\cdot\|$ on M_n, for example, $\|A\|^D = \max_{\|B\|=1} |\operatorname{tr} B^*A| = \max_{\|B\|\leq 1} |\operatorname{tr} B^*A| = \max_{B\neq 0} \frac{|\operatorname{tr} B^*A|}{\|B\|}$.

Exercise. Show that $\|\cdot\|_F^D = \|\cdot\|_F$, that is, the Frobenius norm on M_n is self-dual. *Hint*: $|\langle A, B\rangle_F| \leq \|A\|_F \|B\|_F$ with equality for $A = B$.

Theorem 5.6.39. *Let $\|\cdot\|$ be a norm on M_n. Then*

    (a) $\|\cdot\|$ is self-adjoint if and only if $\|\cdot\|^D$ is self-adjoint
    (b) $\|\cdot\|$ is unitarily invariant if and only if $\|\cdot\|^D$ is unitarily invariant

Proof. In each case, the "only if" implication follows from a computation; the "if" implication follows from the duality theorem (5.5.9(c)).
(a) Suppose that $\|\cdot\|$ is self-adjoint. Then

$$\|A^*\|^D = \max_{B\neq 0} \frac{|\operatorname{tr} B^*A^*|}{\|B\|} = \max_{B\neq 0} \frac{|\operatorname{tr} BA^*|}{\|B^*\|} = \max_{B\neq 0} \frac{|\operatorname{tr} BA^*|}{\|B\|}$$

$$= \max_{B\neq 0} \frac{|\operatorname{tr}(BA^*)^*|}{\|B\|} = \max_{B\neq 0} \frac{|\operatorname{tr} AB^*|}{\|B\|}$$

$$= \max_{B\neq 0} \frac{|\operatorname{tr} B^*A|}{\|B\|} = \|A\|$$

(b) Suppose that $\|\cdot\|$ is unitarily invariant. Then for any unitary $U, V \in M_n$ we have

$$\|UAV\|^D = \max_{B\neq 0} \frac{|\operatorname{tr} B^*UAV|}{\|B\|} = \max_{B\neq 0} \frac{|\operatorname{tr}(U^*BV^*)^*A|}{\|B\|}$$

$$= \max_{C\neq 0} \frac{|\operatorname{tr} C^*A|}{\|UCV\|} = \max_{C\neq 0} \frac{|\operatorname{tr} C^*A|}{\|C\|} = \|A\|$$

$\square$

Exercise. Show that the dual of the (noninduced) matrix norm $\|\cdot\|_1$ on M_n (5.6.0.1) is the norm $\|\cdot\|_\infty$ on M_n (5.6.0.3), that the inequality $\|A^*\|_1 \leq \|A\|_1^D$ is *not* valid for all $A \in M_n$, that $\|\cdot\|_1^D$ is *not* a matrix norm, and that $\|AB\|_1^D \leq \|A^*\|_1 \|B\|_1^D$ for all $A, B \in M_n$. *Hint*: Review the computation of the dual of the vector norm $\|\cdot\|_1$ on $\mathbf{C}^n$ (5.4.15a).

Exercise. Let $\|\cdot\|$ be a norm on M_n and let $A \in M_n$ be given. Explain why there is some $X \in M_n$ such that $\|X\| = 1$ and $|\operatorname{tr} X^*A| = \|A\|^D$. For any $Y \in M_n$ with $\|Y\| = 1$, why is $|\operatorname{tr} Y^*A| \leq \|A\|^D$?

Theorem 5.6.40. *Let $\|\|\cdot\|\|$ be a matrix norm on M_n. Then*

$$\|\|AB\|\|^D \leq \begin{cases} \|\|A^*\|\| \, \|\|B\|\|^D \\ \|\|A\|\|^D \, \|\|B^*\|\| \end{cases}$$

for all $A, B \in M_n$. If $\|\|A^*\|\| \leq \|\|A\|\|^D$ for all $A \in M_n$, then $\|\|\cdot\|\|^D$ is a matrix norm on M_n.

Proof. We prove only the second upper bound. Let $X \in M_n$ be such that $\||X\|| = 1$ and $|\operatorname{tr} X^* AB| = \||AB\||^D$. Use submultiplicativity of $\|| \cdot \||$ to compute

$$\||AB\||^D = |\operatorname{tr} X^* AB| = |\operatorname{tr}(XB^*)^* A| \le \||XB^*\|| \, \||A\||^D$$

$$\le \||X\|| \, \||B^*\|| \, \||A\||^D = \||A\||^D \, \||B^*\||$$

If $\||B^*\|| \le \||B\||^D$ for all $B \in M_n$, then $\||AB\||^D \le \||A\||^D \||B^*\|| \le \||A\||^D \||B\||^D$ for all $A, B \in M_n$. $\qquad\square$

Exercise. Show that the dual of the induced matrix norm $\||\cdot\||_1$ on M_n (5.6.4) is the norm $N_\infty(A) = \sum_{j=1}^n \|a_j\|_\infty$, in which $A = [a_1 \ \ldots \ a_n]$ is partitioned according to its columns. Show that $\||A^*\||_1 \le N_\infty(A)$ for all $A \in M_n$, with equality if rank $A \le 1$. Explain why, on the basis of this inequality and the preceding theorem, $N_\infty(A)$ must be a matrix norm; see (5.6.0.5) for a computational proof of this fact. Why is $N_\infty(\cdot)$ not an induced matrix norm? If $\Lambda = \operatorname{diag}(\lambda_1, \ldots, \lambda_n)$ is diagonal, verify that $\||\Lambda\||_1 = \max_i |\lambda_i|$ and $\||\Lambda\||_1^D = N_\infty(\Lambda) = |\lambda_1| + \cdots + |\lambda_n|$; note that $\||\cdot\||_1$ is induced by an absolute vector norm. *Hint*:

$$\||A\||_1^D = \max_{\||B\||_1 = 1} |\operatorname{tr} B^* A| \le \max_{\||B\||_1 = 1} \sum_{j=1}^n |b_j^* a_j|$$

$$\le \max_{\||B\||_1 = 1} \sum_{j=1}^n \|a_j\|_\infty \|b_j\|_1 \le \sum_{j=1}^n \|a_j\|_\infty \max_{\||B\||_1 = 1} \||B\||_1$$

Describe a linear combination of 0–1 matrices E_{ij} for which equality is achieved, and conclude that $\||\cdot\||_1^D = N_\infty(\cdot)$.

We have seen that the dual of a matrix norm need not be a matrix norm, and that the dual of an induced matrix norm can be a matrix norm that is not induced. The following theorem says that *the dual of an induced matrix norm is always a matrix norm*, which provides a new way to construct matrix norms: Take the dual of any induced matrix norm.

Theorem 5.6.41. *Let $\|| \cdot \||$ be a matrix norm on M_n that is induced by the norm $\|\cdot\|$ on $\mathbb{C}^n$. Then*

(a) $\||A^*\|| = \max\{|\operatorname{tr} B^* A| : \||B\|| = 1 \text{ and } \operatorname{rank} B = 1\}$

(b) $\||A^*\|| \le \||A\||^D$ *for all* $A \in M_n$

(c) $\||A\||^D$ *is a matrix norm*

(d) $\||A^*\|| = \||A\||^D$ *if* rank $A \le 1$

Proof. (a) If $B = xy^*$ for some nonzero $x, y \in \mathbf{C}^n$, then (5.6.30) ensures that $|||B||| = |||xy^*||| = \|x\| \, \|y\|^D$. Compute

$$\max_{\text{rank } B=1} \frac{|\operatorname{tr} B^*A|}{|||B|||} = \max_{x \neq 0 \neq y} \frac{|\operatorname{tr}(yx^*A)|}{\|x\| \, \|y\|^D} = \max_{x \neq 0 \neq y} \frac{|x^*Ay|}{\|x\| \, \|y\|^D}$$

$$= \max_{x \neq 0 \neq y} \frac{|y^*A^*x|}{\|x\| \, \|y\|^D} = \max_{x \neq 0 \neq y} \left| \frac{y^*}{\|y\|^D} A^* \frac{x}{\|x\|} \right|$$

$$= \max_{\|\eta\|^D = \|\xi\| = 1} |\eta^*A^*\xi| = \max_{\|\xi\|=1} \left\| A^*\xi \right\|^{DD}$$

$$= \max_{\|\xi\|=1} \left\| A^*\xi \right\| = |||A^*|||$$

(b) Observe that

$$|||A^*||| = \max_{\text{rank } B=1} \frac{|\operatorname{tr} B^*A|}{|||B|||} \leq \max_{B \neq 0} \frac{|\operatorname{tr} B^*A|}{|||B|||} = |||A|||^D$$

(c) This assertion follows from (b) and (5.6.40).

(d) Suppose that $A = uv^*$ for some nonzero $u, v \in \mathbf{C}^n$. According to (5.6.30), $|||A^*||| = \|v\| \, \|u\|^D$, so we must show that $|||A|||^D = \|v\| \, \|u\|^D$. Compute

$$|||uv^*|||^D = \max_{B \neq 0} \frac{|\operatorname{tr} B^*uv^*|}{|||B|||} = \max_{B \neq 0} \frac{|v^*B^*u|}{|||B|||}$$

$$= \max_{B \neq 0} \frac{|u^*Bv|}{|||B|||} \leq \max_{B \neq 0} \frac{\|u\|^D \|Bv\|}{|||B|||}$$

$$\leq \max_{B \neq 0} \frac{\|u\|^D \, |||B||| \, \|v\|}{|||B|||} = \|u\|^D \|v\|$$

To conclude the proof, we must show that $B = xy^*$ can be chosen such that $|||B||| = 1$ and $|u^*Bv| = \|u\|^D \|v\|$. Invoke (5.5.9(d)) to choose vectors x and y such that (i) $\|x\| = 1$ and $x^*u = \|u\|^D$ ($f(\cdot) = \|\cdot\|^D$ here) and (ii) $\|y\|^D = 1$ and $y^*v = \|v\|$ ($f(\cdot) = \|\cdot\|$ here). Then $|||B||| = \|x\| \, \|y\|^D = 1$ and $|u^*Bv| = |u^*x| \, |y^*v| = \|u\|^D \|v\|$. $\square$

Exercise. Let $\|\cdot\|$ be a norm on M_n and consider its adjoint $\|\cdot\|'$. Explain why the dual of the norm $\|\cdot\|'$ is the function $\nu(A) = \|A^*\|^D$ for each $A \in M_n$. If $||| \cdot |||$ is an induced matrix norm on M_n, explain why $\nu(A) = |||A^*|||^D$ is a matrix norm. What is $\nu(A)$ if $||| \cdot |||$ is the matrix norm $||| \cdot |||_1$? This construction gives yet another way to create new matrix norms: For any induced matrix norm $||| \cdot |||$, take the dual of its adjoint.

Our final theorem is a dual of (5.6.36).

Theorem 5.6.42. *Suppose that an absolute norm $\|\cdot\|$ on $\mathbf{C}^n$ induces the matrix norm $||| \cdot |||$ on M_n, and let $||| \cdot |||^D$ be its dual. Then $||| \cdot |||^D$ is a matrix norm, and for each diagonal matrix $\Lambda = \operatorname{diag}(\lambda_1, \ldots, \lambda_n) \in M_n$, we have $|||\Lambda|||^D = |\lambda_1| + \cdots + |\lambda_n|$.*

Proof. The preceding theorem ensures that $||| \cdot |||^D$ is a matrix norm. Write $\Lambda = \operatorname{diag}(e^{i\theta_1}|\lambda_1|, \ldots, e^{i\theta_n}|\lambda_n|)$ and let $U = \operatorname{diag}(e^{i\theta_1}, \ldots, e^{i\theta_n})$; (5.6.36) ensures that

$\|\|U\|\| = 1$. Then

$$\|\|\Lambda\|\|^D = \max_{\|\|B\|\|=1} |\operatorname{tr} B^*\Lambda| \geq |\operatorname{tr} U^*\Lambda| = |\lambda_1| + \cdots + |\lambda_n|$$

Conversely, write $\Lambda = \lambda_1 E_{11} + \cdots + \lambda_n E_{nn}$, in which each $E_{ii} = e_i e_i^*$ (0.1.7). Then

$$\|\|\Lambda\|\|^D \leq \|\|\lambda_1 E_{11}\|\|^D + \cdots + \|\|\lambda_n E_{nn}\|\|^D$$
$$= |\lambda_1|\,\|\|E_{11}\|\|^D + \cdots + |\lambda_n|\,\|\|E_{nn}\|\|^D$$

so it is sufficient to show that each $\|\|E_{ii}\|\|^D = 1$. Since each E_{ii} has rank one, (5.6.30) and (5.4.13) ensure that $\|\|E_{ii}\|\|^D = \|e_i\|\,\|e_i\|^D \geq 1$. For any vector $x = [x_i] \in \mathbf{C}^n$ and any $i \in \{1, \ldots, n\}$, we have $|x| \geq |x_i e_i|$, so monotonicity of $\|\cdot\|$ ensures that $\|x\| = \|\,|x|\,\| \geq \|\,|x_i e_i|\,\| = |x_i|\,\|e_i\|$. Therefore,

$$\|e_i\|^D = \max_{x \neq 0} \frac{|x^* e_i|}{\|x\|} = \max_{x \neq 0} \frac{|x_i|}{\|x\|} \leq \max_{x \neq 0} \frac{\|x\|}{\|x\|\,\|e_i\|} = \frac{1}{\|e_i\|}$$

so $\|e_i\|\,\|e_i\|^D \leq 1$. We conclude that each $\|e_i\|\,\|e_i\|^D = 1$, as required. $\qquad\square$

Example. The spectral norm on M_n is unitarily invariant and is induced by the Euclidean norm, an absolute norm. Theorem 5.6.39 ensures that $\|\|\cdot\|\|_2^D$ is unitarily invariant, so if $A \in M_n$ and $A = V\Sigma W^*$ is a singular value decomposition, the preceding theorem permits us to compute

$$\|\|A\|\|_2^D = \|\|V\Sigma W^*\|\|_2^D = \|\|\Sigma\|\|_2^D$$
$$= \operatorname{tr} \Sigma = \sigma_1(A) + \cdots + \sigma_n(A) = \|\|A\|\|_{\operatorname{tr}}$$

Theorem 5.6.42 ensures that this new norm, the *trace norm*, is a unitarily invariant matrix norm. This result is more than a little astonishing, as it is far from obvious that the sum of all the singular values is either a subadditive or a submultiplicative function on M_n; see (7.4.7) and (7.4.10).

Problems

5.6.P1 Explain why the l_1-norm on M_n is a matrix norm that is not an induced norm.

5.6.P2 Give an example of a 2-by-2 projection other than I and 0. Show that 0 and 1 are the only possible eigenvalues of a projection. Explain why a projection A is diagonalizable and why $\|\|A\|\| \geq 1$ for any matrix norm $\|\|\cdot\|\|$ if $A \neq 0$.

5.6.P3 If $\|\|\cdot\|\|$ is a matrix norm on M_n, show that $c\|\|\cdot\|\|$ is a matrix norm for all $c \geq 1$. Show, however, that neither $c\|\|\cdot\|\|_1$ nor $c\,\|\cdot\|_2$ is a matrix norm for any $c < 1$.

5.6.P4 In (5.6.1) the same norm is involved in two different ways: Measuring the size of x and measuring the size of Ax. More generally, we might define $\|\|\cdot\|\|_{\alpha,\beta}$ by $\|\|A\|\|_{\alpha,\beta} = \max_{\|x\|_\alpha=1} \|Ax\|_\beta$, in which $\|\cdot\|_\alpha$ and $\|\cdot\|_\beta$ are two (possibly different) vector norms. Is such a function $\|\|\cdot\|\|_{\alpha,\beta}$ a matrix norm? This notion might be used to define a norm on m-by-n matrices, since $\|\cdot\|_\alpha$ may be taken to be a norm on $\mathbf{C}^n$ and $\|\cdot\|_\beta$ may be taken to be a norm on $\mathbf{C}^n$. What properties like those of an induced matrix norm does $\|\|\cdot\|\|_{\alpha,\beta}$ have in this regard?

5.6.P5 Let $\|\| \cdot \|\|_p$ denote the matrix norm on M_n induced by the l_p-norm on $\mathbf{C}^n$, $p \geq 1$. Use (5.4.21) and (5.6.21) to show that

$$\max_{A \neq 0} \frac{\|\|A\|\|_{p_1}}{\|\|A\|\|_{p_2}} = n^{(\min\{p_1, p_2\})^{-1} - (\max\{p_1, p_2\})^{-1}}$$

Deduce that, for all $A \in M_n$ and all $p \geq 1$,

$$n^{\frac{1}{p} - 1} \|\|A\|\|_1 \leq \|\|A\|\|_p \leq n^{1 - \frac{1}{p}} \|\|A\|\|_1$$

$$n^{-\left|\frac{1}{p} - \frac{1}{2}\right|} \|\|A\|\|_2 \leq \|\|A\|\|_p \leq n^{\left|\frac{1}{p} - \frac{1}{2}\right|} \|\|A\|\|_2$$

and

$$n^{-\frac{1}{p}} \|\|A\|\|_\infty \leq \|\|A\|\|_p \leq n^{\frac{1}{p}} \|\|A\|\|_\infty$$

5.6.P6 Verify that axioms (1)–(3) for $\|\| \cdot \|\|$ imply that the same axioms hold for $\|\| \cdot \|\|_S$ in (5.6.7). Thus, (5.6.7) remains valid if "matrix norm" in the hypothesis and conclusion is replaced by "norm on matrices."

5.6.P7 This problem generalizes the construction in (5.6.33.1). Let $N_1(\cdot), \ldots, N_m(\cdot)$ be matrix norms on M_n, let $\|\cdot\|$ be an absolute norm on $\mathbf{C}^n$ such that $\|x\| \geq \|x\|_\infty$ for all $x \in \mathbf{C}^m$, and define $\|\|A\|\| = \left\| [N_1(A) \ \ldots \ N_m(A)]^T \right\|$. (a) Show that $\|e_k\| \geq 1$ for any standard basis vector e_k. (b) Show that $\|\| \cdot \|\|$ is a matrix norm on M_n.

5.6.P8 Show that the nonsingular matrices of M_n are *dense* in M_n; that is, show that every matrix in M_n is the limit of nonsingular matrices. Are the singular matrices dense in M_n?

5.6.P9 Show that the set of norms on $\mathbf{C}^n$ is convex for every $n \geq 1$, but the set of matrix norms on M_n is not convex for any $n \geq 2$. If $N_1(\cdot)$ and $N_2(\cdot)$ are matrix norms on M_n, show that $N(\cdot) = \frac{1}{2}(N_1(\cdot) + N_2(\cdot))$ is a matrix norm if and only if

$$(N_1(A) - N_2(A))(N_1(B) - N_2(B)) \leq 2(N_1(A)N_1(B) - N_1(AB))$$
$$+ 2(N_2(A)N_2(B) - N_2(AB))$$

for all $A, B \in M_n$. See (7.4.10.2) for a proof that the set of unitarily invariant matrix norms is convex.

5.6.P10 Let $\|\cdot\|$ be a given norm on $\mathbf{C}^n$. Partition any $A = [a_1 \ \ldots \ a_n] \in M_n$ according to its columns and define $N_{\|\cdot\|}(A) = \max_{1 \leq i \leq n} \|a_i\|$. (a) Show that $N_{\|\cdot\|}(\cdot)$ is a norm on M_n. (b) Show that $N_{\|\cdot\|}(\cdot)$ is a matrix norm on M_n if and only if $\|x\| \geq \|x\|_1$ for all $x \in \mathbf{C}^n$. (c) For each $i = 1, \ldots, n$ let $d_i(A) = \|a_i\|$ if $a_i \neq 0$ and $d_i(A) = 1$ if $a_i = 0$, and define $D_A = \mathrm{diag}(d_1(A), \ldots, d_n(A))$. Explain why $N_{\|\cdot\|}(AD_A^{-1}) \leq 1$. (d) If $N_{\|\cdot\|}(\cdot)$ is a matrix norm, explain why $\rho(AD_A^{-1}) \leq 1$, $|\det(AD_A^{-1})| \leq 1$, and $|\det A| \leq \det D_A = d_1(A) \cdots d_n(A)$. Conclude that $|\det A| \leq \|a_1\| \cdots \|a_n\|$ if $N_{\|\cdot\|}(\cdot)$ is a matrix norm. (Be careful: What if some $a_i = 0$?) (e) Consider $\|\cdot\| = \|\cdot\|_1$. Explain why $N_{\|\cdot\|_1}(A) = \|\|A\|\|_1$ and conclude that

$$|\det A| \leq \|a_1\|_1 \cdots \|a_n\|_1 \tag{5.6.43}$$

(f) Consider $\|\cdot\| = n \|\cdot\|_\infty$. Explain why $N_{n\|\cdot\|_\infty}(A) = n \|A\|_\infty$ and conclude that $|\det A| \leq n^n \|A\|_\infty^n$. Compare this bound with the one in (2.3.P10). Which is better?

(g) Consider $\|\cdot\| = \|\cdot\|_2$. Explain why $N_{\|\cdot\|_2}(A)$ is *not* a matrix norm, so we cannot use the method in (d) to conclude that

$$|\det A| \leq \|a_1\|_2 \cdots \|a_n\|_2 \tag{5.6.44}$$

Nevertheless, this inequality (Hadamard's inequality) is correct; see (2.1.P23) or (7.8.3). What is going on here? (h) Explain why (5.6.44) is a better bound than (5.6.43); it's the same reason that the method in (d) fails for the Euclidean norm.

5.6.P11 Explain why $\|\|AA^*\|\|_2 = \|\|A^*A\|\|_2 = \|\|A\|\|_2^2$.

5.6.P12 Let $A, B \in M_n$ be given and let $\|\|\cdot\|\|$ be a matrix norm on M_n. Why is $\|\|AB \pm BA\|\| \leq 2\|\|A\|\| \, \|\|B\|\|$? These upper bounds are not very satisfying, since one has a feeling that $AB - BA$ ought to be smaller than $AB + BA$. One can prove a more satisfying upper bound if A and B are positive semidefinite and we use the spectral norm: (a) If A is positive semidefinite, show that $\|\|A - \frac{1}{2}\|\|A\|\|_2 I\|\|_2 = \frac{1}{2}\|\|A\|\|_2$. (b) Let $\alpha, \beta \in \mathbf{C}$ and explain why $\|\|AB - BA\|\| = \|\|(A - \alpha I)(B - \beta I) - (B - \beta I)(A - \alpha I)\|\| \leq 2\|\|A - \alpha I\|\| \, \|\|B - \beta I\|\|$. (c) Let $\alpha = \frac{1}{2}\|\|A\|\|_2$ and $\beta = \frac{1}{2}\|\|B\|\|_2$. Conclude that $\|\|AB - BA\|\|_2 \leq \frac{1}{2}\|\|A\|\|_2 \|\|B\|\|_2$ if A and B are positive semidefinite.

5.6.P13 If $A \in M_n$ is singular, explain why $\|\|I - A\|\| \geq 1$ for every matrix norm $\|\|\cdot\|\|$.

5.6.P14 Let $\|\|\cdot\|\|_\alpha$ and $\|\|\cdot\|\|_\beta$ be given matrix norms on M_n. Under what conditions is the matrix norm $\|\|\cdot\|\| = \max\{\|\|\cdot\|\|_\alpha, \|\|\cdot\|\|_\beta\}$ an induced norm?

5.6.P15 Give an example of a matrix A such that $\rho(A) < \|\|A\|\|$ for every matrix norm $\|\|\cdot\|\|$.

5.6.P16 Let $A = [a_{ij}] \in M_n$ with $n \geq 2$. Explain why the function $\|\|\cdot\|\|$ defined on M_n by $\|\|A\|\| = n \max_{1 \leq i,j \leq n} |a_{ij}|$ is a matrix norm that is not induced.

5.6.P17 Use the idea in (5.6.16) to compute the inverse of the matrix

$$\begin{bmatrix} 1 & -2 & 1 \\ 0 & 1 & 3 \\ 0 & 0 & 1 \end{bmatrix}$$

5.6.P18 Explain how to generalize the method in (5.6.P17) to invert any nonsingular upper triangular matrix $A \in M_n$.

5.6.P19 The spectral radius $\rho(\cdot)$ is a nonnegative, continuous, homogeneous function on M_n that is not a matrix norm, norm, seminorm, or pre-norm on M_n. Give examples to show that (a) $\rho(A) = 0$ is possible for some $A \neq 0$; (b) $\rho(A + B) > \rho(A) + \rho(B)$ is possible; and (c) $\rho(AB) > \rho(A)\rho(B) > 0$ is possible.

5.6.P20 Show that $\|AB\|_2 \leq \|\|A\|\|_2\|B\|_2$ and $\|AB\|_2 \leq \|A\|_2\|\|B\|\|_2$ for all $A, B \in M_n$. Deduce that $\|A\|_2 \leq \sqrt{n}\|\|A\|\|_2$.

5.6.P21 Show that $\|\|A\|\|_2 \leq \|\|A\|\|^{1/2}\|\|A^*\|\|^{1/2}$ for any matrix norm $\|\|\cdot\|\|$ on M_n and all $A \in M_n$. Deduce that $\|\|A\|\|_2 \leq \|\|A\|\|_1^{1/2}\|\|A\|\|_\infty^{1/2}$.

5.6.P22 Show that a matrix norm $\|\|\cdot\|\|$ is unital if and only if $\|\|A\|\|^D \geq |\operatorname{tr} A|$ for all $A \in M_n$.

5.6.P23 Verify that the entries in the following 6-by-6 table give the best constants $C_{\alpha\beta}$ such that $\|\|A\|\|_\alpha \leq C_{\alpha\beta}\|\|A\|\|_\beta$ for all $A \in M_n$. For example, we claim that the Frobenius norm (row 5, omitting the top label row) and the spectral norm (column 2, omitting the left label column) satisfy the inequality $\|A\|_2 \le \sqrt{n}\,\||A\||_2$ for all $A \in M_n$; the constant $\sqrt{n}$ is in position 5,2 in the table. Every norm in the table is a matrix norm.

$\||\cdot\||_\alpha \backslash \||\cdot\||_\beta$	$\||\cdot\||_1$	$\||\cdot\||_2$	$\||\cdot\||_\infty$	$\|\cdot\|_1$	$\|\cdot\|_2$	$n\|\cdot\|_\infty$
$\||\cdot\||_1$	1	$\sqrt{n}$	n	1	$\sqrt{n}$	1
$\||\cdot\||_2$	$\sqrt{n}$	1	$\sqrt{n}$	1	1	1
$\||\cdot\||_\infty$	n	$\sqrt{n}$	1	1	$\sqrt{n}$	1
$\|\cdot\|_1$	n	$n^{3/2}$	n	1	n	n
$\|\cdot\|_2$	$\sqrt{n}$	$\sqrt{n}$	$\sqrt{n}$	1	1	1
$n\|\cdot\|_\infty$	n	n	n	n	n	1

5.6.P24 Show that the bound (5, 2) in (5.6.P23) can be improved to $\|A\|_2 \le (\mathrm{rank}\, A)^{1/2}\,\||A\||_2$.

5.6.P25 Let $A \in M_n$ be the circulant matrix (0.9.6.1) whose first row is $[a_1 \ldots a_n]$, and let $\omega = e^{2\pi i/n}$. Show that

$$|a_1 + \cdots + a_n| \le \max_{\ell=1,\ldots,n} \left| \sum_{k=0}^{n-1} a_{k+1}\omega^{k(\ell-1)} \right| \le \||A\||_2 \le |a_1| + \cdots + |a_n|$$

5.6.P26 If $A \in M_n$ and $\rho(A) < 1$, show that the *Neumann series* $I + A + A^2 + \cdots$ converges to $(I - A)^{-1}$.

5.6.P27 Any polynomial $f(z)$ of degree at least 1 can be written in the form $f(z) = \gamma z^k p(z)$, in which γ is a nonzero constant and

$$p(z) = z^n + a_{n-1}z^{n-1} + a_{n-2}z^{n-2} + \cdots + a_1 z + a_0 \qquad (5.6.45)$$

is a monic polynomial such that $p(0) = a_0 \ne 0$. The roots of $p(z) = 0$ are the nonzero roots of $f(z) = 0$, and it is these roots for which we can give various bounds. Let $C(p) \in M_n$ denote the companion matrix (3.3.12) of the polynomial $p(z)$ in (5.6.45). The eigenvalues of $C(p)$ are the zeroes of the polynomial p, including multiplicities (3.3.14). (a) Use (5.6.9) to show that if $\tilde{z}$ is a root of $p(z) = 0$ and if $\||\cdot\||$ is any matrix norm on M_n, then $|\tilde{z}| \le \||C(p)\||$. In the following, $\tilde{z}$ represents any root of $p(z) = 0$. (b) Use $\|\cdot\|_2$ to show that

$$|\tilde{z}| \le \sqrt{n + |a_0|^2 + |a_1|^2 + \cdots + |a_{n-1}|^2} \qquad (5.6.46)$$

(c) Use $\||\cdot\||_\infty$ to show that

$$|\tilde{z}| \le \max\{|a_0|, 1 + |a_1|, \ldots, 1 + |a_{n-1}|\}$$
$$\le 1 + \max\{|a_0|, |a_1|, \ldots, |a_{n-1}|\} \qquad (5.6.47)$$

This bound on the roots is known as *Cauchy's bound*. (d) Use $\||\cdot\||_1$ to show that

$$|\tilde{z}| \le \max\{1, |a_0| + |a_1| + \cdots + |a_{n-1}|\}$$
$$\le 1 + |a_0| + |a_1| + \cdots + |a_{n-1}| \qquad (5.6.48)$$

which is known as *Montel's bound*. Why is it is poorer than Cauchy's bound? (e) Use $\|\cdot\|_1$ to show that

$$|\tilde{z}| \le (n - 1) + |a_0| + |a_1| + \cdots + |a_{n-1}|$$

which is a poorer bound than Montel's bound for all $n > 2$. (f) Use $n \|\cdot\|_\infty$ to show that

$$|\tilde{z}| \le n \max\{1, |a_0|, |a_1|, \ldots, |a_{n-1}|\}$$

which is a poorer bound than (5.6.48).

5.6.P28 Continuing with the notation of the preceding problem, we seek to improve the bound in (5.6.46). Let $s = |a_0|^2 + |a_1|^2 + \cdots + |a_{n-1}|^2$. Write the companion matrix as $C(p) = S + R$, in which $S = J_n(0)^T$ is the transpose of the n-by-n nilpotent Jordan block and $R = C(p) - J_n(0)$ is a rank-one matrix whose last column is the only nonzero column. (a) Show that $SR^* = RS^* = 0$, $\|\|SS^*\|\|_2 = 1$, and $\|\|RR^*\|\|_2 = \|\|R^*R\|\|_2 = s$. (b) Show that

$$\|\|C(p)\|\|_2^2 = \|\|C(p)C(p)^*\|\|_2 = \|\|(S+R)(S+R)^*\|\|_2$$
$$= \|\|SS^* + RR^*\|\|_2 \le \|\|SS^*\|\|_2 + \|\|RR^*\|\|_2$$

and deduce *Carmichael and Mason's bound*

$$|\tilde{z}| \le \sqrt{1 + |a_0|^2 + |a_1|^2 + \cdots + |a_{n-1}|^2} = \sqrt{s+1} \qquad (5.6.49)$$

which is a better bound than (5.6.46) for all $n \ge 2$. (c) Finally, use the exact value of the largest singular value of $C(p)$ (3.3.16) to obtain the even better bound

$$|\tilde{z}| \le \sqrt{\frac{1}{2}\left(s + 1 + \sqrt{(s+1)^2 - 4|a_0|^2}\right)} = \sigma_1(C(p)) \qquad (5.6.50)$$

Show that $1 \le \sigma_1(C(p)) < \sqrt{s+1}$ (and $\sigma_1(C(p)) = 1$ if and only if $a_1 = \cdots = a_{n-1} = 0$), so the bound (5.6.50) is always better than the bound (5.6.49).

5.6.P29 Apply Montel's bound (5.6.48) to the polynomial

$$q(z) = (z-1)p(z)$$
$$= z^{n+1} + (a_{n-1} - 1)z^n + (a_{n-2} - a_{n-1})z^{n-1} + \cdots + (a_0 - a_1)z + a_0$$

and show that

$$|\tilde{z}| \le \max\{1, |a_0| + |a_0 - a_1| + \cdots + |a_{n-2} - a_{n-1}| + |a_{n-1} - 1|\}$$

Show that the second term in this expression is not less than 1 and deduce another bound of Montel

$$|\tilde{z}| \le |a_0| + |a_0 - a_1| + \cdots + |a_{n-2} - a_{n-1}| + |a_{n-1} - 1|$$

5.6.P30 Use the preceding bound of Montel to prove *Kakeya's theorem*: If $f(z) = a_n z^n + a_{n-1}z^{n-1} + \cdots + a_1 z + a_0$ is a given polynomial with real nonnegative coefficients a_i that are monotone in the sense that $a_n \ge a_{n-1} \ge \cdots \ge a_1 \ge a_0$; then all the roots of $f(z) = 0$ lie in the unit disc; that is, all $|\tilde{z}| \le 1$.

5.6.P31 The preceding four problems all concern upper bounds on the absolute values of the roots of $p(z) = 0$, but they can be used to obtain lower bounds as well. Show that if $p(z)$ is given by (5.6.45) with $a_0 \ne 0$, then

$$q(z) = \frac{1}{a_0}z^n p\left(\frac{1}{z}\right) = z^n + \frac{a_1}{a_0}z^{n-1} + \frac{a_2}{a_0}z^{n-2} + \cdots + \frac{a_{n-1}}{a_0}z + \frac{1}{a_0}$$

is a polynomial of degree n whose zeroes are the reciprocals of the roots of $p(z) = 0$. Use the respective upper bounds on the roots of $q(z) = 0$ to obtain the following lower bounds on the roots $\tilde{z}$ of $p(z) = 0$.

Cauchy:

$$|\tilde{z}| \geq \frac{|a_0|}{\max\{1, |a_0| + |a_{n-1}|, |a_0| + |a_{n-2}|, \ldots, |a_0| + |a_1|\}}$$

$$\geq \frac{|a_0|}{|a_0| + \max\{1, |a_{n-1}|, |a_{n-2}|, \ldots, |a_1|\}}$$

Montel:

$$|\tilde{z}| \geq \frac{|a_0|}{\max\{|a_0|, 1 + |a_1| + |a_2| + \cdots + |a_{n-1}|\}}$$

$$\geq \frac{|a_0|}{1 + |a_0| + |a_1| + \cdots + |a_{n-1}|}$$

Carmichael and Mason:

$$|\tilde{z}| \geq \frac{|a_0|}{\sqrt{1 + |a_0|^2 + |a_1|^2 + \cdots + |a_{n-1}|^2}} = \frac{|a_0|}{\sqrt{s + 1}}$$

in which $s = |a_0|^2 + |a_1|^2 + \cdots + |a_{n-1}|^2$. Finally, use (5.6.9(b)) and the exact value of the smallest singular value of $C(p)$ (3.3.16) to obtain the lower bound

$$|\tilde{z}| \geq \sqrt{\frac{1}{2}\left(s + 1 - \sqrt{(s+1)^2 - 4|a_0|^2}\right)}$$

$$= \frac{\sqrt{2}|a_0|}{\sqrt{s + 1 + \sqrt{(s+1)^2 - 4|a_0|^2}}} = \frac{|a_0|}{\sigma_1(C(p))} \tag{5.6.51}$$

Explain why it is better than the preceding lower bound derived from Carmichael and Mason's bound. Use the two bounds (5.6.50) and (5.6.51) to describe an annulus that contains all of the zeroes of $p(z)$. What is that annulus if $p(z) = z^5 + 1$?

5.6.P32 Consider the polynomial $p(z) = \frac{1}{n!}z^n + \frac{1}{(n-1)!}z^{n-1} + \cdots + \frac{1}{2}z^2 + z + 1$, which is the nth partial sum of the power series for the exponential function e^z. Show that all roots $\tilde{z}$ of $p(z) = 0$ satisfy the inequality $\frac{1}{2} \leq |\tilde{z}| \leq 1 + n!$. Apply Kakeya's theorem to $z^n p(1/z)$ to show that all the roots actually satisfy $|\tilde{z}| \geq 1$.

5.6.P33 Since $\rho(A) = \rho(D^{-1}AD)$ for any nonsingular matrix D, the methods used in (5.6.P27) can be applied to $D^{-1}C(p)D$ to obtain other bounds on the zeroes of the polynomial $p(z)$ in (5.6.45). Make the computationally convenient choice $D = \mathrm{diag}(p_1, \ldots, p_n)$ with all $p_i > 0$ and generalize Cauchy's bound (5.6.47) to

$$|\tilde{z}| \leq \max\left\{|a_0|\frac{p_n}{p_1}, |a_1|\frac{p_{n-1}}{p_1} + \frac{p_{n-1}}{p_n}, |a_2|\frac{p_{n-2}}{p_1} + \frac{p_{n-2}}{p_{n-1}}, \ldots\right.$$

$$\left. \ldots, |a_{n-2}|\frac{p_2}{p_1} + \frac{p_2}{p_3}, |a_{n-1}| + \frac{p_1}{p_2}\right\}, \tag{5.6.52}$$

which is valid for any positive parameters $p_1, p_2, \ldots, p_n$.

5.6.P34 If all the coefficients a_k in (5.6.45) are nonzero, choose the parameters in the preceding problem to be $p_k = p_1/|a_{n-k+1}|$, $k = 2, 3, \ldots, n$ and deduce *Kojima's bound* on the zeroes $\tilde{z}$ of $p(z)$ from (5.6.52):

$$|\tilde{z}| \leq \max\left\{\left|\frac{a_0}{a_1}\right|, 2\left|\frac{a_1}{a_2}\right|, 2\left|\frac{a_2}{a_3}\right|, \ldots, 2\left|\frac{a_{n-2}}{a_{n-1}}\right|, 2|a_{n-1}|\right\} \tag{5.6.53}$$

5.6.P35 Now choose the parameters in (5.6.P33) to be $p_k = r^k$, $k = 1, \ldots, n$ for some $r > 0$ and show that (5.6.52) implies the bound

$$|\tilde{z}| \leq \max\{|a_0|r^{n-1}, |a_1|r^{n-2} + r^{-1}, |a_2|r^{n-3} + r^{-1}, \ldots,$$
$$\ldots, |a_{n-2}|r + r^{-1}, |a_{n-1}| + r^{-1}\} \tag{5.6.54}$$
$$\leq \frac{1}{r} + \max_{0 \leq k \leq n-1}\{|a_k|r^{n-k-1}\} \quad \text{for any } r > 0$$

5.6.P36 If $A \in M_n$, show that the Hermitian matrix $\hat{A} = \begin{bmatrix} 0 & A \\ A^* & 0 \end{bmatrix} \in M_{2n}$ has the same spectral norm as A.

5.6.P37 Show that the spectral norm, unlike the Frobenius norm, is not derived from an inner product on M_n.

5.6.P38 Let $A \in M_n$ be given. Show that there is a matrix norm $\|\|\cdot\|\|$ such that $\|\|A\|\| = \rho(A)$ if and only if every eigenvalue of A of maximum modulus is semisimple, that is, if and only if whenever $J_k(\lambda)$ is a Jordan block of A and $|\lambda| = \rho(A)$, then $k = 1$.

5.6.P39 The result in the preceding problem can be improved if the matrix norm is the spectral norm. A *spectral matrix* is one whose spectral norm and spectral radius are equal. (a) If $U \in M_n$ is unitary and $\alpha \in \mathbf{C}$, explain why αU is spectral. (b) Use (2.3.1) to show that if $A \in M_n$ is not a scalar multiple of a unitary matrix, then A is spectral if and only if there is a unitary $U \in M_n$ such that $U^*AU = \|\|A\|\|_2(B \oplus C)$, in which $B = [b_{ij}]$ is upper triangular, $\|\|B\|\|_2 < 1$, all $|b_{ii}| < 1$, and C is a diagonal unitary matrix. Explain why every maximum-modulus eigenvalue of a spectral matrix is not only a semisimple eigenvalue but also a normal eigenvalue. (c) If $A, B \in M_n$ are spectral, show that $\rho(AB) \leq \rho(A)\rho(B)$.

5.6.P40 (a) Compute the spectral norms of $\begin{bmatrix} 1 & 1 \\ 1 & 1 \end{bmatrix}$ and $\begin{bmatrix} 1 & 1 \\ 1 & -1 \end{bmatrix}$. Conclude that the spectral norm on M_n, although it is induced by an absolute norm on $\mathbf{C}^n$, is not an absolute norm on M_n; for a fundamental reason why it is not, see (7.4.11.1). (b) Compute the spectral norms of $\begin{bmatrix} 1 & 1 \\ -1 & 1 \end{bmatrix}$ and $\begin{bmatrix} 1 & 1 \\ 0 & 1 \end{bmatrix}$. Conclude that setting an entry of a matrix to zero can increase its spectral norm. (c) For $A \in M_n$, show that $\|\|A\|\|_2 \leq \|\||A|\|\|_2$. (d) If $A, B \in M_n$ have real nonnegative entries and $A \leq B$, show that $\|\|A\|\|_2 \leq \|\|B\|\|_2$.

5.6.P41 Let $\|\cdot\|$ be an absolute norm on $\mathbf{C}^n$ and let $\|\|\cdot\|\|$ be the matrix norm on M_n that it induces. Define $N(A) = \|\||A|\|\|$. (a) Show that $\|\|A\|\| \leq N(A)$ for all $A \in M_n$ and there is a vector z with nonnegative entries such that $\|z\| = 1$ and $N(A) = \||A|z\|$. (b) Show that $N(\cdot)$ is an absolute matrix norm on M_n. (c) If $A, B \in M_n$ have real nonnegative entries and $A \leq B$, show that $\|\|A\|\| \leq \|\|B\|\|$. (d) If $\|\|\cdot\|\|$ is the spectral norm $\|\|\cdot\|\|_2$, show that $\|\|A\|\|_2 \leq \|\||A|\|\|_2 \leq \sqrt{\operatorname{rank} A}\|\|A\|\|_2$ for all $A \in M_n$, and $\|\|A\|\|_2 \leq \|\|B\|\|_2$ if A and B have real nonnegative entries and $A \leq B$.

5.6.P42 Although the spectral norm is not absolute, it and every matrix norm that is induced by an absolute vector norm enjoys the weaker monotonicity property established in part (c) of the preceding problem. A norm $\|\cdot\|$ on M_n is *monotone on the positive orthant* if $\|A\| \geq \|B\|$ whenever $A, B \in M_n(\mathbf{R})$ and $A \geq B \geq 0$ (entrywise inequality). Provide details for the following proof that any matrix norm that is induced by a monotone vector norm is monotone on the positive orthant: Let $A, B \in M_n(\mathbf{R})$ and suppose that $A \geq B \geq 0$. Let the matrix norm $\|\|\cdot\|\|$ be induced by a monotone vector norm $\|\cdot\|$. Then $\|\|B\|\| = \max_{x \neq 0} \frac{\|Bx\|}{\|x\|} = \max_{x \neq 0} \frac{\|\|Bx\|\|}{\|x\|} \leq \max_{x \neq 0} \frac{\|B|x|\|}{\|x\|} \leq \max_{x \neq 0} \frac{\|A|x|\|}{\|x\|} \leq \max_{x \neq 0} \frac{\|Ax\|}{\|x\|} = \|\|A\|\|$.

5.6.P43 Let $A \in M_n$ and let $U^*AU = T$ be a unitary upper triangularization (2.3.1). Use the power series definition of e^A to show that $e^T = U^*e^AU$. Deduce that $\det e^A = e^{\operatorname{tr} A}$, so e^A is always nonsingular.

5.6.P44 Suppose that $A = [a_{ij}] \in M_n(\mathbf{R})$ has only integer entries (positive, negative, or zero) and that $K = \max |a_{ij}| = \|A\|_\infty$. Let $\lambda_1, \ldots, \lambda_m$ be the nonzero eigenvalues of A, including multiplicities. Explain why (a) $|\lambda_i| \leq nK$ (an integer) for each $i = 1, \ldots, m$; (b) every nonzero coefficient of the characteristic polynomial $p_A(t)$ is an integer, and therefore has modulus at least one; (c) $p_A(t) = t^{n-m}g_A(t)$, in which $g_A(t)$ is a polynomial of degree m such that $|g_A(0)| = |\lambda_1 \cdots \lambda_m| \geq 1$; (d) $\min_{i=1,\ldots,m} |\lambda_i| \geq 1/(nK)^{m-1} \geq 1/(nK)^{n-1}$; (e) if A is a nonsingular symmetric 4-by-4 matrix whose entries are ± 1 and 0, no entry of A^{-1} has absolute value greater than 64; indeed, no column of A^{-1} has Euclidean norm greater than 64.

5.6.P45 Let $\|\| \cdot \|\|$ be a matrix norm on M_n that is induced by a norm $\| \cdot \|$ on $\mathbf{C}^n$, and suppose that $A = XY^*$, in which $X = [x_1 \ \ldots \ x_k] \in M_{n,k}$ and $Y = [y_1 \ \ldots \ y_k] \in M_{n,k}$. Show that $\|\|A\|\| \leq \sum_{i=1}^k \|x_i\| \ \|y_i\|^D$, with equality if $k = 1$.

5.6.P46 Let $A \in M_n$ be nonsingular and suppose that a matrix norm $\|\| \cdot \|\|$ on M_n is induced by the vector norm $\| \cdot \|$ on $\mathbf{C}^n$. Show that $\|\|A^{-1}\|\| = 1/\min_{\|x\|=1} \|\|Ax\|\|$.

In the following five problems, $\|\| \cdot \|\|$ is a matrix norm on M, $\mathcal{S}_n = \{X : X \in M_n$ and X is singular$\}$, and for a given $A \in M_n$, $\operatorname{dist}_{\|\|\cdot\|\|}(A, \mathcal{S}_n) = \inf\{\|\|A - B\|\| : B \in \mathcal{S}_n\}$ is the distance from A to the set of singular matrices in M_n. We want to show that $\operatorname{dist}_{\|\|\cdot\|\|}(A, \mathcal{S}_n)$ and $\|\|A^{-1}\|\|$ are intimately related.

5.6.P47 If $A, B \in M_n$, A is nonsingular, and B is singular, show that

$$\|\|A - B\|\| \geq 1/\|\|A^{-1}\|\| \tag{5.6.55}$$

Can a nonsingular matrix be closely approximated by a singular matrix?

5.6.P48 (a) Explain why $\mathcal{S}_n$ is a closed set, that is, if $X_i \in \mathcal{S}_n$ for all $i = 1, 2, \ldots$, $B \in M_n$, and $\|\|X_i - B\|\| \to 0$ as $\iota \to \infty$, then $B \in \mathcal{S}_n$. (b) Explain why $\operatorname{dist}_{\|\|\cdot\|\|}(A, \mathcal{S}_n) > 0$ if A is nonsingular and why there is some $B_0 \in \mathcal{S}_n$ such that $\operatorname{dist}_{\|\|\cdot\|\|}(A, \mathcal{S}_n) = \|\|A - B_0\|\|$. (c) If A is nonsingular, explain why $\operatorname{dist}_{\|\|\cdot\|\|}(A, \mathcal{S}_n) \geq \|\|A^{-1}\|\|^{-1}$.

5.6.P49 Suppose that $\|\| \cdot \|\|$ is induced by the norm $\| \cdot \|$ on $\mathbf{C}^n$ and suppose that $A \in M_n$ is nonsingular. Let $x_0, y_0 \in \mathbf{C}^n$ be vectors such that $\|x_0\| = \|y_0\|^D = 1$ and $y_0^*A^{-1}x_0 = \|\|A^{-1}\|\|$; see (5.6.P54). Let $E = -x_0y_0^*/\|\|A^{-1}\|\|$. (a) Show that $\|\|E\|\| = \|\|A^{-1}\|\|^{-1}$. (b) Show that $(A + E)A^{-1}x_0 = 0$, so $A + E \in \mathcal{S}_n$. (c) Conclude that $\operatorname{dist}_{\|\|\cdot\|\|}(A, \mathcal{S}_n) = \|\|A^{-1}\|\|^{-1}$ for every nonsingular A if $\|\| \cdot \|\|$ is an induced matrix norm.

5.6.P50 Suppose that $\|\| \cdot \|\|$ is a matrix norm on M_n that is *not* induced. Then (5.6.26) ensures that there is an *induced* matrix norm $N(\cdot)$ on M_n such that $N(A) \leq \|\|A\|\|$ for all $A \in M_n$. (a) Explain why there is a $\hat{C} \in M_n$ such that $N(\hat{C}) < \|\|\hat{C}\|\|$. (b) Show that there is a *nonsingular* $C \in M_n$ such that $N(C^{-1}) < \|\|C^{-1}\|\|$. (c) Explain why $\operatorname{dist}_{\|\|\cdot\|\|}(C, \mathcal{S}_n) \geq \operatorname{dist}_{N(\cdot)}(C, \mathcal{S}_n) = N(C^{-1})^{-1} > \|\|C^{-1}\|\|^{-1}$.

5.6.P51 Explain why a matrix norm $\|\| \cdot \|\|$ on M_n is induced if and only if $\operatorname{dist}_{\|\|\cdot\|\|}(A, \mathcal{S}_n) = \|\|A^{-1}\|\|^{-1}$ for every nonsingular $A \in M_n$.

5.6.P52 Work out the construction in (5.6.P49) for the spectral norm. If $A \in M_n$ is nonsingular and $A = V\Sigma W^*$ is a singular value decomposition (2.6.3.1), show that one may take x_0 to be the last column of V and y_0 to be the last column of W. Why is σ_n the distance from A (in the norm $\||\cdot\||_2$) to a nearest singular matrix? Show that $A + E = V\hat{\Sigma}W^*$, in which $\hat{\Sigma} = \mathrm{diag}(\sigma_1, \ldots, \sigma_{n-1}, 0)$.

5.6.P53 Work out the construction in (5.6.P49) for the maximum row sum norm and $A = \begin{bmatrix} 1 & 0 \\ 1 & 1/2 \end{bmatrix}$. Show that one may take $x_0 = [-1\ 1]^T$ and $y_0 = [0\ 1]^T$. Why is $1/4$ the distance from A (in the norm $\||\cdot\||_\infty$) to a nearest singular matrix? Show that $A + E = \begin{bmatrix} 1 & 1/4 \\ 1 & 1/4 \end{bmatrix}$.

5.6.P54 The general principle being invoked to ensure existence of the vectors x_0 and y_0 in (5.6.P49) is that the Cartesian product of compact sets is compact. The compact sets involved in this case are the unit balls of the norm $\|\cdot\|$ and its dual. (a) Show that the Cartesian product $\mathbf{C}^n \times \mathbf{C}^n = \{(x, y) : x, y \in \mathbf{C}^n\}$ is a complex normed linear space if we define $(x, y) + (\xi, \eta) = (x + \xi, y + \eta)$, $\alpha(x, y) = (\alpha x, \alpha y)$, and $N((x, y)) = \max\{\|x\|, \|y\|^D\}$. (b) Show that $B_{\|\cdot\|} \times B_{\|\cdot\|^D}$ is a closed subset of $\mathbf{C}^n \times \mathbf{C}^n$ with respect to the norm $N(\cdot)$. (c) Show that $B_{\|\cdot\|} \times B_{\|\cdot\|^D}$ is a bounded subset of $\mathbf{C}^n \times \mathbf{C}^n$ with respect to the norm $N(\cdot)$. (d) Conclude that $B_{\|\cdot\|} \times B_{\|\cdot\|^D}$ is a compact subset of $\mathbf{C}^n \times \mathbf{C}^n$ with respect to the norm $N(\cdot)$. (e) Assume that $A \in M_n$ is nonsingular. The real-valued function $f(x, y) = |y^* A^{-1} x|$ is continuous on $B_{\|\cdot\|} \times B_{\|\cdot\|^D}$, so it achieves its maximum value at some point $(\hat{x}, y_0) \in B_{\|\cdot\|} \times B_{\|\cdot\|^D}$. There is some real θ such that $y_0^* A^{-1}(e^{i\theta}\hat{x}) = |y_0^* A^{-1}\hat{x}|$. Take $x_0 = e^{i\theta}\hat{x}$. Why is $(x_0, y_0) \in B_{\|\cdot\|} \times B_{\|\cdot\|^D}$? Why is $\|x_0\| = \|y_0\|^D = 1$?

5.6.P55 Let $\||\cdot\||$ be a given matrix norm on M_m, and define a function $N(\cdot) : M_{mn} \to \mathbf{R}$ as follows: Partition each $A \in M_{mn}$ as $A = [A_{ij}]_{i,j=1}^n$, in which each block $A_{ij} \in M_m$. Define $N(A) = \max_{1 \le i \le n} \sum_{j=1}^n \||A_{ij}\||$. (a) Show that $N(\cdot)$ is a matrix norm on M_{mn}. (b) What is $N(\cdot)$ if $m = 1$? For an application of matrix norms of this type, see (6.1.P17).

5.6.P56 Let $\||\cdot\||$ be a self-adjoint matrix norm on M_n; for example, it could be a unitarily invariant matrix norm. Show that $\||A\||_2 \le \||A\||$ for every $A \in M_n$.

5.6.P57 Let the eigenvalues of $A \in M_n$ be arranged so that $|\lambda_1| \ge \cdots \ge |\lambda_n|$ and let $\sigma_1 \ge \cdots \ge \sigma_n$ be its ordered singular values. Show that $|\lambda_1 \cdots \lambda_r| \le \sigma_1 \cdots \sigma_r$ for each $r = 1, \ldots, n$ by applying the bound in (5.6.9) to the spectral norm and the compound matrix $C_r(A)$. See (2.6.P33) and (2.3.P12).

5.6.P58 (a) Give an example of matrices $A, B \in M_2$ such that $\|AB\|_2 \ne \|BA\|_2$ (the Frobenius norm). (b) If $A, B \in M_n$, A is normal, and B is Hermitian, show that $\|AB\|_2 = \|BA\|_2$. See (7.3.P43) for a generalization.

Further Readings. Further discussion of the problem of determining bounds between induced norms (5.6.18) is in H. Schneider and G. Strang, Comparison theorems for supremum norms, *Numer. Math.* 4 (1962) 15–20. The bounds in the table in (5.6.P23) are taken from B. J. Stone, Best possible ratios of certain matrix norms, *Numer. Math.* 4 (1962) 114–116, which contains additional bounds and references. For further discussion of the use of matrix norms to locate zeroes of polynomials (5.6.P27 to P35), see M. Fujii and F. Kubo, Operator norms as bounds for roots of algebraic equations, *Proc. Japan Acad.* 49 (1973) 805–808. The entire book Belitskii and Lyubich (1988) is devoted to matrix norms.

5.7 Vector norms on matrices

Although all the axioms for a norm are necessary for a useful notion of "size" for matrices, for some important applications the submultiplicativity axiom for a matrix norm is not necessary. For example, the Gelfand formula (5.6.14) does not require submultiplicativity and is valid for vector norms and even pre-norms. In this section, we discuss vector norms on matrices, that is, norms on the vector space M_n that are not necessarily submultiplicative. We use $G(\cdot)$ to denote a generic vector norm on M_n and begin with some examples of norms on M_n that may or may not be matrix norms.

Example 1. If $G(\cdot)$ is a norm on M_n, and if $S, T \in M_n$ are nonsingular, then

$$G_{S,T}(A) = G(SAT), \qquad A \in M_n \tag{5.7.1}$$

is a norm on M_n. Even if $G(\cdot)$ is a matrix norm, $G_{S,T}(\cdot)$ need not be submultiplicative unless $T = S^{-1}$ (5.6.7).

Exercise. Show that $G_{S,T}(\cdot)$ in (5.7.1) is always a norm on M_n.

Exercise. Let $S = T = \frac{1}{2}I$, let $G(\cdot) = n\|\cdot\|_\infty$, and show that $G_{S,T}(\cdot)$ is not a matrix norm.

Example 2. The *Hadamard product* of two matrices $A = [a_{ij}]$ and $B = [b_{ij}]$ of the same size is their entrywise product $A \circ B = [a_{ij}b_{ij}]$. If $H \in M_n$ has no zero entries and if $G(\cdot)$ is any norm on M_n, then

$$G_H(A) = G(H \circ A), \qquad H \in M_n, |H| > 0 \tag{5.7.2}$$

is a norm on M_n. Even if $G(\cdot)$ is a matrix norm, $G_H(\cdot)$ need not be submultiplicative.

Exercise. Show that $G_H(\cdot)$ in (5.7.2) is always a norm.

Exercise. Show that $G_H(\cdot)$ in (5.7.2) may or may not be a matrix norm, depending on the choice of H. Consider the matrix norm $G(\cdot) = \|\|\cdot\|\|_1$, the matrices

$$H_1 = \begin{bmatrix} 1 & 1 \\ 1 & 1 \end{bmatrix} \quad \text{or} \quad H_2 = \begin{bmatrix} 2 & 1 \\ 1 & 2 \end{bmatrix} \tag{5.7.3}$$

and

$$A = \begin{bmatrix} 0 & 1 \\ 0 & 0 \end{bmatrix}, \quad B = \begin{bmatrix} 0 & 0 \\ 1 & 0 \end{bmatrix}, \quad \text{and} \quad AB \tag{5.7.4}$$

Notice that $G_{H_1}(C) \leq G_{H_2}(C)$ for all $C \in M_2$.

Example 3. The function $G_c(\cdot)$ defined by

$$G_c\left(\begin{bmatrix} a & b \\ c & d \end{bmatrix}\right) = \frac{1}{2}[|a + d| + |a - d| + |b| + |c|] \tag{5.7.5}$$

is a norm on M_2.

Exercise. Show that $G_c(\cdot)$ in (5.7.5) is a norm but not a matrix norm. *Hint*: Consider the matrices (5.7.4).

Example 4. If $A \in M_n$, the set $F(A) = \{x^*Ax : x \in \mathbf{C}^n \text{ and } x^*x = 1\}$ is the *field of values* or *numerical range* of A, and the function

$$r(A) = \max_{\|x\|_2=1} |x^*Ax| = \max\{|z| : z \in F(A)\}$$

is the *numerical radius* of A.

Exercise. Show that $r(\cdot)$ is a norm on M_n. *Hint:* For the positivity axiom (1a), see (4.1.P6). The numerical radius is not a matrix norm, however; see (5.7.P10).

Example 5. The l_∞ norm on M_n is

$$\|A\|_\infty = \max_{1 \le i,j \le n} |a_{ij}| \tag{5.7.7}$$

We saw in (5.6.0.3–4) that $\|\cdot\|_\infty$ is a norm on M_n but not a matrix norm. However, $n\|\cdot\|_\infty$ is a matrix norm.

The preceding examples demonstrate that there are many norms on M_n that are not matrix norms. Some of these norms share some of the properties of matrix norms that follow from submultiplicativity, and some do not. But each norm on M_n is *equivalent* to any matrix norm (in the sense that they have the same convergent sequences); in fact, a somewhat more general result follows immediately from (5.4.4).

Theorem 5.7.8. *Let f be a pre-norm on M_n, that is, a real-valued function on M_n that is positive, homogeneous, and continuous (5.4.4), and let $\|\|\cdot\|\|$ be a matrix norm on M_n. Then there exist finite positive constants C_m and C_M such that*

$$C_m \|\|A\|\| \le f(A) \le C_M \|\|A\|\| \tag{5.7.9}$$

for all $A \in M_n$. In particular, these inequalities are valid if $f(\cdot)$ is a vector norm on M_n.

The bounds (5.7.9) are often useful in extending facts about matrix norms to vector norms on matrices, or, more generally, to pre-norms on matrices. For example, the Gelfand formula (5.6.14) extends in this manner.

Theorem 5.7.10. *If f is a pre-norm on M_n, in particular, if it is a vector norm, then $\lim_{k\to\infty}[f(A^k)]^{1/k} = \rho(A)$ for all $A \in M_n$.*

Proof. Let $\|\|\cdot\|\|$ be a matrix norm on M_n and consider the inequality (5.7.9), which implies that

$$C_m^{1/k} \|\|A^k\|\|^{1/k} \le [f(A^k)]^{1/k} \le C_M^{1/k} \|\|A^k\|\|^{1/k}$$

for all $k = 1, 2, 3, \ldots$. But $C_m^{1/k} \to 1$, $C_M^{1/k} \to 1$, and $\|\|A^k\|\|^{1/k} \to \rho(A)$ as $k \to \infty$ (5.6.14) so we conclude that $\lim_{k\to\infty}[f(A^k)]^{1/k}$ exists and has the asserted value. $\square$

Example 5 illustrates a second sense in which any norm on M_n is equivalent to a matrix norm. A positive scalar multiple of the norm $\|\cdot\|_\infty$ is a matrix norm. This is no accident: Every norm on M_n becomes a matrix norm after multiplication by a suitable positive constant. This fundamental result is a consequence of continuity of a norm function and compactness of its unit sphere.

Theorem 5.7.11. *Let $G(\cdot)$ be a vector norm on M_n and let*

$$c(G) = \max_{G(A)=1=G(B)} G(AB)$$

For a real positive scalar γ, $\gamma G(\cdot)$ is a matrix norm on M_n if and only if $\gamma \geq c(G)$. If $\lvert\lvert\lvert \cdot \rvert\rvert\rvert$ is a matrix norm on M_n, if C_m and C_M are positive constants such that

$$C_m \lvert\lvert\lvert A \rvert\rvert\rvert \leq G(A) \leq C_M \lvert\lvert\lvert A \rvert\rvert\rvert \quad \text{for all } A \in M_n \tag{5.7.11a}$$

and if we set $\gamma_0 = C_M/C_m^2$, then $\gamma_0 G(\cdot)$ is a matrix norm and hence $\gamma_0 \geq c(G)$.

Proof. The value $c(G)$ is the maximum of a positive continuous function over a compact set, so it is finite and positive. For any nonzero $A, B \in M_n$ we have

$$c(G) \geq G\left(\frac{A}{G(A)} \frac{B}{G(B)}\right) = \frac{G(AB)}{G(A)G(B)}$$

that is, $G(AB) \leq c(G)G(A)G(B)$, and hence

$$c(G)G(AB) \leq (c(G)G(A))(c(G)G(B))$$

for all $A, B \in M_n$. Consequently, $c(G)G(\cdot)$ is a matrix norm. If $\gamma > c(G)$, then

$$\gamma G(AB) \leq \frac{\gamma}{c(G)} (c(G)G(A))(c(G)G(B)) = (\gamma G(A))(c(G)G(B))$$

$$\leq (\gamma G(A))(\gamma G(B))$$

so $\gamma G(\cdot)$ is a matrix norm. If $\gamma > 0$, $\gamma < c(G)$, and $\gamma G(\cdot)$ is a matrix norm, then for all A and B such that $G(A) = G(B) = 1$, we have $\gamma G(AB) \leq \gamma G(A)\gamma G(B) = \gamma^2$, so $\max_{G(A)=1=G(B)} G(AB) \leq \gamma < c(G)$, which is a contradiction. Finally, compute

$$\gamma_0 G(AB) \leq \gamma_0 C_M \lvert\lvert\lvert AB \rvert\rvert\rvert \leq \gamma_0 C_M \lvert\lvert\lvert A \rvert\rvert\rvert \, \lvert\lvert\lvert A \rvert\rvert\rvert$$

$$\leq \gamma_0 \frac{C_M}{C_m^2} G(A)G(B) = (\gamma_0 G(A))(\gamma_0 G(B))$$

so $\gamma_0 G(\cdot)$ is a matrix norm. $\qquad\square$

Exercise. Deduce the Gelfand formula for vector norms on M_n from (5.7.11) and (5.6.14).

An important property of any matrix norm $\lvert\lvert\lvert \cdot \rvert\rvert\rvert$ is that it is *spectrally dominant*, that is, $\lvert\lvert\lvert A \rvert\rvert\rvert \geq \rho(A)$ for every $A \in M_n$. It is noteworthy that a vector norm on M_n can be spectrally dominant even if it is not submultiplicative; we now investigate circumstances under which this can occur.

Definition 5.7.12. *A norm $\lVert \cdot \rVert$ on $\mathbf{C}^n$ and a vector norm $G(\cdot)$ on M_n are* compatible *if $\lVert Ax \rVert \leq G(A)\lVert x \rVert$ for all $x \in \mathbf{C}^n$ and all $A \in M_n$. The term* consistent *is sometimes used, and the norm $\lVert \cdot \rVert$ is sometimes said to be* subordinate *to the norm $G(\cdot)$.*

Theorem 5.7.13. *If $\lvert\lvert\lvert \cdot \rvert\rvert\rvert$ is a matrix norm on M_n, then there is a norm on $\mathbf{C}^n$ that is compatible with it. If $\lVert \cdot \rVert$ is a norm on $\mathbf{C}^n$, then there is a matrix norm on M_n that is compatible with it.*

Proof. For any nonzero vector z, the norm $\|\cdot\|_z$ defined in (5.6.27) is compatible with a given matrix norm $\||\cdot\||$: $\|Ax\|_z = \||Axz^*\|| \leq \||A\|| \, \||xz^*\|| = \||A\|| \, \|x\|_z$. Any given norm on $\mathbf{C}^n$ is compatible with the matrix norm on M_n that it induces (5.6.2(b)). $\square$

Theorem 5.7.14. *Let $G(\cdot)$ be a norm on M_n that is compatible with a norm $\|\cdot\|$ on $\mathbf{C}^n$. Then*

$$G(A_1)\cdots G(A_k) \geq \rho(A_1 \cdots A_k)\, \text{for all } A_1, \ldots, A_k \in M_n,\ k = 1, 2, \ldots \quad (5.7.15)$$

In particular, $G(A)$ is spectrally dominant.

Proof. Consider the case $k = 2$ and let $x \in \mathbf{C}^n$ be a nonzero vector such that $A_1 A_2 x = \lambda x$ with $|\lambda| = \rho(A_1 A_2)$. Then

$$\rho(A_1 A_2)\|x\| = \|\lambda x\| = \|A_1 A_2 x\| = \|A_1(A_2 x)\|$$
$$\leq G(A_1)\|A_2 x\| \leq G(A_1)G(A_2)\|x\|$$

Since $\|x\| \neq 0$, we conclude that $\rho(A_1 A_2) \leq G(A_1)G(A_2)$. The general case follows by induction. $\square$

Exercise. Verify the cases $k = 1$ and $k = 3$ in the preceding theorem.

Which vector norms on M_n are compatible with some norm on $\mathbf{C}^n$? The condition (5.7.15) is necessary; to show that it is also sufficient, we need a technical lemma.

Lemma 5.7.16. *Let $G(\cdot)$ be a vector norm on M_n that satisfies (5.7.15). There is a finite positive constant $\gamma(G)$ such that*

$$G(A_1)\cdots G(A_k) \geq \gamma(G)\||A_1 \cdots A_k\||_2$$

for all $A_1, A_2, \ldots, A_k \in M_n$ and all $k = 1, 2, \ldots$

Proof. Let k be a given positive integer, let $A_1, \ldots, A_k \in M_n$ be given, and let $A_1 \cdots A_k = V\Sigma W^*$ be a singular value decomposition (2.6.3). The hypothesis permits us to use (5.7.15) to compute

$$G(V^*)G(A_1)\cdots G(A_k)G(W) \geq \rho(V^* A_1 \cdots A_k W) = \rho(\Sigma) = \||\Sigma\||_2$$
$$= \||V^* A_1 \cdots A_k W\||_2 = \||A_1 \cdots A_k\||_2$$

The final equality follows from unitary invariance of the spectral norm. Since $G(\cdot)$ is a continuous function on the compact set of unitary matrices, $\mu(G) = \max\{G(U) : U \in M_n \text{ is unitary}\}$ is finite and positive. We conclude that

$$G(A_1)\cdots G(A_k) \geq \frac{1}{G(V^*)G(W)}\||A_1 \cdots A_k\||_2$$
$$\geq \mu(G)^{-2}\||A_1 \cdots A_k\||_2$$

$\square$

Theorem 5.7.17. *A vector norm $G(\cdot)$ on M_n is compatible with some norm on $\mathbf{C}^n$ if and only it satisfies the inequality (5.7.15).*

Proof. One implication has already been proved in (5.7.14). To prove the other, we claim it is sufficient to show that there is a matrix norm $\||\cdot\||$ on M_n such that

$G(A) \geq \|\|A\|\|$ for all $A \in M_n$. If such a matrix norm exists, let $\|\cdot\|$ be a norm on $\mathbf{C}^n$ that is compatible with it (5.7.13), and let $x \in \mathbf{C}^n$ and $A \in M_n$ be given. Then $\|Ax\| \leq \|\|A\|\|\|x\| \leq G(A)\|x\|$, so the norm $\|\cdot\|$ is also compatible with $G(\cdot)$.

For a given $A \in M_n$, there are myriad ways to represent it as a product of matrices or as a sum of products of matrices. Define

$$\|\|A\|\| = \inf\left\{\sum_i G(A_{i1})\cdots G(A_{ik_i}) : \sum_i A_{i1}\cdots A_{ik_i} = A, \text{ each } A_{ij} \in M_n\right\}$$

The function $\|\|\cdot\|\|$ is nonnegative and homogeneous. Is it positive? If $\sum_i A_{i1}\cdots A_{ik_i} = A \neq 0$, then (5.7.16) and the triangle inequality for the spectral norm ensure that

$$\sum_i G(A_{i1})\cdots G(A_{ik_i}) \geq \sum_i \gamma(G)\|\|A_{i1}\cdots A_{ik_i}\|\|_2$$
$$\geq \gamma(G)\left\|\left\|\sum_i A_{i1}\cdots A_{ik_i}\right\|\right\|_2 = \gamma(G)\|\|A\|\|_2 > 0$$

so $\|\|\cdot\|\|$ is positive. The triangle inequality and submultiplicativity for $\|\|\cdot\|\|$ follow from its definition as an infimum of sums of products. $\square$

Exercise. Explain carefully why the function $\|\|\cdot\|\|$ constructed in the preceding theorem is subadditive and submultiplicative. *Hint*: If $C = A + B$ or $C = AB$, then every representation of A and B (separately) as sums of products yields a representation of C as a sum of products; of course, not all representations of C as sums of products arise in this way.

Exercise. Let $G(\cdot)$ be a vector norm on M_2 and let $J_2(0)$ be the nilpotent Jordan block of size 2. If $G(\cdot)$ were compatible with a norm $\|\cdot\|$ on $\mathbf{C}^2$, explain why $\|e_1\| = \|J_2(0)e_2\| \leq G(J_2(0))\|e_2\|$ and $\|e_2\| = \|J_2(0)^T e_1\| \leq G(J_2(0)^T)\|e_1\|$, which implies that $\|e_1\| \leq G(J_2(0))G(J_2(0)^T)\|e_1\|$. Conclude that the inequality $G(J_2(0))G(J_2(0)^T) \geq 1$ is a necessary condition for $G(\cdot)$ to be compatible with some vector norm on $\mathbf{C}^2$. Explain why the norm $G_c(\cdot)$ defined in (5.7.5) is not compatible with any norm on $\mathbf{C}^2$.

Exercise. Even though the norm $G_c(\cdot)$ defined in (5.7.5) is not compatible with any norm on $\mathbf{C}^2$, show that it is spectrally dominant. Discuss in light of (5.7.17). *Hint*: Use (1.2.4b) to show that

$$\rho\left(\begin{bmatrix} a & b \\ c & d \end{bmatrix}\right) \leq \frac{1}{2}\{|a - d| + \sqrt{|a + d|^2 + 4|bc|}\} \leq G_c\left(\begin{bmatrix} a & b \\ c & d \end{bmatrix}\right)$$

We have seen that some vector norms on M_n have compatible norms on $\mathbf{C}^n$ and some do not. Those that do are spectrally dominant; those that do not can be either spectrally dominant or not. We have necessary and sufficient conditions for a vector norm on M_n to be compatible with some norm on $\mathbf{C}^n$, and we know that any norm on $\mathbf{C}^n$ is compatible with the submultiplicative norm on M_n that it induces. When is a norm on $\mathbf{C}^n$ compatible with a norm on M_n that is not submultiplicative? Always.

Theorem 5.7.18. *Every norm on $\mathbf{C}^n$ is compatible with a vector norm on M_n that is not a matrix norm.*

Proof. Let $\|\cdot\|$ be a norm on $\mathbf{C}^n$ and let $P \in M_n$ be any permutation matrix with a zero main diagonal, for example, the circulant matrix (0.9.6.2). Let $\|\|\cdot\|\|$ denote the matrix norm on M_n that is induced by $\|\cdot\|$. For any $A = [a_{ij}] \in M_n$ define

$$G(A) = \||A\|| + \||P\|| \; \||P^T\|| \max_{1 \leq i \leq n} |a_{ii}|$$

Then $G(\cdot)$ is a norm on M_n. Moreover, $G(A) \geq \||A\||$ and $\|Ax\| \leq \||A\|| \, \|x\| \leq G(A)\|x\|$ for all $A \in M_n$ and all $x \in \mathbf{C}^n$, so $G(\cdot)$ is compatible with $\|\cdot\|$. However, $G(P) = \||P\||$ and $G(P^T) = \||P^T\||$, so

$$G(PP^T) = G(I) = \||I\|| + \||P\|| \; \||P^T\||$$
$$= 1 + \||P\|| \; \||P^T\|| > G(P)G(P^T)$$

Thus, $G(\cdot)$ is not submultiplicative.   □

Exercise. Let $A = [a_{ij}] \in M_n$ and consider the norm $G(A) = \||A + \mathrm{diag}(a_{11}, \ldots, a_{nn})\||_\infty$. Show that $G(\cdot)$ has the form (5.7.2) (what is H?) and hence is a norm on M_n. Show that $G(\cdot)$ is compatible with the norm $\|\cdot\|_\infty$ on C^n. Let $A = \begin{bmatrix} 0 & 1 \\ 1 & 0 \end{bmatrix}$ and compute $G(A)$ and $G(A^2)$. Explain why $G(\cdot)$ is not submultiplicative.

Our final goal in this section is to present a necessary and sufficient condition for spectral dominance of a norm $G(\cdot)$ on M_n, and we focus on a weak version of submultiplicativity: If for each $A \in M_n$ there is a positive constant γ_A such that $G(A^k) \leq \gamma_A G(A)^k$ for all $k = 1, 2, \ldots$, then $G(A^k)^{1/k} \leq \gamma_A^{1/k} G(A)$ for all $k = 1, 2, \ldots$ and hence (5.7.10) ensures that $\rho(A) \leq G(A)$, that is, $G(\cdot)$ is spectrally dominant. In showing that this sufficient condition is also necessary, subadditivity of $G(\cdot)$ turns out to be crucial.

Exercise. Let $G(\cdot)$ be a norm on M_n. Explain why the following are equivalent:

   (a) For each $A \in M_n$ there is a positive constant γ_A (depending only on A and $G(\cdot)$) such that $G(A^k) \leq \gamma_A G(A)^k$ for all $k = 1, 2, \ldots$.

   (b) For each $A \in M_n$ such that $G(A) = 1$, the sequence $G(A), G(A^2), G(A^3), \ldots$ is bounded.

   (c) For each $A \in M_n$ such that $G(A) = 1$, the entries of all the matrices $A, A^2, A^3, \ldots$ lie in a bounded set.

Exercise. Let $G(\cdot)$ be a norm on M_n, let $S \in M_n$ be nonsingular, and let $G_S(A) = G(SAS^{-1})$ for any $A \in M_n$. Explain why $G(\cdot)$ is spectrally dominant if and only if $G_S(\cdot)$ is spectrally dominant.

Lemma 5.7.19. *Let $G(\cdot)$ be a spectrally dominant norm on M_n, let $A \in M_n$ be given, and let λ be an eigenvalue of A such that $|\lambda| = \rho(A)$. If λ is not semisimple, then $G(A) > \rho(A)$.*

Proof. If $\rho(A) = 0$, then 0 is not a semisimple eigenvalue of A if and only if $A \neq 0$, in which case $G(A) > 0$. We may therefore assume that $\rho(A) \neq 0$. Since we may normalize any maximum-modulus eigenvalue by considering $e^{i\theta} \rho(A)^{-1} A$, we may also assume that $\lambda = 1$ is an eigenvalue of A and $G(A) \geq \rho(A) = 1$. Suppose that 1 is not a semisimple eigenvalue of A, that is, the Jordan canonical form of A is $J_m(1) \oplus B$, in which $m \geq 2$, $B \in M_{n-m}$, and $\rho(B) \leq 1$. We must show that $G(A) > 1$. The preceding exercise permits us to assume that $A = J_m(1) \oplus B$; we must show that $G(A) > 1$. Let $E_m \in M_m$ be the matrix with $m, 1$ entry equal to 1 and all other entries equal to 0. Let $F = E_m \oplus 0_{n-m}$ and write $A = I_m \oplus B + J_m(0) \oplus 0_{n-m}$. For any $\varepsilon > 0$, let $A_\varepsilon = A + \varepsilon F = (I_m + J_m(0) + \varepsilon E_m) \oplus B$. Then $\rho(I_m + J_m(0) + \varepsilon E_m) = 1 + \varepsilon^{1/m} > \rho(B)$; see (1.2.P22). Consequently,

$$1 + \varepsilon^{1/m} = \rho(A_\varepsilon) \leq G(A_\varepsilon) = G(A + \varepsilon F)$$
$$\leq G(A) + G(\varepsilon F) = G(A) + \varepsilon G(F)$$

If $G(A) = 1$, then $\varepsilon^{1/m} \leq \varepsilon G(F)$, or $1 \leq \varepsilon^{\frac{m-1}{m}} G(F)$, which is not possible for all $\varepsilon > 0$ since $\varepsilon^{\frac{m-1}{m}} \to 0$ as $\varepsilon \to 0$. We conclude that $G(A) > 1$. $\quad\square$

Theorem 5.7.20. *A norm $G(\cdot)$ on M_n is spectrally dominant if and only if for each $A \in M_n$ there is a positive constant γ_A (depending only on A and $G(\cdot)$) such that*

$$G(A^k) \leq \gamma_A G(A)^k \quad \text{for all } k = 1, 2, \ldots \tag{5.7.20a}$$

Proof. We need to show only that the stated condition is necessary. Suppose that $A \in M_n$ and $G(A) = 1 \geq \rho(A)$. The preceding lemma ensures that each Jordan block of A has the form $J_m(\lambda)$, in which $|\lambda| \leq 1$, and $m = 1$ if $|\lambda| = 1$. Consequently, either $|\lambda| < 1$ and $J_m(\lambda)^k \to 0$ (5.6.12), or $|\lambda| = 1$, $m = 1$, and $J_m(\lambda)^k = \lambda^k, k = 1, 2, \ldots$ is a bounded sequence. We conclude that the entries of all the matrices $A, A^2, A^3, \ldots$ lie in a bounded set. $\quad\square$

Problems

5.7.P1 Let $G(\cdot)$ be a vector norm on M_n and let $z \in \mathbf{C}^n$ be nonzero. Show that the function $\|x\| = G(xz^*)$ is a norm on $\mathbf{C}^n$. What is this function if $z = e$ or if $z = e_1$?

5.7.P2 Let $G(\cdot)$ be a vector norm on M_n, let $A \in M_n$, and let $\epsilon > 0$. Show that there is a positive constant $K(\epsilon, A)$ such that $(\rho(A) - \epsilon)^k \leq G(A^k) \leq (\rho(A) + \epsilon)^k$ for all $k > K(\epsilon, A)$.

5.7.P3 Let $G(\cdot)$ be a vector norm on M_n and let $A \in M_n$. (a) Use the preceding problem to show that if $\rho(A) < 1$, then $G(A^k) \to 0$ as $k \to \infty$. At what rate? (b) Conversely, if $G(A^k) \to 0$ as $k \to \infty$, show that $\rho(A) < 1$. (c) What can you say about convergence of power series of matrices using vector norms?

5.7.P4 Let $G(\cdot)$ be a vector norm on M_n and define the function $G' : M_n \to \mathbf{R}$ by $G'(A) = \max_{G(B)=1} G(AB)$. Show that $G'(\cdot)$ is a unital matrix norm on M_n. If $G(I) = 1$, show that $G'(A) \geq G(A)$ for all $A \in M_n$.

The next four problems continue the notation and assumptions of (5.7.P4).

5.7.P5 Let $G(\cdot)$ be a matrix norm. Show that $G'(A) \leq G(A)$ for all $A \in M_n$ and if $G(I) = 1$, then $G'(\cdot) = G(\cdot)$.

5.7.P6 Define $G''(A) = \max_{G'(B)=1} G'(AB)$. Show that $G''(\cdot) = G'(\cdot)$.

5.7.P7 If $G(I) = 1$, show that $G(\cdot)$ is a matrix norm if and only if $G'(A) \leq G(A)$ for all $A \in M_n$.

5.7.P8 Show that one can reverse the order in which A and B appear in the definition of $G'(\cdot)$ in (5.7.P4) and thereby obtain another matrix norm; show with an example that it can be different from $\mathcal{G}'(\cdot)$.

5.7.P9 Show that the set of all vector seminorms on $\mathbf{C}^n$ that are compatible with a given norm on M_n is a convex set; in fact, it is a convex cone.

5.7.P10 Show that the numerical radius is not a matrix norm on M_n by considering the matrices (5.7.4) and comparing $r(AB)$ with $r(A)r(B)$.

5.7.P11 (a) Show that $r(J_2(0)) = \frac{1}{2}$. (b) Explain why the numerical radius is not compatible with any norm on $\mathbf{C}^n$. (c) Show that the spectrum of $A \in M_n$ is contained in its field of values. (d) Explain why the numerical radius is spectrally dominant.

5.7.P12 Explain why no norm on $\mathbf{C}^n$ is compatible with the norm $\| \cdot \|_\infty$ on M_n, but that some norm on $\mathbf{C}^n$ is compatible with the norm $n\| \cdot \|_\infty$ on M_n.

5.7.P13 For $A = [a_{ij}] \in M_{m,n}$, let $r_i(A) = [a_{i1} \ \ldots \ a_{in}]^T$ and $c_j(A) = [a_{1j} \ \ldots \ a_{mj}]^T$, and suppose that $\| \cdot \|_\alpha$ and $\| \cdot \|_\beta$ are norms on $\mathbf{C}^n$ and $\mathbf{C}^m$, respectively. Define $G_{\beta,\alpha} : M_{m,n} \to \mathbf{R}$ by

$$G_{\beta,\alpha}(A) = \| [\|r_1(A)\|_\alpha \ \ldots \ \|r_m(A)\|_\alpha]^T \|_\beta$$

and define $G^{\alpha,\beta} : M_{m,n} \to \mathbf{R}$ by

$$G^{\alpha,\beta}(A) = \| [\|c_1(A)\|_\beta \ \ldots \ \|c_n(A)\|_\beta]^T \|_\alpha$$

Show that $G_{\beta,\alpha}(\cdot)$ and $G^{\alpha,\beta}(\cdot)$ are each norms on $M_{m,n}$ but that $G^{\alpha,\beta}(\cdot)$ is not necessarily the same as $G_{\alpha,\beta}(\cdot)$.

5.7.P14 Compare $G_{\beta,\alpha}(\cdot)$ in the preceding problem with the norms $\|| \cdot \||_{\alpha,\beta}$ defined in (5.6.P4), and show by example that even when $m = n$ (and even when $\| \cdot \|_\alpha = \| \cdot \|_\beta$), $G_{\beta,\alpha}(\cdot)$ need not be a matrix norm on M_n.

5.7.P15 Consider the norms in (5.7.P13). (a) If $\| \cdot \|_\alpha = \| \cdot \|_2 = \| \cdot \|_\beta$, what norm is $G_{\beta,\alpha}(\cdot)$? How about $G^{\alpha,\beta}(\cdot)$? (b) If $\| \cdot \|_\alpha = \| \cdot \|_1$ and $\| \cdot \|_\beta = \| \cdot \|_\infty$, what norm is $G_{\beta,\alpha}(\cdot)$? How about $G^{\beta,\alpha}(\cdot)$? What about $G_{\alpha,\beta}(\cdot)$ and $G^{\alpha,\beta}(\cdot)$?

5.7.P16 Let $n \geq 2$ and let $G(\cdot)$ be a seminorm on M_n that is similarity invariant, that is, $G(SAS^{-1}) = G(A)$ for all $A, S \in M_n$ such that S is nonsingular. (a) Show that $G(N) = 0$ for every nilpotent $N \in M_n$ and conclude that $G(\cdot)$ cannot be a norm. (b) Show that $G(A) = n^{-1}G(I_n)| \operatorname{tr} A|$ for all $A \in M_n$.

5.7.P17 If $G(\cdot)$ is a norm on M_n, the *spectral characteristic* of $G(\cdot)$ is $m(G) = \max_{G(A)\leq 1} \rho(A)$. Show that $G(\cdot)$ is spectrally dominant if and only if $m(G) \leq 1$, and show that any norm on M_n may be converted into a spectrally dominant norm via multiplication by a constant; the least such constant is $m(G)$. A norm $G(\cdot)$ on M_n is *minimally spectrally dominant* if $m(G) = 1$.

5.7.P18 If $G(\cdot)$ is a unital norm on M_n, explain why $m(G) \geq 1$. Explain why a spectrally dominant unital norm in M_n must be minimally spectrally dominant. Why is every induced matrix norm minimally spectrally dominant? Why is the numerical radius minimally spectrally dominant?

5.7.P19 Show that the spectral characteristic is a convex function on the cone of norms on M_n and deduce that the set of all spectrally dominant norms on M_n is convex.

5.7.P20 Prove the following assertions about the numerical radius function $r(\cdot)$ on M_n:
(a) $r(\cdot)$ is not unitarily invariant, but it is *unitary similarity invariant*: $r(U^*AU) = r(A)$
whenever $U \in M_n$ is unitary. (b) $r(A) = \max_{\|x\|_2=1} |x^*Ax| \le \max_{\|x\|_2=1} \|Ax\|_2 \|x\|_2 = \|\|A\|\|_2$ for all $A \in M_n$, and $r(A) = \rho(A) = \|\|A\|\|_2$ whenever A is normal. Give an example
of an $A \in M_n$ such that $r(A) < \|\|A\|\|_2$. (c) $r(A) = r(A^*)$ for all $A \in M_n$. (d) $\|\|A\|\|_2 \le 2r(A)$
for all $A \in M_n$. (e) The bounds

$$\frac{1}{2}\|\|A\|\|_2 \le r(A) \le \|\|A\|\|_2 \tag{5.7.21}$$

are sharp.

5.7.P21 Use the inequalities (5.7.21) and (5.7.11) to show that $4r(\cdot)$ is a matrix norm on
M_n. Consider $A = J_2(0)$, A^*, and AA^* to show that $\gamma r(\cdot)$ is not a matrix norm for any
$\gamma \in (0, 4)$.

5.7.P22 Deduce from (5.7.21) and the inequality

$$\frac{1}{\sqrt{n}}\|A\|_2 \le \|\|A\|\|_2 \le \|A\|_2 \tag{5.7.22}$$

in (5.6.P23) that

$$\frac{1}{2\sqrt{n}}\|A\|_2 \le r(A) \le \|A\|_2 \tag{5.7.23}$$

for all $A \in M_n$. Show that the upper bound is sharp. Verify that $A = J_2(0)$ and $A = I$ are
examples of equality in the lower bounds of (5.7.21) and (5.7.22), respectively, and that
$A = E_{11}$ is an example of equality in the upper bounds of (5.7.21) and (5.7.22). Explain
why the upper bound in (5.7.23) must, therefore, be sharp, and give an example of a case of
equality. However, the lower bound in (5.7.23) is not sharp. Why is there a finite maximal
positive constant c_n such that $c_n\|A\|_2 \le r(A)$ for all $A \in M_n$? It is known that $c_n = (2n)^{-1/2}$
for even n and $c_n = (2n-1)^{-1/2}$ for odd n. For even n, the cases of equality are matrices
that are unitarily similar to a direct sum of matrices of the form $\alpha J_2(0)$ in which $|\alpha| = r(A)$;
an additional single 1-by-1 direct summand $[\alpha]$, $|\alpha| = r(A)$, must be included when n is
odd.

5.7.P23 If $x \in \mathbf{C}^n$ and $X = xx^*$ (a Hermitian rank-one matrix), show that $\|X\|_2 = \|x\|_2^2$.
Show that the field of values of $A \in M_n$ is the set of projections (using the Frobenius
inner product) of A onto the set of unit norm rank-one Hermitian matrices. Explain why
$r(A) = \max\{|\langle A, X\rangle_F| : X \text{ is a rank-one Hermitian matrix and } \|X\|_2 = 1\}$ and show that
$r(A) \le \|A\|_2$.

5.7.P24 The numerical radius is related to a natural approximation problem. For a given
$A \in M_n$, suppose that we wish to approximate A as well as possible in the Frobenius norm
(a *least squares* approximation) by a scalar multiple of a Hermitian matrix of rank one. Let
$c \in \mathbf{C}$, $x \in \mathbf{C}^n$, and $\|x\|_2 = 1$. Show that $\|A - cxx^*\|_2^2 \ge \|A\|_2^2 - 2|c\langle A, xx^*\rangle_F| + |c|^2$,
which is minimized for $c = \langle A, \tilde{x}\tilde{x}^*\rangle_F$, in which $\tilde{x}$ is a unit vector for which $r(A) = |\tilde{x}^*A\tilde{x}|$.
Conclude that $\|A - r(A)\tilde{x}\tilde{x}\|_2 \le \|A - cxx^*\|_2$ for all $c \in \mathbf{C}$ and all unit vectors x.

5.7.P25 The numerical radius $r(\cdot)$ is spectrally dominant, so it satisfies the weak power
inequality (5.7.20a). The purpose of this problem is to show that it actually satisfies the
stronger power inequality $r(A^m) \le r(A)^m$ for all $m = 1, 2, \dots$ and all $A \in M_n$.

(a) Why is it sufficient to prove that if $r(A) \leq 1$, then $r(A^m) \leq 1$ for all $m = 1, 2, \ldots$?

(b) Let $m \geq 2$ be a given positive integer, fixed for the rest of the argument, and let $\{w_k\} = \{e^{2\pi i k/m}\}_{k=1}^m$ denote the set of mth roots of unity. Notice that $\{w_k\}$ is a finite multiplicative group and that $\{w_j w_k\}_{k=1}^m = \{w_k\}_{k=1}^m$ for each $j = 1, 2, \ldots, m$. Observe that $1 - z^m = \prod_{k=1}^m (1 - w_k z)$ and show that

$$p(z) = \frac{1}{m} \sum_{j=1}^m \prod_{\substack{k=1 \\ k \neq j}}^m (1 - w_k z) = 1 \quad \text{for all } z \in \mathbf{C}$$

(c) Show that

$$I - A^m = \prod_{k=1}^m (I - w_k A) \quad \text{and} \quad I = \frac{1}{m} \sum_{j=1}^m \prod_{\substack{k=1 \\ k \neq j}}^m (I - w_k A)$$

(d) Let $x \in \mathbf{C}^n$ be any unit vector, $\|x\|_2 = 1$, and let $A \in M_n$. Verify that

$$1 - x^* A^m x = x^*(I - A^m)x = (Ix)^*(I - A^m)x$$

$$= \left(\frac{1}{m} \sum_{j=1}^m \prod_{\substack{k=1 \\ k \neq j}}^m (I - w_k A)x \right)^* \left(\prod_{k=1}^m (I - w_k A)x \right)$$

$$= \frac{1}{m} \sum_{j=1}^m z_j^*[(I - w_j A)z_j], \quad \text{in which } z_j = \prod_{\substack{k=1 \\ k \neq j}}^m (I - w_k A)x$$

$$= \frac{1}{m} \sum_{\substack{j=1 \\ z_j \neq 0}}^m \|z_j\|_2^2 \left(1 - w_j \left(\frac{z_j}{\|z_j\|_2} \right)^* A \left(\frac{z_j}{\|z_j\|_2} \right) \right)$$

(e) Replace A by $e^{i\theta} A$ in the preceding identity to obtain

$$1 - e^{im\theta} x^* A^m x = \frac{1}{m} \sum_{\substack{j=1 \\ z_j \neq 0}}^m \|z_j\|_2^2 \left(1 - e^{i\theta} w_j \left(\frac{z_j}{\|z_j\|_2} \right)^* A \left(\frac{z_j}{\|z_j\|_2} \right) \right)$$

for any real θ. Now suppose that $r(A) \leq 1$, show that the real part of the right-hand side of this identity is nonnegative for any $\theta \in \mathbf{R}$, and deduce that the real part of the left-hand side must also be nonnegative for all $\theta \in \mathbf{R}$. Since θ is arbitrary, argue that $|x^* A^m x| \leq 1$ and hence that $r(A^m) \leq 1$.

5.7.P26 Even though the numerical radius satisfies the power inequality $r(A^m) \leq r(A)^m$ for all $A \in M_n$ and all $m = 1, 2, \ldots$ it does not satisfy the inequalities $r(A^{k+m}) \leq r(A^k)r(A^m)$ for all $A \in M_n$ and all $m, k = 1, 2, \ldots$. Verify this by considering $A = J_4(0)$, $k = 1$, and $m = 2$. Show that $r(A^2) = r(A^3) = \frac{1}{2}$ and $r(A) < 1$.

5.7.P27 Let $P \in M_n$ be a projection, so $P^2 = P$. Assume that $0 \neq P \neq I$. Why is $I - P$ a projection? Why is $\||P|\| \geq 1$ for every matrix norm $\|| \cdot \||$? Use the unitary similarity canonical form of P (3.4.3.3) to show that (a) not only is $\||P\||_2 \geq 1$, but also *every* nonzero singular value of P is greater than or equal to 1; (b) P and $I - P$ have the same singular values that are greater than 1; (c) The fields of values of P and $I - P$ are the same. (d) The numerical radii of P and $I - P$ are the same.

Further Readings. For more discussion of inequalities involving the numerical radius, see M. Goldberg and E. Tadmor, On the numerical radius and its applications, *Linear Algebra Appl.* 42 (1982) 263–284. The proof of the power inequality in (5.7.P25) is taken from C. Pearcy, An elementary proof of the power inequality for the numerical radius, *Michigan Math. J.* 13 (1966) 289–291. See Chapter 1 of Horn and Johnson (1991) for more information about the field of values and numerical radius. Some of the material of this section is from C. R. Johnson, Multiplicativity and compatibility of generalized matrix norms, *Linear Algebra Appl.* 16 (1977) 25–37; Locally compatible generalized matrix norms, *Numer. Math.* 27 (1977) 391–394; and Power inequalities and spectral dominance of generalized matrix norms, *Linear Algebra Appl.* 28 (1979) 117–130, where further results may be found.

5.8 Condition numbers: inverses and linear systems

As an application of matrix and vector norms, we consider the problem of bounding the errors in a computed inverse of a matrix and in a computed solution to a system of linear equations.

If the computations to invert a given nonsingular matrix $A \in M_n$ are performed in floating point arithmetic on a digital computer, there are inevitable and unavoidable errors of rounding and truncation. Furthermore, the entries of A might be the result of an experiment or measurement that is subject to errors; there could be some uncertainty about their values. How do errors in computation and errors in the data affect the entries of a computed matrix inverse?

For many common algorithms, roundoff errors during computation and errors in the data can be modeled in the same way. Let $||| \cdot |||$ be a given matrix norm and suppose that $A \in M_n$ is nonsingular. We wish to compute the inverse of A, but instead, the matrix we have to work with is $B = A + \Delta A$, in which we assume that

$$||| A^{-1} \Delta A ||| < 1 \tag{5.8.0}$$

to ensure that B is nonsingular. Since $B = A(I + A^{-1}\Delta A)$ and $\rho(A^{-1}\Delta A) \le ||| A^{-1}\Delta A ||| < 1$, the assumption (5.8.0) ensures that $-1 \notin \sigma(A^{-1}\Delta A)$, and hence B is nonsingular.

We have $A^{-1}(\Delta A)B^{-1} = A^{-1}(B - A)B^{-1} = A^{-1} - B^{-1}$, so

$$||| A^{-1} - B^{-1} ||| = ||| A^{-1}\Delta A B^{-1} ||| \le ||| A^{-1}\Delta A ||| \; ||| B^{-1} ||| \tag{5.8.1}$$

Since $B^{-1} = A^{-1} - A^{-1}(\Delta A)B^{-1}$, we also have

$$||| B^{-1} ||| \le ||| A^{-1} ||| + ||| A^{-1}\Delta A B^{-1} ||| \le ||| A^{-1} ||| + ||| A^{-1}\Delta A ||| \; ||| B^{-1} |||$$

which is equivalent to the inequality

$$||| B^{-1} ||| = ||| (A + \Delta A)^{-1} ||| \le \frac{||| A^{-1} |||}{1 - ||| A^{-1}\Delta A |||} \tag{5.8.2}$$

Combining (5.8.1) and (5.8.2) produces the bound

$$||| A^{-1} - B^{-1} ||| \le \frac{||| A^{-1} ||| \; ||| A^{-1}\Delta A |||}{1 - ||| A^{-1}\Delta A |||} \le \frac{||| A^{-1} ||| \; ||| A^{-1} ||| \; ||| \Delta A |||}{1 - ||| A^{-1}\Delta A |||}$$

Thus, an upper bound on the relative error in the computed inverse is

$$\frac{|||A^{-1} - (A + \Delta A)^{-1}|||}{|||A^{-1}|||} \leq \frac{|||A^{-1}|||\,|||A|||}{1 - |||A^{-1}\Delta A|||}\frac{|||\Delta A|||}{|||A|||}$$

The quantity

$$\kappa(A) = \begin{cases} |||A^{-1}|||\,|||A||| & \text{if } A \text{ is nonsingular} \\ \infty & \text{if } A \text{ is singular} \end{cases} \tag{5.8.3}$$

is the *condition number for matrix inversion with respect to the matrix norm* $|||\cdot|||$. Notice that $\kappa(A) = |||A^{-1}|||\,|||A||| \geq |||A^{-1}A||| = |||I||| \geq 1$ for any matrix norm. We have proved the bound

$$\frac{|||A^{-1} - (A + \Delta A)^{-1}|||}{|||A^{-1}|||} \leq \frac{\kappa(A)}{1 - |||A^{-1}\Delta A|||}\frac{|||\Delta A|||}{|||A|||} \tag{5.8.4}$$

If we strengthen the assumption (5.8.0) to

$$|||A^{-1}|||\,|||\Delta A||| < 1 \tag{5.8.5}$$

and observe that

$$|||A^{-1}|||\,|||\Delta A||| = |||A^{-1}|||\,|||A|||\frac{|||\Delta A|||}{|||A|||} = \kappa(A)\frac{|||\Delta A|||}{|||A|||}$$

then it follows from (5.8.4) that

$$\frac{|||A^{-1} - (A + \Delta A)^{-1}|||}{|||A^{-1}|||} \leq \frac{\kappa(A)}{1 - \kappa(A)\frac{|||\Delta A|||}{|||A|||}}\frac{|||\Delta A|||}{|||A|||} \tag{5.8.6}$$

This is an upper bound for the relative error in a computed inverse of A as a function of the relative error in the data and the condition number of A. Such a bound is called an *a priori* bound, since it involves only data that are known before any computations are done.

If $|||A^{-1}|||\,|||\Delta A|||$ is not only less than 1, but also *very much* less than 1, the right-hand side (5.8.6) is of the order of $\kappa(A)|||\Delta A|||/|||A|||$, so we have good reason to believe that the relative error in the inverse is of the same order as the relative error in the data, provided that $\kappa(A)$ is not large.

For purposes of inversion, we say that A is *ill conditioned* or *poorly conditioned* if $\kappa(A)$ is large; if $\kappa(A)$ is small (near 1), we say that A is *well conditioned*; if $\kappa(A) = 1$, we say that A is *perfectly conditioned*. Of course, all of these statements about the quality of the conditioning are with respect to a specific matrix norm $|||\cdot|||$.

Exercise. If $A \in M_n$ is nonsingular and the spectral norm is used, explain why $\kappa(A) = \sigma_1(A)/\sigma_n(A)$, the ratio of largest and smallest singular values.

Exercise. If $U, V \in M_n$ are unitary and if a unitarily invariant matrix norm is used in (5.8.3), explain why $\kappa(A) = \kappa(UAV)$: Unitary transformations of a given matrix do not make its conditioning worse. This observation underlies many stable algorithms in numerical linear algebra.

Exercise. Explain why a nonsingular $A \in M_n$ is a scalar multiple of a unitary matrix if and only if, with respect to the spectral norm, $\kappa(A) = 1$.

Exercise. Show that $\kappa(AB) \leq \kappa(A)\kappa(B)$ for any $A, B \in M_n$ with respect to any matrix norm, so we have an upper bound on the growth of the condition number of a matrix that is subjected to a sequence of transformations. What can you say if each of those transformations is unitary?

Similar considerations can be used to give a priori bounds on the accuracy of a solution to a system of linear equations. Suppose that we want to solve a linear system

$$Ax = b, \quad A \in M_n \text{ is nonsingular and } b \in \mathbf{C}^n \text{ is nonzero} \tag{5.8.7}$$

but because of computational errors or uncertainty in the data, we actually solve (exactly) a perturbed system

$$(A + \Delta A)\tilde{x} = b + \Delta b, \quad A, \Delta A \in M_n, b, \Delta b \in \mathbf{C}^n, \tilde{x} = x + \Delta x$$

How close is $\tilde{x}$ to x, that is, how large could Δx be? We can use matrix norms and compatible vector norms to obtain a bound on the relative error of the solution as a function of the relative errors in the data and the condition number of A.

Let a matrix norm $||| \cdot |||$ on M_n and a consistent vector norm $||\cdot||$ on $\mathbf{C}^n$ be given, and assume again that the inequality (5.8.0) is satisfied. Since $Ax = b$, our system is

$$(A + \Delta A)\tilde{x} = (A + \Delta A)(x + Dx) = Ax + (\Delta A)x + (A + \Delta A)\Delta x$$
$$= b + (\Delta A)x + (A + \Delta A)\Delta x = b + \Delta b$$

or

$$(\Delta A)x + (A + \Delta A)\Delta x = \Delta b$$

Therefore, $\Delta x = (A + \Delta A)^{-1}(\Delta b - (\Delta A)x)$ and

$$||\Delta x|| = \left\| (A + \Delta A)^{-1}(\Delta b - \Delta Ax) \right\|$$
$$\leq |||(A + \Delta A)^{-1}||| (||\Delta b|| + ||(\Delta A)x||)$$

Invoking (5.8.2) and compatibility, we have

$$||\Delta x|| \leq \frac{|||A^{-1}|||}{1 - |||A^{-1}\Delta A|||}(||\Delta b|| + |||\Delta A||| \, ||x||)$$

and hence

$$\frac{||\Delta x||}{||x||} \leq \frac{|||A^{-1}||| \, |||A|||}{1 - |||A^{-1}\Delta A|||} \left(\frac{||\Delta b||}{|||A||| \, ||x||} + \frac{|||\Delta A|||}{|||A|||} \right)$$

Using the definition of $\kappa(A)$ and the bound $||b|| = ||Ax|| \leq |||A||| \, ||x||$, we obtain

$$\frac{||\Delta x||}{||x||} \leq \frac{\kappa(A)}{1 - |||A^{-1}\Delta A|||} \left(\frac{||\Delta b||}{||b||} + \frac{|||\Delta A|||}{|||A|||} \right) \tag{5.8.8}$$

If we make the stronger assumption (5.8.6) again, we obtain the weaker but more transparent bound

$$\frac{||\Delta x||}{||x||} \leq \frac{\kappa(A)}{1 - \kappa(A)\frac{|||\Delta A|||}{|||A|||}} \left(\frac{||\Delta b||}{||b||} + \frac{|||\Delta A|||}{|||A|||} \right) \tag{5.8.9}$$

This bound has the same character and consequence as (5.8.6): If the coefficient matrix in the linear system (5.8.7) is well conditioned, then the relative error in the solution is of the same order as the relative errors in the data.

If a computed solution $\hat{x}$ to (5.8.7) is in hand, it can be used in an *a posteriori* bound. Once again, let $||| \cdot |||$ be a matrix norm that is compatible with a vector norm $|| \cdot ||$, let x be the exact solution to (5.6.7), and consider the *residual vector* $r = b - A\hat{x}$. Since $A^{-1}r = A^{-1}(b - A\hat{x}) = A^{-1}b - \hat{x} = x - \hat{x}$, we have the bounds $||x - \hat{x}|| = ||A^{-1}r|| \leq |||A^{-1}||| \, ||r||$ and $||b|| = ||Ax|| \leq |||A||| \, ||x||$, or $1 \leq |||A||| \, ||x||/||b||$. Then

$$||x - \hat{x}|| \leq |||A^{-1}||| \, ||r|| \leq \frac{|||A||| \, ||x||}{||b||} |||A^{-1}||| \, ||r||$$

$$= |||A||| \, |||A^{-1}||| \frac{||r||}{||b||} ||x||$$

so the relative error between the computed solution and the exact solution has the bound

$$\frac{||x - \hat{x}||}{||x||} \leq \kappa(A)\frac{||r||}{||b||} \tag{5.8.10}$$

in which the matrix norm used to compute the condition number $\kappa(A)$ is compatible with the vector norm $|| \cdot ||$. For a well-conditioned problem, the relative error in the solution is of the same order as the relative size of the residual. For an ill-conditioned problem, however, a computed solution that yields a small residual may still be very far from the exact solution.

A common characteristic of matrix norm error bounds is their conservatism: An upper bound may be large even though the actual error is small. However, if a matrix A of moderate size with moderate-size elements has a large condition number, then A^{-1} must have some large entries, and it is well to exercise great caution for the following reason.

If $Ax = b$ and if we set $C = [c_{ij}] = A^{-1}$, then differentiating the identity $x = Cb$ with respect to the entry b_j gives the identities

$$\frac{\partial x_i}{\partial b_j} = c_{ij}, \qquad i, j = 1, \ldots, n \tag{5.8.11}$$

Furthermore, if we consider $C = A^{-1}$ as a function of A, then its entries are just rational functions of the entries of A and hence are differentiable. The identity $CA = I$ means that $\sum_{p=1}^{n} c_{ip}a_{pq} = \delta_{iq}$ for all $i, q = 1, \ldots, n$, and hence

$$\sum_{p=1}^{n}\left(\frac{\partial c_{ip}}{\partial a_{jk}}a_{pq} + \delta_{pq,jk}c_{ip}\right) = \sum_{p=1}^{n}\frac{\partial c_{ip}}{\partial a_{jk}}a_{pq} + \delta_{qk}c_{ij} = 0$$

or

$$\sum_{p=1}^{n}\frac{\partial c_{ip}}{\partial a_{jk}}a_{pk} = -\delta_{qk}c_{ij}, \qquad i, j, k = 1, \ldots, n$$

Now differentiate the identity $x = Cb$ with respect to a_{jk} to obtain

$$\frac{\partial x_i}{\partial a_{jk}} = \sum_{p=1}^{n} \frac{\partial c_{ip}}{\partial a_{jk}} b_p = \sum_{p=1}^{n} \sum_{q=1}^{n} \frac{\partial c_{ip}}{\partial a_{jk}} a_{pq} x_q$$

$$= \sum_{q=1}^{n} \left(\sum_{p=1}^{n} \frac{\partial c_{ip}}{\partial a_{jk}} a_{pq} \right) x_q = \sum_{q=1}^{n} \left(-\delta_{qk} c_{ij} \right) x_q = -c_{ij} x_k$$

which is the identity

$$\frac{\partial x_i}{\partial a_{jk}} = -c_{ij} \sum_{p=1}^{n} c_{kp} b_p \tag{5.8.12}$$

Thus, (5.8.11) and (5.8.12) warn us that if $C = A^{-1}$ has any relatively large entries, then some entry of the solution x may have a large and unavoidable sensitivity to perturbations in some of the entries of b and A.

Problems

5.8.P1 Let $A \in M_n$ be nonsingular and normal. Explain why the condition number for inversion of A with respect to the spectral norm is $\kappa(A) = \rho(A)\rho(A^{-1})$.

5.8.P2 Compute the eigenvalues and the inverse of the normal matrix $A_\epsilon = \begin{bmatrix} 1 & -1 \\ -1 & 1+\epsilon \end{bmatrix}$, $\epsilon > 0$. Show that the ratio of the largest to smallest absolute eigenvalues of A_ϵ is $O(\epsilon^{-1})$ as $\epsilon \to 0$. Use (5.8.P1) to conclude that the condition number of A with respect to the spectral norm is $\kappa(A_\epsilon) = O(\epsilon^{-1})$. Use the exact form of A_ϵ^{-1} to confirm that $\kappa(A_\epsilon) = O(\epsilon^{-1})$ with respect to *any* norm.

5.8.P3 Compute the eigenvalues and the inverse of the matrix $B_\epsilon = \begin{bmatrix} 1 & -1 \\ 1 & -1-\epsilon \end{bmatrix}$, $\epsilon > 0$. Explain why B_ϵ is not normal. Use the exact form of B_ϵ^{-1} to show that the condition number of B_ϵ is $\kappa(B_\epsilon) = O(\epsilon^{-1})$ with respect to any matrix norm, and hence for small ϵ, B_ϵ. However, show that the ratio of the largest to smallest absolute eigenvalues of B is bounded as $\epsilon \to 0$. What about the ratio of the largest and smallest singular values of B_ϵ?

5.8.P4 Show that $\kappa(A) \geq \rho(A)\rho(A^{-1})$ for any nonsingular $A \in M_n$ and any matrix norm. Thus, if the ratio of its absolute largest and smallest eigenvalues is large, A must be ill conditioned for inversion, whether or not it is normal. However, the preceding problem shows that nonnormal matrices can be ill conditioned for inversion even if this ratio is not large.

5.8.P5 The condition number $\kappa(A)$ for inversion depends on the matrix norm used. Show that all condition numbers for inversion are equivalent in the sense that if $\kappa_\alpha(A) = \|\|A^{-1}\|\|_\alpha \|\|A\|\|_\alpha$ and $\kappa_\beta = \|\|A^{-1}\|\|_\beta \|\|A\|\|_\beta$, then there exist finite positive constants $C_{\alpha,\beta}$ and $C_{\beta,\alpha}$ such that

$$C_{\alpha,\beta}\kappa_\alpha(A) \leq \kappa_\beta(A) \leq C_{\beta,\alpha}\kappa_\alpha(A) \quad \text{for all } A \in M_n$$

5.8.P6 Let $\|\| \cdot \|\|$ be a matrix norm on M_n that is induced by a vector norm $\|\cdot\|$ on $\mathbf{C}^n$, and let $A \in M_n$ be nonsingular. (a) Show that the condition number (5.8.3) of A can be computed without using $||| \cdot |||$, as follows:

$$\kappa(A) = \frac{\max\{\|Ax\| : \|x\| = 1\}}{\min\{\|Ax\| : \|x\| = 1\}}$$

(b) Show that $\kappa(A) = 1$ if and only if A is a nonzero scalar multiple of an isometry for the norm $\|\cdot\|$.

5.8.P7 If $\det A$ is small (or large), must $\kappa(A)$ be large?

5.8.P8 Let B_ϵ be the matrix in (5.8.P3) ($\epsilon > 0$) and consider the linear system $B_\epsilon x = [1\ 1]^T$ with exact solution $x = [1\ 0]^T$ and an approximate solution $\hat{x} = [1 + \epsilon^{-1/2}\ \epsilon^{-1/2}]^T$. Show that $\|r\|/\|b\| = O(\epsilon^{1/2})$ as $\epsilon \to 0$ but that the relative error in the solution is $\|x - \hat{x}\|/\|x\| = O(\epsilon^{-1/2})$ as $\epsilon \to 0$. Thus, relatively small residuals can be observed even if the corresponding approximate solution has large errors. Explain how (5.8.10) converts the (small) relative residual into a correct (large) upper bound on the relative error.

5.8.P9 A commonly cited example of an ill-conditioned matrix is the Hilbert matrix (0.9.12). Since H_n is normal, its condition number for inversion with respect to the spectral norm is $\kappa(H_n) = \rho(H_n)\rho(H_n^{-1})$. It is a fact that the condition number of H_n is asymptotically equal to e^{cn}, in which the constant c is approximately 3.5, and it is also a fact that $\rho(H_n) = \pi + O(1/\log n)$ as $n \to \infty$. We have $\kappa(H_3) \sim 5 \times 10^2$, $\kappa(H_6) \sim 1.5 \times 10^7$, and $\kappa(H_8) \sim 1.5 \times 10^{10}$. Explain why H_n is so poorly conditioned even though the entries of H_n are uniformly bounded and $\rho(H_n)$ is not large.

5.8.P10 If the spectral norm is used, show that $\kappa(A^*A) = \kappa(AA^*) = \kappa(A)^2$. Explain why the problem of solving $A^*Ax = y$ may be intrinsically less tractable numerically than the problem of solving $Ax = z$.

5.8.P11 Let $A \in M_n$ be nonsingular. Use (5.6.55) to show that $\kappa(A) \geq |||A|||/|||A - B|||$ for any singular $B \in M_n$. Here, $||| \cdot |||$ is any matrix norm and $\kappa(\cdot)$ is the associated condition number. This lower bound can be useful in showing that A is ill conditioned.

5.8.P12 Let $A = [a_{ij}] \in M_n$ be an upper triangular matrix with all $a_{ii} \neq 0$. Use the preceding problem to show that the condition number of A with respect to the maximum row sum norm has the lower bound $\kappa(A) \geq |||A|||_\infty / \min_{1 \leq i \leq n} |a_{ii}|$.

5.8.P13 See (5.6.P47 to P51). If $A \in M_n$ is nonsingular and $||| \cdot |||$ is a matrix norm on M_n, explain why $\kappa(A) = |||A|||/\text{dist}_{|||\cdot|||}(A, \mathcal{S}_n)$ if $||| \cdot |||$ is induced and $\kappa(A) \geq |||A|||/\text{dist}_{|||\cdot|||}(A, \mathcal{S}_n)$ if $||| \cdot |||$ is *not* induced, with strict inequality for some A.

5.8.P14 Show that the condition number of the companion matrix (3.3.12) with respect to the spectral norm is

$$\kappa(C(p)) = \frac{s + 1 + \sqrt{(s+1)^2 - 4|a_0|^2}}{2|a_0|} \tag{5.8.13}$$

in which $s = |a_0|^2 + |a_1|^2 + \cdots + |a_{n-1}|^2$.

Further Reading. Finding a priori bounds for errors in solutions of linear systems of equations has been a central problem in numerical linear algebra; see Stewart (1973).

CHAPTER 6

Location and Perturbation
of Eigenvalues

6.0 Introduction

The eigenvalues of a diagonal matrix are very easy to locate, and the eigenvalues of a matrix are continuous functions of the entries, so it is natural to ask whether one can say anything useful about the eigenvalues of a matrix that is "nearly diagonal" in the sense that its off-diagonal elements are dominated in some way by the main diagonal entries. Such matrices arise in practice: Large systems of linear equations resulting from numerical discretization of boundary value problems for elliptic partial differential equations are typically of this form.

In differential equations problems involving the long-term stability of an oscillating system, it can be important to know that all of the eigenvalues of a given matrix are in the left half-plane. In statistics or numerical analysis, one may want to show that all the eigenvalues of a given Hermitian matrix are positive. In this chapter, we describe simple criteria that are sufficient to ensure that the eigenvalues of a given matrix are included in sets such as a given half plane, disc, or ray.

All the eigenvalues of a matrix A are located in a disc in the complex plane centered at the origin that has radius $\|\|A\|\|$, in which $\|\| \cdot \|\|$ is any matrix norm. Are there other, smaller, sets that are readily determined and either include or exclude the eigenvalues? We identify several such sets in this chapter.

If a matrix A is subjected to a perturbation $A \rightarrow A + E$, then continuity of the eigenvalues ensures that if the perturbation matrix E is small in some sense, then the eigenvalues should not change too drastically. In this chapter, we explore the behavior of the eigenvalues when the matrix is perturbed, and present some explicit bounds that limit how far the eigenvalues can move after a perturbation of the matrix.

6.1 Geršgorin discs

For any $A \in M_n$, we can always write $A = D + B$, in which $D = \text{diag}(a_{11}, \ldots, a_{nn})$ captures the main diagonal of A, and $B = A - D$ has a zero main diagonal. If we set $A_\epsilon = D + \epsilon B$, then $A_0 = D$ and $A_1 = A$. The eigenvalues of $A_0 = D$ are easy to locate: They are just the points $a_{11}, \ldots, a_{nn}$ in the complex plane, together with their multiplicities. We know that if ϵ is small enough, then the eigenvalues of A_ϵ are located in some small neighborhoods of the points $a_{11}, \ldots, a_{nn}$. The following *Geršgorin disc theorem* makes this observation precise: Some easily computed discs that are centered at the points a_{ii} are guaranteed to contain the eigenvalues of A.

Theorem 6.1.1 (Geršgorin). *Let $A = [a_{ij}] \in M_n$, let*

$$R_i'(A) = \sum_{j \neq i} |a_{ij}|, \quad i = 1, \ldots, n \tag{6.1.1a}$$

denote the deleted absolute row sums *of A, and consider the n Geršgorin discs*

$$\{z \in \mathbf{C} : |z - a_{ii}| \leq R_i'(A)\}, \quad i = 1, \ldots, n$$

The eigenvalues of A are in the union of Geršgorin discs

$$G(A) = \bigcup_{i=1}^{n} \{z \in \mathbf{C} : |z - a_{ii}| \leq R_i'(A)\} \tag{6.1.2}$$

Furthermore, if the union of k of the n discs that comprise $G(A)$ forms a set $G_k(A)$ that is disjoint from the remaining $n - k$ discs, then $G_k(A)$ contains exactly k eigenvalues of A, counted according to their algebraic multiplicities.

Proof. Let λ, x be an eigenpair for A, so $Ax = \lambda x$ and $x = [x_i] \neq 0$. Let $p \in \{1, \ldots, n\}$ be an index such that $|x_p| = \|x\|_\infty = \max_{1 \leq i \leq n} |x_i|$. Then $|x_i| \leq |x_p|$ for all $i = 1, 2, \ldots, n$, and of course $x_p \neq 0$ since $x \neq 0$. Equating the pth entries of both sides of the identity $Ax = \lambda x$ reveals that $\lambda x_p = \sum_{j=1}^{n} a_{pj} x_j$, which we write as

$$x_p(\lambda - a_{pp}) = \sum_{j \neq p} a_{pj} x_j$$

The triangle inequality and our assumption about x_p ensure that

$$|x_p| |\lambda - a_{pp}| = \left| \sum_{j \neq p} a_{pj} x_j \right| \leq \sum_{j \neq p} |a_{pj} x_j| = \sum_{j \neq p} |a_{pj}| |x_j|$$

$$\leq |x_p| \sum_{j \neq p} |a_{pj}| = |x_p| R_p'$$

Since $x_p \neq 0$, we conclude that $|\lambda - a_{pp}| \leq R_p'$; that is, λ is in the disc $\{z \in \mathbf{C} : |z - a_{pp}| \leq R_p'(A)\}$. In particular, λ is in the larger set $G(A)$ defined in (6.1.2).

Now suppose that k of the n discs that comprise $G(A)$ are disjoint from all of the remaining $n - k$ discs. After a suitable permutation similarity of A, we may assume that $G_k(A) = \bigcup_{i=1}^{k} \{z \in \mathbf{C} : |z - a_{ii}| \leq R_i'\}$ is disjoint from $G_k(A)^c = \bigcup_{i=k+1}^{n} \{z \in \mathbf{C} : |z - a_{ii}| \leq R_i'\}$. Write $A = D + B$, in which $D = \text{diag}(a_{11}, \ldots, a_{nn})$ and $B = A - D$. Define $A_\epsilon = D + \epsilon B$, and assume throughout the rest of the argument that $\epsilon \in [0, 1]$. Observe that $A_0 = D$, $A_1 = A$, and $R_i'(A_\epsilon) = R_i'(\epsilon B) = \epsilon R_i'(A)$ for each $i = 1, \ldots, n$.

Consequently, each of the n Geršgorin discs of A_ϵ is contained in the corresponding Geršgorin disc of A. In particular,

$$G_k(A_\epsilon) = \bigcup_{i=1}^{k} \{z \in \mathbf{C} : |z - a_{ii}| \le \epsilon R_i'(A)\}$$

is contained in $G_k(A)$ and is disjoint from $G_k(A)^c$. We know that all the eigenvalues of A_ϵ are contained in $G(A_\epsilon)$, which is contained in $G_k(A) \cup G_k(A)^c$.

Let Γ be a simple closed rectifiable curve in the complex plane that surrounds $G_k(A)$ and is disjoint from $G_k(A)^c$; Γ does not pass through any eigenvalue of any A_ϵ. Let $p_\epsilon(z)$ denote the characteristic polynomial of A_ϵ, so $p_\epsilon(z) \ne 0$ for all $z \in \Gamma$ and all $\epsilon \in [0, 1]$. The zeroes of $p_\epsilon(z)$ are the eigenvalues of A_ϵ, counted according to their algebraic multiplicities, and the coefficients of $p_\epsilon(z) = \det(zI - A_\epsilon) = \det(zI - D - \epsilon B)$ are polynomials in ϵ. The argument principle ensures that the number of zeroes of $p_\epsilon(z)$ inside Γ (that is, in the bounded region whose boundary is Γ) is

$$N(\epsilon) = \frac{1}{2\pi i} \oint_\Gamma \frac{p_\epsilon'(z)}{p_\epsilon(z)} dz$$

The integrand is a rational function of both z and ϵ that is an analytic function of z in a neighborhood of Γ for each $\epsilon \in [0, 1]$. Consequently, the integer-valued function $N(\epsilon)$ is continuous on the interval $[0, 1]$, so it is a constant function there. Since $p_0(t) = (t - a_{11}) \cdots (t - a_{nn})$ has exactly k zeroes inside Γ (namely, the points $a_{11}, \ldots, a_{kk}$), we know that $N(0) = k$. Thus, $N(1) = N(0) = k$ is the number of eigenvalues of A inside Γ. Finally, the first assertion of the theorem ensures that any eigenvalues of A inside Γ are contained in $G_k(A)$, which proves the assertion that exactly k eigenvalues of A lie in $G_k(A)$. $\qquad \square$

The hypothesis for the second assertion in (6.1.1) does not require that the set $G_k(A)$ be connected. If $G_k(A)$ is not connected, it is a union of two or more disjoint unions of discs, to each of which the theorem can be applied again; one obtains in this way a more refined description of the location of the eigenvalues of A. If $G_k(A)$ is connected, no further refinement via (6.1.1) is possible; the best statement we can make is that it contains exactly k eigenvalues of A.

The set $G(A)$ in (6.1.2) is the *Geršgorin set* (for rows) of A; the boundaries of the Geršgorin discs are *Geršgorin circles*. Since A and A^T have the same eigenvalues, one can obtain a Geršgorin disc theorem for *columns* of A by applying the Geršgorin theorem to A^T. The resulting set contains the eigenvalues of A and is determined by the diagonal entries of A and its *deleted absolute column sums*

$$C_j'(A) = \sum_{i \ne j} |a_{ij}|, \quad j = 1, \ldots, n \qquad (6.1.2a)$$

Corollary 6.1.3. *The eigenvalues of* $A = [a_{ij}] \in M_n$ *are in the union of n discs*

$$\bigcup_{j=1}^{n} \{z \in \mathbf{C} : |z - a_{jj}| \le C_j'\} = G(A^T) \qquad (6.1.4)$$

Furthermore, if the union of k of these discs forms a set $\mathcal{G}_k(A)$ that is disjoint from the remaining $n - k$ discs, then $\mathcal{G}_k(A)$ contains exactly k eigenvalues of A, counted according to their algebraic multiplicities.

    Exercise. Explain why the eigenvalues of A are in $G(A) \cap G(A^T)$. Illustrate with the 3-by-3 matrix $A = [a_{ij}]$ with $a_{ij} = i/j$.

The eigenvalues of A are in the two Geršgorin sets (6.1.2) and (6.1.4); in particular, they contain the largest modulus eigenvalue of A. The point in the ith disc in $G(A)$ that is farthest from the origin has modulus $|a_{ii}| + R_i' = \sum_{j=1}^{n} |a_{ij}|$, so the largest of these values is an upper bound for the spectral radius of A. Of course, a similar argument can be made for the absolute column sums.

Corollary 6.1.5. *If $A = [a_{ij}] \in M_n$, then*

$$\rho(A) \leq \min \left\{ \max_i \sum_{j=1}^{n} |a_{ij}|, \max_j \sum_{i=1}^{n} |a_{ij}| \right\}$$

    This result is no surprise, since it says that $\rho(A) \leq |||A|||_\infty$ and $\rho(A) \leq |||A|||_1$; see (5.6.9). But it is interesting to have an essentially geometric derivation of this fact.

    Since $S^{-1}AS$ has the same eigenvalues as A whenever S is nonsingular, we can apply the Geršgorin theorem to $S^{-1}AS$ and thereby obtain additional eigenvalue inclusion sets for A. A particularly convenient choice is $S = D = \text{diag}(p_1, p_2, \ldots, p_n)$ with all $p_i > 0$. Applying the Geršgorin disc theorem to $D^{-1}AD = [p_j a_{ij}/p_i]$ and to its transpose yields the following result.

Corollary 6.1.6. *Let $A = [a_{ij}] \in M_n$ and let $p_1, p_2, \ldots, p_n$ be positive real numbers. The eigenvalues of A are in the union of n discs*

$$\bigcup_{i=1}^{n} \left\{ z \in \mathbf{C} : |z - a_{ii}| \leq \frac{1}{p_i} \sum_{j \neq i} p_j |a_{ij}| \right\} = G(D^{-1}AD)$$

Furthermore, if the union of k of these discs forms a set $G_k(D^{-1}AD)$ that is disjoint from each of the $n - k$ remaining discs, then there are exactly k eigenvalues of A (counting algebraic multiplicities) in $G_k(D^{-1}AD)$. The same is true of the set

$$\bigcup_{j=1}^{n} \left\{ z \in \mathbf{C} : |z - a_{jj}| \leq p_j \sum_{i \neq j} \frac{1}{p_i} |a_{ij}| \right\} = G(DA^T D^{-1})$$

The matrix $A = \begin{bmatrix} 1 & 1 \\ 0 & 2 \end{bmatrix}$ has eigenvalues 1 and 2. A straightforward application of the Geršgorin theorem gives a rather gross estimate for the eigenvalues (Figure 6.1.7a), but the extra parameters in the preceding corollary give enough flexibility to obtain an arbitrarily good estimate of the eigenvalues (Figure 6.1.7b).

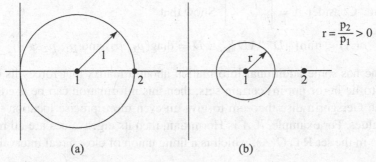

Figure 6.1.7. Geršgorin discs per (6.1.1) and (6.1.6).

Exercise. Consider the matrix

$$A = \begin{bmatrix} 7 & -16 & 8 \\ -16 & 7 & -8 \\ 8 & -8 & -5 \end{bmatrix}$$

Use the Geršgorin theorem to say as much as you can about the location of the eigenvalues of A and the spectral radius of A. Then consider $D^{-1}AD$, with $D = \text{diag}(p_1, p_2, p_3)$. Can you obtain any improvement in your location of the eigenvalues? Finally, compute the actual eigenvalues and comment on how well you did with the estimates.

Exercise. Explain why every eigenvalue of A is in the set $\cap_D G(D^{-1}AD)$, in which the intersection is over all diagonal matrices with positive main diagonal entries.

The idea of introducing free parameters can also be used to obtain a more general form of the estimates (6.1.5) for the spectral radius.

Corollary 6.1.8. *Let* $A = [a_{ij}] \in M_n$. *Then*

$$\rho(A) \leq \min_{p_1,\ldots,p_n>0} \max_{1\leq i\leq n} \frac{1}{p_i} \sum_{j=1}^{n} p_j |a_{ij}|$$

and

$$\rho(A) \leq \min_{p_1,\ldots,p_n>0} \max_{1\leq j\leq n} p_j \sum_{i=1}^{n} \frac{1}{p_i} |a_{ij}|$$

Exercise. Prove the preceding corollary.

Exercise. Let $A = \begin{bmatrix} a & b \\ c & d \end{bmatrix}$ have real positive entries. (a) Compute an explicit diagonal matrix $\tilde{D}$ such that $|||\tilde{D}^{-1}A\tilde{D}|||_\infty = \min_D |||D^{-1}AD|||_\infty$, in which the minimum is taken over all 2-by-2 diagonal matrices D with positive main diagonal entries. (b) Calculate $|||\tilde{D}^{-1}A\tilde{D}|||_\infty$ and $\rho(A)$. Observe that they are equal.

It follows from (8.1.31) that if A is an n-by-n real matrix with positive entries (or, more generally, is nonnegative and irreducible), then the minimum over all D of the maximum row sum of $D^{-1}AD$ is equal to the spectral radius of A. This need not be the case if A has some negative entries.

Exercise. Consider $A = \begin{bmatrix} 1 & 1 \\ -1.5 & 2 \end{bmatrix}$. Show that

$$\rho(A) < \min\{\|\|D^{-1}AD\|\|_\infty : D = \text{diag}(p_1, p_2) \text{ and } p_1, p_2 > 0\}$$

If one has some additional information about a matrix that forces its eigenvalues to lie in (or not in) certain sets, then this information can be used along with the Geršgorin disc theorem to give an even more precise location for the eigenvalues. For example, if A is Hermitian, then its eigenvalues are all real, so they are in the set $\mathbf{R} \cap G(A)$, which is a finite union of closed real intervals.

Exercise. What can you deduce from (6.1.1) about the location of the eigenvalues of a skew-Hermitian matrix? A unitary matrix? A real orthogonal matrix?

Since a square matrix is nonsingular if and only if zero is not in its spectrum, it is of interest to develop conditions that exclude zero from a set known to contain the eigenvalues.

Definition 6.1.9. *A matrix* $A = [a_{ij}] \in M_n$ *is* diagonally dominant *if*

$$|a_{ii}| \geq \sum_{j \neq i} |a_{ij}| = R'_i \quad \text{for all} \quad i = 1, \ldots, n$$

It is strictly diagonally dominant *if*

$$|a_{ii}| > \sum_{j \neq i} |a_{ij}| = R'_i \quad \text{for all} \quad i = 1, \ldots, n$$

From the geometry of the situation it is apparent that zero cannot lie in any closed Geršgorin disc if A is strictly diagonally dominant. Furthermore, if all the main diagonal entries a_{ii} are real and positive, then each of these discs lies in the open right half-plane; if A is Hermitian as well, then its eigenvalues are real, so they must all be real and positive. We summarize these observations in the following theorem, of which part (a) is known as the *Levy–Desplanques theorem*; see (5.6.17).

Theorem 6.1.10. *Let* $A = [a_{ij}] \in M_n$ *be strictly diagonally dominant. Then*

(a) A *is nonsingular*
(b) *if* $a_{ii} > 0$ *for all* $i = 1, \ldots, n$, *then every eigenvalue of* A *has positive real part*
(c) *if* A *is Hermitian and* $a_{ii} > 0$ *for all* $i = 1, \ldots, n$, *then* A *is positive definite*

Exercise. Consider $\begin{bmatrix} 1 & 1 \\ 1 & 1 \end{bmatrix}$ and $\begin{bmatrix} 1 & 1 \\ 1-\epsilon & 1 \end{bmatrix}$. Show that diagonal dominance is not sufficient to guarantee nonsingularity and that strict diagonal dominance is not necessary for nonsingularity.

If we use the extra parameters in (6.1.6) carefully, we can relax slightly the assumption of strict diagonal dominance as a sufficient condition for nonsingularity.

Theorem 6.1.11. *Suppose that* $A = [a_{ij}] \in M_n$ *has nonzero diagonal entries. If* A *is diagonally dominant and* $|a_{ii}| > R'_i$ *for at least* $n - 1$ *values of* $i \in \{1, \ldots, n\}$, *then it is nonsingular.*

Proof. For some k we have $|a_{ii}| > R_i'$ for all $i \neq k$, and $|a_{kk}| \geq R_k'$. If $|a_{kk}| > R_k'$, nonsingularity of A follows from (6.1.10), so we suppose that $|a_{kk}| = R_k' > 0$. In (6.1.6), let $p_i = 1$ for all $i \neq k$ and let $p_k = 1 + \epsilon, \epsilon > 0$. Then

$$\frac{1}{p_k} \sum_{j \neq k} p_j |a_{kj}| = \frac{1}{1 + \epsilon} R_k' < |a_{kk}| \quad \text{for any} \quad \epsilon > 0$$

and

$$\frac{1}{p_i} \sum_{j \neq i} p_j |a_{ij}| = R_i' + \epsilon |a_{ik}| \quad \text{for all} \quad i \neq k$$

Since $R_i' < |a_{ii}|$ for all $i \neq k$, we can choose $\epsilon > 0$ small enough so that $R_i' + \epsilon |a_{ik}| < |a_{ii}|$ for all $i \neq k$. Then (6.1.6) ensures that the point $z = 0$ is excluded from $G(D^{-1}AD)$. It follows that A is nonsingular. $\qquad\square$

The Geršgorin theorem and its variations give inclusion sets for the eigenvalues of A that depend only on the main diagonal entries of A and the *absolute values* of its off-diagonal entries. Using the fact that $S^{-1}AS$ has the same eigenvalues as A led us to (6.1.6) and to the fact that the closed set

$$\bigcap_D G(D^{-1}AD), \quad D = \text{diag}(p_1, \ldots, p_n), \quad \text{all } p_i > 0 \tag{6.1.12}$$

contains the eigenvalues of $A \in M_n$. We might be able to get even smaller inclusion sets for the eigenvalues if we were to admit similarities that are not necessarily diagonal, but if we restrict ourselves just to diagonal similarities and use just the main diagonal entries and the absolute values of the off-diagonal entries, can we somehow do better than (6.1.12)? The answer is no, for the following reason: Let z be any given point on the boundary of the set (6.1.12). Then R. Varga has shown that there exists a matrix $B = [b_{ij}] \in M_n$ such that z is an eigenvalue of B, $b_{ii} = a_{ii}$ for all $i = 1, \ldots, n$, and $|b_{ij}| = |a_{ij}|$ for all $i, j = 1, \ldots, n$.

Problems

6.1.P1 Consider the following iterative algorithm to solve the n-by-n system of linear equations $Ax = y$, in which A and y are given:

(i) Let $B = I - A$ and rewrite the system as $x = Bx + y$.
(ii) Choose an initial vector $x^{(0)}$.
(iii) For $m = 0, 1, 2, \ldots$, calculate $x^{(m+1)} = Bx^{(m)} + y$ and hope that $x^{(m)} \to x$ as $m \to \infty$.

(a) Let $\epsilon^{(m)} = x^{(m)} - x$ and show that $\epsilon^{(m)} = B^m(x^{(0)} - x)$. (b) Conclude that if $\rho(I - A) < 1$, then $x^{(m)} \to x$ as $m \to \infty$ regardless of the choice of the initial approximation $x^{(0)}$. (c) Use the Geršgorin theorem to give a simple explicit condition on A that is sufficient for this algorithm to work.

6.1.P2 Show that $\bigcap_S G(S^{-1}AS) = \sigma(A)$; the intersection is taken over all nonsingular S.

6.1.P3 Let $A = [a_{ij}] = [a_1 \ldots a_n] \in M_n$. Use (6.1.5) to show that

$$|\det A| \leq \prod_{j=1}^{n} \left(\sum_{i=1}^{n} |a_{ij}| \right) = \prod_{j=1}^{n} \|a_j\|_1$$

with a similar inequality for the rows. Compare with the approach in (5.6.P10).

6.1.P4 Let $A \in M_n$ and consider the set $G(A)$ defined in (6.1.2). In the text we showed that the assertion in (6.1.1) that all the eigenvalues of A are in $G(A)$ implies (6.1.10a). Prove the converse implication.

6.1.P5 Suppose that the n Geršgorin discs of $A \in M_n$ are mutually disjoint. (a) If A is real, show that every eigenvalue of A is real. (b) If $A \in M_n$ has real main diagonal entries and its characteristic polynomial has only real coefficients, show that every eigenvalue of A is real.

6.1.P6 If $A = [a_{ij}] \in M_n$ and if $|a_{ii}| > R_i'$ for k different values of i, use properties of principal submatrices of A to show that rank $A \geq k$.

6.1.P7 Suppose that $A \in M_n$ is idempotent but $A \neq I$. Show that A cannot be strictly diagonally dominant (or irreducibly diagonally dominant; see (6.2.25) and (6.2.27)).

6.1.P8 Suppose that $A \in M_n$ is strictly diagonally dominant, that is, $|a_{ii}| > R_i'$ for all $i = 1, \ldots, n$. Show that $|a_{kk}| > C_k'$ for at least one value of $k = 1, \ldots, n$.

6.1.P9 Suppose that $A = [a_{ij}] \in M_n$ is strictly diagonally dominant, and let $D = \mathrm{diag}(a_{11}, \ldots, a_{nn})$. Explain why D is nonsingular and show that $\rho(I - D^{-1}A) < 1$.

6.1.P10 If $A = [a_{ij}] = [a_1 \ldots a_n] \in M_n$, show that rank $A \geq \sum_{i:a_i \neq 0}(|a_{ii}|/\|a_i\|_1)$.

6.1.P11 If $A = [a_{ij}] = [a_1 \ldots a_n] \in M_n$, show that rank $A \geq \sum_{i:a_i \neq 0}(|a_{ii}|^2/\|a_i\|_2^2)$.

6.1.P12 Let $A \in M_n$. Show that $G(A) = G(A^T)$ if A is a Toeplitz matrix or, more generally, if A is persymmetric and all of its main diagonal entries are equal.

6.1.P13 Suppose that $A = [a_{ij}] \in M_n(\mathbf{R})$ is strictly diagonally dominant. Show that $\det A$ has the same sign as the product $a_{11} \cdots a_{nn}$.

6.1.P14 Let $A = [a_{ij}] \in M_n$. (a) If $|a_{ii} - a_{jj}| > R_i' + R_j'$ for some $i, j \in \{1, \ldots, n\}$, explain why the Geršgorin discs of A corresponding to its rows i and j are disjoint. (b) Suppose that $|a_{ii} - a_{jj}| > R_i' + R_j'$ for all distinct $i, j \in \{1, \ldots, n\}$. Explain why A has n distinct eigenvalues. (c) Suppose that A is real and $|a_{ii} - a_{jj}| > R_i' + R_j'$ for all distinct $i, j \in \{1, \ldots, n\}$. Explain why A has n distinct real eigenvalues.

6.1.P15 Suppose that $A = [a_{ij}] \in M_n$ is diagonally dominant. (a) Show that $\rho(A) \leq 2 \max_i |a_{ii}|$. (b) If A is strictly diagonally dominant, explain why $\rho(A) < 2 \max_i |a_{ii}|$.

6.1.P16 This problem explores the maxim: *Gaussian elimination preserves strict diagonal dominance.* Let $n \geq 2$ and suppose that $A \in M_n$ is strictly diagonally dominant. (a) Explain why each of the leading principal submatrices of A is nonsingular. (b) Partition $A = \begin{bmatrix} a & y^T \\ x & B \end{bmatrix}$, in which $x, y \in \mathbf{C}^{n-1}$. Explain why Gaussian elimination on the first column of A produces the matrix $A' = \begin{bmatrix} a & y^T \\ 0 & C \end{bmatrix}$, in which $C = B - a^{-1}xy^T$. Show that C (and hence A') is strictly diagonally dominant. (c) Partition $A = \begin{bmatrix} A_{11} & A_{12} \\ A_{21} & A_{22} \end{bmatrix}$, in which $A_{11} \in M_k$. Let

$C = A_{22} - A_{21}A_{11}^{-1}A_{12}$ be the Schur complement of A_{11} in A. Use part (a) and induction to explain why $\begin{bmatrix} A_{11} & A_{12} \\ 0 & C \end{bmatrix}$ (the result of block Gaussian elimination on the first block column of A) is strictly diagonally dominant.

6.1.P17 This problem explores *block matrix generalizations of Geršgorin's theorem*. Let $|||\cdot|||$ be a given matrix norm on M_m, and consider the matrix norm $N(\cdot)$ on M_{mn} defined in (5.6.P55), whose notation we adopt. For any $A = [A_{ij}]_{i,j=1}^n \in M_{mn}$, define

$$\mathcal{R}_i' = \sum_{j \neq i} |||A_{ij}|||, \quad i = 1, \ldots, n$$

Let $D = A_{11} \oplus \cdots \oplus A_{nn}$. (a) Suppose that $z \in \mathbf{C}$ is not an eigenvalue of D. Explain why $zI - A = (zI - D)(I - (zI - D)^{-1}(A - D))$ and then use the exercise following (5.6.16) to conclude that if $N((zI - D)^{-1}(A - D)) < 1$, then z is not an eigenvalue of A. Show that $N((zI - D)^{-1}(A - D)) \leq \max_{1 \leq i \leq n}(|||(zI - A_{ii})^{-1}|||\, \mathcal{R}_i')$. (b) If $z \in \mathbf{C}$ is not an eigenvalue of any of the matrices $A_{11}, \ldots, A_{nn}$ and if $|||(zI - A_{ii})^{-1}|||^{-1} > \mathcal{R}_i'$ for each $i = 1, \ldots, n$, explain why z is not an eigenvalue of A. (c) Explain why every eigenvalue of A is contained in the set

$$\bigcup_{i=1}^n \sigma(A_{ii}) \cup \bigcup_{i=1}^n \{z \in \mathbf{C} : z \notin \sigma(A_{ii}) \text{ and } |||(zI - A_{ii})^{-1}|||^{-1} \leq \mathcal{R}_i'\} \qquad (6.1.13)$$

(d) We say that $A \in M_{mn}$ is *block strictly diagonally dominant with respect to a matrix norm* $|||\cdot|||$ on M_m if each diagonal block A_{ii} is nonsingular and $|||A_{ii}^{-1}|||^{-1} > \mathcal{R}_i'$ for each $i = 1, \ldots, n$. Use (6.1.13) to show that A is nonsingular if there is a matrix norm on M_m such that A is block strictly diagonally dominant with respect to it. (e) Suppose that $m = 1$. What is block strict diagonal dominance in this case? Show that the sets (6.1.13) and (6.1.2) are the same in this case. Write out the preceding derivation of (6.1.13) in this case and obtain a proof of Geršgorin's theorem that is different from the one in the text. (f) Now suppose that $|||\cdot|||$ is the spectral norm on M_m, each diagonal block A_{ii} is normal, and $\sigma(A_{ii}) = \{\lambda_1^{(i)}, \ldots, \lambda_m^{(i)}\}$ for each $i = 1, \ldots, m$. Show that the eigenvalue inclusion set (6.1.13) is a union of discs

$$\bigcup_{i=1}^n \bigcup_{j=1}^m \{z \in \mathbf{C} : |z - \lambda_j^{(i)}| \leq \sum_{k \neq l} |||A_{ik}|||_2\} \qquad (6.1.14)$$

in this case. What is this set when $m = 1$? (g) Let $m = n = 2$, $A_{11} = A_{22} = \begin{bmatrix} 0 & 1 \\ 1 & 0 \end{bmatrix}$, and $A_{12} = A_{21}^T = \begin{bmatrix} 0 & 0 \\ .5 & 0 \end{bmatrix}$. Explain why $A = [A_{ij}]_{i,j=1}^2$ is not diagonally dominant but is nevertheless nonsingular because it is block diagonally dominant with respect to the maximum column sum matrix norm on M_2. Use (6.1.2) to show that the eigenvalues of A are in $[-1.5, 1.5]$; use (6.1.14) to show that they are in the smaller set $[-1.5, -.5] \cup [.5, 1.5]$. The eigenvalues of A are approximately ± 1.2808 and $\pm .7808$.

6.1.P18 Let $X = [x_1 \ldots x_k] \in M_{n,k}$ have full column rank. Show that there is a nonsingular $R \in M_k$ such that the matrix $Y = [y_{ij}] = [y_1 \ldots y_k] = XR$ has the following property: There are k distinct indices $i_1, \ldots, i_k \in \{1, \ldots, n\}$ such that $y_{i_j j} = \|y_j\|_\infty$ for each $j = 1, \ldots, k$.

6.1.P19 Let λ be an eigenvalue of $A = [a_{ij}] \in M_n$ with geometric multiplicity $k \geq 1$. We claim that there are k distinct indices $i_1, \ldots, i_k \in \{1, \ldots, n\}$ such that $\lambda \in \{z \in \mathbf{C} : |z - a_{i_j i_j}| \leq R'_{i_j}\}$ for each $j = 1, \ldots, k$. Provide details: (a) Let the columns of $X \in M_{n,k}$ be a basis of the λ-eigenspace of A, and let $Y = [y_1 \; \cdots \; y_k] = XR$ have the property described in the preceding problem. (b) $AY = \lambda Y$. (c) Revisit the proof of (6.1.1) and use each eigenpair λ, y_i in the argument. (d) What happens if $k = n$? (e) Why is $\lambda \in \cap_{j=1}^k \{z \in \mathbf{C} : |z - a_{i_j i_j}| \leq R'_{i_j}\}$?

6.1.P20 Let λ be an eigenvalue of $A \in M_n$ with geometric multiplicity at least $k \geq 1$. (a) Show that λ is contained in each union of $n - k + 1$ different Geršgorin discs of A, that is,

$$\lambda \in \bigcup_{j=1}^{n-k+1} \{z \in \mathbf{C} : |z - a_{i_j i_j}| \leq R'_{i_j}(A)\} \tag{6.1.15}$$

for *any* choice of indices $1 \leq i_1 < \cdots < i_{n-k+1} \leq n$. (b) There are $\binom{n}{k-1}$ different possibilities for the union of disks described in (6.1.15). Why is λ in their intersection? (c) Discuss the cases $k = 1$ and $k = n$.

6.1.P21 If a matrix has special structure, hypotheses weaker than those in (6.1.10–11) may be sufficient to ensure nonsingularity. Revisit (2.2.P10) and explain why a circulant matrix $A = [a_{ij}] \in M_n$ is nonsingular if *any one row* is diagonally dominant, that is, if there is some $i \in \{1, \ldots, n\}$ such that $|a_{ii}| > R'_i$.

Notes and Further Readings. The original reference for (6.1.1) is S. Geršgorin, Über die Abgrenzung der Eigenwerte einer Matrix, *Izv. Akad. Nauk. S.S.S.R.* 7 (1931) 749–754; 6.1.P14 is taken from Geršgorin's paper. For a historical perspective on the ideas involved in Geršgorin's theorem, see O. Taussky, A recurring theorem on determinants, *Amer. Math. Monthly* 61 (1949) 672–676. An exposition of a generalization of Geršgorin's theorem is in R. A. Brualdi and S. Mellendorf, Sets in the complex plane containing the eigenvalues of a matrix, *Amer. Math. Monthly* 101 (1994) 975–985. The qualitative assertion in (6.1.10a) has a quantitative version: The smallest singular value of A is bounded from below by $\min_{1 \leq i \leq n} \{|a_{ii}| - \frac{1}{2}(R'_i + C'_i)\}$; see (3.7.17) in Horn and Johnson (1991). R. Varga's book Varga (2004) contains an in-depth discussion of Geršgorin discs, their history, and generalizations. For a proof that the set (6.1.12) has the optimality property stated in the last paragraph of the section, see R. Varga, Minimal Gerschgorin sets, *Pacific J. Math.* 15 (1965) 719–729. The results in (6.1.P18 and P19) about Geršgorin discs and geometric multiplicities are due to F. J. Hall and R. Marsli.

6.2 Geršgorin discs – a closer look

We have seen that strict diagonal dominance is sufficient for nonsingularity but that diagonal dominance is not. Consideration of some 2-by-2 examples suggests the conjecture that diagonal dominance together with strict inequality

$$|a_{ii}| > R'_i = \sum_{j \neq i} |a_{ij}| \text{ for at least one value of } i = 1, \ldots, n \tag{6.2.1}$$

may be sufficient for nonsingularity. Unfortunately, this is not the case, as is shown by the example

$$\begin{bmatrix} 4 & 2 & 1 \\ 0 & 1 & 1 \\ 0 & 1 & 1 \end{bmatrix} \tag{6.2.1a}$$

However, there are useful conditions on a diagonally dominant matrix under which the condition (6.2.1) *is* sufficient to guarantee nonsingularity, and they lead to some very interesting ideas in graph theory. The fundamental observation is that if A is diagonally dominant, then zero cannot be in the *interior* of any Geršgorin disc.

Lemma 6.2.2. *Let $A = [a_{ij}] \in M_n$ and $\lambda \in \mathbf{C}$ be given. Then*

(a) *λ is not in the interior of any Geršgorin disc of A if and only if*

$$|\lambda - a_{ii}| \geq R_i' = \sum_{j \neq i} |a_{ij}| \quad \text{for all} \quad i = 1, \ldots, n \tag{6.2.2a}$$

(b) *if λ is on the boundary of the Geršgorin set $G(A)$ in (6.1.2), then it satisfies the inequalities (6.2.2a)*

(c) *A is diagonally dominant if and only if $\lambda = 0$ satisfies the inequalities (6.2.2a)*

Exercise. Prove the preceding lemma.

Exercise. Consider the point $\lambda = 0$ and the matrix $A = \begin{bmatrix} 1 & 1 \\ 1 & i \end{bmatrix} \oplus \begin{bmatrix} -1 & 1 \\ 1 & -i \end{bmatrix}$. Explain why a point in the interior of $G(A)$ can satisfy the inequalities (6.2.2a).

A careful analysis of the proof of (6.1.1) clarifies what happens if an eigenvalue of A satisfies the inequalities (6.2.2a), in particular, if it is a boundary point of $G(A)$.

Lemma 6.2.3. *Let λ, x be an eigenpair for $A = [a_{ij}] \in M_n$ and suppose that λ satisfies the inequalities (6.2.2a). Then*

(a) *if $p \in \{1, \ldots, n\}$ and $|x_p| = \|x\|_\infty$, then $|\lambda - a_{pp}| = R_p'$; that is, the pth Geršgorin circle of A passes through λ*

(b) *if $p, q \in \{1, \ldots, n\}$, $|x_p| = \|x\|_\infty$, and $a_{pq} \neq 0$, then $|x_q| = \|x\|_\infty$*

Proof. Suppose that $|x_p| = \|x\|_\infty$. Then (6.1.1a) ensures that

$$|\lambda - a_{pp}|\|x\|_\infty = |\lambda - a_{pp}||x_p| = \left| \sum_{j \neq p} a_{pj} x_j \right| \tag{6.2.4}$$

$$\leq \sum_{j \neq p} |a_{pj}||x_j| \leq \sum_{j \neq p} |a_{pj}|\|x\|_\infty = R_p' \|x\|_\infty$$

and hence $|\lambda - a_{pp}| \leq R_p'$. However, the inequalities (6.2.2a) ensure that $|\lambda - a_{pp}| \geq R_p'$, so $|\lambda - a_{pp}| = R_p'$, which is assertion (a). Thus, we have *equality* in both of the inequalities in (6.2.4):

$$|\lambda - a_{pp}|\|x\|_\infty = \sum_{j \neq p} |a_{pj}||x_j| = \sum_{j \neq p} |a_{pj}|\|x\|_\infty = R_p'\|x\|_\infty \tag{6.2.4a}$$

Assertion (b) follows from the center identity in (6.2.4a):

$$\sum_{j \neq p} |a_{pj}|(\|x\|_\infty - |x_j|) = 0$$

Because each summand is nonnegative, it must be zero. Thus, $a_{pq} \neq 0$ implies that $|x_q| = \|x\|_\infty$. $\square$

The preceding lemma looks rather technical, but it has as an immediate consequence the following useful result and its corollary.

Theorem 6.2.5. *Let* $A \in M_n$, *and let* $\lambda, x = [x_i]$ *be an eigenpair of* A *such that* λ *satisfies the inequalities (6.2.2a). If every entry of* A *is nonzero, then*

(a) *every Geršgorin circle of* A *passes through* λ
(b) $|x_i| = \|x\|_\infty$ *for all* $i = 1, \ldots, n$

Exercise. Deduce (6.2.5) from (6.2.3).

Corollary 6.2.6. *Let* $A = [a_{ij}] \in M_n$, *and suppose that every entry of* A *is nonzero. If* A *is diagonally dominant and if there is a* $k \in \{1, \ldots, n\}$ *such that* $|a_{kk}| > R'_k$, *then* A *is nonsingular.*

Proof. Since A is diagonally dominant, $\lambda = 0$ satisfies the inequalities (6.2.2a). The hypothesis ensures that the kth Geršgorin circle does *not* pass through 0, so it follows from the preceding theorem that 0 is not an eigenvalue of A. $\square$

The preceding corollary is both useful and interesting, but we can do much better if we use the information in (6.2.3) more carefully.

Definition 6.2.7. *A matrix* $A = [a_{ij}] \in M_n$ *is said to have property SC if for each pair of distinct integers* $p, q \in \{1, \ldots, n\}$ *there is a sequence of distinct integers* $k_1 = p$, $k_2, \ldots, k_m = q$ *such that each entry* $a_{k_1 k_2}, a_{k_2 k_3}, \ldots, a_{k_{m-1} k_m}$ *is nonzero.*

For example, consider $p = 2$, $q = 1$, and the matrix in (6.2.1a). Then $k_2 = 3$ is the only choice possible. But it is not possible to choose $k_3 = 1$ since the entry in position 3, 1 is zero. Thus, the matrix in (6.2.1a) does not have property *SC*.

Exercise. For $p = 1, q = 2$, and the matrix in (6.2.1a), find a sequence of integers that satisfies the condition stated in (6.2.7).

Using this notion and (6.2.3), we can improve (6.2.5) as follows.

Theorem 6.2.8 (Better theorem). *Let* $A \in M_n$, *and let* $\lambda, x = [x_i]$ *be an eigenpair of* A *such that* λ *satisfies the inequalities (6.2.2a). If* A *has property SC, then*

(a) *every Geršgorin circle passes through* λ
(b) $|x_i| = \|x\|_\infty$ *for all* $i = 1, \ldots, n$

Proof. Let $p \in \{1, \ldots, n\}$ be an index such that $|x_p| = \|x\|_\infty$. Then (6.2.3a) ensures that $|\lambda - a_{pp}| = R'_p$, so the pth Geršgorin circle passes through λ. Let $q \in \{1, \ldots, n\}$ be *any* index such that $q \neq p$. Because A has property *SC*, there is a sequence of distinct indices $k_1 = p$, $k_2, \ldots, k_m = q$ such that each entry $a_{k_1 k_2}, \ldots, a_{k_{m-1} k_m}$

is nonzero. Since $a_{k_1 k_2} \neq 0$, (6.2.3b) ensures that $|x_{k_2}| = \|x\|_\infty$ and (6.2.3a) ensures that $|\lambda - a_{k_2 k_2}| = R'_{k_2}$. Proceeding in this way, we conclude that $|x_{k_i}| = \|x\|_\infty$ and $|\lambda - a_{k_i k_i}| = R'_{k_i}$ for each $i = 2, \ldots, m$. In particular, for $i = m$, we conclude that the qth Geršgorin circle passes through λ and $|x_q| = \|x\|_\infty$. $\square$

Just as in (6.2.6), we can now deduce a useful sufficient condition for non-singularity.

Corollary 6.2.9 (Better corollary). *Suppose that $A = [a_{ij}] \in M_n$ has property SC. If A is diagonally dominant and if there is a $k \in \{1, \ldots, n\}$ such that $|a_{kk}| > R'_k$, then A is nonsingular.*

Exercise. Deduce (6.2.9) from (6.2.8).

What is this strange property *SC*? Notice that it involves only the *locations* of the off-diagonal nonzero entries of A; the main diagonal entries and the values of the nonzero off-diagonal entries are irrelevant. Motivated by this observation, we define two matrices related to A.

Definition 6.2.10. *For any given $A = [a_{ij}] \in M_{m,n}$, define $|A| = [|a_{ij}|]$ and $M(A) = [\mu_{ij}]$, in which $\mu_{ij} = 1$ if $a_{ij} \neq 0$ and $\mu_{ij} = 0$ if $a_{ij} = 0$. The matrix $M(A)$ is the* indicator matrix *of A.*

Exercise. Show that $A \in M_n$ has property *SC* if and only if either (and hence both of) $|A|$ or $M(A)$ has property *SC*.

The sequence of nonzero entries of A that arises in the statement of property *SC* can be summarized visually by certain paths in a graph associated with A.

Definition 6.2.11. *The* directed graph *of $A \in M_n$, denoted by $\Gamma(A)$, is the directed graph on n nodes $P_1, P_2, \ldots, P_n$ such that there is a directed arc in $\Gamma(A)$ from P_i to P_j if and only if $a_{ij} \neq 0$.*

Examples.

$$A_1 = \begin{bmatrix} 1 & 1 \\ 1 & 1 \end{bmatrix}; \qquad \Gamma(A_1) =$$

$$A_2 = \begin{bmatrix} 0 & 1 \\ 1 & 0 \end{bmatrix}; \qquad \Gamma(A_2) =$$

$$A_3 = \begin{bmatrix} 1 & 1 \\ 0 & 0 \end{bmatrix}; \qquad \Gamma(A_3) =$$

$$A_4 = \begin{bmatrix} 4 & 2 & 1 \\ 0 & 1 & 1 \\ 0 & 1 & 1 \end{bmatrix}; \qquad \Gamma(A_4) =$$

Definition 6.2.12. *A directed path γ in a graph Γ is a sequence of arcs $P_{i_1}P_{i_2}$, $P_{i_2}P_{i_3}$, $P_{i_3}P_{i_4}$, ... in Γ. The* ordered list of nodes *in the directed path γ is P_{i_1}, P_{i_2}, ... The* length *of a directed path is the number of arcs in the directed path if this number is finite; otherwise, the directed path is said to have infinite length. A* cycle *(sometimes called a* simple directed cycle*) is a directed path that begins and ends at the same node; this node must occur exactly twice in the ordered list of nodes in the path, and no other node can occur more than once in the list. A cycle of length 1 is a* loop *or* trivial cycle.

Definition 6.2.13. *A directed graph Γ is* strongly connected *if between each pair of distinct nodes P_i, P_j in Γ there is a directed path of finite length that begins at P_i and ends at P_j.*

Theorem 6.2.14. *Let $A \in M_n$. Then A has property SC if and only if the directed graph $\Gamma(A)$ is strongly connected.*

Exercise. Prove the preceding theorem.

Exercise. If each pair of nodes of a directed graph Γ belongs to at least one cycle, explain why Γ is strongly connected. Consider the matrix

$$\begin{bmatrix} 0 & 1 & 0 \\ 1 & 0 & 1 \\ 0 & 1 & 0 \end{bmatrix}$$

and give a counterexample to the converse implication.

There may be more than one directed path between two given nodes of a directed graph, but two such paths with different lengths may not be essentially different; one of them might contain repetitions of one or more subpaths. If one ever visits a given node twice in going along a directed path, then the directed path may be shortened (without changing the end points) by deleting all the intermediate arcs between the first and second visits to the node (the subgraph deleted is, or contains, a cycle).

Observation 6.2.15. *Let Γ be a directed graph on n nodes. If there is a directed path in Γ between two given nodes, then there is a directed path between them that has length not greater than $n - 1$.*

To determine if a given matrix A has property SC, one can check to see if $\Gamma(A)$ is strongly connected. If n is not large or if $M(A)$ has a special structure, then one may be able to inspect $\Gamma(A)$ and ascertain that there is a path between each pair of nodes. Alternatively, the following theorem provides the foundation for a computational algorithm that does not rely on visual inspection.

Theorem 6.2.16. *Let $A \in M_n$, and let P_i and P_j be given nodes of $\Gamma(A)$. The following are equivalent:*

(a) *There is a directed path of length m in $\Gamma(A)$ from P_i to P_j.*

(b) *The i, j entry of $|A|^m$ is nonzero.*

(c) *The i, j entry of $M(A)^m$ is nonzero.*

Proof. We proceed by induction. For $m = 1$ the assertion is trivial. For $m = 2$ we compute

$$(|A|^2)_{ij} = \sum_{k=1}^{n} |A|_{ik} |A|_{kj} = \sum_{k=1}^{n} |a_{ik}||a_{kj}|$$

so that $(|A|^2)_{ij} \neq 0$ if and only if for at least one value of k, both a_{ik} and a_{kj} are nonzero. But this is the case if and only if there is a path of length 2 in $\Gamma(A)$ from P_i to P_j. In general, suppose that the assertion has been proved for $m = q$. Then

$$(|A|^{q+1})_{ij} = \sum_{k=1}^{n} (|A|^q)_{ik} |A|_{kj} = \sum_{k=1}^{n} (|A|^q)_{ik}|a_{kj}| \neq 0$$

if and only if for at least one value of k, both $(|A|^q)_{ik}$ and $|a_{kj}|$ are nonzero. This is the case if and only if there is a path from P_i to P_k of length q and one from P_k to P_j of length 1, that is, if and only if there is a path from P_i to P_j of length $q + 1$.

The same argument works for $M(A)$. $\qquad\square$

Definition 6.2.17. *Let $A = [a_{ij}] \in M_n$. We say that $A \geq 0$ (A is nonnegative) if all its entries a_{ij} are real and nonnegative. We say that $A > 0$ (A is positive) if all its entries a_{ij} are real and positive.*

Corollary 6.2.18. *Let $A \in M_n$. Then $|A|^m > 0$ if and only if from each node P_i to each node P_j in $\Gamma(A)$ there is a directed path in $\Gamma(A)$ of length m. The same is true for $M(A)^m$.*

Corollary 6.2.19. *Let $A \in M_n$. The following are equivalent:*

(a) *A has property SC.*

(b) *$(I + |A|)^{n-1} > 0$.*

(c) *$(I + M(A))^{n-1} > 0$.*

Proof. $(I + |A|)^{n-1} = I + (n-1)|A| + \binom{n-1}{2}|A|^2 + \cdots + |A|^{n-1} > 0$ if and only if for each pair i, j of nodes with $i \neq j$ *at least one* of the matrices $|A|, |A|^2, \ldots, |A|^{n-1}$ has a positive entry in position i, j. But (6.2.16) ensures that this happens if and only if there is a directed path in $\Gamma(A)$ from P_i to P_j. This is equivalent to $\Gamma(A)$ being strongly connected, which is equivalent to A having property SC. $\qquad\square$

Exercise. Prove the assertion involving $M(A)$ in the preceding corollary.

Corollary 6.2.20. *If* $A \in M_n$, $i, j \in \{1, \ldots, n\}$, *and* $i \neq j$, *there is a path in* $\Gamma(A)$ *from* P_i *to* P_j *if and only if the* i, j *entry of* $(I + |A|)^{n-1}$ *is nonzero.*

Exercise. Use the preceding corollary to give an explicit computational test for property SC that involves about $\log_2(n - 1)$ matrix multiplications instead of $n - 2$ matrix multiplications. *Hint*: Consider $(I + |A|)^2$, the square of this matrix, and so on.

We now introduce one more equivalent characterization of property SC. It is based on the fact that strong connectivity of $\Gamma(A)$ is just a topological property of $\Gamma(A)$; it has nothing to do with the labeling assigned to the nodes of $\Gamma(A)$. If we permute the labels of the nodes, the graph stays either strongly connected or not strongly connected. Interchanging the ith and jth rows of A, as well as the ith and jth columns, affects $\Gamma(A)$ by interchanging the labels on nodes P_i and P_j. Conversely, relabeling the nodes of $\Gamma(A)$ corresponds to interchanging some rows and columns of A. Thus, a permutation similarity $A \to P^T A P$ (the result of finitely many interchanges of rows and columns) is equivalent to permuting the labels of the nodes of $\Gamma(A)$.

It is important to know whether some permutation of the rows and columns of A can be found that brings it into the special block form described in the following definition.

Definition 6.2.21. *A matrix* $A \in M_n$ *is reducible if there is a permutation matrix* $P \in M_n$ *such that*

$$P^T A P = \begin{bmatrix} B & C \\ 0_{n-r,r} & D \end{bmatrix} \quad and \ 1 \leq r \leq n-1$$

In the preceding definition, we do not insist that any of the blocks B, C, and D have nonzero entries. We require only that a lower-left $(n - r)$-by-r block of zero entries can be created by some sequence of row and column interchanges. However, we do insist that both of the square matrices B and D have size at least one, so no 1-by-1 matrix is reducible.

Exercise. If $A \in M_n$ is reducible, explain why it has at least $n - 1$ zero entries.

Suppose that we want to solve a system of linear equations $Ax = y$, and suppose that A is reducible. If we write $\tilde{A} = P^T A P = \begin{bmatrix} B & C \\ 0 & D \end{bmatrix}$, we have $Ax = P\tilde{A}P^T x = y$, or $\tilde{A}(P^T x) = P^T y$. Set $P^T x = \tilde{x} = \begin{bmatrix} z \\ \zeta \end{bmatrix}$ (unknown) and $P^T y = \tilde{y} = \begin{bmatrix} w \\ \omega \end{bmatrix}$ (known), in which $z, w \in \mathbf{C}^r$ and $\zeta, \omega \in \mathbf{C}^{n-r}$. Then the system of equations to be solved is equivalent to $\tilde{A}\tilde{x} = \tilde{y} = \begin{bmatrix} B & C \\ 0 & D \end{bmatrix}\begin{bmatrix} z \\ \zeta \end{bmatrix} = \begin{bmatrix} w \\ \omega \end{bmatrix}$, that is, to the two systems $D\zeta = \omega$ and $Bz + C\zeta = w$. If we first solve $D\zeta = \omega$ for ζ, and then solve $Bz = w - C\zeta$ for z, we have *reduced* the original problem to two smaller problems. A linear system with a *reducible* coefficient matrix can be *reduced* to two smaller linear systems.

Definition 6.2.22. *A matrix* $A \in M_n$ *is irreducible if it is not reducible.*

Theorem 6.2.23. *Let* $A \in M_n$. *The following are equivalent:*

   (a) A is irreducible.
   (b) $(I + |A|)^{n-1} > 0$.
   (c) $(I + M(A))^{n-1} > 0$.

Proof. To show that (a) and (b) are equivalent, it suffices to prove that A is reducible if and only if $(I + |A|)^{n-1}$ has a zero entry. Suppose first that A is reducible and that for some permutation matrix P we have $P^T A P = \begin{bmatrix} B & C \\ 0 & D \end{bmatrix} = \tilde{A}$, in which $B \in M_r$, $D \in M_{n-r}$, and $1 \leq r \leq n-1$. Notice that $P^T |A| P = |P^T A P| = |\tilde{A}|$ since the effect of a permutation similarity is only to permute rows and columns; also notice that each of the matrices $|\tilde{A}|^2, |\tilde{A}|^3, \ldots, |\tilde{A}|^{n-1}$ has a lower-left $(n-r)$-by-r zero block. Thus

$$
\begin{aligned}
P^T (I + |A|)^{n-1} P &= (I + P^T |A| P)^{n-1} = (I + |P^T A P|)^{n-1} \\
&= (I + |\tilde{A}|)^{n-1} \\
&= I + (n-1)|\tilde{A}| + \binom{n-1}{2}|\tilde{A}|^2 + \cdots + |\tilde{A}|^{n-1}
\end{aligned}
$$

in which each summand has a lower-left $(n-r)$-by-r zero block. Thus, $(I + |A|)^{n-1}$ is reducible, so it has a zero entry.

Conversely, suppose for some indices $p \neq q$ that the p, q entry of $(I + |A|)^{n-1}$ is 0. Then there is no directed path in $\Gamma(A)$ from P_p to P_q. Define the set of nodes

$$
S_1 = \{P_i : P_i = P_q \text{ or there is a path in } \Gamma(A) \text{ from } P_i \text{ to } P_q\}
$$

and let S_2 be the set of all nodes of $\Gamma(A)$ that are not in S_1. Notice that $S_1 \cup S_2 = \{P_1, \ldots, P_n\}$ and $P_q \in S_1 \neq \varnothing$, so $S_2 \neq \{P_1, \ldots, P_n\}$. If there were a path from some node P_i of S_2 to some node P_j of S_1, then (by definition of S_1) there would be a path from P_i to P_q and so P_i would be in S_1. Thus, there are no paths *from* any node of S_2 *to* any node of S_1. Now relabel the nodes so that $S_1 = \{\tilde{P}_1, \ldots, \tilde{P}_r\}$ and $S_2 = \{\tilde{P}_{r+1}, \ldots, \tilde{P}_n\}$. Let P be the permutation matrix that corresponds to the relabeling. Then

$$
\tilde{A} = P^T A P = \begin{bmatrix} B & C \\ 0 & D \end{bmatrix}, \quad B \in M_r, \ D \in M_{n-r}
$$

and hence A is reducible.

The argument for (a) and (c) is the same. $\qquad \square$

Let us summarize.

Theorem 6.2.24. *Let $A \in M_n$. The following are equivalent:*

(a) *A is irreducible.*
(b) *$(I + |A|)^{n-1} > 0$.*
(c) *$(I + M(A))^{n-1} > 0$.*
(d) *$\Gamma(A)$ is strongly connected.*
(e) *A has property SC.*

Definition 6.2.25. *Let $A \in M_n$. We say that A is* irreducibly diagonally dominant *if*

(a) *A is irreducible*
(b) *A is diagonally dominant, that is, $|a_{ii}| \geq R_i'(A)$ for all $i = 1, \ldots, n$*
(c) *There is an $i \in \{1, \ldots, n\}$ such that $|a_{ii}| > R_i'(A)$*

Exercise. Show by example that a matrix can be irreducible and diagonally dominant without being irreducibly diagonally dominant.

What we have learned about an irreducible matrix and any of its eigenvalues on the boundary of its Geršgorin set can be summarized as follows:

Theorem 6.2.26 (Taussky). *Let $A \in M_n$ be irreducible and suppose that $\lambda \in \mathbb{C}$ satisfies the inequalities (6.2.2a); for example, λ could be a boundary point of the Geršgorin set $G(A)$. If λ is an eigenvalue of A, then every Geršgorin circle of A passes through λ. Equivalently, if some Geršgorin circle of A does not pass through λ, then it is not an eigenvalue of A.*

Corollary 6.2.27 (Taussky). *Let $A = [a_{ij}] \in M_n$ be irreducibly diagonally dominant. Then*

(a) *A is nonsingular*

(b) *if every main diagonal entry of A is real and positive, then every eigenvalue of A has positive real part*

(c) *if A is Hermitian and every main diagonal entry is positive, then every eigenvalue of A is positive, that is, A is positive definite*

Problems

6.2.P1 Let $A \in M_n$ be irreducible and suppose that $n \geq 2$. Show that A does not have a zero row or column.

6.2.P2 Show by an example that the hypothesis of irreducibility in (6.2.28) is necessary.

6.2.P3 Suppose that $A = [a_{ij}] \in M_n$, that $\lambda, x = [x_i]$ is an eigenpair of $|A|$, and that all $x_i > 0$. Let $D = \mathrm{diag}(x_1, \ldots, x_n)$. Explain why every Geršgorin circle of $D^{-1}|A|D$ passes through λ and why $\lambda = \rho(|A|)$. Draw a picture. What can you say about the absolute row sums of $D^{-1}AD$?

6.2.P4 It will be proved in Chapter 8 that a square matrix with positive entries always has a positive eigenvalue and an associated eigenvector with positive entries. Use this fact and the preceding problem to show that $\rho(A) \leq \rho(|A|)$ for all $A \in M_n$.

6.2.P5 Use (6.2.28) to show that Cauchy's bound (5.6.47) on the zeroes $\tilde{z}$ of the polynomial $p(z) = z^n + a_{n-1}z^{n-1} + \cdots + a_1 z + a_0, a_0 \neq 0$ can be improved to $|\tilde{z}| < \max\{|a_0|, |a_1| + 1, |a_2| + 1, \ldots, |a_{n-1}| + 1\}$, provided that not all of the real numbers $|a_0|, |a_1| + 1, |a_2| + 1, \ldots, |a_{n-1}| + 1$ are the same. What improvements can be made in Montel's bound (5.6.48), Carmichael and Mason's bound (5.6.49), and Kojima's bound (5.6.53)?

6.2.P6 Explain why (a) an irreducible upper Hessenberg matrix is unreduced, and give an example of an unreduced upper Hessenberg matrix that is reducible; (b) a Hermitian or symmetric tridiagonal matrix is unreduced if and only if it is irreducible.

6.2.P7 Let $A \in M_n$ be the real symmetric tridiagonal matrix whose main diagonal entries are all 2 and whose superdiagonal entries are all -1. Use (6.2.27) to show that A is positive definite.

6.2.P8 Let $A \in M_n$. We know that $\rho(A) \leq |||A|||_\infty$. If A is irreducible and not all the absolute row sums of A are equal, explain why $\rho(A) < |||A|||_\infty$. Can the assumption of irreducibility be omitted?

6.3 Eigenvalue perturbation theorems

Let $D = \text{diag}(\lambda_1, \ldots, \lambda_n) \in M_n$, let $E = [e_{ij}] \in M_n$, and consider the perturbed matrix $D + E$. Theorem 6.1.1 ensures that the eigenvalues of $D + E$ are contained in the set

$$\bigcup_{i=1}^{n} \{z \in \mathbf{C} : |z - \lambda_i - e_{ii}| \leq R_i'(E) = \sum_{j \neq i} |e_{ij}|\}$$

which is contained in the set

$$\bigcup_{i=1}^{n} \{z \in \mathbf{C} : |z - \lambda_i| \leq R_i(E) = \sum_{j=1}^{n} |e_{ij}|\}$$

Thus, if $\hat{\lambda}$ is an eigenvalue of $D + E$, there is some eigenvalue λ_i of D such that $|\hat{\lambda} - \lambda_i| \leq |\!|\!| E |\!|\!|_\infty$. We can use this bound to obtain a perturbation bound for the eigenvalues of a diagonalizable matrix.

Observation 6.3.1. *Let $A \in M_n$ be diagonalizable, and suppose that $A = S\Lambda S^{-1}$, in which S is nonsingular and Λ is diagonal. Let $E \in M_n$. If $\hat{\lambda}$ is an eigenvalue of $A + E$, there is an eigenvalue λ of A such that*

$$|\hat{\lambda} - \lambda| \leq |\!|\!| S |\!|\!|_\infty \, |\!|\!| E |\!|\!|_\infty \, |\!|\!| S^{-1} |\!|\!|_\infty = \kappa_\infty(S) |\!|\!| E |\!|\!|_\infty$$

in which $\kappa_\infty(\cdot)$ is the condition number with respect to the matrix norm $|\!|\!| \cdot |\!|\!|_\infty$.

Proof. Since $A + E$ and $S^{-1}(A + E)S = \Lambda + S^{-1}ES$ have the same eigenvalues, and since Λ is diagonal, the preceding argument shows that there is some eigenvalue λ of A for which $|\hat{\lambda} - \lambda| \leq |\!|\!| S^{-1}ES |\!|\!|_\infty$. The stated inequality follows from submultiplicativity of the matrix norm $|\!|\!| \cdot |\!|\!|_\infty$. □

The maximum row sum norm is induced by the sum norm on $\mathbf{C}^n$, which is an absolute norm. We can use (5.6.36) to generalize the preceding observation.

Theorem 6.3.2 (Bauer and Fike). *Let $A \in M_n$ be diagonalizable, and suppose that $A = S\Lambda S^{-1}$, in which S is nonsingular and Λ is diagonal. Let $E \in M_n$ and let $|\!|\!| \cdot |\!|\!|$ be a matrix norm on M_n that is induced by an absolute norm on $\mathbf{C}^n$. If $\hat{\lambda}$ is an eigenvalue of $A + E$, there is an eigenvalue λ of A such that*

$$|\hat{\lambda} - \lambda| \leq |\!|\!| S |\!|\!| \, |\!|\!| S^{-1} |\!|\!| \, |\!|\!| E |\!|\!| = \kappa(S) |\!|\!| E |\!|\!| \qquad (6.3.3)$$

in which $\kappa(\cdot)$ is the condition number with respect to the matrix norm $|\!|\!| \cdot |\!|\!|$.

Proof. If $\hat{\lambda}$ is an eigenvalue of $S^{-1}(A + E)S = \Lambda + S^{-1}ES$, then $\hat{\lambda}I - \Lambda - S^{-1}ES$ is singular. If $\hat{\lambda}$ is an eigenvalue of A, the bound (6.3.3) is trivially satisfied. Suppose that $\hat{\lambda}$ is not an eigenvalue of A, so $\hat{\lambda}I - \Lambda$ is nonsingular. In this case, $(\hat{\lambda}I - \Lambda)^{-1}(\hat{\lambda}I - \Lambda - S^{-1}ES) = I - (\hat{\lambda}I - \Lambda)^{-1}S^{-1}ES$ is singular, so (5.6.16) ensures that $|\!|\!| (\hat{\lambda}I - \Lambda)^{-1}S^{-1}ES |\!|\!| \geq 1$. Using (5.6.36), we compute

$$1 \leq |\!|\!| (\hat{\lambda}I - \Lambda)^{-1}S^{-1}ES |\!|\!| \leq |\!|\!| S^{-1}ES |\!|\!| \, |\!|\!| (\hat{\lambda}I - \Lambda)^{-1} |\!|\!|$$

$$= |\!|\!| S^{-1}ES |\!|\!| \max_{1 \leq i \leq n} |\hat{\lambda} - \lambda_i|^{-1} = \frac{|\!|\!| S^{-1}ES |\!|\!|}{\min_{1 \leq i \leq n} |\hat{\lambda} - \lambda_i|}$$

and hence

$$\min_{1 \le i \le n} |\hat{\lambda} - \lambda_i| \le \||S^{-1}ES\|| \le \||S^{-1}\|| \, \||S\|| \, \||E\|| = \kappa(S) \||E\||$$

$\Box$

Exercise. Give an example of a matrix norm that does not satisfy the assumption of the theorem.

Exercise. Explain why a unitary matrix has condition number 1 with respect to the spectral norm.

The condition number $\kappa(\cdot)$ arose in (5.8) in the context of a priori error bounds for computed inverses or solutions of linear equations, but we now see that it arises in the context of a priori error bounds for computed eigenvalues of a diagonalizable matrix. If we think of $\hat{\lambda}$ as an exactly computed eigenvalue of the perturbed matrix $A + E$, then (6.3.3) ensures that the relative error made by using it as an approximation to an eigenvalue λ of A satisfies the inequality

$$\frac{|\hat{\lambda} - \lambda|}{\||E\||} \le \kappa(S)$$

The matrix norm used must be induced by an absolute vector norm, and the columns of S are any independent set of eigenvectors of A. If $\kappa(S)$ is small (near 1), then small perturbations in the data can result in only small changes in the eigenvalues. If $\kappa(S)$ is very large, however, then a computed eigenvalue of $A + E$ might be a poor approximation to an eigenvalue of A.

If A is normal, then one may take S to be unitary, which has condition number 1 with respect to the spectral norm. Thus, normal matrices are well conditioned with respect to eigenvalue computations.

Corollary 6.3.4. *Let $A, E \in M_n$ and suppose that A is normal. If $\hat{\lambda}$ is an eigenvalue of $A + E$, then there is an eigenvalue λ of A such that $|\hat{\lambda} - \lambda| \le \||E\||_2$.*

In the preceding corollary, neither the perturbation matrix E nor the perturbed matrix $A + E$ need be normal. For example, A might be a real symmetric matrix A that is subjected to a real, but not necessarily symmetric, perturbation.

Exercise. Provide details for a proof of (6.3.4).

Exercise. If $A, E \in M_n$ are Hermitian, if $\lambda_1 \le \cdots \le \lambda_n$ are the ordered eigenvalues of A, if $\hat{\lambda}_1 \le \cdots \le \hat{\lambda}_n$ are the ordered eigenvalues of $A + E$, and if $\lambda_1(E) \le \cdots \le \lambda_n(E)$ are the ordered eigenvalues of E, use Weyl's inequalities (4.3.2a,b) to show that

$$\lambda_1(E) \le \hat{\lambda}_k - \lambda_k \le \lambda_n(E) \quad \text{for each } k = 1, \ldots, n \qquad (6.3.4.1)$$

and that $|\hat{\lambda}_k - \lambda_k| \le \rho(E) = \||E\||_2$. Why is this bound better than the one in (6.3.4)? What can you say if all the eigenvalues of E are nonnegative?

It is common in numerical applications to have both A and the perturbing matrix E be real and symmetric. In this case, and in the more general situation in which both

A and $A + E$ are normal, there is a Frobenius norm upper bound for perturbations to all of the eigenvalues.

Theorem 6.3.5 (Hoffman and Wielandt). *Let $A, E \in M_n$, assume that A and $A + E$ are both normal, let $\lambda_1, \dots, \lambda_n$ be the eigenvalues of A in some given order, and let $\hat{\lambda}_1, \dots, \hat{\lambda}_n$ be the eigenvalues of $A + E$ in some given order. There is a permutation $\sigma(\cdot)$ of the integers $1, \dots, n$ such that*

$$\sum_{i=1}^{n} |\hat{\lambda}_{\sigma(i)} - \lambda_i|^2 \leq \|E\|_2^2 = \mathrm{tr}(E^* E) \tag{6.3.6}$$

Proof. Let $\Lambda = \mathrm{diag}(\lambda_1, \dots, \lambda_n)$, let $\hat{\Lambda} = \mathrm{diag}(\hat{\lambda}_1, \dots, \hat{\lambda}_n)$, let $V, W \in M_n$ be unitary matrices such that $A = V \Lambda V^*$ and $A + E = W \hat{\Lambda} W^*$, and let $U = V^* W = [u_{ij}]$. Using unitary invariance of the Frobenius norm, we compute

$$\|E\|_2^2 = \|(A + E) - A\|_2^2 = \|W \hat{\Lambda} W^* - V \Lambda V^*\|_2^2$$
$$= \|V^* W \hat{\Lambda} - \Lambda V^* W\|_2^2 = \|U \hat{\Lambda} - \Lambda U\|_2^2$$
$$= \sum_{i,j=1}^{n} |\hat{\lambda}_i - \lambda_j|^2 |u_{ij}|^2$$

Just as in the proof of (4.3.49), we observe that the matrix $[|u_{ij}|^2]$ is doubly stochastic. Therefore,

$$\|E\|_2^2 = \sum_{i,j=1}^{n} |\hat{\lambda}_i - \lambda_j|^2 |u_{ij}|^2$$

$$\geq \min \{ \sum_{i,j=1}^{n} |\hat{\lambda}_i - \lambda_j|^2 s_{ij} : S = [s_{ij}] \in M_n \text{ is doubly stochastic} \}$$

The function $f(S) = \sum_{i,j=1}^{n} |\hat{\lambda}_i - \lambda_j|^2 s_{ij}$ is a linear function on the compact convex set of doubly stochastic matrices, so (8.7.3) (a corollary of Birkhoff's theorem) ensures that f attains its minimum at a permutation matrix $P = [p_{ij}]$. If P^T corresponds to the permutation $\sigma(\cdot)$ of the integers $1, \dots, n$, we have

$$\|E\|_2^2 \geq \sum_{i,j=1}^{n} |\hat{\lambda}_i - \lambda_j|^2 p_{ij} = \sum_{i=1}^{n} |\hat{\lambda}_{\sigma(i)} - \lambda_i|^2 \qquad \square$$

Theorem 6.3.5 says that eigenvalues of a normal matrix enjoy a strong stability under perturbations, but it does not identify a permutation of the eigenvalues that satisfies the stated inequality. Not every permutation will do, and indeed there is always one for which the inequality in (6.3.6) is reversed; see (6.3.P8). But in the important special case of Hermitian matrices, a natural ordering of the eigenvalues will do.

Corollary 6.3.8. *Let $A, E \in M_n$. Assume that A is Hermitian and $A + E$ is normal, let $\lambda_1, \dots, \lambda_n$ be the eigenvalues of A, arranged in increasing order $\lambda_1 \leq \dots \leq \lambda_n$,* *and let* $\hat{\lambda}_1, \ldots, \hat{\lambda}_n$ *be the eigenvalues of* $A + E$, *ordered so that* $\mathrm{Re}\,\hat{\lambda}_1 \leq \cdots \leq \mathrm{Re}\,\hat{\lambda}_n$. *Then*

$$\sum_{i=1}^{n} |\hat{\lambda}_i - \lambda_i|^2 \leq \|E\|_2^2$$

Proof. The preceding theorem ensures that there is a permutation of the given order for the eigenvalues of $A + E$ such that

$$\sum_{i=1}^{n} |\hat{\lambda}_{\sigma(i)} - \lambda_i|^2 \leq \|E\|_2^2 \tag{6.3.9}$$

If the eigenvalues of $A + E$ in the list $\hat{\lambda}_{\sigma(1)}, \ldots, \hat{\lambda}_{\sigma(n)}$ are already in increasing order of their real parts, there is nothing to prove. If not, two successive eigenvalues in the list are not ordered in this way, say, $\mathrm{Re}\,\hat{\lambda}_{\sigma(k)} > \mathrm{Re}\,\hat{\lambda}_{\sigma(k+1)}$. A computation reveals that

$$|\hat{\lambda}_{\sigma(k)} - \lambda_k|^2 + |\hat{\lambda}_{\sigma(k+1)} - \lambda_{k+1}|^2 = |\hat{\lambda}_{\sigma(k+1)} - \lambda_k|^2 + |\hat{\lambda}_{\sigma(k)} - \lambda_{k+1}|^2 + \Delta(k)$$

in which $\Delta(k) = 2(\lambda_k - \lambda_{k+1})(\mathrm{Re}\,\hat{\lambda}_{\sigma(k+1)} - \mathrm{Re}\,\hat{\lambda}_{\sigma(k)}) \geq 0$. Thus,

$$|\hat{\lambda}_{\sigma(k)} - \lambda_k|^2 + |\hat{\lambda}_{\sigma(k+1)} - \lambda_{k+1}| \geq |\hat{\lambda}_{\sigma(k+1)} - \lambda_k|^2 + |\hat{\lambda}_{\sigma(k)} - \lambda_{k+1}|^2$$

and the two eigenvalues $\hat{\lambda}_{\sigma(k)}$ and $\hat{\lambda}_{\sigma(k+1)}$ can be interchanged without increasing the sum of squared differences in (6.3.9). By a finite sequence of such interchanges (which do not increase the left-hand side of (6.3.9)), the list of eigenvalues $\hat{\lambda}_{\sigma(1)}, \ldots, \hat{\lambda}_{\sigma(n)}$ can be transformed into the list $\hat{\lambda}_1, \hat{\lambda}_2, \ldots, \hat{\lambda}_n$. $\qquad\square$

In an important special case of the preceding corollary, both A and $A + E$ are Hermitian, or even real and symmetric. For a generalization of (6.3.8) in this case, see (7.4.9.3).

Exercise. If $A, E \in M_n$ are Hermitian and if their eigenvalues are arranged in the same (increasing or decreasing) order, explain why

$$\sum_{i=1}^{n} (\lambda_i(A + E) - \lambda_i(A))^2 \leq \|E\|_2^2$$

Exercise. Consider $A = \begin{bmatrix} 0 & 0 \\ 0 & 4 \end{bmatrix}$ and $E = \begin{bmatrix} -1 & -1 \\ 1 & -3 \end{bmatrix}$. Explain why the assertion in (6.3.5) need not be true if one of A and $A + E$ is not normal. *Hint:* $\sum_{i=1}^{2} (\lambda_i(A + E) - \lambda_i(A))^2 = 16$ for any ordering of the eigenvalues.

If A is not diagonalizable, there are no bounds known that are as easy to state as those in (6.3.2). However, there is a simple explicit formula that describes how a *simple* eigenvalue varies when the matrix entries are perturbed. Underlying our formula are basic facts in (1.4.7) and (1.4.12), which we restate here.

Lemma 6.3.10. *Let λ be a simple eigenvalue of $A \in M_n$. Let x and y be, respectively, right and left eigenvectors of A corresponding to λ. Then*

(a) $y^*x \neq 0$

(b) there is a nonsingular $S \in M_n$ such that $S = [x \ S_1]$, $S^{-*} = [\frac{y}{x^*y} \ Z_1]$, $S_1, Z_1 \in M_{n,n-1}$,

$$A = S \begin{bmatrix} \lambda & 0 \\ 0 & A_1 \end{bmatrix} S^{-1} \qquad (6.3.11)$$

and λ is not an eigenvalue of $A_1 \in M_{n-1}$.

Theorem 6.3.12. *Let* $A, E \in M_n$ *and suppose that* λ *is a simple eigenvalue of* A. *Let* x *and* y *be, respectively, right and left eigenvectors of* A *corresponding to* λ. *Then*

(a) *for each given* $\varepsilon > 0$ *there exists a* $\delta > 0$ *such that, for all* $t \in \mathbf{C}$ *such that* $|t| < \delta$, *there is a unique eigenvalue* $\lambda(t)$ *of* $A + tE$ *such that* $|\lambda(t) - \lambda - ty^*Ex/y^*x| \leq |t|\varepsilon$

(b) $\lambda(t)$ *is continuous at* $t = 0$, *and* $\lim_{t \to 0} \lambda(t) = \lambda$

(c) $\lambda(t)$ *is differentiable at* $t = 0$, *and*

$$\left. \frac{d\lambda(t)}{dt} \right|_{t=0} = \frac{y^*Ex}{y^*x} \qquad (6.3.13)$$

Proof. Our strategy is to find a matrix that is similar to $A + tE$, has $\lambda + tx^*Ey/y^*x$ as its 1, 1 entry, and whose Geršgorin disc associated with its first row has radius at most $|t|\varepsilon$ and is disjoint from the other $n - 1$ Geršgorin discs. Let $\mu = \min\{|\lambda - \hat{\lambda}| : \hat{\lambda} \text{ is an} eigenvalue of A and $\hat{\lambda} \neq \lambda\}$; the hypothesis is that $\mu > 0$. Let $\varepsilon \in (0, \mu/7)$.

Using the notation of (6.3.10), let $\eta = y/y^*x$ and perform the similarity

$$S^{-1}(A + tE)S = S^{-1}AS + tS^{-1}ES = \begin{bmatrix} \lambda & 0 \\ 0 & A_1 \end{bmatrix} + t \begin{bmatrix} \eta^*Ex & \eta^*ES_1 \\ Z_1^*Ex & Z_1^*ES_1 \end{bmatrix}$$

$$= \begin{bmatrix} \lambda + t\eta^*Ex & t\eta^*ES_1 \\ tZ_1^*Ex & A_1 + tZ_1^*ES_1 \end{bmatrix}$$

which puts $\lambda + ty^*Ex/y^*x$ in position 1, 1. Now perform a similarity that puts A_1 into an "almost diagonal" upper triangular form described in (2.4.7.2): $A_1 = S_\varepsilon T_\varepsilon S_\varepsilon^{-1}$, in which T_ε is upper triangular, has the eigenvalues of A_1 on its main diagonal, and has deleted absolute row sums at most ε. Let $\mathcal{S}_\varepsilon = [1] \oplus S_\varepsilon$ and compute the similarity

$$\mathcal{S}_\varepsilon^{-1}S^{-1}(A + tE)S\mathcal{S}_\varepsilon = \begin{bmatrix} \lambda + t\eta^*Ex & t\eta^*ES_1\mathcal{S}_\varepsilon \\ t\mathcal{S}_\varepsilon^{-1}Z_1^*Ex & T_\varepsilon + t\mathcal{S}_\varepsilon^{-1}Z_1^*ES_1\mathcal{S}_\varepsilon \end{bmatrix}$$

Now let $\mathcal{R}(r) = [1] \oplus rI_{n-1}$ with $r > 0$ such that $r \|\eta^*ES_1\mathcal{S}_\varepsilon\|_1 < \varepsilon$, and perform a final similarity

$$\mathcal{R}(r)^{-1}\mathcal{S}_\varepsilon^{-1}S^{-1}(A + tE)S\mathcal{S}_\varepsilon\mathcal{R}(r)$$

$$= \begin{bmatrix} \lambda + t\eta^*Ex & tr\eta^*ES_1\mathcal{S}_\varepsilon \\ tr^{-1}\mathcal{S}_\varepsilon^{-1}Z_1^*Ex & T_\varepsilon + t\mathcal{S}_\varepsilon^{-1}Z_1^*ES_1\mathcal{S}_\varepsilon \end{bmatrix} \qquad (6.3.13a)$$

Choose $\delta_1 > 0$ such that $\delta_1|\eta^*Ex| < \varepsilon$, choose $\delta_2 > 0$ such that $\delta_2 \|r^{-1}\mathcal{S}_\varepsilon^{-1}Z_1^*Ex\|_\infty < \varepsilon$, and choose $\delta_3 > 0$ such that $\delta_3 \|\mathcal{S}_\varepsilon^{-1}Z_1^*ES_1\mathcal{S}_\varepsilon\|_\infty < \varepsilon/2$. Let $\delta = \min\{\delta_1, \delta_2, \delta_3, 1\}$ and suppose that $0 < |t| < \delta$. Any main diagonal entry τ of $T_\varepsilon + t\mathcal{S}_\varepsilon^{-1}Z_1^*ES_1\mathcal{S}_\varepsilon$ is at most ε from an eigenvalue $\hat{\lambda}$ of A_1, and every point in the Geršgorin disc around τ is at most 4ε from $\hat{\lambda}$ (ε from the displacement due to a diagonal entry of $t\mathcal{S}_\varepsilon^{-1}Z_1^*ES_1\mathcal{S}_\varepsilon$, ε from an absolute deleted row sum of T_ε, ε from an entry of $tr^{-1}\mathcal{S}_\varepsilon^{-1}Z_1^*Ex$, and ε from a row of $t\mathcal{S}_\varepsilon^{-1}Z_1^*ES_1\mathcal{S}_\varepsilon$). The radius of the Geršgorin disc G_1 associated with the main diagonal entry $\lambda + t\eta^*Ex$ is at most $|t\varepsilon| \le \varepsilon$, and no point in that disc is at a distance greater than $2|t\eta^*Ex| < 2\varepsilon$ from λ. Since $4\varepsilon + 2\varepsilon = 6\varepsilon \le 6\mu/7 < \mu$, it follows that G_1 is disjoint from the Geršgorin discs associated with rows $2, \ldots, n$ of the matrix in (6.3.13a). Theorem 6.1.1 ensures that there is a unique eigenvalue $\lambda(t)$ of $A + tE$ in G_1, so $|\lambda(t) - \lambda - t\eta^*Ex| \le |t|\varepsilon$.

The assertion in (b) follows from observing that

$$|\lambda(t) - \lambda| \le |\lambda(t) - \lambda - t\eta^*Ex| + |t\eta^*Ex| \le |t|\varepsilon + \varepsilon \le 2\varepsilon$$

if $|t| < \delta$. The inequality

$$\left| \frac{\lambda(t) - \lambda}{t} - \eta^*Ex \right| < \varepsilon \quad \text{if } 0 < |t| < \delta$$

implies the assertion in (c). $\qquad\qquad\qquad\qquad\qquad\qquad\qquad\qquad\qquad\qquad$ □

Exercise. Using the assumptions and notation of the preceding theorem, let $x = [x_i]$ and $y = [y_i]$. Explain why

$$\frac{\partial \lambda}{\partial a_{ij}} = \frac{\bar{y}_i x_j}{y^*x} \qquad\qquad (6.3.13b)$$

for all $i, j = 1, \ldots, n$. *Hint*: Let $E = E_{ij}$, the n-by-n matrix whose only nonzero entry is a one in the i, j position.

Exercise. Let $\epsilon > 0$ be given. Consider the matrix $A = \begin{bmatrix} 1 & 1 \\ 0 & 1+\epsilon \end{bmatrix}$, the simple eigenvalue $\lambda = 1$, and the right and left eigenvectors $x = [1 \ 0]^T$ and $y = [\epsilon - 1]^T$. Compute $\partial \lambda/\partial a_{ij}$ for all four pairs i, j. What happens as $\epsilon \to 0$? Conclude that an eigenvalue of a matrix can be very sensitive to certain matrix perturbations if its associated right and left eigenvectors are nearly orthogonal.

The formula (6.3.13) for the derivative of a simple eigenvalue has an analog for singular values; see (7.3.12).

In contrast to the situation for eigen*values*, the eigen*vectors* of a diagonalizable matrix may suffer radical changes with only small perturbations in the entries of the matrix. For example, if $A = \begin{bmatrix} 1 & 0 \\ 0 & 1 \end{bmatrix}$, $E = \begin{bmatrix} \epsilon & \delta \\ 0 & 0 \end{bmatrix}$, and $\epsilon, \delta \ne 0$, the eigenvalues of $A + E$ are $\lambda = 1$ and $1 + \epsilon$ and the respective normalized right eigenvectors are

$$\begin{bmatrix} 1 \\ 0 \end{bmatrix} \quad \text{and} \quad \frac{1}{(\epsilon^2 + \delta^2)^{1/2}} \begin{bmatrix} -\delta \\ \epsilon \end{bmatrix}$$

By choosing the ratio of ϵ to δ appropriately, the second eigenvector can be chosen to point in any direction whatsoever, for ϵ and δ both arbitrarily small.

Our eigenvalue perturbation estimates so far have all been a priori bounds; they do not involve the computed eigenvalues or eigenvectors or any quantity derived from them. Let $\hat{x} \ne 0$ be an "approximate eigenvector" of $A \in M_n$ and let $\hat{\lambda}$ be a corresponding "approximate eigenvalue." If A is diagonalizable, we can use the *residual vector* $r = A\hat{x} - \hat{\lambda}\hat{x}$ to estimate how well $\hat{\lambda}$ approximates an eigenvalue of A.

Theorem 6.3.14. *Let $A \in M_n$ be diagonalizable with $A = S \Lambda S^{-1}$ and $\Lambda = diag(\lambda_1, \ldots, \lambda_n)$. Let $||| \cdot |||$ be a matrix norm on M_n that is induced by an absolute vector norm $\| \cdot \|$ on $\mathbf{C}^n$. Let $\hat{x} \in \mathbf{C}^n$ be nonzero, let $\hat{\lambda} \in \mathbf{C}$, and let $r = A\hat{x} - \hat{\lambda}\hat{x}$.*

(a) There is an eigenvalue λ of A such that

$$|\hat{\lambda} - \lambda| \leq |||S||| \, |||S^{-1}||| \frac{\|r\|}{\|\hat{x}\|} = \kappa(S) \frac{\|r\|}{\|\hat{x}\|} \tag{6.3.15}$$

in which $\kappa(\cdot)$ is the condition number with respect to $||| \cdot |||$.

(b) If A is normal, there is an eigenvalue λ of A such that

$$|\hat{\lambda} - \lambda| \leq \frac{\|r\|_2}{\|\hat{x}\|_2} \tag{6.3.16}$$

Proof. If $\hat{\lambda}$ is an eigenvalue of A, then the asserted bounds are trivially satisfied, so we suppose that it is not an eigenvalue of A. Then $r = A\hat{x} - \hat{\lambda}x = S(\Lambda - \hat{\lambda}I)S^{-1}\hat{x}$ and $\hat{x} = S(\Lambda - \hat{\lambda}I)^{-1}S^{-1}r$. We use (5.6.36) to compute

$$\|\hat{x}\| = \|S(\Lambda - \hat{\lambda}I)^{-1}S^{-1}r\| \leq |||S(\Lambda - \hat{\lambda}I)^{-1}S^{-1}||| \, \|r\|$$
$$\leq |||S||| \, |||S^{-1}||| \, |||(\Lambda - \hat{\lambda}I)^{-1}||| \, \|r\| = \kappa(S)|||(\Lambda - \hat{\lambda}I)^{-1}||| \, \|r\|$$
$$= \kappa(S) \max_{\lambda \in \sigma(A)} |\lambda - \hat{\lambda}|^{-1} \|r\|$$

so that

$$\|\hat{x}\| \min_{\lambda \in \sigma(A)} |\lambda - \hat{\lambda}| \leq \kappa(S)\|r\|$$

The bound (6.3.16) in the normal case is a consequence of the fact that a normal matrix is unitarily diagonalizable, the matrix norm induced by the Euclidean norm is the spectral norm, and a unitary matrix has unit condition number with respect to the spectral norm. □

The preceding result should be contrasted with the a posteriori bound (5.8.10) on the relative error in the solution to a system of linear equations. If the matrix of coefficients of a system of linear equations is ill-conditioned, even if it is normal, a small residual does not ensure a small relative error in the solution. However, (6.3.16) says that if A is normal, and if an approximate eigenpair has a small residual, then the absolute error in the eigenvalue is guaranteed to be small; no condition number appears in the bound.

This pleasant result for the eigenvalues is not matched by a similarly pleasant result for the eigenvectors. Even for a real symmetric matrix, a small residual does not guarantee that an approximate eigenvector is close to an eigenvector.

Exercise. Consider $A = \begin{bmatrix} 1 & \epsilon \\ \epsilon & 1 \end{bmatrix}$ with $\epsilon > 0$. Take $\hat{\lambda} = 1$ and $\hat{x} = [1\,0]^T$, and show that the residual is $r = [0\,\epsilon]^T$. Show that the eigenvectors of A are $[1\,1]^T$ and $[1\,-1]^T$ for all $\epsilon > 0$, so that $\hat{x}$ is not approximately parallel to either of these two vectors no matter how small ϵ is. Show that the eigenvalues of A are $1 + \epsilon$ and $1 - \epsilon$, and verify the bound (6.3.16).

Problems

6.3.P1 Let $A = [a_{ij}] \in M_n$ be normal and let $\lambda_1, \ldots, \lambda_n$ be its eigenvalues. Show that there is a permutation σ of the integers $1, \ldots, n$ such that $\sum_{i=1}^{n} |a_{ii} - \lambda_{\sigma(i)}|^2 \leq \sum_{i \neq j} |a_{ij}|^2$.

6.3.P2 The upper bounds in (6.3.14) involve the norm of the residual vector $r = A\hat{x} - \hat{\lambda}\hat{x}$. For a given $A \in M_n$ and a given nonzero $\hat{x}$, what is an optimal choice of $\hat{\lambda}$? (a) For the Euclidean norm and a given nonzero $\hat{x}$, show that $\|r\|_2 \geq \|A\hat{x} - (\hat{x}^* A\hat{x})\hat{x}\|_2$ for all $\hat{\lambda} \in \mathbf{C}$. (b) If A is normal and y is a unit vector, explain why there is at least one eigenvalue of A in the disc

$$\{z \in \mathbf{C} : |z - y^* A y| \leq (\|Ay\|_2^2 - |y^* A y|^2)^{1/2}\} \tag{6.3.17}$$

(c) If A is Hermitian and y is a unit vector, explain why there is at least one eigenvalue of A in the real interval

$$\{t \in \mathbf{R} : |t - y^* A y| \leq (\|Ay\|_2^2 - (y^* A y)^2)^{1/2}\}$$

6.3.P3 Partition a normal matrix $A \in M_n$ as $A = \begin{bmatrix} B & X \\ Y & C \end{bmatrix}$, in which $B \in M_k$ and $C \in M_{n-k}$. Let β be an eigenvalue of B and let γ be an eigenvalue of C. (a) Use (6.3.14b) to show that there is an eigenvalue of A in the disc $\{z \in \mathbf{C} : |z - \beta| \leq \||Y\||_2\}$ as well as in the disc $\{z \in \mathbf{C} : |z - \gamma| \leq \||X\||_2\}$. (b) If $k = 1$ and $A = \begin{bmatrix} b & x^* \\ y & C \end{bmatrix}$, in which $b \in \mathbf{C}$ and $x, y \in \mathbf{C}^{n-1}$, explain why there is an eigenvalue of A in the disc $\{z \in \mathbf{C} : |z - b| \leq \|x\|_2\}$ as well as in the disc $\{z \in \mathbf{C} : |z - \gamma| \leq \|x\|_2\}$.

6.3.P4 Let $A \in M_n$ be normal, let $\mathcal{S}$ be a given k-dimensional subspace of $\mathbf{C}^n$, and let $\gamma \in \mathbf{C}$ and $\delta > 0$ be given. (a) If $\|Ax - \gamma x\|_2 \leq \delta$ for every unit vector $x \in \mathcal{S}$, show that there are at least k eigenvalues of A in the disc $\{z \in \mathbf{C} : |z - \gamma| \leq \delta\}$. (b) Explain why the assertion in case $k = 1$ of (a) is equivalent to the assertion in (6.3.14b), and specialize your proof in (a) to give an alternative proof of the bound (6.3.16).

6.3.P5 Let t_0 be real and consider the polynomial $p(t) = (t - t_0)^2$. For $\epsilon > 0$, show that the zeroes of the polynomial $p(t) - \epsilon$ are $t_0 \pm \epsilon^{1/2}$. Explain why the ratio of the perturbation in a zero of a polynomial to a perturbation in its coefficients can be unbounded.

6.3.P6 Consider the bound in (6.3.4), which says that for a normal matrix, the ratio of the perturbation in an eigenvalue to a perturbation in the matrix entries is bounded. Since the eigenvalues of the matrix are just the zeroes of its characteristic polynomial, explain how this pleasant situation could be consistent with the conclusion of the preceding problem. The moral is that it is very unwise to attempt to compute the eigenvalues of a matrix (normal or otherwise) by forming its characteristic polynomial and then computing its zeroes: This can turn an inherently well-conditioned problem into an ill-conditioned one!

6.3.P7 Consider $A = \begin{bmatrix} 0 & 1 \\ 0 & 0 \end{bmatrix}$, $E = \begin{bmatrix} 0 & 0 \\ 1 & 0 \end{bmatrix}$, and $A + tE$ for $t > 0$. (a) Does A satisfy the hypotheses of (6.3.12)? (b) Show that the eigenvalues of $A + tE$ are $\pm\sqrt{t}$, and explain why the eigenvalue $\lambda(t) = \sqrt{t}$ is continuous but not differentiable at $t = 0$. (c) Does A satisfy the hypotheses of (6.3.2)? (d) Let λ be an eigenvalue of A and let $\lambda(t)$ be an eigenvalue of $A + tE$. Explain why there is no $c > 0$ such that $|\lambda(t) - \lambda| \leq c \||tE\||$ for all $t > 0$ and contrast with the bound (6.3.3).

6.3.P8 Use the argument in the proof of (6.3.5) to show that, under the hypotheses of the theorem, there is a permutation τ of the integers $1, \ldots, n$ such that $\sum_{i=1}^{n} |\hat{\lambda}_{\tau(i)} - \lambda_i|^2 \geq \|E\|_2^2$.

6.3.P9 In the proof of (6.3.5) we used the fact that if $U = [u_{ij}] \in M_n$ is unitary, then $A = [|u_{ij}|^2]$ is doubly stochastic and *unistochastic* (see (4.3.P10)). Show that the following doubly stochastic matrix is not unistochastic:

$$\begin{bmatrix} \frac{1}{2} & \frac{1}{2} & 0 \\ \frac{1}{2} & 0 & \frac{1}{2} \\ 0 & \frac{1}{2} & \frac{1}{2} \end{bmatrix}$$

6.3.P10 Consider the real symmetric matrix $A(t) = \begin{bmatrix} 0 & t \\ t & 0 \end{bmatrix}$, $t \in \mathbf{R}$. Show that the eigenvalues of $A(t)$ are $\lambda_1(t) = |t|$ and $\lambda_2(t) = -|t|$, neither of which is differentiable at $t = 0$. Does this contradict (6.3.12)? Why?

Further Readings. The first version of (6.3.2) appeared in F. Bauer and C. Fike, Norms and exclusion theorems, *Numer. Math.* 2 (1960) 137–141. The original version of (6.3.5) is in A. J. Hoffman and H. Wielandt, The variation of the spectrum of a normal matrix, *Duke Math. J.* 20 (1953) 37–39. An elementary proof of this result in the real symmetric case is in Wilkinson (1965), pp. 104–109. Theorem 6.3.12 is only the first part of a very interesting story. For a summary of the rest of the tale, see J. Moro, J. V. Burke, and M. L. Overton, On the Lidskii–Vishik–Lyusternik perturbation theory for eigenvalues of matrices with arbitrary Jordan structure, *SIAM J. Matrix Anal. Appl.* 18 (1997) 793–817; for more details, see Baumgärtel (1985), Chatelin (1993), and Kato (1980).

6.4 Other eigenvalue inclusion sets

We have discussed Geršgorin discs in some detail. Many authors, perhaps attracted by the geometrical elegance of the Geršgorin theory, have generalized its ideas and methods to obtain other types of eigenvalue inclusion sets. We now discuss a few of these to give a flavor of what has been done.

The first theorem, due to Ostrowski, gives an eigenvalue inclusion set that is a union of discs, like the Geršgorin set, but the radii of the discs depend on both deleted row and column sums. The row and column versions of Geršgorin's theorem are included in this result, which gives a continuum of eigenvalue inclusion sets that interpolate between (6.1.2) and (6.1.4).

Theorem 6.4.1 (Ostrowski). *Let $A = [a_{ij}] \in M_n$, let $\alpha \in [0, 1]$, and let R_i' and C_i' denote the deleted row and column sums of A:*

$$R_i' = \sum_{j \neq i} |a_{ij}| \quad and \quad C_i' = \sum_{j \neq i} |a_{ij}| \tag{6.4.2}$$

The eigenvalues of A are in the union of n discs

$$\bigcup_{i=1}^{n} \{z \in \mathbf{C} : |z - a_{ii}| \leq R_i'^{\alpha} C_i'^{1-\alpha}\} \tag{6.4.3}$$

Proof. The cases $\alpha = 0$ and $\alpha = 1$ were proved in (6.1.1) and (6.1.3), so we may assume that $0 < \alpha < 1$. Furthermore, we may assume that all $R'_i > 0$, because we may perturb A by inserting a small nonzero entry into any row in which $R'_i = 0$; the resulting matrix has an eigenvalue inclusion set (6.4.4) that is larger than the set for A, and the result follows in the limit as the perturbation goes to zero.

Now suppose that $Ax = \lambda x$ with $x = [x_i] \neq 0$. Then for each $i = 1, 2, \ldots, n$ we have

$$|\lambda - a_{ii}||x_i| = \left| \sum_{j \neq i} a_{ij} x_j \right| \leq \sum_{j \neq i} |a_{ij}||x_j| = \sum_{j \neq i} |a_{ij}|^\alpha (|a_{ij}|^{1-\alpha}|x_j|)$$

$$\leq \left(\sum_{j \neq i} (|a_{ij}|^\alpha)^{1/\alpha} \right)^\alpha \left(\sum_{j \neq i} (|a_{ij}|^{1-\alpha}|x_j|)^{1/(1-\alpha)} \right)^{1-\alpha} \quad (6.4.4)$$

$$= R_i'^\alpha \left(\sum_{j \neq i} |a_{ij}||x_j|^{1/(1-\alpha)} \right)^{1-\alpha}$$

which, since $R'_i > 0$, is equivalent to

$$\frac{|\lambda - a_{ii}|}{R_i'^\alpha}|x_i| \leq \left(\sum_{j=1} |a_{ij}||x_j|^{1/(1-\alpha)} \right)^{1-\alpha}$$

and hence

$$\left[\frac{|\lambda - a_{ii}|}{R_i'^\alpha} \right]^{1/(1-\alpha)} |x_i|^{1/(1-\alpha)} \leq \sum_{j \neq i}^n |a_{ij}||x_j|^{1/(1-\alpha)} \quad (6.4.5)$$

Hölder's inequality (Appendix B) is employed in (6.4.4) with $p = 1/\alpha$ and $q = p/(p-1) = 1/(1-\alpha)$. Now sum (6.4.5) on i to get

$$\sum_{i=1}^n \left(\frac{|\lambda - a_{ii}|}{R_i'^\alpha} \right)^{1/(1-\alpha)} |x_i|^{1/(1-\alpha)} \leq \sum_{i=1}^n \sum_{j \neq i} |a_{ij}||x_j|^{1/(1-\alpha)}$$

$$= \sum_{j=1}^n C'_j |x_j|^{1/(1-\alpha)} \quad (6.4.6)$$

If

$$\left(\frac{|\lambda - a_{ii}|}{R_i'^\alpha} \right)^{1/(1-\alpha)} > C'_i$$

for every i such that $x_i \neq 0$, then (6.4.6) could not be correct. Therefore, there is some $k \in \{1, \ldots, n\}$ such that $x_k \neq 0$ and

$$\left(\frac{|\lambda - a_{kk}|}{R_k'^\alpha} \right)^{1/(1-\alpha)} \leq C'_k$$

It follows that $|\lambda - a_{kk}| \leq R_k'^\alpha C_k'^{1-\alpha}$, so λ is in the set (6.4.3). $\qquad \square$

Exercise. Consider $A = \begin{bmatrix} 1 & 4 \\ 1 & 6 \end{bmatrix}$ and compare its Geršgorin row and column eigenvalue inclusion sets with Ostrowski's set for $\alpha = \frac{1}{2}$. What estimate does Ostrowski's theorem give for the spectral radius of A and how does it compare with the Geršgorin estimates (6.1.5)?

Exercise. What is the Ostrowski version of (6.1.6)?

In the following theorem, due to A. Brauer, the familiar elements of Geršgorin's theorem are also present, but now the rows are taken *two* at a time; the eigenvalue inclusion sets are no longer discs but sets known as *ovals of Cassini*. The proof parallels the proof of Geršgorin's theorem, but selects the *two* largest modulus entries of an eigenvector. Brauer's eigenvalue inclusion set is a subset of Geršgorin's.

Theorem 6.4.7 (Brauer). *Let $A = [a_{ij}] \in M_n$ and assume that $n \geq 2$. The eigenvalues of A are in the union of $n(n-1)/2$ ovals of Cassini*

$$\bigcup_{i \neq j} \{z \in \mathbf{C} : |z - a_{ii}||z - a_{jj}| \leq R'_i R'_j\} \tag{6.4.8}$$

which is contained in the Geršgorin set (6.1.2).

Proof. Let λ be an eigenvalue of A, and suppose that $Ax = \lambda x$ with $x = [x_i] \neq 0$. There is an entry of x that has largest absolute value, say x_p, so $|x_p| \geq |x_i|$ for all $i = 1, \ldots, n$ and $x_p \neq 0$. If all the other entries of x are zero, then the assumption that $Ax = \lambda x$ means that $a_{pp} = \lambda$, which is in the set (6.4.8).

Now suppose that x has at least two nonzero entries, and let x_q be an entry with second largest absolute value; that is, $x_p \neq 0 \neq x_q$ and $|x_p| \geq |x_q| \geq |x_i|$ for all $i \in \{1, \ldots, n\}$, $i \neq p$. It follows from the identity $Ax = \lambda x$ that $x_p(\lambda - a_{pp}) = \sum_{j \neq p} a_{pj} x_j$, and hence

$$|x_p| |\lambda - a_{pp}| = \left| \sum_{j \neq p} a_{pj} x_j \right| \leq \sum_{j \neq p} |a_{pj}||x_j| \leq \sum_{j \neq p} |a_{pj}||x_q| = R'_p |x_q|$$

or

$$|\lambda - a_{pp}| \leq R'_p \frac{|x_q|}{|x_p|} \tag{6.4.9}$$

But we also have $x_q(\lambda - a_{qq}) = \sum_{j \neq q} a_{qj} x_j$, which implies that

$$|x_q| |\lambda - a_{qq}| = \left| \sum_{j \neq q} a_{qj} x_j \right| \leq \sum_{j \neq q} |a_{qj}| |x_j| \leq \sum_{j \neq q} |a_{qj}| |x_p| = R'_q |x_p|$$

or

$$|\lambda - a_{qq}| \leq R'_q \frac{|x_p|}{|x_q|} \tag{6.4.10}$$

Taking the product of (6.4.9) and (6.4.10) permits us to eliminate the unknown ratios of entries of x and obtain the inequality

$$|\lambda - a_{pp}||\lambda - a_{qq}| \leq R'_p \frac{|x_q|}{|x_p|} R'_q \frac{|x_p|}{|x_q|} = R'_p R'_q$$

Thus, the eigenvalue λ lies in the set (6.4.8).

Let $C_{ij} = \{z \in \mathbf{C} : |z - a_{ii}||z - a_{jj}| \leq R'_i R'_j\}$ be the Cassini oval associated with rows $i, j \in \{1, \ldots, n\}$, $i \neq j$. We claim that $C_{ij} \subset G_i \cup G_j$, in which $G_i = \{z \in \mathbf{C} : |z - a_{ii}| \leq R'_i(A)\}$ is the Geršgorin disc associated with row i of A. If $R_i R_j = 0$, our claim is certainly correct, so we may suppose that $R_i R_j > 0$. In this case,

$$C_{ij} = \{z \in \mathbf{C} : \frac{|z - a_{ii}|}{R'_i} \frac{|z - a_{jj}|}{R'_j} \leq 1\}$$

If z is a point that satisfies the inequality $\frac{|z-a_{ii}|}{R'_i} \frac{|z-a_{jj}|}{R'_j} \leq 1$, then not both of the ratios in this product can be greater than one. This means that either $z \in G_i$ or $z \in G_j$, which verifies our claim. It follows that $\cup_{i \neq j} C_{ij} \subset \cup_i G_i$. $\qquad\square$

Exercise. What is the column sum version of Brauer's theorem?

Any theorem about eigenvalue inclusion sets implies (and, indeed, is implied by) a related theorem about nonsingularity: We can use Ostrowski's and Brauer's theorems to formulate conditions that prohibit $z = 0$ from being in the respective eigenvalue inclusion sets.

Corollary 6.4.11. *Let $A = [a_{ij}] \in M_n$ with $n \geq 2$. Each of the following conditions implies that A is nonsingular:*

(a) (Ostrowski) For some $\alpha \in [0, 1]$, $|a_{ii}| > R'^\alpha_i C'^{1-\alpha}_i$ for all $i = 1, \ldots, n$.
(b) (Brauer) $|a_{ii}||a_{jj}| > R'_i R'_j$ for all distinct $i, j = 1, \ldots, n$.

Exercise. Use (6.4.1) and (6.4.7) to prove (6.4.11).

Brauer's eigenvalue inclusion set (6.4.8) is smaller than Geršgorin's (6.1.2), so it is perhaps not surprising that it fails to have the boundary property described in (6.2.8).

Exercise. Verify that the matrix

$$\begin{bmatrix} 1 & 1 & 1 \\ 2 & 4 & 0 \\ 1 & 0 & 2 \end{bmatrix} \tag{6.4.11a}$$

is irreducible and its eigenvalue $\lambda = 0$ is on the boundary of the Brauer set (6.4.8). Explain why the Cassini ovals $|z - 1||z - 2| = 2$ and $|z - 1||z - 4| = 4$ pass through λ, but $|z - 4||z - 2| = 2$ does not.

Brauer's theorem involves products of deleted row sums taken two at a time. An attractive possibility to obtain additional eigenvalue inclusion sets is to take deleted row sums of m rows of $A \in M_n$ and consider a union of sets of the form

$$\bigcup_{i_1, \ldots, i_m \in \mathcal{I}_m} \left\{ z \in \mathbf{C} : \prod_{k=1}^m |z - a_{i_k i_k}| \leq \prod_{k=1}^m R'_{i_k} \right\} \tag{6.4.12}$$

in which $\mathcal{I}_m = \{i_1, \ldots, i_m \in \{1, \ldots, n\} : i_1, \ldots, i_m$ are distinct$\}$. For each m, there are $\binom{n}{m}$ sets of this form: $m = 1$ gives the n Geršgorin discs and $m = 2$ gives Brauer's $n(n - 1)/2$ ovals of Cassini. However, for $m \geq 3$ the sets (6.4.12) need not be eigenvalue inclusion sets for A, as shown by the example

$$A = J_2 \oplus I_2 \qquad (6.4.13)$$

in which $J_2 = \begin{bmatrix} 1 & 1 \\ 1 & 1 \end{bmatrix} \in M_2$. The sets (6.4.12) for this matrix all collapse to the point $z = 1$ if $m = 3$ or $m = 4$.

Exercise. Show that the eigenvalues of the matrix in (6.4.13) are $\lambda = 0, 1, 1$, and 2. Sketch the sets (6.4.12) for $m = 1$, $m = 2$, and $m = 3, 4$. Show that the same phenomenon occurs for all $m \geq 3$ by considering

$$A = J_2 \oplus I_n \in M_{n+2} \qquad (6.4.14)$$

One problem with the set (6.4.12) is that it admits products that contain deleted row sums that are zero. Of course, this does not happen if the matrix A is irreducible; all $R_i' > 0$ in this case. However, even if A is irreducible, the set (6.4.12) still might not be an eigenvalue inclusion set for A. Consider the perturbation of (6.4.13) given by

$$A_\epsilon = \begin{bmatrix} 1 & 1 & \epsilon & \epsilon \\ 1 & 1 & 0 & 0 \\ \epsilon & 0 & 1 & 0 \\ \epsilon & 0 & 0 & 1 \end{bmatrix}, \quad 1 > \epsilon \geq 0 \qquad (6.4.15)$$

The directed graph $\Gamma(A_\epsilon)$ of A_ϵ is

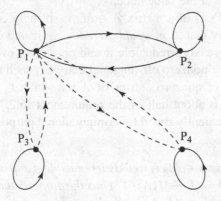

in which the dashed arcs disappear when $\epsilon = 0$.

Exercise. Verify that if $\epsilon \neq 0$, then $\Gamma(A_\epsilon)$ is strongly connected, A_ϵ is irreducible, $R_1' = 1 + 2\epsilon$, $R_2' = 1$, $R_3' = \epsilon$, $R_4' = \epsilon$, and the eigenvalues of A_ϵ are $\lambda_\epsilon = 1, 1$, $1 + (1 + 2\epsilon^2)^{1/2}$, and $1 - (1 + 2\epsilon^2)^{1/2}$.

Since any product of three or more deleted row sums for the matrix (6.4.15) contains at least one factor of ϵ, the sets (6.4.12) cannot be eigenvalue inclusion sets for either $m = 3$ or $m = 4$ if ϵ is positive and sufficiently small.

What property of the matrices (6.4.13) and (6.4.15) allows $m = 1$ and $m = 2$ to be acceptable in (6.4.12) for providing an eigenvalue inclusion set, but not $m = 3$ and

$m = 4$? Observe that the directed graphs in each case contain cycles of length 1 and 2, but not of length 3 or 4.

Motivated by this observation, consider the irreducible 3-by-3 matrix

$$B = \begin{bmatrix} -2 & \frac{1}{6} & -\frac{1}{8} \\ 0 & 1 & -\frac{1}{4} \\ 24 & 0 & 1 \end{bmatrix} \qquad (6.4.15a)$$

which has $\lambda = 0$ as a triple eigenvalue. The deleted row sums of B are $R_1' = 7/24$, $R_2' = 1/4$, and $R_3' = 24$. Brauer's set (6.4.12) is

$$\left\{ z \in \mathbf{C} : |z + 2|\,|z - 1|^2 \le \frac{7}{4} \right\}$$

which does not include λ. The directed graph $\Gamma(B)$ contains cycles of length 1, 2, and 3. Brauer's theorem ensures that the union of the three sets (6.4.12) for $m = 2$ contains λ. However, observe that just *one* of these sets,

$$\{ z \in \mathbf{C} : |z + 2|\,|z - 1| \le 7 \}$$

contains all the eigenvalues of B. This one set corresponds to the nodes of $\Gamma(B)$ that are in the only cycle of length two. This not an accident.

A directed graph Γ is *strongly connected* if from each of its nodes there is a directed path in Γ to *any other* node. We say that Γ is *weakly connected* if from each of its nodes there is a directed path to *some other* node and back, that is, *each node in Γ belongs to some nontrivial cycle*. A *trivial cycle* (or *loop*) is a directed path of length one that begins and ends at the same node.

The directed graph $\Gamma(A)$ of a matrix $A \in M_n$ is strongly connected if and only if A is irreducible. We say that A is *weakly irreducible* if and only if $\Gamma(A)$ is weakly connected. Thus, A is weakly irreducible if and only if for each $i = 1, \ldots, n$ the ith row of A has at least one nonzero off-diagonal entry $a_{i j_i}$ such that there is a sequence $a_{k_1 k_2}, a_{k_2 k_3}, \ldots, a_{k_{m-1} k_m}$ of nonzero entries of A for which $k_1 = j_i$ and $k_m = i$. This cumbersome condition is about half of the requirement (6.2.7) that A have property SC, and it is more conveniently stated for computational purposes in a form analogous to (6.2.23).

Lemma 6.4.16. *A matrix $A \in M_n$ is weakly irreducible if and only if $B = [b_{ij}] = (I + |A|)^{n-1}$ (equivalently, $B = (I + M(A))^{n-1}$) has the property that for each $i = 1, \ldots, n$ there is at some $j \ne i$ such that $b_{ij} b_{ji} \ne 0$, that is, for each $i = 1, \ldots, n$ there is least one nonzero off-diagonal entry b_{ij} such that b_{ji} is nonzero.*

Exercise. Prove Lemma (6.4.16). *Hint*: Use the ideas in (6.2.19).

Exercise. Suppose that $A \in M_n$. Show that A is weakly irreducible if and only if $\Gamma((I + |A|)^{n-1})$ has the property that each of its nodes belongs to a cycle of length 2. What is the corresponding property for irreducible? Which property is weaker? Recall that a cycle is *simple* by definition; only the initial (which is the same as the final) node can appear in the list of nodes more than once.

Exercise. If $A \in M_n$ is weakly irreducible, explain why all $R_i' > 0$ and all $C_i' > 0$.

A *preorder* on a set S is a relation R defined between *all* pairs of points of S such that for any pair of elements $s, t \in S$, either $s R t$ or $t R s$ or both. A preorder must also be reflexive ($s R s$ for every $s \in S$) and transitive (if $s R t$ and $t R u$, then $s R u$). A preorder might not be symmetric ($s R t$ if and only if $t R s$), and it is possible to have $s R t$ and $t R s$ without having $s = t$. A point z in a subset S_0 of S is said to be *a maximal element* of S_0 if $s R z$ for all $s \in S_0$.

Exercise. Let a nonempty set $S \subset \mathbf{C}$ be given. Show that the relation between pairs $z, w \in S$ defined by

$$z R w \quad \text{if and only if } |z| \leq |w|$$

is a preorder on $\mathbf{C}$.

Lemma 6.4.17. *Let S be a nonempty finite set on which a preorder R is defined. Then S contains at least one maximal element.*

Proof. Arrange the elements in any order $s_1, \ldots, s_k$. Set $s = s_1$. If $s_2 R s$, leave s alone; if not, then set $s = s_2$. If $s_3 R s$, leave s alone; if not, then set $s = s_3$. Continue this process with $s_4, \ldots, s_k$. The final value of s is a maximal element. $\square$

If Γ is a directed graph and if P_i is a node of Γ, we define $\Gamma_{\text{out}}(P_i)$ to be the set of nodes different from P_i that can be reached from P_i via some directed path of length 1. Notice that if Γ is weakly connected, then $\Gamma_{\text{out}}(P_i)$ is nonempty for every node $P_i \in \Gamma$.

Let us denote by $C(A)$ the set of nontrivial cycles in the directed graph $\Gamma(A)$. For the matrix (6.4.13), $C(A)$ consists of the single cycle $\gamma = P_1 P_2, P_2 P_1$; for the matrix (6.4.15) there are three nontrivial cycles, all of length 2; for the matrix (6.4.15a) there are two nontrivial cycles, one of length 2 and one of length 3.

Theorem 6.4.18 (Brualdi). *Let $A = [a_{ij}] \in M_n$ and suppose that $n \geq 2$. If A is weakly irreducible, then every eigenvalue of A is contained in the set*

$$\bigcup_{\gamma \in C(A)} \left\{ z \in \mathbf{C} : \prod_{P_i \in \gamma} |z - a_{ii}| \leq \prod_{P_i \in \gamma} R_i' \right\} \qquad (6.4.19)$$

The notation means that if $\gamma = P_{i_1} P_{i_2}, \ldots, P_{i_k} P_{i_{k+1}}$ is a nontrivial cycle with $P_{i_{k+1}} = P_{i_1}$, then the corresponding set in the union (6.4.19) is defined by a product with exactly k factors; the index i takes on the k values $i_1, \ldots, i_k$.

Proof. Weak irreducibility of A ensures that each of its deleted row sums is positive, so if λ is an eigenvalue of A and $\lambda = a_{ii}$ for some $i = 1, \ldots, n$, then λ is in the interior of the set (6.4.19).

For the rest of the argument, we suppose that λ is an eigenvalue of A and $\lambda \neq a_{ii}$ for all $i = 1, \ldots, n$. Let $Ax = \lambda x$ for some nonzero $x = [x_i] \in \mathbf{C}^n$. Define a preorder R on the nodes of Γ by

$$P_i R P_j \quad \text{if and only if } |x_i| \leq |x_j| \qquad (6.4.20)$$

We claim that there exists a cycle γ' in $\Gamma(A)$ with the following three properties:

(a) $\gamma' = P_{i_1}P_{i_2}, P_{i_2}P_{i_3}, \ldots, P_{ik}P_{i_{k+1}}$ is a nontrivial cycle with $k \geq 2$ and $P_{i_{k+1}} = P_{i_1}$.

(b) For each $j = 1, \ldots, k$, the node $P_{i_{j+1}}$ is a maximal node in $\Gamma_{\text{out}}(P_{ij})$; that is, $|x_{i_{j+1}}| \geq |x_m|$ for all m such that $P_m \in \Gamma_{\text{out}}(P_{ij})$.

(c) All $x_{i_j} \neq 0$, $j = 1, \ldots, k$.

$$(6.4.21)$$

If γ' is a cycle that satisfies the conditions (6.4.21), then the identity $Ax = \lambda x$ implies that for every $j = 1, \ldots, k$ we have

$$(\lambda - a_{i_j i_j})x_{i_j} = \sum_{m \neq i_j}^{n} a_{i_j m} x_m = \sum_{P_m \in \Gamma_{\text{out}}(P_{i_j})} a_{i_j m} x_m$$

and hence

$$|\lambda - a_{i_j i_j}||x_{i_j}| = \left| \sum_{P_m \in \Gamma_{\text{out}}(P_{i_j})} a_{i_j m} x_m \right| \leq \sum_{P_m \in \Gamma_{\text{out}}(P_{i_j})} |a_{i_j m}||x_m| \qquad (6.4.22)$$

$$\leq \sum_{P_m \in \Gamma_{\text{out}}(P_{i_j})} |a_{i_j m}||x_{i_{j+1}}| \qquad (6.4.22a)$$

$$= R'_{i_j}|x_{i_{j+1}}|$$

Now take the product of the inequalities (6.4.22) over all the nodes in γ' to obtain

$$\prod_{j=1}^{k} |\lambda - a_{i_j i_j}||x_{i_j}| \leq \prod_{j=1}^{k} R'_{i_j}|x_{i_{j+1}}| \qquad (6.4.23)$$

But

$$\prod_{j=1}^{k} |\lambda - a_{i_j i_j}| = \prod_{P_i \in \gamma'} |\lambda - a_{ii}| \quad \text{and} \quad \prod_{j=1}^{k} R'_{i_j} = \prod_{P_i \in \gamma'} R'_i$$

and since $P_{i_{k+1}} = P_{i_1}$, we also have $x_{i_{k+1}} = x_{i_1}$. Therefore,

$$\prod_{j=1}^{k} |x_{i_j}| = \prod_{j=1}^{k} |x_{i_{j+1}}| \neq 0 \qquad (6.4.24)$$

Thus, dividing (6.4.23) by (6.4.24), we obtain

$$\prod_{P_i \in \gamma'} |\lambda - a_{ii}| \leq \prod_{P_i \in \gamma'} R'_i \qquad (6.4.25)$$

Since γ' is a nontrivial cycle in $\Gamma(A)$, the eigenvalue λ is in the set (6.4.19).

We must now show that there is a cycle γ' that satisfies the conditions (6.4.21). Let i be any index for which $x_i \neq 0$, and note that $\Gamma_{\text{out}}(P_i)$ is nonempty because $\Gamma(A)$ is weakly connected. Since $x_i \neq 0$ and $\lambda - a_{ii} \neq 0$, the identity

$$0 \neq (\lambda - a_{ii})x_i = \sum_{j \neq i} a_{ij}x_j = \sum_{P_j \in \Gamma_{\text{out}}(P_i)} a_{ij}x_j$$

ensures that there is at least one node of $\Gamma_{\text{out}}(P_i)$ (call it P_j), for which the corresponding eigenvector entry x_j is nonzero. Let $P_{i_1} = P_i$, and let P_{i_2} be a maximal node among the nodes in $\Gamma_{\text{out}}(P_{i_1})$, that is, $|x_{i_2}| \geq |x_m|$ for all m such that $P_m \in \Gamma_{\text{out}}(P_{i_1})$. In particular, $|x_{i_2}| \geq |x_j| > 0$.

Suppose that the preceding construction has produced a directed path $P_{i_1}P_{i_2}$, $P_{i_2}P_{i_3}$, $\ldots$, $P_{i_{j-1}}P_{i_j}$ of length $j - 1$ that satisfies conditions (b) and (c) of (6.4.21); we have just done this for $j = 2$. Then the identity

$$0 \neq (\lambda - a_{i_j i_j})x_{i_j} = \sum_{P_m \in \Gamma_{\text{out}}(P_{i_j})} a_{i_j m}x_m$$

ensures that there is at least one node in $\Gamma_{\text{out}}(P_{i_j})$ for which the corresponding eigenvector entry is nonzero. Choose $P_{i_{j+1}}$ to be a maximal node in $\Gamma_{\text{out}}(P_{i_j})$, which ensures that $x_{i_{j+1}} \neq 0$.

There are only finitely many nodes in $\Gamma(A)$, so this construction for $j = 2, 3, \ldots$ eventually produces a first maximal node $P_{i_q} \in \Gamma_{\text{out}}(P_{i_{q-1}})$ that was produced as a node P_{i_p} at some previous step ($2 \leq p + 1 < q$). Then $\gamma' = P_{i_p}P_{i_{p+1}}$, $P_{i_{p+1}}P_{i_{p+2}}$, $\ldots$, $P_{i_{q-1}}P_{i_q}$ is a cycle in $\Gamma(A)$ that satisfies all three conditions in (6.4.21). $\square$

Brualdi's theorem has a sharper form when A is irreducible; it is the generalized Brauer (6.4.7) version of (6.2.26).

Theorem 6.4.26 (Brualdi). *Let $A = [a_{ij}] \in M_n$ be irreducible and suppose that $n \geq 2$. A boundary point λ of the set (6.4.19) can be an eigenvalue of A only if the boundary of each set*

$$\left\{ z \in \mathbf{C} \colon \prod_{P_i \in \gamma} |z - a_{ii}| \leq \prod_{P_i \in \gamma} R_i' \right\} \tag{6.4.27}$$

passes through λ for every nontrivial cycle $\gamma \in C(A)$.

Proof. Since all $R_i' > 0$, if $\lambda = a_{ii}$ for some $i \in \{1, \ldots, n\}$, then λ is not on the boundary of the set (6.4.27). Thus, we may assume that $\lambda \neq a_{ii}$ for all $i = 1, \ldots, n$ and we may continue the argument used in (6.4.18) with the same notation, but with the additional assumption that λ is an eigenvalue of A that lies on the boundary of the set (6.4.19). Just as in the proof of (6.2.3), λ must satisfy the inequality

$$\prod_{P_i \in \gamma} |\lambda - a_{ii}| \geq \prod_{P_i \in \gamma} R_i'$$

for all nontrivial cycles $\gamma \in C(A)$, with equality for at least one $\gamma \in C(A)$. Comparing this inequality with (6.4.25), we see that

$$\prod_{P_i \in \gamma'} |\lambda - a_{ii}| = \prod_{P_i \in \gamma'} R_i' \tag{6.4.28}$$

for the particular cycle γ' constructed in the proof of (6.4.18). Thus, the inequality in (6.4.23) is an equality, as are both of the inequalities in (6.4.22), for all $j \in \{1, , \ldots, k\}$. In particular, the inequality (6.4.22a) is an equality, and hence for each $P_{i_j} \in \gamma'$ and for all m such that $P_m \in \Gamma_{out}(P_{i_j})$, $|x_m| = |x_{i_{j-1}}| = c_{i_{j+1}} = $ constant. This conclusion follows for any cycle that satisfies the conditions (6.4.21).

Now define the set

$$K = \{P_i \in \Gamma(A): |x_m| = c_i = \text{ constant for all } m \text{ such that } P_m \in \Gamma_{out}(P_i)\}$$

We know that K is not empty because all the nodes of γ' are in K. We would like to show that every node of $\Gamma(A)$ is in K.

Suppose that a node P_q of $\Gamma(A)$ is not in K. Because $\Gamma(A)$ is strongly connected, there is at least one directed path in $\Gamma(A)$ from each node of K to this external node P_q. If we select from all such directed paths a path with shortest length, then its first arc must be from a node in K to a node P_f that is not in K. If we use the same preorder on the nodes of $\Gamma(A)$ that we used in the proof of (6.4.18), then we may employ the same construction that we used there: start with the node $P_f = P_{j_1}$, select a maximal node $P_{j_2} \in \Gamma_{out}(P_{j_1})$, select a maximal node $P_{j_3} \in \Gamma_{out}(P_{j_2})$, and so on. At each step, $\Gamma_{out}(P_{j_i})$ is nonempty because $\Gamma(A)$ is weakly (even strongly) connected, and the maximal node satisfies condition (c) of (6.4.21) for the same reason as before.

If, at some step of this construction, we have a choice between selecting a maximal node that is in K or not in K, we choose one that is *not* in K. If, at any step, all the maximal nodes from which we may choose are in K, choose any one of them and then follow a directed path of shortest length (necessarily in K) to a first node that is not in K and resume selecting maximal nodes as before. The definition of K ensures that *any* directed path in K has the property that each of its nodes is a maximal node in Γ_{out} of its predecessor node; this is condition (b) of (6.4.21). Because the complement of K has only finitely many nodes, this construction ultimately produces a first maximal node in the complement of K that was produced as a node at some previous step. The directed path between the first and second occurrences of this node in the construction is a nontrivial directed cycle, which may not be simple because of the way we have forced the path to leave K whenever the construction leads to a node in K. There may be finitely many cycles in the part of the path that lies within K, but they can be pruned off to leave a simple directed cycle γ'', which satisfies the conditions (6.4.21) and contains at least one node that is not in K.

Since the cycle γ'' satisfies the conditions (6.4.21), it can be used in place of the cycle γ' in the proof of (6.4.18). By the argument in the first paragraph of the present proof, we conclude that $|x_m| = c_{j_r} = $ constant for all $P_m \in \Gamma_{out}(P_{j_r})$ for all $P_{j_r} \in \gamma''$. Therefore, every node in γ'' is in K, a contradiction to the conclusion that γ'' contains at least one node that is not in K. This shows that every node of $\Gamma(A)$ is in K.

If γ is *any* nontrivial cycle in $\Gamma(A)$, it automatically satisfies the conditions (6.4.21) because all its nodes are in K. It may therefore be used in place of γ' in the proof of (6.4.18), and hence it may be used in place of γ' in (6.4.28). This is the desired conclusion: The boundary of every set (6.4.27) passes through λ. $\qquad \square$

Corollary 6.4.29. *If $A \in M_n$ and $n \geq 2$, then each of the following conditions ensures that A is nonsingular:*

(a) A is weakly irreducible and

$$\prod_{P_i \in \gamma} |a_{ii}| > \prod_{P_i \in \gamma} R_i'$$

for every nontrivial cycle $\gamma \in C(A)$.

(b) A is irreducible and

$$\prod_{P_i \in \gamma} |a_{ii}| \geq \prod_{P_i \in \gamma} R_i'$$

for every nontrivial cycle $\gamma \in C(A)$, with strict inequality for at least one cycle.

As our final result, we state a strengthened form of Brauer's theorem that locates the eigenvalues of A in a union of possibly fewer ovals of Cassini than are required in (6.4.7). It can dramatically reduce the number of ovals to be considered if A is a sparse (but irreducible) matrix that has many symmetrically located zero entries.

Theorem 6.4.30 (Kolotilina). *Let $n \geq 2$ and let $A = [a_{ij}] \in M_n$ be irreducible. Every eigenvalue of A is contained in the set*

$$\bigcup_{\substack{i \neq j \\ |a_{ij}|+|a_{ji}| \neq 0}} \{z \in \mathbb{C} : |z - a_{ii}||z - a_{jj}| \leq R_i' R_j'\} \tag{6.4.31}$$

The notation means that an oval corresponding to distinct rows i and j appears in the union only if either a_{ij} or a_{ji} is nonzero.

> *Exercise.* Apply the preceding theorem to the matrix (6.4.11a). Which of the three Cassini ovals in (6.4.7) does it permit us to omit? Does the omitted oval contain any eigenvalues of A?

Problems

6.4.P1 Show that if $n \geq 2$ and $A = [a_{ij}]$ satisfies Brauer's condition (6.4.11b) for non-singularity, then $|a_{ii}| > R_i'$ for all but at most one value of $i = 1, \ldots, n$. Thus, Brauer's condition is only slightly weaker than the Levy–Desplanques condition (strict diagonal dominance) in (6.1.10a). How is this related to (6.1.11)?

6.4.P2 Consider $A = \begin{bmatrix} 2 & 3 \\ 1 & 3 \end{bmatrix}$. Show that both conditions (6.4.11) ensure that A is non-singular, but neither (6.1.10a) nor (6.1.11) guarantees nonsingularity. What about the column form of (6.1.11)?

6.4.P3 Show that every irreducible $A \in M_n$ with $n \geq 2$ is weakly irreducible. Give an example of a weakly irreducible matrix that is not irreducible.

6.4.P4 Use the arguments in (6.1.10) and (6.2.6) to provide details for a proof of (6.4.29).

6.4.P5 Show that $A \in M_n$ is weakly irreducible if and only if A is *not* permutation similar to a block triangular matrix, one of whose diagonal blocks is 1-by-1.

6.4.P6 Consider the matrix

$$A = \begin{bmatrix} -2 & 4 & -3 \\ 0 & 1 & -\frac{1}{4} \\ 1 & 0 & 1 \end{bmatrix}$$

(a) Show that $\lambda = 0$ is a triple eigenvalue of A. (b) Show that the set (6.4.12) for $m = 3$ is $\{z \in \mathbf{C} : |z + 2| \, |z - 1|^2 \leq \frac{7}{4}\}$, which does not contain λ. (c) Show that the set (6.4.12) for A^T and $m = 3$ *does* contain λ. (d) Determine the set (6.4.19) for A and show that it contains λ.

6.4.P7 Let $A = [a_{ij}] \in M_n$ and suppose that $a_{ii} = 0$ for each $i = 1, \ldots, n$. Order the set of deleted absolute row sums of A as $R'_{[1]} \geq \cdots \geq R'_{[n]}$. Show that $\rho(A) \leq (R'_{[1]} R'_{[2]})^{1/2}$.

6.4.P8 Although the Brauer set (6.4.8) does not have a boundary eigenvalue property like the one described in (6.2.8), a subset of the Brauer set does have such a property: It is known that if $A = [a_{ij}] \in M_n$ is irreducible and if λ is an eigenvalue of A that is a boundary point of the set

$$\bigcup_{\gamma \in C(A)} \bigcup_{\substack{P_i, P_j \in \gamma \\ P_i \neq P_j}} \{z \in \mathbf{C} : |z - a_{ii}| \, |z - a_{jj}| \leq R'_i R'_j\}$$

then λ is in the set $\{z \in \mathbf{C} : |z - a_{ii}| \, |z - a_{jj}| = R'_i R'_j\}$ for each $\gamma \in C(A)$ and each distinct pair of nodes $P_i, P_j \in \gamma$. (a) Explain why this theorem does not exclude the possibility of $\lambda = 0$ being an eigenvalue of the matrix (6.4.11a). (b) Deduce the following criterion for nonsingularity of A: $|a_{ii}| \, |a_{jj}| \geq R'_i R'_j$ for every pair of distinct nodes $P_i, P_j \in \gamma$ for every $\gamma \in C(A)$; and for at least one $\gamma_0 \in C(A)$, $|a_{ii}| \, |a_{jj}| > R'_i R'_j$ for every pair of distinct nodes $P_i, P_j \in \gamma_0$.

Notes and Further Readings. For more details about eigenvalue inclusion sets and many references to the original literature, see R. Brualdi, Matrices, eigenvalues, and directed graphs, *Linear Multilinear Algebra* 11 (1982) 143–165. Theorem 6.4.7 appears in A. Ostrowski, Über die Determinanten mit überwiegender Hauptdiagonale, *Comment. Math. Helv.* 10 (1937) 69–96; it was independently rediscovered 10 years later and published in A. Brauer, Limits for the characteristic roots of a matrix: II, *Duke Math. J.* 14 (1947) 21–26. For this reason, (6.4.7) is sometimes referred to as the Ostrowski–Brauer theorem. For a proof of (6.4.30), see L. Yu. Kolotilina, Generalizations of the Ostrowski–Brauer theorem, *Linear Algebra Appl.* 364 (2003) 65–80. The theorem in (6.4.P6) is proved in X. Zhang and D. Gu, A note on A. Brauer's theorem, *Linear Algebra Appl.* 196 (1994) 163–174. For a singular value inclusion set that is analogous to (6.4.31), see L. Li, The undirected graph and estimates of matrix singular values, *Linear Algebra Appl.* 285 (1998) 181–188.

CHAPTER 7
Positive Definite and Semidefinite Matrices

7.0 Introduction

A class of Hermitian matrices with a special positivity property arises naturally in many applications. Hermitian (and, in particular, real symmetric) matrices with this positivity property also provide one generalization to matrices of the notion of a positive number. This observation often provides insight into the properties and applications of positive definite matrices. The following examples illustrate several ways in which these special Hermitian matrices arise.

7.0.1 Hessians, minimization, and convexity.
Let f be a smooth real-valued function on some domain $D \subset \mathbf{R}^n$. If $y = [y_i]$ is an interior point of D, then Taylor's theorem says that

$$f(x) = f(y) + \sum_{i=1}^{n} (x_i - y_i) \frac{\partial f}{\partial x_i}\bigg|_y$$

$$+ \frac{1}{2} \sum_{i,j=1}^{n} (x_i - y_i)(x_j - y_j) \frac{\partial^2 f}{\partial x_i \partial x_j}\bigg|_y + \cdots$$

for points $x \in D$ that are near y. If y is a *critical point* of f, then all the first-order partial derivatives vanish at y, and we have

$$f(x) - f(y) = \frac{1}{2} \sum_{i,j=1}^{n} (x_i - y_i)(x_j - y_j) \frac{\partial^2 f}{\partial x_i \, \partial x_j}\bigg|_y + \cdots$$

$$= \frac{1}{2}(x - y)^T H(f; y)(x - y) + \cdots$$

The real n-by-n matrix

$$H(f; y) = \left[\frac{\partial^2 f}{\partial x_i \, \partial x_j}\bigg|_y \right]_{i,j=1}^{n}$$

is the *Hessian* of f at y; equality of the mixed partial derivatives ($\partial^2 f/\partial x_i \partial x_j = \partial^2 f/\partial x_j \partial x_i$) ensures that it is symmetric. If the quadratic form

$$z^T H(f; y)z, \quad z \neq 0, \quad z \in \mathbf{R}^n \qquad (7.0.1.1)$$

is always positive, then y is a *relative minimum* for f. If this quadratic form is always negative, then y is a *relative maximum* for f. If $n = 1$, these criteria are just the usual second derivative test for a relative minimum or a maximum.

If the quadratic form (7.0.1.1) is nonnegative at all points of D (not just at the critical points of f), then f is a *convex function* in D. This is a direct generalization of the familiar situation when $n = 1$.

7.0.2 Covariance matrices. Let $X_1, \ldots, X_n$ be real or complex random variables with finite second moments on some probability space with expectation functional E, and suppose that $\mu_i = E(X_i)$ are the respective means. The *covariance matrix* of the random vector $X = (X_1, \ldots, X_n)^T$ is the matrix $A = [a_{ij}]$ in which

$$a_{ij} = E[(\bar{X}_i - \bar{\mu}_i)(X_j - \mu_j)], \qquad i, j = 1, \ldots, n$$

It is apparent that A is Hermitian. Moreover, for any $z = [z_i] \in \mathbf{C}^n$, we have

$$z^* A z = E\left(\sum_{i,j=1}^{n} \bar{z}_i(\bar{X}_i - \bar{\mu}_i)z_j(X_j - \mu_j)\right)$$

$$= E\left|\sum_{i=1}^{n} z_i(X_i - \mu_i)\right|^2 \geq 0$$

The only properties of the expectation functional that are involved in this observation are its linearity, homogeneity, and nonnegativity; that is, $E[Y] \geq 0$ whenever Y is a nonnegative random variable.

The same observation can be made without recourse to probabilistic language. If one has a family of complex valued functions $f_1, \ldots, f_n$ on the real line, if g is a real-valued function, and if all the integrals

$$a_{ij} = \int_{-\infty}^{\infty} \overline{f_i(x)} f_j(x)g(x)dx, \qquad i, j = 1, \ldots, n$$

are defined and converge, then the matrix $A = [a_{ij}]$ is Hermitian. Moreover,

$$z^* A z = \sum_{i,j=1}^{n} \int_{-\infty}^{\infty} \bar{z}_i \overline{f_i(x)} z_j f_j(x)g(x)\, dx$$

$$= \int_{-\infty}^{\infty} \left|\sum_{i=1}^{n} z_i f_i(x)\right|^2 g(x)\, dx$$

so this quadratic form is nonnegative if the function g is nonnegative.

7.0.3 Algebraic moments of nonnegative functions. Let f be an absolutely integrable real-valued function on the unit interval $[0, 1]$ and consider the numbers

$$a_k = \int_0^1 x^k f(x)\, dx \qquad (7.0.3.1)$$

The sequence $a_0, a_1, a_2, \ldots$ is a *Hausdorff moment sequence*, and it is naturally associated with the real quadratic form

$$\sum_{j,k=0}^n a_{j+k} z_j z_k = \sum_{j,k=0}^n \int_0^1 x^{j+k} z_j z_k f(x) dx = \int_0^1 \left(\sum_{k=0}^n z_k x^k \right)^2 f(x) dx \quad (7.0.3.2)$$

The matrix $A = [a_{i+j}]$ is real symmetric. If the function f is nonnegative, then $z^T A z \geq 0$ for all $z \in \mathbf{R}^{n+1}$ and each $n = 0, 1, 2, \ldots$. A matrix with the structure of A (the elements a_{ij} are a function only of $i + j$) is called a *Hankel matrix*, whether or not its quadratic form is nonnegative; see (0.9.8).

7.0.4 Trigonometric moments of nonnegative functions. Let f be an absolutely integrable real-valued function on $[0, 2\pi]$ and consider the numbers

$$a_k = \int_0^{2\pi} e^{ik\theta} f(\theta) d\theta, \qquad k = 0, \pm 1, \pm 2, \ldots \qquad (7.0.4.1)$$

The sequence $a_0, a_1, a_{-1}, a_2, a_{-2}, \ldots$ is a *Toeplitz moment sequence*, and it is naturally associated with the quadratic form

$$\sum_{j,k=0}^n a_{j-k} z_j \bar{z}_k = \sum_{j,k=0}^n \int_0^{2\pi} e^{i(j-k)\theta} z_j \bar{z}_k f(\theta)\, d\theta$$

$$= \int_0^{2\pi} \left| \sum_{k=0}^n z_k e^{ik\theta} \right|^2 f(\theta)\, d\theta \qquad (7.0.4.2)$$

The matrix $A = [a_{i-j}]$ is Hermitian. If the function f is nonnegative, then $z^* A z \geq 0$ for all $z \in \mathbf{C}^{n+1}$ and each $n = 0, 1, 2, \ldots$. A matrix that has the structure of A (the elements a_{ij} are a function only of $i - j$) is called a *Toeplitz matrix*, whether or not its quadratic form is nonnegative; see (0.9.7). *Bochner's theorem* says that nonnegativity of the quadratic form (7.0.4.2) is both necessary and sufficient for the numbers a_k to be generated by a slight modification of the formula (7.0.4.1) (a nonnegative measure $d\mu$ replaces $f(\theta)\, d\theta$).

7.0.5 Discretization and difference schemes for numerical solution of differential equations. Suppose that we have a two-point boundary value problem of the form

$$-y''(x) + \sigma(x)y(x) = f(x), \qquad 0 \leq x \leq 1$$
$$y(0) = \alpha$$
$$y(1) = \beta$$

in which α and β are given real constants and $f(x)$ and $\sigma(x)$ are given real-valued functions. If we discretize this problem and look only for the values of $y(kh) \equiv y_k$, $k = 0$,

$1, \ldots, n + 1$, and if we use a divided difference approximation to the derivative term

$$y''(kh) \cong \frac{y((k+1)h) - 2y(kh) + y((k-1)h)}{h^2} = \frac{y_{k+1} - 2y_k + y_{k-1}}{h^2}$$

we obtain a system of linear equations

$$\frac{-y_{k+1} + 2y_k - y_{k-1}}{h^2} + \sigma_k y_k = f_k, \quad k = 1, 2, \ldots, n$$
$$y_0 = \alpha$$
$$y_{n+1} = \beta$$

in which $h = 1/(n + 1)$, $y_k = y(kh)$, $\sigma_k = \sigma(kh)$, and $f_k = f(kh)$. The boundary conditions can be incorporated into the first ($k = 1$) and last ($k = n$) equations to give the linear system

$$(2 + h^2 \sigma_1)y_1 - y_2 = h^2 f_1 + \alpha$$

$$-y_{k-1} + (2 + h^2 \sigma_k)y_k - y_{k+1} = h^2 f_k, \qquad k = 2, 3, \ldots, n - 1$$

$$-y_{n-1} + (2 + h^2 \sigma_n)y_n = h^2 f_n + \beta$$

This system can be written as $Ay = w$, in which $y = [y_k] \in \mathbf{R}^n$, $w = [h^2 f_1 + \alpha, h^2 f_2, \ldots, h^2 f_{n-1}, h^2 f_n + \beta]^T \in \mathbf{R}^n$, and $A \in M_n$ is the tridiagonal matrix

$$A = \begin{bmatrix} 2 + h^2 \sigma_1 & -1 & & & \\ -1 & 2 + h^2 \sigma_2 & -1 & & \mathbf{0} \\ & \ddots & \ddots & \ddots & \\ & -1 & 2 + h^2 \sigma_{n-1} & -1 \\ \mathbf{0} & & -1 & 2 + h^2 \sigma_n \end{bmatrix} \quad (7.0.5.1)$$

The matrix A is real, symmetric, and tridiagonal regardless of the values of $\sigma(x)$, but if we want to be able to solve $Ay = w$ for any given right-hand side, then we must impose some condition on $\sigma(x)$ to ensure that A is nonsingular.

The real quadratic form associated with A is

$$x^T A x = \left(x_1^2 + \sum_{i=1}^{n-1} (x_i - x_{i+1})^2 + x_n^2 \right) + h^2 \sum_{i=1}^{n} \sigma_i x_i^2$$

The first group of three terms is nonnegative; it vanishes only if all the entries of x are equal, and equal to zero. If the function σ is nonnegative, then the last sum is nonnegative and

$$x^T A x \geq \left(x_1^2 + \sum_{i=1}^{n-1} (x_i - x_{i+1})^2 + x_n^2 \right) \geq 0 \qquad (7.0.5.2)$$

If A is singular, then there is a nonzero vector $\hat{x} \in \mathbf{R}^n$ such that $A\hat{x} = 0$, and hence $\hat{x}^T A \hat{x} = 0$. But then the central group of terms in (7.0.5.2) vanishes, which implies that $\hat{x} = 0$. Thus, if the function σ is nonnegative, the matrix A is nonsingular and the discretized boundary value problem can be solved for arbitrary boundary conditions α and β.

This is a typical situation in the study of numerical solutions of ordinary or partial differential equations. For computational stability it is desirable to design a discretization of a differential equation problem that leads to a system of linear equations $Ay = w$ in which A is positive definite, and it is usually possible to do so when the differential equations are elliptic.

Matrices with the special positivity property illustrated in these examples are the object of study in this chapter. These matrices arise in many applications: In harmonic analysis, in complex analysis, in the theory of vibrations of mechanical systems, and in other areas of matrix theory such as the singular value decomposition and the solution of linear least squares problems.

Problems

7.0.P1 If the sequence a_k is generated by the formula (7.0.3.1) with a nonnegative function f, show that the quadratic forms

$$\sum_{i,j=1}^{n} a_{i+j+1} z_i z_j \quad \text{and} \quad \sum_{i,j=1}^{n} (a_{i+j} - a_{i+j+1}) z_i z_j, \qquad z = [z_i] \in \mathbf{R}^n$$

are both nonnegative.

7.0.P2 Make a sketch illustrating which diagonals are constant in a Hankel matrix. Do the same for a Toeplitz matrix.

7.0.P3 Show that the matrix A in (7.0.5.1) is always irreducible, and that it is irreducibly diagonally dominant if the function σ is nonnegative. Use (6.2.27) to show that A is nonsingular and that all of its eigenvalues are positive.

Further Readings. For a short survey of facts about real positive definite matrices, see C. R. Johnson, Positive definite matrices, *Amer. Math. Monthly* 77 (1970) 259–264. Other surveys that focus on positive definite matrices and contain numerous references are O. Taussky, Positive definite matrices, pp. 309–319 of *Inequalities*, ed. O. Shisha, Academic Press, New York, 1967; and O. Taussky, Positive definite matrices and their role in the study of the characteristic roots of general matrices, *Adv. Math.* 2 (1968) 175–186. Bhatia (2007) is a book entirely devoted to positive definite matrices; several special topics are treated in depth.

7.1 Definitions and properties

A Hermitian matrix $A \in M_n$ is *positive definite* if

$$x^* A x > 0 \quad \text{for all nonzero } x \in \mathbf{C}^n \tag{7.1.1a}$$

it is *positive semidefinite* if

$$x^* A x \geq 0 \quad \text{for all nonzero } x \in \mathbf{C}^n \tag{7.1.1b}$$

Implicit in these definitions is the fact that if A is Hermitian, then $x^* A x$ is *real* for all $x \in \mathbf{C}^n$; see (4.1.3). Conversely, if $A \in M_n$ and $x^* A x$ is *real* for all $x \in \mathbf{C}^n$, then

A is Hermitian, so assuming that A is Hermitian in the preceding definitions, while customary, is actually superfluous; see (4.1.4). Of course, if A is positive definite, it is also positive semidefinite. In this section, we continue the discussion of these ideas that we began in (4.1).

Exercise. What are the positive definite and positive semidefinite matrices in M_1?

Exercise. Explain why $\begin{bmatrix} 1 & 1 \\ 1 & 1 \end{bmatrix}$ is positive semidefinite but not positive definite.

Exercise. If $A = [a_{ij}] \in M_n$ is positive definite, explain why $\bar{A} = [\bar{a}_{ij}]$, A^T, A^*, and A^{-1} are all positive definite. *Hint*: If $Ay = x$, $x^*A^{-1}x = y^*A^*y$.

The concepts *negative definite* and *negative semidefinite* may be defined for $A \in M_n$ by reversing the inequalities in (7.1.1a) and (7.1.1b) or, equivalently, by saying that $-A$ is positive definite or positive semidefinite, respectively. If A is Hermitian and x^*Ax has both positive and negative values, then A is said to be *indefinite*.

Observation 7.1.2. *Let $A \in M_n$ be Hermitian. If A is positive definite, then all of its principal submatrices are positive definite. If A is positive semidefinite, then all of its principal submatrices are positive semidefinite.*

Proof. Let α be a proper subset of $\{1, \ldots, n\}$ and consider the principal submatrix $A[\alpha]$; see (0.7.1). Let $x \in \mathbf{C}^n$ be a vector such that $x[\alpha] \neq 0$ and $x[\alpha^c] = 0$. Then $x \neq 0$ and $x[\alpha]^*A[\alpha]x[\alpha] = x^*Ax > 0$. Since the nonzero vector $x[\alpha]$ is arbitrary, we conclude that $A[\alpha]$ is positive definite. The second assertion follows in the same way. $\square$

Exercise. Explain why every main diagonal entry of a positive definite (respectively, positive semidefinite) matrix is a positive (respectively, nonnegative) real number.

Observation 7.1.3. *Let $A_1, \ldots, A_k \in M_n$ be positive semidefinite and let $\alpha_1, \ldots, \alpha_k$ be nonnegative real numbers. Then $\sum_{i=1}^{k} \alpha_i A_i$ is positive semidefinite. If there is a $j \in \{1, \ldots, k\}$ such that $\alpha_j > 0$ and A_j is positive definite, then $\sum_{i=1}^{k} \alpha_i A_i$ is positive definite.*

Proof. Let $x \in \mathbf{C}^n$ be nonzero and observe that $x^*(\sum_{i=1}^{k} \alpha_i A_i)x = \sum_{i=1}^{k} \alpha_i(x^*A_i x) \geq 0$ since each $\alpha_i \geq 0$ and each $x^*A_i x \geq 0$. The latter sum is positive if any summand is positive. $\square$

Observation 7.1.4. *Each eigenvalue of a positive definite (respectively, positive semidefinite) matrix is a positive (respectively, nonnegative) real number.*

Proof. Let λ, x be an eigenpair of a positive semidefinite matrix A and calculate $x^*Ax = x^*\lambda x = \lambda x^*x$. Then $\lambda = (x^*Ax)/x^*x \geq 0$ if A is positive semidefinite, and $\lambda > 0$ if A is positive definite. $\square$

Corollary 7.1.5. *Let $A \in M_n$ be positive semidefinite (respectively, positive definite). Then* tr A, det A, *and the principal minors of A are all nonnegative (respectively, positive). Moreover,* tr $A = 0$ *if and only if $A = 0$.*

Proof. The trace of A is the sum of its eigenvalues, which are all nonnegative (respectively, positive); if that sum is zero, then each eigenvalue is zero and hence the diagonalizable matrix A is zero; see (1.3.4). The determinant of A is the product of its eigenvalues, which are all nonnegative. Principal minors are determinants of principal submatrices, so they are products with nonnegative or positive factors; see (7.1.2). $\square$

Exercise. If $A \in M_n$ is negative definite, explain why: The eigenvalues and trace of A are negative; $\det A$ is negative for odd n and positive for even n.

Exercise. Consider the Hermitian matrix $A = \begin{bmatrix} 1 & 0 \\ 0 & -1 \end{bmatrix}$. Exhibit a nonzero vector such that $x^* A x = 0$ but $Ax \neq 0$.

The following result shows that the phenomenon illustrated in the preceding exercise cannot occur for a positive semidefinite matrix.

Observation 7.1.6. *Let $A \in M_n$ be positive semidefinite and let $x \in \mathbf{C}^n$. Then $x^* A x = 0$ if and only if $Ax = 0$.*

Proof. Suppose that $x \neq 0$ and $x^* A x = 0$. Consider the polynomial $p(t) = (tx + Ax)^* A(tx + Ax) = t^2 x^* A x + 2t x^* A^2 x + x^* A^3 x = 2t \|Ax\|_2^2 + x^* A^3 x$. The hypotheses ensure that $p(t) \geq 0$ for all real t. However, if $\|Ax\|_2 \neq 0$ then for sufficiently large negative values of t we would have $p(t) < 0$. We conclude that $\|Ax\|_2 = 0$, so $Ax = 0$. $\square$

Corollary 7.1.7. *A positive semidefinite matrix is positive definite if and only if it is nonsingular.*

Proof. Suppose that $A \in M_n$ is positive semidefinite. The preceding observation ensures that the following statements are equivalent: (a) A is singular; (b) there is a nonzero vector x such that $Ax = 0$; (c) there is a nonzero vector x such that $x^* A x = 0$; (d) A is not positive definite. $\square$

An important property of a positive semidefinite matrix is that it remains positive semidefinite after a *congruence.

Observation 7.1.8. *Let $A \in M_n$ be Hermitian and let $C \in M_{n,m}$.*

(a) *Suppose that A is positive semidefinite. Then $C^* A C$ is positive semidefinite, nullspace $C^* A C = $ nullspace AC, and rank $C^* A C = $ rank AC.*
(b) *Suppose that A is positive definite. Then rank $C^* A C = $ rank C, and $C^* A C$ is positive definite if and only if rank $C = m$.*

Proof. (a) Let $x \in \mathbf{C}^n$. Let $y = Cx$ and observe that $x^* C^* A C x = y^* A y \geq 0$. Thus, $C^* A C$ is positive semidefinite. The remaining assertions follow from (7.1.6): $C^* A C x = 0 \Leftrightarrow x^* C^* A C x = (Cx)^* A(Cx) = 0 \Leftrightarrow A(Cx) = A C x = 0$. Thus, the null spaces of $C^* A C$ and AC are the same, so they have the same rank.
(b) Since A is nonsingular, it follows from (a) that rank $C = $ rank $AC = $ rank $C^* A C$. The preceding corollary ensures that the positive semidefinite matrix $C^* A C \in M_m$ is positive definite if and only if it is nonsingular, which is the case if and only if $m = $ rank $C^* A C$, which is equal to rank C. $\square$

Exercise. If $A \in M_n$ and $C \in M_n$ with $m > n$, show by example that C^*AC may be positive definite even if $A \in M_n$ is positive semidefinite and not positive definite.

The following observation permits many results about positive definite matrices to be extended to positive semidefinite matrices with a continuity argument.

Observation 7.1.9. *Let $A \in M_n$ be Hermitian. Then A is positive semidefinite if and only if there is a sequence of positive definite matrices $A_1, A_2, \ldots$ such that $A_k \to A$ as $k \to \infty$.*

Proof. If A is positive semidefinite, let $A_k = A + k^{-1}I$, $k = 1, 2, \ldots$. Conversely, if $A_k \to A$ as $k \to \infty$ and each A_k is positive definite, then for any nonzero $x \in \mathbf{C}^n$ we have $x^*A_kx > 0$ for each $k = 1, 2, \ldots$, so $\lim_{k \to \infty} x^*A_kx = x^*Ax \geq 0$. $\square$

Positive definite and semidefinite matrices have two perhaps surprising special properties that have profound consequences. We encountered these properties – *row and column inclusion* – in (3.5.3) during our study of LU factorizations: Let $A \in M_n$ be partitioned as $A = \begin{bmatrix} A_{11} & A_{12} \\ A_{21} & A_{22} \end{bmatrix}$, in which $A_{11} \in M_k$. We say that A has the *column inclusion property* if range $A_{12} \subset$ range A_{11} for every $k \in \{1, \ldots, n-1\}$. We say that A has the *row inclusion property* if A^* has the column inclusion property.

Exercise. Partition $A \in M_n$ as $A = \begin{bmatrix} A_{11} & A_{12} \\ A_{21} & A_{22} \end{bmatrix}$, in which $A_{11} \in M_k$ and $k \in \{1, \ldots, n-1\}$. Explain why the following statements are equivalent:

(a) A has the column inclusion property.
(b) For each $k \in \{1, \ldots, n-1\}$, nullspace $A_{11}^* \subset$ nullspace A_{12}^*.
(c) For each $k \in \{1, \ldots, n-1\}$, every column of A_{12} is a linear combination of the columns of A_{11}.
(d) For each $k \in \{1, \ldots, n-1\}$, there is an $X \in M_{k,n-k}$ such that $A_{12} = A_{11}X$.
(e) For each $k \in \{1, \ldots, n-1\}$, rank$[A_{11}\ A_{12}] = $ rank A_{11}.

Exercise. What are the corresponding equivalent statements for the row inclusion property?

Exercise. Suppose that $A \in M_n$ is Hermitian. Explain why A has the column inclusion property if and only if it has the row inclusion property.

Observation 7.1.10. *Every positive semidefinite matrix has the row and column inclusion properties. In particular, if $A = [a_{ij}]$ is positive semidefinite and $a_{kk} = 0$ for some $k \in \{1, \ldots, n\}$, then $a_{ik} = a_{ki} = 0$ for each $i = 1, \ldots, n$.*

Proof. Let $A \in M_n$ be positive semidefinite and partition it as $A = \begin{bmatrix} A_{11} & A_{12} \\ A_{12}^* & A_{22} \end{bmatrix}$, in which $A_{11} \in M_k$ is Hermitian and $k \in \{1, \ldots, n-1\}$. It suffices to show that nullspace $A_{11} \subset$ nullspace A_{12}^*. If A_{11} is nonsingular, there is nothing to prove, so we assume that $\xi \in \mathbf{C}^k$ is nonzero and $\xi^*A_{11} = 0$; we must show that $\xi^*A_{12} = 0$. Let $x = \begin{bmatrix} \xi \\ 0 \end{bmatrix} \in \mathbf{C}^n$. Then $x^*Ax = \xi^*A_{11}\xi = 0$, so (7.1.6) ensures that $x^*A = 0$. Then $0 = x^*A = \xi^*[A_{11}\ A_{12}] = [\xi^*A_{11}\ \xi^*A_{12}] = [0\ \xi^*A_{12}]$, so $\xi^*A_{12} = 0$. For the second assertion, observe that row inclusion ensures that each entry a_{ik} is a scalar multiple of a_{kk}. $\square$

Exercise. Let $A = \begin{bmatrix} A_{11} & A_{12} \\ A_{12}^* & A_{22} \end{bmatrix} \in M_n$ be positive semidefinite. If either $A_{11} = 0$ or $A_{22} = 0$, explain why $A_{12} = 0$.

Exercise. Let $A \in M_n$ be positive semidefinite. Partition $A = [a_1 \ \ldots \ a_n]$ according to its columns, let $\alpha \subset \{1, \ldots, n\}$ be any nonempty index set, and let $j \in \{1, \ldots, n\}$ be any column index. Explain why $a_j[\alpha]$ is in the column space of $A[\alpha]$. *Hint*: Permutation similarity preserves positive definiteness; see (7.1.8).

There is a related, and much larger, class of matrices that also has the row and column inclusion properties. Although the matrices in this class need not be positive definite or even Hermitian, they all have positive semidefinite Hermitian parts; see (4.1.2).

Exercise. Write $A \in M_n$ as $A = H + iK$, in which H and K are Hermitian. If $x \in \mathbb{C}^n$, explain why the following statements are equivalent: (a) $x^*Ax = 0$; (b) $x^*A^*x = 0$; and (c) $x^*Hx = x^*Kx = 0$.

Lemma 7.1.11. *Suppose that $A \in M_n$ has positive semidefinite Hermitian part $H(A) = \frac{1}{2}(A + A^*)$. Then*

(a) nullspace $A \subset$ nullspace $H(A)$ *and* nullspace $A^* \subset$ nullspace $H(A)$
(b) rank $H(A) \leq$ rank A
(c) the following statements are equivalent:
   (i) A *and* $H(A)$ *have the same null space.*
   (ii) A^* *and* $H(A)$ *have the same null space.*
   (iii) rank $A =$ rank $H(A)$.

Proof. (a) Write $A = H + iK$, in which H and K are Hermitian, and let $x \in \mathbb{C}^n$. If either $x^*A = 0$ or $Ax = 0$, then $x^*Ax = 0$, which implies that $x^*Hx = 0$. It follows from (7.1.6) that $Hx = 0$.
(b) Follows from (a).
(c) Equality in the two inclusions in (a) follows from equality in (b).   □

Exercise. If $A \in M_n$ and $H(A)$ is positive definite, explain why A is nonsingular.

Observation 7.1.12. *Let $A \in M_n$ have positive semidefinite Hermitian part $H(A)$. If* rank $A =$ rank $H(A)$, *then A has the row and column inclusion properties.*

Proof. (a) Partition $A \in M_n$ as $A = \begin{bmatrix} A_{11} & A_{12} \\ A_{21} & A_{22} \end{bmatrix}$, in which $A_{11} \in M_k$ and $k \in \{1, \ldots, n-1\}$. Since $H(A)$ is positive semidefinite, the preceding lemma ensures that A, A^*, and $H(A)$ have the same null space. If A_{11} is nonsingular, there is nothing to prove, so assume that A_{11} is singular. First consider the column inclusion property. We assume that $\xi \in \mathbb{C}^k$ is nonzero and $\xi^*A_{11} = 0$; we must show that $\xi^*A_{12} = 0$. Let $x = \begin{bmatrix} \xi \\ 0 \end{bmatrix}$ $\in \mathbb{C}^n$. Then $0 = \xi^*A_{11}\xi = x^*Ax = x^*H(A)x + ix^*K(A)x$, so $x^*H(A)x = 0$. It follows from (7.1.6) that $H(A)x = 0$. Since the null spaces of A^* and $H(A)$ are the same, we have $0 = x^*A = \xi^*[A_{11} \ A_{12}] = [\xi^*A_{11} \ \xi^*A_{12}] = [0 \ \xi^*A_{12}]$, so $\xi^*A_{12} = 0$. The same approach (using equality of the null spaces of A and $H(A)$) shows that A^* has the column inclusion property, so A has the row inclusion property.   □

Exercise. The sufficient condition in the preceding observation is not a necessary condition. Consider $A = \begin{bmatrix} i & 0 \\ 0 & -i \end{bmatrix}$. Explain why A has both the row and column inclusion properties, but rank A > rank $H(A)$.

Exercise. Deduce (7.1.10) from (7.1.12).

Corollary 7.1.13. *If $A \in M_n$ has positive definite Hermitian part, then A has the row and column inclusion properties.*

Proof. If $H(A)$ is nonsingular, then (7.1.11(b)) ensures that rank $A = n = $ rank $H(A)$, so the assertion follows from (7.1.12). $\qquad\square$

Our final observation is a generalization of a fact about real numbers: If a and b are real and nonnegative, then $a + b = 0$ if and only if $a = b = 0$.

Observation 7.1.14. *Let $A, B \in M_n$ be positive semidefinite. Then*

(a) $A + B = 0$ *if and only if* $A = B = 0$
(b) rank$(A + B) > 0$ *if and only if at least one of A or B is nonzero*

Proof. (a) Only the forward implication requires a proof. Let $A = [a_{ij}]$ and $B = [b_{ij}]$ and suppose that $A + B = 0$. Then $a_{ii} + b_{ii} = 0$ for each $i = 1, \ldots, n$ and each summand is real and nonnegative, so each $a_{ii} = b_{ii} = 0$. It follows from the second assertion in (7.1.10) that $A = B = 0$.
(b) rank$(A + B) = 0$ if and only if $A + B = 0$ if and only if $A = B = 0$. $\qquad\square$

Problems

7.1.P1 Let $A = [a_{ij}] \in M_n$ be positive semidefinite. Why is $a_{ii}a_{jj} \geq |a_{ij}|^2$ for all distinct $i, j \in \{1, \ldots, n\}$? If A is positive definite, why is $a_{ii}a_{jj} > |a_{ij}|^2$ for all distinct $i, j \in \{1, \ldots, n\}$? If there is a pair of distinct indices i, j such that $a_{ii}a_{jj} = |a_{ij}|^2$, why is A singular?

7.1.P2 Use the preceding problem to prove the second assertion in (7.1.10): A positive semidefinite matrix has a zero entry on its main diagonal if and only if the entire row and column to which that entry belongs is zero.

7.1.P3 Let $A = [a_{ij}] \in M_n$ be positive semidefinite and have positive main diagonal entries. Show that the matrix $[a_{ij}/\sqrt{a_{ii}a_{jj}}]$ is positive semidefinite, that all its main diagonal entries are $+1$, and that all its entries are bounded by 1 in absolute value. Such a matrix is called a *correlation matrix*.

7.1.P4 If $A = [a_{ij}] \in M_n$ is a correlation matrix, show that $|a_{ij}| \leq 1$ for all $i, j = 1, \ldots, n$. Can equality occur? Can equality occur if A is positive definite?

7.1.P5 Let $A \in M_n$ be Hermitian. If $|\operatorname{tr} A| < \|A\|_2$ (the Frobenius norm), show that A is indefinite.

7.1.P6 Let $A \in M_n$ and $B \in M_m$ be Hermitian. Show that $A \oplus B$ is positive semidefinite if and only both A and B are positive semidefinite. What can you say in the positive definite case?

7.1.P7 A function $f : \mathbf{R} \to \mathbf{C}$ is said to be a *positive definite function* if the matrix $[f(t_i - t_j)] \in M_n$ is positive semidefinite for all choices of points $\{t_1, \ldots, t_n\} \subset \mathbf{R}$ and all $n = 1, 2, \ldots$. If f is a positive definite function, why is $f(-t) = \bar{f}(t)$ for all $t \in \mathbf{R}$? Use (7.1.5) to show that if f is a positive definite function, then (a) $f(0) \geq 0$; (b) f is a bounded function and $|f(t)| \leq f(0)$ for all $t \in \mathbf{R}$; and (c) if f is continuous at 0, then it is continuous everywhere.

7.1.P8 If $f_1, \ldots, f_n$ are positive definite functions and if $a_1, \ldots, a_n$ are nonnegative real numbers, show that $f = a_1 f_1 + \cdots + a_n f_n$ is a positive definite function.

7.1.P9 Show that $f(t) = e^{ist}$ is a positive definite function for each $s \in \mathbf{R}$. Use the preceding problem to show that $f(t) = a_1 e^{is_1 t} + \cdots + a_n e^{is_n t}$ is a positive definite function for any choice of points $s_1, \ldots, s_n \in \mathbf{R}$ and any nonnegative real numbers $a_1, \ldots, a_n$.

7.1.P10 Prove that $\cos t$ is a positive definite function.

7.1.P11 Is $\sin t$ a positive definite function?

7.1.P12 If g is a nonnegative and integrable function on $\mathbf{R}$, show that $f(t) = \int_{-\infty}^{\infty} e^{its} g(s) ds$ is a positive definite function. Explain why the following functions are positive definite:

(a) $f(t) = \frac{\sin(\alpha t)}{\alpha t} = \frac{1}{2\alpha} \int_{-\alpha}^{\alpha} e^{its} ds, \quad \alpha > 0$

(b) $f(t) = e^{-t^2} = \frac{1}{2\sqrt{\pi}} \int_{-\infty}^{\infty} e^{its} e^{-s^2/2} ds$

(c) $f(t) = e^{-|t|} = \frac{1}{\pi} \int_{-\infty}^{\infty} \frac{e^{its}}{1+s^2} ds$

(d) $f(t) = \frac{1+it}{1+t^2} = \frac{1}{1-it} = \int_0^{\infty} e^{its} e^{-s} ds$

Alternative approaches to showing that the functions in (b) and (c) are positive definite are in (7.2.P12 and P14).

7.1.P13 (a) If f is a positive definite function, show that $\bar{f}$ and $\frac{1}{2}(f + \bar{f}) = \operatorname{Re} f$ are positive definite functions. (b) Deduce from the preceding problem that $g(t) = 1/(1+t^2)$ is a positive definite function. (c) Is $h(t) = it/(1+t^2)$ a positive definite function?

7.1.P14 Let $A \in M_n$ be positive semidefinite, and consider the bordered matrix $B = \begin{bmatrix} A & y \\ y^* & \alpha \end{bmatrix}$. If B is positive semidefinite, explain why $y \in \operatorname{range} A$.

7.1.P15 Let f be a positive definite function and suppose that there is a positive real number τ such that $f(\tau) = f(0)$. Use the preceding problem to show that f is periodic with period τ, that is, $f(t) = f(t - \tau)$ for all real t.

7.1.P16 Let $\lambda_1, \ldots, \lambda_n \in \mathbf{C}$ be given and suppose that $\operatorname{Re} \lambda_j > 0$ for all $j = 1, \ldots, n$. Show that $A = [(\lambda_i + \bar{\lambda}_j)^{-1}]_{i,j=1}^n$ is positive semidefinite, and it is positive definite if $\lambda_1, \ldots, \lambda_n$ are distinct. Conclude that the Hankel matrices $A = [(i + j)^{-1}]_{i,j=1}^n$ and $B = [(i + j - 1)^{-1}]_{i,j=1}^n$ are positive definite.

7.1.P17 Let J_n be the n-by-n all-ones matrix; see (0.2.8). Show that $x^* J_n x = |x_1 + \cdots + x_n|^2$ and conclude that J_n is positive semidefinite for all $n = 1, 2, \ldots$.

7.1.P18 (a) Suppose that $0 < \alpha_1 < \cdots < \alpha_n$ and let $A = [\min\{\alpha_i, \alpha_j\}]_{i,j=1}^n$. Show that $A = \alpha_1 J_n + (\alpha_2 - \alpha_1)(0_1 \oplus J_{n-1}) + (\alpha_3 - \alpha_2)(0_2 \oplus J_{n-2}) + \cdots + (\alpha_n - \alpha_{n-1})(0_{n-1} \oplus J_1)$ and prove that A is positive definite. (b) Let $\beta_1, \ldots, \beta_n$ be positive real numbers, not necessarily algebraically ordered and not necessarily distinct. Explain why the *min matrix* $[\min\{\beta_i, \beta_j\}]$ is positive semidefinite and is positive definite if $\beta_i \neq \beta_j$ whenever $i \neq j$.

(c) Show that the *reciprocal max matrix* $[(\max\{\beta_i, \beta_j\})^{-1}]$ is positive semidefinite and is positive definite if $\beta_i \neq \beta_j$ whenever $i \neq j$.

7.1.P19 Use the preceding problem and a limiting argument to show that the kernel $K(s, t) = \min\{s, t\}$ is positive semidefinite on $[0, N]$ for any $N > 0$, that is,

$$\int_0^N \int_0^N \min\{s, t\} \bar{f}(s) f(t) ds \, dt \geq 0 \tag{7.1.15}$$

for all continuous complex-valued function f on $[0, N]$.

7.1.P20 Prove that $\int_0^N \int_0^N \min\{s, t\} \bar{f}(s) f(t) ds \, dt = \int_0^N \left| \int_t^N f(s) ds \right|^2 dt$ for every continuous complex valued function f on $[0, N]$ and use it to give an alternate proof of the assertion in the preceding problem. Why does this proof show that $K(s, t) = \min\{s, t\}$ is positive definite?

7.1.P21 Let $A \in M_n$ have positive semidefinite Hermitian part. (a) Explain why the same is true of SAS^* for any nonsingular $S \in M_n$. (b) Explain why every block in the *congruence canonical form of A has positive semidefinite Hermitian part. (c) Consider the three types of *congruence canonical blocks listed in (4.5.21). Which of these blocks have positive semidefinite Hermitian part? Show that the only Type 0 block with this property is $[0]$; no Type II block has this property; and the only Type I blocks with this property are $[e^{i\theta}]$ with $-\pi/2 \leq \theta \leq \pi/2$ and $-i\begin{bmatrix} 0 & 1 \\ 1 & i \end{bmatrix} = \begin{bmatrix} 0 & -i \\ -i & 1 \end{bmatrix}$. (d) Show that rank A = rank $H(A)$ if and only if the *congruence canonical form of A is the direct sum of a zero matrix and blocks of the form $[e^{i\theta}]$ with $-\pi/2 < \theta < \pi/2$. (e) Explain why $H(A)$ is positive definite if and only if there is a nonsingular $S \in M_n$ such that

$$A = S \operatorname{diag}(e^{i\theta_1}, \ldots, e^{i\theta_n}) S^*, \quad \text{each } \theta_j \in (-\pi/2, \pi/2) \tag{7.1.15}$$

(f) If $H(A)$ is positive definite, show that $H(A^{-1})$ is positive definite. (g) Show that $H(A)$ is positive semidefinite if and only if A is *congruent to a block diagonal matrix of the form

$$(I_p + i\Lambda_p) \oplus (0_q + i\Gamma_q) \oplus (E_{2r} + iF_{2r}) \tag{7.1.16}$$

in which $I_p \in M_p$ is an identity matrix, $\Lambda_p \in M_p$ is real diagonal, $0_q \in M_q$ is a zero matrix, $\Gamma_q = 0_{q_1} \oplus I_{q_2} \oplus (-I_{q_3}) \in M_q$ is an inertia matrix, $E_{2r} = \begin{bmatrix} 0 & 0 \\ 0 & 1 \end{bmatrix} \oplus \cdots \oplus \begin{bmatrix} 0 & 0 \\ 0 & 1 \end{bmatrix} \in M_{2r}$, and $F_{2r} = \begin{bmatrix} 0 & 1 \\ 1 & 0 \end{bmatrix} \oplus \cdots \oplus \begin{bmatrix} 0 & 1 \\ 1 & 0 \end{bmatrix} \in M_{2r}$. Moreover, this block diagonal matrix is unique up to permutations of the diagonal entries of Λ_q. (h) To what form does (7.1.16) reduce if $H(A)$ is positive definite? Explain how this form is consistent with (7.1.15).

7.1.P22 Let $A \in M_n$ have positive semidefinite Hermitian part $H(A)$. We claim that if $H(A^2)$ is positive semidefinite, then rank A = rank $H(A)$, and hence (7.1.12) ensures that A has the row and column inclusion properties. Provide details: (a) Let r = rank $H(A)$. After a suitable unitary similarity, we may assume that $A = \Lambda + iK$, in which $\Lambda = L \oplus 0_{n-r}$, $L \in M_r$ is positive diagonal, and $K = \begin{bmatrix} K_{11} & K_{12} \\ K_{12}^* & K_{22} \end{bmatrix}$ is Hermitian and partitioned conformally to Λ. (b) The lower-right block of A^2 is $-K_{12}^* K_{12} - K_{22}^2$, so $K_{12} = 0$ and $K_{22} = 0$ if $H(A^2)$ is positive semidefinite. (c) $A = (L + iK_{11}) \oplus 0_{n-r}$, so rank A = rank $L = r$.

7.1.P23 Consider $A = \begin{bmatrix} 1 & -2 \\ 2 & 1 \end{bmatrix}$. Show that $H(A)$ is positive definite, so (7.1.13) ensures that A has the row and column inclusion properties even though $H(A^2)$ is not positive semidefinite.

7.1.P24 Let $A = \begin{bmatrix} A_{11} & A_{12} \\ A_{12}^* & A_{22} \end{bmatrix} \in M_n$ be positive semidefinite. Use (7.1.10) to show that rank $A \le$ rank $A_{11} +$ rank A_{22}.

7.1.P25 Let $A \in M_n$ be positive semidefinite, suppose that $n = km$, and partition $A = [A_{ij}]_{i,j=1}^k$ as a k-by-k block matrix in which each block is m-by-m. We claim that the compressed matrix $\mathcal{T} = [\text{tr } A_{ij}]_{i,j=1}^k \in M_k$ is positive semidefinite. Provide details: (a) Let $e_1, \ldots, e_m$ be the standard basis of $\mathbf{C}^m$ and let $e = e_1 + \cdots + e_m$. For a given $p \in \{1, \ldots, m\}$, describe the vector $\text{vec}(e_p e^T) \in \mathbf{C}^{m^2}$; see (0.7.8). (b) For each $p \in \{1, \ldots, m\}$, construct $X_p \in M_{m^2, m}$ as follows: its pth column is $\text{vec}(e_p e^T)$; all other columns are zero. Explain why $\mathcal{T} = \sum_{p=1}^m X_p^* A X_p$. (c) Explain why $\mathcal{T}$ is positive semidefinite; use (7.1.8) and (7.1.3). For a different proof, and other compressions that preserves positive semidefiniteness, see (7.2.P25). The compressed matrix $\mathcal{T}$ is known in the physics literature as a *partial trace* of A.

7.1.P26 Let $A \in M_n$. Suppose that its Hermitian part $H(A)$ is positive semidefinite and rank $A = $ rank $H(A)$. Explain why there are lower triangular matrices $L, L' \in M_n$ and upper triangular matrices $U, U' \in M_n$ such that L and U' are nonsingular and $A = LU = L'U'$.

7.1.P27 Let $A, B \in M_n$ be positive semidefinite and let $\alpha \subset \{1, \ldots, n\}$. (a) Explain why rank $A^k = $ rank A for each $k = 1, 2, \ldots$. (b) Use (7.1.10) to show that rank$(AB)[\alpha] \le \min\{\text{rank } A[\alpha], \text{rank } B[\alpha]\}$ and rank $A^2[\alpha] = $ rank $A[\alpha]$. (c) Explain why rank $A[\alpha] = $ rank $A^2[\alpha] = $ rank $A^4[\alpha] = \cdots = $ rank $A^{2^k}[\alpha] = \cdots$ and show that rank $A[\alpha] = $ rank $A^k[\alpha]$ for every $k = 2, 3, \ldots$.

7.1.P28 This problem is a continuation of (4.5.P21). Let $A \in M_n$ be positive semidefinite and partitioned as $A = \begin{bmatrix} B & C \\ C^* & D \end{bmatrix}$. If B is singular, we cannot form its Schur complement, but the column inclusion property permits us to form a generalized Schur complement. (a) Let $C = BX$ and verify the *congruence

$$\begin{bmatrix} I & 0 \\ -X^* & I \end{bmatrix} \begin{bmatrix} B & C \\ C^* & D \end{bmatrix} \begin{bmatrix} I & -X \\ 0 & I \end{bmatrix} = \begin{bmatrix} B & 0 \\ 0 & D - X^*BX \end{bmatrix}$$

(b) The matrix X whose existence is guaranteed by the column inclusion property need not be unique. However, if $C = BY$, show that $X^*BX = Y^*BY$, so the matrix $\tilde{S} = D - X^*BX$ is well-defined, independent of the choice of X such that $C = BX$. (c) If B is nonsingular, show that $\tilde{S} = S = D - C^*B^{-1}C$, which is the Schur complement of B in A. Consequently, it is reasonable to call $\tilde{S}$ *the generalized Schur complement* of B in A. (d) Explain why $\tilde{S} = D - X^*BX$ is positive semidefinite, and why rank $A = $ rank $B + $ rank $\tilde{S}$. (e) Why may we regard the two statements in (d) as analogs of the identities (4.5.28) in Haynsworth's theorem? For a further development of these ideas, see (7.3.P8).

7.1.P29 Let $A = H_1 + iK_1, B = H_2 + iK_2 \in M_n$, in which H_1, H_2, K_1, and K_2 are Hermitian, and H_1 and H_2 are positive definite. Use the *congruence canonical form (7.1.15) to show that the following are equivalent:

(a) A and B are *congruent.

(b) $A^{-*}A$ and $B^{-*}B$ are similar.

(c) $A^{-*}A$ and $B^{-*}B$ have the same eigenvalues.

(d) $H_1^{-1}K_1$ and $H_2^{-1}K_2$ are similar.

(e) $H_1^{-1}K_1$ and $H_2^{-1}K_2$ have the same eigenvalues.

7.1.P30 Let $A \in M_n$ have positive definite Hermitian part. Show that $A^{-*}A$ is similar to a unitary matrix and that $I + A^{-*}A$ is nonsingular.

7.2 Characterizations and properties

Positive definite and semidefinite matrices can be characterized in many different, sometimes surprising, ways. We have already met the first of these characterizations as (4.1.8).

Theorem 7.2.1. *A Hermitian matrix is positive semidefinite if and only if all of its eigenvalues are nonnegative. It is positive definite if and only if all of its eigenvalues are positive.*

> *Exercise.* Deduce from the preceding theorem that a nonsingular Hermitian matrix $A \in M_n$ is positive definite if and only if A^{-1} is positive definite.

> *Exercise.* Let $A \in M_n$ be positive semidefinite. Use (7.2.1) to show that A is positive definite if and only if rank $A = n$.

Corollary 7.2.2. *If $A \in M_n$ is positive semidefinite, then so is each A^k, $k = 1, 2, \ldots$.*

Proof. If the eigenvalues of A are $\lambda_1, \ldots, \lambda_n$, then the eigenvalues of A^k are $\lambda_1^k, \ldots, \lambda_n^k$. The latter are nonnegative if the former are.   □

Corollary 7.2.3. *Suppose that $A = [a_{ij}] \in M_n$ is Hermitian and strictly diagonally dominant. If $a_{ii} > 0$ for all $i = 1, 2, \ldots, n$, then A is positive definite.*

Proof. This is (6.1.10(c)). The conditions imply that each Geršgorin disc for A lies in the open right half-plane. Since the eigenvalues of a Hermitian matrix are all real, the eigenvalues of A must all be positive.   □

The next characterization is not very practical for verifying positive definiteness computationally, but it can be of theoretical utility.

Corollary 7.2.4. *Let $A \in M_n$ be Hermitian, and let $p_A(t) = a_n t^n + a_{n-1} t^{n-1} + \cdots + a_{n-m} t^{n-m}$ be its characteristic polynomial, in which $a_n = 1$, $a_{n-m} \neq 0$, and $1 \leq m \leq n$. Then A is positive semidefinite if and only if $a_k a_{k+1} < 0$ for each $k = n - m, \ldots, n - 1$.*

Proof. The hypothesis is that the leading coefficients of $p_A(t)$ are nonzero and alternate strictly in sign. If this condition is met, $p_A(t)$ has no negative zeroes, so A has only nonnegative eigenvalues. Conversely, if A is positive semidefinite, denote its positive eigenvalues by $\lambda_1, \ldots, \lambda_m$; its remaining $n - m$ eigenvalues are all zero. An induction argument shows that the signs of the coefficients of the polynomials $(t - \lambda_1)$, $(t - \lambda_1)(t - \lambda_2), \ldots, (t - \lambda_1)(t - \lambda_2) \cdots (t - \lambda_m)$ alternate strictly; multiplying by t^{n-m} gives $p_A(t)$.   □

The following theorem gives a converse to the observation in (7.1.5) about principal minors of a positive semidefinite matrix, with a perhaps surprising statement in the positive definite case.

Theorem 7.2.5 (Sylvester's criterion). *Let $A \in M_n$ be Hermitian.*

(a) *If every principal minor of A (including det A) is nonnegative, then A is positive semidefinite.*

(b) *If every leading (respectively, trailing) principal minor of A is positive (including det A), then A is positive definite.*

(c) *If the first $n - 1$ leading principal minors (respectively, the last $n - 1$ trailing principal minors) of A are positive and det $A \geq 0$, then A is positive semidefinite.*

Proof. (a) Let $r = \text{rank } A$. If $r = 0$, there is nothing to prove, so we suppose that $r \geq 1$. The hypothesis ensures that $E_k(A)$, the sum of all principal minors of size k, is nonnegative for each $k = 1, \ldots, n$. Every Hermitian matrix is rank principal (0.7.6), so some r-by-r principal submatrix of A is nonsingular; it follows that $E_r(A)$ is positive. If $k > r$, every minor of size k is zero, so $E_k = 0$. The representation (1.2.13) for the characteristic polynomial of A is

$$p_A(t) = t^{n-r}(t^r - E_1 t^{r-1} + \cdots + (-1)^{r-1} E_{r-1} t + (-1)^r E_r)$$

in which $E_r > 0$ and the hypothesis implies that E_k is nonnegative for each $k = 1, \ldots, r - 1$. The coefficient sign pattern of the polynomial $p(t)/t^{n-m}$ ensures that it has no zeroes in the interval $(-\infty, 0]$; all of its zeroes must be positive. We conclude that the eigenvalues of A are nonnegative, so it is positive semidefinite.

(b) Let A_k denote the leading principal submatrix $A[\{1, \ldots, k\}]$, $k = 1, \ldots, n$. If $\det A_1 > 0$, then A_1 is positive definite. If $k \in \{1, \ldots, n - 1\}$ and A_k is positive definite, then all of its eigenvalues are positive; the interlacing inequalities (4.3.18) ensure that all the eigenvalues of A_{k+1} are positive except perhaps for its smallest eigenvalue. But the product of *all* the eigenvalues of A_{k+1} is $\det A_{k+1}$, which is positive, so we can conclude that the smallest eigenvalue of A_{k+1} is positive and A_{k+1} is positive definite. It follows by induction that $A_n = A$ is positive definite. The statement about trailing principal minors follows from the statement about leading principal minors and a suitable permutation similarity of A.

(c) The hypothesis and the interlacing argument in (b) ensure that A has at least $n - 1$ positive eigenvalues. If $\det A = 0$, then the remaining eigenvalue is zero, so A is positive semidefinite. $\qquad\square$

Exercise. The leading principal minors of the Hermitian matrix $\begin{bmatrix} 0 & 0 \\ 0 & -1 \end{bmatrix}$ are nonnegative, but it is not positive semidefinite. What is going on here? Does this contradict the preceding theorem?

Every positive real number has a unique positive kth root for each $k = 1, 2, \ldots$. Positive definite matrices have a corresponding property.

Theorem 7.2.6. *Let $A \in M_n$ be Hermitian and positive semidefinite, let $r = \text{rank } A$, and let $k \in \{2, 3, \ldots\}$.*

(a) *There is a unique Hermitian positive semidefinite matrix B such that $B^k = A$.*

(b) *There is a polynomial p with real coefficients such that $B = p(A)$. Consequently, B commutes with any matrix that commutes with A.*

(c) range A = range B, so rank A = rank B.

(d) B is real if A is real.

Proof. Represent $A = U \Lambda U^*$, in which $U = [U_1 \ U_2]$ is unitary, $U_1 \in M_{n,r}$, $\Lambda = \text{diag}(\lambda_1, \ldots, \lambda_r) \oplus 0_{n-r}$, and $\lambda_1, \ldots, \lambda_r$ are positive. Define $B = U \Lambda^{1/k} U^*$, in which $\Lambda^{1/k} = \text{diag}(+\lambda_1^{1/k}, \ldots, +\lambda_r^{1/k}) \oplus 0_{n-r}$ and the unique nonnegative kth root is taken in each case. Then B is Hermitian and positive semidefinite, and $B^k = A$. Observe that range A = range B is the column space of U_1, so rank A = rank $B = r$. If A is real and positive semidefinite, then U may be chosen to be real orthogonal, so our construction produces a real matrix B in this case. It remains to address the questions of uniqueness and commutativity.

Let p be a polynomial such that $p(\lambda_i) = +\lambda_i^{1/k}, i = 1, \ldots, r$ and $p(0) = 0$ if $r < n$; see (0.9.11), which ensures that p has real coefficients. Then $p(\Lambda) = \Lambda^{1/k}$ and $p(A) = p(U \Lambda U^*) = U p(\Lambda) U^* = U \Lambda^{1/k} U^* = B$, which verifies (b). If C is a positive semidefinite Hermitian matrix such that $C^k = A$, then $B = p(A) = p(C^k)$ and hence B commutes with C. Theorem 4.1.6 ensures that there is a unitary V that simultaneously diagonalizes B and C, so $B = V \Lambda_1 V^*$ and $C = V \Lambda_2 V^*$, in which $\Lambda_1, \Lambda_2 \in M_n$ are nonnegative diagonal. Since $B^k = A = C^k$, we deduce that $\Lambda_1^k = \Lambda_2^k$. Uniqueness of the nonnegative kth root of a nonnegative number implies that $\Lambda_1 = (\Lambda_1^k)^{1/k} = (\Lambda_2^k)^{1/k} = \Lambda_2$, so $B = C$. $\square$

We denote the unique positive (semi)definite square root of a positive (semi)definite matrix A by $A^{1/2}$; $A^{1/k}$ denotes the unique positive (semi)definite kth root of A for each $k = 1, 2, \ldots$. See (7.2.P20) for an application of the uniqueness assertion in the preceding theorem.

Exercise. Use the construction in the proof of the preceding theorem to compute $\begin{bmatrix} 5 & 4 \\ 4 & 5 \end{bmatrix}^{1/2}$.

Exercise. If A is positive definite, show that $(A^{1/2})^{-1} = (A^{-1})^{1/2}$.

Theorem 7.2.7. *Let $A \in M_n$ be Hermitian.*

(a) *A is positive semidefinite if and only if there is a $B \in M_{m,n}$ such that $A = B^* B$.*

(b) *If $A = B^* B$ with $B \in M_{m,n}$, and if $x \in \mathbf{C}^n$, then $Ax = 0$ if and only if $Bx = 0$, so nullspace A = nullspace B and rank A = rank B.*

(c) *If $A = B^* B$ with $B \in M_{m,n}$, then A is positive definite if and only if B has full column rank.*

Proof. (a) If $A = B^* B$ for some $B \in M_{m,n}$, then $x^* A x = x^* B^* B x = \|Bx\|_2^2 \geq 0$. The asserted factorization can be achieved, for example, with $B = A^{1/2}$ and $m = n$.
(b) If $Ax = 0$, then $x^* A x = \|Bx\|_2^2 = 0$; if $Bx = 0$ then $Ax = B^* B x = 0 = 0$, so A and B have the same null spaces and hence the same nullity and rank.
(c) The nullity of A is zero if and only if the nullity of B is zero if and only if rank $B = n$. $\square$

See (7.2.P9) for a refinement of the preceding theorem.

Corollary 7.2.8. *A Hermitian matrix A is positive definite if and only if it is* $*$*congruent to the identity.*

Proof. This is simply a restatement of (7.2.7). ∎

Exercise. Let $A \in M_n$ be positive definite and suppose that $A = C^*C$ with $C \in M_n$. Show that there is a unitary $V \in M_n$ such that $C = VA^{1/2}$. *Hint*: Show that $A^{-1/2}C^*CA^{-1/2} = (CA^{-1/2})^*(CA^{-1/2}) = I$.

A factorization $A = B^*B$ of a positive semidefinite matrix can be achieved in various ways. For example, every square matrix C has a QR factorization (2.1.14), so it can be written as $C = QR$, in which Q is unitary, and R is upper triangular, has nonnegative diagonal entries, and has the same rank as C. Then $C^*C = (QR)^*QR = R^*Q^*QR = R^*R$.

Corollary 7.2.9 (Cholesky factorization). *Let $A \in M_n$ be Hermitian. Then A is positive semidefinite (respectively, positive definite) if and only if there is a lower triangular matrix $L \in M_n$ with nonnegative (respectively, positive) diagonal entries such that $A = LL^*$. If A is positive definite, L is unique. If A is real, L may be taken to be real.*

Proof. Let $A^{1/2} = QR$ be a QR factorization and let $L = R^*$. Then $A = A^{1/2}A^{1/2} = (A^{1/2})^*A^{1/2} = R^*Q^*QR = R^*R = LL^*$. The asserted properties of L follow from the properties of R stated in (b) and (e) of (2.1.14). ∎

Let $v_1, \ldots, v_m$ be vectors in an inner product space V with inner product $\langle \cdot, \cdot \rangle$. The *Gram matrix* of the vectors $v_1, \ldots, v_m$ with respect to the inner product $\langle \cdot, \cdot \rangle$ is $G = [\langle v_j, v_i \rangle]_{i,j=1}^m \in M_m$. If $A \in M_n$ is positive semidefinite, partition $A^{1/2} = [v_1 \ \ldots \ v_n]$ according to its columns and notice that $A = A^{1/2}A^{1/2} = (A^{1/2})^*A^{1/2} = [v_i^*v_j] = [\langle v_j, v_i \rangle]_{i,j=1}^n$, in which $\langle \cdot, \cdot \rangle$ is the Euclidean inner product on $\mathbf{C}^n$. Thus, *every positive semidefinite matrix is a Gram matrix.* The following theorem provides another characterization of positive semidefinite matrices by establishing the converse implication in a wider context: a Gram matrix of vectors in *any* inner product space, finite dimensional or not, is positive semidefinite.

Theorem 7.2.10. *Let $v_1, \ldots, v_m$ be vectors in an inner product space V with inner product $\langle \cdot, \cdot \rangle$, and let $G = [\langle v_j, v_i \rangle]_{i,j=1}^m \in M_m$. Then*

(a) *G is Hermitian and positive semidefinite*
(b) *G is positive definite if and only if the vectors $v_1, \ldots, v_m$ are linearly independent*
(c) *$\operatorname{rank} G = \dim \operatorname{span}\{v_1, \ldots, v_m\}$*

Proof. (a) Let $\| \cdot \|$ be the norm derived from the given inner product and let $x = [x_i] \in \mathbf{C}^m$. The properties listed in (5.1.3) ensure that G is Hermitian and

$$x^*Gx = \sum_{i,j=1}^m \langle v_j, v_i \rangle \bar{x}_i x_j = \sum_{i,j=1}^m \langle x_j v_j, x_i v_i \rangle$$

$$= \left\langle \sum_{j=1}^m x_j v_j, \sum_{i=1}^m x_i v_i \right\rangle = \left\| \sum_{i=1}^m x_i v_i \right\|^2 \geq 0 \qquad (7.2.11)$$

so G is positive semidefinite.

(b) The inequality (7.2.11) is an equality if and only if $\sum_{i=1}^{m} x_i v_i = 0$. This cannot happen if $x \neq 0$ and the vectors $v_1, \ldots, v_m$ are linearly independent; in this case G is positive definite. Conversely, if $x^* G x > 0$ whenever $x \neq 0$, then $\left\| \sum_{i=1}^{m} x_i v_i \right\| \neq 0$ whenever $x \neq 0$, which implies that the vectors $v_1, \ldots, v_m$ are linearly independent.

(c) Let $r = \text{rank } G$ and let $d = \dim \text{span}\{v_1, \ldots, v_m\}$. Since G is rank principal, it has a nonsingular, and hence positive definite, principal submatrix of size r. That principal submatrix is the Gram matrix of r of the vectors v_i, so (b) ensures that these vectors are linearly independent. This means that $r \leq d$. On the other hand, d of the vectors v_i are linearly independent, and the Gram matrix of these vectors (positive definite by (b) again) is a principal submatrix of G. This means that $d \leq r$. We conclude that $r = d$. □

Exercise. Let $A \in M_n$ be positive semidefinite and have rank r. Explain why there are vectors $v_1, \ldots, v_n \in \mathbf{C}^n$ such that $\text{rank}[v_1 \ \ldots \ v_n] = r$ and $A = [v_i^* v_j]_{i,j=1}^n$, which is a Gram matrix with respect to the standard inner product. *Hint:* See the discussion preceding (7.2.10).

Exercise. If $A \in M_n$ is the Gram matrix of the vectors $v_1, \ldots, v_n$ in an inner product space, explain why each principal submatrix is the Gram matrix of a list of vectors chosen from the list $v_1, \ldots, v_n$.

Exercise. With (7.2.10) in mind, discuss (7.1.P25) as well as (7.1.P12 and P16). In each case, what is the vector space V, the inner product $\langle \cdot, \cdot \rangle$, the vectors v_i, and the Gram matrix G?

Problems

7.2.P1 Let $A \in M_n$ be Hermitian. Show that A^{2k} is positive semidefinite for all $k = 1, 2, \ldots$ and e^A is positive definite. See the exercises and text following (5.6.15).

7.2.P2 Let $A \in M_n$ be positive semidefinite and let $x \in \mathbf{C}^n$. Show that $x^* A x = \left\| A^{1/2} x \right\|_2^2$ and deduce (7.1.6) from this identity.

7.2.P3 Let $A = [\min\{i, j\}]_{i,j=1}^n$ and let R be the n-by-n upper triangular matrix with entries $+1$ on and above the main diagonal. (a) Show that $A = R^T R$ (an LU factorization of A) and conclude that A is positive definite. (b) Show that R^{-1} is the upper bidiagonal matrix with $+1$ entries on the main diagonal and -1 entries on the first superdiagonal. (c) Show that $A^{-1} = R^{-1} R^{-T} = [\alpha_{ij}]$ is the symmetric tridiagonal matrix with $\alpha_{nn} = 1$, and with $\alpha_{ii} = 2$ and $\alpha_{i,i+1} = -1$ for $i = 1, \ldots, n - 1$.

7.2.P4 If $A \in M_n$ is Hermitian and has positive leading principal minors, show that, in the LDU factorization of A described in (3.5.6b), $U = L^*$ and D is positive diagonal. Use the factorization $A = LDL^*$ to give a proof of (7.2.5b) that does not involve interlacing.

7.2.P5 (a) Verify that $L_1 = \begin{bmatrix} 2 & 0 \\ 1 & \sqrt{3} \end{bmatrix}$ provides a Cholesky factorization (7.2.9) of the positive definite matrix $A_1 = \begin{bmatrix} 4 & 2 \\ 2 & 4 \end{bmatrix}$, and that $4 \cdot 4 \geq 2^2 \cdot (\sqrt{3})^2 = \det A_1$. (b) Let $A = [a_{ij}] \in M_n$ be positive definite and let $A = LL^*$ be a Cholesky factorization. Let $L = [c_{ij}]$, so $c_{ij} = 0$ if $j > i$. Show that $\det A = c_{11}^2 \cdots c_{nn}^2$. Show that each $a_{ii} = |c_{i1}|^2 + \cdots + |c_{i,i-1}|^2 + c_{ii}^2 \geq$

c_{ii}^2, with equality if and only if $c_{ik} = 0$ for each $k = 1, \ldots, i - 1$. Deduce *Hadamard's inequality*: $\det A \leq a_{11} \cdots a_{nn}$, with equality if and only if A is diagonal.

7.2.P6 Let $n \geq 2$, let $A \in M_n$ be Hermitian, and let $B \in M_{n-1}$ be a leading principal submatrix of A. If B is positive semidefinite and rank $B =$ rank A, show that A is positive semidefinite.

7.2.P7 What are necessary and sufficient conditions on the signs of its minors for a Hermitian matrix A to be negative definite (semidefinite)?

7.2.P8 A positive semidefinite or positive definite matrix can have a Hermitian, but not positive semidefinite, square root. It can also have non-Hermitian square roots. Compute the squares of the matrices $\begin{bmatrix} a & b \\ -a^2/b & -a \end{bmatrix}$ for any $a, b \in \mathbf{C}$ with $b \neq 0$, and $\begin{bmatrix} 1 & 1 \\ 0 & -1 \end{bmatrix}$.

7.2.P9 The representation in (7.2.7) can always be achieved with a matrix B that has full row rank and orthogonal rows. (a) Suppose that $A \in M_n$ is positive semidefinite and let $r =$ rank A. Let $A = U \Lambda U^*$, in which U is unitary, $\Lambda = \Lambda_r \oplus 0_{n-r}$, and Λ_r is positive diagonal. Partition $U = [U_1 \ U_2]$ conformally to Λ. Show that $A = B^* B$ in which $B = \Lambda_r^{1/2} U_1^* \in M_{r,n}$ has full row rank and orthogonal rows. (b) Deduce that a rank one positive semidefinite matrix may always be written in the form xx^* for some $x \in \mathbf{C}^n$.

7.2.P10 Let $A \in M_n$. Theorem 4.1.7 says that A is similar to A^* via a *Hermitian* matrix if and only if A is similar to a real matrix. Show that A is similar to A^* via a *Hermitian positive definite* matrix if and only if A is similar to a real diagonal matrix.

7.2.P11 Let $A \in M_n$ be Hermitian. (a) Show that A is positive definite if and only if adj A is positive definite and $\det A > 0$. (b) If n is odd, show that $\text{adj}(-I_n)$ is positive semidefinite, so the determinant condition in (a) cannot be omitted. (c) If A is positive semidefinite, show that adj A is positive semidefinite and $\det A \geq 0$. (d) If adj A is positive semidefinite and $\det A \geq 0$, show by example that A need not be positive semidefinite.

7.2.P12 Let $r \in \mathbf{C}$ be nonzero and consider the symmetric Toeplitz matrix $M(r, n) = [r^{|i-j|}]_{i,j=1}^n \in M_n(\mathbf{R})$, sometimes called a *Markovian matrix*. Evaluate $D_n = \det M(r, n)$ as follows: (a) Let M_{ij} denote the submatrix of $M(r, n)$ obtained by deleting row i and column j. Show that $\det M_{ij} = 0$ whenever $|i - j| \geq 2$ and explain why adj $M(r, n)$ is tridiagonal and symmetric. (b) Show that $D_2 = 1 - r^2$ and use (a) to show that $D_{n+1} = D_n - r^2 D_n = (1 - r^2) D_n = (1 - r^2)^n$ by expanding according to cofactors of the first row. (c) For $n \geq 2$, conclude that $M(r, n)$ is nonsingular for all nonzero complex $r \neq \pm 1$. (d) For $r \in (-1, 1)$ and $n \geq 2$, use (7.2.5) to show that the real symmetric matrix $M(r, n)$ is positive definite. (e) Show that $f(t) = e^{-|t|}$ is a positive definite function on $\mathbf{R}$; see (7.1.P7).

7.2.P13 If $r \neq \pm 1$, explain why the Markovian matrix $M(r, n)$ in the preceding problem has an inverse that is symmetric and tridiagonal. Show that $(1 - r^2) M(r, n)^{-1}$ has the entry $-r$ in every position of the superdiagonal and subdiagonal; it has main diagonal entries $1, 1 + r^2, \ldots, 1 + r^2, 1$.

7.2.P14 Let $r \in \mathbf{C}$ be nonzero and consider the symmetric Toeplitz matrix $G(r, n) = [r^{(i-j)^2}]_{i,j=1}^n \in M_n$, sometimes called a *Gaussian matrix*. Evaluate $D_n = \det G(r, n)$ as follows: (a) For $j = n, n - 1, \ldots, 2$, subtract r^{2j-3} times column $j - 1$ from column j to produce zero entries in positions $(1, 2), \ldots, (1, n)$. If $\min\{i, j\} \geq 2$, the new entry in position i, j is $1 - r^{2(i-1)}$ times the original entry. (b) Repeat this elimination process $n - 2$

times to obtain a lower triangular matrix. (c) Conclude that $D_n = \prod_{k=1}^{n-1}(1 - r^{2k})D_{n-1} = \prod_{k=1}^{n-1}(1 - r^{2k})^{n-k}$. (d) For $n \geq 2$, conclude that $G(r, n)$ is nonsingular for all nonzero $r \in \mathbf{C}$ such that $r \notin \{z \in \mathbf{C} : z^{2k} = 1, k = 1, \ldots, n - 1\}$. (e) For $r \in (-1, 1)$ and $n \geq 2$, use (7.2.5) to show that the real symmetric matrix $G(r, n)$ is positive definite. (f) Show that $f(t) = e^{-t^2}$ is a positive definite function on $\mathbf{R}$.

7.2.P15 Let $A \in M_n$ be positive semidefinite and have ordered eigenvalues $\mu_1 \leq \cdots \leq \mu_n$. Let $z \in \mathbf{C}^n$ be nonzero. (a) Verify the identities $A + zz^* = \begin{bmatrix} A^{1/2} & z \end{bmatrix} \begin{bmatrix} A^{1/2} \\ z^* \end{bmatrix}$ and $\begin{bmatrix} A^{1/2} \\ z^* \end{bmatrix} \begin{bmatrix} A^{1/2} & z \end{bmatrix} = \begin{bmatrix} A & A^{1/2}z \\ z^*A^{1/2} & z^*z \end{bmatrix}$. Denote the latter matrix by B and let $\lambda_1 \leq \lambda_2 \leq \cdots \leq \lambda_n \leq \lambda_{n+1}$ be its ordered eigenvalues. (b) Deduce from (1.3.22) that $\lambda_2 \leq \cdots \leq \lambda_n \leq \lambda_{n+1}$ are the ordered eigenvalues of $A + zz^*$ and $\lambda_1 = 0$. (c) Explain why each of (4.3.9) and (4.3.17) ensures that the interlacing inequalities

$$\lambda_1 \leq \mu_1 \leq \lambda_2 \leq \mu_2 \leq \cdots \leq \mu_{n-1} \leq \lambda_n \leq \mu_n \leq \lambda_{n+1}$$

are valid. (d) Explain why each of (4.3.9) and (4.3.17) implies the other.

7.2.P16 Let $A \in M_n$ be positive definite and not a scalar matrix. Show that the condition number $\kappa(A + tI)$ with respect to the spectral norm is a strictly monotone decreasing convex function of $t \in [0, \infty)$.

7.2.P17 Let $A, B \in M_n$ and suppose that A is positive definite. Show that $C = A + B + B^* + BA^{-1}B^*$ is positive semidefinite. What is C if $n = 1$ and why is it nonnegative in this case?

7.2.P18 If $A \in M_n$ is nonsingular, show that $B = A + A^{-*}$ is nonsingular.

7.2.P19 Let $A \in M_n$ be positive definite and let $x \in \mathbf{C}^n$ be a unit vector. (a) Show that $(x^*Ax)^{-1} \leq x^*A^{-1}x$, with equality if and only if x is an eigenvector of A. (b) If $A = [a_{ij}]$ is a nonsingular correlation matrix and $A^{-1} = [\alpha_{ij}]$, explain why each $\alpha_{ii} \geq 1$, with equality for some $i = p$ if and only if $a_{pj} = a_{jp} = 0$ for all $j = 1, \ldots, n$ such that $j \neq p$.

7.2.P20 Let $A, B \in M_n$ be positive definite. Theorem 4.5.8 says that there is a nonsingular $S \in M_n$ such that $A = SBS^*$. (a) Show that one may choose $S = A^{1/2}B^{-1/2}$, which need not be Hermitian. (b) Show that one may also choose $S = B^{-1/2}(B^{1/2}AB^{1/2})^{1/2}B^{-1/2}$, and this choice is positive definite. (c) Use (7.2.6a) to show that there is a *unique* positive definite S such that $A = SBS^*$.

7.2.P21 Let $A, B \in M_n$ be positive semidefinite. (a) If A and B commute, show that AB is Hermitian and positive semidefinite. (b) Give an example to show that AB need not be Hermitian. (c) Use (1.3.22) to explain why AB and $A^{1/2}BA^{1/2}$ have the same eigenvalues, and the latter matrix has real nonnegative eigenvalues. Why doesn't this reasoning permit us to conclude that AB is diagonalizable? Nevertheless, see (7.6.2b). (d) If A is positive definite, explain why AB is similar to $A^{1/2}BA^{1/2}$, which is similar to a nonnegative diagonal matrix.

7.2.P22 Let $A, G, H \in M_n$ be positive definite and suppose that $GAG = HAH$. We claim that $G = H$. Provide details: (a) Let $X = A^{1/2}G$ and $Y = A^{1/2}H$. Then $X^*X = Y^*Y$. (b) $(YX^{-1})^{-1} = (YX^{-1})^*$, so $YX^{-1} = A^{1/2}GH^{-1}A^{-1/2}$ is unitary. (c) Every eigenvalue of GH^{-1} has modulus one. (d) GH^{-1} is diagonalizable and each of its eigenvalues is positive. (e) Every eigenvalue of GH^{-1} is $+1$ and hence $GH^{-1} = I$.

7.2.P23 Let $A, B \in M_n$ be positive definite. The matrix

$$G(A, B) = A^{1/2}(A^{-1/2}BA^{-1/2})^{1/2}A^{1/2}$$

is the *geometric mean* of A and B. (a) Why is $G(A, B)$ positive definite? (b) If A and B commute, show that $G(A, B) = A^{1/2}B^{1/2} = B^{1/2}A^{1/2} = G(B, A)$. (c) Show that $X = G(A, B)$ is the *unique* solution of the equation $XA^{-1}X = B$. What is X if $n = 1$? (d) Show that $XA^{-1}X = B$ if and only if $XB^{-1}X = A$, and conclude that $G(A, B) = G(B, A)$. (e) Show that $G(A, \bar{A}) = G(A, A^T)$ is real. (f) Show that $G(A, A^{-T}) = G(A, \bar{A}^{-1})$ is complex orthogonal and coninvolutory.

7.2.P24 Let $A \in M_n$ be positive semidefinite. Represent $A = X^*X$ as the Gram matrix of the columns of $X \in M_n$; see (7.2.7a). Let $k \in \{1, \ldots, n\}$. (a) Explain why every principal minor of A of size k is zero if and only if every list of k vectors chosen from the columns of X is linearly dependent. (b) If every principal minor of A of size k is zero, show that rank $A < k$ and every principal minor of A of size $m \geq k$ is zero. (c) Consider $A = \begin{bmatrix} 0 & 1 \\ 1 & 0 \end{bmatrix}$ and explain why the assertion in (b) need not be correct if A is Hermitian but not positive semidefinite.

7.2.P25 Let $A \in M_n$ be positive semidefinite, suppose that $n = km$, and partition $A = [A_{ij}]_{i,j=1}^k$ as a k-by-k block matrix in which each block is m-by-m. Let $C_p(A_{ij})$ denote the pth compound matrix, $p \in \{1, \ldots, m\}$; see (0.8.1). Recall that tr $C_p(A_{ij}) = E_p(A_{ij})$; see (2.3.P12). We claim that the compressed matrices $\mathcal{T} = [\text{tr } A_{ij}]_{i,j=1}^k \in M_k$, $\mathcal{C}_p = [C_p(A_{ij})]_{i,j=1}^k \in M_k$, $\mathcal{E}_p = [E_p(A_{ij})]_{i,j=1}^k$, and $\mathcal{D} = [\det A_{ij}]_{i,j=1}^k$ are all positive semidefinite. Provide details: (a) Let $A = B^*B$, in which $B \in M_n$. Partition $B = [B_1 \ \ldots \ B_k]$ in which each $B_j \in M_{n,m}$. (b) Show that $\mathcal{T} = [\text{tr}(B_i^*B_j)]_{i,j=1}^k$, explain why $\mathcal{T}$ is a Gram matrix, and conclude that it is positive semidefinite. Compare with (7.1.P25). (c) Use the multiplicativity property of the pth compound matrix to explain why $\mathcal{C}_p = [C_p(B_i^*B_j)]_{i,j=1}^k = [C_p(B_i)^*C_p(B_j)]_{i,j=1}^k$ is a Gram matrix and conclude that it is positive semidefinite. (d) Combine the results in (b) and (c) to show that $\mathcal{E}_p$ is positive semidefinite. (e) Observe that $\mathcal{D} = \mathcal{E}_m$ and conclude that it is positive semidefinite.

7.2.P26 Let $A, B \in M_n$ be positive semidefinite. Show that (a) $0 \leq \text{tr } AB \leq |||A|||_2 \text{ tr } B$. (b) $\sqrt{\text{tr } AB} \leq \sqrt{\text{tr } A}\sqrt{\text{tr } B} \leq \frac{1}{2}(\text{tr } A + \text{tr } B)$.

7.2.P27 Let $A_1, \ldots, A_m \in M_n$ be positive semidefinite. Show that $\left\|\sum_{i=1}^m A_i\right\|_2^2 \geq \sum_{i=1}^m \|A_i\|_2^2$.

7.2.P28 Let $A \in M_n$ be positive semidefinite and let $A = B^*B$ be any representation of the form guaranteed in (7.2.7), with $B = [b_1 \ \ldots \ b_n]$. (a) Show that A is a correlation matrix if and only if each b_j is a unit vector. (b) A vector $x = [x_i] \in \mathbb{C}^n$ is *balanced* if $|x_i| \leq \sum_{j \neq i} |x_j|$ for each $i = 1, \ldots, n$. If A is a correlation matrix, show that every vector in its null space is balanced. (c) Show that every main diagonal entry of $A = [a_{ij}]$ is positive if and only if each $b_j \neq 0$. (d) If every main diagonal entry of A is positive, $D = \text{diag}(\sqrt{a_{11}}, \ldots, \sqrt{a_{nn}})$, and $x \in \text{nullspace } A$, show that Dx is a balanced vector.

7.2.P29 Let $n \geq 2$ and let $A = [a_{ij}] \in M_n$ be a correlation matrix. (a) Explain why $|a_{ij}| \leq 1$ for every pair i, j of distinct indices, with strict inequality if A is positive definite. (b) Use (6.1.1) to show that $\sigma(A) \subset [0, n]$. Consider the example $A = J_n$ and explain why no smaller interval contains every eigenvalue of every n-by-n correlation matrix. (c) If A is positive definite, show that $\lambda \in (0, n)$. (d) Suppose that A is tridiagonal. (i) Use (6.1.1) to show that $\lambda \in [0, 3]$. (ii) Use (1.4.P4) to show that $\lambda \in [0, 2]$, that $\lambda \in \sigma(A)$ if and only if $2 - \lambda \in \sigma(A)$, and that $\lambda = 1$ is an eigenvalue of A if n is odd. (iii) Explain why $\lambda = 2$ is an eigenvalue of A if and only if A is singular and conclude that $\sigma(A) \subset (0, 2)$ if A is positive definite.

7.2.P30 Let $A, B \in M_n$ be Hermitian. (a) If A is positive definite, show that AB is similar to a real diagonal matrix. Show that the numbers of positive, negative, and zero eigenvalues of AB and B are the same. (b) Give an example to show that AB need not be diagonalizable if A is positive semidefinite and singular. (c) If A is positive semidefinite, explain why $\mathrm{tr}(AB) = \mathrm{tr}(A^{1/2}BA^{1/2})$ is real.

7.2.P31 Let $A = [a_{ij}] \in M_n(\mathbf{R})$ be symmetric and positive definite, and suppose that $a_{ij} \leq 0$ if $i \neq j$. We claim that A^{-1} has nonnegative entries. Provide details: (a) Order the eigenvalues of A as $0 < \lambda_1 \leq \cdots \leq \lambda_n$ and let $\mu \geq \max\{\lambda_n, \max_{1 \leq i \leq n} a_{ii}\}$. Then $B = \mu I - A$ has nonnegative entries; its eigenvalues are $\mu - \lambda_1 \geq \cdots \geq \mu - \lambda_n \geq 0$. (b) $\rho(B) = \mu - \lambda_1 < \mu$. (c) $A^{-1} = \mu^{-1}(I - \mu^{-1}B)^{-1} = \mu^{-1}\sum_{k=0}^{\infty}\mu^{-k}B^k \geq 0$. For a generalization (same conclusion but a much weaker hypothesis), see (8.3.P15).

7.2.P32 Let $\langle \cdot, \cdot \rangle$ be an inner product on $\mathbf{C}^n$, let $\mathcal{B} = \{e_1, \ldots, e_n\}$ be the standard orthonormal basis for $\mathbf{C}^n$, and let $G \in M_n$ denote the Gram matrix of $\mathcal{B}$ with respect to the inner product $\langle \cdot, \cdot \rangle$. Show that $\langle x, y \rangle = y^*Gx$ for all $x, y \in \mathbf{C}^n$. Conclude that a function $\langle \cdot, \cdot \rangle : \mathbf{C}^n \times \mathbf{C}^n \to \mathbf{C}^n$ is an inner product if and only if there is a positive definite matrix G such that $\langle x, y \rangle = y^*Gx$ for all $x, y \in \mathbf{C}^n$.

7.2.P33 The *Jordan product* of $A, B \in M_n$ is $]A, B[= AB + BA$. The commutator of A and B is $[A, B] = AB - BA$, so the Jordan product is sometimes called the *anticommutator*. Let A and B be Hermitian. (a) Show that $]A, B[$ is Hermitian, so it has real eigenvalues and a real trace. (b) If A and B are positive definite, show that $\mathrm{tr}\,]A, B[> 0$, but consider $A = \begin{bmatrix} 20 & 0 \\ 0 & 1 \end{bmatrix}$ and $B = \begin{bmatrix} 2 & 1 \\ 1 & 2 \end{bmatrix}$ and conclude that $]A, B[$ can have some negative eigenvalues. (c) Show that $[A, B]$ is skew Hermitian, has pure imaginary eigenvalues, and has trace zero.

Problems (7.2.P34 to P36) explore some ideas that arise in studying finite-dimensional quantum systems.

7.2.P34 Let $R \in M_n$ be a Hermitian positive semidefinite matrix such that $\mathrm{tr}\,R = 1$ (a *density matrix*), and define the function $\mathrm{Cov}_R(\cdot, \cdot) : M_n \times M_n \to \mathbf{C}$ by

$$\mathrm{Cov}_R(X, Y) = \mathrm{tr}(RXY^*) - (\mathrm{tr}(RX))(\mathrm{tr}(RY^*))$$

(the *covariance of X and Y in the state R*). (a) Explain why $\|R^{1/2}\|_2 = 1$ (the Frobenius norm), and why $R^{1/2} = R$ if and only if $\mathrm{rank}\,R = 1$ if and only if $R = uu^*$ for some Euclidean unit vector u. If $\mathrm{rank}\,R = 1$, our quantum system is in a *pure state*; if $\mathrm{rank}\,R > 1$ it is in a *mixed state*. The expression $\mathrm{tr}(RX)$ is interpreted as *the average (mean) of X in the state R*. (b) Show that $\mathrm{Cov}_R(X, Y) = \langle R^{1/2}X, R^{1/2}Y \rangle_F - \langle R^{1/2}X, R^{1/2} \rangle_F \langle R^{1/2}, R^{1/2}Y \rangle_F$ (the Frobenius inner product). (c) Show that $\mathrm{Cov}_R(\cdot, \cdot)$ is sesquilinear and $\mathrm{Cov}_R(X, X) \geq 0$, so $\mathrm{Cov}_R(\cdot, \cdot)$ is a semi-inner product on the complex vector space M_n. Show that $\mathrm{Cov}_R(\lambda I, \mu I) = 0$ and $\mathrm{Cov}_R(X - \lambda I, Y - \mu I) = \mathrm{Cov}_R(X, Y)$ for all $\lambda, \mu \in \mathbf{C}$. (d) Define $\mathrm{Var}_R(X) = \mathrm{Cov}_R(X, X)$ (the *variance of X in the state R*). Show that

$$\mathrm{Var}_R(X) = \mathrm{tr}(RXX^*) - |\mathrm{tr}(RX)|^2$$

This variance is interpreted as the average of XX^* minus the absolute square of the average of X (both in the state R). (e) Use the Cauchy–Schwarz inequality (5.1.8) to explain why

$$\text{Var}_R(X)\,\text{Var}_R(Y) \geq |\text{Cov}_R(X, Y)|^2 \tag{7.2.12}$$

7.2.P35 (Continuation; same notation) Let $A, B \in M_n$ be Hermitian (the *observables* of our quantum system). Let $[A, B]$ and $]A, B[$ be, respectively, the commutator and Jordan product of A and B. (a) Show that

$$\text{Cov}_R(A, B) = \text{tr}(RAB) - (\text{tr}(RA))(\text{tr}(RB))$$

in which both averages $\text{tr}(RA)$ and $\text{tr}(RB)$ are real. (b) Show that

$$\text{Im}\,\text{Cov}_R(A, B) = \frac{1}{2i}(\text{tr}(RAB) - \text{tr}(RBA)) = \frac{1}{2i}\,\text{tr}(R[A, B])$$

which reflects the average of the commutator in the state R and is interpreted as a measure of noncommutativity of the observables A and B in the state R. (c) Let $A_0 = A - (\text{tr}(RA))I$ and $B_0 = B - (\text{tr}(RB))I$. Verify that $\text{tr}(RA_0) = \text{tr}(RB_0) = 0$ (mean zero in the state R). Show that $\text{tr}(R\,]A_0, B_0[) = \text{tr}(R\,]A, B[) - 2(\text{tr}(RA))(\text{tr}(RB))$. (d) Show that

$$\text{Re}\,\text{Cov}_R(A, B) = \frac{1}{2}\,\text{tr}(R\,]A_0, B_0[) = \frac{1}{2}(\text{Cov}_R(A, B) + \text{Cov}_R(B, A))$$

(e) Explain why the inequality

$$\text{Var}_R(A)\,\text{Var}_R(B) \geq \frac{1}{4}|\text{tr}(R[A, B])|^2 + \frac{1}{4}(\text{Cov}_R(A, B) + \text{Cov}_R(B, A))^2 \tag{7.2.13}$$

is just an elaboration of the Cauchy–Schwarz inequality (7.2.12) associated with the semi-inner product $\text{Cov}_R(\cdot, \cdot)$. This is the *Schrödinger uncertainty principle*; it implies the weaker inequality

$$\text{Var}_R(A)\,\text{Var}_R(B) \geq \frac{1}{4}|\text{tr}(R[A, B])|^2 \tag{7.2.14}$$

which is the *Heisenberg uncertainty principle*. (f) If $AR = \lambda R$ for some real λ (R is an *eigenstate of* A), show that $\text{Var}_R(A) = 0$. Explain why every density matrix is an eigenstate of a scalar matrix.

7.2.P36 (Continuation; same notation) Define the function $\text{Corr}_R(\cdot, \cdot) : M_n \times M_n \to \mathbf{C}$ by

$$\text{Corr}_R(X, Y) = \text{tr}(RXY^*) - \text{tr}(R^{1/2}XR^{1/2}Y^*)$$

(the *Wigner–Yanase correlation*). Define $I_R(X) = \text{Corr}_R(X, X)$ (the *Wigner-Yanase skew information*),

so

$$I_R(X) = \text{tr}(RXX^*) - \text{tr}(R^{1/2}XR^{1/2}X^*)$$

(a) Show that $I_R(X)$ is real. (b) Show that

$$\text{Corr}_R(X, Y) = \langle R^{1/2}X, R^{1/2}Y \rangle_F - \langle R^{1/2}X, YR^{1/2} \rangle_F$$

and

$$I_R(X) = \|R^{1/2}X\|_2^2 - \langle R^{1/2}X, XR^{1/2} \rangle_F \geq \|R^{1/2}X\|_2^2 - \|R^{1/2}X\|_2 \|XR^{1/2}\|_2$$

(c) Consider $R = \text{diag}(4, 9)$ and show that $I_R(J_2(0)) = -2$. The problem here is that $\left\| J_2(0) R^{1/2} \right\|_2 > \left\| R^{1/2} J_2(0) \right\|_2$. (d) Explain why $\text{Corr}_R(\cdot, \cdot)$ is a sesquilinear form on M_n that is not a semi-inner product on the complex vector space M_n. Show that $\text{Corr}_R(\lambda I, \mu I) = 0$ and $\text{Corr}_R(X - \lambda I, Y - \mu I) = \text{Corr}_R(X, Y)$ for all $\lambda, \mu \in \mathbf{C}$. (e) If $X \in M_n$ is normal, show that $I_R(X) \geq 0$. (f) Show that $\text{Corr}_R(X, Y) = \text{Cov}_R(X, Y)$ if rank $R = 1$; see (7.2.P34(a)). (g) Now let $A, B \in M_n$ be Hermitian, so $I_R(A)$ is real and nonnegative; it is regarded as a measure of the information content of the density matrix R with respect to an observable A. (h) Explain why $\mathcal{H}_n = \{A \in M_n : A = A^*\}$ is a real vector space. (i) Explain why $\text{Re Corr}_R(\cdot, \cdot)$ is a bilinear function on the real vector space $\mathcal{H}_n$ and $\text{Re Corr}_R(A, A) \geq 0$, so $\text{Re Corr}_R(\cdot, \cdot)$ is a semi-inner product on $\mathcal{H}_n$. (j) Show that

$$I_R(A) = \text{tr}(RA^2) - \text{tr}((R^{1/2}A)^2) = -\frac{1}{2} \text{tr}([R^{1/2}, A]^2) \qquad (7.2.15)$$

$$\text{Re Corr}_R(A, B) = \frac{1}{2}(\text{Corr}_R(A, B) + \text{Corr}_R(B, A))$$

$$= \frac{1}{4}(I_R(A + B) - I_R(A - B)) \qquad (7.2.16)$$

and

$$\text{Im Corr}_R(A, B) = \frac{1}{2i}(\text{Corr}_R(A, B) - \text{Corr}_R(B, A))$$

$$= \frac{1}{2i} \text{tr}(R[A, B]) = \text{Im Cov}_R(A, B) \qquad (7.2.17)$$

(k) Use the Cauchy-Schwarz inequality to show that

$$I_R(A)I_r(B) \geq \frac{1}{4}(\text{Corr}_R(A, B) + \text{Corr}_R(B, A))^2$$

$$= \frac{1}{16}(I_R(A + B) - I_R(A - B))^2 \qquad (7.2.18)$$

(l) Consider the function $f : M_m \times M_n \to \mathbf{C}$ defined by $f(X, Y) = \text{tr}(R^{1/2}XR^{12}Y^*)$. Show that f is sesquilinear and $f(X, X) \geq 0$. (m) Explain why $|f(X, I)|^2 \leq f(X, X)f(I, I) = f(X, X)$ and hence $|\text{tr}(RX)|^2 \leq \text{tr}(R^{1/2}XR^{1/2}X^*)$ for all $X \in M_n$. (n) Explain why $(\text{tr}(RA))^2 \leq \text{tr}((R^{1/2}A)^2)$ and deduce that $I_R(A) \leq \text{Var}_R(A)$ for every observable A (skew information is not greater than variance).

7.3 The polar and singular value decompositions

Each complex scalar can be factored as $z = re^{i\theta}$, in which r is real and nonnegative (we may think of it as a 1-by-1 positive semidefinite matrix) and $e^{i\theta}$ has modulus one (we may think of it as a 1-by-1 unitary matrix). The factor $r = |z|$ is always uniquely determined, but the factor $e^{i\theta}$ is uniquely determined only if z is nonzero. The polar decomposition is a matrix analog of this scalar factorization; it is an immediate consequence of the singular value decomposition.

Theorem 7.3.1 (Polar decomposition). *Let $A \in M_{n,m}$.*

(a) *If $n < m$, then $A = PU$, in which $P \in M_n$ is positive semidefinite and $U \in M_{n,m}$ has orthonormal rows. The factor $P = (AA^*)^{1/2}$ is uniquely determined; it is a polynomial in AA^*. The factor U is uniquely determined if* rank $A = n$.

(b) *If $n = m$, then $A = PU = UQ$, in which $P, Q \in M_n$ are positive semidefinite and $U \in M_n$ is unitary. The factors $P = (AA^*)^{1/2}$ and $Q = (A^*A)^{1/2}$ are uniquely determined; P is a polynomial in AA^* and Q is a polynomial in A^*A. The factor U is uniquely determined if A is nonsingular.*

(c) *If $n > m$, then $A = UQ$, in which $Q \in M_m$ is positive semidefinite and $U \in M_{n,m}$ has orthonormal columns. The factor $Q = (A^*A)^{1/2}$ is uniquely determined; it is a polynomial in A^*A. The factor U is uniquely determined if* rank $A = m$.

(d) *If A is real, the factors P, Q, and U in (a), (b), and (c) may be taken to be real.*

Proof. We adopt the notation of (2.6.3), which ensures that there are unitary matrices $V \in M_n$ and $W \in M_m$, and a nonnegative diagonal matrix $\Sigma \in M_{n,m}$ with a special structure, such that $A = V \Sigma W^*$. Let $q = \min\{n, m\}$ and let $\Sigma_q \in M_q$ be the diagonal matrix of singular values defined in (2.6.3.1).

(a) Let $W = [W_1 \ W_2]$, in which $W_1 \in M_{m,n}$. Then $A = V \Sigma W^* = V[\Sigma_n \ 0]W^* = V \Sigma_n W_1^* = (V \Sigma_n V^*)(V W_1^*) = PU$, in which $P = V \Sigma_n V^*$ is positive semidefinite and $U = V W_1^*$ has orthonormal rows. Since $P^2 = V \Sigma_n \Sigma_n V^* = V \Sigma \Sigma^T V^* = (V \Sigma W^*)(W \Sigma^T V^*) = AA^*$, P is uniquely determined as the (polynomial) positive semidefinite square root of AA^*; see (7.2.6). If rank $A = n$, then Σ_n and P are positive definite, so $U = P^{-1}A$ is uniquely determined.

(b) Let $\Sigma = \Sigma_n$. We have $A = V \Sigma W^* = (V \Sigma V^*)(V W^*) = (V W^*)(W \Sigma W^*)$, so if we let $P = V \Sigma V^*$, $Q = W \Sigma W^*$, and $U = V W^*$, then we have factorizations of the required form. Since $P^2 = AA^*$ and $Q^2 = A^*A$, P and Q are uniquely determined as the respective (polynomial) positive semidefinite square roots of AA^* and A^*A. If A is nonsingular, then $U = P^{-1}A = AQ^{-1}$ is uniquely determined.

(c) Apply (a) to A^*.

(d) Corollary 2.6.7 ensures that the unitary factors V and W may be chosen to be real orthogonal if A is real. The matrices P and Q are real if V and W are real. $\square$

Exercise. Let $x \in \mathbf{C}^n = M_{n,1}$ be nonzero. Show that its polar decomposition is $x = up$, in which $p = \|x\|_2 > 0$ and $u = x/\|x\|_2$.

Exercise. Let $A \in M_n$. Use the polar decomposition to show that AA^* is unitarily similar to A^*A. What unitary matrix provides the similarity?

Uniqueness of the positive semidefinite factors in the polar decompositions has many important consequences. One of them (see (7.3.P33)) motivates the following *thin* version of the singular value decomposition, which shows that *any* unitary matrix that diagonalizes A^*A can be used as the right unitary factor in a singular value decomposition of A.

Theorem 7.3.2. *Let $A \in M_{n,m}$, let $q = \min\{n, m\}$, and let $r = $ rank A. Suppose that $A^*A = W\Lambda W^*$, in which $W \in M_m$ is unitary, $\Lambda = \text{diag}(\sigma_1^2, \ldots, \sigma_r^2) \oplus 0_{m-r}$,* and $\sigma_1 \geq \cdots \geq \sigma_r > 0$ *are the ordered positive singular values of A. Let* $\Sigma_r = $ diag$(\sigma_1, \ldots, \sigma_r) \in M_r$ *and define* $\Sigma = \begin{bmatrix} \Sigma_r & 0 \\ 0 & 0 \end{bmatrix} \in M_{n,m}$.

(a) *(thin SVD) Partition* $W = [W_1 \ W_2]$, *in which* $W_1 \in M_{m,r}$. *There is a* $V_1 \in M_{n,r}$ *with orthonormal columns such that* $A = V_1 \Sigma_r W_1^*$.

(b) *There is a unitary* $V \in M_n$ *such that* $A = V \Sigma W^*$.

(c) *If A is real, then the matrices* W, V, *and* V_1 *in (a) and (b) may be taken to be real.*

Proof. (a) Let $D = \Sigma_r \oplus I_{m-r} \in M_m$ and partition $X = AWD^{-1} = [V_1 \ Z] \in M_{n,m}$, in which $V_1 \in M_{n,r}$. Then $X^*X = D^{-1}W^*A^*AWD^{-1} = D^{-1}\Lambda D^{-1} = I_r \oplus 0_{m-r}$, and a block product computation reveals that

$$\begin{bmatrix} I_r & 0 \\ 0 & 0_{m-r} \end{bmatrix} = X^*X = \begin{bmatrix} V_1^* \\ Z^* \end{bmatrix} [V_1 \ z] = \begin{bmatrix} V_1^*V_1 & \star \\ \star & Z^*Z \end{bmatrix}$$

Therefore, $Z = 0$, V_1 has orthonormal columns, and $A = XDW^* = [V_1 \ 0](\Sigma_r \oplus I_{n-r})\begin{bmatrix} W_1^* \\ W_2^* \end{bmatrix} = V_1\Sigma_r W_1^*$.

(b) Let $V = [V_1 \ V_2] \in M_n$ be unitary and observe that $A = V_1\Sigma_r W_1^* = [V_1 \ V_2]\begin{bmatrix} \Sigma_r & 0 \\ 0 & 0 \end{bmatrix}\begin{bmatrix} W_1^* \\ W_2^* \end{bmatrix} = V\Sigma W^*$. (c) If A is real, then $A^*A = A^TA$ is real, so it can be diagonalized by a real orthogonal matrix W. $\square$

Associated with any $A \in M_{n,m}$ is the Hermitian matrix A^*A, whose eigenvalues tell us what the singular values of A are, but the relationship between the eigenvalues of A^*A and the singular values of A is nonlinear. Another Hermitian matrix associated with A has better properties in this regard.

Theorem 7.3.3. *Let* $A \in M_{n,m}$, *let* $q = \min\{n, m\}$, *let* $\sigma_1 \geq \cdots \geq \sigma_q$ *be the ordered singular values of A, and define the Hermitian matrix*

$$\mathcal{A} = \begin{bmatrix} 0 & A \\ A^* & 0 \end{bmatrix} \tag{7.3.4}$$

The ordered eigenvalues of $\mathcal{A}$ are

$$-\sigma_1 \leq \cdots - \sigma_q \leq \underbrace{0 = \cdots = 0}_{|n-m|} \leq \sigma_q \leq \cdots \leq \sigma_1$$

Proof. Suppose that $n \geq m$ and let $A = V\Sigma W^*$ be a singular value decomposition, in which $\Sigma = [\Sigma_m \ 0]^T \in M_{n,m}$. Write the left unitary factor as $V = [V_1 \ V_2] \in M_n$, in which $V_1 \in M_{n,m}$. Let $\hat{V} = V_1/\sqrt{2}$ and $\hat{W} = W/\sqrt{2}$, and define

$$U = \begin{bmatrix} \hat{V} & -\hat{V} & V_2 \\ \hat{W} & \hat{W} & 0_{m,n-m} \end{bmatrix} \in M_{m+n}$$

A calculation reveals that U is unitary and

$$\mathcal{A} = U \begin{bmatrix} \Sigma_m & 0 & 0 \\ 0 & -\Sigma_m & 0 \\ 0 & 0 & 0_{n-m} \end{bmatrix} U^*$$

If $m < n$, consider A^* instead. $\square$

The preceding theorem provides a bridge to link properties of eigenvalues of Hermitian matrices with properties of singular values of arbitrary matrices; see (7.3.P16) for an example of how to use that bridge. The following two corollaries also make use of it. The first is a pair of singular value perturbation results that are obtained from Weyl's inequalities for eigenvalues of Hermitian matrices, and the Hermitian Hoffman–Wielandt theorem; the second is an interlacing theorem that is obtained from Cauchy's interlacing theorem for bordered Hermitian matrices.

Corollary 7.3.5. *Let $A, B \in M_{n,m}$ and let $q = \min\{m, n\}$. Let $\sigma_1(A) \geq \cdots \geq \sigma_q(A)$ and $\sigma_1(B) \geq \cdots \geq \sigma_q(B)$ be the nonincreasingly ordered singular values of A and B, respectively. Then*

(a) $|\sigma_i(A) - \sigma_i(B)| \leq \||A - B\||_2$ *for each* $i = 1, \ldots, q$
(b) $\sum_{i=1}^{q}(\sigma_i(A) - \sigma_i(B))^2 \leq \|A - B\|_2^2$

Proof. (a) Let $E = A - B$ and apply (6.3.4.1) to $\mathcal{A} = \begin{bmatrix} 0 & A \\ A^* & 0 \end{bmatrix}$ and $\mathcal{E} = \begin{bmatrix} 0 & E \\ E^* & 0 \end{bmatrix}$.
(b) Apply (6.3.9) to $\mathcal{A}$ and $\mathcal{E}$; see the exercise following (6.3.8). □

Exercise. Provide details for the proofs of the two parts of the preceding corollary.

Corollary 7.3.6. *Let $A \in M_{n,m}$, let $q = \min\{m, n\}$, and let $\hat{A}$ be a matrix obtained from A by deleting any one of its columns or rows. Let $\sigma_1 \geq \cdots \geq \sigma_q$ and $\hat{\sigma}_1 \geq \cdots \geq \hat{\sigma}_q$ denote the respective ordered singular values of A and $\hat{A}$, in which we define $\hat{\sigma}_q = 0$ if $n \geq m$ and a column is deleted, or if $n \leq m$ and a row is deleted. Then*

$$\sigma_1 \geq \hat{\sigma}_1 \geq \sigma_2 \geq \hat{\sigma}_2 \geq \cdots \geq \sigma_q \geq \hat{\sigma}_q \tag{7.3.7}$$

Proof. Let $\mathcal{A} = \begin{bmatrix} 0 & A \\ A^* & 0 \end{bmatrix}$. Deleting row i from A corresponds to deleting row i and column i from $\mathcal{A}$; deleting column j from A corresponds to deleting row $n + j$ and column $n + j$ from $\mathcal{A}$. Let $\mathcal{A}_d$ denote the result of performing either deletion on $\mathcal{A}$. Then (4.3.17) ensures that the eigenvalues of $\mathcal{A}_d$ interlace the eigenvalues of $\mathcal{A}$. The preceding theorem ensures that the interlacing inequalities between the eigenvalues of $\mathcal{A}$ and $\mathcal{A}_d$ contain the inequalities (7.3.7). □

Exercise. Provide details for a proof of the preceding corollary.

The following analog of the Courant–Fischer theorem provides another example of the close logical relationship between eigenvalues of Hermitian matrices and singular values of arbitrary matrices.

Theorem 7.3.8. *Let $A \in M_{n,m}$, let $q = \min\{n, m\}$, let $\sigma_1(A) \geq \cdots \geq \sigma_q(A)$ be the ordered singular values of A, and let $k \in \{1, \ldots, q\}$. Then*

$$\sigma_k(A) = \min_{\{S:\dim S = m-k+1\}} \max_{\{x:0\neq x\in S\}} \frac{\|Ax\|_2}{\|x\|_2} \tag{7.3.9}$$

and

$$\sigma_k(A) = \max_{\{S:\dim S = k\}} \min_{\{x:0\neq x\in S\}} \frac{\|Ax\|_2}{\|x\|_2} \tag{7.3.10}$$

Proof. These characterizations follow from (4.2.7) and (4.2.8). If $\lambda_1 \leq \lambda_2 \leq \cdots \leq \lambda_m$ are the ordered eigenvalues of the positive semidefinite Hermitian matrix A^*A, then $\sigma_k^2(A) = \lambda_{m-k+1}(A^*A)$, and (4.2.7) ensures that

$$\sigma_k^2(A) = \lambda_{m-k+1}(A^*A) = \min_{\{S:\dim S=m-k+1\}} \max_{\{x:0\neq x\in S\}} \frac{x^*A^*Ax}{x^*x}$$

$$= \min_{\{S:\dim S=m-k+1\}} \max_{\{x:0\neq x\in S\}} \frac{\|Ax\|_2^2}{\|x\|_2^2}$$

The second identity is proved in the same way. $\square$

Exercise. Let $A \in M_n$. Explain why $\|Ax\|_2 \leq \sigma_1(A)\|x\|_2$ for every $x \in \mathbf{C}^n$. *Hint*: (5.6.2b).

Exercise. Let $A, B \in M_n$. Use the preceding theorem to show that $\sigma_k(AB) \leq \sigma_1(A)\sigma_k(B)$ for each $k = 1, \ldots, n$.

The final theorem of this section states a useful basic principle.

Theorem 7.3.11. *Let n, p, and q be positive integers with $p \leq q$. Let $A \in M_{p,n}$ and $B \in M_{q,n}$. Then $A^*A = B^*B$ if and only if there is a $V \in M_{q,p}$ with orthonormal columns such that $B = VA$. If A and B are real, then V may be taken to be real.*

Proof. If $B = VA$, then $B^*B = A^*V^*VA = A^*A$. Conversely, if $A^*A = B^*B$, then use (7.3.2) and its notation to write $A = V_1\Sigma_r W_1^*$ and $B = V_2\Sigma_r W_1^*$, in which $V_1 \in M_{p,r}$ and $V_2 \in M_{q,r}$ have orthonormal columns. Let $\hat{V}_1 = \begin{bmatrix} V_1 \\ 0 \end{bmatrix} \in M_{q,r}$ (no augmentation is needed if $p = q$). Then $\hat{V}_1$ has orthonormal columns, so (2.1.18) ensures that there is a unitary $U \in M_q$ such that $V_2 = U\hat{V}_1$. If we partition $U = [V \ Z]$ with $V \in M_{q,r}$, then $V_2 = U\hat{V}_1 = [V \ Z]\begin{bmatrix} V_1 \\ 0 \end{bmatrix} = VV_1$, so $B = V_2\Sigma_r W_1^* = VV_1\Sigma_r W_1^* = VA$. If A and B are real, (7.3.2) and (2.1.18) ensure that W_1, V_1, and U may be taken to be real. $\square$

In a typical application of the preceding theorem, we are given a matrix X (perhaps with some special structure) and we use some facts about positive semidefinite matrices to factor X^*X as $X^*X = Y^*Y$, in which Y has some special form. If the dimensions match up correctly, we can conclude that $X = VY$ for some matrix V with orthonormal columns; see (7.3.P34) for an example.

Problems

In the following problems, $\sigma_1(X) \geq \cdots \geq \sigma_q(X)$ are the ordered singular values of $X \in M_{m,n}$ and $q = \min\{m, n\}$.

7.3.P1 Explain why the singular values of A are the eigenvalues of the positive semidefinite factors P and Q in the polar decomposition (7.3.1).

7.3.P2 Let $A, B \in M_n$. Let $A = P_1U_1$ and $B = P_2U_2$ be polar decompositions. Show that A and B are unitarily equivalent if and only if P_1 and P_2 are unitarily similar.

7.3.P3 Show that $A \in M_n$ has a zero singular value if and only if it has a zero eigenvalue.

7.3.P4 Let $A \in M_{m,n}$ and let $A = V\Sigma W^*$ be a singular value decomposition in which diag $\Sigma = [\sigma_1 \; \ldots \; \sigma_q]^T$. Partition $V = [v_1 \; \ldots \; v_m]$ and $W = [w_1 \; \ldots \; w_n]$. (a) Show that $A^*v_k = \sigma_k w_k$, $Aw_k = \sigma_k v_k$, and $v_k^* A w_k = \sigma_k$ for each $k = 1, \ldots, q$. The unit vectors w_k are *(right) singular vectors* of A; the unit vectors v_k are *left singular vectors* of A. (b) Let $i \in \{1, \ldots, q\}$. Show that $\max\{\|Ax\|_2 : x \in \text{span}\{w_i, \ldots, w_n\} \text{ and } \|x\|_2 = 1\} = \sigma_i = \min\{\|Ax\|_2 : x \in \text{span}\{w_1, \ldots, w_i\} \text{ and } \|x\|_2 = 1\}$.

7.3.P5 Let $A, E \in M_{m,n}$, let $k \in \{1, \ldots, q\}$, suppose that σ_k is a simple nonzero singular value of A, and let v_k and w_k, respectively, be unit singular vectors such that $Aw_k = \sigma_k v_k$. (a) Explain why σ_k is a simple eigenvalue of $\mathcal{A} = \begin{bmatrix} 0 & A \\ A^* & 0 \end{bmatrix}$ with associated eigenvector $x = \begin{bmatrix} v_k \\ w_k \end{bmatrix}$. (b) Use (6.3.12) to show that

$$\frac{d}{dt}\sigma_k(A + tE)|_{t=0} = \text{Re } v_k^* E w_k \tag{7.3.12}$$

7.3.P6 Let $B \in M_n(\mathbf{R})$, let $A(t) = \begin{bmatrix} B & x \\ y^* & t \end{bmatrix} \in M_{n+1}(\mathbf{R})$ for all $t \in \mathbf{R}$, and suppose that at least one of B, x, y is nonzero. Let $\mu = \max\{\sigma_1(\begin{bmatrix} B \\ y^* \end{bmatrix}), \sigma_1([B \; x])\}$. (a) Explain why $\sigma_1(A(t)) \geq \mu > 0$ for all t and why there is some $t_0 \in \mathbf{R}$ such that $\sigma_1(A(t_0)) = \min\{\sigma_1(A(t)) : t \in \mathbf{R}\} > 0$. (b) If $\sigma_1(A(t_0))$ is not a simple singular value of $A(t_0)$, explain why $\mu = \sigma_1(A(t_0))$. (c) If $\sigma_1(A(t_0))$ is a simple singular value of $A(t_0)$, use (7.3.12) to show that $\mu = \sigma_1(A(t_0))$.

7.3.P7 Let $A \in M_{m,n}$ and let $A = V\Sigma W^*$ be a singular value decomposition. Define $A^\dagger = W\Sigma^\dagger V^*$, in which $\Sigma^\dagger$ is obtained from Σ by first replacing each nonzero singular value with its inverse and then transposing. Show that: (a) $AA^\dagger$ and $A^\dagger A$ are Hermitian; (b) $AA^\dagger A = A$; (c) $A^\dagger A A^\dagger = A^\dagger$; (d) $A^\dagger = A^{-1}$ if A is square and nonsingular; (e) $(A^\dagger)^\dagger = A$; and (f) $A^\dagger$ is uniquely determined by the properties (a)–(c). The matrix $A^\dagger$ is the *Moore–Penrose generalized inverse* of A. Alternatively, write down a singular value decomposition for $A^\dagger$ and show that its three factors are uniquely determined by (a)–(c).

7.3.P8 This problem is a continuation of (7.1.P28), whose notation we adopt. (a) Use the identities (a)-(c) in the preceding problem to show that $X^*BX = C^*B^\dagger C$ for any X such that $C = BX$. (b) Conclude that the positive semidefinite matrix $A = \begin{bmatrix} B & C \\ C^* & D \end{bmatrix}$ is *congruent to $B \oplus (D - C^*B^\dagger C)$. (c) Explain why $D - C^*B^\dagger C = D - C^*B^{-1}C$ if B is nonsingular. (d) Comment on the wisdom of regarding $D - C^*B^\dagger C$ as a *generalized Schur complement* of B in A.

7.3.P9 A *least squares solution* to the linear system $Ax = b$ is a vector x such that $\|x\|_2$ is minimal among all vectors x for which $\|Ax - b\|_2$ is minimal. Show that $x = A^\dagger b$ is the unique least squares solution to $Ax = b$.

7.3.P10 Let $A = V\Sigma W^*$ be a singular value decomposition of $A \in M_{m,n}$ and let $r = \text{rank } A$. Show that (a) the last $n - r$ columns of W are an orthonormal basis for the null space of A; (b) the first r columns of V are an orthonormal basis for the range of A; (c) the last $n - r$ columns of V are an orthonormal basis for the null space of A^*; and (d) the first r columns of W are an orthonormal basis for the range of A^*.

7.3.P11 Let $A \in M_{m,n}$. Show that $\sigma_1(A) = \max\{|x^*Ay| : x \in \mathbf{C}^m, \ y \in \mathbf{C}^n, \text{ and } \|x\|_2 = \|y\|_2 = 1\}$.

7.3.P12 (a) Let $A \in M_{m,n}$ and $B \in M_{p,n}$, let $C = \begin{bmatrix} A \\ B \end{bmatrix} \in M_{m+p,n}$, let rank $C = r$, and let $C = V\Sigma W^*$ be a singular value decomposition. Show that the last $n - r$ columns of W are an orthonormal basis for the intersection of the null spaces of A and B. (b) If $A \in M_{m,n}$ and $B \in M_{m,p}$, how can one use the singular value decomposition to obtain an orthonormal basis for range $A +$ range B?

7.3.P13 Derive the singular value decomposition (2.6.3) from the polar decomposition (7.3.1).

7.3.P14 Let $A \in M_n$. Show that A is diagonalizable by similarity if and only if there is a positive definite Hermitian matrix P such that $P^{-1}AP$ is normal.

7.3.P15 Let $A \in M_{m,n}$. Show that $A^\dagger = \lim_{t \to 0}(A^*(AA^* + tI)^{-1})$.

7.3.P16 Let $A, B \in M_{m,n}$. Two basic inequalities for singular values are

$$\sigma_{i+j-1}(A + B) \leq \sigma_i(A) + \sigma_j(B) \text{ if } 1 \leq i, j \leq q \text{ and } i + j \leq q + 1 \qquad (7.3.13)$$

and

$$\sigma_{i+j-1}(AB^*) \leq \sigma_i(A)\sigma_j(B) \text{ if } 1 \leq i, j \leq q \text{ and } i + j \leq q + 1 \qquad (7.3.14)$$

(a) To prove (7.3.13), let $\mathcal{A}, \mathcal{B} \in M_{m+n}$ be Hermitian block matrices defined as in (7.3.4). Explain the singular value-eigenvalue identity $\sigma_k(A) = \lambda_{m+n-k+1}(\mathcal{A})$, for any $k \in \{1, \dots, q\}$. Derive (7.3.13) from this identity and the Weyl inequalities (4.3.1). (b) Deduce from (7.3.13) that $\sigma_1(A + B) \leq \sigma_1(A) + \sigma_1(B)$. Why is this not surprising? (c) Give an example to show that the inequality $\sigma_i(A + B) \leq \sigma_i(A) + \sigma_i(B)$ need not be valid if $i > 1$. (d) Prove the perturbation bound

$$|\sigma_i(A + B) - \sigma_i(A)| \leq \sigma_1(B) \text{ for any } i \in \{1, \dots, q\} \qquad (7.3.15)$$

(f) Deduce from (7.3.14) that $\sigma_1(AB^*) \leq \sigma_1(A)\sigma_1(B)$. Why is this not surprising? For a proof of (7.3.14) that uses only tools at hand (polar decomposition, subspace intersection, (7.3.P4(b)), and (7.3.8)), see Theorem 3.3.16 in Horn and Johnson (1991).

7.3.P17 Let the eigenvalues of $A \in M_n$ be arranged so that $|\lambda_1(A)| \geq \cdots \geq |\lambda_n(A)|$. (a) A sequence of inequalities due to H. Weyl (1949) describes a multiplicative majorization between the nonincreasingly ordered absolute eigenvalues and singular values of A:

$$|\lambda_1 \cdots \lambda_k| \leq \sigma_1 \cdots \sigma_k \text{ for each } k = 1, \dots, n \qquad (7.3.16)$$

with equality for $k = n$. For a proof, see Theorem 3.3.2 in Horn and Johnson (1991); see (5.6.P57) for a different proof. Explain why Weyl's product inequalities are valid for $k = 1$ and for $k = n$. (b) The multiplicative inequalities (7.3.16) imply the additive inequalities

$$|\lambda_1| + \cdots + |\lambda_k| \leq \sigma_1 + \cdots + \sigma_k \text{ for each } k = 1, \dots, n \qquad (7.3.17)$$

For a proof, see theorem 3.3.13 in Horn and Johnson (1991). Explain why the inequalities (7.3.15) are not (quite) a majorization relationship between the absolute eigenvalues of A and its singular values.

7.3.P18 The case $k = n$ of the inequalities (7.3.17) can be approached with tools at hand, and without relying on the product inequalities (7.3.16). Adopt the notation of the preceding problem and provide details: (a) Let $A = UTU^*$ in which $T = [t_{ij}]$ is upper triangular and each $t_{ii} = \lambda_i$. Explain why there is a diagonal unitary matrix $D = \text{diag}(d_1, \ldots, d_n)$ such that each $d_i t_{ii} = |\lambda_i|$. (b) Let $DT = V\Sigma W^*$, in which $V = [v_1 \ldots v_n]$ and $W = [w_1 \ldots w_n]$ are unitary, $\Sigma = \text{diag}(s_1, \ldots, s_n)$, and $s_1 \geq \cdots \geq s_n \geq 0$. Why is $s_j = \sigma_j$ for each $j = 1, \ldots, n$? (c) Explain why $\sum_j |\lambda_j| = \text{tr } DT = \text{tr}(\sum_j \sigma_j v_j w_j^*) = |\sum_j \sigma_j w_j^* v_j| \leq \sum_j \sigma_j$, with equality if and only if $w_j = e^{i\theta_j} v_j$ for each j such that $\sigma_j \neq 0$. (d) Deduce that there is a diagonal unitary E such that $V\Sigma W^* = V\Sigma EV^*$. (e) Explain why DT is normal. Conclude that T is diagonal and A is normal. (f) Explain why

$$|\lambda_1| + \cdots + |\lambda_n| \leq \sigma_1 + \cdots + \sigma_n \tag{7.3.18}$$

with equality if and only if A is normal. (g) Explain why

$$|\text{tr } A| \leq \sigma_1 + \cdots + \sigma_n \tag{7.3.19}$$

with equality if and only if A is Hermitian and positive semidefinite.

7.3.P19 Let $A, B \in M_n$. (a) Although AB and BA have the same eigenvalues, after considering $\begin{bmatrix} 0 & 1 \\ 0 & 0 \end{bmatrix}$ and $\begin{bmatrix} 0 & 0 \\ 0 & 1 \end{bmatrix}$, explain why AB and BA need not have the same singular values. (b) Why do AB and B^*A^* have the same singular values? (c) If A and B are Hermitian, show that AB and BA have the same singular values. (d) If A and B are normal, show that AB and BA have the same singular values.

7.3.P20 Let $A \in M_{m,n}$ and let $\mathcal{A}$ denote the matrix (7.3.4). Let $v \in \mathbf{C}^n$ and suppose that $Av \neq 0$. Let $u = (Av)/\|Av\|_2$ and let $y = \frac{1}{\sqrt{2}}\begin{bmatrix} u \\ v \end{bmatrix}$. Show that $y^*\mathcal{A}y = \|Av\|_2$ and compute the upper bound (6.3.17) for $\mathcal{A}$ and y. Conclude that there is at least one singular value of A in the real interval

$$\{t \in \mathbf{R} : |t - \|Av\|_2| \leq \frac{1}{\sqrt{2}} \left\| (A^*u - \|Av\|_2 v) \right\|_2$$

$$= \frac{1}{\sqrt{2}\|Av\|_2} \left\| (A^*A - \|Av\|_2^2 I)v \right\|_2 \}$$

7.3.P21 Use (7.3.15) to explain why "small" perturbations to a matrix cannot decrease its rank, but they can increase it. How small is "small"?

7.3.P22 Show that $A, B \in M_{m,n}$ are unitarily equivalent if and only if $\text{tr}((A^*A)^k) = \text{tr}((B^*B)^k)$ for $k - 1, \ldots, n$. If $m = n$, compare this condition with the one in (2.2.8) that is necessary and sufficient to determine whether A and B are unitarily similar.

7.3.P23 Let $A, B \in M_n$. (a) Show that $AA^* = BB^*$ if and only if there is a unitary U such that $A = BU$. (b) If A is nonsingular, $A = BU$, and U is unitary, show that $A\bar{A} = B\bar{B}$ if and only if $A = U^T B$. (c) If A and B are nonsingular, $AA^* = BB^*$, and $A\bar{A} = B\bar{B}$, show that $A^T\bar{A} = B^T\bar{B}$. (d) Consider $x = [1\ 1]^T$, $y = [1\ -1]^T$, $A = \begin{bmatrix} 0 & x^T \\ 0_{2,1} & 0_2 \end{bmatrix}$, and $B = \begin{bmatrix} 0 & y^T \\ 0_{2,1} & 0_2 \end{bmatrix}$. Explain why the implication in (c) need not be valid if the nonsingularity assumption is omitted. (e) Construct block matrices K_A', $K_B' \in M_{4n}$ by replacing the 2, 4 blocks in the matrices (4.4.32) by zero blocks. If A and B are nonsingular, use (4.4.P46) to show that A and B are unitarily congruent if and only if K_A' and K_B' are unitarily similar.

7.3.P24 Let $A, B \in M_{m,n}$. Show that A and B are unitarily equivalent if and only if $\begin{bmatrix} 0 & A \\ A^* & 0 \end{bmatrix}$ and $\begin{bmatrix} 0 & B \\ B^* & 0 \end{bmatrix}$ are (unitarily) similar.

7.3.P25 Let $A = [a_{ij}] \in M_n$ and let $U \in M_n$ be unitary. (a) Show that $|\operatorname{tr}(UA)| \leq \sum_{i,j} |a_{ij}|$. (b) Show that $\sigma_1 + \cdots + \sigma_n \leq \sum_{i,j} |a_{ij}|$ and compare with (2.3.P14).

7.3.P26 Let $A \in M_2$ be Hermitian, positive semidefinite, and nonzero. Let $\tau = +(\operatorname{tr} A + 2\sqrt{\det A})^{1/2}$. (a) Show that

$$A^{1/2} = \tau^{-1}(A + \sqrt{\det A}\, I_2)$$

(b) Use this representation to compute the square root in the exercise following (7.2.6).

7.3.P27 Let $A \in M_2$ be nonzero, let $A = PU$ and $A = VQ$ be polar decompositions in which P and Q are positive semidefinite (and uniquely determined). Let $s = (\|A\|_2^2 + 2|\det A|)^{1/2}$. Show that

$$P = s^{-1}(AA^* + |\det A| I_2) \quad \text{and} \quad Q = s^{-1}(A^*A + |\det A| I_2)$$

Notice that P and Q are real if A is real.

7.3.P28 Let $A \in M_2$ be nonzero and let θ be any real number such that $\det A = e^{i\theta}|\det A|$. Let $Z_\theta = A + e^{i\theta} \operatorname{adj} A^*$ and let $\delta = |\det Z_\theta|$. (a) Show that $\delta = (\sigma_1 + \sigma_2)^2 \neq 0$. (b) Show that

$$U = \delta^{-1/2}(A + e^{i\theta} \operatorname{adj} A^*)$$

is unitary, and that U^*A and AU^* are positive semidefinite. (c) Let P and Q be the positive semidefinite matrices determined in the preceding problem. Explain why $A = PU = UQ$ are polar decompositions. (d) If A is real, explain why U may be chosen to be real.

7.3.P29 Use the two preceding problems to compute left and right polar decompositions for $A = \begin{bmatrix} 0 & -1 \\ 0 & 0 \end{bmatrix}$.

7.3.P30 If $A \in M_n$ is nonsingular, explain why the ordered singular values of A^{-1} are $\sigma_n^{-1} \geq \cdots \geq \sigma_1^{-1}$.

7.3.P31 Explain why $A \in M_n$ is a scalar multiple of a unitary matrix if and only if $\sigma_1 = \cdots = \sigma_n$.

7.3.P32 Let $A \in M_{n,m}$ and suppose that rank $A = r$. (a) Use the thin singular value decomposition in (7.3.2(a)) to provide a full-rank factorization of A as $A = XY^*$, in which $X \in M_{n,r}$, $Y \in M_{m,r}$, and rank $X = $ rank $Y = r$. (b) Let B be a submatrix of A that is the intersection of r linearly independent rows of A and r linearly independent columns of A. Use the full-rank factorization $A = XY^*$ to show that B is nonsingular.

7.3.P33 Let $A \in M_{n,m}$ with $n \geq m$ and let $P = (A^*A)^{1/2} = W\Sigma W^* \in M_m$. Use (7.3.1) to show that there is a $V \in M_{n,m}$ with orthonormal columns such that $A = V\Sigma W^*$. What about the case $n < m$?

7.3.P34 In (7.2.9) we derived the Cholesky decomposition from the QR factorization. Use (7.3.12) to derive the QR factorization from the Cholesky factorization.

7.3.P35 Let $A \in M_n$, and let $A = PU$ be a polar decomposition. Show that A is normal if and only if $PU = UP$.

7.3.P36 Let $A \in M_n$ be normal and suppose that its singular values are distinct. (a) What can you say about the eigenvalues of A? (b) If A^*A is real, show that A is symmetric.

7.3.P37 Let $A, B \in M_n$ be Hermitian and similar: $A = SBS^{-1}$. If $S = UQ$ is a polar decomposition, show that A and B are unitarily similar via U.

7.3.P38 Let $V, W \in M_n$ be unitary and *congruent: $V = SWS^*$. If $S = PU$ is a polar decomposition, show that V and W are unitarily similar via U.

7.3.P39 Let $\Lambda = \mathrm{diag}(\lambda_1, \ldots, \lambda_r)$ and $M = \mathrm{diag}(\mu_1, \ldots, \mu_r)$, in which $|\lambda_i| = |\mu_i| = 1$ for each $i = 1, \ldots, r$. Let $D = \Lambda \oplus 0_{n-r}$ and $E = M \oplus 0_{n-r}$. Suppose that $S = \begin{bmatrix} S_{11} & S_{12} \\ S_{21} & S_{22} \end{bmatrix} \in M_n$ is nonsingular and $D = SES^*$. (a) Show that S_{11} is nonsingular and $\Lambda = S_{11} M S_{11}^*$. (b) Show that there is a permutation matrix $P \in M_r$ such that $D = PMP^T$.

7.3.P40 Let $A \in M_n$. Suppose that rank $A = r$ and suppose that A is *congruent to diagonal matrices $D = \Lambda \oplus 0_{n-r}$ and $E = M \oplus 0_{n-r}$, in which Λ and M are unitary. Show that Λ and M are permutation similar. Refer to the discussion of the *congruence canonical form in (4.5) (in the special case (4.5.24) of *congruence to a diagonal matrix) and explain why this is not surprising.

7.3.P41 Let $A, B \in M_n$ be normal. Show that A is *congruent to B if and only if rank $A =$ rank B and every ray from the origin in the complex plane contains the same number of nonzero eigenvalues of A and B.

7.3.P42 Let $A \in M_{m,n}$, let $q = \min\{m, n\}$ let $\mathcal{A} = \begin{bmatrix} A & 0 \\ 0 & 0 \end{bmatrix} \in M_{r,s}$, and let $t = \min\{r, s\}$. Explain why the singular values of $\mathcal{A}$ are $\sigma_1(A), \ldots, \sigma_q(A)$ together with $t - q$ zero singular values.

7.3.P43 Let $\|\cdot\|$ be a unitarily invariant norm on M_n. If $A, B \in M_n$ are normal, show that $\|AB\| = \|BA\|$. Compare this result with (5.6.P58(b)).

7.3.P44 Let $A \in M_{m,n}$ and let $\hat{A} \in M_{r,s}$ be a submatrix of A obtained by deleting certain of its rows and/or columns. Let $p = m - r + n - s$. For any $X \in M_{k,\ell}$, let $\sigma_1(X) \geq \sigma_2(X) \geq \cdots$ denote its nonincreasingly ordered singular values and define $\sigma_i(X) = 0$ if $i > \min\{k, \ell\}$. Deduce from (7.3.6) that

$$\sigma_i(A) \geq \sigma_i(\hat{A}) \geq \sigma_{i+p}(A), \quad 1 \leq i \leq \min\{r, s\} \tag{7.3.20}$$

7.3.P45 Let $A, B \in M_n$. We claim that that A is unitarily similar to B if and only if there is a nonsingular $S \in M_n$ such that $A = SBS^{-1}$ and $A^* = SB^*S^{-1}$. Provide details: (a) Show that $A(SS^*) = (SS^*)A$. (b) Let $S = PU$ be a polar decomposition, in which the positive definite matrix P is a polynomial in SS^*. Explain why $AP = PA$. (c) Conclude that $B = S^{-1}AS = U^*AU$. (d) Compare this argument with the one used to prove (2.5.21).

Notes and Further Readings. The real case of (7.3.3) was published by C. Jordan in 1874; some authors refer to the Hermitian matrix in (7.3.4) as the *Wielandt matrix.* For additional applications of (7.3.14) and a historical survey, see R. A. Horn and I. Olkin, When does $A^*A = B^*B$ and why does one want to know?, *Amer. Math. Monthly* 103 (1996) 470–482. Problems (7.3.P26 to 28) provide explicit polar decompositions for 2-by-2 matrices. Explicit polar decompositions are also available for companion matrices; see P. van den Driessche and H. K. Wimmer, Explicit polar decompositions of companion matrices, *Electron. J. Linear Algebra* 1 (1996) 64–69. Problem (7.3.P28) and generalizations to matrices of size three or more are discussed in R. A. Horn, G. Piazza, and T. Politi, Explicit polar decompositions of complex matrices, *Electron. J. Linear Algebra* 18 (2009) 693–699.

7.4 Consequences of the polar and singular value decompositions

The polar and the singular value decompositions appear in a host of interesting matrix analytic problems. We sample several of them in this section; more are presented in the problems. Throughout this section, if $X \in M_{m,n}$, we let $q = \min\{m, n\}$ and denote its nonincreasingly ordered singular values by $\sigma_1(X) \geq \cdots \geq \sigma_q(X)$. The matrix $\Sigma(X) = [s_{ij}]$ is the m-by-n diagonal matrix such that $s_{ii} = \sigma_i(X)$ for each $i = 1, \ldots, q$.

7.4.1 von Neumann's trace theorem. The following singular value inequality plays a key role in many matrix approximation problems. Our proof of the core inequality is in (8.7), where we obtain it as an application of Birkhoff's theorem.

Theorem 7.4.1.1 (von Neumann). *Let* $A, B \in M_{m,n}$, *let* $q = \min\{m, n\}$, *and let* $\sigma_1(A) \geq \cdots \geq \sigma_q(A)$ *and* $\sigma_1(B) \geq \cdots \geq \sigma_q(B)$ *denote the non-increasingly ordered singular values of A and B, respectively. Then*

$$\operatorname{Re} \operatorname{tr}(AB^*) \leq \sum_{i=1}^{q} \sigma_i(A)\sigma_i(B) \qquad (7.4.1.2)$$

Proof. If $m = n$, the stated inequality is exactly the assertion of (8.7.6). If $m > n$, augment A and B with a block of zeroes to obtain square matrices, that is, define $\mathcal{A} = [A \ \ 0], \mathcal{B} = [B \ \ 0] \in M_m$. Then $\mathcal{A}\mathcal{B}^* = AB^*$, so it follows from (8.7.6) that $\operatorname{Re} \operatorname{tr}(AB^*) = \operatorname{Re} \operatorname{tr}(\mathcal{A}\mathcal{B}^*) \leq \sum_{i=1}^{m} \sigma_i(\mathcal{A})\sigma_i(\mathcal{B}) = \sum_{i=1}^{n} \sigma_i(A)\sigma_i(B)$. If $m < n$, define $\mathcal{A} = \begin{bmatrix} A \\ 0 \end{bmatrix}, \mathcal{B} = \begin{bmatrix} B \\ 0 \end{bmatrix} \in M_n$. Then $\mathcal{A}\mathcal{B}^* = \begin{bmatrix} AB^* & 0 \\ 0 & 0 \end{bmatrix}$, so it follows from (8.7.6) again that $\operatorname{Re} \operatorname{tr}(AB^*) = \operatorname{Re} \operatorname{tr}(\mathcal{A}\mathcal{B}^*) \leq \sum_{i=1}^{n} \sigma_i(\mathcal{A})\sigma_i(\mathcal{B}) = \sum_{i=1}^{m} \sigma_i(A)\sigma_i(B)$. $\qquad \square$

Corollary 7.4.1.3. *Let* $A, B \in M_{m,n}$, *let* $q = \min\{m, n\}$, *and let* $\sigma_1(A) \geq \cdots \geq \sigma_q(A)$ *and* $\sigma_1(B) \geq \cdots \geq \sigma_q(B)$ *denote the nonincreasingly ordered singular values of A and B, respectively. Then*

(a) $\|A - B\|_2^2 \geq \sum_{i=1}^{q}(\sigma_i(A) - \sigma_i(B))^2$

(b) $\sum_{i=1}^{q} \sigma_i(A) = \begin{cases} \max\limits_{\text{unitary } U \in M_n} \operatorname{Re} \operatorname{tr}(AU) \text{ if } m \leq n \\ \max\limits_{\text{unitary } U \in M_m} \operatorname{Re} \operatorname{tr}(UA) \text{ if } m \geq n \end{cases}$

(c) $\sum_{i=1}^{q} \sigma_i(A)\sigma_i(B) = \max\{\operatorname{Re} \operatorname{tr}(ATB^*U) : T \in M_n \text{ and } U \in M_m \text{ are unitary}\}$

(d) $\sum_{i=1}^{q} \sigma_i(AB^*) \leq \sum_{i=1}^{q} \sigma_i(A)\sigma_i(B)$

Proof. (a) Use the Frobenius inner product and (7.4.1.2) to compute

$$
\begin{aligned}
\|A - B\|_2^2 &= \langle A - B, A - B \rangle_F \\
&= \langle A, A \rangle_F - \langle A, B \rangle_F - \langle B, A \rangle_F + \langle B, B \rangle_F \\
&= \sum_{i=1}^q \sigma_i^2(A) - 2 \operatorname{Re} \operatorname{tr}(AB^*) + \sum_{i=1}^q \sigma_i^2(B) \\
&\geq \sum_{i=1}^q \sigma_i^2(A) - 2 \sum_{i=1}^q \sigma_i(A)\sigma_i(B) + \sum_{i=1}^q \sigma_i^2(B) \\
&= \sum_{i=1}^q (\sigma_i(A) - \sigma_i(B))^2
\end{aligned}
$$

(b) If $m \leq n$, let $\mathcal{A} = \begin{bmatrix} A \\ 0 \end{bmatrix} \in M_n$ and use (7.4.1.2):

$$
\begin{aligned}
\operatorname{Re} \operatorname{tr}(AU) = \operatorname{Re} \operatorname{tr} \mathcal{A}U &\leq \sum_{i=1}^n \sigma_i(\mathcal{A})\sigma_i(U^*) \\
&= \sum_{i=1}^n \sigma_i(\mathcal{A}) = \sum_{i=1}^q \sigma_i(A)
\end{aligned}
$$

If $\Lambda = PU$ is a polar factorization, then $\operatorname{Re} \operatorname{tr}(AU^*) = \operatorname{Re} \operatorname{tr} P = \sum_{i=1}^q \sigma_i(A)$, so the upper bound can be achieved. If $m \geq n$, let $\mathcal{A} = [A\ \ 0] \in M_m$ and use (7.4.1.2) again:

$$
\begin{aligned}
\operatorname{Re} \operatorname{tr}(UA) = \operatorname{Re} \operatorname{tr} U\mathcal{A} &\leq \sum_{i=1}^m \sigma_i(U)\sigma_i(\mathcal{A}^*) \\
&= \sum_{i=1}^m \sigma_i(\mathcal{A}) = \sum_{i=1}^q \sigma_i(A)
\end{aligned}
$$

If $A = UQ$ is a polar factorization, then $\operatorname{Re} \operatorname{tr}(U^*A) = \operatorname{Re} \operatorname{tr} Q = \sum_{i=1}^q \sigma_i(A)$, so the upper bound can be achieved.

(c) Use (7.4.1.2) to compute $\operatorname{Re} \operatorname{tr}(ATB^*U) \leq \sum_{i=1}^q \sigma_i(AT)\sigma_i(U^*B) = \sum_{i=1}^q \sigma_i(A) \sigma_i(B)$ for any unitary T and U. If $A = V_1 \Sigma_1 W_1^*$ and $B = V_2 \Sigma_2 W_2^*$ are singular value decompositions in which the diagonal entries of Σ_1 and Σ_2 are, respectively, $\sigma_1(A) \geq \cdots \geq \sigma_q(A)$ and $\sigma_1(B) \geq \cdots \geq \sigma_q(B)$, then the upper bound is achieved for $T = W_1 W_2^*$ and $U = V_2 V_1^*$; with these choices, $ATB^*U = V_1 \Sigma_1 \Sigma_2^T V_1^*$, whose trace is $\sum_{i=1}^q \sigma_i(A)\sigma_i(B)$.

(d) Let $AB^* = PU$ be a polar decomposition. Use (7.4.1.2) to compute

$$
\sum_{i=1}^q \sigma_i(AB^*) = \operatorname{tr} P = \operatorname{Re} \operatorname{tr}(AB^*U)
$$

$$
\leq \sum_{i=1}^q \sigma_i(A)\sigma_i(U^*B) = \sum_{i=1}^q \sigma_i(A)\sigma_i(B)
$$

$\square$

Part (b) of the preceding corollary ensures that $\operatorname{Re} \operatorname{tr} A \leq \sum_{i=1}^{q} \sigma_i(A)$ for any $A \in M_{m,n}$, but for some applications it is important to identify the case of equality. For example, if A is square, the following theorem shows that $\operatorname{Re} \operatorname{tr} A = \sum_{i=1}^{q} \sigma_i(A)$ if and only if A is positive semidefinite.

Theorem 7.4.1.4. *Let $A = [a_{ij}] \in M_{m,n}$, let $q = \min\{m, n\}$ and $p = \max\{m, n\}$, let $\alpha = \{1, \ldots, q\}$, and let $\sigma_1 \geq \cdots \geq \sigma_q$ be the nonincreasingly ordered singular values of A. Then $\operatorname{Re} \operatorname{tr} A \leq \sum_{i=1}^{q} \sigma_i$, with equality if and only if the leading principal submatrix $A[\alpha]$ is positive semidefinite and A has no nonzero entries outside this principal submatrix.*

Proof. We are concerned only with the case of equality. To prove that the asserted conditions are sufficient, observe that if the principal submatrix $A[\alpha]$ is positive semidefinite, then its eigenvalues are its singular values, which are also the singular values of A since no other entries of A are nonzero; the trace of $A[\alpha]$ is the sum of its eigenvalues, which is the sum of the singular values of A.

Now suppose that $\operatorname{Re} \sum_{i=1}^{q} a_{ii} = \sum_{i=1}^{q} \sigma_i$. If $A = 0$ there is nothing to prove, so let rank $A = r \geq 1$. If necessary, augment A with zero blocks to obtain a square matrix $\mathcal{A} = \begin{bmatrix} A & 0_{m,p-n} \\ 0_{p-m,n} & 0_{p-m,p-n} \end{bmatrix} \in M_p$, which has the same trace and singular values as A. Let $\mathcal{A} = V \Sigma_r W^*$ be a thin singular value decomposition (7.3.2a), in which $V = [v_1 \ \ldots \ v_r] \in M_{p,r}$ and $W = [w_1 \ \ldots \ w_r] \in M_{p,r}$ have orthonormal columns, and $\Sigma_r = \operatorname{diag}(\sigma_1, \ldots, \sigma_r)$. Then

$$\operatorname{Re} \operatorname{tr} \mathcal{A} = \operatorname{Re} \operatorname{tr} A = \operatorname{Re} \sum_{i=1}^{q} a_{ii} = \operatorname{Re} \sum_{i=1}^{p} \sum_{k=1}^{r} v_{ik} \sigma_k \bar{w}_{ik}$$

$$= \operatorname{Re} \sum_{k=1}^{r} \sigma_k \sum_{i=1}^{p} v_{ik} \bar{w}_{ik}$$

$$= \sum_{k=1}^{r} \sigma_k \operatorname{Re}(w_k^* v_k) = \sum_{k=1}^{r} \sigma_k = \sum_{k=1}^{q} \sigma_k$$

It follows that $\operatorname{Re}(w_k^* v_k) = 1$ for each $k = 1, \ldots, r$. Since

$$1 = \operatorname{Re}(w_k^* v_k) \overset{(\gamma)}{\leq} |w_k^* v_k| \overset{(\delta)}{\leq} \|v_k\|_2^2 \|w_k\|_2^2 = 1$$

equality at (δ) and the equality case of the Cauchy–Schwarz inequality ensure that there are scalars d_k such that $v_k = d_k w_k$ for each $k = 1, \ldots, r$; equality at (γ) ensures that each $d_k = 1$. Therefore, $V = W$ and $\mathcal{A} = V \Sigma_r V^*$ is positive semidefinite. It follows that its principal submatrix $A[\alpha]$ is positive semidefinite (7.1.2) and no other entries of $\mathcal{A}$ (and hence of A) are nonzero (7.1.10). $\qquad\square$

Corollary 7.4.1.5. *Let $A, B \in M_{m,n}$, let $q = \min\{m, n\}$, and let $\sigma_1(A) \geq \cdots \geq \sigma_q(A)$ and $\sigma_1(B) \geq \cdots \geq \sigma_q(B)$ denote the nonincreasingly ordered singular values of A and B, respectively. Then*

(a) $\|A - B\|_2^2 \geq \sum_{i=1}^{q} (\sigma_i(A) - \sigma_i(B))^2$, with equality if and only if $\operatorname{Re} \operatorname{tr}(AB^*) = \sum_{i=1}^{q} \sigma_i(A)\sigma_i(B)$

(b) If $\|A - B\|_2^2 = \sum_{i=1}^{q}(\sigma_i(A) - \sigma_i(B))^2$, then both AB^* and B^*A are positive semidefinite, so $\operatorname{tr}(AB^*)$ is real and nonnegative

Proof. (a) The stated inequality is (7.4.1.3a), so we are concerned only with the case of equality, which occurs if and only if the one inequality in the proof of (7.4.1.3a) is an equality, that is, if and only if $\operatorname{Re}\operatorname{tr}(AB^*) = \sum_{i=1}^{q}\sigma_i(A)\sigma_i(B)$.

(b) If $\|A - B\|_2^2 = \sum_{i=1}^{q}(\sigma_i(A) - \sigma_i(B))^2$, the preceding theorem and (7.4.1.3d) ensure that

$$\operatorname{Re}\operatorname{tr}(AB^*) \le \sum_{i=1}^{q}\sigma_i(AB^*) \le \sum_{i=1}^{q}\sigma_i(A)\sigma_i(B) = \operatorname{Re}\operatorname{tr}(AB^*)$$

so $\operatorname{Re}\operatorname{tr}(AB^*) = \sum_{i=1}^{q}\sigma_i(AB^*)$, and the preceding theorem ensures that AB^* is positive semidefinite. Since $\|A - B\|_2^2 = \|B^* - A^*\|_2^2$, it follows that $B^*(A^*)^* = B^*A$ is also positive semidefinite. $\qquad\square$

7.4.2 A nearest singular matrix and a nearest rank-k matrix.

Every matrix that is sufficiently close to a nonsingular matrix A (with respect to some norm) is nonsingular (see the exercise preceding (5.6.17)), but what can we say about the distance from A to the closed set of singular matrices? How can we identify a nearest singular matrix? Is it unique?

Let $A = V\Sigma W^* \in M_n$ be a singular value decomposition in which $\Sigma = \operatorname{diag}(\sigma_1(A), \ldots, \sigma_n(A))$ and $\sigma_n(A) > 0$. If $B \in M_n$ is singular, then $\sigma_n(B) = 0$. The inequality (7.4.1.3a) ensures that

$$\|A - B\|_2^2 > \sum_{i=1}^{n}(\sigma_i(A) - \sigma_i(B))^2$$

$$= \sum_{i=1}^{n-1}(\sigma_i(A) - \sigma_i(B))^2 + \sigma_n^2(A) > \sigma_n^2(A)$$

for every singular $B \in M_n$, so any B such that $\|A - B\|_2^2 = \sigma_n^2(A)$ is a closest singular matrix to A in the Frobenius norm. The singular values of such a matrix are uniquely determined: they must be the $n - 1$ largest singular values of A and one zero. If we let $\Sigma_0 = \operatorname{diag}(\sigma_1(A), \ldots, \sigma_{n-1}(A), 0)$ and take $B_0 = V\Sigma_0 W^*$, then $\|A - B_0\|_2^2 = \sum_{i=1}^{n}(\sigma_i(A) - \sigma_i(B))^2 = \sigma_n^2(A)$ and, as predicted by (7.4.1.5), AB_0^* and B_0^*A are positive semidefinite. The distance in the Frobenius norm from A to B_0 is $\sigma_n(A)$, and no singular matrix can be closer, so $\sigma_n(A)$ is the distance in the Frobenius norm from A to the closed set of singular matrices. We may think of B_0 as a best singular approximation to A in the Frobenius norm.

What about uniqueness? If $\sigma_{n-1}(A) = \sigma_n(A)$, let $\hat{\Sigma}_0 = \operatorname{diag}(\sigma_1(A), \ldots, \sigma_{n-2}(A), 0, \sigma_n(A))$. Then $C_0 = V\hat{\Sigma}_0 W^*$ is singular, $B_0 \ne C_0$, and $\|A - C_0\|_2 = \sigma_{n-1}(A) = \sigma_n(A) = \|A - B_0\|_2$, so a best singular approximation to A is not unique in this case. If $\sigma_{n-1}(A) > \sigma_n(A)$, however, then B_0 is the only singular matrix such that $\|A - B\|_2 = \sigma_n(A)$; see (7.4.P17).

If $A \in M_{m,n}$, $\operatorname{rank} A = r$, and $1 \le k < r$, the same principles can be employed to find a "best rank-k approximation" to A. Let $A = V\Sigma W^*$ be a singular value decomposition in which the diagonal entries of $\Sigma \in M_{m,n}$ are $\sigma_1(A) \geq \cdots \geq \sigma_q(A)$). If $B \in M_{m,n}$ and rank $B = k$, then (7.4.1.3a) ensures that

$$\|A - B\|_2^2 \geq \sum_{i=1}^q (\sigma_i(A) - \sigma_i(B))^2$$

$$= \sum_{i=1}^k (\sigma_i(A) - \sigma_i(B))^2 + \sum_{i=k+1}^q \sigma_i^2(A) \geq \sum_{i=k+1}^q \sigma_i^2(A)$$

so any B such that $\|A - B\|_2^2 = \sum_{i=k+1}^q \sigma_i^2(A)$ is a best rank-k approximation to A. The singular values of such a matrix are again uniquely determined: They must be the k largest singular values of A and $q - k$ zeroes. If we let $\Sigma_0 \in M_{m,n}$ be the diagonal matrix whose diagonal entries are $\sigma_1(A), \ldots, \sigma_k(A)$ and $q - k$ zeroes and take $B_0 = V\Sigma_0 W^*$, then $\|A - B_0\|_2^2 = \sum_{i=k+1}^q \sigma_i^2(A)$, so B_0 is a best rank-k approximation to A; it is unique if and only if $\sigma_{k-1}(A) > \sigma_k(A)$. The distance in the Frobenius norm from A to a closest rank k matrix is $(\sum_{i=k+1}^q \sigma_i^2(A))^{1/2}$.

7.4.3 Least squares solution of a linear system. Let $A \in M_{m,n}$ and $b \in \mathbf{C}^m$ be given, let $m \geq n$, and suppose that rank $A = k$. Consider how one might "solve" the linear system $Ax = b$ using the singular value decomposition $A = V\Sigma W^*$. We want to choose an $x \in \mathbf{C}^n$ so that $\|Ax - b\|_2$ is minimized. Partition $V = [v_1 \ \ldots \ v_m] \in M_m$ and $W = [w_1 \ \ldots \ w_n] \in M_n$ according to their columns. The vector $Ax - b = V\Sigma W^*x - b$ has the same Euclidean norm as the vector $\Sigma W^*x - V^*b$. Let $\xi = W^*x$ and $\beta = V^*b$, so $\xi = [\xi_i] = [w_i^*x]_{i=1}^n$ and $\beta = [\beta_i] = [w_i^*\beta]_{i=1}^m$. The Euclidean norm of the vector

$$\Sigma\xi - \beta = \begin{bmatrix} \sigma_1\xi_1 - \beta_1 & \cdots & \sigma_k\xi_k - \beta_k & \beta_{k+1} & \cdots & \beta_m \end{bmatrix}^T$$

achieves its minimum value $(\sum_{i=k+1}^m |\beta_i|^2)^{1/2}$ if we choose $\xi_i = \sigma_i^{-1}\beta_i = \sigma_i^{-1}w_i^*\beta$ for each $i = 1, \ldots, k$; the Euclidean norm of the vector ξ is minimized if we then choose $\xi_i = 0$ for each $i = k + 1, \ldots, n$. That is, $x = \sum_{i=1}^k (\sigma_i^{-1}v_i^*b)w_i$ is a vector of minimum Euclidean norm such that $\|Ax - b\|_2$ achieves its minimum value of $(\sum_{i=k+1}^m |v_i^*b|^2)^{1/2}$.

Exercise. If $A \in M_{m,n}$ and rank $A = n$, use the preceding analysis to explain why there is an $x \in \mathbf{C}^n$ such that $Ax = b$ if and only if b is orthogonal to nullspace A^*. Explain why this solution x is unique and can be expressed as $x = (A^*A)^{-1}A^*b$.

7.4.4 Approximation by a scalar multiple of a unitary matrix. What is the best least squares approximation to a given $A \in M_n$ by a scalar multiple of a unitary matrix? Invoking (7.4.1.3a), for any unitary $U \in M_n$ and any $c \in \mathbf{C}$, we have

$$\|A - cU\|_2^2 \geq \sum_{i=1}^n (\sigma_i(A) - \sigma_i(cU))^2 = \sum_{i=1}^n (\sigma_i(A) - |c|\sigma_i(U))^2$$

$$= \sum_{i=1}^n (\sigma_i(A) - |c|)^2 = \sum_{i=1}^n \sigma_i^2(A) - 2|c|\sum_{i=1}^n \sigma_i(A) + n|c|^2$$

which is minimized if $|c| = \frac{1}{n} \sum_{i=1}^{n} \sigma_i(A) = \mu$, the mean of the singular values of A. The resulting lower bound is

$$\|A - cU\|_2^2 \geq \sum_{i=1}^{n} \sigma_i^2(A) - n\mu^2 = \sum_{i=1}^{n} (\sigma_i(A) - \mu)^2$$

for any unitary $U \in M_n$. We know that if cU is a minimizer, then $(cU)^*A$ is positive semidefinite. The polar decomposition (7.3.1) suggests that if $A = PU_0$ is a polar decomposition, then its unitary polar factor might be a good candidate. We compute $\operatorname{tr} P = \sum_{i=1}^{n} \sigma_i(A) = n\mu$ and

$$\|PU_0 - \mu U_0\|_2^2 = \|P - \mu I\|_2^2 = \operatorname{tr} P^2 - 2\mu \operatorname{tr} P + n\mu^2$$
$$= \|A\|_2^2 - 2n\mu^2 + n\mu^2 = \|A\|_2^2 - n\mu^2$$

so $(\frac{1}{n} \operatorname{tr} P)U_0$ is a best least squares approximation to A by a scalar multiple of a unitary matrix.

7.4.5 The unitary procrustes problem. Let $A, B \in M_{m,n}$. How well can A be approximated in the Frobenius norm by the "rotation" UB for some unitary $U \in M_m$? This question is known in factor analysis as the *unitary Procrustes problem for A and B*.

For any unitary $U \in M_m$ we have

$$\|A - UB\|_2^2 = \|A\|_2^2 - 2\operatorname{Re}\operatorname{tr}(AB^*U^*) + \|B\|_2^2$$

$$\geq \|A\|_2^2 - 2\sum_{i=1}^{m} \sigma_i(AB^*) + \|B\|_2^2 \qquad (7.4.5.1)$$

with equality if and only if AB^*U^* is positive semidefinite. If $AB^* = PU_0$ is a polar decomposition, then $AB^*U_0^* = P$ is positive semidefinite and $\operatorname{tr}(AB^*U_0^*) = \operatorname{tr} P = \sum_{i=1}^{m} \sigma_i(AB^*)$, so $\|A - UB\|_2^2 = \|A\|_2^2 - 2\operatorname{tr} P + \|B\|_2^2$ achieves the lower bound in (7.4.5.1).

Thus, $U_0 B$ is a best least squares approximation to A by a unitary rotation of B and $\|A - U_0 B\|_2^2 = \|A\|_2^2 - 2\operatorname{tr} P + \|B\|_2^2$.

7.4.6 A two-sided rotation problem. Let $A, B \in M_{m,n}$. How well can A be approximated in the Frobenius norm by a two-sided rotation UBT for some unitary matrices $U \in M_m$ and $T \in M_m$?

For any such unitary U and T, (7.4.1.3(a)) ensures that

$$\|A - UBT\|_2^2 \geq \sum_{i=1}^{q} (\sigma_i(A) - \sigma_i(UBT))^2 = \sum_{i=1}^{q} (\sigma_i(A) - \sigma_i(B))^2 \qquad (7.4.6.1)$$

Let $A = V_1 \Sigma_1 W_1^*$ be a singular value decomposition in which the diagonal entries of Σ_1 are $\sigma_1(A) \geq \cdots \geq \sigma_q(A)$, and let $B = V_2 \Sigma_2 W_2^*$ be a singular value decomposition in which the diagonal entries of Σ_2 are $\sigma_1(B) \geq \cdots \geq \sigma_q(B)$. Let $U_0 = V_1 V_2^*$ and $T_0 = W_2 W_1^*$. Then $\|A - U_0 B T_0\|_2^2 = \|V_1 \Sigma_1 W_1^* - V_1 \Sigma_2 W_1^*\|_2^2 = \|\Sigma_1 - \Sigma_2\|_2^2 = \sum_{i=1}^{m} (\sigma_i(A) - \sigma_i(B))^2$, so $U_0 B T_0$ achieves the lower bound in (7.4.6.1).

7.4.7 Unitarily invariant norms and symmetric gauge functions. If $A \in M_{m,n}$, if $A = V\Sigma W^*$ is a singular value decomposition, and if $\| \cdot \|$ is a unitarily invariant norm, then $\|A\| = \|V\Sigma W^*\| = \|\Sigma\|$, so a unitarily invariant norm of a matrix depends only on its singular values. What can we say about the nature of that dependence?

Suppose that $X = [x_{ij}]$, $Y = [y_{ij}] \in M_{m,n}$ are diagonal and have diagonal entries $x_{ii} = x_i$ and $y_{ii} = y_i$ for $i = 1, \ldots, q$. Let $x = [x_i]$, $y = [y_i] \in \mathbf{C}^q$. Then X^*X is diagonal and has diagonal entries $|x_1|^2, \ldots, |x_q|^2$ (and $n - q$ additional zero entries if $q < n$), so the (not necessarily nonincreasingly ordered) singular values of X are $|x_1|, \ldots, |x_q|$. Define the function $g : \mathbf{C}^q \to \mathbf{R}^+$ by

$$g(x) = g([x_1 \ \cdots \ x_q]^T) = \|X\|$$

The function g inherits certain properties from the norm $\| \cdot \|$:

(a) $g(x) \geq 0$ for all $x \in \mathbf{C}^q$ since $\|X\|$ is always nonnegative.
(b) $g(x) = 0$ if and only if $x = 0$ since $\|X\| = 0$ if and only if $X = 0$.
(c) $g(\alpha x) = |\alpha| g(x)$ for all $x \in \mathbf{C}^q$ and all $\alpha \in \mathbf{C}$ since $\|\alpha X\| = |\alpha| \|X\|$ for all $\alpha \in \mathbf{C}$ and all $X \in M_{m,n}$.
(d) $g(x + y) \leq g(x) + g(y)$ for all $x, y \in \mathbf{C}^q$ since $\|X + Y\| \leq \|X\| + \|Y\|$ for all $X, Y \in M_{m,n}$.

These four properties ensure that g is a *norm* on $\mathbf{C}^q$; it has two additional properties:

(e) g is an absolute norm on $\mathbf{C}^q$ since the matrices X and $|X|$ associated with the vectors $x = [x_i]$ and $|x| = [|x_i|]$ have the same singular values, namely, $|x_1|, \ldots, |x_q|$.
(f) $g(Px) = g(x)$ for all $x \in \mathbf{C}^q$ and every permutation matrix $P \in M_q$ since $\| \cdot \|$ is unitarily invariant. For example, if $q = m \leq n$, then $g(x) = \|X\| = \|PX(P^T \oplus I_{n-m})\| = g(Px)$.

Exercise. If $q = n \leq m$, explain why $g(x) = g(Px)$ for all $x \in \mathbf{C}^n$ and every permutation matrix $P \in M_n$.

Exercise. Explain why the Euclidean norm, the max norm, and the sum norm are the vector norms g associated with, respectively, the Frobenius, spectral, and trace norms according to the preceding prescription. Are these norms permutation-invariant absolute norms?

Definition 7.4.7.1. *A function* $g : \mathbf{C}^q \to \mathbf{R}^+$ *is a* symmetric gauge function *if it is an absolute vector norm such that* $g(x) = g(Px)$ *for every* $x \in \mathbf{C}^q$ *and every permutation matrix* $P \in M_q$.

The preceding discussion shows that every unitarily invariant norm on $M_{m,n}$ determines a symmetric gauge function on $\mathbf{C}^q$. The interesting half of the following theorem says that every unitarily invariant norm is determined by a symmetric gauge function, so there is a one-to-one correspondence between unitarily invariant norms on $M_{m,n}$ and symmetric gauge functions on $\mathbf{C}^q$.

Theorem 7.4.7.2. *Let m and n be given positive integers and let $q = \min\{m, n\}$. For any $A \in M_{m,n}$, let $A = V\Sigma(A)W^*$, in which $V \in M_m$ and $W \in M_n$ are unitary and $\Sigma(A) = [s_{ij}] \in M_{m,n}$ is a nonnegative diagonal matrix whose diagonal entries are* the nonincreasingly ordered singular values of A: $\sigma_1(A) \geq \cdots \geq \sigma_q(A)$. Let $s(A) = [\sigma_1(A) \ldots \sigma_q(A)]^T$.

(a) Let $\| \cdot \|$ be a unitarily invariant norm on $M_{m,n}$. For any $x = [x_i] \in \mathbf{C}^q$, let $X = [x_{ij}] \in M_{m,n}$ be the diagonal matrix such that $x_{ii} = x_i$ for each $i = 1, \ldots, q$. The function $g : \mathbf{C}^q \to \mathbf{R}^+$ defined by $g(x) = \|X\|$ is a symmetric gauge function on $\mathbf{C}^q$.

(b) Let g be a symmetric gauge function on $\mathbf{C}^q$. The function $\| \cdot \| : M_{m,n} \to \mathbf{R}^+$ defined by $\|A\| = g(s(A))$ is a unitarily invariant norm on $M_{m,n}$.

Proof. The assertion in (a) has already been proved, so we have only (b) to deal with. First observe that $\| \cdot \|$ is a well-defined function on $M_{m,n}$ because the singular values of a matrix are uniquely determined. Unitary invariance of the singular values of a matrix ensures that $\|UAV\| = g(s(UAV)) = g(s(A)) = \|A\|$ for all unitary $U \in M_m$ and $V \in M_n$. Because g is a vector norm, we have $\|A\| \geq 0$ for all $A \in M_{m,n}$, with equality if and only if $g(s(A)) = 0$ if and only if $s(A) = 0$ if and only if $A = 0$. Homogeneity follows from observing that $\|cA\| = g(s(cA)) = g(|c|s(A)) = |c|\, g(s(A)) = |c|\, \|A\|$.

Finally, we must show that $\| \cdot \|$ satisfies the triangle inequality. For any given $A, B \in M_{m,n}$, compute

$$\|A + B\| = g(s(A + B)) \overset{(\alpha)}{=} g^{DD}(s(A + B)) \overset{(\beta)}{=} \max_{g^D(y)=1} \operatorname{Re}(y^* s(A + B))$$

$$= \max_{g^D(s(C))=1} \sum_{i=1}^{q} \sigma_i(A + B)\sigma_i(C)$$

$$\overset{(\gamma)}{=} \max_{g^D(s(\Sigma))=1} \max_{T,U \text{ unitary}} \operatorname{Re} \operatorname{tr}((A + B)T\Sigma^* U)$$

$$\leq \max_{g^D(s(\Sigma))=1} \max_{T,U \text{ unitary}} \operatorname{Re} \operatorname{tr}(AT\Sigma^* U)$$

$$+ \max_{g^D(s(\Sigma))=1} \max_{T,U \text{ unitary}} \operatorname{Re} \operatorname{tr}(BT\Sigma^* U)$$

$$\overset{(\gamma)}{=} \max_{g^D(s(\Sigma))=1} \sum_{i=1}^{q} \sigma_i(A)\sigma_i(\Sigma) + \max_{g^D(s(\Sigma))=1} \sum_{i=1}^{q} \sigma_i(B)\sigma_i(\Sigma)$$

$$= \max_{g^D(y)=1} \operatorname{Re}(y^* s(A)) + \max_{g^D(y)=1} \operatorname{Re}(y^* s(B))$$

$$\overset{(\beta)}{=} g^{DD}(s(A)) + g^{DD}(s(B)) \overset{(\alpha)}{=} g(s(A)) + g(s(B))$$

$$= \|A\| + \|B\|$$

We have used (5.5.9(c)) and the hypothesis that g is a norm at the identities labeled (α), (5.5.10) at the identities labeled (β), and (7.4.1.3(c)) at the identities labeled (γ). We have also made use of (4.3.52) and the fact that the vectors $s(A + B)$, $s(A)$, and $s(B)$ have nonnegative entries, so only y vectors with nonnegative entries need to be considered in achieving the respective maxima. $\square$

A familiar example of a family of symmetric gauge functions on $\mathbf{C}^n$ is the family of l_p norms (5.2.4). The unitarily invariant norms on $M_{m,n}$ determined by the l_p norms are known as *Schatten p-norms*.

7.4.8 Ky Fan's dominance theorem. The family of k-norms (5.2.5) are symmetric gauge functions that play a special role in the theory of unitarily invariant norms. The corresponding unitarily invariant norms on $M_{m,n}$ are known as *Ky Fan k-norms*; we denote them by

$$\|A\|_{[k]} = \sigma_1(A) + \cdots + \sigma_k(A), \quad k = 1, \ldots, q = \min\{m, n\} \tag{7.4.8.1}$$

Suppose that $\|\cdot\|$ is a unitarily invariant norm on $M_{m,n}$ and let g be the symmetric gauge function associated with it, as described in the preceding section, whose notation we adopt. Then, for any $A \in M_{m,n}$, we have

$$\|A\| = g(s(A)) = \max_{g^D(y)=1} \mathrm{Re}(y^* s(A))$$

$$= \max_{g^D(s(\Sigma))=1} \sum_{i=1}^{q} \sigma_i(A)\sigma_i(\Sigma) \tag{7.4.8.2}$$

and a summation by parts yields the identity

$$\sum_{i=1}^{q} \sigma_i(A)\sigma_i(\Sigma) = \sigma_1(\Sigma)\sigma_1(A) + \sum_{i=2}^{q-1}\left((\sigma_i(\Sigma) - \sigma_{i+1}(\Sigma))\sum_{j=1}^{i}\sigma_j(A)\right)$$

$$+ \sigma_q(\Sigma)\sum_{j=1}^{q}\sigma_i(A)$$

$$= \sigma_1(\Sigma)\|A\|_{[1]} + \sum_{i=2}^{q-1}(\sigma_i(\Sigma) - \sigma_{i+1}(\Sigma))\|A\|_{[i]} \tag{7.4.8.3}$$

$$+ \sigma_q(\Sigma)\|A\|_{[q]}$$

Observe that $\sigma_1(\Sigma) \geq 0$, each $\sigma_i(\Sigma) - \sigma_{i+1}(\Sigma) \geq 0$, and $\sigma_q(\Sigma) \geq 0$, so if $B \in M_{m,n}$ and $\|A\|_{[k]} \leq \|B\|_{[k]}$ for each $k = 1, \ldots, q$, then

$$\sum_{i=1}^{q} \sigma_i(A)\sigma_i(\Sigma) = \sigma_1(\Sigma)\|A\|_{[1]} + \sum_{i=2}^{q-1}(\sigma_i(\Sigma) - \sigma_{i+1}(\Sigma))\|A\|_{[i]}$$

$$+ \sigma_q(\Sigma)\|A\|_{[q]}$$

$$\leq \sigma_1(\Sigma)\|B\|_{[1]} + \sum_{i=2}^{q-1}(\sigma_i(\Sigma) - \sigma_{i+1}(\Sigma))\|B\|_{[i]}$$

$$+ \sigma_q(\Sigma)\|B\|_{[q]}$$

$$= \sum_{i=1}^{q} \sigma_i(B)\sigma_i(\Sigma)$$

Thus,

$$\|A\| = \max_{g^D(s(\Sigma))=1} \sum_{i=1}^{q} \sigma_i(A)\sigma_i(\Sigma)$$

$$\leq \max_{g^D(s(\Sigma))=1} \sum_{i=1}^{q} \sigma_i(B)\sigma_i(\Sigma) = \|B\|$$

This argument shows that if $\|A\|_{[k]} \leq \|B\|_{[k]}$ for each $k = 1, \ldots, q$, then $\|A\| \leq \|B\|$ for *every* unitarily invariant norm $\|\cdot\|$. Conversely, if $\|A\| \leq \|B\|$ for every unitarily invariant norm $\|\cdot\|$ on $M_{m,n}$, then this inequality must be valid for the Ky Fan k-norms on $M_{m,n}$. We summarize these conclusions in the following theorem.

Theorem 7.4.8.4. *Let $A, B \in M_{m,n}$ be given. Then $\|A\| \leq \|B\|$ for every unitarily invariant norm $\|\cdot\|$ on $M_{m,n}$ if and only if $\|A\|_{[k]} \leq \|B\|_{[k]}$ for each $k = 1, \ldots, q = \min\{m, n\}$.*

7.4.9 Approximation bounds for unitarily invariant norms. The inequalities (7.4.1.3(a)) and (7.3.5(b)) say that for the Frobenius norm and any $A, B \in M_{m,n}$, $\|A - B\|_2 \geq \|\Sigma(A) - \Sigma(B)\|_2$. The inequality (7.3.5(a)) says that for the spectral norm and any $A, B \in M_{m,n}$, $\||A - B\||_2 \geq \||\Sigma(A) - \Sigma(B)\||_2$. In fact, these inequalities are valid for *every* unitarily invariant norm on $M_{m,n}$.

Theorem 7.4.9.1. *Let m and n be given positive integers and let $q = \min\{m, n\}$. For any $A, B \in M_{m,n}$, let $A = V_1 \Sigma(A) W_1^*$ and $B = V_2 \Sigma(B) W_2^*$, in which $V_1, V_2 \in M_m$ and $W_1, W_2 \in M_n$ are unitary, and $\Sigma(A) = [s_{ij}(A)], \Sigma(B) = [s_{ij}(B)] \in M_{m,n}$ are nonnegative diagonal matrices whose diagonal entries $s_{ii}(A) = \sigma_i(A)$ and $s_{ii}(B) = \sigma_i(B)$ are the nonincreasingly ordered singular values of A and B, respectively. Then $\|A - B\| \geq \|\Sigma(A) - \Sigma(B)\|$ for every unitarily invariant norm $\|\cdot\|$ on $M_{m,n}$.*

Proof. Let

$$\mathcal{A} = \begin{bmatrix} 0 & A \\ A^* & 0 \end{bmatrix} \quad \text{and} \quad \mathcal{B} = \begin{bmatrix} 0 & B \\ B^* & 0 \end{bmatrix}$$

According to (7.3.3), the algebraically nonincreasingly ordered eigenvalues of $\mathcal{A}$ are

$$\sigma_1(A) \geq \cdots \geq \sigma_q(A) \geq \underbrace{0 = \cdots = 0}_{|m-n|} \geq -\sigma_q(A) \geq \cdots \geq -\sigma_1(A)$$

with similar expressions for the algebraically nonincreasingly ordered eigenvalues of $\mathcal{B}$ and $\mathcal{A} - \mathcal{B}$. The differences of the respective ordered eigenvalues of $\mathcal{A}$ and $\mathcal{B}$ are $\pm(\sigma_1(A) - \sigma_1(B)), \ldots, \pm(\sigma_q(A) - \sigma_q(B))$ together with $|m - n|$ zeroes. Although it is not clear how to order these values algebraically, the q algebraically largest values are $|\sigma_1(A) - \sigma_1(B)|, \ldots, |\sigma_q(A) - \sigma_q(B)|$. Theorem 4.3.47(b) ensures that $\lambda(\mathcal{A} - \mathcal{B})$ majorizes $\lambda^\downarrow(\mathcal{A}) - \lambda^\downarrow(\mathcal{B})$, that is,

$$\sum_{i=1}^{k} \sigma_i(A - B) \geq \max_{1 \leq i_1 < \cdots < i_k \leq q} \sum_{j=1}^{k} |\sigma_{i_j}(A) - \sigma_{i_j}(B)|, \quad k = 1, \ldots, q$$

Inspection of these inequalities reveals that they are exactly the inequalities

$$\|A - B\|_{[k]} \geq \|\Sigma(A) - \Sigma(B)\|_{[k]}, \quad k = 1, \ldots, q$$

so (7.4.8.4) ensures that $\|A - B\| \geq \|\Sigma(A) - \Sigma(B)\|$ for every unitarily invariant norm $\|\cdot\|$ on $M_{m,n}$. $\qquad\qquad\qquad\square$

One consequence of the preceding theorem is a generalization of the problem of finding a best (in the sense of least squares) rank-k approximation to a given $A \in M_{m,n}$ with rank $A > k$, which we considered in (7.4.2). If $\|\cdot\|$ is a unitarily invariant norm on $M_{m,n}$ and if $B \in M_{m,n}$ and rank $B = k$, then $\sigma_1(B) \geq \cdots \geq \sigma_k(B) > 0 = \sigma_{k+1}(B) = \cdots = \sigma_q(B)$. Using the fact that a unitarily invariant norm on diagonal matrices in $M_{m,n}$ is a monotone norm, we have

$$\begin{aligned}
\|A - B\| &\geq \|\Sigma(A) - \Sigma(B)\| \\
&= \|\mathrm{diag}(\sigma_1(A) - \sigma_1(B), \ldots, \sigma_k(A) - \sigma_k(B), \sigma_{k+1}(A), \ldots, \sigma_q(A))\| \\
&\geq \|\mathrm{diag}(0, \ldots, 0, \sigma_{k+1}(A), \ldots, \sigma_q(A))\| \qquad (7.4.9.2)
\end{aligned}$$

for any $B \in M_{m,n}$ such that rank $B = k$. If $A = V\Sigma(A)W^*$ is a singular value decomposition, we can always attain equality in the inequality (7.4.9.2) with $B = V\Sigma_0 W^*$, in which $\Sigma_0 \in M_{m,n}$ is a nonnegative diagonal matrix with diagonal entries $\sigma_1(A), \ldots, \sigma_k(A)$, and $q - k$ zeroes. Thus, the same matrix that provides a best rank-k approximation to A in the Frobenius norm provides a best approximation in *every* unitarily invariant norm.

Exercise. How closely can $A \in M_{m,n}$ be approximated in the spectral norm by a rank-k matrix?

Another consequence of (7.4.9.1) is a version of (6.3.8) (the Hoffman–Wielandt theorem) that is valid for any unitarily invariant norm. For a Hermitian matrix $H \in M_n$, $\mathrm{diag}\,\lambda^{\downarrow}(H) \in M_n$ is the diagonal matrix whose diagonal entries are the nonincreasingly ordered eigenvalues of H.

Corollary 7.4.9.3 (Mirsky). *Let $A, B \in M_n$ be Hermitian and let $\|\cdot\|$ be a unitarily invariant norm on M_n. Then*

$$\|\mathrm{diag}\,\lambda^{\downarrow}(A) - \mathrm{diag}\,\lambda^{\downarrow}(B)\| \leq \|A - B\| \qquad (7.4.9.4)$$

Proof. Let $\mu \in [0, \infty)$ be such that $A + \mu I$ and $B + \mu I$ are both positive semidefinite. Then

$$\Sigma(A + \mu I) = \mathrm{diag}\,\lambda^{\downarrow}(A + \mu I) = \mathrm{diag}\,\lambda^{\downarrow}(A) + \mu I$$

and $\Sigma(B + \mu I) = \mathrm{diag}\,\lambda^{\downarrow}(B) + \mu I$. Theorem 7.4.9.1 ensures that

$$\begin{aligned}
\|\mathrm{diag}\,\lambda^{\downarrow}(A) - \mathrm{diag}\,\lambda^{\downarrow}(B)\| &= \|(\Sigma(A + \mu I) - \mu I) - (\Sigma(B + \mu I) - \mu I)\| \\
&= \|\Sigma(A + \mu I) - \Sigma(B + \mu I)\| \\
&\leq \|(A + \mu I) - (B + \mu I)\| = \|A - B\|
\end{aligned}$$

$$\qquad\qquad\qquad\qquad\qquad\qquad\qquad\qquad\qquad\qquad\qquad\qquad\qquad\qquad\square$$

Exercise. Let $A, E \in M_n$ be Hermitian. For which unitarily invariant norms are the bounds in (6.3.4) and (6.3.8) consequences of the preceding corollary?

7.4.10 Unitarily invariant matrix norms. Let $\|\cdot\|$ be a unitarily invariant matrix norm on M_n. For any $A \in M_n$, (5.6.34(d)) ensures that $\|A\| \geq \sigma_1(A)$. The following theorem provides a converse to this observation.

Theorem 7.4.10.1. *A unitarily invariant norm $\|\cdot\|$ on M_n is a matrix norm if and only if $\|A\| \geq \sigma_1(A)$ for all $A \in M_n$.*

Proof. Suppose that $\|X\| \geq \sigma_1(X)$ for all $X \in M_n$, and let $A, B \in M_n$ be given. We must show that $\|AB\| \leq \|A\| \, \|B\|$. Let g be the symmetric gauge function determined by $\|\cdot\|$. In the following computation, we adopt the notation of (7.4.7) and use the fact that g is a monotone norm as well as the singular value inequality $\sigma_k(AB) \leq \sigma_1(A)\sigma_k(B)$; see the exercise preceding (7.3.11):

$$
\begin{aligned}
\|AB\| &= g(s(AB)) = g([\sigma_1(AB) \, \sigma_2(AB) \, \ldots \, \sigma_n(AB)]^T) \\
&\leq g([\sigma_1(A)\sigma_1(B) \, \sigma_1(A)\sigma_2(B) \, \ldots \, \sigma_1(A)\sigma_n(B)]^T) \\
&= \sigma_1(A)g([\sigma_1(B) \, \sigma_2(B) \, \ldots \, \sigma_n(B)]^T) \\
&= \sigma_1(A)g(s(B)) = \sigma_1(A) \, \|B\| \\
&\leq \|A\| \, \|B\|
\end{aligned}
$$

$\square$

The theorem ensures that the Ky Fan k-norms and the Schatten p-norms for $p \geq 1$ are unitarily invariant matrix norms on M_n.

Although a convex combination of matrix norms need not be a matrix norm (see (5.6.P9)), a convex combination of unitarily invariant matrix norms is always a unitarily invariant matrix norm.

Corollary 7.4.10.2. *Let $\|\cdot\|_a$ and $\|\cdot\|_b$ be unitarily invariant matrix norms on M_n and let $\alpha \in [0, 1]$. Then $\alpha \|\cdot\|_a + (1 - \alpha) \|\cdot\|_b$ is a unitarily invariant matrix norm on M_n.*

Proof. The convex combination $\alpha \|\cdot\|_a + (1 - \alpha) \|\cdot\|_b$ is a unitarily invariant norm, and since

$$\alpha \|A\|_a + (1 - \alpha) \|A\|_b \geq \alpha\sigma_1(A) + (1 - \alpha)\sigma_1(A) = \sigma_1(A)$$

the preceding theorem ensures that it is a matrix norm. $\square$

7.4.11 Absolute unitarily invariant norms on matrices. The Frobenius norm of a matrix $A = [a_{ij}] \in M_{m,n}$ can be expressed as $\|A\|_2 = (\sigma_1(A)^2 + \cdots + \sigma_n(A)^2)^{1/2}$ and as $\|A\|_2 = (\sum_{i,j} |a_{ij}|^2)^{1/2}$, so it is both unitarily invariant and absolute. Are there other absolute unitarily invariant norms on $M_{m,n}$? The following exercise is a first step toward an answer.

Exercise. Let $\alpha \geq \beta > 0$ be given. Let $a = \frac{1}{2}(\sqrt{\alpha^2 + \beta^2} + \alpha - \beta), b = \sqrt{\alpha\beta/2}$, and $c = \frac{1}{2}(\sqrt{\alpha^2 + \beta^2} - \alpha + \beta)$. Consider the real symmetric matrices $B_\pm = \begin{bmatrix} a & b \\ b & \pm c \end{bmatrix}$. Show that $a, b, c > 0$; the eigenvalues of B_+ are $\sqrt{\alpha^2 + \beta^2}$ and 0;

and the eigenvalues of B_- are α and $-\beta$. Conclude that the singular values of B_+ are $\sqrt{\alpha^2 + \beta^2}$ and 0, while the singular values of B_- are α and β. *Hint*: (1.2.4b), $a + c = \sqrt{\alpha^2 + \beta^2}$, and $a - c = \alpha - \beta$.

Theorem 7.4.11.1. *Let $\|\cdot\|$ be a unitarily invariant norm on $M_{m,n}$. Then $\|\cdot\|$ is an absolute norm if and only if it is a positive scalar multiple of the Frobenius norm.*

Proof. For convenience, suppose that $\|\cdot\|$ is normalized so that $\|E_{11}\| = 1$, in which $E_{11} \in M_{m,n}$ has a 1 in position 1, 1 and zero entries elsewhere. Let $q = \min\{m, n\}$ and let $\sigma_1 \geq \cdots \geq \sigma_q$ denote the singular values of a given $A \in M_{m,n}$. We claim that $\|A\| = \|A\|_2 = (\sigma_1^2 + \cdots + \sigma_q^2)^{1/2}$.

If rank $A = 1$, then $\|A\| = \|\Sigma(A)\| = \|\sigma_1 E_{11}\| = \sigma_1 \|E_{11}\| = \sigma_1 = \|A\|_2$.

If rank $A = 2$, define $A_\pm = \begin{bmatrix} B_\pm & 0 \\ 0 & 0 \end{bmatrix} \in M_{m,n}$, in which $B_\pm = \begin{bmatrix} a & b \\ b & \pm c \end{bmatrix}$, $a = \frac{1}{2}(\sqrt{\sigma_1^2 + \sigma_2^2} + \sigma_1 - \sigma_2)$, $b = \sqrt{\sigma_1 \sigma_2/2}$, and $c = \frac{1}{2}(\sqrt{\sigma_1^2 + \sigma_2^2} - \sigma_1 + \sigma_2)$. The preceding exercise ensures that the singular values of A_- are σ_1, σ_2, and $q - 2$ zeroes, while the singular values of A_+ are $\sqrt{\sigma_1^2 + \sigma_2^2}$ and $q - 1$ zeroes. In the latter case, rank $A_+ = 1$, so $\|A_+\| = \|A_+\|_2 = \sqrt{\sigma_1^2 + \sigma_2^2} = \|A\|_2$. Since A and A_- have the same singular values and $\|\cdot\|$ is unitarily invariant, we have $\|A\| = \|A_-\|$; since $\|\cdot\|$ is absolute, we also have $\|A_-\| = \|A_+\|$. It follows that $\|A\| = \|A_-\| = \|A_+\| = \|A\|_2$.

Proceed by induction on the rank of A. Suppose that rank $A = r \geq 3$ and $\|X\| = \|X\|_2$ for every $X \in M_n$ such that rank $X \leq r - 1$. Define $A_\pm = \begin{bmatrix} B_\pm & 0 \\ 0 & 0 \end{bmatrix} \in M_{m,n}$, in which $B_\pm = \begin{bmatrix} a & b \\ b & \pm c \end{bmatrix} \oplus \text{diag}(\sigma_2, \ldots, \sigma_{r-1})$, $a = \frac{1}{2}(\sqrt{\sigma_1^2 + \sigma_r^2} + \sigma_1 - \sigma_r)$, $b = \sqrt{\sigma_1 \sigma_r/2}$, and $c = \frac{1}{2}(\sqrt{\sigma_1^2 + \sigma_r^2} - \sigma_1 + \sigma_r)$. The preceding exercise ensures that the singular values of A_- are $\sigma_1, \ldots, \sigma_r$ and $q - r$ zeroes, while the singular values of A_+ are $\sqrt{\sigma_1^2 + \sigma_r^2}, \sigma_2, \ldots, \sigma_{r-1}$, and $q - r + 1$ zeroes. The induction hypothesis ensures that $\|A_+\| = \|A_+\|_2 = \|A\|_2$; as in the rank two case, the hypotheses of unitary invariance and absoluteness ensure that $\|A\| = \|A_-\| = \|A_+\| = \|A\|_2$.

If $\|\cdot\|$ is not necessarily normalized, we have shown that $\|\cdot\| / \|E_{11}\| = \|\cdot\|_2$, that is, $\|A\| = \|E_{11}\| \|A\|_2$ for every $A \in M_{m,n}$. $\quad\square$

Exercise. Use the preceding theorem to give a conceptual proof that the spectral norm on M_n is not an absolute norm if $n \geq 2$. Compare with (5.6.P40).

7.4.12 Inequalities of Kantorovich and Wielandt. Let $A \in M_n$ be Hermitian and positive definite; let λ_1 and λ_n be its smallest and largest eigenvalues. Our goals are to show that the following two classical inequalities are equivalent and valid and to explore some of their analytic and geometric consequences.

Kantorovich's inequality is

$$(x^* A x)(x^* A^{-1} x) \leq \frac{(\lambda_1 + \lambda_n)^2}{4\lambda_1 \lambda_n} \|x\|_2^4 \text{ for all } x \in \mathbf{C}^n \qquad (7.4.12.1)$$

and *Wielandt's inequality* is

$$|x^*Ay|^2 \le \left(\frac{\lambda_1 - \lambda_n}{\lambda_1 + \lambda_n}\right)^2 (x^*Ax)(y^*Ay) \text{ for all orthogonal } x, y \in \mathbf{C}^n \quad (7.4.12.2)$$

The inequality (7.4.12.1) is valid if $x = 0$, and (7.4.12.2) is valid if either $x = 0$ or $y = 0$, so in the following discussion we consider only the case $x \ne 0 \ne y$.

Our approach to Kantorovich's inequality begins with the two positive semidefinite matrices $\lambda_n I - A$ and $A - \lambda_1 I$, and the positive definite matrix A^{-1}. These three Hermitian matrices commute, so their product is Hermitian and positive semidefinite. Therefore, for any nonzero vector x, we have

$$0 \le x^*(\lambda_n I - A)(A - \lambda_1 I)A^{-1}x = x^*((\lambda_1 + \lambda_n)I - \lambda_1\lambda_n A^{-1} - A)x$$

and hence

$$x^*Ax + \lambda_1\lambda_n(x^*A^{-1}x) \le (\lambda_1 + \lambda_n)(x^*x) \quad (7.4.12.3)$$

Let $t_0 = \lambda_1\lambda_n(x^*A^{-1}x)$ and rewrite (7.4.12.3) in the equivalent form

$$t_0(x^*Ax) \le t_0(\lambda_1 + \lambda_n)(x^*x) - t_0^2 \quad (7.4.12.4)$$

The function $f(t) = t(\lambda_1 + \lambda_n)(x^*x) - t^2$ is concave and has a critical point at $t = (x^*x)(\lambda_1 + \lambda_n)/2$, where it has a global maximum. Therefore, $f(t_0) \le (x^*x)^2(\lambda_1 + \lambda_n)^2/4$ and it follows from (7.4.12.4) that

$$\lambda_1\lambda_n(x^*A^{-1}x)(x^*Ax) \le \frac{1}{4}(\lambda_1 + \lambda_n)^2(x^*x)^2$$

This is Kantorovich's inequality (7.4.12.1).

We claim that Kantorovich's inequality implies Wielandt's inequality. Consider a 2-by-2 positive definite matrix $B = \begin{bmatrix} a & b \\ b & c \end{bmatrix}$, whose inverse is $B^{-1} = (\det B)^{-1} \operatorname{adj} B = \begin{bmatrix} c/\det B & * \\ * & * \end{bmatrix}$. Let $\mu_1 \le \mu_2$ be the eigenvalues of B. With $x = e_1$ and $A = B$, the inequality (7.4.12.1) is

$$\frac{(\mu_1 + \mu_2)^2}{4\mu_1\mu_2} \ge (e_1^*Be_1)(e^*B^{-1}e_1) = \frac{ac}{ac - |b|^2} = \frac{1}{1 - \frac{|b|^2}{ac}}$$

and a calculation reveals that

$$\frac{|b|^2}{ac} \le \left(\frac{\mu_1 - \mu_2}{\mu_1 + \mu_2}\right)^2 = \left(\frac{1 - \frac{\mu_2}{\mu_1}}{1 + \frac{\mu_2}{\mu_1}}\right)^2 \quad (7.4.12.5)$$

Now let x and y be any pair of orthonormal vectors in $\mathbf{C}^n$ and consider the positive definite 2-by-2 matrix

$$B = [x \ y]^*A[x \ y] = \begin{bmatrix} x^*Ax & x^*Ay \\ y^*Ax & y^*Ay \end{bmatrix}$$

The interlacing inequalities (4.3.38) of the Poincaré separation theorem ensure that the eigenvalues $\mu_1 \le \mu_2$ of B satisfy the inequalities $0 < \lambda_1 \le \mu_1 \le \mu_2 \le \lambda_n$, so $0 < \frac{\mu_2}{\mu_1} \le \frac{\lambda_2}{\lambda_1}$. The inequality (7.4.12.5) and monotonicity of the function

$f(t) = (1 - t)^2/(1 + t)^2$ on $(1, \infty)$ ensure that

$$\frac{|x^*Ay|^2}{(x^*Ax)(y^*Ay)} \leq \left(\frac{1 - \frac{\mu_2}{\mu_1}}{1 + \frac{\mu_2}{\mu_1}}\right)^2 \leq \left(\frac{1 - \frac{\lambda_2}{\lambda_1}}{1 + \frac{\lambda_2}{\lambda_1}}\right)^2 = \left(\frac{\lambda_1 - \lambda_n}{\lambda_1 + \lambda_n}\right)^2$$

This is Wielandt's inequality (7.4.12.2).

Exercise. Explain why both (7.4.12.1) and (7.4.12.2) are satisfied if $x \in \mathbf{C}^n$ is an eigenvector of A. *Hint*: The arithmetic–geometric inequality.

Exercise. If $x \in \mathbf{C}^n$ is not an eigenvector of A, explain why $A^{-1}x - (x^*A^{-1}x)x \neq 0$ and $x - (x^*A^{-1}x)Ax \neq 0$.

Exercise. If $x \in \mathbf{C}^n$ is a unit vector, explain why $(x^*Ax)(x^*A^{-1}x) \geq 1$, with strict inequality if x is not an eigenvector of A, *Hint*: $1 = (x^*x)^2 = (x^*A^{1/2}A^{-1/2}x)^2 \leq \|A^{1/2}x\|_2^2 \|A^{-1/2}x\|_2^2 = (x^*Ax)(x^*A^{-1}x)$, with equality only if $A^{1/2}x = \alpha A^{-1/2}x$, in which case x is an eigenvector of A.

To deduce Kantorovich's inequality from Wielandt's inequality, let $x \in \mathbf{C}^n$ be a unit vector that is not an eigenvector of A, and define $y = A^{-1}x - (x^*A^{-1}x)x$. Then $y \neq 0$ and a computation reveals that $x^*y = 0$, $Ay = x - (x^*A^{-1}x)Ax \neq 0$, $x^*Ay = 1 - (x^*Ax)(x^*A^{-1}x) < 0$, and $y^*Ay = -(x^*A^{-1}x)(x^*Ay)$. Wielandt's inequality in this case is

$$(x^*Ay)^2 \leq -\left(\frac{\lambda_1 - \lambda_n}{\lambda_1 + \lambda_n}\right)^2 (x^*Ax)(x^*A^{-1}x)(x^*Ay)$$

so we have

$$(x^*Ax)(x^*A^{-1}x) - 1 = -x^*Ay \leq \left(\frac{\lambda_1 - \lambda_n}{\lambda_1 + \lambda_n}\right)^2 (x^*Ax)(x^*A^{-1}x)$$

from which it follows that

$$(x^*Ax)(x^*A^{-1}x) \leq \frac{(\lambda_1 + \lambda_n)^2}{4\lambda_1\lambda_n}$$

This is Kantorovich's inequality.

Exercise. Let $u, v \in \mathbf{C}^n$ be orthonormal vectors such that $Au = \lambda_1 u$ and $Av = \lambda_n v$. Let $x = (u + v)/\sqrt{2}$ and $y = (u - v)/\sqrt{2}$. Show that (7.4.12.2) is an equality for the orthonormal vectors x and y, and that (7.4.12.1) is an equality for this unit vector x.

If $B \in M_n$ is nonsingular with singular values $\sigma_1 \geq \cdots \geq \sigma_n > 0$, and if we take $A = B^*B$ in Wielandt's inequality (7.4.12.2), we obtain the inequality

$$|\langle Bx, By \rangle| \leq \left(\frac{\sigma_1^2 - \sigma_n^2}{\sigma_1^2 + \sigma_n^2}\right) \|Bx\| \, \|By\| = \left(\frac{\kappa^2 - 1}{\kappa^2 + 1}\right) \|Bx\| \, \|By\| \quad (7.4.12.6)$$

in which $x, y \in \mathbf{C}^n$ are orthogonal vectors and $\kappa = \sigma_1/\sigma_n$ is the spectral condition number of B. Let $\theta_\kappa \in (0, \pi/2]$ be the unique angle such that $\cos\theta_k = (\kappa^2 - 1)/(\kappa^2 + 1)$.

Exercise. Show that $\sin\theta_\kappa = 2\kappa/(\kappa^2+1)$ and $\cot(\theta_\kappa/2) = \kappa$, with $\theta_\kappa \in (0, \pi/2]$.

If B, x, and y are real, then (7.4.12.6) can be written in the form

$$\cos\theta_{Bx,By} = \frac{|\langle Bx, By\rangle|}{\|Bx\|\,\|By\|} \leq \cos\theta_\kappa \text{ for all orthogonal nonzero } x, y \in \mathbf{R}^n \tag{7.4.12.7}$$

in which $\theta_{Bx,By} \in (0, \pi/2]$ is the angle between the real vectors Bx and By; see (0.6.3.1). This formulation gives us the geometric inequality $0 \leq \theta_\kappa \leq \theta_{Bx,By}$. Moreover, since there are nonzero orthogonal vectors for which (7.4.12.2) – and hence also (7.4.12.7) – is an equality, we have the geometric interpretation that θ_k (determined solely by the spectral condition number of B) is the minimum angle between the real vectors Bx and By as x and y range over all orthonormal pairs of real vectors. If κ is large, then $\frac{\kappa^2-1}{\kappa^2+1} = \frac{1-\kappa^{-2}}{1+\kappa^{-2}}$ is close to 1 and $\theta_\kappa = \cos^{-1}(\frac{1-\kappa^{-2}}{1+\kappa^{-2}})$ is close to zero, and conversely. Thus, κ is large if and only if there is an orthonormal pair of vectors x, y such that Bx and By are nearly parallel.

Exercise. Let $B \in M_n$ be nonsingular, let κ be its spectral condition number, and take $A = B^*B$ in Kantorovich's inequality. Deduce that

$$\|Bx\|_2\,\|B^{-*}x\|_2 \leq \left(\frac{2\kappa}{\kappa^2+1}\right)\|x\|_2^2 \tag{7.4.12.8}$$

and

$$\sin\theta_\kappa\,\|Bx\|_2\,\|B^{-*}x\|_2 \leq \|x\|_2^2 \tag{7.4.12.9}$$

for any $x \in \mathbf{C}^n$.

Problems

7.4.P1 Suppose that $0 < \lambda_1 \leq \cdots \leq \lambda_n$, $\alpha_1, \ldots, \alpha_n$ are nonnegative and $\alpha_1 + \cdots + \alpha_n = 1$. Let $A = (\lambda_1 + \lambda_n)/2$ and $G = \sqrt{\lambda_1\lambda_n}$ (the arithmetic and geometric means of λ_1 and λ_n). Derive the scalar Kantorovich inequality

$$\left(\sum_{i=1}^n \alpha_i\lambda_i\right)\left(\sum_{i=1}^n \alpha_i\lambda_i^{-1}\right) \leq A^2G^{-2} \tag{7.4.12.10}$$

from (7.4.12.1).

7.4.P2 Let $A = [a_{ij}] \in M_n$ be positive definite and have eigenvalues $0 < \lambda_1 \leq \cdots \leq \lambda_n$. We know that $|a_{ij}|^2 < a_{ii}a_{jj}$ for all $i \neq j$; see (7.1.P1). Use (7.4.12.2) to prove the better bound $|a_{ij}|^2 \leq \left(\frac{\lambda_1-\lambda_n}{\lambda_1+\lambda_n}\right)^2 a_{ii}a_{jj}$ for all $i \neq j$.

7.4.P3 Prove the following 2-matrix generalization of (7.4.12.1): Let $B, C \in M_n$ be commuting positive definite matrices with eigenvalues $0 < \lambda_1 \leq \cdots \leq \lambda_n$ and $0 < \mu_1 \leq \cdots \leq \mu_n$, respectively. The *Greub–Rheinboldt inequality* says that

$$(x^*B^2x)(x^*C^2x) \leq \frac{(\lambda_1\mu_1 + \lambda_n\mu_n)^2}{4\lambda_1\lambda_n\mu_1\mu_n}(x^*BCx)^2, \quad \text{any } x \in \mathbf{C}^n \tag{7.4.12.11}$$

which we can express equivalently as

$$\frac{\langle Bx, Cx \rangle}{\|Bx\|_2 \|Cx\|_2} \geq \frac{2\sqrt{\lambda_1 \lambda_n \mu_1 \mu_n}}{\lambda_1 \mu_1 + \lambda_n \mu_n}, \quad \text{any nonzero } x \in \mathbf{C}^n \qquad (7.4.12.12)$$

Unlike Kantorovich's inequality, however, if neither B nor C is a scalar matrix, there need not be a unit vector x for which the Greub–Rheinboldt inequality is an equality. (a) Prove (7.4.12.11). (b) Show that equality is possible in (7.4.12.12) if at least one of B, C is a scalar matrix. (c) For what choice of commuting positive definite matrices B and C does (7.4.12.11) reduce to (7.4.12.1)? (d) If B, C, and x are real, interpret (7.4.12.12) geometrically as a lower bound on the cosine of the smaller angle between the vectors Bx and By, that is, as an upper bound on the smaller angle between Bx and By.

7.4.P4 Let $A \in M_n$ be positive definite and have eigenvalues $0 < \lambda_1 \leq \cdots \leq \lambda_n$. Let $u_1, u_n \in \mathbf{C}^n$ be orthonormal vectors such that $Au_1 = \lambda_1 u_1$ and $Au_n = \lambda_n u_n$, and let $\kappa = \lambda_n / \lambda_1$ be the spectral condition number of A. Deduce from (7.4.12.12) that

$$\frac{\langle x, Ax \rangle}{\|x\|_2 \|Ax\|_2} \geq \frac{2\sqrt{\lambda_1 \lambda_n}}{\lambda_1 + \lambda_n} = \frac{2\sqrt{\kappa}}{\kappa + 1}, \quad \text{any nonzero } x \in \mathbf{C}^n \qquad (7.4.12.13)$$

with equality for the vector $x_0 = \lambda_n^{1/2} u_1 + \lambda_1^{1/2} u_n$. If A and x are real, interpret the preceding inequality geometrically:

$$\cos \theta_{x,Ax} \geq \frac{2\sqrt{\kappa}}{\kappa + 1}, \quad \text{any unit vector } x \in \mathbf{R}^n \qquad (7.4.12.14)$$

in which the lower bound is achieved for the vector x_0. Thus, $0 \leq \theta_{x,Ax} \leq \cos^{-1}(2\kappa^{1/2}(\kappa + 1)^{-1})$ for every unit vector x, with equality in the lower bound for every eigenvector of A and equality in the upper bound for x_0.

7.4.P5 Let $A \in M_n$ be nonsingular, and let κ be its spectral condition number. Use the polar decomposition and the Kantorovich inequality to show that

$$|(x^* Ax)(x^* A^{-1} x)| \leq \frac{1}{4}(\kappa^{1/2} + \kappa^{-1/2})^2 \|x\|_2^4, \quad \text{any } x \in \mathbf{C}^n \qquad (7.4.12.15)$$

with equality for some unit vector x.

7.4.P6 Let κ be the spectral condition number of the positive definite matrix A. Show that the Kantorovich and Wielandt inequalities are, respectively,

$$(x^* Ax)(x^* A^{-1} x) \leq \frac{1}{4}(\kappa^{1/2} + \kappa^{-1/2})^2 \|x\|_2^4, \quad \text{any } x \in \mathbf{C}^n \qquad (7.4.12.16)$$

and

$$|x^* Ay|^2 \leq \left(\frac{\kappa - 1}{\kappa + 1}\right)^2 (x^* Ax)(y^* Ay), \quad \text{all orthogonal } x, y \in \mathbf{C}^n \qquad (7.4.12.17)$$

7.4.P7 Let $A \in M_n$ be nonsingular and Hermitian and have spectral condition number κ. Show that

$$\max_{\|x\|_2 = 1}(\|Ax\|_2 \|A^{-1} x\|_2) = \frac{1}{2}(\kappa + \kappa^{-1})$$

Exhibit a vector x for which the maximum is achieved.

7.4.P8 Let $A \in M_n$ be positive definite, and suppose that all its eigenvalues lie in the interval $[m, M]$, in which $0 < m < M < \infty$. Show that $(x^*Ax)(x^*A^{-1}x) \leq (m + M)^2 \|x\|_2^4 / 4mM$ for all $x \in \mathbf{C}^n$.

7.4.P9 Let $\alpha_1, \ldots, \alpha_n, \beta_1, \ldots, \beta_n$ be (not necessarily ordered) positive real numbers. We know that $\sum_{i=1}^{n} \alpha_i \beta_i \leq \sum_{i=1}^{n} \alpha_i^{\downarrow} \beta_i^{\downarrow}$; see (4.3.54). The Kantorovich inequality permits us to reverse this inequality:

$$\sum_{i=1}^{n} \alpha_i^{\downarrow} \beta_i^{\downarrow} \leq \frac{m + M}{2\sqrt{mM}} \sum_{i=1}^{n} \alpha_i \beta_i \qquad (7.4.12.18)$$

in which $0 < m \leq \alpha_i/\beta_i \leq M < \infty$ for all $i = 1, \ldots, n$. Provide details: (a) Let $A = \mathrm{diag}(\alpha_1/\beta_1, \ldots, \alpha_n/\beta_n)$ and $x = [\sqrt{\alpha_i \beta_i}]_{i=1}^{n}$. Compute $(x^T A x)(x^T A^{-1} x)$ and use (7.4.12.1) to provide an upper bound for it. (b) $(\sum_{i=1}^{n} \alpha_i^{\downarrow} \beta_i^{\downarrow})^2 \leq (\sum_{i=1}^{n} \alpha_i^2)(\sum_{i=1}^{n} \alpha_i^2)$ gives a lower bound.

7.4.P10 Let $x, y \in \mathbf{C}^n$ be nonzero vectors and let $A, B \in M_n$ be positive definite. (a) Show that $|x^*y|^2 \leq (x^*Ax)(y^*A^{-1}y)$, with equality for $x = A^{-1}y$. (b) Conclude that the function $f(A, y) = (y^*A^{-1}y)^{-1}$ has the variational representation

$$f(A, y) = \min_{x^*y \neq 0} \frac{x^*Ax}{|x^*y|^2}$$

(c) Deduce that $f(A + B, y) \geq f(A, y) + f(B, y)$. (d) Let $y = e_i$, the ith standard unit basis vector, and deduce that $\gamma_{ii}^{-1} \geq \alpha_{ii}^{-1} + \beta_{ii}^{-1}$, in which $(A + B)^{-1} = [\gamma_{ij}]$, $A^{-1} = [\alpha_{ij}]$, and $B^{-1} = [\beta_{ij}]$. This is *Bergström's inequality*. (e) Explain why Bergström's inequality can be written in the form

$$\frac{\det(A + B)}{\det(A + B)[\{i\}^c]} \geq \frac{\det A}{\det A[\{i\}^c]} + \frac{\det B}{\det B[\{i\}^c]}, \quad i = 1, \ldots, n \qquad (7.4.12.19)$$

in which the denominators are cofactors of main diagonal entries.

7.4.P11 Let $A \in M_n$ be positive definite, let $x, y \in \mathbf{C}^n$, and let α, β be real and positive. Let $\mathcal{A}_\alpha = \begin{bmatrix} A & x \\ x^* & \alpha \end{bmatrix}$ and $\mathcal{B}_\beta = \begin{bmatrix} B & y \\ y^* & \beta \end{bmatrix}$. (a) Show that $\det \mathcal{A}_\alpha / \det A = \alpha - x^*A^{-1}x$, $\det \mathcal{B}_\alpha / \det B = \beta - y^*B^{-1}y$, and $\det(\mathcal{A}_\alpha + \mathcal{B}_\alpha)/\det(A + B) = \alpha + \beta - (x + y)^*(A + B)^{-1}(x + y)$. (b) Show that

$$\frac{\det(\mathcal{A}_\alpha + \mathcal{B}_\alpha)}{\det(A + B)} - \frac{\det \mathcal{A}_\alpha}{\det A} - \frac{\det \mathcal{B}_\alpha}{\det B}$$
$$= x^*A^{-1}x + y^*B^{-1}y - (x + y)^*(A + B)^{-1}(x + y) \qquad (7.4.12.20)$$

(c) Explain why $\mathcal{A}_\alpha$ and $\mathcal{B}_\beta$ are positive definite for all sufficiently large positive α, β, and use (7.4.12.20) to show that (7.4.12.19) implies the *Berenstein–Veinstein inequality*:

$$x^*A^{-1}x + y^*B^{-1}y \geq (x + y)^*(A + B)^{-1}(x + y) \qquad (7.4.12.21)$$

(d) Use (7.4.12.20) to show that (7.4.12.21) implies (7.4.12.19), and conclude that the Bergström and Berenstein–Veinstein inequalities are equivalent.

7.4.P12 Let $N_1(\cdot)$ and $N_2(\cdot)$ be unitarily invariant norms on $M_{m,n}$. Show that the function $f(A) = N_1(A)/N_2(A)$ is constant on rank-one matrices in $M_{m,n}$. What is the constant?

7.4.P13 For any complex number z and any real number x, we have the inequality $|z - \operatorname{Re} z| \leq |z - x|$. A plausible generalization of this to square matrices $A \in M_n$ is

$$\left\| A - \frac{1}{2}(A + A^*) \right\| \leq \| A - H \| \tag{7.4.12.22}$$

for all Hermitian $H \in M_n$. Prove that this inequality is valid for all unitarily invariant norms $\| \cdot \|$ and, more generally, for all self-adjoint norms. Conclude that the distance (with respect to $\| \cdot \|$) from a given $A \in M_n$ to the closed set of Hermitian matrices in M_n is $\frac{1}{2} \| A - A^* \|$; this is the norm of the skew-Hermitian part of A.

7.4.P14 For any complex number z, $|\operatorname{Re} z| \leq |z|$. Let $\| \cdot \|$ be a unitarily invariant norm and let $A \in M_n$. Show that (a) $\| (A + A^*)/2 \| \leq \| A \|$ (Hermitian part); (b) $\| (A + A^T)/2 \| \leq \| A \|$ (symmetric part); (c) $\| (A + \bar{A})/2 \| \leq \| A \|$ (real part).

7.4.P15 Let $A \in M_n$ and let $\| \cdot \|$ be a unitarily invariant norm on M_n. Use (7.4.9.1) to show that $\| A - U \| \geq \| \Sigma(A) - I \|$ for every unitary $U \in M_n$, with equality if U is the unitary factor in a polar decomposition of A. Conclude that $\| \Sigma(A) - I \|$ is the distance (with respect to $\| \cdot \|$) from A to the compact set of unitary matrices in M_n.

7.4.P16 Let $A \in M_n$ have a singular value decomposition $A = V \Sigma(A) W^*$ and let $\| \cdot \|$ be a unitarily invariant norm on M_n. Show that

$$\| \Sigma(A) - I \| \leq \| A - U \| \leq \| \Sigma(A) + I \| \tag{7.4.12.23}$$

for any unitary $U \in M_n$.

7.4.P17 Let $A \in M_n$ be nonsingular, and let $A = V \Sigma(A) W^*$ be a singular value decomposition in which $\Sigma(A) = \operatorname{diag}(\sigma_1(A), \ldots, \sigma_n(A))$. We showed in (7.4.2) that there are at least two different best singular approximations to A in the Frobenius norm if $\sigma_{n-1}(A) = \sigma_n(A)$. Provide details for the following outline of a proof that if $B \in M_n$ is singular, $\| A - B \|_2 = \sigma_n(A)$, and $\sigma_{n-1}(A) > \sigma_n(A)$, then B is the matrix B_0 constructed in (7.4.2). The condition $\sigma_{n-1}(A) > \sigma_n(A)$ is used only in (d) and (e). (a) Let $\Sigma_0 = \operatorname{diag}(\sigma_1(A), \ldots, \sigma_{n-1}(A), 0)$. Review the discussion in (7.4.2) and explain why the singular values of B must be the same as those of Σ_0, that is, $\Sigma(B) = \Sigma_0$. (b) Explain why AB^* and B^*A are positive semidefinite and $\operatorname{tr}(AB^*) = \sum_{i=1}^{n-1} \sigma_i^2(A)$. (c) Show that there are unitary $X, Y \in M_n$ such that $A = X\Sigma(A)Y^*$, $B = X\Lambda Y^*$, and $\Lambda = \operatorname{diag}(\lambda_1, \ldots, \lambda_n) = P\Sigma_0 P^T$ for some permutation matrix P. (d) Show that $\Lambda = \Sigma_0$. (e) Show that there is a unitary $Z = U \oplus [e^{i\theta}] \in M_n$ such that $X = VZ$, $Y = WZ$, and $Z\Sigma(A) = \Sigma(A)Z$. (f) Explain why $Z\Sigma_0 = \Sigma_0 Z$ and conclude that $B = B_0$.

7.4.P18 Let $\| \cdot \|$ be a unitarily invariant norm on $M_{n,m}$. Show that $\| A \| \leq \| |A| \|$ for all $A \in M_{n,m}$.

Notes and Further Readings. For generalizations of Kantorovich's inequality and references, see A. Clausing, Kantorovich-type inequalities, *Amer. Math. Monthly* 89 (1982) 314–320. For more information about inequalities valid for all unitarily invariant norms, see L. Mirsky, Symmetric gauge functions and unitarily invariant norms, *Quart. J. Math. Oxford* 11 (1960) 50–59, and K. Fan and A. J. Hoffman, Some metric inequalities in the space of matrices, *Proc. Amer. Math. Soc.* 6 (1955) 111–116. As an example of how these results are applied in statistics, and for further references to the statistics literature, see C. R. Rao, Matrix approximations and reduction of dimensionality in multivariate statistical analysis, *Multivariate Analysis–V*, Proceedings of the Fifth International Symposium on Multivariate Analysis, P. R. Krishnaiah, North-Holland, Amsterdam, 1980, pp. 1–22.

7.5 The Schur product theorem

Definition 7.5.1. *If* $A = [a_{ij}] \in M_{m,n}$ *and* $B = [b_{ij}] \in M_{m,n}$, *then the* Hadamard product *(Schur product) of* A *and* B *is the entrywise product matrix* $A \circ B = [a_{ij}b_{ij}] \in M_{m,n}$.

Like the usual matrix product, the Hadamard product distributes over matrix addition: $A \circ (B + C) = (A \circ B) + (A \circ C)$; unlike the usual matrix product, the Hadamard product is commutative: $A \circ B = B \circ A$.

The Hadamard product arises naturally from several different points of view. For example, if f and g are real-valued continuous periodic functions on **R** with period 2π and if

$$a_k = \int_0^{2\pi} e^{ik\theta} f(\theta)d\theta \quad \text{and} \quad b_k = \int_0^{2\pi} e^{ik\theta} g(\theta)d\theta, \quad k = 0, \pm 1, \pm 2, \ldots$$

are their trigonometric moments (Fourier coefficients), then the convolution product

$$h(\theta) = \int_0^{2\pi} f(\theta - t)g(t)dt$$

of f and g has trigonometric moments $c_k = \int_0^{2\pi} e^{ik\theta} h(\theta)d\theta$ that satisfy the identities $c_k = a_k b_k, k = 0, \pm 1, \pm 2, \ldots$. Thus, the Toeplitz matrix of trigonometric moments of h is the Hadamard product of the Toeplitz matrices of trigonometric moments of f and g:

$$[c_{i-j}] = [a_{i-j}] \circ [b_{i-j}]$$

If f and g are both nonnegative real-valued functions, then their convolution is also a nonnegative real-valued function. Therefore, as shown in (7.0.4.1), the matrices $[a_{i-j}]$, $[b_{i-j}]$, and $[c_{i-j}]$ are all positive semidefinite. This is an instance of the Schur product theorem: The Hadamard product of two positive semidefinite matrices is positive semidefinite.

As another example, consider the integral operator

$$K(f) = \int_a^b K(x, y)f(y)dy$$

in which $f \in C[a, b]$ and the kernel $K(x, y)$ is a continuous function on a finite interval $[a, b] \times [a, b]$. Suppose that the kernel $H(x, y)$ satisfies the same conditions, and consider the (pointwise) product kernel $L(x, y) = K(x, y)H(x, y)$ and the associated integral operator

$$L(f) = \int_a^b L(x, y)f(y)\,dy = \int_a^b K(x, y)H(x, y)f(y)dy$$

The linear mapping $f \to K(f)$ is a limit of matrix-vector multiplications (approximate the integral as a finite Riemann sum), and many properties of integral operators can be deduced by taking appropriate limits of results known for matrices. The (pointwise) product of integral kernels leads to an integral operator that is, from this point of view, a continuous analog of the Hadamard product of matrices.

If an integral kernel $K(x, y)$ has the property that

$$\int_a^b \int_a^b K(x, y) f(x) \bar{f}(y) \, dx \, dy \geq 0$$

for all $f \in C[a, b]$, then $K(x, y)$ is said to be a *positive semidefinite kernel*. It is a classical result (Mercer's theorem) that if $K(x, y)$ is a continuous positive semidefinite kernel on a finite interval $[a, b]$, then there exist positive real numbers $\lambda_1, \lambda_2, \ldots$ (known as "eigenvalues") and continuous functions $\phi_1(x), \phi_2(x), \ldots$ (known as "eigenfunctions") such that

$$K(x, y) = \sum_{i=1}^{\infty} \frac{\phi_i(x) \bar{\phi}_i(y)}{\lambda_i} \quad \text{on} \quad [a, b] \times [a, b]$$

and the series converges absolutely and uniformly.

If $K(x, y)$ and $H(x, y)$ are both continuous positive semidefinite kernels on $[a, b]$, then $H(x, y)$ also has an absolutely and uniformly convergent representation

$$H(x, y) = \sum_{i=1}^{\infty} \frac{\psi_i(x) \bar{\psi}_i(y)}{\mu_i} \quad \text{on} \quad [a, b] \times [a, b]$$

with all $\mu_i > 0$. The (pointwise) product kernel $L(x, y) = K(x, y) H(x, y)$ has the representation

$$L(x, y) = \sum_{i,j=1}^{\infty} \frac{\phi_i(x) \psi_j(x) \bar{\phi}_i(y) \bar{\psi}_j(y)}{\lambda_i \mu_j} \quad \text{on} \quad [a, b] \times [a, b]$$

which also converges absolutely and uniformly. Then

$$\int_a^b \int_a^b L(x, y) f(x) \bar{f}(y) \, dx \, dy = \sum_{i,j=1}^{\infty} \frac{1}{\lambda_i \mu_j} \left| \int_a^b \phi_i(x) \psi_j(x) f(x) \, dx \right|^2 \geq 0$$

so $L(x, y)$ is also positive semidefinite. This is another instance of the Schur product theorem.

Exercise. The usual matrix product of two Hermitian matrices is Hermitian if and only if they commute. Show that the Hadamard product of two Hermitian matrices is always Hermitian.

Exercise. Consider $A = \begin{bmatrix} 2 & 1 \\ 1 & 1 \end{bmatrix}$ and $B = \begin{bmatrix} 2 & 1 \\ 1 & 3 \end{bmatrix}$. Show that A, B, and $A \circ B$ are positive definite, but AB is not symmetric, so it is not positive semidefinite. Verify that AB is diagonalizable and has positive eigenvalues. Is this an accident? *Hint:* (7.2.P21).

The sesquilinear form associated with a Hadamard product has a convenient representation as the trace of an ordinary matrix product.

Exercise. Let $A = [a_{ij}]$, $B = [b_{ij}] \in M_n$ be given. Verify that $\sum_{i,j=1}^{n} a_{ij}b_{ij} = \text{tr}(AB^T)$.

Lemma 7.5.2. *Let $A, B \in M_n$ and $x, y \in \mathbf{C}^n$ be given. Let $\text{diag}\, x$ and $\text{diag}\, y$ be the n-by-n diagonal matrices whose respective main diagonal entries are the respective entries of x and y; see (0.9.1). Then*

$$x^*(A \circ B)y = \text{tr}((\text{diag}\, \bar{x})A(\text{diag}\, y)B^T)$$

Proof. Let $A = [a_{ij}]$, $B = [b_{ij}]$, $x = [x_i]$, and $y = [y_i]$. Then $(\text{diag}\, \bar{x})A = [\bar{x}_i a_{ij}]$ and $B\, \text{diag}\, y = [b_{ij}y_j]$. Use the preceding exercise to compute

$$\text{tr}((\text{diag}\, \bar{x})A(\text{diag}\, y)B^T) = \text{tr}(((\text{diag}\, \bar{x})A)(B\, \text{diag}\, y)^T)$$

$$= \sum_{i,j=1}^{n} (\bar{x}_i a_{ij})(b_{ij}y_j) = x^*(A \circ B)y$$

$\square$

Exercise. Let $x, y \in \mathbf{C}^n$ and $A \in M_n$. Show that $(xy^*) \circ A = (\text{diag}\, x)A(\text{diag}\, \bar{y})$.

Exercise. If $A \in M_n$ is Hermitian (in particular, if A is positive semidefinite), explain why $A^T = \bar{A}$.

The first assertion of the following theorem is the *Schur product theorem*.

Theorem 7.5.3. *Let $A, B \in M_n$ be positive semidefinite.*

(a) *$A \circ B$ is positive semidefinite.*
(b) *If A is positive definite and every main diagonal entry of B is positive, then $A \circ B$ is positive definite.*
(c) *If both A and B are positive definite, then $A \circ B$ is positive definite.*

Proof. Let $A = [a_{ij}]$, $B = [b_{ij}]$, and $x = [x_i]$.
(a) Let $C = (\text{diag}\, x)\bar{B}^{1/2}$ and use the preceding lemma to compute

$$x^*(A \circ B)x = \text{tr}((\text{diag}\, \bar{x})A(\text{diag}\, x)\bar{B})$$
$$= \text{tr}(\bar{B}^{1/2}(\text{diag}\, \bar{x})A(\text{diag}\, x)\bar{B}^{1/2}) = \text{tr}(C^*AC)$$

It follows from (7.1.8(a)) that C^*AC is positive semidefinite, so it has nonnegative eigenvalues and a nonnegative trace. Thus, $x^*(A \circ B)x \geq 0$ for all $x \in \mathbf{C}^n$, so $A \circ B$ is positive semidefinite.
(b) Let $\lambda_1 > 0$ be the smallest eigenvalue of A, let $\beta > 0$ be the smallest main diagonal entry of B, and let $x = [x_i] \in \mathbf{C}^n$ be a nonzero vector. Then $A - \lambda_1 I$ has nonnegative eigenvalues, so it is positive semidefinite and hence $(A - \lambda_1 I) \circ B$ is positive semidefinite. Then $0 \leq x^*((A - \lambda_1 I) \circ B)x = x^*(A \circ B)x - \lambda_1 x^*(I \circ B)x$, so

$$x^*(A \circ B)x \geq \lambda_1 x^*(I \circ B)x = \lambda_1 \sum_{i=1}^{n} b_{ii}|x_i|^2 \geq \lambda_1 \beta \, \|x\|_2^2 > 0$$

(c) If B is positive definite, then (7.1.2) ensures that its main diagonal entries are positive; the assertion follows from (b). $\square$

Exercise. For any nonzero $x \in \mathbf{C}^n$, show that the rank-one matrices xx^* and $\bar{x}x^T$ are positive semidefinite.

Theorem 7.5.4 (Moutard). *Let $A = [a_{ij}] \in M_n$. Then A is positive semidefinite if and only if* $\operatorname{tr}(AB^T) = \sum_{i,j=1}^n a_{ij}b_{ij} \geq 0$ *for every positive semidefinite $B = [b_{ij}] \in M_n$.*

Proof. Suppose that A and B are positive semidefinite, and let $e \in \mathbf{C}^n$ be the all-ones vector. Then $\operatorname{diag}(e) = I$ and $\operatorname{tr}(AB^T) = \operatorname{tr}((\operatorname{diag} e)A(\operatorname{diag} e)B^T) = e^*(A \circ B)e$ is nonnegative since $A \circ B$ is positive semidefinite. Conversely, if $\operatorname{tr}(AB^T) \geq 0$ whenever B is positive semidefinite, let $x = [x_i] \in \mathbf{C}^n$, let $B = \bar{x}x^T$, and compute $\operatorname{tr}(AB^T) = \sum_{i,j=1}^n a_{ij}\bar{x}_i x_j = x^*Ax \geq 0$. $\qquad\square$

Application 7.5.5. Let $D \subset \mathbf{R}^n$ be an open bounded set. The real second-order linear differential operator on $C^2(D)$ given by

$$Lu = \sum_{i,j=1}^n a_{ij}(x)\frac{\partial^2 u}{\partial x_i \, \partial x_j} + \sum_{i=1}^n b_i(x)\frac{\partial u}{\partial x_i} + c(x)u \qquad (7.5.6)$$

is said to be *elliptic* in D if the matrix $A(x) = [a_{ij}(x)]$ is positive definite for all $x \in D$. Suppose that $u \in C^2(D)$ satisfies the equation $Lu = 0$ in D. What can we say about the local maxima or minima of the function u in D? If $y \in D$ is a local minimum for u, then $\partial u/\partial x_i = 0$ at y for all $i = 1, \ldots, n$ and the Hessian matrix $[\partial^2 u/\partial x_i \, \partial x_j]$ is positive semidefinite at y. Therefore, $Lu = 0 = \sum_{i,j=1}^n a_{ij}\frac{\partial^2 u}{\partial x_i \, \partial x_j} + cu$, so the preceding theorem ensures that $-cu = \sum_{i,j=1}^n a_{ij}\frac{\partial^2 u}{\partial x_i \, \partial x_j} \geq 0$ at the point y. In particular, $u(y) > 0$ if $c(y) < 0$. A similar argument shows that $u(y) < 0$ at a relative maximum $y \in D$ if $c(y) < 0$. These simple observations are the heart of the following important principle.

Weak minimum principle 7.5.7. Let the operator L defined by (7.5.6) be elliptic in D, and suppose that $c(x) < 0$ in D. If $u \in C^2(D)$ satisfies $Lu = 0$ in D, then u cannot have a negative interior relative minimum or a positive interior relative maximum. If, in addition, u is continuous on the closure of D and u is nonnegative on the boundary of D, then u must be nonnegative everywhere in D.

From the weak minimum principle follows one of the fundamental uniqueness theorems for partial differential equations.

Fejér's uniqueness theorem 7.5.8. Suppose that the operator L defined by (7.5.6) is elliptic, assume that $c(x) < 0$ in D, let f be a given real-valued function on D, and let g be a given real-valued function on ∂D. Then there is at most one solution to the following boundary value problem:

u is twice continuously differentiable in D
$Lu = f$ in D
u is continuous on the closure of D
$u = g$ on ∂D

Proof. If u_1 and u_2 are two solutions to this problem, then the functions $\pm v = u_1 - u_2$ are solutions to the problem $Lv = 0$ in D and $v = 0$ on ∂D. The weak minimum principle says that v and $-v$ are both nonnegative in D, so $v = 0$ in D. $\square$

Exercise. Explain how the weak minimum principle and Fejer's uniqueness theorem apply to the partial differential equation $\sum_{i=1}^{n} \frac{\partial^2 u}{\partial x_i\, \partial x_i} - \lambda u = 0$ in $D \subset \mathbf{R}^n$, in which λ is a positive real parameter.

If $A = [a_{ij}] \in M_n$ is positive semidefinite, then $A \circ A = [a_{ij}^2]$ is also positive semidefinite. It follows from an induction argument that every positive integer Hadamard power $A^{(k)} = [a_{ij}^k]$ is positive semidefinite, $k = 1, 2, \ldots$. Since any non-negative linear combination of positive semidefinite matrices is positive semidefinite, it follows that

$$[p(a_{ij})] = a_0 J_n + a_1 A + a_2 A^{(2)} + \cdots + a_m A^{(m)}$$
$$= [a_0 + a_1 a_{ij} + a_2 a_{ij}^2 + \cdots + a_m a_{ij}^m]$$

is positive semidefinite whenever $p(t) = a_0 + a_1 t + \cdots + a_m t^m$ is a polynomial with nonnegative coefficients; the matrix J_n is the all-ones matrix. More generally, if $f(z) = \sum_{k=0}^{\infty} a_k z^k$ is an analytic function with all $a_k \geq 0$ and radius of convergence $R > 0$, then a limit argument shows that $[f(a_{ij})] \in M_n$ is positive semidefinite if all $|a_{ij}| < R$. An important example is $f(z) = e^z$, whose power series converges for all $z \in \mathbf{C}$ and whose coefficients are all positive: $a_k = 1/k!$. The Hadamard exponential matrix $[e^{a_{ij}}]$ is defined for every $A = [a_{ij}] \in M_n$; it is positive semidefinite whenever A is positive semidefinite, and it fails to be positive definite if and only if A is singular in an especially obvious way.

Theorem 7.5.9. *Let $A = [a_{ij}] \in M_n$ be positive semidefinite.*

(a) *The Hadamard powers $A^{(k)} = [a_{ij}^k]$ are positive semidefinite for all $k = 1, 2, \ldots$; they are positive definite if A is positive definite.*

(b) *Let $f(z) = a_0 + a_1 z + a_2 z^2 + \cdots$ be an analytic function with nonnegative co-efficients and radius of convergence $R > 0$. Then $[f(a_{ij})]$ is positive semidefinite if $|a_{ij}| < R$ for all $i, j \in \{1, \ldots, n\}$; it is positive definite if, in addition, A is positive definite and $a_i > 0$ for some $i \in \{1, 2, \ldots\}$.*

(c) *The Hadamard exponential matrix $[e^{a_{ij}}]$ is positive semidefinite; it is positive definite if and only if no two rows of A are identical.*

Proof. Only the assertions about positive definiteness require justification. The assertion in (a) follows from (7.5.3) and an induction. The assertion in (b) follows from (a). See (7.5.P18 to P21) for a proof of the assertion in (c). $\square$

Problems

7.5.P1 Let $A, B \in M_n$ be positive semidefinite. Provide details for the following outline of an alternative proof that $A \circ B$ is positive semidefinite: (a) There are matrices $X = [x_1 \ \ldots \ x_n]$, $Y = [y_1 \ \ldots \ y_n] \in M_n$ such that $XX^* = A$ and $YY^* = B$; (b) $A = \sum_{i=1}^{n} x_i x_i^*$

and $B = \sum_{i=1}^{n} y_i y_i^*$; (c) $A \circ B = \sum_{i,j=1}^{n}(x_i x_i^*) \circ (y_j y_j^*)$; (d) If $\xi = [\xi_i]$, $\eta = [\eta_i] \in \mathbf{C}^n$, then $(\xi\xi^*) \circ (\eta\eta^*) = (\xi \circ \eta)(\xi \circ \eta)^*$ is a rank-one positive semidefinite matrix.

7.5.P2 Let $A, B \in M_n$. Suppose that $H(A)$ (the Hermitian part of A) is positive definite and B is positive definite. (a) Show that $H(A \circ B)$ is positive definite. (b) Explain why $A \circ B$ has the row and column inclusion properties.

7.5.P3 If $A = [a_{ij}] \in M_n$ is positive semidefinite, show that the matrix $[|a_{ij}|^2]$ is also positive semidefinite.

7.5.P4 Let $A = [a_{ij}] \in M_n$ be positive semidefinite. The preceding problem ensures that $A \circ \bar{A} = [|a_{ij}|^2]$ is positive semidefinite, but what about the Hadamard absolute value matrix $|A| = [|a_{ij}|]$? (a) Suppose that A is positive definite. For $n = 1, 2, 3$, use Sylvester's criterion (7.2.5) to show that $|A|$ is positive definite. Use a limit argument to show that the same conclusion is valid if A is positive semidefinite ($n = 1, 2, 3$ only). (b) Problem (7.1.P10) ensures that $\cos t$ is a positive definite function, so the matrix $C = [\cos(t_i - t_j)]$ is positive semidefinite for all choices of $t_1, \ldots, t_n \in \mathbf{R}$ and for all $n = 1, 2, \ldots$. Let $n = 4$; let $t_1 = 0, t_2 = \pi/4, t_3 = \pi/2$, and $t_4 = 3\pi/4$. Compute $|C|$ and show that it is not positive semidefinite.

7.5.P5 Consider the matrix $|C| \in M_4$ in (7.5.P4). Compute $|C| \circ |C|$ and verify that it is positive semidefinite. Conclude that $B = |C| \circ |C|$ is a positive semidefinite matrix whose nonnegative "Hadamard square root" is not positive semidefinite. Contrast this with the situation for the ordinary square root $B^{1/2}$.

7.5.P6 Consider the matrix

$$A = \begin{bmatrix} 10 & 3 & -2 & 1 \\ 3 & 10 & 0 & 9 \\ -2 & 0 & 10 & 4 \\ 1 & 9 & 4 & 10 \end{bmatrix}$$

Show that A is positive definite but $|A|$ is not positive semidefinite.

7.5.P7 Let $K(x, y)$ be a continuous integral kernel on a finite interval $[a, b]$. Show that $K(x, y)$ is a positive semidefinite kernel if and only if the matrix $[K(x_i, x_j)] \in M_n$ is positive semidefinite for all choices of the points $x_1, \ldots, x_n \in [a, b]$ and all $n = 1, 2, \ldots$.

7.5.P8 Use (7.5.P7) and the Schur product theorem to show that the ordinary (pointwise) product of positive semidefinite integral kernels is positive semidefinite.

7.5.P9 Show that $f \in C(\mathbf{R})$ is a positive definite function if and only if $K(s, t) = f(s - t)$ is a positive semidefinite integral kernel.

7.5.P10 Explain why the product of two positive definite functions is a positive definite function.

7.5.P11 If $A = [a_{ij}] \in M_n$ is positive semidefinite, show that the matrix $[a_{ij}/(i + j)]$ is also positive semidefinite.

7.5.P12 Let $A = [a_{ij}] \in M_n$ be positive semidefinite and suppose that each of its entries is nonzero. Consider the Hadamard inverse matrix $A^{(-1)} = [a_{ij}^{-1}]$. Show that $A^{(-1)}$ is positive semidefinite if and only if rank $A = 1$, that is, if and only if $A = xx^*$ for some $x \in \mathbf{C}^n$ with nonzero entries.

7.5.P13 Let $A, B \in M_n$. Suppose that A is positive definite and B is positive semidefinite. Let $\nu(B)$ denote the number of nonzero main diagonal entries of B. (a) Explain why B is permutation similar to $0_{n-\nu(B)} \oplus C$, in which $C \in M_{\nu(B)}$ is positive semidefinite. (b) Why is $\nu(B) \geq \operatorname{rank} B$? (c) Use (7.5.3) to show that $\operatorname{rank}(A \circ B) \geq \nu(B) \geq \operatorname{rank} B$.

7.5.P14 Let $A \in \check{M}_n$ be positive definite. The matrix $A \circ A^{-T} = A \circ \bar{A}^{-1}$ is known in chemical engineering process control as a *relative gain array*. (a) Explain why $A \circ A^{-T}$ is positive definite, so its smallest eigenvalue $\lambda_{\min}$ is positive. (b) Use the trace identity in (7.5.2) to show that $\lambda_{\min} \geq 1$.

7.5.P15 Provide details for the following outline of a proof that the Hilbert matrix $H_n = [1/(i + j - 1)] \in M_n$ is positive semidefinite: (a) $X = [\xi_{ij}] = [(i - 1)(j - 1)/ij] \in M_n$ is a positive semidefinite matrix and $0 \leq \xi_{ij} < 1$ for all $i, j = 1, \ldots, n$; (b) $Y = [i^{-1}j^{-1}] \in M_n$ is a positive semidefinite matrix with positive diagonal entries; (c) $Z = [1/(1 - \xi_{ij})]$ is positive semidefinite; (e) Use (7.2.5) and (0.9.12.2) to show that H_n is actually positive definite; see (7.5.P22) for a different approach.

7.5.P16 Let $A \in M_n$ be Hermitian. Show that A is positive semidefinite if and only if $A \circ B$ is positive semidefinite for every positive semidefinite $B \in M_n$.

7.5.P17 Let $n_1, \ldots, n_m$ be m given distinct positive integers and let $\gcd(n_i, n_j)$ denote the greatest common divisor of n_i and n_j, $i, j = 1, \ldots, m$. We claim that the *gcd matrix* $G = [\gcd(n_i, n_j)] \in M_m$ is real symmetric positive semidefinite. Provide details: (a) Let $2 \leq p_1 < \cdots < p_d$ be an ordered list of all the distinct prime number factors of all the integers $n_1, \ldots, n_m$. Then, for each $i = 1, \ldots, m$, $n_i = p_1^{\nu(i,1)} \cdots p_d^{\nu(i,d)}$ for a unique set of positive integers $\nu(i, j)$, $i = 1, \ldots, m$, $j = 1, \ldots, d$. (b) $\gcd(n_i, n_j) = p_1^{\min\{\nu(i,1),\nu(j,1)\}} \cdots p_d^{\min\{\nu(i,d),\nu(j,d)\}}$. (c) Each matrix $[\min\{\nu(i, k), \nu(j, k)\}], k = 1, \ldots, d$, is positive semidefinite. (d) Each matrix $G_k = [p_k^{\min\{\nu(i,k),\nu(j,k)\}}]_{i,j=1}^m$ is positive semidefinite. (e) $G = G_1 \circ \cdots \circ G_d$.

The following four problems provide a proof of the assertion in (7.5.9(c)) about positive definiteness of a Hadamard exponential matrix.

7.5.P18 Let $A = [a_{ij}] \in M_n$ be positive semidefinite and let $B_t = [e^{ta_{ij}}]$. Why is B_t positive semidefinite for all $t > 0$? Show that the following are equivalent:

(a) $B_1 = [e^{a_{ij}}]$ is singular.
(b) There is a nonzero $x \in \mathbb{C}^n$ such that $B_t x = 0$ for all $t > 0$.
(c) B_t is singular for all $t > 0$.

The following ideas might be useful: $x \neq 0$ and $x^* B_1 x = 0 \Rightarrow 0 = x^* B_1 x = x^* J_n x + x^* A x + \frac{1}{2!} x^* A^{(2)} x + \cdots \Rightarrow x^* J_n x = 0$ and $x^* A^{(k)} x = 0$ for all $k = 1, 2, \ldots \Rightarrow 0 = x^* B_t x = x^* J_n x + t x^* A x + \frac{t^2}{2!} x^* A^{(2)} x + \cdots$ for all $t > 0 \Rightarrow B_t x = 0$ for all $t > 0$.

7.5.P19 Let $A = \begin{bmatrix} \alpha_1 & \beta \\ \bar{\beta} & \alpha_2 \end{bmatrix} \in M_2$ be positive semidefinite. We know that $B = \begin{bmatrix} e^{\alpha_1} & e^{\beta} \\ e^{\bar{\beta}} & e^{\alpha_2} \end{bmatrix}$ is positive semidefinite. Provide details to show that B is singular if and only if $\alpha_1 = \alpha_2 = \beta$. (a) $\det B = 0 \Rightarrow \alpha_1 + \alpha_2 = 2\operatorname{Re}\beta \Rightarrow (\alpha_1^2 + \alpha_2^2)/2 = 2(\operatorname{Re}\beta)^2 - \alpha_1\alpha_2$. (b) A positive semidefinite $\Rightarrow \alpha_1\alpha_2 \geq |\beta|^2$. (c) The arithmetic–geometric inequality ensures that

$$2(\operatorname{Re}\beta)^2 - \alpha_1\alpha_2 = \frac{\alpha_1^2 + \alpha_2^2}{2} \geq \alpha_1\alpha_2 \geq (\operatorname{Re}\beta)^2 + (\operatorname{Im}\beta)^2 \qquad (7.5.10)$$

(d) $(\operatorname{Re}\beta)^2 \geq \alpha_1\alpha_2 + (\operatorname{Im}\beta)^2 \geq (\operatorname{Re}\beta)^2 + 2(\operatorname{Im}\beta)^2 \Rightarrow \operatorname{Im}\beta = 0$ and $\alpha_1 + \alpha_2 = 2\beta$.
(e) It follows from (7.5.10) and (b) that

$$\beta^2 \geq 2\beta^2 - \alpha_1\alpha_2 = \frac{\alpha_1^2 + \alpha_2^2}{2} \geq \alpha_1\alpha_2 \geq \beta^2$$

so equality in the arithmetic–geometric mean inequality implies that $\alpha_1 = \alpha_2$.

7.5.P20 Let $n \geq 2$ and suppose that $A = [a_{ij}] \in M_n$ is positive semidefinite. If there are distinct $p, q \in \{1, \dots, n\}$ such that $a_{pp} = a_{qq} = a_{pq} = \alpha$, show that the pth and qth rows of A are identical.

7.5.P21 Let $n \geq 2$, let $A = [a_{ij}] \in M_n$ be positive semidefinite, and let $B = [e^{a_{ij}}]$ be the Hadamard exponential of A, which is positive semidefinite. We claim that B *is positive definite if and only if* A *has distinct rows*. Consider the equivalent assertion: B is singular if and only if two rows of A are identical. Sufficiency of the latter condition is apparent, so assume that B is singular and provide details for the following outline of a proof that two rows of A must be identical: (a) Let $B_t = [e^{ta_{ij}}]$. Problem (7.5.P18) ensures that B_t is positive semidefinite and singular for all $t > 0$. Let $D_t = \operatorname{diag}(e^{-ta_{11}/2}, \dots, e^{-ta_{nn}/2})$. Then $C_t = D_t B_t D_t = [e^{-t(a_{ii}+a_{jj}-2a_{ij})/2}]$ is a singular correlation matrix for all $t > 0$. (b) We know that $b_{ii}b_{jj} \geq |b_{ij}|^2$ for all $i, j \in \{1, \dots, n\}$. If $b_{ii}b_{jj} > |b_{ij}|^2$ for all distinct i, j, then $e^{a_{ii}+a_{jj}} > e^{2\operatorname{Re}a_{ij}} \Rightarrow a_{ii} + a_{jj} - 2\operatorname{Re}a_{ij} > 0$ for all distinct $i, j \Rightarrow C_t \to I_n$ as $t \to \infty$, so C_t is nonsingular for all sufficiently large t. This contradiction shows that there must be distinct $p, q \in \{1, \dots, n\}$ such that $b_{pp}b_{qq} = |b_{pq}|^2$, that is, the principal submatrix $\begin{bmatrix} b_{pp} & b_{pq} \\ b_{qp} & b_{qq} \end{bmatrix}$ of B is singular. (c) Problem (7.5.P19) ensures that $a_{pp} = a_{qq} = a_{pq}$ and (7.5.P20) tells us that rows p and q of A are identical.

7.5.P22 Revisit (7.5.P15) and use the ideas in (7.5.P18) to show that the Hilbert matrix is positive definite. (a) $Z = J_n + X + X^{(2)} + X^{(3)} + \cdots$ is positive semidefinite. (b) If $x \in \mathbf{C}^n$ is nonzero and $x^*Zx = 0$, then $x^*J_nx = 0$ and $x^*X^{(k)}x = 0$ for all $k = 1, 2, \dots$. Consequently, $J_nx = 0$ and $X^{(k)}x = 0$ for all $k = 1, 2, \dots$. (c) Let $\alpha_j = (j-1)/j$ for $j = 2, \dots, n$. Explain why $\sum_{i=1}^n x_i = 0$ and $\sum_{i=1}^n \alpha_i^k x_i = 0$ for each $k = 1, 2, \dots$. (d) Why does it follow that $x = 0$? (e) Conclude that H_n is positive definite.

7.5.P23 Let $z_1, \dots, z_n$ be distinct complex numbers. (a) Show that the n-by-n matrices $[e^{z_i\bar{z}_j}]$ and $[\cosh(z_i\bar{z}_j)]$ are positive definite. (b) If $f(z) = (1-z)^{-1}$ and $|z_i| < 1$ for each $i = 1, \dots, n$, show that $[f(z_i\bar{z}_j)]$ is positive definite.

7.5.P24 Let $A, B = [b_{ij}] \in M_n$ be positive semidefinite. (a) Examine the proof of (7.5.3(b)) and explain why $\lambda_{\min}(A \circ B) \geq \lambda_{\min}(A)\min\{b_{ii}\}$, which is useful only if A is positive definite and B has positive main diagonal entries. (b) Show that $\lambda_{\max}(A \circ B) \leq \lambda_{\max}(A)\max\{b_{ii}\}$, which is useful without any further assumptions.

7.5.P25 Let $A \in M_n$ be positive semidefinite, let $z \in \mathbf{C}^n$, let $c \in \mathbf{R}$, and let $e \in \mathbf{R}^n$ be the all-ones vector. Define $B = [b_{ij}] = A + ze^* + ez^* + cJ_n$. (a) Explain why B, while Hermitian, need not be positive semidefinite. (b) Show that the Hadamard exponential matrix $H = [e^{b_{ij}}]$ is positive semidefinite, and that it is positive definite unless A has two equal rows. (c) If $x \in \mathbf{C}^n$ satisfies the condition $x^*e = 0$, show that $x^*Bx \geq 0$. The matrix B is *conditionally positive semidefinite*; the condition is that x must belong to the $(n-1)$-dimensional subspace of $\mathbf{C}^n$ that is orthogonal to the vector e.

Notes and Further Readings. The first systematic study of norm and eigenvalue bounds for entrywise products of matrices seems is in I. Schur, Bemerkungen zur Theorie der beschränkten Bilinearformen mit unedlich vielen Veränderlichen, *J. Reine Angew. Math.* 140 (1911) 1–28; the results in (7.5.3) and (7.5.P24) are Satz VII in this paper. Theorem 7.5.4 was published by Th. Moutard in 1894; L. Fejér recognized in 1918 that it implies the Schur product theorem. J. Hadamard studied term-by-term products of Maclaurin series of analytic functions in 1899. For a short historical survey of the Hadamard product, see section 5.0 of Horn and Johnson (1991). Problem (7.5.P25) describes a class of Hermitian matrices, larger than the class of positive semidefinite matrices, whose Hadamard exponentials are positive semidefinite. See chapter 5 and section 6.3 of Horn and Johnson (1991) for a discussion of this point and a detailed treatment of Hadamard products.

7.6 Simultaneous diagonalizations, products, and convexity

In this section we discuss two different ways to diagonalize a pair of Hermitian matrices and preserve Hermicity. The first, which we address in the following theorem, is useful in dealing with products, while the second is useful in dealing with linear combinations.

Theorem 7.6.1. *Let $A, B \in M_n$ be Hermitian.*

(a) *If A is positive definite, then there is a nonsingular $S \in M_n$ such that $A = SIS^*$ and $B = S^{-*}\Lambda S^{-1}$, in which Λ is real diagonal. The inertias of B and Λ are the same, so Λ is nonnegative diagonal if B is positive semidefinite and Λ is positive diagonal if B is positive definite.*

(b) *If A and B are positive semidefinite and rank $A = r$, then there is a nonsingular $S \in M_n$ such that $A = S(I_r \oplus 0_{n-r})S^*$ and $B = S^{-*}\Lambda S^{-1}$, in which Λ is nonnegative diagonal.*

Proof. (a) Theorem 4.5.7 ensures that there is a nonsingular $T \in M_n$ such that $T^{-1}AT^{-*} = I$. The matrix T^*BT is Hermitian, so there is a unitary $U \in M_n$ such that $U^*(T^*BT)U = \Lambda$ is diagonal. Let $S = TU$. Then $S^{-1}AS^{-*} = U^*T^{-1}AT^{-*}U = U^*IU = I$ and $S^*BS = U^*T^*BTU = \Lambda$. Theorem 4.5.8 tells us that B and Λ have the same inertia.

(b) Using (4.5.7) again, choose a nonsingular $T \in M_n$ such that $T^{-1}AT^{-*} = I_r \oplus 0_{n-1}$ and partition $T^*BT = \begin{bmatrix} B_{11} & B_{12} \\ B_{12}^* & B_{22} \end{bmatrix}$ conformally to it. Since T^*BT is positive semidefinite, (7.1.10) ensures that there is an $X \in M_{n-r}$ such that $B_{12} = B_{11}X$. Let $R = \begin{bmatrix} I_r & -X \\ 0 & I_{n-r} \end{bmatrix}$ and compute $R^*(T^*BT)R = B_{11} \oplus (B_{22} - X^*B_{11}X)$; see (7.1.P28). There are unitary matrices $U_1 \in M_r$ and $U_2 \in M_{n-r}$ such that $U_1^*B_{11}U_1 = \Lambda_1$ and $U_2^*(B_{22} - X^*B_{11}X)U_2 = \Lambda_2$ are real diagonal. Let $U = U_1 \oplus U_2$, $\Lambda = \Lambda_1 \oplus \Lambda_2$, and $S = TRU$. A computation reveals that $S^{-1}AS^{-*} = I_r \oplus 0_{n-r}$ and $S^*BS = \Lambda$. □

Exercise. In the proof of part (a) of the preceding theorem, explain why one possible choice of T is the matrix $A^{1/2}$, so $S = A^{1/2}U$, in which U is any unitary matrix such that $A^{1/2}BA^{1/2} = U\Lambda U^*$ is a spectral decomposition. If A and B are real, use this observation to show that S may be chosen to be real.

The preceding theorem has an immediate application to some questions about matrix products.

Corollary 7.6.2. *Let* $A, B \in M_n$ *be Hermitian.*

(a) *If* A *is positive definite, then* AB *is diagonalizable and has real eigenvalues. If, in addition,* B *is positive definite or positive semidefinite, then* Λ *has positive or nonnegative eigenvalues, respectively.*

(b) *If* A *and* B *are positive semidefinite, then* AB *is diagonalizable and has nonnegative eigenvalues.*

Proof. (a) Use part (a) of the preceding theorem to represent $A = SS^*$ and $B = S^{-*}\Lambda S^{-1}$. Then $AB = SS^*S^{-*}\Lambda S^{-1} = S\Lambda S^{-1}$.

(b) Use part (b) of the preceding theorem to represent $A = S(I_r \oplus 0_{n-r})S^*$ and $B = S^{-*}\Lambda S^{-1}$. Then $AB = S(I_r \oplus 0_{n-r})S^*S^{-*}\Lambda S^{-1} = S(\Lambda_1 \oplus 0_{n-r})$, in which $\Lambda = \Lambda_1 \oplus \Lambda_2$ is partitioned conformally to $I_r \oplus 0_{n-r}$. $\qquad\square$

One case not covered in the corollary requires a different approach. The example $A = \begin{bmatrix} 1 & 0 \\ 0 & 0 \end{bmatrix}$ and $B = \begin{bmatrix} 0 & 1 \\ 1 & 0 \end{bmatrix}$ shows that the product of a positive semidefinite matrix and a Hermitian matrix need not be diagonalizable. The next theorem shows that this example is typical: AB is always quasi-diagonalizable, and any 2-by-2 blocks in its Jordan canonical form are nilpotent.

Theorem 7.6.3. *Let* $A, B \in M_n$ *be Hermitian, and assume that* A *is positive semidefinite and singular. Then* AB *is similar to* $\Lambda \oplus N$, *in which* Λ *is real diagonal and* $N = J_2(0) \oplus \cdots \oplus J_2(0)$ *is a direct sum of 2-by-2 nilpotent blocks; either direct summand* Λ *or* N *may be absent.*

Proof. Choose a nonsingular S such that $S^{-1}AS^{-*} = I_r \oplus 0_{n-r}$ and partition $S^*BS = [B_{ij}]$ conformally to $I_r \oplus 0_{n-r}$. Then $S^{-1}ABS = (S^{-1}AS^{-*})(S^*BS) = \begin{bmatrix} B_{11} & B_{12} \\ 0 & 0 \end{bmatrix}$ and $B_{11} \in M_r$ is Hermitian. If B_{11} is nonsingular, then (2.4.6.1) ensures that $\begin{bmatrix} B_{11} & B_{12} \\ 0 & 0 \end{bmatrix}$ is similar to $B_{11} \oplus 0_{n-r}$, which is diagonalizable. If rank $B_{11} = p < r$, then B_{11} is similar to $D \oplus 0_{r-p}$, in which $D \in M_p(\mathbf{R})$ is nonsingular real diagonal. Partition $B_{12} = \begin{bmatrix} C_1 \\ C_2 \end{bmatrix}$, in which $C_1 \in M_{p,n-r}$, and use (2.4.6.1) again:

$$\begin{bmatrix} D & 0 & C_1 \\ 0 & 0 & C_2 \\ 0 & 0 & 0 \end{bmatrix} \text{ is similar to } \begin{bmatrix} D & 0 & 0 \\ 0 & 0 & C_2 \\ 0 & 0 & 0 \end{bmatrix} = D \oplus \begin{bmatrix} 0 & C_2 \\ 0 & 0 \end{bmatrix}$$

Finally, observe that $\begin{bmatrix} 0 & C_2 \\ 0 & 0 \end{bmatrix}$ is nilpotent and has a zero square, so its Jordan canonical form is a direct sum of a zero matrix and rank C_2 copies of $J_2(0)$. $\qquad\square$

We now turn to a second way to diagonalize a pair of Hermitian matrices and preserve Hermicity. Proofs of the following assertions are parallel to the arguments used to prove (7.6.1).

Theorem 7.6.4. *Let $A, B \in M_n$ be Hermitian.*

(a) *If A is positive definite, then there is a nonsingular $S \in M_n$ such that $A = SIS^*$ and $B = S\Lambda S^*$, in which Λ is real diagonal. The inertias of B and Λ are the same, so Λ is nonnegative diagonal if B is positive semidefinite; it is positive diagonal if B is positive definite. The main diagonal entries of Λ are the eigenvalues of the diagonalizable matrix $A^{-1}B$.*

(b) *If A and B are positive semidefinite and rank $A = r$, then there is a nonsingular $S \in M_n$ such that $A = S(I_r \oplus 0_{n-r})S^*$ and $B = S\Lambda S^*$, in which Λ is nonnegative diagonal; rank $B = $ rank Λ.*

Proof. (a) Choose a nonsingular $T \in M_n$ such that $T^{-1}AT^{-*} = I$. Choose a unitary $U \in M_n$ such that $U^*(T^{-1}BT^{-*})U = \Lambda$ is diagonal. Let $S = TU$. Then $S^{-1}AS^{-*} = U^*T^{-1}AT^{-*}U = U^*IU = I$ and $S^{-1}BS^{-*} = U^*T^{-1}BT^{-*}U = \Lambda$. The final assertion follows from a computation: $A^{-1}B = (S^{-*}S^{-1})(S\Lambda S^*) = S^{-*}\Lambda S^*$.

(b) Choose a nonsingular $T \in M_n$ such that $T^{-1}AT^{-*} = I_r \oplus 0_{n-1}$ and partition $T^{-1}BT^{-*} = \begin{bmatrix} B_{11} & B_{12} \\ B_{12}^* & B_{22} \end{bmatrix}$ conformally to it. Let $X \in M_{n-r}$ be such that $B_{12} = B_{11}X$. Let $R = \begin{bmatrix} I_r & -X \\ 0 & I_{n-r} \end{bmatrix}$ and compute $R^*(T^{-1}BT^{-*})R = B_{11} \oplus (B_{22} - X^*B_{11}X)$. There are unitary matrices $U_1 \in M_r$ and $U_2 \in M_{n-r}$ such that $U_1^*B_{11}U_1 = \Lambda_1$ and $U_2^*(B_{22} - X^*B_{11}X)U_2 = \Lambda_2$ are real diagonal. Let $U = U_1 \oplus U_2$, $\Lambda = \Lambda_1 \oplus \Lambda_2$, and $S = TRU$. Then $S^{-1}AS^{-*} = I_r \oplus 0_{n-r}$ and $S^{-1}BS^{-*} = \Lambda$.   □

Exercise. In the proof of part (a) of the preceding theorem, explain why one possible choice of T is the matrix $A^{1/2}$, so $S = A^{1/2}U$, in which U is any unitary matrix such that $A^{-1/2}BA^{-1/2} = U\Lambda U^*$ is a spectral decomposition. What can you say if A and B are real? Use (7.2.9) to describe a possible choice of T that is lower triangular. What is the corresponding choice of U? What is S?

A variant of the preceding results can be proved with similar methods.

Theorem 7.6.5. *Let $A, B \in M_n$. If A is positive definite and B is complex symmetric, then there is a nonsingular $S \in M_n$ such that $A = SIS^*$ and $B = S\Lambda S^T$, in which Λ is nonnegative diagonal. The main diagonal entries of Λ^2 are the eigenvalues of the diagonalizable matrix $A^{-1}B\bar{A}^{-1}\bar{B}$.*

Proof. Choose a nonsingular matrix $R \in M_n$ such that $R^{-1}AR^{-*} = I$. Use (4.4.4(c)) to choose a unitary $U \in M_n$ such that $U^*R^{-1}BR^{-T}\bar{U} = \Lambda$ is nonnegative diagonal. Let $S = RU$. Then $S^{-1}AS^{-*} = U^*R^{-1}AR^{-*}U = U^*IU = I$ and $S^{-1}BS^{-T} = U^*R^{-1}BR^{-T}\bar{U} = \Lambda$. A computation verifies the second assertion: $A^{-1}B\bar{A}^{-1}\bar{B} = (S^{-*}S^{-1})(S\Lambda S^T)(S^{-T}\bar{S}^{-1})(\bar{S}\Lambda S^*) = S^{-*}\Lambda^2 S^*$.   □

This result has applications to complex function theory: The Grunsky inequalities for univalent analytic functions discussed in (4.4) are inequalities between quadratic forms generated by a positive definite Hermitian matrix and a complex symmetric matrix.

The following result is an application of (7.6.4(a)).

Theorem 7.6.6. *The function* $f(A) = \log \det A$ *is a strictly concave function on the convex set of positive definite Hermitian matrices in* M_n.

Proof. Let $A, B \in M_n$ be positive definite. We must show that

$$\log \det(\alpha A + (1 - \alpha)B) \geq \alpha \log \det A + (1 - \alpha) \log \det B \qquad (7.6.7)$$

for all $\alpha \in (0, 1)$, with equality if and only if $A = B$. Use (7.6.4(a)) to write $A = SIS^*$ and $B = S\Lambda S^*$ for some nonsingular $S \in M_n$ and $\Lambda = \text{diag}(\lambda_1, \ldots, \lambda_n)$ with each $\lambda_i > 0$. Then

$$f(\alpha A + (1 - \alpha)B) = f(S(\alpha I + (1 - \alpha)\Lambda)S^*) = f(SS^*) + f(\alpha I + (1 - \alpha)\Lambda)$$
$$= f(A) + f(\alpha I + (1 - \alpha)\Lambda)$$

and

$$\alpha f(A) + (1 - \alpha)f(B) = \alpha f(A) + (1 - \alpha)f(S\Lambda S^*)$$
$$= \alpha f(A) + (1 - \alpha)(f(SS^*) + f(\Lambda))$$
$$= \alpha f(A) + (1 - \alpha)f(A) + (1 - \alpha)f(\Lambda)$$
$$= f(A) + (1 - \alpha)f(\Lambda)$$

It suffices to show that $f(\alpha I + (1 - \alpha)\Lambda) \geq (1 - \alpha)f(\Lambda)$ for all $\alpha \in (0, 1)$. This follows from strict concavity of the logarithm function:

$$f(\alpha I + (1 - \alpha)\Lambda) = \log \prod_{i=1}^{n}(\alpha + (1 - \alpha)\lambda_i) = \sum_{i=1}^{n} \log(\alpha + (1 - \alpha)\lambda_i)$$

$$\geq \sum_{i=1}^{n}(\alpha \log 1 + (1 - \alpha) \log \lambda_i)$$

$$= (1 - \alpha)\sum_{i=1}^{n} \log \lambda_i = (1 - \alpha) \log \prod_{i=1}^{n} \lambda_i$$

$$= (1 - \alpha)\log \det \Lambda = (1 - \alpha)f(\Lambda)$$

This inequality is an equality if and only if every $\lambda_i = 1$, if and only if $\Lambda = I$ if and only if $B = SIS^* = A$. $\qquad \square$

Theorem 7.6.6 is often used in the following form, which is obtained by exponentiating the inequality (7.6.7).

Corollary 7.6.8. *Let* $A, B \in M_n$ *be positive definite and let* $0 < \alpha < 1$. *Then*

$$\det(\alpha A + (1 - \alpha)B) \geq (\det A)^{\alpha}(\det B)^{1-\alpha} \qquad (7.6.9a)$$

with equality if and only if $A = B$.

Exercise. If $A, B \in M_n$ are positive definite, explain why

$$\det\left(\frac{A + B}{2}\right) \geq \sqrt{\det AB} \qquad (7.6.9b)$$

with equality if and only if $A = B$. This inequality may be thought of as an arithmetic–geometric mean inequality for the determinant.

Another application of (7.6.4(a)) draws on the same ideas as (7.6.6).

Theorem 7.6.10. *The function $f(A) = \operatorname{tr} A^{-1}$ is a strictly convex function on the convex set of positive definite Hermitian matrices in M_n.*

Proof. Let $A, B \in M_n$ be positive definite. We must show that

$$\operatorname{tr}(\alpha A + (1 - \alpha)B)^{-1} \le \alpha \operatorname{tr} A^{-1} + (1 - \alpha) \operatorname{tr} B^{-1}$$

for all $\alpha \in (0, 1)$, with equality if and only if $A = B$. Use (7.6.4(a)) to write $A = SIS^*$ and $B = S\Lambda S^*$ for some nonsingular $S \in M_n$ and $\Lambda = \operatorname{diag}(\lambda_1, \ldots, \lambda_n)$ with each $\lambda_i > 0$. Let $s_1, \ldots, s_n$ be the (necessarily positive) main diagonal entries of the positive definite matrix $S^{-1}S^{-*}$. Then

$$
\begin{aligned}
\operatorname{tr}(\alpha A + (1 - \alpha)B)^{-1} &= \operatorname{tr}(\alpha SS^* + (1 - \alpha)S\Lambda S^*)^{-1} \\
&= \operatorname{tr}(S^{-*}(\alpha I + (1 - \alpha)\Lambda)^{-1}S^{-1}) \\
&= \operatorname{tr}((\alpha I + (1 - \alpha)\Lambda)^{-1}S^{-1}S^{-*}) \\
&= \sum_{i=1}^{n}(\alpha + (1 - \alpha)\lambda_i)^{-1}s_i \\
&\le \sum_{i=1}^{n}(\alpha s_i + (1 - \alpha)\lambda_i^{-1}s_i) \\
&= \operatorname{tr}(\alpha S^{-1}S^{-*} + (1 - \alpha)S^{-1}\Lambda^{-1}S^{-*}) \\
&= \alpha \operatorname{tr} A^{-1} + (1 - \alpha) \operatorname{tr} B^{-1}
\end{aligned}
$$

with equality if and only if every $\lambda_i = 1$, if and only if $A = B$. $\square$

For a stronger result, see (7.7.P14).

Problems

7.6.P1 Let $A \in M_n$. Show that the following statements are equivalent:

(a) A is similar to a Hermitian matrix.
(b) A is diagonalizable and has real eigenvalues.
(c) $A = HK$, in which $H, K \in M_n$ are Hermitian and at least one factor is positive definite.
(d) A is similar to A^* via a positive definite similarity.

Compare with (4.1.7).

7.6.P2 Let $A, B \in M_n$ be Hermitian. If there are real numbers α and β such that $\alpha A + \beta B$ is positive definite, show that there is a nonsingular $S \in M_n$ such that $S^{-1}AS^{-*}$ and $S^{-1}BS^{-*}$ are real diagonal.

7.6.P3 Show that the matrix $A^{-1}B\bar{A}^{-1}\bar{B}$ in (7.6.5) is similar to a matrix that is *congruent to the positive definite matrix A^{-T}.

7.6.P4 Use the following idea to prove (7.6.1(b)): Let $S \in M_n$ be nonsingular and such that $A + B = S(I_m \oplus 0_{n-m})S^*$. Partition $S^{-1}AS^{-*} = [A_{ij}]$ and $S^{-1}BS^{-*} = [B_{ij}]$ conformally to $I_m \oplus 0_{n-m}$. Then (7.1.14) ensures that $A_{22} + B_{22} = 0 \Rightarrow A_{22} = B_{22} = 0$, so $A_{12} = B_{12} = 0$. $A_{11} + B_{11} = I_m \Rightarrow A_{11}$ and B_{11} commute, so they are simultaneously unitarily diagonalizable.

7.6.P5 Consider the two real quadratic forms $5x_1^2 - 2x_1x_2 + x_2^2$ and $x_1^2 + 2x_1x_2 - x_2^2$. (a) Explain *why* there is a nonsingular change of variables $x \to S\xi$ that transforms the given quadratic forms to $\alpha_1\xi_1^2 + \alpha_2\xi_2^2$ and $\beta_1\xi_1^2 - \beta_2\xi_2^2$, respectively, with positive coefficients $\alpha_1, \alpha_2, \beta_1, \beta_2$. (b) Explain *how* you could determine an S that effects this transformation.

7.6.P6 Let $A, B \in M_n$ be positive semidefinite. Show that $\det(A + B) \geq \det A + \det B$, with equality if and only if either $A + B$ is singular or $A = 0$ or $B = 0$.

7.6.P7 Let $A, B \in M_n$ be Hermitian and suppose that A is positive definite. Use (7.6.4) to show that $A + B$ is positive definite if and only if every eigenvalue of $A^{-1}B$ is greater than -1.

7.6.P8 Let $C \in M_n$ be Hermitian, and write $C = A + iB$ with $A, B \in M_n(\mathbf{R})$. If C is positive definite, we claim that $|\det B| < \det A$ and $\det C \leq \det A$. Provide details: (a) Verify that A is symmetric and B is skew symmetric, so the eigenvalues of B are pure imaginary and occur in conjugate pairs. (b) Show that C is positive definite if and only if A is positive definite and every eigenvalue of $iA^{-1}B$ is greater than -1. (c) If A is positive definite, show that the eigenvalues of $iA^{-1}B$ are either zero or occur in $\pm$ pairs. (d) If C is positive definite then A is positive definite, every eigenvalue of $iA^{-1}B$ lies in the interval $(-1, 1)$, and the eigenvalues of $iA^{-1}B$ are either zero or occur in $\pm$ pairs. (e) If C is positive definite, conclude that $|\det iA^{-1}B| < 1$ and hence $|\det B| < \det A$; this is an inequality of H. P. Robertson. (f) If C is positive definite, then $\det C = \det A \det(I + iA^{-1}B)$. Why is $0 < \det(I + iA^{-1}B) \leq 1$, with equality if and only if $B = 0$? (g) If C is positive definite, conclude that $\det C \leq \det A$, with equality if and only if $B = 0$; this is an inequality of O. Taussky. See (7.8.19) and (7.8.24) for generalizations of these inequalities.

7.6.P9 In (4.1.7) we found that $A \in M_n$ is the product of two Hermitian matrices if and only if A is similar to a real matrix. Use (7.6.1(a)) to show that $A \in M_n$ is the product of two positive definite Hermitian matrices if and only if A is diagonalizable and has positive eigenvalues.

7.6.P10 If $A, B, C \in M_n$ are positive definite, we know that AB is positive definite if and only if it is Hermitian ($AB = BA$). Show that $S = ABC$ is positive definite if and only if it is Hermitian ($ABC = CBA$). Show that $\text{tr}(AB)$ is always positive, and give an example to show that $\text{tr}(ABC)$ can be negative.

7.6.P11 Provide the details for the following alternative proof of the result in the preceding problem, with the same notation and assumptions: Let $S(\alpha) = ((1 - \alpha)C + \alpha A)BC$ for all $\alpha \in [0, 1]$, and assume that $S(1)$ is Hermitian. (a) Why is $S(\alpha)$ Hermitian for all $\alpha \in [0, 1]$? (b) Why is $S(\alpha)$ nonsingular for all $\alpha \in [0, 1]$? (c) The eigenvalues of $S(\alpha)$ depend continuously on α, all the eigenvalues of $S(0)$ are positive, and every eigenvalue of $S(\alpha)$ is nonzero, for all $\alpha \in [0, 1]$. Conclude that every eigenvalue of $S(1)$ is positive.

7.6.P12 Let $A, B \in M_n$ be Hermitian and assume that A is positive semidefinite. Theorem 7.6.3 says that AB is similar to a direct sum of a positive diagonal matrix $D_+ \in M_\pi(\mathbf{R})$, a negative diagonal matrix $D_- \in M_\nu(\mathbf{R})$, a zero matrix, and s copies of $J_2(0)$, the

2-by-2 nilpotent Jordan block. (a) Examine the proof of (7.6.3) and show that $\pi \leq i_+(B)$, $\nu \leq i_-(B)$, and $s = \text{rank}(AB) - \text{rank}((AB)^2)$; see (4.5.6).

7.6.P13 Let $A, B \in M_n(\mathbf{R})$ be symmetric and positive definite, and let $x(t) = [x_1(t) \ldots x_n(t)]^T$. Use (7.6.2(a)) or (7.6.4(a)) to show that every solution $x(t)$ of $Ax''(t) = -Bx(t)$ is bounded on $(-\infty, \infty)$.

7.6.P14 Let $A, B \in M_n$ be Hermitian. (a) If A is positive definite, or if A and B are positive semidefinite, show that $\rho(AB) = 0$ if and only if $AB = 0$. (b) If A is positive semidefinite and $\rho(AB) = 0$, is $AB = 0$?

7.6.P15 Let $A, B \in M_n$ be positive semidefinite and nonzero. Show that $\text{tr}\, AB = \left\| A^{1/2} B^{1/2} \right\|_2^2 \geq 0$ with equality if and only if $AB = 0$.

7.6.P16 Let $A, B \in M_n$ be Hermitian and suppose that A is positive semidefinite. Show that AB is similar to a real diagonal matrix if and only if $\text{rank}(AB) = \text{rank}(AB)^2$.

7.6.P17 Let $\mathcal{S} \subset M_n$ be a compact convex set of positive semidefinite matrices that contains at least one positive definite matrix. (a) Explain why $\mu = \sup\{\det A : A \in \mathcal{S}\}$ is positive and finite, and why there is a matrix $Q \in \mathcal{S}$ such that $\mu = \det Q$. (b) Use (7.6.9b) to show that if $A_1, A_2 \in \mathcal{S}$ and $\det A_1 = \det A_2 = \mu$, then $A_1 = A_2$. (c) Conclude that there is a *unique* matrix $Q \in \mathcal{S}$ such that $\det Q = \max\{\det A : A \in \mathcal{S}\}$.

The following 10 problems explore properties of the *ellipsoid* $\mathcal{E}(A) = \{x \in \mathbf{F}^n : x^* A x \leq 1\}$ associated with a positive definite matrix $A \in M_n(\mathbf{F})$ ($\mathbf{F} = \mathbf{R}$ or $\mathbf{C}$).

7.6.P18 Let $A \in M_n(\mathbf{F})$ be positive definite. Show that $\mathcal{E}(A)$ is the unit ball of the norm $\nu_A(x) = \left\| A^{1/2} x \right\|_2$, so it is a convex set.

7.6.P19 (Continuation; same notation) Let $A \in M_n(\mathbf{R})$ be positive definite. Show that the volume of $\mathcal{E}(A)$ is $\text{vol}(\mathcal{E}(A)) = c_n / \sqrt{\det A}$, in which $c_n = \pi^{n/2} \Gamma(1 + n/2)$ is the volume of the Euclidean unit ball in $\mathbf{R}^n$; c_n satisfies the recurrence $c_1 = 2$, $c_2 = \pi$, $c_n = \frac{2\pi}{n} c_{n-2}$ for $n = 3, 4, \ldots$. Observe that larger determinants correspond to smaller volumes.

7.6.P20 (Continuation; same notation) Let $\|\cdot\|$ be a norm on $\mathbf{F}^n$ and define $E(\|\cdot\|) = \{B \in M_n(\mathbf{F}) : B$ is positive semidefinite and $x^* B x \leq \|x\|^2$ for all $x \in \mathbf{F}^n\}$. Let $A \in M_n(\mathbf{F})$ be positive definite and consider the norm $\nu_A(x) = \left\| A^{1/2} x \right\|_2$ defined in (7.6.P18); its unit ball is $\mathcal{E}(A)$. Show that (a) the unit ball of $\|\cdot\|$ is contained in $\mathcal{E}(B)$ for every positive definite $B \in E(\|\cdot\|)$; (b) there is an $\varepsilon > 0$ such that $\varepsilon \nu_A(x) = \nu_{\varepsilon^2 A}(x) \leq \|x\|$ for all $x \in \mathbf{F}^n$, so $\varepsilon^2 A \in E(\|\cdot\|)$; (c) $\mathcal{E}(\varepsilon^2 A)$ contains the unit ball of $\|\cdot\|$; (d) the set $E(\|\cdot\|)$ is convex, compact (closed and bounded; any norm on $M_n(\mathbf{F})$), and contains a positive definite matrix; (e) there is a unique positive definite matrix $Q \in E(\|\cdot\|)$ with maximum determinant; (f) if $\mathbf{F} = \mathbf{R}$, $Q \in E(\|\cdot\|)$ is the unique positive definite matrix such that $\text{vol}(\mathcal{E}(Q)) = \min\{\text{vol}(\mathcal{E}(B)) : B$ is positive definite and $\mathcal{E}(B)$ contains the unit ball of $\|\cdot\|\}$.

The ellipsoid $\mathcal{E}(Q)$ is the *Loewner ellipsoid associated with the norm* $\|\cdot\|$. The positive definite matrix $L = Q^{1/2}$ is the *Loewner–John matrix associated with the norm* $\|\cdot\|$. The unit ball of $\|\cdot\|$ is contained in (and must touch) the unit ball of $\nu_Q(\cdot) = \|L \cdot\|_2$, so $\|Lx\|_2 \leq \|x\|$ for all $x \in \mathbf{F}^n$ with equality for some nonzero $x_0 \in \mathbf{F}^n$. Moreover, $\|L(cx_0)\|_2 = \|cx_0\|$ for all $c \in \mathbf{F}$ with modulus one.

7.6.P21 (Continuation; same notation) Let $\mathcal{F}_{\|\cdot\|} = \{A \in M_n(\mathbf{F}) : A$ is an isometry for $\|\cdot\|\}$ be the isometry group of $\|\cdot\|$; see (5.4.P11). Let $A \in \mathcal{F}_{\|\cdot\|}$ (so $|\det A| = 1$), let L be the Loewner–John matrix associated with $\|\cdot\|$, and let $Q = L^2$. Show that

(a) $x^*A^*QAx = (Ax)^*Q(Ax) \le \|Ax\|^2 = \|x\|^2$, so $A^*QA \in E(\|\cdot\|)$; (b) $\det(A^*QA) = |\det A|^2 \det Q = \det Q$, so

$$A^*QA = Q \qquad (7.6.11)$$

(c) it follows from (7.6.11) that

$$LAL^{-1} \text{ is unitary if } \mathbf{F} = \mathbf{C}; \text{ it is real orthogonal if } \mathbf{F} = \mathbf{R} \qquad (7.6.12)$$

(d) $\mathcal{F}$ is a bounded multiplicative matrix group; each member of $\mathcal{F}$ is similar, via the same positive definite matrix L, to a unitary matrix in $M_n(\mathbf{F})$; each member of $\mathcal{F}$ is diagonalizable; $L\mathcal{F}L^{-1}$ is a subgroup of the unitary group in $M_n(\mathbf{F})$.

7.6.P22 (Continuation; same notation) Let $\mathcal{G} \subset M_n(\mathbf{F})$ be a bounded multiplicative group of matrices. We claim that there is a norm $\|\cdot\|_{\mathcal{G}}$ on $\mathbf{F}^n$ such that every member of $\mathcal{G}$ is an isometry of $\|\cdot\|_{\mathcal{G}}$. Provide details: (a) Let $\|\cdot\|$ be any norm on $\mathbf{F}^n$. Then $\|x\|_{\mathcal{G}} = \sup\{\|Bx\| : B \in \mathcal{G}\}$ defines a norm on $\mathbf{F}^n$. (b) If $A \in \mathcal{G}$, then $\|Ax\|_{\mathcal{G}} = \sup\{\|BAx\| : B \in \mathcal{G}\} = \sup\{\|Cx\| : C \in \mathcal{G}\} = \|x\|_{\mathcal{G}}$.

7.6.P23 (Continuation; same notation) Let $\mathcal{G} \subset M_n(\mathbf{F})$ be a bounded multiplicative group of matrices. Deduce from the preceding two problems that the Loewner–John matrix L associated with the norm $\|\cdot\|_{\mathcal{G}}$ is a positive definite matrix such that LAL^{-1} is unitary for every $A \in \mathcal{G}$. This result is a strong form of *Auerbach's theorem*: A bounded multiplicative group of complex (respectively, real) matrices is similar to a group of unitary (respectively, real orthogonal) matrices.

7.6.P24 (Continuation; same notation) Let $\|\cdot\|$ be an absolute norm on $\mathbf{F}^n$. Show that its Loewner–John matrix is positive diagonal.

7.6.P25 (Continuation; same notation) Let $\|\cdot\|$ be an absolute norm on $\mathbf{F}^n$ that is permutation-invariant (that is, $\|\cdot\|$ is a symmetric gauge function) and let L be its Loewner–John matrix. Show that $L = \alpha I$, in which $\alpha = \min\{\|x\| : \|x\|_2 = 1\}$.

7.6.P26 (Continuation; same notation) Show that the Loewner–John matrix L associated with the l_p norms on $\mathbf{F}^n$ is αI, in which $\alpha = 1$ if $1 \le p \le 2$ and $\alpha = n^{(p-2)/2p}$ if $p \ge 2$.

7.6.P27 Let $\|\cdot\|$ be a norm on $\mathbf{R}^n$. (a) Show that $\mathrm{vol}\,\mathcal{E}(Q) \ge c_n / \prod_{i=1}^n \|e_i\|$, in which the vectors e_i are the standard basis vectors in $\mathbf{R}^n$ and c_n is the volume of the Euclidean unit ball in $\mathbf{R}^n$. (b) Conclude that $\mathrm{vol}\,\mathcal{E}(Q) \ge c_n$ if $\prod_{i=1}^n \|e_i\| \le 1$, which is the case if $\|\cdot\|$ is a standardized norm. (c) Confirm that $\mathrm{vol}\,\mathcal{E}(Q) \ge \pi$ if $n = 2$ and $\|\cdot\| = \|\cdot\|_p$, an l_p norm on $\mathbf{R}^2$ with $1 \le p \le \infty$. What is $\mathcal{E}(Q)$ if $1 \le p \le 2$? if $p = \infty$?

7.6.P28 Deduce (7.6.4(a)) from the more general result in (4.5.17(a)).

7.6.P29 Deduce (7.6.5) from the more general result in (4.5.17(c)).

Further Readings. For a variety of results about products of matrices from various positivity classes, see C. S. Ballantine and C. R. Johnson, Accretive matrix products, *Linear Multilinear Algebra* 3 (1975) 169–185. For a proof of the assertion in (7.6.P12(c)), see R. A. Horn and Y. P. Hong, The Jordan canonical form of a product of a Hermitian and a positive semidefinite matrix, *Linear Algebra Appl.* 147 (1991) 373–386. Problems (7.6.P18 to P25) are adapted from E. Deutsch and H. Schneider, Bounded groups and norm-Hermitian matrices, *Linear Algebra Appl.* 9 (1974) 9–27, which contains other interesting applications of the Loewner–John matrix.

7.7 The Loewner partial order and block matrices

The natural analogy between Hermitian matrices and real numbers, with positive semidefinite matrices as the analogs of nonnegative real numbers, suggests an order relation among Hermitian matrices.

Definition 7.7.1. *Let* $A, B \in M_n$. *We write* $A \succeq B$ *if* A *and* B *are Hermitian and* $A - B$ *is positive semidefinite; we write* $A \succ B$ *if* A *and* B *are Hermitian and* $A - B$ *is positive definite.*

Observe that $A \succeq 0$ if A is Hermitian and positive semidefinite; $A \succ 0$ if A is Hermitian and positive definite.

Exercise. Show that $A \succeq B$ and $B \succeq A$ if and only if $A = B$.

Exercise. Show that the relation $\succeq$ is transitive and reflexive, but that it is not a total order if $n > 1$: If $n > 1$, there are always Hermitian matrices $A, B \in M_n$ such that neither $A \succeq B$ nor $B \succeq A$.

The preceding exercise shows that the order relation defined in (7.7.1) is a *partial order*. It is often referred to as the *Loewner partial order*.

Exercise. If $A \succeq B$ and $C \succeq 0$, explain why $A \circ C \succeq B \circ C$. *Hint*: $A - B \succeq 0$ and (7.5.3a).

Exercise. If $A \in M_n$ is Hermitian with smallest and largest eigenvalues $\lambda_{\min}(A)$ and $\lambda_{\max}(A)$, respectively, explain why $\lambda_{\max}(A)I \succeq A \succeq \lambda_{\min}(A)I$. *Hint*: (4.2.2(c)).

Exercise. Let $A \in M_n$ be Hermitian. Explain why: $I \succeq A$ if and only if $\lambda_{\max}(A) \le 1$; $I \succ A$, if and only if $\lambda_{\max}(A) < 1$.

A partial order on a real linear space can be defined by identifying a special closed convex cone and saying that one element of the linear space is "greater" than another if their difference lies in the special cone. For the Loewner partial order, the elements of the real linear space are n-by-n Hermitian matrices and the elements of the closed convex cone are positive semidefinite matrices. For this partial order and matrices of size one, the real linear space is $\mathbf{R}$ and the closed convex cone is $[0, \infty)$: This gives the usual order relation on the real numbers.

A different partial order provides the setting for the following chapter: The real linear space is $M_n(\mathbf{R})$ and the closed convex cone consists of real matrices with nonnegative entries.

Exercise. Show by example that if $A \succeq B$ and if $A \ne B$, it does not follow that $A \succ B$.

Let $X \in M_{n,m}$. Then $\sigma_1(X) = \lambda_{\max}(XX^*)^{1/2} = \lambda_{\max}(X^*X)^{1/2} = \sigma_1(X^*)$ is the largest singular value (the spectral norm) of X. We say that X is a *contraction* if $\sigma_1(X) \le 1$; it is a *strict contraction* if $\sigma_1(X) < 1$.

The following properties of the Loewner partial order generalize familiar facts about real numbers.

Theorem 7.7.2. *Let* $A, B \in M_n$ *be Hermitian and let* $S \in M_{n,m}$. *Then*

(a) *if* $A \succeq B$, *then* $S^*AS \succeq S^*BS$

(b) *if* rank $S = m$, *then* $A \succ B$ *implies* $S^*AS \succ S^*BS$

(c) *if* $m = n$ *and* $S \in M_n$ *is nonsingular, then* $A \succ B$ *if and only if* $S^*AS \succ S^*BS$; $A \succeq B$ *if and only if* $S^*AS \succeq S^*BS$

(d) $I_m \succ S^*S$ *(respectively,* $I_n \succ SS^*$ *) if and only if* S *is a strict contraction;* $I_m \succeq S^*S$ *(respectively,* $I_n \succeq SS^*$ *) if and only if* S *is a contraction*

Proof. (a) If $(A - B) \succeq 0$, then (7.1.8(a)) ensures that $S^*(A - B)S = S^*AS - S^*BS \succeq 0$.

(b) This assertion follows in the same way from (7.1.8(b)).

(c) If $S^*AS \succ S^*BS$, then $S^{-*}(S^*AS)S^{-1} = A \succ B = S^{-*}(S^*BS)S^{-1}$, and likewise for the remaining assertion.

(d) $I_m \succeq S^*S$ if and only if $1 \geq \lambda_{\max}(S^*S) = \sigma_1(S)^2$, and likewise for the remaining assertions. $\qquad\square$

Theorem 7.7.3. *Let* $A, B \in M_n$ *be Hermitian and suppose that* A *is positive definite.*

(a) *If* B *is positive semidefinite, then* $A \succeq B$ *(respectively,* $A \succ B$ *) if and only if* $\rho(A^{-1}B) \leq 1$ *(respectively,* $\rho(A^{-1}B) < 1$ *) if and only if there is a positive semidefinite contraction (respectively, strict contraction)* X *such that* $B = A^{1/2}XA^{1/2}$.

(b) $A^2 \succeq B^2$ *(respectively,* $A^2 \succ B^2$ *) if and only if* $\sigma_1(A^{-1}B) \leq 1$ *(respectively,* $\sigma_1(A^{-1}B) < 1$ *) if and only if there is a contraction (respectively, strict contraction)* X *such that* $B = AX = X^*A$.

Proof. (a) It follows from (a) and (d) of the preceding theorem that $A \succeq B$ if and only if $I = A^{-1/2}AA^{-1/2} \succeq A^{-1/2}BA^{-1/2}$ if and only if $1 \geq \sigma_1(A^{-1/2}BA^{-1/2})$. But $A^{-1/2}BA^{-1/2}$ is positive semidefinite, so

$$\sigma_1(A^{-1/2}BA^{-1/2}) = \lambda_{\max}(A^{-1/2}BA^{-1/2})$$
$$= \lambda_{\max}(A^{-1/2}A^{-1/2}B) = \lambda_{\max}(A^{-1}B)$$

Finally, (7.6.2(a)) ensures that $A^{-1}B$ has real nonnegative eigenvalues, so $\lambda_{\max}(A^{-1}B) = \rho(A^{-1}B)$. Conversely, if $B = A^{1/2}XA^{1/2}$ and X is a positive semidefinite contraction, then $A^{-1}B = A^{-1}A^{1/2}XA^{1/2} = A^{-1/2}XA^{1/2}$ is similar to a positive semidefinite contraction, so $\rho(A^{-1}B) = \rho(X) = \sigma_1(X) \leq 1$. The argument is the same if $A \succ B$, but in this case $I = A^{-1/2}AA^{-1/2} \succ A^{-1/2}BA^{-1/2}$, so $1 > \sigma_1(A^{-1/2}BA^{-1/2})$.

(b) Since B^2 is positive semidefinite, $A^2 \succeq B^2$ if and only if $I \succeq A^{-1}B^2A^{-1}$ if and only if $1 \geq \sigma_1(A^{-1}B^2A^{-1}) = \lambda_{\max}((A^{-1}B)(A^{-1}B)^*) = \sigma_1(A^{-1}B)^2$. Let $X = A^{-1}B$. Then X is a contraction and $B = AX$ is Hermitian, so $AX = X^*A$. Conversely, if X is a contraction and $B = AX = X^*A$, then $B^2 = AXX^*A$, so $A^2 - B^2 = A(I - XX^*)A$ is positive semidefinite. $\qquad\square$

Exercise. Suppose that A is positive definite and B is positive semidefinite. If $\sigma_1(A^{-1}B) \leq 1$, explain why both $A \succeq B$ and $A^2 \succeq B^2$. *Hint:* $\sigma_1(X) \geq \rho(X)$ for all $X \in M_n$; see (5.6.9).

Corollary 7.7.4. *Let $A, B \in M_n$ be Hermitian. Let $\lambda_1(A) \leq \cdots \leq \lambda_n(A)$ and $\lambda_1(B) \leq \cdots \leq \lambda_n(B)$ be the ordered eigenvalues of A and B, respectively.*

(a) *If $A \succ 0$ and $B \succ 0$, then $A \succeq B$ if and only if $B^{-1} \succeq A^{-1}$.*

(b) *If $A \succ 0$, $B \succeq 0$, and $A \succeq B$, then $A^{1/2} \succeq B^{1/2}$.*

(c) *If $A \succeq B$, then $\lambda_i(A) \geq \lambda_i(B)$ for each $i = 1, \ldots, n$.*

(d) *If $A \succeq B$, then $\operatorname{tr} A \geq \operatorname{tr} B$ with equality if and only if $A = B$.*

(e) *If $A \succeq B \succeq 0$, then $\det A \geq \det B \geq 0$.*

Proof. (a) The preceding theorem ensures that $A \succeq B$ if and only if $\rho(A^{-1}B) = \rho(BA^{-1}) \leq 1$ if and only if $B^{-1} \succeq A^{-1}$.

(b) Let $X = A^{-1/2}B^{1/2}$. If $A \succeq B$, then $1 \geq \rho(A^{-1}B) = \rho(A^{-1/2}B^{1/2}B^{1/2}A^{-1/2}) = \rho((A^{-1/2}B^{1/2})(A^{-1/2}B^{1/2})^*) = \sigma_1(A^{-1/2}B^{1/2})^2 \geq \rho(A^{-1/2}B^{1/2})^2$. The criterion in (7.7.3(a)) ensures that $A^{1/2} \succeq B^{1/2}$.

(c) $A = B + (A - B)$ and $A - B \succeq 0$; the asserted eigenvalue inequalities follow from (4.3.12).

(d) The asserted inequality follows from (c). Since $A - B \succeq 0$, (7.1.5) ensures that $\operatorname{tr}(A - B) = 0$ if and only if $A - B = 0$.

(e) Use (c) to compute $\det A = \prod_{i=1}^{n} \lambda_i(A) \geq \prod_{i=1}^{n} \lambda_i(B) = \det B \geq 0$. $\square$

Exercise. Let $A = \begin{bmatrix} 3 & 1 \\ 1 & 2 \end{bmatrix}$ and $B = \begin{bmatrix} 2 & 0 \\ 0 & 1 \end{bmatrix}$. Show that $A \succ B \succ 0$ but $A^2 - B^2$ is not positive semidefinite. Thus, the implications in (7.7.4b) cannot be reversed.

Exercise. If $A \succ B$ in each of the assertions in the preceding corollary, how can the respective conclusions be strengthened?

If a partitioned Hermitian matrix $H = \begin{bmatrix} A & B \\ B^* & C \end{bmatrix}$ has a nonsingular leading principal submatrix A, the fundamental identity for the Schur complement (0.8.5.3) is the nonsingular *congruence

$$\begin{bmatrix} I & 0 \\ Y^* & I \end{bmatrix}\begin{bmatrix} A & B \\ B^* & C \end{bmatrix}\begin{bmatrix} I & Y \\ 0 & I \end{bmatrix} = \begin{bmatrix} A & 0 \\ 0 & C - B^*A^{-1}B \end{bmatrix} \qquad (7.7.5)$$

in which $Y = -A^{-1}B$. This identity and (7.7.2(c)) reveal that H is positive definite if and only if both A and its Schur complement $C - B^*A^{-1}B$ are positive definite; it is positive semidefinite if and only if $A \succ 0$ and $C - B^*A^{-1}B \succeq 0$. This observation has a host of pleasant consequences. We prepare to present some of them by establishing the following lemma.

Lemma 7.7.6. *Let $X \in M_{p,q}$ and let $K = \begin{bmatrix} I_p & X \\ X^* & I_q \end{bmatrix} \in M_{p+q}$. Then*

(a) *K is positive definite if and only if X is a strict contraction*

(b) *K is positive semidefinite if and only if X is a contraction*

Proof. The identity (7.7.5) and (7.7.2(d)) ensure that $\begin{bmatrix} I_p & X \\ X^* & I_q \end{bmatrix} \succ 0$ if and only if $I_q - X^*X \succ 0$ if and only if $\sigma_1(X) < 1$. The assertion in (b) follows in a similar fashion. $\square$

Theorem 7.7.7. *Let $H = \begin{bmatrix} A & B \\ B^* & C \end{bmatrix} \in M_{p+q}$ be Hermitian with $A \in M_p$ and $C \in M_q$. The following are equivalent:*

(a) H is positive definite.

(b) A is positive definite and $C - B^* A^{-1} B$ is positive definite.

(c) A and C are positive definite and $\rho(B^* A^{-1} B C^{-1}) < 1$.

(d) A and C are positive definite and $\sigma_1(A^{-1/2} B C^{-1/2}) < 1$.

(e) A and C are positive definite and there is a strict contraction $X \in M_{p,q}$ such that
$B = A^{1/2} X C^{1/2}$.

Proof. (a) $\Leftrightarrow$ (b): Demonstrated in our discussion of (7.7.5).

(b) $\Rightarrow$ (c): Follows from (7.7.3(a)).

(c) $\Leftrightarrow$ (d): Let $X = A^{-1/2} B C^{-1/2}$. Then $1 > \rho(B^* A^{-1} B C^{-1}) = \rho(C^{-1/2} B^* A^{-1} B C^{-1/2}) = \rho(X^* X) = \sigma_1(X)^2$.

(d) $\Rightarrow$ (e): For $X = A^{-1/2} B C^{-1/2}$, $\sigma_1(X) < 1$ and $B = A^{1/2} X C^{1/2}$.

(e) $\Rightarrow$ (a): Let $B = A^{1/2} X C^{1/2}$, in which $X \in M_{p,q}$ and $\sigma_1(X) < 1$. Let $S = A^{1/2} \oplus C^{1/2}$. The preceding lemma ensures that $\begin{bmatrix} I_p & X \\ X^* & I_q \end{bmatrix} \succ 0$, so $H = S^* \begin{bmatrix} I_p & X \\ X^* & I_q \end{bmatrix} S \succ 0$. $\qquad\square$

We use the following lemma to obtain a version of the preceding theorem in the positive semidefinite case.

Lemma 7.7.8. *Let $A \in M_n$ be positive semidefinite and singular, and let $A_k = A + k^{-1} I_n$ for each $k = 1, 2, \ldots$. Let $X_k \in M_{m,n}$ be a contraction for each $k = 1, 2, \ldots$. Then*

(a) *each A_k is positive definite and $\lim_{k \to \infty} A_k^{1/2} = A^{1/2}$*

(b) *there is a sequence of positive integers $k_i \to \infty$ as $i \to \infty$ such that $X = \lim_{i \to \infty} X_{k_i}$ exists and is a contraction*

Proof. (a) Let $r = \operatorname{rank} A$, let $\lambda_1, \ldots, \lambda_r$ be the positive eigenvalues of A, and let $A = U(\operatorname{diag}(\lambda_1, \ldots, \lambda_r) \oplus 0_{n-r}) U^*$ be a spectral decomposition. Then continuity of the square root function on $[0, \infty)$ ensures that $A_k^{1/2} = U(\operatorname{diag}((\lambda_1 + k^{-1})^{1/2}, \ldots, (\lambda_r + k^{-1})^{1/2}) \oplus k^{-1/2} I_{n-r}) U^* \to U(\operatorname{diag}(\lambda_1^{1/2}, \ldots, \lambda_r^{1/2}) \oplus 0_{n-r}) U^* = A^{1/2}$ as $k \to \infty$.

(b) If $B = [b_{ij}] = [b_1 \ \ldots \ b_n] \in M_{m,n}$ is a contraction, then for any entry of B, we have

$$|b_{ij}|^2 \le \|b_j\|_2^2 = \|Be_j\|_2^2 = e_j^T B^* B e_j \le \lambda_{\max}(B^* B) \|e_j\|_2^2 = \sigma_1(B)^2 \le 1$$

The given sequence of contractions is therefore a bounded sequence in $M_{m,n}$, so it contains a convergent subsequence $X_{k_i} \to X$ as $i \to \infty$. Theorem 2.6.4 ensures that $\sigma_1(X) = \lim_{i \to \infty} \sigma_1(X_{k_i}) \le 1$, so X is a contraction. $\qquad\square$

Theorem 7.7.9. *Let $H = \begin{bmatrix} A & B \\ B^* & C \end{bmatrix} \in M_{p+q}$ be Hermitian, with $A \in M_p$ and $C \in M_q$. The following two statements are equivalent:*

(a) *H is positive semidefinite.*

(b) *A and C are positive semidefinite, and there is a contraction $X \in M_{p,q}$ such that $B = A^{1/2} X C^{1/2}$.*

If H is positive semidefinite, we may choose the contraction in (b) to be

$$X = \lim_{i \to \infty} (A + k_i^{-1} I_p)^{-1/2} B (C + k_i^{-1} I_q)^{-1/2} \qquad (7.7.9.1)$$

for some sequence of positive integers $k_i \to \infty$ as $i \to \infty$.

If H is positive semidefinite and A and C are nonsingular, then $X = A^{-1/2} B C^{-1/2}$ and the following statements are equivalent:

(c) *A and C are positive definite and $\rho(B^* A^{-1} B C^{-1}) \leq 1$.*

(d) *A and C are positive definite and $A^{-1/2} B C^{-1/2}$ is a contraction.*

(e) *A and C are positive definite and $C - B^* A^{-1} B$ is positive semidefinite.*

Proof. (a) $\Rightarrow$ (b): Consider $H_k = H + k^{-1} I_n$ for each $k = 1, 2, \ldots$. Then H_k, $A_k = A + k^{-1} I_p$, and $C_k = C + k^{-1} I_q$ are positive definite for each $k = 1, 2, \ldots$, so (7.7.7(e)) ensures that there is a contraction $X_k \in M_{p,q}$ such that $B = A_k^{1/2} X_k C_k^{1/2}$ for each $k = 1, 2, \ldots$. The preceding lemma tells us that there is a sequence $k_i \to \infty$ such that $X = \lim_{i \to \infty} X_{k_i}$ is a contraction, $\lim_{i \to \infty} A_{k_i}^{1/2} = A^{1/2}$, $\lim_{i \to \infty} C_{k_i}^{1/2} = C^{1/2}$, and $B = \lim_{i \to \infty} A_{k_i}^{1/2} X_{k_i} C_{k_i}^{1/2} = A^{1/2} X C^{1/2}$.

(b) $\Rightarrow$ (a): If $B = A^{1/2} X C^{1/2}$ and X is a contraction, let $S = A^{1/2} \oplus C^{1/2}$. Then (7.1.8(b)) and (7.7.6) ensure that $H = \begin{bmatrix} A & B \\ B^* & C \end{bmatrix} = S \begin{bmatrix} I_p & X \\ X^* & I_q \end{bmatrix} S^*$ is positive semidefinite.

(b) $\Rightarrow$ (c) $\Rightarrow$ (d) $\Rightarrow$ (e) $\Rightarrow$ (a): Proceed as in the corresponding implications in (7.7.7). $\qquad\square$

The characterization (7.7.9.1) has an important consequence: If

$$(A + \varepsilon I_p)^{-1/2} B (C + \varepsilon I_q)^{-1/2}$$

is Hermitian, skew Hermitian, symmetric, skew symmetric, positive semidefinite, or real for each sufficiently small $\varepsilon > 0$, then a contraction X such that $B = A^{1/2} X C^{1/2}$ may be chosen to have the same property.

Corollary 7.7.10. *Let $A, C \in M_p$ be Hermitian.*

(a) *If $\begin{bmatrix} A & I_p \\ I_p & C \end{bmatrix} \succeq 0$, then $A \succ 0$, $C \succ 0$, $A \succeq C^{-1}$, and $C \succeq A^{-1}$.*

(b) *If $A \succ 0$, $C \succ 0$, and either $A \succeq C^{-1}$ or $C \succeq A^{-1}$, then $\begin{bmatrix} A & I_p \\ I_p & C \end{bmatrix} \succeq 0$.*

(c) *If $A \succ 0$, then $\begin{bmatrix} A & I_p \\ I_p & A^{-1} \end{bmatrix} \succeq 0$.*

(d) *If $A \succ 0$, then the following are equivalent: (i) $\begin{bmatrix} A & I_p \\ I_p & A \end{bmatrix} \succeq 0$; (ii) $A \succeq A^{-1}$; (iii) $A \succeq I \succeq A^{-1}$.*

Proof. (a) The hypothesis ensures that $A \succeq 0$ and $C \succeq 0$. Theorem 7.7.9(b) ensures that there is a contraction X such that $I = A^{1/2} X C^{1/2}$, so both $A^{1/2}$ and $C^{1/2}$ (and hence both A and C) are nonsingular. Then $A \succeq C^{-1}$ and $C \succeq A^{-1}$ follow from (7.7.9(e)) and (7.7.4(a)).

(b) and (c) follow from (7.7.5).

(d) (i) $\Rightarrow$ (ii) and (iii) $\Rightarrow$ (i) follow from (7.7.5). (ii) $\Rightarrow$ (iii) follows from (7.7.3(a)): $A \succeq A^{-1} \Rightarrow \rho(A^{-2}) \leq 1 \Rightarrow \rho(A^{-1}) \leq 1 \Rightarrow A \succeq I$ and $I \succeq A^{-1}$. $\qquad\square$

The inequalities in the following theorem arise in complex function theory and harmonic analysis; they are best understood as facts about partitioned positive semidefinite matrices.

Theorem 7.7.11. *Let $A \in M_p$ and $C \in M_q$ be positive semidefinite and let $B \in M_{p,q}$. The following four statements are equivalent:*

(a) $(x^*Ax)(y^*Cy) \geq |x^*By|^2$ *for all $x \in \mathbf{C}^p$ and all $y \in \mathbf{C}^q$.*

(b) $x^*Ax + y^*Cy \geq 2|x^*By|$ *for all $x \in \mathbf{C}^p$ and all $y \in \mathbf{C}^q$.*

(c) $H = \begin{bmatrix} A & B \\ B^* & C \end{bmatrix}$ *is positive semidefinite.*

(d) *There is a contraction $X \in M_{p,q}$ such that $B = A^{1/2}XC^{1/2}$.*

If A and C are positive definite, then the following statement is equivalent to (c):

(e) $\rho(B^*A^{-1}BC^{-1}) \leq 1$.

Proof. (a) $\Rightarrow$ (b): This implication follows from the arithmetic–geometric mean inequality: $\frac{1}{2}(x^*Ax + y^*Cy) \geq (x^*Ax)^{1/2}(y^*Cy)^{1/2} \geq |x^*By|$.

(b) $\Rightarrow$ (c): Let $z = [x^* \ y^*]^*$ and compute

$$z^*Hz = x^*Ax + y^*Cy + 2\operatorname{Re}(x^*By) \geq x^*Ax + y^*Cy - 2|x^*By|$$

which (b) ensures is nonnegative.

(c) $\Rightarrow$ (d): This implication is in (7.7.9).

(d) $\Rightarrow$ (a): If $B = A^{1/2}XC^{1/2}$ and X is a contraction, then the Cauchy–Schwarz inequality and (7.3.9) permit us to compute

$$|x^*By|^2 = |x^*A^{1/2}XC^{1/2}y|^2 = |(A^{1/2}x)^*(XC^{1/2}y)|^2$$
$$\leq \left\|A^{1/2}x\right\|_2^2 \left\|XC^{1/2}y\right\|_2^2 \leq \left\|A^{1/2}x\right\|_2^2 \sigma_1(X) \left\|C^{1/2}y\right\|_2^2$$
$$\leq \left\|A^{1/2}x\right\|_2^2 \left\|C^{1/2}y\right\|_2^2 = (x^*Ax)(y^*Cy)$$

(e) $\Leftrightarrow$ (c): This equivalence is in (7.7.7).   □

The following special case of the preceding theorem introduces a generalization of the notion of positive definiteness: $x^*Ax \geq |x^*Bx|$ for all x, rather than $x^*Ax \geq 0$ for all x; for a different generalization see (7.7.P16).

Corollary 7.7.12. *Let $A \in M_n$ be positive semidefinite and let $B \in M_n$ be Hermitian. The following four statements are equivalent:*

(a) $x^*Ax \geq |x^*Bx|$ *for all $x \in \mathbf{C}^n$.*

(b) $x^*Ax + y^*Ay \geq 2|x^*By|$ *for all $x, y \in \mathbf{C}^n$.*

(c) $H = \begin{bmatrix} A & B \\ B & A \end{bmatrix}$ *is positive semidefinite.*

(d) *There is a Hermitian contraction $X \in M_n$ such that $B = A^{1/2}XA^{1/2}$.*

If A is positive definite, then the following statement is equivalent to (a):

(e) $\rho(A^{-1}B) \leq 1$.

Proof. (a) $\Rightarrow$ (b): Let $x, y \in \mathbf{C}^n$ and use the inequality in (a), the triangle inequality, and Hermicity of B to compute

$$
\begin{aligned}
2(x^*Ax + y^*Ay) = (x+y)^*A(x+y) + (x-y)^*A(x-y) \\
\geq |(x+y)^*B(x+y)| + |-(x-y)^*A(x-y)| \\
\geq |(x+y)^*B(x+y) - (x-y)^*A(x-y)| = 4|x^*By|
\end{aligned}
$$

(b) $\Rightarrow$ (c) $\Rightarrow$ (d) $\Rightarrow$ (a): Let $A = C$ and $B = B^*$ in the preceding theorem. Since $(A + \varepsilon I)^{-1/2}B(A + \varepsilon I)^{-1/2}$ is Hermitian for each $\varepsilon > 0$, (7.7.9) ensures that we may take X to be Hermitian.

(e) $\Leftrightarrow$ (c): In the preceding theorem, (e) becomes $\rho(A^{-1}B)^2 \leq 1$ if $A = C$ and $B = B^*$. $\square$

The next corollary provides a version of the basic representation (7.7.3(a)) in the positive semidefinite case.

Corollary 7.7.13. *Let* $A, B \in M_n$ *be positive semidefinite. The following statements are equivalent:*

- *(a)* $A \succeq B$.
- *(b)* $\begin{bmatrix} A & B \\ B & A \end{bmatrix} \succeq 0$.
- *(c) There is a positive semidefinite contraction* $X \in M_n$ *such that* $B = A^{1/2}XA^{1/2}$.

Proof. Since $A \succeq B$ if and only if $x^*Ax \geq x^*Bx$ for all $x \in \mathbf{C}^n$, the asserted equivalences follow from (7.7.12) in the special case in which the Hermitian matrix B is positive semidefinite, *provided* we can choose X to be positive semidefinite instead of merely Hermitian. But $(A + \varepsilon I)^{-1/2}B(A + \varepsilon I)^{-1/2}$ is positive semidefinite for each $\varepsilon > 0$, so (7.7.9.1) ensures that we may take X to be positive semidefinite. $\square$

Corollary 7.7.14. *Let* $A, B, C, D \in M_n$ *be Hermitian and suppose that* A *and* C *are positive semidefinite. If* $x^*Ax \geq |x^*Bx|$ *and* $x^*Cx \geq |x^*Dx|$ *for all* $x \in \mathbf{C}^n$, *then* $x^*(A \circ C)x \geq |x^*(B \circ D)x|$ *for all* $x \in \mathbf{C}^n$.

Proof. The hypotheses and the implication (a) $\Rightarrow$ (c) in (7.7.12) ensure that $\begin{bmatrix} A & B \\ B & A \end{bmatrix} \succeq 0$ and $\begin{bmatrix} C & D \\ D & C \end{bmatrix} \succeq 0$, so (7.5.3) tells us that $\begin{bmatrix} A & B \\ B & A \end{bmatrix} \circ \begin{bmatrix} C & D \\ D & C \end{bmatrix} = \begin{bmatrix} A \circ C & B \circ D \\ B \circ D & A \circ C \end{bmatrix} \succeq 0$. The conclusion now follows from the implication (c) $\Rightarrow$ (a) in (7.7.12). $\square$

The inverse of a positive definite matrix is positive definite, and any principal submatrix of a positive definite matrix is positive definite. If we apply these two operations successively in both orders, we obtain matrices that obey an interesting inequality.

Theorem 7.7.15. *Let* $H \in M_n$ *be positive definite and let* $\alpha \subset \{1, \ldots, n\}$. *Then* $H^{-1}[\alpha] \succeq (H[\alpha])^{-1}$.

Proof. Since a permutation congruence of a positive definite matrix is positive definite, we may assume that $H = \begin{bmatrix} A & B \\ B^* & C \end{bmatrix}$, $\alpha = \{1, \ldots, k\}$, and $H[\alpha] = A$. The identity (0.7.3.1) ensures that $H^{-1}[\alpha] = (A - BC^{-1}B^*)^{-1} = (A - B^*C^{-1}B)^{-1} \succ 0$.

The inequality $A \succeq A - B^*C^{-1}B \succ 0$ and (7.7.4(a)) tell us that $(A - B^*C^{-1}B)^{-1} \succeq A^{-1}$, so $H^{-1}[\alpha] = (A - B^*C^{-1}B)^{-1} \succeq A^{-1} = (H[\alpha])^{-1}$. $\qquad \square$

Exercise. Let $A = [a_{ij}] \in M_n$ be positive definite and let $A^{-1} = [\alpha_{ij}]$. Deduce from the preceding theorem that $\alpha_{ii} \geq 1/a_{ii}$ for each $i = 1, \ldots, n$.

Several 2-by-2 block matrices that are positive (semi)definite can be constructed using (7.7.9(b)).

Theorem 7.7.16. *Let $A \in M_n$ be positive definite. The following matrices are positive semidefinite and singular:*

(a) $\begin{bmatrix} A & X \\ X^* & X^*A^{-1}X \end{bmatrix}$ *for any $X \in M_{n,m}$.*

(b) $\begin{bmatrix} A & I_n \\ I_n & A^{-1} \end{bmatrix}$.

(c) $\begin{bmatrix} A & A \\ A & A \end{bmatrix}$.

Proof. Use (7.7.9). In each case, we verify that for $\begin{bmatrix} A & B \\ B^* & C \end{bmatrix}$ we have A nonsingular and $C = B^*A^{-1}B$.
(a) $X^*A^{-1}X - X^*A^{-1}X = 0$.
(b) Take $X = I_n$ in (a).
(c) Take $X = A$ in (a). $\qquad \square$

Exercise. Show that $\begin{bmatrix} A & A \\ A & A \end{bmatrix} \succeq 0$ if $A \succeq 0$. *Hint*: Consider $S\begin{bmatrix} I & I \\ I & I \end{bmatrix}S^*$ and $S = A^{1/2} \oplus A^{1/2}$.

Many inequalities for Hadamard products of positive semidefinite matrices can be derived by considering Hadamard products of suitable positive semidefinite 2-by-2 block matrices; (7.5.3) ensures that these products are positive semidefinite. The following theorem is an example of this technique.

Theorem 7.7.17. *Let $A, B \in M_n$ be positive definite. Then*

(a) $A^{-1} \circ B^{-1} \succeq (A \circ B)^{-1}$
(b) $A^{-1} \circ A^{-1} \succeq (A \circ A)^{-1}$
(c) $A^{-1} \circ A \succeq I \succeq (A^{-1} \circ A)^{-1}$

Proof. (a) The preceding theorem and the Schur product theorem ensure that

$$\begin{bmatrix} A & I_n \\ I_n & A^{-1} \end{bmatrix} \circ \begin{bmatrix} B & I_n \\ I_n & B^{-1} \end{bmatrix} = \begin{bmatrix} A \circ B & I_n \\ I_n & A^{-1} \circ B^{-1} \end{bmatrix}$$

is positive semidefinite. Therefore, (7.5.3(c)) and (7.7.10(a)) ensure that $A^{-1} \circ B^{-1} \succeq (A \circ B)^{-1}$.
(b) Set $B = A$ in (a).
(c) Set $B = A^{-1}$ in (a) and invoke (7.7.10(d)). $\qquad \square$

Our final result, a consequence of the preceding theorem, provides a lower bound that gives a quantitative version of (7.5.3(a)). For a different lower bound, see (7.5.P24).

Theorem 7.7.18. *Let* $A, B \in M_n$ *be positive definite. Then* $\lambda_{\min}(A \circ B) \geq \max\{\lambda_{\min}(AB), \lambda_{\min}(AB^T)\}.$

Proof. Since $\lambda_{\min}(B^{1/2}AB^{1/2}) = \lambda_{\min}(AB)$, we have $B^{1/2}AB^{1/2} \succeq \lambda_{\min}(AB)I$, which is equivalent to $A \succeq \lambda_{\min}(AB)B^{-1}$. It follows from (7.7.17(c)) that $A \circ B \succeq \lambda_{\min}(AB)(B^{-1} \circ B) \succeq \lambda_{\min}(AB)I$ and hence $\lambda_{\min}(A \circ B) \geq \lambda_{\min}(AB)$. Likewise, $\lambda_{\min}((B^{1/2})^T A (B^{1/2})^T) = \lambda_{\min}(AB^T)$, so $A \succeq \lambda_{\min}(AB^T)B^{-T}$ and (7.5.P14) ensures that $A \circ B \succeq \lambda_{\min}(AB^T)(B^{-T} \circ B) \succeq \lambda_{\min}(AB^T)I$. It follows that $\lambda_{\min}(A \circ B) \geq \lambda_{\min}(AB^T)$. $\qquad\square$

Problems

7.7.P1 Consider $\begin{bmatrix} 4 & 0 \\ 0 & 2 \end{bmatrix}$ and $\begin{bmatrix} 1 & 0 \\ 0 & 3 \end{bmatrix}$, and explain why the implication in (7.7.4(c)) cannot be reversed. However, if $A, B \in M_n$ are Hermitian; if $A = U\Lambda U^*$ and $B = VMV^*$ are spectral decompositions in which $\Lambda = \mathrm{diag}(\lambda_1, \dots, \lambda_n)$, $M = \mathrm{diag}(\mu_1, \dots, \mu_n)$, $\lambda_1 \leq \cdots \leq \lambda_n$, and $\mu_1 \leq \cdots \leq \mu_n$; and if $\Lambda \succeq M$, show that there is a unitary W such that $W^*AW \succeq B$; in fact, we may take $W = UV^*$.

7.7.P2 Let $A_1, A_2, B_1, B_2 \in M_n$ be Hermitian. If $A_1 \succeq B_1$ and $A_2 \succeq B_2$, show that $A_1 + A_2 \succeq B_1 + B_2$.

7.7.P3 The assertion in (7.7.4(b)) can be improved. Use (7.7.8) to show that if $A \succeq B \succeq 0$, then $A^{1/2} \succeq B^{1/2}$.

7.7.P4 Let $A, B, C, D \in M_n$ be Hermitian. Suppose that $A \succeq B \succeq 0$ and $C \succeq D \succeq 0$. Show that $A \circ C \succeq B \circ D \succeq 0$.

7.7.P5 Let $A, B \in M_n$ be Hermitian. If $A \succeq B$ and $\alpha \subset \{1, \dots, n\}$, show that $A[\alpha] \succeq B[\alpha]$.

7.7.P6 Let $A, B \in M_n$ be positive semidefinite. If $A \succeq B$, show that range $B \subseteq$ range A.

7.7.P7 Let $A \in M_n$ and let $y \in \mathbf{C}^n$ be nonzero. Show that there is an $x \in \mathbf{C}^n$ such that $\|x\|_2 \leq 1$ and $Ax = y$ if and only if $AA^* \succeq yy^*$.

7.7.P8 Let $H = \begin{bmatrix} A & B \\ B^* & C \end{bmatrix} \in M_n$ be positive definite, let $A \in M_k$, and let $\alpha = \{1, \dots, k\}$. Examine the proof of (7.7.15) and explain why $H^{-1}[\alpha] \succ (H[\alpha])^{-1}$ if and only if B has full column rank.

7.7.P9 Suppose that $H = \begin{bmatrix} A & B \\ B^* & C \end{bmatrix} \succeq 0$. Show that $\min\{\mathrm{rank}\, A, \mathrm{rank}\, C\} \geq \mathrm{rank}\, B$. In particular, A and C are positive definite if B is square and nonsingular.

7.7.P10 Let $A \in M_n$ be positive definite, and let $x, y \in \mathbf{C}^n$. Show that $(x^*Ax)(y^*A^{-1}y) \geq |x^*y|^2$.

7.7.P11 If $H = \begin{bmatrix} A & B \\ B^* & C \end{bmatrix}$ is positive semidefinite and $A, C \in M_p$, show that $(\det A)(\det C) \geq |\det B|^2$. What can you say if H is positive definite?

7.7.P12 Let $A, B \in M_n$ and let $Z = \begin{bmatrix} I & A \\ B^* & I \end{bmatrix}$. Verify that $ZZ^* = \begin{bmatrix} I + AA^* & A + B \\ (A+B)^* & I + B^*B \end{bmatrix}$ and show that $|\det(A + B)|^2 \leq (\det(I + AA^*))(\det(I + BB^*))$.

7.7.P13 Let $A, B \in M_n$ be Hermitian and let $\alpha \in (0, 1)$. Verify that $\alpha A^2 + (1 - \alpha)B^2 \succeq (\alpha A + (1 - \alpha)B)^2 + \alpha(1 - \alpha)(A - B)^2 \succeq (\alpha A + (1 - \alpha)B)^2$. Conclude that $f(t) = t^2$ is strictly convex on Hermitian matrices.

7.7.P14 Let $A, B \in M_n$ be positive definite and let $\alpha \in (0, 1)$. (a) Show that $\alpha A^{-1} + (1 - \alpha)B^{-1} \succeq (\alpha A + (1 - \alpha)B)^{-1}$, with equality if and only if $A = B$. Thus, $f(t) = t^{-1}$ is strictly convex on positive definite matrices.

7.7.P15 Let $A = [a_{ij}] \in M_n$ be positive semidefinite, let $B = [b_{ij}] \in M_n$ be Hermitian, and suppose that $x^*Ax \geq |x^*Bx|$ for all $x \in \mathbf{C}^n$. (a) Show that $x^*[a_{ij}^k]x \geq |x^*[b_{ij}^k]x|$ for all $x \in \mathbf{C}^n$ and all $k = 1, 2, \ldots$. (b) Show that $x^*[e^{a_{ij}}]x \geq |x^*[e^{b_{ij}}]x|$ for all $x \in \mathbf{C}^n$.

7.7.P16 Let $A \in M_n$ be positive semidefinite and let $B \in M_n$ be symmetric. Show that the following statements are equivalent:

(a) $x^*Ax \geq |x^T Bx|$ for all $x \in \mathbf{C}^n$.
(b) $x^*Ax + y^*Ay \geq 2|x^T By|$ for all $x, y \in \mathbf{C}^n$.
(c) $x^*\bar{A}x + y^*Ay \geq 2|x^*By|$ for all $x, y \in \mathbf{C}^n$.
(d) $H = \begin{bmatrix} \bar{A} & B \\ \bar{B} & A \end{bmatrix}$ is positive semidefinite.
(e) There is a symmetric contraction $X \in M_n$ such that $B = \bar{A}^{1/2}XA^{1/2}$.

If A is positive definite, then the following statements are each equivalent to (c):

(f) $\rho(\bar{B}\bar{A}^{-1}BA) \leq 1$.
(g) $\sigma_1(\bar{A}^{-1/2}BA^{-1/2}) \leq 1$.

7.7.P17 Let $A, B, C, D \in M_n$. Suppose that A and C are positive semidefinite, while B and D are symmetric. If $x^*Ax \geq |x^T Bx|$ and $x^*Cx \geq |x^T Dx|$ for all $x \in \mathbf{C}^n$, show that $x^*(A \circ C)x \geq |x^T(B \circ D)x|$ for all $x \in \mathbf{C}^n$.

7.7.P18 Let $A = [a_{ij}] \in M_n$ be positive semidefinite, let $B = [b_{ij}] \in M_n$ be symmetric, and suppose that $x^*Ax \geq |x^T Bx|$ for all $x \in \mathbf{C}^n$. (a) Show that $x^*[a_{ij}^k]x \geq |x^T[b_{ij}^k]x|$ for all $x \in \mathbf{C}^n$ and all $k = 1, 2, \ldots$. (b) Show that $x^*[e^{a_{ij}}]x \geq |x^T[e^{b_{ij}}]x|$ for all $x \in \mathbf{C}^n$.

7.7.P19 Let f be a complex analytic function on the unit disc that is normalized so that $f(0) = 0$ and $f'(0) = 1$. Consider the Grunsky inequalities (4.4.1). Explain why

$$\sum_{i,j=1}^n \frac{x_i \bar{x}_j}{1 - z_i \bar{z}_j} \geq \left| \sum_{i,j=1}^n x_i x_j \left(\frac{z_i z_j}{f(z_i)f(z_j)} \frac{f(z_i) - f(z_j)}{z_i - z_j} \right)^{\pm 1} \right| \tag{7.7.19}$$

for all $z_1, \ldots, z_n \in \mathbf{C}$ with $|z_i| < 1$, all $x_1, \ldots, x_n \in \mathbf{C}$, and all $n = 1, 2, \ldots$ if and only if f is one-to-one.

7.7.P20 (a) Explain why the contractions X in (7.7.7(e)), (7.7.9(b)), and (7.7.11(d)) can be expressed as $X = (A^\dagger)^{1/2}B(C^\dagger)^{1/2}$ (Moore–Penrose inverses), with similar expressions in (7.7.12(d)) and in (7.7.P16(d)). (b) Revisit (7.3.P8) and explain why the Hermitian block matrix H in (7.7.9) is positive semidefinite if and only if A is positive semidefinite and $C \succeq B^*A^\dagger B$.

7.7.P21 Let $A \in M_n$ be positive definite and let $B \in M_n$ be positive semidefinite. (a) Show that there is a positive scalar c such that $cA \succeq B$. (b) Show that the smallest such scalar is $c = \rho(A^{-1}B)$. (c) Show that the smallest positive scalar c such that $cA \circ X \succeq X$ for all $X \succeq 0$ is $c = e^T A^{-1}e$ (the all-ones vector).

7.7.P22 Let $A \in M_n$ be positive definite and let $A = A_1 + i A_2$, in which A_1 and A_2 are real. Show that A_1 is real symmetric and positive definite, while A_2 is real skew symmetric.

7.7.P23 Let $A \in M_n$ be positive definite. We claim that $\operatorname{Re} A^{-1} \succeq (\operatorname{Re} A)^{-1}$ and range $\operatorname{Im} A \subset$ range $\operatorname{Re} A$. Refer to (1.3.P20) and provide details: Let $A = A_1 + i A_2$ and $A^{-1} = B_1 + i B_2$, in which A_1, A_2, B_1, B_2 are real. Let $H = \begin{bmatrix} A_1 & A_2 \\ -A_2 & A_1 \end{bmatrix}$ and $K =$. $\begin{bmatrix} B_1 & B_2 \\ -B_2 & B_1 \end{bmatrix}$. (a) H is unitarily similar to $A \oplus \bar{A}$, so A is positive definite if and only if H is positive definite. (b) $H^{-1} = K$. (c) Let $\alpha = \{1, \ldots, n\}$. Then $\operatorname{Re} A^{-1} = B_1 = H^{-1}[\alpha] \succeq (H[\alpha])^{-1} = A_1^{-1} = (\operatorname{Re} A)^{-1}$. (d) H has the column inclusion property.

7.7.P24 (Continuation; same notation) (a) What inequality do you obtain if you choose $\alpha = \{j\}$ in the preceding problem ($j \in \{1, \ldots, n\}$)? (b) Show that there is a real skew-symmetric strict contraction X such that $\operatorname{Im} A = (\operatorname{Re} A)^{1/2} X (\operatorname{Re} A)^{1/2}$. (c) Show that $\det(\operatorname{Re} A) > |\det(\operatorname{Im} A)|$. Why is this inequality not very interesting if n is odd?

7.7.P25 Let $A \in M_n$ be Hermitian and let $A = A_1 + i A_2$, in which A_1 and A_2 are real. Show that A is positive definite if and only if A_1 is positive definite and $\rho(A_1^{-1} A_2) < 1$.

7.7.P26 Let $C_1, \ldots, C_k \in M_n$ be positive definite. Let $E = (\operatorname{Re} C_1) \circ \cdots \circ (\operatorname{Re} C_k)$ and $F = (\operatorname{Im} C_1) \circ \cdots \circ (\operatorname{Im} C_k)$. Show that $\det E > |\det F|$. Why is $E + iF$ positive definite if k is odd? What goes wrong when k is even?

7.7.P27 Let $A, B \in M_{m,n}$. We claim that $\sigma_1(A \circ B) \leq \sigma_1(A) \sigma_1(B)$. Provide details: Assume that $A \neq 0 \neq B$. Let $X = A/\sigma_1(A)$ and $Y = B/\sigma_1(B)$. Then $\begin{bmatrix} I_m & X \\ X^* & I_n \end{bmatrix} \circ \begin{bmatrix} I_m & Y \\ Y^* & I_n \end{bmatrix} = \begin{bmatrix} I_m & X \circ Y \\ (X \circ Y)^* & I_n \end{bmatrix}$ is positive semidefinite, so $X \circ Y$ is a contraction.

7.7.P28 Let $A \in M_n$ be positive definite. Show that the bordered matrix $\begin{bmatrix} A & x \\ x^* & a \end{bmatrix} \in M_{n+1}$ is positive definite if and only if $a > x^* A^{-1} x$.

7.7.P29 Let $A \in M_n$ be positive definite. Show that $A^{-1} \circ \cdots \circ A^{-1} \succeq (A \circ \cdots \circ A)^{-1}$ (same number of factors in each Hadamard product).

7.7.P30 Let $A \in M_n$ be nonsingular. Show that $(A^{-T} \circ A)e = e$, in which e is the all-ones vector. If A is positive definite, explain why $A^{-T} \circ A \succ I$ is not possible, although (7.7.18) ensures that $A^{-T} \circ A \succeq I$.

7.7.P31 Let $A \in M_n$ be Hermitian and nonsingular, and let $S \subset \mathbf{C}^n$ be a subspace. Suppose that A is positive definite on S, that is, $x^* A x > 0$ for all nonzero $x \in S$. Which of the two following assumptions is necessary and sufficient for A to be positive definite (on $\mathbf{C}^n$)? (a) A is positive definite on $S^\perp$, or (b) A^{-1} is positive definite on $S^\perp$? Before attempting a proof, consider the example $A = \begin{bmatrix} 1 & 2 \\ 2 & 1 \end{bmatrix}$ and $S = \operatorname{span}\{e_1\}$.

7.7.P32 Let $A, B \in M_n$ be positive definite. Show that $A \succeq B \Leftrightarrow \begin{bmatrix} I & B^{1/2} \\ B^{1/2} & A \end{bmatrix} \succeq 0 \Leftrightarrow \begin{bmatrix} B^{-1} & I \\ I & A \end{bmatrix} \succeq 0$.

7.7.P33 Let $A_i, B_i \in M_n$ be positive definite for each $i = 1, \ldots, k$ and let $\alpha_i \geq 0$ for each $i = 1, \ldots, k$. Suppose that each $A_i \succeq B_i$ and $\alpha_1 + \cdots + \alpha_k = 1$. (a) Show that $\sum_{i=1}^k \alpha_i A_i \succeq (\sum_{i=1}^k \alpha_i B_i^{1/2})^2$ and $\sum_{i=1}^k \alpha_i A_i \succeq (\sum_{i=1}^k \alpha_i B_i^{-1})^{-1}$. (b) What do the inequalities in (a) say if $n = 1$, $k = 2$, and $\alpha_1 = \alpha_2$? Prove these scalar inequalities directly.

7.7.P34 Let $A \in M_n$. Show that (a) $A A^* \succeq A^* A$ if and only if A is normal; (b) $(A A^*)^{1/2} \succeq (A^* A)^{1/2}$ if and only if A is normal.

7.7.P35 Let $A \in M_n$ be nonsingular. (a) Show that $\begin{bmatrix} (AA^*)^{1/2} & A \\ A^* & (A^*A)^{1/2} \end{bmatrix}$ is positive semidefinite and singular. (b) Show that $\begin{bmatrix} (AA^*)^{1/2} & A \\ A^* & (AA^*)^{1/2} \end{bmatrix}$ is positive semidefinite if and only if A is normal.

7.7.P36 Let $A, B \in M_n$ be Hermitian and let $H = \begin{bmatrix} A & B \\ B & A \end{bmatrix}$. (a) Revisit (1.3.P19) and explain why $H \succeq 0$ if and only if $A \pm B \succeq 0$. (b) Deduce the equivalence of (7.7.12(a)) and (7.7.12(c)) from (a).

7.7.P37 Let $A, B \in M_n$ be positive definite. Show that $A \circ B^{-1} + A^{-1} \circ B \succeq 2I$.

7.7.P38 Let $X \in M_n$ be Hermitian. Show that X is a contraction if and only if $I \succeq X^2$.

7.7.P39 Let $A, B \in M_n$. Show that $\begin{bmatrix} I & A \\ A^* & A^*A \end{bmatrix} \succeq 0$ and deduce that $A^*A \circ B^*B \succeq (A \circ B)^*(A \circ B)$.

7.7.P40 Let $A \in M_p$ be positive definite, let $B \in M_q$ be positive semidefinite, and suppose that $H = \begin{bmatrix} A & B \\ B^* & C \end{bmatrix} \in M_{p+q}$ is positive semidefinite. Denote the Schur complement of A in H by $S_H(A) = C - B^*A^{-1}B$. We claim that

$$S_H(A) = \max \left\{ E \in M_q : E = E^* \text{ and } H \succeq \begin{bmatrix} 0 & 0 \\ 0 & E \end{bmatrix} \right\} \tag{7.7.20}$$

The "max" is with respect to the Loewner partial order. Provide details: (a) Use the *congruence in (7.7.5) to show that $H - \begin{bmatrix} 0 & 0 \\ 0 & E \end{bmatrix}$ is *congruent to $A \oplus (S_H(A) - E)$. (b) $H - \begin{bmatrix} 0 & 0 \\ 0 & E \end{bmatrix} \succeq 0$ if and only if A is positive definite and $S_H(A) - E \succeq 0$. (c) The "max" is achieved for $E = S_H(A)$, which is positive semidefinite.

7.7.P41 (Continuation; same notation) Let $H_1 = \begin{bmatrix} A_1 & B_1 \\ B_1^* & C_1 \end{bmatrix}$, $H_2 = \begin{bmatrix} A_2 & B_2 \\ B_2^* & C_2 \end{bmatrix} \in M_{p+q}$ be positive semidefinite, and suppose that $A_1, A_2 \in M_p$ are positive definite. Use the variational characterization (7.7.20) of the Schur complement to prove the following.

(a) Monotonicity of the Schur complement: If $H_1 \succeq H_2$, then $S_{H_1}(A_1) \succeq S_{H_2}(A_2)$.

(b) Concavity of the Schur complement: $S_{H_1+H_2}(A_1 + A_2) \succeq S_{H_1}(A_1) + S_{H_2}(A_2)$.

(c) Explain why $H_1 \circ H_2 \succeq \begin{bmatrix} 0 & 0 \\ 0 & S_{H_1}(A_1) \end{bmatrix} \circ H_2 = \begin{bmatrix} 0 & 0 \\ 0 & S_{H_1}(A_1) \circ C_2 \end{bmatrix} \succeq \begin{bmatrix} 0 & 0 \\ 0 & S_{H_1}(A_1) \circ S_{H_2}(A_2) \end{bmatrix}$ and conclude that

$$S_{H_1 \circ H_2}(A_1 \circ A_2) \succeq S_{H_1}(A_1) \circ C_2 \succeq S_{H_1}(A_1) \circ S_{H_2}(A_2)$$

7.7.P42 Let $A \in M_n$ and suppose that $H(A) = \frac{1}{2}(A + A^*)$ (the Hermitian part of A) is positive definite. Use the *congruence canonical form (7.1.15) to show that $H(A)^{-1} \succeq H(A^{-1}) \succ 0$.

7.7.P43 Let $A, B \in M_n$ be positive definite and suppose that $A \succeq B$. Prove that $\det(A + B) \geq \det A + n(\det A)^{\frac{n-1}{n}}(\det B)^{\frac{1}{n}} \geq \det A + n \det B$. Compare with (7.6.P6).

7.7.P44 Let $A, B \in M_n$ be positive definite and suppose that $A \succeq B$. Provide details to show *from the definition* that $B^{-1} \succeq A^{-1}$: (a) Let $x, y \in \mathbf{C}^n$ be nonzero. We must show that $x^*B^{-1}x \geq x^*A^{-1}x$. (b) $(y - B^{-1}x)^*B(y - B^{-1}x) \geq 0 \Rightarrow 2\operatorname{Re} y^*x - y^*Ay \geq 2\operatorname{Re} y^*x - y^*By$. (c) Now let $y = A^{-1}x$.

7.7.P45 Let $A, B \in M_n$ be positive definite and let $H = \begin{bmatrix} B^{-1} & I \\ I & A \end{bmatrix}$. Compute H/B^{-1} and H/A. Explain why $A \succeq B$ if and only if $H \succeq 0$ if and only if $B^{-1} \succeq A^{-1}$.

Notes and Further Readings. In 1934, C. Loewner (K. Löwner) characterized functions of matrices that are monotone with respect to his eponymous partial order: $A \succeq B \Rightarrow f(A) \succeq f(B)$. He discovered that f is a monotone matrix function if and only if its *difference quotient kernel* $L_f(s, t) = (f(s) - f(t))/(s - t)$ is positive semidefinite. For example, (7.7.4) shows that the functions $f(t) = -t^{-1}$ and $f(t) = t^{1/2}$ are monotone on positive definite matrices, while the exercise following (7.7.4) shows that the monotone real-valued function $f(t) = t^2$ is not monotone on positive definite matrices. The following table of functions, difference quotient kernels, and associated matrices (each $x_i \in (0, \infty)$) illustrates Loewner's theory:

$$f(t) = -t^{-1} \quad L_f = \frac{1}{st} \quad [\xi_i \xi_j]_{i,j=1}^n \succeq 0$$
$$f(t) = \sqrt{t} \quad L_f = \frac{1}{\sqrt{s}+\sqrt{t}} \quad [(\xi_i + \xi_j)^{-1}]_{i,j=1}^n \succeq 0 \ (7.1.P16)$$
$$f(t) = t^2 \quad L_f = s + t \quad [\xi_i + \xi_j]_{i,j=1}^n \text{ is indefinite } (1.3.25)$$

There is also a theory of *convex matrix functions*: $\alpha f(A) + (1 - \alpha)f(B) \succeq f(\alpha A + (1 - \alpha)B)$ for all $\alpha \in (0, 1)$. Problems (7.7.P13 and P14) study the strictly convex functions $f(t) = t^2$ (on Hermitian matrices) and $f(t) = t^{-1}$ (on positive definite matrices). The functions $f(t) = -t^{1/2}$ and $f(t) = t^{-1/2}$ are known to be strictly convex on positive definite matrices. See section 6.6 in Horn and Johnson (1991), Bhatia (1997), and Donoghue (1974) for more information about monotone and convex matrix functions. For more information about the ideas in (7.7.11–13) and (7.7.P15 to P18), see C. H. FitzGerald and R. A. Horn, On the structure of Hermitian–symmetric inequalities, *J. London Math. Soc.* 15 (1977) 419–430. For further references related to (7.7.15) and (7.7.17), see C. R. Johnson, Partitioned and Hadamard product matrix inequalities, *J. Research NBS* 83 (1978) 585–591.

7.8 Inequalities involving positive definite matrices

Positive definite matrices are involved in a rich variety of inequalities involving determinants, eigenvalues, diagonal entries, and other quantities. In this section, we examine some of these inequalities.

The fundamental determinant inequality for positive definite matrices is Hadamard's inequality. Many other inequalities are either equivalent to it or generalizations of it.

Theorem 7.8.1 (Hadamard's inequality). *Let* $A = [a_{ij}] \in M_n$ *be positive definite. Then*

$$\det A \leq a_{11} \cdots a_{nn} \tag{7.8.2}$$

with equality if and only if A is diagonal.

Proof. Since A is positive definite, it has positive main diagonal entries and is diagonally *congruent to a correlation matrix. Let $D = \text{diag}(a_{11}^{1/2}, \ldots, a_{nn}^{1/2})$ and define $C = D^{-1}AD^{-1}$, which is also positive definite; it has unit diagonal entries, so $\text{tr } C = n$. Let $\lambda_1, \ldots, \lambda_n$ be the (necessarily positive) eigenvalues of C. Then the arithmetic–geometric mean inequality ensures that

$$\det C = \lambda_1 \cdots \lambda_n \leq \left(\frac{1}{n}(\lambda_1 + \cdots + \lambda_n)\right)^n = \left(\frac{1}{n}\text{tr } C\right)^n = 1$$

with equality if and only if each $\lambda_i = 1$. Since C is Hermitian and hence diagonalizable, each $\lambda_i = 1$ if and only if $C = I$. Thus,

$$\det A = \det(DCD) = (\det C)(\det D)^2$$
$$= (\det C)(a_{11} \cdots a_{nn}) \leq a_{11} \cdots a_{nn}$$

with equality if and only if $A = DCD = D^2 = \text{diag}(a_{11}, \dots, a_{nn})$. $\square$

For any nonsingular $A \in M_n(\mathbf{R})$, $|\det A|$ is the volume of the real n-dimensional parallelepiped whose edges are given by the columns of A. This volume is largest if the edges are orthogonal, and in this case the volume is the product of the lengths of the edges. The following version of Hadamard's inequality is an algebraic statement of this geometric inequality; it is valid even for complex square matrices.

Corollary 7.8.3 (Hadamard's inequality). *Let $B \in M_n$ be nonsingular and partition $B = [b_1 \ \dots \ b_n]$ and $B^* = [\beta_1 \ \dots \ \beta_n]$ according to their columns. Then*

$$|\det B| \leq \|b_1\|_2 \cdots \|b_n\|_2 \quad \text{and} \quad |\det B| \leq \|\beta_1\|_2 \cdots \|\beta_n\|_2 \qquad (7.8.4)$$

The respective inequalities in (7.8.4) are equalities if and only if the columns (respectively, rows) of B are orthogonal.

Proof. Apply (7.8.2) to the positive definite matrix $A = B^*B$: $\det A = |\det B|^2$, and the main diagonal entries of A are $\|b_1\|_2^2, \dots, \|b_n\|_2^2$. The columns of B are orthogonal if and only if A is diagonal. The second inequality in (7.8.4) follows from applying the first to B^*. $\square$

Exercise. We have deduced (7.8.4) from (7.8.2). Now show that (7.8.4) implies (7.8.2). *Hint*: If A is positive definite, use (7.2.7) to write $A = B^*B$ (any such B will do). Apply (7.8.4) to B and square.

Hadamard's inequality makes a statement about certain principal submatrices of a positive definite matrix. We now turn to three other inequalities that make statements of the same type. Each is equivalent to Hadamard's inequality.

Theorem 7.8.5 (Fischer's inequality). *Suppose that the partitioned Hermitian matrix*

$$H = \begin{bmatrix} A & B \\ B^* & C \end{bmatrix} \in M_{p+q}, \quad A \in M_p \text{ and } C \in M_q$$

is positive definite. Then

$$\det H \leq (\det A)(\det C) \qquad (7.8.6)$$

Proof. Let $A = U\Lambda U^*$ and $C = V\Gamma V^*$ be spectral decompositions, in which U and V are unitary and $\Lambda = \text{diag}(\lambda_1, \dots, \lambda_p)$ and $\Gamma = \text{diag}(\gamma_1, \dots, \gamma_q)$ are positive diagonal. Let $W = U \oplus V$ and compute

$$W^*HW = \begin{bmatrix} \Lambda & U^*BV \\ V^*B^*U & \Gamma \end{bmatrix}$$

Hadamard's inequality (7.8.2) ensures that

$$\det H = \det(W^*HW) \leq (\lambda_1 \cdots \lambda_p)(\gamma_1 \cdots \gamma_q) = (\det A)(\det C)   \square$$

Exercise. We have deduced (7.8.6) from (7.8.2). Use (7.8.6) to prove by induction that if $H = [H_{ij}]_{i,j=1}^{k}$ is partitioned as a k-by-k block matrix in which each diagonal block $H_{ii} \in M_{n_i}$, then

$$\det H \leq (\det H_{11}) \cdots (\det H_{kk}) \tag{7.8.7}$$

Now explain why (7.8.2) follows from (7.8.7), and conclude that Fischer's inequality is equivalent to Hadamard's inequality.

The Fischer and Hadamard inequalities involve determinants of disjoint principal submatrices. Hadamard's inequality is $\det A \leq (\det A[\{1\}]) \cdots (\det A[\{n\}])$, while Fischer's inequality involves a pair of complementary principal submatrices: $\det A \leq (\det A[\alpha])(\det A[\alpha^c])$, in which $\alpha \subset \{1, \ldots, n\}$ and we observe the convention that $\det A[\varnothing] = 1$. An inequality due to Koteljanskiĭ (often called the *Hadamard–Fischer inequality*) embraces the noncomplementary case: The principal submatrices are permitted to overlap. We adopt the convention that $\det A[\alpha] = 1$ if the index set α is empty. The following lemma is a consequence of Fischer's inequality.

Lemma 7.8.8. *Let $B \in M_m$ be positive definite. Let $\alpha, \beta \subset \{1, \ldots, m\}$. Suppose that α^c and β^c are nonempty and disjoint, and $\alpha \cup \beta = \{1, \ldots, m\}$. Then $\det B[\alpha^c \cup \beta^c] \leq (\det B[\alpha^c])(\det B[\beta^c])$.*

Proof. There is no loss of generality to assume that $\beta^c = \{1, \ldots, k\}, \alpha^c = \{j, \ldots, m\}$, and $1 < k < j < m$. Then $A[\alpha^c]$ and $A[\beta^c]$ are complementary principal submatrices of $A[\alpha^c \cup \beta^c]$, so (7.8.6) ensures that $\det A[\alpha^c \cup \beta^c] \leq (\det A[\alpha^c])(\det A[\beta^c])$. $\qquad\square$

Theorem 7.8.9 (Koteljanskiĭ's inequality). *Let $A \in M_n$ be positive definite and let $\alpha, \beta \subset \{1, \ldots, n\}$. Then*

$$(\det A[\alpha \cup \beta])(\det A[\alpha \cap \beta]) \leq (\det A[\alpha])(\det A[\beta]) \tag{7.8.10}$$

Proof. There is no loss of generality to assume that $\alpha \cup \beta = \{1, \ldots, n\}$ (if not, work within the principal submatrix $A[\alpha \cup \beta]$). We may also assume that $\alpha \cap \beta$ is nonempty (if it is empty, then $\beta = \alpha^c$ and (7.8.10) reduces to (7.8.6)). Finally, we may assume that both α^c and β^c are nonempty (if α^c is empty then $\alpha = \{1, \ldots, n\}$ and (7.8.10) is trivial). These three assumptions ensure that α^c and β^c are disjoint and nonempty. Our strategy is to use Jacobi's identity (0.8.4.2) to express $\det A[\alpha \cap \beta]$ in a form to which we can apply the preceding lemma, and then use Jacobi's identity again. Compute

$$\frac{\det A[\alpha \cap \beta]}{\det A} = \det A^{-1}[(\alpha \cap \beta)^c] = \det A^{-1}[\alpha^c \cup \beta^c]$$
$$\leq (\det A^{-1}[\alpha^c])(\det A^{-1}[\beta^c])$$
$$= \frac{\det A[\alpha]}{\det A} \frac{\det A[\beta]}{\det A}$$

and hence $(\det A)(\det A[\alpha \cap \beta]) \leq (\det A[\alpha])(\det A[\beta])$. $\qquad\square$

Exercise. We have deduced (7.8.10) from (7.8.6) (via (7.8.9)). Show that (7.8.10) implies (7.8.6), and conclude that Koteljanskiĭ's inequality is equivalent to Hadamard's inequality.

Another equivalent version of Hadamard's inequality is due to Szász. For each $k \in \{1, \ldots, n\}$, let $P_k(A)$ denote the product of the $\binom{n}{k}$ k-by-k principal minors of $A \in M_n$. Notice that $P_n(A) = \det A$ and $P_1(A) = a_{11} \cdots a_{nn}$, so Hadamard's inequality (7.8.2) may be restated as $P_n(A) \le P_1(A)$.

Theorem 7.8.11 (Szász's inequality). *Let $A \in M_n$ be positive definite. Then*

$$P_{k+1}(A)^{\binom{n-1}{k}^{-1}} \le P_k(A)^{\binom{n-1}{k-1}^{-1}} \quad \text{for each} \quad k = 1, \ldots, n-1 \qquad (7.8.12)$$

Proof. The identity $A^{-1} = (\det A)^{-1} \operatorname{adj} A$ reminds us that each diagonal entry of A^{-1} is the ratio of a principal minor of A of size $n-1$ and $\det A$. Thus, an application of (7.8.2) to the positive definite matrix A^{-1} gives the inequality

$$\frac{1}{\det A} = \det A^{-1} \le \frac{P_{n-1}(A)}{(\det A)^n}$$

and hence $P_n(A)^{n-1} = (\det A)^{n-1} \le P_{n-1}(A)$. It follows that

$$P_n(A) \le P_{n-1}(A)^{1/(n-1)} = P_{n-1}(A)^{\binom{n-1}{n-2}^{-1}} \qquad (7.8.13)$$

This is the case $k = n - 1$ of Szász's family of inequalities. The remaining cases may be derived inductively. For example, to obtain the next case, apply (7.8.13) to each principal submatrix of A of size $n - 1$. Since each principal submatrix of A of size $n - 2$ occurs twice as a principal submatrix of some principal submatrix of size $n - 1$, we obtain the inequality $P_{n-1}(A)^{n-2} \le P_{n-2}(A)^2$, which implies that

$$P_{n-1}(A)^{\binom{n-1}{n-2}^{-1}} = P_{n-1}(A)^{\frac{1}{n-1}} \le P_{n-2}(A)^{\frac{2}{(n-1)(n-2)}}$$
$$= (P_{n-2}(A)^2)^{\frac{1}{(n-1)(n-2)}} = P_{n-2}(A)^{\binom{n-1}{n-3}^{-1}}$$

This is the case $k = n - 2$ of Szász's inequalities. The remaining cases follow in the same way. $\qquad \square$

Exercise. We have deduced (7.8.12) from (7.8.2). Use (7.8.12) to show that

$$a_{11} \cdots a_{nn} = P_1(A) \ge P_2(A)^{\binom{n-1}{2}^{-1}} \ge \cdots \ge P_{k+1}(A)^{\binom{n-1}{k}^{-1}}$$
$$\ge \cdots \ge P_n(A)^{\binom{n-1}{n-1}^{-1}} = \det A \qquad (7.8.14)$$

in which $k = 2, \ldots, n - 1$. Conclude that Szász's inequality is a refinement of Hadamard's inequality that is equivalent to it.

Lemma 7.8.15. *Let $A = [a_{ij}] \in M_n$ be positive semidefinite and partitioned as $A = \begin{bmatrix} a_{11} & x^* \\ x & A_{22} \end{bmatrix}$, in which $A_{22} \in M_{n-1}$. Define*

$$\alpha(A) = \begin{cases} \frac{\det A}{\det A_{22}} & \text{if } A_{22} \text{ is positive definite} \\ 0 & \text{otherwise} \end{cases}$$

Then $\tilde{A} = \begin{bmatrix} a_{11} - \alpha(A) & x^* \\ x & A_{22} \end{bmatrix}$ is positive semidefinite.

Proof. There is nothing to prove if A is singular, so assume that A is positive definite. Apply Sylvester's criterion (7.2.5) to the trailing principal minors. The trailing minors $\det(\tilde{A}[\{k, \ldots, n\}]) = \det A[\{k, \ldots, n\}]$ are positive for each $k = 2, \ldots, n$; $\det \tilde{A} = \det A - \alpha(A) \det A_{22} = \det A - \det A = 0$. $\qquad\square$

Exercise. Prove Hadamard's inequality (7.8.2) by induction using (7.8.15).

Hadamard's inequality (7.8.2) may be stated as

$$\underbrace{1 \cdots 1}_{n \text{ times}} \det A \leq \det(I \circ A)$$

The following theorem is a substantial generalization of this observation.

Theorem 7.8.16 (Oppenheim-Schur inequalities). *Let* $A = [a_{ij}], B = [b_{ij}] \in M_n$ *be positive semidefinite. Then*

$$\max\{a_{11} \cdots a_{nn} \det B, b_{11} \cdots b_{nn} \det A\} \leq \det(A \circ B) \qquad (7.8.17)$$

and

$$a_{11} \cdots a_{nn} \det B + b_{11} \cdots b_{nn} \det A \leq \det(A \circ B) + \det(AB) \qquad (7.8.18)$$

Proof. We continue to use the notation in (7.8.15) and begin with (7.8.17), which one verifies is correct for $n = 1$. We proceed by induction on the dimension, let $n \geq 2$, and assume that (7.8.17) is correct for matrices of sizes at most $n - 1$. Since $\tilde{A}$ is positive semidefinite, $\tilde{A} \circ B$ is positive semidefinite and

$$0 \leq \det(\tilde{A} \circ B) = \det(A \circ B) - \det\begin{bmatrix} \alpha(A)b_{11} & 0 \\ * & A_{22} \circ B_{22} \end{bmatrix}$$
$$= \det(A \circ B) - \alpha(A)b_{11} \det(A_{22} \circ B_{22})$$

The induction hypothesis ensures that

$$\det(A \circ B) \geq \alpha(A)b_{11}(b_{22} \cdots b_{nn} \det A_{22}) = b_{11}b_{22} \cdots b_{nn} \det A$$

The other inequality in (7.8.17) follows from the first and the identity $A \circ B = B \circ A$.

Now consider (7.8.18), which one verifies is correct for $n - 1$. We proceed again by induction, let $n \geq 2$, and assume that (7.8.18) is correct for matrices of sizes at most $n - 1$. Apply (7.8.17) to $\tilde{A} \circ B$:

$$(a_{11} - \alpha(A))a_{22} \cdots a_{nn} \det B \leq \det(\tilde{A} \circ B)$$
$$= \det(A \circ B) - \alpha(A)b_{11} \det(A_{22} \circ B_{22})$$

Now apply the induction hypothesis to $\det(A_{22} \circ B_{22})$:

$$\det(A \circ B) \geq (a_{11} - \alpha(A))a_{22} \cdots a_{nn} \det B + \alpha(A)b_{11} \det(A_{22} \circ B_{22})$$
$$\geq (a_{11} - \alpha(A))a_{22} \cdots a_{nn} \det B$$
$$\quad + \alpha(A)b_{11}(a_{22} \cdots a_{nn} \det B_{22} + b_{22} \cdots b_{nn} \det A_{22}$$
$$\quad - \det(A_{22}B_{22}))$$

Rearrange this inequality as follows:

$$\det(A \circ B) + \det(AB) - a_{11} \cdots a_{nn} \det B - b_{11} \cdots b_{nn} \det A$$
$$\geq \det(AB) - \alpha(A)a_{22} \cdots a_{nn} \det B$$
$$+ \alpha(A)b_{11}(a_{22} \cdots a_{nn} \det B_{22} - \det(A_{22}B_{22}))$$
$$= \alpha(A)(a_{22} \cdots a_{nn} - \det A_{22})(b_{11} \det B_{22} - \det B)$$

Finally, observe that (7.8.2) ensures that $a_{22} \cdots a_{nn} - \det A_{22} \geq 0$ and (7.8.6) ensures that $b_{11} \det B_{22} - \det B \geq 0$, so

$$\det(A \circ B) + \det(AB) - a_{11} \cdots a_{nn} \det B - b_{11} \cdots b_{nn} \det A \geq 0$$

This inequality is a restatement of (7.8.18).   □

Exercise. If $A, B \in M_n$ are positive definite, deduce from the preceding theorem that

$$(\det A)(\det B) \leq \det(A \circ B)$$

and hence $\det(A \circ A^{-1}) \geq 1$. Compare this inequality with (7.7.17(c)).

Exercise. If $A, B \in M_n$ are positive definite, explain why

$$(\det A)(\det B) \leq a_{11} \cdots a_{nn} \det B \leq \det(A \circ B) \leq a_{11} \cdots a_{nn} b_{11} \cdots b_{nn}$$

A determinant inequality of a rather different sort applies to matrices with positive definite Hermitian part. It follows from the inequality $|z| \geq |\text{Re} z|$ for complex numbers.

Theorem 7.8.19 (Ostrowski–Taussky inequality). *Let $H, K \in M_n$ be Hermitian and let $A = H + iK$. If H is positive definite, then*

$$\det H \leq |\det(H + iK)| = |\det A| \tag{7.8.20}$$

with equality if and only if $K = 0$, that is, if and only if A is Hermitian.

Proof. We have $A = H(I + iH^{-1}K)$, so (7.8.20) is equivalent to the inequality $|\det(I + iH^{-1}K)| \geq 1$. Corollary 7.6.2(a) ensures that $H^{-1}K$ is diagonalizable and has real eigenvalues $\lambda_1, \ldots, \lambda_n$. Then

$$|\det(I + H^{-1}K)| = \prod_{j=1}^{n} |1 + i\lambda_j|$$

and it suffices to note that $|1 + i\lambda|^2 = 1 + \lambda^2 \geq 1$ for any real number λ, with equality if and only if $\lambda = 0$. Therefore, the inequality (7.8.20) is an equality if and only if $H^{-1}K = 0$ if and only if $K = 0$ if and only if $A = H$.   □

The following inequality involving the sum of two positive definite matrices is a consequence of a classical scalar inequality.

Theorem 7.8.21 (Minkowski's determinant inequality). *Let $A, B \in M_n$ be positive definite. Then*

$$(\det A)^{1/n} + (\det B)^{1/n} \leq (\det(A + B))^{1/n} \tag{7.8.22}$$

with equality if and only if $A = cB$ for some $c > 0$.

Proof. Theorem 7.6.4 ensures that there is a nonsingular $S \in M_n$ such that $A = SIS^*$ and $B = S\Lambda S^*$, in which $\Lambda = \text{diag}(\lambda_1, \ldots, \lambda_n)$ is positive diagonal. The assertion (7.8.22) is that

$$
\begin{aligned}
(\det SS^*)^{1/n} + (\det S\Lambda S^*)^{1/n} &= |\det S|^{2/n} + |\det S|^{2/n}(\det \Lambda)^{1/n} \\
&= |\det S|^{2/n}(1 + (\det \Lambda)^{1/n}) \\
&\leq |\det S|^{2/n}(\det(I + \Lambda))^{1/n} \\
&= (\det(SS^* + S\Lambda S^*))^{1/n}
\end{aligned}
$$

so we must prove that $1 + (\det \Lambda)^{1/n} \leq (\det(I + \Lambda))^{1/n}$, that is,

$$
1 + \left(\prod_{i=1}^{n} \lambda_j \right)^{1/n} \leq \left(\prod_{i=1}^{n} (1 + \lambda_i) \right)^{1/n} \tag{7.8.23}
$$

This inequality is a special case of Minkowski's product inequality (B10), which is an equality if and only if $\lambda_1 = \cdots = \lambda_n = c > 0$ if and only if $A = cB$. $\square$

Exercise. Let $A, B \in M_n$ be positive definite. Derive the inequality $\det(A + B) \geq \det A + \det B$ from (7.8.22). Compare with (7.4.P6).

The inequality (7.8.20) can be improved for matrices of size two or larger. For $n = 1$, the inequality in the following theorem is $\text{Re } z + |\text{Im } z| \leq |z|$, which is false if $\text{Im } z \neq 0$.

Theorem 7.8.24. *Let $n \geq 2$, let $H, K \in M_n$ be Hermitian, and let $A = H + iK$. If H is positive definite, then*

$$
\det H + |\det K| \leq |\det(H + iK)| = |\det A| \tag{7.8.25}
$$

If $n = 2$, the inequality (7.8.25) is an equality if and only if $K = cH$ for some $c \in \mathbf{R}$; if $n \geq 3$, it is an equality if and only if $K = 0$, that is, if and only if A is Hermitian.

Proof. Proceed as in the proof of (7.8.19). We have $A = H(I + iH^{-1}K)$, so $|\det A| = (\det H)|\det(I + iH^{-1}K)|$ and $\det H + |\det K| = (\det H)(1 + |\det(H^{-1}K)|)$. Since $H^{-1}K$ is similar to a real diagonal matrix $\Lambda = \text{diag}(\lambda_1, \ldots, \lambda_n)$, we must prove the inequality

$$
|\det(I + iH^{-1}K)| = \prod_{j=1}^{n} |1 + i\lambda_j| \geq 1 + \prod_{j=1}^{n} |\lambda_j| = \det I + |\det(H^{-1}K)|
$$

Each $|1 + i\lambda_j|^2 = 1 + \lambda_j^2$, so it suffices to prove the equivalent inequality

$$
\prod_{j=1}^{n} (1 + \lambda_j^2) \geq (1 + \prod_{j=1}^{n} |\lambda_j|)^2, \quad n \geq 2 \text{ and each } \lambda_j \in \mathbf{R} \tag{7.8.26}
$$

If $n = 2$, the arithmetic–geometric inequality ensures that

$$
\begin{aligned}
(1 + \lambda_1^2)(1 + \lambda_2^2) &= 1 + \lambda_1^2 + \lambda_2^2 + \lambda_1^2 \lambda_2^2 \\
&\geq 1 + 2|\lambda_1 \lambda_2| + \lambda_1^2 \lambda_2^2 = (1 + |\lambda_1 \lambda_2|)^2
\end{aligned}
$$

with equality if and only if $\lambda_1 = \lambda_2 = c$, if and only if $K = cH$. Now suppose that $n \geq 3$ and compute

$$
\prod_{j=1}^{n}(1 + \lambda_j^2) = 1 + \prod_{j=1}^{n}\lambda_j^2 + \prod_{j=1}^{n-1}\lambda_j^2 + \sum_{j=1}^{n}\lambda_j^2 + \text{ nonnegative terms}
$$

$$
\geq 1 + \prod_{j=1}^{n-1}\lambda_j^2 + \lambda_n^2 + \prod_{j=1}^{n}\lambda_j^2 + \sum_{j=1}^{n-1}\lambda_j^2
$$

$$
\geq 1 + (\prod_{j=1}^{n-1}\lambda_j^2 + \lambda_n^2) + \prod_{j=1}^{n}\lambda_j^2
$$

$$
\geq 1 + 2(\prod_{j=1}^{n-1}|\lambda_j|)|\lambda_n| + \prod_{j=1}^{n}\lambda_j^2 = (1 + \prod_{j=1}^{n}|\lambda_j|)^2
$$

The first inequality in this computation is the result of discarding a sum of nonnegative terms, the second inequality results from discarding the sum $\sum_{j=1}^{n-1}\lambda_j^2$, and the final inequality is an application of the arithmetic–geometric mean inequality. Consequently, if $n \geq 3$ and the inequality (7.8.26) is an equality, then $\lambda_1 = \cdots = \lambda_{n-1} = 0$ and $\lambda_n = \prod_{j=1}^{n-1}\lambda_j$, which is zero. Conversely, if $\lambda_1 = \cdots = \lambda_n = 0$, then (7.8.26) is an equality. If $n \geq 3$, we conclude that the inequality (7.8.26) is an equality if and only if $\Lambda = 0$ if and only if $K = 0$ if and only if A is Hermitian. $\qquad \square$

Exercise. Provide details for a proof by induction that (7.8.26) is an equality for some $n \geq 3$ if and only if $K = 0$.

Exercise. If $K = cH$, verify that (7.8.25) is an equality if $n = 2$. What goes wrong if $n > 2$?

The inequality (7.8.20) can be strengthened by combining elements of the proofs of (7.8.21) and (7.8.24).

Theorem 7.8.27 (Fan's determinant inequality). *Let $H, K \in M_n$ be Hermitian and let $A = H + iK$. If H is positive definite, then*

$$
(\det H)^{2/n} + |\det K|^{2/n} \leq |\det(H + iK)|^{2/n} = |\det A|^{2/n} \tag{7.8.28}
$$

with equality if and only if all of the eigenvalues of $H^{-1}K$ (necessarily real) have the same absolute value.

Proof. Theorem 7.6.4 ensures that there is a nonsingular $S \in M_n$ such that $H = SIS^*$ and $K = S\Lambda S^*$, in which $\Lambda = \text{diag}(\lambda_1, \ldots, \lambda_n)$ is real diagonal; its diagonal entries are the eigenvalues of the diagonalizable matrix $H^{-1}K$. The inequality (7.8.28) is equivalent to the inequality

$$
|\det(I + i\Lambda)|^{2/n} \geq 1 + |\det \Lambda|^{2/n}
$$

Since $(\prod_{j=1}^{n} |1 + i\lambda_j|)^{2/n} = (\prod_{j=1}^{n}(1 + \lambda_j^2))^{1/n}$ and $|\det \Lambda|^{2/n} = (\prod_{j=1}^{n} \lambda_j^2)^{1/n}$, we must prove that

$$1 + \left(\prod_{j=1}^{n} \lambda_j^2\right)^{1/n} \leq \left(\prod_{j=1}^{n}(1 + \lambda_j^2)\right)^{1/n}.$$

which is another special case of Minkowski's inequality (B10). The case of equality is $\lambda_1^2 = \cdots = \lambda_n^2 = c \geq 0$. We have $c = 0$ if and only if $K = S\Lambda S^* = 0$. $\qquad \square$

Exercise. Under the hypotheses of the preceding theorem, explain why all of the eigenvalues of $H^{-1}K$ have the same absolute value if and only if $(H^{-1}K)^2 = cI$ for some $c \geq 0$.

Exercise. If there is a real number γ such that $K = \gamma H$, show that the inequality (7.8.28) is an equality. This is not the only case of equality, of course.

Exercise. Derive the inequality (7.8.20) from each of (7.8.25) and (7.8.28).

Problems

7.8.P1 Let $A, B \in M_n$ be positive semidefinite. Use (7.8.17) to show that $A \circ B$ is positive definite if A is positive definite and B has positive diagonal entries; this is (7.5.3(b)).

7.8.P2 Let $A = [A_{ij}]_{i,j=1}^{n} \in M_{nk}$, with each $A_{ij} \in M_k$. Prove the following block generalization of Hadamard's inequality (7.8.2):

$$|\det A| \leq \left(\prod_{i=1}^{n}\left(\sum_{j=1}^{n} \||A_{ij}\||_2^2\right)\right)^{k/2}$$

What is this inequality if $k = 1$? if $n = 1$?

7.8.P3 Let $A, B \in M_n$ be positive definite. Show that the following statements are equivalent:

(a) $A \circ B = AB$.
(b) $\det(A \circ B) = \det(AB)$.
(c) A and B are positive diagonal matrices.

7.8.P4 Let $A \in M_n$ be positive definite and let $f(A) = (\det A)^{1/n}$. (a) Show that

$$f(A) = \min\{\frac{1}{n} \operatorname{tr}(AB) : B \text{ is positive definite and } \det B = 1\} \qquad (7.8.29)$$

(b) Deduce that $f(A)$ is a concave function on the convex set of positive definite matrices.
(c) Derive (7.8.21) from (b).

7.8.P5 Suppose that $H_+ = \begin{bmatrix} A & B \\ B^* & C \end{bmatrix}$ is positive definite. (a) Show that $H_- = \begin{bmatrix} A & -B \\ -B^* & C \end{bmatrix}$ is positive definite. (b) Apply Minkowski's inequality (7.8.22) to the two positive definite matrices $H_\pm$ and deduce Fischer's inequality (7.8.6).

7.8.P6 Let $H = \begin{bmatrix} A & B \\ B^* & C \end{bmatrix} \in M_n$ be positive definite. Let $H = LL^*$ be a Cholesky factorization (7.2.9), in which $L = \begin{bmatrix} L_{11} & 0 \\ L_{21} & L_{22} \end{bmatrix}$, so $A = L_{11}L_{11}^*$ and $C = L_{22}L_{22}^* + L_{21}L_{21}^*$. Use these representations to prove Fischer's inequality (7.8.6).

7.8.P7 Let $A = [a_{ij}] \in M_3(\mathbf{R})$. If all $|a_{ij}| \leq 1$, we claim that $|\det A| \leq 3\sqrt{3}$ and that this bound is never attained. Provide details:

$$\frac{\partial}{\partial a_{ij}}(\det A) = (-1)^{i+j} \det A[\{i\}^c, \{j\}^c] \quad \text{and} \quad \frac{\partial^2}{\partial a_{ij}^2}(\det A) = 0$$

If $\det A[\{i\}^c, \{j\}^c] = 0$, then $\det A$ is independent of the value of a_{ij}, which may therefore be taken to be ± 1. If $\det A[\{i\}^c, \{j\}^c] \neq 0$, then $\det A$ does not have a relative maximum or minimum with respect to a_{ij} if $-1 < a_{ij} < 1$. Thus, $|\det A|$ achieves its maximum value within the given constraints when all $a_{ij} = \pm 1$. There are only finitely many such matrices for $n = 3$. What is the result for general $n > 3$? If A has complex entries, use the maximum principle (the maximum modulus theorem) for analytic functions to show that $|\det A|$ cannot have a maximum in the interior of the set $\{A \in M_n : \text{all } |a_{ij}| \leq 1\}$.

7.8.P8 Let $A = [a_{ij}] \in M_n$. (a) Use Hadamard's inequality to show that $|\det A| \leq \|A\|_\infty^n \, n^{n/2}$, a famous inequality in the theory of Fredholm integral equations. (b) Consider the characteristic polynomial (1.2.10a) of A and show that $|a_{n-k}| \leq \binom{n}{k} \|A\|_\infty^k k^{k/2}$ for each $k = 1, \ldots, n$.

7.8.P9 Let $A \in M_n$ be positive definite. Show that $\det A = \min\{\prod_{i=1}^n v_i^* A v_i : v_1, \ldots, v_n \in C^n \text{ are orthonormal}\}$.

7.8.P10 Let $A \in M_n$ be positive definite and let $u_1, \ldots, u_n \in \mathbf{C}^n$ be orthonormal. Use the preceding problem to show that $u_1, \ldots, u_n$ are eigenvectors of A and $u_1^* A u_1, \ldots, u_n^* A u_n$ are eigenvalues of A if and only if $\det A = \prod_{i=1}^n u_i^* A u_i$.

7.8.P11 Let $A = \begin{bmatrix} A_{11} & A_{12} \\ A_{12}^* & A_{22} \end{bmatrix}$ be positive definite. Prove the *reverse Fischer inequality* for Schur complements: $\det(A/A_{11}) \det(A/A_{22}) \leq \det A$; see (0.8.5).

7.8.P12 Let $A = [a_{ij}] \in M_n$ be positive definite. Partition $A = \begin{bmatrix} A_{11} & x \\ x^* & a_{nn} \end{bmatrix}$, in which $A_{11} \in M_{n-1}$. Use the Cauchy expansion (0.8.5.10) or the Schur complement to show that $\det A = (a_{nn} - x^* A_{11}^{-1} x) \det A_{11} \leq a_{nn} \det A_{11}$, with equality if and only if $x = 0$. Use this observation to give a proof by induction of Hadamard's inequality (7.8.2) and its case of equality.

7.8.P13 Let $A = [a_{ij}] \in M_n$ be positive definite and have eigenvalues $\lambda_1, \ldots, \lambda_n$. Consider the kth elementary symmetric function $S_k(t_1, \ldots, t_n)$ defined in (1.2.14). Observe that $S_1(\lambda_1, \ldots, \lambda_n) = \text{tr } A = S_1(a_{11}, \ldots, a_{nn})$ and that Hadamard's inequality (7.8.2) may be restated as $S_n(\lambda_1, \ldots, \lambda_n) \leq S_n(a_{11}, \ldots, a_{nn})$. Use (1.2.16) and (7.8.2) to show that $S_k(\lambda_1, \ldots, \lambda_n) \leq S_k(a_{11}, \ldots, a_{nn})$ for each $k = 1, \ldots, n$.

Versions of the basic determinant inequalities (7.8.20, 25, 28) are valid for nonsingular normalizable matrices; these inequalities can all be obtained with a unified approach. Let $A = H + iK \in M_n$, in which H and K are Hermitian. Suppose that A is nonsingular and is diagonalizable by *congruence, that is, there is a nonsingular $S \in M_n$ and a diagonal unitary matrix $D = \text{diag}(e^{i\theta_1}, \ldots, e^{i\theta_n})$ such that $A = SDS^*$; see (4.5.24) and (4.5.P37). In this case, we have $H = S\Gamma S^*$ and $K = S\Sigma S^*$, in which $\Gamma = \text{diag}(\cos \theta_1, \ldots, \cos \theta_n)$ and $\Sigma = \text{diag}(\sin \theta_1, \ldots, \sin \theta_n)$. We use this notation in the following seven problems.

7.8.P14 If $C \in M_n$ has positive definite Hermitian part, show that C is diagonalizable by *congruence. Give an example of a matrix that is diagonalizable by *congruence but does not have positive definite Hermitian part.

7.8.P15 Verify that $|\det H| = |\det S|^2 |\det \Gamma| = |\det S|^2 \prod_{j=1}^{n} |\cos \theta_j|$, $|\det K| = |\det S|^2 |\det \Gamma| = |\det S|^2 \prod_{j=1}^{n} |\sin \theta_j|$, and $|\det A| = |\det S|^2$.

7.8.P16 (Analog of (7.8.20)) We claim that $|\det H| \leq |\det A|$. Explain why, to prove this claim, it suffices to show that $\prod_{j=1}^{n} |\cos \theta_j| \leq 1$, and do so.

7.8.P17 (Analog of (7.8.25)) We claim that $|\det H| + |\det K| \leq |\det A|$ if $n \geq 2$. Explain why, to prove this claim, it suffices to show that $\prod_{j=1}^{n} |\cos \theta_j| + \prod_{j=1}^{n} |\sin \theta_j| \leq 1$ if $n \geq 2$, and do so. What goes wrong for $n = 1$?

7.8.P18 (Analog of (7.8.28)) We claim that $|\det H|^{2/n} + |\det K|^{2/n} \leq |\det A|^{2/n}$. Explain why, to prove this claim, it suffices to show that $(\prod_{j=1}^{n} \cos^2 \theta_j)^{1/n} + (\prod_{j=1}^{n} \sin^2 \theta_j)^{1/n} \leq 1$, and do so.

7.8.P19 (Another approach to (7.8.22)): If H and K are positive definite, we claim that $(\det H)^{1/n} + (\det K)^{1/n} \leq (\det(H + K))^{1/n}$. Explain why, to prove this claim, it suffices to show that $(\prod_{j=1}^{n} \cos \theta_j)^{1/n} + (\prod_{j=1}^{n} \sin \theta_j)^{1/n} \leq (\prod_{j=1}^{n} (\cos \theta_j + \sin \theta_j))^{1/n}$ under the assumption that each $\theta_j \in (0, \pi/2)$. Do so.

7.8.P20 Deduce the inequalities (7.8.20), (7.8.25), and (7.8.28) from (7.8.P16, P17 and P18).

7.8.P21 Let $A = [a_{ij}]$, $B = [b_{ij}] \in M_n$ be positive definite. When do the Oppenheim–Schur inequalities in (7.8.16) become equalities? The equality case in Hadamard's inequality (7.8.2) plays a key role in answering this question. (a) Explain why

$$\det(A \circ B) \geq a_{11} \cdots a_{nn} \det B + (b_{11} \cdots b_{nn} - \det B) \det A$$
$$\geq a_{11} \cdots a_{nn} \det B \geq \det(AB)$$

(b) Show that $\det(A \circ B) = a_{11} \cdots a_{nn} \det B$ if and only if B is diagonal, and $\det(A \circ B) = \det(AB)$ if and only if both A and B are diagonal. (c) If $n = 2$, show that the inequality (7.8.18) is always an equality. (d) If $n \geq 3$, it is known that (7.8.18) is an equality if and only if there is a permutation matrix $P \in M_n$ such that $PAP^T = A_1 \oplus A_2$ and $PBP^T = B_1 \oplus B_2$, in which $A_1, B_1 \in M_2$ and both $A_2, B_2 \in M_{n-2}$ are diagonal. Explain why this condition is sufficient for equality in (7.8.18).

Further Reading. For a proof that the condition stated in (7.8.P21(d)) is necessary if (7.8.18) is an equality, see A. Oppenheim, Inequalities connected with definite Hermitian forms, *J. London Math. Soc.* 5 (1930) 114–119. The equality cases for positive semidefinite A and B are discussed in X. D. Zhang and C. X. Ding, The equality case for the inequalities of Oppenheim and Schur for positive semi-definite matrices, *Czech. Math. J.* 59 (2009) 197–206.

CHAPTER 8

Positive and Nonnegative Matrices

8.0 Introduction

Suppose that there are $n \geq 2$ cities $C_1, \ldots, C_n$ among which migration takes place as follows: Simultaneously at 8:00 A.M. each day a constant fraction a_{ij} of the current population of city j moves to city i for all distinct $i, j \in \{1, \ldots, n\}$; the fraction a_{jj} of the current population of city j remains in city j. Thus, if we denote the population of city i on day m by $p_i^{(m)}$, we have the recursive relation

$$p_i^{(m+1)} = a_{i1}p_1^{(m)} + \cdots + a_{in}p_n^{(m)}, \quad i = 1, \ldots, n, \quad m = 0, 1, \ldots$$

between the population distributions on days m and $m + 1$. If we denote the n-by-n matrix of migration coefficients by $A = [a_{ij}]$ and the population distribution vector on day m by $p^{(m)} = [p_i^{(m)}]$, then

$$p^{(m+1)} = Ap^{(m)} = AAp^{(m-1)} = \cdots = A^{m+1}p^{(0)}, \quad m = 0, 1, \ldots$$

in which $p^{(0)}$ is the initial population distribution. Observe that $0 \leq a_{ij} \leq 1$ for all $i, j \in \{1, \ldots, n\}$, and $\sum_{i=1}^{n} a_{ij} = 1$ for each $j = 1, \ldots, n$.

To make sensible long-range plans for city services and capital investment, government officials wish to know how the population will be distributed among the cities far into the future; that is, they want to know about the asymptotic behavior of $p^{(m)}$ for large m. But since $p^{(m)} = A^m p^{(0)}$, we are led to investigate the asymptotic behavior of A^m.

As an example, let us consider in detail the case of two cities. We have $a_{11} + a_{21} = 1 = a_{12} + a_{22}$, so if we denote $a_{21} = \alpha$ and $a_{12} = \beta$, we have

$$A = \begin{bmatrix} 1 - \alpha & \beta \\ \alpha & 1 - \beta \end{bmatrix}$$

and we are interested in A^m for large m. If A were diagonalizable, we could compute A^m explicitly. Thus, we begin by computing the eigenvalues of A: $\lambda_2 = 1$ and $\lambda_1 = 1 - \alpha - \beta$. Since $0 \leq \alpha, \beta \leq 1$, we have $\lambda_2 = 1 \geq |\lambda_1| = |1 - \alpha - \beta|$, so $1 = |\lambda_2| = \rho(A)$ and the spectral radius of A is an eigenvalue of A. Moreover, except in the trivial case

$\alpha = \beta = 0$ (in which case A is reducible), we see that $\lambda_2 = \rho(A)$ is a simple eigenvalue of A.

If $\alpha + \beta \neq 0$, the respective eigenvectors are $x = [\beta \ \alpha]^T$ (for $\lambda_2 = 1$) and $z = [1 \ -1]^T$ (for λ_1), so in this case A is diagonalizable and $A = S \Lambda S^{-1}$, in which

$$\Lambda = \begin{bmatrix} 1 & 0 \\ 0 & 1 - \alpha - \beta \end{bmatrix}, S = \begin{bmatrix} \beta & 1 \\ \alpha & -1 \end{bmatrix}, \text{ and } S^{-1} = \frac{1}{\alpha + \beta} \begin{bmatrix} 1 & 1 \\ \alpha & -\beta \end{bmatrix}$$

The entries of the eigenvector x are nonnegative; they are positive if A is irreducible.

If α and β are not both 1, then $|\lambda_1| = |1 - \alpha - \beta| < 1$ and so $\lambda_1^m \to 0$ as $m \to \infty$. Thus, in this case we have

$$\lim_{m \to \infty} A^m = S \left(\lim_{m \to \infty} \Lambda^m \right) S^{-1} = S \begin{bmatrix} 1 & 0 \\ 0 & 0 \end{bmatrix} S^{-1} = \frac{1}{\alpha + \beta} \begin{bmatrix} \beta & \beta \\ \alpha & \alpha \end{bmatrix}$$

and so the equilibrium population distribution

$$\lim_{m \to \infty} p^{(m)} = \frac{1}{\alpha + \beta} \begin{bmatrix} \beta & \beta \\ \alpha & \alpha \end{bmatrix} \begin{bmatrix} p_1^{(0)} \\ p_2^{(0)} \end{bmatrix} = \frac{p_1^{(0)} + p_2^{(0)}}{\alpha + \beta} \begin{bmatrix} \beta \\ \alpha \end{bmatrix}$$

does not depend on the initial distribution. The matrices A^m approach a limit whose columns are proportional to the eigenvector x that is associated with the eigenvalue 1 (the spectral radius of A), and the limiting population distribution is proportional to this same eigenvector.

The two exceptional cases are easily analyzed individually. If $\alpha = \beta = 0$, then $A = I$, $\lim_{m \to \infty} A^m = I$, and $\lim_{m \to \infty} p^{(m)} = p^{(0)}$, so the limiting distribution depends on the initial distribution.

If $\alpha = \beta = 1$, then $A = \begin{bmatrix} 0 & 1 \\ 1 & 0 \end{bmatrix}$, and the two cities exchange their entire populations on successive days. The powers of A do not approach a limit and neither does the population distribution if the initial population distribution is unequal. However, there is a sense in which an "average equilibrium" is attained, namely,

$$\lim_{m \to \infty} \frac{1}{m} \sum_{k=1}^{m} A^k = \begin{bmatrix} .5 & .5 \\ .5 & .5 \end{bmatrix} \quad \text{and} \quad \lim_{m \to \infty} \frac{1}{m} \sum_{k=1}^{m} p^{(k)} = \frac{p_1^{(0)} + p_2^{(0)}}{2} \begin{bmatrix} 1 \\ 1 \end{bmatrix}$$

Exercise. Verify that these two assertions about limits are correct.

In summary, we found the following in this example:

1. The spectral radius $\rho(A)$ is an eigenvalue of A; it is not just the absolute value of an eigenvalue.
2. The eigenvector x associated with the eigenvalue $\rho(A)$ can be taken to have nonnegative entries, which are positive if A is irreducible.
3. If every entry of A is positive, then $\rho(A)$ is a simple eigenvalue that is strictly larger than the modulus of any other eigenvalue.
4. If every entry of A is positive, then $\lim_{m \to \infty} (A/\rho(A))^m$ exists and is a rank-one matrix, each of whose columns is proportional to the eigenvector x.
5. Even if some entry of A is zero, $\lim_{m \to \infty} (1/m) \sum_{k=1}^{m} (A/\rho(A))^k$ exists.

These conclusions are generally true for $n \geq 2$, but it is not possible to analyze the general case with simple direct methods. New tools are required and are developed in the rest of this chapter.

Problems

8.0.P1 Show that the matrix $A = \begin{bmatrix} 1 & 1 \\ 0 & 1 \end{bmatrix}$ has spectral radius 1 and the sequence $A, A^2, A^3, \ldots$ is unbounded.

8.0.P2 Consider the matrix $A_\epsilon = \begin{bmatrix} (1+\epsilon)^{-1} & (1+\epsilon)^{-1} \\ \epsilon^2(1+\epsilon)^{-1} & (1+\epsilon)^{-1} \end{bmatrix}$, $\epsilon > 0$. (a) Show that $\lambda_2 = 1$ is a simple eigenvalue of A_ϵ, that $\rho(A) = \lambda_2 = 1$, and that $1 > |\lambda_1|$. (b) Show that $x = (1+\epsilon)^{-1}[1 \; \epsilon]^T$ and $y = (1+\epsilon)(2\epsilon)^{-1}[\epsilon \; 1]^T$ are eigenvectors of A_ϵ and A_ϵ^T, respectively, corresponding to the eigenvalue $\lambda = 1$. (c) Calculate A_ϵ^m explicitly, $m = 1, 2, \ldots$. (d) Show that $\lim_{m \to \infty} A_\epsilon^m = \frac{1}{2}\begin{bmatrix} 1 & \epsilon^{-1} \\ \epsilon & 1 \end{bmatrix}$. (e) Calculate xy^T and comment. (f) What happens if $\epsilon \to 0$?

8.0.P3 If an n-by-n matrix of intercity migration coefficients is irreducible, what can you say about the freedom of travel of the populace?

Further Readings. For a wealth of information about properties of positive and nonnegative matrices as well as many references to the theoretical and applied literature, see Berman and Plemmons (1994) and Seneta (1973). The book Varga (2000) contains a summary of results about nonnegative matrices, with special emphasis on applications to numerical analysis.

8.1 Inequalities and generalities

Let $A = [a_{ij}] \in M_{m,n}$ and $B = [b_{ij}] \in M_{m,n}$ and define $|A| = [|a_{ij}|]$ (entrywise absolute value). If A and B have real entries, we write

$$A \geq 0 \text{ if all } a_{ij} \geq 0, \text{ and } A > 0 \text{ if all } a_{ij} > 0$$

$$A \geq B \text{ if } A - B \geq 0, \text{ and } A > B \text{ if } A - B > 0$$

The reversed relations $\leq$ and $<$ are defined similarly. If $A \geq 0$, we say that A is a *nonnegative* matrix, and if $A > 0$, we say that A is a *positive* matrix. The following simple facts follow immediately from the definitions.

Exercise. Let $A, B \in M_{m,n}$. Show that

(8.1.1) $|A| \geq 0$ and $|A| = 0$ if and only if $A = 0$
(8.1.2) $|aA| = |a||A|$ for all $a \in \mathbf{C}$
(8.1.3) $|A + B| \leq |A| + |B|$
(8.1.4) $A \geq 0$ and $A \neq 0 \Rightarrow A > 0$ only if $m = n = 1$
(8.1.5) if $A \geq 0$, $B \geq 0$, and $a, b \geq 0$, then $aA + bB \geq 0$
(8.1.6) if $A \geq B$ and $C \geq D$, then $A + C \geq B + D$
(8.1.7) if $A \geq B$ and $B \geq C$, then $A \geq C$

Proposition 8.1.8. *Let* $A = [a_{ij}] \in M_n$ *and* $x = [x_i] \in \mathbb{C}^n$ *be given.*

(a) $|Ax| \leq |A|\,|x|$.

(b) *Suppose that* A *is nonnegative and has a positive row. If* $|Ax| = A|x|$, *then there is a real* $\theta \in [0, 2\pi)$ *such that* $e^{-i\theta}x = |x|$.

(c) *Suppose that* x *is positive. If* $Ax = |A|x$, *then* $A = |A|$, *so* A *is nonnegative.*

Proof. (a) The assertion follows from the triangle inequality:

$$|Ax|_k = |\sum_j a_{kj}x_j| \leq \sum_j |a_{kj}x_j| = \sum_j |a_{kj}|\,|x_j| = (|A|\,|x|)_k \qquad (8.1.8.1)$$

for each $k = 1, \ldots, n$.

(b) The hypothesis is that $A \geq 0$, $a_{k1}, \ldots, a_{kn}$ are all positive, and $|Ax| = A|x|$. Then $|Ax|_k = |\sum_j a_{kj}x_j| = \sum_j a_{kj}|x_j| = (A|x|)_k$. This is a case of equality in the triangle inequality (8.1.8.1), so there is a $\theta \in \mathbb{R}$ such that $e^{-i\theta}a_{kj}x_j = a_{kj}|x_j|$ for each $j = 1, \ldots, n$; see Appendix A. Since each a_{kj} is positive, it follows that $e^{-i\theta}x_j = |x_j|$ for each $j = 1, \ldots, n$, that is, $e^{-i\theta}x = |x|$.

(c) We have $|A|x = \text{Re}(|A|x) = \text{Re}(Ax) = (\text{Re}\,A)x$, so $(|A| - \text{Re}\,A)x = 0$. But $|A| - \text{Re}\,A \geq 0$ and $x > 0$, so (8.1.1) ensures that $|A| = \text{Re}\,A$. Then $A = |A| \geq 0$. $\qquad \square$

Exercise. Let $A, B, C, D \in M_n$, let $x, y \in \mathbb{C}^n$, and let $m \in \{1, 2, \ldots\}$. Show that

(8.1.9) $|AB| \leq |A|\,|B|$

(8.1.10) $|A^m| \leq |A|^m$

(8.1.11) if $0 \leq A \leq B$ and $0 \leq C \leq D$, then $0 \leq AC \leq BD$

(8.1.12) if $0 \leq A \leq B$, then $0 \leq A^m \leq B^m$

(8.1.13) if $A \geq 0$, then $A^m \geq 0$; if $A > 0$, then $A^m > 0$

(8.1.14) if $A > 0$, $x \geq 0$, and $x \neq 0$, then $Ax > 0$

(8.1.15) if $A \geq 0$, $x > 0$, and $Ax = 0$, then $A = 0$

(8.1.16) if $|A| \leq |B|$, then $\|A\|_2 \leq \|B\|_2$

(8.1.17) $\|A\|_2 = \|\,|A|\,\|_2$

Of course, the assertions (8.1.16–17) hold for any absolute vector norm on matrices, not just the Frobenius norm. Our first application of these facts is to an inequality for the spectral radius.

Theorem 8.1.18. *Let* $A, B \in M_n$ *and suppose that* B *is nonnegative. If* $|A| \leq B$, *then* $\rho(A) \leq \rho(|A|) \leq \rho(B)$.

Proof. Invoking (8.1.10) and (8.1.12), we have $|A^m| \leq |A|^m \leq B^m$ for each $m = 1, 2, \ldots$. Thus, (8.1.16) and (8.1.17) ensure that

$$\|A^m\|_2 \leq \|\,|A|^m\,\|_2 \leq \|B^m\|_2 \text{ and } \|A^m\|_2^{1/m} \leq \|\,|A|^m\,\|_2^{1/m} \leq \|B^m\|_2^{1/m}$$

for each $m = 1, 2, \ldots$. If we now let $m \to \infty$ and apply the Gelfand formula (5.6.14), we deduce that $\rho(A) \leq \rho(|A|) \leq \rho(B)$. $\qquad \square$

Corollary 8.1.19. *Let* $A, B \in M_n$ *be nonnegative. If* $0 \leq A \leq B$, *then* $\rho(A) \leq \rho(B)$.

Corollary 8.1.20. *Let $A = [a_{ij}] \in M_n$ be nonnegative.*

(a) If $\tilde{A}$ is principal submatrix of A, then $\rho(\tilde{A}) \leq \rho(A)$.
(b) $\max_{i=1,\ldots,n} a_{ii} \leq \rho(A)$.
(c) $\rho(A) > 0$ if any main diagonal entry of A is positive.

Proof. (a) If $r = n$, there is nothing to prove. Suppose that $1 \leq r < n$, let $\tilde{A}$ be an r-by-r principal square submatrix of A, and let P be a permutation matrix such that $PAP^T = \begin{bmatrix} \tilde{A} & B \\ C & D \end{bmatrix}$. The preceding theorem ensures that

$$\rho(\tilde{A}) = \rho(\tilde{A} \oplus 0_{n-r}) = \rho\left(\begin{bmatrix} \tilde{A} & 0 \\ 0 & 0 \end{bmatrix}\right)$$
$$\leq \rho\left(\begin{bmatrix} \tilde{A} & B \\ C & D \end{bmatrix}\right) = \rho(PAP^T) = \rho(A)$$

(b) Take $r = 1$ to see that $a_{ii} \leq \rho(A)$ for all $i = 1, \ldots, n$.
(c) $\rho(A) \geq \max_{i=1,\ldots,n} a_{ii} > 0$. $\square$

Exercise. The hypothesis that A is nonnegative is essential for the inequalities in (8.1.20). Consider $A = \begin{bmatrix} 1 & 1 \\ -1 & -1 \end{bmatrix}$. Is $1 \leq \rho(A)$?

Exercise. If $A > 0$, why is $\rho(A) > 0$?

Since we shall soon have rather good upper bounds on the spectral radius of a nonnegative matrix, (8.1.18) will be useful in obtaining upper bounds on the spectral radius of an arbitrary matrix.

Lemma 8.1.21. *Let $A = [a_{ij}] \in M_n$ be nonnegative. Then $\rho(A) \leq |||A|||_\infty = \max_{1 \leq i \leq n} \sum_{j=1}^n a_{ij}$ and $\rho(A) \leq |||A|||_1 = \max_{1 \leq j \leq n} \sum_{i=1}^n a_{ij}$. If all the row sums of A are equal, then $\rho(A) = |||A|||_\infty$; if all the column sums of A are equal, then $\rho(A) = |||A|||_1$.*

Proof. We know that $|\lambda| \leq \rho(A) \leq |||A|||$ for any eigenvalue λ of A and any matrix norm $||| \cdot |||$. If all the row sums of A are equal, then $e = [1 \ldots 1]^T$ is an eigenvector of A with eigenvalue $\lambda = |||A|||_\infty$ and so $|||A|||_\infty = \lambda \leq \rho(A) = |||A|||_\infty$. The statement for column sums follows from applying the same argument to A^T. $\square$

The *largest* row sum of a nonnegative matrix is an *upper* bound on its spectral radius; it may seem surprising that the *smallest* row sum is a *lower* bound.

Theorem 8.1.22. *Let $A = [a_{ij}] \in M_n$ be nonnegative. Then*

$$\min_{1 \leq i \leq n} \sum_{j=1}^n a_{ij} \leq \rho(A) \leq \max_{1 \leq i \leq n} \sum_{j=1}^n a_{ij} \tag{8.1.23}$$

and

$$\min_{1 \leq j \leq n} \sum_{i=1}^n a_{ij} \leq \rho(A) \leq \max_{1 \leq j \leq n} \sum_{i=1}^n a_{ij} \tag{8.1.24}$$

Proof. Let $\alpha = \min_{1 \leq i \leq n} \sum_{j=1}^{n} a_{ij}$. If $\alpha = 0$, let $B = 0$. If $\alpha > 0$, define $B = [b_{ij}]$ by letting each $b_{ij} = \alpha a_{ij}(\sum_{k=1}^{n} a_{ik})^{-1}$. Then $A \geq B \geq 0$ and $\sum_{j=1}^{n} b_{ij} = \alpha$ for all $i = 1, \ldots, n$. The preceding lemma ensures that $\rho(B) = \alpha$, and (8.1.19) tells us that $\rho(B) \leq \rho(A)$. The upper bound in (8.1.23) is the norm bound in (8.1.21). The column sum bounds follow from applying the row sum bounds to A^T.   □

Corollary 8.1.25. *Let $A = [a_{ij}] \in M_n$. If A is nonnegative and either $\sum_{j=1}^{n} a_{ij} > 0$ for all $i = 1, \ldots, n$ or $\sum_{i=1}^{n} a_{ij} > 0$ for all $j = 1, \ldots, n$, then $\rho(A) > 0$. In particular, $\rho(A) > 0$ if $n \geq 2$ and A is irreducible and nonnegative.*

Exercise. Let $A, B \in M_n$ be nonnegative and suppose that $n \geq 2$. Suppose that A is irreducible and B is positive. Explain why A cannot have a zero row or a zero column, but all its main diagonal entries can be zero. Why is AB positive?

We can generalize the preceding theorem by introducing some free parameters. If $A \geq 0$, $S = \mathrm{diag}(x_1, \ldots, x_n)$, and all $x_i > 0$, then $S^{-1}AS = [a_{ij}x_i^{-1}x_j] \geq 0$ and $\rho(A) = \rho(S^{-1}AS)$. Applying (8.1.22) to $S^{-1}AS$ yields the following result.

Theorem 8.1.26. *Let $A = [a_{ij}] \in M_n$ be nonnegative. Then for any positive vector $x = [x_i] \in \mathbf{R}^n$ we have*

$$\min_{1 \leq i \leq n} \frac{1}{x_i} \sum_{j=1}^{n} a_{ij}x_j \leq \rho(A) \leq \max_{1 \leq i \leq n} \frac{1}{x_i} \sum_{j=1}^{n} a_{ij}x_j \qquad (8.1.27)$$

and

$$\min_{1 \leq j \leq n} x_j \sum_{i=1}^{n} \frac{a_{ij}}{x_i} \leq \rho(A) \leq \max_{1 \leq j \leq n} x_j \sum_{i=1}^{n} \frac{a_{ij}}{x_i} \qquad (8.1.28)$$

Corollary 8.1.29. *Let $A = [a_{ij}] \in M_n$ be nonnegative and let $x = [x_i] \in \mathbf{R}^n$ be a positive vector. If $\alpha, \beta \geq 0$ are such that $\alpha x \leq Ax \leq \beta x$, then $\alpha \leq \rho(A) \leq \beta$. If $\alpha x < Ax$, then $\alpha < \rho(A)$; if $Ax < \beta x$, then $\rho(A) < \beta$.*

Proof. If $\alpha x \leq Ax$, then $\alpha x_i \leq (Ax)_i$ and $\alpha \leq \min_{1 \leq i \leq n} x_i^{-1} \sum_{j=1}^{n} a_{ij}x_j$, so the preceding theorem ensures that $\alpha \leq \rho(A)$. If $\alpha x < Ax$, then there is some $\alpha' > \alpha$ such that $\alpha x < \alpha' x \leq Ax$. In this event, $\rho(A) \geq \alpha' > \alpha$. The upper bounds can be verified in a similar fashion.   □

Corollary 8.1.30. *Let $A \in M_n$ be nonnegative. If x is a positive eigenvector of A, then $\rho(A), x$ is an eigenpair for A; that is, if $A \geq 0$, $x > 0$, and $Ax = \lambda x$, then $\lambda = \rho(A)$.*

Proof. If $x > 0$ and $Ax = \lambda x$, then $\lambda \geq 0$ and $\lambda x \leq Ax \leq \lambda x$. But then (8.1.29) ensures that $\lambda \leq \rho(A) \leq \lambda$.   □

Corollary 8.1.31. *Let $A = [a_{ij}] \in M_n$ be nonnegative. If A has a positive eigenvector, then*

$$\rho(A) = \max_{x > 0} \min_{1 \leq i \leq n} \frac{1}{x_i} \sum_{j=1}^{n} a_{ij}x_j = \min_{x > 0} \max_{1 \leq i \leq n} \frac{1}{x_i} \sum_{j=1}^{n} a_{ij}x_j \qquad (8.1.32)$$

Exercise. Prove the preceding corollary. *Hint*: Use the positive eigenvector x in (8.1.27).

Corollary 8.1.33. *Let $A = [a_{ij}] \in M_n$ be nonnegative and write $A^m = [a_{ij}^{(m)}]$. If A has a positive eigenvector $x = [x_i]$, then for all $m = 1, 2, \ldots$ and for all $i = 1, \ldots, n$ we have*

$$\sum_{j=1}^n a_{ij}^{(m)} \leq \left(\frac{\max_{1 \leq k \leq n} x_k}{\min_{1 \leq k \leq n} x_k} \right) \rho(A)^m \qquad (8.1.34a)$$

and

$$\left(\frac{\min_{1 \leq k \leq n} x_k}{\max_{1 \leq k \leq n} x_k} \right) \rho(A)^m \leq \sum_{j=1}^n a_{ij}^{(m)} \qquad (8.1.34b)$$

If $\rho(A) > 0$, then the entries of $[\rho(A)^{-1} A]^m$ are uniformly bounded for $m = 1, 2, \ldots$.

Proof. Let $x = [x_i]$ be a positive eigenvector of A. Then (8.1.30) ensures that $Ax = \rho(A)x$, so $A^m x = \rho(A)^m x$ for each $m = 1, 2, \ldots$. Since $A^m \geq 0$, for any $i = 1, \ldots, n$ we have

$$\rho(A)^m \max_{1 \leq k \leq n} x_k \geq \rho(A^m)x_i = (A^m x)_i = \sum_{j=1}^n a_{ij}^{(m)} x_j$$

$$\geq \left(\min_{1 \leq k \leq n} x_k \right) \sum_{j=1}^n a_{ij}^{(m)}$$

Since $\min_{1 \leq k \leq n} x_k > 0$, the asserted upper bound on $\sum_{j=1}^n a_{ij}^{(m)}$ follows:

$$\rho(A)^m \frac{\max_{1 \leq k \leq n} x_k}{\min_{1 \leq k \leq n} x_k} \geq \sum_{j=1}^n a_{ij}^{(m)}$$

The asserted lower bound follows in a similar fashion. $\qquad \square$

Problems

8.1.P1 If $A \in M_n$ is nonnegative and if A^k is positive for some positive integer k, explain why $\rho(A) > 0$.

8.1.P2 Give an example of a 2-by-2 matrix A such that $A \geq 0$, A is not positive, and $A^2 > 0$.

8.1.P3 Suppose that $A \in M_n$ is nonnegative and nonzero. If A has a positive eigenvector, explain why $\rho(A) > 0$.

8.1.P4 Let $A \in M_n$. Corollary 5.6.13 ensures that for each $\varepsilon > 0$ there is a nonnegative matrix $C(A, \varepsilon)$ such that $|A^m| \leq (\rho(A) + \varepsilon)^m C(A, \varepsilon)$ for all $m = 1, 2, \ldots$. If we assume that A is nonnegative and has a positive eigenvector, explain why there is a nonnegative matrix $C(A)$ such that $|A^m| \leq \rho(A)^m C(A)$ for all $m = 1, 2, \ldots$. Consider $A = \begin{bmatrix} 1 & 1 \\ 0 & 1 \end{bmatrix}$ and explain why the assumption about a positive eigenvector cannot be omitted.

8.1.P5 If $A \in M_n$ is nonnegative and has a positive eigenvector, show that A is diagonally similar to a nonnegative matrix, all of whose row sums are equal. Equal to what?

8.1.P6 Give an example to show that a reducible nonnegative matrix can have a positive eigenvector.

8.1.P7 Let $A = [a_{ij}] \in M_n$ be nonnegative and let $x = [x_i] \in \mathbf{R}^n$ be a positive vector. (a) Explain why (8.1.27) can be restated as

$$\min_{1 \le i \le n} \frac{(Ax)_i}{x_i} \le \rho(A) \le \max_{1 \le i \le n} \frac{(Ax)_i}{x_i} \tag{8.1.29}$$

(b) Show that the choice $x = e$ (the all-ones vector) in (8.1.29) gives the bounds in (8.1.23). (c) If A has positive row sums $R_i = (Ae)_i$, $i = 1, \ldots, n$, show that the choice $x = Ae$ in (8.1.29) leads to the improved bounds

$$\min_{1 \le i \le n} R_i \le \min_{1 \le i \le n} \frac{1}{R_i} \sum_{j=1}^n a_{ij} R_j \le \rho(A) \le \max_{1 \le i \le n} \frac{1}{R_i} \sum_{j=1}^n a_{ij} R_j \le \max_{1 \le i \le n} R_i$$

8.1.P8 Let $A, B \in M_n$ be nonnegative and suppose that $A \ge B \ge 0$. Show that $\|\|A\|\|_2 \ge \|\|B\|\|_2$.

8.1.P9 Let $A \in M_n$. Use (8.1.18) to show that $\|\|A\|\|_2 \le \|\||A|\|\|_2$.

8.1.P10 Let $A = [A_{ij}]_{i,j=1}^k \in M_n$, in which each $A_{ij} \in M_{n_i, n_j}$ and $n_1 + \cdots + n_k = n$. Let $G(\cdot)$ be a given vector norm on all M_{n_i, n_j}, $1 \le i, j \le k$, that is compatible with a given norm $\|\cdot\|$ on all $\mathbf{C}^{n_i}$, $1 \le i \le k$ (5.7.12). Let $\mathcal{A} = [G(A_{ij})] \in M_k$. (a) Show that $\rho(A) \le \rho(\mathcal{A})$. (b) Give examples of some vector norms $G(\cdot)$ on matrices for which the bound $\rho(A) \le \rho(\mathcal{A})$ is valid and explain why. (c) Explain why (5.6.9(a)) and (8.1.18) are special cases of (a); describe the partition, $G(\cdot)$, and $\|\cdot\|$ in each case.

8.1.P11 Let $A = [a_{ij}] \in M_n$ be nonnegative, let σ be a given permutation of $\{1, \ldots, n\}$, and let $\gamma = a_{1\sigma(1)} a_{2\sigma(2)} \cdots a_{n\sigma(n)}$. Show that $\rho(A) \ge \gamma^{1/n}$. This inequality is interesting only if there is a σ such that $\gamma > 0$, which is the case if and only if $\Gamma(A)$ contains a cycle of length n.

8.2 Positive matrices

The theory of nonnegative matrices assumes its simplest and most elegant form for positive matrices, and it is for this case that Oskar Perron published the fundamental discoveries in 1907. In developing this theory, we begin with some remarkable properties of eigenvectors that are associated with eigenvalues of maximum modulus.

Lemma 8.2.1. *Let $A \in M_n$ be positive. If λ, x is an eigenpair of A and $|\lambda| = \rho(A)$, then $|x| > 0$ and $A|x| = \rho(A)|x|$.*

Proof. The hypotheses ensure that $z = A|x| > 0$ (8.1.14). We have $z = A|x| \ge |Ax| = |\lambda x| = |\lambda| |x| = \rho(A)|x|$, so $y = z - \rho(A)|x| \ge 0$. If $y = 0$, then $\rho(A)|x| = A|x| > 0$, so $\rho(A) > 0$ and $|x| > 0$. If, however, $y \ne 0$, (8.1.14) again ensures that $0 < Ay = Az - \rho(A)A|x| = Az - \rho(A)z$, in which case $Az > \rho(A)z$. It follows from (8.1.29) that $\rho(A) > \rho(A)$, which is not possible. We conclude that $y = 0$. $\square$

From this technical result we now deduce a basic fact about positive matrices.

Theorem 8.2.2. *If $A \in \mathbf{M}_n$ is positive, there are positive vectors x and y such that $Ax = \rho(A)x$ and $y^T A = \rho(A)y^T$.*

Proof. There is an eigenpair λ, x of A with $|\lambda| = \rho(A)$. The preceding lemma ensures that $\rho(A)$, $|x|$ is also an eigenpair of A and $|x| > 0$. The assertion about y follows from considering A^T. $\qquad\square$

Exercise. If $A \in M_n$ and $A > 0$, use (8.1.31) and the preceding theorem to explain why

$$\rho(A) = \max_{x>0} \min_i \frac{1}{x_i} \sum_{j=1}^n a_{ij} x_j = \min_{x>0} \max_i \frac{1}{x_i} \sum_{j=1}^n a_{ij} x_j \qquad (8.2.2a)$$

After strengthening the conclusion of (8.2.1), we will be able to show that the only maximum modulus eigenvalue of a positive matrix is its spectral radius.

Lemma 8.2.3. *Let $A \in M_n$ be positive. If λ, x is an eigenpair of A and $|\lambda| = \rho(A)$, then there is a $\theta \in \mathbf{R}$ such that $e^{-i\theta} x = |x| > 0$.*

Proof. The hypothesis is that $x \in \mathbf{C}^n$ is nonzero and $|Ax| = |\lambda x| = \rho(A)|x|$; (8.2.1) ensures that $A|x| = \rho(A)|x|$ and $|x| > 0$. Since $|Ax| = \rho(A)|x| = A|x|$ and some (in fact, every) row of A is positive, (8.1.8b) ensures that there is a $\theta \in \mathbf{R}$ such that $e^{-i\theta} x = |x|$. $\qquad\square$

Theorem 8.2.4. *Let $A \in M_n$ be positive. If λ is an eigenvalue of A and $\lambda \neq \rho(A)$, then $|\lambda| < \rho(A)$.*

Proof. Let λ, x be an eigenpair of A, so $|\lambda| \leq \rho(A)$. If $|\lambda| = \rho(A)$, (8.2.3) ensures that $w = e^{-i\theta} x > 0$ for some $\theta \in \mathbf{R}$. Since $Aw = \lambda w$ and $w > 0$, it follows from (8.1.30) that $\lambda = \rho(A)$. $\qquad\square$

If A is positive, we now know that $\rho(A)$ is its eigenvalue of strictly largest modulus. What can be said about the geometric or algebraic multiplicity of $\rho(A)$?

Theorem 8.2.5. *If $A \in M_n$ is positive, then the geometric multiplicity of $\rho(A)$ as an eigenvalue of A is 1.*

Proof. Suppose that $w, z \in \mathbf{C}^n$ are nonzero vectors such that $Aw = \rho(A)w$ and $Az = \rho(A)z$. Then $w = \alpha z$ for some $\alpha \in \mathbf{C}$. Lemma 8.2.3 ensures that there are real numbers θ_1 and θ_2 such that $p = [p_j] = e^{-i\theta_1} z > 0$ and $q = [q_j] = e^{-i\theta_2} w > 0$. Let $\beta = \min_{1 \leq i \leq n} q_i p_i^{-1}$ and let $r = q - \beta p$. Notice that $r \geq 0$ and at least one entry of r is zero. If $r \neq 0$, then $0 < Ar = Aq - \beta Ap = \rho(A)q - \beta\rho(A)p = \rho(A)(q - \beta p) = \rho(A)r$, so $\rho(A)r > 0$ and $r > 0$, which is a contradiction. We conclude that $r = 0$, $q = \beta p$, and $w = \beta e^{i(\theta_2 - \theta_1)} z$. $\qquad\square$

Corollary 8.2.6. *Let $A \in M_n$ be positive. There is a unique vector $x = [x_i] \in \mathbf{C}^n$ such that $Ax = \rho(A)x$ and $\sum_i x_i = 1$. Such a vector must be positive.*

Exercise. Prove (8.2.6).

The unique normalized eigenvector characterized in (8.2.6) is the *Perron vector* of A, sometimes called the *right Perron vector*; $\rho(A)$ is the *Perron root* of A. Of course, the matrix A^T is positive if A is positive, so all the preceding results about eigenvectors of A apply to A^T as well. An eigenvector $y = [y_i]$ of A^T corresponding to the eigenvalue $\rho(A)$ and normalized so that $\sum_i x_i y_i = 1$ is positive and unique; it is the *left Perron vector* of A.

> *Exercise.* If $A \in M_n$ is positive, explain carefully why any nonzero vector y such that $y^T A = \rho(A) y^T$ can be normalized as described in the preceding sentence, and why, after this normalization, it is positive and unique.

Our final result about the spectral radius of a positive matrix is that its algebraic multiplicity is also 1. Consequently, the powers of a positive matrix have a very special asymptotic behavior.

Theorem 8.2.7. *Let $A \in M_n$ be positive. The algebraic multiplicity of $\rho(A)$ as an eigenvalue of A is 1. If x and y are the right and left Perron vectors of A, then $\lim_{m \to \infty} (\rho(A)^{-1} A)^m = xy^T$, which is a positive rank-one matrix.*

Proof. We know that $\rho(A) > 0$, and that x and y are positive vectors such that $Ax = \rho(A)x$, $y^T A = \rho(A) y^T$, and $y^* x = y^T x = 1$. Theorem 1.4.12b ensures that $\rho(A)$ has algebraic multiplicity 1, and (1.4.7b) tells us that there is a nonsingular $S = [x \; S_1]$ such that $S^{-*} = [y \; Z_1]$ and $A = S([\rho(A)] \oplus B)S^{-1}$. Since $\rho(A)$ is a simple eigenvalue of A that is its eigenvalue of strictly largest modulus, $\rho(B) < \rho(A)$, that is, $\rho(\rho(A)^{-1} B) < 1$. Theorem 5.6.12 ensures that

$$\left(\frac{1}{\rho(A)} A \right)^m = S \begin{bmatrix} 1 & 0 \\ 0 & (\rho(A)^{-1} B)^m \end{bmatrix} S^{-1} \tag{8.2.7a}$$

$$\to [x \; S_1] \begin{bmatrix} 1 & 0 \\ 0 & 0_{n-1} \end{bmatrix} \begin{bmatrix} y^T \\ Z_1^T \end{bmatrix} = xy^T \text{ as } m \to \infty \qquad \square$$

We now summarize the principal results obtained in this section for positive matrices.

Theorem 8.2.8 (Perron). *Let $A \in M_n$ be positive. Then*

(a) $\rho(A) > 0$

(b) $\rho(A)$ *is an algebraically simple eigenvalue of A*

(c) *there is a unique real vector $x = [x_i]$ such that $Ax = \rho(A)x$ and $x_1 + \cdots + x_n = 1$; this vector is positive*

(d) *there is a unique real vector $y = [y_i]$ such that $y^T A = \rho(A) y^T$ and $x_1 y_1 + \cdots + x_n y_n = 1$; this vector is positive*

(e) $|\lambda| < \rho(A)$ *for every eigenvalue λ of A such that $\lambda \neq \rho(A)$*

(f) $(\rho(A)^{-1} A)^m \to xy^T$ *as $m \to \infty$*

Perron's theorem has many applications, one of which exhibits an eigenvalue inclusion set for any square complex matrix. This inclusion set is determined by the spectral radius and main diagonal entries of a dominating nonnegative matrix.

Theorem 8.2.9 (Fan). *Let $A = [a_{ij}] \in M_n$. Suppose that $B = [b_{ij}] \in M_n$ is nonnegative and $b_{ij} \geq |a_{ij}|$ for all $i \neq j$. Then every eigenvalue of A is in the union of n discs*

$$\bigcup_{i=1}^{n} \{z \in \mathbf{C} : |z - a_{ii}| \leq \rho(B) - b_{ii}\} \tag{8.2.9a}$$

In particular, A is nonsingular if $|a_{ii}| > \rho(B) - b_{ii}$ for all $i = 1, \ldots, n$.

Proof. First, assume that $B > 0$. Theorem 8.2.8 ensures that there is a positive vector x such that $Bx = \rho(B)x$, and hence

$$\sum_{j \neq i} |a_{ij}|x_j \leq \sum_{j \neq i} b_{ij}x_j = \rho(B)x_i - b_{ii}x_i \text{ for each } i = 1, \ldots, n$$

Thus, we have

$$\frac{1}{x_i} \sum_{j \neq i} |a_{ij}|x_j \leq \rho(B) - b_{ii} \text{ for each } i = 1, \ldots, n$$

The result follows from (6.1.6) with $p_i = x_i$.

If some entry of B is zero, consider $B_\epsilon = B + \epsilon J_n$ for $\epsilon > 0$. Then $b_{ij} + \epsilon > |a_{ij}|$ for all $i \neq j$, so Ky Fan's eigenvalue inclusion set with respect to B_ϵ is a union of n disks of the form $\{z \in \mathbf{C} : |z - a_{ii}| \leq \rho(B_\epsilon) - (b_{ii} + \epsilon)\}$. The assertion for a nonnegative B now follows from observing that $\rho(B_\epsilon) - (b_{ii} + \epsilon) \to \rho(B) - b_{ii}$ as $\epsilon \to 0$.

If $|a_{ii}| > \rho(B) - b_{ii}$ for all $i = 1, \ldots, n$, then $z = 0$ is not in the set (8.2.9a). $\qquad\square$

Part (f) of (8.2.8) guarantees that a certain limit exists. The proof of (8.2.7) and the bounds in (5.6.13) give an upper bound on the rate of convergence:

$$\left\| (\rho(A)^{-1}A)^m - xy^T \right\|_\infty = \left\| S \begin{bmatrix} 1 & 0 \\ 0 & (\rho(A)^{-1}B)^m \end{bmatrix} S^{-1} \right\|_\infty \leq Cr^m \qquad (8.2.10)$$

in which r is any real number in the open interval $(|\lambda_{n-1}|/\rho(A), 1)$, C is a positive constant that depends on r and the positive matrix A, and $|\lambda_{n-1}| = \max\{|\lambda| : \lambda \in \sigma(A)$ and $\lambda \neq \rho(A)\}$ is the modulus of a second-largest-modulus eigenvalue of A, sometimes called a *secondary eigenvalue*. It is known that

$$\frac{|\lambda_{n-1}|}{\rho(A)} \leq \frac{1 - \kappa^2}{1 + \kappa^2} \qquad (8.2.11)$$

in which $\kappa = \min\{a_{ij} : i, j = 1, \ldots, n\}/\max\{a_{ij} : i, j = 1, \ldots, n\}$. This upper bound is easy to compute and can be used as the rate parameter r in (8.2.10).

Problems

8.2.P1 If $A \in M_n$ is positive, describe in detail the asymptotic behavior of A^m as $m \to \infty$.

8.2.P2 The second exercise following (6.1.8) involves a 2-by-2 positive matrix. Discuss that exercise in light of the exercise following (8.2.2).

8.2.P3 Apply the results derived in this section (which make no assumption about diagonalizability) to the matrix $A = \begin{bmatrix} 1-\alpha & \beta \\ \alpha & 1-\beta \end{bmatrix}$, $0 < \alpha, \beta < 1$, and compare with the conclusions reached in (8.0). Use (8.2.8(b)) to explain why the eigenvalues of A must be distinct.

8.2.P4 Consider the general intercity migration model with $n > 2$ cities as described in (8.0). If all the migration coefficients a_{ij} are positive, what is the asymptotic behavior of the population distribution $p^{(m)}$ as $m \to \infty$?

8.2.P5 Let $A, B \in M_n$ and suppose that $A > B > 0$. Use the "min max" characterization of $\rho(B)$ (8.2.2a) to show that $\rho(A) > \rho(B)$.

8.2.P6 If $A \in M_n$ is positive and if $x = [x_i]$ is its Perron vector, explain why $\rho(A) = \sum_{i,j=1}^n a_{ij}x_j$.

8.2.P7 Let $n \geq 2$ and let $A \in M_n$ be nonsingular. If A is positive, show that A^{-1} cannot be nonnegative. If A is nonnegative, show that A^{-1} is nonnegative only if A has exactly one nonzero entry in each column. How is such a matrix related to a permutation matrix?

8.2.P8 Let $A \in M_n$ be positive. Let x and y be positive vectors (not necessarily the Perron vectors) such that $Ax = \rho(A)x$ and $A^T y = \rho(A)y$. Explain why $(\rho(A)^{-1}A)^m \to (y^T x)^{-1}xy^T$.

8.2.P9 Let $A \in M_n$ be positive and let $x = [x_i]$ be the Perron vector of A. (a) Suppose that *either* $\min_i \sum_{j=1}^n a_{ij} = \rho(A)$ *or* $\max_i \sum_{j=1}^n a_{ij} = \rho(A)$. Show that $x_1 = \cdots = x_n$ and deduce that *every* row sum of A is equal to $\rho(A)$. (b) Consider the two basic inequalities (8.1.23). Explain why either both inequalities are strict or both are equalities. Moreover, either every row sum of A is the same or both inequalities are strict. What about the pairs of inequalities (8.1.24), (8.1.27), and (8.1.28)?

8.2.P10 Suppose that $A = [a_{ij}] \in M_n$ is both positive and symmetric and has exactly one positive eigenvalue. Show that $a_{ij} \geq \sqrt{a_{ii}a_{jj}} \geq \min\{a_{ii}, a_{jj}\}$ for all $i, j = 1, \ldots, n$.

8.2.P11 Let $A \in M_n$ be positive, let $\rho(A)$ be its spectral radius, and let $x = [x_i]$ and y be positive vectors such that $Ax = \rho(A)x$ and $y^T A = \rho(A)y^T$. We know that $\rho(A)$ has geometric multiplicity 1. Let $D = \mathrm{diag}(x_1, \ldots, x_n)$, let $B = D^{-1}AD$, and let $p_B(t) = p_A(t)$ be its characteristic polynomial. Provide details for the following alternative argument to show that (i) $\rho(A)$ has algebraic multiplicity 1, and (ii) $\mathrm{adj}(\rho(A)I - A) = \gamma xy^T$ for some $\gamma > 0$. (a) B is positive and has the same eigenvalues as A. (b) Every row sum of B is equal to $\rho(A) = \rho(B)$. (c) $p_B(\rho(B)) = 0$, and to show that $\rho(B)$ is a *simple* eigenvalue, it suffices to show that $p'_B(t)|_{t=\rho(B)} \neq 0$. (d) $p'_B(t) = \mathrm{tr}\,\mathrm{adj}(tI - B) = \sum_i p_{B_i}(t)$ (0.8.10.2), in which $B_i = B[\{i\}^c]$ is a principal submatrix of B of size $n - 1$. (e) Every row sum of each B_i is strictly less than $\rho(B)$, so each $\rho(B_i) < \rho(B)$. (f) Each $p_{B_i}(t)$ has its largest real zero at $\rho(B_i)$ and $p_{B_i}(t) \to +\infty$ as $t \to \infty$, so each $p_{B_i}(\rho(B)) > 0$. (g) $p'_B(t)|_{t=\rho(B)} > 0$. (h) $\mathrm{adj}(\rho(A)I - A) = \gamma xy^T$ (1.4.11). (i) $p'_B(t)|_{t=\rho(B)} = \mathrm{tr}\,\mathrm{adj}(\rho(A)I - A) = \gamma y^T x \Rightarrow \gamma > 0$. (j) If the eigenvalues of A other than $\rho(A)$ are $\lambda_2, \ldots, \lambda_n$, explain why $\gamma = (\rho(A) - \lambda_2) \cdots (\rho(A) - \lambda_n)/y^T x$ and use this representation to explain why $\gamma > 0$.

8.2.P12 Provide details for the following proof that the spectral radius of a positive matrix $A \in M_n$ is an eigenvalue with algebraic multiplicity 1. (a) Define $B = D^{-1}AD$ as in the preceding problem, so B is positive and has the same eigenvalues as A, and every row sum of B is equal to $\rho(A)$. (b) For the maximum row sum matrix norm, we have $\|\|B\|\|_1 = \rho(B)$. (c) (5.6.P38) and (8.2.5).

8.2.P13 Let $A \in M_n$ be positive. Show that $\rho(A) = \lim_{m\to\infty}(\mathrm{tr}\,A^m)^{1/m}$.

8.2.P14 Let $A \in M_n$ be positive. Explain why (a) $\mathrm{adj}(\rho(A)I - A)$ is positive; (b) each column of $\mathrm{adj}(\rho(A)I - A)$ is a positive multiple of the Perron vector of A; (c) each row of $\mathrm{adj}(\rho(A)I - A)$ is a positive multiple of the left Perron vector of A. If $\rho(A)$ is known, these observations provide an algorithm to compute the left and right Perron vectors of A without solving any linear equations.

8.2.P15 Let $A, B \in M_n(\mathbf{R})$. Suppose that $0 \leq A \leq B$ but $A \neq B$, so some (nonnegative) entry of A is strictly less than the corresponding entry of B. Then (8.1.9) ensures that $\rho(A) \leq \rho(B)$. (a) Consider $A = \begin{bmatrix} 0 & 1 \\ 0 & 0 \end{bmatrix}$ and $B = \begin{bmatrix} 0 & 2 \\ 0 & 0 \end{bmatrix}$ to show that $\rho(A) = \rho(B)$ is possible. (b) If B is positive, however, use (8.2.8) to show that $\rho(A) < \rho(B)$.

8.2.P16 Let $A \in M_n$ be positive, and let $x \in \mathbf{R}^n$ be a nonnegative and nonzero eigenvector of A. Refer to (1.4.P6) and the principle of biorthogonality to explain why x cannot be an eigenvector associated with any eigenvalue of A other than $\lambda = \rho(A)$. Why must it be a positive vector?

Further Readings. For many different bounds on the ratio $|\lambda_{n-1}|/\rho(A)$, including the bound (8.2.11), see U. Rothblum and C. Tan, Upper bounds on the maximum modulus of subdominant eigenvalues of nonnegative matrices, *Linear Algebra Appl.* 66 (1985) 45–86.

8.3 Nonnegative matrices

What parts of the theory developed in the preceding section can be generalized (perhaps by a suitable limit argument) to nonnegative matrices that are not positive? The only results in Perron's theorem that generalize by taking limits are contained in the following theorem.

Theorem 8.3.1. *If $A \in M_n$ is nonnegative, then $\rho(A)$ is an eigenvalue of A and there is a nonnegative nonzero vector x such that $Ax = \rho(A)x$.*

Proof. For any $\epsilon > 0$, define $A(\epsilon) = A + \epsilon J_n$. Let $x(\epsilon) = [x(\epsilon)_i]$ be the Perron vector of $A(\epsilon)$, so $x(\epsilon) > 0$ and $\sum_{i=1}^{n} x(\epsilon)_i = 1$. Since the set of vectors $\{x(\epsilon) : \epsilon > 0\}$ is contained in the compact set $\{x : x \in \mathbf{C}^n, \|x\|_1 \leq 1\}$, there is a monotone decreasing sequence $\epsilon_1 \geq \epsilon_2 \geq \cdots$ with $\lim_{k \to \infty} \epsilon_k = 0$ such that $\lim_{k \to \infty} x(\epsilon_k) = x$ exists. Since $x(\epsilon_k) > 0$ and $\|x(\epsilon_k)\|_1 = 1$ for all $k = 1, 2, \ldots$, the limit vector $x = \lim_{k \to \infty} x(\epsilon_k)$ must be nonnegative and nonzero (indeed, $\|x\|_1 = 1$). Theorem 8.1.18 ensures that $\rho(A(\epsilon_k)) \geq \rho(A(\epsilon_{k+1})) \geq \cdots \geq \rho(A)$ for all $k = 1, 2, \ldots$, so $\rho = \lim_{k \to \infty} \rho(A(\epsilon_k))$ exists and $\rho \geq \rho(A)$. However, $x \neq 0$ and

$$Ax = \lim_{k \to \infty} A(\epsilon_k) x(\epsilon_k) = \lim_{k \to \infty} \rho(A(\epsilon_k)) x(\epsilon_k)$$
$$= \lim_{k \to \infty} \rho(A(\epsilon_k)) \lim_{k \to \infty} x(\epsilon_k) = \rho x$$

so ρ is an eigenvalue of A. It follows that $\rho \leq \rho(A)$, so $\rho = \rho(A)$. $\quad\square$

There is a generalization of the "max min" part of the variational characterization (8.1.32) of the spectral radius to nonnegative matrices and vectors. To approach it, we prove the half of (8.1.29) that remains correct for nonnegative matrices and vectors.

Theorem 8.3.2. *Let $A \in M_n$ be nonnegative, and let $x \in \mathbf{R}^n$ be nonnegative and nonzero. If $\alpha \in \mathbf{R}$ and $Ax \geq \alpha x$, then $\rho(A) \geq \alpha$.*

Proof. Let $A = [a_{ij}]$, let $\epsilon > 0$, and define $A(\epsilon) = A + \epsilon J_n > 0$. Then $A(\epsilon)$ has a positive left Perron vector $y(\epsilon)$: $y(\epsilon)^T A(\epsilon) = \rho(A(\epsilon))y(\epsilon)^T$. We are given that $Ax - \alpha x \geq 0$, so $A(\epsilon)x - \alpha x > Ax - \alpha x \geq 0$ and hence $y(\epsilon)^T (A(\epsilon)x - \alpha x) = (\rho(A(\epsilon)) - \alpha)y(\epsilon)^T x > 0$. Since $y(\epsilon)^T x > 0$, we have $\rho(A(\epsilon)) - \alpha > 0$ for all $\epsilon > 0$. But $\rho(A(\epsilon)) \to \rho(A)$ as $\epsilon \to 0$, so we conclude that $\rho(A) \geq \alpha$. $\quad\square$

Corollary 8.3.3. *If $A \in M_n$ is nonnegative, then*

$$\rho(A) = \max_{\substack{x \geq 0 \\ x \neq 0}} \min_{\substack{1 \leq i \leq n \\ x_i \neq 0}} \frac{1}{x_i} \sum_{j=1}^{n} a_{ij} x_j \qquad (8.3.3a)$$

Proof. Let x be any nonzero nonnegative vector and let $\alpha = \min_{x_i \neq 0} \sum_j a_{ij} x_j / x_i$. Then $Ax \geq \alpha x$, so the preceding theorem ensures that $\rho(A) \geq \alpha$, and hence

$$\rho(A) \geq \max_{\substack{x \geq 0 \\ x \neq 0}} \min_{\substack{1 \leq i \leq n \\ x_i \neq 0}} \frac{1}{x_i} \sum_{j=1}^{n} a_{ij} x_j$$

Now use (8.3.1) to choose a nonzero nonnegative x such that $Ax = \rho(A)x$, which shows that equality can be attained with $\alpha = \rho(A)$. $\square$

Exercise. Consider $A = \begin{bmatrix} 1 & 0 \\ 0 & 2 \end{bmatrix}$ and $x = \begin{bmatrix} 1 \\ 0 \end{bmatrix}$. Explain why the implication $Ax \geq \alpha x \Rightarrow \rho(A) \geq \alpha$ in (8.1.29) need not be correct if the nonnegative vector x is not positive. Show that the "min max" characterization in (8.1.32) need not be correct for nonnegative matrices.

The matrix in the preceding exercise has no positive left or right eigenvector. A nonnegative matrix with a positive left or right eigenvector has some special properties.

Theorem 8.3.4. *Let $A \in M_n$ be nonnegative. Suppose that there is a positive vector x and a nonnegative real number λ such that either $Ax = \lambda x$ or $x^T A = \lambda x^T$. Then $\lambda = \rho(A)$.*

Proof. Suppose that $x = [x_i] \in \mathbf{R}^n$ and $Ax = \lambda x$. Let $D = \mathrm{diag}(x_1, \ldots, x_n)$ and define $B = D^{-1}AD$, which has the same eigenvalues as A. Then $Be = D^{-1}ADe = D^{-1}Ax = \lambda D^{-1}x = \lambda e$, so every row sum of the nonnegative matrix B is equal to λ. It follows from (8.1.21) that $\rho(B) = \lambda$. If $x^T A = \lambda x^T$, apply this argument to A^T. $\square$

Exercise. Let $A \in M_n$ be nonnegative. Explain why (a) every column (respectively, row) sum of A is equal to one if and only if $e^T A = e^T$ (respectively, $Ae = e$); (b) if $e^T A = e^T$ then $e^T A^m = e^T$ (respectively, if $Ae = e$ then $A^m e = e$) for each $m = 2, 3, \ldots$; and (c) under either assumption in (b), every entry of A^m is between zero and one for every $m = 1, 2, \ldots$, so A is power bounded.

Theorem 8.3.5. *Suppose that $A \in M_n$ is nonnegative and has a positive left eigenvector.*

(a) *If $x \in \mathbf{R}^n$ is nonzero and $Ax \geq \rho(A)x$, then x is an eigenvector of A corresponding to the eigenvalue $\rho(A)$.*

(b) *If $A \neq 0$, then $\rho(A) > 0$ and every eigenvalue λ of A such that $|\lambda| = \rho(A)$ is semisimple, that is, every Jordan block of A corresponding to a maximum-modulus eigenvalue is one-by-one.*

Proof. Let y be a positive left eigenvector of A. The preceding theorem ensures that $A^T y = \rho(A)y$.

(a) We know that $x \neq 0$ and $Ax - \rho(A)x \geq 0$. We need to show that $Ax - \rho(A)x = 0$. If $Ax - \rho(A)x \neq 0$, then $y^T(Ax - \rho(A)x) > 0$. However, $y^T(Ax - \rho(A)x) = \rho(A)y^Tx - \rho(A)y^Tx = 0$, which is a contradiction.

(b) Since y is positive and A is nonzero and nonnegative, some entry of y^TA is positive. Consequently, the identity $y^TA = \rho(A)y^T$ ensures that $\rho(A) > 0$. Let $D = \mathrm{diag}(y_1, \ldots, y_n)$ and let $B = \rho(A)^{-1}DAD^{-1}$. It suffices to show that every eigenvalue of B with unit modulus is semisimple. Compute $e^TB = \rho(A)^{-1}e^TDAD^{-1} = \rho(A)^{-1}y^TAD^{-1} = \rho(A)^{-1}\rho(A)yD^{-1} = e^T$. The preceding exercise ensures that every column sum of the nonnegative matrix B is one, so B is power bounded and the assertion follows from (3.2.5.2). $\qquad\square$

Exercise. Restate and prove the preceding theorem under the assumption that A has a positive right eigenvector corresponding to the eigenvalue $\rho(A)$.

Exercise. Give an example to show that the hypothesis that A is nonnegative cannot be omitted in the preceding theorem. *Hint*: $A = \begin{bmatrix} 1 & -1 \\ 2 & -2 \end{bmatrix}$ and $x = e$.

If $A \in M_n$ is nonnegative, its eigenvalue $\rho(A)$ is called the *Perron root* of A. Because an eigenvector (even if normalized) associated with the Perron root of a nonnegative matrix need not be uniquely determined, there is no well-determined notion of "the Perron vector" for a nonnegative matrix. For example, every nonzero nonnegative vector is an eigenvector of the nonnegative matrix $A = I$ associated with the Perron root $\rho(A) = 1$.

Problems

8.3.P1 Show by examples that the items from (8.2.11) that are not included in (8.3.1) are not generally true of all nonnegative matrices.

8.3.P2 If $A \in M_n$ is nonnegative and A^k is positive for some $k \geq 1$, show that A has a positive eigenvector.

8.3.P3 If $A = [a_{ij}] \in M_n$ is nonnegative and tridiagonal, show that all the eigenvalues of A are real.

8.3.P4 Show by example that the following generalization of (8.1.30) is false: If $A \in M_n$ is nonnegative and has a nonnegative eigenvector x, then $Ax = \rho(A)x$.

8.3.P5 Consider $A = \begin{bmatrix} 0 & 1 \\ 0 & 1 \end{bmatrix}$ and $x = [1 \ 2]^T$. Explain why (8.3.5) need not be correct if we omit the assumption that A has a positive left eigenvector.

8.3.P6 Let $A \in M_n$ be nonnegative and nonzero. (a) If A commutes with a positive matrix B, show that the left and right Perron vectors of B are, respectively, left and right eigenvectors of A associated with the eigenvalue $\rho(A)$. (b) Compare and contrast the result in (a) with the information in (1.3.19). (c) If A has positive left and right eigenvectors, show that there is a positive matrix that commutes with A.

8.3.P7 Suppose that $A \in M_n$ is nonnegative. (a) If A has a nonnegative eigenvector with $r \geq 1$ positive entries and $n - r$ zero entries, show that there is a permutation matrix P such that $P^TAP = \begin{bmatrix} B & C \\ 0 & D \end{bmatrix}$ is nonnegative, $B \in M_r$, $D \in M_{n-r}$, and B has a positive eigenvector. If $r < n$, conclude that A is reducible. (b) Explain why A is irreducible if and only if *all* of its nonnegative eigenvectors are positive.

8.3.P8 Let $A \in M_n$ be nonnegative. Use the preceding problem to show that either A is irreducible or there is a permutation matrix P such that

$$P^T A P = \begin{bmatrix} A_1 & & \bigstar \\ & \ddots & \\ \mathbf{0} & & A_k \end{bmatrix} \tag{8.3.6}$$

is block upper triangular, and each diagonal block is irreducible (possibly a 1-by-1 zero matrix). This is an *irreducible normal form* (*Frobenius normal form*) of A. Observe that $\sigma(A) = \sigma(A_1) \cup \cdots \cup \sigma(A_k)$ (including multiplicities), so the eigenvalues of a nonnegative matrix are zero (with arbitrary multiplicity) together with the spectra of finitely many nonnegative nonzero irreducible matrices; see (8.4.6) for the special properties of their spectra. An irreducible normal form of A is not necessarily unique.

8.3.P9 A matrix $A = [a_{ij}] \in M_n(\mathbf{R})$ whose *off-diagonal* entries are all nonnegative is said to be *essentially nonnegative*. If A is essentially nonnegative, explain why there is some $\lambda > 0$ such that $\lambda I + A \geq 0$. Use this observation and (8.3.1) to show that if $A \in M_n$ is essentially nonnegative, then A has a real eigenvalue $r(A)$ (often called the *dominant eigenvalue* of A) with the property that $r(A) \geq \mathrm{Re}\, \lambda_i$ for every eigenvalue λ_i of A. Show that $r(A)$ need not be the eigenvalue of A with largest modulus, but if A is nonnegative, then $r(A) = \rho(A)$.

8.3.P10 Let $A \in M_n$ be nonnegative and consider the real symmetric nonnegative matrix $H(A) = \frac{1}{2}(A + A^T)$. Show that $\rho(A) \leq \lambda_{\max}(H(A))$.

8.3.P11 Suppose that $A \in M_n$ is nonnegative. (a) Explain why its characteristic polynomial can be factored as $p_A(t) = (t - \rho(A))g(t)$, in which $g(t) = t^{n-1} + \gamma_1 t^{n-2} + \gamma_2 t^{n-3} + \cdots$ and $\gamma_1 = \rho(A) - \mathrm{tr}\, A$. Thus, $\gamma_1 = 0$ if and only if $\mathrm{tr}\, A = \rho(A)$. (b) If $n = 3$ and $\mathrm{tr}\, A = \rho(A) > 0$, explain why the eigenvalues of A are $\rho(A)$ and $\pm\sqrt{\det A/\rho(A)}$, which are either real or pure imaginary. (c) *Magic squares* are n-by-n positive matrices whose entries are distinct integers between 1 and n^2; all row sums and column sums, and the sums of the entries in the main diagonal and counterdiagonal are equal. If $A \in M_n$ is a magic square, explain why $\rho(A) = \frac{1}{2}n(n^2 + 1)$ is an eigenvalue of A and $p_A(t) = (t - \rho(A))(t^{n-1} + \gamma_2 t^{n-3} + \cdots)$.

8.3.P12 Let $A \in M_n$ be nonnegative. We claim that $\mathrm{adj}(\rho(A)I - A)$ is nonnegative. Provide details: (a) If r is real and $r > \rho(A)$, show that $\det(rI - A) > 0$. (b) If $r > \rho(A)$, show that $(rI - A)^{-1}$ is positive. (c) If $r > \rho(A)$, deduce from (a) and (b) that $\mathrm{adj}(rI - A) > 0$. (d) Conclude that $\mathrm{adj}(\rho(A)I - A) \geq 0$.

8.3.P13 Let $A \in M_n$ be nonnegative. (a) If $\rho(A)$ has geometric multiplicity greater than 1, explain why $\mathrm{adj}(\rho(A)I - A) = 0$. (b) If $\rho(A)$ has algebraic multiplicity greater than 1, $\mathrm{adj}(\rho(A)I - A)$ can be nonzero, but why is every main diagonal entry of $\mathrm{adj}(\rho(A)I - A)$ zero?

8.3.P14 Let $A \in M_n$ be nonnegative. Explain why (a) $\rho(A)$ can have geometric multiplicity greater than 1, but only if every minor of $\rho(A)I - A$ is zero; (b) $\rho(A)$ can have algebraic multiplicity greater than 1, but only if every principal minor of $\rho(A)I - A$ is zero.

8.3.P15 Let $A = [a_{ij}] \in M_n(\mathbf{R})$. Suppose that $a_{ij} \leq 0$ for all $i \neq j$, and suppose that every real eigenvalue of A is positive; such a matrix is called an *M-matrix*. Provide details to show that A^{-1} is nonnegative: (a) Let $\mu = \max a_{ii}$, so $\mu > 0$; (b) $B = \mu I - A$ is nonnegative and $\rho(B)$ is an eigenvalue of B; (c) $\mu - \rho(B)$ is an eigenvalue of A, so $\mu > \rho(B)$; (d) $A^{-1} = \mu^{-1} \sum_{k=0}^{\infty} \mu^{-k} B^k \geq 0$. For a special case of this problem, see (7.2.P31).

8.3.P16 Let $A, B \in M_n(\mathbf{R})$. (a) Show that A is nonsingular and A^{-1} is nonnegative if and only if whenever $x, y \in \mathbf{R}^n$ and $Ax \geq Ay$, then $x \geq y$. (b) A is said to be a *monotone matrix* if it satisfies either of the equivalent conditions in (a). If A and B are monotone matrices, show that AB is a monotone matrix. (c) Explain why every M-matrix is a monotone matrix.

Further Readings. See C. R. Johnson, R. B. Kellogg, and A. B. Stephens, Complex eigenvalues of a nonnegative matrix with a specified graph II, *Linear Multilinear Algebra* 7 (1979) 129–143, and C. R. Johnson, Row stochastic matrices similar to doubly stochastic matrices, *Linear Multilinear Algebra* 10 (1981) 113–130, for results about the eigenvalue possibilities for nonnegative matrices. The book Bapat and Raghavan (1997) is a comprehensive reference for results about nonnegative matrices. See section 2.5 of Horn and Johnson (1991) for 18 equivalent characterizations of M-matrices.

8.4 Irreducible nonnegative matrices

It is a useful heuristic principle that results about matrices with no zero entries can often be generalized to irreducible matrices. We have seen one instance of this principle in the extensions of the basic Geršgorin theorem in Chapter 6, and we now exhibit another. The basic idea has already been established in (6.2.24); we restate the relevant portion here.

Lemma 8.4.1. *Let $A \in M_n$ be nonnegative. Then A is irreducible if and only if $(I + A)^{n-1} > 0$.*

Exercise. Explain why $A \in M_n$ is irreducible if and only if A^T is irreducible.

We also need the following two lemmas.

Lemma 8.4.2. *Let $\lambda_1, \ldots, \lambda_n$ be the eigenvalues of $A \in M_n$. Then $\lambda_1 + 1, \ldots, \lambda_n + 1$ are the eigenvalues of $I + A$ and $\rho(I + A) \leq \rho(A) + 1$. If A is nonnegative, then $\rho(I + A) = \rho(A) + 1$.*

Proof. The first assertion is a consequence of (2.4.2). We have $\rho(I + A) = \max_{1 \leq i \leq n} |\lambda_i + 1| \leq \max_{1 \leq i \leq n} |\lambda_i| + 1 = \rho(A) + 1$. However, (8.3.1) ensures that $\rho(A) + 1$ is an eigenvalue of $I + A$ if $A \geq 0$, so $\rho(I + A) = \rho(A) + 1$ in this case. □

Lemma 8.4.3. *If $A \in M_n$ is nonnegative and A^m is positive for some $m \geq 1$, then $\rho(A)$ is the only maximum-modulus eigenvalue of A; it is positive and algebraically simple.*

Proof. Let $\lambda_1, \ldots, \lambda_n$ be the eigenvalues of A. Then $\lambda_1^m, \ldots, \lambda_n^m$ are the eigenvalues of A^m. Theorem 8.2.8 ensures that exactly one of $\lambda_1^m, \ldots, \lambda_n^m$ is equal to $\rho(A^m) = \rho(A)^m$, which is positive; all the rest have modulus strictly less than $\rho(A^m)$. Consequently, $n - 1$ of $\lambda_1, \ldots, \lambda_n$ are strictly less than $\rho(A)$ in modulus; (8.3.1) ensures that $\rho(A)$ is the remaining eigenvalue. $\square$

Now we investigate how much of Perron's theorem generalizes to nonnegative irreducible matrices. The name of Frobenius is associated with generalizations of Perron's results about positive matrices to nonnegative matrices.

Theorem 8.4.4 (Perron–Frobenius). *Let $A \in M_n$ be irreducible and nonnegative, and suppose that $n \geq 2$. Then*

(a) $\rho(A) > 0$

(b) $\rho(A)$ *is an algebraically simple eigenvalue of A*

(c) *there is a unique real vector $x = [x_i]$ such that $Ax = \rho(A)x$ and $x_1 + \cdots + x_n = 1$; this vector is positive*

(d) *there is a unique real vector $y = [y_i]$ such that $y^T A = \rho(A)y^T$ and $x_1 y_1 + \cdots + x_n y_n = 1$; this vector is positive*

Proof. (a) Corollary 8.1.25 shows that $\rho(A) > 0$ under conditions even weaker than irreducibility.

(b) If $\rho(A)$ is a multiple eigenvalue of A, then (8.4.2) ensures that $\rho(A) + 1 = \rho(I + A)$ is a multiple eigenvalue of $I + A$ and hence $(1 + \rho(A))^{n-1} = \rho((I + A)^{n-1})$ is a multiple eigenvalue of the positive matrix $(I + A)^{n-1}$, which contradicts (8.2.8(b)).

(c) Theorem 8.3.1 ensures that there is a nonnegative nonzero vector x such that $Ax = \rho(A)x$. Then $(I + A)^{n-1}x = (\rho(A) + 1)^{n-1}x$, and since $(I + A)^{n-1}$ is positive (8.4.1), it follows from (8.1.14) that $(I + A)^{n-1}x$, and hence also $x = (\rho(A) + 1)^{1-n}(I + A)^{n-1}x$, is positive. If we impose the normalization $e^T x = 1$, then (b) ensures that x is unique.

(d) This follows by applying (c) to A^T. $\square$

The preceding theorem ensures that the left and right eigenspaces of an irreducible nonnegative matrix A associated with its Perron root are one-dimensional. The vector x in (8.4.4(c)) is the *(right) Perron vector* of A; the vector y in (8.4.4(d)) is its *left Perron vector*.

Theorem 8.4.4(c–d) ensure that the results in (8.1.30–33) and (8.3.4–5) apply to irreducible nonnegative matrices. Of particular importance is the variational characterization (8.1.32) of the spectral radius. These observations are crucial in the following extension of (8.1.18).

Theorem 8.4.5. *Let $A, B \in M_n$. Suppose that A is nonnegative and irreducible, and $A \geq |B|$. Let $\lambda = e^{i\varphi}\rho(B)$ be a given maximum-modulus eigenvalue of B. If $\rho(A) = \rho(B)$, then there is a diagonal unitary matrix $D \in M_n$ such that $B = e^{i\varphi}DAD^{-1}$.*

Proof. Let x be a nonzero vector such that $Bx = \lambda x$, and let $\rho = \rho(A) = \rho(B)$. Then

$$\rho|x| = |\lambda x| = |Bx| \leq |B|\,|x| \overset{(\alpha)}{\leq} A|x| \tag{8.4.5a}$$

Theorem 8.3.5 and the inequality $A|x| \geq \rho|x|$ imply that $A|x| = \rho|x|$, and (8.4.4) ensures that $|x|$ is positive. Equality in the inequality (α) in (8.4.5a) tells us that $(A - |B|)x = 0$; since x is positive and $A - |B| \geq 0$, (8.1.1) ensures that $A = |B|$. Let D be the unique diagonal unitary matrix such that $x = D|x|$. The identity $Bx = \lambda x = e^{i\varphi}\rho x$ is equivalent to the identity $BD|x| = e^{i\varphi}\rho D|x|$, or $e^{-i\varphi}D^{-1}BDx = \rho|x| = A|x| = |B||x|$. If we let $C = e^{-i\varphi}D^{-1}BD$, we have $C|x| = |C||x|$, so (8.1.8(c)) ensures that $C = |C| = |B| = A$. Thus, $B = e^{i\varphi}DAD^{-1}$. $\qquad\square$

If A is positive, Perron's theorem ensures that $\rho(A)$ is the unique eigenvalue of A of largest modulus. If A is nonnegative but not positive, it may have eigenvalues of maximum modulus other than $\rho(A)$. However, if A is also irreducible, then these eigenvalues (in fact, *all* of its eigenvalues) occur in a regular pattern.

Corollary 8.4.6. *Let $A \in M_n$ be irreducible and nonnegative, and suppose that it has exactly k distinct eigenvalues of maximum modulus. Then*

  (a) *A is similar to $e^{2\pi i p/k}A$ for each $p = 0, 1, \ldots, k-1$*
  (b) *if $J_{m_1}(\lambda) \oplus \cdots \oplus J_{m_\ell}(\lambda)$ is a direct summand of the Jordan canonical form of A, and if $p \in \{1, \ldots, k-1\}$, then $J_{m_1}(e^{2\pi i p/k}\lambda) \oplus \cdots \oplus J_{m_\ell}(e^{2\pi i p/k}\lambda)$ is also a direct summand of the Jordan canonical form of A*
  (c) *the maximum-modulus eigenvalues of A are $e^{2\pi i p/k}\rho(A)$, $p = 0, 1, \ldots, k-1$, and each has algebraic multiplicity 1*

Proof. If $k = 1$, there is nothing to prove, so assume that $k \geq 2$. Let $\lambda_p = e^{i\varphi_p}\rho(A)$, $p = 0, 1, \ldots, k-1$, be the distinct maximum-modulus eigenvalues of A, in which $0 = \varphi_0 < \varphi_1 < \varphi_2 < \cdots < \varphi_{k-1} < 2\pi$. Let $\mathcal{S} = \{\varphi_0 = 0, \varphi_1, \varphi_2, \ldots, \varphi_{k-1}\}$, which is the set of (exactly k) distinct arguments of the maximum-modulus eigenvalues of A. Since A is real, its eigenvalues occur in conjugate pairs, so $\varphi_{k-1} = 2\pi - \varphi_1$, $\varphi_{k-2} = 2\pi - \varphi_2$, etc., that is, for each $\varphi_p \in \mathcal{S}$, the element $\varphi_{k-p} \in \mathcal{S}$ is such that $\varphi_{k-p} + \varphi_p = 0 \pmod{2\pi}$.

Now apply the preceding theorem with $B = A$ and $\lambda = e^{i\varphi_p}\rho(A)$ for any $p = 0, 1, \ldots, k-1$. We find that $A = B = e^{i\varphi_p}D_pAD_p^{-1} = D_p(e^{i\varphi_p}A)D_p^{-1}$, that is, A is similar to $e^{i\varphi_p}A$ for any $p = 0, 1, \ldots, k-1$. Therefore, if $J_{m_1}(\lambda) \oplus \cdots \oplus J_{m_\ell}(\lambda)$ is a direct summand of the Jordan canonical form of A, then $J_{m_1}(e^{i\varphi_p}\lambda) \oplus \cdots \oplus J_{m_\ell}(e^{i\varphi_p}\lambda)$ is also a direct summand. If we apply this observation to the part of the Jordan canonical form of A associated with any maximum-modulus eigenvalue $\lambda = e^{i\varphi_q}\rho(A)$ and take $p = k - q$, we find that

$$J_{m_1}(e^{i(\varphi_{k-p}+\varphi_p)}\rho(A) \oplus \cdots \oplus J_{m_\ell}(e^{i(\varphi_{k-p}+\varphi_p)}\rho(A))$$
$$= J_{m_1}(\rho(A)) \oplus \cdots \oplus J_{m_\ell}(\rho(A))$$

is a direct summand of the Jordan canonical form of A. However, (8.4.4(b)) ensures that $\ell = 1$ and $m_1 = 1$, that is, each maximum-modulus eigenvalue is simple.

Since A is similar to $e^{i\varphi_p}A$ as well as to $e^{i\varphi_q}A$, it follows that A is similar to $e^{i(\varphi_p+\varphi_q)}A$ for any $p, q \in \{0, 1, \ldots, k-1\}$. That is, for each pair of elements $\varphi_p, \varphi_p \in \mathcal{S}, \varphi_p + \varphi_p \pmod{2\pi}$ is also in $\mathcal{S}$. By induction, we can conclude that $r\varphi_1 = \varphi_1 + \cdots + \varphi_1 \pmod{2\pi}$ is in the finite set $\mathcal{S}$ for all $r = 1, 2, \ldots$. The $k + 1$ elements $\varphi_1, 2\varphi_1, \ldots, k\varphi_1$,

$(k + 1)\varphi_1$ of S cannot all be distinct, so there are positive integers $r > s \geq 1$ such that $r\varphi_1 = s\varphi_1 \pmod{2\pi}$, in which case $1 < (r - s) \leq k$. It follows that $(r - s)\varphi_1 = 0 \pmod{2\pi}$, that is, $e^{i(r-s)\varphi_1} = 1$, so $e^{i\varphi_1}$ is a root of unity. Let p (necessarily in $\{1, \ldots, k\}$) be the smallest positive integer such that $e^{ip\varphi_1} = 1$. Choose any $\varphi_m \in S$. Divide the interval $[0, 2\pi)$ into p half-open subintervals $[0, \varphi_1), [\varphi_1, 2\varphi_1), \ldots, [(p - 1)\varphi_1, 2\pi)$. Since φ_m is in one of these subintervals, there is some integer q with $0 \leq q \leq p - 1$ such that $q\varphi_1 \leq \varphi_m < (q + 1)\varphi_1$; that is, $0 \leq \varphi_m - q\varphi_1 < \varphi_1$. It follows that $\varphi_m - q\varphi_1 = 0$ since $\varphi_m - q\varphi_1 \in S$ and φ_1 is the smallest nonzero element of S. We conclude that *each* element φ_m is *some* positive integer multiple of φ_1. If $p < k$, there would be fewer than k distinct elements in the set $\{0, \varphi_1, 2\varphi_1, \ldots\}$, which we have just shown to contain every point in S. We conclude that $p = k$, $\varphi_m = 2\pi m/k$ for each $m = 0, 1, \ldots, k - 1$, and the maximum-modulus eigenvalues of A are $\rho(A), e^{2\pi i/k}\rho(A), \ldots, e^{2\pi i(k-1)/k}\rho(A)$. □

Suppose that $A \in M_n$ is irreducible and nonnegative, and has k eigenvalues of maximum modulus. The preceding theorem ensures that the number of eigenvalues of A (including multiplicities) on any circle $\{z \in \mathbf{C} : |z| = r > 0\}$ is a nonnegative integer multiple of k, possibly a zero multiple. Thus, k must be a divisor of the number of nonzero eigenvalues of A.

Exercise. If $A \in M_n$ is nonnegative, explain why $\operatorname{tr} A^m \geq 0$ for each $m = 1, 2, \ldots$.

Exercise. Can an irreducible nonnegative matrix $A \in M_3$ with spectral radius 1 have eigenvalues 1, i, and $-i$? Can it have those eigenvalues if we drop the requirement that it be irreducible? *Hint*: Apply the preceding corollary; consider $\operatorname{tr} A^2$.

Corollary 8.4.7. *Suppose that $A \in M_n$ is irreducible and nonnegative. If A has $k > 1$ eigenvalues of maximum modulus, then every main diagonal entry of A is zero. Moreover, every main diagonal entry of A^m is zero for each positive integer m that is not divisible by k.*

Proof. Let $\varphi = 2\pi/k$. Corollary 8.4.6a ensures that A is similar to $e^{i\varphi}A$, so A^m is similar to $e^{im\varphi}A$ for each $m = 1, 2, 3, \ldots$ and $\operatorname{tr} A^m = e^{im\varphi}\operatorname{tr} A^m$. Since $e^{im\varphi}$ is real and positive only if m is an integer multiple of k, this is impossible if A^m has any positive main diagonal entry and m is not divisible by k. □

Exercise. Suppose that $A \in M_n$ is irreducible and nonnegative. To ensure that $\rho(A)$ is the only eigenvalue of A that has maximum modulus, explain why it is *sufficient* for A to have at least one nonzero main diagonal entry. Consider the matrix

$$\begin{bmatrix} 0 & 1 & 1 \\ 1 & 0 & 1 \\ 1 & 1 & 0 \end{bmatrix}$$

and explain why this sufficient condition is not *necessary*. Can you find a 2-by-2 example?

The statement in (8.4.7) can be made more precise: If $A \in M_n$ is irreducible and non-negative, and has $k > 1$ eigenvalues of maximum modulus, then there is a permutation matrix P such that

$$PAP^T = \begin{bmatrix} 0 & A_{12} & & 0 \\ \vdots & 0 & \ddots & \\ 0 & & \ddots & A_{k-1,k} \\ A_{k,1} & 0 & \cdots & 0 \end{bmatrix} \tag{8.4.8}$$

in which the main diagonal zero blocks are square; see theorem 1.8.3 in Bapat and Raghavan (1997).

Problems

8.4.P1 Show by examples that the items in (8.2.11) that are not included in (8.4.4) are not generally true of irreducible nonnegative matrices.

8.4.P2 Give an example of an $A \in M_n$ such that $\rho(I + A) \neq \rho(A) + 1$. Give a condition on A that is necessary and sufficient to have $\rho(I + A) = \rho(A) + 1$. Why is this condition satisfied if A is nonnegative.

8.4.P3 Irreducibility is a sufficient but not necessary condition for a nonnegative matrix to have a positive eigenvector. Consider $\begin{bmatrix} 1 & 1 \\ 0 & 0 \end{bmatrix}$ and $\begin{bmatrix} 1 & 0 \\ 1 & 0 \end{bmatrix}$ to show that a reducible nonnegative matrix may or may not have a positive eigenvector.

8.4.P4 If $n \geq 2$ and if $A \in M_n$ is irreducible and nonnegative, show that the entries of the matrices $(\rho(A)^{-1}A)^m$ are uniformly bounded as $m \to \infty$.

8.4.P5 If $A, B \in M_n$, then AB and BA have the same eigenvalues. Consider $\begin{bmatrix} 0 & 1 \\ 0 & 1 \end{bmatrix}$ and $\begin{bmatrix} 0 & 0 \\ 1 & 1 \end{bmatrix}$. Explain why (a) even if A and B are nonnegative, AB can be irreducible while BA is reducible; (b) an irreducible matrix can be similar (even unitarily similar) to a reducible matrix.

8.4.P6 Show that the assertion in (8.3.P6(a)) remains correct if the hypothesis that B is positive is replaced by the weaker hypothesis that B is irreducible and nonnegative.

8.4.P7 Show that the companion matrix of the polynomial $t^k - 1 = 0$ is an example of a k-by-k nonnegative matrix with k eigenvalues of maximum modulus. Sketch the location of these eigenvalues in the complex plane.

8.4.P8 Let p, q, and r be given positive integers. Construct a nonnegative matrix of size $p + q + r$ whose eigenvalues of maximum modulus are all of the pth, qth, and rth roots of unity.

8.4.P9 An irreducible nonnegative matrix is said to be *cyclic of index k* if it has $k \geq 1$ eigenvalues of maximum modulus. Discuss the aptness of this term.

8.4.P10 If $A \in M_n$ is cyclic of index $k \geq 1$, explain why its characteristic polynomial is $p_A(t) = t^r(t^k - \rho(A)^k)(t^k - \mu_2^k) \cdots (t^k - \mu_m^k)$ for some nonnegative integers r and m and some complex numbers μ_i with $|\mu_i| < \rho(A)$, $i = 2, \ldots, m$. Comment on the pattern of zero and nonzero coefficients in $p_A(t)$ and give a criterion for A to have only one eigenvalue of maximum modulus based on the form of the characteristic polynomial.

8.4.P11 Let $n > 1$ be a prime number. If $A \in M_n$ is irreducible, nonnegative, and nonsingular, explain why either $\rho(A)$ is the only eigenvalue of A of maximum modulus or all the eigenvalues of A have maximum modulus.

8.4.P12 Let $p(t)$ be a polynomial of the form (3.3.11) in which $a_0 \neq 0$, and let $\tilde{p}(t) = t^n - |a_{n-1}| t^{n-1} - \cdots - |a_1| t - |a_0|$. Show that $\tilde{p}(t)$ has a simple positive zero r that is not exceeded by the modulus of any zero of either $\tilde{p}(t)$ or $p(t)$. What can you say about the zeroes of $\tilde{p}(t)$ if it has exactly $k > 1$ zeroes of modulus r?

8.4.P13 Let $A = [a_{ij}] \in M_n$ be irreducible and nonnegative, and let $x = [x_i]$ and $y = [y_i]$ be its right and left Perron vectors, respectively. (a) Explain why $\rho(A)$ is a differentiable function of a_{ij} for each $i, j \in \{1, \dots, n\}$ and why $\partial \rho(A)/\partial a_{ij} = x_i y_j$ for each $i, j \in \{1, \dots, n\}$. (b) Why is $\partial \rho(A)/\partial a_{ij} > 0$ for all $i, j \in \{1, \dots, n\}$?

8.4.P14 Let $A, B \in M_n$ be nonnegative and suppose that A is irreducible. (a) Use the preceding problem to show that $\rho(A + B) > \rho(A)$ if $B \neq 0$. (b) Explain why $A + B$ is irreducible and use (8.4.5) to show that $\rho(A + B) > \rho(A)$ if $B \neq 0$.

8.4.P15 Let $A \in M_n$ be nonnegative. (a) If A is irreducible, explain why any nonnegative eigenvector of A is a positive scalar multiple of the Perron vector of A. (b) If A has a linearly independent pair of nonnegative eigenvectors, explain why A must be reducible.

8.4.P16 Let $A \in M_n$ be nonnegative. (a) Explain why A is irreducible if and only if there is a polynomial $p(t)$ such that every entry of $p(A)$ is nonzero. (b) If $p(t)$ is a polynomial of degree d or less such that every entry of $p(A)$ is nonzero, explain why $(I + A)^d > 0$. (c) Suppose that the minimal polynomial of A has degree m. Show that A is irreducible if and only if $(I + A)^{m-1} > 0$.

8.4.P17 Let $A \in M_n$ be nonnegative and consider the problem of finding a best rank-one approximation to A in the sense of least squares: If either AA^T or $A^T A$ is irreducible, find an $X \in M_n$ such that $\|A - X\|_2 = \min\{\|A - Y\|_2 : Y \in M_n \text{ and rank } Y = 1\}$. Show that such an X is nonnegative, unique, and given by $X = \sqrt{\rho(AA^T)} v w^T$, in which $v, w \in \mathbf{R}^n$ are positive unit eigenvectors of AA^T and $A^T A$ associated with the eigenvalue $\rho(AA^T)$.

8.4.P18 Find a best rank-one least squares approximation to each of the matrices $\begin{bmatrix} 1 & 1 \\ 1 & 1 \end{bmatrix}$, $\begin{bmatrix} 1 & 1 \\ 0 & 1 \end{bmatrix}$, and $\begin{bmatrix} 0 & 0 \\ 1 & 1 \end{bmatrix}$. Explain why a best rank-one least squares approximation to $I \in M_n$ is not unique if $n > 1$.

8.4.P19 Show that all the assertions of (8.2.P9) are correct under the weaker hypothesis that A is irreducible and nonnegative.

8.4.P20 Let $n \geq 2$ and $A \in M_n(\mathbf{R})$ be given. (a) Explain why every negative eigenvalue of A^2 has even algebraic and geometric multiplicities. (b) Compute the squares of the matrices $\begin{bmatrix} 0 & 2 \\ -1 & -\frac{1}{2} \end{bmatrix}$,

$$
\begin{bmatrix} 0 & -1 & 1 \\ 0 & 0 & 1 \\ 0 & -1 & 0 \end{bmatrix}, \text{ and } \begin{bmatrix} 0 & 1 & \cdots & 1 & n \\ -1 & 0 & \cdots & 0 & 1 \\ \vdots & \vdots & \ddots & \vdots & \vdots \\ -1 & 0 & \cdots & 0 & 1 \\ -1 & -1 & \cdots & -1 & -\frac{n-1}{n} \end{bmatrix} \in M_n
$$

Which of these matrices shows that $A^2 \leq 0$ and $A^2 \neq 0$ is possible, but with some zero entries? Which shows (and for what n?) that A^2 can have no zero entries and only one positive entry? (c) If $A^2 \leq 0$, explain why A^2 must be reducible. (d) If $n > 2$ and at least $n^2 - n + 2$ of the entries of A^2 are negative, explain why A^2 has at least one positive entry.

8.4.P21 Let $A \in M_n$ be irreducible and nonnegative. (a) If there is a nonnegative nonzero vector x and a positive scalar α such that $Ax \leq \alpha x$, show that x is positive. (b) Deduce from (a) that any nonnegative eigenvector of A is positive, and is a positive scalar multiple of the Perron vector of A.

8.4.P22 Let $x_1, \ldots, x_{n+2}$ be given unit vectors in $\mathbf{R}^n$, and let $G = [x_i^T x_j] \in M_{n+2}(\mathbf{R})$ be their Gram matrix. (a) If $I - G$ is nonnegative, show that it (and hence also G) must be reducible. (b) Explain why there are at most $n + 1$ vectors in $\mathbf{R}^n$ such that the angle between any two of them is greater than $\pi/2$.

8.4.P23 Let $A \in M_n$ be irreducible and nonnegative, and let x and y, respectively, be its right and left Perron vectors. Explain why $\operatorname{adj}(\rho(A)I - A)$ is a positive scalar multiple of the rank-one positive matrix xy^T.

8.4.P24 Let $A \in M_n$ be nonnegative and suppose that $\rho(A) > 0$. If λ is a maximum-modulus eigenvalue of A, use (8.3.6) and (8.4.6) to show that $\lambda/\rho(A) = e^{i\theta}$ is a root of unity and $e^{ip\theta}\rho(A)$ is an eigenvalue of A for each $p = 0, 1, \ldots, k - 1$. Illustrate by examples that these need not be the *only* maximum-modulus eigenvalues of A, and they need not be simple.

8.4.P25 The matrix $A_1 = \begin{bmatrix} 0 & 1 \\ 1 & 0 \end{bmatrix}$ shows that the result in (8.2.P13) need not be correct for nonnegative matrices that are not positive. Explain why $\lim_{m \to \infty} (\operatorname{tr} A_1^m)^{1/m}$ does not exist; nevertheless, $\limsup_{m \to \infty} (\operatorname{tr} A_1^m)^{1/m} = 1 = \rho(A_1)$. Provide details for the following steps to show that this limit result is correct for any nonnegative matrix A: If $\rho(A) = 0$, explain why $\lim_{m \to \infty} (\operatorname{tr} A^m)^{1/m} = \rho(A)$. Now assume that $\rho(A) > 0$. (a) $\operatorname{tr} A^m = |\sum_{i=1}^n \lambda_i(A^m)| \leq |\sum_{i=1}^n \sigma_i(A^m)| = \||A^m\||_{\operatorname{tr}}$, in which the λ_i are eigenvalues, the σ_i are singular values, and $\||\cdot\||_{\operatorname{tr}}$ is the trace norm; see the example following (5.6.42). (b) $\limsup_{m \to \infty} (\operatorname{tr} A^m)^{1/m} \leq \limsup_{m \to \infty} \||A^m\||_{\operatorname{tr}}^{1/m} = \rho(A)$. (c) Consider an irreducible normal form (8.3.6) of A, and let $A_{i_1} \in M_{n_1}, \ldots, A_{i_g} \in M_{n_g}$ be all of the diagonal blocks in (8.3.6) such that $\rho(A_i) = \rho(A)$. If A_{i_ℓ} has exactly k_ℓ eigenvalues with modulus $\rho(A)$, (8.4.6) ensures that $\rho(A)^{k_\ell}$ is an eigenvalue of $A_{i_\ell}^{k_\ell}$ with multiplicity k_ℓ and all other eigenvalues have strictly smaller modulus. Then $\operatorname{tr} A_{i_\ell}^{pk_\ell} = \rho(A)^{pk_\ell}(k_\ell + o(1))$ for $p = 1, 2, \ldots$, in which $o(1)$ is a quantity that tends to zero as $p \to \infty$. Construct a sequence of positive integers $m_j \to \infty$ such that $\operatorname{tr} A_{i_1}^{m_j} + \cdots + \operatorname{tr} A_{i_g}^{m_j} = (k_1 + \cdots + k_g + o(1))\rho(A)^{m_j}$ for $j = 1, 2, \ldots$. (d) $\operatorname{tr} A^{m_j} \geq (k_1 + \cdots + k_g + o(1))\rho(A)^{m_j}$ for each $j = 1, 2, \ldots$. (e) For any given $\varepsilon \in (0, 1/2)$, $\limsup_{m \to \infty} (\operatorname{tr} A^m)^{1/m} \geq \lim_{m \to \infty} (k_1 + \cdots + k_g - \varepsilon)^{1/m} \rho(A)$. (e) Conclude that if $A \in M_n$ is nonnegative, then

$$\limsup_{m \to \infty} (\operatorname{tr} A^m)^{1/m} = \rho(A) \tag{8.4.9}$$

8.4.P26 Let $A \in M_n$ be nonnegative and let $x = [x_i]$ and $y = [y_i]$ be nonnegative vectors such that $Ax = \rho(A)x$ and $y^T A = \rho(A)y^T$. We claim that A is irreducible if and only if no principal minor of $\rho(A)I - A$ is zero if and only if $\operatorname{adj}(\rho(A)I - A) = cxy^T$ is positive. Provide details: (a) If A is irreducible, then (8.4.4) ensures that $\rho(A)$ is simple, and $y = [y_i]$

and $x = [x_i]$ are positive. Then $\mathrm{adj}(\rho(A)I - A) = cxy^T$ is nonnegative and nonzero; its main diagonal entries are $cx_1 y_1, \ldots cx_n y_n$, which are nonzero, so they are positive and equal to the principal minors of $\rho(A)I - A$. Then $c > 0$ and $cxy^T = \mathrm{adj}(\rho(A)I - A) > 0$. (b) Conversely, if no principal minor of $\rho(A)I - A$ is zero, then $\rho(A)$ is simple, $\mathrm{adj}(\rho(A)I - A) = cxy^T$ has a positive main diagonal, and each of x, y, and c is positive. Problem (8.3.P7) ensures that A is irreducible.

8.5 Primitive matrices

An examination of the proof of (8.2.7) reveals that it is valid for an irreducible nonnegative matrix provided that we make one additional assumption: There are no eigenvalues of maximum modulus other than the spectral radius. This property is so important that it motivates a definition.

Definition 8.5.0. *A nonnegative matrix $A \in M_n$ is* primitive *if it is irreducible and has only one nonzero eigenvalue of maximum modulus.*

The notion of primitivity is due to Frobenius (1912).

Theorem 8.5.1. *If $A \in M_n$ is nonnegative and primitive, and if x and y are, respectively, the right and left Perron vectors of A, then $\lim_{m \to \infty}(\rho(A)^{-1}A)^m = xy^T$, which is a positive rank-one matrix.*

Proof. We have in hand all the ingredients required in the proof of (8.2.7): $\rho(A)$ is a simple eigenvalue with positive associated right and left eigenvectors x and y such that $x^T y = 1$. We can perform the factorization (8.2.7a), in which every eigenvalue of B has modulus strictly less than $\rho(A)$, so $\lim_{m \to \infty}(\rho(A)^{-1}B)^m = 0$. $\qquad\square$

We have now generalized all of Perron's theorem from the class of positive matrices to the class of primitive nonnegative matrices. But how, in practice, can one test a given irreducible nonnegative matrix for primitivity without computing its maximum-modulus eigenvalues? The following characterization of primitivity, while not itself a computationally effective test, leads to several useful criteria.

Theorem 8.5.2. *If $A \in M_n$ is nonnegative, then A is primitive if and only if $A^m > 0$ for some $m \geq 1$.*

Proof. If A^m is positive, there is a directed path of length m between every pair of nodes of the directed graph $\Gamma(A)$ of A, so $\Gamma(A)$ is strongly connected and A is irreducible. In addition, (8.4.3) ensures that there are no maximum-modulus eigenvalues of A other than $\rho(A)$, which is algebraically simple. Conversely, if A is primitive, then $\lim_{m \to \infty}(\rho(A)^{-1}A)^m = xy^T > 0$, so there is some m such that $(\rho(A)^{-1}A)^m > 0$. $\qquad\square$

Exercise. If $A \in M_n$ is nonnegative and irreducible, and if $A^m > 0$, explain why $A^p > 0$ for all $p = m + 1, m + 2, \ldots$.

The characterization in the preceding theorem, and the information in (8.4.6) about the maximum-modulus eigenvalues of nonnegative irreducible matrices, provide a graph-theoretical criterion for primitivity.

Theorem 8.5.3. *Let $A \in M_n$ be irreducible and nonnegative, and let $P_1, \ldots, P_n$ be the nodes of the directed graph $\Gamma(A)$. Let $L_i = \{k_1^{(i)}, k_2^{(i)}, \ldots\}$ be the set of lengths of all directed paths in $\Gamma(A)$ that both start and end at the node P_i, $i = 1, \ldots, n$. Let g_i be the greatest common divisor of all the lengths in L_i. Then A is primitive if and only if $g_1 = \cdots = g_n = 1$.*

Proof. Irreducibility of A implies that no set L_i is empty: For each i and for any $j \neq i$, there is a path in $\Gamma(A)$ that joins P_i to P_j; there is also a path in $\Gamma(A)$ that joins P_j to P_i. If A is primitive, then (8.5.2) ensures that there is some $m \geq 1$ such that $A^m > 0$, and hence $A^k > 0$ for all $k \geq m$. But then $m + p \in L_i$ for each integer $p \geq 1$ and each $i = 1, \ldots, n$, so $g_i = 1$ for all $i = 1, \ldots, n$.

Suppose that $A = [a_{ij}]$ is not primitive and has exactly $k > 1$ eigenvalues of maximum modulus. Corollary 8.4.7 ensures that A^m has a zero main diagonal for every m that is not an integral multiple of k; for each such m, there is no directed path in $\Gamma(A)$ that both starts and ends at any node of $\Gamma(A)$. Thus, $L_i \subset \{k, 2k, 3k, \ldots\}$, and hence $g_i \geq k > 1$ for each $i = 1, \ldots, n$. $\quad\square$

A theorem of Romanovsky provides additional insight into the preceding result: If $A \in M_n$ is irreducible and nonnegative, then $g_1 = g_2 = \cdots = g_n = k$ is the number of maximum-modulus eigenvalues of A.

The following result is useful in many situations; in particular, it shows that an irreducible nonnegative matrix with positive main diagonal must be primitive.

Lemma 8.5.4. *If $A \in M_n$ is irreducible and nonnegative, and if all its main diagonal entries are positive, then $A^{n-1} > 0$, so A is primitive.*

Proof. If every main diagonal entry of A is positive, let $\alpha = \min\{a_{11}, \ldots, a_{nn}\} > 0$ and define $B = A - \mathrm{diag}(a_{11}, \ldots, a_{nn})$. Then B is nonnegative and irreducible (because A is irreducible), and $A \geq \alpha I + B = \alpha(I + (1/\alpha)B)$. Then (8.4.1) ensures that $A^{n-1} \geq \alpha^{n-1}(I + (1/\alpha)B)^{n-1} > 0$. $\quad\square$

Exercise. If $A \in M_n$ is nonnegative has positive diagonal entries, and if the i, j entry of A^m is positive, explain why the i, j entry of A^{m+p} is positive for each integer $p \geq 1$.

Although an irreducible nonnegative matrix may have a reducible power, all powers of a nonnegative primitive matrix are primitive.

Lemma 8.5.5. *Let $A \in M_n$ be nonnegative and primitive. Then A^m is nonnegative and primitive for every integer $m \geq 1$.*

Proof. Since all sufficiently large powers of A are positive, the same is true for A^m for any m. If A^m were reducible, then A^{mp} would be reducible for all $p = 2, 3, \ldots$, and hence these matrices cannot be positive. This contradiction shows that no power of A can be reducible. $\quad\square$

The characterization in (8.5.2) is not a computationally effective test for primitivity since no upper bound on the powers to be computed is given. The following theorem provides a finite (but discouragingly large) upper bound.

Theorem 8.5.6. *Let $A \in M_n$ be nonnegative. If A is primitive, then $A^k > 0$ for some positive integer $k \leq (n-1)n^n$.*

Proof. Because A is irreducible, there is a directed path from the node P_1 in $\Gamma(A)$ back to itself; let k_1 be the shortest such path, so that $k_1 \leq n$. The matrix A^{k_1} has a positive entry in its 1,1 position, and any power of A^{k_1} also has a positive 1,1 entry. Primitivity of A and (8.5.5) ensure that A^{k_1} is irreducible, so there is a directed path from the node P_2 in $\Gamma(A^{k_1})$ back to itself; let $k_2 \leq n$ be the length of the shortest such path. The matrix $(A^{k_1})^{k_2} = A^{k_1 k_2}$ has positive 1,1 and 2,2 entries. Continue this process down the main diagonal to obtain a matrix $A^{k_1 \cdots k_n}$ (with each $k_1 \leq n$) that is irreducible and has positive diagonal entries. Lemma 8.5.4 ensures that $(A^{k_1 \cdots k_n})^{n-1} > 0$. Finally, observe that $k_1 \cdots k_n(n-1) \leq n^n(n-1)$. $\qquad\qquad\square$

If $A \in M_n$ is nonnegative and primitive, the least k such that $A^k > 0$ is the *index of primitivity* of A, which we denote by $\gamma(A)$. We know that $\gamma(A) \leq n^n(n-1)$, and that $\gamma(A) \leq n-1$ if the main diagonal of A is positive. The following theorem gives an upper bound that is much smaller than the former bound, and only twice as large as the latter one if A has only one positive diagonal entry. If there is at least one cycle in $\Gamma(A)$ that has length s, and if no cycle in $\Gamma(A)$ has length less than s, we say that *the shortest cycle in $\Gamma(A)$ has length s.*

Theorem 8.5.7. *Let $A \in M_n$ be nonnegative and primitive, and suppose that the shortest cycle in $\Gamma(A)$ has length s. Then $\gamma(A) \leq n + s(n-2)$, that is, $A^{n+s(n-2)} > 0$.*

Proof. Because A is irreducible, every node in $\Gamma(A)$ is contained in a cycle, and any shortest cycle has length at most n. We may assume that the distinct nodes in a shortest cycle are $P_1, P_2, \ldots, P_s$. Notice that $n + s(n-2) = n - s + s(n-1)$ and consider $A^{n-s+s(n-1)} = A^{n-s}(A^s)^{n-1}$. Partition $A^{n-s} = \begin{bmatrix} X_{11} & X_{12} \\ X_{21} & X_{22} \end{bmatrix}$ with $X_{11} \in M_s$ and $X_{22} \in M_{n-s}$. Because the nodes $P_1, \ldots, P_s$ comprise a cycle in $\Gamma(A)$, for each positive integer m and any $i \in \{1, \ldots, s\}$, there is a directed path in $\Gamma(A)$ of length m from P_i to *some* P_j with $j \in \{1, \ldots, s\}$. In particular, taking $m = n - s$, each row of X_{11} must contain at least one positive entry. For each $i \in \{s+1, \ldots, n\}$ there is a directed path in $\Gamma(A)$ of length $r \leq n - s$ (the number of nodes not in the cycle) from P_i (not in the cycle) to *some* node in the cycle. If $r < n - s$, one can go an additional $n - s - r$ steps around the cycle to obtain a directed path in $\Gamma(A)$ of length exactly $n - s$ from P_i to *some* node in the cycle. It follows that there is at least one nonzero entry in each row of X_{21}.

Now partition $(A^s)^{n-1} = \begin{bmatrix} Y_{11} & Y_{12} \\ Y_{21} & Y_{22} \end{bmatrix}$ with $Y_{11} \in M_s$ and $Y_{22} \in M_{n-s}$. Because $P_1, \ldots, P_s$ comprise a cycle in $\Gamma(A)$, there is a loop at each node $P_1, \ldots, P_s$ in $\Gamma(A^s)$. Since A is primitive, A^s is also primitive, and hence it is irreducible. Therefore, for any $i, j \in \{1, \ldots, n\}$ there is a directed path in $\Gamma(A^s)$ of length at most $n - 1$ from P_i to P_j. By first going a sufficient number of times around the loop at P_i, we can always construct such a path that has length exactly $n - 1$. It follows that $Y_{11} > 0$ and $Y_{12} > 0$.

To complete the argument, we compute

$$A^{n-s}(A^s)^{n-1} = \begin{bmatrix} X_{11} & X_{12} \\ X_{21} & X_{22} \end{bmatrix} \begin{bmatrix} Y_{11} & Y_{12} \\ Y_{21} & Y_{22} \end{bmatrix} \geq \begin{bmatrix} X_{11}Y_{11} & X_{11}Y_{12} \\ X_{21}Y_{11} & X_{21}Y_{12} \end{bmatrix}$$

and use (8.1.14) to conclude that $A^{n-s}(A^s)^{n-1} > 0$. $\qquad\square$

One consequence of (8.5.7) is a celebrated result of H. Wielandt, which gives a sharp upper bound for the index of primitivity.

Corollary 8.5.8 (Wielandt). *Let $A \in M_n$ be nonnegative. Then A is primitive if and only if $A^{n^2-2n+2} > 0$.*

Proof. If some power of A is positive, then A is primitive, so only the converse implication is of interest. If $n = 1$, the result is trivial, so assume that $n > 1$. If A is primitive, then it is irreducible and there are cycles in $\Gamma(A)$. If the shortest cycle in $\Gamma(A)$ has length n, then the length of every cycle in $\Gamma(A)$ is a multiple of n and (8.5.3) tells us that A cannot be primitive. Thus, the length of the shortest cycle in $\Gamma(A)$ is $n - 1$ or less, so (8.5.7) tells us that $\gamma(A) \leq n + s(n-2) \leq n + (n-1)(n-2) = n^2 - 2n + 2$. $\qquad\square$

Wielandt gave an example (see (8.5.P4)) to show that the bound $\gamma(A) \leq n^2 - 2n + 2$ is best possible for matrices that have a zero main diagonal.

We know that if A has a positive main diagonal, then it is primitive if and only if $A^{n-1} > 0$. The following result of Holladay and Varga uses the ideas employed in the proof of (8.5.7) to provide a bound on the index of primitivity if some, but perhaps not all, of the main diagonal entries are positive.

Theorem 8.5.9. *Let $A \in M_n$ be irreducible and nonnegative, and suppose that A has d positive main diagonal entries, $1 \leq d \leq n$. Then $A^{2n-d-1} > 0$; that is, $\gamma(A) \leq 2n - d - 1$.*

Proof. Under the stated hypotheses, A must be primitive, and $\Gamma(A)$ has d cycles with (minimum) length one. We may assume that $P_1, \ldots, P_d$ are the nodes in $\Gamma(A)$ that have loops. Consider $A^{2n-d-1} = A^{n-d}(A^1)^{n-1}$ and partition $A^{n-d} = \begin{bmatrix} X_{11} & X_{12} \\ X_{21} & X_{22} \end{bmatrix}$ and $A^{n-1} = \begin{bmatrix} Y_{11} & Y_{12} \\ Y_{21} & Y_{22} \end{bmatrix}$, in which $X_{11}, Y_{11} \in M_d$ and $X_{22}, Y_{22} \in M_{n-d}$. The argument in the proof of (8.5.7) shows that each row of the blocks X_{11} and X_{21} contains at least one nonzero entry, the blocks Y_{11} and Y_{12} are positive, and $A^{n-d}A^{n-1}$ is positive. $\qquad\square$

Exercise. Show that $A = \begin{bmatrix} 0 & 1 \\ 1 & 1 \end{bmatrix}$ is primitive. What are its eigenvalues? Compute the bounds on $\gamma(A)$ given by (8.5.7) and (8.5.9). What is the exact value of $\gamma(A)$?

If one wishes to verify that a given nonnegative matrix is primitive, one could check that it is irreducible and that Wielandt's condition (8.5.8) is satisfied. Matrices arising in practice frequently have a special structure that makes it easy to see whether the associated directed graph is strongly connected. Furthermore, if a nonnegative matrix is irreducible and some main diagonal entry is positive, then it must be primitive. However, if the matrix with a zero main diagonal is large and its entries have no special structure, then it may be necessary to use (8.4.1) or (8.5.8) to check for irreducibility or primitivity. In either case, a useful strategy is to square the matrix repeatedly until the resulting power exceeds the critical value ($n - 1$ or $n^2 - 2n + 2$, respectively). For example, if $n = 10$, then calculation of $(I + A)^2$, $(I + A)^4$, $(I + A)^8$, and $(I + A)^{16}$ is sufficient to verify irreducibility; this is four matrix multiplications instead of the eight required by a direct application of (8.4.1). Similarly, calculation of A^2, A^4, A^8, A^{16}, A^{32}, A^{64}, and A^{128} is sufficient to verify primitivity when $n = 10$; this is 7 matrix multiplications instead of 81. We are making implicit use of (8.5.P3) in these considerations.

Problems

8.5.P1 One sometimes encounters an alternative definition of primitivity: A nonnegative square matrix A is primitive if there is a positive integer m such that $A^m > 0$. Is this definition consistent with (8.5.0)?

8.5.P2 If $A \in M_n$ is nonnegative and primitive and $A^m = [a_{ij}^{(m)}]$, show that $\lim_{m \to \infty}(a_{ij}^{(m)})^{1/m} = \rho(A)$ for all $i, j = 1, \ldots, n$.

8.5.P3 If $A \in M_n$ is nonnegative and primitive, we know that A^m is primitive for any positive integer m. However, if $A, B \in M_n$ are nonnegative and primitive, give an example to show that AB need not be primitive.

8.5.P4 For $n \geq 3$, Wielandt's matrix $A = [a_{ij}] \in M_n$ has $a_{1,2} = a_{2,3} = \cdots = a_{n-1,n} = a_{n,1} = a_{n,2} = 1$; all other entries are zero. Construct $\Gamma(A)$ and use it to show that A is irreducible and primitive. Show that the 1, 1 entry of A^{n^2-2n+1} is zero and $A^{n^2-2n+2} > 0$.

8.5.P5 Let $A \in M_n$ be nonnegative and irreducible. Explain why A is primitive if at least one main diagonal entry is positive. Show that this sufficient condition is necessary for $n = 2$ but not for $n \geq 3$.

8.5.P6 Provide details for a proof of (8.5.9).

8.5.P7 Discuss the computational shortcuts suggested at the end of this section.

8.5.P8 If $A \in M_n$ is idempotent, then $A = \lim_{m \to \infty} A^m$. If A is nonnegative, irreducible, and idempotent, show that A is a rank-one positive matrix.

8.5.P9 Let $A \in M_n$ be nonnegative. Give an example to show that $\lim_{m \to \infty}(\rho(A)^{-1}A)^m$ can exist even if A is not primitive. Indeed, A can be reducible and can have multiple eigenvalues of maximum modulus.

8.5.P10 Prove the following partial converse of (8.5.1): If $A \in M_n$ is nonnegative and irreducible, and if $\lim_{m \to \infty}(\rho(A)^{-1}A)^m$ exists, then A is primitive.

8.5.P11 Show that $A = \begin{bmatrix} 0 & 1 \\ 1 & 0 \end{bmatrix}$ is irreducible but A^2 is reducible. Does this contradict (8.5.5)?

8.5.P12 Give an example of an irreducible nonnegative matrix $A \in M_n$ such that $\lim_{m \to \infty}(\rho(A)^{-1}A)^m$ does not exist.

8.5.P13 If $\epsilon > 0$ and if $A \in M_n$ is nonnegative and irreducible, prove that $A + \epsilon I$ is primitive. Conclude that every nonnegative irreducible matrix is a limit of nonnegative primitive matrices.

8.5.P14 A nonnegative matrix $A = [a_{ij}]$ is *combinatorially symmetric* provided that $a_{ij} > 0$ if and only if $a_{ji} > 0$ for all $i, j = 1, \ldots, n$. If A is combinatorially symmetric and primitive, show that $A^{2n-2} > 0$. Can you strengthen the bound for $\gamma(A)$, given more information about the cycle structure of $\Gamma(A)$?

8.5.P15 If n is prime and $A \in M_n$ is nonnegative, irreducible, and nonsingular, show that either (a) A is primitive or (b) A is similar to the companion matrix of $x^n - \rho(A)^n = 0$, so all of its eigenvalues have maximum modulus.

8.5.P16 One way to compute the Perron vector and spectral radius of a nonnegative matrix $A \in M_n$ is the power method:

$$x^{(0)} \quad \text{is an arbitrary positive vector,} \qquad \sum_{i=1}^{n} x_i^{(0)} = 1$$

$$y^{(m+1)} = Ax^{(m)} \quad \text{for all} \quad m = 0, 1, 2, \ldots$$

$$x^{(m+1)} = \frac{y^{(m+1)}}{\sum_{i=1}^{n} y_i^{(m+1)}} \quad \text{for all} \quad m = 0, 1, 2, \ldots$$

If A is primitive, show that the sequence of vectors $x^{(m)}$ converges to the (right) Perron vector of A and that the sequence of numbers $\sum_{i=1}^{n} y_i^{(m+1)}$ converges to the Perron root of A. What is the rate of convergence? Is the hypothesis of primitivity necessary?

8.5.P17 If $A \in M_n$ is nonnegative, show that primitivity of A depends only on the *location* of the zero entries and not on the magnitudes of the nonzero entries.

8.5.P18 If $A \in M_n$ is nonnegative, irreducible, and symmetric, show that A is primitive if and only if $A + \rho(A)I$ is nonsingular. In particular, this condition is met if A is positive semidefinite. Symmetric nonnegative matrices with 0s and 1s as entries arise naturally as adjacency matrices of undirected graphs.

8.5.P19 Calculate the eigenvalues and eigenvectors of each of the following matrices and categorize them according to the key concepts of the chapter (nonnegative, irreducible, primitive, positive, and so forth): $\begin{bmatrix} 1 & 1 \\ 1 & 1 \end{bmatrix}, \begin{bmatrix} 0 & 1 \\ 1 & 1 \end{bmatrix}, \begin{bmatrix} 1 & 0 \\ 1 & 1 \end{bmatrix}, \begin{bmatrix} 1 & 0 \\ 0 & 1 \end{bmatrix}, \begin{bmatrix} 0 & 1 \\ 1 & 0 \end{bmatrix}, \begin{bmatrix} 1 & 0 \\ 0 & 0 \end{bmatrix}, \begin{bmatrix} 0 & 1 \\ 0 & 0 \end{bmatrix}, \begin{bmatrix} 0 & 0 \\ 0 & 0 \end{bmatrix}.$

8.5.P20 In the proof of (8.5.7), explain why each *column* of X_{11} and X_{12} contains at least one nonzero entry, and why $Y_{21} > 0$.

Further Reading. For a proof of Romanovsky's theorem, see V. Romanovsky, Recherches sur les chaînes de Markoff, *Acta Math.* 66 (1936) 147–251. For a nonnegative primitive matrix $A \in M_n$, Wielandt's theorem says that $A^{(n-1)^2+1} > 0$; for a proof that $A^{(m-1)^2+1} > 0$, in which m is the degree of the minimal polynomial of A, see J. Shen, Proof of a conjecture about the exponent of primitive matrices, *Linear Algebra Appl.* 216 (1995) 185–203.

8.6 A general limit theorem

Even if a nonnegative matrix A is irreducible, the normalized powers of A need have no limit, as the example $A = \begin{bmatrix} 0 & 1 \\ 1 & 0 \end{bmatrix}$ shows. Nevertheless, there is a precise sense in which, *on the average*, this limit does exist.

Exercise. Let $\theta \in (0, 2\pi)$. Show that $(1 - e^{i\theta}) \sum_{m=1}^{N} e^{im\theta} = e^{i\theta} - e^{i(N+1)\theta}$ and conclude that

$$\frac{1}{N} \sum_{m=1}^{N} e^{im\theta} = \frac{e^{i\theta} - e^{i(N+1)\theta}}{N(1 - e^{i\theta})} \to 0 \text{ as } N \to \infty$$

Exercise. Let $B \in M_n$ and suppose that $\rho(B) < 1$. Show that $(I - B) \sum_{m=1}^{N} B^m = B - B^{N+1}$ and conclude that

$$\frac{1}{N} \sum_{m=1}^{N} B^m = \frac{1}{N}(B - B^{N+1})(I - B)^{-1} \to 0 \text{ as } N \to \infty$$

Theorem 8.6.1. *Let $A \in M_n$ be irreducible and nonnegative, let $n \geq 2$, and let x and y, respectively, be the right and left Perron vectors of A. Then*

$$\lim_{N \to \infty} \frac{1}{N} \sum_{m=1}^{N} (\rho(A)^{-1} A)^m = xy^T \tag{8.6.2}$$

Moreover, there exists a finite positive constant $C = C(A)$ such that

$$\left\| \frac{1}{N} \sum_{m=1}^{N} (\rho(A)^{-1} A)^m - xy^T \right\|_{\infty} \leq \frac{C}{N} \tag{8.6.3}$$

for all $N = 1, 2, \ldots$.

Proof. If A is primitive, $\rho(A)^{-1} A$ can be factored as in (8.2.7a), in which x is the first column of S and y is the first column of S^{-1}. We have

$$\frac{1}{N} \sum_{m=1}^{N} (\rho(A)^{-1} A)^m = S \begin{bmatrix} 1 & 0 \\ 0 & \frac{1}{N} \sum_{m=1}^{N} B^m \end{bmatrix} S^{-1}$$

in which $\rho(B) < 1$, so the preceding exercise ensures that

$$\frac{1}{N} \sum_{m=1}^{N} (\rho(A)^{-1} A)^m \to S \begin{bmatrix} 1 & 0 \\ 0 & 0_{n-1} \end{bmatrix} S^{-1} = xy^T$$

Now suppose that A has exactly $k > 1$ eigenvalues of maximum modulus and let $\theta = 2\pi/k$. Corollary 8.4.6(c) ensures that the maximum-modulus eigenvalues of $\rho(A)^{-1} A$ are $1, e^{i\theta}, e^{2i\theta}, \ldots, e^{(k-1)i\theta}$, and each is a simple eigenvalue. Thus, there is a nonsingular $S \in M_n$ such that x is its first column, y is the first column of S^{-1}, and

$$\rho(A)^{-1} A = S([1] \oplus [e^{i\theta}] \oplus \cdots \oplus [e^{(k-1)i\theta}] \oplus B)S^{-1}$$

in which $B \in M_{n-k}$ and $\rho(B) < 1$. The preceding exercises ensure that

$$\frac{1}{N} \sum_{m=1}^{N} (\rho(A)^{-1} A)^m = S([1] \oplus [\lambda_{1,N}] \oplus \cdots \oplus [\lambda_{k-1,N}] \oplus \mathcal{B}_N)S^{-1}$$

in which

$$
\begin{aligned}
\lambda_{1,N} &= \frac{e^{i\theta} - e^{i(N+1)\theta}}{N(1-e^{i\theta})} \to 0 \text{ as } N \to \infty \\
\vdots \quad &\qquad \vdots \qquad\qquad\qquad\qquad \vdots \\
\lambda_{k-1,N} &= \frac{e^{i(k-1)\theta} - e^{i(N+1)(k-1)\theta}}{N(1-e^{i(k-1)\theta})} \to 0 \text{ as } N \to \infty \\
\mathcal{B}_N &= \frac{1}{N}(B - B^{N+1})(I-B)^{-1} \to 0 \text{ as } N \to \infty
\end{aligned}
\tag{8.6.4}
$$

Therefore,

$$
\frac{1}{N}\sum_{m=1}^{N}(\rho(A)^{-1}A)^m \to S([1] \oplus [0] \oplus \cdots \oplus [0] \oplus 0_{n-k})S^{-1} = xy^T
$$

The bound in (8.6.3) is revealed in the representation

$$
\begin{aligned}
&\frac{1}{N}\sum_{m=1}^{N}(\rho(A)^{-1}A)^m - xy^T \\
&= S([1] \oplus [\lambda_{1,N}] \oplus \cdots \oplus [\lambda_{k-1,N}] \oplus \mathcal{B}_N)S^{-1} - S([1] \oplus 0_{n-1})S^{-1} \\
&= S([0] \oplus [\lambda_{1,N}] \oplus \cdots \oplus [\lambda_{k-1,N}] \oplus \mathcal{B}_N)S^{-1} \\
&= \frac{1}{N}S\left([0] \oplus [N\lambda_{1,N}] \oplus \cdots \oplus [N\lambda_{k-1,N}] \oplus N\mathcal{B}_N\right)S^{-1}
\end{aligned}
\tag{8.6.5}
$$

The identities (8.6.4) ensure that the matrix factor in (8.6.5) is bounded as $N \to \infty$. $\square$

Problems

8.6.P1 Let $A = \begin{bmatrix} 0 & 1 \\ 1 & 0 \end{bmatrix}$. Compute the direct sum in the matrix factor in (8.6.5). Explain why the bound in (8.6.1) cannot be improved.

8.6.P2 Suppose that $A \in M_n$ is irreducible and nonnegative, let $n \geq 2$, and write $A^m = [a_{ij}^{(m)}]$ for $m = 1, 2, \ldots$. For each pair i, j, show that $a_{ij}^{(m)} > 0$ for infinitely many values of m. Give an example to show that there may also be infinitely many values of m for which $a_{ij}^{(m)} = 0$.

8.7 Stochastic and doubly stochastic matrices

A nonnegative matrix $A \in M_n$ with the property that $Ae = e$, that is, all its row sums are $+1$, is said to be a *(row) stochastic matrix*; each row may be thought of as a discrete probability distribution on a sample space with n points. A *column stochastic matrix* is the transpose of a row stochastic matrix, that is, $e^T A = e^T$; such matrices arose in the intercity population migration model discussed in (8.0). Stochastic matrices also arise in the study of Markov chains and in a variety of modeling problems in economics and operations research.

The defining identity $Ae = e$ and (8.3.4) tell us that $+1$ is not only an eigenvalue of a stochastic matrix, but also its spectral radius.

The set of stochastic matrices in M_n is a compact set (its entries all lie in the closed real interval $[0, 1]$) that is also convex: If A and B are stochastic and $\alpha \in [0, 1]$, then

$$(\alpha A + (1 - \alpha)B)e = \alpha Ae + (1 - \alpha)Be = \alpha e + (1 - \alpha)e = e$$

Thus, the stochastic matrices in M_n form an easily recognized family of nonnegative matrices with a particular positive eigenvector in common. Nonnegative matrices with a positive eigenvector have many special properties (for example, see (8.1.30), (8.1.31), (8.1.33), (8.3.4)), which therefore are possessed by all stochastic matrices.

Exercise. Explain why an n-by-n stochastic matrix has at least n nonzero entries.

A stochastic matrix $A \in M_n$ such that A^T is also stochastic is said to be *doubly stochastic*; all row and column sums are $+1$. The set of doubly stochastic matrices in M_n is the intersection of two compact convex sets, so it is compact and convex. A nonnegative matrix $A \in M_n$ is doubly stochastic if and only if both $Ae = e$ and $e^T A = e^T$.

We encountered two special classes of doubly stochastic matrices in (4.3.49) and (6.3.5): Orthostochastic or unistochastic matrices of the form $A = [|u_{ij}|^2]$, in which $U = [u_{ij}] \in M_n$ is either real orthogonal or unitary.

Another special class of doubly stochastic matrices is the set (group) of permutation matrices. The permutation matrices are the fundamental and prototypical doubly stochastic matrices, for Birkhoff's theorem says that any doubly stochastic matrix is a convex combination of finitely many permutation matrices.

Exercise. Suppose that $n \geq 2$, $A = [a_{ij}] \in M_n$ is doubly stochastic, and some $a_{ii} = 1$. Explain why (a) $a_{ki} = a_{ik} = 0$ for all $k \in \{1, \ldots, n\}$ such that $k \neq i$; (b) A is permutation similar to $[1] \oplus B$, in which B is doubly stochastic; (c) the main diagonal entries of B are obtained from the main diagonal entries of A by removing one entry equal to $+1$; and (d) the characteristic polynomials of A and B are related by the identity $p_A(t) = (t - 1)p_B(t)$.

In preparation for our proof of Birkhoff's theorem, we establish the following lemma.

Lemma 8.7.1. *Let $A = [a_{ij}] \in M_n$ be a doubly stochastic matrix that is not the identity matrix. There is a permutation σ of $\{1, \ldots, n\}$ that is not the identity permutation and is such that $a_{1\sigma(1)} \cdots a_{n\sigma(n)} > 0$.*

Proof. Suppose that every permutation σ of $\{1, \ldots, n\}$ that is not the identity permutation σ_0 has the property that $a_{1\sigma(1)} \cdots a_{n\sigma(n)} = 0$. This assumption and (0.3.2.1) permit us to compute the characteristic polynomial of A:

$$p_A(t) = \det(tI - A) = \prod_{i=1}^{n}(t - a_{ii}) + \sum_{\sigma \neq \sigma_0}\left(\operatorname{sgn}\sigma \prod_{i=1}^{n}(-a_{i\sigma(i)})\right)$$

$$= \prod_{i=1}^{n}(t - a_{ii})$$

It follows that the main diagonal entries of A are its eigenvalues. Since $+1$ is an eigenvalue of A, *at least one* of its main diagonal entries is $+1$. The preceding exercise ensures that A is permutation similar to $[1] \oplus B$, in which $B = [b_{ij}] \in M_{n-1}$ is doubly stochastic; its main diagonal entries are obtained from the main diagonal entries of A by omitting one $+1$ entry; $+1$ is an eigenvalue of B; and $p_B(t) = p_A(t)/(t-1) = \prod_{i=1}^{n-1}(t - b_{ii})$. Applying the preceding argument to B shows that some $b_{ii} = 1$, so *at least two* main diagonal entries of A are $+1$. Continuing in this way, after at most $n-1$ steps, we conclude that *every* main diagonal entry of A is $+1$, so $A = I$. This contradiction shows that some product $a_{1\sigma(1)} \cdots a_{n\sigma(n)}$ must be positive. $\qquad\square$

In our proof of Birkhoff's theorem, we use the preceding lemma to extract a positive multiple of a permutation matrix from a given doubly stochastic matrix A. We do this extraction in such a way as to create a new doubly stochastic matrix that has at least one more zero entry than A. Iterating this extraction process leads in finitely many steps to a representation of A as a finite convex combination of permutation matrices.

Exercise. Let $A \in M_n$ be doubly stochastic. If A has exactly n positive entries, explain why it is a permutation matrix. If A is not a permutation matrix, explain why it has at most $n^2 - n - 1$ zero entries.

Theorem 8.7.2 (Birkhoff). *A matrix $A \in M_n$ is doubly stochastic if and only if there are permutation matrices $P_1, \ldots, P_N \in M_n$ and positive scalars $t_1, \ldots, t_N \in \mathbf{R}$ such that $t_1 + \cdots + t_N = 1$ and*

$$A = t_1 P_1 + \cdots + t_N P_N \tag{8.7.3}$$

Moreover, $N \le n^2 - n + 1$.

Proof. The sufficiency of the representation (8.7.3) is clear; we prove its necessity by exhibiting an algorithm that constructs it in finitely many steps.

If A is a permutation matrix, there is nothing to prove. If not, the preceding lemma ensures that there is a nonidentity permutation σ of $\{1, \ldots, n\}$ such that $a_{1\sigma(1)} \cdots a_{n\sigma(n)} > 0$. Let $\alpha_1 = \min\{a_{1\sigma(1)}, \ldots, a_{n\sigma(n)}\}$ and define the permutation matrix $P_1 = [p_{ij}] \in M_n$ by $p_{i\sigma(i)} = 1$ for each $i = 1, \ldots, n$. If $\alpha_1 = 1$, then A is a permutation matrix, so $0 < \alpha_1 < 1$. Let $A_1 = (1 - \alpha_1)^{-1}(A - \alpha_1 P_1)$ and check that A_1 is doubly stochastic and has at least one more zero entry than A, and $A = (1 - \alpha_1)A_1 + \alpha_1 P_1$. If A_1 is a permutation matrix, we stop because we have achieved a representation of the form (8.7.3) with two summands. If not, we repeat this argument and find $\alpha_2 \in (0, 1)$ and a permutation matrix P_2 such that $A_2 = (1 - \alpha_2)^{-1}(A_1 - \alpha_2 P_2)$ is doubly stochastic and has at least one more zero entry than A_1, and

$$A = (1 - \alpha_1)(1 - \alpha_2)A_2 + (1 - \alpha_1)\alpha_2 P_2 + \alpha_1 P_1$$

If A_2 is a permutation matrix, we stop because we have achieved a representation of the form (8.7.3) with three summands. If not, we iterate until we are forced to stop because some A_k is a permutation matrix, at which point we have a representation of the form (8.7.3) with $k + 1$ summands. Since A_k has at least k zero entries, and since a doubly stochastic matrix has at most $n^2 - n$ zero entries, we can have at most $n^2 - n$ iterations and at most $n^2 - n + 1$ summands in (8.7.3). $\qquad\square$

The following corollary is an important consequence of Birkhoff's theorem that has many applications; for example, it was a key element in our proof of the

Hoffman–Wielandt theorem (6.3.5). It assures us that if we want to find the maximum value of a convex function on the set of doubly stochastic matrices, it suffices to restrict our attention to permutation matrices. See Appendix B for a discussion of convex functions, convex sets, and extreme points.

Corollary 8.7.4. *The maximum (respectively, minimum) of a convex (respectively, concave) real-valued function on the set of doubly stochastic n-by-n matrices is attained at a permutation matrix.*

Proof. Let f be a convex real-valued function on the set of n-by-n doubly stochastic matrices, let A be a doubly stochastic matrix at which f attains its maximum value, represent $A = t_1 P_1 + \cdots + t_N P_N$ as a convex combination of permutation matrices, and let k be an index such that $f(P_k) = \max\{f(P_i) : i = 1, \ldots, n\}$. Then

$$f(A) = f(t_1 P_1 + \cdots + t_N P_N) \leq t_1 f(P_1) + \cdots + t_N f(P_N)$$
$$\leq t_1 f(P_k) + \cdots + t_N f(P_k) = (t_1 + \cdots + t_N) f(P_k) = f(P_k)$$

Since f achieves its maximum at A, we have $f(A) = f(P_k)$. A similar argument validates the assertion about the minimum of a concave function.   □

A nonnegative matrix $A \in M_n$ is *doubly substochastic* if $Ae \leq e$ and $e^T A \leq e^T$, that is, all row and column sums are at most one. The following lemma shows that any doubly substochastic matrix is dominated by a doubly stochastic matrix.

Lemma 8.7.5. *Let $A \in M_n$ be doubly substochastic. There is a doubly stochastic matrix $S \in M_n$ such that $A \leq S$.*

Proof. For any doubly substochastic matrix $S \in M_n$, let $N(S)$ denote the number of row sums and column sums of S that are less than one, that is, the number of entries of the vectors Ae and $A^T e$ whose entries are less than one.

Let $A \in M_n$ be doubly substochastic. Then A is doubly stochastic if and only if $N(A) = 0$. If $N(A) > 0$ and we can show that there is always a doubly substochastic $C \in M_n$ such that both $A \leq C$ and $N(C) < N(A)$, then the assertion of the lemma would follow from a finite induction.

Let $A = [a_{ij}] \in M_n$ be doubly substochastic and suppose that $N(A) > 0$. Since the sum of the row sums of A equals the sum of its column sums, there must be a row (say, row i) and a column (say, column j) whose sums are each less than one. Modify A by increasing the entry a_{ij} until either the ith row sum or the jth column sum (or both) equals one; let C be this modified matrix. Then C is doubly substochastic, $A \leq C$, and $N(C) < N(A)$.   □

In our final result, we deduce the square case of von Neumann's trace theorem (7.4.1.1) from the preceding lemma and (8.7.4).

Exercise. Let $U = [u_{ij}]$, $V = [v_{ij}] \in M_n$ be unitary and let $S = [|u_{ij} v_{ji}|]$. Show that S is doubly substochastic. *Hint*: The Cauchy–Schwarz inequality.

Theorem 8.7.6 (von Neumann). *Let the ordered singular values of* $A, B \in M_n$ *be* $\sigma_1(A) \geq \cdots \geq \sigma_n(A)$ *and* $\sigma_1(B) \geq \cdots \geq \sigma_n(B)$. *Then*

$$\operatorname{Re} \operatorname{tr}(AB) \leq \sum_{i=1}^{n} \sigma_i(A) \sigma_i(B)$$

Proof. Let $A = V_1 \Sigma_A W_1^*$ and $B = V_2 \Sigma_B W_2^*$ be singular value decompositions, in which $V_1, W_1, V_2, W_2 \in M_n$ are unitary, $\Sigma_A = \operatorname{diag}(\sigma_1(A), \ldots, \sigma_n(A))$, and $\Sigma_B = \operatorname{diag}(\sigma_1(B), \ldots, \sigma_n(B))$. Let $U = W_1^* V_2 = [u_{ij}]$ and $V = W_2^* V_1 = [v_{ij}]$. Then

$$\operatorname{Re} \operatorname{tr}(AB) = \operatorname{Re} \operatorname{tr}(V_1 \Sigma_A W_1^* V_2 \Sigma_B W_2^*)$$

$$= \operatorname{Re} \operatorname{tr}(\Sigma_A W_1^* V_2 \Sigma_B W_2^* V_1) = \operatorname{Re} \operatorname{tr}(\Sigma_A U \Sigma_B V)$$

$$= \operatorname{Re} \sum_{i,j=1}^{n} \sigma_i(A) \sigma_j(B) u_{ij} v_{ji} = \sum_{i,j=1}^{n} \sigma_i(A) \sigma_j(B) \operatorname{Re}(u_{ij} v_{ji})$$

$$\leq \sum_{i,j=1}^{n} \sigma_i(A) \sigma_j(B) |u_{ij} v_{ji}|$$

The preceding exercise tells us that the matrix $[|u_{ij} v_{ji}|]$ is doubly substochastic, and (8.7.5) ensures that there is a doubly stochastic matrix C such that $[|u_{ij} v_{ji}|] \leq C = [c_{ij}]$. Therefore,

$$\operatorname{Re} \operatorname{tr}(AB) \leq \sum_{i,j=1}^{n} \sigma_i(A) \sigma_j(B) c_{ij}$$

$$\leq \max\{ \sum_{i,j=1}^{n} \sigma_i(A) \sigma_j(B) s_{ij} : S = [s_{ij}] \text{ is doubly stochastic}\}$$

The function $f(S) = \sum_{i,j=1}^{n} \sigma_i(A) \sigma_j(B) s_{ij}$ is a linear (and therefore convex) function on the set of doubly stochastic matrices, so (8.1.4) tells us that it attains its maximum at a permutation matrix $P = [p_{ij}]$. If π is the permutation of $\{1, \ldots, n\}$ such that $p_{ij} = 1$ if and only if $j = \pi(i)$, then

$$\operatorname{Re} \operatorname{tr}(AB) \leq \sum_{i,j=1}^{n} \sigma_i(A) \sigma_j(B) p_{ij} = \sum_{i=1}^{n} \sigma_i(A) \sigma_{\pi(i)}(B)$$

$$\leq \sum_{i=1}^{n} \sigma_i(A) \sigma_i(B)$$

The final inequality follows from (4.3.52). □

Problems

8.7.P1 Show that the sets of stochastic and doubly stochastic matrices in M_n each constitute a semigroup under matrix multiplication; that is, if $A, B \in M_n$ are (doubly) stochastic, then AB is (doubly) stochastic.

8.7.P2 Let $A \in M_n$ be stochastic and let λ be any eigenvalue of A such that $|\lambda| = 1$; $\lambda = 1$ is one such eigenvalue, but there could be others. Show that A is power bounded and conclude that λ is a semisimple eigenvalue of A.

8.7.P3 Let $A \in M_n$ be a nonnegative nonzero matrix that has a positive eigenvector $x = [x_i]$, and let $D = \mathrm{diag}(x_1, \ldots, x_n)$. Show that $\rho(A)^{-1}D^{-1}AD$ is stochastic. This observation permits many questions about nonnegative matrices with a positive eigenvector to be reduced to questions about stochastic matrices.

8.7.P4 Let $A \in M_n$ be a nonnegative nonzero matrix that has a positive eigenvector. Explain why every maximum-modulus eigenvalue of A is semisimple.

8.7.P5 Let $A \in M_n$ be doubly stochastic and let $\sigma_1(A)$ be its largest singular value. Show in two ways that $\sigma_1(A) = \rho(A) = 1$, that is, *a doubly stochastic matrix is a spectral matrix*. (a) Use the representation (8.7.3). (b) Use (2.6.P29).

8.7.P6 Let $\|\| \cdot \|\|$ be the matrix norm on M_n induced by a permutation-invariant norm $\|\cdot\|$ on $\mathbf{R}^n$. Show that $\|\|A\|\| = 1$ for every doubly stochastic matrix in M_n.

8.7.P7 Show that any permutation matrix is an extreme point of the convex set of doubly stochastic matrices. What more can you say if A is a permutation matrix?

8.7.P8 Explain why a matrix is an extreme point of the compact convex set of doubly stochastic n-by-n matrices *if and only if* it is a permutation matrix.

8.7.P9 Let $A \in M_n$ be doubly stochastic. (a) Show that A cannot have exactly $n + 1$ positive entries. (b) If A is not a permutation matrix, explain why it has at most $n^2 - n - 2$ zero entries.

8.7.P10 Show that a 2-by-2 doubly stochastic matrix is symmetric and has equal diagonal entries.

8.7.P11 Show that the representation (8.7.3) need not be unique.

8.7.P12 Let $A \in M_n$ be doubly stochastic, symmetric, and positive semidefinite; let $A^{1/2}$ be its positive semidefinite square root. (a) Show that $A^{1/2}e = e$, so all the row (and column) sums of $A^{1/2}$ are $+1$. (b) Although $A^{1/2}$ need not be nonnegative, show that it *is* nonnegative (and hence doubly stochastic) if $n = 2$.

8.7.P13 Let $\|\| \cdot \|\|$ be a unitarily invariant matrix norm on M_n. Show that $\|\|A\|\| \leq \|\|I\|\|$ for every doubly stochastic matrix $A \in M_n$.

8.7.P14 If $A \in M_n$ is doubly stochastic and reducible, show that A is permutation similar to a matrix of the form $A_1 \oplus A_2$, in which both A_1 and A_2 are doubly stochastic.

8.7.P15 The upper bound in (8.7.2) can be reduced from $n^2 - n + 1$ to $(n^2 - n + 1) - (n - 1) = n^2 - 2n + 2$. Provide details: (a) Every n-by-n doubly stochastic matrix $S = [s_{ij}]$ is a solution of the $2n - 1$ linear equations $\sum_{k=1}^n s_{ik} = 1$, $i = 1, \ldots, n$ and $\sum_{k=1}^n s_{ki} = 1$, $i = 1, \ldots, n - 1$. (b) Write these equations in the form $A \,\mathrm{vec}\, S = e \in \mathbf{R}^{2n-1}$; see (0.7.7). (c) $A \in M_{2n-1,n^2}$ has full row rank. (d) $\dim \mathrm{nullspace}\, A = n^2 - 2n + 1$. (e) The set of doubly stochastic n-by-n matrices may be thought of as a convex polyhedron in $\mathbf{R}^{n^2-2n+1}$. (f) (8.7.3) and Carathéodory's theorem (see Appendix B) imply that any doubly stochastic n-by-n matrix is a convex combination of at most $n^2 - 2n + 2$ permutation matrices.

Notes and Further Readings. Theorem 8.7.2 appears in G. Birkhoff, Tres observaciones sobre el álgebra lineal, *Univ. Nac. Tucumán Rev. Ser. A* 5 (1946) 147–150. In 1916,

D. Kőnig published a theorem that is equivalent to (8.7.2) for matrices with nonnegative rational entries; see p. 239 of the 1936 book Kőnig (1936) or p. 381 of Kőnig (1990). Consequently, (8.7.2) is sometimes referred to as the Birkhoff–Kőnig theorem. The only nonconstructive part of the proof of (8.7.2) is the identification of a particular permutation σ of $\{1, \ldots, n\}$ such that $a_{1\sigma(1)} \cdots a_{n\sigma(n)} > 0$. A constructive algorithm to identify such a permutation is described on pp. 64–65 of Bapat and Raghavan (1997); this book provides a comprehensive discussion of doubly stochastic matrices (see chapter 2) and many references to the literature.

APPENDIX A
Complex Numbers

A *complex number* has the form $z = a + ib$, in which a and b are real numbers and i is a formal symbol that satisfies the relation $i^2 = -1$. The real number a is the *real part* of z and is denoted by $\operatorname{Re} z$; the real number b is the *imaginary part* of z and is denoted by $\operatorname{Im} z$. The *complex conjugate* of the complex number $z = a + ib$ is $\bar{z} = a - ib$. Addition and multiplication of complex numbers $z_1 = a_1 + ib_1$ and $z_2 = a_2 + ib_2$ are defined by

$$z_1 + z_2 = (a_1 + a_2) + i(b_1 + b_2), \quad z_1 z_2 = a_1 a_2 - b_1 b_2 + i(a_1 b_2 + a_2 b_1)$$

Addition is the result of adding real parts and imaginary parts separately; multiplication is the result of algebraic expansion together with the relation $i^2 = -1$. The additive inverse of $z = a + ib$ is $-z = -a + i(-b)$, and, as long as $z \neq 0 = 0 + i0$, the multiplicative inverse of z is

$$\frac{1}{z} = \frac{a - ib}{a^2 + b^2} = \frac{a}{a^2 + b^2} + i\left(\frac{-b}{a^2 + b^2}\right)$$

Subtraction and division of complex numbers z_1 and z_2 are defined by

$$z_1 - z_2 = z_1 + (-z_2), \quad \frac{z_1}{z_2} = z_1\left(\frac{1}{z_2}\right) = \frac{z_1 \bar{z}_2}{z_2 \bar{z}_2}$$

The set of all complex numbers is denoted by $\mathbf{C}$; the operations of addition and multiplication are commutative, and $\mathbf{C}$ constitutes a field under these operations, with $0 = 0 + i0$ as the additive identity and $1 = 1 + i0$ as the multiplicative identity. The real numbers $\mathbf{R}$ form a subfield of $\mathbf{C}$. The *modulus* (or *absolute value*) of z, denoted $|z|$, is the nonnegative real number $|z| = +(z\bar{z})^{1/2} = +((\operatorname{Re} z)^2 + (\operatorname{Im} z)^2)^{1/2}$; $|z| = 0$ if and only if $z = 0$. If $z_2 \neq 0$, the quotient z_1/z_2 is $(1/|z_2|^2)z_1\bar{z}_2$. The operations of multiplication and complex conjugation commute ($\overline{z_1 z_2} = \bar{z}_1 \bar{z}_2$), $\overline{(\bar{z})} = z$, $\operatorname{Re}(z_1 + z_2) = \operatorname{Re} z_1 + \operatorname{Re} z_2$, and $\operatorname{Im}(z_1 + z_2) = \operatorname{Im} z_1 + \operatorname{Im} z_2$. We have $\operatorname{Re} z = \frac{1}{2}(z + \bar{z})$ and $\operatorname{Im} z = \frac{1}{2i}(z - \bar{z})$. The real numbers are those $z \in \mathbf{C}$ such that $\operatorname{Im} z = 0$, or equivalently, such that $z = \bar{z}$. For any $z \in \mathbf{C}$, $\operatorname{Re} z \leq |z|$, with equality if and only if z is real and nonnegative, in which case $z = |z|$.

Geometrically, the complex numbers $\mathbf{C}$ may be thought of as a Cartesian (coordinate) plane with origin at 0, a "real axis" ("x-axis"), and an "imaginary axis" ("y-axis"). A complex number $z = a + ib$ may be identified with the real ordered pair (a, b) (rectangular coordinates) that specifies a point in the Cartesian plane. The *real axis* is $\{z : \operatorname{Im} z = 0\}$, and the *imaginary axis* is $\{z : \operatorname{Re} z = 0\}$. The projection of $z \in \mathbf{C}$ onto the real axis (imaginary axis) is $\operatorname{Re} z$ ($i \operatorname{Im} z$). Complex conjugation is reflection across the real axis, and $|z|$ is the Euclidean distance of z from the origin. The open (closed) *right half-plane* of $\mathbf{C}$ is $\{z \in \mathbf{C} : \operatorname{Re} z > (\geq) 0\}$, and the open (closed) *upper half-plane* of $\mathbf{C}$ is $\{z \in \mathbf{C} : \operatorname{Im} z > (\geq) 0\}$. The *unit disc* of $\mathbf{C}$ is $\{z \in \mathbf{C} : |z| \leq 1\}$, and the disc about $a \in \mathbf{C}$ of radius r is $\{z \in \mathbf{C} : |z - a| \leq r\}$.

The complex plane may also be described by *polar coordinates*: the position of $z \in \mathbf{C}$ is specified by the radius $r = |z|$ of the circle about the origin on which z lies, and the angle θ, measured counterclockwise from the real line, of the ray from the origin on which z lies. The polar coordinates of z are (r, θ). The angle $\theta = \arg z$ is the *argument* of z. The notation $z = re^{i\theta}$ is used, in which $e^{i\theta} = \cos\theta + i\sin\theta$. We have $|e^{i\theta}| = +(\cos^2\theta + i\sin^2\theta)^{1/2} = 1$, $(e^{i\theta})^{-1} = e^{-i\theta}$, and $|e^{i\theta}z| = |z|$. Since $e^{i\theta} = e^{i(\theta \pm 2n\pi)}$, $n = 1, 2, \ldots$, $\arg z$ is only determined $\operatorname{mod} 2\pi$. The transformation from polar to rectangular coordinates for $z = a + ib = re^{i\theta}$ is $a = r\cos\theta$ and $b = r\sin\theta$. The transformation from rectangular to polar coordinates is $r = |z| = (a^2 + b^2)^{1/2}$ and, if $r \neq 0$, $\theta = \arcsin\frac{b}{r} = \arg z$ (take $0 \leq \theta < 2\pi$ for the *principal value* of the argument). The unit disc in $\mathbf{C}$ is $\{re^{i\theta} : 0 \leq r \leq 1 \text{ and } 0 \leq \theta < 2\pi\}$. For each $z \in \mathbf{C}$, there is a real number θ such that $e^{-i\theta}z = |z|$: If $z \neq 0$, take $\theta = \arg z$; if $z = 0$, any real θ will do. For a given $z \in \mathbf{C}$, $e^{-i\theta}z = |z|$ if and only if z lies on the ray $\{re^{i\theta} : r \geq 0\}$.

For any given complex numbers $z_1, \ldots, z_m$ the *triangle inequality* is $|z_1 + \cdots + z_m| \leq |z_1| + \cdots + |z_m|$. To prove this basic inequality and identify the case of equality, let θ be a real number such that $e^{-i\theta}(z_1 + \cdots + z_m) = |z_1 + \cdots + z_m|$. Then

$$|z_1 + \cdots + z_m| = \operatorname{Re}|z_1 + \cdots + z_m| = \operatorname{Re}(e^{-i\theta}(z_1 + \cdots + z_m))$$
$$= \operatorname{Re}(e^{-i\theta}z_1) + \cdots + \operatorname{Re}(e^{-i\theta}z_m)$$
$$\leq |e^{-i\theta}z_1| + \cdots + |e^{-i\theta}z_m| = |z_1| + \cdots + |z_m|$$

with equality if and only if $\operatorname{Re}(e^{-i\theta}z_k) = |e^{-i\theta}z_k|$ for each $k = 1, \ldots, m$, that is, if and only if each $z_k = e^{i\theta}|z_k|$ lies on the same ray $\{re^{i\theta} : r \geq 0\}$.

Convex Sets and Functions

Let V be a vector space over a field that contains the real numbers. A *convex combination* of a selection $v_1, \ldots, v_k \in V$ of elements of V is a linear combination whose coefficients are real, nonnegative, and sum to 1:

$$\alpha_1 v_1 + \cdots + \alpha_k v_k; \quad \alpha_1, \ldots, \alpha_k \geq 0, \quad \sum_{i=i}^{k} \alpha_i = 1$$

A subset K of V is said to be *convex* if any convex combination of any selection of elements from K lies in K. Equivalently, K is convex if all convex combinations of *pairs* of points in K are again in K. Geometrically, this may be interpreted as saying that the line segment joining any two points of K must lie in K; that is, K has no "dents" or "holes." A convex set K for which $\alpha x \in K$ whenever $\alpha > 0$ and $x \in K$ is called a *convex cone* (equivalently, positive linear combinations from K are in K). It is straightforward to verify that both the set sum and the intersection of two convex sets (respectively, convex cones) is again a convex set (respectively, convex cone).

Now let V be a real or complex vector space with a given norm, so one can speak of open, closed, and compact sets in V. An *extreme point* of a closed convex set K is a point $z \in K$ that may be written as a convex combination of points from K in only a trivial way; that is, $z = \alpha x + (1 - \alpha)y$, $0 < \alpha < 1$, $x, y \in K$, implies $x = y = z$. A closed convex set may have a finite number of extreme points (for example, a polyhedron), infinitely many extreme points (for example, a closed disc), or no extreme points (for example, the closed upper half-plane in $\mathbf{R}^2$). A compact convex set always has extreme points, however. The *convex hull* of a set S of points in V, denoted $\mathrm{Co}(S)$, is the set of all convex combinations of all selections of points from S, or, equivalently, the "smallest" convex set (intersection of all convex sets) containing S. The Krein–Milman theorem says that a compact convex set is the closure of the convex hull of its extreme points. A compact convex set is said to be *finitely generated* if it has finitely many extreme points; the extreme points are the *generators* of the convex set. A theorem of Carathéodory (sometimes called the Carathéodory–Steinitz theorem) says that any point in the convex hull of a set $S \subset \mathbf{R}^n$ is a convex combination of at most $n + 1$ points of S.

Now suppose that V is a real inner product space with inner product $\langle \cdot, \cdot \rangle$. The *separating hyperplane theorem* states that if K_1, $K_2 \subseteq V$ are two given nonempty nonintersecting convex sets with K_1 closed and K_2 compact, then there exists a hyperplane H in V such that K_1 lies in one of the closed half-spaces determined by H while K_2 lies in the other; that is, H separates K_1 and K_2. A *hyperplane H* in V is a translation of the orthogonal complement of a one-dimensional subspace of $V : H = \{x \in V : \langle x - p, q \rangle = 0\}$ for given vectors $p, q \in V, q \neq 0$. The hyperplane H determines two open *half-spaces*: $H^+ = \{x \in V : \langle x - p, q \rangle > 0\}$, $H^- = \{x \in V : \langle x - p, q \rangle < 0\}$. The sets $H_0^+ = H^+ \cup H$ and $H_0^- = H^- \cup H$ are the closed half-spaces determined by H. Thus, separation means that $K_1 \subseteq H_0^+$ and $K_2 \subseteq H_0^-$ for some vectors p, q. There are various ways to strengthen the separation conclusion by making various additional assumptions about the two convex sets. For example, if the closures of K_1 and K_2 do not intersect, then the separation may be taken to be strict; that is, $K_1 \subseteq H^+$, $K_2 \subseteq H^-$. The closure of the convex hull of any bounded set $S \subset V$ can be obtained as the intersection of all closed half-spaces that contain S.

If V is the vector space $\mathbf{C}^n$ with complex inner product $\langle \cdot, \cdot \rangle$, hyperplanes and half-spaces are defined similarly, *except* that $\mathbf{C}^n$ must be identified with $\mathbf{R}^{2n}$ and $\langle \cdot, \cdot \rangle$ must be replaced with the real inner product $\mathrm{Re}\langle \cdot, \cdot \rangle$ as follows: Identify $x + iy \in \mathbf{C}^n$ with $\left[\begin{smallmatrix} x \\ y \end{smallmatrix}\right] \in \mathbf{R}^{2n}$, and note that $\mathrm{Re}\langle x_1 + iy_1, x_2 + iy_2 \rangle = \langle x_1, x_2 \rangle + \langle y_1, y_2 \rangle$ by conjugate linearity of the complex inner product. Then $\langle x_1, x_2 \rangle + \langle y_1, y_2 \rangle$ is the (real) inner product of $\left[\begin{smallmatrix} x_1 \\ y_1 \end{smallmatrix}\right]$ and $\left[\begin{smallmatrix} x_2 \\ y_2 \end{smallmatrix}\right]$, and hyperplanes and half-spaces defined in $\mathbf{R}^{2n}$ have the appropriate geometric interpretation in $\mathbf{C}^n$.

A real valued function f defined on a convex set $K \subseteq V$ is said to be *convex* if

$$f(\alpha x + (1 - \alpha)y) \leq \alpha f(x) + (1 - \alpha)f(y) \tag{B1}$$

for all $0 < \alpha < 1$ and all $x, y \in K, y \neq x$. If the inequality (B1) is always strict, then f is said to be *strictly convex*. If the inequality (B1) is reversed for all $0 < \alpha < 1$ and all $x, y \in K, y \neq x$, then f is said to be *concave* (or *strictly concave* if it is reversed and always strict). Equivalently, a concave (respectively, strictly concave) function is the negative of a convex (respectively strictly convex) function. Geometrically, the chord joining any two function values $f(x)$ and $f(y)$ lies above (respectively, below) the graph of a convex (respectively, concave) function. A linear function is both convex and concave. If $V = \mathbf{R}^n$ and K is an open set, the Hessian

$$H(x) \equiv \left[\frac{\partial^2 f}{\partial x_i \, \partial x_j}(x) \right]$$

which is a symmetric matrix in $M_n(\mathbf{R})$, exists almost everywhere in K for a bounded convex function f and is necessarily positive semidefinite for points in K at which it exists. It is positive definite in the strictly convex case. Conversely, a function whose Hessian is positive semidefinite (respectively, positive definite) throughout a convex set is convex (respectively, strictly convex). Similarly, negative definiteness corresponds to concavity.

Optimization of convex and concave functions has some pleasant properties. On a compact convex set the maximum (respectively, minimum) of a convex (respectively, concave) function is attained at an extreme point. On the other hand, on a convex set, the set of points at which the minimum (respectively, maximum) of a convex (respectively, concave) function is attained is convex and any local minimum (respectively, maximum) is a global minimum (respectively, maximum). For example, a strictly convex function attains a minimum at at most one point of a convex set, and a critical point is necessarily a minimum.

Convex combinations of real numbers obey some simple and frequently useful inequalities. If $x_1, \ldots, x_k$ are given real numbers, then

$$\min_{1 \leq i \leq k} x_i \leq \sum_{i=1}^{k} a_i x_i \leq \max_{i \leq i \leq k} x_i \tag{B2}$$

for any convex combination: $\alpha_1, \alpha_2, \ldots, \alpha_k \geq 0$ and $\alpha_1 + \cdots + \alpha_k = 1$.

Consideration of certain convex functions $f(\cdot)$ of one variable on an interval leads to various classical inequalities. One can use induction to show that the defining two-point inequality (B1) on the interval implies an n-point inequality

$$f\left(\sum_{i=1}^{n} \alpha_i x_i\right) \leq \sum_{i=i}^{n} \alpha_i f(x_i), \quad n = 2, 3, \ldots \tag{B3}$$

whenever $\alpha_i \geq 0$, $a_1 + \cdots + \alpha_n = 1$, and all x_i are in the interval.

Application of (B3) to the strictly convex function $f(x) = -\log x$ over the interval $(0, \infty)$ leads to the *weighted arithmetic–geometric mean inequality*

$$\sum_{i=1}^{n} \alpha_i x_i \geq \prod_{i=1}^{n} x_i^{a_i}, \quad x_i \geq 0 \tag{B4}$$

which contains the *arithmetic–geometric mean inequality*

$$\frac{1}{n} \sum_{i=i}^{n} x_i \geq \left(\prod_{i=1}^{n} x_i\right)^{1/n}, \quad x_i \geq 0 \tag{B5}$$

when all $\alpha_i = 1/n$. The inequality is an equality if and only if all x_i are equal.

Application of (B3) to the strictly convex function $f(x) = x^p$, $p > 1$, over the interval $(0, \infty)$ leads to *Hölder's inequality*

$$\sum_{i=1}^{n} x_i y_i \leq \left(\sum_{i=1}^{n} x_i^p\right)^{1/p} \left(\sum_{i=1}^{n} y_i^q\right)^{1/q}, \quad x_i, y_i > 0, p > 1, \frac{1}{p} + \frac{1}{q} = 1 \tag{B6}$$

Hölder's inequality is an equality if and only if the vectors $[x_i^p]$ and $[y_i^q]$ are linearly dependent. If we take $p = q = 2$, we obtain a version of the *Cauchy–Schwarz inequality*

$$\sum_{i=1}^{n} x_i y_i \leq \left(\sum_{i=1}^{n} x_i^2\right)^{1/2} \left(\sum_{i=1}^{n} y_i^2\right)^{1/2}, \quad x_i, y_i \in \mathbf{R} \tag{B7}$$

which is an equality if and only if the vectors $[x_i]$ and $[y_i]$ are linearly dependent. As a limiting case of Hölder's inequality we obtain

$$\sum_{i=1}^{n} x_i y_i \leq \left(\sum_{i=1}^{n} x_i\right) \max_{1 \leq i \leq n} y_i, \quad x_i, y_i \geq 0 \tag{B8}$$

From Hölder's inequality one can deduce *Minkowski's sum inequality*

$$\left(\sum_{i=1}^{n}(x_i + y_i)^p\right)^{1/p} \leq \left(\sum_{i=1}^{n}x_i^p\right)^{1/p} + \left(\sum_{i=1}^{n}y_i^p\right)^{1/p}, \quad x_i, y_i \geq 0, p > 1 \quad \text{(B9)}$$

which is an equality if and only if the vectors $[x_i]$ and $[y_i]$ are linearly dependent.

Minkowski's product inequality

$$\left(\prod_{i=1}^{n}(x_i + y_i)\right)^{1/n} \geq \left(\prod_{i=1}^{n}x_i\right)^{1/n} + \left(\prod_{i=1}^{n}y_i\right)^{1/n} \quad x_i, y_i \geq 0 \quad \text{(B10)}$$

is a consequence of the arithmetic–geometric mean inequality. The inequality (B10) is an equality if and only if the vectors $[x_i]$ and $[y_i]$ are linearly dependent.

Jensen's inequality says that

$$\left(\sum_{i=1}^{n}x_i^{p_1}\right)^{1/p_1} > \left(\sum_{i=1}^{n}x_i^{p_2}\right)^{1/p_2} \quad x_i, y_i > 0, 0 < p_1 < p_2 \quad \text{(B11)}$$

It follows from the computation

$$\frac{\left(\sum_{i=1}^{n}x_i^{p_2}\right)^{1/p_2}}{\left(\sum_{i=1}^{n}x_i^{p_1}\right)^{1/p_1}} = \left(\sum_{i=1}^{n}\left(\frac{x_i^{p_1}}{\sum_{j=1}^{n}x_j^{p_1}}\right)^{p_2/p_1}\right)^{1/p_2}$$

$$< \left(\sum_{i=1}^{n}\frac{x_i^{p_1}}{\sum_{j=1}^{n}x_j^{p_1}}\right)^{1/p_2} = \left(\frac{\sum_{i=1}^{n}x_i^{p_1}}{\sum_{j=1}^{n}x_j^{p_1}}\right)^{1/p_2} = 1$$

Further Readings. For more information about convex sets and geometry see Valentine (1964). For more about convex functions and inequalities, see Boas (1972). Proofs of the classical inequalities B4–B11 can be found in Beckenbach and Bellman (1965) and Hardy, Littlewood, and Pólya (1959).

The Fundamental Theorem
of Algebra

One historical motivation for introducing the complex numbers $\mathbf{C}$ was that polynomials with real coefficients might not have real zeroes. For example, a calculation reveals that $\{1 + i, 1 - i\}$ are zeroes of the polynomial $p(t) = t^2 - 2t + 2$, which has no real zeroes. All zeroes of any polynomial with real coefficients are, however, contained in $\mathbf{C}$. In fact, all zeroes of all polynomials with complex coefficients are in $\mathbf{C}$. Thus, $\mathbf{C}$ is an *algebraically closed field*: There is no field $\mathbf{F}$ such that $\mathbf{C}$ is a subfield of $\mathbf{F}$, and such that there is a polynomial with coefficients from $\mathbf{C}$ and with a zero in $\mathbf{F}$ that is not in $\mathbf{C}$.

The *fundamental theorem of algebra* states that any polynomial p with complex coefficients and of degree at least 1 has at least one zero in $\mathbf{C}$. Using synthetic division, if $p(z) = 0$, then $t - z$ divides $p(t)$; that is, $p(t) = (t - z)q(t)$, in which $q(t)$ is a polynomial with complex coefficients, whose degree is 1 smaller than that of p. The zeroes of p are z, together with the zeroes of q. The following theorem is a consequence of the fundamental theorem of algebra.

Theorem. *A polynomial of degree $n \geq 1$ with complex coefficients has, counting multiplicities, exactly n zeroes among the complex numbers.*

The multiplicity of a zero z of a polynomial p is the largest integer k for which $(t - z)^k$ divides $p(t)$. If a zero z has multiplicity k, then it is counted k times toward the number n of zeroes of p. It follows that a polynomial with complex coefficients may always be factored into a product of linear factors over the complex numbers.

If a polynomial p with *real* coefficients has some nonreal complex zeroes, they must occur in *conjugate pairs*, since, if $0 = p(z)$, then $0 = \bar{0} = \overline{p(z)} = p(\bar{z})$. The observation that $(x - z)(x - \bar{z}) = x^2 - 2\text{Re}(z)x + |z|^2$ ensures that any real polynomial may be factored into a product of powers of linear and quadratic factors over the reals; each irreducible quadratic factor corresponds to a conjugate pair of complex roots.

Further Reading. For an elementary proof of the fundamental theorem of algebra, see Childs (1979).

Continuity of Polynomial Zeroes and Matrix Eigenvalues

It is an important fact, typically proved using complex analysis, that the n zeroes of a polynomial of degree $n \geq 1$ with complex coefficients depend continuously on the coefficients.

For $z \in \mathbf{C}^n$, let $f(z) = [f_1(z) \; \ldots \; f_m(z)]^T$, in which $f_i : \mathbf{C}^n \to \mathbf{C}$, $i = 1, \ldots, m$. The function $f : \mathbf{C}^n \to \mathbf{C}^m$ is *continuous* at z if each f_i is continuous at z, $i = 1, \ldots, m$. The function $f_i : \mathbf{C}^n \to \mathbf{C}$ is continuous at z if, for a given vector norm $\| \cdot \|$ on $\mathbf{C}^n$ and each $\epsilon > 0$, there is a $\delta > 0$ such that if $\| z - \zeta \| < \delta$, then $|f_i(z) - f_i(\zeta)| < \epsilon$.

One might be tempted to describe continuous dependence of the zeroes of a polynomial on its coefficients by requiring continuity of the function $f : \mathbf{C}^n \to \mathbf{C}^n$ that takes the n coefficients (all but the leading 1) of a monic polynomial of degree n to the n zeroes of the polynomial. There is a problem, however: There is no obvious way to define this function, since there is no natural way to define an ordering among the n zeroes. As a quantitative statement of the continuous dependence of the zeroes on the coefficients of a polynomial, we offer the following.

Theorem D1. *Let $p(t) = t^n + a_1 t^{n-1} + \cdots + a_{n-1} t + a_n$ and $q(t) = t^n + b_1 t^{n-1} + \cdots + b_{n-1} t + b_n$ be polynomials of degree $n \geq 1$ with complex coefficients. Let $\lambda_1, \ldots, \lambda_n$ be the zeroes of p in some order and let $\mu_1, \ldots, \mu_n$ be the zeroes of q in some order (counting multiplicities in both cases). Define*

$$\gamma = 2 \max_{1 \leq k \leq n} \{ |a_k|^{1/k}, |b_k|^{1/k} \}$$

Then there exists a permutation τ of $\{1, \ldots, n\}$ such that

$$\max_{1 \leq j \leq n} |\lambda_j - \mu_{\tau(j)}| \leq 2^{\frac{2n-1}{n}} \left(\sum_{k=1}^{n} |a_k - b_k| \gamma^{n-k} \right)^{1/n}$$

In the same spirit, the following explicit bounds ensure continuity of matrix eigenvalues.

Theorem D2. *Let $A, B \in M_n$ be given. Let $\lambda_1, \ldots, \lambda_n$ be the eigenvalues of A in some order and let $\mu_1, \ldots, \mu_n$ be the eigenvalues of B in some order (counting multiplicities*

563

in both cases). Then there exists a permutation τ of $\{1, \ldots, n\}$ such that

$$\max_{1 \leq j \leq n} |\lambda_j - \mu_{\tau(j)}| \leq 2^{\frac{2n-1}{n}} (\|\|A\|\|_2 + \|\|B\|\|_2)^{\frac{n-1}{n}} \|\|A - B\|\|_2^{\frac{1}{n}}$$

Further Reading. The two theorems quoted here are in R. Bhatia, L. Elsner, and G. Krause, Bounds on the variation of the roots of a polynomial and the eigenvalues of a matrix, *Linear Algebra Appl.* 142 (1990) 195–209.

Continuity, Compactness, and Weierstrass's Theorem

Let V be a finite-dimensional real or complex vector space with a given norm $\|\cdot\|$. The *closed ball of radius* ϵ about $x \in V$ is $B_\epsilon(x) = \{y \in V : \| y - x \| \le \epsilon\}$; the corresponding *open ball* is $B_\epsilon(x) = \{y \in V : \| y - x \| < \epsilon\}$. A set $S \subseteq V$ is *open* if, for each $x \in S$, there is an $\epsilon > 0$ such that $B_\epsilon(x) \subseteq S$. A set $S \subseteq V$ is *closed* if the complement of S in V is open. A set $S \subseteq V$ is *bounded* if there is an $r > 0$ such that $S \subseteq B_r(0)$. Equivalently, a set $S \subseteq V$ is closed if and only if the limit of any convergent (with respect to $\|\cdot\|$) sequence of points in S is itself in S; S is bounded if and only if it is contained in some ball of finite radius. A set $S \subseteq V$ is *compact* if it is both closed and bounded.

For a given set $S \subseteq V$ and a given real-valued function f defined on S, $\inf_{x \in S} f(x)$ and $\sup_{x \in S} f(x)$ need not be finite, and even if they are, there may or may not be points $x_{\min}$ and $x_{\max}$ in S such that $f(x_{\min}) = \inf_{x \in S} f(x)$ and $f(x_{\max}) = \sup_{x \in S} f(x)$, that is, f need not attain a maximum or minimum value on S. However, under certain circumstances, we may be sure that f attains both maximum and minimum values on S.

A real- or complex-valued function f defined on a given set $S \subseteq V$ is *continuous at* $x_0 \in S$ if, for each $\epsilon > 0$, there is a $\delta > 0$ such that $|f(x) - f(x_0)| < \epsilon$ whenever $x \in S$ and $\|x - x_0\| < \delta$; f is *continuous on* S if it is continuous at each point of S; f is *uniformly continuous on* S if, for each $\epsilon > 0$, there is a $\delta > 0$ such that $|f(x) - f(y)| < \epsilon$ whenever $x, y \in S$ and $\|x - y\| < \delta$.

Theorem (Weierstrass). *Let S be a compact subset of a finite-dimensional real or complex vector space V with a given norm $\|\cdot\|$, and let $f : S \to \mathbf{R}$ be a continuous function. There exists a point $x_{\min} \in S$ such that*

$$f(x_{\min}) \le f(x) \quad \text{for all} \quad x \in S$$

and a point $x_{\max} \in S$ such that

$$f(x) \le f(x_{\max}) \quad \text{for all} \quad x \in S$$

That is, f attains its minimum and maximum values on S.

Of course, the values $\max_{x \in S} f(x)$ and $\min_{x \in S} f(x)$ in Weierstrass's theorem may each be attained at more than one point of S.

If either of the key hypotheses (compact S and continuous f) of Weierstrass's theorem is violated, the conclusion may fail. The assumption that S is a subset of a finite-dimensional normed linear space is not essential, however. With suitable definitions of *compact* and *continuous*, Weierstrass's theorem is valid for a continuous real-valued function on a compact subset of a general topological space.

Further Readings. For more information about analysis and normed linear spaces, see chapter 2 of Kreyszig (1978) or chapter 3 of Conway (1990).

APPENDIX F

Canonical Pairs

Any square complex matrix A can be represented uniquely as $A = S(A) + C(A)$, in which $S(A) = \frac{1}{2}(A + A^T)$ is symmetric and $C(A) = \frac{1}{2}(A - A^T)$ is skew symmetric; it can also be represented uniquely as $A = H(A) + iK(A)$, in which both $H(A) = \frac{1}{2}(A + A^*)$ and $K(A) = \frac{1}{2i}(A - A^*)$ are Hermitian. A simultaneous congruence of $S(A)$ and $C(A)$ corresponds to a congruence of A; a simultaneous *congruence of $H(A)$ and $K(A)$ corresponds to a *congruence of A.

Canonical forms for the pairs $(S(A), C(A))$ and $(H(A), K(A))$ (called *canonical pairs*) can be derived from congruence and *congruence canonical forms for A. In describing the canonical pairs, we use the following k-by-k matrices:

$$
M_k = \begin{bmatrix} 0 & 1 & & 0 \\ 1 & 0 & \ddots & \\ & \ddots & \ddots & 1 \\ 0 & & 1 & 0 \end{bmatrix}, \quad
N_k = \begin{bmatrix} 0 & 1 & & 0 \\ -1 & 0 & \ddots & \\ & \ddots & \ddots & 1 \\ 0 & & -1 & 0 \end{bmatrix}
$$

$$
X_k = \begin{bmatrix} 0 & & & & (-1)^{k+1} \\ & & & \iddots & 0 \\ & & -1 & \iddots & \\ & 1 & 0 & & \\ -1 & 0 & & & \\ 1 & 0 & & & 0 \end{bmatrix}
$$

$$
Y_k = \begin{bmatrix} 0 & & & & & 0 \\ & & & & \iddots & (-1)^k \\ & & & 0 & \iddots & \\ & & 0 & 1 & & \\ & 0 & -1 & & & \\ 0 & 1 & & & & 0 \end{bmatrix}
$$

We also use a 2-parameter version of the type I matrix (4.5.19)

$$\Delta_k(a, b) = \begin{bmatrix} 0 & & & a \\ & & \ddots & b \\ & a & \ddots & \\ a & b & & 0 \end{bmatrix}, \qquad a, b \in \mathbf{C}$$

A *matrix pair* (A, B) is an ordered pair of square matrices of the same size. The *direct sum* of two matrix pairs is $(A_1, A_2) \oplus (B_1, B_2) = (A_1 \oplus B_1, \ A_2 \oplus B_2)$, in which $A_1, A_2 \in M_p$ and $B_1, B_2 \in M_q$. The *skew sum* of two square matrices of the same size is

$$[A \setminus B] = \begin{bmatrix} 0 & B \\ A & 0 \end{bmatrix}$$

Matrix pairs (A_1, A_2) and (B_1, B_2) are said to be *simultaneously congruent* (respectively, *simultaneously *congruent*) if there is a nonsingular matrix R such that $A_1 = R^T B_1 R$ and $A_2 = R^T B_2 R$ (respectively, $A_1 = R^* B_1 R$ and $A_2 = R^* B_2 R$). This transformation is a *simultaneous congruence* (respectively, a *simultaneous *congruence*) of a matrix pair via R.

The following theorem lists the canonical pairs that can occur and their associations with the three types of *congruence and congruence canonical matrices listed in (4.5.21) and (4.5.25).

Theorem F1. *(a) Each pair (S, C) consisting of a symmetric complex matrix S and a skew-symmetric complex matrix C of the same size is simultaneously congruent to a direct sum of pairs, determined uniquely up to permutation of summands, of the following three types, each associated with the indicated congruence canonical matrix type (4.5.25) for $A = S + C$:*

Type 0: $J_n(0)$	(M_n, N_n)
Type I: Γ_n	(X_n, Y_n) if n is odd,
	(Y_n, X_n) if n is even
Type II: $H_{2n}(\mu)$	$([J_n(\mu + 1) \setminus J_n(\mu + 1)^T],$
	$[J_n(\mu - 1) \setminus - J_n(\mu - 1)^T])$
$0 \neq \mu \neq (-1)^{n+1}$ and μ is determined up to replacement by μ^{-1}	

The Type II pair can be replaced by two alternative pairs

Type II: $H_{2n}(\mu)$	$([I_n \setminus I_n], [J_n(\nu) \setminus - J_n(\nu)^T])$
$0 \neq \mu \neq -1$	$\nu \neq 0$ if n is odd, $\nu \neq \pm 1$,
$\mu \neq 1$ if n is odd	ν is determined up to replacement by $-\nu$
Type II: $H_{2n}(-1)$	$([J_n(0) \setminus J_n(0)^T], [I_n \setminus - I_n])$
n is odd	n is odd

in which

$$\nu = \frac{\mu - 1}{\mu + 1}$$

(b) Each pair (H, K) of Hermitian matrices of the same size is simultaneously *congruent to a direct sum of pairs, determined uniquely up to permutation of summands, of the following four types, each associated with the indicated *congruence canonical matrix type (4.5.21) for $A = H + iK$:

Type 0: $J_n(0)$	(M_n, iN_n)
Type I: $\lambda \Delta_n$	$\pm (\Delta_n(1, 0), \Delta_n(c, 1))$
$\|\lambda\| = 1, \ \lambda^2 \neq -1$	$c \in \mathbf{R}$
Type I: $\lambda \Delta_n$	$\pm (\Delta_n(0, 1), \Delta_n(1, 0))$
$\lambda^2 = -1$	
Type II: $H_{2n}(\mu)$	$([I_n \setminus I_n], [J_n(a + ib) \setminus J_n(a + ib)^*])$
$\|\mu\| > 1$	$a, b \in \mathbf{R}, \ a + bi \neq i, \ b > 0$

in which

$$a = \frac{2 \operatorname{Im} \mu}{|1 + \mu|^2}, \qquad b = \frac{|\mu|^2 - 1}{|1 + \mu|^2}, \qquad c = \frac{\operatorname{Im} \lambda}{\operatorname{Re} \lambda}$$

Further Readings. The canonical pairs presented are taken from R. A. Horn and V. V. Sergeichuk, Canonical forms for complex matrix congruence and *congruence, *Linear Algebra Appl.* 416 (2006) 1010–1032. For alternative versions of canonical pairs, see P. Lancaster and L. Rodman, Canonical forms for Hermitian matrix pairs under strict equivalence and congruence, *SIAM Review* 47 (2005) 407–443, P. Lancaster and L. Rodman, Canonical forms for symmetric/skew-symmetric real matrix pairs under strict equivalence and congruence, *Linear Algebra Appl.* 406 (2005) 1–76, R. C. Thompson, Pencils of complex and real symmetric and skew matrices, *Linear Algebra Appl.* 147 (1991) 323–371, and V. V. Sergeichuk, Classification problems for systems of forms and linear mappings, *Math. USSR-Izv.* 31 (1988) 481–501.

References

Aitken, A. C. 1956. *Determinants and Matrices*. 9th ed. Oliver and Boyd, Edinburgh.

Axler, S. 1996. *Linear Algebra Done Right*. Springer, New York.

Bapat, R. B. 1993. *Linear Algebra and Linear Models*. Hindustan Book Agency, Delhi.

Bapat, R. B., and T. E. S. Raghavan. 1997. *Nonnegative Matrices and Applications*. Cambridge University Press, Cambridge.

Barnett, S. 1975. *Introduction to Mathematical Control Theory*. Clarendon Press, Oxford.

Barnett, S. 1979. *Matrix Methods for Engineers and Scientists*. McGraw-Hill, London.

Barnett, S. 1983. *Polynomials and Linear Control Systems*. Dekker, New York.

Barnett, S. 1990. *Matrices: Methods and Applications*. Clarendon Press, Oxford.

Barnett, S., and C. Storey. 1970. *Matrix Methods in Stability Theory*. Barnes and Noble, New York.

Baumgärtel, H. 1985. *Analytic Perturbation Theory for Matrices and Operators*. Birkhäuser, Basel.

Beckenbach, E. F., and R. Bellman. 1965. *Inequalities*. Springer, New York.

Belitskii, G. R., and Yu. I. Lyubich. 1988. *Matrix Norms and Their Applications*. Birkhäuser, Basel.

Bellman, R. 1997. *Introduction to Matrix Analysis*. 2nd ed. SIAM, Philadelphia.

Berman, A. M. Neumann, and R. J. Stern. 1989. *Nonnegative Matrices in Dynamic Systems*. Wiley, New York.

Berman, A., and R. J. Plemmons. 1994. *Nonnegative Matrices in the Mathematical Sciences*. SIAM, Philadelphia.

Bernstein, D. 2009. *Matrix Mathematics*. 2nd ed. Princeton University Press, Princeton.

Bhatia, R. 1987. *Perturbation Bounds for Matrix Eigenvalues*. Longman Scientific and Technical, Essex.

Bhatia, R. 1997. *Matrix Analysis*. Springer, New York.

Bhatia, R. 2007. *Positive Definite Matrices*. Princeton University Press, Princeton.

Boas, R. P. Jr. 1972. *A Primer of Real Functions*. 2nd ed. Carus Mathematical Monographs 13. Mathematical Association of America, Washington, D.C.

Bonsall, F. F., and J. Duncan. 1971. *Numerical Ranges of Operators on Normed Spaces and of Elements of Normed Algebras*. Cambridge University Press, Cambridge.

Bonsall, F. F., and J. Duncan. 1973. *Numerical Ranges II*. Cambridge University Press, Cambridge.

Brualdi, R. A., and H. J. Ryser. 1991. *Combinatorial Matrix Theory*. Cambridge University Press, New York.

Brualdi, R. A., and B. L. Shader. 1995. *Matrices of Sign-Solvable Linear Systems*. Cambridge University Press, New York.

Campbell, S. L., and C. D. Meyer. 1991. *Generalized Inverses of Linear Transformations*. Dover, Mineola.

Carlson, D., C. Johnson, D. Lay, and D. Porter, eds. 2002. *Linear Algebra Gems*. Mathematical Association of America, Washington, D.C.

Chatelin, F. 1993. *Eigenvalues of Matrices*. John Wiley, New York.

Childs, L. 1979. *A Concrete Introduction to Higher Algebra*. Springer, Berlin.

Clark, J., K. C. O'Meara, and C. I. Vinsonhaler. 2011. *Advanced Topics in Linear Algebra. Weaving Matrix Problems Through the Weyr Form*. Oxford University Press, Oxford.

Conway, J. 1990. *A Course in Functional Analysis*. 2nd ed. Springer, New York.

Courant, R., and D. Hilbert. 1953. *Methods of Mathematical Physics. Vol. I*, Interscience, New York (originally Julius Springer, Berlin, 1937).

Crilly, T. 2006. *Arthur Cayley: Mathematician Laureate of the Victorian Age*. The Johns Hopkins University Press, Baltimore.

Cullen, C. G. 1966. *Matrices and Linear Transformations*. Addison-Wesley, Reading.

Donoghue, W. F. Jr. 1974. *Monotone Matrix Functions and Analytic Continuation*. Springer, Berlin.

Faddeeva, V. N., Trans. C. D. Benster. 1959. *Computational Methods of Linear Algebra*. Dover, New York.

Fallatt, S. M., and C. R. Johnson. 2011. *Totally Nonnegative Matrices*. Princeton University Press, Princeton.

Fan, Ky. 1959. *Convex Sets and Their Applications*. Lecture Notes, Applied Mathematics Division, Argonne National Laboratory, Summer.

Fiedler, M. 1975. *Spectral Properties of Some Classes of Matrices*. Lecture Notes, Report No. 75.01R. Chalmers University of Technology and the University of Göteborg.

Fiedler, M. 1986. *Special Matrices and Their Applications in Numerical Mathematics*. Martinus Nijhoff, Dordrecht.

Franklin, J. 1968. *Matrix Theory*. Prentice-Hall, Englewood Cliffs.

Gantmacher, F. R. 1959. *The Theory of Matrices*. 2 vols. Chelsea, New York.

Gantmacher, F. R. 1959. *Applications of the Theory of Matrices*. Interscience, New York.

Gantmacher, F. R., and M. G. Krein. 1960. *Oszillationsmatrizen, Oszillationskerne, und kleine Schwingungen mechanische Systeme*. Akademie, Berlin.

Glazman, I. M., and Ju. Il Ljubic. 2006. *Finite-Dimensional Linear Analysis*. Dover, Mineola.

Gohberg, I., P. Lancaster, and L. Rodman. 1982. *Matrix Polynomials*. Academic Press, New York.

Gohberg, I., P. Lancaster, and L. Rodman. 1983. *Matrices and Indefinite Scalar Products*. Birkhäuser, Boston.

Gohberg, I., P. Lancaster, and L. Rodman. 2006. *Invariant Subspaces of Matrices with Applications*. John Wiley, New York, 1986; SIAM, Philadelphia.

Gohberg, I. C., and M. G. Krein. 1969. *Introduction to the Theory of Linear Nonselfadjoint Operators*. American Mathematical Society, Providence.

Golan, J. S. 2004. *The Linear Algebra a Beginning Graduate Student Ought to Know*. Kluwer, Dordrecht.

Graham, A. 1981. *Kronecker Products and Matrix Calculus with Applications*. Horwood, Chichester.

Graybill, F. A. 1983. *Matrices with Applications to Statistics*. 2nd ed. Wadsworth, Belmont.

Greub, W. H. 1978. *Multilinear Algebra*. 2nd ed. Springer, New York.

Golub, G., and C. VanLoan. 1996. *Matrix Computations*. 3rd ed. The Johns Hopkins University Press, Baltimore.

Halmos, P. R. 1958. *Finite-Dimensional Vector Spaces*. Van Nostrand, Princeton.

Halmos, P. R. 1967. *A Hilbert Space Problem Book*. Van Nostrand, Princeton.

Hardy, G. H., J. E. Littlewood, and G. Pólya. 1959. *Inequalities*. Cambridge University Press, Cambridge (first ed. 1934).

Higham, N. J. 2008. *Functions of Matrices*. SIAM, Philadelphia.

Hirsch, M. W., and S. Smale. 1974. *Differential Equations, Dynamical Systems, and Linear Algebra*. Academic Press, New York.

Hogben, L. ed. 2007. *Handbook of Linear Algebra*. Chapman and Hall, Boca Raton.

Horn, R. A., and C. R. Johnson. 1991. *Topics in Matrix Analysis*. Cambridge University Press, Cambridge.

Hoffman, K., and R. Kunze. 1971. *Linear Algebra*. 2nd ed. Prentice-Hall, Englewood Cliffs.

Householder, A. S. 1964. *The Theory of Matrices in Numerical Analysis*. Blaisdell, New York.

Householder, A. S. 1972. *Lectures on Numerical Algebra*. Mathematical Association of America, Buffalo.

Jacobson, N. 1943. *The Theory of Rings*. American Mathematical Society, New York.

Kaplansky, I. 1974. *Linear Algebra and Geometry: A Second Course*. Dover, Mineola.

Karlin, S. 1960. *Total Positivity*. Stanford University Press, Stanford.

Kato, T. 1980. *Perturbation Theory for Linear Operators*, Springer, Berlin.

Kellogg, R. B. 1971. *Topics in Matrix Theory*. Lecture Notes, Report No. 71.04, Chalmers Institute of Technology and the University of Göteberg.

Kőnig, D. 1936. *Theorie der endlichen und unendlichen Graphen*. Akademische Verlagsgesellschaft, Leipzig.

Kőnig, D. 1990. *Theory of Finite and Infinite Graphs*. Birkhäuser, Boston, (translated by R. McCoart with commentary by W. T. Tutte).

Kowalsky, H. 1969. *Lineare Algebra*. 4th ed. deGruyter, Berlin.

Kreyszig, E. 1978. *Introductory Functional Analysis with Applications*. John Wiley, New York.

Lancaster, P. 1969. *Theory of Matrices*. Academic Press, New York.

Lang, S. 1987. *Linear Algebra*. 3rd ed. Springer, New York.

Lancaster, P., and M. Tismenetsky. 1985. *The Theory of Matrices with Applications*. 2nd ed. Academic Press, New York.

Lawson, C., and R. Hanson. 1974. *Solving Least Squares Problems*. Prentice-Hall, Englewood Cliffs.

Lax, P. D. 2007. *Linear Algebra and Its Applications*. 2nd ed. John Wiley, New York.

MacDuffee, C. C. 1946. *The Theory of Matrices*. Chelsea, New York.

Marcus, M. 1973–1975. *Finite Dimensional Multilinear Algebra*. 2 vols. Dekker, New York.

Marcus, M., and H. Minc. 1964. *A Survey of Matrix Theory and Matrix Inequalities*. Allyn and Bacon, Boston.

Marshall, A. W., and I. Olkin. 1979. *Inequalities: Theory of Majorization and Its Applications*. Academic Press, New York.

Minc, H. 1988. *Nonnegative Matrices*. John Wiley, New York.

Mirsky, L. 1963. *An Introduction to Linear Algebra*. Clarendon Press, Oxford.

Muir, T. 1930. *The Theory of Determinants in the Historical Order of Development*. 4 vols. Macmillan, London, 1906, 1911, 1920, 1923; Dover, New York, 1966. *Contributions to the History of Determinants, 1900–1920*. Blackie, London.

Newman, M. 1972. *Integral Matrices*. Academic Press, New York.

Noble, B., and J. W. Daniel. 1988. *Applied Linear Algebra*, 3rd ed. Prentice-Hall, Englewood Cliffs.

Ortega, J. M. 1987. *Matrix Theory: A Second Course*. Plenum Press, New York.

Parlett, B. 1998. *The Symmetric Eigenvalue Problem*. SIAM, Philadelphia.

Parshall, K. H. 2006. *James Joseph Sylvester: Jewish Mathematician in a Victorian World*. The Johns Hopkins Press, Baltimore.

Perlis, S. 1952. *Theory of Matrices*. Addison-Wesley, Reading.

Radjavi, H., and P. Rosenthal. 2000. *Simultaneous Triangularization*. Springer, New York.

Rogers, G. S. 1980. *Matrix Derivatives*. Lecture Notes in Statistics, vol. 2. Dekker, New York.

Roman, S. 2010. *Advanced Linear Algebra*. 3rd ed. Graduate Texts in Mathematics. Springer, New York.

Rudin, W. 1976. *Principles of Mathematical Analysis*. 3rd ed. McGraw-Hill, New York.

Seneta, E. 1973. *Nonnegative Matrices*. John Wiley, New York.

Serre, D. 2002. *Matrices: Theory and Applications*. Springer, New York.

Stewart, G. W. 1973. *Introduction to Matrix Computations*. Academic Press, New York.

Stewart, G. W., and J.-G. Sun. 1990. *Matrix Perturbation Theory*. Academic Press, New York.

Strutt, J. W. Baron Rayleigh. 1894. *The Theory of Sound*. Dover, New York, 1945. 1st ed. Macmillan, London, 1877; 2nd ed., revised and enlarged, Macmillan, London.

Suprunenko, D. A., and R. I. Tyshkevich. 1968. *Commutative Matrices*. Academic Press, New York.

Todd, J. ed. 1962. *Survey of Numerical Analysis*. McGraw-Hill, New York.

Turnbull, H. W. 1950. *The Theory of Determinants, Matrices and Invariants*. Blackie, London.

Turnbull, H. W., and A. C. Aitken. 1962. *An Introduction to the Theory of Canonical Matrices*. Blackie, London, 1932; 2nd ed. Blackie, London, 1945; 3rd ed. Dover, Mineola.

Valentine, F. A. 1964. *Convex Sets*. McGraw-Hill, New York.

Varga, R. S. 2000. *Matrix Iterative Analysis*. 2nd ed. Springer, New York.

Varga. R. S. 2004. *Geršgorin and His Circles*. Springer, New York.

Wedderburn, J. H. M. 1934. *Lectures on Matrices*. American Mathematical Society Colloquium Publications XVII. American Mathematical Society, New York.

Wielandt, H. 1996. *Topics in the Analytic Theory of Matrices*. Lecture Notes prepared by R. Meyer. Department of Mathematics, University of Wisconsin, Madison, 1967. Published as pp. 271–352 in *Helmut Wielandt, Mathematische Werke, Mathematical Works. Volume 2: Linear Algebra and Analysis*. Edited by Bertram Huppert and Hans Schneider. Walter de Gruyter, Berlin.

Wilkinson, J. H. 1965. *The Algebraic Eigenvalue Problem*. Clarendon Press, Oxford.

Zhan, X. 2002. *Matrix Inequalities*. Springer, Berlin.

Zhang, F. 2009. *Linear Algebra: Challenging Problems for Students*. 2nd ed. The Johns Hopkins University Press, Baltimore.

Zhang, F. 2011. *Matrix Theory: Basic Results and Techniques*. 2nd ed. Springer, New York.

Zhang, F. ed. 2005. *The Schur Complement and its Applications*. Springer, New York.

Zurmühl, R., and S. Falk. 1986. *Matrizen und ihre Anwendungen, 2: Numerische Methoden*. Springer, Heidelberg.

Zurmühl, R., and S. Falk. 2011. *Matrizen und ihre Anwendungen, 1: Grundlagen*. Springer, Heidelberg.

Notation

$\mathbf{R}$	the real numbers
$\mathbf{R}^n$	real vector space of real n-vectors, $M_{n,1}(\mathbf{R})$
$\mathbf{C}$	the complex numbers
$\mathbf{C}^n$	complex vector space of complex n-vectors, $M_{n,1}$
$\mathbf{F}$	a field
$\mathbf{F}^n$	vector space (over $\mathbf{F}$) of n-vectors with entries from $\mathbf{F}$
$M_{m,n}(\mathbf{F})$	m-by-n matrices with entries from $\mathbf{F}$
$M_{m,n}$	m-by-n complex matrices; same as $M_{m,n}(\mathbf{C})$
M_n	n-by-n complex matrices; same as $M_{n,n}(\mathbf{C})$
A, B, C, etc.	matrices; $A = [a_{ij}]$, etc.
x, y, z, etc.	column vectors; $x = [x_i]$, etc.
I_n	identity matrix in M_n; I if size is clear from context
$0_{m,n}$	zero matrix in $M_{m,n}$; 0 if size is clear from context
$\bar{A}$	matrix of complex conjugates of entries of A
A^T	transpose of A
A^*	conjugate transpose of A; same as $\bar{A}^T$
A^{-1}	inverse of $A \in M_n$
A^{-T}	inverse of transpose of $A \in M_n$
A^{-*}	inverse of conjugate transpose of $A \in M_n$
$|A|$	$[|a_{ij}|]$ for $A = [a_{ij}] \in M_n$
adj A	adjugate of $A \in M_n(\mathbf{F})$; transposed cofactor matrix
$A^{\dagger}$	Moore-Penrose inverse of $A \in M_{m,n}$
A^D	Drazin inverse of $A \in M_n$
$\mathcal{B}$	a basis of a vector space
e_i	ith standard basis vector
e	an all-ones vector; the base of natural logarithms
$[v]_{\mathcal{B}}$	$\mathcal{B}$-coordinate representation of a vector v
$_{\mathcal{B}_1}[T]_{\mathcal{B}_2}$	$\mathcal{B}_1$-$\mathcal{B}_2$ basis representation of linear transformation T
$\binom{n}{k}$	binomial coefficient, $n!/(k!(n-k)!)$
$p_A(\cdot)$	characteristic polynomial of $A \in M_n$

$\kappa(A)$	condition number of A (with respect to some norm)
$\det A$	determinant of $A \in M_n$
$\oplus$	direct sum
$\Gamma(A)$	directed graph of A
$\|\cdot\|^D$	dual norm of a norm $\|\cdot\|$
$f^D(\cdot)$	dual norm of a pre-norm $f(\cdot)$
λ	usually denotes an eigenvalue
$[\lambda_i(A)]$	vector of eigenvalues of $A \in M_n$
$n!$	factorial: $n(n-1)\cdots 2 \cdot 1$
$G_k(A)$	kth Geršgorin disc
$G(A)$	Geršgorin region, union of Geršgorin discs
$GL(n, \mathbf{F})$	group of nonsingular matrices in $M_n(\mathbf{F})$
$A \circ B$	Hadamard product of $A, B \in M_{m,n}(\mathbf{F})$
$\gamma(A)$	index of primitivity of a primitive $A \in M_n(\mathbf{R})$
$M(A)$	indicator matrix of $A \in M_n$
$J_k(\lambda)$	k-by-k Jordan block with eigenvalue λ
$q_A(\cdot)$	minimal polynomial of $A \in M_n$
$\|\cdot\|_1$	l_1 norm (sum norm)
$\|\cdot\|_2$	l_2 norm (Euclidean norm); Frobenius norm
$\|\cdot\|_\infty$	l_∞ norm (max norm)
$\|\cdot\|_p$	l_p norm
$\|x\|_{[k]}$	k-norm on vectors
$\|\|A\|\|_1$	max column sum matrix norm
$\|\|A\|\|_2$	spectral matrix norm; largest singular value
$\|\|A\|\|_\infty$	max row sum matrix norm
$N_\infty(A)$	sum of max norms of columns of A (matrix norm)
$\|\|A\|\|_{tr}$	trace norm (matrix norm)
$\|A\|_{[k]}$	Ky Fan k-norm (matrix norm)
$r(A)$	numerical radius (vector norm on matrices)
$\perp$	orthogonal complement
$\text{per } A$	permanent of $A \in M_n$
$\text{rank } A$	rank of $A \in M_{m,n}$
$\text{sgn } \tau$	signum of a permutation τ
σ	usually denotes a singular value
$\sigma_1(A)$	largest singular value of A; spectral norm
$[\sigma_i(A)]$	vector of singular values of A
$\text{span}(\mathcal{S})$	span of a set $\mathcal{S}$ of vectors
$\text{range } A$	span of the columns of A; the column space of A
$\text{nullspace } A$	subspace of solutions of $Ax = 0$
$\rho(A)$	spectral radius of $A \in M_n$
$\sigma(A)$	spectrum; eigenvalues of A, with multiplicities
$A[\alpha, \beta]$	submatrix of A determined by index sets α and β
$A[\alpha]$	principal submatrix of A determined by index set α
$\text{tr } A$	trace of $A = [a_{ij}]$; $\sum_i a_{ii}$
$E_k(A)$	sum of principal minors of A of size k
$S_k(A)$	kth elementary symmetric function of eigenvalues

A/A_{11}	Schur complement of A_{11} in A
$C_r(A)$	rth compound matrix of A
Λ, Σ	typically, diagonal matrices
$[A, B]$	commutator of $A, B \in M_n$; $AB - BA$
$]A, B[$	Jordan product of $A, B \in M_n$; $AB + BA$
$C(a, b)$	$\begin{bmatrix} a & b \\ -b & a \end{bmatrix}$, real Jordan canonical form block
$w(A, \lambda)$	Weyr characteristic: $(w_1(A, \lambda), \ldots, w_q(A, \lambda))$
$x_i(A)^{\downarrow}$	nonincreasingly ordered entries of a real vector x
$x_i(A)^{\uparrow}$	nondecreasingly ordered entries of a real vector x
Δ_k	Type I (symmetric) *congruence canonical block
$H_{2k}(\mu)$	Type II congruence and *congruence canonical block
Γ_k	Type I (real) congruence canonical block
$\langle A, B \rangle_F$	Frobenius inner product
$B_{\|\cdot\|}(r; x)$	norm ball of radius r around the point x
$F(A)$	field of values of $A \in M_n$
$R_i'(A)$	deleted row sum
$C_i'(A)$	deleted column sum
$A \succeq B$	Loewner partial order on Hermitian matrices
$H(A)$	Hermitian part of $A \in M_n$

Hints for Problems

Section 1.0

1.0.P1 $f(x) = x^T A x$ is a continuous function on the compact set $\{x \in \mathbf{R}^n : x^T x = 1\}$.

Section 1.1

1.1.P7 Let λ, x be an eigenvalue–eigenvector pair of A. Then $x^* A x = \lambda x^* x$. But $x^* x > 0$ and $\overline{x^* A x} = x^* A^* x = x^* A x$ is real.

1.1.P13 $Ax = \lambda x \Rightarrow (\det A)x = (\text{adj } A)Ax = \lambda(\text{adj } A)x$; $\lambda \neq 0 \Rightarrow (\text{adj } A)x = (\lambda^{-1} \det A)x$. If $\lambda = 0$, then either adj $A = 0$ (if rank $A < n - 1$) or adj $A = \alpha x y^T$ (if rank $A = n - 1$) (0.8.2).

Section 1.2

1.2.P10 Any non-real complex zeroes of a polynomial with real coefficients occur in conjugate pairs; $p_A(t)$ has real coefficients if $A \in M_n(\mathbf{R})$.

1.2.P11 Let $\mathcal{B}$ be a basis for V and consider $[T]_{\mathcal{B}}$.

1.2.P14 Evaluate $\det(tI - A)$ by cofactors along the first column, and then use cofactors along the first row in the next step.

1.2.P21 Example 1.2.8.

1.2.P23 Look at the last two terms of $p_A(t)$; (1.2.13).

Section 1.3

1.3.P4 Review the proof of (1.3.12) and show that B and A are simultaneously diagonalizable. See (0.9.11) and explain why there is a (Lagrange interpolating) polynomial $p(t)$ of degree at most $n - 1$ such that the eigenvalues of B are $p(\alpha_1), \ldots, p(\alpha_n)$.

1.3.P7 If $A^2 = B$ and $Ax = \lambda x$, show that $\lambda^4 x = A^4 x = B^2 x = 0$ and explain why both eigenvalues of A are zero. Then tr $A = 0$ so $A = \begin{bmatrix} a & b \\ c & -a \end{bmatrix}$. Also, $\det A = 0$, so $a^2 + bc = 0$. Then $A^2 = ?$

579

1.3.P10 If some linear combination is zero, say, $0 = \sum_{i=1}^{k}\sum_{j=1}^{n_i} c_{ij}x_j^{(i)} = \sum_{i=1}^{k} y^{(i)}$, use (1.3.8) to show that each $y^{(i)} = 0$.

1.3.P13 Consider $\begin{bmatrix} 0 & 0 \\ 0 & 0 \end{bmatrix}$ and $\begin{bmatrix} 0 & 1 \\ 0 & 0 \end{bmatrix}$.

1.3.P16 $p_A(t) = t^{n-r}(t^r - t^{r-1}\operatorname{tr} B + \cdots \pm \det B)$, so $E_r(A) \neq 0$. (1.2.13) Consider $\begin{bmatrix} 1 & i \\ i & -1 \end{bmatrix}$.

1.3.P17 Let $S = C + iD$ with $C = (S + \bar{S})/2$ and $D = (S - \bar{S})/(2i)$. Then $AS = SB$ and $A\bar{S} = \bar{S}B$ imply that $AC = CB$ and $AD = DB$. Proceed as in the proof of (1.3.28).

1.3.P19 (b) See (0.9.10).

1.3.P21 (o), (n), and (k).

1.3.P23 Consider a similarity of $\mathcal{A}$ via $\begin{bmatrix} I_n & X \\ 0 & I_m \end{bmatrix}$.

1.3.P26 (b) $\tilde{A}_{pq} = \sum_{i,j=1}^{n} P_{pi}A_{ij}P_{qj}^T = \sum_{i,j=1}^{n} \left(e_p^T A_{ij} e_q\right) \varepsilon_i \varepsilon_j^T$.

1.3.P33 (b) (1.3.8). (e) The exercise preceding (1.2.10).

1.3.P35 (a) If $\mathcal{A}$ is reducible, use (1.3.17) to give an example of an $A \in M_n$ such that $A \notin \mathcal{A}$. (b) Every subspace is $\mathcal{A}$-invariant if $\mathcal{A} = \{0\}$. (f) If not, let z be a nonzero vector that is orthogonal to the subspace $\mathcal{A}^*x$, that is, $(A^*x)^*z = x^*Az = 0$ for all $A \in \mathcal{A}$. Use (d) to choose an $A \in \mathcal{A}$ such that $Az = x$. (g) If $d = \min\{\operatorname{rank} A : A \in \mathcal{A} \text{ and } A \neq 0\} > 1$, choose any $A_d \in \mathcal{A}$ with $\operatorname{rank} A_d = d$. Choose distinct i, j such that the vectors Ae_i and Ae_j are linearly independent (pair of columns of A_d), so $A_d e_i \neq 0$ and $A_d e_j \neq \lambda A_d e_i$ for all $\lambda \in \mathbf{C}$. Choose $B \in \mathcal{A}$ such that $B(A_d e_i) = e_j$. Then $A_d B A_d e_i \neq \lambda A_d e_i$ for all $\lambda \in \mathbf{C}$. The range of A_d is $A_d B$-invariant ($A_d B(A_d x) = A_d(B A_d x)$) so it contains an eigenvector of $A_d B$ (1.3.18). Thus, there is an x such that $A_d x \neq 0$ and for some $\lambda_0 \in \mathbf{C}$, $(A_d B - \lambda_0 I)(A_d x) = 0$. Hence, $A_d B A_d - \lambda_0 A_d \in \mathcal{A}$, $A_d B A_d - \lambda_0 A_d \neq 0$, and $\operatorname{rank}(A_d B A_d - \lambda A_d) < d$. This contradiction implies that $d = 1$. (h) For any given nonzero $\eta, \zeta \in \mathbf{C}^n$, choose $A, B \in \mathcal{A}$ such that $\eta = Ay$ and $\zeta = B^*z$. Then $\eta\zeta^* = A(yz^*)B \in \mathcal{A}$.

1.3.P40 (b) If $B = \sum_{i=1}^{N} \alpha_i A_i = 0$, then $0 =]A_i, B[= \alpha_i A_i^2$.

1.3.P41 (a) Consider $D_2 D_1 A D_2 D_2^{-1}$.

Section 1.4

1.4.P6 (a) The principle of biorthogonality.

1.4.P7 Assume without loss of generality that $\lambda_n = 1$ and let $y^{(1)}, \ldots, y^{(n)}$ be linearly independent eigenvectors associated with $\lambda_1, \ldots, \lambda_n$. Write $x^{(0)}$ (uniquely: Why?) as $x^{(0)} = \alpha_1 y^{(1)} + \cdots + \alpha_n y^{(n)}$, with $\alpha_n \neq 0$. Then $x^{(k)} = c_k(\alpha_1 \lambda_1^k y^{(1)} + \cdots + \alpha_{n-1}\lambda_{n-1}^k y^{(n-1)} + \alpha_n y^{(n)})$ for some scalar $c_k \neq 0$. Since $|\lambda_i|^k \to 0$, $i = 1, \ldots, n-1$, $x^{(k)}$ converges to a scalar multiple of $y^{(n)}$.

1.4.P9 (b) Consider $S^{-1}AS$ with $S = \begin{bmatrix} I & 0 \\ z^T & 1 \end{bmatrix}$.

1.4.P11 Why is the list consisting of the first $n - 1$ columns of $A - \lambda I$ linearly independent?

1.4.P13 (a) See (1.2.16).

1.4.P15 $(A - \lambda I + \kappa z w^*)u = 0 \Rightarrow \kappa(w^*u)y^*z = 0 \Rightarrow w^*u = 0 \Rightarrow u = \alpha x \Rightarrow w^*x = 0$.

1.4.P16 Let $\lambda, x = [x_i]_{i=1}^n$ be an eigenvalue–eigenvector pair for A. Then $(A - \lambda I)x = 0 \Rightarrow cx_{k-1} + (a - \lambda)x_k + bx_{k+1} = 0 \Rightarrow x_{k+1} + \frac{a-\lambda}{b}x_k + \frac{c}{b}x_{k-1} = 0, k = 1, \ldots, n$, a second order difference equation with boundary conditions $x_0 = x_{n+1} = 0$, and indicial equation $t^2 + \frac{a-\lambda}{b}t + \frac{c}{b} = 0$ with roots r_1 and r_2. The general solution of the difference equation is (a) $x_k = \alpha r_1^k + \beta r_2^k$ if $r_1 \neq r_2$, or (b) $x_k = \alpha r_1^k + k\beta r_1^k$ if $r_1 = r_2$; α and β are determined by the boundary conditions. In either case, $r_1 r_2 = c/b$ (so $r_1 \neq 0 \neq r_2$) and $r_1 + r_2 = -(a - \lambda)/b$ (so $\lambda = a + b(r_1 + r_2)$). If $r_1 = r_2$, then $0 = x_0 = x_{n+1} \Rightarrow x = 0$. Thus, $r_1 \neq r_2$ and $x_k = \alpha r_1^k + \beta r_2^k$, so $0 = x_0 = \alpha + \beta$ and $0 = x_{n+1} = \alpha(r_1^{n+1} - r_2^{n+1}) \Rightarrow (r_1/r_2)^{n+1} = 1 \Rightarrow r_1/r_2 = e^{\frac{2\pi i \kappa}{n+1}}$ for some $\kappa \in \{1, \ldots, n\}$. Since $r_1 r_2 = c/b$ we have $r_1 = \pm\sqrt{c/b}\, e^{\frac{\pi i \kappa}{n+1}}$ and $r_2 = \pm\sqrt{c/b}\, e^{\frac{-\pi i \kappa}{n+1}}$ (same choice of signs). Thus, the eigenvalues of A are $\{a + b(r_1 + r_2) : \kappa = 1, \ldots, n\} = \{a + 2\sqrt{bc}\cos(\frac{\pi\kappa}{n+1}) : \kappa = 1, \ldots, n\}$ (for a fixed choice of sign for the square root).

Section 2.1

2.1.P2 Use (2.1.4(g)) and (1.1.P1).

2.1.P14 $U^{-1} = U^T = U^*$.

2.1.P16 (g) (1.3.23) the eigenvalues of $[w \; x]^T[-(w^T w)^{-1}w \; y] \in M_2(\mathbf{R})$ are $\frac{1}{2}(x^T y - 1 \pm i(1 - (x^T y)^2)^{1/2})$.

2.1.P22 (0.2.7).

2.1.P24 (b) $\det B \leq \sqrt{27}$.

2.1.P25 $C_r(U^*)$ versus $C_r(U^{-1})$ (0.8.1).

2.1.P26 (a) Examine the proof of (2.1.14). (b) The exercise following (2.1.10).

2.1.P28 (a) Apply the construction in the preceding problem to the first column of A (with $k = n - 1$), then apply it (with $k = n - 2$) to the second column of the transformed matrix.

2.1.P29 The exercise following (2.1.10) or (2.1.P21).

Section 2.2

2.2.P8 $\text{tr}\, C = ?; \begin{bmatrix} 0 & b \\ a & 0 \end{bmatrix}^2 = ?$ Write $A = UBU^*$, in which $B = [b_{ij}]$ has zero main diagonal entries. Then write $B = B_L + B_R$, in which $B_L = [\beta_{ij}]$, $\beta_{ij} = b_{ij}$ if $i \geq j$ and $\beta_{ij} = 0$ if $j > i$.

2.2.P9 Write $A = UBU^*$, in which $B = [b_{ij}]$ has zero main diagonal entries. Then write $B = B_L + B_R$, in which $B_L = [\beta_{ij}]$; $\beta_{ij} = b_{ij}$ if $i > j$ and $\beta_{ij} = 0$ if $j > i$.

2.2.P10 If $\lambda_\ell = 0$, then $|a_i| \leq \sum_{j \neq i} |a_j|$.

Section 2.3

2.3.P6 If $\Delta_1, \Delta_2 \in M_n$ are both upper triangular, what is the main diagonal of $\Delta_1\Delta_2 - \Delta_2\Delta_1$?

2.3.P8 When is $\begin{bmatrix} a & b \\ 0 & c \end{bmatrix}$ complex orthogonal?

2.3.P11 If $T \in M_n$ is strictly upper triangular, what does T^2 look like? T^3? T^{n-1}? T^n?

2.3.P12 (a) (0.8.1) and (2.1.P25). $C_r(UTU^*) = C_r(U)C_r(T)C_r(T)^*$.

2.3.P13 (2.3.4).

2.3.P14 (2.3.1). $\sum_i |\lambda_i| = \text{tr}(DT) = \text{tr}(DU^*AU)$ for a suitable diagonal unitary D.

Section 2.4

2.4.P10 (2.4.P9).

2.4.P11 (c) (2.4.P2).

2.4.P12 (b) Explain why we may assume that A and B are upper triangular and C is strictly upper triangular. (c) If $A = S\Lambda S^{-1}$, $\Lambda = \lambda_1 I_{n_1} \oplus \cdots \oplus \lambda_d I_{n_d}$, and $\lambda_i \neq \lambda_j$ if $i \neq j$, let $\mathcal{C} = S^{-1}CS$ and $\mathcal{B} = S^{-1}BS$. If Λ commutes with $\mathcal{C}$, then $\mathcal{C}$ is block diagonal conformal to Λ. But $\mathcal{C} = \Lambda\mathcal{B} - \mathcal{B}\Lambda$ has zero diagonal blocks, so $\mathcal{C} = 0$. (f) Either $C = 0$ or rank $C = 1$; invoke Laffey's theorem in the preceding problem. (g) A, B, and C are nilpotent, but $B^2 - C^2$ is not; $A + B$ is even nonsingular. (h) Consider

$$A = \begin{bmatrix} e^{4i\theta} & 0 & 0 \\ 0 & e^{2i\theta} & 0 \\ 0 & 0 & e^{2i\theta} \end{bmatrix} \quad \text{and} \quad B = \begin{bmatrix} 0 & 1 & 0 \\ 0 & 0 & 1 \\ 1 & 0 & 0 \end{bmatrix}, \quad \theta = \frac{\pi}{3}$$

Show that $C^3 = 0$ and some eigenvalue of AB is not of the form $e^{ik\theta}$.

2.4.P14 Let the columns of $U_2 \in M_{n,n-r}$ be an orthonormal basis for the null space of A^* and let $U = [U_1 \ U_2]$ be unitary. Then $U^*AU = \begin{bmatrix} U_1^*AU_1 & U_1^*AU_2 \\ 0 & 0 \end{bmatrix}$. Let $V \in M_r$ be unitary and such that $U_1^*AU_1 = V\Delta V^*$ and Δ is upper triangular. Let $W = V \oplus I_{n-r}$. Consider $(UW)^*A(UW)$.

2.4.P15 (c) What is $p_{A,I}(1, t)$ and why is $p_{A,I}(I, A) = 0$?

2.4.P19 If A and B are simultaneously upper triangularizable and $p(s, t)$ is any polynomial in two noncommuting variables, then $p(A, B)(AB - BA)$ is nilpotent. (2.4.8.7). Explain why $p(A_{ii}, B_{ii})(A_{ii}B_{ii} - B_{ii}A_{ii})$ is nilpotent for $i = 1, 2$. (b) Consider $p(A_1, \ldots, A_m)(A_i A_j - A_j A_i)$.

2.4.P22 (a) $K_m = [\sum_{k=1}^d \nu_k \mu_k^{i+j-2}]_{i,j=1}^m = \sum_{k=1}^d \nu_k v_k^{(m)}(v_k^{(m)})^T$. (c) rank $V_m = $ rank $D = d \Rightarrow$ rank $K_m \leq d$; K_d nonsingular $\Rightarrow$ rank $K_m \geq d$.

2.4.P25 (b) (2.4.P10).

2.4.P27 Take $p(t) = p_{CB}(t)$ in the preceding problem.

2.4.P28 (0.4.6(e)).

2.4.P31 What is $p_A(t)$ in this case?

2.4.P33 As in the proof of (2.4.6.1), argue that $A = SDS^{-1}$, in which $D = A_{11} \oplus A_{22}$ and $S = \begin{bmatrix} I_k & S_{12} \\ 0 & I_{n-k} \end{bmatrix}$. Let $C = S^{-1}BS$. Then $C^p = D$, so D commutes with C, which must be block diagonal conformal with D, and $B = SCS^{-1}$.

2.4.P35 If $\mathbf{F} = \mathbf{C}$, consider $U = \text{diag}(e^{i\theta_1}, \ldots, e^{i\theta_n})$ with $0 \leq \theta_1 < \cdots < \theta_n < 2\pi$. If $\mathbf{F} = \mathbf{R}$, consider $U = \text{diag}(\pm 1, \ldots, \pm 1)$, in which $n - 1$ of the entries have the same sign and the remaining entry has the opposite sign. In either case, consider variations on $U = \begin{bmatrix} 0 & 1 \\ 1 & 0 \end{bmatrix} \oplus I_{n-2}$.

Section 2.5

2.5.P18 Consider $A = \begin{bmatrix} 0 & 1 \\ 2 & 0 \end{bmatrix}$ and A^2.

2.5.P20 $\|(A - \lambda I)x\|_2 = \|(A - \lambda I)^*x\|_2$.

2.5.P25 (2.5.16).

2.5.P26 (h) These are all classical polynomial interpolation problems. Look carefully at (0.9.11.4).

2.5.P27 (a) (2.5.16). $(AB^*)A = A(B^*A)$. (b) Let $B = A\bar{A}$. $\Rightarrow$: $BA = A\bar{B} \Rightarrow B^*A = A\bar{B}^*$ (2.5.16). $\Leftarrow$: $BB^* = A(\overline{AA^*A^T}) = A(\overline{A^T A^* A})$ and $B^*B = (AA^*A^T)\bar{A}$.

2.5.P28 (b) $(AB)A = A(BA)$.

2.5.P31 Use (2.5.8).

2.5.P33 Let $U \in M_n$ be a unitary matrix that simultaneously diagonalizes every member of $\mathcal{F}$, let $B = U\text{diag}(1, 2, \ldots, n)U^*$, let $A_\alpha = U\Lambda_\alpha U^*$ with $\Lambda_\alpha = \text{diag}(\lambda_1^{(\alpha)}, \ldots, \lambda_n^{(\alpha)})$, and take $p_\alpha(t)$ to be the Lagrange interpolation polynomial such that $p_\alpha(k) = \lambda_k^{(\alpha)}$, $k = 1, 2, \ldots, n$.

2.5.P34 (b) (2.4.11.1).

2.5.P35 (a) If $xx^* = yy^*$ and $x_k \neq 0$, then $x_j = \overline{(y_k/x_k)}y_j$ for all $j = 1, \ldots; n$. (b) $AA^* = A^*A \Leftrightarrow uu^* = vv^*$, in which $u = x/\|x\|_2$ and $v = y/\|y\|_2$.

2.5.P37 (b) (1.3.P14). (d) (2.5.P35).

2.5.P38 (b) (2.4.P12).

2.5.P43 (a) Use the defect from normality in (2.5.P42).

2.5.P44 (b) $\text{tr}(A^2 B^2) = \text{tr}((AB)(AB)^*)$.

2.5.P46 If $A \in M_n(\mathbf{R})$ and $T = U^*AU$ is upper triangular, then $\bar{T} = U^T A\bar{U}$ is unitarily similar to A and hence to T, so the sets of main diagonal entries of T and $\bar{T}$ are identical.

2.5.P49 If $A = S\Lambda S^{-1}$, let $S = RQ$ be an RQ factorization. Then $R^{-1}AR$ is normal.

2.5.P55 (2.5.P15), and (2.4.P10).

2.5.P56 (2.5.P26).

2.5.P57 Use (2.5.17). Which blocks in (2.5.17.1) are symmetric or skew symmetric?

2.5.P60 $x^T(ne_j - e) = nx_j$; use the Cauchy–Schwarz inequality (0.6.3).

2.5.P61 (c) The preceding problem and $\Delta(A)$ in (2.5.P42).

2.5.P64 (1.3.P39).

2.5.P65 (0.8.1).

2.5.P67 Consider $\Lambda = \Lambda_r \oplus 0_{n-r}$ and $B = [B_{ij}]_{i,j=1}^2$, in which $\Lambda_r \in M_r$ is nonsingular and B is partitioned conformally to Λ. Then $\Lambda B = 0 \Rightarrow B_{11} = 0$ and $B_{12} = 0 \Rightarrow B_{21} = 0 \Rightarrow B\Lambda = 0$.

2.5.P68 The hypothesis is that A is unitarily similar to $A_r \oplus 0_{n-r}$, in which A_r is nonsingular.

2.5.P69 (a) Compare blocks in the respective identities $M_A W = W M_B$, starting in block position $(k, 1)$. Work to the right until reaching block position $(k, k - 1)$. Move up to block position $(k - 1, 1)$ and work to the right until reaching block position $(k - 1, k - 2)$. Repeat this process, moving up one block row at a time until reaching block position $(2, 1)$. (b) (2.5.2).

2.5.P70 (a) Use the preceding problem. (c) (2.2.8).

2.5.P74 Consider $A = \begin{bmatrix} 0 & i \\ i+1 & 0 \end{bmatrix}$, $B = A^T$, $X = \begin{bmatrix} i & 0 \\ 0 & i+1 \end{bmatrix}$.

2.5.P75 (b) (2.5.16).

2.5.P76 (b) See the discussion preceding (2.5.2).

Section 2.6

2.6.P3 Let $n \leq m$ (if $n > m$, consider B^* and A^*). After a common unitary equivalence $(A = V\Sigma W^*, A \to V^* A W, B \to V^* B W)$ we may assume that $A = \Sigma$. If ΣB^* and $B^* \Sigma$ are normal, then $\Sigma\Sigma^T B = B\Sigma^T\Sigma$ (2.5.P27). Let $\Sigma = [\Sigma_q \ 0]$, $\Sigma_q = s_1 I_{n_1} \oplus \cdots \oplus s_d I_{n_d}$, $s_1 > \cdots > s_d \geq 0$, $B = [B_1 \ B_2]$ with $B_1 \in M_n$, and $B_1 = [B_{ij}]_{i,j=1}^d$ conformal to Σ_q. Then $\Sigma\Sigma^T B = B\Sigma^T\Sigma \Rightarrow \Sigma_q^2 B_1 = B_1 \Sigma_q^2$ and $\Sigma_q^2 B_2 = 0$; moreover, $\Sigma_q B$ is normal. If $s_d > 0$, then $B_2 = 0$, $B_1 = B_{11} \oplus \cdots \oplus B_{dd}$, and each B_{ii} is normal. If $s_d = 0$, then $B_2 = \begin{bmatrix} 0 \\ C \end{bmatrix}$, in which C is n_d-by-$(m - n)$, $B_1 = B_{11} \oplus \cdots \oplus B_{d-1,d-1} \oplus B_{dd}$, and each $B_{11}, \ldots, B_{d-1,d-1}$ is normal; replace each normal B_{ii} with its spectral decomposition and replace $[B_{dd} \ C]$ with its singular value decomposition.

2.6.P8 Let $A = V\Sigma W^*$. Then rank $AB =$ rank $\Sigma W^* B$, and $\Sigma W^* B$ has at most rank A nonzero rows.

2.6.P9 Let $D = \Sigma_1 \oplus I_{n-r}$, show that $(AWD^{-1})^*(AWD^{-1}) = I_r \oplus 0_{n-r}$, and conclude that $AWD^{-1} = [V_1 \ 0_{n,n-r}]$, in which V_1 has orthonormal columns. Let $V = [V_1 \ V_2]$ be unitary.

2.6.P10 The preceding problem.

2.6.P11 (2.5.5) and (2.6.P9).

2.6.P13 It suffices to consider only the case in which $A = \Sigma$.

2.6.P14 (b) A is normal if and only if Σ^2 commutes with $W^* V$.

2.6.P15 (2.5.P42).

2.6.P16 (2.6.P3).

2.6.P19 Examine the block diagonal entries of the identities $U^* U = I$ and $UU^* = I$. For example, $U_{11} U_{11}^* + U_{12} U_{12}^* = I_k \Rightarrow U_{11} U_{11}^* = I_k - U_{12} U_{12}^* \Rightarrow \sigma_i^2(U_{11}) = 1 - \sigma_{k-i+1}^2(U_{12})$.

2.6.P21 (compare with (2.6.P3)) If $A\bar{B} = AB^*$ is normal, then $(A\bar{B})^T = \bar{B}A = B^* A$ is normal. We may take $A = \Sigma$ (use (2.6.6a) to write $A = U\Sigma U^T$, so $\Sigma\bar{B}$ is normal and $\bar{B} = U^* B\bar{U}$ is symmetric). If $\Sigma\bar{B}$ and $\bar{B}\Sigma$ are normal and B is symmetric, then $\Sigma^2 B = B\Sigma^2$ (2.5.P27). Write $\Sigma = s_1 I_{n_1} \oplus \cdots \oplus s_d I_{n_d}$, in which $s_1 > \cdots > s_d \geq 0$ and partition $B = [B_{ij}]_{i,j=1}^d$ conformally to Σ. Then $\Sigma^2 B = B\Sigma^2 \Rightarrow B = B_{11} \oplus \cdots \oplus B_{dd}$ and each

B_{ii} is symmetric; B_{ii} is also normal if $s_i > 0$. If $s_i > 0$, replace B_{ii} with $Q_i \Lambda_i Q_i^T$, in which Q_i is real orthogonal and Λ_i is diagonal (2.5.P57); if $s_d = 0$, replace B_{dd} with the special singular value decomposition in (2.6.6(a)).

2.6.P25 Let $A = V \Sigma W^*$ be a singular value decomposition with $\Sigma = \Sigma_r \oplus 0_{n-r}$, write $A = V \Sigma (W^* V) V^*$, and partition $W^* V = \begin{bmatrix} K & L \\ M & N \end{bmatrix}$.

2.6.P26 (a) $L = 0 \Rightarrow M = 0 \Rightarrow K$ is unitary. (c) Proceed as in (2.6.P23(c)).

2.6.P28 (a) Let $A = V \Sigma W^*$ be a singular value decomposition with $\Sigma = \Sigma_1 \oplus 0_{n-r}$ and partition $V = [V_1 \ V_2]$, $W = [W_1 \ W_2]$ with $V_1, W_1 \in M_{n,r}$. Then $W_1 = V_1 U$ for some unitary $U \in M_r$ (why?) ensures that $A = V \begin{bmatrix} \Sigma_1 U & 0 \\ 0 & 0 \end{bmatrix} V^*$. (b) (1.3.P16).

2.6.P29 (2.5.P38).

2.6.P31 (b) (1.3.P19).

2.6.P32 Consider $\begin{bmatrix} 0 & I_n \\ I_n & 0 \end{bmatrix} \mathcal{A}$ and use (2.6.P7).

2.6.P33 (0.8.1) $C_r(V \Sigma W^*) = C_r(V) C_r(\Sigma) C_r(W)^*$.

2.6.P35 Apply (2.6.9) to the matrix $A - (\frac{1}{n} \operatorname{tr} A) I$.

Section 2.7

2.7.P1 Permute and partition.

2.7.P2 Suppose that $m = n$. Let $A = V \Sigma W^*$ be a singular value decomposition in which $\Sigma = C \oplus I_{n-\nu}$ and $C = \operatorname{diag}(c_1, \ldots, c_\nu)$ with each $c_j \in [0, 1)$. Let $S = \operatorname{diag}((1 - c_1^2)^{1/2}, \ldots, (1 - c_\nu^2)^{1/2})$ and consider

$$(V_1 \oplus V) \begin{bmatrix} C & S & 0 \\ -S & C & 0 \\ 0 & 0 & I_{n-\nu} \end{bmatrix} (W_1 \oplus W)^* \in M_{n+\nu}$$

in which $V_1, W_1 \in M_\nu$ are arbitrary unitary matrices. If $m \neq n$, pad A with a zero block to obtain a square contraction of size $\max\{m, n\}$ that has $\nu + |m - n|$ singular values that are strictly less than one.

2.7.P6 (b) If $u_{12} \neq 0$ and $u_{12}/u_{21} = e^{i\phi}$, consider a similarity via $D = \operatorname{diag}(1, e^{i\phi/2})$.

Section 3.1

3.1.P12 (e) Both say that there are exactly k blocks $J_\ell(\lambda)$ of size $\ell = p$ in the Jordan canonical form of A.

3.1.P16 (3.1.18).

3.1.P17 (3.1.P16).

3.1.P18 (b) (3.1.P17) and (1.3.22).

3.1.P19 (c) (3.1.18).

3.1.P20 (1.3.P16).

3.1.P22 (a) Show that there is a positive diagonal D such that DAD^{-1} is symmetric and apply (3.1.P21). (b) Perturb A and use continuity.

3.1.P23 (a) Proceed as in (3.1.P22); choose a positive diagonal D such that DAD^{-1} is Hermitian. (c) Consider iA; (1.4.P4).

3.1.P24 (d) Consider the directed graphs of A and B (6.2).

3.1.P26 Use (2.5.P19) to explain why it suffices to show that if $x \in \mathbf{C}^n$ and $A^2 x = 0$, then $Ax = 0$. $A^2 x = 0 \Rightarrow 0 = \|A^2 x\|_2^2 = x^* A^* A^* A A x = x^* A^* A A^* A x = \|A^* A x\|_2^2 \Rightarrow 0 = x^* A^* A x = \|Ax\|_2^2$.

3.1.P27 $A = S \Lambda S^{-1} = QR\Lambda R^{-1} Q^* \Rightarrow R\Lambda R^{-1}$ is normal and upper triangular.

3.1.P30 The preceding problem.

Section 3.2

3.2.P2 Why does it suffice to consider the case in which A is a Jordan matrix? Suppose that $A = J_k(\lambda) \oplus J_\ell(\lambda) \oplus J$, in which J is either empty or is a Jordan matrix and $k, \ell \geq 1$. For any polynomial $p(t)$ the leading $k + \ell$ diagonal entries of $p(A)$ are all equal to $p(\lambda)$. But $-I_k \oplus I_\ell \oplus I_{n-k-\ell}$ commutes with A.

3.2.P3 (1.3.P20(f)).

3.2.P15 Use the two preceding problems.

3.2.P19 (1.4.7).

3.2.P20 (a) (3.2.11.1).

3.2.P24 (a) Write $A^T = SAS^{-1}$ (3.2.3.1). Then $D = AB - B(SAS^{-1}) \Rightarrow DS = A(BS) - (BS)A$. Also, $AD = DA^T \Rightarrow A(DS) = (DS)A$. Invoke Jacobson's lemma to conclude that DS is nilpotent. (b) Let $A = S\Lambda S^{-1}$ with $\Lambda = \lambda_1 I_{n_1} \oplus \cdots \oplus \lambda_d I_{n_d}$; let $\mathcal{D} = S^{-1} DS$ and $\mathcal{B} = S^{-1} B S^{-T}$. Then $\Lambda \mathcal{D} = \mathcal{D}\Lambda$ and $\mathcal{D} = \Lambda \mathcal{B} - \mathcal{B}\Lambda$. Conclude that $\mathcal{D}$ and $\mathcal{B}$ are block diagonal conformal to Λ and $\mathcal{D} = 0$. (c) $D^2 = ABD - BA^T D = A(BD) - (BD)A$. Also, $AD^2 = (AD)D = DA^T D = D^2 A$. Invoke Jacobson's lemma to conclude that D^2 is nilpotent. (d) Let $A = SJS^{-1}$ with $J = J_{n_1}(\lambda_1) \oplus \cdots \oplus J_{n_d}(\lambda_d)$; let $\mathcal{D} = S^{-1} DS$ and $\mathcal{B} = S^{-1} B S^{-T}$. Then $J\mathcal{D} = \mathcal{D}J^T$ and $\mathcal{D} = J\mathcal{B} - \mathcal{B}J^T$. Conclude that $\mathcal{D} = \mathcal{D}_1 \oplus \cdots \oplus \mathcal{D}_d$ and $\mathcal{B} = \mathcal{B}_1 \oplus \cdots \oplus \mathcal{B}_d$ are block diagonal conformal to J, $J_i \mathcal{D}_i = \mathcal{D}_i J_i^T$, and $\mathcal{D}_i = J_i \mathcal{B}_i - \mathcal{B}_i J_i^T$ for each $i = 1, \ldots, d$ $(J_i := J_{n_i}(\lambda_i))$. Let $J_i^T = S_i J_i S_i^{-1}$. Then $J_i(\mathcal{D}_i S_i) = (\mathcal{D}_i S_i) J_i$, and $(\mathcal{D}_i S_i) = J_i(\mathcal{B}_i S_i) - (\mathcal{B}_i S_i)J_i$. Invoke Jacobson's lemma.

3.2.P26 Show that $(B - B^T)A = 0$ and explain why rank$(B - B^T) \leq 1$. See (2.6.P27).

3.2.P27 (c) See (2.6.P12).

3.2.P29 (a) tr $C = 0$ or use Jacobson's Lemma.

3.2.P31 Modify the argument in the preceding problem. (c′) There is a nonsingular $X \in M_n$ such that $XA = A^T X$. (d′) There is a nonsingular $Y \in M_k$ such that $YA_{11}^T = A_{11}Y$. Let $C = Y \oplus 0_{n-k}$. (f′) Let $B = CX$. Then $AB = BA$.

3.2.P32 See (2.4.P12(c)) for the diagonalizable case. Explain why it suffices to consider the case $A = J_2(\lambda)$. Invoke (3.2.4.2) and argue that C is strictly upper triangular; explain why strict upper triangularity of $AB - BA$ implies that B is strictly upper triangular. Jacobson's lemma can be useful in these arguments.

3.2.P33 Use (2.3.1) to reduce (via a unitary similarity) to the case in which A is upper triangular. Then B is a polynomial in A as well as a polynomial in A^*, so it is both upper and lower triangular.

3.2.P34 (a) Each of $\bar{A}$ and A^T is a polynomial in A, so they commute.

3.2.P36 (a) (3.2.5.2).

Section 3.3

3.3.P3 Show that $t^2 - t = t(t - 1)$ annihilates A.

3.3.P10 Use cofactors to compute the determinant.

3.3.P13 (1.2.20).

3.3.P19 Suppose that $C^{n-1} \neq 0$ and use (3.2.4.2).

3.3.P20 (d) What is column $n - 1$ of AB?

3.3.P22 (3.2.P4).

3.3.P26 (f) $(\lambda_i I - A)q_i(A) = q_A(A)$.

3.3.P28 $K^2 = I$, so there are three possibilities for the minimal polynomial of K.

3.3.P29 Let $K = SDS^{-1}$ with $D = I_m \oplus (-I_{n-m})$ and let $\mathcal{A} = S^{-1}AS = [A_{ij}]_{i,j=1}^2$. Then $K A$ is similar to $D\mathcal{A}$ and $\mathcal{A} = D\mathcal{A}D \Rightarrow A_{12} = 0$ and $A_{22} = 0$.

3.3.P30 Let $T = iI_m \oplus I_{n-m}$ and compute $T\mathcal{B}T^{-1}$.

3.3.P31 Use (3.3.4).

3.3.P33 Apply Schur's inequality (2.3.2a) to the companion matrix of p.

Section 3.4

3.4.P1 What is the real Jordan canonical form of A?

3.4.P4 (a) Use (3.4.2.10) and the identities (3.4.2.12). (b) $w_i(A, \lambda_j)^2 \geq w_i(A, \lambda_j)$.

3.4.P5 Use (3.4.3.1) as in (3.4.3.3); the unitary Weyr form of A is again block 2-by-2, but now F_{12} has full column rank.

3.4.P8 (c) Suppose that W has exactly p diagonal blocks (that is, $w_p(J, 0) > 0$ and $w_{p+1}(J, 0) = 0$). Where do the first p rows and columns of $P^T W P$ come from? What does the leading p-by-p principal submatrix of $P^T W P$ look like? If $w_p(J, 0) > 1$, what does $P^T W P[\{p + 1, \ldots, 2p\}]$ look like? What happens if $w_p(J, 0) = 1$?

Section 4.1

4.1.P3 If $A = SBS^{-1}$, show that $A = U\Lambda U^*$ and $B = V\Lambda V^*$ with U and V unitary, so $U^*AU = \Lambda = V^*BV$.

4.1.P6 If $x^*Ax = 0$ for all $x \in \mathbf{C}^n$, (4.1.4) ensures that A is Hermitian. Let x be any eigenvector of A; why must its corresponding eigenvalue be zero?

4.1.P9 Let $A = [a_{ij}]$ and $B = [b_{ij}]$. Use $x = e_i$ and $y = e_j$ to show that $|a_{ij}| = |b_{ij}|$ for all $i, j = 1, \ldots, n$. Let $x = e_i, y = se_j + te_k$ to show that $|sa_{ij} + ta_{ik}|^2 = |sb_{ij} + tb_{ik}|^2$ and hence $\text{Re}(s\bar{t}[a_{ij}\bar{a}_{ik} - b_{ij}\bar{b}_{ik}]) = 0$ for all $s, t \in \mathbf{C}$. Deduce that $a_{ij}/b_{ij} = a_{ik}/b_{ik}$ if $b_{ij}b_{ik} \neq 0$.

4.1.P12 $A = \begin{bmatrix} 0 & 1 \\ 0 & 0 \end{bmatrix}$.

4.1.P13 (a) (2.4.P2). (b) (4.1.P12).

4.1.P18 (2.5.P44).

4.1.P19 $x = (I - A)x + Ax$ is a sum of vectors in the null space and range of A. If the null space is orthogonal to the range, then $x^*Ax = ((I - A)x + Ax)^*Ax = x^*(A^*A)x$ is real.

4.1.P20 Show that $(A - A^*)^3 = 0$; $A - A^*$ is normal.

4.1.P25 (0.8.1) and (2.3.P12).

Section 4.2

4.2.P3 Take $x = e_i$.

4.2.P7 If S_1 and S_2 are subspaces of $\mathbf{C}^n$ with dim $S_1 = p$ and dim $S_2 = q$, let the columns of $X \in M_{n,p}$ and $Y \in M_{n,q}$, respectively, be bases for S_1 and S_2, respectively, and let $Z = [X \ Y]$. Explain why $S_1 + S_2 = \operatorname{range} Z$, so rank $Z = \dim(S_1 + S_2)$. Since rank $Z +$ nullity $Z = p + q$, it suffices to show that nullity $Z = \dim(\operatorname{range} X \cap \operatorname{range} Y)$.

4.2.P8 $x^*(A + B)x = x^*Ax + x^*Bx \geq x^*Ax$.

Section 4.3

4.3.P5 Use interlacing.

4.3.P6 Consider

$$\begin{bmatrix} 0 & i & 1 \\ -i & 0 & 1 \\ 1 & 1 & 0 \end{bmatrix}$$

4.3.P8 Use (4.3.48), (4.3.49), or the definition.

4.3.P9 The preceding problem; $x = e$ and $x = e_i$.

4.3.P12 $\sum_{i,j=1}^m |b_{ij}|^2 \leq \sum_{i,j=1}^m |b_{ij}|^2 + \sum_{i,j} |c_{ij}|^2 \leq \sum_{i=1}^m R_i^2 \leq \sum_{i=1}^m \sigma_i^2$, so $\sum_{i=1}^m \sigma_i^2 = \sum_{i,j=1}^m |b_{ij}|^2 \Rightarrow \sum_{i,j} |c_{ij}|^2 = 0$.

4.3.P14 Use the quasilinearization (4.3.39).

4.3.P15 (b) (2.1.14): $X = VR$ and det $X^*X = \det R^*R$.

4.3.P16 (a) $A = U^*\Lambda U$ with a unitary $U = [u_1 \ u_2] = [u_{ij}] \in M_2$ and a diagonal $\Lambda = \operatorname{diag}(\lambda_1, \lambda_2) = \lambda_1 I + (\lambda_2 - \lambda_1)E_{22}$. Compute $a_{12} = u_1^*\Lambda u_2 = (\lambda_2 - \lambda_1)\bar{u}_{21}u_{22}$. (b) Interlacing.

4.3.P17 (a) What is the nth entry of $Ax = \lambda x$? (f) $p_n(\lambda) = p_{n-1}(\lambda) = 0 \Rightarrow p_{n-1}(\lambda) = p_{n-2}(\lambda) = 0 \Rightarrow \cdots \Rightarrow p_0(\lambda) = 0$.

4.3.P19 Compute tr$(A + zz^*)$.

4.3.P20 (1.2.P13).

4.3.P21 Eigenvalue interlacing.

4.3.P22 Why is rank $A \leq 2$? If $a_i \neq a_j$, why does the principal submatrix $\begin{bmatrix} 2a_i & a_i + a_j \\ a_i + a_j & 2a_j \end{bmatrix}$ have one positive and one negative eigenvalue? Compare with (1.3.25).

4.3.P23 Consider $A = \text{diag } x$ and $B = \text{diag } y$.

4.3.P25 (a) What are the entries of AA^*? (b) Apply (4.3.P17) to AA^*.

4.3.P29 Use (4.3.39).

4.3.P30 Let U_{ii} be unitary and such that each $U_{ii}^* A_{ii} U_{ii} = \Lambda_i$ is diagonal. Let $U = U_{11} \oplus \cdots \oplus U_{kk}$. Why is $d(A)$ majorized by $\lambda(A_{11} \oplus \cdots \oplus A_{kk}) = d(U^* A U)$? Why is $d(U^* A U)$ majorized by $\lambda(A)$?

Section 4.4
4.4.P1 (4.4.4c) with $S = U\Sigma^{1/2}$.

4.4.P4 If $A = Q\Lambda Q^T$ with Λ real diagonal and Q real orthogonal, write $\Lambda = \Sigma D^2$ and let $U = QD$. When can all the factors in the factorization $A = U\Sigma U^T$ be taken to be real?

4.4.P5 Partition $W = [W_{ij}] = U^* V$ conformally to Σ. $\Sigma \bar{W} = W\Sigma \Rightarrow s_i W_{ij} = s_j \bar{W}_{ij} \Rightarrow$ W_{ii} is real if $s_i \neq 0$ and $(s_i - s_j) \text{tr } W_{ij} W_{ij}^* = 0$.

4.4.P6 Consider the sign of $r(A) = (\text{tr } A\bar{A})^2 - 4|\det A|^2$, the discriminant of $p_{A\bar{A}}(t)$, and the fact that $\text{tr } A\bar{A} < 0$ if $A\bar{A}$ has two negative eigenvalues. The product of the eigenvalues is $|\det A|^2$.

4.4.P9 (b) (4.4.P1).

4.4.P19 Represent A as in (4.4.10). Use Schur's inequality, (4.4.11a,b), and (4.4.12a,b) to show that $\text{tr } AA^* \geq \text{tr } \Delta\Delta^* + \sum_{j=1}^{q/2} \text{tr } \Gamma_{jj} \Gamma_{jj}^* \geq \text{tr } \Delta\bar{\Delta} + \text{tr } \Gamma\bar{\Gamma} = \text{tr } A\bar{A}$.

4.4.P22 (3.3.P28).

4.4.P23 (0.9.8).

4.4.P24 (a) Use (1.4.12(a)); alternatively, use (2.4.11.1(b)). (b) Use (1.4.12(b)).

4.4.P26 If Q is complex orthogonal and $A = QBQ^T$, then every word $W(A, A^*) = W(A, A^T)$ is similar (via Q) to $W(B, B^T) = W(B, B^*)$. Use (2.2.6) and (2.5.21).

4.4.P28 $\det(I + X) = \prod_{i=1}^n (1 + \lambda_i(X))$ and (4.4.13).

4.4.P29 (b) (0.8.5.1), the preceding problem, and a continuity argument. (e) (3.4.1.7). (f) (3.2.P30).

4.4.P30 $SAS^{-1} = SBS^T S^{-T} CS^{-1}$.

4.4.P31 (4.4.25).

4.4.P32 (3.2.3.1) and (3.2.11).

4.4.P33 (b) Show that the negation of each condition implies the negation of the other; use (3.1.P12).

4.4.P35 (2.5.3) and (2.5.14).

4.4.P38 (d) (1.3.P19).

4.4.P39 $A = B \Lambda B^{-1} = B^{-T} \Lambda B \Rightarrow B^T B \Lambda = \Lambda B^T B \Rightarrow S \Lambda = \Lambda S \Rightarrow A = Q \Lambda Q^T.$

4.4.P44 (a) What does $A \bar{A} = A A^*$ say about the blocks in (4.4.31)?

4.4.P45 (4.4.22).

4.4.P46 (b) (2.5.P69 and P70).

4.4.P47 (a) Follow the hint in (2.5.P69(a)).

4.4.P48 (2.5.10).

Section 4.5

4.5.P2 (4.4.19).

4.5.P4 Use (1.3.2) and the proof of (4.5.17(a)). Let $S A_1^{-1} A_i S^{-1}$ be real diagonal for all $i = 1, \ldots, k$. Let $B_i = S^{-*} A_i S^{-1}$ and show that $\{B_i\}$ is a commuting family of Hermitian matrices. There is a unitary U such that $U B_i U^*$ is diagonal for all $i = 2, \ldots, k$; $T = US$ provides the required *congruence.

4.5.P10 (4.5.P9(c)). $A = U(B \oplus 0_{n-r})U^* \Rightarrow A = U_1 B U_1^*$. Let $P \in M_n$ be a permutation matrix such that $P U_1 = \begin{bmatrix} X \\ Y \end{bmatrix}$ and $X \in M_r$ is nonsingular. Then $P A P^T = ?$

4.5.P11 (a) Why is 0_{n-r} the singular part of A with respect to *congruence, and why is Λ a regular part of A? What is the *cosquare of Λ? (b) Diagonalize A by unitary *congruence, construct a *congruence to a direct sum of canonical blocks, and invoke uniqueness in (4.5.21).

4.5.P13 What is the Jordan canonical form (respectively, *congruence canonical form) of a unitary matrix?

4.5.P14 (b) If $Ax = 0$, then $H^{-1} K x = i x$. (c) (4.5.23).

4.5.P18 What are the eigenvalues of $A^{-1} B$?

4.5.P20 (d) Consider the continuous function $a_{n-r}(A)$ on $\mathcal{S}$. Why does it have constant sign on $\mathcal{S}$? If $A, B \in \mathcal{S}$ have different inertias, let $f : [0, 1] \to \mathcal{S}$ be a continuous function with $f(0) = A$ and $f(1) = B$ and consider $g(t) = a_{n-r}(f(t))$.

4.5.P22 If the inequalities (4.3.18) are not satisfied, let k be the *smallest* index such that either (a) $\lambda_k(A) > \lambda_k(B)$ or (b) $\lambda_k(B) > \lambda_{k+1}(A)$. Suppose that (a) is the case, so $\lambda_k(A) > \lambda_k(B) \geq \lambda_{k-1}(B) \geq \lambda_{k-1}(A)$. Let $\alpha \in (\lambda_k(B), \lambda_k(A))$. Why is $B - \alpha I$ nonsingular? Use Haynsworth's theorem to show that $i_-(A - \alpha I) \geq i_-(B - \alpha I)$. Why is $\lambda_k(A) - \alpha > 0 > \lambda_{k-1}(A) - \alpha$ and $\lambda_k(B) - \alpha < 0$? Why does $A - \alpha I$ have $k - 1$ negative eigenvalues? Why does $B - \alpha I$ have at least k negative eigenvalues. Consider case (b).

4.5.P26 Use (4.5.27) and the exercise preceding (4.5.26). The cosquares of $H_{2k}(\mu)$ and $H_{2k}(\bar{\mu})$ are similar if and only if either $\mu = \bar{\mu}$ or $\mu = \bar{\mu}^{-1}$.

4.5.P31 If A is nonsingular and *congruent to a real matrix, then $A^{-*} A$ is similar to a real matrix.

4.5.P33 Use (3.2.3) for $J_k(0)$ and (4.5.27) for Γ_k and $H_{2k}(\mu)$.

4.5.P34 Use (3.2.3) for $J_k(0)$. If $J_k(\mu)^T = S J_k(\mu) S^{-1}$ and $S = \begin{bmatrix} 0_n & S^* \\ S^{-1} & 0_n \end{bmatrix}$, then $S H_{2k}(\mu)^T S^* = H_{2k}(\mu)$.

4.5.P36 Use (4.5.7) to reduce to the case in which $A = I_{i_+} \oplus (-I_{i_-}) \oplus 0_{i_0}$ and $i_+ i_- > 0$. Let $\alpha = \{1, \ldots, i_+\}$, $\beta = \{i_+ + 1, \ldots, i_+ + i_-\}$, and $\gamma = \{i_+ + i_- + 1, \ldots, n\}$. Choose vectors x such that $x^* A x = 0$ and that permit one to deduce from $x^* B x = 0$ that certain entries of B are zero and others are proportional to the corresponding entries of A. Adjustable scalars c and d in the following choices of x have modulus 1. (1) Take $i \in \alpha$ and $j \in \beta$. Choose $x = e_1 + c e_j$. Then $x^* B x = b_{11} + b_{jj} + 2 \operatorname{Re} c b_{1j} = 0$. Choose c to get $b_{11} + b_{jj} \pm |b_{1j}| = 0$, so $b_{jj} = -b_{11}$. Now choose $x = e_i + c e_j$. Then $x^* B x = b_{ii} + b_{jj} + 2 \operatorname{Re} c b_{ij} = b_{ii} - b_{11} + 2 \operatorname{Re} c b_{ij} = 0$ and $b_{ii} - b_{11} \pm |b_{ij}| = 0$, so $b_{ij} = 0$ and $b_{ii} = b_{11}$. Conclude that $B[\alpha, \beta] = 0$, $\operatorname{diag} B[\alpha] = [b_{11} \; \ldots \; b_{11}]^T$, and $\operatorname{diag} B[\beta] = -[b_{11} \; \ldots \; b_{11}]^T$. (2) Take $i \in \alpha$, $j \in \beta$, and $k \in \gamma$. Choose $x = e_k$. Then $x^* B x = b_{kk} = 0$. Now choose $x = e_i + c e_j + d e_k$. Then $x^* B x = b_{ii} + 2 \operatorname{Re} c b_{ij} + 2 \operatorname{Re} d b_{ik} + b_{jj} + 2 \operatorname{Re} \bar{c} d b_{jk} + b_{kk} = 2 \operatorname{Re} d b_{ik} + 2 \operatorname{Re} \bar{c} d b_{jk} = 0$, so $|b_{ik}| = \pm |b_{jk}|$. Conclude that $B[\alpha, \gamma] = 0$, $B[\beta, \gamma] = 0$, and $\operatorname{diag} B[\gamma] = 0$. (3) If $|\gamma| > 1$, take $i, j \in \gamma$, $i \neq j$, and $x = e_i + c e_j$. Then $x^* B x = b_{ii} + b_{jj} + 2 \operatorname{Re} c b_{ij} = 2 \operatorname{Re} c b_{ij} = 0$, so $b_{ij} = 0$. Conclude that $B[\gamma] = 0$. (4) If $|\alpha| > 1$, take $i, j \in \alpha$, $i \neq j$, $k \in \beta$, and $x = 3 e_i + 4 c e_j + 5 e_k$. Then $x^* B x = 24 \operatorname{Re} c b_{ij} = 0$, so $b_{ij} = 0$. Conclude that $B[\alpha] = b_{11} I_{i_+}$. (5) If $|\beta| > 1$, take $i \in \alpha$, $j, k \in \beta$, $j \neq k$, and $x = 5 e_i + 3 e_j + 4 c e_k$. Conclude that $b_{jk} = 0$ and $B[\beta] = -b_{11} I_{i_-}$.

Section 4.6

4.6.P3 If $c_1, \ldots, c_k \in \mathbf{C}$ and $z = c_1 x_1 + \cdots + c_k x_k = 0$, then $A \bar{z} = \lambda (\overline{c_1} x_1 + \cdots + \overline{c_k} x_k) = 0$, so $(\operatorname{Re} c_1) x_1 + \cdots + (\operatorname{Re} c_k) x_k = 0$ and $(\operatorname{Im} c_1) x_1 + \cdots + (\operatorname{Im} c_k) x_k = 0$.

4.6.P6 (4.4.4(c)).

4.6.P7 Represent A as in (4.6.9).

4.6.P9 Why must $A \bar{A}$ have at least one nonnegative eigenvalue?

4.6.P14 If $\mu \neq 0$, consider $H_{2k}(\mu)^{-1} H_{2k}(\mu) \overline{H_{2k}(\mu)}$.

4.6.P16 (a) $A \bar{A} z_j = ?$ (b) Look at the Type I blocks. (c) Linear independence over $\mathbf{C}$ always implies linear independence over $\mathbf{R}$. (d) In this case, though not in general, linear independence over $\mathbf{R}$ implies linear independence over $\mathbf{C}$.

4.6.P17 Consider the linear system $A \bar{x}_1 = X b_1$ involving an unknown vector b_1. Why is this linear system consistent (0.4.2)? Why does it have a *unique* solution?

4.6.P18 (f) g is equal to half the geometric multiplicity of σ^2 as an eigenvalue of $R_2(A)^2$, which is equal to the geometric multiplicity of σ^2 as an eigenvalue of $A \bar{A}$.

4.6.P19 (1.3.P21), parts (m) and (c), together with (4.6.18).

4.6.P21 (4.6.11), (4.6.7), (4.6.P17 and P20).

4.6.P24 (b) (1.3.P19).

4.6.P26 $D^2 \Delta = \Delta \bar{\Delta} \Delta = \Delta (\overline{\bar{\Delta} \Delta}) = \Delta D^2$.

4.6.P27 (a) Start with (4.6.3). If A is unitarily congruent to Δ as in the preceding problem, then normality ensures that Δ is block diagonal and two cases must be considered: $\Delta_j \overline{\Delta_j} = 0$ or $\Delta_j \overline{\Delta_j} = \lambda I_{n_j}$ with $\lambda > 0$.

Section 5.1

5.1.P4 (a) Use the inner product to express each of the four terms in (5.1.9).

5.1.P10 $\|\vec{0}\| = \|0\vec{0}\| = |0|\|\vec{0}\| \Rightarrow \|\vec{0}\| = 0$ ($\vec{0}$ denotes the zero vector; 0 is the zero scalar).
$0 = \|\vec{0}\| = \|x - x\| \leq \|x\| + \|-x\| = 2\|x\| \Rightarrow$ (1).

5.1.P14 Since $\sum_{i=1}^{n}(x_i - \mu) = 0$, $(x_j - \mu)^2 = (\sum_{i \neq j}(x_i - \mu))^2 \leq (n-1)\sum_{i \neq j}(x_i - \mu)^2 = n(n-1)\sigma^2 - (n-1)(x_j - \mu)^2$.

Section 5.2

5.2.P6 (2.1.13).

5.2.P7 (a) $\|x\| \left\| \frac{x}{\|x\|} - \frac{y}{\|y\|} \right\| \leq \|x\| \left\| \frac{x}{\|x\|} - \frac{y}{\|x\|} \right\| + \|x\| \left\| \frac{y}{\|x\|} - \frac{y}{\|y\|} \right\| = \|x - y\| + |\|y\| - \|x\||$
$\leq 2\|x - y\|$. (c) $\|x\|\|y\| \left\| \frac{x}{\|x\|} - \frac{y}{\|y\|} \right\|^2 = 2\|x\|\|y\| - 2\operatorname{Re}\langle x, y \rangle = 2\|x\|\|y\| - (\|x\|^2 + \|y\|^2 - \|x - y\|^2) = \|x - y\|^2 - (\|x\| - \|y\|)^2$ and hence $4\|x - y\|^2 - (\|x\| + \|y\|)^2$
$\left\| \frac{x}{\|x\|} - \frac{y}{\|y\|} \right\|^2 = \frac{(\|x\| - \|y\|)^2}{\|x\|\|y\|}((\|x\| + \|y\|)^2 - \|x - y\|^2) \geq 0$.

5.2.P9 Consider the unit vector $u(t) = (b - a)^{-1/2}$, $\lambda = (b - a)^{1/2}(\alpha + \beta)/2$, and $\mu = (b - a)^{1/2}(\gamma + \delta)/2$.

5.2.P11 Without loss of generality, assume that $\|x\| \leq \|y\|$ and compute $\|x + y\| = \left\| \frac{\|x\|}{\|x\|}x + \frac{\|x\|}{\|y\|}y + (1 - \frac{\|x\|}{\|y\|})y \right\| \leq \|x\| \left\| \frac{x}{\|x\|} + \frac{y}{\|y\|} \right\| + (1 - \frac{\|x\|}{\|y\|})\|y\| = \|x\|(\left\| \frac{x}{\|x\|} + \frac{y}{\|y\|} \right\| - 1) + \|y\| = \|x\| + \|y\| + \|x\|(\left\| \frac{x}{\|x\|} + \frac{y}{\|y\|} \right\| - 2)$. For the lower bound, consider $\|x + y\| = \left\| \frac{\|y\|}{\|y\|}y + \frac{\|y\|}{\|x\|}x - (\frac{\|y\|}{\|x\|} - 1)x \right\| \geq \|y\| \left\| \frac{\|y\|}{\|y\|}y + \frac{\|y\|}{\|x\|}x \right\| - (\frac{\|y\|}{\|x\|} - 1)\|x\|$.

5.2.P12 (5.2.16) ensures that $\max\{\|x\|, \|y\|\} \left\| \frac{x}{\|x\|} - \frac{y}{\|y\|} \right\| \leq \|x - y\| - \|x\| - \|y\| + 2\max\{\|x\|, \|y\|\} = \|x - y\| + |\|x\| - \|y\||$.

5.2.P13 (a) Why is $\langle H, iK \rangle_F + \langle iK, H \rangle_F = 0$?

Section 5.3

5.3.P2 Consider $y = e_2$, $z = e_1$.

Section 5.4

5.4.P3 Jensen's inequality (Appendix B) and Hölder's inequality:

$$\|x\|_{p_1}^{p_1} = \sum_{i=1}^{n} |x_i|^{p_1} \leq \left(\sum_{i=1}^{n} \left(|x_i|^{p_1} \right)^{\frac{p_2}{p_1}} \right)^{\frac{p_1}{p_2}} \left(\sum_{i=1}^{n} \left(1^{\frac{p_2}{p_2 - p_1}} \right) \right)^{\frac{p_2 - p_1}{p_2}}$$
$$= n^{\frac{p_2 - p_1}{p_2}} \|x\|_{p_2}^{p_1}$$

5.4.P4 Consider $f(x) = 1/\|x\|_\alpha$ on the unit sphere S of $\|\cdot\|_\beta$. If f is unbounded on S, there is a sequence $\{x_N\} \subset S$ with $\|x_N\|_\alpha < 1/N$ and $\|x_N\|_\beta = 1$ for all $N = 1, 2, \ldots$, which contradicts equivalence of $\|\cdot\|_\alpha$ and $\|\cdot\|_\beta$.

5.4.P8 If $1 < k < n$ reduce to: for $y_1 \geq \cdots \geq y_n \geq 0$, maximize $x_1 y_1 + \cdots + x_n y_n$ subject to $x_1 \geq \cdots \geq x_k \geq 0$, $x_k = x_{k+1} = \cdots = x_n$, $x_1 + \cdots + x_k = 1$. Let $x_i = x_k + t_i$, $i = 1, \ldots, k - 1$ and maximize $f(x_k, t_1, \ldots, t_{k-1}) = x_k(y_1 + \cdots + y_n) + t_1 y_1 + \cdots +$

$t_{k-1} y_{k-1}$ over $\mathcal{S} = \{x_k, t_1, \ldots, t_{k-1} : kx_k + t_1 + \cdots + t_{k-1} = 1$ and $x_k, t_1, \ldots, t_{k-1} \geq 0\}$. The maximum of f is achieved at an extreme point of $\mathcal{S}$: $x_k = \frac{1}{k}$ and $t_1 = \cdots = t_{k-1} = 0$, or $x_k = 0$ and some $t_i = 1$ and all other $t_j = 0$.

5.4.P9 (5.4.13) and (5.4.14) with $A = \begin{bmatrix} 1 & -1 \\ 0 & 1 \end{bmatrix}$ and $\|x\| = \|Ax\|_1$.

5.4.P12 $\|A^* y\|^D = \max_{\|x\|=1} |y^* Ax| = \max_{\|A^{-1}z\|=1} |y^* z| = \max_{\|z\|=1} |y^* z| = \|y\|^D$.

5.4.P13 (5.4.P11(e)). Consider an isometry $A = [a_{ij}]$ for l_p with $1 \leq p < 2 < q \leq \infty$; otherwise, consider A^* and l_q. For each standard basis vector e_j, $\|Ae_j\|_p = \|e_j\|_p = 1$, so (a) $\sum_{i=1}^n |a_{ij}|^p = 1$ for each $j = 1, \ldots, n$; (b) $|a_{ij}| \leq 1$ for all $i, j = 1, \ldots, n$; and (c) $\sum_{i,j=1}^n |a_{ij}|^p = n$. Consideration of A^* and q shows that $\sum_{i,j=1}^n |a_{ij}|^q = n$ as well. However, $|a_{ij}|^q \leq |a_{ij}|^p$ with equality if and only if $a_{ij} = 0$ or 1. Thus, each column of A contains exactly one nonzero entry, and it has unit modulus; nonsingularity ensures that no row contains more than one nonzero entry.

5.4.P18 Consider the continuous function $\det Y$, in which $Y = [y_1 \ \cdots \ y_n]$.

Section 5.5

5.5.P8 (5.4.16) and (5.5.14).

5.5.P10 (5.4.22) and the duality theorem.

5.5.P11
$$\|[x_1 \ \ldots \ x_{k-1} \ \alpha x_k \ x_{k+1} \ \ldots \ x_n]^T\|$$
$$= \|(1-\alpha)[x_1 \ \ldots \ x_{k-1} \ 0 \ x_{k+1} \ \ldots \ x_n]^T + \alpha x\|$$
$$\leq (1-\alpha)\|[x_1 \ \ldots \ x_{k-1} \ 0 \ x_{k+1} \ \ldots \ x_n]^T\| + \alpha\|x\|$$
$$\leq (1-\alpha)\|x\| + \alpha\|x\| = \|x\|$$

Section 5.6

5.6.P7 Suppose that $\max_k N_k(A) = N_j(A)$. $\|\|AB\|\| = \|[N_1(AB) \ \ldots \ N_m(AB)]^T\| \leq \|[N_1(A)N_1(B) \ \ldots \ N_m(AB)N_m(B)]^T\| \leq (\max_k N_k(A))\|\|B\|\| \leq N_j(A)\|e_j\| \ \|\|B\|\| \leq \|\|A\|\| \ \|\|B\|\|$.

5.6.P9 Consider $N_1(\cdot) = \|\cdot\|_1$, $N_2(\cdot) = \|\cdot\|_2$, $A = \begin{bmatrix} 0 & 1 \\ 0 & 1 \end{bmatrix}$, and $B = A^T$.

5.6.P10 (a) $\Rightarrow$: For a given x, choose $\theta_1, \ldots, \theta_n$ so that $|\sum_{j=1}^n e^{i\theta_j} x_j| = \|x\|_1$. Let $A = [e^{i\theta_1} e_1 \ \ldots \ e^{i\theta_n} e_1]$ and $B = [x \ \ldots \ x]$, so that $\|x\|_1 \|e_1\| = N_{\|\cdot\|}(AB) \leq N_{\|\cdot\|}(A)N_{\|\cdot\|}(B) = \|e_1\| \ \|x\|$. $\Leftarrow$: Let $A = [a_1 \ \ldots \ a_n]$ and $B = [b_1 \ \ldots \ b_n] = [b_{ij}]$. Then $N_{\|\cdot\|}(AB) = \max_j \|Ab_j\| = \max_j \|\sum_i a_i b_{ij}\| \leq \max_j \sum_i |b_{ij}| \ \|a_i\| \leq \|\|B\|\|_1 N_{\|\cdot\|}(B) \leq N_{\|\cdot\|}(A)N_{\|\cdot\|}(B)$. (d) $|\det B| \leq \rho(B)^n$. Why? (g) To conclude that $|\det(AD_A^{-1})| \leq 1$, it is *sufficient* to know that $\rho(AD_A^{-1}) \leq 1$, but this condition is not necessary. Consider $A = \begin{bmatrix} 1 & 1 \\ 1 & 2 \end{bmatrix}$, for which $\rho(AD_A^{-1}) \sim 1.37$ but nevertheless $1 = |\det A| \leq \|a_1\|_2 \|a_2\|_2 = \sqrt{10} < \|a_1\|_1 \|a_2\|_1 = 6$.

5.6.P12 (a) How are the eigenvalues and singular values of a Hermitian matrix related? Why do the eigenvalues of the Hermitian matrix $A - \frac{1}{2}\|\|A\|\|_2 I$ lie in the real interval $[-\frac{1}{2}\|\|A\|\|_2, \frac{1}{2}\|\|A\|\|_2]$?

5.6.P17 Only three terms in the series are nonzero.

5.6.P18 Choose a diagonal matrix D such that DA has all 1s on the main diagonal.

5.6.P19 (c) Consider $\begin{bmatrix} 0 & 1 \\ 0 & 0 \end{bmatrix}$, $\begin{bmatrix} 0 & 0 \\ 1 & 0 \end{bmatrix}$, $\begin{bmatrix} 0 & 1 \\ 1 & 0 \end{bmatrix}$, and $\begin{bmatrix} 1 & 1 \\ 0 & 1 \end{bmatrix}$.

5.6.P20 The Frobenius norm is unitarily invariant and monotone; $\|\Sigma C\|_2^2 = \sum_{i,j} |\sigma_i c_{ij}|^2 \leq \sigma_1 \sum_{i,j} |c_{ij}|^2$.

5.6.P21 $\rho(AA^*) \leq \|\|AA^*\|\|$.

5.6.P22 $\Rightarrow$: $\|\|A\|\|^D = \max_{B \neq 0} \frac{|\operatorname{tr} B^* A|}{\|\|B\|\|} \geq \frac{|\operatorname{tr} I^* A|}{\|\|I\|\|}$. $\Leftarrow$: $\|\|I\|\| = \max_{B \neq 0} \frac{|\operatorname{tr} B^* I|}{\|\|B\|\|^D} \leq \max_{B \neq 0} \frac{\|B\|\|^D}{\|B\|\|^D}$, but $\|\|I\|\| \geq 1$ for any matrix norm.

5.6.P23 Hint (i, j) in the following list pertains to establishing the (i, j) entry in the 6-by-6 table of constants in (5.6.P23). The matrix given in each case is one for which the inequality $\|\|A\|\|_\alpha \leq C_{\alpha\beta} \|\|A\|\|_\beta$ is an equality. The "equality" matrices are all in M_n: I is the identity matrix; $J = ee^T$ has all entries 1; $A_1 = ee_1^T$ has all 1s in its first column and all other entries are 0; $E_{11} = e_1 e_1^T$ has only its 1,1 entry 1 and all other entries are 0.

$(1, 2)$ follows from $(2, 1)$ by (5.6.21)

$(1, 3)$ $\|\|A\|\|_1 \leq \|A\|_1 \leq n \|\|A\|\|_\infty$; A_1

$(1, 4)$ A_1

$(1, 5)$ $(\max_{1 \leq j \leq n} \sum_{i=1}^n |a_{ij}|)^2 \leq \sum_{j=1}^n (\sum_{i=1}^n |a_{ij}|)^2 \leq (\sum_{j=1}^n 1)(\sum_{i=1}^n |a_{ij}|^2)$ (Cauchy–Schwarz inequality); A_1

$(1, 6)$ $\max_{1 \leq j \leq n} \sum_{i=1}^n |a_{ij}| \leq n \max_{1 \leq i, j \leq n} |a_{ij}|$; J

$(2, 1)$ follows from $(2, 5)$ and $(5, 1)$; A_1^*

$(2, 3)$ follows from $(2, 5)$ and $(5, 3)$; A_1

$(2, 4)$ follows from $(2, 5)$ and $(5, 4)$; A_2

$(2, 5)$ $\sigma_1(A) \leq (\sum_{i=1}^n \sigma_i^2(A))^{1/2} = \|A\|_2$; A_1

$(2, 6)$ follows from $(2, 5)$ and $(5, 6)$; J

$(3, 1)$ follows from $(1, 3)$ by (5.6.21); A_1^*

$(3, 2)$ follows from $(2, 3)$ by (5.6.21); A_1^*

$(3, 4)$ A_1^*

$(3, 5)$ similar to $(1, 5)$; A_1^*

$(3, 6)$ similar to $(1, 6)$; J

$(4, 1)$ $\sum_{j=1}^n \sum_{i=1}^n |a_{ij}| \leq n \max_{1 \leq j \leq n} \sum_{i=1}^n |a_{ij}|$; I

$(4, 2)$ follows from $(4, 5)$ and $(5, 2)$; consider the Fourier matrix, (2.2.P10).

$(4, 3)$ similar to $(4, 1)$; I

$(4, 5)$ $(\sum_{i,j=1}^n |a_{ij}|)^2 = \sum_{i,j,p,q=1}^n |a_{ij}||a_{pq}| \leq \frac{1}{2} \sum_{i,j,p,q=1}^n (|a_{ij}|^2 + |a_{pq}|^2)$ (arithmetic–geometric mean inequality); J

$(4, 6)$ $\sum_{i,j=1}^n |a_{ij}| \leq n^2 \max_{1 \leq i, j \leq n} |a_{ij}|$; J

$(5, 1)$ $\sum_{j=1}^n \sum_{i=1}^n |a_{ij}|^2 \leq \sum_{j=1}^n (\sum_{i=1}^n |a_{ij}|)^2 \leq n(\max_{1 \leq j \leq n} \sum_{i=1}^n |a_{ij}|)^2$; I

$(5, 2)$ $\sum_{i,j=1}^n |a_{ij}|^2 = \operatorname{tr} A^* A = \sum_{i=1}^n \sigma_i^2(A) \leq n\sigma_1^2(A)$; I

$(5, 3)$ similar to $(5, 1)$; I

$(5, 4)$ $\sum_{i,j=1}^n |a_{ij}|^2 \leq (\sum_{i,j=1}^n |a_{ij}|)^2$; E_{11}

$(5, 6)$ $\sum_{i,j=1}^n |a_{ij}|^2 \leq n^2 \max_{1 \leq i, j \leq n} |a_{ij}|^2$; J

$(6, 1)$ $\max_{1 \leq i, j \leq n} |a_{ij}| \leq \max_{1 \leq j \leq n} \sum_{i=1}^n |a_{ij}|$; I

$(6, 2)$ $\max_{1 \leq i, j \leq n} |a_{ij}|^2 \leq \max_{1 \leq i \leq n} \sum_{j=1}^n |a_{ij}|^2 = \max_{1 \leq i \leq n} (A^*A)_{ii} \leq \rho(A^*A)$, (4.2.P3); I

$(6, 3)$ similar to $(6, 1)$; I

$(6, 4)$ $\max_{1 \leq i, j \leq n} |a_{ij}| \leq \sum_{i,j=1}^n |a_{ij}|$; E_{11}

$(6, 5)$ $\max_{1 \leq i, j \leq n} |a_{ij}|^2 \leq \sum_{i,j=1}^n |a_{ij}|^2$; E_{11}

5.6.P24 rank A = number of nonzero singular values of A and $\|A\|_2 = (\sigma_1^2 + \cdots + \sigma_n^2)^{1/2}$.

5.6.P25 Use (2.2.9) and (0.9.6.3); C_n is unitary.

5.6.P26 If $A \in M_n$ and $\rho(A) < 1$, show that the *Neumann series* $I + A + A^2 + \cdots$ converges to $(I - A)^{-1}$.

5.6.P36 $\||\hat{A}\||_2 = \rho(\hat{A}^*\hat{A})^{1/2}$.

5.6.P37 Consider $\begin{bmatrix} 1 & 0 \\ 1 & 0 \end{bmatrix}, \begin{bmatrix} 0 & 1 \\ 0 & 0 \end{bmatrix}$, and the parallelogram identity.

5.6.P38 $\Rightarrow$: If $A \neq 0$, let $B = A/\||A\||$, so $\||B^m\||$ is bounded as $m \to \infty$. If $J_k(e^{i\theta})$ is a Jordan block of B and $k > 1$, then $J_k(e^{i\theta})^m$ is not bounded as $m \to \infty$. $\Leftarrow$: If A is not a scalar matrix, use (3.1.21) and consider the matrix norm $\||X\|| = \||S(\varepsilon)^{-1} X S(\varepsilon)\||_\infty$ with $0 < \varepsilon < \max\{\rho(A) - |\lambda| : \lambda \in \sigma(A)\}$.

5.6.P39 (b) Choose a Schur triangularization $A/\rho(A) = UTU^*$ in which the eigenvalues with modulus strictly less than one appear first in the main diagonal of T, followed by the eigenvalues with modulus equal to one. The spectral norm of T is one, so every column of T has Euclidean norm at most one. Consider columns of T containing a main diagonal entry with modulus one. (c) See the exercise following (5.6.9).

5.6.P40 (c) $\|Ax\|_2 = \|\,|Ax|\,\|_2 \le \|\,|A|\,|x|\,\|_2$. (d) $\|Ax\|_2 = \|\,|Ax|\,\|_2 \le \|A|x|\,\|_2 \le \|B|x|\,\|_2$.

5.6.P41 (a) If $\|x\| = 1$ and $\||A\|| = \|Ax\|$, then $\|Ax\| = \|\,|Ax|\,\| \le \|\,|A|\,|x|\,\| \le \max_{\|y\|=1} \|\,|A|y\| = N(A) \le \max_{\|y\|=1} \|\,|A|\,|y|\,\|$. (b) If $z \ge 0$ and $N(AB) = \|\,|AB|z\|$, then $N(AB) \le \|\,|A|\,|B|z\| \le N(A)\|\,|B|z\| \le N(A)N(B)$. (c) Let $z \ge 0$ be such that $\|z\| = 1$ and $N(A) = \|\,|A|z\| = \|Az\| = \||A\||$. Then $\|Az\| \le \|Bz\| \le \||B\||$. (d) $\|\,|A|\,\|_2 \le \||A\||_2$.

5.6.P43 If $\lambda_1, \ldots, \lambda_n$ are the main diagonal entries of T, then $e^{\lambda_1}, \ldots, e^{\lambda_n}$ are the main diagonal entries of e^T.

5.6.P44 Consider the matrix norm $n \|\cdot\|_\infty$.

5.6.P47 $B = A - (A - B) = A(I - A^{-1}(A - B))$ is singular, so $\||A^{-1}(A - B)\|| \ge 1$.

5.6.P48 (c) (5.6.55).

5.6.P56 $\||A\||_2^2 = \rho(A^*A) \le \||A^*A\|| \le \||A\||^2$.

5.6.P58 (b) $(AB)^*(AB) = (A^*B)^*(A^*B)$ and $\|\cdot\|_2$ is self-adjoint.

Section 5.7

5.7.P3 Consider $G(A^k x e^T)$ if $Ax = \lambda x$ and $x \neq 0$. (c) What can you say about convergence of power series of matrices using vector norms?

5.7.P11 (b) A previous exercise shows that $r(J_2(0))r(J_2(0)^T) \ge 1$ is necessary for compatibility.

5.7.P16 (a) N is similar to $2N$. (b) If $B \in M_n$ has zero main diagonal, then it is a linear combination of nilpotent matrices so $G(B) = 0$. Use (2.2.3) and (5.1.2) to show that $G(A) = G((n^{-1} \operatorname{tr} A)I_n + B) = n^{-1}|\operatorname{tr} A|G(I_n)$.

5.7.P19 For given norms $G_1(\cdot)$ and $G_2(\cdot)$ on M_n and $\alpha \in [0, 1]$, we must show that $m(\alpha G_1 + (1 - \alpha)G_2) \leq \alpha m(G_1) + (1 - \alpha)m(G_2)$, that is,

$$\max_{A \neq 0} \frac{\rho(A)}{\alpha G_1(A) + (1 - \alpha)G_2(A)} \leq \max_{A \neq 0} \frac{\alpha \rho(A)}{G_1(A)} + \max_{A \neq 0} \frac{(1 - \alpha)\rho(A)}{G_2(A)}$$

Explain why it is sufficient to show that $(\alpha a + (1 - \alpha)b)^{-1} \leq \alpha/a + (1 - \alpha)/b$ if $a, b > 0$, which is just convexity of the function $f(x) = x^{-1}$.

5.7.P20 (d) $A = H(A) + iK(A)$, in which $H(A)$ and $K(A)$ are Hermitian (0.2.5), $H(A) = (A + A^*)/2$, and $K(A) = (A - A^*)/2i$. Compute $\|\|A\|\|_2 \leq \|\|H(A)\|\|_2 + \|\|K(A)\|\|_2 = r(H(A)) + r(K(A)) \leq r(A) + r(A^*) = 2r(A)$. (e) Consider suitable n-by-n versions of E_{11} and $J_2(0)$.

5.7.P23 Use the Cauchy–Schwarz inequality or (5.6.41(a)).

5.7.P25 (b) Notice that $p(z)$ is a polynomial of degree at most $m - 1$, and that $p(z) = \frac{1}{m} \sum_{j=1}^{m}(1 - z^m)/(1 - w_j z)$, so that $p(z) = p(w_1 z) = \cdots = p(w_m z)$ for all $z \in \mathbf{C}$. Hence $p(z) = \text{constant} = p(0) = 1$.

Section 5.8

5.8.P7 Consider $A = \lambda I \in M_n$.

5.8.P14 (3.3.17).

Section 6.1

6.1.P3 If some column a_j is zero, there is nothing to prove. If all $a_j \neq 0$, let $B = A \, \text{diag}(\|a_1\|_1, \ldots, \|a_n\|_1)^{-1}$. Then (6.1.5) ensures that $\rho(B) \leq 1$, so $|\det B| \leq 1$.

6.1.P4 Apply (6.1.10a) to the matrix $\lambda I - A$.

6.1.P6 Apply (6.1.10) to a principal submatrix of A and use (0.4.4d).

6.1.P9 Use Corollary 6.1.5.

6.1.P10 rank $A = \text{rank}(AD)$ for any nonsingular diagonal matrix D, so it suffices to assume that all $a_{ii} \geq 0$ and all $\|a_i\|_1$ are either zero or 1. In this case, all the eigenvalues of A lie in the unit disc, and one must show that rank $A \geq \sum_{i=1}^{n} a_{ii}$. Explain why $\sum_{i=1}^{n} a_{ii} = \sum_{i=1}^{n} \lambda_i \leq \sum_{i=1}^{n} |\lambda_i| \leq$ number of nonzero eigenvalues of $A \leq$ rank A.

6.1.P11 As in (6.1.P10), it suffices to consider the case in which all $\|a_i\|_2$ are zero or 1; in this case we must show that rank $A \geq \sum_{i=1}^{n} |a_{ii}|^2 = \sum_{i=1}^{n} |e_i^* a_i|^2$, in which $e_1, \ldots, e_n$ are the standard orthonormal basis vectors in $\mathbf{C}^n$. If A has rank k, choose orthonormal vectors $v_1, \ldots, v_k \in \mathbf{C}^n$ such that $\text{span}\{v_1, \ldots, v_k\} = \text{span}\{a_1, \ldots, a_n\}$. Then $a_i = \sum_{j=1}^{k}(v_j^* a_i)v_j$, so $e_i^* a_i = \sum_{j=1}^{k}(v_j^* a_i)(e_i^* v_j)$ and $\sum_{i=1}^{n} |e_i^* a_i|^2 \leq \sum_{i=1}^{n} \left(\left(\sum_{j=1}^{k} |v_j^* a_i|^2\right)\left(\sum_{j=1}^{k} |e_i^* v_j|^2\right)\right) = \sum_{j=1}^{k} \sum_{i=1}^{n} |e_i^* v_j|^2 = \sum_{j=1}^{k} 1$.

6.1.P13 Choose a diagonal real orthogonal matrix D such that the main diagonal entries of DA are all positive; use (6.1.10(b)).

6.1.P16 (b) $|a^{-1}| \sum_j |y_j| < 1$, $\sum_{j \neq i} |b_{ij}| < |b_{ii}| - |x_i|$, and $\sum_{j \neq i} |c_{ij}| = \sum_{j \neq i} |b_{ij} - a^{-1}x_i y_j| \leq \sum_{j \neq i} |b_{ij}| + |x_i|(|a^{-1}| \sum_j |y_j|)$. (c) Uniqueness of the Schur complement (0.8.5(a)).

6.1.P18 Choose a permutation matrix P_1 such that a largest-modulus entry of the first column of $P_1 X$ is in position 1, 1. Let $R_1 = \begin{bmatrix} e^{i\theta} & z^* \\ 0 & I_{k-1} \end{bmatrix}$ and choose θ and z such that $P_1 X R_1 = \begin{bmatrix} \|x_1\|_\infty & 0 \\ * & X_2 \end{bmatrix}$. Now consider $X_2 \in M_{n-1,k-1}$ and deflate sequentially, as in the proof of (2.3.1), to obtain a permutation matrix $P = P_{k-1} \cdots P_1$ and a nonsingular upper triangular matrix $R = R_1 \cdots R_{k-1}$ such that PXR is an upper triangular matrix whose diagonal entries are the max norms of their respective columns. Let $Y = XR$.

6.1.P20 The preceding problem.

Section 6.2
6.2.P4 Start with the case in which A has no zero entries, then argue by continuity.

6.2.P5 Show that the companion matrix $C(p)$ in (3.3.12) is irreducible if $a_0 \neq 0$.

6.2.P8 (6.2.26) and (6.2.1a).

Section 6.3
6.3.P1 Apply (6.3.5) to $B = \mathrm{diag}(a_{11}, \ldots, a_{nn})$ and $E = A - B$.

6.3.P3 (a) Let $\xi \in \mathbf{C}^k$ be a unit vector such that $B\xi = \beta\xi$, let $x = \begin{bmatrix} \xi \\ 0 \end{bmatrix}$, and consider the residual vector $Ax - \beta x$. (b) (2.5.P37).

6.3.P4 If $\lambda_1, \ldots, \lambda_n$ are the eigenvalues of A, why are $|\lambda_1 - \gamma|^2, \ldots, |\lambda_n - \gamma|^2$ the eigenvalues of the positive semidefinite matrix $B = (A - \gamma I)^*(A - \gamma I)$? The hypothesis is that $x^* B x \leq \delta^2$ for every unit vector $x \in \mathcal{S}$. Apply (4.2.10(b)).

Section 6.4
6.4.P7 (6.4.7).

Section 7.1
7.1.P1 (7.1.2) and (7.1.5).

7.1.P3 Perform a congruence by a suitable diagonal matrix.

7.1.P5 If A is positive semidefinite, show that $(\mathrm{tr}\, A)^2 \geq \mathrm{tr}\, A^2$.

7.1.P7 $\det[f(t_i - t_j)]_{i,j=1}^n \geq 0$. For (a), consider $n = 1$; for (b), consider $n = 2$; and for (c), consider $n = 3$.

7.1.P10 $\cos t = (e^{it} + e^{-it})/2$.

7.1.P14 (7.1.10).

7.1.P15 Consider $[f(t_i - t_j)]$ with $t_1 = 0$, $t_2 = -\tau$, and $t_3 = -t$.

7.1.P16 Consider $\int_0^\infty \left| \sum_{k=1}^n x_k e^{-\lambda_k s} \right|^2 ds$ with $x = [x_i] \in \mathbf{C}^n$. If $f(s) = \sum_{k=1}^n x_k e^{-\lambda_k s} = 0$ for all $s > 0$, then $f(0) = 0$, $f'(0) = 0$, $\ldots$, $f^{(n-1)}(0) = 0$, a system of linear equations for the entries of x.

7.1.P18 (a) The preceding problem. (b) Algebraically ordering the numbers corresponds to a permutation similarity of the min matrix. Consider the representation in (a), in which some of the summands are zero. (c) Consider $[\min\{\beta_i^{-1}, \beta_j^{-1}\}]$.

7.1.P19 Express the integral as the limit of Riemann sums over partitions of $[0, N]$ with equally spaced points.

7.1.P20 Express the double integral as an iterated integral and integrate by parts.

7.1.P21 (f) $A^{-1} = (S^{-*})D^{-1}(S^{-*})^*$. What does D^{-1} look like? Alternatively, observe that A^{-*} is *congruent to $A = A^{-*}AA^{-1}$ and $H(A^{-*}) = H(A^{-1})$. (g) The direct summands correspond to *congruences of the various possibilities for the Toeplitz decompositions of the Type 0 and Type I blocks in the *congruence canonical form of A.

7.1.P24 rank $A \leq$ rank $[\begin{smallmatrix} A_{11} & A_{12} \end{smallmatrix}]$ + rank $[\begin{smallmatrix} A_{12}^* & A_{22} \end{smallmatrix}]$ = rank $[\begin{smallmatrix} A_{11} & A_{11}X \end{smallmatrix}]$ + rank $[\begin{smallmatrix} A_{22}Y & A_{22} \end{smallmatrix}]$.

7.1.P26 (7.1.2) and (3.5.3).

7.1.P27 (a) $A = \begin{bmatrix} A_{11} & A_{11}X \\ \star & \star \end{bmatrix}$ and $B = \begin{bmatrix} B_{11} & \star \\ Y^*B_{11} & \star \end{bmatrix}$, so $AB = \begin{bmatrix} A_{11}B_{11} + A_{11}XY^*B_{11} & \star \\ \star & \star \end{bmatrix}$ and $(AB)[\alpha] = A_{11}(I + XY^*)B_{11}$. If $A = B$, then $X = Y$ and $I + XX^*$ is positive definite. (c) (a) $\Rightarrow$ rank $A^{2^k}[\alpha] = \text{rank}(A^{2^k-k}A^k) \leq \text{rank } A^k[\alpha] = \text{rank}(A^{k-1}A)[\alpha] \leq \text{rank } A[\alpha]$.

7.1.P28 (b) $X^*B = Y^*B$, so $X^*BX = Y^*BX = Y^*BY$. (c) Choose $X = B^{-1}C$.

7.1.P29 (a) This means that $A = S_1 D S_1^*$ and $B = S_2 D S_2^*$; $D = \text{diag}(e^{i\theta_1}, \ldots, e^{i\theta_n})$, each $\theta_j \in (-\pi/2, \pi/2)$; $D = \Gamma + i\Sigma$, $\Gamma = \text{diag}(\cos\theta_1, \ldots, \cos\theta_n)$, $\Sigma = \text{diag}(\sin\theta_1, \ldots, \sin\theta_n)$. (b) Similar to D^2, which has a unique square root with eigenvalues in the open right half-plane. (d) Similar to $T = \text{diag}(\tan\theta_1, \ldots, \tan\theta_n)$.

Section 7.2

7.2.P6 Interlacing. If the smallest eigenvalue of A is negative, it has more nonzero eigenvalues than B.

7.2.P10 If $A^* = BAB^{-1}$ and $B = B^*$ is positive definite, why is $B^{-1/2}A^*B^{1/2} = B^{1/2}AB^{-1/2}$? If $A = S\Lambda S^{-1}$, then $A = BA^*B^{-1}$ with $B = SS^*$.

7.2.P11 (a) If A is nonsingular, then adj(adj A) = $(\det A)^{n+2}A$. (c) (2.5.P47) or consider $A_\epsilon = A + \epsilon I, \epsilon > 0$. (d) Consider an $A \in M_n$ with $n \geq 3$ and rank $A \leq n - 2$.

7.2.P12 (a) If $i = 1$ and $j > 2$, the first column of M_{1j} is a multiple of the second column. (e) For $s = \frac{i}{m}$ and $t = \frac{j}{m}$, $f(s - t) = e^{-\frac{1}{m}|i-j|} = r^{|i-j|}$ in which $r = e^{-\frac{1}{m}} \in (0, 1)$. Use continuity and a limiting argument.

7.2.P13 (7.2.P12(a)). Use $M(r, n)M(r, n)^{-1} = M(r, n)^{-1}M(r, n) = I$ to determine the entries of $M(r, n)^{-1}$.

7.2.P14 (f) Proceed as in (7.2.P12(e)).

7.2.P15 To reduce to the positive semidefinite case, replace A by $A + cI_n$ and B by $B + cI_{n+1}$ with a sufficiently large positive c, as in the proof of (4.3.17).

7.2.P17 Let $D = A^{1/2} + BA^{-1/2}$ and compute DD^*.

7.2.P18 $B = A(I + (A^*A)^{-1})$.

7.2.P19 (a) Consider $\|x\|_2^2 = (A^{1/2}x)^*(A^{-1/2}x)$ and use the Cauchy–Schwarz inequality. (b) Let $x = e_i$ in (a).

7.2.P20 (c) If $A = SBS^*$, then $B^{1/2}AB^{1/2} = (B^{1/2}SB^{1/2})^2$.

7.2.P21 (a) (4.1.6).

7.2.P22 (d) (7.2.P21).

7.2.P23 (c) The preceding problem. (e) $GA^{-1}G = \bar{A} \Rightarrow \bar{A} = \bar{G}A^{-1}\bar{G}$. (f) $GA^{-1}G = A^{-T} \Rightarrow A^{-T} = G^{-T}A^{-1}G^{-T}$.

7.2.P24 (b) (7.2.10).

7.2.P26 The Cauchy–Schwarz inequality for the Frobenius inner product, the table in (5.4.P3), and the arithmetic–geometric mean inequality.

7.2.P27 $\text{tr}((\sum_{i=1}^{m} A_i)^*(\sum_{i=1}^{m} A_i)) = \sum_{i,j=1}^{m} \text{tr}(A_i A_j) \geq \sum_{i=1}^{m} \text{tr } A_i^2$.

7.2.P28 (b) $x_i = -\sum_{j \neq i} b_i^* b_j x_j = 0$ for each $i = 1, \ldots, n$.

7.2.P29 (a) (7.1.P1).

7.2.P30 (a) $A^{-1/2}ABA^{1/2} = A^{1/2}BA^{1/2}$; (4.5.8). (b) $A = \begin{bmatrix} 1 & 0 \\ 0 & 0 \end{bmatrix}$ and $C = \begin{bmatrix} 0 & 1 \\ 1 & 0 \end{bmatrix}$.

7.2.P33 (b) (7.2.P30(c)).

7.2.P34 (c) Use the Cauchy Schwarz inequality for the Frobenius inner product.

7.2.P36 (a) X^*RX and $R^{1/4}XR^{1/2}X^*R^{1/4}$ are positive semidefinite. (e) (5.6.P58). (l) $R^{1/4}XR^{1/2}X^*R^{1/4}$ is positive semidefinite.

Section 7.3
7.3.P6 (c) Consider $\frac{d}{dt}\sigma_1(A(t_0) + tE)|_{t=0}$, in which $E = \begin{bmatrix} 0_n & 0 \\ 0 & 1 \end{bmatrix}$.

7.3.P7 (f) If $X, Y \in M_{m,n}$ (in their role as $A^\dagger$) satisfy (a)–(c), then $X = X(AX)^* = XX^*A^* = X(AX)^*(AY)^* = XAY = (XA)^*(YA)^*Y = A^*Y^*Y = (YA)^*Y = Y$. Alternatively, write down a singular value decomposition for $A^\dagger$ and show that its three factors are uniquely determined by (a)–(c).

7.3.P11 $|x^*Ay| \leq \|x\|_2\|Ay\|_2$.

7.3.P13 Apply the spectral theorem to P.

7.3.P14 If $A = S\Lambda S^{-1}$, let $S = PU$.

7.3.P16 (c) $A = \begin{bmatrix} 1 & 0 \\ 0 & 0 \end{bmatrix}$ and $B = \begin{bmatrix} 0 & 0 \\ 0 & 1 \end{bmatrix}$. (d) $\sigma_i(A) = \sigma_i((A+B) - B) \leq \sigma_i(A+B) + \sigma_1(B)$.

7.3.P19 (d) If $\Lambda, M \in M_n$ are diagonal and U is unitary, $\Lambda U M$ and $\bar{\Lambda}U\bar{M}$ are diagonally unitarily equivalent.

7.3.P22 (2.4.P9).

7.3.P23 (a) (7.3.1).

7.3.P25 (a) $\text{tr}(UA) = \sum_{i,j} u_{ij}a_{ji}$. (b) $\text{tr } \Sigma = \text{tr}(V^*AW) = \text{tr}(WV^*\Sigma)$.

7.3.P26 (a) Use $p_A(A) = 0$ to show that $(A^{1/2})^2 = A$.

7.3.P28 (c) Represent $A = V\Sigma W^*$ and consider separately the cases (i) A is nonsingular ($\sigma_1 \geq \sigma_2 > 0$) and (ii) A is singular ($\sigma_1 > \sigma_2 = 0$).

7.3.P34 If $A \in M_{m,n}$ with $m \geq n$, we have $A^*A = R^*R$, in which $R \in M_n$ is upper triangular.

7.3.P35 If A is normal, $P = Q$ in (7.3.1).

7.3.P36 Consider a polar factorization $A = PU$ and use the preceding problem.

7.3.P37 $A = SBS^{-1} = S^{-*}BS^* \Rightarrow B$ commutes with S^*S and hence with Q.

7.3.P38 $V = SWS^* \Rightarrow P^{-1}V = (UWU^*)P \Rightarrow V = UWU^*$; uniqueness of the unitary factor in (7.3.1(b)).

7.3.P39 (b) The preceding problem.

7.3.P40 The preceding problem.

7.3.P41 The preceding problem.

7.3.P43 (7.3.P35); $\|AB\| = \|PUQV\| = \|PQ\| = \|QP\|$.

Section 7.4

7.4.P3 (a) Since $B = U\Lambda U^*$ and $C = UMU^*$ for some unitary $U \in M_n$, write the asserted inequality using $y = U^*x$ and then using $z = (\Lambda M)^{1/2}y$. Then apply (7.4.12.1) with $B = \Lambda M^{-1}$ to show that the asserted inequality is valid (and is sharp) with a constant of the form $\lambda_1 \lambda_n \mu_j \mu_k / (\lambda_1 \mu_j + \lambda_n \mu_k)^2$ for some choice of indices $1 \leq j \neq k \leq n$. Show that the least constant of this form occurs for $j = 1$ and $k = n$.

7.4.P4 Take $B = A^{1/2}$ and $C = I$.

7.4.P5 $|x^*Ax|^2 = |x^*PUx|^2 = |(P^{1/2}x)^*(P^{1/2}Ux)| \leq (x^*Px)((Ux)^*P(Ux))$ so

$$|(x^*Ax)(x^*A^{-1}x)| \leq (x^*Px)(x^*P^{-1}x)((Ux)^*P(Ux))((Ux)^*P^{-1}(Ux))$$

Use (7.4.12.1) twice.

7.4.P10 (a) $x^*y = (A^{1/2}x)^*(A^{-1/2}y)$.

7.4.P11 (a) (0.8.5.10). (d) Write $A = \begin{bmatrix} A_n & \xi \\ \xi^* & a_{nn} \end{bmatrix}$ and $B = \begin{bmatrix} B_n & \eta \\ \eta^* & b_{nn} \end{bmatrix}$. Put $A_n \to A$, $B_n \to B$, $A \to \mathcal{A}_\alpha$, $B \to \mathcal{B}_\beta$, $\xi \to x$, and $\eta \to y$ in the left-hand side of (7.4.12.20), which is nonnegative by assumption. This proves (7.4.12.19) for $i = n$. Permute for the general case.

7.4.P13 $A - \frac{1}{2}(A + A^*) = \frac{1}{2}(A - H) + \frac{1}{2}(H - A^*)$, so $\| A - \frac{1}{2}(A + A^*) \| \leq \frac{1}{2} \| A - H \| + \frac{1}{2} \| H - A^* \|$.

7.4.P16 See the preceding problem for the lower bound. For the upper bound use (7.3.P16) to show that $\sigma_i(A + (-U)) \leq \sigma_i(A) + 1$, which means that $\|A - U\|_{[k]} \leq \|\Sigma(A) + I\|_{[k]}$ for each $k = 1, \ldots, n$. Invoke (7.4.8.4).

7.4.P17 (b) (7.4.1.7). (c) (2.6.P4). (d) Use the condition (4.3.52a) for the case of equality in (4.3.51); $w_i = \sigma_i(A)$, $y_i = \lambda_i$, and $x_i = \sigma_i(B)$ $(= \sigma_i(A)$ for $i = 1, \ldots, n - 1$ and $x_n = 0)$. Suppose that A has d distinct singular values $s_1 > \cdots > s_{d-1} = \sigma_{n-1}(A) > s_d = \sigma_n(A)$ with respective multiplicities $n_1, \ldots, n_d$, with $n_d = 1$. Then $(\sigma_{n-1}(A) - \sigma_n(A))(\sum_{i=1}^{n-1} \sigma_i(B) - \sum_{i=1}^{n-1} \lambda_i) = 0 \Rightarrow \lambda_n = 0$. Now work from the top down through the distinct singular values. $(\sigma_{n_1}(A) - \sigma_{n_1+1}(A))(\sum_{i=1}^{n_1} \sigma_i(B) - \sum_{i=1}^{n_1} \lambda_i) = (s_1 - s_2)(n_1 s_1 - \sum_{i=1}^{n_1} \lambda_i) = 0 \Rightarrow \lambda_1 = \cdots = \lambda_{n_1} = s_1$ since all $\lambda_i \leq s_1$. If $d > 2$, $(\sigma_{n_1+n_2}(A) - \sigma_{n_1+n_2+1}(A))(\sum_{i=1}^{n_1+n_2} \sigma_i(B) - \sum_{i=1}^{n_1+n_2} \lambda_i) = (s_2 - s_3)(n_2 s_2 - \sum_{i=n_1+1}^{n_1+n_2} \lambda_i) = 0 \Rightarrow \lambda_{n_1+1} = \cdots = \lambda_{n_1+n_2} = s_2$. (e) (2.6.5).

7.4.P18 Show that $\sigma_k(A) \le \sigma_k(|A|)$ for all $k = 1, \ldots, n$ by observing that $\|Ax\|_2 \le \| |Ax| \|_2 \le \| |A| |x| \|_2$ and then invoking (7.3.8). Then use (7.4.8.4).

Section 7.5

7.5.P1 (a) (7.2.7).

7.5.P3 Consider $A \circ \bar{A}$.

7.5.P7 To show that the matrix condition is sufficient, consider a Riemann sum approximation to the integral

$$\int_a^b \int_a^b K(x, y) f(x) \bar{f}(y) dx\, dy \cong \sum_{i,j=1}^n K(x_i, y_j) f(x_i) \bar{f}(x_j) \Delta x_i \Delta x_j$$

To show that the matrix condition is necessary, consider a function $f(x) = \sum_{i=1}^n a_i \delta_\epsilon(x - x_i)$, in which $\delta_\epsilon(x)$ is an "approximate delta function," which is continuous and nonnegative, vanishes identically outside the interval $[-\epsilon, \epsilon]$, and satisfies $\int_{-\infty}^\infty \delta_\epsilon(x) dx = 1$. Now let $\epsilon \to 0$.

7.5.P11 (7.1.P16).

7.5.P12 Apply the inequality in (7.1.P1) to both A and $A^{(-1)}$. Conclude that every principal minor of A of size two is zero and use (7.2.P24).

7.5.P14 (a) Let $x = [x_i]$ be a unit vector and let $B = A^{1/2}(\operatorname{diag} x) A^{-1/2}$. The eigenvalues of B are x_i, so $\|B\|_F^2 \ge \|x\|_2^2 = 1$. Compute $x^*(A \circ A^{-1})x = \operatorname{tr}((\operatorname{diag} \bar{x}) A (\operatorname{diag} x) A^{-1}) = \operatorname{tr}(B^*B) - \|D\|_F^2$. (4.2.2(c)).

7.5.P15 (c) (7.5.9(b)) with $f(t) = 1/(1 - t)$.

7.5.P17 (c) (7.1.P18). (d) (7.5.9(b)) with $f_k(t) = e^{t \ln p_k}$.

7.5.P18 tA is positive semidefinite if $t > 0$.

7.5.P20 There is nothing to show if $n = 2$, so let $n \ge 3$ and $\alpha = a_{pp}$. After a permutation, we may assume that $p = 1$ and $q = 2$. The leading 2-by-2 principal submatrix of A is $P = \begin{bmatrix} \alpha & \alpha \\ \alpha & \alpha \end{bmatrix}$, so (7.1.10) ensures that $\begin{bmatrix} a_{1j} \\ a_{2j} \end{bmatrix}$ is in the range of P for each $j = 3, \ldots, n$.

7.5.P22 (d) (0.9.11).

7.5.P23 (b) (7.5.P21) and the ideas in (7.5.P22).

7.5.P25 (b) $H = \alpha(X \circ Y)$, in which $\alpha > 0$, X is positive semidefinite or positive definite, and Y is a rank-one positive semidefinite matrix with positive main diagonal entries.

Section 7.6

7.6.P1 (d) If $A^* = S^{-1}AS$, show that AS is Hermitian and use (7.6.4).

7.6.P5 The exercise following (7.6.4).

7.6.P6 (7.6.4).

7.6.P7 $A + B = A(I + A^{-1}B)$.

7.6.P8 (b) The preceding problem.

7.6.P10 Write $S = (AB)C = EC$, in which E has positive eigenvalues. If S is Hermitian, explain why $E = SC^{-1}$ has the same number of positive eigenvalues as S, and conclude that S is positive definite. Consider $\begin{bmatrix} 10 & 0 \\ 0 & 1 \end{bmatrix}$, $\begin{bmatrix} 1 & -1 \\ -1 & 2 \end{bmatrix}$, and $\begin{bmatrix} 3 & 5 \\ 5 & 10 \end{bmatrix}$.

7.6.P12 (a) Interlacing: B_{11} is a principal submatrix of $S^{-*}BS^{-1}$.

7.6.P15 The preceding problem.

7.6.P16 (7.6.3).

7.6.P17 $\mu \geq \det((A_1 + A_2)/2) \geq (\det A_1 \det A_2)^{1/2} = \mu$.

7.6.P18 (5.2.6).

7.6.P19 The volume of $\mathcal{E}(A)$ is

$$\text{vol}(\mathcal{E}(A)) = \int_{x \in \mathcal{E}(A)} dV(x) = \int_{\|y\|_2 \leq 1} |\det J(y)| \, dV(y)$$

in which $J(y) = [\partial x_i / \partial y_y]$ is the Jacobian matrix for the change of variables $x \to y = A^{1/2}x$.

7.6.P20 (b) (5.4.4). (e) (7.6.P17).

7.6.P21 (b) (7.6.P20). (c) $(LAL^{-1})^*(LAL^{-1}) = ?$

7.6.P24 Let U_k be a diagonal matrix whose diagonal entry in position k is $+1$ and all the other diagonal entries are -1. Then U_k is an isometry for $\|\cdot\|$ and (7.6.11) implies that U_k commutes with Q. Consider $k = 1, 2, \ldots$ in succession and invoke (2.4.4.3).

7.6.P25 The preceding problem ensures that L is positive diagonal. $P^T L^2 P = L^2$ for every permutation matrix P. The unit ball of $\|x\|$ is contained in (and must touch) $\mathcal{E}(\alpha^2 I)$, so $\|x/\|x\|\|_2 \leq \alpha^{-1}$ for all $x \neq 0$.

7.6.P26 (5.4.21).

7.6.P27 (a) If $Q = [q_{ij}]$, then $e_i Q e_i \leq \|e_i\|^2$, so Hadamard's inequality ensures that $\prod_{i=1}^n \|e_i\|^2 \geq \prod_{i=1}^n q_{ii} \geq \det Q$.

Section 7.7

7.7.P6 (7.7.3(a)) and (7.2.6(c)).

7.7.P7 The preceding problem and (7.7.2) if A is nonsingular. If A is singular, use its singular value decomposition to reduce to the nonsingular case.

7.7.P9 $B = A^{1/2}XC^{1.2}$.

7.7.P10 Use (7.7.11) and (7.7.16), or invoke the Cauchy–Schwarz inequality: $x^*y = (A^{-1}x)^*(Ay)$.

7.7.P14 (a) (7.6.4).

7.7.P15 (a) (7.7.14). (b) Power series and the triangle inequality.

7.7.P16 Proceed as in the proof of (7.7.12). Write (b) as $\overline{x^*Ax} + y^*Ay \geq 2|x^T By|$ and let $x \to \bar{x}$. Invoke (7.7.9) with $A \to \bar{A}$ and $B^* = \bar{B}$. $(\bar{A} + \varepsilon I)^{-1/2}B(A + \varepsilon I)^{-1/2}$ is symmetric for each $\varepsilon > 0$.

7.7.P17 Use (c) in the preceding problem.

7.7.P19 The preceding problem.

7.7.P21 We need $cA \succeq J = ee^T$.

7.7.P22 Consider x^*Ax for real x.

7.7.P25 See (b) in the preceding problem.

7.7.P30 Look at the cofactor representation for the elements of A^{-1}.

7.7.P31 (7.7.15).

7.7.P33 (a) The preceding problem.

7.7.P34 (b) (7.7.4(d)).

7.7.P35 Let $A = PU = UQ$ be polar decompositions and explain why $A^*(AA^*)^{-1/2}A = (U^*P)P^{-1}(UQ) = Q = (A^*A)^{1/2}$. (b) The preceding problem.

7.7.P37 Consider (7.7.11(b)), $\begin{bmatrix} A & I \\ I & A^{-1} \end{bmatrix}$, and $\begin{bmatrix} B^{-1} & I \\ I & B \end{bmatrix}$.

7.7.P41 (a) $H_2 - \begin{bmatrix} 0 & 0 \\ 0 & S_{H_2}(A_2) \end{bmatrix} \succeq 0$ and $H_1 \succeq H_2 \Rightarrow H_1 - \begin{bmatrix} 0 & 0 \\ 0 & S_{H_2}(A_2) \end{bmatrix} \succeq 0$, and hence $S_{H_1}(A_1) \succeq S_{H_2}(A_2)$. (b) $H_1 - \begin{bmatrix} 0 & 0 \\ 0 & S_{H_1}(A_1) \end{bmatrix} \succeq 0$ and $H_2 - \begin{bmatrix} 0 & 0 \\ 0 & S_{H_2}(A_2) \end{bmatrix} \succeq 0 \Rightarrow H_1 + H_2 - \begin{bmatrix} 0 & 0 \\ 0 & S_{H_1}(A_1) + S_{H_2}(A_2) \end{bmatrix} \succeq 0 \Rightarrow S_{H_1+H_2}(A_1 + A_2) \succeq S_{H_1}(A_1) + S_{H_2}(A_2)$. (c) (7.7.P4).

7.7.P42 If $z \in \mathbf{C}$ and $\text{Re } z > 0$, you must show that $(\text{Re } z)^{-1} \geq \text{Re}(z^{-1})$.

7.7.P43 $\det(A + B) = (\det A)\det(I + A^{-1}B)$ and $\rho(A^{-1}B) \leq 1$.

Section 7.8

7.8.P2 Apply Fischer's inequality to A^*A and use $|\det B| \leq \|B\|_2^k$ if $B \in M_k$.

7.8.P3 (7.8.18): $(a_{11} \cdots a_{nn} - \det A)\det B + (b_{11} \cdots b_{nn} - \det B)\det A \leq 0$.

7.8.P4 (a) (7.6.2b) and the arithmetic–geometric mean inequality: $\sum \lambda_i(AB) \geq n(\prod \lambda_i(AB))^{1/n}$.

7.8.P6 (7.7.4(e)).

7.8.P9 Let $V = [v_1 \ldots v_n] \in M_n$ and apply (7.8.2) to V^*AV.

7.8.P11 $\det A = (\det A_{11})\det(A/A_{11})$ and $A_{11} \succeq A/A_{22}$.

7.8.P13 $E_k(\lambda_1, \ldots, \lambda_n)$ is a sum of principal minors of size k, each of which is bounded from above by a product of k distinct diagonal entries.

7.8.P17 $\prod_{j=1}^n |\cos \theta_j| + \prod_{j=1}^n |\sin \theta_j| \leq |\cos \theta_1 \cos \theta_2| + |\sin \theta_1 \sin \theta_2| \leq (\cos^2 \theta_1 + \sin^2 \theta_1)^{1/2}(\cos^2 \theta_2 + \sin^2 \theta_2)^{1/2}$.

7.8.P18 Minkowski's inequality (B10): $(\prod_{j=1}^n \cos^2 \theta_j)^{1/n} + (\prod_{j=1}^n \sin^2 \theta_j)^{1/n} \leq (\prod_{j=1}^n (\cos^2 \theta_j + \sin^2 \theta_j))^{1/n}$.

7.8.P19 Minkowski's inequality (B10) again.

Section 8.0

8.0.P2 (f) Set $B_\epsilon = (1 + \epsilon)A_\epsilon$ and proceed as in the text to diagonalize B_ϵ.

Section 8.1

8.1.P5 Consider the remarks preceding (8.1.26).

8.1.P8 $\|\|A\|\|_2^2 = \rho(A^T A)$. Why is $A^T A \geq B^T B$?

8.1.P9 $\rho(A^* A) \leq \rho(|A^* A|)$.

8.1.P10 (a) If $|A| > 0$, let $x > 0$ be such that $|A|x = \rho(|A|)x$, and partition $x^T = [x_1^T \ \ldots \ x_k^T]^T$, in which each $x_i \in \mathbf{R}^{n_i}$. Let $\xi_i = \|x_i\|$ and let $\xi = [\xi_i] \in \mathbf{R}^k$. Explain why $\rho(|A|)\xi_i = \left\|\sum_j |A_{ij}|x_j\right\| \leq \sum_j G(|A_{ij}|)\xi_j$ for each $i = 1, \ldots, k$, $\mathcal{A}\xi \geq \rho(|A|)\xi$, and $\rho(|A|) \leq \rho(\mathcal{A})$ (8.1.29). If A has some zero entries, consider $A + \varepsilon J_n$ and use a continuity argument.

8.1.P11 The main diagonal entries of A^n are $\sum a_{i,i_2} a_{i_2,i_3} \cdots a_{i_{n-1},n} a_{i_n,i}$, $i = 1, \ldots, n$, in which the sum is over all integers $i_2, \ldots, i_n \in \{1, \ldots, n\}$. Why is γ one of the summands? (8.1.20).

Section 8.2

8.2.P1 There are three cases: $A^m \to 0$, A^m diverges, and A^m converges to a positive matrix. Characterize and analyze each case.

8.2.P5 Let x be the Perron vector of A so that $Ax > Bx$.

8.2.P9 (a) If $\min_i \sum_{j=1}^n a_{ij} = \rho(A)$, let p be any index such that $x_p = \min_i x_i$ and explain why $\rho(A)x_p = \sum_{j=1}^n a_{pj}x_j \geq \sum_{j=1}^n a_{pj}x_p \geq \rho(A)x_p$ and hence $x_i = x_p$ for all $i = 1, \ldots, n$.

8.2.P10 Interlacing and (8.2.8) ensure that each 2-by-2 principal submatrix has exactly one positive eigenvalue.

8.2.P11 (j) (1.4.P13). Any non-real eigenvalues occur in conjugate pairs.

8.2.P13 For any given $\varepsilon > 0$, why is there an N such that $1 - \varepsilon < \text{tr}(\frac{1}{\rho(A)}A^m) = 1 + r_1^m + \cdots + r_n^m < 1 + \varepsilon$ for all $m > N$ $(r_k = \lambda_k(A)/\rho(A))$?

8.2.P14 (c) (8.2.P11).

8.2.P15 (b) If any zero entry of A is zero, increase it slightly to obtain a positive matrix A' such that $0 \leq A' \leq B$ but $A' \neq B$. Let y be the left Perron vector of A' and let x be the right Perron vector of B. Then $\rho(B')y^T x = y^T B' x < y^T A x = \rho(A)y^T x$.

Section 8.3

8.3.P1 Consider $\begin{bmatrix} 1 & 1 \\ 0 & 1 \end{bmatrix}$, $\begin{bmatrix} 1 & 0 \\ 0 & 1 \end{bmatrix}$, and $\begin{bmatrix} 0 & 1 \\ 1 & 0 \end{bmatrix}$.

8.3.P2 If $x > 0$ and $A^k x = \rho(A^k)x$, then $A^k(Ax) = \rho(A^k)(Ax)$. Invoke (8.2.5) and (8.3.4).

8.3.P3 If all the sub- and superdiagonal entries are positive, then there is a positive diagonal matrix D such that $D^{-1}AD$ is symmetric. Now use a limit argument.

8.3.P6 (b) $BAx = ABx = \rho(B)Ax$, (8.2.6), and (8.3.4). Alternatively, $B = S([\rho(B)] \oplus B_1)S^{-1}$ and $A = S([\lambda] \oplus A_1)S^{-1}$, in which the first column of S is the Perron vector of B and $\rho(B)$ is not an eigenvalue of B_1.

8.3.P9 The eigenvalues of $\lambda I + A$ are $\lambda + \lambda_i$.

8.3.P10 (8.3.1): $\rho(A) \le \max\{y^T Ay : y \in \mathbf{R}^n, y \ge 0, \text{ and } \|y\|_2 = 1\} \le \max\{y^T H(A)y : y \in \mathbf{R}^n \text{ and } \|y\|_2 = 1\}$.

8.3.P12 (a) Let $\lambda \in \sigma(A)$. If λ is real, then $r - \lambda \ge 0$. If λ is not real, then $\bar{\lambda} \in \sigma(A)$ and $(r - \lambda)(r - \bar{\lambda}) > 0$. (b) (5.6.16). (c) Continuity.

8.3.P13 (1.2.13), (1.2.15), and the preceding problem. $\rho(A)$ not simple $\Rightarrow S_{n-1}(\rho(A)I - A) = 0 \Rightarrow E_{n-1}(\rho(A)I - A) = 0 \Rightarrow \text{tr adj}(\rho(A)I - A) = 0 \Rightarrow$ each main diagonal entry is zero.

8.3.P16 (a) The hypothesis is that $Ax \ge 0 \Rightarrow x \ge 0$. If $Ax = 0$, then $A(-x) = 0$ so $x \ge 0$ and $-x \ge 0$. If z is a column of A^{-1}, then Az is a column of I_n, which is nonnegative.

Section 8.4

8.4.P6 (8.4.1).

8.4.16 (c) $(1 + t)^{n-1} = q_A(t)h(t) + r(t)$, in which $r(t)$ has degree at most $m - 1$.

8.4.17 Use the characterization of a best rank-one approximation given in (7.4.1).

8.4.19 The hint for (8.2.P9) permits us to conclude that $x_j = x_p$ for all j such that $a_{pj} > 0$. Let $q \ne p$ and let $k_1 = p, k_2, k_3, \ldots, k_m = q$ be a sequence of distinct indices such that each entry $a_{k_1 k_2}, a_{k_2 k_3}, \ldots, a_{k_{m-1} k_m}$ is positive. Explain why $x_p = x_{k_2} = \cdots = x_{k_{m-1}} = x_q$.

8.4.20 (c) If A^2 is irreducible, nonpositive, and nonzero, consider the multiplicity of its negative eigenvalue $-\rho(A^2)$. (d) How few zero entries can a reducible matrix have?

8.4.21 (a) $(I + A)x \le (\alpha + 1)x \Rightarrow (I + A)^{n-1}x \le (\alpha + 1)^{n-1}x$.

8.4.22 (a) rank $G \le n \Rightarrow \lambda = 1$ is an eigenvalue of $I - G$ with multiplicity at least two.

8.4.24 Let $A_1 = \begin{bmatrix} 0 & 1 \\ 1 & 0 \end{bmatrix}$ and let A_2 be the companion matrix of $p(t) = t^3 - 1$. Consider $A_1 \oplus A_1$ and $A_1 \oplus A_2$.

8.4.26 (a) (1.4.11) and (8.3.P12 and P13). (b) (8.3.P14).

Section 8.5

8.5.P3 Consider $\begin{bmatrix} 1 & 1 \\ 1 & 0 \end{bmatrix}$ and $\begin{bmatrix} 0 & 1 \\ 1 & 1 \end{bmatrix}$.

8.5.P4 Think of A as a linear transformation acting on the standard basis $\{e_1, \ldots, e_n\}$. Then $A : e_i \to ? A^{n-1} : e_i \to ? A^{(n-1)(n-1)} : e_1 \to ?$

8.5.P10 If $|\mu| = \rho(A)$, $\mu \ne \rho(A)$, $z \ne 0$, and $Az = \mu z$, then $(\rho(A)^{-1}A)^m z \to ?$

8.5.P14 Consider A^2 and use (8.5.5) and (8.5.6). Use (8.5.8).

Section 8.7

8.7.P2 The preceding problem and (3.2.5.2).

8.7.P4 The preceding problems.

8.7.P5 $\||A\||_2 \leq \sum_i \alpha_i \||P_i\||_2$.

8.7.P6 $Ae = e \Rightarrow \||A\|| \geq 1$. $\||P\|| = 1$ for every permutation matrix $\Rightarrow \||A\|| \leq 1$.

8.7.P7 If $A = \alpha_1 B + \alpha_2 C$ with $\alpha_1, \alpha_2 \in (0, 1)$, $\alpha_1 + \alpha_2 = 1$, and A, B, C are doubly stochastic, then every entry of B and C in the same position as a zero entry of A must be zero.

8.7.P8 (8.7.2).

8.7.P9 If A has $n + 1$ positive entries, some row contains at least two positive entries, so two different columns of A contain at least two positive entries each. At most $n - 3$ positive entries are contained in the remaining $n - 2$ columns.

8.7.P13 Use the Fan dominance theorem and (8.7.3).

8.7.P14 If A is permutation similar to the block matrix in (6.2.21), what are the column sums of the block B? What must the sum of its row sums be?

Index

Printed in the United States
by Baker & Taylor Publisher Services